GENERAL
ZOOLOGY

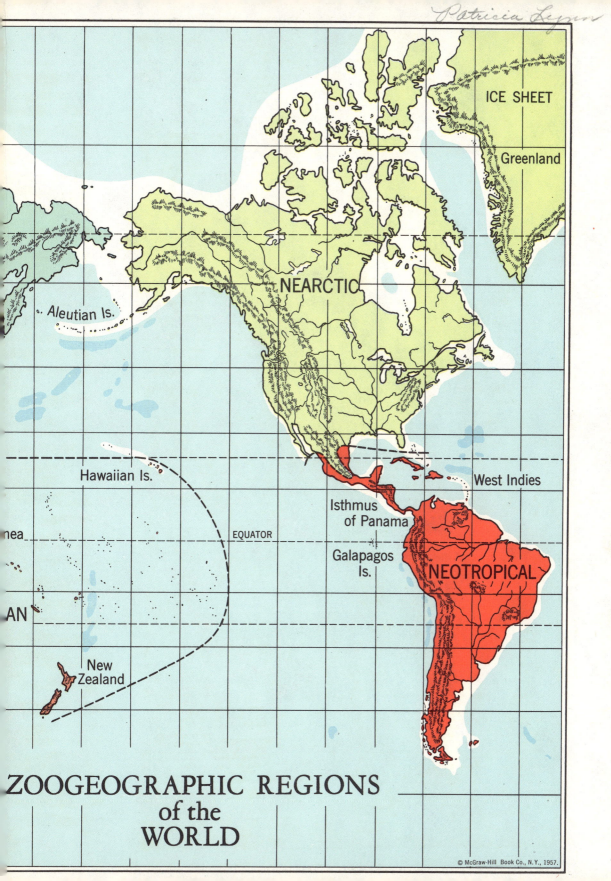

ICE SHEET

Greenland

NEARCTIC

Aleutian Is.

Hawaiian Is.

West Indies

Isthmus
of Panama

nea

EQUATOR

Galapagos
Is.

NEOTROPICAL

AN

New
Zealand

ZOOGEOGRAPHIC REGIONS
of the
WORLD

© McGraw-Hill Book Co., N.Y., 1957.

GENERAL ZOOLOGY
Sixth Edition

Tracy I. Storer

Late Professor of Zoology
University of California, Davis

Robert C. Stebbins

Professor of Zoology
Department of Zoology and
Museum of Vertebrate Zoology
University of California, Berkeley

Robert L. Usinger

Late Professor of Entomology
University of California, Berkeley

James W. Nybakken

Professor of Biological Sciences
California State University at Hayward and
the Moss Landing Marine Laboratories

McGraw-Hill Book Company

New York / St. Louis / San Francisco / Auckland / Bogotá / Düsseldorf / Johannesburg /
London / Madrid / Mexico / Montreal / New Delhi / Panama / Paris / São Paulo /
Singapore / Sydney / Tokyo / Toronto

Library of Congress Cataloging in Publication Data

Main entry under title:

General zoology.

    Includes index.
    1. Zoology. I. Storer, Tracy Irwin, date
QL47.2.G46 1979   591   78-12558
ISBN 0-07-061780-5

GENERAL ZOOLOGY

1234567890   DODO   7832109

This book was set in Palatino by Progressive Typographers.
The editors were James E. Vastyan, Janet Wagner, and Stephen Wagley;
the designer was Beverly G. Haw, A Good Thing, Inc.;
the production supervisor was Dennis J. Conroy.
New drawings were done by J & R Services, Inc.
R. R. Donnelley & Sons Company was printer and binder.

CONTENTS

Preface

This text is a general introduction to zoology at the college and university level. The subject matter is arranged to facilitate readings with either lectures or laboratory work and for reference use. Part I deals with general principles of animal biology. The first chapters deal with the finer structures of the body and their organization into special systems that carry on essential life processes. Succeeding chapters consider the more general phases of animal existence—reproduction, heredity, distribution, and evolution.

In Part II, following an introduction to classifying and naming of animals, each chapter describes the structure and some aspects of the physiology of typical animals belonging to each major group—from the protozoans to humans. The accounts of natural history and reproduction show how structure and function serve in the life histories of these and related animals. Some interrelations between animals and their environments are discussed, and there is frequent mention of the importance of other animals to humans. The broader relationships among animals are dealt with further in the general chapters on distribution and evolution.

The synopses of classification, a special feature of the book, have been prepared from recent publications with the help of specialists and organized on a uniform pattern. They show the extent, diversity, and relationships of animals constituting each of the larger groups and something of their mode of occurrence. They may also aid in identification of specimens as to phylum, class, or order. Common or notable representatives are mentioned by scientific and common names, using North American examples principally. The geologic ranges of groups and various fossil representatives (marked †) are mentioned.

The references at the ends of chapters direct the reader to further information. Some "classic" references—by Mendel, Darwin, and others—are cited so that the reader may become acquainted with a few of the major contributions to biologic thought. Comprehensive works in foreign languages are included. The lists include significant new books and other important recent literature.

Technical words in the text are distinguished by different typefaces, as follows: many anatomic and other terms, where emphasis is useful, **humerus, monohybrid;** scientific names of genera or species, *Rana pipiens, Mus;* names of families and higher systematic groups, CULICIDAE, PROTOZOA.

Throughout the summaries of classification, taxonomic levels are distinguished typographically.

This sixth edition of *General Zoology* contains the most extensive changes that the text has undergone since it was first published. We felt that such changes were necessary not only to update the book with respect to current advances in the field, but also to correct some long-standing criticisms concerning the organization of the material and the sequence of presentation of various subjects. Perhaps the most obvious change to long-time users of the book is the absence of the traditional chapter on the frog (Chapter 2 in previous editions) and the incorporation of the essential material into Chapter 31 on the Amphibia. The decision to undertake this change was a difficult one made only after many reviewers and users suggested it. It is our belief that by delaying discussion of the frog until later in the book and in the context of other amphibians the usefulness of the book will be enhanced and the sequence of the discussion topics will be improved.

Another major organizational change has been to rearrange the chapters of Part II, the Animal Kingdom, such that those phyla which are most closely related are treated together and that the progression is from more primitive to more advanced. In order to effect these changes we have integrated the ctenophores into the chapter with the cnidarians (coelenterates), placed all the pseudocoelomates together

with the nematodes in a single chapter, and moved the echinoderm chapter to a position near the chordates. Furthermore, the arthropod chapters have been altered such that phylogenetically similar subgroups are treated together and the more primitive forms are treated before the more advanced. Finally, the chapter on miscellaneous phyla has been reorganized to place the similar phyla next to each other and each is preceded by a discussion of general features of each group. The result of all this reorganization has been to make the groupings of the various animal phyla conform much more closely to the accepted phyletic groupings of major invertebrate textbooks.

Perhaps more important than the reorganization is the addition to each animal-phyla chapter of a new section outlining the relationships and phylogeny of the group and another section describing the general anatomical and physiological functioning of the group. We felt that such sections were necessary to give a clear understanding of what the relationships were and the evidence on which they are based, and to give the student an awareness that all the major phyla have important functional differences which may not be apparent in a simple listing of their anatomical parts. We feel that these additions will make it easier to understand the great animal diversity in the world.

Yet another significant change has been to place the chapter on chemistry (Chapter 2) before the chapter on cells. This provides a more logical progression from the molecular basis of life to the cellular basis. In addition, the cell chapter has been almost completely rewritten to update and expand the sections on cell organelles and their function and those on mitosis and cell division.

Additional new sections have been added to this edition. They include: consideration of the relevance of the study of zoology to modern life (Chapter 1); expanded coverage of protein structure and enzyme action (Chapter 2); discussions of the kingdoms of organisms, the origin of the metazoa, and the role of embryological features in classification (Chapter 14); the regulation of body water balance by the kidneys in vertebrates (Chapter 7); osmoregulation in fishes (Chapters 29 and 30); how hormones act on target cells, including a discussion of the action of cyclic AMP (Chapter 8); expanded sections on sensory structure and function and on transmission of nerve impulses, including the events occurring at synaptic junctions (Chapter 9); new sections on the process of fertilization, and on early stages of embryonic development in mammals (Chapter 10); new information on the nature of genes, and on gene mapping in viruses; a section on genetic engineering, including a discussion of recombinant DNA research (Chapter 11); expanded discussion of continental drift (Chapter 12); natural selection, adaptation, genetic polymorphism, and speciation (Chapter 13); and a new section on molecular evolution (Chapter 13).

In the chapters on vertebrates information has been added on hearing in amphibians and reptiles, snake locomotion, and neuroendocrine control of migration in birds. The sections on bird flight and bird voice have been expanded, and new information has been added on hormonal control of reproduction in humans. Of special importance has been the addition of a section on evolution to chapters on vertebrates. The section gives an overview of evolutionary advances made by each group, in an effort to identify the selective value of outstanding features of the vertebrate classes and to relate these features to mode of life and evolutionary origin.

In the discussions of function, adaptation, and evolution that characterize so many parts of the book, we sometimes refer to a given structure as *for* a particular purpose. This has been done to reduce the awkward wording that is often required to skirt the pitfalls of teleology. We do not imply purposeful action on the part of the organism.

Terminology has been updated in this edition. The most significant change here is the adoption of the term Cnidaria to replace the old term Coelenterata. This adoption is in keeping with the usage now found in most major textbooks on invertebrate zoology.

The illustrations have been enhanced by new photographs. Many of the photographs show representatives of the various animal phyla, but new transmission and scanning electron micrographs of cell organelles have been added as well. Finally, some new line drawings have been prepared to accompany the new sections, a number of them by Robert Stebbins.

In addition to the many people who have assisted in previous editions of this book, we would particularly like to acknowledge those who helped on this edition. We thank Richard Mariscal, Ron Larson, and Gary McDonald for their fine photographs. We want to thank the many instructors who provided information about their classroom experiences with the fifth edition. And we wish to acknowledge the perceptive comments of the following people, who read the entire manuscript for the sixth edition: Troy L. Best, Howard D. Booth, George Butterstein, James C. Greene, and Judith A. Theile. Thanks also go to persons who provided information and advice in their areas of expertise: James D. Yarbrough (cell physiology); Thomas Dietz (human physiology); Robert Macey (muscle contraction); Curtis Williams (endocrinology); Richard Steinhardt (nerve transmission); Atonie W. Blackler (reproduction and development); Charles and Sharon Nicoll (human reproduction); Robert Tamarin (genetics); Peter F. Brussard (ecology); Troy L. Best (evolution); Alice M. Brues (human evolution); William A. Brueske (protozoans); Barbara N. Burkett (invertebrates); T. Michael Peters (terrestrial arthropods); John McCosker and Richard Rosenblatt (fishes); David and Marvalee Wake (amphibians); Ned K. Johnson (birds); and Peter Dalby (mammals). Anne Fetzer and Kathleen Roberts helped with library work. Anna Rose Stebbins helped with editorial matters.

Robert C. Stebbins
James W. Nybakken

GENERAL ANIMAL BIOLOGY

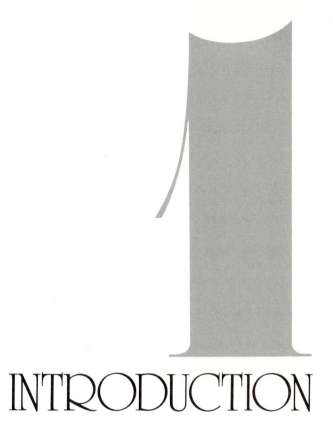

INTRODUCTION

Animals differ from one another in size, structure, manner of life, and other features. Man has acquired enough detailed knowledge about animals to fill a large library, but much more needs to be learned, and there are many unanswered questions.

What is life? In what ways are the various kinds of animals alike or unlike in structure, internal processes, and modes of life? How do animals carry on their activities? How are the many kinds related to one another? How have they evolved? In what ways does the human being resemble and differ from other living things? How are humans affected by animals, and how have their activities influenced those about them? The answers to many of these questions are provided by the **science of zoology** (Greek *zoön*, animal + *logos*, discourse), which deals with animal life.

1-1 Diversity of life The world contains an enormous number of living things. No one knows exactly how many different types (species) of organisms inhabit our planet, but even if one excludes the microscopic forms and the plants, estimates exceed 1 million. Some feel that there may be as many as 2 million species of animals alone. Even more species, now extinct, have lived during past geological time. Some animal species are abundant in terms of individuals, while many others are rare or uncommon. Some form of life inhabits nearly every conceivable area of the earth. Hence, organisms are found in such unlikely places as the deepest trenches of the ocean, in hot springs, and in the ice and snow of the Arctic and Antarctic. All bodies of water and nearly all land surfaces have characteristic living creatures inhabiting them. Living creatures do not exist separately in a vacuum but always with other organisms in a physical environment; thus each different environment has a characteristic set of species. These characteristic assemblages of organisms interact with each other and with the physical environment in which they exist. The total of all these complex interactions constitutes what has been called the "web of life" or the "balance of nature" but which zoologists more often designate as the discipline of ecology (Chap. 12).

Human beings are also animals and, as such, also exist with other organisms in a physical environment. Although humans have used their intelligence and tool-making ability to create environments—cities, houses, space capsules—which insulate them to a greater or lesser extent from contact and interaction with the physical and biological environment of the planet, they cannot escape entirely. They are still dependent on certain organisms to provide food, they still harbor parasites and disease organisms, and they usually interact with other animals or at least with other human beings. They are still subject to the principles of zoology in that their life processes are similar to those of most other animals.

Despite the enormous numbers of different animals in the world, the basic processes which function to ensure maintenance of life are similar among all. All animals are thus organized and operate according to certain biological and physical laws. These laws and concepts serve as a framework for organizing the first section of this text. All animals are constructed around certain chemical elements and compounds and operate according to certain physical laws, the most important of which have to do with energy. These basic laws of chemistry and physics are covered in Chap. 2. The basic building block of any animal, the cell, is covered in Chap. 3. Chapters 4 to 11 discuss in detail the various essential life processes which are shared by each living organism. Chapter 12 details the principles which govern the organization of animals into interacting groups, and Chap. 13 discusses the continuity of life through time, indicating how change occurs through time.

1-2 Science Science (Latin *scientia*, knowledge) is exact knowledge or tested and verified human ex-perience. It is our way of carefully exploring our environment, the material universe. The raw materials of science are **facts,** the real state of things. Science seeks facts to demonstrate the natural orderly relationships among phenomena; it is self-testing and it avoids myth, legend, or bias (prejudice). Simple facts—that fire is hot, water is wet, etc.—may be determined by direct observation, but even these gain precision by the use of scientific instruments, permitting observations by one person to be compared with those of others. In many fields of science progress is dependent upon the instruments that are available, and the development of a new tool such as the electron microscope or the cyclotron opens up new subjects and methods not even suspected earlier.

The records of science are accumulated facts or **data** (sing., *datum*). Qualitative data deal with different kinds of things, subjective estimates of degree, and quantitative data with dimensions, weights, or other facts that can be expressed in numerical terms.

A **scientist** is a person of inquiring mind, curious about natural phenomena, who asks questions and seeks answers supported by dependable evidence. Absolute honesty in thought and action are basic to the **scientific method,** which is the making of careful observations and experiments, then using the data obtained to formulate general principles. The scientific method begins with some observations that arouse speculation as to their meaning. For example, it is commonly noted that moths are attracted to a light. To the scientist this suggests that a principle is involved in the relation between cause (light) and effect (attraction). The scientist sets up a temporary working explanation, or **hypothesis**—that moths react positively to light—then plans experiments to test the hypothesis. Various kinds of moths are exposed to lights of different intensity and wavelength. If the experiments support the hypothesis, the scientist can formulate a more definite **theory** to explain the observed facts. In this case the theory might state that certain kinds of moths are attracted to light in the blue portion of the spectrum but are repelled by longer wavelengths that appear to us as yellow or red. Such a theory then becomes the basis for extensive trials over a period of years. As a result of such scientific study and deduction, electrical manufacturers have produced yellow lights that are considered to be "insectproof," not attractive to insects. Moths, destructive to agricultural crops can be sampled and even lured to their death by "light traps" with white or blue lamps. Finally, through repeated proof as to cause and effect, a theory may be restated as a general **principle** or **law**—but even this is never beyond criticism. New facts may be discovered that require the principle to be restated or discarded. Thus the scientific method accepts no knowledge as completely fixed or infallible but constantly seeks added evidence to test and to formulate basic principles of nature. Useful observations in zoology can be made by anyone who uses the scientific method. The thrill of discovering new facts is a rewarding experience.

1-3 Specialized fields of science   The average person has more or less **common knowledge** about animals. A child learns that dogs bark and are covered with hair, that birds fly and are covered with feathers, and many similar facts. Much detailed information of this sort is known to primitive peoples who live in the wild and to farmers whose livelihood depends on an understanding of nature. Other people may develop a special interest in **natural history** and study the habits of birds or other common animals. The science of zoology includes these kinds of information together with all other knowledge about animals, both popular and technical. With **botany,** the science of plants, and **microbiology,** the study of bacteria, viruses, and monerans, it forms **biology** (Gr. *bios,* life), the science of living things. Biology in turn is one of the natural sciences that have to do with the phenomena of nature. Some other natural sciences are geology—earth structure; mineralogy—substances in the earth's crust; physiography—surface features of the earth; and meteorology—weather and climate. The natural sciences may be contrasted with the **physical sciences**—physics dealing with the properties of matter and energy and chemistry with the constitution and transformation of matter. Applied science is directed at the solution of practical problems such as many of those in agriculture or engineering, whereas "pure" or basic science has no such immediate objective. Yet many economic and medical problems have been resolved by using the findings of basic scientific studies.

In ancient times one man such as Aristotle (384–322 B.C.) might master the entire field of science, and only a century ago Louis Agassiz (1807–1873) and some others knew and could teach all the natural sciences. With the subsequent great increase in knowledge this is no longer possible, and scientific fields have of necessity been divided and redivided. A scientist now must specialize in one field or related parts of a few. Although unfortunate in some respects, specialization has made for much more rapid advance in both science and industry. A scientist has, therefore, the problem of working in a particular branch of knowledge yet endeavoring to develop a broad vision of the world about him.

Scientific publications are the means of communication between scientists throughout the world. They have grown progressively until about a million separate articles now are published each year. Any one scientist finds difficulty in keeping abreast of current publications and to do so must rely in part on special abstracting journals, reviews, and digests.

Some principle subdivsions of zoology and the chapters where they are discussed are:

Morphology (Gr. *morphe,* form): structure as a whole (Chaps. 4–9, 15–35)

Histology (Gr. *histos,* tissue): microstructure of tissues (Chap. 3)

Cytology (Gr. *kytos,* hollow): structures and functions within cells (Chap. 3)

Physiology (Gr. *physis,* nature): living processes or functions within animals (Chaps. 2–10, 15–35)

Nutrition (L. *nutrio,* feed): use and conversion of food substances (Chap. 5)

Embryology (Gr. *en*, in + *bryo,* swell): growth and development of the new individual within the egg or mother (uterus) (Chap. 10)

Genetics (Gr. *genesis,* origin): heredity and variation (Chaps. 11, 13)

Parasitology (Gr. *para,* beside + *sitos,* food): study of animals that live and subsist on or in other animals (Chaps. 12, 15–26)

Natural history: life and behavior of animals in their natural surroundings (Chaps. 13, 15–36)

Ethology: behavior of animals (Chap. 12)

Ecology (Gr. *oikos,* house): relations of animals to their environments (Chaps. 12, 15–36)

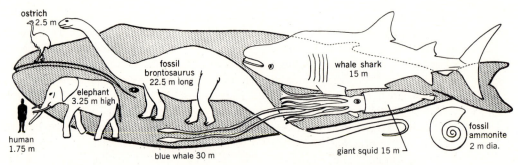

Figure 1-1 The largest animals as compared with humans. (*Adapted from C. R. Knight.*)

Zoogeography (Gr. *zoön*, animal + geography): distribution of animals in space (Chap. 12)

Paleontology (Gr. *palaios*, ancient + *ont*, being): fossil animals and their distribution in time (Chaps. 12, 13, 15–35)

Evolution (L. *e*, out + *volvo*, roll): origin and differentiation of animal life (Chap. 13)

Biochemistry (Gr. *bios*, life): study of the chemical compounds and processes involved in living organisms (Chap. 2)

Taxonomy (systematics) (Gr. *taxis*, arrangement + *nomos*, law): classification of animals and principles thereof (Chaps. 14, 15–35)

Zoology is also divided for the study of particular animal groups:

Entomology: study of insects: further subdivided into insect taxonomy, economic entomology, etc. (Chap. 25)

Mammalogy: study of mammals (Chap. 34)

Helminthology (Gr. *helmins*, worm): study of various worms (Chaps. 18, 19, 23)

1-4 Living things   The average person can distinguish most kinds of living organisms from nonliving or inorganic matter—a tree or bird or worm from a rock or some chemical substance—but this is not easy with some lower forms of life. The seed of a plant or the egg of an insect seems inert, but each, when placed under suitable conditions, will soon reveal its living nature.

Six important characteristic features of living things serve to separate them from the nonliving:

1 **Metabolism**   Within living organisms, a complex series of essential chemical processes, collectively termed **metabolism,** are constantly in progress. Important in metabolism are such common activities as taking in and digesting food; assimilating the digested food into the body; respiration, which is the process of releasing the energy of the assimilated food; and excretion, which is the removal of waste materials produced during the release of energy. Nonliving things are not capable of enacting these processes by taking in foreign materials, transforming them to produce energy, and then eliminating waste products. Admittedly, some nonliving materials release energy. For example, radioactive materials give off energy as they decay into other elements. But nonliving things do not derive energy by taking in or transforming foreign matter.

2 **Growth**   All living organisms grow by developing new parts between or within older ones. Thus, growth occurs by addition from within. This is growth by **intussusception,** and it is a defining feature for living things. Nonliving things do grow, but growth is always by addition of material to the outside, not the inside, as in a crystal (Fig. 1-3).

3 **Irritability**   Living things react to changes in their environment, a capacity which is generally termed **irritability.** Response to environmental changes may take any of several forms, and the degree of response is not always proportional to

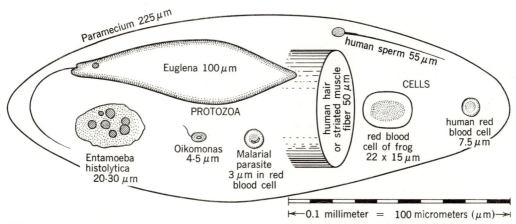

Figure 1-2   Some of the smallest animals and protistans and some animal cells, all included within the outline of a *Paramecium*.   Magnified about 550 times.

the magnitude of the stimulus. The organism is usually not permanently altered by the stimulus. Nonliving things do not respond to stimuli in a similar way, for if they react, as in the expansion of a metal when heated, there is a definite quantitative relation between the stimulus (heat) and the effect produced (expansion).

4 Reproduction   Each kind of living organism has the ability to reproduce itself in kind.

5 Form and size   Each type of living organism usually has a definite form and characteristic size (Figs. 1-1, 1-2). Nonliving materials often vary in both size and form; mineral crystals are quite constant in form but vary in size.

6 Chemical composition   Living organisms are composed mainly of four chemical elements: carbon (C), hydrogen (H), oxygen (O), and nitrogen (N) in various but definite proportions; these four occur with lesser amounts of other elements. These elements, when bound with one or more atoms of carbon, form complex organic molecules, often of great molecular weight; together they make the living substance or **protoplasm** of the plant or animal. The same and other chemical elements occur in the much smaller molecules that compose the nonliving minerals, rocks, and soil.

Viruses are, in some ways, intermediate between living and nonliving things (some have been crystallized); but they can grow only in living cells.

1-5 Physical and chemical basis of life   The fantastic diversity of life on this planet rests on a surprisingly few of the 92 naturally occurring chemical elements. Atoms of the low-molecular-weight elements carbon, hydrogen, nitrogen, and oxygen make up more than 95 percent of the living material (protoplasm) of the planet. In various chemical combinations these four elements form the essential framework of the bodies of all animals, from simple inorganic compounds such as water (H_2O) through small organic compounds such as amino acids, sugars, and fats up to large complex macromolecules of protein, starch, and nucleic acids.

Perhaps 30 or so other elements are among the minor constituents of living matter. These minor constituents are minor in terms of total amounts present but are still vital to the functioning of certain living systems. Examples of such important elements are calcium, needed in vertebrate bone and mollusk shell construction; phosphorus, vital in all living energy relations and protein structure; and iron and copper, needed to carry oxygen in vertebrate and invertebrate respiratory systems. Many of these minor elements are integral parts of the large and complex organic molecules.

All animals require energy in the form of food in order to sustain life processes. They must obtain this energy ultimately from plants or other autotrophic organisms (such as certain bacteria), as they cannot manufacture their own foods from simple inorganic

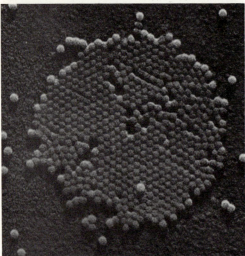

Figure 1-3 Nonliving versus living things. *Top.* Crystals of quartz (long, white) and of iron pyrites. About natural size. *Bottom.* Crystallized form of the poliomyelitis virus which, when in cells of animals or humans, "lives" and multiplies. ×121,000. (*Electron microscope photo from W. M. Stanley.*)

components. Certain laws of physics govern energy relationships, and they are as valid for living systems as for nonliving. These laws are embodied in the field of thermodynamics, which deals with energy and its transformation.

Thus, because the elements of which an animal body is composed are chemical and obey certain chemical laws, and because animals exist only by using energy, which operates under certain rigid

physical laws, both chemistry and physics are significant in a study of life and life processes. As we shall see in later chapters, energy in particular is important because it acts at all levels, from the molecular to the whole ecosystem.

1-6 Animals versus plants Most macroscopic organisms (those which may be seen clearly with the unaided eye) can easily be referred to either the Plant or the Animal Kingdom. This is not true for many microscopic organisms or for the group of large organisms known as fungi (mushrooms and relatives). Hence, scientists now classify organisms into either four or five kingdoms: ANIMALIA (animals), PLANTAE (plants), FUNGI (fungi and slime molds), PROTISTA, and MONERA. The last two kingdoms include very small organisms such as bacteria and amoebas (Sec. 15-3). By far the larger number of organisms, however, fall into either the Plant or Animal Kingdom. Some important differences between plants and animals follow.

1 **Form and structure** The animal body form is rather invariable, the organs are mostly internal, growth usually produces changes in proportions with age, the cell membranes are delicate, and the body fluids contain sodium chloride (NaCl). In plants the form often is variable, organs are added externally, the cells commonly are within thick cellulose walls, and sodium chloride usually is toxic. Most growth is at the ends of organs and often continues throughout life, but each kind of plant has a normal growth limit.

2 **Metabolism** Animals require complex organic materials as food, obtained by eating plants or other animals. These foods are broken down (digested) and reorganized chemically within the body. Oxygen (O_2) is usually needed for respiration. The end products of metabolism are mainly carbon dioxide (CO_2), water (H_2O), and urea ($NH_2)_2CO$. Plants also carry on metabolism, but in addition they use water, carbon dioxide from the air, and inorganic chemicals obtained in solution from the soil. By photosynthesis—the action of sunlight on the green pigment known as chlorophyll—these simple materials are formed into var-

ious organic compounds, and oxygen is released as a by-product (Fig. 12-1).

3 **Nervous system and movement** Most animals have a nervous system and respond quickly to stimuli; plants have no such system and react slowly. Animals commonly can move about or move parts of the body but certain kinds become fixed early in life (sponges, sea anemones, oysters, barnacles), and other fixed forms (hydroids, bryozoans) are of plantlike form.

1-7 Animal life of the world About one million kinds, or species, of living animals are known, and new sorts are constantly being discovered. Some are enormously abundant, others are present in moderate numbers, and still others are rare. For convenience in study and to indicate relationships between the different kinds, the Animal Kingdom is

Figure 1-4 Plants versus animals and protistans. *Above.* Venus's flytrap, a plant that closes its leaf blades to catch insects for food. *Left.* Euglena, a one-celled organism "claimed" by both botanists and zoologists (Sec. 15-10). *Right.* Colonial hydroid, a marine and animal of plantlike form (Sec. 17-13).

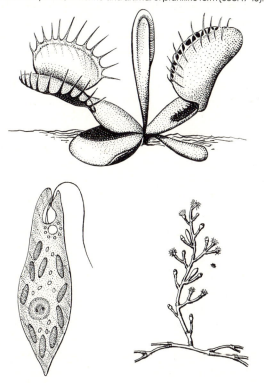

divided into various groups, large and small (Chap. 14).

Every kind of animal has special requirements for life, determined by its structure and its needs as to food, shelter, and reproduction. Different parts of the earth are covered with fresh water or salt water or with soil or rocks of many types. Tropical regions receive more solar heat than those near the poles, and the amount of moisture in the atmosphere or precipitated as rain or snow varies locally. Therefore the physical environment on different parts of the earth is diverse. This influences the kinds of plants that grow on the land, and the plant cover in turn influences the types of animals that can live in any one place. In consequence, the numbers and kinds vary widely on different parts of the earth (Chap. 12).

Each animal species is influenced by the physical environment and by the other plants and animals that make up its biological environment. Most animals are affected by predators, diseases, and competitors. The total of all these interactions comprises the "web of life," or the "balance of nature," a dynamic complex of physical and biological forces that acts on every living organism, including human beings.

1-8 Relation of animals to humans Early humans probably lived much as do primitive peoples today, by gathering wild seeds and fruits and hunting animals (Chap. 35). Such activity, however, requires large land areas to sustain each human, hence groups of humans existing in that way are low in number. Modern human societies with their large populations can only be sustained by intensive agriculture in the form of cultivated grains (wheat, corn, and rice) and domesticated animals (cattle, sheep, swine, and poultry). Both primitive and civilized peoples rely on animals in fresh water and salt water—fishes, oysters, crabs, and others—for additional supplies of animal protein. However, the aquatic harvest declines where overexploited.

The wool of sheep and pelts of fur-bearing mammals provide clothing, bird feathers are used in pillows and quilts, hides supply leather and glue, hair is made into felt, and the glands and other internal organs yield many medicinal preparations. Honey, beeswax, and natural sponges are other useful animal products. The livestock and meat-packing industries, commercial fisheries, the fur trade, and beekeeping are animal industries that provide profitable employment for thousands of persons.

Animals used in studies in both field and laboratory have contributed immensely to humans' understanding of themselves and their environment as well as to their enjoyment of nature. Much practical knowledge has been gained by these studies in the fields of medicine, physiology, embryology, genetics, behavior, and animal husbandry.

Large predatory mammals are rarely any longer dangerous to humans in civilized countries, but they kill some domestic livestock and wild animals. The insects and rodents that feed upon crop plants, grasses, herbs, or trees take a heavy toll that necessitates large expenditures for control. Other insects and the "house" rats and mice damage stored foods and property. Some insects, spiders, scorpions, and snakes are dangerously venomous. The many kinds of parasites—protozoans, worms, insects, and ticks—bring illness and death to humans, their domestic livestock, and useful wild species. Disease factors carried by animals and protozoans have exercised a dominant role in the history of humankind; examples are the protozoan parasites of malaria and the virus of yellow fever, both carried by mosquitoes; the bacteria of plague, transmitted by fleas; and typhus, spread by lice and fleas.

1-9 Why study zoology   What is the relevance of this science to students today? In an increasingly complex, interrelated, and technological world, these are important questions which deserve an answer by way of an introduction to this book. In these latter decades of the twentieth century, humans are facing several converging problems which could bring about a most serious crisis. These problems include increasing population, increasing pollution, decreasing food and energy for the human population, and decreasing diversity of life forms. All these problems are intimately related to zoology. The study of populations, how they function, and how they are controlled is one aspect of the discipline of ecology (Chap. 12). Energy and its apportionment and conservation in living animals are major aspects of several disciplines of zoology, including molecular biology or cell physiology (Chap. 2) and most other aspects of physiology (Chaps. 4–10), while transfer of energy on a population level is one aspect of ecology. Diversity of life, that is, the great variety of animal forms present on this planet and their successful adaptation to diverse habitats, is the topic of Part II of this book, which is designed to acquaint the reader with the spectrum of animal life inhabiting this earth and which may be important in the structuring of stable life-support systems of this planet.

All the major problems facing human beings in the latter quarter of this century are biological: the population explosion, the food and energy shortages, and pollution. They cannot be solved without an adequate knowledge of the principles which govern animal life upon this planet, namely, zoology. One cannot expect, for example, to come to grips with the human population problem unless one understands the principles underlying the growth, maintenance, and decline of animal populations, a part of ecology, for humans are also subject to the same rules. Similarly, one cannot expect to understand either the basis of the current food crisis or its resolution without a knowledge of the different types of foods, why they are needed, and how they serve in living animal systems, another discipline of zoology (physiology). Finally, in a time when animal species all over the world are increasingly subject to extinction and diminishing population sizes, it is important to understand something about this diversity if anything is to be preserved. Why study zoology? The future of the human race may depend upon it!

1-10 The study of zoology   Many students have some general knowledge about animals, gained from everyday experiences in country or city, or from museums and zoos. To expand this background, the beginning course in zoology provides a general account of the Animal Kingdom, the structure of representatives of different groups, the bodily (physiological) processes, and the ways that animals live. To this is added an understanding of how animals grow and reproduce, the principles of inheritance (heredity), the distribution of animal life over the earth today and in past geologic time, and finally how the existing kinds of animals came into being—the theory of organic evolution. Further work in zoology will consider other aspects of the subject and in greater detail.

The requirements for a successful study of zoology are few: (1) ability to observe carefully and to report accurately that which is seen; (2) absolute honesty in all work—a prime requirement in all branches of science; (3) clear thinking to arrive at dependable deductions or inferences from observations; and (4) a judicial attitude to appraise the relative values of conflicting evidence and to arrive at appropriate conclusions—but with a willingness to abandon or alter such conclusions in the presence of evidence pointing in another direction. Skill in attaining all these requirements may be gained even in an elementary course.

References

The vast literature of zoology includes thousands of separate books and hundreds of periodicals, both technical and popular. The items listed here and in other chapters are a few selected works in which the student will find elementary and advanced materials on various aspects of zoology. There are many elementary textbooks of zoology and animal biology that deal with parts or all of the Animal Kingdom and with general biological phenomena and principles.

The following works include systematic accounts of the Animal Kingdom:

Grzimek, H. C. Bernhard (editor). 1970. Grzimek's animal life encyclopedia. New York, Van Nostrand Reinhold. 13 vols. Translated from German. *Most recent attempt at an encyclopedic treatment; readable, well illustrated.*

Harmer, S. F., and **A. E. Shipley** (editors). 1895–1909. The Cambridge natural history. London, Macmillan & Co., Ltd. 10 vols. Reprinted 1960. New York, Stechert-Hafner, Inc. PROTOZOA to MAMMALIA.

Lankester, E. R. (editor). 1900–1909. A treatise on zoology. London, A. & C. Black, Ltd. 8 vols. *Incomplete; omits annelids, insects, land vertebrates, etc.* Reprinted. New York, Stechert-Hafner, Inc.

Moore, R. C. (editor). 1952–   . Treatise on invertebrate paleontology. Lawrence, Kansas Geological Society and The University of Kansas Press. *Detailed coverage including fine synopses of the biology of various groups. Multivolume and incomplete.*

Parker, T. J., and **W. A. Haswell.** 1897. A textbook of zoology. London, Macmillan & Co., Ltd. 1940. Vol. 1. 6th ed. Invertebrates. Rev. by Otto Lowenstein. xxxii + 770 pp., 733 figs. 1962. Vol. 2. 7th ed. Vertebrates. Rev. by A. J. Marshall. xxiii + 952 pp., 659 figs.

Romer, A. S. 1966. 3d ed. Vertebrate paleontology. Chicago, University of Chicago Press. viii + 468 pp.

Some important comprehensive works in foreign languages are:

Bronn, H. G., and others. 1859–   . Klassen und Ordnungen des Tierreichs. Leipzig. Akademische Verlagsgesellschaft Geest & Portig K.-G. "6 vols." (actually about 50, still incomplete).

Grassé, Pierre-P. (editor). 1948–   . Traité de zoologie, anatomie, systématique, biologie. Paris, Masson et Cie. 17 vols. planned, 13 issued to date.

Kükenthal, W., and **T. Krumbach** (editors). 1925–   . Handbuch der Zoologie, eine Naturgeschichte der Stamme des Tierreichs. Berlin, Walter de Gruyter & Co. 8 vols. (actually 14 or more, still incomplete).

The following deal with invertebrates:

Barnes, R. D. 1974. 3d ed. Invertebrate zoology. Philadelphia, W. B. Saunders Company. xii + 870 pp., illus.

Barrington, E. J. W. 1967. Invertebrate structure and function. Boston, Houghton Mifflin Company. x + 549 pp., illus.

Bayer, F. M., and **H. B. Owre.** 1968. The free living lower invertebrates. New York, The Macmillan Company. viii + 229 pp., 271 figs.

Beklemishev, V. W. 1969. Principles of comparative anatomy of invertebrates. 2 vols. Chicago, University of Chicago Press.

Bullough, W. S. 1950. Practical invertebrate anatomy. London, Macmillan & Co., Ltd. xi + 463 pp., 168 figs. PROTOZOA to CEPHALOCHORDATA.

Carthy, J. D. 1958. The behaviour of the invertebrates. London, George Allen & Unwin, Ltd. xiii + 380 pp., 148 figs. *Experimental.*

Dales, R. Phillip (editor). 1969. Practical invertebrate zoology. Seattle, University of Washington Press. xii + 356 pp., 140 figs.

Daugherty, E. C. (editor). 1963. The lower Metazoa: Comparative biology and physiology. Berkeley, University of California Press. xi + 478 pp., illus.

Edmondson, W. T. (editor). 1959. Ward and Whipple, Fresh-water biology. 2d ed. New York, John Wiley & Sons, Inc. 20 + 1,248 pp., illus. FUNGI, ALGAE, PROTOZOANS TO ARTHROPODS; *keys and descriptions.*

Florkin, M., and **B. Scheer** (editors). 1967–1975. Chemical zoology. 9 vols. New York, Academic Press, Inc.

Covers various aspects of physiology of most invertebrate phyla.

Fretter, V., and **A. Graham.** 1976. A functional anatomy of invertebrates. New York, Academic Press, Inc. vi + 589 pp., 212 figs.

Gardiner, M. 1972. The biology of invertebrates. New York, McGraw-Hill Book Company. 945 pp., illus.

Giese, A, and **J. Pearse** (editors). Reproduction of marine invertebrates. 1974. Vol. I. Acoelomate and pseudocoelomate metazoans. xi + 546 pp., illus. 1975. Vol. II. Entoprocts and lesser coelomates. xiii + 344 pp., illus. 1975. Vol. III. Annelids, echiurans. 343 pp. New York, Academic Press, Inc.

Gosner, K. L. 1971. Guide to identification of marine and estuarine invertebrates. New York, Interscience Publishers. 693 pp. (Atlantic Coast).

Hadži, Jovan. 1963. The evolution of the Metazoa. New York, Pergamon Press. xii + 499 pp., 62 figs.

Hyman, L. H. 1940. The invertebrates: Protozoa through Ctenophora. Vol. 1, xii + 726 pp., 221 figs. 1951. Platyhelminthes and Rhynchocoela. Vol. 2, vii + 550 pp., 208 figs. 1951. Acanthocephala, Aschelminthes, and Entoprocta. Vol. 3, vii + 572 pp., 223 figs. 1955. Echinodermata. Vol. 4, vii + 763 pp., 280 figs. 1959. Smaller coelomate groups. Vol. 5, viii + 783 pp., 241 figs. 1967. Mollusca I. Vol. 6, vii + 832 pp., 249 figs. New York, McGraw-Hill Book Company. *Modern, general; other volumes planned.*

Jagersten, G. 1972. Evolution of the metazoan life cycle. New York, Academic Press, Inc. 280 pp.

Kaestner, A. Invertebrate zoology. 1967. Vol. 1, 597 pp. 1968. Vol. 2, 472 pp. 1970. Vol. 3, 523 pp. Translated from 2d Ger. edition by H. W. Levi and L. R. Levi, New York, Wiley Interscience.

MacGinitie, G. E., and **Nettie MacGinitie.** 1968. Natural history of marine animals. 2d ed. New York, McGraw-Hill Book Company. xii + 523 pp., 286 figs.

Meglitsch, Paul. 1972. Invertebrate zoology. 2d ed. New York, Oxford University Press. xviii + 834 pp., illus.

Ricketts, E. F., and **Jack Calvin.** 1952. Between Pacific tides. 4th ed. Rev. by Joel W. Hedgpeth. Stanford, Calif., Stanford University Press. xiv + 614 pp., 22 pls., 302 figs. *Marine invertebrates, arranged by habitats.*

Russell-Hunter, W. D. 1968. A biology of the lower invertebrates. New York, The Macmillan Company. x + 181 pp., illus.

————. 1969. A biology of the higher invertebrates. New York, The Macmillan Company. xi + 244 pp., illus.

Smith, R. I., and **J. Carlton** (editors). 1975. Light's manual, Intertidal invertebrates of the central California coast. Berkeley, University of California Press. xvii + 716 pp., illus., 156 pls.

The following deal with vertebrates:

Alexander, R. M. 1975. The chordates. London, Cambridge University Press. vi + 480 pp.

Blair, W. F., A. P. Blair, P. Brodkorb, F. R. Cagel, and **G. A. Moore.** 1968. 2d ed. Vertebrates of the United States. New York, McGraw-Hill Book Company. ix + 616 pp. *A systematic account with keys to species.*

Hildebrand, M. 1974. Analysis of vertebrate structure. New York, John Wiley & Sons. xv + 710 pp.

McCauley, W. J. 1971. Vertebrate physiology. Philadelphia, W. B. Saunders Company. xiv + 422 pp.

Orr, R. T. 1976. 4th ed. Vertebrate biology. Philadelphia, W. B. Saunders Company. viii + 472 pp.

Romer, A. S. 1959. The vertebrate story. Chicago, University of Chicago Press. vii + 437 pp.

Romer, A. S., and **T. S. Parsons.** 1977. 5th ed. The vertebrate body. Philadelphia, W. B. Saunders Company. viii + 624 pp.

Wake, M. H. (editor). 1979. Hyman's comparative vertebrate anatomy. Chicago, University of Chicago Press.

Wessells, N. K. (editor). 1974. Vertebrate structures and functions. San Francisco, W. H. Freeman and Company. 440 pp. *Readings from Scientific American, 1955 to 1974.*

Young, J. Z. 1962. 2d ed. The life of vertebrates. New York, Oxford University Press. xv + 820 pp.

A few books on the history of zoology are:

Asimov, Isaac. 1964. A short history of biology. Garden City, N.Y., Natural History Press. ix + 189 pp., 7 figs.

Gabriel, M. L., and **Seymour Fogel.** 1955. Great experiments in biology. Englewood Cliffs, N.J., Prentice-Hall, Inc., xiii + 317 pp. *Excerpts from notable original works by Hooke, Pasteur, Mendel, and others.*

Ley, Willy. 1968. Dawn of zoology. Englewood Cliffs, N.J., Prentice-Hall, Inc. viii + 280 pp., illus.

Locy, W. A. 1925. The growth of biology. New York, Henry Holt & Co. xiv + 481 pp., 140 figs.

Nordenskiöld, E. 1924. The history of biology. New York, Alfred A. Knopf, Inc. x + 629 + xv pp., illus.

Peattie, D. C. 1936. Green laurels: The lives and achievements of the great naturalists. New York, Simon & Schuster, Inc., xxiii + 368 pp., illus.

Singer, Charles. 1959. A history of biology to about the year 1900: A general introduction to the story of living things. 3d and rev. ed. New York, Abelard-Schuman, Limited, xxxv + 579 pp., 194 figs.

Two index periodicals for finding original articles are:

Biological Abstracts. 1926– . Baltimore, Union of American Biological Societies. *Comprehensive listing of new scientific literature in biology with abstract of each article; indexed by author and subject.*

Zoological Record. 1864– . London, Zoological Society of London and British Museum. *Annual volumes listing new books and papers in zoology; indexed by author, subject matter, and systematic position of the animals discussed.*

For definitions of zoological terms not in glossary of this book see:

Abercrombie, Michael, C. J. Hickman, and **M. L.**

Johnson. 1962. A dictionary of biology. Chicago, Aldine Publishing Company, 254 pp.

Gray, Peter. 1967. The dictionary of the biological sciences. New York, Reinhold Publishing Corporation. xx + 602 pp.

Jaeger, E. C. 1955. A source-book of biological names and terms. Springfield, Ill., Charles C Thomas, Publishers, Inc. xxxv + 323 pp.

Pennak, R. W. 1964. Collegiate dictionary of zoology. New York, The Ronald Press Company. 583 pp.

2

SOME CHEMISTRY OF LIFE

In order to stay alive, animals, and the cells which constitute them, are constantly carrying on chemical changes. These chemical changes and interactions are called **reactions.** Chemical reactions in living cells and tissues are responsible for activities such as the production of secretions in gland cells, the release of energy for the movement of muscles, and the digestion of complex food molecules. If most reactions are stopped, the cells and the animal will die in a few minutes. In order to understand animals it is thus necessary to first cover some fundamentals of chemistry. Study of chemical compounds and their reactions in the cells and fluids of living organisms is the field of **biochemistry,** which seeks ultimately to understand the chemical phenomena we call **life.** A recent branch of biochemistry called **molecular biology** is rapidly accumulating knowledge on some of the detailed chemical aspects of life, as in the case of DNA and RNA (Secs. 2-28, 11-17).

The present chapter discusses some general physical and chemical principles, the basic components of living matter, and a few chemical aspects of cell metabolism. Other biochemical substances and metabolic processes are described in Chaps. 4 to 9.

Physical properties

2-1 Matter, weight, and gravity The substance of the universe, the earth, and living organisms is termed **matter.** Under different environmental conditions of temperature and pressure, any particular kind of matter may be in one of three physical states

—solid, liquid, or gas. Water, a common type of matter, may be variously solid ice, fluid water, or water vapor. Animal shells and skeletons are mostly solids, the blood plasma and much of the content of body cells are fluid, and gases are present in lungs or dissolved in body fluids. Almost any animal comprises matter in three states.

The **mass,** or quantity of matter in any object or body, is one of its basic attributes. Certain forces attract any two bodies of matter, the degree of attraction being dependent upon their masses and distance apart. The attraction between the earth and that of any animal or other object on or near its surface is termed **gravity,** and the value of this force is its **weight.**

The force of gravity keeps animals against the surface of the earth or any solid object on which they may be. It acts more rapidly in air than in a heavier medium such as water, where resistance to movement is greater. The weight of any given animal would be less on the moon (small mass) but much greater on Jupiter (larger mass). The volume relation of weight of any object in reference to some standard (such as water) is termed its **specific gravity.** That of a gas is low, whereas that of metals such as iron or gold is high. Among animals specific gravity, and particularly the surface volume relationships, determine their habits and influence the types of environments in which they can live. Bats, birds, and insects are able to fly because of their extensive wing surfaces, and some aquatic invertebrates swim and float readily because they have much surface in relation to weight. The effective specific gravity of any aquatic animal is less than that of a comparable land dweller, because the former is buoyed up by the weight of water it displaces.

Because of another property or force, **inertia,** a body at rest tends to remain so, and one in motion tends to continue in motion. Inertia is directly related to mass. A child's wagon requires less force to start into motion (overcoming inertia) than an automobile, but the wagon meets more surface resistance to motion and tends to stop sooner than the heavier vehicle. The same is true of animals. An insect has less inertia than a bear; hence it can start and stop more quickly. In the absence of gravity and friction with the air, water, or ground, a body once set in motion would continue on indefinitely, but on the earth resistance of the surroundings eventually overcomes the inertia of movement. Any animal, large or small, must exert propulsive power to remain in motion.

2-2 Cohesion and adhesion For particles of matter of submicroscopic size (molecules; see Sec. 2-4) other forces operate: that of **cohesion** tends to keep particles of the same kind together and that of **adhesion** those of different kinds. Cohesion of molecules at the surface of a body of water (or other fluid) produces an elastic skinlike effect termed **surface tension** that tends to make the surface minimum in extent. This tension has an appreciable elastic strength; it will support a clean needle laid on the surface. Water striders and other insects can "walk" on the surface film because their feet are covered by a nonwettable wax that does not break the cohesive force. Surface tension rounds up rainwater as drops, and microscopic amounts of oil within animal cells are formed into spherical droplets by this force. Adhesion and surface tension are responsible for the rise of fluid in a fine capillary tube. An insect that falls with its wings on the surface film of a pool may be unable to rise again because of adhesion between its wings and the water. Cohesion and adhesion keep the various microscopic parts of the body attached to each other.

All the phenomena named—gravity, inertia, cohesion, surface tension, and adhesion—are involved in the structural makeup and bodily processes of animals at both gross and microscopic levels.

2-3 Energy Another basic component of the universe is **energy,** "the capacity to do work." All activities of organisms involve energy; examples are the movements of animals, the digestion and use of food, and the transmission of nerve impulses. Energy may be manifested in several ways: *motion,* such as the flight of an insect; *heat,* an increase in temperature (due to random movements of particles within matter); *chemical change,* or reaction, as in the digestion of food; *electric current,* flow of impulses along the course of a nerve; and *light,* transmission of photons. All these forms, which are more or less interconvertible, are termed **kinetic energy,** the energy of motion (Gr. *kinein,* to move). A second kind is **potential energy,** the energy of position. An upraised hand or foot has potential energy, but as it swings to throw or kick a ball, this is converted to the kinetic energy of motion.

Two basic laws govern all energy conversion: The **first law of thermodynamics** states that in any closed system the total quantity of energy remains unchanged. In an animal the total received in food is expended in movements, digestion, and other bodily processes or lost as heat radiated into the environment. None has actually been "lost" to the system of which the animal is a part. The **second law of thermodynamics** holds that heat is the end form of all energy transformations and that all forms of energy may be entirely transformed into heat, but that heat may never be transformed completely into the other forms. The energy received by an animal is variously converted in its body, but all that is involved in motion, friction, chemical conversions, or even nerve impulses finally becomes heat that is transferred to its environment.

The energy in the world is almost all derived from the sun. Solar radiation is responsible for the development and growth of plants, upon which in turn virtually all animals depend (Chap. 12).

Chemical principles

2-4 Structure of matter   In everyday experience we learn to recognize some of the thousands of kinds of matter or substances to which names are given—water, iron, sugar, etc. Mere inspection, however, will not show whether any particular substance is pure—of one kind—or a mixture of two or more. Ordinary water, for example, usually contains both oxygen (a gas) and salts (solids), in solution. If we are to learn the actual properties of water alone, it must be rid of other substances. The science of chemistry deals with the structure and composition of substances and with the reactions that these materials undergo.

Chemical research has shown that each kind of pure substance consists of ultramicroscopic struc-

tural units called **molecules.** In turn each molecule is made up of one or more **chemical elements.** An element is a material which cannot be broken down into simpler form by ordinary chemical means. The particles of an element are termed **atoms,** which are basic building blocks, and all atoms of an element are similar. A molecule of water consists of two atoms of hydrogen and one of oxygen. For convenience in stating chemical facts and describing chemical reactions the names of elements are represented by symbols: H for hydrogen, O for oxygen, and C for carbon; some others are shown in Fig. 2-2. The formula for the water molecule is therefore H_2O, that of the gas oxygen is O_2, and that of common table sugar is $C_{12}H_{22}O_{11}$. In all, 92 naturally occurring chemical elements have been identified, named, and studied. An additional 11 have been made synthetically in research laboratories, making a total of 103.

By indirect methods we have learned that atoms, in turn, are composed of even smaller particles. No one has been able to see the ultraminute molecules, atoms, or smallest components; but many careful physical experiments and calculations have made it possible to determine their weight, learn their electric charges, and compute their speed. From these and other data the structural makeup of molecules and atoms has been visualized, and models of many have been made.

2-5 Atoms   The atom is considered as having a spherical outline with a central **nucleus** around which are smaller units, one or more, called **electrons,** each revolving in an orbit, or shell; most atoms have more than one shell (Fig. 2-1). The makeup of an

Figure 2-1 Model of presumed structure of a helium atom.

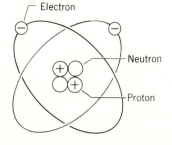

atom thus roughly resembles our solar system with its central sun (nucleus) and revolving planets (electrons). In both there is a vast amount of space between the components. If an atom were enlarged to a sphere 30 m (100 ft) in diameter, the nucleus would be perhaps 1 cm (½ in) through. Around the nucleus electrons would be whirling so fast as to be a faint blur.

The nucleus is composed of **protons,** each of which bears a single positive charge, and also **neutrons** that are uncharged. For every positively charged proton in the nucleus there is an electron, negatively charged, in one of the shells. The entire atom, therefore, is neutral, as the positive and negative charges are equal.

Atoms of chemical elements differ from one another in the number of electrons, protons, and neutrons each contains. The elements can be arranged in a **periodic table** according to the number of **protons** each contains. Hydrogen has 1 proton; therefore its **atomic number** is 1; helium has 2; sodium has 11; and so on. The **atomic weight** is an arbitrary number

Figure 2-2 First part of the periodic table diagramming the structure of atoms. The central number represents the nucleus and its net positive charge—the atomic number. Small black dots represent electrons, negatively charged, in their respective orbits. The atoms shown include those of elements common (C, H, O, N) or essential (Na, P, etc.) in living matter; still others are present in minute amounts as trace elements (Fe, Si, etc.). Iron (Fe) is misplaced here because five kinds of atoms are omitted after calcium (Ca).

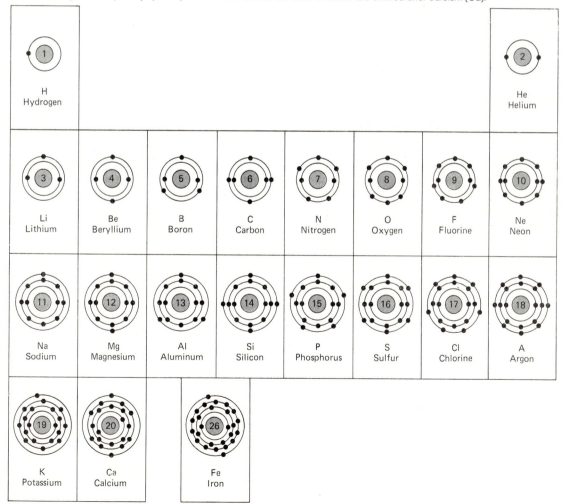

assigned to each kind of atom with reference to carbon (12) as a standard. It is approximately equal to the sum of the number of protons and neutrons in the nucleus. The electrons are practically weightless. Sample atomic weights are hydrogen, 1; carbon, 12; sodium, 23; uranium, 238.

2-6 Isotopes   All atoms of an element have the same atomic number but not the same atomic weight, because some contain more neutrons than others. Those of any one element that differ in mass are called **isotopes.** Certain kinds of isotopes release electrons or other electromagnetic radiation and hence are radioactive. Some of these can be produced artifically, and some occur in nature, resulting from bombardment of the atoms by cosmic rays. For example, carbon 14 (^{14}C) resembles its "parent" ^{12}C but is radioactive. It can be incorporated into a carbon-containing substance that is fed or injected into an animal. Then the ^{14}C can be traced into various parts of the body or into new compounds by a device that records radioactivity (Geiger counter). Research using isotopes is revealing some of the most intimate details of chemical processes in organisms.

The atmosphere contains mostly ^{12}C but also a little ^{14}C, and live plants incorporate the two into their substance in that ratio. When a tree is chopped down, the ratio begins to decline as the ^{14}C atoms decay. (The half-life of ^{14}C is 5,580 years; one-half is lost in that time.) In a relic of wood or charcoal found in nature the ^{12}C/^{14}C ratio provides a measure of age of the relic. This is much used in dating archaeologic specimens.

2-7 Electrons and bonds   Different kinds of atoms have one to seven concentric shells. Each shell can contain only a fixed maximum number of electrons, 2 in the innermost but 8 in the outermost shell (except palladium). An atom is unreactive only when all its shells are complete. Helium (2 electrons and neon (2 + 8), for example, are inert gases that do not react with other atoms. Most kinds of atoms, however, have fewer than 8 electrons in the outermost shell and are potentially reactive—able to enter chemical reactions and become joined to other atoms. Thus, sodium (2 + 8 + 1) and chlorine (2 + 8 + 7), both unstable, can join by an **electron transfer reaction** to form a stable molecule, sodium chloride (Fig. 2-3). The sodium atoms loses 1 electron, whereby its second shell (8 electrons) becomes the outermost and the chlorine gains 1 to make its outermost shell complete (8 electrons).

Atoms become joined together in units called **molecules** by means of **chemical bonds** that involve the transfer or sharing of electrons. An ionic bond results when one or more electrons are transferred from one atom to another as in the sodium chloride reaction just described. In a covalent bond the electrons are shared as in the case of hydrogen, where each atom has one electron, and when two atoms are joined, they share the two electrons, and thereby the shell has its maximum number; this is the state in hydrogen gas (H_2). Bonds may be designated symbolically by a dash(—) or a dot(·), or they may be omitted (NaCl).

The **valence,** or combining capacity of an atom, depends on how many electrons are present in its outermost shell in relation to the number required to "fill" or "empty" that shell (Fig. 2-2). For example, the valence of hydrogen is 1; that of carbon is 4. Others are N, 3; O, 2; Na, 1; Cl, 1; Ca, 2; P, 3 (or 5), and so on.

2-8 Ions, electrolytes, and compounds   When the outer orbit contains fewer than half the total number of electrons that it can hold, it may lose one or more; if more than half, it may gain electrons. A change in the number of electrons changes the electrical nature of the atom—in gaining electrons it becomes negative, but in losing any it becomes positive. An atom thus changed is termed an **ion;** with an excess of electrons it becomes an **anion** (having a negative charge; in an electric field it moves toward the anode, or positive pole); with a deficit it becomes a **cation** (going to the cathode, or negative pole).

A substance formed by the joining of two or more different kinds of atoms is a **compound** (Fig. 2-3). The combination of water with a chemical compound dissolved in it is called a **solution.** A compound which dissociates into anions and cations when dissolved in water forms a solution which will

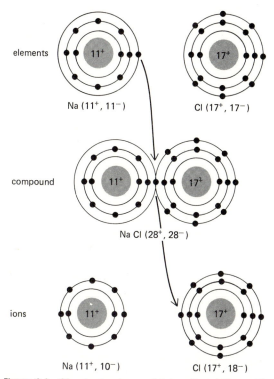

elements

Na (11⁺, 11⁻)   Cl (17⁺, 17⁻)

compound

Na Cl (28⁺, 28⁻)

ions

Na (11⁺, 10⁻)   Cl (17⁺, 18⁻)

Figure 2-3   Chemical union and later dissociation. *Elements.* Each with like number of positive and negative charges. Sodium (Na), only one electron in outermost orbit; chlorine (Cl), lacking one in outermost orbit. *Compound.* Sodium chloride (NaCl, table salt); single electron shared by both. *Ions.* When dissolved in water, some of the compound dissociates into ions, each with a complete outer orbit; sodium ion has net + charge of 1, chlorine a net − charge of 1.

conduct an electric current. Hence any chemical compound which will dissociate into ions in water is called an **electrolyte.**

Any compound that releases H^+ ions (protons) when dissolved in water is called an **acid.** Common examples are hydrochloric acid, acetic acid (in vinegar), and lactic acid (in sour milk). A **base,** or **alkali,** is a compound that in solution releases OH^- ions or accepts hydrogen ions. Caustic soda (NaOH) and ammonia water (NH_4OH) are household examples. Either acids or alkalies, when concentrated, are severe irritants and will "burn" the skin and the delicate coverings of the eyes and mouth.

The strength of an acid or base is indicated by the relative number of hydrogen (H^+) or hydroxyl (OH^-) ions present in a given solution. This is represented

by an exponential value termed the *pH,* where a value of 7 is neutral and increasing values to 14 indicate stronger bases, while decreasing values to 0 indicate stronger acids. (Fig. 2-4). Most body fluids are close to the neutral value, the pH of human blood being about 7.3, or very slightly alkaline.

2-9 Chemical reactions   Mixing an acid and a base produces a **salt.** The H^+ and OH^- ions combine to form water (H_2O), and the remaining ions join as the new compound, the salt. When, for example, hydrochloric acid (HCl) and sodium hydroxide (NaOH) are mixed in solution, the result is the salt sodium chloride (NaCl) and water. The metallic sodium ion has replaced the H^+ ion of the acid. The process of recombination is a **chemical reaction** and can be expressed in symbols as a **chemical equation,** thus:

$$HCl + NaOH \longrightarrow NaCl + H_2O$$

Acid   Base   Salt   Water

In this case the HCl and NaOH are the **reactants** and NaCl and H_2O are the **products** of the chemical reaction. The arrow indicates the course of the reaction. If the reaction is reversible, as is true of many biologic reactions in living organisms, a dual symbol ⇌ is used.

A chemical reaction presumably is based on collisions between particles (atoms, ions, or molecules) of two kinds. The rate of reaction depends upon (1) the nature of the particles, some reacting more readily than others; (2) the concentration, since the potential number of collisions increases if more are present; and (3) the temperature, because higher temperature makes particles move faster, collisions are more frequent and more violent, and therefore the collisions are more likely to cause interaction. A

Figure 2-4   The pH range.

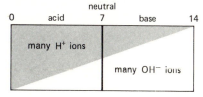

neutral

0   acid   7   base   14

many H^+ ions

many OH^- ions

fourth effect is that of a catalyst, or enzyme (Sec. 2-19), which makes the reaction speedier by bringing the particles closer together or reducing the amount of energy necessary to make the reaction go.

2-10 Mixtures   When a substance is mixed with any fluid the result is variously a solution, suspension, or colloid. In a solution the molecules, or ions of dissolved substance (the solute), soon become evenly distributed throughout those of the liquid (the solvent). Many acids, bases, salts, and other compounds (e.g., sugars) form true solutions in which the solute soon disappears from view and the solvent becomes clear. One liquid may be dissolved in another, such as alcohol in water, and a gas may be dissolved in a fluid, as oxygen is in water. If, however, the dispersed particles are of larger size (groups of molecules), a **suspension** results; mixing clay or flour in water yields a cloudy product; if allowed to stand undisturbed it will slowly clear as the particles settle to the bottom. An **emulsion** is a mixture of a fluid and fine droplets of another liquid; milk containing droplets of cream (butterfat) and mayonnaise (oil, vinegar, raw egg) are examples.

A **colloid** (Gr. *Kolla*, glue) results when the particles are of intermediate size—too large to enter solution but too small to settle out. Glue is a colloid of animal gelatin in water, the particles remain suspended indefinitely. The water is termed the **matrix** (continuous or external phase), and the other material is the **inclusion** (dispersed or internal phase). Colloid particles measure 1/10,000 to 1/1,000,000 mm in diameter; they are larger than most chemical molecules but cannot be seen with an ordinary microscope. Such division of matter into minute particles results in an enormous increase in ratio of surface to volume. Thus a solid cubic centimeter of any substance has a surface of 6 cm² (about 1 in²), but when dispersed as particles 1/100,000 mm in diameter the total surface is about 6000 m², or 1⅜ acres! The large surfaces provided by colloid dispersion in living matter are important for the chemical changes constantly in progress there. Colloids do not diffuse through membranes (Sec. 2-11) and when dried usually are masses of indefinite form; by contrast, crystalloids (e.g., salt, sugar) diffuse readily and when dried produce crystals of regular and characteristic structure. A colloid system may be either a semirigid **gel** or a more fluid **sol** (like gelatin in water when cold or warmed). In living matter which is largely colloidal these states may interchange during metabolic processes.

2-11 Diffusion and osmosis   Molecules in any kind of matter are constantly in motion, and the differences between states of matter—solid, liquid, or gas—result from the relative degree of motion possible. In a solid such as iron or brick the field of motion is extremely small. When more movement is possible, the substance is a liquid, and when the limit of motion is even greater, the result is a gas. In a fluid or gas the molecules move out in all directions until they are evenly distributed throughout the available space. Under a high-power microscope a suspension of minute particles in fluid shows a dancing **brownian movement** of the particles resulting from bombardment of the particles by molecules of the suspension medium.

Net movement of molecules from places of higher to lower concentration is termed **diffusion.** If an odoriferous gas (e.g., hydrogen sulfide) is released in one corner of a room, it diffuses and can be smelled in any part of the air. When a solid such as sugar or salt is immersed in water, it quickly **dissolves,** and molecules of the compound are later uniformly spread throughout the water, as one can confirm by withdrawing a drop in a pipette from any portion and tasting it.

The forces that repel one molecule from another result in a **diffusion pressure** proportional to the number of molecules present per unit volume of space. If two gases are enclosed in a container, each becomes diffused equally, and the total pressure is the sum of the two partial pressures. In like manner there is diffusion pressure in a solution when a quantity of any substance is dissolved in a fluid.

If a vessel containing water is divided in two by a metal partition, then sugar may be dissolved in one compartment and salt in the other, but the two will not mix. If, however, the partition is of collodion, cellophane, or parchment, sugar will diffuse through from the first compartment to the second, and salt in the opposite direction. The thin sheet acts as a **permeable membrane** having submicroscopic pores that permit the molecules of sugar and of salt to pass. Many of the finer structures in animal bodies are surrounded by **semipermeable membranes** that are selective in their action. Such membranes regulate the movement of food substances, respiratory gases, other essential materials, and wastes between body parts. Some membranes permit passage of larger molecules than others, and the rate of passage varies with the kind of membrane and the kinds and amounts of material on the two sides.

When unequal concentrations of dissolved substances occur on the opposite sides of a semipermeable membrane, the resulting differences in diffusion pressure bring an exchange of water and of the dissolved substances through the membrane until there is equilibrium (equal diffusion pressure) on the two sides. The diffusion of water through a semipermeable membrane is termed **osmosis.** A fluid is said to be **isotonic** when it contains the same amount of water and dissolved substance (i.e., solute) as the solution or cell to which it is compared. Solutions used to immerse living cells or tissues for study are made isotonic with the natural fluids that surrounded them in the body, as to the kinds and amounts of the principal salts (0.9% NaCl for mam-

malian blood or tissues, etc.). A **hypertonic** solution contains less water and more dissolved substances than that in the solution with which it is being compared, and a **hypotonic** solution contains more water and less dissolved substances.

Two experiments with artificial semipermeable membranes (collodion or cellophane) will demonstrate (Fig. 2-5) diffusion and osmosis. The end of a thistle tube is covered by a semipermeable membrane and inverted in a beaker; the thistle tube contains 10% salt solution (NaCl; molecular weight, 58), and the beaker contains only pure water (A). Some salt will diffuse through the membrane from the thistle tube to the beaker and some water from the beaker to the thistle tube until an equilibrium occurs with equal parts of salt and water in each (B). When a solution of hemoglobin is placed in the thistle tube (C), however, water will move from the beaker to the thistle tube by osmosis, so that the level fluid rises in the tube and lowers in the beaker (D). This results because hemoglobin molecules are too large (molecular weight, 63,000 to 68,000) to pass through pores in the membrane. These experiments show processes involved in transfer of materials across membranes of living cells in animal bodies. Diffusion and osmosis are fundamental in the physiologic processes of animals and their cells, including food absorption and utilization, respiration, and excretion (Secs. 5-10, 7-1, 7-11, 7-13).

Figure 2-5 Simple diffusion and osmosis. *Left. A.* End of thistle tube containing 10% salt solution is covered by a permeable membrane and inverted in a beaker of pure water. *B.* Salt diffuses out through the membrane, and water diffuses in, until solution is of same strength on both sides (equilibrium). *Right. C.* Hemoglobin solution in tube, pure water in beaker. *D.* Molecules of hemoglobin are too large to pass pores in *semi*permeable membrane, but water diffuses inward, diluting hemoglobin solution; level of fluid becomes higher in tube and lower in beaker.

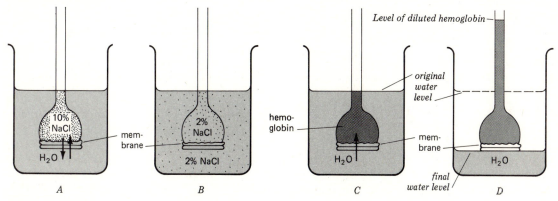

2-12 Buffers   Cells can function only within rather close physical and chemical limits, including (1) temperatures between about 0°C (32°F) and 40 or 45°C (104 or 113°F); (2) presence of oxygen gas within certain pressures; (3) definite and limited concentrations of salts; and (4) a delicate balance between H^+ and OH^- ions—the **acid-base equilibrium** (regulation of pH). This equilibrium is maintained by **buffers,** which are combinations of certain salts and weak acids that react with strong acids and strong bases to produce weak acids or bases, salts and water. Blood, for example, contains carbonate buffers made up of the salts sodium and potassium bicarbonate ($NaHCO_3$ and $KHCO_3$) and of the weak acid carbonic acid (H_2CO_3). If a strong acid such as hydrochloric (HCl) enters the blood, the salt of the buffer converts it to a weak acid which cannot lower the pH as much as HCl can:

$$NaHCO_3 + HCl \longrightarrow NaCl + H_2CO_3$$

Sodium bicarbonate   Hydrochloric acid   Sodium chloride   Carbonic acid

On the other hand, if a strong base such as sodium hydroxide (NaOH) enters the blood, the carbonic acid member of the buffer will neutralize it:

$$H_2CO_3 + NaOH \longrightarrow NaHCO_3 + H_2O$$

Carbonic acid   Sodium hydroxide   Sodium carbonate   Water

Seawater is buffered by bicarbonate to a pH of about 8.1, one of its favorable features as an environment for animals.

Chemical components of living matter

About 30 of the 92 elements are present in animal cells and tissues, but four of them provide most of the living soft parts by percentage: oxygen (62 percent), carbon (20), hydrogen (10), and nitrogen (3). Others are calcium (2.5), phosphorus (1.14), chlorine (0.16), sulfur (0.14), potassium (0.11), sodium (0.10), magnesium (0.07), iodine (0.014), and iron (0.010). The remainder are "trace elements," present in very

small amounts. No single element is peculiar to living creatures.

Chemical processes in the body result from interactions of the ions, atoms, and molecules that form the substance of the animal. As detailed in later chapters, there are nervous and chemical controls operating constantly in the functions of the body to maintain a relatively steady internal environment. When the normal pattern becomes deranged, illness or death ensues. "Life" may be regarded as an enormously complex system of interacting physical and chemical processes that are intricately and delicately balanced.

2-13 Water, gases, and salts   Different animals and their parts contain from 5 to 95 percent of water (H_2O), an average being 65 to 75 percent. It is more abundant in young cells or animals than in older ones and in the lower aquatic animals than in the higher terrestrial types. Water (1) is the best solvent for inorganic substances and is a good solvent for many organic compounds; (2) favors the dissociation of electrolytes dissolved in it; (3) has high surface tension; and (4) has a large capacity to absorb heat. All the necessary life processes are dependent upon these characteristics of water, and the life of organisms on this earth would be impossible without it. Watery body fluids (lymph and blood plasma), with both inorganic and organic content, surround cells in the animal body and transport materials within it. Terrestrial animals usually have a suitable body covering and other means to restrict undue loss of water.

Oxygen (O_2) and carbon dioxide (CO_2) are present as **gases** in the respiratory organs of air breathers and dissolved in tissue fluids or blood generally. The mineral or **inorganic salts** are conspicuous in skeletons or shells, often as calcium carbonate, $CaCO_3$, calcium phosphate, $Ca_3(PO_4)_2$, or silicon dioxide, SiO_2—secreted by specific cells or tissues. Sodium chloride (NaCl) and other salts are present in small concentrations; these ions are important in many activities of cells.

2-14 Organic compounds   Substances that contain carbon (C) in combination with hydrogen or/and oxygen are called **organic compounds.** Fully a million or more kinds now are known. Originally it was thought that they were produced only in living organisms, but chemists have learned to synthesize a large variety, including many not found in nature.

The element carbon is somewhat different from the elements discussed thus far. It needs 4 electrons to complete its outer ring or shell, but instead of giving up 4 or obtaining 4 and becoming electrically charged, as the anions and cations of electrolytes or salts do, it obtains these electrons by sharing them with other atoms. Hence, when placed in water, most carbon compounds do not dissociate into ions as do electrolytes or salts nor do they conduct electricity. Furthermore, many are not even soluble in water. Carbon, with 4 shared electrons or bonds, may share these electrons with other carbon atoms, thus forming chains or rings (Fig. 2-6). The free bonds become joined by hydrogen (H), oxygen (O), hydroxyl (OH), nitrogen (N), phosphorus (P), and sometimes sulfur (S) or other ions or elements. Compounds in the animal body and the changes they undergo often are complex.

The two basic types are (1) the straight-chain hydrocarbons such as methane (CH_4), ethane (C_2H_6), propane (C_3H_8), or ethene (C_2H_4); and (2) the ring, or aromatic, compounds such as benzene (C_6H_6) (Fig. 2-6). Within any organic molecule, whether straight-chain or aromatic, only certain parts of the molecule take part in the chemical reactions, the rest being relatively inert (often abbreviated **R**). These active parts of the organic molecules are termed **functional groups.** These groups have characteristic arrangements of bonds and elements which bestow distinctive properties on the organic compound of which they are a part. Examples of such groups are alcohol (—CHOH), aldehyde (—COH), acid or carboxyl (—COOH), ketone (C=O), ester (—COO—), and amine (—CH_2NH_2); also chloride (—Cl), phosphate (≡PO_4), and a methyl residue (—CH_3).

Thus from CH_4, methane (or marsh gas), by substitution of functional groups for hydrogen atoms, we can obtain $CHCl_3$, chloroform (an anesthetic), CCl_4, carbon tetrachloride (a cleaning fluid), or CH_3OH, methanol or methyl (wood) alcohol. If the last compound is oxidized (2H removed), there results HCHO, formaldehyde, a gas that in aqueous

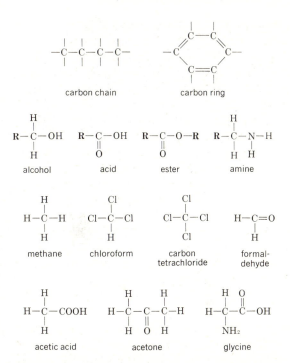

Figure 2-6 Structural formulas of simple organic materials. *Top.* Carbon chain and 6-carbon ring. *Center.* Carbon residue (**R**) joined to functional groups. *Bottom.* Simple compounds.

solution (formalin) is used to preserve biologic specimens. The methyl residue (CH_3) joined to a carboxyl group (—COOH) forms acetic acid (CH_3COOH), present in vinegar. Isopropyl alcohol (C_3H_7OH), by removal of 2H atoms, becomes a ketone called acetone (C_3H_6O), a useful organic solvent. Reacting methyl alcohol (CH_3OH) with butyric acid (C_3H_7COOH) yields the methyl ester of butyric acid ($C_3H_7COOCH_3$), found in pineapple. Simple esters provide the principal flavors in many fruits, and complex esters make up vegetable oils and animal fats. Organic amines derive essentially from ammonia (NH_3) by replacing 1 to 3H atoms with a hydrocarbon residue; an example is NH_2CH_3. The related amino acids contain both amine (—NH_2) and carboxyl (—COOH) groups, and hence may act as either base or acid. Glycine (NH_2CH_2COOH) is the simplest amino acid (Sec. 2-18). In the ring compounds of carbon, the position where a functional group or residue (one or more) is attached to the ring influences the chemical character of the resulting compound.

These examples are among the most elementary in organic chemistry. The reactions involved can be deduced easily from the structural formulas in Fig. 2-6. Compounds in animal bodies and the changes they undergo in physiologic processes often are far more complex.

Many organic compounds may be subdivided by a process termed **hydrolysis** (Gr. *hydro, water + lysis,* to loosen). Upon adding a water molecule (H_2O), a compound (AB) splits into two parts:

$$A\text{-}B + H_2O \rightleftharpoons A\text{-}H + B\text{-}OH$$

Reversal of hydrolysis (bottom arrow above) is a reaction termed **dehydration.** During digestion complex organic foods are divided into simpler compounds by hydrolysis. By the reverse process two-substances are combined to form a more complex molecule, with loss of H_2O. In the animal body many essential compounds are thus synthesized from simpler materials. All these reactions are controlled by special types of organic substances called enzymes (Sec. 2-19).

2-15 Carbohydrates   These are compounds of carbon, hydrogen, and oxygen, generally in the ratio of 1 carbon atom to 2 hydrogen atoms to 1 oxygen atom. They are the simplest and most abundant organic substances in nature—the sugars, starches, and celluloses. They are the primary products of photosynthesis in green plants (which also synthesize fats and proteins). Carbohydrates make up about three-fourths of the solids in plants. They provide much of the immediate or ultimate food for animals and are much used by humans (food, fabrics, wood, paper). Many are sugars: mono-, di-, or polysaccharides. A monosaccharide (simple sugar) cannot be divided by hydrolysis into simpler sugars. The most common example is glucose ($C_6H_{12}O_6$), which is important in energy pathways in animals. Another is ribose ($C_5H_{10}O_5$), present in nucleic acids (Sec. 2-20). Other monosaccharides have from 3 to 10 carbon atoms. The 6-carbon, or hexose, sugar **glucose** ($C_6H_{12}O_6$) is the end product in digestion of

carbohydrates, is regularly present (about 0.1 percent as blood sugar) in circulating blood, and is the source of energy for muscle contraction and other bodily activities. Other important 6-carbon sugars are fructose and galactose. Disaccharides, or double sugars, are composed of 2 monosaccharide units joined by dehydration. Sucrose, or table sugar ($C_{12}H_{22}O_{11}$), a disaccharide, is composed of 1 molecule each of glucose and fructose (Fig. 2-8):

$$C_6H_{12}O_6 + C_6H_{12}O_6 \underset{H_2O}{\overset{H_2O}{\rightleftharpoons}} C_{12}H_{22}O_{11}$$

Another double sugar, lactose (glucose + galactose), is present in the milk of mammals used by their suckling young. Maltose is a disaccharide formed from 2 glucose units.

Polysaccharides are large molecules composed of many simple sugar units joined together by dehydration. The commonest are all polymers of glucose. The glucose polymer (multiple molecule) typical in animals is **glycogen** ($C_6H_{10}O_5)_m$. *m* stands for any number of moderate size. It is commonly stored in vertebrate liver and can be reconverted into glucose for transport by the blood. Plants store their reserves as **starch** (*m* very large); huge long-chain

Figure 2-7 Three ways of picturing the glucose molecule. *Chemical formula.* A shorthand description. *Structural formula.* Relative positions of the atoms and the bonds between them. *Model.* "Architecture," showing spatial relations.

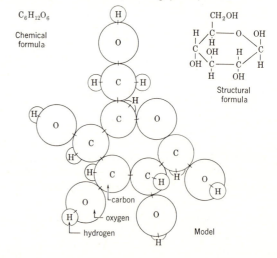

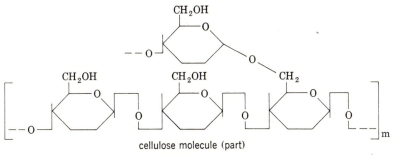

β-D-glucose

β-D-ribose

sucrose

fructose

glycogen molecule (part)

cellulose molecule (part)

Figure 2-8 Structural formulas of some common carbohydrates. β-D-glucose, β-D-ribose, and fructose are simple sugars. Sucrose is a disaccharide, and glycogen and cellulose are polysaccharides. The mode of attachment of glucose units in glycogen (└─O─┘) is an α-linkage and that in cellulose (⌐O⌐) a β-linkage. Asterisk indicates that changing H and OH around gives an α-linkage; m indicates that unit is repeated.

molecules are formed, and these accumulate into starch grains of visible size, as seen in potato or cereal. When starch or glycogen is eaten by animals, it is hydrolyzed during digestion back into glucose units. The essential relationship between glucose and glycogen or starch is thus:

$$mC_6H_{12}O_6 \rightleftharpoons (C_6H_{10}O_5)_m + mH_2O$$

Glucose   Glycogen or starch

Cellulose, another glucose polymer, is the fibrous support material in plants. It can be digested by only a few animals but is readily broken down by microorganisms in the rumen of cattle, etc. (Fig. 2-8).

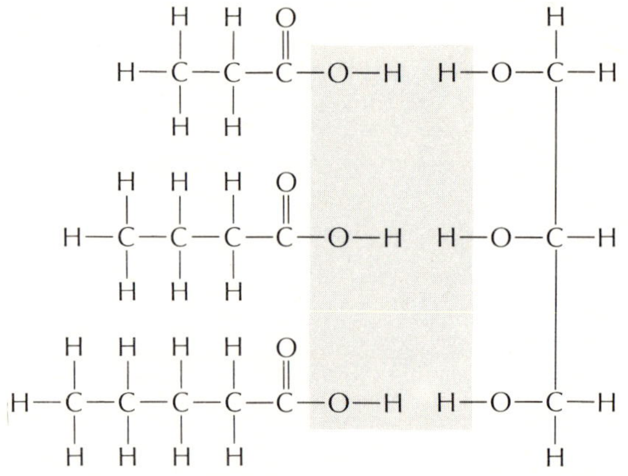

Figure 2-9 Structure of a fat. Shaded area encloses symbols for three water molecules lost when fatty acid and glycerol combine.

2-16 Lipids   The fats, oils, waxes, steroids, and related substances contain carbon and hydrogen with less oxygen than in carbohydrates. They are greasy and soluble in organic liquids such as ether, chloroform, or benzene but rarely in water. Some, such as olive oil or cod-liver oil, are fluid at ordinary temperatures. Others are solid, such as the fat under the skin of various mammals or fats that have been processed into butter, lard, or tallow. A few are firm waxes like those in the human ear and in the combs of honeybees.

A true fat is composed of 1 molecule of glycerol (glycerin) $C_3H_5(OH)_3$ joined to three fatty acids with loss of 3 water molecules (Fig. 2-9).

Fatty acids contain the carboxyl group (—COOH), usually joined to a long, straight-chain carbon residue. Examples are butyric acid $CH_3(CH_2)_2COOH$ in milk and butter, and stearic acid $CH_3(CH_2)_{16}COOH$ in animal fats. Saturated fatty acids have no double bonds between adjacent carbon atoms, while unsaturated fatty acids have one or more double bonds. Foods which are advertised as having polyunsaturated fats or oils are thus composed of fatty acids with many double bonds. Fats may be decomposed (saponified) by alkali (NaOH, etc.) to yield glycerin and soap. Digestive processes effect a comparable breakdown of fats in food. After absorption the components are variously synthesized into substances needed in the body.

As food, fats of plant or animal origin yield twice as much energy per gram as do carbohydrates or proteins. In the body, carbohydrate may be converted into fat (hence excess sugars and starches may be fattening); more rarely fat is transformed into carbohydrate. Many mammals and birds accumulate fat under the skin and between the internal organs that insulates against heat loss; the blubber of whales and seals is a conspicuous example. Fat is also a fuel reserve in fasting, hibernation, or migration, and oil in the liver of sharks and some bony fishes serves similarly. Cell and nuclear membranes (Secs. 3-3, 3-5) and the sheaths of certain nerve fibers (Sec. 3-17) have high lipid content.

Beeswax, lanolin, and other waxes contain fatty acids and an alcohol other than glycerol; they are of high molecular weight. Phospholipids are fatty substances containing also phosphorus and nitrogen that are important in cell membranes; lecithin in egg yolk is another example.

2-17 Sterols   These compounds are lipids which have complex molecules that include three 6-carbon rings, one 5-carbon ring, and at least one OH radical. Cholesterol, ecdysone, calciferol (vitamin D), most sex hormones, and certain cancer-producing substances are in this group (Fig. 2-10).

2-18 Proteins   Most abundant of organic materials in animal protoplasm are the **proteins.** Be-

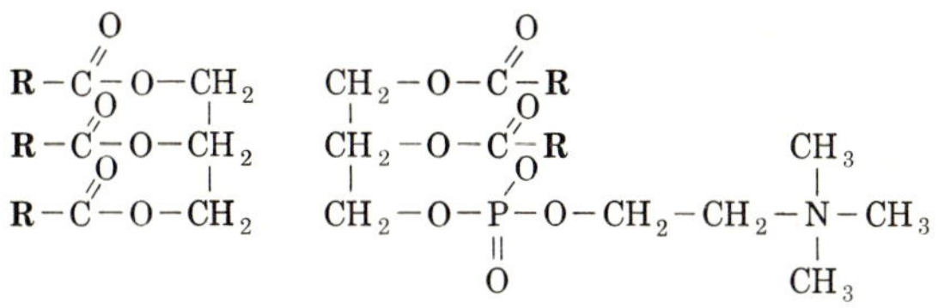

$$R-\overset{\overset{O}{\|}}{C}-O-CH_2$$

triglyceride   lecithin

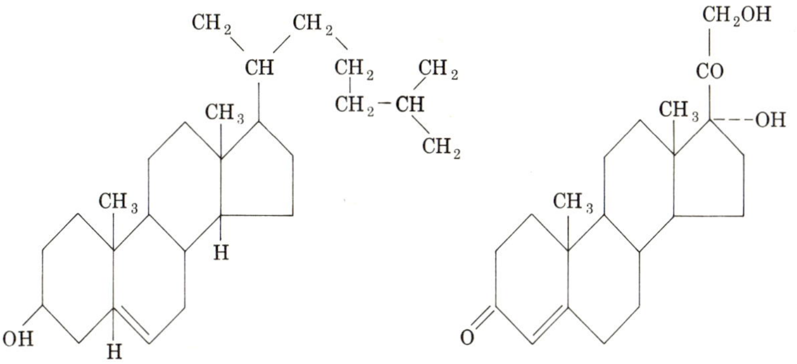

cholesterol   cortisone

Figure 2-10 Structural formulas of some lipids and sterols. *Top.* A simple lipid (triglyceride) and lecithin, a phospholipid. **R** stands for the remainder of the long carbon chain of the fatty acid. *Bottom.* Two sterols, one a hormone (cortisone).

sides carbon, hydrogen, oxygen, and nitrogen (12 to 19 percent), they contain sulfur, and some have iron, iodine, phosphorus, or other elements. Protein molecules are complex and relatively huge, with molecular weights from about 6000 to 10,000,000. Proteins are almost infinite in variety; those in any one individual and its parts usually differ from those in all others. In the human body, 70 percent of which is water, about half the remaining 30 percent is protein (muscle, over 33 percent; collagen in cartilage and bone, 20 percent; skin, 10 percent; etc.). Many functionally important bodily components are proteins, including the enzymes (Sec. 2-19), antibodies (Sec. 6-7), and some of the hormones (Chap. 8). Protein molecules are constantly being built up and broken down in plant and animal cells. Biochemists are learning their makeup and how to synthesize smaller proteinlike molecules in the laboratory.

The basic components of proteins are **amino acids,** each of which contains a basic amino group (—NH₂), an acidic carboxyl group (—COOH), and a side chain (**R**) attached to a carbon atom. All have a common formula (NH₂·CHR·COOH), but the **R** component is different in each. For examples see Fig. 2-11. Glycine (the left structure in Fig. 2-11B) is

the simplest; its **R** is one H atom. Alanine has a CH₃ residue instead of H; arginine contains three CH₂ residues; cystine, present in hair and nails, contains SH groups; and tyrosine, which includes a phenolic ring, is a precursor for the hormones thyroxine and adrenaline and also for black pigment (melanin).

Twenty kinds of amino acids enter the makeup of proteins during biologic synthesis. This takes place on the ribosomes under direction of messenger RNA, which in turn is under genetic control (Sec. 2-28). Several modifications of the 20 amino acids may take place subsequently, so that 23 (or more) kinds of amino acids are found in plants, and animals synthesize some. For normal health, 8 are required in man and 10 in laboratory rats. The protein requirements for human beings are largely a matter of obtaining enough of the several essential amino acids.

All protein molecules are long chains of amino acids joined by **peptide linkages** (—CO·NH—), the amino group (—NH₂) of one becoming attached to the carboxyl group (—COOH) of another by dehydration under enzyme control. In this process a molecule of water is released thus:

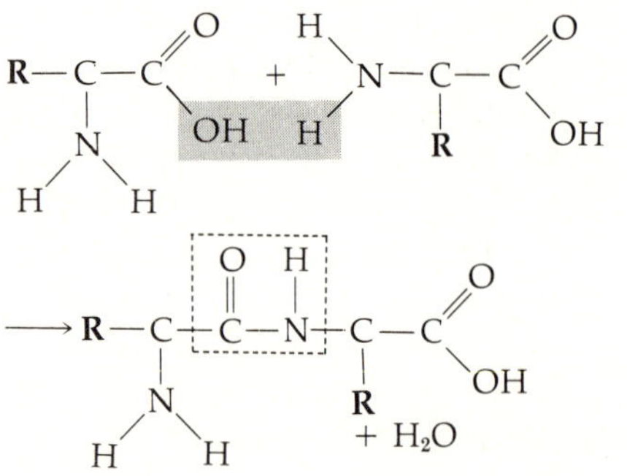

Two amino acids so joined form a dipeptide and many polypeptides.

Long polypeptide chains are termed **proteins.** Proteins and the properties associated with them are the result of more than just the sequence of amino acids. For this reason several different levels of organization are recognized. The kind and sequence of amino acids in a protein constitutes the **primary structure** of the molecule. Because of the peculiarities of the peptide bonding and the three-dimensional

Figure 2-11 Amino acids and peptide chains. *Top.* Four amino acids; the parts (NH₂·CHR·COOH) are the same in each but **R** is different. *Center.* Two amino acids about to be joined by peptide linkage (---) by removal of water (OH + H); the **R** components are shown in boldface (**H, CH₃**). *Bottom.* Part of a polypeptide. The **R** will differ according to the amino acids present. Bonds at ends connect to other components not shown.

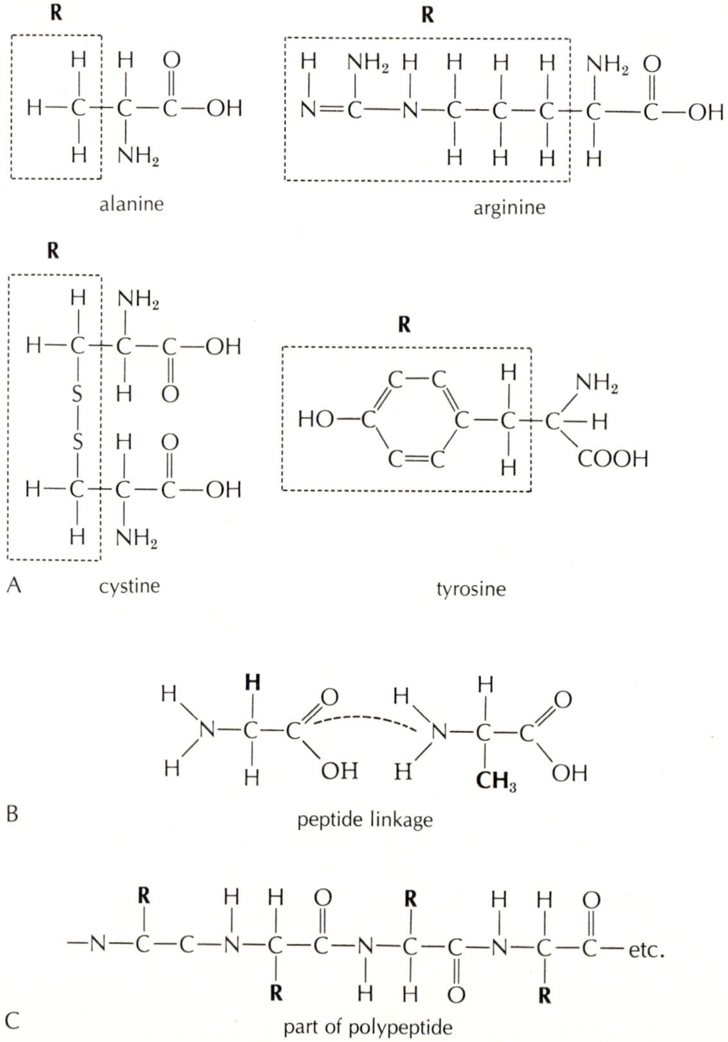

arrangement of the joined amino acids, the polypeptide chains in proteins tend to assume a helical or coiled structure like that of a spring. This is the **alpha helix** (Fig. 2-12). This spiral form is maintained by weak hydrogen bonds between the carboxyl (—COOH) group of one amino acid and the amino (—NH$_2$) group of an adjacent one in the helix. This is the **secondary structure** of the protein. The helical strands of the protein may themselves be bent and folded in space. Such bending and folding is characteristic of certain proteins and confers certain properties on the molecule. Such three-dimensional forms are the **tertiary structure** of the protein. Tertiary structure is maintained by strong and weak bonds between certain amino acids in the folded chains. Bonds which maintain this structure include disulfide bonds, covalent bonds, and hydrogen bonds.

The exact amino acid sequence of a protein was first reported in 1954 by Sanger and coworkers, for the molecule of insulin. The structures of many other proteins now have been determined.

The bodies of animals contain a great variety of proteins—soluble forms in blood and tissue fluids, gelatinous components in soft aquatic creatures, contractile substance in muscle, fibrous materials in cartilage and the organic foundation of bones, and insoluble kinds, (keratins) in cuticle, feathers, hair, nails or claws, and horns. Animals of many kinds eat proteins and digest them into the component amino acids. After being absorbed, these are used to synthesize the proteins characteristic of their own bodies. Some protein may be deaminated and the residues used for other purposes.

Because proteins differ in composition and structure, no two species of living organism contain exactly the same types; this **protein specificity** leads to complications when protein from one species is introduced (other than as food) into the body of another. Such **foreign proteins** may cause illness and death. Proteins of some plant pollens inhaled from the air often produce allergy in man. Transfusion of blood, if not of compatible type between donor and recipient (Sec. 6-8) can lead to shock and death.

Simple proteins consist entirely of amino acids. Examples are the water-soluble albumins of blood, milk, and egg white, the globulins of seeds and blood soluble in salt solutions, and the insoluble albuminoids of connective tissue and keratin. **Conjugated proteins** include simple proteins joined to another nonprotein substance. The nonprotein moiety is called the **prosthetic group.** Examples of such proteins include the hemoprotein (chromoprotein) in red blood cells and the phosphoprotein casein in milk and egg yolk; nucleoproteins are proteins attached to nucleic acids (Sec. 2-20).

Heat, pressure, acidity, and other conditions may alter the chain structure of proteins by breaking some bonds or by changing the three-dimensional configuration of the molecule, a change called **denaturation.** If the effect is slight, recovery may occur, but if severe, the change will be irreversible, as when fluid egg albumen is boiled and becomes solid or when meat (largely protein) is cooked. Denaturation of a protein leads to loss of most or all of its chemical activity.

2-19 Enzymes   Many chemical reactions that proceed slowly if at all are speeded by the action of a **catalyst,** a substance that accelerates the process without itself undergoing permanent change or being used up. Inorganic hydrogen and oxygen gases, for example, when brought together do not react, but in the presence of finely divided (colloidal) platinum as a catalyst they combine to make water—

Figure 2-12 The alpha helix. Shown diagrammatically is a portion of a polypeptide chain illustrating the secondary helical structure of a protein molecule. The dotted lines indicate stabilizing hydrogen bonds. (*After Weiss.*)

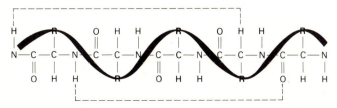

often with explosive violence. Catalysts are used much by industry, as in the cracking of petroleum and many other processes. Complex organic catalysts, the **enzymes,** are universal in living matter, and life is dependent on them. They promote a multitude of chemical reactions which would otherwise proceed too slowly to maintain vital processes. A sugar such as glucose, exposed to oxygen in the air, undergoes scant change, but in the cells of a living animal it is oxidized rapidly to yield energy. Chemically enzymes comprise a protein plus a prosthetic group, or **coenzyme.** The latter often includes a vitamin of the B group. Coenzyme A (CoA), involved in the Krebs cycle (Sec. 2-25), includes pantothenic acid.

Enzymes are involved in respiration, digestion, secretion and excretion, muscle contraction, and nerve conduction. For centuries before their chemical nature and mode of action were learned, enzymes were used in such practical processes as the fermentation of crushed grapes to produce wine (changing sugars to alcohol), of milk to make cheese, and of bread dough to make it rise.

Hundreds of enzymes are known, some only by the reactions they produce. The same kinds often are present in many species of plants and animals. Dozens have been purified and many crystallized. Molecular weights range from 12,000 upward. Some include metallic ions (Fe, Cu, Mg, Co, etc). The enzyme ribonuclease has been found to consist of 124 amino acids of 17 (or 19) kinds.

Enzymes of glycolysis (Sec. 2-24) are freely dispersed in the fluids within cells, and those involved in aerobic respiration (Krebs cycle, Sec. 2-25) are in the mitochondria (Sec. 3-4). In the digestive tract enzymes (Sec. 5-5) are produced by secretory granules in certain cells and passed into the cavity for extracellular action on food components. Protein-splitting digestive enzymes (proteases) are formed in cells as inactive **zymogens** and require another substance, an **activator,** to become functional. The inactive trypsinogen, secreted in the pancreas, becomes the active protease trypsin only after it has entered the small intestine and become activated there by the protein enterokinase from cells in the intestinal wall.

Most enzymes are named by adding the suffix -ase to the name of the substance, or **substrate,** acted upon or to the reaction that is promoted. Thus the enzyme that splits milk sugar, or lactose, into two simpler sugars is called lactase, and the enzymes catalyzing the removal of hydrogen are termed dehydrogenases. In general, enzymes are specific, each acting on a single substrate (urease acts only on urea), but some will work on different substrates of related structure.

Enzymes usually are water-soluble and are precipitated by strong alcohol or concentrated solutions of metallic salts. They are more active at temperatures of 20 to 40°C and lose their catalytic power when heated to 50 to 60°C or higher. This probably is the reason that most animals die at such temperatures. The inactivity of cold-blooded animals in winter is presumably due to their enzymes having reduced activity at low temperatures. Most enzymes behave like proteins and do not diffuse through semipermeable membranes, so they are retained at the sites where their work is needed.

Different kinds of enzymes act on carbohydrates, fats, proteins, phosphoric acid esters, and many molecules of intermediate stages at all levels of metabolism. Some serve to subdivide complex materials received in food and others to build, from simpler components, the organic substances essential in body structure and metabolism. Hydrolytic enzymes (e.g., carbohydrases) that act to divide by adding water and splitting a molecule into two or more smaller ones (Sec. 2-14) usually act on certain linkages and thus are substrate-specific. Other types (phosphorylases), which function in energy production, are active only in the presence of high-energy phosphate bonds (Sec. 2-23). Lipases that act on fats are able to deal with various fatty substrates, except the phospholipids, which are degraded by phospholipases. Enzymes serving in oxidation and reduction—oxidases, catalases, and dehydrogenases (Sec. 2-22)—are specific, each kind acting on a particular substrate or linkage.

The reactions induced by most enzymes are reversible, depending upon the amount of substrate and reaction products present (law of mass action). A unit of enzyme may catalyze 10,000 to 1,000,000 units of substrate; thus pepsin will digest 50,000 times its weight of boiled egg white in 2 h. The rate of action

of some is unbelievably rapid—a molecule of cata-
lase will reduce 5 million molecules/min of hydrogen
peroxide (H_2O_2) to water and oxygen. Each enzyme
does best at a particular temperature and pH value:
pepsin at pH 1.2 to 1.8 (HCl) in the stomach, but
trypsin at 6.8 to 7.5 (mildly alkaline) in the intestine.
Some enzymes work in a "team," or "chain," pattern, one after another on successive stages in the
buildup or subdivision of a sequence of compounds;
nine are involved in the splitting of glucose to pyruvic acid (Fig. 2-18) and as many or more in the Krebs
respiration cycle (Fig. 2-19). Similar teams of enzymes are required to manufacture amino acids in
stepwise procedure.

Enzyme function is not completely understood,
but enough is now known to outline the probable
course of events in an enzyme-catalyzed event. All
enzymes have certain regions (loci) upon the molecule called **active sites.** These are areas of specific
molecular and/or spatial configuration which are
complementary to portions of the substrate molecule
upon which the enzyme acts. In a reaction the enzyme (EN) first binds to the substrate (SUB) at this
active site to form a temporary enzyme-substrate
complex (EN-SUB). During this temporary union the
substrate molecule is activated and then cleaves to
release the enzyme in its original form and the produce or products (PR).

$$EN + SUB \rightleftharpoons (EN\text{-}SUB) \rightleftharpoons EN + PR$$

These reactions are reversible (as indicated by the
dual set of arrows), and whether they go from left to
right or vice versa depends on several factors such as
the concentration of the substrate and products, temperature, and enzyme concentration. Generally the
reaction continues to move in one direction (to the
products) as long as substrate concentration is high
and product concentration is low (or the product is
removed as it is formed).

Current theory of enzymatic specificity suggests
that the active sites on enzymes have a three-dimensional configuration such that only substrate molecules which fit that configuration are accepted. Linkage between the enzyme and substrate is thus
analogous to a **lock and key** relationship (Fig. 2-13).
This classic theory of enzyme substrate specificity

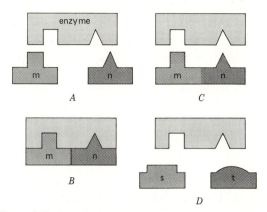

**Figure 2-13 Simplified diagram of enzyme and substrate
interaction.** *A–C.* Two substrate components (*m, n*) fit together on the enzyme, becoming joined (as by dehydration),
and separate as a unit molecule. Other *m* and *n* components
then may follow the same sequence. Alternatively the *mn* compound may be divided by the *C* to *A* sequence. *D.* Two other
nonfitting components (*s,t*) cannot be joined by an enzyme of
the type shown.

does not, however, explain all aspects of enzymatic
action and hence has been modified by addition of
the **induced fit theory.** According to this hypothesis,
when the substrate and enzyme are fitted together
there is a further interaction between the amino
acids of the active site and the substrate which can
further change the configuration of the active site.
This is apparently done to bring certain chemical
groups necessary for catalytic activity into more precise alignment with the substrate. High specificity of
enzymes is thus explained on the basis of three-dimensional spatial complementarity (lock and key)
coupled with correct alignment of reactive groups
(induced fit).

Since the enzyme-catalyzed reaction depends, at
least in part, on a three-dimensional lock-and-key
fit, it is possible for other molecules which are not the
specific substrate but which have the same or a similar three-dimensional configuration to occupy the
site and shut off the system.

Normal action by many kinds of enzymes is essential for the well-being of any organism; disturbance of one or a few may result in illness or death.
Thus, if a person inhales cyanide (HCN), it will combine with metallic ions in certain enzymes of the
body, upsetting their usual function, and the individual is poisoned. Again, certain kinds of bacteria

require para-aminobenzoic acid ($NH_2 \cdot C_6H_4 \cdot COOH$) for their growth. If a person ill from infection by such bacteria is given the sulfa drug sulfanilamide ($NH_2 \cdot C_6H_4 \cdot SO_2NH_2$)—a chemically and spatially related substance—the bacteria will decline, because the alternative substrate (serving as an **inhibitor**) is accepted by their enzymes but does not provide a product required for their needs.

2-20 Nucleic acids These large, complex organic molecules have fundamental roles in protein synthesis and heredity. First identified in 1870 in the sperm of salmon, their chemical makeup and basic biological significance have been demonstrated by many scientists. Two classes are known: **deoxyribose nucleic acid (DNA)**, present only in cell nuclei, and **ribose nucleic acid (RNA)**, found in both nuclei and cytoplasm, especially on ribosomes (Fig. 3-3). Many kinds of each are known, differing primarily in their base sequences (Sec. 2-28).

Nucleic acids are polymers composed of three types of molecules: a 5-carbon sugar, deoxyribose or ribose; nitrogenous organic bases, purines and pyrimidines (ring compounds with 2 or more atoms of nitrogen); and a phosphate group (Fig. 2-14). These parts are joined as follows: sugar + base + phosphate = nucleotide; many nucleotides joined (by polymerization) = nucleic acid; and nucleic acid + protein = nucleoprotein. In DNA and RNA the nucleotides are joined to one another in long chains through linkages between the sugars and phosphate groups, which form the backbone of the chain. The bases stick out from the chain as free side arms. In DNA the purine bases are adenine (A) and guanine (G), and the pyrimidines are cytosine (C) and thymine (T). In RNA thymine is replaced by uracil (U), and the sugar is ribose.

One important discovery about the DNA molecule is that it exists not as a single nucleotide chain but as two chains which are complementary to each other and linked together by hydrogen bonds between adjacent bases. Complementarity is maintained in the molecule by having a certain base hydrogen bonded only to one other base. This is known as base pairing. Thus T always joins to A and G to C. Why does this base pairing occur? Apparently be-

Ribose
(*absent in deoxyribose)

Phosphoric acid

Adenine

Thymine
(*only H in uracil)

Guanine
PURINES

Cytosine
PYRIMIDINES

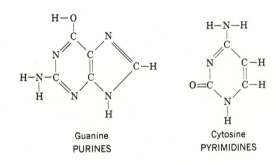

adenine
(nitrogenous base)

ribose
(sugar)

phosphate

Adenylic acid, a nucleotide

Figure 2-14 Components of nucleic acids.

cause of the coiling of the two DNA strands, there remains only a certain distance between the two strands. This distance is such that the two purines (A and G) cannot fit because they are too big and the two pyrimidines (C and T) are too small to allow hydrogen bonding. Thus only a pyrimidine-to-purine spacing is possible. Final specificity is due to incompatibility of hydrogen bonds except between A and T (three each) and G and C (two each). Whenever DNA is separated into its component parts, the number of adenine and thymine molecules is always equal, and the same is true for guanine and cytosine. What this base pairing means is that if the sequence of base pairs in one strand of the two is known, the other is automatically known. This is of great significance for genetics.

The huge DNA molecule resembles a ladder, the sides of alternate sugar and phosphate molecules and the rungs of nitrogenous bases between the sugar molecules, each pair of opposing bases joined centrally by hydrogen bonds (Fig. 2-15). Finally, the DNA molecule is twisted in a compact double helix, or spiral.[1]

DNA molecules from different animal species differ mainly in the sequence of bases. DNA in chromosomes is the genetic material (*genes*) that controls heredity (Sec. 11-18), being capable of reproducing itself in kind (duplication of chromosomes); it also produces RNA, which directs synthesis of proteins (Sec. 2-28). Chromatin of cell nuclei (Sec. 3-5) is rich in DNA, so it is readily stained by basic dyes. DNA is capable of "manufacturing" a duplicate. In this process the two DNA strands unwind or uncoil. Two new complementary strands are then formed by the action of the enzyme **DNA polymerase,** which lines up the appropriate nucleotides to match the complementary ones on each of the original strands and links them together.

Cellular metabolism

All living processes require supplies of chemical substances and energy. Because the many reactions involved in synthesis and breakdown of materials all take place within living cells of plants and animals, the subject of cellular energy relationships is called **cellular metabolism.** During recent decades studies by an ever-increasing number of biochemists and physiologists have revealed many of the fascinating details. Refined research methods such as use of radioactive isotopes, chromatography, and microchemical analyses have made known many of the chemical components, enzymes, and processes involved. Since molecules and their reactions are invisible, most of the evidence is indirect, based on complex physical and chemical tests. There is reasonable proof for some parts, but others remain in the realm of hypothesis. This section outlines the simpler aspects of cell metabolism.

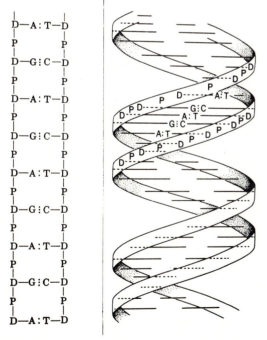

Figure 2-15 *Left.* **Diagram of part of DNA double chain.** D, deoxyribose sugar; P, phosphoric acid; A, adenine; G, guanine; C, cytosine; T, thymine. In RNA the D is replaced by R (ribose) and T by U (uracil). Dots(:) are hydrogen bonds between nitrogenous bases. *Right.* **Part of helix (diagrammatic).** Actually the components are closely compacted.

[1] *The model of the DNA molecule was proposed in 1953 by James D. Watson and Francis H. C. Crick of Cambridge University. Maurice H. F. Wilkins of Kings College, London, clarified knowledge of the DNA fiber. In 1962 the three jointly received a Nobel prize for their spectacular contributions.*

2-21 Metabolism and energy   In essence, the entire living world depends on a series of reversible reactions summarized by the photosynthetic equation:

Sun energy in → $C_6H_{12}O_6$
Glucose
(energy stored)

$CO_2 + H_2O$ → Energy out as ATP

Photosynthesis

Cell respiration

The energy derives from sunlight. Plants transform light energy by the photosynthetic action of chlorophyll into chemical energy and store it in the bonds of organic compounds. In this process carbon dioxide is taken in and oxygen is released. Chemical bonds in the organic molecules represent stored or potential energy. Some of this is used by the plants themselves in synthesizing other of their own necessary components. The energy needed by animals is obtained either directly by feeding on plants or secondarily by eating other plant-consuming animals. Animals use glucose (and other organic compounds) and oxygen—by respiration—to derive most of their energy, giving off CO_2 and H_2O as by-products. The processes in both plants and animals actually are complex, and various other intermediate compounds are involved; the reactions are all enzyme-controlled and rapid.

The syntheses of compounds in organisms are called **endergonic** reactions, because they require energy from a source outside the reacting substances, as in photosynthesis. All other life processes are **exergonic,** because they liberate energy from potential sources in physiologic fuels. Maintenance and repair, muscular contraction, secretion, and other physiologic oxidations are all of the latter type. Such oxidations are the basis of life; death results when they cease.

2-22 Oxidation and reduction   In the burning of coal, fat, or sugar in the presence of oxygen, there are released CO_2, H_2O, and energy as heat—but the reaction occurs at high temperature, and more importantly, the energy is rendered unusable. By con-

trast, oxidation of physiologic fuels such as carbohydrate, lipid, or protein in an animal proceeds at moderate temperature (usually 40°C or lower). The by-products again are CO_2, H_2O, and energy, but relatively little of the energy is lost as heat, and much of it is stored or made available for vital processes. This controlled oxidation results from the passage of the food molecule along a chain of enzymes in which each catalyzes a part of the total oxidation. This sequence usually is divided into two parts: (1) the original molecule is broken down into 2-carbon units, and (2) these are oxidized to carbon dioxide and water. The route, or enzyme chain, which is followed is different in the first part depending on the type of food molecule involved (i.e., whether carbohydrate, lipid, or protein). Once the initial molecule has been broken into 2-carbon units, however, the second phase of the sequence is the same for all. Amino acids are used as energy sources if the animal has not enough carbohydrate and lipid; if used, they may enter the sequence in several places (Sec. 2-27). **Oxidation** of a substance is the loss or transfer of one or more electrons (e^-) to another material. The latter undergoes **reduction** because it gains an electron. Other kinds of reactions also are oxidation, including the removal of hydrogen (even when the hydrogen combines with something other than oxygen).

2-23 ATP and the energy cycle   In many vital processes of plants and animals phosphorus (P) is an essential constituent in either inorganic or organic combination. Human bones contain about 65 percent calcium hydroxyapatite, which is a complex compound of calcium and phosphate. A normal adult excretes 3 to 4 g of phosphoric acid daily, which must be replaced by the food. The calcium-phosphorus ratio in the diet of grazing animals influences their rate of gain. At the cellular level, simple fermentation of sugar by yeast is speeded by adding phosphate, which becomes organically combined in the reaction. Many enzyme-controlled reactions need phosphate compounds in synthesizing body components, in providing energy for muscle contraction, and in other activities. Special enzymes control the addition or removal (phos-

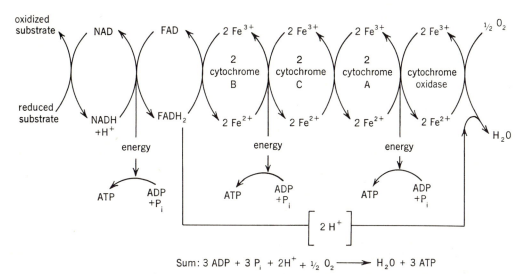

Figure 2-16 The electron transport system. The Fe^{3+} and Fe^{2+} represent the oxidized and reduced states of iron, the active atom of the cytochrome enzymes. Energy is released at the points indicated and immediately trapped in the high-energy bonds of ATP. Of the two hydrogen ions removed from the substrate, one is incorporated in the NAD molecule (NADH) and the other remains free in the surrounding medium as H^+, its electron having been incorporated in NADH. $FADH_2$ releases hydrogen as H^+ ions, the electrons being further transferred via the cytochromes. The H^+ ions are not transferred but remain in the medium until oxygen is activated by the transferred electrons, at which time they join to form water in the presence of cytochrome oxidase. P_i stands for inorganic phosphate (PO_4).

phorylation, dephosphorylation) of phosphoric acid groups.

The central role in the energy cycle in all animals is played by an organic molecule called **adenosine triphosphate (ATP)**. All energy released by respiratory oxidation which is to be made available for use is stored in this molecule or the closely related ADP (adenosine diphosphate), in special energy-rich phosphate bonds. The energy so stored is then available for all energy-requiring reactions of the animal. The ATP molecule (Fig. 2-17) comprises adenine ($C_5H_4N_5$, a nitrogenous base), the sugar ribose ($C_4H_8O_3$), and three phosphoric acid groups ($—HPO_3$), two attached by special high-energy bonds. ADP contains two acid groups and one high-energy bond.

The relationship between ATP and ADP can be expressed by the following reversible equation:

$$ADP + P_i + energy \rightleftharpoons ATP$$

Here P_i refers to inorganic phosphate. The cell can convert ATP back to ADP and thereby produce a controlled release of energy needed for the many biological processes. These processes include the en-

ergy required to initiate the oxidation of glucose and other molecules and the energy needed for running and replicating the cell, producing secretions, and contracting muscle.

In the animal body, ATP is produced in one of two ways, which will become easier to understand if the reader reviews the following explanation (Secs.

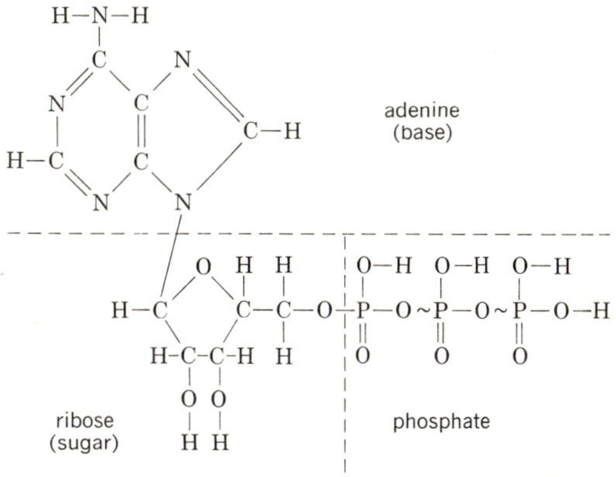

Figure 2-17 Adenosine triphosphate (ATP). The symbol $\sim$ indicates an energy-rich bond. Loss of the outermost $—PO_3H_2$ gives the diphosphate (ADP), with release of much energy; loss of the second phosphate unit leaves the monophosphate (AMP), yielding further energy.

2-24 to 2-27). In the first method, called substrate-linked phosphorylation, ATP is formed directly when ADP reacts with an organic molecule containing a phosphate group held in a high-energy bond. The phosphate group and the high-energy bond are transferred to ADP to form ATP.

A second, or indirect, way, is the process that occurs when pairs of hydrogen ions and electrons which have been removed by enzymatic action (dehydrogenase enzymes) from some organic molecule undergoing oxidation are passed through a special system of linked enzymes known as the **electron transport system.** The enzymes in this system are oxidases called cytochromes. They contain iron and effect oxidation by transfer of electrons (Fig. 2-16). By gaining an electron, the iron of each cytochrome is reduced ($Fe^{3+} + e^- \rightarrow Fe^{2+}$). Then, as it gives up an electron to the next cytochrome in the chain, it is itself oxidized and regenerated to Fe^{3+}. The final cytochrome of the chain transfers its electron to oxygen, which then unites with the hydrogen removed by the dehydrogenases to form water. This entire process generates energy, and this energy is trapped in high-energy phosphate bonds. ADP and phosphate interact with this cytochrome system, forming ATP and trapping that energy for later use by the animal body. In this second method, the phosphate comes from an inorganic compound. Not surprisingly, it is called inorganic phosphate and is abbreviated as P_i. This method of energy trapping is termed **oxidative phosphorylation.**

For each pair of electrons processed completely through this cytochrome system, a maximum of 3 ATP molecules may be generated. In certain cases, however, the hydrogen ions removed from the organic molecules do not pass through the entire system of cytochromes. Instead, they enter part way down the chain. As a result, only 2 ATP molecules are formed during the transfer of electrons. Such a "short-circuiting" occurs once in the Krebs cycle (between succinic and fumaric acid) and results in the formation of a lesser number of ATP molecules than would be expected (Sec. 2-25).

Synthesis of compounds by cells usually requires the presence of ATP or ADP. The linkage of simple sugars to produce disaccharides or starch requires the presence of the sugar phosphate in some cases but only the enzyme (glycosidase) and inorganic phosphate (H_3PO_4) in others. Proteins are formed from amino acids by action of RNA and enzymes but without need of phosphorus.

2-24 Glycolysis, or anaerobic respiration The most common energy source for animals is the carbohydrate glucose or one of its storage polymers such as glycogen. The potential energy of glucose is locked in the bonds between the C, H, and O atoms in the molecule, and about 690,000 calories (cal)/mole (180 g) can be freed upon degradation according to the following equation:

$$C_6H_{12}O_6 + 6O_2 \longrightarrow 6H_2O + 6CO_2 + 690,000 \text{ cal}$$

In the living cell this energy is released in stepwise fashion, each step being regulated by enzymes. The stepwise sequence can be divided into two parts. The first part is anaerobic and is called **glycolysis.** The enzymes for this part are located in the cytoplasm. The end product is a keto organic acid, pyruvic acid. This part recovers only a fraction of the energy. The second part is called the **Krebs cycle** and is aerobic. In this latter part degradation to CO_2 and H_2O is completed and much energy is released (Sec. 2-25). For glycolysis the starting compound is glucose phosphate. Obtaining glucose phosphate from glucose requires the expenditure of one high-energy phosphate bond from ATP. Glycogen, however, produces glucose phosphate directly by reacting with inorganic phosphate, thus saving one high-energy bond.

The subsequent events leading from glucose phosphate to pyruvic acid involve at least six steps, each catalyzed by a specific enzyme (Fig. 2-18). It is important to note that in the second step the original glucose phosphate molecule reacts again with ATP to produce a sugar with two phosphate groups. This is fructose diphosphate. This molecule then splits into two slightly different 3-carbon compounds which are interconvertible. From this point each of the two 3-carbon molecules passes through the enzyme sequence in such a way that two molecules of pyruvic acid are derived for each initial glucose phosphate molecule.

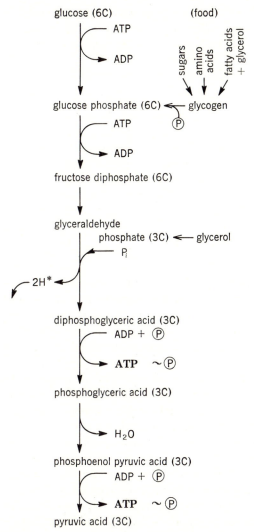

glucose (6C)   (food)

|-- ATP

|--> ADP

sugars amino acids fatty acids + glycerol

glucose phosphate (6C) <-- glycogen
                        ℗

|-- ATP

|--> ADP

fructose diphosphate (6C)

glyceraldehyde
   phosphate (3C) <-- glycerol
            |-- P$_i$

2H* <--

diphosphoglyceric acid (3C)
            |-- ADP + ℗
            |--> **ATP** ~℗

phosphoglyceric acid (3C)

            |--> H$_2$O

phosphoenol pyruvic acid (3C)
            |-- ADP + ℗
            |--> **ATP** ~℗

pyruvic acid (3C)

Figure 2-18 Glycolysis. Starting with the products of digestion (glucose, fatty acids, glycerol, and amino acids) or with stored glycogen, the process of glycolysis proceeds through a series of intermediate compounds to end with pyruvic acid. The number of carbon atoms (C) in each stage is indicated. Points of direct energy production are marked by the symbol ~℗. Derived products are shown in boldface (**ATP**). Splitting of fructose disphosphae results in two molecules of glyceraldehyde phosphate, so that all products thenceforth are doubled—but only one sequence is shown. 2H* atoms go to the cytochrome system in Fig. 2-16. P$_i$ = inorganic phosphate.

After formation of the 3-carbon molecules, each reacts with phosphoric acid to produce a diphosphate compound and is also dehydrogenated (=oxidation) by the action of a special dehydrogenase. In this reaction one phosphate bond in the diphos-

phate compound becomes a high-energy bond; in the next step it is transferred to ADP to form ATP. ATP has trapped some of the energy in the original glucose molecule. The remaining steps in the sequence involve the internal rearrangement of the 3-carbon molecules with the loss of water to make the remaining phosphate bond into a high-energy bond. In the next step this high-energy bond is transferred to ATP, trapping again a fraction of the original energy of the glucose molecule. The remaining molecule is a 3-carbon compound called pyruvic acid.

Formation of pyruvic acid ends the glycolytic sequence. If the sequence was initiated with free glucose, a study would show that 2 ATP molecules were used to yield fructose diphosphate and that 4 ATP molecules were generated in passing from fructose diphosphate to pyruvic acid (2 ATP molecules for each 3-carbon compound), for a net gain of only 2 ATP molecules. If started with glycogen, it would be found that only 1 ATP molecule per glucose unit was used to obtain fructose diphosphate, and the net gain is 3 ATP molecules for each glucose unit. The only loss from the original glucose molecule is 4 hydrogen atoms which are held by by NADH + H$^+$.

The fate of pyruvic acid depends on whether or not oxygen is available. If oxygen is not available or the second half of the sequence is somehow blocked, pyruvic acid becomes the hydrogen acceptor, and the NADH + H$^+$ transfers the 4 hydrogen atoms to pyruvic acid to form 2 lactic acid molecules, and 2 ATP molecules are the net gain.

Usually the second sequence is not blocked and oxygen is available. In this case pyruvic acid is further transformed, and the pair of hydrogen ions held by NADH + H$^+$ is passed on to the cytochrome system (Fig. 2-16), where each pair, in passing to oxygen, generates 3 additional ATP molecules, or a total of 6, making total of 8 net ATPs formed from each glucose molecule during glycolysis.

2-25 Aerobic respiration (the Krebs cycle) When pyruvic acid has been formed, only a small part of the total energy in a glucose molecule has been liberated and trapped in the high-energy bonds of ATP. The final sequence, the citric acid, or **Krebs cycle,** releases most of the remaining energy in the glucose molecule in another series of stepwise enzyme-controlled reactions. The end products are carbon dioxide and water. These reactions occur in the mitochondria of cells.

Before entering the Krebs cycle, the 3-carbon pyruvic acid must first be split into a 2-carbon unit through loss of carbon dioxide (decarboxylation). In this process a special enzyme called **coenzyme A** enters the reaction. It has the vitamin thiamine as a functional group. In the reaction one carbon atom is split off and the remaining 2-carbon or acetyl unit is attached to the coenzyme A molecule to form **acetyl coenzyme A (CoA).** Hydrogen is also removed in this sequence and transferred to the electron transport system, where 3 ATP molecules are generated. Acetyl coenzyme A is a central molecule in energy production in cells because it is the molecule that enters the Krebs cycle. Also it is the end product of the oxidation of fatty acids and some amino acids, and they enter the Krebs cycle at this point (Sec. 2-26; Fig. 2-20).

The first step in the Krebs cycle is the reaction of acetyl coenzyme A with the 4-carbon acid oxaloacetic acid and water to form the 6-carbon citric acid and regenerate coenzyme A. Citric acid then undergoes an internal rearrangement through loss of water at one site and addition at another to form isocitric acid. This acid is then dehydrogenated to form oxalosuccinic acid, which is decarboxylated to yield α-ketoglutaric acid, a 5-carbon acid structurally similar to the amino acid glutamic acid (which can be transformed to α-ketoglutaric acid by loss of the amino group). The α-ketoglutaric acid then undergoes a series of reactions in which it loses hydrogen, is decarboxylated, and joins coenzyme A to form succinyl coenzyme A, which reacts with ADP to form ATP and succinic acid. Succinic acid subsequently loses hydrogen to form fumaric acid, which in turn reacts with water to form malic acid. In the final step malic acid loses hydrogen to generate the starting compound, oxaloacetic acid. Thus the pyrivuc acid molecule has been completely oxidized to carbon dioxide and water during one turn of the cycle.

One turn of the cycle generates 3 molecules of carbon dioxide, uses 3 of water (Fig. 2-19), and pro-

duces 10 hydrogen atoms and 10 electrons (e^-). Only a single ATP molecule is generated directly in the reaction going from α-ketoglutaric acid to succinic acid. Since 2 molecules of pyruvic acid are generated for each glucose molecule, the aerobic cycle must make two turns to completely oxidize all pyruvic acid. Hence the overall equation for this section is

$$2 \text{ Pyruvic acid} + 10O_2 + 30 \text{ ADP} + 30 \text{ phosphate} \longrightarrow 6CO_2 + 4H_2O + 30 \text{ ATP}$$

This shows oxygen being used and 2 molecules of water generated but not used. It also indicates the net generation of 30 high-energy bonds, whereas the cycle generates but 2 directly. How is this discrepancy resolved? The answer lies in the 10 pairs of hydrogen atoms and electrons removed in the cycle. They are transferred to the electron transport system of enzymes, where in passage along the chain 3 ATP molecules are generated for each pair of electrons transferred if the starting point is NADH + H$^+$, but only 2 ATP molecules if the starting point is FADH$_2$ (Fig. 2-16). At the end of the chain the electrons are transferred to oxygen, activating it to react with the 2 hydrogen atoms to produce 1 molecule of water. Since 10 pairs of hydrogen atoms are produced and transferred, 10 water molecules result and 10 oxygen molecules are consumed. Six water molecules are used in the cycle, leaving a net gain of 4, as shown in the chemical equation. The reason for 30 ATP molecules rather than 32 as would be expected (10 pairs of hydrogen molecules transferred = 30 ATP molecules plus 2 produced directly, or a total of 32) is that one pair of hydrogen molecules enters the electron transport system at FADH$_2$, not NADH + H$^+$, thereby generating but 2 ATP molecules for each turn of the cycle (Fig. 2-16). The total energy production, assuming glucose as the initial substrate, is then the sum of the total of ATP molecules generated in glycolysis and in the Krebs cycle. Glycolysis produces 10 ATP molecules, 4 directly and 6 through hydrogen atoms transferred via the electron transport system. However, 2 ATP molecules were used to produce glucose phosphate and fructose diphosphate, so the net gain is 8. From both glycolysis and the Krebs

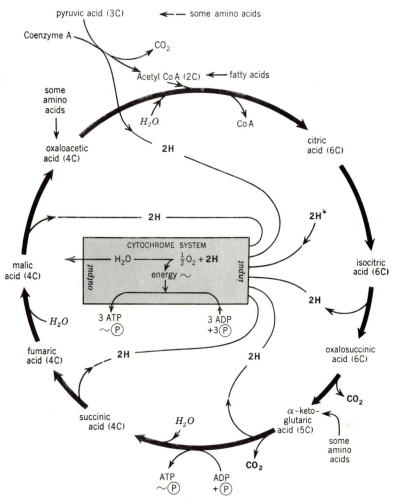

Figure 2-19 Krebs cycle. Pyruvic acid (derived by glycolysis; Fig. 2-18) is oxidized to CO_2 and H_2O to yield energy. Most of the energy is obtained by passage of the H atoms through the cytochrome system to join with O and form water. For every 2H transferred, 3 ATP molecules are generated, each with a high-energy bond (* = 2H from glycolysis). Points of energy production are marked by the symbol ~; trapping of this energy in ATP is indicated by ~Ⓟ. The number of carbon atoms (C) in each compound is shown. Places where fatty acids and amino acids may enter the cycle are marked.

cycle, therefore, there is a net gain of 38 ATP molecules (Table 2-1).

Burning 1 mole (= 1 gram-molecule, 180 g) of glucose yields 690 kilocalories (kcal), and forming 1 mole of ATP from ADP requires about 8 kcal (reversibly available as cell energy output). The cellular efficiency is 8 × 38/690, or about 44 percent—in contrast with 30 percent for a steam engine. The rest is lost as heat. The cell, therefore, is a relatively efficient machine.

2-26 Respiration of fats Carbohydrates such as glucose are not the only source of energy in the cell. Fats and lipids also may be used to produce energy; indeed, complete oxidation of a fat produces more energy than a carbohydrate.

To derive energy from a fat in a cell it must first be broken into its component fatty acids and glycerol. Glycerol may enter the energy production sequence in glycolysis at the point where the original 6-carbon molecule (fructose diphosphate) is split into two 3-

Table 2-1
SUMMARY OF CELLULAR RESPIRATION

Reaction	ATP Production		
	Via Removal of ~$\textcircled{P}$ From Organic Compound	Via Cytochrome System	
		No. of H+ Pairs	ATP Generated
Glycolysis (anaerobic, cytoplasm):			
Glucose + 2ADP + 2P$_i$ → 2 pyruvic acid + 2ATP + 4H+*	2ATP		
Intermediate (aerobic):			
4H+ (from glycolysis)* + O$_2$ + 6ADP + 6P$_i$ → 6ATP + 2H$_2$O		2	6ATP
2 Pyruvic acid + 2CoA + 6ADP + 6P$_i$ + O$_2$ →		2	6ATP
2 acetyl CoA + 2CO$_2$ + 6ATP + 2H$_2$O			
Krebs cycle (aerobic in mitochondria):			
2 Acetyl CoA + 2 oxaloacetic acid + 24ADP + 24P$_i$ + 4O$_2$ + 6H$_2$O →			
2CoA + 2 oxaloacetic acid + 4CO$_2$ + 24ATP + 8H$_2$O	2ATP	8	22ATP
Totals			
Glucose + 6O$_2$ + 6H$_2$O + 38ADP + 38P$_i$ →	4ATP	12	34ATP
6CO$_2$ + 12H$_2$O + 38ATP			

* The 2 pairs of H from glycolysis (anaerobic) cannot generate 6ATP unless transferred to the electron transport system, which is an aerobic system, hence the separation above.
SOURCE: Adapted from Daniel J. Simmons.

carbon molecules (Fig. 2-18). However, the initial stages of the breakdown of fatty acids are different from those of glucose. Splitting fatty acids into 2-carbon units is termed **β-oxidation,** another sequence of enzyme-catalyzed reactions (Fig. 2-20).

The initial step is the reaction of the fatty acid with acetyl coenzyme A, using energy derived from ATP. The activated fatty acid then is dehydrogenated (oxidized) to form a double bond between the second and third carbons of the chain. Next, water is added to remove the double bond and form an alcohol (C—OH) at carbon 3 (the β position). Another dehydrogenation follows, producing a ketone group (C═O) on carbon 3. The final step is reaction with another molecule of coenzyme A, which results in splitting the fatty acid into two fragments, the shorter of which is acetyl coenzyme A, which may now enter the Krebs cycle at the point noted previously (Fig. 2-19). The remainder of the molecule may now repeat the cycle until completely divided into 2-carbon units (Fig. 2-20).

For each 2-carbon unit produced, 4 hydrogen atoms are removed and transferred to oxygen via the electron transport system (Fig. 2-16). This produces 5 ATP molecules. (One transfer is via FAD, thus by-passing NADH + H+, and yields 2, not 3, ATP molecules.) One ATP molecule is necessary to activate the fatty acid molecule, but once activated, the fatty acid molecule can repeat the cycle without further activation. Hence the first 2-carbon unit produces 4 ATP molecules and all the rest 5 for each 2-carbon unit produced. Fatty acids consist of long carbon chains (those in animal fats, such as palmitic and stearic acids, have 16 and 18 carbon atoms, respectively), so that complete splitting of them to acetyl coenzyme A molecules will produce many ATP molecules, 39 for palmitic acid and 44 for stearic acid. For this reason, fats are a better source of energy than carbohydrates.

Once the fatty acid has reached the acetyl coenzyme A stage, it can enter the Krebs cycle. Fats not oxidized are stored; they may also be converted into carbohydrates and stored as glycogen.

2-27 Amino acid oxidation Proteins are broken down into their component amino acids in the course of digestion and absorbed into the body as amino acids. Most amino acids serve in the synthesis of new body protein, but they may enter the aforementioned energy production pathways (Fig. 2-21).

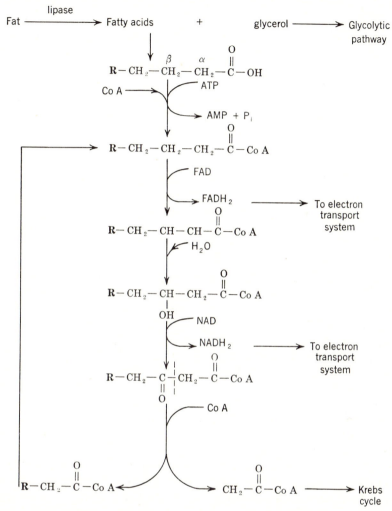

Figure 2-20 β-Oxidation of fatty acids. In this mechanism the fatty acid molecule is split at the second, or β-, carbon atom to generate acetyl CoA, which may then enter the Krebs cycle directly. The remaining molecule is the original fatty acid less 2 carbon atoms. This unit then is recycled until it is completely broken into 2-unit molecules. P_i = inorganic phosphate; **R** = remainder of a fatty acid chain.

First, however, the —NH_2 group must be removed by an enzyme-catalyzed reaction called **deamination.** The end products are ammonia (NH_3) and a particular organic acid called an α-keto acid.

$$R-\underset{\underset{NH_2}{|}}{CH}-COOH \xrightarrow[\text{deamination}]{\text{oxygen}}$$

Amino acid

$$R-\underset{\underset{O}{\|}}{C}-COOH + NH_3$$

α-Keto acid   Ammonia

Ammonia is a very poisonous substance and is excreted by the animal, sometimes after combination with another substance to render it less toxic (Sec. 7-13).

The specific α-keto acids produced by deamination vary with the amino acid undergoing reaction. Some keto acids produced are identical with those involved in the energy cycle and hence may enter the cycle directly. In this way alanine becomes pyruvic acid, glutamic acid becomes α-ketoglutaric acid, and aspartic acid becomes oxaloacetic acid. Others re-

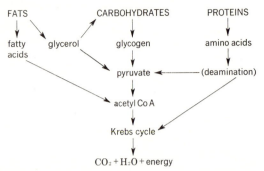

Figure 2-21 **Essential metabolic paths in use of carbohydrates, fats, and proteins (see Fig. 2-19, Krebs cycle).**

quire additional transformations to yield an acid that can enter the cycle.

2-28 DNA, RNA, and protein synthesis About 20 kinds of amino acids are components of proteins. Each type of protein is a distinctive sequence of amino acids. There are many different proteins in any one animal and additional kinds in other animals, so their variety is almost unlimited. During growth and life there is constant replacement of proteins (and other organic substances) within the cells of an individual.

According to current hypotheses, based on much critical research, the primary function of the nucleic acids DNA and RNA is to control the details of protein synthesis. The DNA molecules occur in chromosomes and compose the functional units of the chromosomes, the genes. Each DNA molecule consists of a double-stranded sequence of pyrimidine and purine bases joined to each other through ribose sugar and phosphate groups. The number of base units in a DNA molecule is not known for certain, but estimates range up to hundreds of thousands. The number also varies among different organisms. The sequences of these bases represent a code that is the genetic code that determines the inherited characteristics of the animal. The code is written in an "alphabet" consisting of four "letters" which are the four bases adenine, guanine, cytosine, and thymine. Since the DNA molecules are very long, with many base sequences, the number of "words" which may be produced by the four-letter alphabet is almost limitless. Each "word" specifies a particu-

lar amino acid. The sequence of the words on the DNA strand determines the sequences of the amino acids in the proteins manufactured by the cell. Different genes consist of segments of DNA molecules, with different base sequences yielding different proteins. The total array of proteins which can be produced by a cell thus is determined by the number of genes or ultimately by the complement of base sequences in DNA.

Since DNA is in the nucleus and protein synthesis occurs in the cytoplasm, how does the DNA template for production of protein molecules function? The answer is found in RNA. At least three different types are known: **messenger RNA** (mRNA), **transfer RNA** (tRNA) and **ribosomal RNA** (rRNA). Each of these types functions in a different manner. Messenger RNA is formed in the nucleus on the DNA template from nucleotides. The DNA double strand "unzips" to permit translation of its base sequences in a manner similar to that in which it undergoes DNA replication. In this case only one strand of the DNA is employed as a template to produce a given mRNA molecule. If both DNAs were used, because of the complementarity principle, the two mRNA molecules would produce complementary proteins and chaos would result.

It must be remembered that in nucleic acids certain bases pair only with certain other specific bases: A pairs with T or U and C with G. The sequence of bases on the DNA strand therefore determines the sequence of bases in the mRNA strand formed on it. The DNA strand, in a photographic sense, constructs a "negative" of itself in the mRNA strand. This is catalyzed by the enzyme **RNA polymerase.** The mRNA strand then moves from the nucleus into the cytoplasm, where it usually attaches to or associates with a ribosome. **Ribosomes** are macromolecular units consisting of several proteins and rRNA and are composed of two subunits, a large and a small, which fit together. Ribosomes cluster together on mRNA strands, forming **polysomes.** The ribosomes then each move along the mRNA strand transcribing the "message" held in the mRNA into a polypeptide chain. How is this done?

As noted, about 20 amino acids are present in animal proteins, and the sequence of bases along the mRNA molecule serves as the means for directing the alignment of these amino acids in protein synthesis. How many bases, therefore, are necessary, as a unit, to code a single amino acid and no other? The four bases acting singly (A, G, C, U) can code only four amino acids, which are not enough; similarly the 16 possible doublet combinations (GA, CU, etc.) would fall short of the necessary number of guiding elements. If, however, three bases are used in triplet combinations, then 64 combinations of A, G, C, and U are possible (GGA, GAA, AGG, etc.). Through the use of artificial RNA molecules containing known sequences of bases, it has been found that the assumption that three bases are needed to code a single amino acid is correct. Each amino acid thus is coded by a three-base sequence, or **triplet.** This was first discovered when an artificial RNA consisting only of U was added to a mixture of amino acids in a reaction tube. The resulting peptide consisted only of phenylalanine. Thus it could be concluded that the triplet code for the amino acid phenylalanine was UUU. Subsequent work along these lines has led to the identification of the base triplets for all the amino acids. The codes for all amino acids are now known. Most amino acids are coded by more than one triplet.

The joining of the proper amino acids on the mRNA template requires the last type of RNA, tRNA, also called soluble RNA (sRNA). Each molecule of tRNA is comparatively short, and both ends are functionally important. One end is the same in all tRNA types. It consists of the same three-base sequence, adenine, cytosine, and cytosine (CCA). It is to this "carrier" end that amino acids attach. Each molecule of tRNA is made specific for a single amino acid by a "recognition" site elsewhere on the molecule. The specificity is spelled out in a three-base sequence which is essentially the inverse of the code for that amino acid in mRNA. In phenylalanine, for example, since the mRNA code is UUU, the specific tRNA triplet is AAA; for tyrosine the mRNA code is AUU, and the specific tRNA molecule has UAA on the recognition site.

For the amino acid to attach to its specific tRNA molecule, it must first be activated by reacting with ATP in the presence of an enzyme to form an enzyme-amino acid-AMP complex. (AMP is an abbreviation for adenosine monophosphate. It is formed when ATP loses both of its high-energy-bond

phosphates.) This complex then reacts with the proper tRNA, setting free AMP and the enzyme and attaching the amino acid to the tRNA (Fig. 2-22).

The amino acid-tRNA molecules then are ready to be formed into a protein. Each tRNA-amino acid molecule, as it arrives at the ribosome on the mRNA strand, will be able to bond only to a sequence of three bases on the mRNA which is the inverse of the triplet on its recognition site. Thus, as in the above example, the AAA triplet of the tRNA-phenylalanine complex will be able to bond only to a sequence of UUU in mRNA; similarly a tRNA with a recognition triplet of GAU can bond only to CUA in mRNA. As the tRNA molecules with their amino acids arrive at the ribosomes, they are guided into a sequence of sites on the ribosome where the amino acids are first properly aligned, then formed into a peptide bond with the preceding amino acids, freeing the tRNA, and then, as the ribosome moves down the mRNA molecule, they are moved out to make room for the next amino acid. The final step is an enzyme-catalyzed formation of peptide bonds between adjacent amino acids to form the protein. The tRNA molecules are freed as the peptide bonding progresses so that they may function again in further protein synthesis. Termination of the peptide chain comes when the ribosome in its movement down the mRNA chain encounters a termination code (UAA, UAG, or UGA) or the end of the mRNA strand. At this point the ribosome is released and the completed protein freed. The protein which has been formed has been built according to instructions determined by the sequence of bases in the original DNA molecule.

2-29 Regulation of protein synthesis It would seem apparent that since each cell in an animal body receives the same complement of genes (DNA), each cell should have the potential to make all proteins characteristic of that animal. Yet not all cells produce all the proteins; different cells produce different proteins, and similar cells produce unlike proteins at different times. There must, therefore, be some mechanism operating within the cell which not only turns genes on and off but also regulates the amount

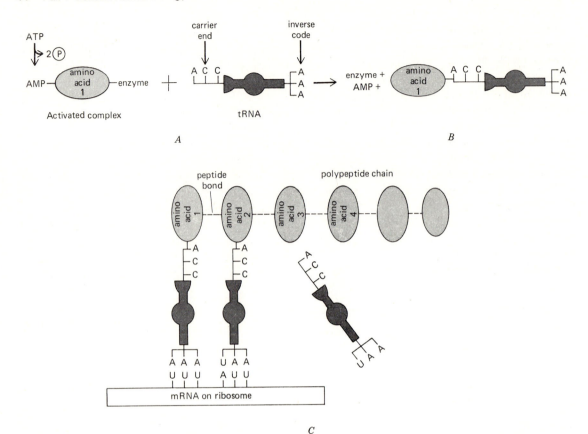

Figure 2-22 Amino acid activation and protein synthesis. Each amino acid is first activated by reacting with an enzyme and ATP to produce an activated complex. The activated complex then reacts with a tRNA molecule which is specific for that amino acid. The tRNA–amino acid complex is then base-paired to the corresponding base on the mRNA molecule on a ribosome. This aligns the amino acids in the sequence that the protein requires. The amino acids are then joined by peptide bonds, the peptide chain is released, and the specific tRNA molecules are free to react with further amino acids. A, C, and U stand for the bases adenine, cytosine, and uracil.

of protein which they produce at any one time. In other words, something must act to **repress** or **induce** protein formation. It has also been demonstrated that in certain organisms when a specific substrate for an enzyme is given, the organism can rapidly produce the enzyme where little existed before. Since all enzymes are proteins, there must also be an explanation for this **induction** of enzyme production.

A hypothesis by Jacob and Monod suggests a mechanism whereby repression and induction of protein synthesis might operate in the cell at the level of the gene. This is the **operon hypothesis.**

The operon hypothesis proposes the existence of two special types of genes, regulators and operators,

in addition to the structural genes responsible for normal protein synthesis. The series of structural genes that produce proteins (enzymes) necessary for any given metabolic function presumably are located next to each other on a chromosome; it is also assumed that associated with each set is a special operator gene. The operator gene must be functioning if the associated structural genes are to be active. The whole unit, structural genes and operator gene, is termed an **operon.** The regulator genes may or may not be adjacent to the operon which they regulate. Regulator genes produce special proteins which act in one of two ways to control the production of protein by the structural genes.

In **repression** the protein produced by a regulator

gene reacts with the end product of the metabolic sequence catalyzed by the enzymes produced by the structural genes. This regulator-end product complex then inhibits the operator gene for the structural proteins; in other words, the operator gene is turned off. As a result the structural genes cease functioning, and the metabolic process ceases. This repression continues as long as the end product is above a certain level, that necessary for formation of the complex. When it falls below that level, the complex disappears, the operator gene again works, and structural genes again produce enzymes which renew the metabolic process. When the end product again builds up, the complex is again formed, and the genes are turned off (Fig. 2-23).

In **induction** the protein produced by the regula-

tor gene itself prevents the operator gene from functioning and the structural genes from producing enzymes for the metabolic sequence. When the substrate for the enzymes produced by the structural genes is introduced into the system, the substrate combines with the regulator gene protein. This releases the inhibition of the operator gene allowing the enzymes for the reaction to be produced and the reaction to continue. This then uses up the substrate. When it is gone, inhibition would again turn off the enzyme production (Fig. 2-24).

Such interaction between regulator genes, operator genes, substrates, and end products saves the cell much energy and enables it to synthesize enzymes only when needed. This is of importance for survival, because a single cell could not accommodate all

Figure 2-23 Repression. (A) Enzymes 1 and 2 are needed to convert substrate into product. (B, C) When there is an excess of product, it unites with the repressor protein controlled by the regulator gene. The repressor product complex turns the operator gene off, stopping further synthesis of mRNA and thus of enzymes 1 and 2. (D) As the cell uses up the product, the P-R complex is depleted, the operon is no longer repressed, and more product is manufactured.

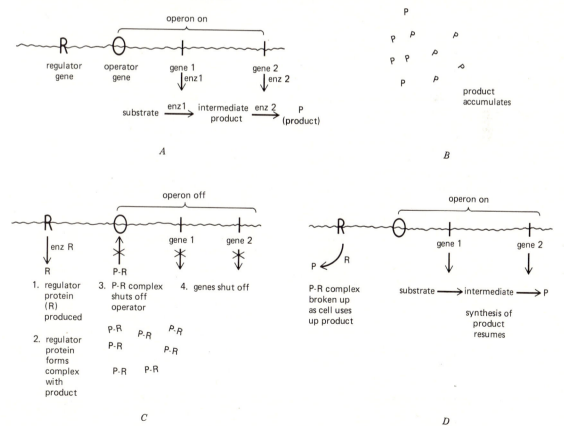

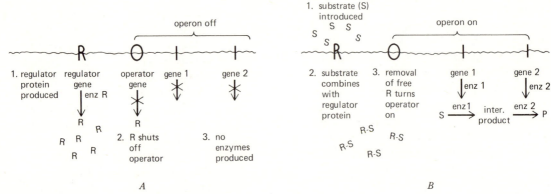

Figure 2-24 Induction. (*A*) In the absence of substrate, repressor protein keeps the operon shut off. (*B*) When substrate becomes available, it combines with the repressor protein, preventing the repressor from shutting off the operon. Synthesis of product starts up.

the enzymes and end products which its full complement of genes might produce if all were active at one time.

The inferences just stated on protein synthesis are derived from studies on certain bacteriophages and bacteria but are believed to represent the situation in higher plants and animals.

The organism is thus an intricate complex of chemical compounds that undergo orderly sequences of enzyme-controlled changes to produce the phenomena we call life. It is simple in fundamental structure, complex in operation, ageless in continuity, and infinite in variety. Most impressive is the principle that simple combinations of four variables outline a genetic code basic for making the proteins essential to survival of all living creatures and their evolution.

References

Baldwin, Ernest. 1962. The nature of biochemistry. New York, Cambridge University Press. xiii + 111 pp., 18 figs. *Simple summary of main subjects.*
_____. 1963. Dynamic aspects of biochemistry, 4th ed.

New York, Cambridge University Press. xxiii + 554 pp., 49 figs., many tables and diagrams.

Barry, J. M. 1964. Molecular biology. Englewood Cliffs, N.J., Prentice-Hall, Inc. 144 pp.

Conn, E. E., and **P. K. Stumpf.** 1976. Outlines of biochemistry. 4th ed. New York, John Wiley & Sons, Inc. vii + 629 pp., illus.

Giese, A. C. 1973. Cell physiology. 4th ed. Philadelphia, W. B. Saunders Company. xxii + 741 pp. illus. *Much physical and biochemical detail.*

Johnson, W. H., and **W. C. Steere** (editors). 1962. This is life: Essays in modern biology. New York, Holt, Rinehart & Winston, Inc. xii + 354 pp., illus. *Plants, animals, bacteria, and viruses.*

Lehninger, Albert L. 1975. Biochemistry. 2d ed. New York, Worth Publishers, Inc. xxiii + 1104 pp., illus.

Speakman, J. C. 1965. Molecules. New York, McGraw-Hill Book Company. vii + 145 pp.

Walsh, E. O'F. 1961. An introduction to biochemistry. London, English Universities Press, Ltd. ix + 454 pp., illus.

White, Abraham, Philip Handler, Emil Smith, Robert Hill, and **I. Lehman.** 1978. Principles of biochemistry. 6th ed. New York, McGraw-Hill Book Company. xi + 1492 pp., illus.

Scientific American in recent years has included many articles, well illustrated, on cells and cellular components and their biochemical processes. Reprints of most of these articles are available in some bookstores.

CELLS AND TISSUES

The basic structural and functional unit of all animals is the cell. Within these minute units the complex biochemical and physical processes which allow life to continue are carried out. Cells are organized into tissues and tissues into organs and organ systems. The latter are discussed in Chaps. 4 to 9, while the physical and chemical reactions occurring within cells have been considered in Chap. 2. The present chapter deals with the organization and function of the various cell components and the types of tissue present in animal bodies.

All the many different parts of the body, large and small, combine to act in a coordinated manner and perform the processes essential to life. Unlike inanimate chemical systems, the living organism is an open system in that there is continual input and output of matter to provide energy and also the ma-

terials for growth and maintenance. The internal environment of each part, however, is in a balanced "steady state" that changes but slightly (Sec. 7-16).

Cells and protoplasm

3-1 History The finer structure of living matter remained unknown until after invention of the compound microscope. In 1665 Robert Hooke reported that thin slices of cork and other plant materials contained minute partitions separating cavities that he named **cells.** Then for nearly 150 years little was added to the subject. In 1808 Mirbel concluded that plants are formed of "membraneous cellular tissue," and Lamarck in 1809 stated that any living body must have its parts of cellular tissue or formed by such tissue. Robert Brown in 1833 described the nucleus as a central feature in plant cells. In 1838 M. J. Schleiden put forth the thesis that cells were the unit of structure in plants, and the next year Theodor Schwann applied this thesis to animals. The latter author indeed wrote of the combination of molecules to form the cell (structure) and the chemical changes which they undergo (metabolism). In 1840 Purkinje named the cell contents **protoplasm.** Thus attention shifted from the cell wall to the far more important contents. Until recently, the term *protoplasm* was used to describe the living substance within a cell. Now, with our knowledge of the complex subcellular organelle system in a cell, the term has ceased to have widespread usage and is more often used as a general term to encompass the whole array of organelles and compounds within a cell. The modern cell theory states that *all animals and plants are composed of cells and cell products.* In all organisms the cell is the fundamental unit, structural and physiological, and there is constant exchange of matter within and between cells in the process of living. In multicellular animals the cells are integrated for proper functioning between body parts, whereas in unicellular animals the cell and organism are one. A multicellular animal generally starts life as a single cell that divides repeatedly to form its body (Chap. 10).

3-2 Study of cells   The study of cell structure and activity is called **cytology** (cell biology). Since most animal cells are minute, the unit of measurement is 1/1000 mm, a **micrometer** (symbol μm); for the finest details the **nanometer** (symbol nm) is most often used (1000 nm = 1 μm; 1 nm = 10 Å). The size and form of cells vary, but each kind in a given species of animal is fairly uniform (Fig. 3-1). Among the smallest known animal cells are the spherical blood cells of the mouse deer (*Tragulus*), with a mean diameter of 1.5 μm. Human red blood cells average 7.5 μm, and many other cells are from 10 to 50 μm. The largest cells are the yolk of bird and shark eggs, about 30 mm (a little over 1 in) in diameter in the chicken (Fig. 10-7).

The human eye can only resolve (distinguish) two points about 0.1 mm (0.004 in) apart, and living

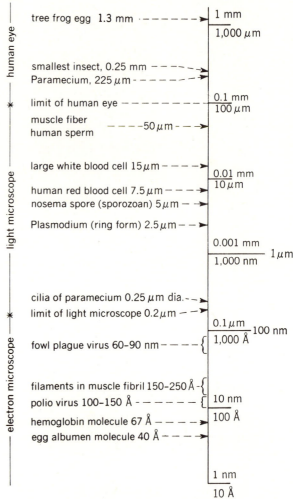

Figure 3-1 Relative sizes of animal and protistan cells and some parts of cells. Each major scale division is $\frac{1}{10}$ of that above. 1 millimeter (mm) = 1000 micrometers (μm; formerly microns); 1 micrometer = 1000 nanometers (nm; formerly millimicrons); 1 nanometer = 10 angstrom units (Å). Visual microscope magnifies about 10 to 2000 times; electron microscope, about 5000 to 100,000 times or more.

cells are essentially transparent under ordinary light. Therefore cell details must be made to have more contrast, and some kind of microscope must be used to examine them (Fig. 3-2). For much cytologic study with the light microscope, cells or cell groups (tissues) are killed and fixed by chemical reagents or frozen, cut into thin sections (10 μm or less), mounted on glass microscope slides, and stained to

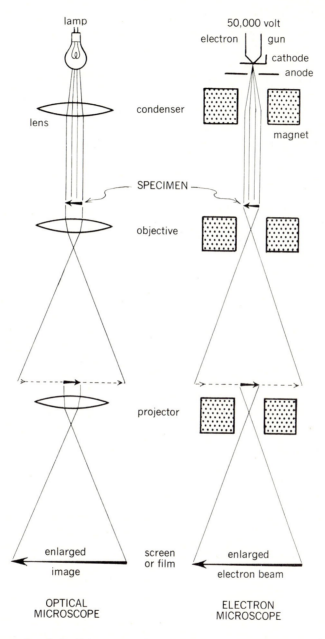

lamp

50,000 volt

electron | gun

cathode

anode

condenser

lens

magnet

SPECIMEN

objective

projector

enlarged
image

screen
or film

enlarged
electron beam

OPTICAL
MICROSCOPE

ELECTRON
MICROSCOPE

Figure 3-2 *Optical microscope* **(inverted).** Light beam from lamp concentrated by condenser lens to illuminate specimen; image focused and magnified by objective and again by projector (eyepiece); enlarged image seen on viewing screen or photographed on film. *Electron microscope.* Narrow beam of high-speed electrons from cathode passes through hole in anode and is concentrated by doughnut-shaped magnetic coil (condenser) to pass through specimen; emerging electron pattern focused and magnified (second magnet) further enlarged (third magnet); "image" either seen on fluorescent screen or recorded on film. Entire instrument enclosed in high vacuum, because electrons are scattered in air.

differentiate the internal details. Care is necessary that these procedures do not alter the original structure. Living cells may be treated with "vital dyes" that stain and render parts more visible without injury. Highly delicate apparatus for microdissection permits removing or adding parts or the injection of reagents for special studies. Isolated cells or bits of tissue may be kept alive and cultured for days (or years) in special chambers provided with nutrient solutions.

The best optical microscope, with a central beam of concentrated light, will resolve parts down to about 0.2 μm. In a dark-field microscope the object is illuminated only from the sides (no central light), and structures within cells appear bright against a dark background. The phase-contrast microscope takes advantage of slight differences in refractive index between cell structures by converting these to striking brightness differences in the visible image of the cell. This is accomplished by artificially shifting the phase of a portion of the light which has passed through the specimen. The resolution of fine detail in transparent living cells studied by this method is comparable to that obtainable by ordinary microscopy only with fixed and stained material.

The image on any of these types of microscope can be seen by the human eye or recorded on photographic film. Far greater magnification and resolution, revealing the ultrastructure of cellular details, is possible with the electron microscope. Instead of light, a high-voltage stream of electrons, emitted by a heated cathode, is focused in a narrow beam by electromagnets—in a manner analogous to the focusing of light by glass lenses. The study sections must be very thin (0.1–0.05 μm = 1000–500 Å) and supported on an even thinner layer of collodion. All parts of the instrument are enclosed in a vacuum chamber. A nonvisible image results from differences in electron dispersion as the beam passes through the object, according to the thickness, density, and chemical (atomic) composition of its parts. The image can be made visible on a fluorescent screen or may be recorded photographically (as a micrograph). The light microscope has an upper limit of magnification of 1500 to 2000 times, whereas that of the transmission electron microscope is 100,000 times or more; and the micrograph then can be enlarged 10 times.

The ultimate means of cellular analysis is x-ray diffraction. A parallel beam of x-rays is passed through an object (nerve, hair fiber, etc.), which spreads (diffracts) the rays according to its structure. An image of the diffracted pattern (rings, concentric bands, or spots) is recorded photographically. By measuring the distances between repeating parts and using certain mathematical calculations, the structure and size of molecules, in nanometers, can be learned!

Cell structure

Cells are of two different types. **Prokaryotic** cells are those which lack a well-defined nucleus and most cell organelles. They are characteristic of the phylum Monera (Chap. 14) and are not considered here. All animal cells are **eukaryotic,** meaning that they have definite nuclei as well as the cell organelles.

Studies of cells formerly dealt mainly with their physical features as seen in thin stained sections. In recent years new methods of study and new tools of research such as the electron microscope and freeze-etching techniques have been devised which have allowed us to learn the subcellular structure as well as the reactions constantly in progress in every living cell. The tiny cell is an amazing unit of complex organelles in which many chemical substances undergo a wide variety of interaction and change—synthesis of new materials, use of food and energy to provide for movement, secretion, or other activities, and rendering of waste products into forms not harmful. Any cell is at least as intricate as an entire petroleum refinery that receives the mixture of hydrocarbons in petroleum, refines and modifies some for fuel and lubricants, and synthesizes many new and different organic compounds to serve various purposes in our modern everyday life. The following section outlines the subcellular structure of a cell.

3-3 Cell membranes Thanks to the electron microscope and sophisticated biochemical work, we now know considerably more about the intracellular organization of a typical animal cell than we did 15 to 20 years ago. The animal cell (Figs. 3-3, 3-4) is bounded externally by a **cell membrane,** or **plasma**

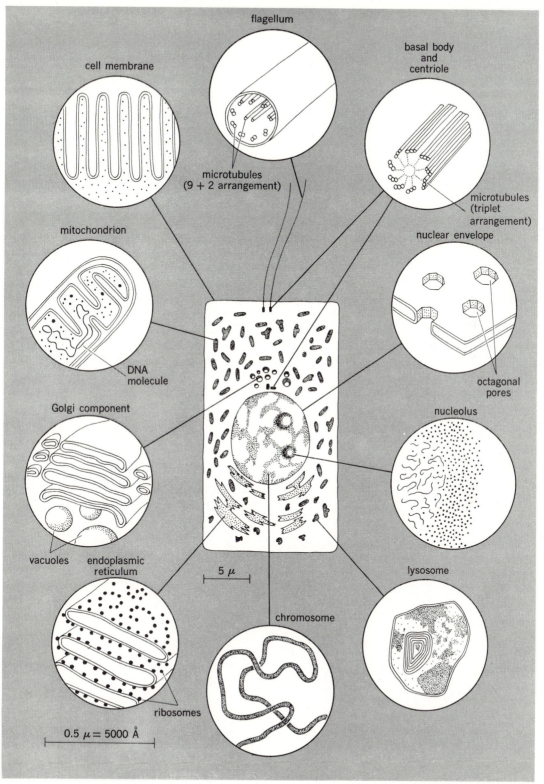

Figure 3-3 Schematic diagram of an animal cell. Not all parts shown will be present or evident in any one cell, either living or fixed and stained.

membrane. This is a structure which separates the cell from the surrounding environment and acts as a selective filter for materials attempting to pass into or out of the cell. Typically it is about 8 nm thick. Biochemical studies reveal that the membrane is composed of two types of molecules, proteins and lipids. The lipids are of the special class called phospholipids. In good transmission-electron micrographs the membrane appears as a pair of parallel black lines enclosing a central lighter area (Fig. 3-6). It was first suggested that the structure of this three-part membrane was one in which layers of lipid (the light central area) were sandwiched between inner and outer layers of protein (the inner and outer black bands). The phospholipids of the center band are molecules which are asymmetrical in that they carry an electric charge at one end which is hydrophilic, while the other end is uncharged and hydrophobic. Given these biochemical features, it was also soon suggested that the lipid inner layer was not single but a double layer in which the uncharged ends of the lipids abutted each other in the center

while the charged ends faced to the outside and inside, where oppositely charged protein molecules attracted them, giving the necessary stability (Fig. 3-5). The above tripartite arrangement of the plasma membrane has been termed the **unit membrane** model. More recent work on the cell membrane using freeze-etch techniques has suggested that the structure may not be as simple as that proposed by the unit membrane model. In particular, it has been demonstrated that the protein molecules penetrate into the lipid layers, often entirely through them, and that they are not continuous on the surface of the lipid (Fig. 3-7). This has led to a second, more recent, model of the structure of the membrane, known as the **fluid mosaic** model. In this model the large protein molecules occur as isolated, various-size globules embedded in the lipid layer or closely associated with the inner and outer surfaces (Fig. 3-5).

Cell membranes exist to regulate the movement of material in and out of the cell. But how do materials cross a seemingly solid barrier? Some ma-

FIGURE 3-4   Transmission electron micrograph of a cell and components: rat liver cell ×10,175. Note the large number of mitochondria. M, mitochondria; E, endoplasmic reticulum; N, nucleus; C, chromatin; Nc, nucleolus. (*Courtesy of Dr. John Belton, California State University, Hayward, Calif.*)

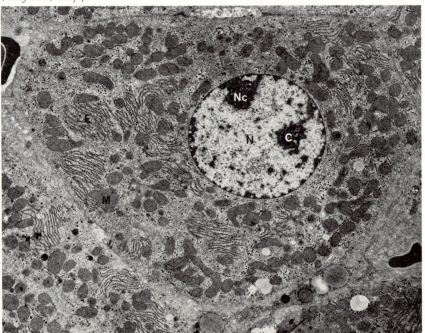

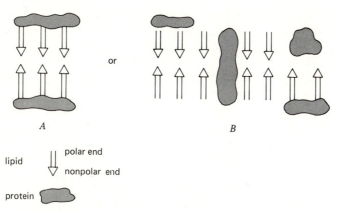

Figure 3-5 **Diagrammatic suggested structures of the cell membrane.** *A.* Unit membrane model. *B.* Fluid mosaic model.

terials, such as water and certain small molecules, pass readily through membranes as through a sieve, suggesting the presence of pores. Although pores have not been observed under the electron microscope, small ones may exist. Other materials are lipid-soluble and diffuse through the lipid layers, but many other molecules are too large to pass through pores and are not soluble in lipid. It is thought that they pass through the membrane by a process of **facilitated diffusion** in which the molecule seeking passage becomes associated with a carrier, possibly a membrane protein, which transports it across the membrane (Fig. 3-7). In all the above cases substances are impelled by diffusion (moving from an area of high concentration to one of low concentration), and no energy is required. Substances may, however, move against a concentration gradient, a process requiring energy. Such movement occurs by **active transport,** a process resembling faciliated diffusion in that it involves association of molecules to be transported with membrane protein carriers. The only difference is that materials are moved from an area of low concentration to a high one.

For those molecules, such as large proteins, which cannot cross membranes via any of these transport mechanisms, the final mechanism seems to be what is called **pinocytosis.** This is similar to phagocytosis, a process in which cells engulf particles. Here the molecule comes to rest on the surface of the membrane and the membrane folds over it, enclos-

ing it in a vacuole or vesicle which moves into the cell.

3-4 Mitochondria Universally present within animal cells (except mature red blood cells) are spherical to rod-shaped structures from 0.2 to 7 μm in size called **mitochondria** (sing. mitochondrion). The number per cell varies with the cell type, sperm cells having as few as 25 and liver cells up to 1000. Structurally each mitochondrion consists of a smooth outer membrane surrounding a complexly folded inner membrane and separated from the latter by a narrow space. The complex folding and projections of the inner membrane are termed **cristae** (Fig. 3-6). Between the cristae is the matrix of the mitochondrion. The folding of the cristae provides increased surface area for enzyme activity. The number of cristae varies according to the activity of the mitochondrion.

Mitochondria occupy an extremely important position in the animal cell because they are the energy-converting organelles. The inner membrane system of the mitochondria is the location of the enzyme systems for Krebs cycle and the electron transport system (Secs. 2-23, 2-25). It is only here that these systems are found, hence all energy production for the cell occurs here.

Mitochondria also contain DNA, the only occurrence of DNA outside the nucleus, and RNA as well as ribosomes. Hence the mitochondrion also con-

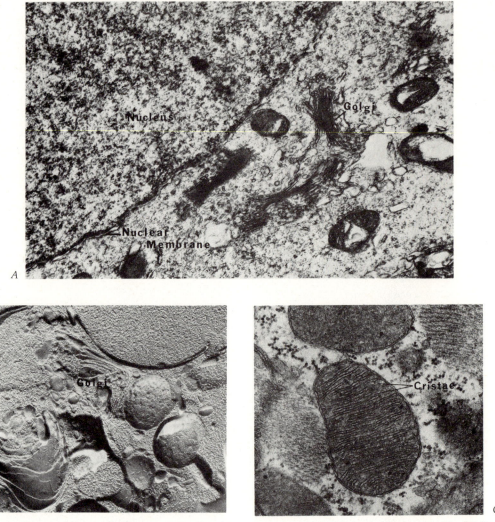

Figure 3-6 A. Transmission electron micrograph of a section of a mouse oocyte cell, ×17,020, showing the structure of the Golgi complex and the nuclear membrane. B. Freeze-fracture preparation of a rat lung cell showing the Golgi complex. ×47,520. C. Enlarged transmission electron micrograph showing the structure of a single mitochondrion; note the cristae. ×52,500. (*Courtesy of Dr. John Belton, California State University, Hayward, Calif.*)

tains the biochemical machinery to manufacture proteins. There is some evidence that certain mitochondrial proteins are produced there.

3-5 The nucleus The most prominent structure visible in an animal cell in a low-power light microscope is the nucleus. With few exceptions, all animal cells have a single nucleus. Those which lack them, such as mammalian red blood cells, have short

life spans. A few have more than one nucleus. Usually it is a rounded structure separated from the rest of the cell by a double membrane called the **nuclear envelope** (Fig. 3-7). The outer membrane is often continuous with the membrane system of the endoplasmic reticulum. The nuclear envelope is penetrated by pores; these may facilitate passage of large organic molecules between the nucleus and the rest of the cell, but this has not been conclusively established (Fig. 3-7). The interior of the nucleus contains

Figure 3-7   Freeze-fracture micrographs of the structure of the nuclear envelope of a cell. *A.* Nuclear envelope showing the large nuclear pores. $\times 27,600$. *B.* Enlarged view of the nuclear envelope showing nuclear pores on the inner part of the membrane. Note the particles showing on the face of the membrane. They are probably proteins of the membrane structure. $\times 108,000$. NP, nuclear pore; Mit, mitochondrion; OM, split through the middle of the outer nuclear membrane. (*Courtesy of Dr. John Belton, California State University, Hayward, Calif.*)

a network of dark-staining fine threads called **chromatin** (Fig. 3-4). Chromatin consists of the DNA-containing structures within the nucleus, which, during cell division, coil up and condense into the identifiable units called **chromosomes** (Fig. 3-9). Also in the nucleus are one or more dense spherical objects called **nucleoli** (sing. nucleolus). Each nucleolus is composed of specialized areas of certain chromosomes called nucleolar organizers, ribosomal RNA, and molecules of ribonucleoprotein, which is a combination of RNA and protein and a probable precursor of ribosomes. The nucleolus is thus concerned with the construction of ribosomes. These ultimately leave the nucleus and are organized on the endoplasmic reticulum as part of the protein-producing apparatus of the cell. The nucleolus disappears at mitosis and is largest in the interphase cell.

The nucleus functions as the control center of the cell, initiating and monitoring all functions. In the chromatin it also contains the genetic material responsible for the continuity of successive generations of cells.

3-6 Endoplasmic reticulum Between the nuclear envelope and the outer plasma membrane of the cell there is a network of membranes forming what is called the endoplasmic reticulum (Fig. 3-4). Two types are recognized, one with ribosomes attached to the membrane surfaces, called **rough endoplasmic reticulum,** and one without, called **smooth endoplasmic reticulum.** The endoplasmic reticulum is continuous with the nuclear envelope. The ribosomes bound onto the rough endoplasmic reticulum are the sites of protein synthesis in the cell. Proteins formed on these ribosomes are believed to be able to pass to the outside via the cavity of the endoplasmic reticulum. Thus rough endoplasmic reticulum is important not only in the protein synthesis but also as a means of communication, channeling products both to the outside of the cell and, via the membrane network, to other parts of the cell. Smooth endoplasmic reticulum, which often has a tubular appearance under the electron microscope, seems to be important in synthesizing and secreting certain steroid hormones, enzymes of carbohydrate metabolism, and enzymes of lipid synthesis. Both rough and

smooth endoplasmic reticulum seem to be continuous within a cell and may be part of a single membrane system in which smooth endoplasmic reticulum may be derived from rough endoplasmic reticulum.

3-7 Golgi complex A final system of membranes found in most animal cells consists of a series of from 3 to 20 parallel flattened sacs closely stacked together to form what is called the **Golgi complex** (Fig. 3-6). The ends of the sacs of the Golgi complex tend to bud off or pinch off various vesicles, which then move to the plasma membrane.

The function of the Golgi apparatus appears to be one of packaging and moving to the outside materials produced in the cell—a role in cell secretion. The major material moved is protein produced by the endoplasmic reticulum, and there is a close relationship between the endoplasmic reticulum and the Golgi complex. In the secretory process, proteins (enzymes) for secretion are produced by the ribosomes of the endoplasmic reticulum, pass into the cavity of the endoplasmic reticulum membrane system, and travel into the Golgi complex via an assumed fusion of the Golgi complex and the endoplasmic reticulum. Once in the Golgi complex the proteins are concentrated in the ends of the flattened sacs. These then pinch off and move to the cell membrane, where the vesicles fuse, discharging the contents to the outside.

3-8 Other organelles In addition to the above major cell organelles, a number of others are often found. **Lysosomes** are small bodies surrounded by a single membrane and containing the enzymes (lysozymes) which degrade material in the cell. These bodies act as waste-disposal units, digesting and removing foreign material brought into the cell from outside or removing cellular organelles which are not needed. Lysosomes are really only one type of a whole class of small single-membrane-bounded organelles called **microbodies,** which contain specific enzymes and which probably originate in the endoplasmic reticulum.

Microtubules are tiny cylindrical elements of animal cells about 20 to 25 nm in diameter formed by molecules of the protein tubulin. Microtubules form the spindle fibers of dividing cells, function as cellular skeletal elements in some protozoans, and form the major formed elements of cilia, flagella, and centrioles (Fig. 3-18). **Centrioles** appears in some animal cells and have a structure similar to cilia and flagella. They were formerly thought to be vital in spindle formation in mitosis, but this appears now not to be the case. They are prominent only in mitotic cells and would seem to function somehow in mitosis.

Microfilaments are smaller than microtubules, ranging in diameter from 4 to 7 nm. Biochemically they are constructed of protein molecules similar to those constituting the contractile elements of muscle (Sec. 4-11). Microfilaments appear to be of major importance in providing motive force for cell contraction, amoeboid movement, and possibly intracellular transport.

The cell cycle

Like whole organisms, cells also have a cycle. They are born, live for a while, and then reproduce. This cycle can be conveniently divided into two parts, a period of growth and preparation for reproduction called interphase and a reproductive period called mitosis. Growth in multicellular organisms is chiefly by multiplication of cells. Cells multiply chiefly by **mitosis,** a rather complex process that involves, importantly, an equal division of the nuclear chromatin in kind and amount (Figs. 3-10, 3-11). Cell division by mitosis is common to all animals and all plants. Mitosis is active during embryonic development, in growth, in repair of injury, and in replacement of body covering at molting. It is also the process involved in malignant growths (tumors, cancer).

3-9 Interphase   As seen in living cells, the cell cycle is a continuous dynamic process in which mitosis is the most dramatic event and also the one that represents the smallest time frame in a complete cell cycle (Fig. 3-8). The longest part of the cycle is **interphase,** the period between the end of one mitosis (telophase) and the initiation of another (prophase). A number of events occur during interphase, the

most significant being the replication of the genetic material (DNA). Recent cytochemical studies have shown that this interphase period may be divided into three parts. The first, termed the **G_1 phase,** immediately follows the end of a previous mitotic division (Fig. 3-8). This first phase is the longest of the three interphase periods and is characterized by growth. The newly produced daughter cell increases in size and undergoes internal chemical changes which somehow prepare it for DNA replication (Sec. 2-20). The duration of the G_1 phase is variable, depending on the type of cell and its level of nutrition. It may be as short as a few hours or as long as several days.

Immediately following G_1 comes the **S phase.** It is in this period that DNA replication or synthesis occurs in the cell. When the replication has finished, this phase of the cell cycle is completed. The period between the completion of DNA synthesis and the beginning of active mitosis (prophase) constitutes the final or **G_2 phase** of interphase. The duration of the S and G_2 phases is again variable, depending on the cell, but both are shorter than G_1 and are usually

Figure 3-8 The cell cycle (diagrammatic), showing relative time spent in each phase.   2N, normal DNA content; 4N, double DNA content.

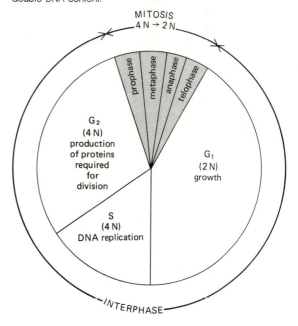

measured in hours. During G_2 certain proteins are produced which are necessary to cell division, and mitosis is arrested in this phase if protein synthesis is stopped. Although DNA synthesis is the most obvious event of this interphase cycle, there are undoubtedly a great many other internal chemical changes in the cell which signal the initiation or the cessation of certain events. This can be demonstrated by fusing cells in one phase with earlier and later phases. Thus, for example, fusion of S-phase cells with G_1 cells causes G_1 cells to begin to make DNA before they would normally do so. This suggests some chemical trigger in S-phase cells which is absent from G_1 cells and about which we are currently ignorant. Similarly, fusion of interphase G_2 cells with cells in active mitosis results in premature condensation of chromosomes.

A final event in interphase occurs with the replication of the centrioles and their movement to a position such that there is a pair at either end of the nucleus. Simultaneously, as the centrioles separate, microtubules begin to appear radiating from the center, or **centrosome,** area where the two centrioles are located. These microtubules will form the spindle, with the centrioles at each pole.

3-10 Mitosis   The function of mitosis is to physically divide the cell into two daughter cells and ensure that each has exactly the same complement of genetic material (DNA). As with interphase, mitosis is also subdivided into separate phases (Fig. 3-8). Whereas in interphase the major changes are internal chemical changes which are not manifest in the appearance of the cell or its organelles (except for size), the changes associated with mitosis affect the cell organelles and are marked by definite morphological changes.

Prophase is the initial stage of mitosis (Fig. 3-10). It is marked by condensation of the dark-staining chromatin granules of the nucleus into visible long threads called chromosomes. Each thread consists of two intertwined parallel longitudinal subunits called **chromatids.** During the remainder of prophase these chromosomes continue to condense into shorter, stubbier units. As they become shorter and fatter, a small, less condensed, lighter area called a **centromere** (kinetochore) becomes visible on each (Fig. 3-9). Nucleoli become dispersed, the material becoming associated with certain chromosomes, and the nuclear envelope breaks down, freeing the chromosomes. Before the breakdown of the nuclear enve-

Figure 3-9   The human chromosomes grouped according to size, position of centromere (white ring), and possession of satellites.   Autosomes (1–22) and sex chromosomes (X, Y). Each chromosome appears distinctly double owing to treatment with colchicine. (Male, diploid, metaphase, × 3,000) *(After Tjio and Puck, 1958, Proc. Natl. Acad. Sci.)*

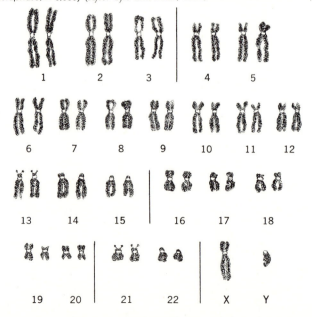

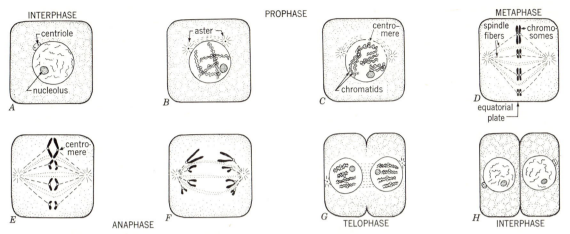

Figure 3-10 Mitosis: division of nuclear materials and cytoplasm of one cell into two. Diagrammatic.

lope, the formation of the **spindle** has been completed in the region of the nucleus. The spindle is a structure consisting of bundles of microtubules derived from the protein tubulin. One set of microtubules runs from pole to pole in the spindle, while another runs from one pole to attach to each chromosome by the centromere. Thus disintegration of the nuclear envelope places the chromosomes on the spindle, where each is attached to a spindle fiber.

The next stage of mitosis is termed **metaphase.** Here the chromosomes become aligned midway between the two poles of the spindle so that they lie in a plane called the **equatorial plate.** At this stage the chromosomes have reached maximum thickness.

In the next, or **anaphase,** stage, the aligned double chromosomes begin to separate. The sister chromatids pull apart on the spindle fibers and head for opposite poles (Fig. 3-10). The genetic material in the chromosomes, which was doubled during the S phase of interphase, is now halved so that each daughter cell has the same complement of DNA as did the G_1 cell.

The last phase, **telophase,** begins when the groups of daughter chromosomes end their polar movement and the centromeres reach the spindle poles. A new nuclear envelope forms and surrounds each set of daughter chromosomes; a new nucleolus forms. Finally, the original cell body itself divides, forming a new membrane between the two sets of daughter chromosomes, a process called **cytoki-**

nesis. The spindle then disintegrates and mitosis is over.

The duration of a complete mitotic cycle depends on the species of animal, the age of the individual, the particular tissue involved, the temperature, etc. The total time of a mitosis has been determined as 9 min for *Drosophila* cleavage and 67 to 205 min for chick mesenchyme cells. In both cases the longest stage was the prophase.

3-11 Significance of mitosis The duplication of the chromatin in interphase, followed by mitosis in which each daughter cell receives exactly the same kind and amount of genetic material as that in the parent, is of great significance in that it ensures a continuous succession of identical cells. If, during the growth process, successive cells received genetic material in a random way so that each was differently organized, growth and cell differentiation into tissues and organs in metazoan (multicellular) animals could not be effected and chaos would result. Mitosis is different from **meiosis** because in meiosis (Sec. 10-5) the genetic material of the parent is halved in each daughter cell. Such division of genetic material occurs only in reproductive tissue.

3-12 Amitosis Direct nuclear division (amitosis) occurs in some cells, including certain ciliates

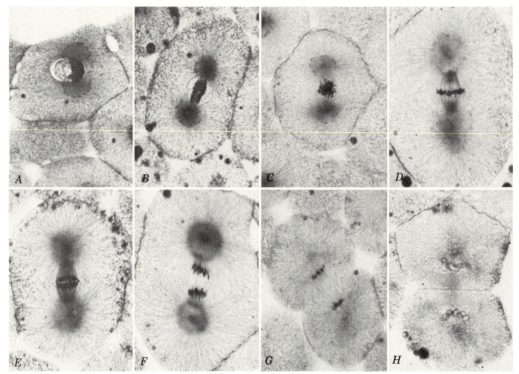

Figure 3-11 Mitosis in egg (blastula) of whitefish. PROPHASE. *A.* Centrosome divides. *B, C.* Centrosomes at opposite poles, chromosomes becoming evident, nuclear membrane disappears. METAPHASE. *D, E.* Chromosomes centered on equator of spindle; in *E,* each divides into two. ANAPHASE. *F, G.* Chromosomes move toward poles, spindle lengthens, centrosomes less evident. TELOPHASE. *H.* Nuclear membrane forming around chromosomes, cytoplasm of the two cells separated by membrane between. (*Photomicrographs by Dr. Hans Ris.*) Compare Fig. 3-10.

and suctorians. The nucleus separates into two parts without forming spindles or chromosomes. In higher animals many cases of cells undergoing amitotic division have been reported under pathologic conditions.

Tissues

The parts of any multicellular animal consist of different kinds of cells. Those of similar structure and function are arranged in groups or layers known as **tissues,** hence multicellular animals (METAZOA) are "tissue animals." In each tissue the cells are essentially alike, being of characteristic size, form, and arrangement, and they are specialized or differentiated, both structurally and physiologically, to perform some particular function such as protection, digestion, or contraction; thus there is a division of

labor between different tissues. **Histology,** or microscopic anatomy, is the study of the structure and arrangement of tissues in organs, in contrast to gross anatomy, which deals with organs and organ systems by dissection (Fig. 3-12).

The cells in a multicellular animal may be divided into (1) **somatic cells,** or **body cells** (and their products), constituting the individual animal throughout its life, and (2) **germ cells** having to do only with reproduction and continuance of the species (Chap. 10). There are five major groups of somatic tissues: (1) epithelial, or covering; (2) connective, or supporting; (3) vascular, or circulatory; (4) muscular, or contractile; and (5) nervous.

3-13 Epithelial tissues These cover the body, outside and inside, as in the skin and the lining of the digestive tract (Fig. 3-13). The cells are compactly

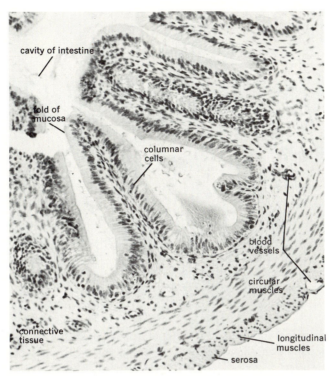

Figure 3-12 Section of frog intestine (duodenum). Photomicrograph shows how several kinds of cells and tissues combine to form an organ.

placed, bonded together by intercellular cement for strength, and often supported below on a basement membrane. Structurally the cells may be (1) squamous, or flat, (2) cuboidal, (3) columnar, (4) ciliated, or (5) flagellated. The tissue may be either (6) simple, with the cells in one layer, or (7) stratified, with multiple layers. A special case is the pseudostratified: nuclei occur in two or more levels, thus appearing to be stratified, but all cells are attached to the basement membrane (Fig. 3-13). Functionally an epithelial tissue may be protective, glandular (secretory), sensory, or be involved in absorption.

Simple squamous epithelium is made up of thin flat cells, like tiles in a floor; such cells form the peritoneum that lines the body cavity and the endothelium of the inner surface of blood vessels in vertebrates. Stratified squamous epithelium forms the outer layers of the human skin (Fig. 4-1) and lines the mouth and anterior portion of the nasal cavities. Cuboidal epithelium, with cubelike cells, is present in salivary glands, kidney tubules, and the thyroid

gland. Columnar epithelium consists of cells taller than wide, with their long sides adjacent; this type lines the stomach and intestine of vertebrates.

A **ciliated cell** bears on its exposed surface numerous short hairlike extensions known as **cilia** (Fig. 3-18). These beat in one direction, the adjacent cilia acting in unison, so that small particles or materials on the surface are moved along. Cuboidal ciliated epithelium lines the sperm ducts of earthworms and other animals, and columnar ciliated epithelium lines the earthworm's intestine and the air passages (trachea, etc.) of land vertebrates. The embryos and young larvae of many aquatic animals are covered with ciliated cells by which they are able to swim about. A **flagellated cell** has one or more slender, whiplike organelles, or **flagella,** on the exposed surface; such cells line the digestive cavities of hydras and collar layers of sponges.

Protective epithelium guards animals from external injury and from infection. It is one-layered on many invertebrates but stratified on land verte-

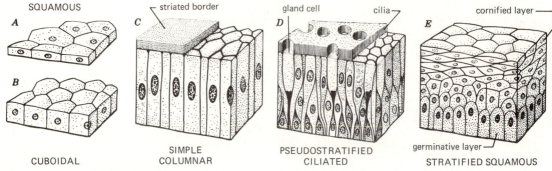

SQUAMOUS

A

B

CUBOIDAL

C ← striated border

SIMPLE
COLUMNAR

gland cell   cilia

D

PSEUDOSTRATIFIED
CILIATED

cornified layer

E

germinative layer

STRATIFIED SQUAMOUS

Figure 3-13   Types of epithelial tissue.

brates. In the latter case, the basal columnar cells (stratum germinativum) produce successive layers of cells by mitosis; these pass outward, flatten, and lose their soft texture, to become cornified or "horny," as they reach the surface (Fig. 3-13*E*). The epithelium on the earthworm secretes a thin homogeneous **cuticle** over its entire exterior surface, and the body covering on arthropods is similarly produced. Nails, claws, hair, and feathers are produced by special groups of epithelial cells.

Glandular epithelium (Fig. 3-14) is specialized for secreting products necessary for use by an animal. Individual gland cells of columnar type (goblet cells) that secrete mucus occur on the exterior of earthworms and in the intestinal epithelium of vertebrates. The multicellular salivary and sebaceous glands of humans and various other mammals are lined with cuboidal cells. The secretions may be either sticky (mucous) or watery (serous).

Epithelial cells specialized to receive certain kinds of external stimuli are called **sensory cells.** Ex-

amples are those in the epidermis of earthworms (Fig. 22-2) and on the tongue and in the nasal passages of humans (Figs. 9-12, 9-13).

3-14 Connective and supportive tissues   These serve to bind the other tissues and organs together and to support the body. They derive from embryonic mesenchyme cells with fine protoplasmic processes. Tissues of this group later become diverse in form; some produce fibers and other intercellular substance, the cells becoming less conspicuous. The supportive tissues include reticular, fibrous, and adipose types together with cartilage, bone, and pigment (Fig. 3-15).

Reticular tissue is a framework of stellate reticular cells and an abundant network of fine reticular fibers which they secrete; it makes the framework of lymph glands, red bone marrow, the spleen, and other organs. **Fibrous connective tissue** consists of scattered cells, rounded or branched in form, with

Figure 3-14   Types of glandular tissue.

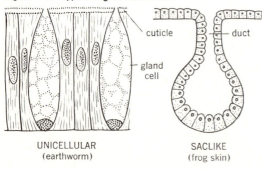

UNICELLULAR
(earthworm)

cuticle — duct

gland
cell

SACLIKE
(frog skin)

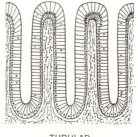

TUBULAR
(human intestine).

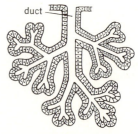

duct

COMPOUND ALVEOLAR
(salivary gland)

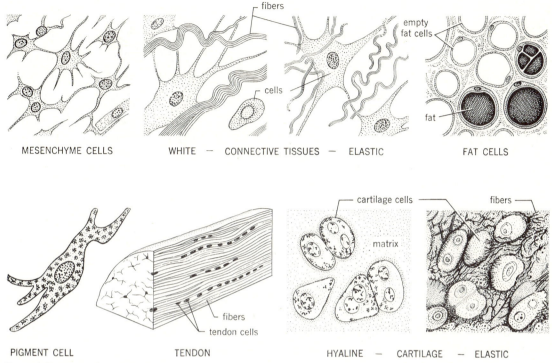

MESENCHYME CELLS   WHITE — CONNECTIVE TISSUES — ELASTIC   FAT CELLS

PIGMENT CELL   TENDON   HYALINE — CARTILAGE — ELASTIC

Figure 3-15 Types of supportive tissue.

the intercellular spaces occupied by delicate fibers. Two major cell types are present, **fibroblasts,** which produce fibers, and **macrophages,** which are protective phagocytes. The **white** (collagenous) **fibers** are made of many fine parallel fibrils, pale in color and often wavy in outline, forming bundles that are crossed or interlaced and occasionally branched; they occur commonly in tendons and around muscles and nerves. **Elastic fibers** are sharply defined and straight, bent, or branched; they bind the skin to the underlying muscles, attach many other tissues and organs to one another, and are present in the lungs, in walls of the larger blood vessels, and elsewhere. Both kinds of fiber occur in the wall of the intestine and in the deeper part (dermis) of vertebrate skin. In **adipose** or **fat tissue** the cells are rounded or polygonal, with thin layers of cytoplasm and the nucleus at one side; they contain droplets of fat, which may form larger globules. Fat is usually dissolved out in prepared microscopic sections, leaving only a framework of cell outlines.

Cartilage (gristle) is a firm yet elastic matrix (chondrin) secreted by small groups of rounded cartilage cells embedded within it and covered by a thin, fibrous perichondrium. **Hyaline cartilage** is bluish white, translucent, and homogeneous under the light microscope. It has, however, a significant proportion of very fine (10 to 20-nm diameter) collagen fibers embedded in the abundant matrix; it covers joint surfaces and rib ends and is present in the nose and in the tracheal rings. It is the skeletal cartilage in the embryos of all vertebrates and in the adults of sharks and rays. It may become impregnated with calcareous salts but as such does not become bone. **Elastic cartilage** containing some yellow fibers is present in the external ears of mammals and in the eustachian tubes. **Fibrocartilage,** the most resistant type, is largely of fibers, with fewer cells and less matrix. It occurs in the pads between the vertebrae of mammals, in the pubic symphysis, and about joints subject to severe strains.

True **bone** or **osseous tissue** occurs only in the

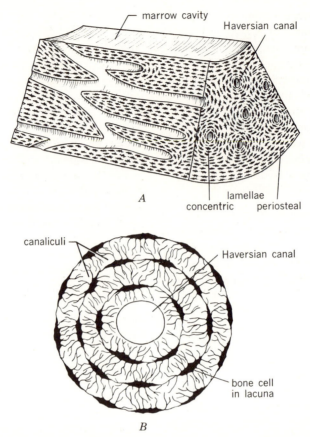

Figure 3-16 Structure of bone (enlarged, diagrammatic). *A*. Sector of long bone in longitudinal and cross section. *B*. Three concentric lamellae around a Haversian canal as seen in a thinly ground cross section. Compare Fig. 4-3.

skeletons of bony fishes and land vertebrates (Fig. 3-16); it is unlike the limy skeletons of invertebrates. Bone is a dense organic matrix (chiefly collagen) with mineral deposits, largely calcium hydroxyapatite, $Ca_{10}(PO_4)_6(OH)_2$ and calcium carbonate. Bone develops either as replacement for previously existing cartilage (cartilage bone) or follows embryonic mesenchymal cells (membrane bone). Both types are produced by **bone cells** called osteoblasts, which later mature into osteocytes. The cells become separate but retain many minute cytoplasmic connections with one another and with blood vessels. Bone is, therefore, a living tissue that may be resorbed in part or changed in composition. During the life of an individual the proportion of mineral gradually increases and the organic material decreases, so that bones are resilient in early youth and brittle in old age.

A bone (Fig. 4-3) is covered by thin fibrous **periosteum,** to which muscles and tendons attach. Within the periosteum are bone cells that function in growth and repair. The mineral substance is deposited in thin layers, or **lamellae.** Those beneath the periosteum are parallel to the surface. Inside, only in mammalian long bones, are many small tubular **concentric lamellae,** forming cylindrical **haversian systems,** the wall of each being of several such lamellae with a central **haversian canal.** The systems are mainly longitudinal but they cross-connect, providing channels for blood vessels and nerves to pass from the periosteum to the interior marrow cavity of a bone. Individual bone cells occupy small spaces, or **lacunae,** between the lamellae; these connect to one another by many fine radiating canals (canaliculi) occupied by the cytoplasmic processes. In flat bones such as those of the skull and in the ends of long bones, the interior lacks regular systems and is more spongy. Cross sections made by sawing such bones show that bone fibers are arranged like beams in arches and trusses to resist compression from the exterior. A slice of bone ground microscopically thin will show the lacunae and canaliculi, which then become filled with air and appear black by refraction. The central cavity in a long bone is filled with soft, spongy **yellow marrow** (containing much fat); the ends and spaces in other bones contain **red marrow,** in which blood cells are produced.

3-15 Vascular or circulatory tissues   The blood and lymph that serve to transport and distribute materials in the body consist of a fluid **plasma** containing free cells, or **corpuscles** (Fig. 3-17; Table 6-1). Colorless **white blood cells,** or **leukocytes,** are present in all animals with body fluids. Several different types of leukocyte are recognized (Fig. 3-17). **Neutrophils** are active phagocytes "policing" the body by engulfing bacteria and other foreign materials. **Lymphocytes** are also defensive, not as phagocytes but through intimate involvement with the immune responses of the animal. Some are capable of being transformed into special antibody-producing cells synthesizing gamma globulins. **Monocytes** are capa-

Figure 3-17 Human blood cells. Erythrocyte about 7.5 μm in diameter. Nuclei of leukocytes are dark (Table 6-1).

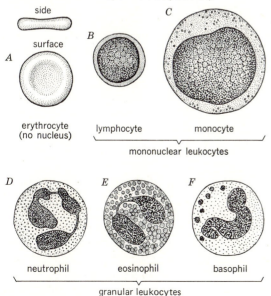

ble of leaving the circulatory system to become phagocytes in tissues. The role of **eosinophils** and **basophils** is poorly understood, but they may play roles in the immune response in the animal. Vertebrate blood also contains **red blood cells,** or **erythrocytes,** colored red by a pigment, **hemoglobin,** that serves for transport of oxygen. Those in most mammals are nonnucleated, biconcave, and usually round, but in other vertebrates they are nucleated, biconvex, and oval. The fluid plasma transports most materials carried in the bloodstream; it is colorless in vertebrates, but that of some invertebrates is colored either blue or red by dissolved respiratory pigment (hemocyanin, hemoglobin, etc.).

3-16 Muscular, or contractile, tissues   Contractile proteins are present throughout the Animal Kingdom. As slender fibrils, they provide the movement of cilia, flagella, muscles, and even chromosomes during mitosis. They also appear to be involved in amoeboid movement (Sec. 15-4). **Flagella** occur in some protozoans, on the collar cells of sponges, and on internal cells of hydras (Chaps. 15 to 17). External **cilia** serve for the locomotion of ciliate protozoans, ctenophores, rotifers, gastrotrichs, flatworms, and the larvae of many invertebrates. Cilia occur on the tentacles of bryozoans, phoronids, brachiopods, some marine worms, and certain cnidarians and on the exterior of starfishes and the gills of mollusks. They line parts of the respiratory and genital tracts of vertebrates, the intestines of mollusks and earthworms, the feeding groove (endostyle) of lower chordates, and the excretory organs of many invertebrates.

Cilia and flagella are shown by the electron microscope to be similar in structure. Both consist of a basal body with or without "rootlets," a basal plate, and a main stalk (Fig. 3-18). The stalk has two central filaments surrounded by nine double filaments. A membrane continuous with that of the cell encloses the filaments. Flagella of some protozoans (*Euglena,* etc.) have fine branches (mastigonemes).

Movements in most animals are produced by long, slender **muscle cells** (Fig. 3-19) that contain minute fibers or myofibrils. When stimulated, they shorten in length or contract, thus drawing together the parts to which the muscles are attached.

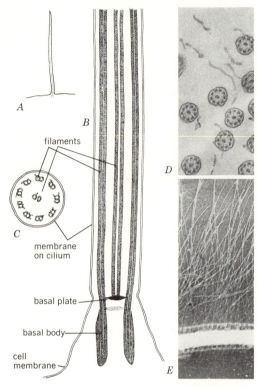

filaments

membrane
on cilium

basal plate

basal body

cell
membrane

**Figure 3-18 Structure of cilia (*Paramecium*) and flagella
(*Euglena*).** *A.* Entire cillum. *B.* Longitudinal section. *C.* Cross
section. Electron micrographs. *D.* Cross section of cilia. ×19,600.
E. Part of a flagellum; shaft (white, at bottom) and fine mastigonemes (above), × 10,800. (*After Dorothy Pitelka.*)

brates is attached to the skeleton and hence is called
skeletal muscle; being under conscious control, it is
also termed **voluntary muscle.**

    Nonstriated or **smooth muscle** consists of deli-
cate spindle-shaped cells, each with one central oval
nucleus and homogeneous fibrils; the cells are arranged in layers, or sheets, held by fibrous connective tissue. Such muscle is found in the internal
organs or viscera of the vertebrate body, as in the
walls of the digestive tract, blood vessels, respiratory
passages, and urinary and genital organs; hence it is
also called **visceral muscle;** not being under conscious control, it is also termed **involuntary muscle.**
In some lower invertebrates the contractile and cytoplasmic portions of a muscle cell are distinct
parts, as may be seen in nematodes (Fig. 19-2*B*).

    Nonstriated muscle is usually capable of slow but
prolonged contraction; in mollusks, it forms the vol-

Figure 3-19 Types of muscle cells and tissues.

Smooth or involuntary

Striated or skeletal

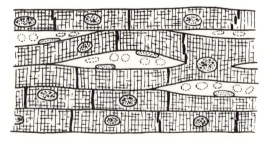

Cardiac or heart

In **striated muscle** the fibrils have alternate dark
and light crossbands of different structure (Fig.
4-11), producing a distinctly crossbanded, or
striated, appearance; the dark bands shorten and
broaden upon contraction. The cells are cylindrical,
scarcely 50 μm in diameter, but some measure 2.5 cm
(1 in) or more in length. Each cell is surrounded by a
delicate membrane (sarcolemma) and contains several to many long nuclei. The vertebrates have
groups of striated muscle cells surrounded by connective tissue sheaths to form **muscles** of various
shapes. These sheaths either attach to the periosteum on bones or gather to form tendons by which
the muscles are attached to the skeleton (Fig. 4-3).
The simultaneous contraction of many fibers causes
a muscle to shorten and bulge, as easily seen in the
biceps of the upper arm. Striated muscle in verte-

untary muscles of the body. Striated muscle can contract rapidly but intermittently and requires frequent rest periods; it occurs in the wing muscles of the swiftest flying insects, in the bodies and viscera of arthropods generally, and throughout the bodies of all vertebrates.

The heart muscle of vertebrates is called **cardiac muscle;** it has delicate cross striations, and the fibers are branched to form an interconnecting network. Cardiac muscle is striated yet involuntary; throughout the life of an individual its only rest period is between successive contractions of the heart.

3-17 Nervous tissues Nervous systems are composed of nerve cells, or **neurons.** The neurons are of varied form (Fig. 3-20) in the systems of different animals and in the several parts of any one system. The individual neuron usually has a large cell body, a conspicuous nucleus, and two or more extensions. The process that transmits stimuli to the cell body is

the **dendrite,** and that carrying impulses away from it is the **axon.** In a large animal an individual neuron may be a meter or more long. Bipolar cells each have one dendrite and one axon; multipolar cells each have multiple dendrites and a single axon. The dendrites are often short and commonly much branched (like a tree) near the cell body, whereas the axon may be short or long and is unbranched save for an occasional collateral fiber. A group of nerve cell bodies, with their conspicuous nuclei, when outside the central nervous system is termed a **ganglion** (pl. ganglia).

A group of fibers or processes bound together by connective tissue is a **nerve.** The central nervous system of animals consists of an aggregation of nerve cells and fibers. Among these are the **neuroglia** (or glia) **cells** of several types that seem to serve as delicate packing to hold neurons apart and may also aid in nutrition of neurons. Nerve fibers are sheathed by special cells called **Schwann cells.** When these cells do not produce additional material and thus lie flat

Figure 3-20 Types of nerve cells. In upper drawing of myelinated nerve fiber, myelin sheath is greatly enlarged relative to diameter of axon. *1–4.* Cross section of myelinated nerve fiber showing stages in growth of Schwann cell, which wraps round and round the axon to form hundreds of adjacent membranes.

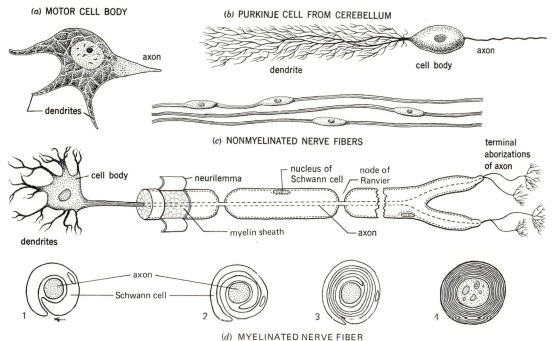

(a) MOTOR CELL BODY

(b) PURKINJE CELL FROM CEREBELLUM

axon

dendrite

cell body

axon

dendrites

(c) NONMYELINATED NERVE FIBERS

terminal aborizations of axon

cell body

neurilemma

nucleus of Schwann cell

node of Ranvier

dendrites

myelin sheath

axon

axon

Schwann cell

1   2   3   4

(d) MYELINATED NERVE FIBER

against the nerve fiber, the nerve is called **nonmyelinated.** When the Schwann cells elaborate a thick lipid coating around the nerve, thus giving it a white appearance, the nerve is termed **myelinated.** Myelinated nerves are constricted at intervals, called **nodes of Ranvier,** which mark the end of one Schwann cell and the beginning of another. The lipid insulation of the nerve fibers permits a greater current flow, thus speeding nerve transmission. The Schwann cell sheath seems to play an important role in the regeneration of damaged nerve fibers. Nonmyelinated fibers are common among invertebrates, and among vertebrates they are found in the sympathetic system and in certain fiber tracts of the spinal cord (internally) and brain (externally). In nerves and on the outside of the spinal cord the myelinated fibers give those parts a whitish appearance.

Organ systems

Every animal, small or large, must carry on a variety of essential functions (Fig. 3-21). Basically these may be reduced to growth, maintenance, and reproduction; all other functions serve these major needs. Actually the bodily operations are complex. In the various groups of the Animal Kingdom, from lowest to highest, there is progressive increase in bodily complexity to carry on these functions. A series of **organ systems** has evolved to serve the various needs. The systems and their principal functions are as follows:

1 **Body covering, or integument**—protection from the environment
2 **Skeletal system**—support (and protection) of the body
3 **Muscular system**—movement and locomotion
4 **Digestive system**—reception and preparation of food; egestion of waste
5 **Circulatory system**—transport of materials
6 **Respiratory system**—exchange of oxygen and carbon dioxide
7 **Excretory system**—disposal of organic wastes and excess fluid
8 **Endocrine glands or system**—regulation of internal processes and adjustments to exterior environment
9 **Nervous system (and sense organs)**—regulation of internal processes and adjustments to exterior environment
10 **Reproductive system**—production of new individuals

Most of the invertebrates and all the vertebrates have the systems just mentioned. In some cases functions are performed without special structural parts being present. Cnidarians, for example, lack respiratory and excretory organs, and flatworms and roundworms have no circulatory or respiratory organs. An organ that continues in use maintains its efficiency, but if unused it tends to degenerate. In sedentary animals and many parasites various

Figure 3-21 Diagram of the essential functions in an animal.

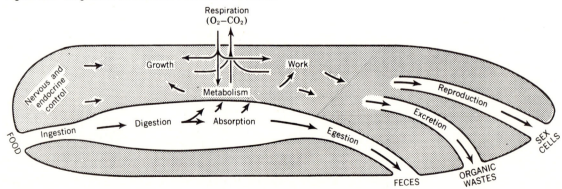

organs have disappeared. Thus the tapeworm, which absorbs nutriment directly from its host, has no digestive tract, and insects such as fleas, lice, and others of burrowing or parasitic habits have no wings.

References

Bloom, William, and **D. W. Fawcett.** 1975. A textbook of histology. 10th ed. Philadelphia, W. B. Saunders Company. xv + 1033 pp., illus.

Brachet, Jean, and **Alfred E. Mirsky.** 1959–1964. The cell, biochemistry, physiology, morphology. New York, Academic Press, Inc. 6 vols.

Copenhaver, W. M., R. P. Bunge, and **M. B. Bung.** 1971. Bailey's textbook of histology. 16th ed. Baltimore, The Williams & Wilkins Company. xv + 745 pp., illus.

De Robertis, E. P. D., F. A. Saez, and **E. M. F. De Robertis, Jr.** 1975. Cell biology. 6th ed. Philadelphia, W. B. Saunders Company. xv + 615 pp., illus.

Di Fiore, M. S. H. 1974. An atlas of human histology. 4th ed. Philadelphia, Lea & Febiger. 252 pp., 114 color pls., 207 figs. *Large, clear, labeled illustrations of tissues.*

Fawcett, Don W. 1966. An atlas of fine structure: The cell, its organelles and inclusions. Philadelphia, W. B. Saunders Company. viii–448 pp., 240 figs.

Langley, L. L. 1968. Cell function. 2d ed. New York, Reinhold Book Corporation. xiii + 364 pp., illus.

Loewy, Ariel, and **Philip Siekevitz.** 1969. Cell structure and function. 2d ed. New York, Holt, Rinehart and Winston, Inc. 516 pp., illus.

Mazia, Daniel, and **Albert Tyler** (editors). 1963. The general physiology of cell specialization. New York, McGraw-Hill Book Company. xiv + 434 pp., 230 illus. *Papers of a symposium.*

Novikoff, Alex B., and **Eric Holtzman.** 1976. Cells and organelles. 2d ed. New York, Holt, Rinehart and Winston, Inc. xii + 400 pp., illus.

Rhodin, J. A. G. 1963. An atlas of ultrastructure. Philadelphia, W. B. Saunders Company. xiv + 222 pp., *82 large plates, each of 1 to 4 electron micrographs showing cellular details; many references.*

Stern, Herbert, and **D. L. Nanney.** 1965. The biology of cells. New York, John Wiley & Sons, Inc. 548 pp., illus.

Weiss, L., and **R. O. Greep** (editors). 1977. Histology. 4th ed. New York, McGraw-Hill Book Company. xv + 1209 pp., illus.

Welsch, U., and **V. Storch.** 1976. Comparative animal cytology and histology. Seattle, University of Washington Press. xiv + 343 pp., 174 figs.

Windle, W. F. 1976. Textbook of histology. 5th ed. New York, McGraw-Hill Book Company. xi + 561 pp., illus.

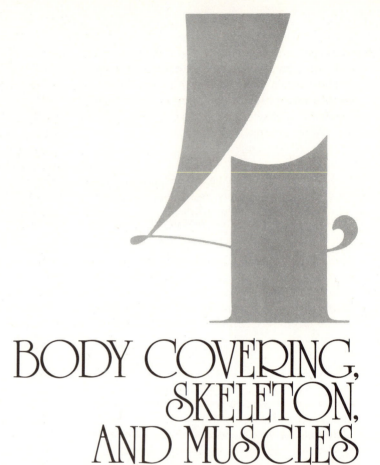

BODY COVERING, SKELETON, AND MUSCLES

In all but the lowest animals the external body covering, the supporting framework or skeleton, and the muscles that serve for movements and locomotion are variously interrelated. The kind of covering, the type of skeleton, and the arrangement of muscles in each group depend on its ancestry, the sort of environment it inhabits, and its mode of life. Among invertebrates the functions of protection and support are often combined in a firm external skeleton with the muscles inside. The most elaborate designs are among the insects and other arthropods, which have jointed body segments and appendages with many individual muscles attached to inward projections of the covering skeleton. By contrast, the vertebrates, almost from their beginning, have had a separate body covering or integument and an internal jointed framework or skeleton with muscles

on the outer surfaces; both hinge and ball-and-socket joints are present. The muscles of vertebrates, by their size and arrangement, are largely responsible for bodily shape.

Body covering

The integument, interposed between the internal and external environment of the animal, performs a great variety of functions and gives rise to such diverse structures as shell, arthropod cuticle, scutes, hair, feathers, and horn. In all animals it protects against injury, abrasion, and wear and against the invasion of disease organisms. It also responds to temperature and to mechanical, chemical, and other stimuli. Other functions include regulation of body water and temperature, respiration, nutrient uptake, and protection against ultraviolet radiation; integumentary glands act in species and individual responses through release of repelling or attracting secretions; pigments function in concealment, warning, and recognition.

4-1 Invertebrates Many protozoans (e.g., *Amoeba*) are covered only by a delicate cell membrane, whereas others (*Paramecium*) also have a firm elastic pellicle. Most multicellular animals are covered by a tissue, the **epidermis.** On many soft-bodied aquatic invertebrates or those of moist environments on land, such as cnidarians, flatworms, and slugs, this is a single layer of cells. The epidermis on many worms and on arthropods secretes an external, noncellular **cuticle** as additional covering; this is delicate on earthworms but resistant on flukes, tapeworms, and roundworms; it forms the exoskeleton of arthropods. On snails and some other animals the epidermis secretes an exoskeleton of shell (Fig. 4-2). The cuticle on arthropods is toughened by **chitin** (Sec. 23-3), a carbohydrate, and usually rendered waterproof by a waxy nonchitinous outer layer of lipoprotein (cuticulin). In lobsters and crabs the cuticle is hardened by deposits of calcium carbonate and in many insects by sclerotization (involving cross-linkages between protein molecules). At the time of molting, enzymes secreted by the epidermis dissolve the inner portion of the old cuticle, the dissolved materials are absorbed, and the outer portion is shed (Sec. 24-18). The desiccation-resistant cuticle, along with other adaptations to life in air, enables the insects, spiders, and relatives to inhabit dry environments.

4-2 Vertebrates The **integument** consists of an outer **epidermis** over an underlying **dermis** that contains blood vessels, nerves, connective tissue, and pigment. The epidermis is derived from ectoderm and the dermis from mesoderm (Sec. 10-14). The epidermis, subjected to abrasion and wear, is replaced by mitotic cell division in its basal layer, the **stratum germinativum.** In many animals the outermost cells

become toughened and **cornified** through accumulation of granules of **keratin,** an insoluble protein that is resistant to wear and to chemical disintegration.

On fishes the epidermis is thin and glandular and closely applied to scales (usually of bone) embedded in the dermis. The glands secrete a mucus that coats the body and protects against disease and injury. On sharks and rays the scales are covered with enamel and project through the skin (Fig. 29-4). Such scales in the mouth region probably gave rise to the first vertebrate teeth (Sec. 29-7). Bony dermal plates were common in early vertebrates (ostracoderms) and provided the membrane bones, developed directly from sheets of embryonic cells, that form important outer elements of the vertebrate skull. Such plates were also prevalent in ancient amphibians.

The land vertebrates (amphibians, reptiles, birds, and mammals) have a **stratified epidermis** of several cell layers (Fig. 3-13) with the outermost portion cornified. Amphibian skin, important in respiration, is glandular and moist. On reptiles, birds, and mammals the cornified part is dry and tougher, more resistant to abrasion and water loss. In reptiles it is thickened into scales, sometimes underlaid with bony scutes. Reptile-like scales are also found on the legs of birds and the tails of rodents.

Birds are covered by **feathers** and most mammals by **hair;** these are nonliving cornified products of the epidermis that conserve body heat, protect against abrasion, smooth contours, and provide streamlining. Feathers form the broad surfaces of wings and tail used in flight. Other cornified epidermal derivatives are the horns of cattle and sheep (but not the calcareous antlers of deer), the claws, nails, hoofs, and horny pads on the feet of various land vertebrates, the beak and leg coverings on birds, the outer scutes on the shells of turtles, and the rattle of rattlesnakes. The skin of mammals contains **sweat glands,** important in cooling the body, and **sebaceous glands,** which secrete a fatty, oily substance that keeps the skin and hair pliable and reduces the rate of evaporation of water. In many mammals fat deposits in the dermis further contribute to insulation; subcutaneous fat is abundant in seals, whales (as blubber), and other forms that live in cold waters. Pigment is scattered throughout the skin, being concentrated in the epidermis in mammals and often

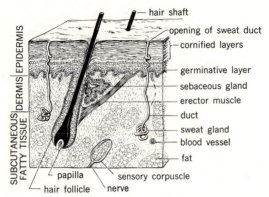

SUBCUTANEOUS FATTY TISSUE | DERMIS | EPIDERMIS

- hair shaft
- opening of sweat duct
- cornified layers
- germinative layer
- sebaceous gland
- erector muscle
- duct
- sweat gland
- blood vessel
- fat
- papilla
- hair follicle
- sensory corpuscle
- nerve

Figure 4-1 Section through human skin. Enlarged and diagrammatic.

enclosed in chromatophores in fish, amphibians, and certain invertebrates.

The human skin (Fig. 4-1) resembles that of other mammals but is scantily haired and thin on most parts. Scarcity of human hair suggests that humans originated in a warm environment. As in some other mammals, evaporation of perspiration secreted by the sweat glands helps cool the body.

The outer epidermis is shed periodically—in one piece in amphibians, snakes, and some lizards, in fragments in other lizards, and by gradual sloughing in crocodilians, most turtles, and mammals. Feathers and hair are molted at intervals and replaced by new coverings.

Skeletal systems

Most animals have a firm framework, or **skeleton,** that gives physical support and protection for the body and often provides surfaces for the attachment of muscles. A skeleton is not absolutely necessary, however, since many aquatic invertebrates and a few land animals have none. Parts of the skeleton in arthropods and vertebrates form jointed appendages that serve as levers for locomotion. In such cases there is integration of structure and function between the skeletal parts and muscles to make their interaction more efficient.

The skeleton may be a shell or other **external** covering (*exoskeleton*), as on corals, mollusks, and arthropods, or **internal** (*endoskeleton*), as with vertebrates (Fig. 4-2). It is **rigid** on corals, many mollusks, and other forms but variously **jointed** and movable in echinoderms, arthropods, and vertebrates. Exo-

Figure 4-2 Animals and protistans with skeletons (diagrammatic). *A, B.* Protozoans. *A.* Radiolarian, framework of strontium sulfate. *B.* Foraminiferan, limy shell. *C.* Sponge, many minute protein, limy, or glassy spicules. *D.* Coral, solid calcareous (limy) cup with partitions. *E.* Rotifer, firm "glassy" cuticle. *F.* Brachiopod, two limy shells. *G.* Echinoderm, internal jointed skeleton of limy plates. *H.* Mollusk, limy shell. *I.* Crustacean, complete exoskeleton with chitin. *J.* Vertebrate, skull, vertebrae, limb girdles, and limb skeleton of bone.

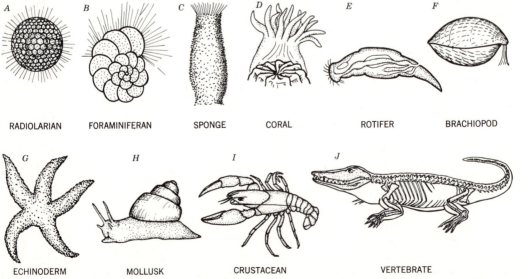

RADIOLARIAN   FORAMINIFERAN   SPONGE   CORAL   ROTIFER   BRACHIOPOD

ECHINODERM   MOLLUSK   CRUSTACEAN   VERTEBRATE

skeletons serving as defensive armor were present on fossil animals such as the trilobites, primitive fishlike ostracoderms, early amphibians (labyrinthodonts), and some ancient reptiles (dinosaurs); they occur also on living brachiopods, most mollusks, barnacles, some fishes, the turtles and tortoises, and the armadillo.

An exoskeleton limits the ultimate size of an animal and may become so heavy that the organism cannot move about because the internal muscles are not large and powerful enough to move the heavy framework. The internal skeleton of a vertebrate involves far less limitation, and some vertebrates have attained huge size; these include the brontosaurs and other fossil reptiles and the living elephants and rhinoceroses. Certain sharks and whales, whose weight is partly supported by the water, are even larger (Fig. 1-1).

4-3 Invertebrate skeletons Some protozoans (Sarcodina, Mastigophora) secrete or otherwise form skeletons of calcareous, siliceous, or organic substances, often of intricate pattern. Sponges secrete microscopic internal rods (spicules) or fibers of the same kinds of materials. The skeletons of corals, brachiopods, echinoderms, and mollusks are mainly lime ($CaCO_3$) and are retained throughout the life of the individual, growing at the margins and becoming thicker with age. All arthropods—crustaceans,

insects, etc.—are covered completely by jointed exoskeletons of organic materials containing chitin (Sec. 23-3). These are flexible at the joints between body segments and appendages but more rigid elsewhere. On crabs and related crustaceans the external covering is reinforced by internal deposits of limy salts. The appendages number one pair per body segment (somite) or fewer and are variously developed as sensory antennae, jaws or other mouth parts, and legs for walking or swimming. Since the skeletons of arthropods when once hardened cannot expand, these animals undergo a complete molt of the old covering at intervals to permit growth; the body enlarges immediately after a molt, before the new covering has hardened. The secreted tubes in which some aquatic worms live, the cases built of bottom debris by some protozoans and certain insect larvae, and the empty snail shells used by hermit crabs all serve as protective exoskeletons.

4-4 Vertebrate skeletons The internal skeleton has a common basic pattern, with the fundamental features seen in a frog (Sec. 31-6). From the cyclostomes to the mammals a progressive sequence may be traced, although there are many differences in the size and form of component parts and in the presence or absence of certain elements. The essential features in a land vertebrate are given in Table 4-1. The skeleton supports the body, provides for attach-

Table 4-1
GENERAL DIVISIONS OF THE SKELETON IN A LAND VERTEBRATE

Axial skeleton (median)			Appendicular skeleton (lateral, paired)	
Skull	Vertebral Column	Thoracic Basket	Pectoral (anterior)	Pelvic (posterior)
Cranium (brain box) Sense capsules (nose, eye, ear) Visceral arches (jaws, hyoid, larynx)	Vertebrae: Cervical (neck) Thoracic (chest) Lumbar (lower back) Sacral (hip) Caudal (tail)	Ribs (paired; bony or cartilaginous) Sternum (breastbone)	Shoulder girdle: Scapula (dorsal) Clavicle (anterior) Coracoid (posterior)	Hip girdle: Ilium Pubis Ischium
			Forelimb: Humerus (upper arm) Radius and ulna (forearm) Carpals (wrist) Metacarpals (palm) Phalanges (fingers)	Hind limb: Femur (thigh) Tibia and fibula (shank) Tarsals (ankle) Metatarsals (sole) Phalanges (toes)

ment of muscles, and houses and protects the brain and nerve cord. In all but the cyclostomes it includes framework for the jaws and the paired fins or limbs. The skeleton is of **cartilage** in adult cyclostomes and the cartilaginous fishes (sharks and rays) and in embryos of all higher vertebrates, but in the adults of bony fishes to mammals (Fig. 4-5) it is largely of **bone** with cartilage over joint surfaces and in a few other places.

Each bone in the body is built on good "engineering" principles in both gross and microscopic structure. Those subject to heavy stresses are reinforced within, and where stout muscles or tendons attach, the exterior is roughened. Long bones are tubes—a given amount of material is more rigid when formed as a tube than as a solid rod. Also thin layers give greater strength than a solid mass, as in plywood; hence the lamellae of bone (Fig. 3-16) afford further stiffness.

Membrane or dermal bones, such as flat bones in the skull, grow as long as the junctions between any two are cartilaginous but cease growing when bony sutures develop between them. The **long bones**—limbs, fingers, centra of vertebrae—are composed of a central **diaphysis** with a cap, or **epiphysis,** at each end; these start as three centers of ossification in the embryo. A layer of epiphyseal cartilage connects each epiphysis to the central (diaphysis) portion.

Growth in length (Fig. 4-3) is by extension of the epiphyseal cartilage; that cartilage in contact with the bone is converted to bony tissue. Increase in length is possible so long as the junction remains cartilaginous. A long bone increases in diameter by deposit of bony tissue on the exterior; at the same time some bony substance is removed from the walls of the interior cavity. Repair of fracture in any bone is by much the same process. Bone grows and is restructured by the action of osteoblasts and osteoclasts (bone-building and bone-resorbing cells) under hormonal control. Bone growth is promoted by the growth hormone (somatotropin) of the anterior pituitary (Sec. 8-12), which accelerates protein synthesis in the epiphyseal plates and increases the rate of osteoblast formation. Parathyroid hormone promotes bone resorption and thyroid calcitonin inhibits it (Secs. 8-5, 8-6). The two hormones interact to maintain a constant level of calcium in the blood. Calcium salts are absorbed from food. When the blood calcium declines, bones are "robbed" of calcium, as may occur during pregnancy or with deficient diet.

4-5 Vertebral column In all chordates the first skeletal element to appear in the embryo is a slender unsegmented and gelatinous rod, the **notochord,**

Figure 4-3 Structure of a long bone and a joint; human elbow region in lengthwise section. The rounded end (*trochlea*) of the humerus fits the (*semilunar*) notch in the olecranon (base of ulna) to make a hinge joint, enclosed in the joint (*articular*) capsule. The ulna, a typical long bone, has a central tubular shaft (*diaphysis*) with a cap (*epiphysis*) at each end. Growth in length occurs in cartilage areas between shaft and ends. The ends are covered with smooth articular cartilage.

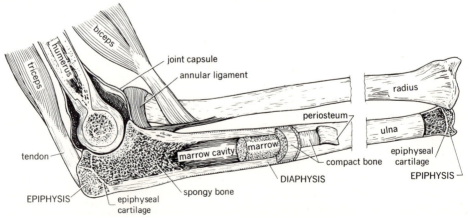

that extends along the dorsal body axis between the digestive tract and the nerve cord. It persists thus throughout life in amphioxus and cyclostomes, but in fishes and other vertebrates is later surrounded and supplanted by the backbone, or **vertebral column,** of separate vertebrae (Fig. 4-4). The spoollike **centrum** of each vertebra has a dorsal **neural arch** which encloses the nerve cord. In the tails of fishes each vertebra also has a ventral **hemal arch** around the main artery and vein; this arch is spread open in the body or trunk region to form riblike structures shielding the internal organs. In land vertebrates the centrum bears a pair of **transverse processes** as points of attachment for the true ribs of these animals (rare or absent in frogs). At either end of the centrum are two **articular processes** by which one vertebra may move slightly on those directly before and behind. The vertebral column of fishes comprises only **trunk** and **tail** regions, but in salamanders, reptiles, birds, and mammals there are five regions: neck, or **cervical;** chest, or **thoracic,** with ribs; lower back, or **lumbar;** pelvic, or **sacral,** joining the hind limb girdle; and tail, or **caudal.** The caudal vertebrae are few in humans and birds. Long-bodied swimming vertebrates have numerous vertebrae, all much alike, as seen in eels and similar fishes, in some salamanders (*Amphiuma*), in some fossil reptiles, and in whales. The living snakes that, in effect, "swim" on land also have many vertebrae (Fig. 32-9). An additional set of articular processes on each vertebra stiffens the intervertebral connections and reduces rotational movement around the long axis of the body, an aid in control when the slender body is elevated. The ribs of land vertebrates usually join ventrally to a breastbone, or **sternum,** but this is lacking in snakes. The sternum of birds has a large median keel for attachment of the stout flight muscles (Fig. 33-5).

4-6 Skull This structure, which frames the vertebrate head, begins in the embryo as cartilage and consists of (1) the **cranium,** or brain box, which houses and protects the brain; (2) three pairs of **sense capsules** for the organs of smell, sight, and hearing; and (3) the **visceral skeleton,** which is a series of paired arches providing the jaws, support for the tongue (hyoid apparatus), and supports for the gill region. In sharks and rays the skull elements remain more or less cartilaginous and the upper jaw is not fused to the cranium. In bony fishes and higher forms the capsules and upper jaw become more completely united to the brain case and the cartilaginous cranium is replaced by numerous bones. These are chiefly membrane bones derived originally from dermal scales and formed directly from mesenchyme (Sec. 3-14), without a transitional cartilage stage. Membrane bones also form the lower jaw of fishes, amphibians, and reptiles. The dentary, a single such bone, is a diagnostic mammalian characteristic. In land vertebrates parts of the visceral arches are put to other uses (Fig. 13-7).

Both the general form and the detailed structure

Figure 4-4 Vertebrae. *A.* Tail region of bony fish; both neural and hemal arches. *B.* Trunk region of bony fish. *C.* Human lumbar vertebra. *D.* Part of vertebral column (human lumbar region) showing how the vertebrae articulate with one another, the pads between the centra, and the openings (foramina) for spinal nerves connecting to nerve cord.

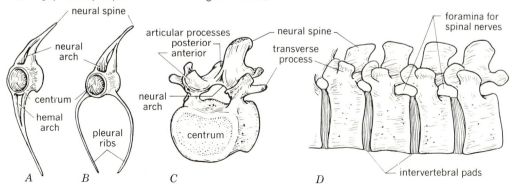

of the skull in adults of various vertebrates are diverse, and a comparative study from fishes to mammals shows many differences, including a great reduction in the number of bones; yet there is an underlying continuity in pattern throughout the entire series.

4-7 Limbs   Cyclostomes have no lateral appendages, but the sharks and bony fishes have two pairs, the **pectoral** and **pelvic fins,** with skeletal elements consisting of **fin rays.** Each pair is supported on a framework, or girdle (Secs. 29-5, 30-5). Land vertebrates have two pairs of **limbs** in place of the fins, and these are supported by the **pectoral** and **pelvic girdles,** respectively. Each limb characteristically ends in five **toes,** or digits. The component bones of the girdles and limbs are homologous from amphibians to mammals, although variously modified in adaptation to special modes of life (Fig. 13-2). Loss of

Figure 4-5 The human skeleton. (Typically of the vertebrate pattern; compare Figs. 31-5, 34-2.)

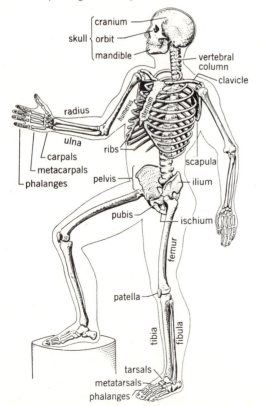

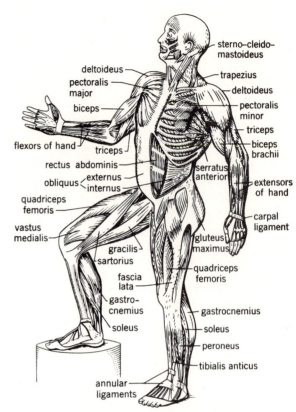

Figure 4-6 Superficial muscles of the human body. Pectoralis major and external oblique removed on left side.

digits, fusion of other bones, and reduction or complete loss of fins, limbs, and girdles have occurred in various vertebrates. Some salamanders have only four or fewer toes on each foot, and no living bird has more than three "fingers" or four toes. Reduction in number of toes occurs in many mammals, the horse being an extreme case, with only one functional toe on each foot (Fig. 34-6). The radius and ulna and the tibia and fibula are fused in many species that have slight rotational movement of the limbs. The limbs and digits are greatly reduced in some salamanders and lizards and are absent in a few lizards and all snakes. Some skinks lack forelimbs (*Rhodona*). Whales, sirens, and the mole lizard (*Bipes*) have no hind limbs, and among fishes the eels lack pelvic fins. Vestiges of the limbs or girdles in whales, boas, and other limbless vertebrates indicate that these animals have descended from ancestors with limbs.

Muscular systems

Movement in animals is primarily by means of intracellular contractile proteins arranged as filaments or **fibrils.** Many cells contain such contractile elements, but in most animals the contractions that bring about changes in shape or form and locomotion are produced by **muscular tissues,** consisting of special cells containing **bundles** of such fibrils. Many multicellular animals that are capable of locomotion have opposed sets of muscles to perform these movements. The most important contractile proteins are **actin** and **myosin,** found throughout the Animal Kingdom from protozoans to vertebrates. The energy for contraction is derived from ATP (Sec. 2-23).

4-8 Invertebrates   Simple protozoans such as *Amoeba* engage in **amoeboid movements** and can contract or extend the one-celled body in any direction (Sec. 15-4). Other protozoans move by means of contractile fibrils contained in cilia and flagella (Sec. 3-16), and in some, such fibrils move mouth parts (*Epidinium*) or basal stalks (*Vorticella*). The fibrillar mechanism is thought to be similar to that of striated muscle (Sec. 4-11).

Both smooth and striated muscle are present, the latter occurring from cnidarians to arthropods. The body wall of cnidarians contains T-shaped epitheliomuscular cells with contractile fibers in the basal part; these cells lie in opposed sets (Fig. 17-2) by which the body can be reduced in either length or diameter. Flatworms usually have muscle fibers in three directions—longitudinal, transverse, and dorsoventral (Fig. 18-3B); contraction of those in any one plane will force the soft but cell-filled body to extend in the other planes, much as the human tongue can be moved. In roundworms the muscle cells somewhat resemble those of cnidarians, but all are aligned against the body wall and parallel to the main body axis (Fig. 19-2). The alternate contraction of fibers along opposite sides of the body enables the worm to bend and straighten, but it can neither twist freely nor extend the body in length. In earthworms the body wall includes two layers of muscles, an outer circular and an inner longitudinal layer (Fig. 22-4). Contraction of the outer layer causes the fluid-filled body to lengthen, and action of the longitudinal muscles shortens it. The mollusks, crustaceans, insects, and other arthropods are the only invertebrates that depart from the layer arrangement of muscles; they have many separate muscles, varied in size, arrangement, and attachments, that move the body segments and the parts of the jointed legs and other appendages. These muscles are fastened to the internal surfaces of the exoskeleton and act over hinge joints between adjacent parts (Fig. 4-7). A caterpillar may have 2000 separate muscles, compared with some 600 muscles in human beings.

4-9 Vertebrates   The muscles that move the skeleton are an obvious feature of vertebrates. However, amoeboid, ciliary, and flagellar movements, a heritage from an invertebrate past, are also present. Movements of mesenchyme and leukocytes are amoeboid, cilia line parts of the respiratory, digestive, genital, and excretory tracts, and a flagellum drives the vertebrate spermatozoon.

Figure 4-7 *A.* Internal segmental muscles of an arthropod, connected to hardened, jointed, telescoping parts of exoskeleton. *B.* Internal muscles in leg of an insect. *C, D.* Hind leg of frog showing two pairs of opposed muscles: extensor versus flexor. In *C* the leg is folded, or *flexed,* by contraction (thickening), bringing the origin and insertion of each muscle closer together. In *D* the leg is *extended* by contraction of the opposed muscles. (*Adapted partly from Guyer, Animal biology, Harper & Brothers.*)

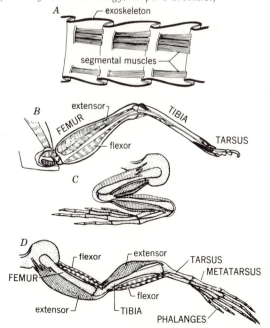

In the fishlike forms—cyclostomes to permanently gilled amphibians—and the limbless reptiles, the muscles are predominantly segmental (Fig. 32-3). They alternate with the vertebrae and provide the undulating movements by which the animals travel. In land vertebrates, from frogs to mammals, the nonsegmental muscles that move the limbs and head are larger and more important (Fig. 4-6). In the higher forms some segmental muscles persist, however, between the vertebrae and the ribs and in the rectus abdominis muscles of the ventral body wall. In locomotion each pair of opposed muscles shows a rhythm of alternate activity. If more than one such pair is involved, their action shows a regular sequence. The primitive pattern, as in an eel, snake, or other slender animal, is a series of waves of contraction passing alternately along either side of the body. The movement of limbs in land forms is also alternate, in both vertebrates (Fig. 4-8) and insects.

4-10 Muscle and nerve In a living animal the contraction (shortening) of a muscle results from impulses passing from the central nervous system along a nerve. A nerve contains numerous fibers (neurons), each of which makes contact with several

Figure 4-8 Locomotion in a toad, a four-footed animal. Beginning at the upper left, the 15 figures show the normal pattern of limb movement; 1 to 4, sequence in flexing the limbs. (*After J. Gray, 1939.*)

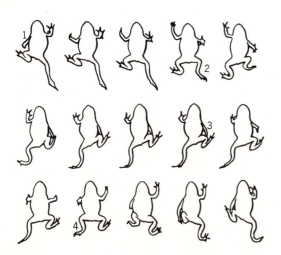

muscle cells. Contraction in response to nerve impulses may be demonstrated with a preparation of the sciatic nerve and gastrocnemius muscle dissected out together from a live frog (Fig. 4-9A). One end of the muscle is attached immovably, and the opposite end is tied to a lever that will magnify and record any change in length of the muscle. Impulses in the form of brief electric shocks are applied to the nerve. Beginning with a shock too weak to produce any result and gradually increasing the intensity, a threshold of stimulation is reached at which slight contraction results. Further increase will produce a greater contraction, but soon a point is reached when still stronger impulses have no further effect. However, if all fibers but one in the nerve are cut and the remaining one is stimulated at increasing intensities, nothing happens until the **threshold** is reached, when the response of those muscle cells connected to the neuron is at once maximal. This is the **all-or-none effect.** The graded increase in action of an entire nerve-muscle preparation results from different muscle cells having slightly different thresholds. Maximum contraction of an entire muscle occurs when every muscle cell is stimulated above threshold.

An individual muscle contraction follows a characteristic pattern (Fig. 4-9B) lasting about 0.1 s. The interval between the initial stimulus and shortening, about 0.01 s, is known as the **latent period.** Although no obvious mechanical change occurs during this phase, reactions are taking place within the muscle which liberate the energy required for contraction, and some relaxation occurs. The second phase, the **period of contraction,** lasts about 0.04 s. Finally, in the **period of relaxation,** about 0.05 s, the muscle returns to its original length and physiologic state. When individual shocks are well spaced in time, the muscle relaxes completely to its original length between them. If the shocks are repeated at close intervals, it does not. With still higher frequency of stimulus there is no relaxation but a smoothly maintained contraction, termed **tetanus.** Normal movements in the entire animal result from tetanic contractions. In humans, at least when they are awake, some muscle cells in every muscle are in tetanus, and the muscles feel firm. This maintained tension, or **tonus,** resistance to stretching, keeps the body in

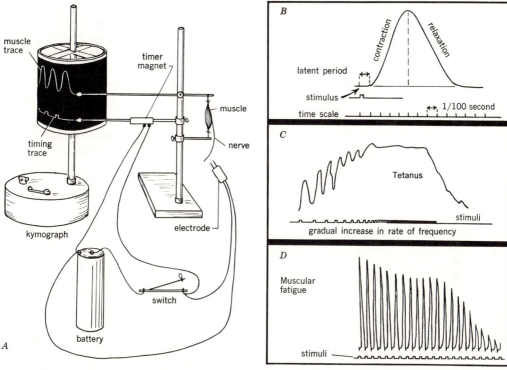

Figure 4-9 Contraction of voluntary muscle (Sec. 4-10). *A.* Nerve-muscle preparation attached to kymograph for recording contraction when nerve is stimulated by electric impulse from battery. *B.* Diagram of normal contraction and relaxation after a single stimulus. *C.* Tetanus (maintained contraction) with increasing frequency of stimuli. *D.* Fatigue resulting from repeated stimuli over a long period. (*B–D from kymograph records.*)

normal condition—while seated, standing, or at work. Tonus continues without fatigue because while some muscle cells are contracted, others are relaxed and resting. The tone of muscles is one indication of health; in some paralyses, such as that of poliomyelitis, the muscles become flabby (atonic), whereas in other conditions they are tightly contracted (tetanic).

The muscle cells in a skeletal muscle are in groups called **motor units;** each group is controlled by one motor neuron, and each unit evidently follows the **all-or-none law:** its muscle cells either contract fully or not at all. No muscle is ever completely at rest; some units (5 percent or so) are always contracted, even during the "complete relaxation" of rest or sleep, thereby generating some heat. If motor stimuli to nerve end plates are experimentally blocked (by the drug curare), all muscles become relaxed, the body is limp, and the temperature declines, since no heat is then produced by muscular contraction.

4-11 Muscular contraction Many striated muscles can contract with extreme speed (as in an insect wing) and do so repeatedly for a time. In a 100-yard dash the runner's leg muscles may contract 30 times within 10 s; during and at the end of the race the runner has an "oxygen debt" that is repaid by quick, heavy breathing for several minutes. When muscles contract, oxygen is used, carbon dioxide is given off, the glycogen content is reduced, heat is produced (see Krebs cycle, Sec. 2-25, Fig. 4-10), and if insufficient oxygen is supplied to the muscle, lactic acid accumulates and the sensation of fatigue is produced.

Special studies by x-ray diffraction and electron

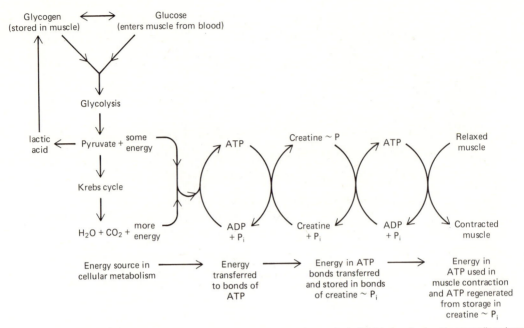

Figure 4-10 Muscle-energy relationships. The immediate source of energy for muscles is ATP. It is resynthesized from creatine phosphate (creatine $\sim P_i$ where $\sim$ means a high-energy bond and P_i inorganic phosphate). Creatine phosphate is in turn resynthesized from the energy derived in part from glycolysis (anaerobic regeneration) but mainly from the Krebs cycle (aerobic regeneration). This energy is furnished in the form of ATP. Part of the energy obtained in the Krebs cycle is used to convert the excess lactic acid or pyruvate to glycogen.

micrographs help to visualize the process of contraction. A muscle is composed of many fibers (muscle cells). The cells are 10 to 100 μm in diameter and, in some muscles, may reach lengths of over 30 cm. Each muscle cell contains a bundle of **myofibrils** about 1 to 2 μm in diameter and numbering several hundred to several thousand. Each myofibril, in turn, is an aggregation of slender parallel strands, or **myofilaments.** A myofibril shows a repeating pattern of distinct bands (Fig. 4-11). Broad, dark A bands alternate with pale I bands, giving striated muscle its characteristic banded appearance. Each I band is crossed by a narrow dark Z line and each A band by a slightly pale H zone. The region between two Z lines is called a **sarcomere,** the basic unit of contraction. When muscle contracts, the I bands and H zones narrow, but the width of the A bands remains essentially the same. The banding and changes in appearance are explained by the structure, arrangement, and movement of the myofilaments within the myofibril (Figs. 4-11, 4-12). The myofilaments are of two kinds, **thick** (110 to 150 Å) and **thin** (50 Å), composed, respec-

tively, of the proteins myosin and actin. Consider their arrangement within a single sarcomere. The thick filaments, of uniform length and equidistant from one another, form a bundle in the dark A-band region and are held in register by linkages at the M line (Fig. 4-11). Thin filaments, attached in the plane of the Z line, project inward from each end of the sarcomere and penetrate the A-band bundle. Penetration is slight to moderate when the muscle is relaxed. The H zone, between the ends of the thin filaments, is wide. Upon contraction, the Z lines come closer together and the thin filaments penetrate farther into the A band, reducing the width of the H zone. Upon extreme contraction, thick and thin filaments may overlap completely, the thin ones touching and crumpling in the center of the A band and the thick ones abutting on the Z lines. I and H zones disappear. Throughout contraction, the length of the filaments remains essentially constant; it is the amount of overlap that changes.

What brings about the movement of the two kinds of filaments relative to one another? The **slid-**

Figure 4-11 Electron micrographs of striated fibers from frog sartorius muscle (compare Sec. 4-11). *Above.* Longitudinal section, × 30,000. I band (pale) of thin filaments only; A band (dark) of thick filaments, still darker at ends where thick and thin filaments overlap. H zone with only thick filaments is between ends of thin filaments. The region between the two Z lines is a sarcomere. The M line is a zone of linkage between thick filaments which maintains their parallel arrangement. *Below.* Cross section × 74,000. I band at right, thin filaments as fine dots; H zone at left of thicker filaments as larger dots; region of overlap with filaments of both sizes. (*Original micrographs by Dr. Lee D. Peachey, Columbia University.*)

ing filament theory of muscular contraction, proposed independently by the English physiologists A. F. Huxley and H. E. Huxley, holds that thick filaments have cross bridges that are able to swing back and forth and hook into receptor sites on the thin filaments. The bridges move the latter in rachetlike fashion toward one another. The power stroke of the bridges in the two halves of the thick filaments is thus toward one another, thereby moving the thin filaments toward the center of the A band. The positioning of the two kinds of filament is such that each

thin one receives cross bridges from three thick ones (Fig. 4-12), thus many bridges impinge on each thin filament.

What are the molecular characteristics of the filaments and bridges that make muscular movement possible? The thick filaments contain molecules of myosin, the thin ones molecules of actin. Myosin molecules have a globular "head" and slender tail (Fig. 4-12). They are oriented tail to tail in the two halves of the thick filament, and the heads project to the sides, forming the cross bridges (Fig. 4-12). Actin

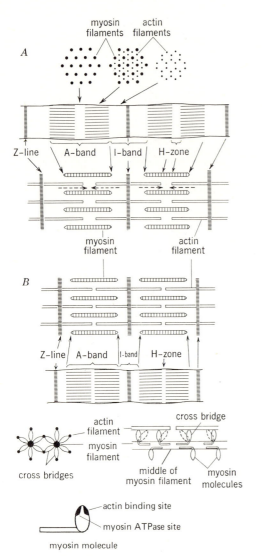

Figure 4-12 Changes in a myofibril as it contracts and actin and myosin myofilaments slide past one another (see also Fig. 4-11). *A. Restiong myofibril. Above.* Myofilaments seen in cross section of a myofibril at three levels indicated by arrows. *Below.* Schematic representation of position of myofilaments. *B.* Myofibril contracted. *Above.* Myofilaments have changed position but not length. *Below.* Presumed mechanism of myofilament movement. *Left.* Myofilaments in cross section showing cross bridges formed by globular heads of myosin molecules. *Right.* Lateral view, showing ratchetlike movement of heads of myosin molecules which are oriented in opposite directions in the two halves of the myosin myofilaments. Dotted arrows in *A* indicate direction of movement of actin myofilaments. (*After H. E. Huxley.*)

molecules are globular and arranged in two intertwined chains. When a myosin bridge moves, it attaches to an active site on the actin filament, pulls the filament a short distance, then releases it and attaches to the next active site. The splitting of ATP provides the energy for this movement. Attachment and detachment of the myosin bridges occurs as follows: The globular head of the myosin molecule contains an actin-binding site and a myosin ATPase site. The former binds the cross bridge to the actin molecule and the latter brings about the splitting of ATP required for movement. Prior to contact with actin the ATPase site splits ATP slowly, but upon contact its activity greatly increases. Detachment of the bridge is achieved by binding a molecule of ATP to myosin (in the presence of magnesium). This reaction returns the bridge to its former state and readies it for attachment to a new actin site.

What initiates and halts muscular activity? In the resting muscle, two **regulator** proteins, **troponin** and **tropomyosin,** associated with the thin filaments, prevent actin from combining with myosin and no contraction occurs. Their inhibitory effect is removed by calcium. Calcium binds to the troponin molecule, producing a change which is transmitted through the tropomyosin molecule, making it possible for actin to combine with myosin. The calcium ions come from storage areas in the **sarcoplasmic reticulum** (Fig. 4-13), which extends among the myofibrils and communicates with the extracellular environment by means of small tubules. (The reticulum is similar to the endoplasmic reticulum of other cells.) At the time of muscle contraction an action potential causes depolarization of the tubules, which quickly spreads into the muscle cell, causing the stored calcium ions to be released to the cytoplasm. Upon cessation of the electrical activity, release of calcium stops and the ions are actively transported back to the storage areas. Inhibition by the regulator proteins is restored.

What are the mechanisms involved in the coupling of the nerve signal with muscular contraction? Skeletal muscle is innervated by motor neurons whose cell bodies lie in the brain or spinal cord. At the site of the muscle, each motor axon branches extensively, and each branch serves a single muscle

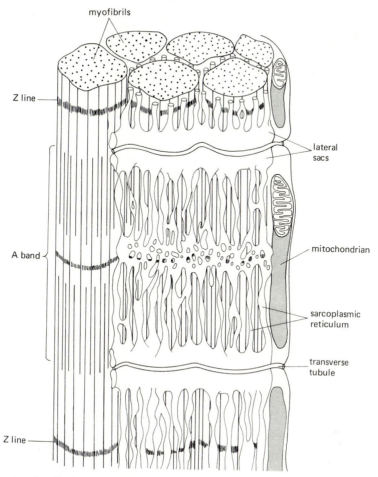

myofibrils

Z line

A band

lateral sacs

mitochondrian

sarcoplasmic reticulum

transverse tubule

Z line

Figure 4-13 The sarcoplasmic reticulum of skeletal muscle. *(From Vander et al., Human physiology, McGraw-Hill, 1975).*

fiber. A given branch terminates in a depression on the muscle membrane called the **motor end plate;** the junction between nerve and muscle is called the **neuromuscular junction.** When an action potential arrives at the junction, it depolarizes the nerve cell membrane and **acetylcholine** (ACh) is released into the space between the nerve and muscle membranes. Release of ACh causes depolarization of the muscle end plate and spread of membrane excitation, which initiates the events of contraction. A given nerve signal is brief. The molecules of ACh last only about 5 ms before they are destroyed by an enzyme, **acetylcholinesterase,** released by the muscle membrane,

and the motor end plate returns to its resting condition.

Muscle fibers and nerve cells have a number of characteristics in common. Both have an electrically excitable cell membrane in which action potentials can be propagated, a characteristic not found in other cells of the body. The neuromuscular junction in skeletal muscle resembles in structure and function a synaptic junction between two nerve fibers (Sec. 9-2). In both cases an action potential is invoked in the receiving cell by neural release of a chemical transmitter, followed by a flux of ions. A stimulus threshold must be reached before the receiving cell will respond (all-or-none response), and there is a brief refractory period following stimulation.

4-12 Energy for muscular contraction   The energy for muscle contraction and other biologic processes is supplied ultimately by glucose ($C_6H_{12}O_6$) derived from absorbed food and carried in the bloodstream. When stored (in liver, muscle, etc.) it is converted into glycogen ($C_6H_{10}O_5)_x$. Then, in essence, glycogen is oxidized to carbon dioxide and water:

$$C_6H_{10}O_5 + 6O_2 \xrightarrow[enzymes]{} 6CO_2 + 5H_2O + energy$$

Formerly it was thought that glycogen served rather directly for the release of energy, because some of it disappears from muscle with repeated strong contractions. Actually, its conversion involves a series of intermediate reactions (see Sec. 2-24; Figs. 2-18, 2-19).

In addition, other substances in muscle are involved in the energy-contraction cycle, including recovery. A resting muscle contains (1) adenosine triphosphate (ATP), (2) creatine phosphate (CP), and (3) glycogen. With contraction each undergoes chemical change, and *each reaction releases much energy*. Chemical analyses show (Fig. 4-10) that

1   ATP is changed to adenosine diphosphate (ADP).
2   Creatine phosphate (CP) is broken down into creatine, giving its inorganic phosphate (P_i) to ADP.
3   Glycogen is converted into lactic acid.

The energy of the three reactions above is utilized as follows:

1   Breakdown of ATP provides the actual energy for muscular contraction.
2   Division of CP supplies the energy to resynthesize ATP.
3   The several reactions converting glycogen to lactic acid afford energy for re-forming ATP or CP from C and P_i.

4   Oxygen reacts with about one-fifth of the lactic acid to provide energy for reconverting the remaining four-fifths to glycogen.

Splitting of both ATP and CP is by hydrolysis (addition of water) and involves no reaction with oxygen; hence it can proceed under anaerobic conditions (absence of oxygen). In contrast, the reconversion of lactic acid to glycogen is aerobic, involving use of oxygen. The "oxygen debt" is built up by the breakdown of glucose to lactic acid when oxygen is in low supply. Lactic acid accumulates and must be removed by oxidation, via the citric acid cycle (Sec. 2-25; Fig. 4-10), at a later time when oxygen becomes available.

Both CP and ATP are present and active in most or all living cells. If ATP is applied to actomyosin threads or to muscle appropriately prepared, the threads or muscle contracts. Actomyosin ATPase is the enzyme that catalyzes the breakdown of ATP to ADP and inorganic phosphate. All reactions in muscles—breakdown or synthesis—are by enzymes specific for each reaction.

Muscles contract to perform work; about 30 percent of the energy used serves this purpose, the balance producing heat. About four-fifths of all bodily heat derives from this source.

Fatigue—the inability to continue contraction—results from accumulation of lactic acid and depletion of glycogen and ATP. Evidently the motor end plates on muscles are rendered inactive by the lactic acid. (If formation of lactic acid is prevented by use of a poison, iodoacetic acid, muscles continue to contract.)

References

The following works, which discuss the animal or human body, organ systems, and physiology, are useful as references for Chaps. 4 to 9. Many excellent, well-illustrated articles on special subjects in these fields appear in *Scientific American* and are available as reprints. Some are cited in the reference lists. The *Annual Review of Physiology* (Palo Alto, Calif., Annual Reviews, Inc.) also should be consulted. It covers a great variety of topics in physiology and includes information on both invertebrates and vertebrates.

Chapman, C. B., and **J. H. Mitchell.** 1965. The physiology of exercise. *Sci. Am.,* vol. 212, no. 5, pp. 88–96.

Cohen, C. 1975. The protein switch of muscle contraction. *Sci. Am.,* vol. 233, no. 5, pp. 36–45. *The contraction of muscle is "turned on" when calcium ions trigger changes in two proteins, tropomyosin and troponin.*

Ganong, W. F. 1977. Review of medical physiology. 8th ed. Los Altos, Calif., Lange Medical Publications.

Gordon, M. S., and others. 1968. Animal function: Principles and adaptations. New York, The Macmillan Company. xvi + 560 pp. *Animal function as it is related to survival in natural environments.*

————, **G. A. Bartholomew, A. D. Grinnell, C. B. Jorgensen,** and **F. N. White.** 1977. Animal physiology: Principles and adaptations. 3d ed. New York, The MacMillan Company. xx + 699 pp.

Gray, J. 1968. Animal locomotion. London, Weidenfeld and Nicolson, Ltd. xi + 479 pp. *Mechanics of animal locomotion—vertebrates and invertebrates.*

Guyton, A. C. 1977. Basic human physiology: Normal function and mechanisms of disease. Philadelphia, W. B. Saunders Company. xi + 931 pp.

Hayashi, T. 1961. How cells move *Sci. Am.,* vol. 205, no. 3, pp. 184–204. *The beating of cilia, movement of an amoeba, and muscular contraction all appear to share an underlying molecular unity.*

Huddart, H. 1975. The comparative structure and function of muscle. Elmsford, N.Y., Pergamon Press, Inc. viii + 397 pp. *Muscle structure and electrical and mechanical activity discussed. Includes information on visceral, skeletal, and cardiac muscle of invertebrates and vertebrates.*

Huxley, H. E. 1965. The mechanism of muscular contraction. *Sci. Am.,* vol. 213, no. 6, pp. 18–27. *The sliding-filament theory of muscular contraction.*

Katz, B. 1966. Nerve, muscle and synapse. New York, McGraw-Hill Book Company. ix + 193 pp.

McLean, F. C. 1955. Bone. *Sci. Am.,* vol. 192, no. 2, pp. 84–91. *The skeleton grows and renews itself while providing rigid support, and it functions in maintaining the level of calcium in the blood.*

Merton, P. A. 1972. How we control the contraction of our muscles. *Sci. Am.,* vol. 226, no. 5, pp. 30–37. *A servomechanism resembling that which controls power-assisted steering in an automobile drives voluntary muscular movements.*

Montagna, W. 1956. The structure and function of the skin. New York, Academic Press, Inc. 356 pp. *A general text that integrates anatomy and function.*

————. 1959. Comparative anatomy. New York, John Wiley & Sons, Inc. x + 397 pp. *Comparative anatomy of the vertebrates for college sophomores.*

————. 1965. The skin. *Sci. Am.,* vol. 212, no. 2, pp. 56–66.

Porter, K. R., and **C. Franzini-Armstrong.** 1965. The sarcoplasmic reticulum. *Sci. Am.,* vol. 212, no. 3, pp. 72–80.

Prosser, C. D., and **F. A. Brown, Jr.** 1973. Comparative animal physiology. 3d ed. Philadelphia, W. B. Saunders Company. xxii + 966 pp. *Includes both vertebrates and invertebrates.*

Romer, A. S. 1962. The vertebrate body. 3d ed. Philadelphia, W. B. Saunders Company. viii + 627 pp., 407 figs. *Structure and function.*

Satir, P. 1974. How cilia move. *Sci. Am.,* vol. 231, no. 4, pp. 44–52. *Cilia are molecular machines powered by ATP.*

Scheer, B. T. 1948. Comparative physiology. New York, John Wiley & Sons, Inc. x + 563 pp., 73 figs. *Protozoa to vertebrates.*

————. 1963. Animal physiology. New York, John Wiley & Sons, Inc. xi + 409 pp., illus. *Includes biochemical materials.*

Smith, D. S. 1965. The flight muscles of insects. *Sci. Am.,* vol. 212, no. 6, pp. 76–88. *The wings of some insects beat hundreds of times per second; their muscles elucidate general muscle function.*

Vander, A. J., J. H. Sherman, and **D. S. Luciano.** 1975. Human physiology: The mechanisms of body function. 2d ed. New York, McGraw-Hill Book Company. vii + 614 pp. *Many excellent illustrations.*

Yapp, W. B. 1960. An introduction to animal physiology. 2d ed. New York, Oxford University Press. xix + 423 pp., 49 figs. *Includes both invertebrates and vertebrates.*

DIGESTIVE SYSTEMS
AND METABOLISM

Green plants build their tissues from inorganic materials by the photosynthetic process, using energy from the sun (Chaps. 1, 12). The food of animals is obtained by eating plants or other animals. It serves two purposes, as a fuel to supply bodily energy and as a source of materials for growth and repair. After being obtained (feeding), it is broken down into simpler chemical substances (digestion) and then is taken into the cells and tissues of the body (absorption), where it is utilized (metabolism).

5-1 Feeding Animals differ widely in their food habits. Some insects feed on the tissues or juices of a single species of plant or the blood of one kind of animal, but most animals take several or many kinds of food. Cattle, deer, rodents, and insects that eat plants are **herbivorous;** cats, sharks, flesh flies, and many marine animals whose food consists entirely or largely of other animals are **carnivorous;** humans, bears, rats, and others that eat various plant and animal materials are general feeders, or **omnivorous;** vultures and some insects that eat dead animals are **scavengers;** and certain frogs, lizards, birds, and mammals that feed primarily on insects are **insectivorous.** Paramecia and certain other protozoans, some sea anemones, certain fishes, and tadpoles, that feed on small particles, living or dead, such as plankton, are termed **microphagous** feeders. In contrast, most higher animals, including humans, that use larger materials are **macrophagous** feeders. A few animals feed on fluids, like the mosquitoes and ticks that suck blood and the aphids that pump in plant juices.

The digestive mechanism in various animals (Fig. 5-1) differs in general form, structural details, and physiologic processes according to the nature of the food, manner of life, and other factors. All means for taking and using food are essentially alike in that materials from the external environment are brought into intimate contact with internal membranous surfaces where digestion and absorption can take place.

5-2 Invertebrates   Many protozoans have no permanent structures for taking or digesting food. An amoeba pushes out lobes (pseudopodia) at any part of its one-celled body to surround an item of food; the latter is taken into a fluid-filled **food vacuole** in the cytoplasm for digestion (Fig. 15-4). In paramecia and other ciliate protozoans a permanent external **oral groove** lined by beating cilia

carries food particles to a definite "cell mouth," where they pass into food vacuoles and are digested (Fig. 15-22). Many animals, from protozoans to lower chordates, draw food to the mouth in a current of water by use of cilia.

The microscopic food of sponges is captured by and digested in flagellated **collar cells** that line certain interior canals of the animal; digestion is thus **intracellular,** as in protozoans. In intracellular digestion, small food particles that are engulfed by **phagocytosis** at the cell surface become surrounded by fluid-filled vacuoles into which digestive enzymes, probably carried chiefly by lysosomes, are released. Cnidarians have a definite **mouth** connected to a saclike **digestive** (gastrovascular) **cavity** within the body that is lined by a tissue layer of special digestive cells (Fig. 17-2). The flatworms (except tapeworms) have a mouth and a branched **digestive tract**

Figure 5-1 Types of digestive systems in animals and protistans; diagrammatic. *A*. Amoeba, food enters at any place on cell surface. *B*. Paramecium, with definite cell mouth. *C*. Hydra, mouth and saclike digestive cavity. *D*. Planaria, mouth and branched digestive tract but no anus. *E*. Earthworm, tubular digestive tract having specialized sections, complete with terminal mouth and anus. *F*. Vertebrate, complete and partly coiled tract with specialized parts and digestive glands, vent at base of tail.

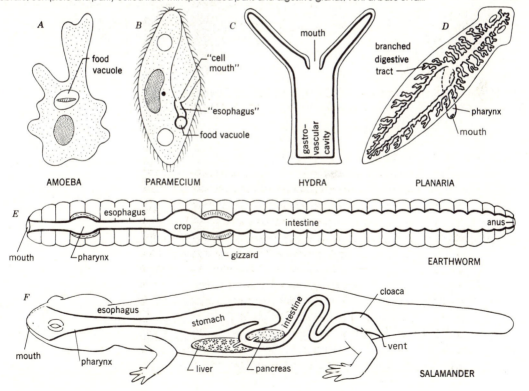

(gastrovascular cavity) extending to all parts of the body (Fig. 18-1). In both the latter groups the tract is **coelenteronic** (incomplete) in that foods enter and undigested residues pass out the same opening, the mouth. In the coelenterates and flatworms, food that has entered the digestive tract is acted upon by enzymes secreted from gland cells in the interior lining. This is **extracellular** digestion, in a digestive cavity, such as occurs in all higher animals; some partly digested food, however, is taken into cells lining the cavity for **intracellular** digestion.

In most other invertebrates the digestive tract is essentially a tube within the body. It opens to the outside (mouth, anus) and is separated from the interior body spaces by selectively permeable membranes. It is termed **enteronic** (complete) because food enters the mouth and passes through various organs for storage, digestion, or absorption and any residues pass out the anus at the opposite end of the system. The parts differ in structure in animals belonging to various groups. (Chaps. 19 to 26), but the names applied to them give some indication of the function of each part. An earthworm, for example, has a **mouth** with fleshy lips to grasp food, a muscular **pharynx** that sucks in the food and lubricates it by mucous secretions, a slender **esophagus** to carry food on to the dilated **crop** for storage, a muscular-walled **gizzard** where food is ground against particles of sand, and a long **intestine** with pouchlike lateral extensions providing a large surface for absorption of digested portions (Fig. 22-3). Undigested residues pass out the anus at the posterior end of the body. Jaws with teeth occur in the mouths of some other annelid worms, in squids and octopuses, in sea urchins, and in many arthropods. The mouth in most mollusks has a radula (Fig. 21-16) bearing many fine horny teeth that serve to rasp off particles of food. The mouth parts of arthropods are modified appendages; those of insects are adapted for either chewing or sucking (Chap. 25; Table 25-2).

5-3 Vertebrates[1]   The digestive system of almost every vertebrate has the following essential parts (Figs. 5-2, 5-3): (1) The **mouth** and **mouth cavity** commonly have **teeth** to grasp, tear, or chew food and a **tongue** (fishes excepted) that may help in capturing or manipulating it; in most land vertebrates the **salivary glands** secrete saliva to lubricate the food and start digestion. (2) The **pharynx** contains gill slits in fishes and some aquatic amphibians but has no direct digestive function. (3) The **esophagus** (gullet) is a flexible tube carrying food past the region of the heart and lungs. (4) The **stomach** is a large pouch where food is stored and some digestion occurs. (5) The **small intestine,** a long, slender, coiled tube, is the principal region for digestion and absorption. (6) The **large intestine (colon)** is the portion where water and salt absorption occurs, some cellulose is partly digested by bacteria, and undigested residues are formed into relatively dry masses (feces) for expulsion through (7) the **cloaca,** which ends with (8) the **anus** or **vent.** In addition, the cloaca is an exit for excretory wastes and sex cells in sharks, amphibians, reptiles, and birds, but these enter it by separate openings; the cloaca is absent in most mammals.

[1] *For a comparison of the digestive and other organ systems in the various classes of vertebrates, see the figures on "general structure" in Chaps. 28 to 34.*

Figure 5-2 Diagram of structure and activities in the digestive tract of a vertebrate. Wavy lines indicate glandular areas.

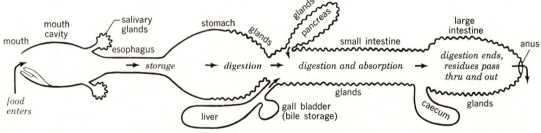

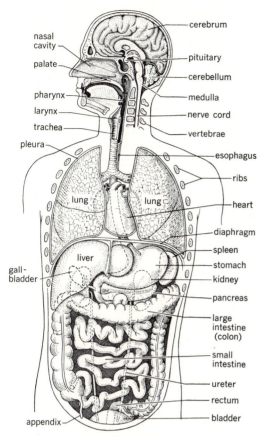

Figure 5-3 **The human digestive system.**

All vertebrates have two large digestive glands, the **liver** and **pancreas,** connected by ducts to the upper part of the small intestine.

Typically all vertebrates have **teeth** in both upper and lower jaws except the living birds, turtles, and a few specialized forms in other classes. Most fishes, amphibians, and reptiles have simple, slender, conical teeth attached to the bone surfaces (Chaps. 29 to 32). In the absence of flexible lips, their teeth serve mainly to grasp the food. Some birds peck or tear food with their beaks, but many swallow it intact. Among mammals, however, the teeth of an individual usually are of several kinds, differentiated for cutting, piercing, shearing, or grinding according to the materials used as food (Secs. 34-8, 34-19). Structurally a tooth has a hard outside enamel, a filling of softer dentine, and a central living pulp supplied

with blood vessels and nerves (Fig. 5-4). Teeth of mammals and some reptiles have the root of each set in a socket of the bony jaw.

5-4 Food and digestion The plant and animal foods taken by animals consist of proteins, carbohydrates, and fats, together with vitamins, minerals, and water. The water and inorganic salts can be absorbed from the digestive tract without change, but the organic materials must be altered before they can be utilized. Some foods are subjected only to chemical alteration, as with the microscopic organisms taken as food by protozoans and other small animals, the fluids of plants sucked up by bees and aphids, the blood pumped in by parasitic worms, leeches, or insects, and the larger prey taken by cnidarians and starfishes. Many other animals have the capacity to reduce food physically. This must be done before chemical digestion can proceed effectively. It is accomplished by teeth in the mouth or elsewhere (pharynx of some fishes, stomach of crayfishes) and by grinding in the gizzard of earthworms and birds. Flesh eaters such as the sharks, large fishes, snakes, hawks, owls, cats, and others bolt down their food intact or in large pieces, and its physical reduction is accomplished by muscular and chemical action in the stomach. Other fishes and the herbivorous mammals that eat plant materials chew their food thoroughly before it can be digested. Insects and

Figure 5-4 **Enlarged section of human tooth in the jaw.** Compare Fig. 34-13.

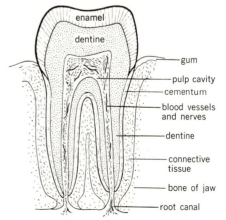

many land vertebrates have salivary glands that provide secretions to moisten the food while it is being chewed and swallowed.

5-5 Digestive enzymes   The chemical aspects of digestion involve the reduction of complex organic substances in the food into simpler molecules that can be passed through cells of the digestive epithelium to enter the fluids and cells of the body. Proteins are reduced to amino acids, fats to fatty acids and glycerol, and carbohydrates to simple sugars (*monosaccharides*) such as glucose. These changes are performed by the digestive enzymes (Sec. 2-19).

These enzymes are produced by all animals from protozoans to mammals, but not the same number or kinds of enzymes are present in every sort of animal. The food in a vacuole within a protozoan changes gradually in form and size as it is acted upon by enzymes. The reaction of the vacuole changes from acid to alkaline during the process, as can be shown by indicator dyes. The cytoplasm therefore has the ability to secrete enzymes and also substances to change the acidity (pH) of fluid in the vacuole. Among lower invertebrates, enzymes are secreted by cells in some or all parts of the digestive tract, but in higher animals only by glands or cells in certain portions of it. In the vertebrates, some are produced in the salivary glands, others regularly in the stomach and small intestine, and most by the pancreas (Fig. 5-2).

5-6 The digestive process in humans (Fig. 5-9) The taking of food into the mouth cavity is a joint action of the lips, tongue, and teeth. The flexible lips are delicately sensitive to the physical character and temperature of the food but not to taste. The tongue, having muscles in three planes, has great mobility in handling the food; on its surface the taste buds (Fig. 9-12) are concentrated. The teeth cut and grind the food.

The food is lubricated by **saliva** secreted by three pairs of **salivary glands,** the submaxillary, sublingual, and parotid (Fig. 5-5). About 1000 mL (1 L; a little over a quart) of saliva is produced per day, mostly at mealtimes. Secretion is a reflex act (Sec. 9-10) stimulated by savory tastes in the food or even by the sight or smell of food that literally "makes your mouth water." The intensity of the stimulus seemingly is related to the water content of the food —dry bread in the mouth stimulates a copious flow, wet bread much less, and water none at all.

Saliva contains a protein, mucin, serving as a lubricant, and an enzyme, **salivary amylase (ptyalin);** the latter, in the normal alkaline medium of the mouth, hydrolyzes starches first to polysaccharide fragments and then to the disaccharide maltose (malt sugar). The action is more rapid on cooked starch but at best is slight because food is in the mouth only a short time. Chewing aids starch digestion by breaking up the food, mixing in the enzyme, and lengthening the time of exposure to amylase. Starch in an average meal requires about an hour for complete

Figure 5-5   *A.* Food in mouth prior to swallowing; respiratory path with epiglottis elevated and glottis open. *B.* Closure of nasal passage and trachea during swallowing.

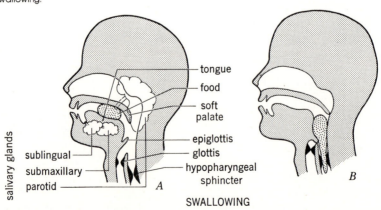

tongue
food
soft palate
epiglottis
glottis
hypopharyngeal sphincter

salivary glands
sublingual
submaxillary
parotid

A
B

SWALLOWING

digestion. The action of amylase continues within the food mass in the stomach until penetrated by the acid gastric juice. About 50 percent of starch digestion is completed by salivary amylase during passage through the stomach.

After a mouthful of food has been ground by the teeth and mixed with saliva, the tongue, by voluntary action, moves it backward into the pharynx and there presses it into a compact bolus. The remainder of the process of swallowing is involuntary, guided by a sequence of reflex movements. Respiration is inhibited, the larynx is raised, and the glottis closes. The soft palate rises to close the nasal cavity. As the bolus of food passes backward, it forces the epiglottis downward to cover the closed glottis, and the hypopharyngeal sphincter opens (Fig. 5-5). Failure in any of these reflexes results in "swallowing the wrong way"—the bolus of food enters the larynx, and choking is followed by a convulsive cough to remove the obstacle. The normal passage of a bolus through the esophagus results from a slow wave of muscular contraction down its walls until the food passes the gastroesophageal sphincter (cardiac valve) and enters the stomach.

All the movements and kneading of food in the digestive tract below the pharynx are by slow rhythmic contraction and relaxation of involuntary muscles, longitudinal and circular, in the wall of the tract. This process is known as **peristalsis.** By alternate action of the muscles, the diameter of the tract at any one place is first enlarged, then reduced. In the stomach the alternate action kneads and mixes the food with gastric secretions. In the intestine this movement, long continued, serves to divide and redivide the contents, to mix it thoroughly, to bring new portions against the inner wall, and to move the contents slowly along.

5-7 The stomach   In the stomach the food is given both physical and chemical treatment, then is passed, a little at a time, to the small intestine. Storage is mainly in the upper part (fundus) of the stomach, and muscular action occurs chiefly in the middle (body) and lower (antral or pyloric) portions. The lower portion ends with the pyloric valve, a circular muscle at the junction with the intestine.

The stomach's secretion has an antiseptic effect on bacteria in the food and it partly digests proteins. Yet surgical removal of the human stomach is not necessarily fatal, because most of the food is digested in the small intestine.

The stomach wall contains gastric glands (an estimated 35 million) that secrete the gastric juice. Pioneer studies on gastric juice were made by William Beaumont (1785–1853) on a man—Alexis St. Martin —whose stomach had been partly shot away. The wound healed so that a permanent fistula (external opening) remained. Over several years Beaumont took and analyzed samples of gastric juice and tested its action on various foods; this provided the first understanding of the digestive process.

The gastric secretions include **mucin,** which further lubricates the food mass, **hydrochloric acid** (about 0.2 percent) released by the stomach's **parietal cells,** and enzymes. The acidity of gastric juice (pH about 1.0) is well known from the unpleasant experience of vomiting. To produce the acid the stomach must secrete hydrogen ions against a large concentration gradient. The concentration of ions in the stomach may be 3 million times that in the blood. The mechanism is unknown, but the parietal cells have an extensive secretory surface of numerous intracellular channels associated with many large mitochondria that provide energy for ion transport.

Of the gastric enzymes, **pepsin,** released by the stomach's **chief cells,** splits proteins (to polypeptides such as proteoses and peptones). **Rennin** causes the casein in milk to coagulate. It occurs chiefly in young animals. (Rennin extracted from the stomachs of calves is used to form the curd in cheese making.) An average person secretes an estimated 2000 to 3000 mL gastric juice daily. The mixture of partly digested food particles and secreted fluids that accumulate in the stomach is known as **chyme.**

Since the stomach secretes acid and proteolytic enzymes, why does it not digest itself? The following factors appear to be responsible by protecting the gastric epithelium: (1) a mildly alkaline mucus, (2) low permeability to the uptake of hydrogen ions, (3) tight, "leakproof" intercellular junctions, and (4) rapid cellular replacement.

5-8 The intestines   The small intestine is a slender tube about 7.6 m (25 ft) long. The first 25 cm (10 in) or so are the duodenum, the long central part is the jejunum, and the remaining 120 to 150 cm (4 or 5 ft) the ileum. When food in semifluid state (chyme) has passed through the pyloric valve into the duodenum, it stimulates secretion of pancreatic juice and fluid from tubular glands in the much-folded wall there and elsewhere in the intestine (Fig. 5-6). The pancreatic secretion contains several enzymes, among which are **trypsin,**[1] **chymotrypsin,** and **carboxypeptidase,** which break down proteins and polypeptides to amino acids; **amylase,** which breaks down polysaccharides into a mixture of glucose and

maltose, completing the action begun by salivary amylase; and **lipase,** which breaks down fats into fatty acids and glycerol; other enzymes break down nucleic acids, etc. Bicarbonates, secreted by pancreatic duct cells, neutralize the hydrochloric acid in the chyme, making the intestinal contents mildly alkaline. This prevents ulcerative damage to the intestine and creates an environment in which the pancreatic enzymes can act. The daily secretion of pancreatic juice may be 500 to 1000 mL.

A third fluid, **bile** (not an enzyme), is added from the liver through the bile duct. This greenish-yellow liquid contains bile salts that facilitate digestion by physically reducing fats to small droplets (emulsification). Bile is secreted by the liver at a rate of 250 to 1,000 mL/day and flows into the duodenum at or near the site of the pancreatic duct. A small amount of bile

[1] *Actually secreted by the pancreas as inactive trypsinogen; becomes active protease trypsin only after it has entered the small intestine and become activated there by the protein enterokinase from cells in the intestinal wall.*

Figure 5-6   Structure of the small intestine (duodenum).   *A.* Cross section. *B.* Diagrammatic longitudinal section, enlarged. *C.* Microvilli on surface of intestinal epithelial cell, × 7000.

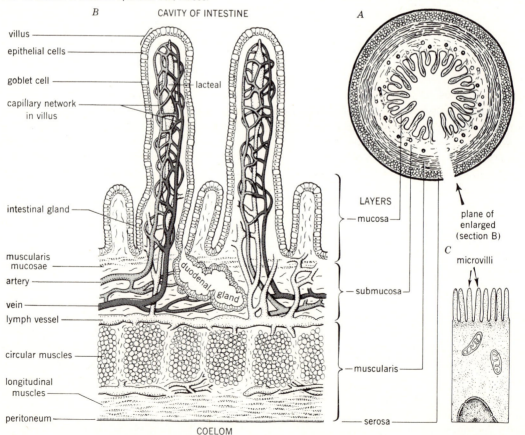

(about 33 mL) is stored in the gallbladder. If the flow of bile is interrupted mechanically, as by gallstones or infection of the bile duct, certain bile pigments accumulate in the blood and body tissues and produce jaundice, with yellowing of the skin. The pigments are breakdown products of hemoglobin from worn out red cells. They give urine and feces their characteristic yellowish to brownish color.

The **liver,** the largest "gland" in the body, besides secreting bile, performs several other functions related to the digestive tract and other parts of the body (Fig. 5-10). Briefly, it (1) stores glucose (as glycogen) and provides a regulated supply to the body as needed; (2) serves in protein synthesis, regulating the concentration of amino acids in the blood; (3) functions in the breakdown and disposal of waste nitrogenous products, converts ammonia (toxic product of protein metabolism) to urea, and helps dispose of other toxic substances; (4) forms a substance (antianemia factor) aiding in red blood cell production—but also breaks down old red cells; (5) produces heparin, a blood anticoagulant; (6) stores vitamins; and (7) is involved in hormone metabolism.

Glands of the mucosa of the small intestine secrete acid-neutralizing bicarbonate and large amounts of mucus but essentially no enzymes. Cellular activity is great, the entire intestinal epithelium being replaced approximately every 36 h.

The **large intestine,** or **colon,** serves principally to store and concentrate undigested and indigestible residues and to dispose of them by way of the rectum and anus. It conserves water by absorption from the mass and actively takes up sodium and small amounts of vitamins (H and K) synthesized by intestinal bacteria. Food residues, together with bacteria, mucus, and dead cells from the intestinal wall, become the **feces** that are expelled at intervals. Food usually passes from the mouth to the end of the small intestine in about $4\frac{1}{2}$ h, but residues may continue for a longer time, even beyond 24 h, in the colon. During this time there is much bacterial action. Bacteria that survive the stomach acidity multiply rapidly, and some bring about more or less putrefaction, especially in the colon, where various gases (nitrogen, H_2S, CO_2) result. Bacteria constitute up to 50 percent of the dry weight of the feces.

5-9 Control of digestion The muscular and secretory activities of the gastrointestinal tract are influenced by (1) autonomic nerves (chiefly the vagus and sympathetic fibers), (2) internal nerve plexuses within the walls of the tract itself, and (3) hormones secreted by gastrointestinal glands. Control is chiefly automatic, involving reflex responses to physical and chemical characteristics of food. Short reflexes are initiated by receptors in the tract itself and employ neural plexuses in the walls of the tract. Much peristaltic mobility is of this type. Longer reflexes may be triggered by receptors both inside and outside the tract; the central nervous system and higher centers may be involved. Thus food entering the mouth stimulates salivary secretion by activating taste-bud cells from which impulses pass along sensory nerves to a salivary center in the medulla. From there motor fibers in the autonomic nervous system (especially parasympathetic fibers) carry impulses to the salivary glands (Fig. 9-9). Simultaneously gastric juice is secreted in the stomach, as was shown by Ivan Pavlov (1849–1936): the esophagus in a dog was severed and the two cut ends brought out to open on the skin of the neck region; when food was placed in the animal's mouth, its stomach secreted gastric juice although no food reached the stomach. The exocrine secretion of the pancreas and the secretion of bile also are stimulated by the presence of food in the mouth. Emotions may greatly influence the activities of the digestive tract. For example, sadness and fear tend to inhibit gastric mobility and anger to increase it.

The gastrointestinal endocrine glands are stimulated both by neural and hormonal signals and directly by the chemistry of the gut contents. In a normal animal, when food enters the stomach, it stimulates the release of **gastrin** (a hormone) by stomach cells, and this brings about the secretion of more gastric juice. The flow of pancreatic juice is also under hormonal control (Fig. 5-7). The chyme, upon entering the duodenum, stimulates the release of hormones (Chap. 8) from cells in the intestinal wall. They are carried in the blood to the pancreas. Acid in the chyme causes the release of **secretin,** which promotes the secretion of the alkaline pancreatic bicarbonate. (Bicarbonate production is also stimulated by the release of stomach gastrin.) Fats and peptide

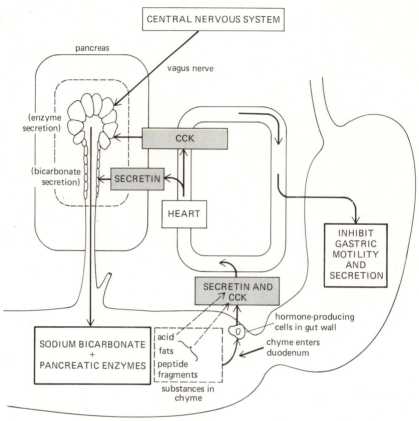

Figure 5-7 Pathways involved in the secretion of pancreatic juice. *(After Vander et al., Human physiology, McGraw-Hill, 1975.)*

fragments cause the release of **cholecystokinin** (CCK, pancreozymin), which stimulates the flow of the pancreatic enzymes. (Gastrin also promotes the secretion of these enzymes.) CCK causes the gallbladder to contract and discharge bile. Pancreatic juice, being alkaline, neutralizes the acid gastric juice, thereby removing the stimulus to its own secretion. Secretin and CCK, along with neural reflexes, inhibit gastric motility and secretion, allowing adequate time for digestion of the chyme before more enters the duodenum. Thus each stage in the digestive process influences others and results in a coordinated procedure.

5-10 Absorption The small intestine is the principal area for absorption, the process that fulfills the purpose of digestion. Through the intestinal wall the

chemical substances derived from food enter the bloodstream and are distributed to the body to be built into living tissues or used for energy. The absorbing surface is increased about 600 times over that of a flat-surface tube by folds in the intestinal mucosa and by 4 million or more minute, mobile projections, or **villi,** which bear still smaller projections, **microvilli,** on the surface of their epithelial cells (Fig. 5-6). Each villus contains blood capillaries and a central **lacteal vessel.** The walls of the cells in the intestinal epithelium are selectively permeable.

The end products of protein and carbohydrate digestion (amino acids, simple sugars) are actively transported (moved against a concentration gradient) through cells of the intestinal mucosa and released into blood capillaries connecting to the hepatic portal vein, whence they are carried to the liver. End products of fat digestion enter the lacteals, which connect to the lymphatic system. Lymph vessels from the intestine join those from elsewhere in the body to form a large trunk, the thoracic duct, that enters the venous system close to the heart (Chap. 6; Fig. 6-7).

5-11 Metabolism When the products of digestion reach their ultimate destinations by way of the blood, they are variously (1) broken down chemically to supply energy (catabolism) and stored as ATP until energy is needed by the cells; (2) stored as glycogen—animal starch—or as depot fat; or (3) built into new cellular structures (anabolism). Synthesis and breakdown are going on simultaneously in every living cell. The two processes are in dynamic equilibrium, with one or the other dominating at various times or places (Figs. 5-9, 5-10). The catabolic processes using glucose to provide energy are discussed in Sec. 2-24.

The liver plays a central role in this process, receiving glucose in blood from the intestine and con-

verting it into glycogen. Glycogen is (1) stored in the liver for use between meals, when it is reconverted into glucose; and (2) released into the bloodstream gradually, thereby maintaining a rather constant glucose (blood sugar) level of about 0.1 percent. The blood sugar level is largely controlled by the hormones insulin and glucagon, formed by the islets of Langerhans in the pancreas (Sec. 8-9). Most of the lactic acid produced in the muscles is also carried by the blood to the liver, where it is converted to liver glycogen. Thus there is a constant circulation of carbohydrate within the body (Fig. 5-10).

When carbohydrates are taken in excess, they can be converted into fats by conversion to 2-carbon units via glycolysis and a process similar to reversal of the β-oxidation pathway to form fatty acids (Fig. 2-9). Fats are stored in fat cells among the muscles, beneath the skin, and elsewhere in the body. In nonmammalian vertebrates and some invertebrates (i.e., insects) they may be stored in special structures, the **fat bodies,** commonly located in the abdominal cavity. Fats are also formed from fatty acids

Figure 5-8 Flow chart of input and output of the body.

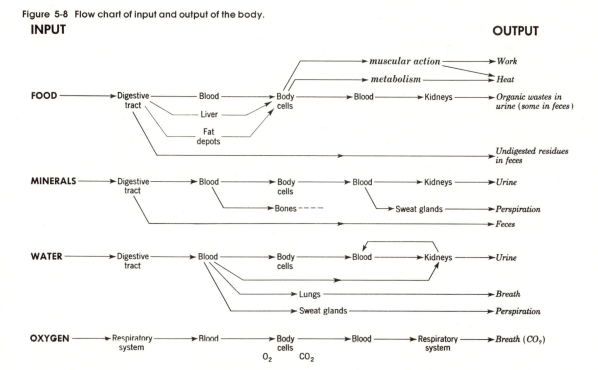

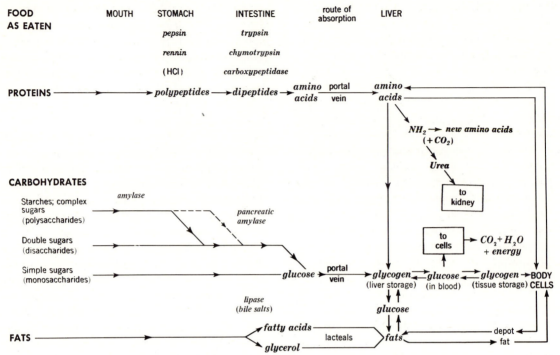

Figure 5-9 History of food in the body—from the mouth through the processes of digestion and distribution, including some functions of the liver. Enzymes named in small lightface italic. (*Adapted from Weisz, 1954.*)

and glycerol absorbed by the lacteals and transported by way of the lymphatic system.

5-12 Utilization   The metabolic alteration of food in the body to yield energy is an oxidative process, analogous to combustion of a fuel but much more complex (Sec. 2-24). The heat of combustion of

Figure 5-10 Outline of carbohydrate use.

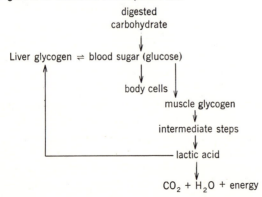

any compound is recorded in calories (cal)—1 cal is the heat needed to raise 1 g of water 1°C at 15°C. In metabolism the kilocalorie is used: 1 kcal = 1000 cal. Laboratory determinations show the fuel value of the three main classes of foodstuffs in kilocalories per gram to be as follows: carbohydrate, 4.1; fat, 9.3; protein, 5.5 (but in an animal the protein value is 4.1, because urea and some other products of protein metabolism are not oxidized). Fat thus is a fuel of high energy value in the animal body.

The amount of energy needed to maintain vital functions in an animal at rest when no food is being digested or absorbed is termed the **basal metabolic rate.** For a 25-year-old human adult male weighing 70 kg (155 lb) at 37°C it is about 1 kcal/kg/h, or 1700 kcal/day (70 × 24 = 1680). In active life the daily energy requirements in kilocalories at age 25 are about as follows: for the average worker, male 3200 and female 2300; sedentary worker, male 2400 and female 1700; heavy worker, male 4000 and female 2800. An estimated 10 percent of caloric value is lost in preparation and cooking or is not absorbed, for which allowance should be made in determining optimal caloric intake. Other variables are introduced by differences in functioning of the thyroid gland (Sec. 8-4), by age, by bodily proportions, and by the environmental temperature.

The percentage of food actually absorbed of the total amount ingested is an indication of the degree or efficiency of utilization. This varies widely, depending on the composition of the foodstuff and the specific needs of each kind of animal. Meat and other materials of animal origin are almost completely utilized, 95 percent or more being absorbed, whereas foods from plant sources are less completely used, depending on the amount of indigestible material. Seeds usually yield higher percentages of usable food than leaf or stem materials.

A daily account of the intake and output for each group of substances shows that the animal body is in a state of balance. Demands are greater in growing individuals or in those which are very active, as compared with sedentary individuals. The balance is maintained by the selective utilization of various substances in the diet. It is upset if any essential and irreplaceable element is lacking and also if the total of foodstuffs is not sufficient for the minimum requirements of the organisms. A diet, therefore, should be adequate in both kinds and amounts of food.

5-13 Kinds of food The preceding discussion shows that the amount of food is not the only criterion for an adequate diet. A balanced diet is a mixture of foods containing all the substances essential for the development, growth, and maintenance of the individual. Foremost among these are the carbohydrates, fats, and proteins. Furthermore, the proteins must include most of the amino acids to provide the body with the building materials to synthesize its own proteins.

Certain additional substances, the **inorganic elements** (minerals) and the **vitamins,** are essential in a balanced diet. Most diets contain adequate amounts of these substances, so their need was not suspected until recent years. Experiments have shown that minute amounts of iron, copper, zinc, manganese, cobalt, and iodine are essential **trace elements** for animals; a number of vitamins are necessary for normal health, growth, and reproduction.

Iodine is an example of an essential trace element. For several hundred years it has been known that iodine deficiency produces in humans a disease called goiter, with a tumorlike bulge in the neck region (Sec. 8-4). Goiter results from malfunction of the thyroid gland, which, in the absence of iodine, cannot produce the essential metabolism-regulating hormone, thyroxine. Iodine makes up 65 percent by weight of the thyroxine molecule, but not more than one part per million is needed in the bloodstream, because the thyroid gland can accumulate and store iodine until the necessary level is reached. The required minute amounts of iodine are now generally provided, in areas where it does not occur naturally, in food or water by use of "iodized salt."

Another deficiency disease is anemia, caused by insufficient iron. Most (66 percent) of this element in the body is contained in the hemoglobin of the blood; additional iron is found in the liver, spleen, and bone marrow, where red blood cells are formed.

Addition of minute amounts of certain antibiotics to the feed of poultry and domestic mammals has been found to increase markedly the rate of growth, evidently by reducing pathogenic organisms in the digestive tract. This advantage may be lost, however, if resistant pathogens invade the animals.

5-14 Vitamins (Table 5-1) These are organic substances, mostly of plant origin, required for normal growth in animals. Their existence was suspected late in the nineteenth century when it was found that a diet of pure carbohydrates, fats, and proteins would not support life. Eijkman (1858–1930) was the first to produce a disease of dietary origin, polyneuritis, in fowls (Fig. 5-11) by feeding them exclusively on polished rice. The resemblance between this condition and a human illness, beriberi, led him to cure many cases in Java with an extract of rice bran. This and other nutritional factors later were named vitamins; the polyneuritis factor was called vitamin B_1. In the last 50 years most of the vitamins have been isolated, crystallized, characterized chemically, and synthesized. Vitamins are effective in minute amounts. The average human being needs only 1 to 2 mg of vitamin B_1 daily, usually supplied by whole-grain bread or cereal. Well-balanced

Table 5-1
THE VITAMINS AND THEIR CHARACTERISTICS

Name, Formula, and Solubility	Important Sources	Functions	Result of Deficiency or Absence (in humans, except as noted)
Lipid-soluble vitamins **A** ($C_{20}H_{30}O$), antixerophthalmic	Plant form (carotene, $C_{40}H_{56}$) in green leaves, carrots, etc.; is changed in liver to animal form ($C_{20}H_{30}O$), present in fish-liver oil (shark); both forms in egg yolk, butter, milk	Maintains integrity of epithelial tissues, especially mucous membranes; needed as part of visual purple in retina of eye	Xerophthalmia (dry cornea, no tear secretion), phrynoderma (toad skin), night blindness, growth retardation, nutritional roup (hoarseness) in birds
D ($C_{28}H_{44}O$), antirachitic	Fish-liver oils, especially tuna, less in cod; beef fat; also exposure of skin to ultraviolet radiation	Regulates metabolism of calcium and phosphorus; promotes absorption of calcium in intestine; needed for normal growth and mineralization of bones	Rickets in young (bones soft, yielding, often deformed); osteomalacia (soft bones), especially in women of Orient
E, or tocopherol ($C_{29}H_{50}O_2$), antisterility	Green leaves, wheat-germ oil and other vegetable fats, meat, milk	Antioxidative; maintains integrity of membranes	Sterility in male fowls and rats, degeneration of testes with failure of spermatogenesis, embryonic growth disturbances, suckling paralysis and muscular dystrophy in young animals
K ($C_{31}H_{46}O_2$), antihemorrhagic	Green leaves, also certain bacteria, such as those of intestinal flora	Essential to production of prothrombin in liver; necessary for blood clotting	Blood fails to clot
Water-soluble vitamins **B** complex Thiamine (B_1) ($C_{12}H_{17}ON_4S$), antineuritic	Yeast, germ of cereals (especially wheat, peanuts, other leguminous seeds), roots, egg yolk, liver, lean meat	Needed for carbohydrate metabolism; thiamine pyrophosphate an essential coenzyme in pyruvate metabolism (stimulates root growth in plants)	On diet high in polished rice beriberi (nerve inflammation); loss of appetite, with loss of tone and reduced motility in digestive tract; cessation of growth; polyneuritis (nerve inflammation) in birds
Riboflavin (B_2) ($C_{17}H_{20}O_6N_4$)	Green leaves, milk, eggs, liver, yeast	Essential for growth; forms prosthetic group of FAD enzymes concerned with intermediate metabolism of food and electron transport system	Cheilosis (inflammation and cracking at corners of mouth), digestive disturbances, "yellow liver" of dogs, curled-toe paralysis of chicks, cataract
Nicotinic acid, or niacin ($C_6H_5O_2N$), antipellagric	Green leaves, wheat germ, egg yolk, meat, liver, yeast	Forms active group of nicotinamide adenine dinucleotide, which functions in dehydrogenation reactions	Pellagra (Fig. 5-12) in humans and monkeys, swine pellagra in pigs, blacktongue in dogs, perosis in birds
Folic acid ($C_{19}H_{19}O_6N_7$)	Green leaves, liver, soybeans, yeast, egg yolk	Essential for growth and formation of blood cells; coenzyme involved in transfer of single-carbon units in metabolism	Anemia, hemorrhage from kidneys, and sprue (defective intestinal absorption) in humans; nutritional cytopenia (reduction in cellular elements of blood) in monkeys; slow growth and anemia in chicks and rats
Pyridoxine (B_6) ($C_8H_{12}O_3N$)	Yeast, cereal grains, meat, eggs, milk, liver	Present in tissues as pyridoxal phosphate, which serves as coenzyme in transamination and decarboxylation of amino acids	Anemia in dogs and pigs; dermatitis in rats; paralysis (and death) in pigs, rats, and chicks; growth retardation

Table 5-1 (*continued*)
THE VITAMINS AND THEIR CHARACTERISTICS

Name, Formula, and Solubility	Important Sources	Functions	Result of Deficiency or Absence (in humans, except as noted)
B complex Pantothenic acid ($C_9H_{17}O_5N$)	Yeast, cane molasses, peanuts, egg yolks, milk, liver	Forms coenzyme A, which catalyzes transfer of various carboxylated groups and functions in carbohydrate and lipid metabolism	Dermatitis in chicks and rats, graying of fur in black rats, "goose-stepping" and nerve degeneration in pigs
Biotin (vitamin H) ($C_{10}H_{16}O_3N_2S$)	Yeast, cereal grains, cane molasses, egg yolk, liver, vegetables, fresh fruits	Essential for growth; functions in CO_2 fixation and fatty acid oxidation and synthesis	Dermatitis with thickening of skin in rats and chicks, perosis in birds
Cyanocobalamin (B_{12}) ($C_{63}H_{90}N_{14}O_{14}PCo$)	Liver, fish, meat, milk, egg yolk, oysters, bacteria and fermentations of *Streptomyces*; synthesized only by bacteria	Formation of blood cells, growth; coenzyme involved in transfer of methyl groups and in nucleic acid metabolism	Pernicious anemia, slow growth in young animals, wasting disease in ruminants
C, or ascorbic acid ($C_6H_8O_6$)	Citrus fruits, tomatoes, vegetables; also produced by animals (except primates and guinea pigs)	Maintains integrity of capillary walls; involved in formation of "intercellular cement"	Scurvy (bleeding in mucous membranes, under skin, and into joints) in humans and guinea pigs

Figure 5-11 Vitamin B₁ deficiency — polyneuritis (nerve inflammation). *A*. Pigeon fed 12 to 24 days on polished rice lacking vitamin B₁. *B*. Same bird, completely normal, a few hours after receiving B₁ concentrate or food high in B₁ content. (*After Harris, Vitamins, J. & A. Churchill, Ltd.*)

A

B

diets provide enough of all vitamins for normal health. Much modern advertising is misleading in claiming that people on normal diets need to take vitamin concentrates regularly.

Eating habits and some methods of preparing food can result in vitamin deficiencies. White bread and canned foods are commonly deficient; the heat of cooking destroys some vitamins, and others are lost when the water is drained off after cooking. Raw fruits and vegetables are helpful in the diet because they retain their vitamins. However, prolonged storage may lead to loss of vitamins.

The restricted diets of some people lead to deficiency diseases (Fig. 5-12). Scurvy (vitamin-C deficiency) was common on long sea voyages in past centuries; British sea captains who later learned to carry limes as a food accessory to prevent scurvy were known as "limeys." Beriberi is prevalent among Oriental peoples who live largely on polished rice, and pellagra among persons on diets largely of corn that is deficient in niacin.

Many vitamins are unrelated chemically, but each evidently regulates one or more bodily processes. Most act as parts of enzymes, or as coenzymes, or influence enzyme systems within cells. Thiamine

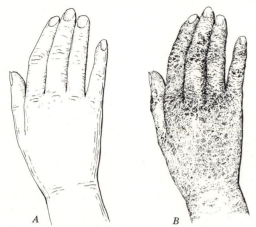

Figure 5-12 Pellagra. *A*. Normal hand. *B*. Hand of person on diet lacking niacin of vitamin B complex: skin thickened, sloughing, cracking, and with extra pigment. (*After Harris, Vitamins, J. & A. Churchill, Ltd.*)

(B_1), for example, may unite with phosphoric acid to form a pyrophosphate which acts as a coenzyme in the breakdown of pyruvic acid with removal of CO_2. Thiamine deficiency results in an accumulation of pyruvic acid in the tissues, with consequent nerve and muscle pain (beriberi).

Vitamin D is actually a hormone, **calciferol,** synthesized in many vertebrates by action of ultraviolet radiation in sunlight on 7-dehydrocholesterol in the skin. The hormone is carried in the blood to sites of bone formation and promotes calcification. When it is absent, bone deformities reflecting the deficiency disease called rickets result. In the absence of normal calcification the bones become flexible and their ability to support the body is seriously impaired. Characteristic symptoms are bowlegs, knock-knees, misshapen boxlike skull, and rib nodules. Fishes, unlike birds and mammals, synthesize calciferol enzymatically without ultraviolet light, which is filtered out by water. Fish-liver oil, high in vitamin D, is an excellent antirachitic medicine, because calciferol is highly stable. Rickets chiefly affects children and animals in parts of the world where the winter is especially long or sunlight is frequently intercepted by smoke and fog; it also occurs in cities, where exposure to sunlight is reduced by air pollutants, shadows of tall buildings, and indoor living.

5-15 Other digestive processes The digestive mechanisms and processes among animals vary in many details. In some birds and insects the lower end of the esophagus is dilated into a crop for temporary storage of food. Most birds also have a stomach of two parts, a slender, soft glandular proventriculus and a larger thick-walled, muscular gizzard, lined by hardened secretion, where food is ground up by grit swallowed for the purpose. The cattle, deer, and other cud-chewing mammals (ruminants) have a stomach of four compartments; the first three are specializations of the esophagus and have rough cornified linings where food is abraded and also worked on by bacteria (Fig. 34-14). Rodents, horses, and some other herbivores have a large, thin-walled caecum at the junction of the small and large intestines; there is some bacterial digestion of cellulose in this chamber. Humans have a short caecum (Fig. 5-3), to which the vermiform appendix attaches (the latter seems to have little function and may become infected, requiring surgical removal).

Microorganisms are essential for the digestion of cellulose by certain termites and a few other wood-eating invertebrates. If the bacteria and protozoans normally in the digestive system of a termite are eliminated (by high temperature), the insect starves, because it cannot produce enzymes to digest the cellulose in its diet of wood. Bloodsucking animals commonly have a special anticoagulant in the saliva that keeps the blood in a fluid state during ingestion. In scale insects and their relatives, which suck plant fluids, the terminal part of the intestine is bent back into a loop fitting into a dilated pouch of the esophagus; this "filter chamber" eliminates excess water from the highly diluted foodstuffs.

A few animals and protistans partly digest their food outside the body. The protozoan *Vampyrella* secretes an enzyme, cellulase, to dissolve the cell walls of an alga, *Spirogyra*, on which it feeds. A starfish extrudes its stomach to envelop and digest large prey. Spiders and the larvae of some beetles (*Dytiscus, Lampyris,* etc.) inject a protease into their prey—insects, tadpoles, slugs, and snails—that predigests parts of these animals, the softened food then being ingested.

Water and inorganic salts also are necessary in the diet, since they are essential elements of biologic structures (Sec. 2-14). Aquatic species obtain both from their environment. Land dwellers may take them in part or entirely with their food, but many drink to obtain an adequate amount of water. Some desert species depend entirely on water in food and do not drink even if water is offered. The normal oxidation of food in all animals yields oxidative, or "metabolic," water (Fig. 2-16). For insects living in dry wood or cereals, this is the main source of water.

References

Brooks, F. P. 1970. Control of gastrointestinal function. New York, The Macmillan Company. x + 222 pp. *Gastrointestinal physiology of significance in clinical medicine. Well illustrated.*

Davenport, H. W. 1977. Physiology of the digestive tract. 4th ed. Chicago, Year Book Medical Publishers, Inc. x + 263 pp.

————. 1972. Why the stomach does not digest itself. *Sci. Am.,* vol. 226, no. 1, pp. 86–93.

Jennings, J. B. 1972. Feeding, digestion and assimilation in animals. 2d ed. London, Macmillan & Co., Ltd. vi + 244 pp. *General introduction to the study of animal nutrition; a comparative approach.*

Kretchmer, N. 1972. Lactose and lactase. *Sci. Am.,* vol. 227, no. 4, pp. 70–78. *Lactase is the enzyme that breaks down lactose (milk sugar). For want of lactase many adults cannot digest milk.*

Levine, R., and **M. S. Goldstein.** 1958. The action of Insulin. *Sci. Am.,* vol. 198, no. 5, pp. 99–106. *Information on how insulin speeds sugar metabolism.*

Loomis, W. F. 1970. Rickets. *Sci. Am.,* vol. 223, no. 6, pp. 76–91. *Insufficient exposure to sunlight is the cause of this first air-pollution disease of smoky cities.*

McDonald, P., R. A. Edwards, and **J. F. D. Greenhalgh.** 1973. Animal nutrition. 2d ed. Edinburgh, Oliver & Boyd Ltd. viii + 479 pp. *An introduction to the study of nutrition in farm animals.*

Neurath, H. 1964. Protein-digesting enzymes. *Sci. Am.,* vol. 211, no. 6, pp. 68–79.

Trowell, H. C. 1954. Kwashiorkor. *Sci. Am.,* vol. 191, no. 6, pp. 46–50. *The most severe and common human nutritional disorder, caused by a deficiency of protein in the diet.*

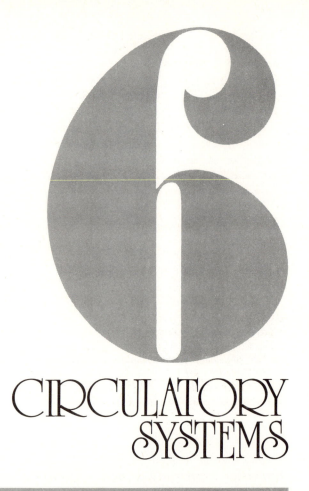

CIRCULATORY SYSTEMS

The life processes of animals and protistans require that nutrients and oxygen be available for metabolism and that wastes be removed promptly. In protozoans these interchanges are aided by streaming movements (cyclosis) of the cytoplasm within the one-celled body, and in the simple multicellular types, the exchanges occur by diffusion between the cells and adjacent body parts. More complex animals, with organs and tissues well removed from the exterior or gut, have a **circulatory system** for internal transport (Fig. 6-1). Its essential parts are (1) the **blood,** consisting of fluid plasma and free cells or blood corpuscles; (2) the **heart** (or an equivalent structure), with muscular walls that contract periodically to pump the blood through the body; and (3) a system of tubular **blood vessels** through which the fluid is moved. The system is **closed** in nemerteans,

holothurians, cephalopods, annelids, and vertebrates, where the vessels convey blood from the heart in various blood vessels and capillary beds among the tissues and back to the heart. Most mollusks and arthropods have an **open** (lacunar) system, blood being pumped from the heart through blood vessels to various organs but returning partly or entirely through body spaces—the **hemocoel**—to the heart.

6-1 Invertebrates Sponges, cnidarians, and flatworms have no circulatory system. Simple diffusion serves to carry digested food, respiratory gases, and wastes between various parts of their bodies. In nematodes, rotifers, and entoprocts body fluid is circulated in a pseudocoel. Most echinoderms have a

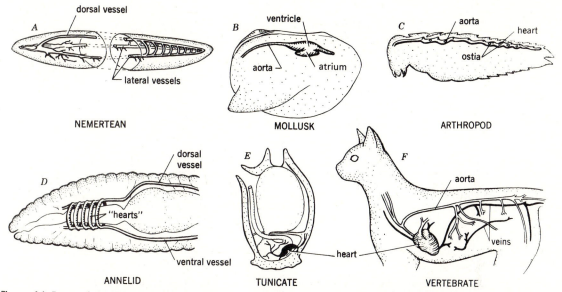

Figure 6-1 Types of circulatory systems in animals. *A.* Nemertean (ribbon worm), simple lengthwise dorsal and lateral vessels with cross connectives. *B.* Mollusk (bivalve), dorsal heart with atrium (1 or 2) and ventricle, anterior and posterior aortas; blood returns through body spaces (hemocoel) — open system. *C.* Arthropod (insect), dorsal tubular heart and aorta; blood returns through body spaces (hemocoel) — open system. *D.* Annelid (earthworm), dorsal and ventral vessels (and others) with cross connectives — closed system. *E.* Tunicate (sea squirt), a heart and aortas, vessels obscure; blood flow reverses. *F.* Vertebrate (mammal), chambered heart, definite aorta, arteries, and veins, with connections to respiratory organs — closed system.

complex coelom divided into several fluid-filled tube systems serving as a circulatory system. The coelomic fluid and the intercellular fluid are variously interconnected, but the ambulacral fluid is separate. Nemerteans have one dorsal and two lateral blood vessels with many cross connections. Pulsations in the walls of these serve to circulate the blood, which contains red blood cells resembling those of vertebrates.

The blood of most invertebrates has relatively few free cells in the plasma as compared with vertebrates. Usually there are amoeboid corpuscles resembling white blood cells, some phagocytic (Sec. 6-4) and some aiding in transport of food or other substances. In insects many of the cells cling to organs and become common in the plasma only after bodily injury or during molt (then there may be up to 30,000 to 70,000 per cubic millimeter in some species). If a respiratory pigment (Sec. 7-9) is present to carry oxygen, it often is dissolved in the plasma.

The heart of invertebrates is always dorsal to the digestive tract (Fig. 6-2). In mollusks it is short, lies

within a thin pericardial sac, and consists of one or two thin-walled atria that receive blood from the body and deliver it to a single muscular-walled **ventricle.** The latter contracts to force the blood through the vessels, or arteries, into tissue spaces, and from these spaces the blood reaches various organs (Fig. 21-20). Insects and many other arthropods have a slender dorsal tube as a heart, with segmentally placed lateral openings (ostia) that receive blood from the body spaces (hemocoel) and pump it through a median aorta, whence it flows through body spaces to organs and tissues (Figs. 24-3, 25-5). Many insects have accessory hearts which serve as boosters to propel blood through the antennae, wings, and legs.

The earthworm has a closed system with several lengthwise vessels through the body and paired transverse connecting vessels in most body segments. The circulation is produced by contractions of the middorsal vessel and by five pairs of lateral hearts in anterior body segments (Fig. 22-3; Table 22-1). In tunicates the heart is a valveless tube lying

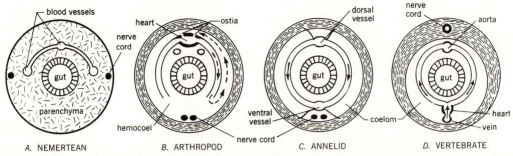

Figure 6-2 Diagrammatic cross sections to show relations of circulatory system to body spaces and organs. Arrows show path of blood flow. *A.* Nemertean. Mere system of vessels, no heart, body filled with packing tissue (parenchyma). *B.* Arthropod or mollusk. An open system. Dorsal heart (and aortas) from which blood escapes into body spaces — the hemocoel — and returns to the heart (some mollusks collect blood from the hemocoel in veins). Coelom reduced to small cavities (in pericardial sac, gonads, etc.). *C.* Annelid. Closed system of vessels containing the blood; body spaces (between gut and body wall); a true coelom. *D.* Vertebrate. Closed system; heart ventral; a true coelom.

in the pericardium, and the path of blood flow reverses at short intervals.

The hearts of vertebrates and mollusks are myogenic, the beat originating in the heart muscle itself, whereas the hearts of crustaceans and *Limulus* are neurogenic, the beat originating with nerve ganglion cells.

6-2 Vertebrates[1] The general features of circulation in vertebrates are shown in Fig. 6-3. Differences among fishes, amphibians, and birds are depicted in Fig. 6-6.

In all vertebrates the blood comprises (1) a nearly colorless plasma; (2) white blood cells, or leukocytes, of several kinds; (3) red cells, or erythrocytes, colored by the contained hemoglobin, which serves to transport oxygen; and (4) smaller cells, the platelets, or thrombocytes. The **plasma** carries dissolved foods, wastes, internal secretions, and some gases. Human blood plasma consists of about 92 percent water plus proteins and other organic compounds and about 0.9

[1] *The various functions of the circulatory system and the general structure of the blood, heart, and vessels of vertebrates are much as described for the frog (Sec. 31-8). Human blood cells are shown in Fig. 3-17, and their characteristics are given in Table 6-1; the structure and action of the human heart are illustrated in Fig. 6-4, and the principal vessels of the human circulatory system in Fig. 6-7; the microanatomy of arteries, capillaries, and veins is shown in Fig. 6-5; progressive differences in the heart and aortic arches of vertebrates are illustrated in Fig. 13-4.*

percent inorganic salts, chiefly sodium chloride; in health, these vary but slightly in amount. A physiologic salt solution containing the same kinds and amounts of these salts can be used to dilute blood without injuring the cells. An average person has 4.5 to 5.5 L (5 to 6 qt) of blood; about 60 percent is plasma.

6-3 Erythrocytes The **red cells** are nucleated in nearly all nonmammalian vertebrates. In mammals they are nonnucleated, biconcave, and circular (oval in camels). Mammalian red cells are nucleated, however, during their development. The red cells average about 4.5 million per cubic millimeter in a woman and 5.0 million in a man. The total number in a human being is about 30 trillion (3×10^{13}); each may live to 120 days and in a lifetime may make 170,000 circuits in the bloodstream. Red cells are more numerous in infants and in persons living at high altitudes; their numbers are altered in some diseases, being reduced in anemia. Red cells are produced chiefly in red bone marrow, and an excess supply is often stored in the spleen. Old cells are destroyed chiefly in the spleen, whence much of the hemoglobin passes to the liver; the pigment of hemoglobin is excreted in the bile, and its iron content is largely returned to the marrow.

6-4 Leukocytes The several kinds of **white cells** have their principal activities in the tissues, and those seen in the bloodstream are usually on their

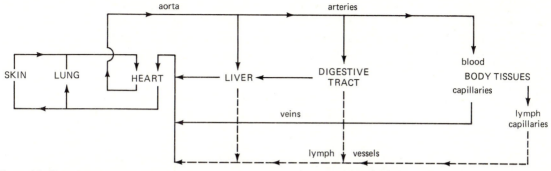

Figure 6-3 Plan of the circulation (flow chart) in a frog or other land vertebrate (the latter lack skin respiratory flow), with both blood vessels (solid lines) and lymph vessels (broken lines); arrows indicate the direction of flow.

Table 6-1
AVERAGE CHARACTERISTICS OF HUMAN BLOOD CELLS

Kind of Cell and Average Number per m³ of Blood	Structure; Color with Wright's Blood Stain;* Diameter	Source	Function
Erythrocytes (red blood cells): 5,000,000 (in males) 4,500,000 (in females)	Nonnucleated, circular, biconcave; orange-buff; 7.5 to 7.7 μm (8.6 μm in fresh blood)	Endothelium of capillaries in bone marrow	Transport oxygen; remain in blood vessels
Leukocytes (white blood cells): 5000 to 9000	Colorless in life		
1. GRANULOCYTES	Nucleus of lobes joined by thread, stains dark lilac; cytoplasm granular, pale blue; 10 to 12 μm	Reticuloendothelial cells outside capillaries of bone marrow	Amoeboid; can leave blood vessels and enter tissues; defend against infection, remove foreign matter
a. Neutrophils, 65 to 75%	Granules stain weakly		Protect against bacterial invasion
b. Eosinophils, 2 to 5%	Granules few, eosin (red)		Respond to inflammation
c. Basophils, 0.5%	Granules large, deep blue		Function unknown
2. LYMPHOCYTES, 20 to 25%	Nucleus single, large, round, deep blue; scant cytoplasm, clear blue; 6 to 10 μm	Lymphoid tissue, spleen, thymus, and lymph glands	Nonmotile; produce antibodies
3. MONOCYTES, 2 to 6%	Nucleus single, large, round, deep blue; much cytoplasm, muddy blue; 12 to 15 μm	Spleen and bone marrow	Motile; phagocytic
Platelets: About 250,000	Small, refractile, no nucleus; dark blue to lilac; 2 to 4 μm	Cytoplasmic fragments of megakaryocytes in bone marrow	Provide substance needed in clotting

* A stain containing two special kinds of dye, methylene blue and eosin, together with sodium bicarbonate and methyl alcohol.

way from their origins in the marrow, spleen, or lymphoid structures to the tissues or to their death. The lymphocytes, with a single rounded nucleus, are commonest in lymph vessels and in the lymph nodes along those vessels; the granulocytes, with lobed nuclei, are in the bloodstream and also around tissues. The estimated life of a white cell is 12 to 13 days. Most white cells of both types can perform amoeboid movements and can crawl out between the endothelial cells lining capillaries into spaces between tissue cells. There many of them can act as **phagocytes** (Gr. *phagos,* eat) to protect the body by consuming bacteria that invade wounds. Others are involved in the development of antibodies for immune defense. In an acute infection, such as appendicitis or pneumonia, the neutrophils and small lymphocytes increase markedly; the total leukocyte count will rise from a normal count of about 5000 to 9000 and up to 20,000 or 30,000 per cubic millimeter to battle the infection. The whitish pus of an infected area consists of dead leukocytes, tissue cells, and blood serum.

The blood platelets, or **thrombocytes,** are an essential element of the blood. They are more or less disk-shaped, much smaller than red cells, and without nuclei. Blood platelets exposed to an anticoagulant and stored in a blood-bank bottle do not survive. In human beings there are more than a trillion blood platelets, and each has a life span of 8 to 10 days. When a blood vessel is injured, they clump and disintegrate, releasing thromboplastin, which initiates the clotting process (Sec. 6-6). The platelets are rich in ATP but lose this energy-rich phosphate during clotting.

6-5 Functions of the blood   The blood performs a variety of duties for the many parts of the body. Some were listed previously, but all may be mentioned here to show the importance of this fluid circulating medium. It serves to carry

1   Oxygen and carbon dioxide between respiratory organs and tissues (Secs. 7-9, 7-10)

2   Water and digested foods from the digestive tract to other organs (Sec. 5-10)

3   Stored foods from one organ or tissue to another as needed (Sec. 5-11)

4   Organic wastes, excess minerals in solution, and water to the excretory organs (Sec. 7-13)

5   Hormones from the glands where produced to the places of use (Chap. 8)

6   Antibodies for immune defense

Besides these miscellaneous transport functions, the blood acts to regulate the pH of tissues within narrow limits by means of buffers such as phosphates and bicarbonates; it is slightly alkaline, with a pH relatively constant at 7.4. Its role in maintaining water balance between the tissues and excretory structures is all-important, and it does this at such a rate that the water content of the blood does not vary appreciably in a normal individual. In the warm-blooded birds and mammals the blood, by differential distribution between the internal organs and the body surface, serves to maintain the temperature of the entire body within close limits. Finally, the blood is the defense mechanism against foreign organisms, having therefore a major role in maintaining normal health and opposing the effects of infection.

6-6 Clotting   When a blood vessel is cut, the issuing blood is soon stopped by a protective **clot.** Within seconds, platelets stick to the edges of the cut and form a plug. The material called **thromboplastin** (thrombokinase), actually several substances which are released from injured tissues and disintegrating blood platelets (thrombocytes), forms from factors present in the blood plasma. In combination with calcium ions always present in the plasma, it acts upon **prothrombin,** a blood protein, to produce **thrombin.** (Prothrombin is produced in the liver, and vitamin K is required.) Thrombin converts a soluble blood protein, **fibrinogen,** into **fibrin.** Fibrin becomes a mass of fine fibers entangling corpuscles to form the clot. Diagrammatically the process is as follows:

$$\text{Thromboplastin}$$
$$\downarrow$$
$$\text{Prothrombin} + \text{calcium} \longrightarrow \text{thrombin}$$
$$\text{Thrombin} + \text{fibrinogen} \longrightarrow \text{fibrin}$$
$$\text{Fibrin} + \text{blood cells} \longrightarrow \text{blood clot}$$

The fluid residue from a clot is the blood serum. The substance **heparin** prevents the formation of thrombin in blood. Decrease in the number of platelets lengthens the average clotting time of 3 min in humans. Blood withdrawn for transfusion or for laboratory use is kept from clotting by addition of heparin or sodium citrate (which removes the calcium ions). Anticoagulants are secreted by leeches, some bloodsucking arthropods, cyclostomes, and vampire bats to keep the blood they take from other animals for food in a fluid state during feeding.

In certain human males known as hemophiliacs ("bleeders"), clotting is delayed or fails; a slight cut or tooth extraction may result in death by loss of blood (hemorrhage). This condition is caused by a sex-linked hereditary disease (hemophilia) transmitted by females but manifested chiefly in males; among females, only those rare individuals homozygous for the defect experience the disease. The excessive bleeding is due to absence of one of several thromboplastic factors necessary to convert prothrombin into thrombin.

6-7 Antibodies When foreign matter (antigen), i.e., bacteria, enters the blood of an animal, a specialized protein, or **antibody,** usually is formed in lymphoid tissue. It is capable of combining chemically with the specific antigen that caused its production. Thus, if a small (sublethal) dose of rattlesnake venom is injected into a pigeon, the bird's plasma, after several days, will contain antibodies capable of neutralizing subsequent injections. The venom has served as an **antigen,** stimulating some tissues to produce an antibody which is carried mainly in the blood plasma. Bacteria and other organisms may serve as antigens. Antibodies can inactivate some viruses, neutralize bacterial toxins, and aid phagocytosis by leukocytes. With the aid of certain activated plasma proteins, some invading organisms are killed outright. Recovery from any germ-caused disease results from the production of antibodies, which often confer a degree of **immunity,** transient or permanent, to further attack by that organism. Humans and other mammals can be rendered immune to certain diseases by injecting the dead or attenuated organisms (**vaccine**) of a particular disease or the immune serum (**antitoxin**) from a horse or other animal that has previously been immunized. Examples are vaccines for smallpox and typhoid

fever and antitoxins for diphtheria, tetanus, and snakebite. Antibodies occur in the globulin fraction of blood plasma. The injecting of gamma globulin, obtained from whole blood of donors, may increase the immunity of the recipient to various diseases, including measles.

6-8 Human blood groups If red blood cells from one person are mixed with blood plasma of another individual, the cells remain separate in some cases but become clumped, or agglutinated, in others. This is a matter of great practical importance when blood from a healthy donor is sought for transfusion into a sick or wounded person; should clumping occur, the patient may die instead of being helped. The blood of donor and patient must be compatible. Extensive tests show that there are two types of **antigens** (agglutinogens). They are polysaccharides called A and B, and they coat the surface of the red cells. The plasma may contain one or two kinds of **antibodies** (agglutinins), known as *a* (*anti-A*) and *b* (*anti-B*). Thus there are four **blood groups** among human beings: group O, having neither antigen A nor antigen B and having antibodies *a* and *b*; group A, antigen A and antibody *b*; group B, antigen B and antibody *a*; and group AB, antigens A and B but no antibodies. The results of mixing cells of any one group with plasma of another are summarized in Table 6-2. A person of group O has no antigens in the red cells and hence can give blood to persons of any

Table 6-2
COMPATIBILITY OF HUMAN BLOOD GROUPS

				Donor blood group			
				O	A	B	AB
				Antigen in red cells			
				None	A	B	AB
Recipient blood group	O	Antibodies in serum	a, b	−	+	+	+
	A		b	−	−	+	+
	B		a	−	+	−	+
	AB		None	−	−	−	−

−, compatible; no agglutination
+, not compatible; agglutinates

group (universal donor). A person of group AB, having no antibodies in the serum, can receive blood of any group (universal recipient) but can give blood only to others of the same group. Blood-group characteristics are inherited and remain constant throughout life.

Anthropoid apes have the same blood groups as humans, but those of monkeys and lower mammals differ from those of humans.

Other antigens (M, N) are present in human red cells. These result in three additional blood types: those persons who have only the M antigen; those with only the N antigen; and those who have both M and N. However, no clumping of cells occurs when these are mixed by blood transfusion. The frequencies of the types vary in different populations, M types being very numerous among American Indians and low among Australian aborigines, whereas the reverse is true for N types. The MN blood types provide evidence in addition to the ABO blood groups for determining cases of doubtful parentage. A parent with M-type blood, for example, could not be the father of a child with N-type blood.

6-9 The Rh factor   About 85 percent of Caucasians have another antigen in their red blood cells, and their blood is known as Rh+ (Rh-positive); those lacking this substance are termed Rh− (Rh-negative), the difference being due to the presence or absence of a carbohydrate on the red corpuscles. The Rh factor is named for the rhesus monkey, in which it was first found. If Rh+ blood is repeatedly transfused into an Rh− individual, the antigen stimulates production of anti-Rh agglutinin. This is called isoimmunization, since both the antigen (Rh) and the antibody (anti-Rh) are in the same species.

Rh− individuals receiving blood from Rh+ donors show no reaction at first but later become isoimmunized, and if they are then transfused with Rh+ blood, a severe reaction occurs which is usually fatal. The anti-Rh agglutinins cause hemolysis of the Rh+ transfused blood.

Another serious problem occurs when an Rh− mother repeatedly bears an Rh+ fetus (that received the Rh+ factor from the father) and becomes immunized by Rh+ fetal erythrocytes that have entered the maternal circulation. Then in later pregnancies the maternal anti-Rh agglutinins cross the placenta, enter the fetal circulation, and hemolyze the fetal red cells, commonly with fatal results. This disease (erythroblastosis) in the fetus or newborn results in loss of about 1 pregnancy in 50 among Caucasians. Once so immunized, such a mother is unlikely to bear a living Rh+ child unless the blood of the fetus or newborn child is replaced immediately after birth.

6-10 Capillaries and the lymphatic system   The blood in the capillaries carries out the ultimate function of the circulatory system—the exchange of nutrients and metabolic end products. This exchange occurs in the tissue fluid, or lymph, which collects in minute interstitial channels between the capillaries and cells. The lymph is essentially a plasma filtrate, a fluid which originates from seepage, or percolation, of water and solutes through capillary walls. Outflow of plasma is promoted at the arterial end of the capillaries by hydrostatic pressure from heart action. The lymphatic pressure is close to zero. Plasma proteins remain in the capillaries because their large size prevents them from diffusing outward. Why is there not a continual loss of plasma from the blood? The hydrostatic force is countered by osmotic pressure, which causes water to enter the capillaries. Water concentration in the lymph is greater than in the blood because of the proteins retained in the capillaries. Water thus moves in and carries with it dissolved solutes to which the capillary walls are permeable. Toward the venous end of the capillaries the situation is reversed. Hydrostatic pressure may be half that at the arterial end and is less than osmotic pressure. There is, therefore, a net movement of water and solutes inward. Solutes (sodium, chloride, glucose, etc.) have nearly identical concentrations in the plasma and tissue fluid. The capillary walls are highly permeable to them, and the slight changes in concentration caused by cellular activity have little effect close to the capillaries.

Solutes, in addition to being carried by the osmotic movement of water, move by passive diffusion in directions determined by their concentration gradients. For example, as glucose is used by active muscle cells, its lymph concentration falls below that in the blood. Glucose in the blood then diffuses into the lymph. Diffusion also occurs between cells and the lymph, but cells differ from capillaries in that they have the capacity for active transport.

The lymph thus plays a vital role in transport between cells. It is primarily extracellular but is returned to the blood as described above and through the lymphatic system. Between cells and capillaries there are fine, thin-walled **lymph vessels** with valves. Most lymph vessels are so delicate that they are not seen in anatomic preparations. They become larger in the thorax and there unite to form the **thoracic duct,** which empties into the venous system near the heart (Fig. 6-7). The lymphatic system carries fluid in only one direction, from the tissues to the blood and heart (Fig. 6-3). The fluid is moved by the massaging action of muscles on the lymph vessels and by changes in thoracic pressure due to breathing. The lymph valves prevent backflow. More than half the protein circulating in the blood is lost from the capillaries in a day and returned to the bloodstream by way of the lymphatic system. The lymphatic system is essential in maintaining blood volume. Also it is the main route by which fats, absorbed by the lacteals in the intestine, are transported to the bloodstream via the thoracic duct. Cholesterol likewise reaches the bloodstream from the tissues by this system. Scattered along the system are many **lymph nodes.** Besides producing lymphocytes, the nodes defend the body from infection by intercepting disease organisms. Thus infections may be accompanied by swelling of lymph nodes in the affected area.

The **spleen** is a part of the lymphatic and circulatory systems, capable of acting as a reservoir to hold a fifth to a third of all blood cells; it serves to regulate the volume of blood cells elsewhere in circulation. In addition it produces white cells (lymphocytes) and destroys old red cells.

The **thymus** provides lymphocytes with immunologic competence and thus aids the body in combating infections (Sec. 8-7).

6-11 The heart The entire circulatory system in any vertebrate is composed of the heart, the blood vessels (arteries, arterioles, capillaries, and veins), and the lymphatic channels and nodes. The heart comprises a series of chambers with slight or heavy muscular walls that receive blood from the veins and

pump it through the arteries. In the two-chambered heart of fishes (one atrium, one ventricle) all blood passing through the heart is unoxygenated; amphibians and most reptiles have 2 atria that receive blood from the body and lungs, respectively, and a single ventricle; in birds and mammals the four-chambered heart (two atria, two ventricles) is really a dual structure, the right side pumping only from the body to the lungs (pulmonary circuit) and the left side from the lungs to the body (systemic circuit) (Figs. 6-4 and 13-4).

Figure 6-4 *Above.* The mammalian (human) heart opened in frontal plane, ventral view. *Below.* Its mode of action (*A–D*). Arrows indicate paths of blood flow. Heavy stipple, unoxygenated blood; light stipple, oxygenated blood. *A.* Atria filling from veins. *B.* Blood entering relaxed ventricles. *C.* Atria contracting. *D.* Ventricles contracting, valves closed, blood forced into aorta and pulmonary arteries. *E.* Location of sinoatrial (s.a.) and atrioventricular (a.v.) nodes. Arrows indicate spread of control. The aorta and pulmonary artery actually emerge from the dorsal side (rear) but are shown here to aid in tracing the flow of blood. (*Modified from Best and Taylor, The human body and its functions, Henry Holt and Company, Inc.*)

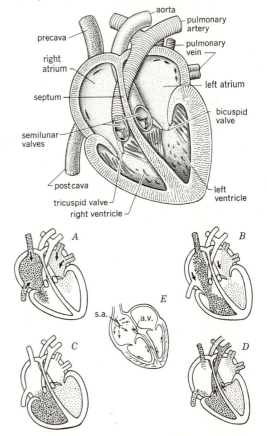

Heart (cardiac) muscle has an innate power of rhythmic contraction from the time it is first formed in the embryo. This remarkable independence can be seen readily in a frog. A heart, removed from the body, will contract for some minutes, and if it is immersed in physiologic saline solution that supplies essential elements and nutrients (NaCl, KCl, $CaCl_2$, etc., and glucose), beating may continue for several days. The primary regulation of heartbeat rests with groups of specialized cardiac cells. The **pacemaker** (sinoatrial node, in the wall of the right atrium) initiates contraction of the atria. After a brief delay it stimulates a second center (atrioventricular node, in the septum between atria), which, in turn, affects a band of fibers, the **bundle of His,** on the walls of the ventricles to start simultaneous contraction there. If living heart tissue is minced and treated with the enzyme trypsin and the individual cells separated, they can be kept alive in a nutrient medium (tissue culture). About 1 in 100 of these cells beats independently and continuously. After several days of growth the cells come in contact and clusters beat at the same rate and in unison. Thus there are two types of cell in the heart, the muscle cells that make up the bulk of the heart and the pacemaker cells that generate the rhythmic heartbeat. The beat also is under nervous control. Stimulation of the sinoatrial node via parasympathetic fibers (in vagus nerve) or local application of acetylcholine slows the heartbeat, but stimulation via sympathetic fibers or application of norepinephrine accelerates it.

The sequence of heart action in humans is as follows: The thin-walled atria fill with blood from the great veins, then contract, moving the blood into the ventricles. As the latter begin to contract, the bicuspid and tricuspid valves are closed by the increasing pressure of the blood. The semilunar valves are still closed, so the blood is blocked in all directions and the ventricle pressure rises. When pressure in the ventricles exceeds that in the arteries, the semilunar valves open and blood spurts into the arterial system. Then the cycle is repeated. The only rest for heart muscle throughout life is in the short intervals while the chambers are filling.

The heart in a quiet normal person contracts about 72 times per minute and pumps about 60 mL of blood per beat. The total volume of blood is about 6000 mL, and all of it could pass through the heart in 100 beats. During a life of 70 years the heart may contract 3 billion times and move nearly 200 million liters of blood. In some small birds and mammals the heart beats 200 to 400 times per minute.

Blood moves from the heart in a series of spurts, which in an exposed artery, as on the wrist or temple, may be felt as the **pulse,** greatest as the heart is contracted (systole) and least when filling (diastole). The pressure declines at progressive distances from the heart because of frictional losses, especially in the arterioles, and the returning flow in the veins is virtually smooth. Typical human pressures, in millimeters of mercury (mmHg), are: arteries, 120/80 (systolic/diastolic); capillaries, 30/10; veins, 10/0. Pressure is influenced by rate of heartbeat, degree of dilation or constriction of vessels, and other factors.

By listening with a stethoscope, the experienced physician can learn much from sounds related to opening and closing of the several valves in the heart. Recording the minute voltages resulting from phases of heart action as an electrocardiogram (ECG) is another means used in searching for abnormalities.

The ever-active heart has its own vessels to supply oxygen and nutrients. From the aorta, just beyond the semilunar valves, the **coronary arteries** run on the heart surface and into its muscles, distributing blood to capillaries there. Blood returns in the **coronary veins** that enter a coronary sinus delivering to the right atrium. From 7 to 10 percent of the aortic output goes to the coronary circulation, testifying to the great needs of the heart. Any stoppage (occlusion) of coronary vessels is usually attended by severe pain (angina pectoris) in the left thorax and arm; this is a serious condition, and fatal if left unattended.

6-12 Aortic arches   (Fig. 13-4) Six pairs of aortic arches appear in the embryos of all vertebrates, proceeding from a ventral aorta at the anterior end of the heart and passing between the gill slits that develop in the sides of the pharynx. The first and second pairs soon disappear. In adult fishes, arches 3 to 6 lead to the gills, where respiratory exchanges occur, and then all join above to form the dorsal aorta; similarly, arches 4 to 6 supply the gills in permanently gilled salamanders. Among land vertebrates, arch 3 forms the common carotid artery on each side, arch 5 disappears, and arch 6 becomes the pulmonary artery. In frogs, toads, and reptiles the two parts of arch 4 become the systemic arches to the dorsal aorta, in birds only the right half persists, and in mammals only the left half, the opposite arch in each forming a subclavian artery.

6-13 Blood vessels   The heart and all vessels are lined throughout with a glassy-smooth endothelium. The walls of the aorta and larger arteries contain heavy layers of elastic tissue and muscle fibers (Fig. 6-5), but the small arterioles are covered by smooth muscle fibers only. The capillary walls, where all exchanges of nutrients and gases, as well as removal of wastes, occur between the bloodstream and tissues, have no muscle fibers but possess many contractile (Rouget) cells on their outer surfaces. Frog muscle shows about 400 capillaries per square millimeter in cross section, and 1 mL of blood in passing has contact with about 2700 mm² of capillary surface. In a dog the comparable figures (by Krogh) are 2600 capillaries and 5600 mm². Capillaries in human skeletal muscles may have a total length of 96,000 km (60,000 miles) and a surface area of about $\frac{1}{2}$ hectare ($1\frac{1}{2}$ acres).

Veins are thin-walled, with connective tissue fibers but few muscles; unlike the arteries, they collapse when empty. The walls of all blood vessels are elastic, and vasomotor nerve fibers control the muscle fibers, causing arterioles to contract so as to alter the amount of blood passing to any region. The veins are provided with a series of valves that aid in maintaining the flow of blood back to the heart. The human circulatory system helps to control the body temperature by regulating the loss of heat. Excess heat acts through a nerve center in the medulla to permit dilation of superficial blood vessels in the skin, where heat may be lost; chilling results in contraction of such vessels.

6-14 Blood circulation in vertebrates   The paths of blood circulation are similar in principle in all vertebrates but differ in details depending on the complexity of the heart (one or two atria and ventricles) and the type of respiration (gills or lungs) (Fig. 6-6). The circulation of the blood in humans was first demonstrated by William Harvey (1578–1657), an English physician, early in the seventeenth century. Harvey tied a ligature above the elbow and saw enlargements at the location of valves in the veins of the forearm. When he held his finger on a vein and pressed the blood out above that point, he noticed that the vein remained empty. (One can repeat this experiment on one's own arm.) By this and other experiments he deduced that blood flows along the veins toward the heart. Harvey reasoned that the blood must enter the extremities through the arteries and pass somehow to the veins, but it remained for the Italian anatomist, Marcello Malpighi, to discover the capillaries 33 years later.

In humans, the **path of circulation** is essentially as follows: The blood arriving from various parts of the body passes into the precaval and postcaval veins to enter the right atrium; it is relatively poor in oxygen, dark red (stippled, Fig. 6-7), and carries carbon dioxide. From the right atrium it flows through the tricuspid valve (Fig. 6-4) to the right ventricle, and thence, as the result of a strong contraction of the ventricle (systole), through the semilunar valve and along the pulmonary artery to the lungs.

In the lungs the blood courses through many small capillaries in membranes covering the alveoli (Fig. 7-3C), where it gives up its carbon dioxide (Sec. 7-10) and becomes reoxygenated (Fig. 6-6). Thence it flows into larger vessels and to the pulmonary veins entering the left atrium. Through the bi-

Figure 6-5 Structure of blood vessels (not to scale). An artery has a thicker muscular layer than a vein, and a vein is usually larger than its corresponding artery. The wall of a capillary consists only of endothelium.

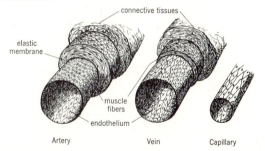

connective tissues

elastic membrane

muscle fibers

endothelium

Artery   Vein   Capillary

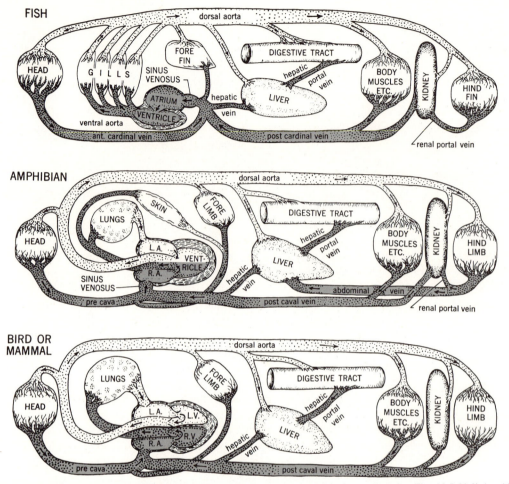

Figure 6-6 Paths of blood circulation in vertebrates. Fish, one atrium and one ventricle; four aortic arches (Figs. 29-5, 30-3). Amphibian, two atria and one ventricle (Fig. 31-6). Bird or mammal, two atria and two separate ventricles (Figs. 33-8, 34-4). Arrows indicate paths of blood flow. Gray, unoxygenated blood; white, oxygenated blood. (Compare Fig. 13-4.)

cuspid valve it reaches the left ventricle, where, by powerful muscular contraction (systole), it is forced into the aorta, the largest and most stout-walled vessel in the body. The aorta divides first into several large, thick-walled arteries that in turn branch and subdivide to supply all parts of the body. The blood travels along the arteries to microscopic arterioles and thence to the nonmuscular capillaries in the tissues.

The most direct route for return of blood to the heart is through the systemic part of the venous system. Capillaries join to form venules, and these merge to form veins, finally collecting in the two great veins, precaval and postcaval.

In addition to the complete cycle outlined, there are several vital side paths. Arterial blood in the abdomen enters a system of capillaries lining the walls of the intestines where digested food is absorbed; then the blood passes into the **portal vein** (a vein connecting two systems of capillaries) to the liver. There it spreads through another system of capillaries, where food substances may be stored in liver cells and other essential processes take place as described in Sec. 5-11. To complete this important side route, blood from the liver gathers in the hepatic vein and flows to the postcaval vein.

An equally essential path takes arterial blood through a double system of capillaries in the kidneys

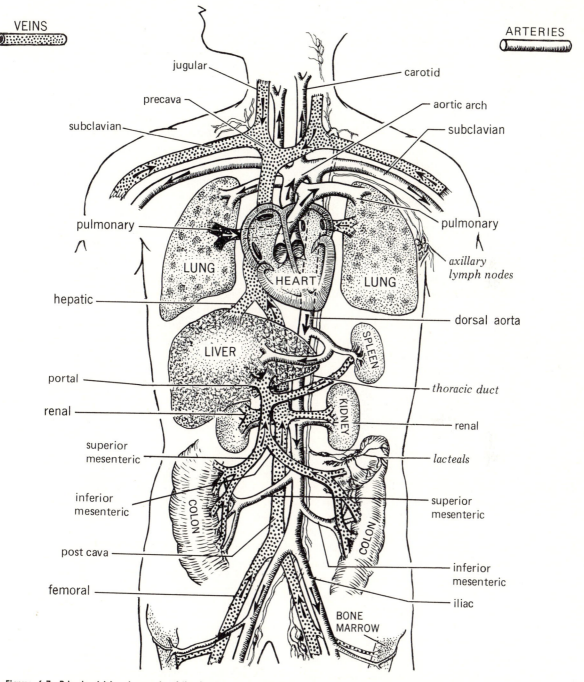

VEINS

ARTERIES

jugular

carotid

precava

aortic arch

subclavian

subclavian

pulmonary

pulmonary

axillary
lymph nodes

LUNG

HEART

LUNG

hepatic

dorsal aorta

LIVER

SPLEEN

portal

thoracic duct

renal

KIDNEY

renal

superior
mesenteric

lacteals

inferior
mesenteric

superior
mesenteric

COLON

COLON

post cava

inferior
mesenteric

femoral

iliac

BONE
MARROW

Figure 6-7 Principal blood vessels of the human circulatory system in relation to internal organs. Stomach, small intestine, bladder, and sex organs omitted. Arrows indicate paths of blood flow; coarse stippling (veins), blood low in oxygen; unstippled (arteries), blood richer in oxygen. Of the pulmonary vessels, the artery (unstippled) carries unoxygenated blood to the lung and the vein (stippled) returns oxygenated blood to the heart. The thoracic duct of the lymphatic system and a few lymph nodes are also shown and labeled in italic.

(Sec. 7-12; Fig. 7-10), whence it returns to the heart. The kidneys are the major organs which control metabolite levels of the blood. During the passage of the blood through them, excess water and wastes are removed to maintain a relatively uniform condition (steady state) in the body as a whole.

In the heart of a human fetus the partition between the two atria is incomplete, and also there is a connection (ductus arteriosus) between the pulmonary artery and aorta. These both serve to shunt most of the blood around the developing but nonfunctional lungs. (Oxygen is obtained through the placenta.) At birth, when air breathing begins, a flap closes the wall between the atria, and the ductus arteriosus is constricted. Occasionally the septum between the ventricles is incomplete at birth and much of the blood bypasses the lungs, resulting in a "blue baby," deficient in oxygen in the blood, a condition which usually can be corrected by surgery.

The heart and circulatory system are so important in many bodily activities that any disturbance in their functions is serious. High blood pressure (hypertension) puts added strain on both the heart and vessels and may end in rupture of an artery, extensive hemorrhage, and death. "Hardening of the arteries" (arteriosclerosis) is another circulatory ailment, found particularly in older persons. A common and serious disease is coronary thrombosis—closure or stoppage by a spasm or clot of some of the coronary vessels supplying blood to the heart muscle itself. The blood supply to the injured area is not completely restored, and a local deficiency in circulation results. The remaining coronary vessels have an added load, and the heart becomes less efficient. Corrective surgery is sometimes successful.

Other activities in the circulatory system include formation of red cells in the bone marrow, the service of the spleen in storing blood, and that of the spleen and liver in destroying old red cells.

6-15 Integration of the circulatory system   The heart and blood vessels are controlled by the nervous system and also by certain substances in the blood. The circulatory system is sensitive to slight changes in the body, and its performance is complex because so many organs and functions are involved. Mild

exercise, for example, brings adjustments in heartbeat, blood pressure, and distribution of blood. The muscular activity requires oxygen and produces carbon dioxide. By chemical and nervous stimulation, the oxygen demand increases arterial pressure, making for a greater flow of blood. The rate of heartbeat quickens from a reflex stimulated by increased pressure in the right atrium. Other reflexes stimulate vessel constrictor centers in the medulla (Fig. 9-7), decreasing blood flow to inactive areas and hastening it where needed. Meanwhile the hypothalamus is activated and epinephrine is secreted, resulting in constriction of blood vessels in the skin and viscera and dilation of those in the muscles. All these changes tend in one direction and, unless checked, would lead to excessive heartbeat and a blood pressure so high it would endanger fine blood vessels of the brain. But pressure receptors and chemoreceptors in the aortic arch and carotid sinuses, acting through the medulla, cause a relaxation of arterial muscles and decrease in heart rate. The various parts of the circulatory system thus are integrated and operate with sensitive checks and balances.

References

(See also references at end of Chap. 4.)

Adolph, E. F. 1967. The heart's pacemaker. *Sci. Am.*, vol. 216, no. 3, pp. 32–37. *A group of specialized cells initiates beats of the heart and contributes to its control.*

Berne, R. M., and **M. N. Levy.** 1967. Cardiovascular physiology. St. Louis, The C. V. Mosby Company. ix + 254 pp., illus.

Capra, J. D., and **A. B. Edmundson.** 1977. The antibody combining site. *Sci. Am.*, vol. 236, no. 1, pp. 50–59. *When an antigen combines with an antibody, it does so at a site that fits it as a lock fits a key.*

Clarke, C. A. 1968. The prevention of "rhesus" babies. *Sci. Am.*, vol. 219, no. 5, pp. 46–52. *Experiments in breeding butterflies led fortuitously to a solution of the Rh problem.*

Fox, H. M. 1950. Blood pigments. *Sci. Am.*, vol. 182, no. 3, pp. 20–22. *A comparison of the major respiratory pigments—hemoglobin, hemocyanin, and chlorocruorin.*

Guyton, A. C. 1963. Circulatory physiology: Cardiac output and its regulation. Philadelphia, W. B. Saunders Company. viii + 468 pp., 199 figs.

_____. 1977. Basic human physiology. 2d ed. Philadelphia, W. B. Saunders Company. xi + 931 pp.

Laki, K. 1962. The clotting of fibrinogen. *Sci. Am.*, vol. 206, no. 3, pp. 60–66.

Mayerson, H. S. 1963. The lymphatic system. *Sci. Am.*, vol. 208, no. 6, pp. 80–90. *The "second" circulation of the body and its role in maintaining the body's steady state.*

McKusick, V. A. 1965. The royal hemophilia. *Sci. Am.*, vol. 213, no. 2, pp. 88–95. *A defective X chromosome passed by Queen Victoria has plagued European royalty for three generations.*

Perutz, M. F. 1964. The hemoglobin molecule. *Sci. Am.*, vol. 211, no. 5, pp. 64–76. *Its 10,000 atoms are arranged in four helical, twisting chains.*

Ponder, E. 1957. The red blood cell. *Sci. Am.*, vol. 196, no. 1, pp. 95–102. *Its effectiveness in oxygen transport is elucidated by study of the molecular structure of its surface.*

Scholander, P. F. 1957. "The wonderful net." *Sci. Am.*, vol. 196, no. 4, pp. 96–107. *The way animals conserve heat and oxygen by a countercurrent arrangement of blood vessels.*

Wiggers, C. J. 1957. The heart. *Sci. Am.*, vol. 196, no. 5, pp. 74–87. *A survey of this intricately woven muscle with special reference to its operation as a pump.*

Wolf, A. V. 1958. Body water. *Sci. Am.*, vol. 199, no. 5, pp. 125–132. *It diffuses freely through tissues but its total amount is closely regulated.*

Wood, J. E. 1968. The venous system. *Sci. Am.*, vol. 218, no. 1, pp. 86–96. *The veins are a reservoir for the blood, not merely passive conduits.*

Zucker, M. B. 1961. Blood platelets. *Sci. Am.*, vol. 204, no. 2, pp. 58–64. *Minute disks in the bloodstream that plug breaks in blood vessels and are constituents of blood clots.*

Zweifach, B. W. 1959. The microcirculation of the blood. *Sci. Am.*, vol. 200, no. 1, pp. 54–60. *The anatomy and function of capillaries are described.*

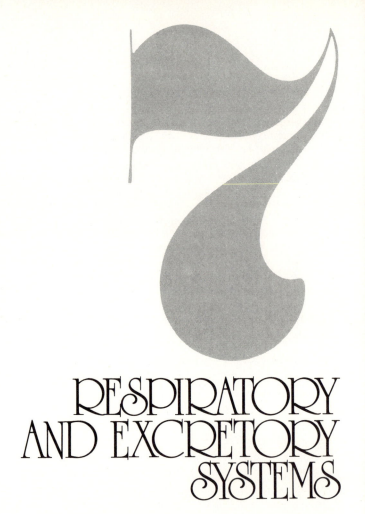

RESPIRATORY AND EXCRETORY SYSTEMS

Respiratory systems

All animals need oxygen for the metabolism in their cells and must dispose of the resulting carbon dioxide. The exchange of these gases is termed **respiration.** Some animals can exist for months on fats or other foods stored in their bodies, many can live a shorter time without water, but few survive long without oxygen, since little is stored in the body. Most animals obtain the oxygen from their environment. The air contains 21 percent oxygen, but water (at 15°C) holds only 0.7 percent or less; the oxygen in the water molecule (H_2O) is not available for respiration.

Animal life undoubtedly originated in the sea,

where a host of animal types still live, deriving their oxygen from that dissolved in the water. In the course of geologic time various animals have become terrestrial and therefore air-breathing. This change required major adaptive modifications, including new methods of respiration.

7-1 Oxygen and carbon dioxide Ordinary respiration in different animals is performed by various **respiratory organs** or **systems,** such as the body covering, gills, lungs, or tracheae. These structures are unlike in appearance but fundamentally the same in function (Figs. 7-1, 7-2); each comprises a moist permeable membrane through which molecules of

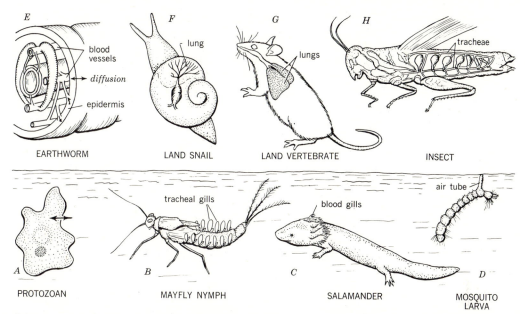

Figure 7-1 Types of respiratory mechanism in animals and protistans. In water (*below*). *A*. Protozoan, diffusion through cell wall. *B*. Mayfly nymph (insect), tracheal gills. *C*. Salamander, blood gills. *D*. Mosquito larva, aquatic, with tube for breathing free air. In air (*above*). *E*. Earthworm, diffusion through moist body wall to blood vessels. *F*. Land snail, moist lung inside body. *G*. Land vertebrate, pair of moist lungs inside body. *H*. Insect, system of air ducts (tracheae) throughout body.

oxygen and carbon dioxide diffuse readily. In accordance with the laws of gases each gas acts independently of others. When a difference in diffusion pressure (Sec. 2-11) exists on the two sides of a membrane, more molecules pass toward the region of lesser pressure than in the opposite direction. The partial pressure of oxygen in the air or water is greater than within an animal body, where it is constantly being used up, so oxygen tends to enter any suitable membrane surface. The partial pressure of carbon dioxide is greater within the animal, so it tends to pass outward. These exchanges occur simultaneously. In many small animals the exchange of gases is direct, from air or water through membranes to tissue cells; but it is more complex in larger species and those with dry or nonpermeable exteriors. In the latter, respiration consists of two stages: **external respiration,** the exchange between environment and respiratory organs; and **internal respiration,** the exchange between body fluids and tissue cells. A third stage, **cellular respiration** (the utilization of oxygen in the cells and release of carbon dioxide), is discussed as part of metabolism (Secs. 2-22, 5-11).

The term **respiration** is normally associated with free oxygen. But some intestinal parasites and some mud-inhabiting invertebrates live in an environment where there is little or no oxygen. These animals may obtain energy anaerobically in the absence of free oxygen through glycolysis (Sec. 2-24).

7-2 Respiratory mechanisms Animals obtain oxygen in five ways: (1) from water or air through a moist surface directly into the body (amoeba, flatworm); (2) from air or water through the thin, moist body wall to blood vessels (earthworms, etc.); (3) from air through spiracles or from water through tracheal gills to a system of air ducts or tracheae to the tissues (insects); (4) from water through moist gill surfaces to blood vessels (fishes, amphibians); and (5) from air through moist lung surfaces to blood vessels (land snails, land vertebrates). In all methods diffusion through a moist surface is involved.

7-3 Diffusion Many aquatic organisms obtain air directly from their environment. In a protozoan,

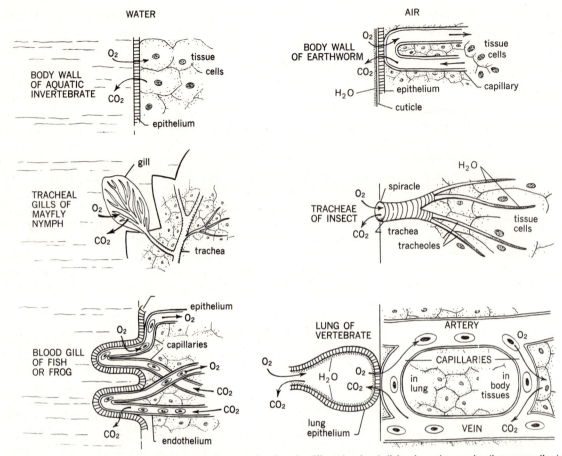

Figure 7-2 Equivalent nature of various respiratory mechanisms in different animals living in water or air; diagrammatic. In air respiratory surfaces are kept moist by body fluids high in water content (stippled areas labeled H_2O).

gaseous exchange takes place through the cell membrane to and from the surrounding water. In sponges, cnidarians, and other lower softbodied invertebrates, the gases diffuse through epithelial cells and thence to those deeper in the body. Some endoparasites are bathed in the body fluids of their hosts, from which they absorb oxygen and to which they give up carbon dioxide. Terrestrial flatworms can live in damp places where respiratory diffusion is possible through the moist epidermis. Direct diffusion would be inadequate for large animals, because their internal organs are distant from the outer surface. Terrestrial species must conserve bodily water

and cannot afford a large, moist exterior for respiration by diffusion.

7-4 Tracheae Insects, centipedes, millipedes, some arachnids, and *Peripatus* have fine tubes branching inward from the body surface to all internal organs. These are called **tracheae.** They develop as ingrowths of the body wall and are lined with chitin. Each ends in microscopic tracheal cells that extend as intracellular **tracheoles,** sometimes forming a capillary network in the tissues. The end of a tracheole is filled with fluid through which oxygen and carbon dioxide diffuse to and from the adjacent tissue cells. Diffusion of gases through the tracheal system is aided by movements of the thoracic and abdominal segments of the body. The efficiency of this system depends on the extremely rapid diffusion rate of oxygen in air as compared with that in water and on the relatively small size of tracheate animals. In many insects the exterior openings, or **spiracles,** of the tracheal system have valves which can close to limit loss of water.

Larvae of dragonflies, mayflies, stoneflies, and some other tracheate arthropods are aquatic. They obtain oxygen by diffusion from the water, either through the cuticle to the tracheal system or through specialized **tracheal gills** which provide more surface for diffusion. Adult water beetles and bugs swim to the surface and take down a silvery bubble of air under the wings. Oxygen from the bubble diffuses into the tracheal system, and the bubble supply is replaced by diffusion from the water.

7-5 Blood gills In most higher animals respiration is aided by a blood transport system. Its simplest form is seen in the earthworm, where oxygen diffuses through the body wall into superficial blood vessels and then passes from the blood to the tissue cells. In frogs the moist skin and the lining of the mouth cavity serve similarly. Many aquatic animals, however, have a more efficient mechanism—**blood gills**—of many slender filaments covered by delicate epidermis and containing capillary networks. The O_2–CO_2 exchange occurs between the surrounding water and the blood within the gills. Free, dissolved oxygen diffuses inward from the water. Cold water holds more oxygen than warm water, and the "white water" of fast-flowing streams has more than the still water of ponds or stagnant swamps. The small amount of oxygen dissolved in water limits the density of animal populations, but aquatic plants partly offset this by releasing oxygen during photosynthesis.

The gill filaments of salamander larvae (Fig. 31-10) and those of some marine worms are merely exposed to the water, but tube-dwelling annelids, aquatic crustaceans such as the crayfish (Fig. 24-4), and many aquatic mollusks have special means to force water over the gills. The gills of fishes and tadpoles are in chambers at the sides of the pharynx, and water taken into the mouth is forced out over the filaments (Fig. 30-6).

7-6 Lungs All land vertebrates, including most amphibians and all reptiles, birds, and mammals, have **lungs.** A lung is a chamber lined by moist epithelium, underlaid by a network of blood capillaries, where atmospheric air can be used. Basically, a lung is like a blood gill but is invaginated rather than evaginated. Its walls are partitioned in varying degree to form compartments called **alveoli.** The alveoli are microscopic chambers, open to pulmonary air flow and surrounded by many blood capillaries (Fig. 7-3C), where respiratory exchanges occur. In some amphibians (*Necturus*), the lungs have little vascularization, are saclike, lack alveoli, and function primarily as hydrostatic organs; in frogs, vascularization is more extensive and alveoli are large, opening widely into the lung cavity; in reptiles, alveoli are more numerous and the trachea is subdivided into **bronchi;** in birds and mammals, vascularization, bronchial subdivision, and numbers of alveoli reach their peak. The overall evolutionary trend has been an increase in vascularization and respiratory surface with a rise in metabolic activity. Bird lungs are dense, and air flows through them rather than in and out as in other lunged vertebrates (Sec. 33-11). En route through the respiratory system, it passes to and from a series of thin-walled **air sacs** (Fig. 33-9); these occupy spaces between internal organs and around or in some bones. The air sacs serve mainly to dissipate excess body heat and to shunt air into the lungs. Some mollusks and arachnids have lungs that function by diffusion (Figs. 21-16, 23-5).

7-7 The human respiratory system The mouth and nose communicate with the lungs through a series of special structures. The **glottis** is an opening in the floor of the pharynx, protected above by a lid, or **epiglottis,** and supported by a cartilaginous framework, the **larynx.** The latter connects to a flexible tube, the **trachea,** or windpipe, that extends into the thorax and forks into two **bronchi,** one to each **lung** (Fig. 7-3).

In the nose the entering air is filtered by hairs and scroll-like turbinal bones or conchae covered with a thick layer of mucous membrane; it is also warmed and moistened. The mouth serves as an alternate route for air, and the pharynx is a passage for air from either the nose or mouth to the larynx (Fig. 5-5).

The larynx, or voice organ, is in the front part of the neck. It is broad above, triangular in shape, and consists of nine cartilages moved by muscles; it contains two folds of mucous membrane with embedded fibrous, elastic ligaments, the **vocal cords.**

The voice is produced by air forced from the lungs to vibrate the vocal cords, and the positioning of the cords is changed to produce various kinds of sounds. In singing a high note, for example, the cords are drawn taut and close together. The sound waves so created pass through the pharynx, mouth, and nasal cavities, which act as resonating chambers; these parts, together with the tongue and lips, are important in speech. The size of the larynx varies

in different individuals; at the time of puberty it grows more rapidly in males than in females, resulting in change to a deeper and lower voice.

The trachea and bronchi are reinforced against collapse by rings of cartilage. The human lungs have 300 million or more alveoli; the entire inner surface is estimated as 40 to 80 m² (about 48 to 96 yd²), some 50 times the area of the skin.

The substance of the lungs is porous and spongy, and contains large numbers of elastic fibers. The right lung is larger, broader, and slightly shorter than the left, owing to the asymmetric positions of heart and liver. The **diaphragm** is a dome-shaped muscular partition that separates the thorax, containing the heart and lungs, from the abdominal cavity. Under normal conditions the lungs occupy fully the thoracic cavity, each lung in its own airtight chamber. The lungs are covered with a membrane, the **visceral pleura;** a similar membrane, the **parietal pleura,** lines the chest cavity; a lubricating **intrapleural fluid** fills the space between. The **intrapleural pressure** is lower than that of the atmosphere, thus

Figure 7-3 Human respiratory system. *A.* Larynx, trachea, and lungs in ventral view; left lung opened. *B.* Part of a bronchus, with cartilages; small blood vessels adjacent. *C.* Bronchiole with alveoli and capillaries; diagrammatic. (Dark vessels carry unoxygenated blood.)

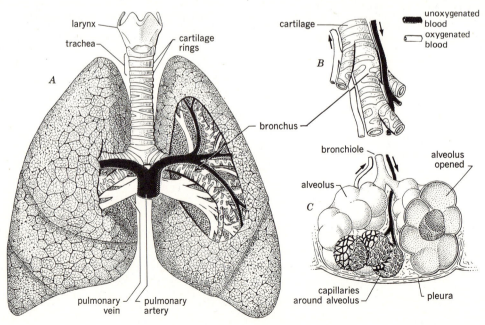

air in the alveoli keeps the lungs expanded against the thoracic walls. If air enters a pleural cavity, by accident or in the treatment of pulmonary disease, the lung collapses.

7-8 Human breathing   This essential process involves muscular movements, partly voluntary, that alter the volume of the thoracic cavity and thereby that of the lungs. At inspiration the ribs are raised and the diaphragm is flattened so that the thoracic space is enlarged. Pressure in the lungs is reduced, whereupon external air (at atmospheric pressure) moves down the trachea and into the lungs. The elastic framework of the lungs is stretched. Expiration results in the reverse process: by relaxation of muscles controlling the ribs and diaphragm, the lungs are compressed, the elastic tissue contracts, and air is forced outward (Fig. 7-4C). The lungs move freely, lubricated by the pleural fluid.

The alveoli are lined with a watery film which, because of surface tension, acts much like stretched rubber; the alveoli tend to contract. Considerable energy would be required to distend them, and respiration would be exhausting were it not for a phos-

pholipoprotein complex, **pulmonary surfactant.** This substance, secreted by the alveolar cells, intersperses with the water molecules in the alveolar film and reduces their cohesiveness and thus the surface tension.

With each quiet breathing movement about one-tenth of the total amount of air that can be moved by maximal respiration (the vital capacity) flows in and out of the lungs (Fig. 7-5). Such movements average about 10 to 20 per minute, and total ventilation is about 8 to 10 L in that period. In violent exercise the frequency may rise to 50 per minute and the ventilation to 20+ L. Atmospheric air averages 21 percent oxygen and expired air 16 percent, a net loss of about 5 percent oxygen and gain of 4.3 percent carbon dioxide. Nitrogen (79 percent) is inert and has no part in breathing. In the lungs there is a gradient, owing to diffusion through the residual air, so that the alveolar air contains only 14 percent oxygen. The basic rate of breathing is controlled mainly by a respiratory center in the medulla of the brain, which sends out rhythmic stimuli. The center contains **inspiratory** and **expiratory neurons.** The former send impulses via motor fibers to the intercostal muscles and via the phrenic nerve to the diaphragm, causing them to

Figure 7-4 Mechanism of breathing. *A*. The lung and its enclosing thoracic wall can be compared to two balloons, one inside the other, with fluid between (gray areas). The inner balloon (the lung) is filled with air. *B*. When the outer balloon is expanded (lifting of chest wall and lowering of diaphragm), expansion of the fluid space causes the fluid pressure to decrease. Fluid pressure then is less than internal air pressure, and the wall of the inner balloon expands. As the inner balloon enlarges, its air pressure decreases and atmospheric air moves in. *(Adapted from Vander et al., Human Physiology, McGraw-Hill, 1975.) C*. Upon inspiration, contraction of intercostal muscles moves ribs upward and outward; diaphragm contracts (moves downward).

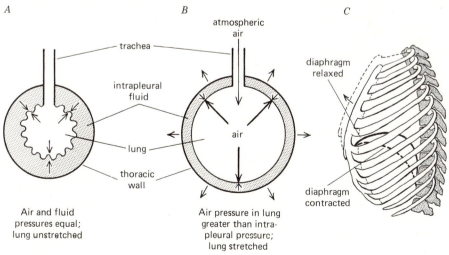

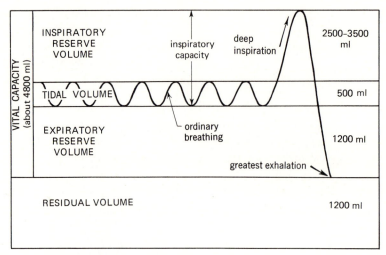

Figure 7-5 Human lung volumes and capacities. Ordinary respiration exchanges only the tidal air (about 10 percent); deep breathing flushes most of the lung (about 4500 mL, or 75 percent); the residual air changes only by diffusion. (*Winton and Bayliss, Human physiology, J. & A. Churchill, Ltd.*)

contract. Expiration occurs when the inspiratory neurons complete their inherent cycle of firing and when they are inhibited by expiratory neurons and by impulses from lung stretch receptors acting via the brain. In quiet breathing, expiration involves passive relaxation of the respiratory muscles and recoil of the elastic lungs, but when exhalation is forceful, there is active muscular contraction.

It has been generally accepted that the rate of breathing is controlled primarily by changes in the concentration of CO_2 in the blood, which affect the medullary respiratory center, a slight increase in CO_2 above normal causing an increase in ventilation and a lowering a decrease. However, there is now evidence that CO_2 itself is not responsible but rather associated changes in hydrogen ion concentration (Sec. 7-10) acting on the respiratory center via the cerebrospinal fluid. The brain is highly sensitive to deviations in H+ concentration. The control of breathing at rest may be aimed primarily at H+ regulation. Respiratory rate is also influenced by other factors. The amount of O_2 dissolved in the blood is monitored by the carotid and aortic bodies located, respectively, at the bifurcation of the common carotid artery and on the arch of the aorta. These chemoreceptors have neural connections with the medullary respiratory center. A fall in the plasma level of O_2 causes increased ventilation. The breathing rate increases during vigorous exercise because of the greater production of carbon dioxide in muscular metabolism. The breath may be held for a limited time, but as the carbon dioxide increases in the blood, the stimulus becomes too strong to be resisted. The breathing rate is also subjected to nervous control of other sorts, as seen in emotional states of anger or excitement. At any instant respiration is controlled by a variety of factors.

If breathing stops because of suffocation, death soon follows unless breathing movements are restored artificially by mouth-to-mouth resuscitation or alternately applying and releasing pressure on the ribs to simulate breathing and bring about gaseous exchange. The lungs of a newborn baby are inflated with the first breath after interruption of the placental circulation.

The air at high altitudes is rarefied (lower barometric pressure), and a given volume contains less oxygen than at sea level. This affects normal respiratory needs of humans and animals. A mountain climber or a person in an unpressurized airplane must use a tank of oxygen and face mask to obtain adequate oxygen. Planes operating above 3000 m (10,000 ft) usually have the air pressure inside raised (pressurized) to ease passengers' breathing.

A person who enters the water without an artificial supply of air can hold the breath and remain submerged about 2 min, then must come up to breathe and repay the oxygen debt incurred. For continued submergence "skin divers" use a tank of compressed air (scuba gear) and face mask. The air provides oxygen and also serves to keep the lungs inflated. The maximum safe limit for experienced divers is about 36 to 46 m (120 to 150 ft). The water pressure increases the amount of oxygen and nitrogen absorbed from the breathing mixture, which in excess can cause loss of consciousness. At normal atmospheric pressure only about 1 L of nitrogen is dissolved in the entire body. The amount increases greatly with increasing depth, and at 7 atmospheres may become seven times that at the surface. A high concentration dissolved in the body fluids acts as an anesthetic on the central nervous system. At depths of more than 60 m (200 ft) a diver may fall asleep. If a diver is brought suddenly to the surface, the nitrogen expands rapidly within the body fluids, because the external pressure on the body is no longer great enough to keep the nitrogen dissolved. The expanding gas forms bubbles causing painful decompression sickness, or "the bends." In severe cases there is mechanical damage to nervous tissues, resulting in mental disorders and paralysis.

7-9 Respiratory function of the blood   After oxygen has crossed the alveolar membrane, it must be delivered to the tissue cells where needed. In human beings and most higher animals this transport is accomplished by the blood. Human blood plasma carries in solution only about 2 percent of the total oxygen. The remainder is transported by the red **hemoglobin,** a conjugated protein, with which it enters into combination in the erythrocytes, or red corpuscles. The small size and flattened shape of these cells ensure rapid diffusion of gases throughout their interior. The process is as follows: After diffusion into the alveolar capillaries, the oxygen unites with the hemoglobin because the tension of the gas is lower in the blood than in the alveoli. The combined **oxyhemoglobin** then travels in the circulation to the tissues, where oxygen tension is lower than in the arterial blood. There oxygen is freed to

diffuse to the cells, and the deoxygenated hemoglobin returns to the lungs in the venous blood.

In addition to diffusion, several other factors influence the extent to which hemoglobin binds oxygen: (1) An increase in hydrogen ion concentration causes it to have less affinity for O_2, and a decrease, the reverse. The H+ concentration is lower in the lung capillaries than in the systemic venous blood. This favors O_2 pickup. In the tissue capillaries the reverse is true. The acidity of actively metabolizing tissues is high and causes the release of more O_2 from hemoglobin than would occur because of diffusion alone. (2) A rise in temperature also reduces hemoglobin's affinity for O_2, and a fall, the opposite. The elevated temperature of metabolically active tissues contributes to O_2 release. (3) DPG (2,3 = diphosphoglycerate) is produced during glycolysis in red cells. An increase in this substance, triggered by a decrease in the tissue oxygen supply, enhances unloading of oxygen by hemoglobin.

The total oxygen capacity of the human blood averages about 1200 mL, and 100 to 350 mL of oxygen passes into the tissues at each circuit. In an hour the body at rest uses about 15 L (4 gal) of oxygen and in strenuous exercise up to 280 L (75 gal).

Hemoglobin itself is a so-called "respiratory pigment," formed by the union of red **heme,** which contains iron, with a colorless protein, **globin.** The unique feature of hemoglobin is that a given amount may combine with different amounts of oxygen, depending on the tension of the gas in contact with the system. Thus the reaction is reversible and may be represented by the generalized equation $Hb_4 + 4O_2 \rightleftharpoons 4HbO_2$, where Hb stands for a molecule of hemoglobin. HbO_2 is oxyhemoglobin, of bright-red color, in contrast to the dark color of unoxygenated hemoglobin. All vertebrates (except icefish) have hemoglobin, but it is sporadic in other animal groups. The respiratory pigment in most mollusks and arthropods is **hemocyanin** (with copper instead of iron) and that of some polychaete tube worms is **chlorocruorin** (containing iron); these pigments are dissolved in the blood plasma. When oxygenated, hemocyanin is blue rather than red and chlorocruorin is green.

Myoglobin is a hemoglobinlike protein in skeletal and cardiac tissue. It apparently facilitates transport of oxygen throughout muscle cells and acts as a reservoir of oxygen for sudden needs. The red muscles of diving birds and mammals contain high concentrations, required for the stringent metabolic demands of diving.

7-10 Carbon dioxide   Respiration involves the exchange of two gases, and it might be inferred that disposal of carbon dioxide is a reversal of the inward flow of oxygen. Actually the mechanism of carbon dioxide transport is somewhat different. When the blood reaches the tissue capillaries it gives off oxygen and picks up carbon dioxide. Carbon dioxide diffuses into the blood because, as a consequence of metabolic activities, its concentration in the extracellular fluid is higher than that of the plasma. It combines with water to form carbonic acid (H_2CO_3). In the blood most of the H_2CO_3 (about 67 percent) is rapidly ionized to bicarbonate (HCO_3^-) and hydrogen ions (H^+). The conversion takes place chiefly within the red cells, where the process is catalyzed by the enzyme **carbonic anhydrase.** Thus CO_2 + $H_2O \underset{\text{carbonic anhydrase}}{\overset{}{\rightleftharpoons}} H_2CO_3 \rightleftharpoons HCO_3^- + H^+$. Considerable CO_2 (about 25 percent) acts directly with hemoglobin (Hb) to form **carbamino compounds** ($HbCO_2$). Thus $CO_2 + Hb \rightleftharpoons HbCO_2$. The remaining 8 percent is dissolved in the plasma and in the red cells. The reactions are reversible. Therefore when the blood reaches the alveolar capillaries, the reactions shown progress to the left, because the concentration of CO_2 in the blood is much higher than that in the alveoli. Carbon dioxide picked up by the tissue capillaries is thereby released to the lungs and the exterior. The bicarbonate ions (HCO_3^-) circulating in the blood are an important buffer, aiding in maintaining the acid—base equilibrium of the blood at a pH of about 7.4 by combining with free hydrogen ions to produce carbonic acid ($HCO_3^- + H^+ \rightleftharpoons H_2CO_3$). Free hydrogen ions contribute to the acidity of body fluids and must be held within narrow limits. The buffering process is facilitated by carbonic anhydrase acting as described above.

Excretory systems

Excretion is the process of ridding the body of wastes resulting from metabolism. The protoplasm and fluids of an animal or protozoan make up a delicately balanced physicochemical system, and it is the function of the excretory system (Fig. 7-6) to maintain this constant internal environment. Excess water, gases, salts, and organic materials, including metabolic wastes, are excreted whereas substances essential for normal functions are conserved.

An important function of excretion is to dispose of waste nitrogenous materials. A second function is to regulate water balance. During digestion the nitrogen-bearing proteins are split into amino acids and absorbed. Some then go to body cells for building new proteins. In the vertebrates others pass to the liver (Fig. 5-9) and lose their amino group through deamination or transamination reactions. The amino group is then formed into urea, uric acid, or ammonia and eliminated by the kidneys. Among mammals, turtles, and amphibians the end product is mostly urea; in land reptiles (also some snails and many insects) the nitrogenous waste is mainly insoluble uric acid. Marine teleost fishes excrete up to a third of the nitrogen as trimethylamine oxide. Many invertebrates and aquatic vertebrates eliminate the nitrogen as ammonia; this compound is highly toxic but readily disposed of, because an excess of water is present. A few animals excrete amino acids directly, and spiders excrete nitrogen as guanine.

7-11 Excretion in invertebrates   The simplest-appearing method of excretion is to pass wastes through the cell membrane into the surrounding water, as occurs in many protozoans. *Amoeba, Paramecium,* and various other freshwater protozoans have one or more **contractile vacuoles** that accumulate excess water from within the cytoplasm and periodically discharge to the exterior so as to maintain the normal fluid balance within the cell. Ammonia is the chief excretory product. Excretions of sponges and cnidarians diffuse from body cells into the epidermis and thence into the water.

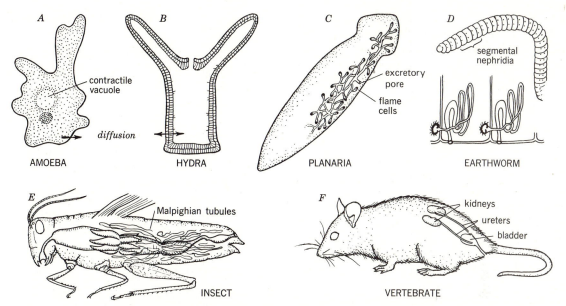

Figure 7-6 Types of excretory mechanisms in animals and protistans. *A.* Amoeba, contractile vacuole and diffusion from cell surface. *B.* Hydra, diffusion from cells. *C.* Planaria, many flame cells connecting to ducts ending in common excretory pore(s). *D.* Earthworm, two nephridia in each body somite, emptying separately through body wall. *E.* Grasshopper, series of fine Malpighian tubules connected to end of midgut. *F.* Vertebrate, two kidneys with ducts to a single bladder discharging to exterior.

Among insects and a few other arthropods the principal excretory organs are slender **Malpighian tubules** (Fig. 7-7*A, B*) attached to the anterior end of the hindgut and closed at their free ends; these tubules collect wastes from the body fluids and discharge them into the hindgut. Both urates and some carbon dioxide are received in solution from the blood; water and other materials are resorbed in the lower parts of the tubules. The final excretions, including uric acid crystals, carbonates, oxalates, and sometimes urea or ammonia, pass out with the feces. The fat body of insects is also a depository for organic wastes and is the chief excretory mechanism in springtails (Collembola), which lack Malpighian tubules. The exoskeleton renders excretory service in some invertebrates, including insects, since nitrogenous materials deposited in it are eliminated when the animal molts. The white pigment in wings of cabbage butterflies, formed from uric acid, is clearly an excretory product.

Flatworms, ribbon worms, and some other inver-

tebrates have **flame cells** (protonephridia, with inner ends closed; Fig. 7-7*C*) scattered among the body cells from which wastes are drawn to pass out in a branched system of ducts (Fig. 18-2). The commonest excretory organs in many animals are tubular structures, the **nephridia** and **coelomoducts.** Primitively these were arranged one pair to a body somite, but they have been variously modified in the course of evolution. Nephridia in annelids are ectodermal in origin. In the earthworm, each somite contains a pair of **nephridia** (metanephridia, inner ends open; Fig. 7-7*D*). The inner end of each has a ciliated funnel, or **nephrostome,** draining from the coelom, and around the long tubule are blood vessels whence wastes are also drawn; the tubule ends externally as a small ventral **nephridiopore** (Fig. 22-6). Mollusks and rotifers have one or two pairs of nephridiumlike organs that drain from the body or blood (Fig. 21-20); nephridia also occur in the chordate amphioxus (Fig. 27-9).

In some annelids and arthropods and in the chor-

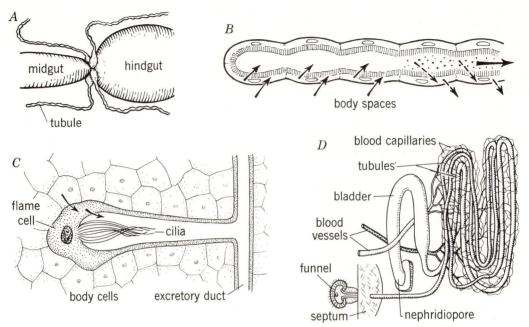

Figure 7-7 *A*. Malpighian tubules of insect attached to gut. *B*. Section of one tubule showing entry of excretions from body spaces (→), resorption routes for water and some other materials (⇢), and exit route of wastes (→). *C*. Planarian flame cell draws fluid wastes from surrounding body cells; bundle of cilia drives fluid into excretory duct. *D*. Earthworm nephridium receives fluid wastes from coelom through funnel and also by diffusion from surrounding blood capillaries.

dates, the principal excretory organs are coelomoducts, mesodermal in origin, probably derived from genital ducts but now variously modified to remove wastes from the body cavity. Crustaceans have two pairs, the renal ("antennal" or "green") and maxillary glands; each has an end sac with a duct opening at the base of an appendage. Only rarely are both developed in the same stage of a single species.

Spiders have coxal glands in the cephalothorax derived from coelomoducts.

7-12 The vertebrate kidney (Figs. 7-8, 7-9; Table 7-1) The principal excretory organs in a vertebrate are two **kidneys.** In general each consists of a mass of coelomoducts opening into a collecting duct. They

Figure 7-8 Basic patterns of vertebrate excretory systems in relation to circulatory system and coelom; diagrammatic (Table 7-1). *Pronephros.* Segmental. Ciliated ducts gather fluid waste from coelom; a knot (glomus) of blood capillaries adjacent. *Mesonephros.* Segmental. Some with open ciliated ducts, others without; branch from duct around knot of blood capillaries forms a glomerulus. *Metanephros.* Nonsegmental. Concentrated groups of glomeruli draining to one large duct; no opening to coelom (compare Fig. 7-10).

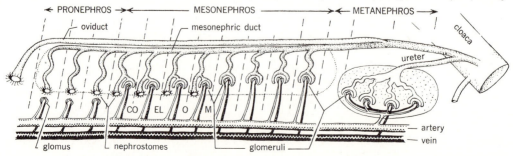

are short and posterior in all but the fishes and sala-manders, where they extend dorsally along most of the body cavity. The kidneys of lower vertebrates—cyclostomes to amphibians—and the embryonic kidneys of the higher groups develop segmentally, a pair per body somite (pronephros, mesonephros); some tubules have nephrostomes opening to the coelom, thus somewhat resembling the nephridia of earthworms. The adult kidneys of reptiles, birds, and mammals (metanephros) are nonsegmental and drain wastes only from the blood.

From each kidney, of whatever type, a common collecting duct, the **ureter,** carries the waste pos-teriorly. In amphibians, reptiles, and birds the two ureters discharge into the **cloaca,** to which a **urinary bladder** connects in amphibians and some reptiles. The waste, or **urine,** is always fluid except in reptiles and birds, where the semisolid excretions (uric acid) are voided as a white paste (guano) with the feces. In most mammals the ureters connect directly to the bladder, whence a median duct, the **urethra,** dis-charges to the exterior, passing through the penis in the male. The interrelated excretory and reproduc-tive systems of vertebrates are commonly termed the **urogenital system.**

Each human kidney (Fig. 7-10) consists of an inner **medulla** and an outer **cortex.** Each contains about a million minute excretory units, or **nephrons.** A nephron is made up of (1) a globular double-walled **Bowman capsule** around a clump of capillaries, or **glomerulus;** and (2) a **tubule** surrounded by blood capillaries. The tubule consists of (1) a proximal con-voluted portion, (2) the loop of Henle, with descend-ing and ascending limbs, and (3) a distal convoluted part. The nephron empties into a collecting duct. The Bowman capsule is about 0.2 mm in diameter, and

Figure 7-9 Urogenital systems of vertebrates; only one side or half of each is shown. For the male mammal the primitive (em-bryonic) positions of testis and ductus deferens are shown by broken lines.

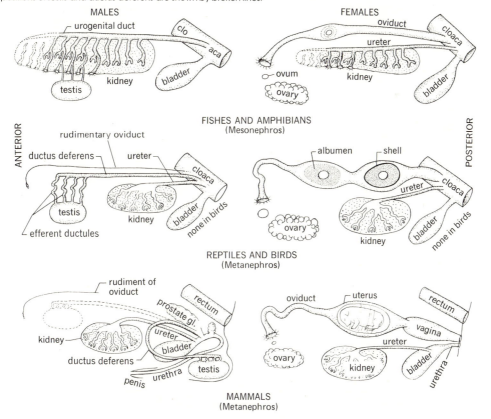

Table 7-1
TYPES OF KIDNEY IN VERTEBRATES

Kidney type	Embryonic history and adult structure	In fishes and amphibians	In reptiles, birds, and mammals
Pronephros, or head kidney	First to appear in embryo; develops segmentally, far forward in body cavity; each unit with a nephrostome opening from the coelom; no glomeruli	Functions in larva; disappears in adult	Appears transiently in embryo and soon disappears
Mesonephros, or midkidney	Develops segmentally in middle part of body cavity; some nephrostomes open to coelom, but excretion chiefly by glomeruli	Becomes functional kidney of adult	Appears after pronephros; functions during embryonic life, disappearing before hatching or birth; duct persists as ductus deferens in male
Metanephros, or hind kidney	Last to develop; not segmental; posterior in body cavity; no nephrostomes; many glomeruli; all excretion from bloodstream	Not developed	Last to appear; becomes functional kidney after hatching or birth

the tubule is 0.015 to 0.025 mm in diameter by 50 to 60 mm (2 to 2½ in) long. The 2 million nephrons of a human being, end to end, would extend for nearly 80 km (50 mi). All the collecting ducts discharge into a central cavity (pelvis) of the kidney that connects to the ureter.

7-13 Kidney function The first step in urine formation is *filtration.* Wastes and other materials are brought in the bloodstream by the renal arteries and arterioles to the glomerulus. According to the generally accepted theory of kidney function, protein-free fluid passes from the capillaries of the glomerulus through the Bowman capsule. This process takes place because of the high pressure in these capillaries, caused by the difference in size of arterioles leading to and from the glomerulus. The fluid in the capsule has the same percentage composition as

blood plasma minus the proteins and cells to which the membranes are impermeable. The second step is differential *resorption* by cells of the tubules—mostly in the proximal convoluted tubule but also in the loop of Henle and the distal convoluted tubule. Resorption may involve work, with the use of oxygen and the expenditure of energy. This is illustrated by materials such as glucose that are present in the blood and in the glomerular filtrate but not in normal urine. Such substances are resorbed completely and in a direction opposite from that expected by simple diffusion.

The most remarkable feature of resorption by the tubules is its selectivity. For example, about 1140 g (2½ lb) of salt (NaCl) passes from the glomeruli into the tubules each day—but normally only 4 to 8 g (0.14 to 0.28 oz) of this leaves the body in the urine. The rest is resorbed into the bloodstream. Urea, on the

Table 7-2
CONCENTRATING ACTION OF THE HUMAN KIDNEY: SOME VALUES IN URINE AND IN BLOOD

	Concentration (percent)							
	Water	Sodium (Na)	Chloride (Cl)	Potassium (K)	Phosphate (PO_4)	Sulfate (SO_4)	Uric acid	Urea
In blood plasma	92	0.30	0.37	0.02	0.009	0.002	0.004	0.03
In urine	95	0.35	0.60	0.15	0.150	0.180	0.050	2.00
Ratio		1	2	7	16	90	12	60

other hand, is constantly being removed; it is about half (30 g daily) of all solids in the urine, where it is in much higher concentration than in the blood plasma (Table 7-2).

Unused amino acids derived from proteins are not stored. Instead they are deaminated—the amino group (NH_2) is removed by an enzyme (deaminase) —and converted to ammonia, a rather toxic material. In the liver of mammals the ornithine cycle combines ammonia with carbon dioxide to yield the less toxic urea which is excreted by the kidney. Three amino

acids in the liver serve in a repeating cycle, aided by the enzyme arginase as shown in Fig. 7-11. The overall reaction is thus $2NH_3 + CO_2 \rightarrow CO(NH_2)_2 + H_2O$. The residue from deaminated amino acids is converted to fat or carbohydrate for further use.

Materials like glucose, sodium, and calcium are called **high-threshold** substances because they are resorbed in considerable quantities, little or none being left to pass out in normal urine; those resorbed in small quantities (urea, uric acid, etc.) are termed **low-threshold** substances. In addition to glomerular

Figure 7-10 The human excretory system. *A*. Entire system, ventral view. *B*. One kidney in median section. *C*. Relations of Bowman corpuscles, tubules, and blood vessels. *D*. One Bowman corpuscle and adjacent tubule (shown also in cross section) — solid arrows show flow of blood, broken arrows the excretory path. *C* and *D* diagrammatic and much enlarged.

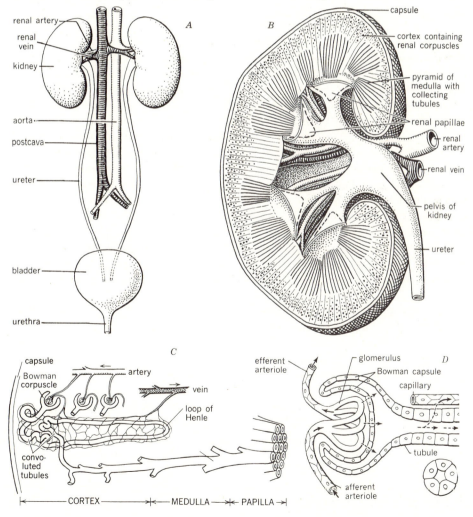

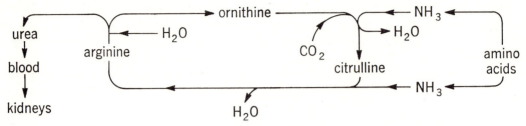

Figure 7-11 Ornithine cycle.

filtration and resorption, there is some direct tubular secretion of waste products difficult to metabolize in the body.

The capacity of the human kidneys is truly remarkable. They comprise scarcely $\frac{1}{200}$ (0.5 percent) of the total body weight yet receive 20 percent of the blood volume pumped by the heart. About 1600 L (425 gal) of blood flows through the kidneys each day, but only about 170 L (45 gal) of fluid is filtered, and of this 168 L is resorbed; thus only 1 or 2 L passes off as urine. In other words, an amount equal to the entire volume of blood is filtered by the tubules about 30 to 36 times per day.

Urinary output is controlled in several ways. Filtration, the diffusion of fluids from blood plasma through the glomerular membranes to the capsular space, is directly influenced by blood pressure. This, in turn, is affected by epinephrine from the adrenal medulla which constricts the blood vessels of the glomeruli. Another hormone, *aldosterone,* from the adrenal cortex, stimulates resorption of sodium and chloride in the kidney tubules and the elimination of potassium. The kidney itself influences its release. With a decrease in sodium intake, specialized cells in the renal arterioles secrete **renin,** which causes a plasma protein angiotensinogen, synthesized by the liver, to split off **angiotensin.** Angiotensin strongly promotes the secretion of aldosterone, which then increases the distal tubular resorption of sodium.

The resorption of water is one of the most basic processes of kidney function. About 80 percent of the water in the tubular fluid is probably recaptured by direct osmosis into the capillaries and thence passes into the venous system as a result of attraction by colloids in the blood. Additional water is taken up in the distal part of the tubule and collecting duct, a

process controlled by the antidiuretic hormone (ADH) of the posterior lobe of the pituitary gland (Sec. 8-12).

The mechanism by which the kidneys regulate body water balance under the influence of ADH is as follows: There is a strong osmotic gradient in the extracellular fluid of the kidney extending from the cortex to the inner portion of the medulla, from a region of low to one of high osmolarity (Fig. 7-12). The gradient is established and maintained by a countercurrent system operating in the hairpin portion of the loops of Henle, which concentrate chloride (Cl^-) and sodium (Na^+) ions in the extracellular fluid around the tubules. This gradient is crucial in the regulation of urine concentration. To understand its formation, let us consider first what happens in the descending limb in the presence of an *existing* extracellular gradient. The walls of this limb are permeable to diffusion of Cl^-, Na^+, and water. Urinary fluid enters the limb from the proximal convoluted tubule at an osmolarity value about that of the blood plasma. As the fluid passes along the tubule it enters zones of increasing osmotic concentration. Water diffuses out of the tubule and NaCl diffuses in, each moving toward its region of lower concentration. Tubular fluid volume is reduced. The fluid in the bottom of the descending limb finally assumes essentially the same osmotic concentration as that of the extracellular fluid (Fig. 7-12). What is the source of the high concentration of extracellular NaCl? We follow the movement of the fluid further. It passes around the bend in Henle's loop and moves along the ascending limb. In contrast to the descending limb, the walls of the ascending limb are relatively impermeable to the diffusion of both water and NaCl. The urine does not assume the osmolarity of the sur-

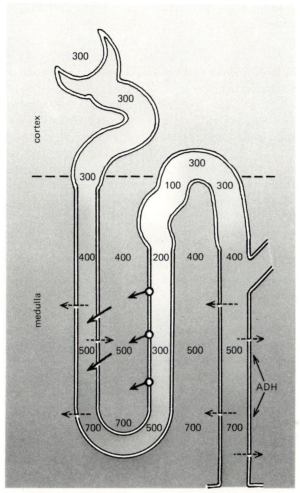

Figure 7-12 Mechanism for controlling the concentration of urine in the mammalian kidney. The osmotic concentration gradient in the extracellular fluid and variation in the concentration of solutes in the tubular fluid of the nephron are shown by graded tones (the darker the tone, the greater the concentration). ←- , diffusion of water; ←ᴏ active transport of chloride; ←, passive diffusion of chloride; 100–700, osmolarity values. Water resorption occurs from the collecting duct under the influence of the antidiuretic hormone (ADH).

rounding fluids. Instead, the tubule walls actively transport NaCl[1] outward. Some of the transported NaCl diffuses into the descending limb, and a cyclic flow of NaCl is established from the ascending to the descending limb, augmented by constantly added new NaCl entering the latter from Bowman's capsule. By this countercurrent, multiplying effect, the osmotic concentration of the extracellular fluid can be built up to high levels.

[1] *It appears that it is the chloride ions that are actively transported and that the sodium ions follow passively.*

What part does the osmotic gradient play in the control of the water content of urine? The final concentration of urine occurs in the collecting ducts. They course through the medulla parallel to the loops of Henle. The urinary fluid entering a collecting duct has equilibrated with the osmotic concentration of the extracellular fluid and circulating blood plasma (Fig. 7-12). As the fluid passes along the duct, it enters zones of increasing osmotic concentration. The duct walls are subject to great variations in permeability to water, depending on the action of ADH. As permeability is increased by ADH, water diffuses out of the duct into the extracellular fluid along the osmotic gradient. Under strong action by ADH, the fluid at the end of the duct may be isosmotic with the extracellular fluid at the tip of the medulla.

There is a correlation between the length of the loops of Henle and the maximal urine concentration achieved by various species. The loops are short in species that live in moist environments and long in desert forms such as kangaroo rats and other desert rodents.

Normal kidney function is essential to health, and any irregularity or disease in the kidney is serious. Certain salts, especially oxalates, may crystallize to form kidney stones in the pelvis of the kidney that sometimes require removal by surgical procedure. The content of the urine may be altered by other abnormal conditions. Urinalysis, therefore, may give useful clues to the general state of bodily function, healthy or otherwise. Abnormal constituents of the urine may be albumin, excess glucose, acetone bodies, cell casts, pus, blood, or bile pigments. It is a remarkable fact that removal of one kidney and even part of the second kidney does not entirely block the total excretory process in humans.

An artificial kidney has been devised and is now available in many hospitals for cases of acute kidney failure or blood poisoning. Blood is diverted from an artery through cellophane tubing set in a circulating bath fluid and thence back to a vein. The cellophane has pores of about the same size as the glomerular capillaries so that substances will diffuse in or out depending on the concentration of each in the bath and in the blood. By adjusting the concentration of substances in the bath, it is possible to add or to remove elements from the blood as desired.

7-14 The bladder and urination   The urine forms at a fairly constant rate, about 1 mL/min. It passes down the ureters to accumulate in the urinary bladder. It is expelled at intervals through the urethra. The bladder is a hollow, pear-shaped organ low in the abdominal cavity. Like the stomach, it can adapt to change in volume without altering the internal pressure. The smooth muscle of the bladder wall accommodates to increase in volume until about 300

mL of urine has accumulated; then a sensation of fullness develops. The desire to urinate may, however, be suppressed by voluntary control of the external urethral sphincter until the total content is 700 to 800 mL. Urination, or emptying of the bladder, is controlled by several reflex mechanisms which involve stretching followed by contraction of the bladder wall with simultaneous relaxation of the sphincter. Even small amounts of urine can be passed by straining, which increases the pressure in the abdomen and compresses the bladder. When the urine reaches the urethra, urination continues by reflex action even though pressure is discontinued.

7-15 Other means of excretion   In higher animals, including humans, some wastes are eliminated by other organs. Metabolic CO_2 is disposed of by routes described in Sec. 7-10. Water is eliminated through the lungs, up to 240 mL (8 oz)/day in human beings. Some other excretory products are voided with the feces, including wastes of heavy metals (iron and calcium), bile pigments excreted by the liver during the breakdown of hemoglobin, and some water.

The human skin, by its sweat glands (Fig. 4-1), also serves for the elimination of water, together with salts, traces of CO_2, and some nitrogenous waste. The loss through perspiration is usually small, but during active sweating as much as 12 L (3 gal) of water may be lost in a day. Under such circumstances salt (NaCl) needed in the body's economy must be replaced by eating salty foods or salt tablets; also the supply of vitamin C must be adequate. Perspiration is only incidentally excretory, its primary function being temperature regulation. It is also influenced by fright and emergency situations, in which a "cold sweat" occurs.

Regulation in the body

7-16 Homeostasis   The famous French physiologist Claude Bernard said, "All the vital mechanisms, varied as they are, have only one object, that of preserving constant the conditions of life in the internal environment." All living organisms maintain a more or less **steady internal state,** known as homeostasis, regardless of extremes in their external environment. In general, the degree to which a particular group has achieved independence of its environment is a measure of its evolutionary progression; some protozoans are affected by nearly every factor in the medium surrounding them, whereas humans are more independent, by one means or another.

Reflex activity of the nervous system and hormones of the endocrine system are the bases of all steady-state control. Every part of the animal body during all stages of growth and reproduction is under their influence. The situation of even the simplest animal is so intricate and little understood that regulatory processes generally are considered piecemeal, in terms of a few readily measurable criteria, rather than as an integrated whole. Osmotic pressure, hydrogen ion concentration, and temperature are three of these criteria, and each is intimately connected with water.

7-17 Osmoregulation   Water is taken in with food and also to some extent by absorption in aquatic forms. It is the universal solvent and carrier, and no organism can be independent of this essential fluid. Because of its property of diffusing across membranes, water is the vehicle for maintenance of the steady state.

The body fluids of all animals are remarkably similar in salt content and resemble that of seawater, suggesting that life originated in the sea at some remote time. Regulation of osmotic pressure (Sec. 2-11) is simple for most soft-bodied marine invertebrates, because their body fluids are in equilibrium (isotonic) with the surrounding water; pressures are the same inside and outside. Fresh water contains only about $\frac{1}{100}$ the salt concentration of seawater. Body fluids of freshwater animals and protistans have a higher salt content than the surrounding medium, and water tends to diffuse inward; the excess is disposed of in various ways. Protozoans dispose of it by means of the contractile vacuole, which usually is absent in marine forms. Other freshwater animals have nephridia, Malpighian tubules, or kid-

neys to excrete the surplus. This work of excretion requires energy, as evidenced by the higher respiratory rate of freshwater animals as compared with closely related marine species.

Aquatic vertebrates maintain osmotic balance in different ways, depending on their evolutionary history. Among bony fishes freshwater fish have blood with higher salt content than the surrounding medium. Their methods of osmoregulation are discussed in Sec. 30-20.

Cormorants, gulls, and other marine birds drink seawater for their internal needs. The water is absorbed by the gut wall, and the excess of salts passes in the bloodstream to a pair of salt-secreting glands located near the eyes, with ducts to the nostrils. Marine and some terrestrial reptiles also excrete salt by nasal glands (Sec. 32-24).

The body not only regulates water but also maintains various ions in a steady state. Sodium, potassium, chloride, and sulfate ions, for example, are usually in different concentrations in the body fluids and in the surrounding medium. This is due to differential excretion, as with magnesium in decapod crustaceans, or differential absorption, as in frogs, where the skin absorbs sodium and chloride ions but not potassium or calcium. In frogs, salt absorption is controlled by hormones from the anterior pituitary.

7-18 Water regulation of land animals   Terrestrial animals experience the hazard of desiccation, and many of them have an impermeable cuticle to resist loss of water from the body surface. Also, means have developed whereby water is resorbed by the kidneys (occasionally by cells in the rectum, cloaca, or bladder).

In humans the excretion or retention of water depends on the state of hydration of the body as a whole. Excessive sweating decreases the volume of fluid that passes out in the urine, whereas drinking quantities of fluid increases the urinary output. Water balance is controlled to a certain extent by thirst, which varies remarkably with the state of hydration, and by kidney action, which is influenced by the antidiuretic hormone (ADH) (Sec. 7-13). In the absence of this hormone, resorption by the kidney tubules decreases. The control mechanism is automatic, because an increase of osmotic pressure in the blood causes increased secretion of the hormone. This, in turn, stimulates resorption and thus conserves water. Alcohol inhibits secretion of the antidiuretic hormone and thus has a dehydrating effect. Caffeine is a diuretic, increasing the glomerular filtration rate and reducing resorption of water by the tubule cells.

7-19 Regulation of pH   The hydrogen ion concentration of most body fluids varies, but it is usually between pH 7 and 8. The regulatory mechanism by the blood, in this case, is the buffering action of such inorganic ions as phosphates and carbonates. Ion pairs, such as HPO_4^{2-} and $H_2PO_4^-$ or CO_3^{2-} and HCO_3^-, act as buffers by combining with excess H^+. For example, the CO_3^{2-} forms HCO_3^-, resulting in a decrease of carbonate, an increase in bicarbonate, and the reduction in concentration of free H^+ ions. When the blood becomes too alkaline, this reaction is reversed. A considerable amount of acid or base thus may be absorbed without altering the pH of body fluids.

7-20 Heat regulation   The metabolism in an animal produces heat (which can be measured and stated in calories; see Sec. 5-12). The body temperature at any given time, however, is a function of the heat produced, received, conserved, and lost. In most animals metabolism is low and the body temperature does not differ much from that of the environment. Such animals are called cold-blooded, or ectothermal, although actually their bodies may be comparatively warm or cold, following fluctuations in outside temperature. Many reptiles maintain their body temperature at "preferred" levels by varying the amount of exposure to the sun. Regulation in such cases is often by behavior. In summer many desert reptiles are active by night and seek shelter during the peak of daytime heat. Other nonregulators have life cycles such that they are in the egg, pupa, or some other resistant stage during winter.

By contrast, birds and mammals are warmblooded, or even-temperatured (endothermal); to maintain this condition their energy production goes up as the outside temperature goes down, and thus the body temperature remains nearly constant. The normal human oral temperature is about 37°C (98.6°F). This varies within a few degrees for various reasons, of which the most important is infection. Regulation is effected by the hypothalamus, which acts as a thermostat.

In cold weather metabolism is increased through muscular activity, including the involuntary act of shivering, and some of the energy produced is in the form of heat. During warm weather, excess heat is lost in two ways: Blood vessels in the skin are dilated so that heat is taken to the surface more rapidly, and the activity of the sweat glands is increased. The actual loss of heat is largely through radiation from the body surface and utilization of heat in the process of evaporating water. Excessive humidity hinders evaporation and is the cause of discomfort on hot, moist days. Clothing produces no heat but holds a layer of warm moist air between it and the skin and thus reduces the loss of heat by evaporation and radiation. Evaporation through increased respiration is an important means of regulating temperature in animals such as dogs that do not perspire.

7-21 Organizational levels   The so-called steady state is not a single, static condition but a dynamic equilibrium of many systems which change in successive stages of development. Furthermore, it exists at various levels. There is the cellular steady state by which individual cells are maintained in equilibrium with their environment; there is regulation at the tissue level, at the organ level, and finally at the level of the whole organism. Regulation and the steady state are at the basis of life.

References

(See also references at end of Chaps. 4 and 6.)

Bently, P. J. 1971. Endocrines and osmoregulation: A comparative account of the regulation of water and salt in vertebrates. New York, Springer-Verlag. xvi + 300 pp.

Clements, J. A. 1962. Surface tension in the lungs. *Sci. Am.*, vol. 207, no. 6, pp. 120–130. *A soaplike agent (sur-*

factant) coats the inner surface of the lungs and appears to prevent their collapse.

Comroe, J. H., Jr. 1974. Physiology of respiration. Chicago, Year Book Medical Publishers, Inc. xii + 316 pp., 110 figs. *An introductory text for medical students.*

————. 1966. The lung. *Sci. Am.*, vol. 214, no. 2, pp. 56–68. *Anatomy and physiology of the human lung.*

Fenn, W. O. 1960. The mechanism of breathing. *Sci. Am.*, vol. 202, no. 1, pp. 138–148. *A description of the human respiratory system.*

Langley, L. L. 1966. Homeostasis. London, Chapman & Hall, Ltd. 114 pp.

Negus, V. 1965. The biology of respiration. Baltimore, The Williams & Wilkins Company. xi + 228 pp., 154 figs. *Comparative.*

Potts, W. T. W., and **G. Parry.** 1964. Osmotic and ionic regulation in animals. New York, Pergamon Press. 423 pp.

Smith, H. W. 1953. The kidney. *Sci. Am.*, vol. 188, no. 1, pp. 40–48. *It not only filters wastes from the blood but also regulates the volume and contents of body fluids.*

————. 1961. From fish to philosopher. Garden City, N.Y., Anchor Books, Doubleday & Company, Inc. 293 pp. Paper.

Solomon, A. K. 1962. Pumps in the living cell. *Sci. Am.*, vol. 207, no. 2, pp. 100–108. *The mechanism of excretion of sodium.*

Williams, C. M. 1953. Insect breathing. *Sci. Am.*, vol. 188, no. 2, pp. 28–32. *The tracheal system of insects provides each cell with oxygen from a private conduit.*

THE ENDOCRINE GLANDS

Glands are cells or groups of cells specialized in structure and function to produce substances needed in bodily processes; these are synthesized from ingredients obtained in the blood, lymph, or other surrounding fluids. Most glands discharge their products through ducts and are called glands of external secretion (*exocrine*). The salivary glands and liver, for example, have ducts carrying secretions to parts of the digestive tract, whereas the mammary and sweat glands discharge through openings on the body surface. In addition there are other glands without ducts whose secretions are carried by the bloodstream to various parts of the body. These are the **endocrine glands,** and their products are **hormones.** They include (1) **epithelial endocrine glands** (derived at an early stage from embryonic germ layers), (2) endocrine organs developed from **transformed nerve gan-** glia (adrenal medulla, insect corpora cardiaca) and (3) **neurosecretory cells,** often in **neurohemal organs** (posterior pituitary, invertebrate sinus gland). Neurosecretory cells are nerve cells that produce hormones. Their secretion is transported along the cell's axon to its swollen terminals where it is stored or released into the circulatory system. A neurohemal organ is a special grouping of neurosecretory axons and their associated blood vessels. All three types of endocrine structure are found in both invertebrates and vertebrates, although not necessarily together in the same taxonomic group. In invertebrates epithelial endocrine glands are known only in cephalopods and arthropods. Minute amounts of endocrine substances exercise profound regulatory control over many bodily functions, stimulating or inhibiting the development, growth, and activities of various tissues, affecting metabolism, and influencing the behavior of the individual.

8-1 Control by hormones Most or all endocrine activities are interrelated and influence one another. They are so closely integrated with activities of the nervous system that we may speak of many body functions as being under **neuroendocrine control.** The exact ways that hormones influence physiologic processes are often unknown. However, several basic mechanisms suggest how they may affect their target cells. One involves the action of steroid hormones on the process of genetic transcription. The hormone passes through the cell's plasma membrane and upon entering the cytoplasm is bound to a protein which is called its cytoplasmic receptor. The hormone–cytoplasmic receptor combination then passes into the nucleus, where it combines with a specific nuclear receptor. There are specific receptors for specific hormones. The hormone–nuclear receptor complex causes the transcription of particular kinds of messenger RNA (Sec. 2-28), and thus brings about the synthesis of specific enzymes or other proteins. It perhaps does so by removing the repressor on RNA polymerase.

Another mechanism involves interaction of the hormone with a specific receptor in the cell membrane which activates (or sometimes inhibits) an enzyme, **adenylate cyclase,** widely present in animal cell membranes. The hormone itself does not enter the cell. Activation of adenylate cyclase catalyzes the production of adenosine 3′,5′-monophosphate (or cyclic AMP) from ATP within the cell, producing a second messenger (the hormone itself is the first). Increase in intracellular cyclic AMP, in general, causes an increase in protein synthesis, the nature of which depends on the specialization or programming of the target cell. Enzymes and other substances may be produced, preexisting enzymes may be activated, or the permeabilities of cell membranes may be altered. Not all hormones activate adenylate cyclase. Some produce their effect by inhibiting it and causing a decrease in cyclic AMP. Furthermore, various steroid hormones and perhaps others do not act via cyclic AMP.

Some hormones bring about prompt responses,

whereas others act on a long-term basis, coordinating growth of organs and metabolic concentrates in blood and other tissues. Hormones are transformed or destroyed, so their effects are not permanent unless a continuing supply is available. The fundamental substances of which they are formed appeared early in biochemical evolution and occur in all multicellular organisms.

The chemical structure of several hormones is now known. Those of the adrenal cortex, gonads, placenta, and insect ecdysial glands are steroids, derived from cholesterol; thyroxine is an amino acid; and insulin and growth hormone are proteins. Oxytocin and vasopressin (antidiuretic hormone) are peptides, each containing nine amino acids in only two of which they differ, yet they have considerably different physiologic properties (Sec. 8-12).

8-2 Invertebrate hormones Hormones occur in flatworms, annelid worms, mollusks, echinoderms, arthropods, and other invertebrates. They influence growth and regeneration, molting, reproduction, metamorphosis, color change, metabolism, and other processes.

Crustaceans (Fig. 8-1) have epithelial endocrine glands consisting of paired **Y organs** in the head and a pair of **androgenic glands** near the sperm ducts. They also have neurosecretory cells scattered throughout the central nervous system. Certain of these are organized in neurohemal organs which include the **sinus glands** in the eyestalks and the **postcommissural** and **pericardial organs.** A major concentration of neurosecretory cells occurs in the **X organ(s),** located in ganglia of the eyestalk. Their axons extend to the sinus gland, where their neurosecretory material reaches the blood. This gland also receives neurosecretory axons from the brain. A variety of hormones are present in the eyestalk.

An important mechanism in the control of molting in crustaceans involves interaction between the sinus glands and the cephalic Y organs. The former produce a molt-inhibiting hormone which acts during the intermolt period to restrain the action of the Y organs; the latter produce a molt-stimulating hormone. Restraint is maintained until just prior to molting, whereupon the Y organs are activated.

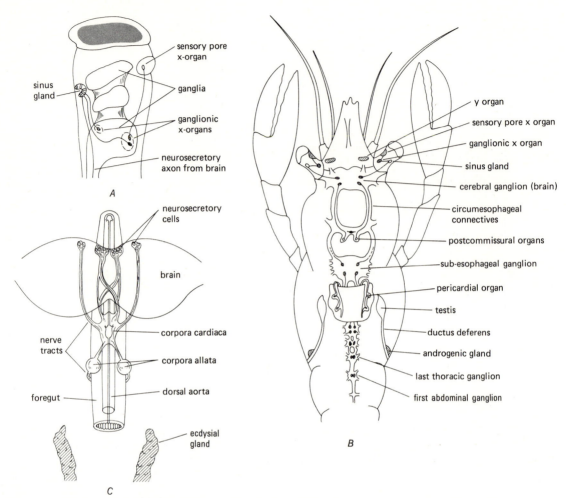

Figure 8-1 Endocrine systems of crustaceans and insects. *A.* Eyestalk of crustacean. *B.* Generalized male crustacean. *C.* Insect endocrine system. Neurosecretory cells shown in black; epithelial endocrine glands crosshatched; neurohemal organs shown in white (sinus gland, postcommissural organs, pericardial organ, corpora cardiaca, corpora allata); epithelial endocrine glands are Y organs, androgenic glands, ecdysial glands. (*A and B after Gorbman and Bern, A textbook of comparative endocrinology, John Wiley & Sons, Inc. 1962, and C after Highnam and Hill, Comparative endocrinology of invertebrates. Edward Arnold Ltd., 1977.*)

Eyestalk hormones also influence development, reproduction, and movement of pigment in retinal and somatic cells. Retinal pigment shifts help adapt the eyes to light and darkness. Somatic pigments—white, red, and yellow (also black, brown, and blue)—can be variously dispersed or concentrated. Such color changes are important in concealment, thermoregulation, and display and are mediated through the eyes. The postcommissural organs also influence color change. Chromatophorotropins appear to be present throughout the central nervous system of crustaceans, and the chromatophores are evidently under multihormonal control.

The secretion of the androgenic glands is essential for development of the gonads. In the amphipod *Orchestia*, early removal of these glands in genetic males causes the primary germ cells to produce oocytes. Transplantation of the glands into females causes transformation of the ovaries into testes and some structural masculinization.

Hormones in insects influence molting, reproduction, diapause, metabolism, excretion, and other processes. The epithelial endocrine glands of insects are paired (or fused) **corpora allata** in the head and **ecdysial (or thoracic) glands** (analogous to the Y organs of crustaceans) in the thorax or head. Neurosecretory structures include (1) the **corpora cardiaca** posterior to the corpora allata and associated with the dorsal aorta, (2) **neurosecretory cells** of the brain (pars intercerebralis), and (3) **ventral ganglia** of the central nervous system. The neurosecretory axons extend from the brain to the corpora cardiaca. The latter is an important neurohemal organ producing its own hormones but serving mainly to store and release neurosecretory hormones from the brain. The corpora allata secrete **juvenile hormone (neotenin),** which promotes growth and differentiation of larval structures and functions to retain juvenile characteristics between molts and to promote the development of a subsequent larval instar (Sec. 25-31). The ecdysial glands produce **ecdysone** (a steroid), the molting hormone. Molting and maturation in insects involves (1) stimulation of the central nervous system, causing neurosecretory cells of the pars intercerebralis to secrete activation or ecdysial gland–stimulating hormone, **ecdysiotropin;** (2) axon transport of the hormone to the corpora cardiaca, where it is released to the blood; (3) blood transport of the hormone to the ecdysial glands, causing the secretion of ecdysone, which initiates the molting process. Ecdysone causes the epidermis to form new cuticle and a molting fluid that loosens the old cuticle. Following the molt, secretion of juvenile hormone by the corpora allata maintains the larval instar until the next molt. After the last molt the corpora allata become temporarily inactive and the adult stage is produced. In holometabolous insects, removal of the corpora allata in a young larva is followed by premature pupation, metamorphosis, and emergence of a miniature adult. In hemimetabolous insects (grasshoppers, etc.), neotenin is present progressively later (or in smaller amounts) during each instar, becoming inactive except in adult females involved in egg development, when the adult stage is reached. Continuing high levels are maintained in holometabolous insects (LEPIDOPTERA) until the last

instar, when the level falls, and during pupation the hormone is inactive. During or shortly after the last molt the ecdysial glands disappear, and molting normally does not occur in adults. Experiments on hemimetabolous bloodsucking bugs (*Rhodnius*) have demonstrated that the brain is necessary for molting to occur. These insects grow by a series of molts, then metamorphose to the winged, sexually mature stage. They usually feed only once in each instar, gorging themselves when opportunity affords and molting at a fixed interval after the meal. When the body is distended, stretch receptors in the body wall send stimuli to the brain. There, usually after 8 days, activation of the pars intercerebralis occurs and molting is initiated. Decapitation experiments (the insects can survive for a time without the head) show that the brain and a time element are involved. Molting will not occur if the head is removed after the animal feeds and before 8 days have elapsed, the time required for activation of the pars intercerebralis. If decapitated after 8 days, however, the bug will molt several weeks later. If, following a blood meal, one bug is decapitated before and the other after the 8-day period and the two are joined surgically so that their bloodstreams mingle, both will molt.

Diapause (Sec. 25-14), a period of arrested growth and development in insects, is under endocrine control. Diapause is an adaptation aiding these animals to survive adverse conditions. Physiologic changes occur which permit the insect to withstand winter cold, desiccation, or lack of food. In pupae and larvae it is invoked by failure of the cerebral neurosecretory cells to produce ecdysiotropin; in adult insects it may be associated with relative inactivity of the corpora allata, sometimes aided by inactivity of the cerebral neurosecretory cells. In some insects embryonic diapause occurs. It is under the control of an inhibitory hormone secreted by the female which affects her developing oocytes. In the silk moth (*Bombyx*) the hormone is released by neurosecretory cells in the suboesophageal ganglion, under brain control. According to the conditions under which they were reared, female *Bombyx* will deposit diapause or nondiapause eggs. In the former, embryonic development is interrupted at an early stage; in the latter, development is direct. Whether *Bombyx* pupae will be diapause producers or not depends on the conditions of temperature and photoperiod under which the eggs are incubated. Eggs that develop in spring under conditions of low temperature and short photoperiod produce adults that lay nondiapause eggs. Their eggs directly produce a second generation that takes advantage of the favorable conditions of summer. Eggs that develop under the high temperatures and long photoperiods of summer produce adults that lay diapause eggs, adapted to withstand the unfavorable conditions of winter. Many insects respond to particular photoperiods (short or long) which determine the onset of diapause. The photoperiodic response is often temperature-dependent. In most insects in temperate regions the optimal temperature for diapause development is 0 to 7°C (32 to 45°F).

In the moth (*Hyalophora cecropia*), metamorphosis of the overwintering pupa into an adult results from the effect of the brain-activation hormone ecdysiotropin on the ecdysial glands (Fig. 8-2). In nature, the cold of winter is necessary to end the diapause that precedes metamorphosis. During pupal diapause, both the brain and the ecdysial glands are inactive. With termination of diapause—which can be induced experimentally by chilling the pupa—the brain becomes activated, neurosecretion is produced, and the ecdysial glands secrete.

Experiments demonstrating these endocrine activities are as follows: (1) A normal pupa does not transform if kept overwinter at room temperature but does so after being stored at 5°C (41°F). (2) If a chilled and an unchilled larva are joined surgically (parabiosis) so that their bloodstreams mingle, both will transform; hormone from the one circulates in the other. (3) If a chilled pupa is dissected into two parts, head-thorax and abdomen, the first transforms into normal foreparts of an adult but the second does not. If, however, brain and ecdysial glands are then implanted in the second, it becomes a normal abdomen, which may lay eggs! (4) If the brain is removed from several chilled larvae, the larvae then grafted to one another in a chain, and a brain transplanted into the first, the whole series will transform in sequence.

Both the cerebral neurosecretory cells and the corpora allata have been implicated in oocyte growth and vitellogenesis in insects, but oogenesis and spermatogenesis seem to proceed independently of hormonal influence.

8-3 Hormones of vertebrates[1]   In most vertebrates, including humans (Fig. 8-3), the endocrine glands include the pituitary, pineal, thyroid, ultimo-

---

[1] *See also control of secretory activities in digestion (Sec. 5-9), iodine and the thyroid (Sec. 5-13), regulation of excretion (Sec. 7-13), endocrines and reproduction (Sec. 10-21), and endocrine effects in birds (Sec. 33-14, 33-21), mammals (Sec. 34-27), and humans (Sec. 35-13).*

Figure 8-2   Endocrine control of metamorphosis in larvae of the cecropia moth (*Hyalophora cecropia*).   *A*. Ligatures tied behind head and thorax before ecdysiotropin (ET) from brain is secreted; no metamorphosis. *B*. Ligatures tied after ET has circulated throughout body but before the ecdysial (thoracic) glands could be activated; head and thorax transform. *C*. Ligatures applied still later; ecdysial gland now activated, ET throughout body, metamorphosis complete. *D*. Position of corpus allatum and ecdysial glands in head of larva; diagrammatic. *E*. Brain and ecdysial glands from chilled larva implanted in isolated abdomen of pupa. *F*. Hormones produce metamorphosis to adult form, and eggs laid! Stipple, larva; lines, pupa; hairy, adult. (*After C. M. Williams, 1948, 1952, Biol. Bull.*)

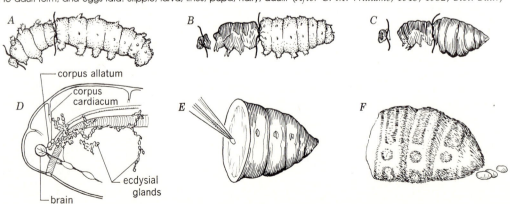

branchial bodies, parathyroids, islets of Langerhans, adrenals, gonads, parts of the gastric and intestinal mucosa, and, in some mammals, the placenta. Epithelial endocrine glands predominate among vertebrates. However the pituitary and adrenal glands are composite, for the posterior pituitary is a neurohemal organ (Sec. 8-12), and the adrenal medulla represents a transformed nerve ganglion. The location, structure, and functions of the endocrine glands are enough alike throughout the vertebrates to suggest a homologous series, differing as to details.

Knowledge of vertebrate hormone function has advanced enormously through experimental research since about 1920. Endocrine functions are studied by (1) removing the glands from either young or adult animals, (2) implanting them into subjects of various ages, (3) feeding the gland substance or an extract of it, (4) injecting the extract into the body, (5) observing individuals with diseased glands, and (6) chemical synthesis of molecular homologues. Deficiency of a particular secretion is denoted by the prefix *hypo-* and an excess by *hyper-*. With the thyroid, for example, a scant supply is referred to as hypothyroidism and an excess is called hyperthyroidism. The science of endocrinology has important applications in human medicine and some bearing on the production of domesticated animals. Also it is essential to understanding animal physiology and behavior. The structure and function of the endocrines are described here with particular reference to human beings.

Figure 8-3 The human endocrine glands. *Left.* Position in body. *Center.* Form of each (not to same scale). *Right.* Enlarged details of cellular structure.

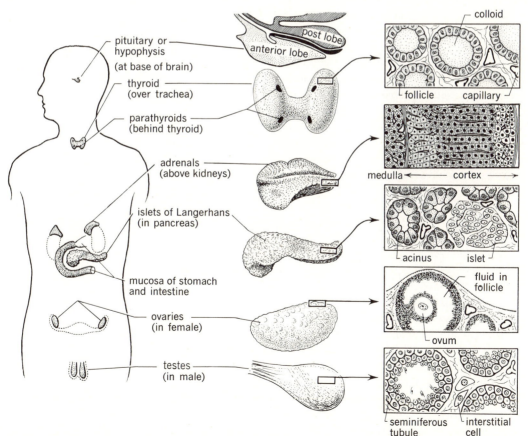

8-4 Thyroid This gland, of two lobes joined by an isthmus, is of epithelial origin. It lies on either side of the trachea below the larynx. It consists of many spherical closed sacs or follicles of microscopic size that are lined with cuboidal cells and surrounded by blood vessels and nerves. The follicles are filled with a colloid containing the hormones **thyroxine** ($C_{15}H_{11}O_4NI_4$, about 65 percent iodine by weight) and **triiodothyronine,** which regulate the general metabolism of the body (calorigenic action) as well as its growth and sexual development. Normal function produces a varying amount of thyroxine depending on age, sex, and other factors. Control is by a thyroid-stimulating hormone (TSH) from the pituitary, regulated by the hypothalamus. The thyroid in turn influences other endocrine glands, including the adrenal cortex and the gonads. With decreased production of thyroxine the subject is less active, is listless, and feels cold because of lowered general metabolism. This condition may be overcome by a daily dose of a few milligrams of thyroid extract taken orally. Overgrowth or overfunction in humans speeds up bodily activities **(basal metabolism),** increasing both heat production and heartbeat. This condition is usually relieved by removing part of the gland surgically. Extreme cases of hyperthyroidism often produce nervous excitability and exophthalmic goiter, or protrusion of the eyeballs. If the thyroid is removed in a young animal, skeletal growth is arrested and sexual maturity fails.

A deficiency of iodine in the soil and water occurs in flooded or glaciated areas and in other places far from the sea, such as the Great Lakes region, the Alps, and the Himalayas. Chronic enlargement of the thyroid, known as **goiter,** often occurs to compensate for the deficiency. If the deficiency is severe, **cretinism** results among children. Cretins are dwarfed in size, with thick puffy skin and coarse facial features; the basal metabolism is decreased, and the sexual organs do not develop; mental growth is seriously retarded; many are imbeciles or idiots, and deaf-mutes are common. Comparable deficiency in adults causes **myxedema,** characterized by thick puffy skin, scant dry hair, lowered metabolism, disturbance of sexual function, and mental lethargy. Treatment with thyroid extract, if commenced early, stimulates young cretins to normal development and commonly restores myxedematous adults. That is why, in regions of deficiency, iodine is added to the diet as iodized salt to prevent the occurrence of such defects.

The thyroid is important in amphibian metamorphosis (Sec. 31-24) but evidently not in that of fishes. Amphibian larval tissues differ in sensitivity to thyroxine and the time at which they are receptive to it. Prolonged low levels may cause only partial metamorphosis, and large amounts may produce deformities leading to death. The increase in thyroxine which characterizes normal development starts slowly; a slight increase occurs during early metamorphosis and a major peak at its climax.

Many experiments have tested the effect of thyroid secretion on amphibian larvae. If the thyroid is removed from tadpoles, they do not transform. Normal development is restored, however, if thyroid tissue is implanted or fed or if an extract of it is injected. Metamorphosis can be accelerated by such procedures, resulting in frogs that are dwarfed in size at the time of transformation.

Amphibian tissues are locally responsive to thyroxine and do not require prior conditioning for response. An implant on one side of the larval mouth causes only that side to assume adult features; one in the tail fin causes resorption only at the site of the implant.

Neoteny (prolongation of larval life) (Sec. 31-24), which occurs naturally in some salamander larvae and to a lesser degree in tadpoles, can usually be terminated by administering thyroxine. Permanently larval amphibians such as the mud puppy and hellbender may respond only slightly. Environmental factors, such as shortage of iodine in food and water or low temperatures, may be responsible for some kinds of neoteny; genetic factors are also sometimes involved.

8-5 Parathyroids These epithelial glands develop as outgrowths of visceral pouches, appearing first in amphibians. They are often four in number in mammals. In humans they are small oval structures lying behind or partly embedded in the thyroid. Their secretion, a polypeptide, **parathormone,** regulates the concentrations of calcium and phosphate in the blood plasma and affects the metabolism of these materials in the body. Secretory activity of the parathyroids is in direct response to calcium levels in the blood. The hormone mobilizes calcium by (1) activating the osteoclasts, which make available calcium from the skeleton, (2) enhancing calcium absorption by the intestine, and (3) increasing its resorption by the kidneys. The first two depend on the presence of vitamin D. Removal of the parathyroids is followed by muscular twitchings and spasmodic contractions of increasing violence, leading to severe convulsions (**parathyroid tetany**) and death. Injection of parathyroid extract stops these effects. Overproduction of parathyroid secretion removes abnormal amounts of calcium from the bones and raises the level of calcium in the blood. Parathyroid glands are unknown in fishes.

8-6 Ultimobranchial bodies   This tissue develops from the last pair of visceral pouches, but its secretory cells are thought to originate from the neural crest. It has yet to be found in cyclostomes. In mammals the ultimobranchial cells migrate into the thyroid, giving rise to **parafollicular** or **C cells.** The C cells secrete a polypeptide, **calcitonin,** which lowers the calcium level of the blood and antagonizes calcium resorption from bone by acting on the osteoclasts. It acts in opposition to parathyroid hormone and perhaps protects the skeleton against excessive resorption. The ultimobranchial bodies are absent in adult mammals.

8-7 Thymus   The thymus develops from visceral pouches. In birds and mammals it is located beneath the sternum in the upper part of the chest. It is most active in early life (in humans, during fetal and preadolescent life). The action of the thymus is not fully understood. It is uncertain whether it is an endocrine gland. It evidently produces a protein substance, **thymosin,** which acts upon lymphocytes that enter the organ and perhaps also on those circulating in the blood, endowing them with immunologic competence. Such lymphocytes are responsible for the cellular immunity manifested in the rejection of tissue grafts and in the response of cells to viruses and invasive bacteria. In thymectomized mice,

grafts of foreign tissue are less likely to be rejected and the capacity to combat infection declines.

8-8 Gastric and intestinal mucosa   Several hormones are produced in cells lining the stomach and small intestine that control secretion of digestive enzymes. Secretin (Fig. 5-7) and cholecystokinin (pancreozymin) from the wall of the duodenum stimulate secretion of pancreatic juice, and gastrin, from the mucosa of the stomach, serves similarly for gastric juice (Sec. 5-9).

8-9 Islets of Langerhans   Within the substance of the pancreas, in addition to the glandular tissue providing digestive enzymes that pass through the pancreatic duct to the intestine, there are many small groups of cells, the **islets of Langerhans,** of different form and staining reaction, not connected to the duct. The cells are of three kinds, alpha$_2$ cells that produce **glucagon,** beta cells that produce **insulin,** and alpha$_1$ cells that evidently secrete gastrin. Insulin (Sec. 5-11) and glucagon interact to control the energy-releasing carbohydrate metabolism of body tissues. Insulin has a stimulating influence, promoting the "burning" and storage of body "fuels." It facilitates the movement of glucose (sugar) into muscle cells, the liver, and adipose and other tissues. It favors glycogen formation and storage in the muscles, liver, and adipose tissue and the formation of fat, and it promotes the uptake of amino acids and their incorporation into tissue protein. Release of insulin is directly stimulated by an increase in the glucose level of the blood. Glucagon, on the other hand, promotes the release of liver glucose, inhibits glycogen formation, and activates the formation of carbohydrates in the liver from amino acids, thus tending to protect against an undesirable fall in blood sugar level (hypoglycemia).

Disease in the islets or removal of the pancreas is followed by an increase of sugar in the blood and urine, a condition known as **diabetes mellitus.** Lack of insulin decreases the capacity of the muscles to use glucose. Copious urine is produced to keep the excess sugar in solution. Dehydration, loss of electrolytes, and circulatory failure may ensue. Protein and fat stores are drawn upon as sources of energy. Formerly diabetes was fatal in children and young adults and a considerable cause of death in older people, but its effects can now largely be prevented by orally taken medicine, diet, and/or daily injection of insulin. In some diseases an excess of insulin is produced, resulting in a drastic reduction in the level of blood sugar. This is comparable to the condition caused by injecting an overdose of insulin into a diabetic. The resulting insulin shock is not unlike some forms of drunkenness and can be overcome by eating sugar and thus raising the level of sugar in the blood. For medical purposes insulin is extracted from the pancreas of cattle and sheep obtained in slaughterhouses. It may soon be produced in quantity by insulin-producing bacteria created by recombinant-DNA research.

8-10 Adrenals These two small glands lie adjacent to the anterior, or upper, end of the kidneys and have an unusually rich blood supply. Each consists of an outer **cortex** and inner **medulla,** of different microscopic structure and embryonic origin. In frogs the adrenals lie along the ventral surface of the kidney, and in sharks the cortex and medulla are separate structures. The medulla (chromaffin tissue) is derived from neural-crest cells and is closely related in embryonic origin and function to the sympathetic nervous system, by which it is controlled. It produces epinephrine and norepinephrine (noradrenaline). Epinephrine acts to support sudden metabolic needs of the body under conditions of emergency. It increases cardiac output and dilates the blood vessels of the muscles, shunting blood where it is needed during exertion. The hormone relaxes smooth muscle of the bronchioles of the lung (hence relieves attacks of asthma) and retards muscular movement in the intestine; it also hastens the transformation of glycogen into glucose. One or two parts per billion of epinephrine are normal in the human bloodstream, but under emotional stress, such as fear or anger, additional amounts are suddenly secreted and blood is shifted from the viscera to the muscles and the brain, so that the individual is ready for "fight or flight." Norepinephrine causes vasoconstriction and confers muscle tone throughout the circulatory system. It also has an effect similar to, but weaker than, that of epinephrine on the blood sugar level and smooth muscle contraction. Unlike removal of other endocrine glands, removal of the adrenal medulla by surgical procedure, which stops the secretion of its hormones, often results in no significant disturbance in experimental animals. Possibly the autonomic nervous system can take over in the absence of these hormones in the protected environment of the laboratory. The situation is different under natural conditions.

The cortex, or outer part, of the adrenal produces several endocrine substances, all steroids. Three classes are recognized: (1) mineralocorticoids, affecting the metabolism of water and electrolytes and the serum concentrations of sodium and potassium; (2) glucocorticoids, influencing metabolism of carbohydrates and proteins; and (3) androgens, affecting sexual development. The more important of these hormones are the mineralocorticoid aldosterone and the glucocorticoids cortisol and corticosterone. Cortisol (or hydrocortisone) is beneficial in treating some types of arthritis. The general effect of these steroids is to help the body cope with cold, other environmental demands, or infection and to aid in regulation of carbohydrate metabolism and electrolyte balance.

Complete removal of both adrenals is followed by death in 10 to 15 days. Earlier symptoms are loss of appetite, vomiting, weakness and prostration, reduction in bodily temperature and metabolism, and loss of water and sodium chloride from the blood. Destruction of the adrenal cortex (Addison's disease) in humans results in bronzing of the skin, gradual decline, and finally death. Excessive androgen production in females causes masculinization.

Perhaps the most important but least understood aspect of adrenal physiology is the response of these glands to stress. The adrenal cortex of animals subjected to stress often enlarges. In addition to physical factors, two "social" ones may be involved—population density and the position of the animal in the hierarchy of dominance. As population density increases, encounters between individuals become more frequent and stresses mount. The top animal in a dominance (or **peck order**) hierarchy apparently experiences less stress than the subordinate individuals.

8-11 Gonads, or sex glands The testes of the male and ovaries of the female are the **gonads,** or **primary sex organs.** The sperm ducts, associated glands, and penis of the male and the oviducts, uterus, and vagina in the female are the **accessory sex organs;** these are related in various ways to reproduction (Chap. 10). External differences between the sexes, or **secondary sexual characteristics,** appear in many animals upon attaining sexual maturity. The gonads, besides producing eggs and sperm, secrete hormones that affect the accessory organs and sexual characteristics. These hormones are the androgens, estrogens, and progesterone, closely related steroid compounds. The former predominate in males, but some androgen is secreted by the ovary. Estrogens in some species are secreted by the testes and adrenocortical tissue. (They are also found in plants.) Other endocrine glands, especially the pituitary and thyroid, also influence the sexual structures and functions.

The thick neck, deeper voice, and belligerent manner of bulls, the antlers of male deer, and the larger comb, wattles, and spurs and crowing habits of roosters are some familiar secondary sexual characteristics. Castration, or removal of the gonads, before sexual maturity, produces striking changes in the form and temperament of these animals. The steer (castrated bull) has a smaller neck and more cowlike voice and is docile; the castrated deer grows no antlers; and the capon (castrated rooster) has smaller comb and spurs and does not crow. In all such castrates the secondary characters are lacking, the accessory sex organs are reduced, sexual behavior is slight or absent, and the individuals tend to accumulate fat.

The hormone of the testis responsible for these changes is **testosterone** ($C_{19}H_{28}O_2$), evidently produced by the **Leydig** or **interstitial cells** between the seminiferous tubules. If testosterone is injected into a castrated individual, the accessory sex organs enlarge, the secondary sexual characters develop, and the behavior becomes that of a normal (uncastrated) animal.

Follicles of the ovary produce a female sex hormone, **estradiol** (and related estrogenic hormones). It is one of the essential factors in the regulation of the estrous cycle (Sec. 35-13) and in the development of secondary sexual characteristics that occurs with the onset of puberty. It is responsible for the phenomenon of estrus, or "heat," in female mammals. Removing the ovaries from an immature female prevents her from becoming sexually mature; the accessory sex organs remain infantile, and the sex instincts are not shown. Injecting estradiol into a castrated female corrects these effects. If injected into a normal (uncastrated) but immature female, sexual maturity is quickly brought about; the accessory organs develop, but the ovaries remain infantile.

The accessory reproductive organs of females, especially after estrus, are controlled by another ovarian hormone, **progesterone** (*progestin*). This is produced by the corpus luteum that forms from a graafian follicle of the ovary after discharge of the ovum and, late in pregnancy, by the placenta. Progesterone, together with estradiol, prepares the uterus for receiving a fertilized ovum. Both hormones, directly or indirectly, induce enlargement of the mammary glands for their subsequent function; later the lactogenic hormone of the pituitary stimulates milk secretion. Progesterone inhibits further ovulation by preventing the secretion of the hypothalamic follicle-stimulating and luteinizing releasing factors (Sec. 35-13). A third ovarian hormone, **relaxin,** also produced by the placenta, facilitates birth by relaxing the pubic ligament, thus permitting the pelvic opening to expand.

8-12 Pituitary At the base of the brain is the **pituitary gland** (hypophysis) (Fig. 8-4), formed during embryonic development of the adenohypophysis, derived from a pouch (Rathke's pocket) on the roof of the mouth, and the neurohypophysis (also called the posterior or neural lobe) from the infundibulum of the brain. Typically the adenohypophysis has three parts: (1) the pars distalis (anterior lobe, or anterior pituitary), (2) the pars intermedia, and (3) the pars tuberalis. The first two are epithelial endocrine glands; the third provides support for the portal circulation that conveys neurosecretory hormones to the anterior lobe, which is not directly connected with the brain. The neurohypophysis is a neurohemal organ with neurosecretory cell bodies located in the hypothalamus. The entire human

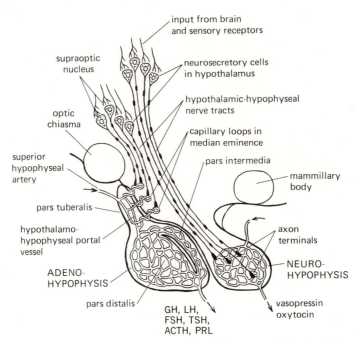

Figure 8-4 Relations between the neurosecretory cells in the hypothalamus and the circulatory system of the pituitary gland in vertebrates.

pituitary weighs only about 0.5 g, but it has an enormous influence on the growth and functioning of the entire body.

In mammals the posterior lobe stores and releases two chemically similar hormones that are produced by the neurosecretory cells in the hypothalamus. **Vasopressin,** or **antidiuretic hormone** (ADH), increases absorption of water in the kidney tubules (Sec. 7-18) and causes constriction of the smooth muscle of the arterioles. **Oxytocin** causes contraction of smooth muscle in the uterus during birth and is used medically to induce labor; it also causes ejection of milk from the mammary glands (Sec. 8-14). The antidiuretic hormone of amphibians is **arginine vasotocin** (Sec. 31-12), found in all vertebrate classes except mammals. The hormones of the posterior lobe can act rapidly, within seconds of their release.

The intermediate lobe secretes **melanophore-stimulating hormone** (MSH), or **intermedin,** which disperses the pigment in the melanophores of amphibians and other lower vertebrates (Sec. 31-12). Although the chemistry of the hormone is known

almost entirely from mammals, its function in higher vertebrates is unclear.

The anterior lobe produces at least six hormones (polypeptides and proteins), each produced in a characteristic cell type. Five of these are tropic hormones that influence the activity of particular target organs, mostly other endocrine glands. The sixth, growth-stimulating hormone, controls metabolic activities of the tissues generally. Four of the anterior lobe hormones (two gonadotropins, adrenocorticotropin, and thyrotropin) are involved in "negative" feedback loops, a rise in blood level of the hormone causing a decrease in hormonal output and a decline causing an increase in output. Somatotropin and prolactin are not involved in feedback relationships. The principal anterior pituitary hormones and their effects are as follows:

1 Growth-stimulating Excessive secretion of this hormone (somatotropin) or overgrowth of the gland causes **gigantism.** If this occurs during early youth, it results in lengthening of the long bones; human giants 2.3 to 2.6 m (7 to 8 ft) tall are produced by extreme overfunction. An excess later in life causes **acromegaly,** in which the forehead, nose, and lower jaw become massive and the facial skin is thick and coarse. Deficiency of this hormone results in **dwarfing,** the individual retaining the proportions of a child.

2 **Gonadotropic** In mammals, **follicle-stimulating hormone** (FSH) promotes the development of ovarian follicles and mature sperm. The **luteinizing hormone** (LH) evokes ovulation, formation of corpora lutea, and production and release of estrogens and progesterone. Both FSH and LH are glycoproteins and are closely related to thyrotropin. When gonadotropins are injected into immature females they cause precocious sexual maturity within a few days; overdoses in rats cause a doubling or trebling of the number of eggs released from the ovaries at one time. In male mammals they stimulate growth of both the seminiferous tubules and the interstitial tissue. Implantation of pituitary gland into amphibians results in rapid maturing and laying of eggs. Removal of the pituitary is followed by atrophy of the gonads and the accessory sex organs.

3 **Lactogenic** This hormone, prolactin, or luteotropic hormone (LTH), has many functions. It is involved in (1) sodium retention in fishes and influences the upstream migration of anadromous species seeking their spawning sites; (2) amphibian metamorphosis; (3) water-seeking behavior of young newts; (4) growth of lizards; (5) crop-sac "milk" production by pigeons; (6) brood patch formation and broodiness in birds; (7) secretion of progesterone, which stimulates and maintains uterine development during pregnancy in mammals; and (8) mammary gland development and lactation. The manifold growth-promoting effects of prolactin suggest that it shares a common molecular ancestry with somatotropin.

4 **Adrenocorticotropic** Growth and secretory activity of the adrenal glands are stimulated. One of these products is the adrenocorticotropic hormone (ACTH), which in turn stimulates the secretion of other hormones, including cortisol.

5 **Thyrotropic** The growth and secretory activity of the thyroid gland are regulated by this hormone of the anterior pituitary.

The pituitary gland is under the control of neurosecretory cells in the hypothalamus. Axons serving the anterior lobe terminate in the median eminence, where their hormones enter the hypothalamo-hypophyseal portal system and are carried by the blood to pituitary target cells (Fig. 8-4). Since there appear to be specific neurosecretory hormones that cause the release or inhibition of hormones produced by the various anterior pituitary cell types, these hormones are called **releasing or inhibitory factors.** Axons that serve the posterior lobe travel down the infundibular stalk and terminate on capillaries in the posterior lobe.

8-13 Pineal The roof of the diencephalon bears the pineal body (epiphysis), long of uncertain function. It secretes **melatonin,** which concentrates the pigment of the melanophores of fishes and amphibians, inhibits gonadal development (antigonadotropic action) and is involved in the regulation of circadian rhythms (daily physiological and behavioral changes) (Sec. 9-17). An enzyme, **hydroxyindole-O-methyltransferase,** participates in the synthesis of melatonin. In mammals the enzyme's activity is promoted by norepinephrine released by sympathetic nerve fibers that extend to the pineal secretory cells from the superior cervical ganglia. The pineal appears to play an important part in the regulation of the reproductive cycle. In the rat, exposure to light suppresses the synthesis and release of melatonin (a gonad inhibitor) by the pineal; thus continuous lighting causes persistent estrus. Messages about environmental light reach the pineal via the retina, inferior accessory optic tract, medial forebrain bundle, and sympathetic nervous system. In birds, in contrast to mammals, neither the retina nor the sympathetic nervous system are required for environmental control of melatonin formation; the pineal possibly responds directly to light or is influenced by extraretinal photoreceptors of unknown location in the brain. In ectothermal vertebrates, photoreceptors of the median or third eye (found in lampreys, frogs, and lizards) or in the pineal itself appear to provide information on changes in environmental light important in circadian activity and seasonal regulation of reproduction. Removing the median (or parietal) eye in lizards causes increased exposure to light and heat, with accompanying increase in thyroid activity, and activation of the gonads.

8-14 The endocrine glands as a system Neurosecretory cells probably formed the original mechanism of hormonal action. They are at the root of nearly all endocrine responses. Cnidarians and certain other primitive invertebrates apparently have only this form of endocrine control. In them, stimulation of neurosecretory cells causes release of a hormone that acts directly in one-step fashion on target structures. In the more advanced invertebrates the one-step method of control continues, but two-step control is introduced with the appearance of epithelial endocrine glands: the neurosecretory cell stimulates an epithelial endocrine gland, which in turn produces a hormone that acts on the target. In vertebrates further complexity, a three-step sequence, is present: the pituitary, stimulated by cerebral neurosecretory cells in the hypothalamus, produces tropins that stimulate other epithelial endocrine glands, which in turn act on target organs. Two- and three-step control predominate in vertebrates, although one-step control continues. An example of the latter is the suckling reflex, in which an exogenous tactile stimulus releases oxytocin from hypothalamic neurosecretory cells and quickly causes ejection of milk by the mammary glands. Two- and three-step control of endocrine function make possible more effective introduction of feedback mechanisms between the glands and neurosecretory systems (Sec. 8-12). The activity of the endocrine glands is closely integrated with that of the nervous system, in vertebrates importantly involving action of the hypothalamus and autonomic nervous system. The latter may directly innervate endocrine glands or act through the hypothalamus. Sensory receptors provide information on changes in the internal and external environment (chemistry of body fluids, changes in environmental lighting, temperature, etc.), and this information is converted to chemical signals by the nerves and glands, bringing about coordination of the activities of organs and tissues.

References

Bagnara, J. T., and **M. E. Hadley.** 1973. Chromatophores and color change. Englewood Cliffs, N.J., Prentice-Hall, Inc. ix + 202 pp. *Animal pigments and the control of chromatophores.*

Barrington, E. J. W. 1975. An introduction to general and comparative endocrinology. London, Oxford University Press. x + 281 pp., 242 figs. *Both invertebrates and vertebrates.*

Bently, P. J. 1976. Comparative vertebrate endocrinology. New York, Cambridge University Press. xi + 415 pp.

Constantinides, P. C., and **N. Carey.** 1949. The alarm reaction. *Sci. Am.,* vol. 180, no. 3, pp. 20–23. *Physiologic changes in animals subjected to stress, with comments on implications for the harassed human animal.*

Etkin, W., and **L. I. Gilbert** (editors). 1968. Metamorphoses: A problem in developmental biology. New York, Appleton Century Crofts. xi + 459 pp. *A number of authors discuss metamorphoses, including endocrine controls, in crustaceans, insects, lower chordates, and amphibians.*

Funkenstein, D. H. 1955. The physiology of fear and anger. *Sci. Am.,* vol. 192, no. 5, pp. 74–80. *Different secretions of the adrenal glands are evidently associated with outwardly and inwardly directed anger in man.*

Guillemin, R., and **R. Burgus.** 1972. The hormones of the hypothalamus. *Sci. Am.,* vol. 227, no. 5, pp. 24–33. *Isolation of hormones ("releasing factors") from the hypothalamus that regulate the pituitary gland.*

Highnam, K. C., and **L. Hill.** 1977. The comparative endocrinology of invertebrates. 2d ed. London, William Clowes and Sons, Ltd. ix + 357 pp.

Li, C. H. 1963. The ACTH molecule. *Sci. Am.,* vol. 209, no. 1, pp. 46–53. *The adrenocorticotropic hormone of the pituitary gland is important in health and disease. Study of its molecule shows how its function is related to structure.*

Pastan, I. 1972. Cyclic AMP. *Sci. Am.,* vol. 227, no. 2, pp. 97–105. *Adenosine monophosphate is a "second messenger" between a hormone and its effects within the cell.*

Pike, J. E. 1971. Prostaglandins. *Sci. Am.,* vol. 225, no. 5, pp. 84–92. *Hormonelike substances that affect a wide range of physiologic processes from contraction of the uterus to secretion from the stomach wall.*

Rasmussen, H., and **M. M. Pechet.** 1970. Calcitonin. *Sci. Am.,* vol. 223, no. 4, pp. 42–50. *A recently discovered thyroid hormone that acts indirectly to control the level of calcium in the blood.*

Thompson, E. O. P. 1955. The insulin molecule. *Sci. Am.*, vol. 192, no. 5, pp. 36–41. *The first protein molecule to have its complete chemical structure described.*

Tombes, A. S. 1970. Introduction to invertebrate endocrinology. New York, Academic Press, Inc. 217 pp.

Turner, C. D., and **J. T. Bagnara.** 1976. General endocrinology. 6th ed. Philadelphia, W. B. Saunders Company, x + 596 pp., illus.

Wilkins, L. 1960. The thyroid gland. *Sci. Am.*, vol. 202, no. 3, pp. 119–129. *Its hormones determine the rate of metabolism. Malfunction can have grave consequences, particularly in childhood.*

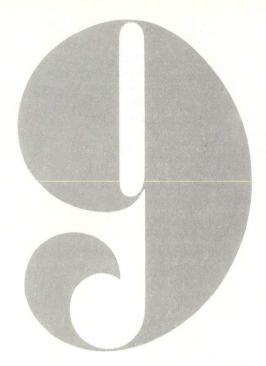

NERVOUS SYSTEMS AND SENSE ORGANS

Nervous systems

All cells are excitable, or irritable. Because of this, every organism is sensitive to changes or stimuli from both its external and its internal environments; to these it responds or reacts in various ways. Every type of organic response, from the simplest action of an amoeba to the most complex human bodily function or mental process, results from this fundamental characteristic of excitability. To perceive stimuli, to transmit these to various body parts, and to effect responses, most animals have a **nervous system** (Fig. 9-1). This system (together with endocrine glands in some) serves also to coordinate and integrate the functions of cells, tissues, and organ systems so that they act harmoniously as a unit.

9-1 Stimulus and response Any physical or chemical change capable of exciting an organism or its parts is a **stimulus.** Common external stimuli derive from temperature, moisture, light, gravity, contact, pressure, oxygen supply, salt concentrations, and odors (chemical emanations). Internal stimuli result from the quantity of food, water, oxygen, or wastes in the body and from fatigue, pain, disease, or other conditions. Some stimuli act directly upon cells or tissues and elicit a direct response (e.g, sunburn), but most animals have various kinds of specialized receptors (sense organs) to receive stimuli.

A **receptor** is a cell or organ having a special sensitivity to some particular kind or kinds of stimulus, as the eye to light and the ear to sound. **Exteroceptors** such as the eye receive stimuli from the external en-

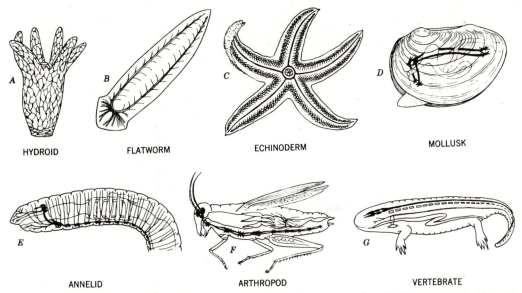

Figure 9-1 Nervous systems in animals (heavy black). *A*. Hydroid, nerve net throughout body. *B*. Flatworm, ganglia in "head" region, two nerve cords. *C*. Echinoderm, nerve ring around mouth, median nerve in each arm. *D*. Bivalve mollusk, three pairs of ganglia and connectives. *E*. Annelid, a "brain" of ganglia in anterior end, double solid ventral nerve cord, segmental ganglia and nerves. *F*. Arthropod, similar to earthworm. *G*. Vertebrate, brain in head, single hollow dorsal nerve cord with paired segmental nerves.

vironment and **interoceptors** from within the body, as with hunger or thirst. (See also proprioceptors, Sec. 9-13.) A stimulus causes the receptor to generate nerve impulses which travel along nerves to the central nervous system; the latter integrates the sensory information and then initiates impulses which excite terminal structures, or **effectors** (muscles, glands), to bring about responses.

Some stimuli are gradual and response is slow, as in the chilling that precedes a sneeze; others are abrupt and produce a quick response, like the jab of a pin. Beyond a certain minimum there may be no quantitative relation between the intensity of a stimulus and the kind or magnitude of the responses that it produces (the all-or-none effect); this depends upon the kinds of cells or organs excited and their physiologic condition. Several weak stimuli in rapid succession may bring a response although each individually is too slight to do so; this is called the summation effect. Upon being excited, muscles contract and thus produce movements, while gland cells pour forth the secretions that were previously synthesized within them.

9-2 Neurons and nerves Nervous systems are composed of nerve cells, or **neurons,** with cell processes known as **dendrites** and **axons.** The neurons are of varied form (Fig. 3-20) in the systems of different animals and in the several parts of any one nervous system. The neuron is the structural and functional unit of the nervous system, which consists chiefly of neurons in orderly arrangement. They constitute about 10 percent of the cells in the human nervous system. The remainder are **glia cells,** which are not electrically excitable but which support the neurons physically, probably also sustain them metabolically, and are believed to participate actively in brain function. Between any two neurons related in function there is a close association, or **synapse;** this is a "physiologic valve" that passes nerve impulses in only one direction, from the axon of one neuron to the dendrite or cell body of the other. A **nerve** consists of one to many neurons (axons or dendrites) bound together by connective tissue and including blood vessels to supply nutrients and oxygen. A large nerve contains many fibers—like the wires in an electric cable.

The **nerve impulse,** or **action potential,** that passes along a nerve fiber involves both chemical and electrical change. The impulse travels at a uniform speed with the same intensity throughout. An electrical wave accompanies it (Fig. 9-2). To understand the propagation of the impulse, let us consider first the neuron at rest. The resting nerve fiber is electrically polarized. It has a **resting potential.** The outside of its selectively permeable membrane is relatively positive, the inside is negative. What causes the polarization, and how is it maintained? The overall numbers of positive and negative ions are about the same outside and inside the cell, but the concentrations of some ions differ greatly. There are about 10 to 15 times more sodium (Na^+)

ions in the extracellular fluid than inside the neuron. Potassium (K^+) ions, however, are about 30 times greater within than without. Movement of both by diffusion and active transport and permeability characteristics of the cell membrane can account for the potential difference. Na^+ ions tend to diffuse inward and K^+ ions outward in the direction of their concentration gradients. The membrane of a resting cell, however, is less permeable to Na^+ than K^+; this and the concentration difference causes Na^+ to enter more slowly than K^+ leaves. The resulting charge separation across the membrane leads to a 60- to 90-millivolt, inside negative, potential. The differences in the concentration gradients would gradually disappear were it not for an **ion pump** of carrier mole-

Figure 9-2 *A.* **Changes in voltage across a nerve membrane during an action potential spike.** Depolarizing and repolarizing phases are indicated. The refractory period, during which the neuron cannot be stimulated, occurs during repolarization. Note brief period of hyperpolarization. *B.* **Upper diagram.** Resting nerve fiber showing its electrical polarization. **(1)–(7).** Time-lapse sequence showing movements of positive charge at nerve cell membrane that account for self-propagating wave of depolarization (and accompanying repolarization) that spreads away from site of stimulus. Horizontal line represents nerve cell membrane, outer surface toward top. (1) Stimulus increases permeability of membrane and Na^+ diffusion into cell sharply increases, causing internal potential to become positive (2). Since unlike charges attract and like charges repel, current (movement of positive charge) flows *away* from stimulated region through intracellular fluid and *toward* stimulated region in extracellular fluid (3). (4) The initial depolarization, as a result of local current flow triggers an action potential in adjacent areas and the process described repeats itself (5)–(7), on to the end of the membrane.

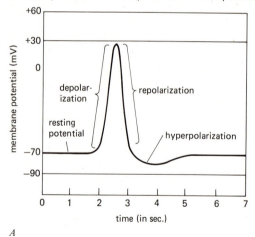

A

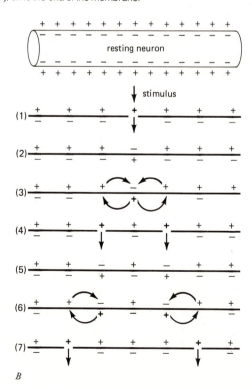

B

cules (probably phosphate-containing protein) within the cell. By active transport, the pump moves Na^+ ions to the cell surface, where they are expelled as fast as they "leak" in and K^+ ions are taken from the surface into the cell. Such ion transport requires the expenditure of cellular energy derived from ATP.

How is the nerve impulse transmitted? If a sufficient mechanical or electrical stimulus is applied to the cell membrane, it becomes **depolarized** at the place stimulated, and a **self-propagating** wave of depolarization (and accompanying repolarization) spreads outward in the membrane (Fig. 9-2). The cell produces a current pulse of its own that amplifies the original stimulus. Upon depolarization, the membrane potential suddenly changes from about -70 to $+30$ mV, then rapidly returns to the resting level. This is the **action potential.** The voltage changes can be shown by inserting a fine electrode into the neuron and another into the fluid surrounding it and connecting the two to a voltmeter. As the impulse passes, the voltmeter registers an abrupt peak, or spike, then a slower decline. Following the peak there is a **refractory period** (1 to 5 ms) during which the depolarized fiber cannot respond to another stimulus (Fig. 9-2A).

What causes the sudden change in potential? The stimulus alters the cell membrane structure, increasing its permeability to Na^+ ions. The rate of Na^+ diffusion into the cell sharply increases. This additional positive charge to the cell begins the process of ·:olarization of the membrane. As the shift in charge begins there is a still greater increase in Na^+ permeability accompanied by additional movement of Na^+ ions into the cell and, in feedback fashion, still further depolarization. When voltage finally declines to nerve impulse threshold level, Na^+ ions enter in such quantity that the internal potential becomes positive. It is then that the rapidly rising part of the action potential occurs (Fig. 9-2A).

The action potential lasts only about 1 ms. The return to the resting level, or **repolarization,** is achieved by (1) a rapid decrease in the entry of Na^+ (as the membrane becomes more positive inside) and (2) an accompanying increase in K^+ permeability, which causes increasing numbers of K^+ ions to move out of the cell, down their concentration gradient. When the former predominance of the K^+ diffu-

sion rate over that of Na^+ is regained, polarization is restored. There may even be a slight **hyperpolarization** of the membrane potential (Fig. 9-2A) as the K^+ ion diffusion briefly overshoots its resting-level rate.

During the first part of the refractory period, the neuron cannot respond to another stimulus of any intensity (this corresponds to the period of Na^+ permeability changes). There then follows a relative refractory period (often lasting about 5 to 10 ms and corresponding roughly with the increase in K^+ permeability), during which a strong stimulus (usually considerably greater than threshold level) can initiate a nerve impulse. A neuron has an **all-or-none response.** A stimulus must be at threshold level for the cell to fire.

How is the nerve signal able to spread from the site of stimulation? The initial action potential triggers, by local current flow, a new potential at adjacent membrane sites, setting in motion the Na^+–K^+ flux. Each new action potential site, in turn, does likewise. Current flow in the nerve fiber is thus in all directions, away from the point of stimulation. However, since action potentials are usually initiated at the dendritic end, the impulse travels in only one direction, toward the tip of the axon (Fig. 9-2B).

Nerve impulses vary in speed. An impulse travels some 30 m/s (98 ft/s) in a frog and up to 120 m/s (392 ft/s) in some mammalian fibers. Transmission is slower in nonmyelinated fibers than in those with a myelin sheath and slower in fibers of small diameter. In myelinated fibers, currents can spread faster from one node (Sec. 3-17) to the next because of the insulating effect of the myelin. Therefore the nerve impulse can propagate itself faster. Nerve impulse propagation in some myelinated fibers may reach over 400 km/h (250 mi/h).

The synapse is the site where electrical activity is transmitted from one nerve cell to another—from a **presynaptic** to a **postsynaptic** neuron. Each postsynaptic neuron has large numbers (often thousands) of synaptic junctions on the surface of its dendrites and cell body, received from hundreds of converging presynaptic neurons. Some of the presynaptic neurons are **excitatory,** some **inhibitory.** Whether the postsynaptic neuron will fire depends on the number of synapses active at a given time and on how many are excitatory or inhibitory.

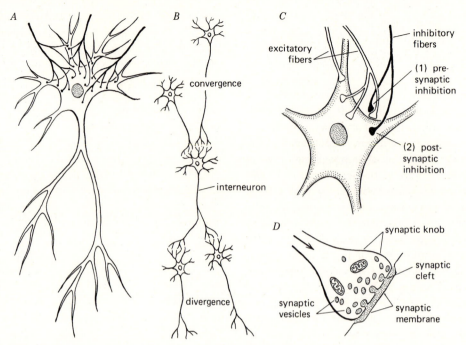

Figure 9-3 Synaptic connections between neurons. *A*. Many axonal branches synapse on dendrites and nerve cell body of postsynaptic neuron. *B*. Intercalation of an interneuron increases versatility of neural response. *C*. Mechanism of neural inhibition by (1) pre- and (2) postsynaptic inhibitory fibers. *D*. Release of transmitter substance into the synaptic cleft by an axonal knob upon arrival of an action potential.

At the ends of the axon branches are **synaptic knobs** or disks containing small membrane-bound vesicles of chemicals (Fig. 9-3). A narrow **synaptic cleft,** about 200 Å wide and of uniform width, separates each knob from the immediately underlying cell surface, the subsynaptic membrane. Upon arrival of a nerve impulse the vesicles, filled with transmitter molecules (a neurohumor), release their contents into the synaptic cleft. If the amount is sufficient, an action potential ensues in the postsynaptic neuron, or, in the case of inhibitory synapses, inhibition occurs. **Acetylcholine** is produced in many synapses, including those at the neuromuscular junction and between pre- and postsynaptic fibers of the autonomic nervous system. In at least some sympathetic synapses the transmitter substance is epinephrine or norepinephrine. The increased amounts of acetylcholine would continue to stimulate the next neuron but for the fact that an enzyme, **acetylcholinesterase,** in the motor end-plate membrane quickly hydrolyzes acetylcholine and returns it to its former level. There is a short delay in passage of the nerve impulse at each synapse.

Activity of a single synapse is seldom sufficient to change the membrane potential of the postsynaptic neuron; the combined action of a number of synapses is usually required or increased frequency of arrival of action potentials. The frequency rate in some neurons may reach about 1000 impulses per second. Postsynaptic activity is also influenced by the location of synaptic junctions. The depolarization threshold of the postsynaptic neuron is lower at the base of its axon than on its cell body and dendrites.

Inhibition of a postsynaptic neuron may occur through direct activation of inhibitory synapses or indirectly by means of presynaptic connections involving synaptic junctions between one neuron and the synaptic knobs of another (Fig. 9-3). There is an interplay between the excitatory and inhibitory synapses which may result in postsynaptic activity depending on which kind of synaptic input predominates.

9-3 Kinds of neurons Sensory or **afferent neurons** are those which conduct impulses from receptors to or toward the central nervous system, and **motor** or **efferent neurons** conduct from the central nervous system to various effectors. Still others, **interneurons** (association or connector neurons), in the brain and spinal cord, join variously between sensory and motor neurons and are of great importance in providing versatility in the routing of nerve impulses (Fig. 9-3). In human beings they constitute about 99 percent of all nerve cells, and thousands may be involved in the complicated neural process of memory or language.

Some nerves contain only sensory fibers, others only motor fibers, and many are mixed nerves including both types. A **ganglion** is a unit containing the cell bodies of a few or many neurons, and the equivalent of ganglia in the brain are known as **nuclei,** or **centers.** In all but the lowest animals the nervous system may be termed a receptor-adjustor-effector system or a sensory-neuromuscular mechanism.

9-4 Invertebrate nervous systems (Fig. 9-1) In sponges the cells about the openings (oscula) in the body wall contract slowly if touched, but these reactions seem to be local responses without true propagation to nearby cells. There are no definite nerve cells or structures. *Hydra* and other cnidarians have a diffuse **nerve net** around the body in or under the epithelium, but no central ganglion. The net is composed of nerve cells (protoneurons) separated by synaptic junctions or joined to one another by cytoplasmic processes. They connect both to receptors (modified epithelial cells) in the epidermis and to the bases of epitheliomuscular cells that contract slowly to alter the body shape. A nerve impulse may spread widely over the net because the synapses, in contrast to those of higher animals, pass impulses either way. Nerve nets also occur in ctenophores,

echinoderms, enteropneusts, and ascidians and even in the gut wall (as nerve plexuses) and on the blood vessels of vertebrates.

In bilaterally symmetrical animals the nervous system is linear, usually comprising one or more pairs of **ganglia** or a **brain** in the anterior end joined to one or more **nerve cords** that extend posteriorly through the body. The nerve cords of invertebrates are mostly ventral and solid, and nerves pass from the ganglia and cords to various organs. Flatworms have two anterior ganglia, with nerves to the head region, and two separated nerve cords which are joined by cross connectives. In mollusks, annelids, and arthropods the paired anterior ganglia (supraesophageal, subesophageal) lie above and below the esophagus and are joined by connectives. The more specialized mollusks lack ventral nerve cords but have large ganglia joined by connectives in the head, foot, and viscera. In annelid worms and the more primitive arthropods, the two ventral nerve cords have a pair of ganglia and a pair or more of nerves in each body segment. In the higher crustaceans, insects, and arachnoids the ventral ganglia are concentrated anteriorly. Starfishes and other echinoderms have a radially arranged nervous system in keeping with their symmetry.

Giant nerve fibers occur in several phyla. Those consisting of single cells that give rise to a large axon are found in nemerteans, cestodes, polychaetes, balanoglossids, and some lower vertebrates. Multicellular giant fibers forming a syncytium are common in annelids, crustaceans, and cephalopod mollusks. In the squid a large nerve fiber in the mantle wall is nearly 1 mm (0.04 in) in diameter and 20 cm (8 in) long. Giant fibers provide more rapid conduction of impulses than the system of synapses and hence are useful in escape or startle reactions.

9-5 Vertebrate nervous systems In all vertebrates the nervous system has a comparable embryonic origin (Sec. 10-17) and is always single, hollow, and dorsal to the digestive tract. It consists of (1) the **central nervous system,** with large anterior brain (Figs. 9-4, 9-6) connected to a spinal or nerve cord, and (2) the **peripheral nervous system** of 10 or 12 pairs of

cranial nerves from the brain, a pair of **spinal nerves** from the cord for each primitive body segment, and the **autonomic nervous system** (Fig. 9-9).

9-6 Brain The parts of the brain are in linear arrangement in fishes and amphibians, and general brain regions and structures are easily seen (Fig. 9-4). These are (1) the two **olfactory lobes** with nerves to the nasal chambers; (2) two **cerebral hemispheres** closely joined to the preceding and also attached to (3) the median **diencephalon.** Behind this are two rounded **optic lobes,** supported on (4) the **midbrain** below and followed by (5) the **cerebellum;** this is over (6) the thin-roofed **medulla oblongata,** which tapers to join the spinal cord. The diencephalon has a

slender dorsal **pineal body,** or **epiphysis.** Below the diencephalon is the **optic chiasma** (crossing of the optic nerves) followed by the **infundibulum,** with the **hypophysis,** or **pituitary gland,** at its posterior end.

The **cavities** within the brain are the first and second **ventricles** in the cerebral hemispheres (Fig. 9-4); these connect to a third ventricle in the diencephalon. From the latter the aqueduct of Sylvius leads to the fourth ventricle located in the medulla. The fourth ventricle is continuous with a minute central canal through the spinal cord. Circulating cerebrospinal fluid fills the ventricles and other cavities and surrounds the brain. Metabolic exchanges for the brain are performed by arteries and veins over its surface and by two dense networks of blood ves-

Figure 9-4 Vertebrate brain. *A.* Early embryonic stage showing three primary divisions and developing optic vesicle. *B.* More advanced stage in dorsal view. *C.* Frontal section showing brain ventricles; I and II are lateral ventricles.

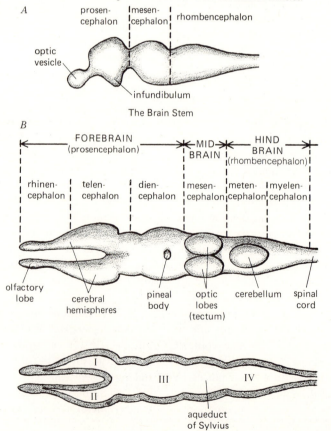

sels, the **anterior choroid plexus** in the roof of the diencephalon and the **posterior choroid plexus** in the roof of the medulla. These plexuses form the cerebrospinal fluid. The brain and spinal cord are surrounded by three membranes (Fig. 9-8), a thicker **dura mater** adhering to the enclosing bones, a delicate **pia mater** close over the nervous tissue itself, and a thin arachnoid mater in the space between. **Cranial nerves** (Fig. 9-7) extend from various parts of the brain to sense organs, muscles, and other structures (Table 9-1).

The brain has several features that resemble those of the nerve cord. As with the cord, it contains aggregations of gray and white matter, the former where nerve cell bodies are concentrated and the latter dominated by myelinated fiber tracts. In the brain stem (see below), as in the cord, the sensory and motor areas tend to be separated, respectively, into dorsal and ventral gray columns (Fig. 9-10), but in the brain these columns tend to break up into brain centers or nuclei. The cranial nerves, most of them homologous with the spinal nerves, also reflect this dorsoventral relationship. Their connections and functions are given in Table 9-1.

Development of an anterior brain is correlated with bilateral symmetry and forward locomotion

Table 9-1
THE PAIRED CRANIAL NERVES OF VERTEBRATES

Number and name of nerve	Origin (in brain)	Distribution (external connections)	Function (chiefly as in humans)
I Olfactory	Olfactory lobe (or bulb)	Olfactory epithelium in nasal cavity	Sensory: smell
II Optic	Optic lobe on midbrain	Retina of eye	Sensory: sight
III Oculomotor	Floor of midbrain	Eye: 4 muscles of eyeball; also iris, lens, upper lid	Motor: movements of eyeball, iris, lens, and eyelid
IV Trochlear	Floor of midbrain (emerges dorsally)	Eye: superior oblique muscle of eyeball	Motor: rotation of eyeball
V Trigeminal	Side of medulla	Top and sides of head, face, jaws, and teeth	Sensory: forehead, scalp, upper eyelid, side of nose, teeth; motor: movement of tongue and of muscles used in chewing
VI Abducens	Side of medulla	Eye: external rectus muscle of eyeball	Motor: rotation of eyeball
VII Facial	Side and floor of medulla	Tongue (anterior $\frac{2}{3}$), muscles of face, of mastication, and of neck	Sensory: taste; motor: facial expression, chewing, movement of neck
VIII Auditory (acoustic)	Side of medulla	Inner ear: (1) organ of Corti in cochlea; (2) semicircular canals	Sensory: (1) hearing, (2) equilibrium
IX Glossopharyngeal	Side of medulla	Tongue (posterior $\frac{1}{3}$), mucous membrane, and muscles of pharynx	Sensory: taste and touch; motor: movements in pharynx
X Vagus (pneumogastric)	Side and floor of medulla	Pharynx, vocal cords, lungs, heart, esophagus, stomach, and intestine	Sensory: vocal cords, lungs; motor: pharynx, vocal cords, lungs, esophagus, stomach, heart; inhibits heartbeat
XI* Spinal accessory	Floor of medulla	Muscles of palate, larynx, vocal cords, and neck	Motor: muscles of pharynx, larynx, and neck
XII* Hypoglossal	Floor of medulla	Muscles of tongue (and neck)	Motor: movements of tongue

* Nos. XI and XII are lacking in amphibians, fishes, and cyclostomes.

and the location of the primary somatic sense organs of sight, smell, and hearing in the head. In the course of evolution a great increase occurred in sensory and motor neurons and especially their connectors, the interneurons. The placement of an interneuron in the simple reflex arc (Sec. 9-10) greatly increases the possible responses to a sensory or motor stimulus.

Early in the development of the vertebrate embryo the brain becomes subdivided into three parts, the **prosencephalon** (forebrain), the **mesencephalon** (midbrain), and the **rhombencephalon** (hindbrain). These primary divisions constitute the **brain stem** (Fig. 9-4*A*). The first and last become further subdivided, yielding five divisions. The prosencephalon differentiates into the (1) rhinencephalon ("nose" brain), (2) telencephalon, and (3) diencephalon. The rhombencephalon gives rise to the (4) metencephalon and (5) myelencephalon. Certain basic features of brain structure and function have changed little in the course of evolution, and they reside primarily in the brain stem. They will be discussed first.

The **medulla,** derived from the rhombencephalon, is a reflex coordinating center for information received from the muscles and joints and from the lateral-line organs (fishes and amphibians) and balance organs of the ear. Its floor contains a large part of the **reticular formation** which extends throughout the brain stem (Fig. 9-5). This is a network of small, many-branched neurons that receive and integrate information from the cord and many parts of the brain. It contains long-established reflex centers for swallowing, vomiting, cardiovascular control, and

respiration. Parts of it are involved in the processes of arousal and sleep. The cerebellum (discussed below) is associated with the anterior roof of the medulla.

The **tectum** is a thickened area of gray matter in the roof of the mesencephalon, above the aqueduct of Sylvius (Fig. 9-5). In fishes and amphibians it is an important sensory and motor projection area and coordinating center in which motor activity is initiated. The tectum originally developed in connection with visual reception, and it contains optic lobes, which vary in size, depending on the importance of vision. The lobes are large in birds, some reptiles, and bony fishes. In mammals the tectum has been reduced largely to a visual and auditory reflex center.

The cerebral hemispheres are an outgrowth of the telencephalon. At first they were evidently primarily olfactory in function (as in cyclostomes), and they continue so in amphibians but become differentiated. A lateral strip, the **paleopallium** ("old" brain), retains olfactory function in later evolutionary stages. A lower part contains the **basal nuclei.** In amphibians the basal nuclei form a correlation center receiving olfactory stimuli and sending messages to the brain stem. The center receives sensory input from the thalamus for correlation with olfactory information. In mammals it is involved in voluntary control of movements. The **corpus striatum** (Fig. 9-5), like the tectum, is an important sensory and motor projection area and an integrating and coordinating center. It is involved in the control of instinctive behavior and is well developed in reptiles

Figure 9-5 Median section of mature, primitive vertebrate brain, showing basic structures.

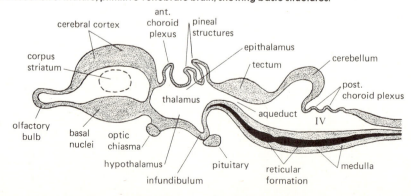

and birds, animals with stereotyped responses. In the former it rivals the tectum in importance and in the latter is a dominant center.

The **diencephalon** contains three thalamic areas. The first, the **thalamus,** is an integrating area for sensory input. It is also a relay and coordinating center between the corpus striatum, cerebral cortex, and motor columns. It contains a significant portion of the reticular system. The second, the **epithalamus** above, contains the habenular nuclei, which receive fibers from the olfactory area and transmit to motor areas of the brain stem. The habenular apparatus receives information from the light-responsive pineal (Sec. 8-13), which is involved in the regulation of circadian rhythms and perhaps reproductive cycles. The third, the **hypothalamus** below, is concerned with the regulation of the internal environment. In contrast with the somatic sensory and motor functions of the tectum, it is a visceral sensory and motor correlation area and integrates basic behavioral pathways requiring correlation of autonomic, endocrine, and somatic functions. It contains centers for thermoregulation, control of water balance, and sleep and is involved in neurosecretory control of the pituitary (Sec. 8-12). Fiber tracts in the walls of the diencephalon connect the thalamic areas.

The foregoing brain functions are pervasive among vertebrates. We now consider some later developments. With increasing complexity of movement, enlargement of the cerebellum occurred. Its size tends to reflect the degree of coordination required for locomotion and maintenance of posture. It does not in itself initiate movements, except for adjustments in posture, but rather influences other parts of the brain responsible for motor activity. The cerebellum and associated **pons** (containing major fiber tracts and a portion of the reticular formation; Fig. 9-7) originate in the metencephalon. The cerebellum receives input from the lateral-line organs in fishes and amphibians, from the semicircular canals (vestibular system), and from proprioceptors in joints, tendons, and muscles. It sends messages to the midbrain and thence to motor areas. In mammals, major fiber tracts connect it with the **cerebral cortex,** for these animals exercise direct cortical control over motor responses; in turn, stimuli are sent from the cortex by way of the pons to the cerebellum for processing.

The most conspicuous achievement in brain evolution has been the expansion of the cerebrum and cerebral cortex. Their great increase appears to have been favored by land life. Terrestrial environments presented new opportunities (and difficulties) for sensory and locomotor development, thus placing a

Figure 9-6 The brains of representative vertebrates, showing progressive increase, especially in the cerebral hemispheres and cerebellum. *Olfactory lobes,* clear; *cerebrum,* lightly stippled; *optic tracts and lobes,* coarsely stippled; *base of midbrain,* wavy lines; *cerebellum,* vertical lines; *medulla oblongata,* horizontal dashes; *pituitary body,* black. Stubs of cranial nerves are outlined.

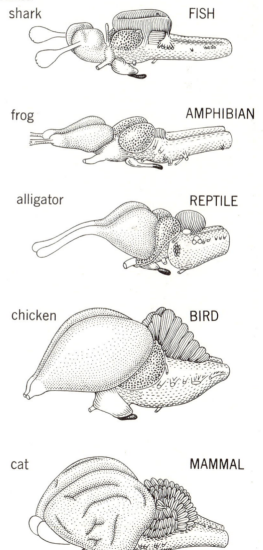

shark — FISH
frog — AMPHIBIAN
alligator — REPTILE
chicken — BIRD
cat — MAMMAL

premium on sensory awareness. It is perhaps no accident that the site for cortical expansion was the olfactory part of the brain. Complex sensory, motor, and correlative activity are required to identify and find the source of odors (Fig. 9-6).

Primitively the gray matter of the cerebral hemispheres was internal and undifferentiated. Later it moved toward the surface, and basal nuclei to the interior. Major connections were established between the latter and the brain stem. In advanced reptiles a new brain area, the **neopallium** appears, destined to become the major integrating and correlative center in mammals. It grows larger and more elaborate and reaches its peak of development in the higher primates, including humans. Major fiber tracts are established between it and the brain stem and between the neopallial areas of the two hemispheres via the corpus callosum. Each hemisphere controls the opposite side of the body, and in humans the left hemisphere is usually dominant. The ancient paleopallium becomes restricted to the floor of the hemispheres. To accommodate the increasing cortical concentration of gray matter, the cortex becomes extensively folded, and in humans, although only about 5 mm ($\frac{1}{4}$ in) thick, contains some 14 billion neurons and their synapses. Sensory information, formerly assembled in the midbrain tectum, now goes to the neopallium via the thalamus, and the midbrain becomes primarily a reflex center. The cortex becomes a major projection area for motor and sensory functions. In lower vertebrates the projection areas are spread out and most of the cortical surface has a direct sensory or motor function. In humans, however, much of the cortex is "unassigned," and the sensory and motor areas are discrete and localized on the two sides of the central fissure (Fig. 9-7). The size of the area for a given function reflects the extent and refinement of control required. Thus both sensory and motor areas for finger and lip movements are large, whereas those for the limbs are much smaller. The sensory and motor regions which preside over corresponding parts of the body are opposite one another, and there is close communication between them. With increasing cortical control, direct paths are established with the motor columns in the cord and cerebellum.

The extensive uncommitted "silent" areas of the brain seem to be largely **association areas,** but their function is unclear. Four major ones are recognized (Fig. 9-7), interconnecting cortical regions. They provide for refinement of control over both the autonomic and somatic systems. In humans they apparently contain the circuitry involved in awareness of sensations and actions, memory, judgment, and reason. Memory pertaining to sensory experience may be stored in association areas. For example, the receptive area for sight provides awareness of the colors of a painting, but the adjacent association area is required for its recognition as a landscape. Among animals, increasing size of the cortex correlates with increasing ability to learn. There is, however, no exact correlation between brain size and intelligence (Sec. 35-8). All these mental attributes are also influenced by other parts of the central nervous system. Destruction of cortical areas other than motor, sensory, or language centers does not always result in obvious behavioral changes.

In the advanced vertebrates the primitive functions of the brain stem persist little changed. The roles of the hypothalamus and reticular system are quite stereotyped. We are generally not conscious of their functioning. An important part of our emotional life appears to center in the hypothalamus and other parts of the **limbic system**—an interconnected series of brain structures including the hypothalamus, thalamus, frontal-lobe cortex, temporal lobe, and their interconnecting pathways.

9-7 Spinal cord and nerves   (Fig. 9-8) The spinal cord extends posteriorly from the medulla and within the neural arches of the vertebrae. The outer **white matter** is made up of bundles of myelinated fibers connecting the various parts of the brain, the nuclei of spinal nerves, and interneurons. The inner **gray matter** contains interneurons and the nuclei of motor neurons. Spinal nerves emerge from the cord, between the vertebrae. Each nerve has two roots. The **sensory** or **dorsal root** carries nerve impulses from parts of the body *into* the spinal cord, and on this root is an enlargement, or **ganglion,** containing nerve cells. The **motor** or **ventral root** comprises fibers that transmit directive impulses outward *from* the cord to the tissues. The two roots of each join

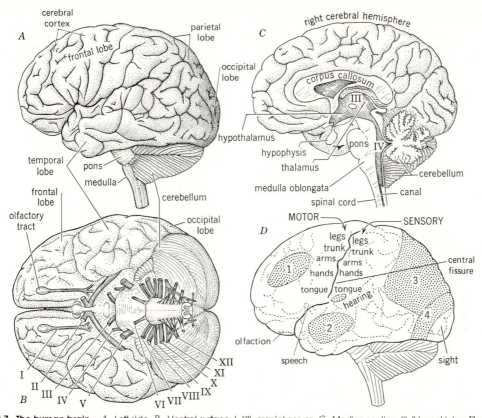

Figure 9-7 The human brain. *A*. Left side. *B*. Ventral surface; I–XII, cranial nerves. *C*. Median section; III, IV, ventricles. *D*. Left side, showing localization of certain functions on the surface of the cerebral cortex; association areas (stippled) are (1) frontal lobe, (2) temporal or auditory, (3) parietal lobe, and (4) visual.

Figure 9-8 The human spinal cord, spinal nerves, and sympathetic nervous system in relation to the vertebrae and the membranes (meninges) about the cord.

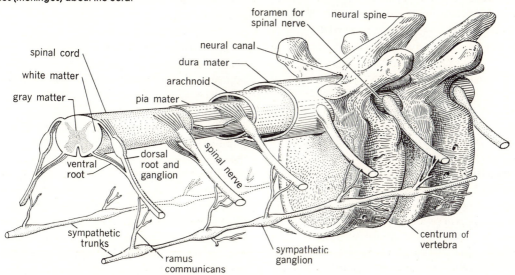

outside the cord as a nerve that extends to a definite part of the body or limbs. If the dorsal root of a spinal nerve is cut, any sensory impulses from the entering fibers fail to reach the cord and brain. Destruction of the ventral root blocks all motor control by fibers in that nerve. The ventral roots are variously injured or destroyed in poliomyelitis, leading to impaired muscular function.

9-8 Autonomic nervous system   (Fig. 9-9) The somatic nerves (cranial, spinal) connect mainly to skeletal muscles and direct the adjustment of an animal to its surroundings. By contrast, the **autonomic ner-**

vous system, of ganglia and fibers connecting to all smooth muscles, glands, and the viscera, deals with the internal environment of the body. It controls routine functions such as the rate of metabolism, heartbeat, and activity of the digestive tract, and maintains the constancy (homeostasis) of components in the blood, lymph, and tissue fluids. In birds and mammals, it closely regulates the body temperature by increasing metabolism and fluffing out the feathers or fur in cold weather and by promoting loss of heat in a warm environment.

The thoracolumbar portion (or **sympathetic system**) of the autonomic nervous system includes two lengthwise chains of connected ganglia along the trunk vertebrae and aorta. Efferent fibers from the spinal cord pass into spinal nerves to enter the sympathetic ganglia as preganglionic fibers. Upon leaving, as postganglionic fibers, those of each group unite as a **plexus,** then distribute to various organs, as with the nerves from the celiac plexus to the stomach, liver, etc. Afferent sympathetic fibers pass directly from organs to the dorsal roots of spinal nerves and into the spinal cord. Still other fibers connect to the erector muscles of the hairs and to sweat glands and small blood vessels. The craniosacral portion (or **parasympathetic system**) of the autonomic nervous system includes fibers in certain cranial nerves, to the iris of the eye (III), the glands and mucous membranes of the mouth (V, VII), and the heart, lungs, stomach, and upper small intestine (X, or vagus); and other fibers from sacral nerves connect to organs in the lower abdomen.

Most visceral organs and some others are innervated by both systems, and the two have more or less opposite effects. The parasympathetic promotes secretion of saliva and digestive juices, increases muscular activity of the intestine, constricts bronchioles in the lungs, slows the heartbeat, and constricts the pupil and adjusts the eye for near vision. In contrast, the sympathetic increases heartbeat, slows gastrointestinal action, dilates the bronchioles, etc. By increasing secretion of epinephrine from the adrenal glands (Sec. 8-10) it also mobilizes bodily resources for emergencies—fright, flight or fight, and injury. The epinephrine constricts blood vessels of the skin and viscera, dilates those of the heart and skeletal muscle, releases glucose from the liver for muscle

Figure 9-9 The human autonomic nervous system and its connections with the central nervous system and the internal organs; diagrammatic and simplified. *Sympathetic trunk and main ganglia* (celiac, etc.), heavy stipple; *sympathetic nerves,* broken lines; *parasympathetic nerves,* heavy solid lines; *cranial nerves,* III, VII, (IX), X; *spinal nerves* numbered for each region of spinal cord.

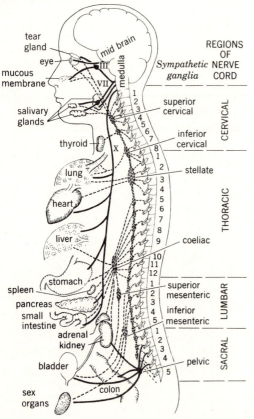

metabolism, and accelerates the clotting time of the blood.

9-9 Sensory pathways   After an impulse reaches the spinal cord, it follows specific sensory pathways to higher centers. These differ for various types of stimuli. Injury to the spinal cord results in loss of sensitivity on the afflicted side in the case of touch and pressure but on the opposite side with pain and temperature. In the first case, the afferent neurons from sense organs for touch and pressure enter the spinal cord and immediately turn upward on the same side to the medulla oblongata. There the impulse is transmitted across a synapse and activates a secondary neuron, axons of which cross the medulla and turn up to end in the thalamus. From there the impulse is transmitted to the cerebral cortex for analysis and integration.

Stimuli for pain and temperature enter the spinal cord through the dorsal roots and into afferent neurons ending in the dorsal horn of the gray matter. Passing a synapse, an impulse enters an interneuron that crosses the spinal cord at the same level and then turns upward on the side opposite the original sensory receptor to ascend past the medulla directly to the thalamus. Thence the impulse goes to the cerebral cortex. In both cases there is a crossing over between sensory receptor and cerebral cortex, but at different levels. Thus injury to one side of the cerebral cortex, as in cerebral hemorrhage, results in loss of sensation on the opposite side of the body.

9-10 Types of response   All patterns of response are an outgrowth of the interaction of heredity and environment. Some responses are largely inherited and some are largely learned, but all contain both genetic and nongenetic elements. Certain responses in animals can be classified readily, but others cannot because they differ from one another in degree and not in kind. Among lower animals many are chiefly invariable, having a strong genetic component, whereas in higher forms the variable responses predominate. The amoeba exhibits many essentially fixed responses; yet it can learn in a simple way. Human behavior is highly variable, but human beings have many constant and involuntary responses.

The essentially unvarying type of response by which an animal orients itself toward or away from a given stimulus is termed a **taxis.** A fish that heads into a current so that the two sides of its body are stimulated equally by flowing water exhibits positive **rheotaxis** (Gr. *rheos,* current), and an insect that climbs directly upward in opposition to gravity is said to show **negative geotaxis** (Gr. *geos,* earth). The moth that flies directly to a light is **photopositive,** or shows positive phototaxis (Gr. *photos,* light), whereas a cockroach that scuttles for cover when spotted by a light at night is **photonegative** (negative phototaxis). These several types of response are considered to depend upon reflexes.

The simplest neural response is a **monosynaptic reflex** which involves only an afferent and efferent neuron and a single synapse. Muscle-stretch reflexes are the only reflexes of this type in humans. When the human leg is bent and suspended freely and the knee tendon is tapped, the leg jerks forward. This knee-jerk reflex is an automatic, unlearned, and involuntary response to the stimulus. The **reflex arc** (Fig. 9-10A) (1) involves a receptor excited by the stimulus and (2) induces a nerve impulse in the dendrite of a sensory (afferent) neuron, which passes through the nerve cell body (in the dorsal root ganglion) and along the sensory axon into the gray matter of the spinal cord. There the impulse (3) crosses a synapse to (4) a second conductor, the motor (efferent) neuron, and continues out its axon in the ventral root to (5) the end organ in contact with an effector; if the latter is a muscle, it is excited to contract. All other human reflexes are **polysynaptic** having one or more interneurons between the sensory and motor pathways. Other simple reflexes are winking of the eyelids when an object is thrust before the eyes and sudden secretion of tears by the tear glands when a bit of dust lodges on the cornea. A reflex may or may not evoke a conscious sensation.

Few if any reflexes in vertebrates are really simple. The majority are polysynaptic. **Allied reflexes** are combined to produce a harmonious effect, such as the muscular movements in a person when walking or in an earthworm or caterpillar when crawling. These actions may be modified or inhibited through

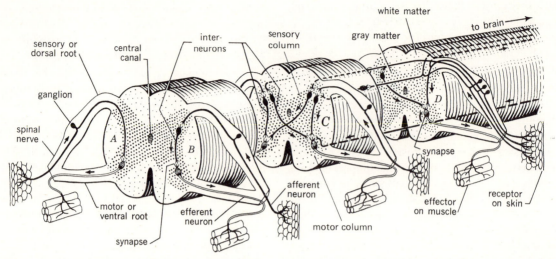

Figure 9-10 Simplified stereogram of the vertebrate spinal cord and nerves to show relations of neurons involved in reflex arcs. *Afferent neurons,* solid; *efferent neurons,* outlined; *interneurons,* broken and solid lines; *receptors,* as on skin; *effectors,* as on muscles Arrows show paths of nerve impulses. Each nerve contains many fibers. *A.* Simple reflex arc. *B.* Reflex arc with one interneuron. *C.* Reflex arc with cross connections. *D.* Reflex arc with cross connections and also others to and from the brain.

interneurons extending to other portions of the nerve cord and to the brain (Fig. 9-10*D*). **Chain reflexes** act in sequence, the response of one becoming the stimulus for the next. A frog reacts to a nearby moving insect by opening its mouth and flipping its tongue forward and back; the captured prey in the mouth stimulates receptors there to bring about closing of the mouth and to start swallowing reflexes in the pharynx and esophagus.

By training, a reflex may be conditioned to follow upon some environmental stimulus other than the original one that evoked it. In a dog the sight of food induces reflexly a flow of saliva, whereas the ringing of a bell does not. The Russian physiologist Pavlov rang a bell whenever food was offered to a dog. After some repetitions the mere sound of the bell, without food, induced salivary secretion in the animal. This Pavlov termed a **conditioned reflex.** Many human acts become conditioned reflexes, often of complex character. Repeated performance of a particular act or procedure becomes a **habit** by some more or less enduring change in the mode of response to a stimulus. Many of the routine activities of human beings are thus reduced to habits so that they are induced by particular stimuli without the intervention of con-

scious control. Dressing oneself, for example, is first learned as a conscious procedure but grows to be a habit. In time it becomes a series of chain and conditioned reflexes, mostly automatic—but not invariable; one may absentmindedly pick up or put on the wrong garment or shoe.

An **instinct** is a complex, largely unlearned pattern of behavior usually involving groups of chain reflexes. Compared with a reflex it is more elaborate but more adaptable. Most instincts serve to maintain the individual or the species. In many kinds of animals the choice and means of obtaining food are instinctive acts throughout life, whereas instincts pertaining to reproduction are manifested only when the individuals become sexually mature. Among animals that live more than 1 year the reproductive instincts are active only during the breeding season. The migrations of birds and fishes and the manner of making nests and caring for young among insects and vertebrates are governed entirely or largely by instinct. The mud dauber wasp exemplifies a complex cycle of instincts. Each female, without previous experience or learning, makes a tube of mud; then just before it is sealed, she captures spiders and paralyzes them with her sting, lays an egg on each, and seals them in the tube. The wasp larvae hatch and feed on the living prey, and when mature the young wasps cut their way out. The parent female never sees her offspring, but her instinctive behavior, and later that of the young, serve to maintain the species. In social insects such as the honeybee (Sec. 25-15), each caste has separate instincts that function together for the well-being of the colony.

A short-lived spider does not have enough time to learn how to spin a complicated web by trial and error. This instinctive act undoubtedly evolved by natural selection over a long period. Higher vertebrates have a long time to learn patterns of behavior from their parents: mammals in particular have a more or less prolonged attachment to the mother while nursing. Some types of learning that have been studied experimentally are (1) classic conditioning, as in the Pavlov salivating dog experiment; (2) instrumental learning, in which a rat or other animal, through trial and error, learns to manipulate mechanical devices to obtain food; and (3) perceptual learning, in which an animal bypasses the trial-and-error stage and comes to the correct answer after analyzing the problem. A tortoise, for example, is stopped by a barrier, while a dog observes the barrier and then walks around it.

Imprinting is a special type of learning involving the interplay of internal and external factors. An animal is most readily imprinted at a particular stage of development by stimuli with certain well-defined properties. Not just any stimulus will do. A young duck, for example, follows the first moving object it sees shortly after hatching. This is usually its mother, but other animals and even inanimate objects have been substituted. Acoustic imprinting may also occur in birds, in some even prior to hatching. Homing migrations of salmon may depend on imprinting during early life, whereupon they are then able to recognize the unique chemical characteristics of the stream where they were hatched. **Intelligence,** or the ability to learn, is best developed among the higher vertebrates and human beings, in whom the brain contains a greater number and more intricate arrangement of conduction paths and larger numbers of interneurons in the cerebral cortex. All the neurons to be present in an animal are developed early in the life of an individual, but new associations and pathways are established throughout life according to the kinds and intensity of stimuli received and the patterns of behavior developed.

Sense organs

Even at the earliest evolutionary stages sensory mechanisms must have responded, in varying degree, to the earth's basic chemical, mechanical, and electromagnetic energy forms that exist throughout the universe. In multicellular organisms there are often special sensory structures, or **sense organs** (Fig. 9-11), for this purpose. They are located so as to meet the environment, tending to be on the surface and around the body in sessile animals but more numerous anteriorly in bilaterally symmetrical species. The sense organs may be grouped according to the energy form to which they respond: (1) **chemoreceptors** are sensitive to changes in the chemical environment and include the senses of taste and smell; (2) **mechanoreceptors** are responsive to mechanical stimuli, usually involving stretching or compression of a receptor membrane, and include the senses of touch, pain, proprioception (muscle and joint sense), equilibrium, the lateral-line sense of fishes and amphibians (Sec. 30-12), and hearing; (3) **photoreceptors** are sensitive to light waves (a narrow segment of the electromagnetic spectrum); (4) **thermoreceptors** to temperature; and (5) **electroreceptors,** in a few fishes (Sec. 30-15), to electric signals. The receptors convert specific sensory information into action potentials.

In general, as animals have entered more complex environments, their organs of external sense have become more elaborate. With the adoption of land life, opportunities for sensing at a distance were greatly increased, and vision, olfaction, and hearing have reached high levels of refinement. The senses, however, are seldom equally developed within a species. Compared with humans, the dog has a more delicate sense of smell, the cat hears sound of higher pitch, the eagle has keener sight, and the honeybee responds to light farther into the violet but not so far into the red.

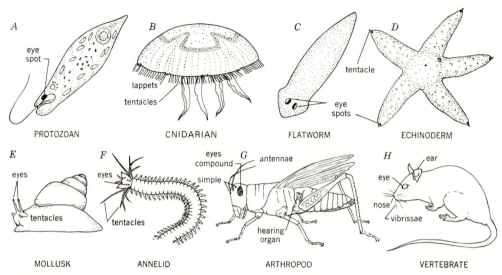

Figure 9-11 Sensory devices and sense organs in various animals and protistans. *A*. Protozoan (*Euglena*), eyespot. *B*. Cnidarian (jellyfish), lappets and tentacles. *C*. Flatworm (*Dugesia*), eyespots. *D*. Echinoderm (starfish), tentacle and eyespot on end of each arm. *E*. Mollusk (land snail), eyes and tentacles on head. *F*. Annelid worm (sandworm), eyes and tentacles on head. *G*. Arthropod (grasshopper), both compound and simple eyes and antennae on head; hearing organ on thorax. *H*. Vertebrate (mammal), eyes, ears, and nose, also sensory hairs (vibrissae) on tip of head.

The human "special senses" are **taste** for certain substances in solution; **smell** for volatile chemicals and gases in the air; **touch,** including contact, pressure, heat, and cold; **hearing** for vibrations in air, water, and solids; and **sight** for light waves.

Chemoreception

All cells respond to chemical changes in their internal and external environments. Unicellular organisms show chemotactic responses to concentrations of chemicals in their surroundings, moving toward food and away from toxic substances. The cells of multicellular organisms respond to CO_2 and O_2 levels, hormones, etc. in body fluids. Early in the evolution of multicellular life, chemicals integrated the activities of cells and continue to do so in the most advanced animals in the neurosecretory properties of nerves and the action of the endocrine glands. Receptors for sensing chemical information in the external environment (tastes, odors, irritating substances) also have a long evolutionary history. A general chemical sensitivity to irritating substances is present over all the body in amphibians and fishes

and many other aquatic animals and in the human mouth and nasal passages.

9-11 Taste and smell   Taste is the perception of dissolved substances by **taste buds.** These contain groups of slender modified epithelial cells with fine hairlike microvilli. The "hairs" are clustered within a small external pore (Fig. 9-12). Taste buds are usually in or about the mouth but occur over the body and fins in fish and on the tarsi of certain insects. A given taste-bud receptor cell may be innervated by more than one sensory neuron. Since the neurons involved in taste have different firing rates in response to different substances, a taste-bud's response appears to depend on the relative activity of the neurons that innervate it. Human taste distinguishes sweets, salts, acids, and bitter (alkaloidal) substances and differs in sensitivity to concentrations of the various materials that can be detected. Other vertebrates and many invertebrates appear to taste the same substances we do with comparable thresholds. The olfactory organs of fishes and other aquatic animals respond like taste buds to substances dissolved in the water.

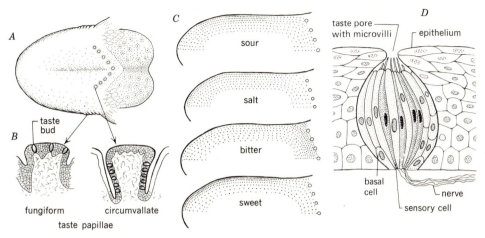

Figure 9-12 The mechanism of taste in humans. *A.* Dorsal surface of tongue. *B.* Two types of taste papillae in section and enlarged. *C.* Relative sensitivity on different parts of the tongue for the four tastes. *D.* Section of a taste bud; enlarged and diagrammatic. (*Partly after Parker, Smell, taste and allied senses in the vertebrates, J. B. Lippincott Company.*)

Besides their ordinary service, the taste buds may help in maintaining the constancy of the body's internal environment. Rats suffering from experimental dietary or endocrine deficiencies, when offered a choice of foods or solutions, chose those containing the needed substances.

Smell, or olfaction—"taste at a distance"—depends in humans on neurons with directly exposed tips that lie in the mucous membrane high in the nasal cavity (Fig. 9-13). Eddy currents of air carry volatile substances directly to these cell endings, which have much greater sensitivity than taste buds. Vast numbers (thousands) of odors can be detected.

The olfactory neuron receptor is enlarged, somewhat rod-shaped, and contains up to 20 motile cilialike filaments that are bathed by mucus on the surface of the nasal epithelium; its axon passes directly to the brain. Since the olfactory epithelium is off the main air-stream through the nose, sniffing aids in detecting odors.

Action potentials appear to be triggered by odorant molecules of a particular configuration and polarity that fit into receptor sites on the olfactory filaments. Humans can detect oil of peppermint at 0.024 mg/L of air and artificial musk at 0.0004 mg/L. Much of our "taste" for food depends upon smell, as shown

Figure 9-13 The mechanism of smell (olfaction) in humans; diagrammatic. *A.* Section of the head showing olfactory epithelium on lateral wall of right nasal cavity. *B.* Transverse section of nasal cavity. *C.* Enlarged microscopic section of the olfactory epithelium. (*Adapted from Parker, Smell, taste and allied senses in the vertebrates, J. B. Lippincott Company.*)

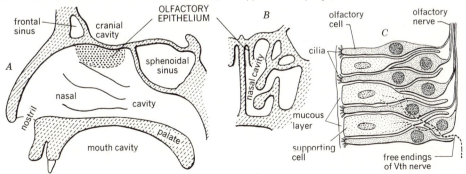

by the fact that when a cold congests the nasal membranes, all food tastes much alike. The sense of smell is vastly more sensitive among many wild mammals and insects and serves them variously in finding food and mates and sometimes in avoiding predators. In some moths the odor of a female may attract a male from many miles downwind.

Mechanoreception

All cells react to mechanical deformation of their surfaces. The simplest mechanoreceptors are free, non-myelinated nerve terminals. Usually, however, the terminal is coupled with hairlike structures or stretch membranes which move and amplify the stimulus. "Hair" cells are present in crustacean statocysts, in the lateral-line system of fishes (Sec. 30-12) and amphibians, and in vertebrate equilibrium and auditory organs. The "hairs" make the cell more responsive to mechanical forces—movements of fluids, gravity, changes in rate of movement, and vibrations. Deformation evidently causes an increase in membrane permeability to certain ions that depolarize the cell.

Vertebrate hair cells are remarkably uniform in structure. Usually 20 to 50 "hairs" project from the cell surface. One of these, the **kinocilium,** is constructed like a typical cilium with a central pair of fibers surrounded by nine doublets (Fig. 3-18). It is located at one side of the hair cluster. Depolarization

occurs when the hairs are deflected toward the kinocilium. Hair-cell excitation is then transmitted chemically to an associated neuron.

9-12 Touch Tactile receptors are common on the tentacles of cnidarians and annelid worms and the antennae of arthropods; the latter commonly have tactile hairs on the body (Fig. 25-4D). On vertebrates, tactile receptors occur over most of the exterior surface. Their morphologic variation is great. Some are **free nerve endings,** but most are special bulbs, or **corpuscles,** that contain the sensory nerve terminations (Fig. 9-14), or the terminals are highly branched and closely associated with hairs. Movement of the hair triggers an action potential. In humans, tactile receptors are most sensitive and closely spaced on the face, lips, and palmar surfaces of the fingers. At the fingertip, a pressure of only 3 g/mm^2 is detected and two points 2 to 3 mm apart will receive separate sensations, whereas on the back of the body the minima are around 50 g/mm^2 and between 60 and 70 mm.

Meissner's corpuscles (Fig. 9-14), receptors of delicate tactile sense, lie near the skin surface. Deep touch is registered by **Pacinian corpuscles,** deeper in the skin and associated with tendons, joints, muscles, and visceral mesenteries. The Pacinian unmyelinated nerve terminal is enclosed by onionlike layers of connective tissue that deform under pressure. Pain receptors are free nerve endings that

Figure 9-14 Receptors and effectors, the end organs related to sensory and motor nerves. *A*. Free sensory nerve endings in cornea of eye. *B*. Meissner corpuscle (sensory) under human epidermis. *C*. Nerve endings on gland cell in pancreas. *D*. Motor fibers on muscles in frog. *E*. Motor end plates on muscle fibers in rabbit. (*Adapted from Cajal, Histology, The Williams & Wilkins Company.*)

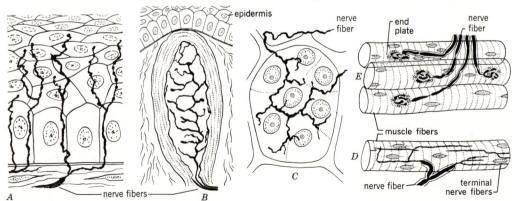

are excited by injury or intense stimulation. Injury appears to release a proteolytic enzyme that breaks down plasma globulins to form certain polypeptides which excite the pain terminals.

9-13 Proprioceptors   A number of "sense organs" in muscles, tendons, connective tissues, and skeletal tissues do not produce well-defined sensations but help coordinate the position of the limbs and are generally concerned with the so-called "kinesthetic sense." (The Pacinian corpuscles are among them.) These are the **proprioceptors.** The simple act of touching a particular part of the body when the eyes are shut is learned by practice and involves unconscious memory of the exact tensions and displacements of muscles required to place the finger in the right spot. The same mechanism is involved in more complicated acts which, once learned, are carried out unconsciously, as, for example, walking, piano playing, skating, riding a bicycle, or typewriting. To a certain extent the proprioceptors are responsible for maintenance of posture.

Muscle **stretch receptors** are important in sensing the amount of stretch and extent of contraction of muscles. They are modified muscle fibers containing a central spindle in which are invaginated sensory nerve endings. Increase in length of the fiber causes an increase in firing of the nerve, and shortening causes a decrease or cessation of firing. A stretch reflex occurs when the receptor is excited and activates (monosynaptically) the motor neurons of adjacent muscle fibers, causing contraction that reverses the stretch. Through interneuronal connections, inhibition of motor neurons of antagonistic muscles occurs.

9-14 Equilibrium   A **statocyst** (Fig. 9-15) is a small organ of equilibrium in which a particle rests among hairlike projections on sensory cells. A change in position of the animal brings the particle, or **statolith,** against one or another of the receptors, which transmits an impulse indicating the body position in respect to gravity. In mollusks the statolith is a small limy concretion, whereas in the crayfish (Sec. 24-14) a particle of sand serves this purpose. Some aquatic animals have hydrostatic organs

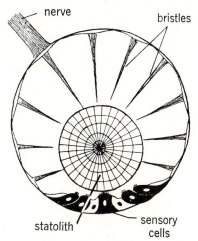

Figure 9-15 Statocyst of a mollusk (enlarged section).

which aid in equilibration by their sensitivity to small differences in pressure. Air bladders in many fishes and air bubbles in some aquatic insects serve this purpose. The inner ear of most vertebrates has a utriculus, a sacculus, and three fluid-filled semicircular canals (Fig. 9-16A) in three separate planes. The first two structures are static balancing organs which give information on the position of the head; the semicircular canals provide information on movements of the head and body. On each canal is a swelling, or **ampulla,** a statocystlike organ of equilibrium, that contains one to many limy particles (otoliths) over sensory "hair cells." Tilting the head or moving the body shifts the otoliths or causes movement of fluid in one or more canals. In turn, these stimulate the hair cells that connect to nerves and bring about reflex muscular movements whereby the body is righted. If a cat is inverted and dropped, the mechanism of equilibration acts to rotate the body so that the animal lands on its feet.

9-15 Hearing   The organ of hearing in most mammals (Fig. 9-16) has an external sound-collecting appendage, or **pinna,** around a tubular **external auditory canal.** A protective wax is secreted by glands along this canal. At the end of the canal sound waves act to set the **eardrum** (*tympanum*, or *tympanic membrane*) into vibration. These movements are amplified and transmitted by three **auditory ossicles**

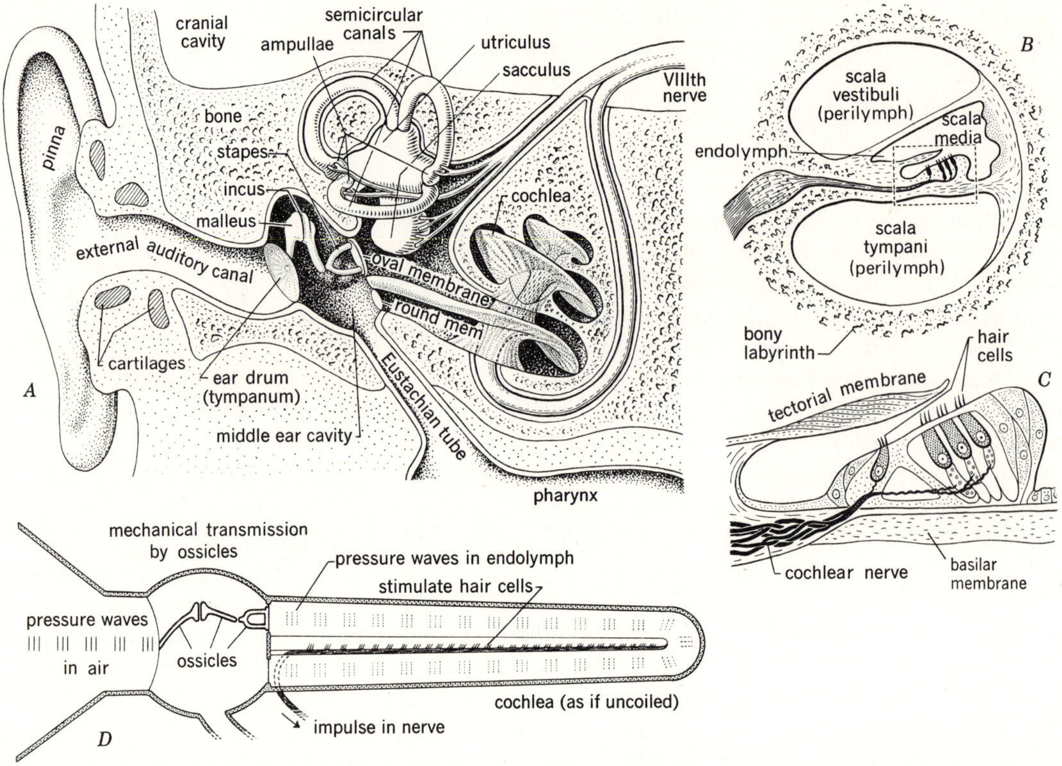

Figure 9-16 The mechanism of hearing and equilibrium in humans. *A.* General structure of the ear; parts within skull enlarged. *B.* Cross section of one part of cochlea (area of *C* in dotted lines). *C.* Enlarged section through spiral organ of Corti with sensory hair cells. *D.* Diagram of sound transmission from the air to an impulse in the auditory nerve.

(*malleus, incus, stapes*) to the oval membrane and produce vibrations in the fluid filling the spiral **cochlea** of the inner ear. In the human ear the tympanum is about twenty times the area of the oval membrane, so that much more sound energy is directed to the latter than could be collected by it alone. The effect also is enhanced by lever action of the ossicles. This provides the pressure necessary to move the cochlear fluid. Within the cochlea is the **organ of Corti,** consisting of a **basilar membrane** containing fibers of different lengths and rows of **hair cells,** the tips of which are embedded in the overhanging **tectorial membrane.** Vibrations transmitted to the perilymph of the scala vestibuli at the oval membrane (Fig. 9-16) cause displacement of the scala media, housing the organ of Corti. The basilar membrane moves and a shearing force is produced on the hairs embedded in the tectorial membrane. The round window allows

the force of vibration to be dissipated by bowing outward into the middle ear cavity. The basilar membrane is narrow and stiff near the oval membrane but becomes wider and more flexible distally. Compressional waves in the perilymph set up a **traveling wave** that moves the length of the basilar membrane, and the membrane moves with greatest amplitude where its elasticity is appropriate for a given frequency, stimulating the hair cells at that site. The narrow basal part of the membrane resonates best with high-frequency tones. The low-frequency vibrations travel farther, and thus the more distant part is most responsive to low tones. Impulses from excited cochlear nerve fibers are transmitted by the auditory nerve, and the brain recognizes the sound frequency.

Loudness of sounds depends on the number of hair cells stimulated. Location is determined by the difference in arrival time of a sound component and its relative intensity at each ear.

Lower land vertebrates lack a pinna, the cochlea is very short or is represented by a small outgrowth (*lagena*), and there is often only one auditory bone (*stapes* or *columella*). Some species, however, (certain reptiles, for example) have an additional element, the extrastapes. The human ear responds to sound frequencies from 20 to 20,000 hertz (Hz), with greatest sensitivity in the 1000- to 4000-Hz range. Bats can hear frequencies up to 200 kHz. Loud music and jet and revved-up motorcycle sounds can permanently damage the auditory hair cells. Sound receptors among invertebrates occur mainly in certain insects (Fig. 9-11).

Photoreception

Animal light detection (and plant photosynthesis) makes use of a narrow segment of the total electromagnetic spectrum that reaches the earth. The human eye is sensitive from violet to red, from about 400 to 700 nanometers (nm), and is most acute at around 500 nm (yellow). Some insects (e.g., bees) see into the ultraviolet, and there are indications that a few animals may see into the infrared.

9-16 Light and sight Photoreceptors sensitive to light are present in earthworms (Fig. 22-11), and there are "eyespots" on various cnidarians and some mollusks. From such simple structures various types of eyes have developed (Fig. 9-17). Among arthropods there are both simple and compound eyes (Fig. 24-7; Secs. 24-14, 25-11). The cephalopod mollusks have eyes much like those of vertebrates, but they are derived differently.

The eye is the most complex vertebrate sense organ (Fig. 9-18). Its structure is analogous to that of a camera, having a transparent biconvex **lens** which focuses images of external objects on the photosensitive interior, as on a photographic film. The lens is encircled by radiating muscle fibers, the **ciliary muscle,** that attaches to a ligament that borders the lens. Contraction of this muscle (in mammals) releases tension on the lens, and it rounds up through its own elasticity for focusing on close objects. The outer

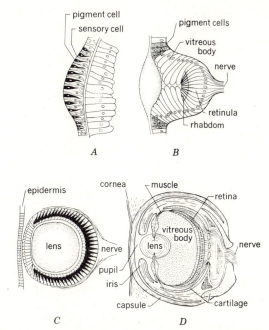

Figure 9-17 Eyes of some invertebrates, seen in median section. *A*. Medusa (cnidarian). *B*. Beetle larva (insect). *C*. Snail (mollusk). *D*. Cuttlefish (mollusk). (*Altered from Claus, Grobben, and Kuhn.*)

sclerotic coat of connective tissue forms a supporting case for the eye, with the cornea as a transparent front. The next or **choroid coat** includes blood vessels and much black pigment to exclude all light except that entering through the lens. The innermost layer, or **retina,** contains the **photoreceptors,** the **rods** and **cones** that connect through interneurons with ganglion cells whose axons form the optic nerve (second cranial). Inside the cornea, part of the choroid coat is specialized as the **iris,** a pigmented disk with a central opening, or pupil, through which light enters the spherical lens just behind. The pupil contracts or dilates reflexly to regulate the amount of light entering the eye. The space before the lens contains a watery **aqueous humor,** and that behind, a jellylike **vitreous humor,** both serving to maintain the form of the eyeball. The eye is moved by six muscles attached to its outer surface. There is remarkable consistency in eye structure (muscles, nerves, etc.) throughout the vertebrates.

The rods and cones are modified cilia, and their outer segments contain large numbers of flat disks

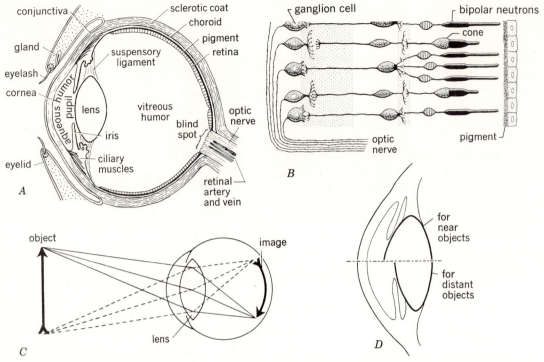

Figure 9-18 The mechanism of sight in humans. *A.* Median vertical section of the eye. *B.* Enlarged diagram of the structure of the retina. Light enters retina from left. *C.* The lens serves to form an image (reduced and inverted) on the retina in the way a camera lens produces an image on the photographic film. *D.* Changes in shape of the lens (accommodation) to focus on near and distant objects.

on which photosensitive pigments are located. The rods (about 125 million in a human eye) form colorless sensations in dim light, and the cones (about 6.5 million) are active in strong light and are sensitive to both white and colors. The rods are more than twice as sensitive as cones, but their resolution is poor, perhaps because many of them are served by a single ganglion cell. They are concentrated at the periphery of the retina, which explains why looking to the side of an object in dim light improves its resolution. Some nocturnal animals have a layer of guanine or other material in the back of the eye which reflects light back through the retina and provides further stimulation of the photoreceptors.

There are three kinds of cones, which respond to red, green, and blue light and, in combination, to intermediate colors. Their populations are interspersed. The cones are centrally located, and the cone-rich and rod-free **fovea centralis,** aligned with cornea and lens, is the region of greatest retinal acuity. Some cones are served by single ganglion cells

and thus have a direct route to the brain. Nerve circuits in the retina and visual cortex compare levels of excitation in the cone populations and color perception results.

The photopigments in the rods and cones are acted upon by light and trigger nerve impulses. A photopigment consists of a small **chromophore** molecule (retinaldehyde or retinal) attached to a large colorless protein, **opsin.** The chromophore, a variant of vitamin A, is a carotenoid derived from plants.

The opsin, different for each kind of photoreceptor, determines the type of sensitivity of the photopigment—whether responding to all light (rods) or to red, green, or blue. Light falling on the photopigment alters the chromophore, which then separates from opsin. In restoration the chromophore is rearranged and the process reversed. The photopigment in the rods is visual purple, or **rhodopsin,** which is bleached by light and must be present for vision in dim light. After exposure to bright light, 30 to 40 min is needed for dark adaptation (vision in very dim light). Severe vitamin A deficiency, which prevents normal chromophore formation, interferes with dark adaptation, producing night blindness. The light-sensitive pigment of cones is **iodopsin,** similar to rhodopsin.

Acuity of vision is improved in many diurnal animals by yellow or red oil droplets in the photoreceptors, yellow lenses, or yellow pigment in the fovea (primates), which, like amber-colored sunglasses, cut down on scatter light and chromatic aberration. Focusing on close objects in mammals, birds, and lizards is accomplished by rounding up the lens through action of the ciliary muscle, whereas in most fishes, amphibians, and snakes the lens is moved forward for close vision. The eyes of many vertebrates move independently but can, when required, function in binocular vision, both eyes focusing on an object in part of the visual field. Such vision facilitates detection of movement in the line of sight toward or away from the viewer and enables humans and some animals to judge distances accurately.

Color vision is present in diurnal teleost fishes, birds, lizards, some snakes, frogs, primates, squirrels, and domestic cats. Surprisingly, however, color vision is poor or lacking in most mammals.

9-17 Extraretinal photoreceptors Retinal photoreceptors are concerned primarily with pattern vision, or the perception of images. Some animals have extraretinal photoreceptive regions in brain or skin outside the retinas of the "true" eyes. In ectothermal vertebrates, extraretinal photoreceptors have been found in the pineal body (Sec. 8-13) and in the median, or pineal, eye in tadpoles, frogs, some lizards, and the tuatara, where they closely resemble in structure the photoreceptors of the eyes. In experiments, extraretinal photoreceptors have been implicated in (1) adjusting circadian locomotor rhythms to changes in daylight (salamanders and frogs), (2) synchronizing reproductive cycles to seasonal changes (birds and lizards), (3) rhythmic daily color changes in tadpoles and fishes, and (4) directional orientation movements (dependent on light cues) in frogs and salamanders. Extraretinal photoreceptors thus appear to be concerned primarily

with physiologic and behavioral responses to non-visual light stimuli.

References

(See also references at end of Chap. 4.)

Amoore, J. E., J. W. Johnston, Jr., and **M. Rubin.** 1964. The stereochemical theory of odor. *Sci. Am.,* vol. 210, no. 2, pp. 42–49. *The sense of smell may be based on the shape of molecules and nerve receptors; seven primary odors are distinguished, each by an appropriately shaped receptor at the olfactory nerve endings.*

Baker, P. F. 1966. The nerve axon. *Sci. Am.,* vol. 214, no. 3, pp. 74–82. *Nerve impulse conduction studied by removing the contents of the giant axon of the squid and replacing them with various solutions.*

Bullock, T. H., and **G. A. Horridge.** 1965. Structure and function in the nervous system of invertebrates. San Francisco, W. H. Freeman and Company. 2 vols. xxvii + 1722 pp., illus.

DiCara, L. V. 1970. Learning in the autonomic nervous system. *Sci. Am.,* vol. 222, no. 1, pp. 30–39. *Behavioral responses long considered involuntary (heartbeat and intestinal contraction) can be influenced by learning.*

Dowling, J. E. 1966. Night blindness. *Sci. Am.,* vol. 215, no. 4, pp. 78–84. *The role of vitamin A in vision.*

Eccles, J. 1965. The synapse. *Sci. Am.,* vol. 212, no. 1, pp. 56–66. *How one nerve cell transmits the nerve impulse to another cell.*

Heimer, L. 1971. Pathways in the brain. *Sci. Am.,* vol. 225, no. 1, pp. 48–60. *Connections among brain cells and brain circuitry revealed by new techniques of staining and microscopy.*

Hendricks, S. B. 1968. How light interacts with living matter. *Sci. Am.,* vol. 219, no. 3, pp. 174–186. *Light activates photosynthesis, vision, and photoperiodism through the mediation of specific pigments.*

Hodgson, E. S. 1961. Taste receptors. *Sci. Am.,* vol. 204, no. 5, pp. 135–144. *The mechanism of taste reception is clarified by study of the hairs on the proboscis of the blowfly.*

Hubel, D. H. 1963. The visual cortex of the brain. *Sci. Am.,* vol. 209, no. 2, pp. 54–62. *The visual cortex is studied to determine how it transforms images on the retina into vision.*

Kalmus, H. 1958. The chemical senses. *Sci. Am.,* vol. 198, no. 4, pp. 97–106. *Experiments with the senses of smell and taste.*

Kandel, E. 1960. Nerve cells and behavior. *Sci. Am.,*

vol. 223, no. 1, pp. 57–70. *Exploration in primitive sea animals of the possibility of studying memory and learning at the level of nerve cells and their interconnections.*

Katz, B. 1961. How cells communicate. *Sci. Am.*, vol. 205, no. 3, pp. 209–220. *Cells communicate by "chemical messengers" and the impulses of the nervous system.*

———. 1966. Nerve, muscle, and synapse. New York, McGraw-Hill Book Company. Paper. ix + 193 pp.

Lentz, T. L. 1968. Primitive nervous systems. New Haven, Conn., Yale University Press. viii + 148 pp.

Llinás, R. R. 1975. The cortex of the cerebellum. *Sci. Am.*, vol. 232, no. 1, pp. 56–71. *The pattern of nerve cell connections has been determined in detail and related to function.*

Livingston, W. K. 1953. What is pain? *Sci. Am.*, vol. 188, no. 3, pp. 59–66. *The perception of injury involves a subtle blend of physiologic and psychologic factors.*

Luria, A. R. 1970. The functional organization of the brain. *Sci. Am.*, vol. 222, no. 3, pp. 66–78. *Injuries to the brain provide clues to the functions of speech and writing.*

MacNichol, E. F., Jr. 1964. Three-pigment color vision. *Sci. Am.*, vol. 211, no. 6, pp. 48–56. *In vertebrate retinas, color is discriminated by three pigments segregated in three kinds of cone receptor.*

McCashland, B. W. 1968. Animal coordinating mechanisms. Dubuque, Iowa, Wm. C. Brown Company. Publishers. Paper. 118 pp.

McEwen, B. S. 1976. Interactions between hormones and nerve tissue. *Sci. Am.*, vol. 235, no. 1, pp. 48–58. *Steroid hormones secreted by the gonads and adrenals are traced to target cells in the brain. In the newborn animal the sex hormones help to lay down brain circuits that control later behavior.*

Miller, W. H., F. Ratliff, and **H. K. Hartline.** 1961. How cells receive stimuli, *Sci. Am.*, vol. 205, no. 3, pp. 222–238. *The properties of cells highly specialized to monitor*

changes in the external and internal environments are elucidated by studies of visual receptors in the horseshoe crab.

Neisser, U. 1968. The processes of vision. *Sci. Am.*, vol. 219, no. 3, pp. 204–214. *The role of eye and brain in visual perception and visual memory.*

Petras, J. M., and **C. R. Noback** (editors). 1969. Comparative and evolutionary aspects of the vertebrate central nervous system. Ann. N.Y. Acad. Sci., vol. 167, pp. 1–513.

Schneider, D. 1974. The sex-attractant receptor of moths. *Sci. Am.*, vol. 231, no. 1, pp. 28–35. *Receptor cells on the silk moth's antennae can detect one molecule of attractant.*

Snider, R. S. 1958. The cerebellum. *Sci. Am.*, vol. 199, no. 2, pp. 84–90. *This long-mysterious region of the brain is evidently a monitor of communication between brain and body.*

Stent, G. S. 1972. Cellular communication. *Sci. Am.*, vol. 227, no. 3, pp. 42–51. *Cells communicate by means of chemical messengers and nerve impulses.*

Walls, G. L. 1942. The vertebrate eye and its adaptive radiation. Bloomfield Hills, Mich., Cranbrook Institute of Science Bulletin 19. xiv + 785 pp. Reprinted 1963. New York, Hafner Publishing Company, Inc.

Werblin, F. S. 1973. The control of sensitivity in the retina. *Sci. Am.*, vol. 228, no. 1, pp. 70–79. *Neuron interactions in the retina help explain how the eye maintains contrast over a wide range of illumination.*

Wilson, V. J. 1966. Inhibition in the central nervous system. *Sci. Am.*, vol. 214, no. 5, pp. 102–110. *When some muscles contract others must be inhibited. In vertebrates inhibition is achieved by inhibitory neurons in brain and spinal cord.*

Young, R. W. 1970. Visual cells. *Sci. Am.*, vol. 223, no. 1, pp. 80–91. *The rods and cones are highly sensitive, yet durable, and they constantly renew themselves, but in different ways.*

10

REPRODUCTION AND DEVELOPMENT

The ability to produce new living individuals is a basic characteristic of all organisms. Early biologists understood correctly how the higher animals reproduce, but for centuries it was believed that many forms of life arose from nonliving materials by **spontaneous generation**—worms and tadpoles from mud and flies from the carcasses of dead animals. These erroneous ideas were gradually abandoned after Francesco Redi (1626?–1697) showed in 1668 that maggots and flies are produced from meat only if living flies have laid eggs on such material. Yet only a century ago it was thought that bacteria and other microorganisms could develop spontaneously. Then in 1861, Louis Pasteur (1822–1895) cultured bacteria in flasks with a long S-shaped neck serving as a trap for airborne organisms, but *unstoppered*. Once the bacteria in a flask were killed by heat, the medium

within remained without life. The principle of **sterilization** that he demonstrated is the basis for destroying microorganisms by heat or chemicals. It is used in present day surgery and medicine, in preserving food by canning, in the keeping of *pasteur*ized milk, and in other aspects of modern life.

All reliable evidence indicates that new life comes only from preexisting life (*omne vivum ex vivo*); this is the process of **biogenesis,** or **reproduction.**

Reproduction

10-1 Asexual reproduction This occurs (Fig. 10-1) when new individuals derive from one "parent" without the intervention of an egg; no special repro-

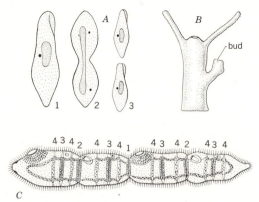

Figure 10-1   Types of asexual reproduction.   *A.* Fission in *Paramecium*, *B.* Budding in *Hydra*. *C.* Fragmentation in a flatworm, *Microstomum;* numbers indicate the sequence of fission planes that will divide the animal into 16 parts, each later able to become an individual. (*After von Graff.*)

ductive structures are involved. It occurs in many plants through the production of spores, bulbs, and shoots, and in animals it is most common in aquatic sessile or drifting forms such as cnidarians, sponges, and tunicates and in polychaetes and planarian worms. It is the predominant form of reproduction in the protists. Asexual reproduction may be by unicellular fission and sporulation or by multicellular budding or fragmentation.

Protozoans such as *Paramecium* multiply by **fission,** in which an individual divides into two halves, usually equal, after which each grows to the original form. The nucleus divides, and then the cytoplasm. Multiple fission, or **sporulation,** occurs in the sporozoans (*Plasmodium,* causative agent of malaria, etc.), where the nucleus divides repeatedly and then the cytoplasm subdivides, so that a part of it surrounds each of the many daughter nuclei (Chap. 15). **Budding** is a type of reproduction in which a new individual arises as an outgrowth, or **bud,** on an older animal; it grows to the form and size of the latter. Budding of sponges, cnidarians, bryozoans, and tunicates results in colonies of many individuals. Freshwater sponges also produce internal buds, or **gemmules** (Sec. 16-9), each of several cells, within a common dense covering. These escape, and later each gemmule produces a new individual. Bryozoans have internal buds known as **statoblasts**

which develop into new individuals. **Fragmentation** occurs in some flatworms (TURBELLARIA) and ribbon worms (NEMERTINEA), an individual breaking into two or more parts, each capable of growing to be a complete animal. Some sea anemones reproduce from fragments at the base of the stalk.

The processes of asexual reproduction are closely related to those of **regeneration.** Regenerative growth is common, in some degree, to most organisms, as in the capacity to replace parts lost by injury. Young animals and species low in the evolutionary scale usually have greater regenerative powers than older or higher animals. A cutting of willow or geranium in moist soil will grow into a complete plant, and pieces of some hydroid cnidarians, if put in sand under seawater will form complete animals. When a flatworm (*Dugesia*) is cut into pieces, each will usually regenerate to form a complete but smaller individual (Fig. 10-2). Sea stars and other echinoderms regenerate lost arms or other parts. Appendages of crabs and other crustaceans and the tails of some salamanders and lizards may be cast off under stress, a process termed **autotomy** (Gr. *auto,* self + *toma,* to cut). Then the animal regenerates the lost part. Regeneration may involve either a local part or a complete reorganization of tissues to produce organs and other structures anew.

Figure 10-2   Regeneration in a flatworm, *Dugesia.*   Portions cut from an entire worm (indicated by broken lines) gradually regenerated (dark stipple) to form entire small worms. (*After Stempell.*)

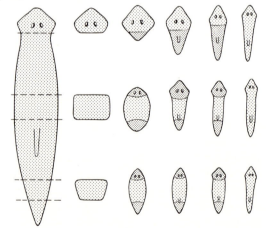

10-2 Sexual reproduction Most animals and plants reproduce by a process in which new individuals develop from sex cells, or **gametes,** produced by the parents. This is **sexual reproduction.** Typically two cells of different kind (male and female) join to create a new individual.

Protozoans have reproductive processes resembling the sexual phenomena of higher animals (Sec.15-25) In the **conjugation** of ciliates (*Paramecium,* etc.), two individuals of similar appearance but evidently differing biochemically, fuse together, exchange micronuclear materials, and then separate to continue fission. Among sporozoans (*Plasmodium,* etc.), two kinds of individuals (macrogametes and microgametes) are formed at certain stages; these fuse permanently in pairs to continue the life cycle. In the colonial flagellate *Volvox,* the same colony or different colonies yield two kinds of free individuals that combine in pairs, one of each, and give rise to new colonies.

In multicellular animals, **sex** is the total of all structural and functional characteristics that distinguish **male** ($\male$) and **female** ($\female$). Both produce free **sex cells,** or **germ cells.** Those of males are minute and known as sperm, or **spermatozoa;** the female releases somewhat larger **eggs,** or **ova** (sing. *ovum*). Besides the necessary differences in reproductive organs, individuals of the two sexes may differ in external or internal form, in physiology, and in behavior.

The germ cells are produced in organs known as **gonads,** the sperm in **testes** (sing. *testis*), or spermaries, and the ova in **ovaries.** These are the **primary sex organs,** and the only sex structures present in cnidarians. Most animals have **accessory sex organs** —ducts, glands, and other organs associated with the gonads to form a **reproductive system** that aids in the reproductive process (Fig. 10-3).

If both male and female systems are in one individual, as in flatworms and earthworms, the animal is termed **monoecious.** In nematodes, arthropods, various other invertebrates and practically all vertebrates, each individual is either male or female; the sexes are **separate,** and such animals are **dioecious.** The term **hermaphrodite** is applied to monoecious species and also to occasional abnormal individuals of dioecious species that contain both male and fe-

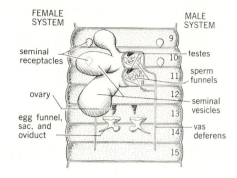

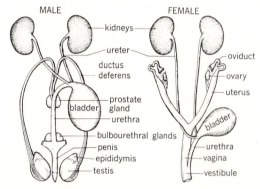

Figure 10-3 Reproductive systems. *Above.* Earthworm (monoecious), both systems in one individual. *Below.* Cat (dioecious), male and female systems in separate individuals. Excretory system also included.

male systems. Some animals (some gastropods and oysters) change sex during the course of their life. If the sperm are produced before the eggs, the condition is known as **protandry;** if eggs are produced first, the condition is known as **protogyny;** in tapeworms and certain snails, self-fertilization occurs.

Sexual reproduction arose early in evolution and now predominates. It made possible almost infinite pooling and reassorting of the genes from pairs of individuals in their descendants and thus much more genetic variability than could occur in asexual reproduction. Variability increases the capacity of a population to adjust through natural selection (Sec. 13-17) to environmental changes.

10-3 Reproductive systems There is variety in the structural details of the reproductive systems of

Table 10-1
THE ESSENTIAL FEATURES OF REPRODUCTIVE SYSTEMS

Sex	Gonads	Sex ducts	Accessory glands	Sperm storage	Terminal organs
Male (♂)	Testis (sper-matozoa)	Efferent ductules, ductus deferens	Sperm-activating and nurturing	Seminal vesicle	Penis
Female (♀)	Ovary (ova)	Oviduct (+ uterus)	Albumen, yolk, shell	Seminal receptacle	Vagina (+ ovipositor)

different animals, but all have a basic similarity in pattern, even between the two sexes (Table 10-1). In the **male reproductive system** of vertebrates the spermatozoa are produced in a series of compartments, or tubules, of the **testis** (Fig. 10-4A). From there, in most species, they travel through small ducts, or efferent ductules, to a larger sperm duct, or **ductus deferens** (vas deferens). The lower end of the latter often is enlarged as a **seminal vesicle** which stores sperm (not in mammals) or augments the seminal fluid. The ductus deferens opens into the

cloaca in lower vertebrates but into the urethra in most mammals, where the spermatozoa emerge through the copulatory organ, or **penis.** Copulatory organs of a different kind are found in certain fishes and amphibians and in most reptiles (Sec. 29-15, 32-19). In forms with a mesonephric kidney (Figs. 7-8, 7-9) (some fishes and amphibians), the wolffian duct that drains the kidney serves also as the sperm duct. In higher vertebrates, however, this becomes exclusively the ductus deferens. **Accessory glands** providing secretions to activate the sperm or for other purposes are sometimes present along the sperm duct. In mammals these are the prostate and bulbo-urethral glands.

The **female reproductive system** produces the ova as individual cells in the **ovary.** In some animals each ovum during its growth is surrounded by follicle cells, as in the frog. The mammalian egg grows in a special **graafian follicle** that enlarges as the egg matures and finally ruptures to release it (Fig. 10-4B). Fluid in the follicle contains hormones important in reproduction (Sec. 35-13). Ovaries of mammals and some other animals are solid, those in a frog are saclike, and those of insects comprise several tubular ovarioles. The ovum acquires its yolk while still in the ovary. Mature eggs are released from the ovary and moved down the conducting tube, or **oviduct,** by the action of muscles in its walls or by cilia lining its interior. **Glands** in the wall of the oviduct may surround the egg with various materials such as the albumen (egg white) and shell in reptiles, birds, and some other animals. The lower part of the duct is enlarged as a temporary storage reservoir for eggs in some species. This part is expanded as the **uterus** in mammals and other animals that retain the eggs for development inside the body. In most vertebrates

Figure 10-4 The gonads of mammals. *A*. Section of a testis. *B*. Small section of an ovary, much enlarged, with several graafian follicles.

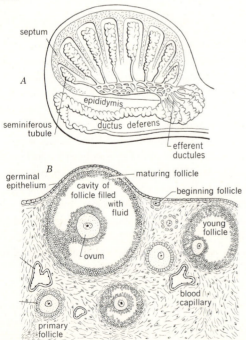

the female reproductive tract opens into the cloaca. In mammals, the terminal part of the tract is usually specialized as the **vagina** to receive the male's penis; some have also a **seminal receptacle** for storage of sperm received.

10-4 Origin of sex cells   There is much interest in the ultimate origin of the sex cells because they give rise to new individuals and serve to transmit hereditary characteristics between successive generations. The doctrine of germinal continuity proposed by August Weismann (1834–1914) held that the sex cells, or **germ plasm,** constituted a substance apart from external influences and from the body, or **somatoplasm.** For each new generation the parent germ plasm produces both the body (soma) and the germ plasm of new individuals. The continuity of germ plasm is clear in some invertebrates (*Ascaris, Sagitta, Miastor*), where one cell in the early cleavage of the egg can be traced as the ultimate origin of the future sex cells. In vertebrates the primary germ cells start at distant sites, then migrate or are transported to the germinal ridges where gonads later develop. In the frog, chick, and mammal they start in the embryonic endoderm and migrate by amoeboid movement or are transported in blood vessels. In frogs they appear in the yolk endoderm, and in mammalian embryos they arise in the yolk sac stalk, the allantois, and the gut, then move to the germinal ridges. There is a direct continuity in the chromosomes by which each cell of the new animal contains the hereditary mechanism established in the egg (**zygote**) as it starts to develop.

The gonads appear during embryonic development in some animals but do not enlarge until the individuals approach sexual maturity; in others they form at the latter stage. Early germ cells in the gonads multiply by mitosis like somatic cells (epithelium); every chromosome replicates, and the two parts separate longitudinally so that each daughter cell receives an identical set of chromosomes. The nucleus in each contains a dual set of chromosomes, the **diploid number,** $2n$ (except in some haploid male bees, etc.). In every pair, one chromosome has been derived from the male parent and the other from the female parent of the individual in question. The two members of each pair are termed **homologous chromosomes.** They are alike in size and shape but may differ in genetic content (Chap. 11). At the approach of sexual maturity the germ cells multiply rapidly; they are then known as **spermatogonia** in the male and as **oogonia** in the female. Before they are able to participate in reproduction, however, their physical and physiologic characteristics must change.

10-5 Maturation and meiosis   The process by which spermatogonia become spermatozoa and oogonia become ova is known as **gametogenesis,** and the resulting matured cells are called **gametes** (Figs. 10-5 to 10-7). The accompanying nuclear changes are termed **meiosis.** Meiosis halves the number of chromosomes, thereby preventing them from doubling with each generation. The gametes of the male and female differ in form, size, and physiology, but the meiotic changes in their nuclei are comparable.

Meiosis consists of two nuclear divisions that follow one another, known as the **first and second meiotic divisions.** These differ from mitosis in two important features: (1) The final number of chromosomes in a gamete is only half (or the **haploid number,** n) of that present in a spermatogonium or oogonium (or somatic cell), and the set of chromosomes in a matured gamete includes but one member from each homologous pair that was present in the unmatured cells. (2) There is a random assortment in this reduction, so that each gamete receives either *one or the other* member of each pair of chromosomes. When two gametes of opposite sex later join in fertilization, the chromosome number in the new individual will be that of the species ($2n$). The manner of division and segregation of the chromosomes during maturation, together with the random meeting of eggs and sperm in fertilization, affords a logical basis for many of the observed phenomena of inheritance on the premise that the chromosomes are the bearers of the determiners, or genes, for hereditary characters. The random sorting provides for variation in the combinations of characters that will appear in different individuals of the new generation (Chap. 11).

The basic chromosome events in meiosis are the same in both sexes (Fig. 10-5). In both the first and

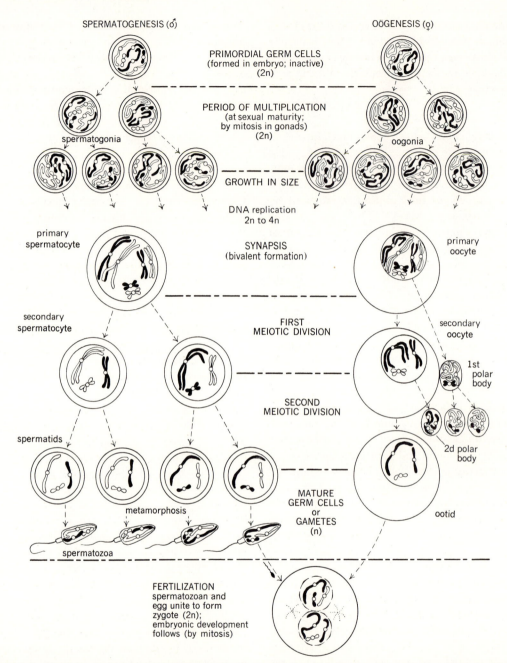

SPERMATOGENESIS (♂)

OÖGENESIS (♀)

PRIMORDIAL GERM CELLS
(formed in embryo; inactive)
(2n)

PERIOD OF MULTIPLICATION
(at sexual maturity;
by mitosis in gonads)
(2n)

spermatogonia

oogonia

GROWTH IN SIZE

DNA replication
2n to 4n

primary
spermatocyte

SYNAPSIS
(bivalent formation)

primary
oocyte

secondary
spermatocyte

FIRST
MEIOTIC DIVISION

secondary
oocyte

1st
polar
body

SECOND
MEIOTIC DIVISION

spermatids

metamorphosis

2d polar
body

MATURE
GERM CELLS
or
GAMETES
(n)

ootid

spermatozoa

FERTILIZATION
spermatozoan and
egg unite to form
zygote (2n);
embryonic development
follows (by mitosis)

Figure 10-5 Formation of sperm and eggs; diagrammatic. The process is similar in the two sexes as to nuclear divisions and chromosomes but differs as to the cytoplasm (*left,* male; *right,* female). The number of chromosomes is shown for each stage. The species is assumed to have six chromosomes (diploid number); chromosomes derived from the previous generation are shown as white (maternal) and black (paternal), respectively. Crossing-over is shown in one pair of chromosomes in the primary spermatocyte. The change from the diploid to the haploid number of chromosomes occurs at the first meiotic division. Compare Fig. 3-10 (mitosis) for details of phases in division.

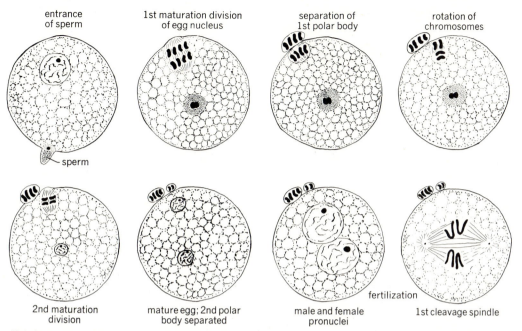

entrance of sperm

1st maturation division of egg nucleus

separation of 1st polar body

rotation of chromosomes

sperm

2nd maturation division

mature egg; 2nd polar body separated

male and female pronuclei

fertilization

1st cleavage spindle

Figure 10-6 Maturation of the ovum, sperm entrance, junction of pronuclei, and preparation for first cleavage in the roundworm, *Ascaris megalocephala.* (The first polar body actually is carried off the surface of the egg by the elevation of the fertilization layer.)

second meiotic divisions prophase to telophase stages are recognized, as in mitosis (Sec. 3-10). The reduction of the number of chromosomes from diploid to haploid occurs during the first meiotic division. In *prophase I,* (1) the diffuse threads of chromatin that make up the interphase chromosomes thicken and the chromosomes become evident as discrete structures. They appear long and threadlike, with granules spaced like beads on a string. DNA replication is complete, and the chromosomes consist of a pair of identical partners, the **sister chromatids,** attached by the centromere. (2) The chromosomes of each homologous pair, called a **bivalent,** come to lie more or less parallel to one another. Since each homologous chromosome consists of two sister chromatids, each bivalent contains four chromatids and is called **tetrad** ($4n$). The centromere of each chromosome is undivided at this time and is shared by sister chromatids. (3) The chromosome pairs shorten and thicken and twist about one another in **synapsis. Crossing-over** (Sec. 11-16) may occur. A piece of one chromatid of the paternally derived chromosome is exchanged for a homologous piece of one chromatid

of the maternally derived chromosome. This produces chromosomes that contain a mixture of paternal and maternal genes. (4) The homologous chromosomes separate but remain attached at specific points called **chiasmata,** where presumably crossing-over of genes has occurred (Fig. 11-18; Sec. 11-16). The chromatids of each chromosome are now distinctly visible but are no longer identical, having exchanged parts. The centromeres remain undivided. (5) The chromosomes thicken and the chiasmata vanish. Nucleolus and nuclear membrane disappear and the tetrads move to the equatorial plate. In *metaphase I,* the spindle forms with the bivalents arranged at random on the equatorial plate, and the centromeres of the two homologous chromosomes orient toward opposite poles. In *anaphase 1,* one maternal pair goes to one pole of the spindle and the paternal pair to the opposite. This is **segregation.** From different bivalents the manner of segregation is by chance—free assortment. Maternal chromatids from one bivalent may migrate on the spindle with maternal *or* paternal chromatids from other bivalents. Thus some maternal and some paternal pairs of chromatids go to one pole, and those of opposite origin to the other pole. Chromosome reassortment is of great importance in contributing to genetic variability. In a species such as the human, with 23 pairs of homologous chromosomes, the chance of two sperm having identical chromosomes is 2^{23}. In *telophase 1* and *interphase,* the cell may or may not divide. In eukaryotic cells often no cell wall or nuclear membrane forms, and the chromosomes may enter directly into the second meiotic division.

In the second meiotic division a spindle again forms and the chromosomes take an equatorial position. The centromeres divide and there is separation of the sister chromatids, which now go as distinct chromosomes to opposite poles of the cell. The cytoplasm divides and the chromosomes elongate. The result of the two meiotic divisions is the formation of four cells, each containing n chromosomes, the haploid number. From each pair of homologous chromosomes any one cell contains one representative of each pair.

10-6 Spermatogenesis   As a male matures sexually, spermatogonia in the testis multiply by mitosis until many are present; then spermatogenesis begins. Each spermatogonium increases in size to become a **primary spermatocyte.** Each primary spermatocyte then undergoes the first meiotic division ($2n$ to n) and gives rise to two **secondary spermatocytes** (each with n chromosomes but duplicated). These in turn divide (second meiotic division) to form **spermatids,** each with the haploid number of chromosomes.

Following the second division, each undergoes **maturation.** Much of the cytoplasm is lost and the nucleus compacts to a small (densely staining) head containing the mass of haploid DNA and the **acrosome,** usually at the sperm tip, with lysins (enzymes) for dissolving the egg membrane. Behind the head is the midpiece, which includes two centrioles and tightly coiled mitochondria (or a mitochondrion), which evidently supply the energy for the newly formed flagellar tail that provides motility. A sperm is like a torpedo, with a range limited by its packaged energy supply. This is the mature male gamete, or **spermatozoan.** Both maturation and metamorphosis are usually completed before the sperm escapes from the testis. (Fig. 10-7).

10-7 Oogenesis   The gonads of females produce fewer sex cells than those of males. The egg acquires most of its physical features before meiosis begins. In the ovary the oogonia become **primary oocytes,** often enlarging with the addition of yolk. Synapsis, bivalent formation, and chromosome reduction occur as in the male, but the division spindle forms near the cell margin. As a consequence, at the first meiotic division virtually all the cytoplasm remains with one nucleus to form the **secondary oocyte;** the other nucleus passes out on the surface of the cell as a microscopic **first polar body.** Likewise, in the second meiotic division, the cytoplasm with one nucleus forms the **ootid,** and the other nucleus is discarded as the **second polar body.** The haploid chromosome content of an ootid results from chance assortment, as with a spermatid. With slight change in nuclear

Figure 10-7   The gametes of several animals.   Size of ova in millimeters. Spermatozoa, greatly enlarged, but not to same scale. (*Mostly after Retzius.*)

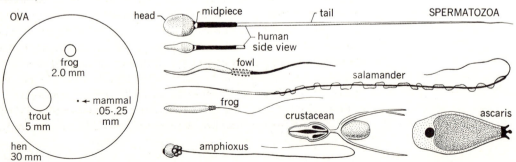

position the ootid becomes a female gamete, or **ovum.** Thus each oogonium yields but one ovum; yet the nuclear divisions that produce the ovum and polar bodies are equivalent to those by which four spermatozoa are derived (Fig. 10-5). In different species, meiosis either occurs in the ovary or after the egg is set free or requires that a sperm must penetrate the egg cytoplasm before it will be completed.

During the maturation process, the egg packages materials needed for embryonic development. Early embryonic stages, unguided by nuclear genes will depend on these stored materials. In addition to accumulating yolk, nuclear products increase as necessary for the many mitotic events to follow. Large amounts of DNA and RNA are assembled. **Lampbrush chromosomes** are observed, seen especially well in salamander oocytes. They are thought to represent active synthesis of mRNA by DNA templates and are named for their puffed–out loops of the double helix of DNA. The oocyte nucleolus is also active, producing ribosomal RNA. As maturation is completed the chromosomes contract and move near the egg surface. Upon fertilization, the first proteins are built from the stored RNA and amino acid reserves. It is thought that messenger RNA templates are freed to act at that time, perhaps through removal of a "masking" protein. The process may depend upon a reaction of the egg's outer layer, the **cortex.**

The egg is prepared in another way to influence embryogenesis. All animal eggs exhibit **polarity,** a differentiation along an anteroposterior axis extending from the animal to vegetal poles. Polarity establishes the direction of future cleavage and subsequent differentiation of the embryo. It arises during the growth and differentiation of the oocyte and may be a property of the cortical layer (Sec. 10-15), since it persists after displacement of the egg's visible components by centrifuging. In echinoderms and mollusks it is dictated by the point of attachment of the oocyte to the ovarian wall, where supplies enter. This point becomes the vegetal pole. In many animals including mammals, however, there is no obvious ovarian influence on polarity.

10-8 Gametes The gametes of various animals differ in form and size, and those of the two sexes in any one species are unlike. The ovum (any covering or shell being disregarded) is spherical or oval and nonmotile and may contain yolk to nourish the newly developing individual. The largest ova are those of some sharks (180 by 140 mm; 5½ by 7 in). The human ovum is only about 0.15 mm in diameter, and some invertebrate eggs are even smaller. Spermatozoa are motile and able to swim in fluid. While usually threadlike, some are amoeboid or of other shapes. Their size is usually microscopic; human spermatozoa are 50 to 70 μm long, but the sperm of a toad (*Discoglossus*) measure 2000 μm (2 mm). The sperm cell is a minute fraction of the egg in volume; with human gametes the ratio is about 1:195,000. Enough human ova to provide the present world population of 4 billion people could be put in a top hat and the sperm to fertilize them in a thimble!

10-9 Fertilization The union of a mature spermatozoan and an ovum is known as **fertilization,** and the resulting cell is a **zygote.** The joining of two haploid (n) nuclei yields a zygote with the diploid number ($2n$) of the species. Fertilization involves the physical entry of the sperm and also physiologic processes in both egg and sperm. Fertilization is irreversible and usually species-specific; only in exceptional cases will "foreign" sperm fertilize an egg. Fertilization stimulates the egg to active cleavage and development; it also provides for combining hereditary characteristics from both parents. In different species sperm penetrate the egg at various stages during maturation, but fusion of the egg and sperm nuclei occurs only after oogenesis is completed. In some invertebrates fertilization occurs at the young primary oocyte stage, in most mammals at the second metaphase, and in cnidarians and echinoderms at the female pronucleus stage. Substances are produced on the surfaces of the gametes that facilitate their union. For example, in sea urchins **fertilizin** (a glycoprotein) is released from the jelly coat of the egg. It interacts with a substance on the surface of the sperm called **antifertilizin** (an acidic protein) in a manner suggesting an antigen-antibody reaction. Fertilizin seems to activate sperm and to aid in their attachment to the surface of the egg. It also tends to screen out foreign sperm. Comparable reactions have been demonstrated in other animals. Many kinds of sperm produce **sperm lysins** (lysosomelike enzymes) which aid in penetrating the egg surface. They are present in the **acrosomal** vesicle at the tip of the sperm head or elsewhere. (Fig. 10-8).

Studies of fertilization have centered on the sea urchin egg, but findings apply to many other animals as well. Following fertilization, a number of important changes occur within the egg. (1) There is a rapid shift in egg membrane potential, evidently accompanied by changes in permeability. (2) A prompt and extensive reorganization of the egg cortex takes place, called the **cortical reaction.** A color change spreads quickly over the egg's surface, and a **fertilization layer** (or membrane) forms. Formation

of this layer involves fusion of membranes that surround **cortical granules** in the egg cortex with the egg's tightly adhering vitelline membrane (Fig. 10-9). The membrane then lifts slightly, and the contents of the granules are deployed in several ways: one portion fuses as a film on the undersurface of the vitelline membrane, there contributing to sperm blockage; another remains at the egg's surface, imbibes water, and forms a **hyaline layer** (chiefly mucopolysaccharide); the third contributes to the fluid in the perivitelline space between the hyaline and fertilization layers. Although elevation of the fertilization layer usually blocks additional sperm, in some eggs, particularly those with much yolk, **polyspermy** may occur. (3) A **fertilization cone,** a protru-

Figure 10-8 Sea urchin (*Arbacia*) sperm penetrating surface of egg. (*Courtesy of E. Anderson.*)

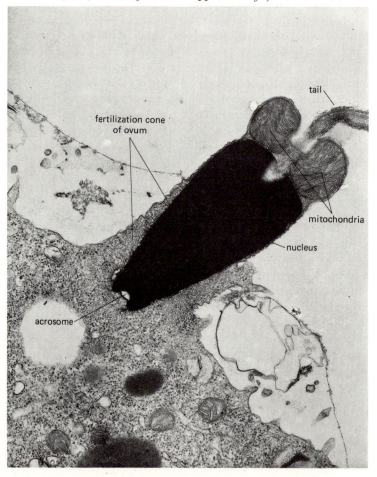

sion of egg cytoplasm, engulfs the sperm. (4) The egg is metabolically activated; its respiration rate increases. Blocks to energy-releasing systems and protein synthesis are removed; DNA synthesis is stimulated. (5) The male and female **pronuclei** are drawn together and fuse, and cleavage begins. The sperm contributes the centriole (Sec. 3-9) needed for spindle formation; the egg centriole ceases to function.

A few monoecious, or hermaphroditic, animals may be self-fertilizing; but **cross-fertilization,** the combining of gametes from two different individuals, is the general rule. A monoecious animal commonly produces its eggs and sperm at different times; if formed simultaneously, they are usually self-sterile.

In **external fertilization,** (1) the eggs and sperm are shed freely into open water (many invertebrates, some marine fishes); (2) the male and female are close together, as over a nest, when eggs and sperm are extruded (lampreys, trout); or (3) there is simultaneous extrusion of eggs and sperm by clasped pairs (frogs and toads). In **internal fertilization,** (4) the male places sperm packets, or spermatophores, on the bottom of a pond or stream or in moist places on land, and one or more are taken into the seminal receptacle of the female (salamanders), or spermatophores are deposited in the mantle cavity of the female (cephalopods); (5) by copulation sperm are transferred into the female's vagina, later to fertilize eggs in her reproductive tract (nematodes, most arthropods, some mollusks, fishes, and amphibians, all reptiles, birds, and mammals). Many water dwellers practice internal fertilization, and it is necessary for nearly all terrestrial species, because sperm can travel only in a fluid medium. In vertebrates having internal fertilization, the sperm enter the oviduct, where the eggs are usually fertilized.

10-10 Special types of sexual reproduction Development of an egg without the entrance of a sperm is known as **parthenogenesis.** It is most common in some invertebrates such as rotifers (Fig. 19-17), plant lice and thrips, some beetles, many ants, bees, and wasps, and some crustaceans; males are unknown in some species of thrips and rotifers. In wild vertebrates parthenogenesis is known in the European

rock lizards (*Lacerta*), some species of North American whiptail lizards (*Cnemidophorus*), and others. Aphids are parthenogenetic during spring and summer; then both sexes are produced. These mate, and the females lay fertilized eggs from which females hatch the next spring to continue parthenogenesis (Fig. 25-32). The queen honeybee produces fertilized eggs (using stored sperm) that develop into females, either workers or queens, but she also lays unfertilized eggs that yield haploid (n) males, or drones (Sec. 25-18).

Larvae of the gallfly (*Miastor*) produce eggs that develop parthenogenetically to yield other larvae. In the liver flukes (TREMATODA), one larval stage, the sporocyst, produces unfertilized eggs that give rise to another stage known as the redia (Sec. 18-10). Such parthenogenesis among larvae is termed *paedogenesis.*

Mature eggs of some sea urchins, frogs, and other animals that normally are fertilized may be induced to develop by **artificial parthenogenesis,** as demonstrated by Jacques Loeb in 1900. The stimuli employed included shaking, heat, dilute organic acids, and hypertonic solutions (water with higher than normal concentrations of salts). By pricking thousands of frog eggs with needles, Loeb induced development in many eggs, obtained over 200 tadpoles, and reared about half (both sexes) through or beyond metamorphosis. The frogs were diploid in chromosome number because their eggs had not undergone meiosis (see Sec. 10-5). Means for inducing artificial parthenogenesis are diverse, but all successful methods achieve the same result—activation of the egg.

Polyembryony is the production of two or more individuals from one egg by separate development from cells formed early in embryonic growth. This occurs in cases of human identical twins, also in some armadillos (four young per egg), in bryozoans, and in some parasitic HYMENOPTERA. In the last-named group, some eggs produce single embryos and others yield from 8 to about 1000.

10-11 Reproduction in general Most species have definite **reproductive seasons.** In temperate and cold regions these are usually in spring or summer, when food is abundant and other conditions for survival of the offspring are favorable. Many invertebrates do not reproduce until the environmental temperature reaches a certain minimum. Other species are influenced by the kind of food available. With some birds and mammals the increase in exposure to daylight with increasing day length acts on the gonads through the pituitary gland to induce breeding.

In some animals, favorable conditions for mating may not coincide with those best for development and production of young. Thus a delay may occur between mating and fertilization or development of the early embryo. Sperm may be stored in the reproductive tract, sometimes in special structures, as in some salamanders, in birds, and in bats. Female kangaroos mate just after giving birth and while suckling the young. The resulting embryo, in the blastocyst stage (Sec. 10-18), lies dormant and free in the uterine cavity, ready to develop further when the demands of infant care subside, there is sudden loss of the young, or an abrupt extension of reproductive effort is required. Kangaroos can have a blastocyte in dispause, an infant in the pouch, and a suckling outside the pouch.

Most animals are **oviparous;** the females release or lay eggs (Fig. 10-9) from which the young later hatch out. Many aquatic invertebrates, most insects, fishes, amphibians, and reptiles, and all birds are of this sort. Some of them are **ovoviviparous,** producing eggs with much yolk that develop within the oviducts ("uterus") of the female. Certain insects, sharks, the coelacanth (*Latimeria*), lizards, and snakes are examples. The mammals and some other animals (*Peripatus, Salpa*) are **viviparous,** producing small eggs that are retained and nourished in the female's uterus.

The **number of eggs** produced during the reproductive season by each female is inversely proportional to the average chance for survival to maturity of any one offspring. If the hazards are great, the number is large. Some parasites produce millions of eggs, the ocean sunfish (*Mola mola*) up to 28 million, the codfish up to 9 million a year, and a brook trout to over 5000; a quail averages 14, a robin 3 to 5, a deer or sheep 2 or 1, and the horse only 1. Some species produce several batches of eggs or broods of young. The **rate of development** until hatching is rather constant in birds and mammals, but with other animals it varies according to environmental temperature or other factors. The approximate time required is characteristic for each species, ranging from a few hours or days for some invertebrates to months for large mammals—9 months in humans and 20 months in elephants.

Many animals have special **breeding habits** that make for greater success in reproduction. These include courting performances that bring the sexes together for mating, nests to provide protection for eggs, and parental care of eggs and young. Eggs are carried on the body or in brood pouches by females of some crustaceans, insects, and spiders and by one

Figure 10-9 Protective coverings of some animal eggs.

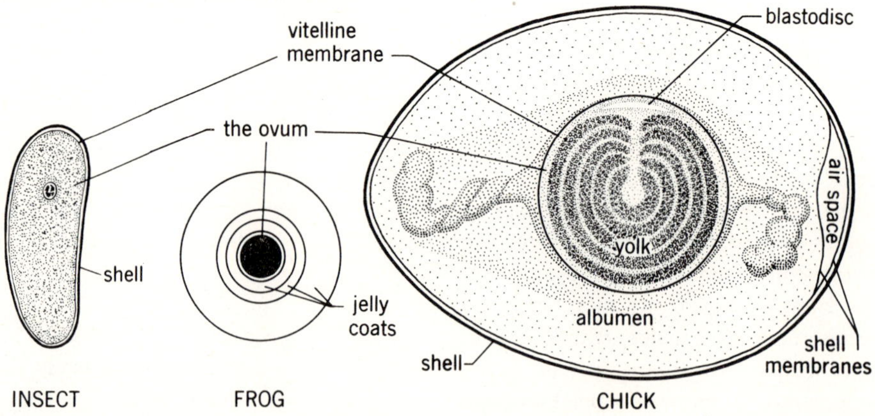

or the other sex in certain fishes and amphibians. Birds incubate their eggs, and heat from the parent's body causes the eggs to develop at a uniform rate. In mammals, development of the young within the mother's uterus achieves a similar result. Young of some ants, bees, and wasps are provided with food in the nest, the termites and social bees feed and tend the young in their colonial nests, the nestlings of birds are fed with food gathered by the parents, and young mammals are nourished by milk secreted from the mammary glands of their mothers.

10-12 Sex ratio   The numerical relation between the sexes in a species is the sex ratio. Theoretically there should be equal numbers of male- and female-producing gametes. Actually there may be a differential either in their production or in mortality among embryos or later stages. In human beings, males experience greater mortality both prenatally and postnatally. The ratio is important in livestock production, and there would be decided practical benefit if sex could be controlled to produce more females.

Development

The starting point for the production of a new individual by sexual reproduction is the fertilized egg, or **zygote.** Repeated mitotic divisions result in many cells that differentiate to form the tissues and organs of the developing individual, or **embryo.** The science that deals with the subject is **embryology.** The following account outlines the early development of a frog (Fig. 10-11) and mentions some features of development in salamanders, birds, and mammals.

10-13 Cleavage and blastula formation   Soon after an egg is fertilized, the single-celled zygote becomes two cells, the two divide into four, and so on. This process of **cleavage** partitions the egg substance into an increasing number of smaller cells, or **blastomeres,** each with an equal number of chromosomes. The cleavage divisions are unusual because they are not separated by periods of cellular growth; the zygote mass is the same at the end of cleavage as at the

start. The size distribution of the cells and the kind of division are a function of the amount and distribution of the yolk. Cleavage is termed **holoblastic** when an entire egg cell divides, as in the frog, and **meroblastic** when only part of the cell divides, as in the chick (Sec. 14-6; Fig. 10-10).

As holoblastic cleavage occurs, as in amphibians, the cells become arranged in the form of a hollow ball, or **blastula,** within which a central cavity, the **blastocoel,** appears (Fig. 10-10). Two major regions are evident, an upper **animal hemisphere,** or pole of small dark cells with little yolk, and an opposite **vegetal hemisphere** below, of larger, pale-colored cells rich in yolk granules. Between them is a **marginal zone** of medium-sized cells.

10-14 Gastrula   As cells continue to proliferate, an inrolling, or **involution,** begins, directly below the center of the gray crescent, a lightly pigmented area of the egg cortex important in guiding early stages of development (Sec. 10-15). A slit forms, and the fold above it is the **dorsal lip of the blastopore,** derived from the crescent. The cells of the animal hemisphere move inward more rapidly than the larger, fewer yolky cells of the vegetal hemisphere. The slit extends laterally and downward and eventually forms a ring of involuting tissue, the **blastopore,** which moves over the yolk-laden vegetal hemisphere and increasingly encloses it. This process of overgrowth is called **epiboly.** Once inside the embryo, the involuting cells move away from the blastopore and form the walls of an enlarging chamber, the **archenteron** (or gastrocoel), the cavity of the primitive gut. The infolding process is called **invagination.** As the archenteron enlarges, the blastocoel is gradually obliterated. Epiboly and invagination are accomplished without change in the overall mass of the embryo, indicating that growth plays no part in the process of gastrulation. When complete the gastrula consists of (1) an outer layer of **ectoderm** from cells of the animal hemisphere, (2) an inner layer of **endoderm** (entoderm) from cells of the vegetal hemisphere, and (3) between these a third layer, **mesoderm.** Since the latter at this stage includes the presumptive notochord, it may be called the **chordamesoderm.** Most of the mesoderm invaginates by

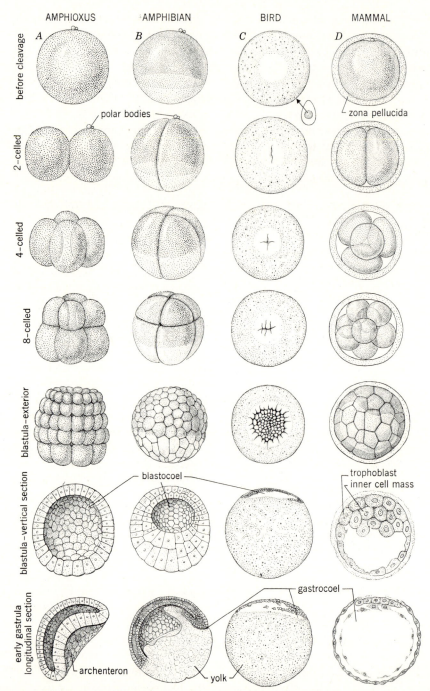

Figure 10-10 Sample stages of cleavage and gastrulation in eggs of chordates. *A.* Amphioxus, cleavage holoblastic, little yolk; egg diameter 0.1 mm. (*After Hatschek.*) *B.* Amphibian (frog), modified holoblastic cleavage, much yolk; diameter 2 mm. (*Various sources.*) *C.* Bird, meroblastic discoid cleavage in small blastodisc on large yolk mass; diameter 30 mm. (*After Blount; and Patten, Early embryology of the chick, McGraw-Hill Book Company.*) *D.* Mammal, cleavage holoblastic, practically no yolk; an outer trophoblast and an inner cell mass formed in blastula; gastrula formed by migration of endoderm cells from inner cell mass (involution); egg surrounded during early cleavage by zona pellucida, which later disappears. (*After Gregory; and Patten, Embryology of the Pig, McGraw-Hill Book Company.*)

rolling over the lateral and ventral lips of the blasto-pore. However, the portion giving rise to the noto-chord moves inward over the dorsal lip and is pre-ceded by the **prechordal plate** mesoderm of the head. These are the **germ layers** from which various tissues and organs will form. The ectoderm will produce the external covering of the body, the nervous system, and the sense organs; the endoderm provides the lining of the digestive tract, its glands, and asso-ciated structures; and the mesoderm gives rise to the supportive tissues, muscles, lining of the body cav-ity, and other parts (Table 10-2).

10-15 Body symmetry and axes The anteropos-terior axis of the embryo is typically related to the axis of polarity of the egg, determined in the ovary (Sec. 10-7). In amphibians (also many mollusks and annelids), dorsoventral and right and left are estab-lished simultaneously at fertilization. The point of

---

Table 10-2
GENERAL OUTLINE OF EMBRYONIC DIFFERENTIATION IN A VERTEBRATE

Early stages	Germ layer formation	Embryonic divisions	Tissues and organs of the adult

Diagram content:

Early stages:
Matured Sperm — Egg → Zygote (fertilized egg) → Cleavage → Blastula → Gastrula

Germ layer formation: Ectoderm, Mesoderm, Endoderm

Embryonic divisions and Tissues and organs of the adult:

Ectoderm →
- Epidermis and epidermal struc-tures (hair, glands—sweat, mammary, sebaceous, etc.), enamel of teeth, lining of mouth cavity, nose, and cloaca (part)
- Receptors
- Neural folds and plate → Nervous system
- Notochord → Notochord; later surrounded by vertebrae

Mesoderm →
- Mesenchyme → Skeleton; other supportive and connective tissues; muscles of head / Circulatory system (heart, vessels, blood) / Dermis (part)
- Epimere (somite) → Dermis (part) / Voluntary muscles
- Mesomere (intermediate) → Excretory system; some genital structures (ductus deferens)
- Hypomere (lateral plate) → Peritoneum and mesenteries / Involuntary muscles of diges-tive and reproductive tracts / Most of reproductive system
- Germ cells

Endoderm →
- Primitive gut → Epithelium of digestive tract, except mouth and anal canal / Pharynx / Middle ear / Thyroid, thymus, parathyroids / Lining of respiratory system / Liver and pancreas / Lining of bladder, urethra

sperm penetration sets the plane of bilateral symmetry and, in most cases, also the plane of the first cleavage furrow. The **gray crescent** (Sec. 10-14, Fig. 10-11) forms middorsally with its midpoint on a meridian (in relation to the anteroposterior axis of the egg) 180° from the point of fertilization. The meridian of sperm penetration thus is midventral. The plane of symmetry lies between the two. The crescent forms upon rotation of the egg cortex. Rotation is indicated by a shift of the egg's cortical pigment relative to the endoplasm. Rotation is toward the vegetal pole on the ventral side, where the sperm penetrated. The crescent is of particular importance in guiding later development. In amniotes (Fig. 10-14), the axis of symmetry is not established until the end of cleavage.

At the end of gastrulation, when all the endoderm is inside, the original egg axis has rotated about 90° (see Fig. 10-11A, G). The former lower end of the axis is then at the completed blastopore, marking the posterior end of the future animal. The chordamesoderm cells indicate the dorsal region; and shortly after gastrulation the paired neural folds on the surface, forward from the blastopore, provide an external indication of the dorsal surface.

10-16 The embryo   After gastrulation major **differentiation** of the embryo begins. From the three germ layers there are outpocketings, inpocketings, thickenings, divisions, and other changes that lead to establishment of the organs and organ systems.

Figure 10-11   Early embryology of the frog.   Long arrow indicates egg axis. *A*. Sperm at surface of egg. *B*. Entrance path of sperm (in plane of page) bisects gray crescent and determines plane of first cleavage. *C*. Late blastula. *D*. Blastopore forming, gastrulation begins. *E*. Processes in gastrulation. *F*, *G*. Gastrulation continues, rotation of egg on axis, anteroposterior and dorsoventral relations established. *H*, *I*. Gastrula completed. *J*. Beginning of organ systems. *D-H*. Longitudinal sections. *I*, *J*. Cross sections. Compare Figs. 10-7, 10-10, 31-15. (*A-C, E-G redrawn by permission from W. C. Curtis and M. C. Guthrie, Textbook of general zoology, John Wiley & Sons, Inc.; D, H-J redrawn from H. Spemann, Embryonic development and induction, Yale University Press.*)

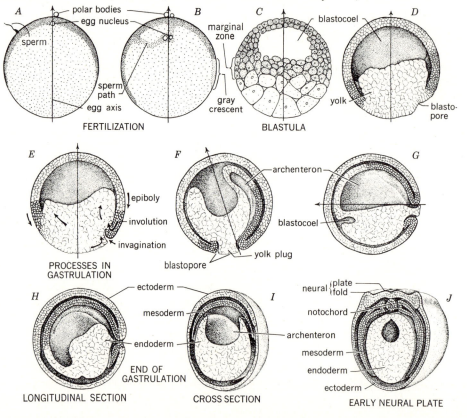

The nervous system starts dorsally as a pair of **neural folds.** The ectoderm between these sinks down and the folds come together to form a **neural tube** (canal), enlarged at the anterior end to become the **brain.** On either side, between the neural tube and ectoderm, a line of cells forms the **neural crests** that will produce the dorsal or sensory roots of spinal nerves to grow into the cord. Motor roots later grow out ventrally from the cord. The neural crests also contribute sympathetic ganglia, the Schwann cells of nerve fibers, pigment cells, and important cartilage elements of the branchial complex.

The early brain has three primary vesicles, the **fore-, mid-,** and **hindbrain.** The forebrain produces the cerebral hemispheres and diencephalon, and from the hindbrain the cerebellum and medulla oblongata are derived. A rounded **optic vesicle** grows laterally on either side of the forebrain and stimulates the ectoderm on the side of the head region to form a thickened **lens vesicle** that subsequently produces the lens of the eye. Meanwhile, the outer surface of each optic vesicle becomes concave by invagination and forms the **retina.**

The endoderm of the primitive gut becomes the inner lining of the **digestive tract.** Anteriorly, at the future pharynx, three outpocketings of the tract on either side meet three corresponding inpocketings from the side of the neck; these break through to form the **gill slits.** A single ventral outpocket, behind the pharynx, forms the **liver bud** that becomes the liver and bile duct. An inpocketing of ectoderm (*stomodeum*) forms ventrally on the head region, and a similar one (*proctodeum*) at the posterior end. In later embryonic life these break through to join the endoderm of the digestive tract, the stomodeum becoming the **mouth cavity** and the proctodeum becoming the anal canal, both lined by ectoderm. During larval life a ventral outpocket of the pharynx grows posteriorly and divides into two lobes. The anterior part gives rise to the larynx and trachea and the lobes to the lungs.

During gastrulation the mesoderm grows inward and penetrates between the ectoderm and endoderm. Cells in its middorsal part become arranged as a solid rod, the **notochord,** between the nerve tube and primitive gut, to serve as a supporting body axis. Prospective mesoderm at either side of the notochord grows down as a curved plate between the ectoderm and endoderm, and the two meet ventrally under the yolk-laden cells. The thin lower part (*hypomere*) of each plate splits into two layers. The outer is applied to the ectoderm and becomes the **parietal peritoneum,** the inner surrounds the gut (and other organs later) to make the **visceral peritoneum** (and smooth muscle of the gut), and the space between the layers is the body cavity, or **coelom.** The uppermost mesoderm (*epimere*) at either side of the nerve tube and notochord forms a lengthwise series of segmental blocks, or **somites.** Each somite differentiates into three parts: a thin outer part (*dermatome*) becomes the **dermis** of the skin, a thick inner part (*myotome*) later gives rise to the **voluntary muscles,** and nearest the notochord a scattering of cells (*sclerotome*) grow about the neural tube and notochord to form the **vertebrae** (axial skeleton), first of cartilage but later replaced by bone. Between the ventral plates and the somites a third portion (*mesomere*) is the forerunner of the **excretory system** and parts of the reproductive system.

Further details of embryonic development are too many to follow here. After some days (depending on the species of frog and the water temperature) the embryo escapes from its gelatinous egg covering to hatch as a tadpole, or **larva.** Shortly it begins to feed and grow. Development continues for some months, and then the larva **metamorphoses** into a frog. (Fig. 31-16).

10-17 Development of the chick   (Figs. 10-10C, 10-12) Eggs of birds, reptiles, and many fishes (also monotreme mammals) contain so much yolk that cleavage of the entire mass is impossible. The process begins in a small **blastodisc** area at the animal pole. By superficial (meroblastic) cleavage a plate of cells is formed that corresponds to the spherical blastula of a frog and is called the **blastoderm.** It becomes three or four cells thick. Its upper and lower layers are known, respectively, as the **epiblast** and **hypoblast.** A median **primitive streak** (Fig. 10-12) develops at the posterior end of the blastoderm and foreshadows the anteroposterior axis of the future embryo. The streak corresponds to the blastopore and its anterior end to the dorsal lip. A groove appears within it, as cells of the epiblast and perhaps also proliferating cells within the streak move downward and laterally

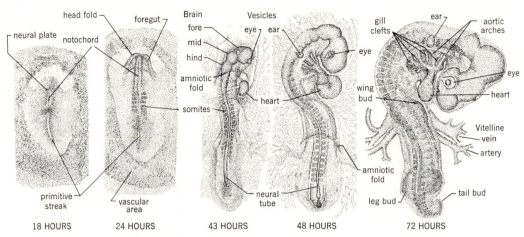

Figure 10-12 The chick embryo during the first 72 h of development (×8). Gastrulation (Fig. 10-10C) probably occurs before the egg is laid. At 21 h of incubation the first somite appears; after about 30 h, at the 9-somite stage, the heart begins beating and food material is brought by vitelline veins from the yolk surface (not shown). During the second day the anterior end turns so that the left side of the embryo lies toward the yolk; this turning is completed on the fourth day. At 48 h the amniotic fold covers the anterior two-thirds of the embryo. (*After Patten, Early embryology of the chick, McGraw-Hill Book Company.*)

to form mesoderm and endoderm. Endoderm is also constructed from cells detached from the blastoderm. It spreads as a flat sheet over the surface of the yolk mass. From the anterior part of the primitive streak, chordamesoderm moves laterally and forward between the ectoderm (derived from epiblast) and endoderm, but the posterior half of the streak yields only extraembryonic mesoderm. The presumptive notochord comes to lie on the anteroposterior axis in front of the streak. During early stages, much of the mesoderm is in the form of **primary mesenchyme** (mesoblast; Table 10-2), a loosely connected network of cells. The mesoderm eventually splits into two sheets with the future coelomic cavity between. Somites develop in the mesoderm at each side of the notochord. **Neural folds** appear in the ectoderm overlying the notochord and grow together to form the neural tube. Meanwhile, a **head fold** appears at the anterior end of the primitive streak; this contains the first evidence of the future brain and foregut. The subsequent development of the chick embryo has much in common with that of the frog as to the way in which organs become established. Gill pouches and **gill slits** appear in the first few days but soon close. The endoderm and mesoderm on all sides spread around the yolk to form a **yolk sac** (Fig. 10-14) that is enclosed by the ventral body wall just before hatching.

10-18 Development of mammals (Figs. 10-10D, 10-13) The eggs of the lowest living mammals (MONO-TREMATA) resemble those of reptiles and of birds in being of some size (1.3 to 1.8 cm; ½ to ¾ in. in platypus), with much yolk and are laid. Eggs of all the higher, or placental, mammals are minute (about 0.1 mm in diameter), virtually yolkless, and retained within the female's body for development. The entire egg divides, and a fluid-filled cavity appears within the globular mass of dividing cells to form the blastula, or **blastocyst.** The cavity is enclosed by a single layer of cells, the **trophoblast;** to one side is a clump of cells, the **inner cell mass** (Fig. 10-10). The blastocyst implants in the uterine lining, and the trophoblast, now a syncytium, breaches maternal blood vessels to begin the formation of the placenta (Fig. 10-13). During implantation, gastrulation occurs. A portion of the inner cell mass next to the blastocoel splits off to form the endoderm; the remaining cells form ectoderm. As in the chick, a primitive streak forms and mesoderm grows out from it between the other two germ layers; a notochord appears. The embryo develops anterior to the primitive streak. Its further development closely resembles that of the chick, including the formation of gill pouches and slits (Fig. 13-5) that soon close. A yolk-sac vestige, a cavity enclosed by endoderm but containing no yolk (Fig. 10-13), forms. This has been

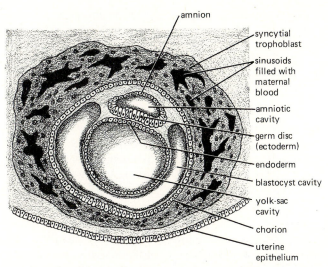

Figure 10-13   **Implanted human embryo (blastocyst), entirely enclosed within maternal tissue (×135).**   Sinusoids filled with maternal blood are present within the trophoblast syncytium. The embryo develops in the area of the germ disc.

taken to mean that the mammals are derived from a stock (the reptiles) with large-yolked eggs and in acquiring other means for developing and nourishing the embryo have not lost all features of their former mode of development.

Early development in mammals occurs much more slowly than in most lower vertebrates or invertebrates. The rat reaches the eight-cell stage in the oviduct on the third day, has blastula in the uterus on the fourth to seventh days, enters the primitive-streak stage on the eighth day, and develops somites I to VI (comparable with a 25-h chick) on the tenth day. After the embryonic membranes and placenta are formed, growth is rapid and birth occurs about 3 weeks after fertilization.

10-19 Embryonic membranes   (Fig. 10-14) Depending on the environment in which development takes place and the mode of life of the organism, special structures may be required to protect the embryo and to supply its metabolic needs during em-

Figure 10-14   **Three stages in development of the embryonic membranes of the chick.**   Note that each membrane is composed of two layers of embryonic tissue. Diagrammatic longitudinal sections; shell, shell membranes, and albumin omitted; compare Figs. 10-10, 10-12. The presence of the amnion characterizes the reptiles, birds, and mammals—the amniotes.

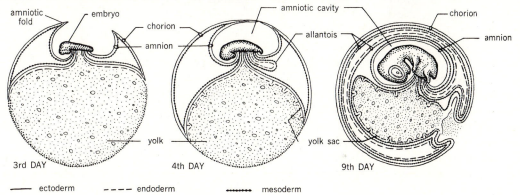

bryonic growth. In most vertebrates, for example, nutrients are enclosed in a yolk-sac membrane outside the body. Such extraembryonic life-support systems reach peak development in the vertebrate **land** (cleidoic) **egg,** a structure which has made possible development on land outside the maternal body. Its principal features are abundant yolk and a liquid interior, enclosed in a more or less impervious shell. A crucial step was the evolution of the **amnion,** which provided the embryo with a built-in natal pond. The amnion forms a closed sac about the embryo and is filled with a watery **amniotic fluid** that keeps the embryo moist and protects it against shock or adhesions. The structure is found in reptiles, birds, and mammals, the **amniotes.** Additional embryonic membranes are the **chorion, yolk sac,** and **allantois.** The chorion participates in the construction of a respiratory membrane beneath the shell and contributes to the formation of the placenta in mammals. Its origin is closely associated with that of the amnion (Fig. 10-14). The yolk sac becomes highly vascular and supplies nutrients. It is gradually incorporated into the digestive tract. The allantois grows out from the hindgut to lie against the chorion, forming the **chorioallantois.** It brings blood vessels that serve in respiration and excretion. All these membranes are ruptured and discarded at birth.

10-20 Placenta and umbilical cord Vascular processes, or villi, of the chorioallantois become embedded in the uterine surface, and the resulting joint embryonic-maternal structure is called the **placenta** (Fig. 10-15). Nutrients and oxygen then pass from maternal blood vessels through several intervening cell layers to the blood of the embryo, and carbon dioxide and excretory wastes move in the opposite direction; but there is no direct connection between maternal and embryonic circulations. The degree of intimacy between maternal and embryonic structures and the form of placenta vary in different mammals. The placenta may be **diffuse,** with scattered villi (pig and cow); **discoid,** or disklike (bats, rodents, and humans); or **zonary,** a cylindrical band (cats and other carnivores). The placenta is complex; some lining tissues of the uterus disappear, and the chorionic villi with their embryonic blood vessels are bathed in maternal blood (Fig. 35-4). Any developing embryo that has acquired characteristic mammalian form is called a **fetus.** From the ventral surface of its abdomen a soft, flexible **umbilical cord** extends, conveying two arteries and a vein that connect to embryonic capillaries in the placenta. When the fetus has completed its growth, **birth** (*parturition*) occurs. The vagina dilates, and slow rhythmic contractions of the uterus gradually force the fetus to the

Figure 10-15 *Left,* Entire reproductive tract of female cat; uterus opened to show the position of one embryo of the six present. *Right.* Cat embryo with zonary placenta cut and part of embryonic membranes removed to show relations of umbilical cord, blood vessels, and membranes.

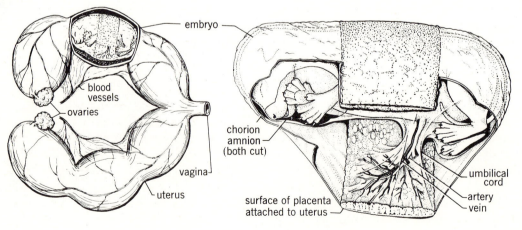

exterior. The amnion either is ruptured in this process or is quickly torn by the mother, so that the newborn young may breathe air; soon the umbilical cord is severed. The placenta either passes out with the fetus or descends later, as the afterbirth, and is usually eaten by the parent. Placentation also occurs in some viviparous fishes and reptiles. In the European skink (*Chalcides chalcides),* for example, interlocking folds develop in the oviduct between maternal tissues and the chorioallantois of the embryo. The placenta is absent in monotremes and usually also in marsupials.

10-21 Role of hormones in reproduction   Hormones (Chap. 8) regulate the sequence and timing of events in reproduction. This is best understood in vertebrates, especially mammals, where several hormones of the anterior pituitary are highly important, together with others originating in the reproductive organs. The subject is discussed in Sec. 35-13.

10-22 Ideas on development   Aristotle and the Romans recognized that mating of a male and female was necessary for the production of offspring in humans, the domestic animals, and some animals in nature, but the causal relation of copulation to birth of young is still unknown to some primitive peoples. Spermatozoa were discovered by Hamm and van Leeuwenhoek in 1677 and soon were thought to have a bearing on development. In 1824 Prevost and Dumas showed that if seminal fluid of frogs was filtered to remove the sperm before the fluid was poured over unfertilized eggs, the eggs failed to develop. Penetration of the egg by sperm and union of the sperm and egg nuclei in fertilization were first observed in sea urchin eggs by Oskar Hertwig and by Fol, independently in 1875. Meanwhile, embryologists had two contrasting theories to explain development. The **preformation theory,** supported by Malpighi, Bonnet, and other seventeenth- and eighteenth-century naturalists, assumed that either the egg or the sperm contained a "germ," completely preformed but minute and invisible, which expanded or enlarged to visible size and form during develop-

ment. Some artists even figured the "homunculus" (little man) presumed to occur in the head of a human sperm. The opposing **theory of epigenesis** assumed that the egg lacked internal organization and some outside force (*vis essentialis*) was responsible for its development. The careful studies of Wolff (1759) on the developing chick dealt a severe blow to the theory of preformation. Modern embryology partakes of both concepts. The oocyte contains preformed information in the DNA nucleotide sequence, but the embryo arises epigenetically from constituents of a largely unordered cytoplasm under the direction of the genetic material. The problem of development becomes in large part one of information retrieval from the DNA information storage bank.

10-23 Factors in differentiation   A bewildering array of complexities is bound up in the construction of a multicellular organism. Many have not yet been unraveled; others, such as the molecular control of protein synthesis, have been at least partly revealed (Sec. 2-29). Even with the theories explaining the latter it is not possible to explain how cells migrate to certain areas in the embryo, how they organize into definite tissues and organs, and how a balance is maintained so that one tissue or organ does not begin to usurp most of the space in the embryo. In other words, no tissue, organ, or group of cells of an organism can develop separately from others. Tissue differentiation and control of growth are brought about to a considerable extent through information exchange between cells via chemical messages that channel the growth of all parts of the embryo. Several processes contribute to differentiation and control. They start with the egg.

The egg is organized, both before and after fertilization. The degree of organization differs among species. Egg organization influences the cleavage pattern and may predetermine embryonic structures that will derive from different parts of the egg. Cleavage may be determinate or indeterminate or may fall into some intermediate category. In **determinate** cleavage, found in eggs of ctenophores, some mollusks, annelids, and tunicates, separation of blastomeres or cell groups during cleavage will result in each producing only that part of the embryo which it would form in an undisturbed egg. This indicates that the material in the egg is organized prior to cleavage and that the early blastomeres receive fixed informational content. Such eggs are called **mosaic.** By contrast, in **indeterminate** cleavage, found in the eggs of some jellyfishes, echinoderms, and vertebrates, the individual blastomeres may be separated, up to the four-celled stage, and each will produce a complete but proportionately smaller embryo; eggs of amphioxus will do likewise in the two-celled stage; and if the blastomeres at the two-celled stage of a salamander are carefully separated, two complete embryos may result if cleavage is bilateral. In these cases each blastomere, if parted, can produce a separate and complete individual. Each cell is **equipotent** (or totipotent) and **regulative,** capable of regulating its own development independently or producing part of an embryo along with other blastomeres. However, after three or four cleavages the fate of the blastomeres becomes fixed.

An egg cortex feature, the gray crescent, is of great importance in differentiation. Hans Spemann (1869–1941), working with newt eggs, demonstrated that in its absence normal development failed to occur. Curtis (1962) pursued the matter further. In **Xenopus,** a small patch of cortex was removed before fertilization, also after fertilization but prior to gray-crescent formation, without interfering with normal development. Once the gray crescent was formed, however, this piece of cortex was endowed with special properties and its removal halted development prior to gastrulation.

Evidence of presumptive organization in the blastula and of the time or stage in which a given feature will appear has been derived for amphibians by vital staining[1] and by transplanting small areas of the embryo. By many studies of this type, it has been possible to map out the fates of specific areas on the surface of the blastula, as indicated in Fig. 10-16.

[1] *The movement and fate of local portions of the embryo may be determined by spotting the surface with vital dyes that stain but do not injure the cells and then tracing the subsequent location of these color spots.*

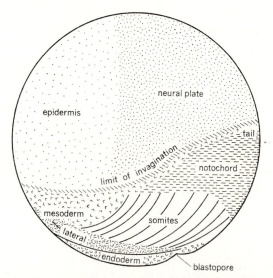

Figure 10-16 Localization of formative areas on the blastula of a salamander (*Triturus*) at the start of gastrulation. Epidermis and neural plate will derive from ectoderm; lateral plate, somites, and notochord from mesoderm. (*Modified from Vogt.*)

Between the two-celled stage (when each blastomere may produce a separate embryo) and the end of gastrulation, cells for each major region of the early embryo have become differentiated (Fig. 10-16).

Prior to gastrulation, differentiation proceeds largely under cortical influence and draws upon materials packaged in the egg (Sec. 10-7). When gastrulation begins, the process requires genetic information from nuclear genes. This is indicated by a great increase in messenger RNA and failure of artificially activated eggs to gastrulate in the absence of the nucleus.

As cell populations increase, groups of proliferating cells influence one another. Through **cooperative interaction** (*critical mass concept*) a great many cells are capable of producing what a few cannot. If, for example, a large piece of an embryo is removed and cultured separately, it will continue to develop normally and produce the tissues expected from that area. However, if the same piece is cut up into many small pieces and cultured, the more specialized tissues will not develop, though the tissue lives. There is apparently a critical mass needed to permit further development of tissue.

One tissue may induce a specific developmental response in another. A tissue that has developed early may evoke a response in nearby undifferentiated tissue. Inductive determination of the fate of tissues in embryonic development was ascribed to **organizers** by Spemann. The **inducing principle** is thought to be a protein that in some way affects the DNA-synthetic mechanism of the reacting cells.

In **synergistic inductions,** two or more developing tissues mutually interact to cause each of them to develop differently than if they were separate. In a mouse embryo, for example, the secretory and collecting tubules of the potential kidney begin development in two different tissue masses. If development is to progress normally, the two must be intimately associated. This has been demonstrated by separating the masses in culture media; the two sets of tubes then never develop. In the process of **inhibition,** one cell or tissue mass impedes the growth of another cell or tissue.

The dorsal lip of the blastopore is an especially potent **inductor** and the response of reacting tissues is quite specific. If at the beginning of gastrulation a piece of presumptive epidermis is planted in the dorsal lip, it will be carried in and will contribute to the production of muscle segments, gut wall, or other organs, depending upon where it arrives; in any case, it contributes to either mesoderm or endoderm derivatives and not to ectoderm. Presumptive lateral mesoderm transplanted into the ectoderm of a gastrula, will become epidermis (ectoderm). If, however, part of the dorsal lip is transplanted under lateral, or belly, ectoderm, it will induce the formation there of a secondary embryo on the host embryo. The notochord and other dorsal lip derivatives come from the transplant, but the nervous tissue arises from the local host ectoderm that otherwise would have become epidermis (Fig. 10-17).

An example of induction in later stages is to be seen in the development of the lens for an eye only in the presence of an optic vesicle. If the vesicle is removed, no lens develops, and if the vesicle is transplanted from the head to elsewhere in the body, a lens may develop wherever an optic vesicle influences ectoderm. Although inductors have a persistent capacity to induce, the competency of target

A

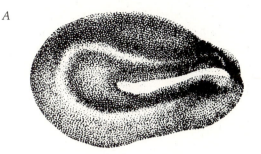

B

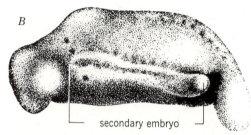

secondary embryo

Figure 10-17 Effect of an implanted organizer on early development of the salamander (*Triturus taeniatus*). *A*. Lateral view showing secondary neural plate induced by implant of a piece of blastoporal lip from another species (*T. cistatus*). White stripe is tissue from *T. cristatus* donor; the rest of the plate has been induced. Host's neural plate, dorsal in position, does not show. *B*. Later stage with secondary embryo showing ear vesicles, nerve tube, two rows of somites, and tail bud. (*After Spemann and Mangold, 1924.*)

tissues to respond waxes and wanes. The developing optic cup is directive and determinative, but the responding tissue is capable of a morphologic response only for a limited time.

Evidence of time sequence in organization has been obtained by exchanging transplants of presumptive epidermis and neural plate (both ectodermal) between embryos of two salamanders that have eggs of different color. Dark presumptive epidermis from *Triturus taeniatus* transplanted to the presumptive neural plate region of the lightcolored *T. cristatus* becomes neural plate if moved during the early gastrula stage but continues as dark epidermis, even in the brain, if transplanted in the late gastrula. The reverse transplant produced opposite results. In the early transplant the region determined the nature of the subsequent differentiation, whereas in the later one the material that was transplanted was already "determined" as to the kind of tissue it was to produce. (Fig. 10-17).

10-24 Gene expression In embryonic development somatic cells specialize to form the tissues and organ systems of the body. Such specialization is ultimately determined by genetic information contained in their nuclei. Yet all these nuclei were derived from a common source, the zygote nucleus, and, with rare exceptions (somatic cell chromosome changes), all receive identical genetic information.[1] What, then, controls cellular differentiation? It appears that differentiation is initiated by an effect of variable egg cytoplasm upon these genetically identical nuclei. If the cytoplasm induces changes in the nuclei that lead to an irreversible restriction of developmental capacity, this should be detected by nuclear transfer experiments. In such experiments a cell nucleus from an embryo at the blastula or more advanced stage is injected into a newly laid unfertilized egg which has had its nucleus removed or destroyed. The development that follows shows the extent to which the transplanted nucleus in the new environment can promote normal development and therefore indicates the range of genetic information it carries. Nuclei from the resulting embryo (all descendants of the original transplanted nucleus) can in turn be injected into enucleated eggs and their development followed. Thus a number of embryos can be obtained from a single original somatic cell nucleus, and it can be determined whether abnormalities in transplant embryos are due to nuclear changes, nonnuclear factors, or an interaction between the two.

There have been many nuclear transplant experiments in amphibians, using *Rana* and *Xenopus*. In *Rana,* if early-stage nuclei from the blastula or gastrula are employed, a high percentage of transplants develop into normal tadpoles, so no irreversible nuclear changes appear to have occurred. As increasingly later stages are used, however, resulting transplants become progressively more abnormal. Similar

results have been obtained with *Xenopus,* indicating that as cells differentiate their nuclei are less able to support normal development after transplantation. However, even transplants of nuclei of skin cells of adult frogs have resulted in larvae that are nearly normal. It is of interest that in some cases a nucleus improves in its capacity to direct activity in a new cytoplasmic environment after repeated transplantation in that environment.

These studies indicate that nuclei of differentiating somatic cells do not generally undergo loss or impairment of their genetic content. Rather, it appears that the specialized information provided by somatic cell nuclei results from inactivation of parts of their genetic material and the influence of the cellular environment in which they are found at a given time.

References

Allen, R. D. 1959. The moment of fertilization. *Sci. Am.,* vol. 201, no. 1, pp. 124–134. *What happens in the egg following fertilization that prevents entry of other sperm.*

Austin, C. R., and **R. V. Short** (editors). 1972. Embryonic and fetal development. New York, Cambridge University Press. viii + 158 pp. *Mammalian development.*

Balinsky, B. I. 1970. An introduction to embryology, 3d ed. Philadelphia, W. B. Saunders Company. xviii + 725 pp., 457 figs.

Beermann, W., et al. 1966. Cell differentiation and morphogenesis. Amsterdam, North-Holland Publishing Company; New York, John Wiley & Sons, Inc. viii + 209 pp. *Information on both plant and animal differentiation. A number of contributors.*

Bell, Eugene (editor). 1967. Molecular and cellular aspects of development. New York, Harper & Row, Publishers, Incorporated. xi + 525 pp.

Berrill, N. J., and **Gerald Karp.** 1976. Development. McGraw-Hill Book Company. viii + 566 pp. *Embryology from the point of view of developmental biology.*

Bonner, James. 1965. The molecular biology of development. New York, Oxford University Press. 185 pp., 18 figs.

Davidson, E. H. 1968. Gene activity in early development. New York, Academic Press, Inc. xi + 375 pp.

DeHaan, R. L., and **H. Ursprung** (editors). 1965. Organ-

---

[1] *In* Ascaris *and certain insects there is actual loss of chromosome material in somatic cell lines.*

ogenesis. New York, Holt, Rinehart and Winston, Inc. 816 pp., illus.

de Terra, Noël. 1974. Cortical control of cell division. *Sci.*, vol. 184, no. 4136, pp. 530–539.

Ebert, J. D. 1965. Interacting systems in development. New York, Holt, Rinehart, and Winston, Inc. viii + 227 pp.

Edwards, R. G., and **R. E. Fowler.** 1970. Human embryos in the laboratory. *Sci. Am.*, vol. 223, no. 6, pp. 44–54. *The culture of human and other mammalian eggs and embryos.*

Epel, D. 1977. The program of fertilization. *Sci. Am.*, vol. 237, no. 5, pp. 128–138. *The fusion of egg and sperm triggers transient changes in the concentration of ions that prevents fusion of additional sperm and initiates development of the embryo.*

Fischberg, M., and **A. W. Blackler.** 1961. How cells specialize. *Sci. Am.*, vol. 205, no. 3, pp. 124–140. *Early steps in differentiation of the embryo are programmed into the egg before fertilization.*

Gray, G. W. 1957. The organizer, *Sci. Am.*, vol. 197, no. 5, pp. 79–88. *History of embryologic studies on the organizing qualities of the blastopore lip of amphibian embryos.*

Gurdon, J. B. 1968. Transplanted nuclei and cell differentiation. *Sci. Am.*, vol. 219, no. 6, pp. 24–35. *Information on how genes are controlled during embryonic development.*

Hamilton, W. J., J. D. Boyd, and **H. W. Mossman.** 1962. Human embryology, 3d ed. Baltimore, The Williams & Wilkins Company. viii + 493 pp., 510 figs. *Includes excellent chapter on comparative vertebrate development.*

Holliday, R., and **J. E. Pugh.** 1975. DNA modification mechanisms and gene activity during development. *Sci.*, vol. 187, no. 4173, pp. 226–232.

Lash, Jay, and **J. R. Whittaker** (editors). 1974. Concepts of development. Stamford, Conn. Sinauer Associates, Inc. x + 469 pp. *Animal development for more advanced students. Many contributors.*

Lillie, F. R. 1952. The development of the chick. 3d ed.

Rev. by H. L. Hamilton. New York, Holt, Rinehart, and Winston, Inc. xi + 624 pp., illus. *The American classic.*

Moscona, A. A. 1961. How cells associate. *Sci. Am.*, vol. 205, no. 3, pp. 142–162. *Describes the forces that promote cellular aggregation and binding.*

Patten, B. M., and **B. M. Carlson.** 1974. Foundations of embryology. 3d ed. New York, McGraw-Hill Book Company. xix + 650 pp.

Rugh, Roberts. 1951. The frog: Its reproduction and development. New York, McGraw-Hill Book Company. x + 336 pp., illus.

————. 1962. Experimental embryology: Techniques and procedures. 3d ed. Minneapolis, Burgess Publishing Company. ix + 501 pp., illus.

Saunders, J. W. 1970. Patterns and principles of animal development. New York, The Macmillan Company. xiii + 282 pp., illus.

Spemann, Hans. 1938. Embryonic development and induction. New Haven, Conn., Yale University Press. xii + 401 pp., 192 figs. *Experiments on amphibian embryos; a classic work.*

Torrey, T. W. 1962. Morphogenesis of the vertebrates. New York, John Wiley & Sons, Inc. viii + 600 pp. illus. *Comparative anatomy and embryology integrated.*

Waddington, C. H. 1966. Principles of development and differentiation. New York, The Macmillan Company Paper. x + 115 pp.

Wessells, N. K. 1971. How living cells change shape. *Sci. Am.*, vol. 225, no. 4, pp. 76–82. *How microtubules and microfilaments make cell movement possible.*

Whittaker, J. R. 1968. Cellular differentiation. Belmont, Calif., Dickenson Publishing Company., Inc. 111 pp.

Willier, B. H., and **J. M. Oppenheimer.** 1974. 2d ed. Foundations of experimental embryology. Hafner Publishing Company, Inc. xxiv + 277 pp. *Classic articles by distinguished developmental biologists.*

Witschi, E. 1956. Development of vertebrates. Philadelphia, W. B. Saunders Company. xvi + 588 pp., 370 figs.

11
HEREDITY AND GENETICS

"Like tends to beget like," in animals and plants. Yet offspring usually differ among themselves and from their parents in varying degree. The ancient Greeks knew that blue-eyed parents have blue-eyed children, that baldness and cross-eyes follow in successive generations, and that certain eye defects run in particular families. This passage of characters from one generation to another is called **inheritance,** or **heredity.**

Heredity deals with characteristics based on descent, the occurrence in living organisms of qualities, either expressed or hidden, that are derived from their ancestors. It involves physical and physiologic characteristics, instincts, and even psychologic features in higher animals and humans. Differences among individuals of a species are called **variations.** These are of two types: **environmentally induced variations** due mainly to food, temperature, or other external factors; and **hereditary variations** that appear in certain offspring with little or no reference to the environment. The science of **genetics** seeks to account for the resemblances and differences due to heredity, their source, and their development.

Genetics has helped the study of evolution, embryology, and other sciences. Much knowledge of human heredity has been obtained, parts of which have practical application. It has a cultural value in dispelling many faulty beliefs regarding inheritance. Genetics aids agriculture by improving the form, yield, resistance to disease, and other features of domesticated animals and cultivated plants.

Little of value was learned about heredity until the eighteenth century, when knowledge developed about sexuality in plants and plant hybridization. Kölreuter (1733–1806) and others produced various fertile hybrids by artificial pollination and described the characters of both parent and hybrid plants, but they had no clear understanding of the hereditary process.

11-1 Mendel   The person who first made decisive experiments in heredity and who formulated the basic laws of genetics was Gregor Johann Mendel (1822–1884), a monk in the Augustinian monastery at Brünn, Austria. He saw clearly that all earlier hybridizers had failed to discover any general laws of inheritance because they had neither traced individual characters through successive generations nor kept complete numerical records of their results. Mendel therefore planned careful experiments to overcome these difficulties. He chose the garden pea (*Pisum*) for study and spent two years in selecting races with distinctive and contrasting characters and in making certain that each original stock was pure. During the next six years he made many crosses by artificial pollination and carried each through three or more generations. Mendel kept count of all plants and seeds of each kind produced, analyzed the results, and from them deduced the two most important fundamental laws of heredity. His report of 1866 in an obscure periodical was brought to worldwide attention only in 1900 by three other investigators, De Vries, Correns, and von Tschermak, who had independently reached similar conclusions.

Discovery of Mendel's findings gave great impetus to the study of heredity, and an enormous amount of careful work has since been done by many geneticists and molecular biologists working with various microorganisms and the higher plants and animals. A major advance came with the use of the fruit fly. *Drosophila melanogaster*, by T. H. Morgan and others. This small insect (Fig. 11-1) is reared easily in bottles provided with food, or culture medium, for the larvae; a pair will produce 200 or more offspring, with successive generations every 10 to 14 days. There followed the impressive discoveries of the physical chemists, in particular those of Wilkins, Watson, and Crick, which provided a molecular basis for an understanding of genetics. Attention has

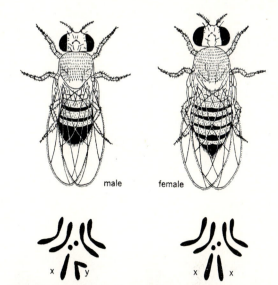

Figure 11-1 The fruit fly, *Drosophila melanogaster,* **used for studies in genetics.** Much enlarged. Diploid sets of chromosomes shown, with sex chromosomes labeled (XY, XX). Male fly smaller, with abdomen marked by three black bands, the last extending beneath the rounded posterior end. Female larger, abdomen swollen but pointed, with five black bands not joined ventrally.

turned increasingly to microorganisms—molds and bacteria (and bacterial viruses)—as sources of genetic information. They reproduce rapidly and can be grown easily in the laboratory under controlled conditions. The bacterium *Escherichia coli* is especially popular. Such abundant material has made it possible to investigate problems requiring experimentation with many individuals and generations and to apply mathematical analysis to the results.

11-2 Monohybrid cross   This is a cross in which the parents differ in one pair of alternative characters. When a true-breeding black guinea pig and a white individual are mated (Fig. 11-2), all the (hybrid) individuals of the next (F_1)[1] generation are black, no matter which parent is black and which is white. These black hybrids, when crossed among themselves, yield in the following (F_2) generation a population of which, *on the average,* three-fourths

---

[1] *The parental generation is designated as P_1 the first generation of offspring as the F_1 (first filial), the second as the F_2, and so on.*

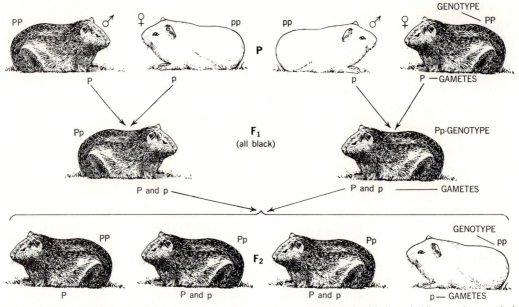

Figure 11-2 Monohybrid cross. Inheritance of coat color in guinea pigs when pure-breeding black and white parents are mated.

are black, like the black grandparent, and one-fourth are white, like the white grandparent. The character of white coat disappears in the F_1 and reappears unchanged in the F_2. If the F_2 white animals are then crossed among themselves, only white individuals result (F_3), whereas among the black F_2 animals some (one-third) produce only blacks and the others (two-thirds) produce both blacks and whites, as in the F_2 generation. Thus, as Mendel said, when two contrasting characters are brought together in a cross, one is usually **dominant** (expressed, or evident) in the next (F_1) generation, and the other is **recessive** (latent, or receding from view). In the following (F_2) generation these two characters are **segregated,** in an average 3:1 ratio. We may, therefore, state in modern terms **Mendel's first law:** *The factors for a pair of characters are segregated.* In the example (Fig. 11-2), all the gametes of the pure-breeding black males carry the factor *P* (black) and those of the white females the factor *p* (white). These factors reappear unchanged in the gametes of the F_1 hybrids, uninfluenced by their association in the hybrids. Each gamete is *pure,* containing only one or the other of the two factors. The factors responsible for a pair of alternative, or contrasting, characters, such as Mendel studied, are termed **alleles.**

11-3 Mechanism of heredity Mendel was the first to differentiate between the actual visible character and that "something" which caused its production. Obviously the character cannot be present in the germ cells that join in fertilization to produce a new individual, but something representing it and responsible for its production is there. This **factor** for the character is the **gene** (Sec. 11-18). A gene is the unit of inheritance that is transmitted in a gamete and controls the development of a character, by interaction with the other genes, the cytoplasm, and the environment.

Since the gametes are the only cells that participate in the formation of new individuals of the next generation, the mechanism of heredity must be sought in them. It will be recalled (Chap. 10) that (1) in meiosis each cell destined to become a gamete receives, by random sorting, either one or the other member of each pair of homologous chromosomes; (2) at fertilization there is a random meeting of an egg and a sperm; (3) the chromosomes of the egg nucleus and sperm nucleus join in the zygote; and (4) during embryonic growth and later, each cell in the new individual receives by mitosis an equal and like number of chromosomes derived from the zygote. As inheritance in sexual reproduction involves the

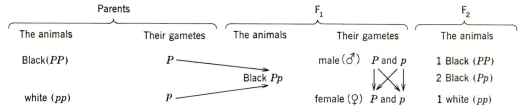

Figure 11-3 Inheritance in a monohybrid cross.

transmission of characteristics from both parents, the chromosomes must be the means by which this occurs. Certain experiments have shown that (1) an egg deprived of its nucleus and then fertilized by a sperm produces an individual with only paternal characteristics; (2) mature eggs caused to develop by artificial parthenogenesis result in animals with only maternal characteristics; and (3) some cytoplasm may be removed from certain eggs, and yet if the egg nucleus is fertilized by a sperm, the resulting individual contains characteristics of both parents. Thus hereditary transmission of characteristics is dependent on the nucleus rather than the cytoplasm in the vast majority of cases.

The evidence is now clear from studies in molecular biology that the chromosomes are indeed the chief structures that bear the hereditary material and that most genes are located on the chromosomes.

11-4 *Explanation of the Mendelian ratio* Hybrid individuals result from combining two gametes of different heredity, and a process of segregation must take place in the germ cells of such hybrids to account for the sorting out of genes to produce characters in the F_2 generation. Mendel realized this but knew nothing of the actual mechanism. The reduction divisions during maturation and the random union of egg and sperm, both discovered since his time, explain this segregation.

If we represent the dominant gene for black (or pigmented) coat in guinea pigs by P and the recessive gene for white by p, then the genetic formulas for the parents (which contain the diploid number of chromosomes) are PP and pp and for their respective (haploid) gametes are P and p. When the gametes of the parents join in fertilization, all the resulting offspring of the F_1 generation will be Pp and black. When the F_1 forms gametes, each sex will produce both P and p gametes in a 1:1 ratio. The possible combinations for the F_2 will be PP (black), Pp and pP (black), and pp (white), giving an average ratio of 3 blacks to 1 white individual. The events in this cross are outlined in Fig. 11-3.

Individual animals such as the parents that are "pure," containing like genes for any one character (PP or pp), are termed **homozygous.** The F_1 hybrid, which appears black and yet contains a gene for white (Pp), is said to be **heterozygous;** it contains two kinds of genes of an allelic pair. The entire genetic constitution of an individual, both expressed and latent, makes up its **genotype,** and its appearance, or the assemblage of characters that are expressed, or evident, constitutes its **phenotype.** Both the parents and two of the F_2 (PP and pp) are homozygous, but the F_1 hybrid and two of the F_2 (Pp) are heterozygous; the F_1 hybrid has a black phenotype, and its genotype is Pp. In the F_2, the whites are homozygous recessives; of the blacks, one-third are homozygous dominants yielding only blacks, and two-thirds are heterozygous. The analysis may be restated (Fig. 11-4) with phenotypes, genotypes, and

Figure 11-4 Genetics of a monohybrid cross.

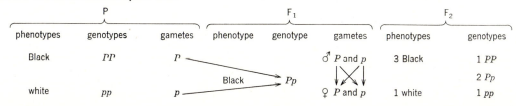

gametes bracketed for the parental, first filial, and second filial generations respectively.

Many other characters have been found to follow Mendel's first law.

11-5 Law of probability   Random sorting of chromosomes and the meeting of eggs and sperm are matters of chance (Sec. 11-3), so that in any cross the actual numbers of individuals showing the contrasting characteristics rarely correspond to the theoretical ratio expected. Thus in one monohybrid mating Mendel obtained in the F_2 generation 7324 pea seeds —5474 round (dominant) and 1850 wrinkled (recessive), a ratio of 2.96:1 rather than the expected 3:1. In another experiment (8023 seeds) the ratio was 3.01:1.

Differences between the actual results and the theoretical ratio in any given case are explainable by the **law of probability,** which deals with the chance that any one event will happen rather than its alternative. **Probability** is defined as the fraction or ratio of instances of one event divided by the total of all possible events. This may be illustrated by tossing a coin. In any single toss the chance, or probability, that either a head or a tail will come up is even ($\frac{1}{2}:\frac{1}{2}$). Sometimes a head will appear two, three, or more times in sequence, but at each toss the probability

that either a head or a tail will come up remains the same ($\frac{1}{2}:\frac{1}{2}$); however, the chances of deviating from the expected ratio decrease with the accumulation of successive tosses. Eventually, with many trials, tails will appear as frequently as heads, realizing the expected ratio. If 10 coins are tossed together, the most frequent result will be 5 heads and 5 tails. The combinations of 4 heads and 6 tails, 4 tails and 6 heads, 3 heads and 7 tails, etc., occur less often, and either 10 heads or 10 tails is rare. In 100 actual trials with 10 coins at once, the most frequent number of heads (and tails) was 5, as expected (Fig. 11-5). Lower and higher numbers decreased successively to 0 and 10.[1] The resulting graph is not entirely symmetrical but would be smoothed out if the number of trials were increased.

The "normal" curve of distribution represents the results expected from an infinite number of trials. With this curve as the theoretical basis one may predict, by appropriate statistical calculations, the probability (but never the certainty) of occurrence of a particular event or character.

11-6 Codominance   With some characters the F_1 hybrid is intermediate between parents. Such incomplete dominance occurs in the Andalusian fowl, where the mating of a black and a splashed-white produces blue offspring in the F_1, and the F_2 results are in the ratio of 1 black, 2 blue, 1 splashed-white

Figure 11-5   Results of tossing 10 coins 100 times (.) and the resulting graph (broken line), together with the normal curve of distribution (solid line).

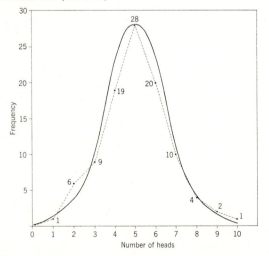

---

[1] *Human inheritance of sex generally follows a 50:50 ratio as in the tossing of coins, yet occasionally large families have all sons or all daughters, representing the extremes of the normal curve. The height of human beings, however, follows a different pattern, being influenced by several genetic factors rather than a single pair as in a Mendelian monohybrid cross. It varies continuously rather than falling into separate categories such as tall, medium, and short. If the heights of 1000 adult white male Americans chosen at random are plotted, the resulting graph conforms closely to a normal curve, ranging from about 140 to 215 cm (55 to 85 in), with the greatest frequency near 173 cm (68 in). A nonrandom sample including also adult females, children, or members of other races might be skewed to the left.*

Standard textbooks in biologic statistics provide formulas for testing the "goodness of fit" of actual counts or measurements to theoretical ratios, for the size of sample necessary to predict that a given event will occur in 99 of 100 cases, and other ways for determining the mathematical reliability of experimental results with biologic materials.

Figure 11-6 Codominance of a Mendelian character in the Andalusian fowl. The "blue" fowls are closely stippled. (*Modified from Hesse-Doflein.*)

(Figs. 11-6, 11-7). The black and splashed-white individuals are homozygous; the blue are heterozygous, as diagrammed in Fig. 11-7. Mendel's first law obviously applies here as well as in cases of complete dominance, but the heterozygous individuals, being blue, are easily recognized.

11-7 Testcross Offspring that show the dominant character in a cross are alike phenotypically but may be either heterozygous or homozygous for that character. To determine their genotype the **testcross** is used, mating the dominant phenotype with a pure recessive individual. If, with guinea pigs, the black individual under test is homozygous (PP), all the offspring from a testcross with the recessive (pp) will

be black (Pp); if, however, it is heterozygous (Pp), the offspring will be about equally black (Pp) and white (pp) (Fig. 11-8). If the testcross is performed with a pure recessive grandparent, it is called a **backcross**. In practical genetics the testcross is used as a rapid means for "purifying" (rendering homozygous) desirable stocks.

11-8 Dihybrid cross When parents differ in two pairs of characters, the F_1 offspring are termed **dihybrids;** in such crosses Mendel found that each pair of characters is inherited independently of the other. This may be illustrated (Fig. 11-9) in the case of the guinea pig, where black, or pigmented, coat (P) is dominant to white (p) and rough, or rosetted, coat

Figure 11-7 Pattern of inheritance in codominance.

P			F_1			F_2	
phenotypes	genotypes	gametes	phenotypes	genotype	gametes	phenotypes	genotypes
Black	BB	B			♂ B and b	1 Black	BB
			Blue	Bb		2 Blue	Bb
splashed white	bb	b			♀ B and b	1 splashed white	bb

Testcross

phenotypes	genotypes	gametes

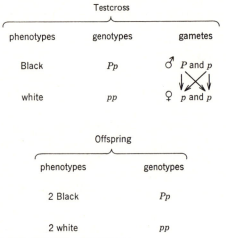

Black Pp ♂ P and p

white pp ♀ p and p

Offspring

phenotypes	genotypes
2 Black	Pp
2 white	pp

Figure 11-8 Genetics of a testcross.

(R) is dominant to smooth (r). The mating of a black rough animal to a smooth white one is outlined in Fig. 11-10.

Four kinds of gametes are formed by each sex; the Punnett checkerboard[1] (Table 11-1) will show the 16 possible matings producing the F_2 generation. The F_2

[1] *Named for its originator, R. C. Punnett.*

phenotypes include 9 black rough (any with PR traits) like the dominant parent and 1 white smooth (pr) like the recessive parent. Two new combinations have appeared, in which there are 3 black smooth (Pr) and 3 white rough (pR). The 9:3:3:1 ratio is characteristic of a dihybrid cross (Table 11-2). There are $9 + 3 = 12$ black to $3 + 1 = 4$ white offspring, and $9 + 3 = 12$ rough to $3 + 1 = 4$ smooth individuals. The ratio of 3 dominant to 1 recessive in each case follows Mendel's first law, as in a monohybrid cross. Each pair of genes has been transmitted independently of the other. The chances for a guinea pig to be black or white are independent of its chances to be rough or smooth. The same 3:1 ratio is expected for each pair of characters acting separately. This illustrates **Mendel's second law:** *When races differ from each other in two (or more) pairs of factors, the inheritance of one pair of factors is independent of that of the other(s).*

The four phenotypes involve nine different genotypes, as shown in the analysis of the F_2 population; $PPRR$, $PPrr$, $ppRR$, and $pprr$ are homozygous, and the other five are heterozygous. The appearance of new phenotypes and genotypes in dihybrid (and multihybrid) crosses is a practical means to obtain

Figure 11-9 Dihybrid cross, with guinea pigs differing in two independent pairs of Mendelian characters of the hair—color and arrangement.

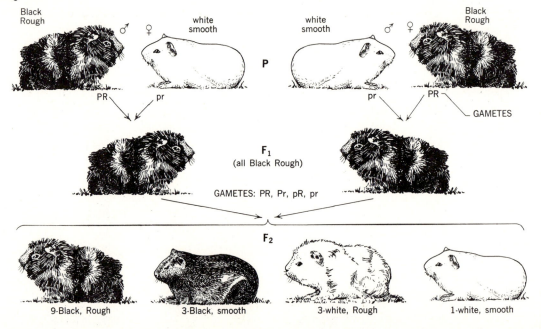

Black Rough ♂   ♀ white smooth   white smooth ♂   ♀ Black Rough

P

PR   pr   pr   PR   GAMETES

F_1 (all Black Rough)

GAMETES: PR, Pr, pR, pr

F_2

9-Black, Rough   3-Black, smooth   3-white, Rough   1-white, smooth

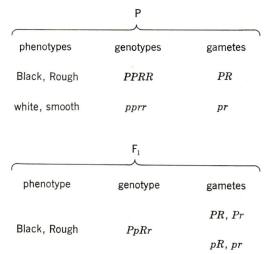

P		
phenotypes	genotypes	gametes
Black, Rough	*PPRR*	*PR*
white, smooth	*pprr*	*pr*

F₁		
phenotype	genotype	gametes
		PR, Pr
Black, Rough	*PpRr*	
		pR, pr

Figure 11-10 Matings in a dihybrid cross.

Table 11-2
ANALYSIS OF THE F₂ OFFSPRING FROM A DIHYBRID CROSS

Phenotypes			
9 Black Rough	3 Black smooth	3 white Rough	1 white smooth

Genotypes			
1-PPRR 2-PPRr 2-PpRR 4-PpRr	1-PPrr 2-Pprr	1-ppRR 2-ppRr	1-pprr

strains of animals or plants with combinations of characters different from those of the parents. In this cross, there are two new types of homozygous individuals—black smooth (*PPrr*) and white rough (*ppRR*).

11-9 Special types of inheritance   Besides the simple types of Mendelian inheritance already described, many others of more complex nature have been discovered that involve the interaction of two or more factors.

For example, the form of a comb differs in several

breeds of domestic fowls, but each remains true within its breed (Fig. 11-11). The Wyandotte has a low, regular **rose** comb with papillae, the Brahma a narrow, higher, three-ridged **pea** comb, and the Leghorn and others an upright blade, or **single** comb. When fowls with rose (or pea) and single comb are crossed, the former is dominant, and the F₂ averages 3 rose (or pea) to 1 single. If rose is crossed with pea, however, the F₁ hybrid bears a **walnut** comb (resembling half a walnut meat) and the F₂ gives 9 walnut, 3 rose, 3 pea, and 1 single (Fig. 11-12). The results differ from ordinary dihybrids in that the F₁ resembles neither parent, and two other types appear in

Figure 11-11 Comb characters of male fowls. (*Adapted from Punnett, Mendelism, The Macmillan Company.*)

Table 11-1
POSSIBLE MATINGS FOR THE F₂ GENERATION IN A DIHYBRID CROSS

		Male gametes			
		PR	Pr	pR	pr
Female gametes	PR	PPRR Black Rough	PPRr Black Rough	PpRR Black Rough	PpRr Black Rough
	Pr	PPRr Black Rough	PPrr Black smooth	PpRr Black Rough	Pprr Black smooth
	pR	PpRR Black Rough	PpRr Black Rough	ppRR white Rough	ppRr white Rough
	pr	PpRr Black Rough	Pprr Black smooth	ppRr white Rough	pprr white smooth

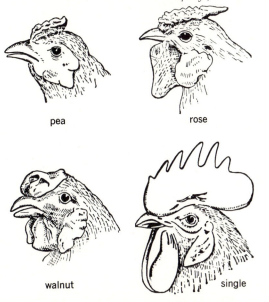

pea

rose

walnut

single

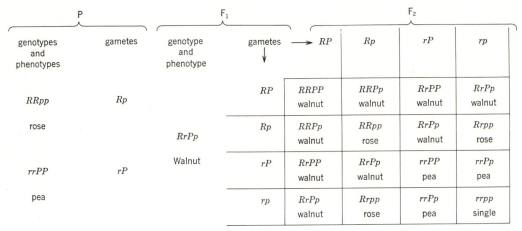

Figure 11-12 Inheritance of comb characters in fowls.

the F_2. This is a case of **interaction of factors** in a cross involving two pairs of factors that affect one structure, the comb. If the dominant gene for rose is represented by R and its allele by r and those for pea by P and p, respectively, the results are as follows: single comb is a double recessive ($rrpp$), rose comb contains either one or two R genes but only recessive p genes, pea comb has either one or two P genes but only recessive r genes, and walnut appears whenever at least one each of R and P are present.

Another modified two-factor ratio occurs in the color of Duroc-Jersey swine, where both factors are necessary for full color (red), either one alone yields partial color (sandy), and the absence of both yields colorless (white) animals.

11-10 Additive factors Some characteristics are the result of two or more genes acting in an **additive** manner; skin color in humans is a good example. There probably are three or four pairs of genes involved, but for simplicity it can be assumed that black pigmentation is determined by only two, B and B'. Then a person with the darkest possible color would have the genotype $BB\ B'B'$, and one with the palest skin would be $bb\ b'b'$. A mating between such individuals would result in offspring with an intermediate brown color, $Bb\ B'b'$ (Fig. 11-13). If two people with the latter genotype mate, their offspring would average the number of alleles for dark pigmentation indicated in the checkerboard. Of 16 children, the expectation would be 1 with black skin

Figure 11-13 Inheritance of human skin color. (Figures in checkerboard indicate the number of genes for dark color in each individual.)

P genotypes and phenotypes	gametes	F₁ genotype and phenotype	gametes	F₂ BB'	Bb'	bB'	bb'
$bbb'b'$ WHITE	bb'		BB'	4	3	3	2
		$BbB'b'$ BROWN	Bb'	3	2	2	1
			bB'	3	2	2	1
$BBB'B'$ BLACK	BB'		bb'	2	1	1	0

(4 dominant alleles, *BB B'B'*), 4 with dark brown skin (3 dominants), 6 with intermediate brown skin (2 dominants), 4 with light brown skin (1 dominant), and 1 with very light skin (no dominant allele, *bb b'b'*).

Descendants of the Africans earlier brought to America interbred so thoroughly with European (and other pale-skinned) groups that the contemporary black population has all possible shades of color. Fairly light-skinned parents, therefore, may have children with quite dark skins, or vice versa.

11-11 Lethal factors Various species of plants and animals carry **lethal factors** which, when homozygous, stop development at some stage, and the individual dies. Their presence is usually detected by an abnormal ratio in the offspring. A conspicuous case is that of the yellow race of the house mouse, *Mus musculus,* which never breeds true. If a yellow mouse is mated to some nonyellow, half the young are yellow and half are nonyellow, a ratio to be expected from mating a heterozygous animal (yellow) with a homozygous recessive (any nonyellow such as agouti). If two yellows are mated, the young average 2 yellow to 1 nonyellow, whereas the expected ratio among the young would be 1 homozygous yellow to 2 heterozygous yellow to 1 nonyellow, but the homozygous yellow dies as an embryo. The "creeper fowl" with short, crooked legs behaves genetically like the yellow mouse. Other lethals are known in *Drosophila,* cattle, sheep, hogs, and horses, and some human defects are due to such factors.

11-12 Miscellaneous effects All the examples discussed previously involve pairs of alternative factors. In many other cases more than two alternative factors affect the same character; these are called **multiple alleles.** Thus, in the domestic rabbit, among various color forms, there are the normal wild type, the complete albino with solidly white coat and pink eyes, and the Himalayan albino with pink eyes and a white coat except for black or dark brown on the ears, nose, and feet. The albino is a simple recessive to the wild type, and the Himalayan albino is likewise recessive to the wild type. When Himalayan and pure albino are crossed, however, all the F_1 off-

spring are Himalayan, and the F_2 yields 3 Himalayan to 1 pure albino; there is no reversion to the wild type. Obviously, Himalayan and albino are alleles of one another, and both are alleles of the wild color. Other instances of multiple alleles are known for coat color in mice, guinea pigs, and rats. In *Drosophila* at least 14 alleles for eye color have been found, from white and ivory through buff and apricot to the wild-type red. Several series are known in plants, especially snapdragons and maize. No more than two alleles of a series can occur in any particular individual.

Still more complex types of inheritance include modifying factors, multiple effects of a single gene, and those in which either the external environment or conditions within the animal change the manner in which a gene regulates the development of a character. An instance of the last kind occurs with dwarfism in the house mouse, where a recessive gene halts growth at about 2 weeks of age and the animals are all sterile; the anterior lobe of the pituitary gland and the cortex of the adrenal are deficient. Continued implantation of anterior pituitary in such young causes resumption of growth; the animals may reach normal size, and males may become fertile. Thus hormones may exercise an influence on the manner in which genes act in the formation of characters.

An example of the effect of external conditions on expression of a character is seen with fur color in the Himalayan rabbit. This species is typically white with dark extremities (nose, ears, feet). Its coloration is due to the Himalayan gene, the expression of which is temperature-dependent. The typical color pattern occurs because the extremities are normally cooler than the rest of the body during fur growth. If, however, the patches of black fur are removed and the extremities are kept warm, white fur grows back. If the patches of white fur are then removed and the skin is chilled, black fur appears. A similar phenomenon occurs in the Siamese cat. Thus modifying genes and environmental factors can influence the extent to which a gene is expressed in the phenotype.

Inheritance usually follows the classic Mendelian genetic pattern, but a few cases are known in which self-duplicating cytoplasmic particles are transmitted as hereditary units. Sensitivity to carbon dioxide seems to be such a case in *Drosophila*. If females from a CO_2-sensitive strain are crossed with wild-type males, the offspring are CO_2-sensitive. In the reciprocal cross with wild-type females and CO_2-sensitive males, the F_1 flies are not sensitive to CO_2. From extensive trials it is clear that CO_2 sensitivity is transmitted in the cytoplasm of the egg, independent of the chromosomes.

11-13 Sex and heredity No factor mentioned previously has any relation to sex; either the male or the female may carry one factor and the other parent its alternative. The situation is different for some other characters, including sex itself. Somatic cells and spermatogonia in male animals contain a pair of homologous chromosomes in which one member may be smaller than the other and sometimes of different shape. These are the **sex chromosomes,** the larger often being the **X chromosome** and the other the **Y chromosome;** in some species the latter is lacking. In *Drosophila* the Y chromosome, counting its long and short arms together, is the longer of the two (Fig. 11-1). A pair of X chromosomes is present in females. Thus a male may be designated as XY (or XO) and a female as XX. All other parts of strictly homologous chromosomes are termed **autosomes.**

Human cells, except gametes, contain 46 chromosomes—the 2 sex chromosomes and 22 pairs of homologous autosomes (Fig. 3-9). During maturation the sex chromosomes segregate freely, like other chromosomes, so that an ovum contains 22 autosomes and an X chromosome, and a sperm has 22 autosomes and either an X or a Y chromosome.

If A is used to represent one haploid set of autosomes, then the genetics of sex in human beings can be diagrammed as in Fig. 11-14.

Thus sex is determined by the kind of sperm that fertilizes an egg. As approximately equal numbers of the two sexes appear in the offspring of most animals, it is reasonable to assume that X sperm and Y sperm are formed in equal numbers and that either kind has an equal chance of fertilizing an egg. The results are the same in species (many insects) that lack the Y chromosome, save that the female has an even number of chromosomes and the male one less. Among birds and moths the situation is the reverse of that in other animals, there being two kinds of eggs and one of sperm; males are therefore XX (sometimes designated as ZZ), and females are XY (or ZW). The Y chromosome evidently carries no known genes for specific characters, yet in *Drosophila* species where it normally occurs, males without a Y chromosome are sterile.

Prior to the discovery of sex chromosomes, many ingenious theories had been proposed to explain sex. Various schemes have been tried to control the sex of offspring, especially in domesticated animals and humans, but none is effective. In certain frogs and some invertebrates, alteration of environmental conditions affects the sex ratio, the proportion of males to females among offspring (Sec. 10-12).

Some abnormalities of sex are known. Occasional fruit flies, bees, and other insects are **gynandromorphs,** with one part of the body showing male and the other female characteristics. Intersexes are individuals arrested at an intermediate stage of sexual development and may show traits of both sexes. Intersexes have been seen in gypsy moths reared experimentally and also in pigs. In rare cases **sex reversal** has been observed among vertebrates, especially poultry (seldom in humans), when an individual originally female has become a male. It is common, however, in certain invertebrates.

11-14 Sex-linked inheritance The X (or Z) chromosome carries genes for **sex-linked** characters, the inheritance of which is therefore related to sex determination. In the fruit fly *Drosophila*, where eye color depends on some sex-linked genes (Fig. 11-15), the normal red eye color is dominant over white eye. If a homozygous red-eyed female and a white-eyed male are mated, all the F_1 flies are red-eyed; when the latter are intercrossed, the F_2 yields an average of 2 red-eyed females to 1 red-eyed male to 1 white-eyed

Figure 11-14 Genetics of human sex.

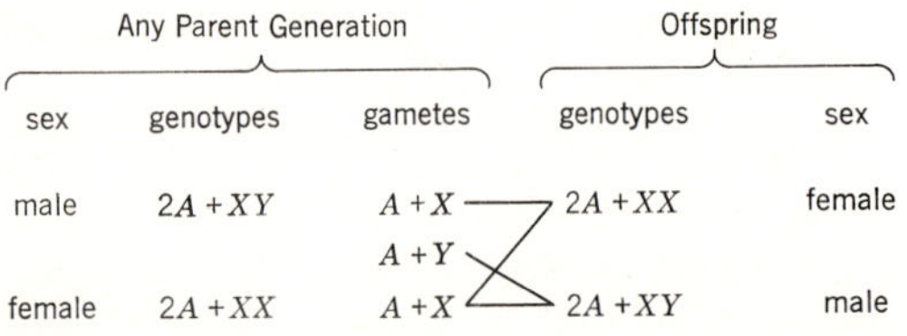

	Any Parent Generation			Offspring	
sex	genotypes	gametes		genotypes	sex
male	$2A + XY$	$A + X$		$2A + XX$	female
		$A + Y$			
female	$2A + XX$	$A + X$		$2A + XY$	male

Red-eyed female
White-eyed male
White-eyed female
Red-eyed male
P
F1
F2

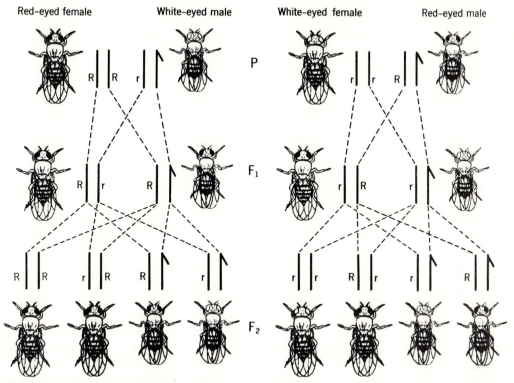

Figure 11-15 Sex determination and sex-linked inheritance of eye color in *Drosophila*. The sex chromosomes are represented by vertical bars with symbols (R, r) for genes of eye color.

male. If X_R represents the gene for red eye and X_r that for white eye, the results of this cross are as in Fig. 11-16. In the P generation each matured ovum carries an X chromosome with the gene for red eye, half the sperm carry an X chromosome with a gene for white eye, and the other sperm contain a Y chromosome with no gene for eye color. In the F_1 generation two kinds of eggs are produced, with the gene for either red eye or white eye, and of the sperm half have a gene (on the X chromosome) for red eye and half have no gene for eye color (Y chromosome). Four kinds of zygotes are thus possible. In the F_2, half the females are homozygous for red eye and half are heterozygous, and the males are red- and white-eyed in equal numbers.

A somewhat different result is obtained in the opposite or **reciprocal cross** with a homozygous white-eyed female and a red-eyed male. In the F_1 the

Figure 11-16 Eye color in *Drosophila*. Homozygous red-eyed female mated to white-eyed male.

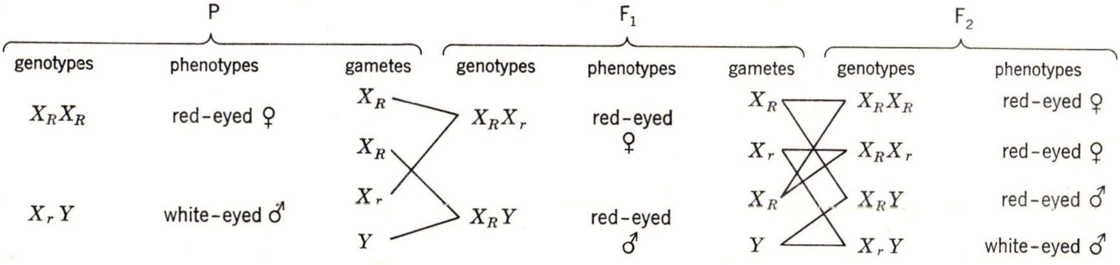

P
F1
F2
genotypes
phenotypes
gametes
XRXR
red-eyed ♀
XrY
white-eyed ♂
XR
XR
Xr
Y
genotypes
phenotypes
gametes
XRXr
red-eyed ♀
XRY
red-eyed ♂
XR
Xr
XR
Y
genotypes
phenotypes
XRXR
red-eyed ♀
XRXr
red-eyed ♀
XRY
red-eyed ♂
XrY
white-eyed ♂

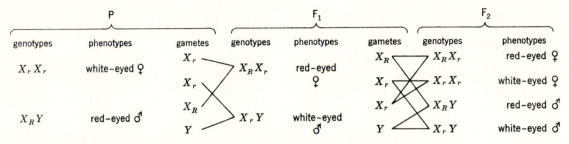

	P			F$_1$			F$_2$	
genotypes	phenotypes	gametes	genotypes	phenotypes	gametes	genotypes	phenotypes	
$X_r X_r$	white-eyed ♀	X_r	$X_R X_r$	red-eyed ♀	X_R	$X_R X_r$	red-eyed ♀	
		X_r			X_r	$X_r X_r$	white-eyed ♀	
		X_R			X_r	$X_R Y$	red-eyed ♂	
$X_R Y$	red-eyed ♂	Y	$X_r Y$	white-eyed ♂	Y	$X_r Y$	white-eyed ♂	

Figure 11-17 **Inheritance of eye color in** *Drosophila* **from white-eyed female parent.**

males are white-eyed and the females red-eyed. In the F$_2$ there result approximately equal numbers of red-eyed females, white-eyed females, red-eyed males, and white-eyed males (Fig. 11-17).

About 150 sex-linked genes have been found in *Drosophila* (Fig. 11-19, chromosome I), and many sex-linked characters are known in other animals and in humans; barred plumage in poultry and red-green color blindness in man (Sec. 11-24) are common examples.

11-15 Linkage Independent assortment (Mendel's second law) was evident in the examples discussed earlier in this chapter, but in the preceding paragraphs some types of inheritance were described that do not involve independent assortment. Many characters of animals tend to be inherited together. The number of pairs of Mendelian factors present in any animal far exceeds the number of chromosome pairs, so each chromosome must carry many genes. There are perhaps some 90,000 genes in humans. Characters that tend to be inherited together are said to be **linked.** Studies of linkage relations in various animals and plants show that the genes occur in **linkage groups,** the members of each group being linked to one another in varying degree while a pair in one linkage group assorts or combines independently with pairs in other linkage groups. When the linkage relations of many genes in a species are known, it is found that there are as many groups of linked genes as chromosome pairs. Many hundreds of linked genes are known in *Drosophila melanogaster;* there are 4 linkage groups (3 large, 1 small) and 4 pairs of chromosomes (3 long, 1 short). *Drosophila willistoni* has 3 linkage groups and 3 chromosome pairs, and *D. virilis* has 6 of each. Among plants, maize has 10 of each, and the garden pea has 7 of each. This evidence is highly important in showing that the genes are contained in the chromosomes.

Figure 11-18 **Crossing-over, the exchange of genes between homologous chromosomes.** *A , a ; B , b* represent pairs of allelic genes in homologous chromosomes that pair during synapsis. *A.* Synapsis of duplicated chromosomes to form tetrad. *B.* Two nonsister chromatids bend across one another. *C.* Chromatids break at point of contact and exchange parts. *D* and *E.* Double crossover.

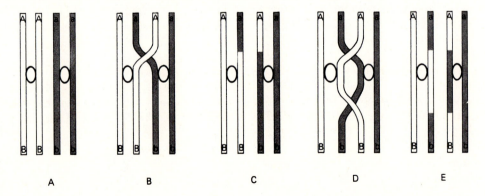

A B C D E

11-16 Crossing-over If the genes for two characters are in one chromosome and the latter remains intact through inheritance, their linkage will be complete and the two characters will occur together; but this is not always the case (Fig. 11-18). The characters separate in a certain number of cases, the percentage of separations varying between different characters, although it is usually constant between any two. Thus short ear and dilute coat color in mice are linked in over 99 percent of individuals, but the percentage is variously lower with many other characteristics. Chromosomes that have exchanged parts have undergone a **crossover,** and the segments involved are recognized by the genes that mark them. During the maturation division the two chromosomes of a homologous pair lie close together or intertwined at synapsis. At this time two nonsister chromatids may bend across each other and break at the point of contact, exchanging homologous parts. This is the physical basis of crossing-over. Often crossing-over may occur simultaneously at several points between two chromatids (Fig. 11-18).

If the genes in a chromosome are in linear arrangement, then two that are far apart will have their linkage transposed by an exchange at any point between them, whereas if they are close together the chance of crossing-over is much less. Assuming that the frequency of crossing-over indicates the relative distance between genes on a chromosome, T. H. Morgan and his coworkers constructed **chromosome maps** for known genes in each linkage group of *Drosophila*. In a mating involving any two linked characters, 1 percent of crossover is taken to represent 1 unit of distance between their respective genes.

Thus the three characters of yellow body, white eye, and ruby eye are all sex-linked and therefore pertain to chromosome I (X). In matings between yellow body and white eye, crossover occurs in 1.5 percent of individuals, between white eye and ruby eye in 6.0 percent, and between yellow body and ruby eye in 7.5 percent; hence the sequence is yellow-white-ruby, with 1.5 units between the genes for yellow and white and 6.0 units between yellow and ruby (1.5 + 6.0 = 7.5). Many crosses between various linked characters provide data for constructing chromosome maps (Fig. 11-19).

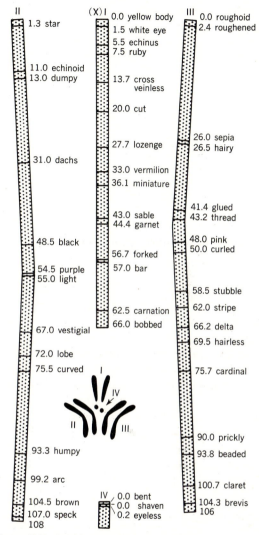

Figure 11-19 Chromosome maps for *Drosophila.* One chromosome of each pair is diagrammed, with the relative positions for some of the many genes "mapped" by linkage studies. Numbers indicate relative distance of each form end of chromosome. Inset shows chromosomes of female diploid cell.

11-17 Giant chromosomes The relatively huge chromosomes in the salivary glands of the larvae of *Drosophila* and other flies are permanently paired as homologous chromosomes. When stained, they show many dark-colored transverse bands, some wide and others narrow (Fig. 11-20). The bands contain much DNA and may be the site of genes, as indicated in cases of deletion. Where a character has disappeared from a laboratory stock of *Drosophila*, part of one chromosome is missing at the position for the gene of that character in the chromosome map. This is important evidence associating the genes with the chromosomes. Each giant chromosome consists of a large number (up to a thousand) of partly replicated, closely aligned chromosomes that have arisen by repeated cycles of DNA unaccompanied by cell division (the centromeric DNA, however, remains unreplicated). The dark bands are probably made up of many lampbrush loops (Figs. 11-20, 11-21), parallel in register, arising from the DNA strands. Gene activity is indicated by local swellings involving one or several adjacent dark bands. At these sites DNA unfolds into open loops and RNA is actively transcribed. The collective groups of loops are called **puffs.** The sequence of appearance of puffs relates to gene function at specific developmental stages of the fly larva.

11-18 The nature of the gene The evidence on characters, genes, chromosomes, and linkage was summarized by T. H. Morgan[1] thus: (1) the characters of an individual are referable to paired genes in the germinal material, held together in a definite

[1] *1926. Theory of the gene. New Haven, Conn., Yale University Press.*

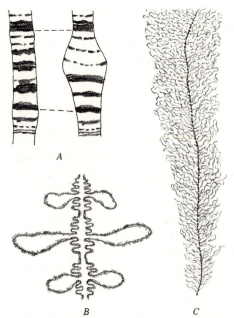

Figure 11-21 *A*. Puffing in a salivary gland chromosome of a fly larva. *B*. Lampbrush loops from oocyte of salamander, representing lateral extensions of chromatids arising from a DNA strand. Transcription of RNA and formation of protein along loop causes fuzzy appearance. *C*. Enlarged portion of loop, showing sites of attachment of RNA polymerase molecules (black dots) involved in elaboration of RNA chains which are attached to the DNA template. Note increasing length of RNA chains, indicating different temporal stages in polymerase activity. (Diagrammatic.)

Figure 11-20 Giant chromosomes from salivary glands of *Drosophila.* Photomicrograph, × 500. Each chromosome is paired and contains many (perhaps 500 to 1000) chromonemata strands, unseparated. L and R refer to left and right arms of chromosomes. Numbers as in Fig. 11-19. (*Courtesy of B. P. Kaufmann.*)

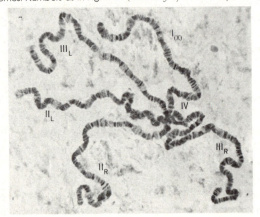

number of linkage groups; (2) at maturation each pair of genes separates (Mendel's first law), so that each gamete contains only one set; (3) the genes in different linkage groups assort independently (Mendel's second law); (4) orderly interchange or crossing-over occurs at times between genes in corresponding (homologous) linkage groups; and (5) the frequency of crossing-over furnishes evidence of the linear arrangement of genes in each linkage group and their position relative to one another. The presence of a particular gene is recognized by some phenotypic characteristic that is produced in an organism.

Far more is now known about genes than in Morgan's day. The gene, once thought to be an indivisible unit of heredity, is actually a linear chain of discrete units, the nucleotides, arranged in pairs (Sec. 2-20) in the double DNA molecule. Work with rapidly reproducing microorganisms—molds, yeasts, bacteria, and viruses—has revealed that each gene contains a number of different sites at which mutations (Sec. 11-19) occur and between which crossing-over can occur. Particularly useful have been studies of the virus T4, parasitic in bacteria. Detailed mapping of two of its rIIA and rIIB genes, which influence the length of the viral life cycle, has been carried out. Some 1000 to 1500 different mutable sites (places where mutations may occur) have been found.

The genes of a number of other viruses and bacteria have been mapped; the procedure, now carried out at the level of the gene itself, resembles that used by Morgan in mapping gene loci (position of the genes) on the chromosomes of *Drosophila*. In the light of these findings, a gene may be defined (J. Watson) as a discrete chromosomal region responsible for a specific cellular product such as the amino acid sequence of a protein or a polypeptide chain. It consists of a linear series of potentially mutable units, each of which can exist in several alternative forms and between which crossing-over can occur. It is now known that the earlier concept of a one gene–one character relationship was inadequate. All complex characters are under the control of many genes.

Evidence that genes are responsible for the synthesis of complex molecules was obtained by Beadle and Tatum, working on the mold *Neurospora*. They studied the organism's ability to synthesize amino acids and vitamins. The wild-type mold lived successfully in agar tubes of *simple* medium containing only sugar, salts, and the vitamin biotin. It could synthesize all the amino acids and other organic compounds needed for growth and reproduction. At the time of sporulation, the wild mold was exposed to x-rays and single irradiated spores were placed on a *complete* medium consisting of a broad spectrum of vitamins, all the amino acids, and the ingredients of the simple medium. The spores grew successfully. But when spores produced by these cultures were transferred to simple medium and to complete medium, all those on the complete medium grew but some on the simple medium failed to do so. The deficient spores were then placed on simple medium rich in vitamins but without amino acids. The vitamin supplement failed to support growth. Irradiation had evidently caused mutations that made it

impossible for the mold to grow without an outside supply of amino acids. Twenty culture tubes of simple medium were then prepared, each with a different amino acid. To these were added the deficient spores. All spores failed to grow except those in the tube containing arginine. Several different mutants were found. One would grow only when arginine was added to the simple medium; another when either arginine or citrulline was added; and a third when arginine, citrulline, or ornithine was added. The three amino acids have a similar molecular structure.

A gene-biochemical relationship is evident in two abnormalities in human waste metabolism. Phenylalanine, an amino acid in many proteins, when excreted is converted by a series of enzymes eventually to carbon dioxide and water by a normal person who has both P and A genes (Fig. 11-22). One stage in the conversion is alkapton. An individual suffering from alkaptonuria (only recessive genes, *aa*) lacks the enzyme to oxidize alkapton (homogentisic acid), and this intermediate product is excreted. The urine on standing turns dark. Later, deposits of alkapton darken the cartilage of the ears and eyes (sclera), and finally arthritis develops. In instances of phenylketonuria (*pp*), phenylalanine accumulates in the blood, cerebrospinal fluid, and urine (whence it is excreted as phenylpyruvic acid). The subjects usually are feebleminded, and the hair pigment becomes lighter. These examples are conspicuous because absence of a dominant gene (P or A) results in complete loss of an essential metabolic stage and af-

Figure 11-22 Routes for disposal of phenylalanine in urine. P, p and A, a are two pairs of alleles.

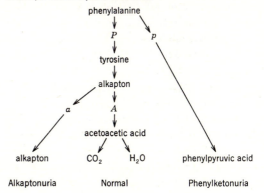

fects other bodily processes. Many biochemical defects in metabolism may be quantitative rather than qualitative as to gene expression and hence are difficult to identify.

11-19 Mutations At times, new characters appear in the offspring of animals and plants that prove to be heritable; these are **mutations,** caused by a change in the genetic mechanism giving rise to the characters concerned. A mutation is a random and stable change in the information-transmitting apparatus of the cell, which then is perpetuated by DNA replication (Sec. 2-20) until changed by another random event. The frequency of mutation can be influenced by tissue and environmental factors. Mutations may be classified as (1) gene mutations and (2) chromosome mutations. They occur in both **somatic** and **germ** cells, but it is only in the latter that the "memory" of the change is transmitted from generation to generation. All have the same effect—an alteration in the nucleotide sequences or amount of DNA and thus the genetic instructions that govern the activities of the cell. The total gene complement of a cell is its **genome.**

Gene or **point** mutations are the most common. They are errors in replication that occur prior to chromosomal duplication. One or more nucleotides may be substituted for others.

The smallest portion of a gene that can produce a mutational effect is a **muton,** and the smallest muton is a single nucleotide pair. Mutons evidently code for individual amino acids. Change in one nucleotide pair can have a significant effect on transcription and coding of a specific protein or enzyme and thus on the activities of the cell. Adjoining mutons often act together, forming larger functional units called **cistrons.** A gene may contain one or more cistrons. Cistrons appear to code for polypeptide chains. Some genes are more likely to mutate than others, thus certain alleles (all the genes situated at a particular locus) are far more common than others.

Reverse mutations may occur. In the T4 bacterial virus at the rII gene sites, it was found that most mutations occurred at single sites and that restoration to the original forms by a reverse mutation could occur. Other mutations, however, resulted in physical deletion of sections of the gene and reverse mutations were highly unlikely. The difference between a gene's mutation rate in one direction and that in the reverse direction is its **mutation pressure.**

Chromosome mutations are alterations in the structure and number of chromosomes. A segment of a chromosome may be (1) lost **(deficiency** or **deletion)** or (2) "read" twice at replication **(duplication).** The normal gene sequence may be altered by (1) exchange of sections between two or more nonhomologous chromosomes **(translocation)** or (2) a 180° rotation of a chromosome section, reversing the linear order of genes in that part **(inversion).** In translocations, genes go into a new linkage group and subsequently assort independently of their own alleles. In changes affecting chromosome number, one set of parental chromosomes may be absent **(haploidy),** or the diploid set may be doubled or tripled **(polyploidy).** Individual chromosomes may be lost or gained. A genome minus one chromosome is called a **monosomic.** In a **polysomic** genome one kind of chromosome may be represented three times **(trisomy)** or four times instead of twice, the diploid number. Human examples of trisomy are known, several of which cause infant death (Sec. 11-26). Polysomic genomes result from **nondisjunction**— failure of paired homologous chromosomes to separate at synapse, thus producing a gamete with one or more extra chromosomes. Chromosome fusions and fissions also occur and are common evolutionary phenomena. Although they alter the number of chromosomes, they do not change the amount of hereditary material.

Polyploids are produced in the following way: Omission of the normal reduction division during meiosis results in germ cells with the diploid instead of the haploid number of chromosomes. When such diploid gametes unite, doubling of the normal diploid number occurs, forming a tetraploid; a triploid is produced by union of a diploid with a normal haploid gamete. Polyploidy is more common in plants than in animals and has been an important mode of speciation in some taxa. A number of plant groups contain species with double, triple, quadruple, etc. the basic chromosome number of the presumed parental stock.

Mutations can be induced artificially by, for instance, radiation (x-rays and ultraviolet light), application of certain mutagenic chemicals (mustard gas, etc.), and high temperature. X-rays applied to *Drosophila*, maize, and barley may increase the mutation rate two hundredfold in some cases, in proportion to the dosage. These mutations prove heritable, like natural ones. It has been suggested that in nature the frequency of mutation has varied in response to major environmental changes such as rising earth temperature and cosmic-ray penetration during reversals of the earth's magnetic field.

A historic case of a mutation was the sudden appearance in a true-breeding stock of red-eyed *Drosophila* of one white-eyed individual; when the latter was bred, this character proved to be heritable, and since it continued in succeeding generations, it became evident that the change was permanent. Over a thousand mutations have been observed in *Drosophila* (Fig. 11-23) and many in other animals and in plants. Most mutations in *Drosophila* are abnormali-

ties, recessive defects, or lethals such as would not survive in nature. But distinctive mutations among domestic animals and plants have been preserved by selective breeding. Examples are the polled (hornless) Hereford cattle, the short-legged (Ancon) sheep, and the short-legged dachshund and long-legged greyhound among dogs. The rate of mutation varies widely in different animals and for different genes; recent calculations indicate one mutation per 10,000 to 1 million cell divisions. Most mutant genes are recessive, but some dominants have appeared.

Because of the wider use of and therefore greater exposure of people to high-frequency radiation, induced mutations are of practical importance. X-ray technicians, workers in atomic-energy plants, and others are threatened with genetic damage because the effects of radiation are cumulative; continued exposure to small amounts over several years may be as serious for an individual as receiving an equivalent dose in a few minutes. Mutations due to irradiation may not be evident for several generations, because they usually are masked by dominant alleles and are seen only as homozygous recessives. The general level of radioactivity is increasing throughout the world in the atomic age. The social and biologic implications of this situation are scarcely perceived, but knowledge is sorely needed.

Figure 11-23 Some mutations of wings and eyes in *Drosophila* used in study of crossing-over and linkage (*After Morgan et al.*)

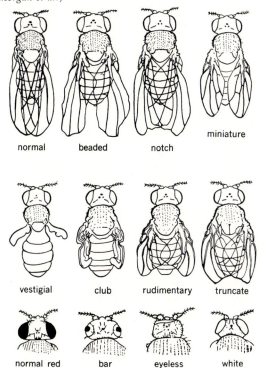

normal beaded notch miniature

vestigial club rudimentary truncate

normal red bar eyeless white

11-20 Inbreeding and outbreeding The mating of closely related individuals is called **inbreeding;** this may involve brother-sister matings or mating of less closely related individuals. Many races of domesticated animals and plants have been derived or improved by inbreeding. Close marriages in human society, as between cousins, are usually banned for fear that defective children may result. Inbreeding tends to produce homozygous stocks. Since most genes for defects are recessive, it provides more opportunity for defective characters to appear. The results will depend on the genetic constitution of the stocks that are inbred. Guinea pigs carried through 23 generations of brother-sister matings by Sewall Wright and others ended with a stock reduced in size, fertility, and resistance to tuberculosis. Helen D. King, however, carried similar matings in white rats through 25 generations, using only the most vigorous individuals in each, and at the end the inbred animals compared favorably with crossbred controls. Similar experiments through more generations with *Drosophila* have ended with normal stocks. Inbreeding with selection can produce satisfactory or improved stocks, combining desirable dominant characteristics in the homozygous condition.

Outcrossing, which is the mating of individuals not closely related, often results in **hybrid vigor,** the offspring exceeding their parents in vigor and size. Outcrossing tends to produce heterozygous individuals in which any defective features are masked by dominant normal characters. Most human marriages are outcrosses, the human population is heterozygous, and defects are relatively rare. People in an isolated community are more likely to be inbred and hence show more defects than in the general population.

11-21 Artificial selection   Selective breeding and the perpetuation of mutations have produced many breeds of domesticated animals that differ markedly in physical, physiologic, and psychologic characteristics from their wild ancestors. Choice among the resulting offspring and discarding of those with unwanted features have yielded many useful forms.

The wild jungle fowl of India (*Gallus bankiva*) is small and slender, with variegated coloration, and the hens lay only 12 to 24 eggs per year. Continued selection has produced from it many domestic breeds of poultry that differ in size, coloration, and egg production. Bantams are only about 25 cm (10 in) high and weigh about 680 g ($1\frac{1}{2}$ lb), whereas meat fowls such as Plymouth Rocks and Rhode Island Reds are up to 40 cm (16 in) tall and weigh $2\frac{1}{2}$ to $3\frac{1}{2}$ kg (6 to 8 lb). White Leghorn hens now average 120 eggs per year, and in some highly selected flocks each produces 200 or more.

Beef cattle (Hereford, shorthorn, and Aberdeen Angus) are selected for shape that yields desirable cuts of meat. Dairy breeds such as the Holstein-Friesian and Jersey have been developed for high milk production and high butterfat, respectively.

The greatest variety perpetuated under domestication is among dogs (derived from wolves of Eurasia), over 100 breeds being recognized. These range from tiny Pekingese to great wolfhounds and from short-legged dachshunds and bulldogs to long-legged greyhounds. The coat color varies from white through shades of yellow, brown, and blue to black, and the character of the coat from nearly hairless in the Chihuahua to the dense sleek coat of the collie and curly hair of the Irish water spaniel.

Artificial selection has perpetuated peculiar mutations that would not survive in nature, such as the pouters and fantails among pigeons or the bulldogs and Pekingese dogs. If any domesticated animals succeed in the wild, they generally revert to resemble the ancestral species. This results from natural selection for genotypes resembling those previously successful in the wild environment. Domestic rabbits released on islands in time produce offspring like the European gray rabbit from which they were derived. Feral swine come to resemble the European wild boar from which they arose.

11-22 Twins   Since the sum total of characteristics in any one individual results from the action and interaction of many genes, the various offspring from one pair of parents will differ. When a female produces more than one young at a birth, each usually develops from a separate egg and the young are genetically different, but if the early blastomeres of one egg part and each gives rise to an individual, all will have the same genetic makeup. The first condition produces **fraternal twins** (triplets, etc.) of the same or of different sex; they are no more alike than the other children of any one family. The second condition yields **identical twins,** always of the same sex, with like physical and physiologic characteristics and having virtually the same mental traits and abilities. In the armadillo, four young of identical kind regularly result from partitioning a single egg.

11-23 Population genetics   Earlier in this chapter heredity was discussed in terms of mating selected pairs of homozygous parents having certain characteristics and determining the kinds of offspring expected in successive generations. Plant and animal geneticists seek to produce organisms with certain desired characteristics, such as larger size or greater resistance to disease. To do this, they make selected matings, first of parental stocks and later from among the progeny in successive generations. With populations in nature it is natural selection that determines which individuals will breed successfully.

A **population** is the total of living individuals of one species in an area. Provided there is no influence that would differentially affect gene frequencies, a population may have any proportion between members of a pair of alleles. The human races are natural populations, and many human marriages are nonselective for most genetic characteristics. To study a natural population, a part, or **sample,** is taken at random.

Population genetics involves application of Mendelian principles to populations, especially as to gene frequencies and the proportions of genotypes under various mating systems. Consider the case of a pair of autosomal alleles, D (dominant) and d (recessive) of equal occurrence. Matings between homozygous parents have the following outcomes:

Parent genotype	DD	dd
Gametes	D	d
F_1 genotypes	Dd	Dd
Gametes	D d	D d
F_2 genotypes	$DD + Dd + dD + dd$, or	
	$\frac{1}{4}DD + \frac{1}{2}Dd + \frac{1}{4}dd$	

Thereafter random matings between the three genotypes of such a population will maintain them in the same proportions in succeeding generations (Table 11-3).

This principle is embodied in the **Hardy-Weinberg law,** which states that *in a given population with random mating and no selection between genotypes, the relative frequency of the genotypes* (in this case DD, Dd, and dd) *tends to remain constant from generation to generation.* The relation of gene frequency to frequency of genotype follows the binomial expression $(p + q)^n$, the basic Hardy-Weinberg formula, where p = the frequency of a given gene, q = that of its allele, and $p + q = 1$. Using the allelic pair considered above, let p represent the frequency of D genes and q the frequency of d genes. Because the genes must be either D or d, $p + q = 1$. (It is customary to use p to designate the dominant gene when present). Each sex produces gametes of each type in the same

Table 11-3
RESULTS OF RANDOM MATINGS

Parents		Offspring DD	Dd	dd
$\frac{1}{4}DD \times$	$\begin{cases}\frac{1}{4}DD\\\frac{1}{2}Dd\\\frac{1}{4}dd\end{cases}$	$\frac{1}{16}$ $\frac{1}{16}$	$\frac{1}{16}$ $\frac{1}{16}$	
$\frac{1}{2}Dd \times$	$\begin{cases}\frac{1}{4}DD\\\frac{1}{2}Dd\\\frac{1}{4}dd\end{cases}$	$\frac{1}{16}$ $\frac{1}{16}$	$\frac{1}{16}$ $\frac{1}{8}$ $\frac{1}{16}$	$\frac{1}{16}$ $\frac{1}{16}$
$\frac{1}{4}dd \times$	$\begin{cases}\frac{1}{4}DD\\\frac{1}{2}Dd\\\frac{1}{4}dd\end{cases}$	$\frac{1}{16}$	$\frac{1}{16}$	$\frac{1}{16}$ $\frac{1}{16}$
Ratio of offspring		$\frac{1}{4}DD + \frac{1}{2}Dd + \frac{1}{4}dd$		

frequency (Fig. 11-24). Genotypic proportions are derived by expansion of the binomial to the second power, thus $(p + q)^2 = p^2 + 2pq + q^2$. In the example above, the proportion of the DD genotype is represented by p^2, Dd by $2pq$ and dd by q^2. This is the familiar Mendelian monohybrid ratio 1:2:1.

In applying the Hardy-Weinberg law, it is assumed that (1) the population is so large that errors in sampling are unimportant; (2) there is no migration of individuals into or out of the population; (3) mutations do not occur or are so rare as to be negligible; (4) gametes carrying the two alleles are produced in equal numbers; (5) matings are at random; and (6) all genotypes (dominant homozygous, heterozygous, or recessive) are equal in survival rate—that is, no artificial or natural selection occurs. For example, among humans there are many people who detect the substance phenylthiocarbamide as intensely bitter (tasters), whereas others find it essentially tasteless (nontasters). This trait is evidently due to a pair of simple Mendelian alleles—T, taster (dominant) and t, nontaster (recessive). Assume that in a given population, 84 percent of persons tested are found to be tasters and the rest (16 percent) nontasters. Because of the presence of the dominant allele T, the nontasters must be homozygous (tt) for the recessive trait. Knowing the proportion of the tt genotype, we can determine the ratio of T and t gametes and the other genotypes in the population. The Punnett

Figure 11-24 **Use of the Punnet checkerboard to determine the frequency of genotypes when the proportion of a recessive phenotype in the population is known in a cross involving a single heterozygous allele.** The example pertains to human detection by taste of the substance phenylthiocarbamide. T, taster; t, nontaster; see text for explanation of circled numbers.

checkerboard (Fig. 11-24A) is useful in showing the steps in this procedure graphically.

The observed proportion of the *tt* genotype in the population was 0.16 (16 percent). Enter this figure in the square for the homozygous recessive (Fig. 11-24B ①). All other values can now be readily determined. The proportion of *t* gametes in the male and female gene pools must be 0.4 (Fig. 11-24B ②), because when multiplied they must equal 0.16 *tt*. If 0.4 gametes of each sex contain the *t* gene, then the remaining 0.6 must carry the *T* gene (Fig. 11-24B ③), because together they must equal 1 (100 percent); $(p + q) = 1$. Once the frequency of the *t* and *T* gametes has been determined, the proportion of remaining genotypes (*TT* and *Tt*) can be obtained by multiplication (Fig. 11-24B ④). The proportions of the F_1 genotypes thus are $0.36\,TT + 0.48\,Tt + 0.16\,tt$.

What happens in the F_2 generation? The proportion of gametes and the genotypic ratios remain the same, as predicted by the Hardy-Weinberg law. The 16 percent F_1 *tt* individuals produce only *t* gametes, but in addition, 24 percent (half) of the gametes produced by the *Tt* individuals are *t*, yielding a total of 40 percent *t* gametes. Likewise, the 36 percent F_1 *TT* individuals produce only *T* gametes, but these are augmented by 24 percent (half) of the *T* gametes from the *Tt* genotypes, giving a total of 60 percent. The F_1 0.4*t* to 0.6*T* gametic ratio is thus maintained in the F_2, and this gametic frequency will again produce genotypes in the proportion found in the F_1 generation. The process described will continue generation after generation unless the equilibrium is altered in ways mentioned earlier.

It is possible also to analyze more complex types of heredity. For example, the proportions of the MN blood groups in humans differ widely in geographically separate populations. In this case one gene (*M*) produces antigen M and its allele (*m*) the antigen N. Neither gene is dominant, so the heterozygote (*Mm*) produces both antigens, forming the MN group, whereas the homozygous genotypes *MM* and *mm* result in groups M and N, respectively. Results from a few of the many studies on this characteristic are shown in Table 11-4. The percentages of each type actually found and those expected according to the formulas used above are so close as to demonstrate again that a simple pair of Mendelian alleles is involved. In this example there are strikingly different proportions between the alleles in separate populations.

Although populations in equilibrium are rare in nature (perhaps they do not exist), the Hardy-Weinberg law tells us what happens at a given gene locus

Table 11-4
DISTRIBUTION OF MN BLOOD GROUPS

Population	% observed/% calculated		
	MM MM	MN Mm	NN mm
Australian aborigines (730)*	3.0/3.17	29.6/29.3	67.4/67.6
Caucasians, United States (6129)	29.2/29.2	49.6/49.7	21.3/21.2
Eskimos, Greenland (569)	83.5/83.4	15.6/15.9	0.9/0.8

* Number of individuals tested in parentheses.

in large random-mating populations in the absence of evolutionary change. In small or isolated populations, chance matings may lead to imbalance of genotypic ratios. The law thus provides an important reference point in studying evolutionary processes and also reveals the mechanism by which genetic variation in a population is conserved. It explains why rare traits persist and why dominant genes do not increase in proportion to their alleles, as once thought. The conservation of uncommon genes is of great importance in the evolutionary process. Their presence greatly increases the range of genetic variability and a population's ability to adjust, through natural selection, to environmental change.

Human inheritance

The genetic characteristics of human beings are passed from generation to generation like those of other animals. Details are sometimes difficult to work out for humans because of the long intervals between generations and the scarcity of records that specify the characteristics of many individuals. By tracing backward through family histories in which the peculiarities of many members are listed, it has been possible to determine the manner of inheritance for some physical features and physiologic characteristics.

11-24 Color blindness The inability to distinguish red from green is a sex-linked recessive character in humans (Table 11-5). Normal males neither have nor transmit the defect; carrier females enjoy normal vision but, being heterozygous for the character, may have color-blind children; color-blind males and females both transmit the defect. In the United States about 8 percent of men and 0.5 percent of women are color-blind.

11-25 Blood groups The importance of blood groups in transfusion was described in Sec. 6-8. The heredity of these groups is based on three codominant alleles for antigen production: antigen A, gene I^A; antigen B, gene I^B; no antigen, gene I^O. Neither I^A nor I^B is dominant over the other, but both are dominant over I^O. The genotypes and blood groups of individuals are as follows:

Genotype	$I^O I^O$	$I^A I^A$ or $I^A I^O$	$I^B I^B$ or $I^B I^O$	$I^A I^B$
Blood group	O	A	B	AB

The relations possible between parents and offspring are shown in Table 11-6. The blood group to which an individual belongs is useful in some but not all medicolegal cases involving parentage, as when two babies are accidentally interchanged in a

Table 11-5
INHERITANCE OF COLOR VISION (RED-GREEN DISCRIMINATION)* IN HUMANS

Parents → ↓	gametes ↓→	Normal ♂ XY		Color-blind ♂ **X**Y	
		X	Y	**X**	Y
			Children		
Normal ♀ XX	X	XX ♀ Normal	XY ♂ Normal	**X**X ♀ Carrier	XY ♂ Normal
Carrier ♀ **X**X	X	XX ♀ Normal	XY ♂ Normal	**X**X ♀ Carrier	XY ♂ Normal
	X	**X**X ♀ Carrier	**X**Y ♂ **Color-blind**	**XX** ♀ **Color-blind**	**X**Y ♂ **Color-blind**
Color-blind ♀ **XX**	**X**	**X**X ♀ Carrier	**X**Y ♂ **Color-blind**	**XX** ♀ **Color-blind**	**X**Y ♂ **Color-blind**

* The gene for this factor, being sex-linked, is carried on the X chromosome. Color blindness (**X**) is recessive to normal vision (X). The heterozygous "carrier" female has normal vision.

Table 11-6
HEREDITY OF HUMAN BLOOD GROUPS

Parents	Children Possible	Children Not possible	Child	Other parent known to be	One parent May be	One parent Cannot be
O × O	O	A, B, AB		O	A, B	
O × A			O	A	O, B	AB
A × A	O, A	B, AB		B	O, A	
O × B				O		
B × B	O, B	A, AB	A	B	A, AB	O, B
A × B	O, A B, AB	—	B	O A	B, AB	O, A
O × AB	A, B	O, AB				
A × AB				A	B, AB	O, A
B × AB	A, B, AB	O	AB	B	A, AB	O, B
AB × AB				AB	A, B, AB	O

hospital or when a woman claims a certain man as the father of a child. For example, if two parents, both of group O, are given a baby of group A, there obviously has been a mistake; likewise a couple of O × AB could not be parents to a group O baby. An A × B couple, however, could not make a valid claim against any particular baby.

11-26 Medical genetics Various human diseases are known to be inherited, such as hemophilia and diabetes (Table 11-7). In **sickle-cell anemia** the precise biochemical change has been discovered. Under low oxygen conditions (in vitro or in vivo) the red blood cells of affected people become sickle-shaped. Homozygous recessives suffer from anemia that is usually fatal, but persons with the single gene difference in the heterozygous condition show no ill effects.

The hemoglobin molecule is a tetramer, containing four polypeptide chains-, two identical α chains and two identical β chains, totaling some 600 amino

Table 11-7
SOME MENDELIAN HEREDITARY HUMAN CHARACTERS

Character	Dominant	Recessive
Normal:		
Skin pigment	Normal	None (albinism)
Eye (iris) color	Brown	Blue
Abnormal:		
Fingers and toes	Short	Normal
	Webbed	Normal
	Extra digits	Normal, 5 each
Nervous function	Huntington's chorea*	Normal
Phenylalanine metabolism	Normal	Phenylketonuric idiocy†
Eye	Opaque lens (hereditary cataract)	Normal
	Internal pressure (glaucoma)	Normal
Ear	Normal hearing	Deaf-mutism
Sex-linked:		
Color vision	Normal	Color-blind
Blood clotting	Normal	Hemophilia

* Degenerative disease of the nervous system leading to death at middle age.
† Caused by recessive gene that blocks metabolism of the amino acid phenylalanine. This substance then accumulates to toxic levels and impairs structural and mental development. The recessive homozygote fails to reproduce.

acids. In sickle hemoglobin the α chains are identical with those found in normal hemoglobin, but 1 out of 146 amino acids in each β chain, a glutamic acid, is replaced by lysine. This small but important change has influenced the human population of Central Africa for a curious reason. Persons who are heterozygous for sickle-cell anemia show an increased resistance to malaria. This selective advantage results in a higher incidence of sickle-cell anemia in regions where malaria is common.

In most human beings the XX-XY chromosome pattern results in individuals normal as to sex. In normal females the nucleus of cells in the epidermis, mouth lining, and elsewhere shows a heavily staining clump of chromatin, rare or absent in males. It is called **sex chromatin** or a **Barr body,** after its discoverer. A Barr body appears to be an inactivated X chromosome. In female mammals it is evidently normal for only one of the two X chromosomes to be active in a cell at any one time. The single X chromosome of the male is rarely so affected. This discovery has made possible determination of the sex of human embryos as early as 3 weeks of age, by examining cells from the amniotic fluid. It has also aided in the study of sexual abnormalities resulting from deviations in the sex chromosome pattern. In chromosome complements with two or more X's, the number of Barr bodies is one less than the number of chromosomes. In **Klinefelter's syndrome** the sex chromosome pattern is XXY, and a single Barr body is found in cell nuclei. The individual is phenotypically male but has very small testes, usually no mature sperm, and may have feminine proportions (enlarged breasts) and be mentally retarded. In **Turner's syndrome** only a single X chromosome is present (XO). The individual is phenotypically female but with infantile mammary glands and vestigial gonads or none; stature may be small, and there is often mental retardation.

Down's syndrome (mongolism) is a condition producing offspring that are physically and mentally retarded, with abnormalities of the face, eyelids, tongue, and other organs. Presence of an eyelid fold, mistakenly thought to resemble that of Orientals, is the basis for the alternative name. About 0.15 percent of births in Caucasian populations show the defect, but many patients with mongolism die early. Its incidence increases with age of the mother

(11 $\times$ at 35 to 39 years and 100 $\times$ at 45 + years), but age of the father has no effect. In twin births both are affected if identical, but only one may show the defect if nonidentical; this indicates a genetic rather than a physiologic basis. Studies of tissue cultures of fibroblast and bone marrow cells in Down's syndrome reveal a condition of trisomy (Sec. 11-19). There are 47 chromosomes, one autosome (No. 21) being triplicate as a result of nondisjunction—failure of paired chromosomes to separate, presumably in the maternal oocyte. This genetic imbalance evidently acts during embryonic development to produce the condition.

11-27 Genetic manipulation   **Eugenics** seeks to improve the human race by applying the principles of genetics. It is fraught with difficulties. Eugenic movements got under way in the early 1900s, soon after the rediscovery of Mendel's laws. Efforts to promote selective breeding among humans historically have been characterized by racism, elitism, and even genocide. It now appears unlikely, on both scientific and sociologic grounds, that a selective breeding approach at the population level could produce desirable results or be successful in a democracy. The human population is large, genetically polymorphic, mobile, and resistant to interference with personal freedoms. Most individuals are a constellation of good and bad traits by anyone's standards. Negative eugenics—efforts to weed out undesirable characteristics—have not been successful. These factors militate against population eugenics, even if agreement could be reached as to the desired genotypes. Other approaches, such as improving educational opportunities and the environment, are more likely to work. There has, however, been a revival of interest in eugenics in recent years, set forth now on a firmer biological basis. On a more directly personal and immediate basis, eugenics has an important role to play in medical genetics and genetic counseling, to reduce the risk, in particular, of various heritable diseases. Opportunities for a positive form of eugenics are also now available through sperm storage (by freezing), perhaps soon to be extended to ova. Semen cryobanks make possible selective use of DNA.

Genetic engineering, a variety of techniques for

manipulating DNA within living cells, is a rapidly growing field. Much of it has been done in microorganisms and lower vertebrates. Some of the procedures may eventually be applicable to humans. For example, bacterial viruses sometimes carry small sections of host DNA from one bacterium to another, where it may be permanently incorporated into the bacterial DNA. It might be possible in humans to insert new genes into host cells, using a harmless virus. Cloning of selected donor genotypes may be possible through nuclear transplant techniques involving implantation of a somatic cell nucleus into an enucleate activated egg (Sec. 10-24). If successful in humans, a series of individuals of identical genotype could be produced. **Recombinant-DNA** research involves manipulation of nucleotide sequences in DNA molecules. "Restriction" enzymes, isolated from bacteria, are used to cut the molecules of DNA into segments. A segment can then be attached to a suitable "carrier" to be inserted into an appropriate host, where it can be propagated and where it may possibly function. In work on bacteria, foreign DNA is linked to plasmid DNA removed from *Escherichia coli*. The recombinant DNA (donor + plasmid DNA) is then returned to the *E. coli* cell, where it replicates in the process of cell multiplication. In this way an indefinite number of copies of the inserted nucleotide sequence can be made. Since nucleotide sequences from many organisms can be thus replicated in vast numbers, the technology carries with it the possibility of both great benefits and risks. On the positive side it (1) provides a method for obtaining a great amount of information on the nucleotide sequences in the DNA of a particular organism, (2) makes possible production of desired cellular products, some of which may be of pharmaceutical and commercial value (e.g., insulin), and (3) allows for genetic infusions from their wild progenitors (declining throughout the world) to strengthen stocks of domestic plants and animals. Negative aspects are possible biohazards resulting from deliberate or accidental release of artificially created new and virulent pathogens that might affect humans or other organisms. *E. coli* is widespread and is an abundant organism in the human intestinal tract. As with advances in nuclear science, molecular biology and genetic engineering are providing the human race with

awesome powers. In such cases the scientific enterprise cannot be divorced from matters of human welfare. Great wisdom will be required in the use of these new technologies.

References

Crick, F. H. C. 1962. The genetic code. *Sci. Am.*, vol. 207, no. 4, pp. 66–74. *How do genes determine the order of amino acids in a protein? It appears that each amino acid is specified by a triplet of nucleic acid bases and that triplets are read in simple sequence.*

_____. 1966. The genetic code: III. *Sci. Am.*, vol. 215, no. 4, pp. 55–62. *Confirmation of how the four-letter language of molecules of nucleic acid controls the 20-letter language of the proteins.*

Dunn, L. C., and **T. Dobzhansky.** 1952. Heredity, race and society. Rev. ed. New York, Mentor Books, New American Library, Inc. 143 pp.

Fiddes, J. C. 1977. The nucleotide sequence of a viral DNA. *Sci. Am.* vol. 237, no. 6, pp. 54–67. *The nucleotide order has been established in full for the 5375-nucleotide DNA of the bacterial virus ϕ X174.*

Ford, E. B. 1971. Ecological genetics. 3d ed. John Wiley & Sons, Inc. xx + 410 pp., illus.

Gardner, E. J. 1972. Principles of genetics. 4th ed. John Wiley & Sons, Inc. xi + 527pp.

Grobstein, C. 1977. The recombinant-DNA debate. *Sci. Am.*, vol. 237, no. 1, pp. 22–33. *The gene-splicing method and the debate over its biohazards.*

Hurwitz, J., and **J. J. Furth.** 1962. Messenger RNA. *Sci. Am.*, vol. 206, no. 2, pp. 41–49. *This nucleic acid acts as a messenger to carry instructions from genes to sites of protein manufacture.*

Lerner, I. M., and **W. J. Libby.** 1976. Heredity, evolution, and society. San Francisco, W. H. Freeman and Company xx + 431 pp.

Mazia, D. 1974. The cell cycle. *Sci. Am.*, vol. 230, no. 1, pp. 54–64. *What happens in the living cell between divisions; four stages are now recognized.*

Mendel, Gregor. 1866. Versuche an Pflanzen-Hybriden [Experiments in plant hybridization] *Verhandlungen der Naturf.* Verein, Brünn, vol. 4, pp. 3–47. The original paper.

Moody, P. A. 1975. Genetics of man. 2d ed. New York, W. W. Norton & Co., Inc., ix + 507 pp.

Peters, J. A. (editor). 1959. Classic papers in genetics. Englewood Cliffs, N.J., Prentice-Hall, Inc. vi + 282 pp., illus. *Historic landmarks from Mendel (1865) to Benzer (1955).*

Srb, A. M., R. D. Owen, and **R. S. Edgar.** 1965. General genetics. 2d ed. San Francisco, W. H. Freeman and Company x + 557 pp.

Stein, G. S., J. S. Stein, and **L. J. Kleinsmith.** 1975. Chromosomal proteins and gene regulation. *Sci. Am.*, vol. 232, no. 2, pp. 46–57. *Apparently the histone proteins keep genes turned off and the nonhistone proteins selectively turn them on.*

Stent, G. S. 1971. Molecular genetics: An introductory narrative. San Francisco, W. H. Freeman and Company xvi + 650 pp., illus.

Stern, Curt. 1973. Principles of human genetics. 3d ed. San Francisco, W. H. Freeman and Company x + 891 pp. *Includes chapters on medical genetics and genetic counseling.*

———, and **E. R. Sherwood.** 1966. The origin of genetics: A Mendel source book. San Francisco, W. H. Freeman and Company xii + 179 pp. *Clarification of events surrounding Gregor Mendel's historic work; basic papers of Mendel and others in English translation.*

Strickberger, M. W. 1968. Genetics. New York, The Macmillan Company x + 868 pp.

Suskind, S. R., and **P. E. Hartman** (editors). Foundations of modern genetics. Englewood Cliffs, N.J., Prentice-Hall, Inc. 8 vols.

Watson, J. D. 1976. Molecular biology of the gene. 3d ed. Menlo Park, Calif., W. A. Benjamin, Inc. xxiv + 739 pp. *Includes excellent and concise chapter summaries.*

White, M. J. D. 1954. Animal cytology and evolution. 2d ed. New York, Cambridge University Press xiv + 454 pp., 147 figs.

12

ANIMAL ECOLOGY AND DISTRIBUTION

Every living organism, from the simplest bacterium to the ponderous elephant, has a distinctive mode of life that depends upon its structure and physiology and also upon the kind of environment that it occupies. Physical and biologic factors act to make a wide variety of environments on different parts of the earth. Conditions are rather constant in some tropical lands and seas, but over much of the earth the temperature, moisture relations, and sunlight change markedly with the seasons. Collectively these influences are known as **climate.** The life cycle of each species is closely adjusted to the climatic conditions of its environment. No animal lives entirely to itself; on the contrary, each is part of an integrated living **community** that includes others of its kind, many different sorts of animals, and plants of few or many types. **Ecology** (Gr. *oikos,* house) is the scien-

tific study of the interrelationships between organisms and their environments; **distribution** is the study of their occurrence in space and time.

For convenience in organizing subject matter, textbooks often treat ecology and evolution as separate topics, as is done here. It is important to emphasize the artificial nature of such separation and the close interrelationship that exists between the two disciplines. The relationships between organisms and their environments—all the physical and biological factors affecting them and influenced by them —are the result of natural selection, thus all ecologic phenomena have an evolutionary explanation. Evolutionary ecology concerns itself with those features of a species or population that have survival value (are evolutionary adaptations) and are not merely the consequences of population dynamics.

Ecology

The physical
environment

Animals and plants are affected by various physical and chemical factors, the most important being (1) sunlight, (2) temperature, (3) water, (4) physical substrate, (5) gravity, (6) pressure, and (7) gases and minerals. Each can be measured and its effects on animals observed but all are interrelated and none acts independently. Sunlight provides the radiant energy used by plants in photosynthesis, but it also warms animal environments and raises the temperature of water, leading to evaporation (and eventually to precipitation of rain or snow). Temperature controls the speed of all chemical reactions, including the biochemical reactions within living organisms.

Water is the solvent for soil minerals essential to plants, is a requirement in animal bodies, and is the medium in which many animals live.

12-1 Sunlight   Almost all the energy used by organisms is derived from the sun (Fig. 12-1). Energy may be transformed from one type to another but is never created or destroyed (Sec. 2-3). Green plants absorb the radiant energy in sunlight and, by the photosynthetic action of chlorophyll in their cells, produce carbohydrates from carbon dioxide and water; they also synthesize proteins and fats. The energy stored in these compounds is the ultimate source used by all animals. It is passed from organism to organism and is the sole source of energy for maintaining and operating all living systems. Energy relations underlie all physical and biotic processes on the earth and make possible the activities of organisms.

Figure 12-1   The chemical cycles of carbon dioxide, oxygen, nitrogen, and minerals in nature.   Arrows indicate the paths of movement of materials from the air (CO_2, O_2, N) and soil (minerals) to and from plants and animals.

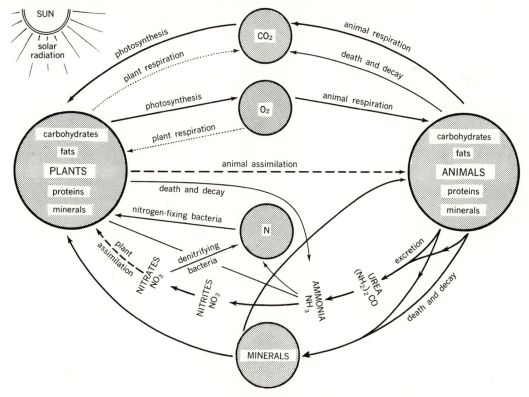

12-2 Temperature   The range of temperatures in the universe covers thousands of degrees, but most life on earth can exist only within a range from zero to 50°C (32 to 122°F) or less. Heat tolerance is affected by moisture and really depends on the evaporating power of the air or the percentage of water vapor in relation to saturation at any given temperature. In the dry air of a desert, for example, a temperature of 32°C (90°F) is not uncomfortable, but the same temperature coupled with high relative humidity is difficult to tolerate.

Temperature influences the growth, fruiting, and survival of the plants upon which various animals depend for food. A prolonged cold spring delays the development of grasses and leaves upon which many insects, rodents, and grazing animals forage and may determine their survival. Unseasonable weather at blossoming time may reduce the subsequent crops of berries or seeds on which various birds feed, forcing them to wander elsewhere for food or starve.

Reptiles, amphibians, fishes, insects, and all other invertebrates have little or no internal regulation of body temperature, although some species achieve a measure of control over body temperature by behavioral means. The rate of chemical processes in their metabolism and hence their growth and activities are affected directly by environmental temperatures, being speeded up by warmth and slowed by cold. Each species has thermal limits; all are killed by prolonged exposure to freezing temperatures or excessive heat. If freezing weather occurs after their eggs or larvae have begun to develop, many are killed and their populations are reduced. Some insects have eggs, larvae, or pupae that overwinter as resting stages with lowered metabolism, beneath the ground surface, within plants, or in the bottoms of ponds and streams, to escape being frozen. Certain insects trapped in ice may survive because the water in their bodies is so full of solutes that it does not freeze.

Reptiles and amphibians must hibernate in the ground or water to escape being frozen in regions that experience low winter temperatures. Some reptiles of arid regions that are abroad by day in the spring become nocturnal in summer to avoid high temperatures. Some **aestivate,** suspending activity in summer. Most freshwater fishes are inactive in cold weather, and each species has an upper limit of heat tolerance. Since temperature changes are slower and not so extreme in the ocean, marine organisms are less affected by seasonal changes of climate; yet many kinds of saltwater fishes migrate north and south seasonally.

The birds and mammals, having insulated bodies and closely regulated temperatures, are much less affected directly by change of environmental temperature, but excessive winter cold or summer heat may impose stress and reduce their food supply. Many birds that summer in Arctic and temperate regions travel, or **migrate,** to warmer regions for the winter to obtain appropriate food. Birds, elk, and deer that summer in high mountains, as in western North America, migrate to lower levels for the same reason.

Ground squirrels and some insectivorous bats enter a winter rest, or **hibernation,** when their food of the warmer seasons becomes unavailable. In hibernation the body temperature drops to about that of the animal's shelter, the heartbeat and respiration become very slow, and the reduced metabolism is supplied mostly by fat stored in the body before entering hibernation.

12-3 Water   There is a constant interchange of water among air, land, and sea and between living organisms and their environments (Secs. 2-13, 7-17). In addition, water profoundly influences the environments of organisms. The water cycle (Fig. 12-2) involves evaporation, cloud formation, precipitation, surface water runoff, and percolation through the soil. Water stores vast quantities of heat, and because its specific heat is so great (requiring 1 cal to raise 1 g of water 1°C at 15°C), any large mass is slow to warm up in the spring and slow to cool in the fall. Water is heaviest at 4°C (39.2°F); it expands on cooling below this point and changes to ice at 0°C (32°F). The force of expansion is so great that rocks are split when water in crevices freezes; this is a mechanism in soil formation. (The cracking of the engine block iron in an automobile when the jacket water freezes is a common example of this power.) The fact that ice floats, being lighter than water, is important to organisms. But for this, ice would form at the bottom of

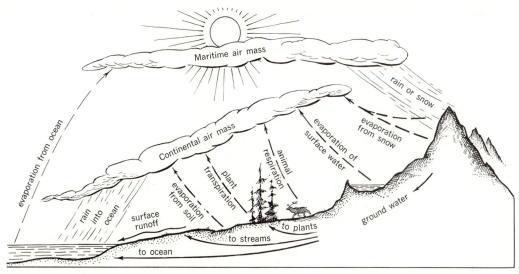

Figure 12-2 The water cycle. Constant interchange of water among air, land, and sea produces various daily and seasonal alterations in the environments of plants and animals.

lakes, and most large bodies of water would have permanent masses of ice in their depths. Instead, water sinks as it cools to 4°C and warmer water rises, creating convection currents. These bring about a spring and autumn turnover in temperate lakes and protect organisms from temperature extremes, since those beneath the ice in deep lakes are never much below 4°C.

Many aquatic environments in eastern North America are relatively stable because water is retained by freezing in winter and that lost by evaporation in summer is replaced by frequent rains. In the Western states, however, lakes fluctuate widely with changes in rainfall or snowfall. Many small ones and some of the larger dry out completely at times, killing all fishes and other strictly aquatic species. The bordering marshes that shelter frogs, turtles, ducks, muskrats, and others also are destroyed. Abrupt changes in the flow of streams alter the conditions for their inhabitants. Severe floods modify the character of the bottom, covering gravel beds with silt, and the rushing waters may actually destroy many creatures. The reduction of a creek to scant flow or scattered pools exposes the aquatic animals to attack by predators of the land and also permits the waters to become overheated. Certain amphibians, insects, and invertebrates breed in transient rain pools, and if the rains are scant or unseasonal, the animals may lack

spawning places, or the pools may dry out, killing the young before growth is completed.

Land animals are affected by the moisture content of the air, or **relative humidity** (percentage of water vapor in relation to saturation at any given temperature). Some are suited to deserts of low humidity, others exist only where the atmosphere is virtually saturated, and many live at intermediate humidities. For small animals the **microclimate** of the restricted places where they feed or find shelter is all-important, usually having a lower temperature and higher humidity than the general climate of the region in which they occur. In arid regions many small creatures remain in seclusion by day, else they would soon die of desiccation. They venture out at night when temperatures are lower and the humidity greater, especially close to the ground. In areas having frequent summer rains or where fields and gardens are irrigated, the humidity near the ground allows small invertebrates to be active during the daytime.

All animals that inhabit soil are affected by change in its moisture content. Earthworms live close to the surface in damp soil during warm weather but go deeper as the surface layer dries, and the same occurs with many insect larvae. In turn, the moles that feed on such animals work in shallower or deeper layers as necessary.

12-4 Pressure   Animals of the land experience differences in atmospheric pressure at elevations above sea level, because the density of the air (hence available oxygen) decreases with altitude. People living in high mountains have larger hearts and more red blood cells to compensate for the lesser oxygen supply, and mountain climbers from low altitudes use oxygen tanks to aid their breathing. Flight of some birds is thought to be less easy in highly rarefied air. At any one place, whatever its altitude, changes in air pressure (barometric reading) are associated with changes in weather. Strong winds—resulting from regional differences in pressure—affect animals variously; birds and insects fly less easily and may be blown to new places, and the rapid moving air has a drying effect on plants and many animals.

In deep lakes and the sea, water pressure increases regularly with depth (1 atmosphere, or 15 lb/10 m), yet the Danish Galathea Expedition dredged nearly 100 species of invertebrates near the Philippine Islands at a depth of about 10,500 m, where the pressure was about 1 ton/cm². Such animals can live because the pressure within their bodies is the same as that outside. Many plankton forms and some fishes are at depths of 400 m by day but rise to the surface at night, gradually experiencing a fortyfold change in pressure. Human skin divers use air tanks both to supply oxygen and to keep up the pressure in their lungs when they are in the water.

12-5 Chemical cycles in nature   The elements that form the bodies of plants and animals are all derived from the environment, and there is a constant interchange, or **circulation,** of these, incident to the life and death of organisms (Fig. 12-1).

Carbon (C) is a constituent of all organic compounds in organisms (Sec. 2-14). From the carbon dioxide (CO_2) in air or water carbon is synthesized into the molecules of carbohydrates, and these, together with proteins and fats, make up the tissues of plants. The plants are eaten by certain animals, and after digestion and absorption in the latter (Chap. 5), the compounds of carbon become reorganized as similar compounds in animals. In turn, these materials pass through other animals. Destructive me-

tabolism in animals yields carbon dioxide as a respiratory waste that returns to the air or water. Huge amounts of carbon exist in the air (6 tons/acre as CO_2), in fossil fuels (coal, oil), and in limestone rocks (carbonates).

Oxygen (O_2) is taken directly from the air or as dissolved in water (Fig. 12-1) to serve oxidative processes in animal bodies. It returns later to the environment, either combined with carbon as carbon dioxide or with hydrogen as water. From the carbon dioxide used by plants the oxygen is released to the environment, but plants also use some oxygen in respiration. A "balanced aquarium" contains animals and plants in such quantities that their mutual needs and outputs of oxygen and carbon dioxide are balanced.

Atmospheric **nitrogen** (N) can be used directly only by nitrogen-fixing bacteria and certain algae in soil or in root nodules of some legumes, which combine it into nitrates (NO_3). Plants use nitrates to form vegetable proteins. These either return by decay to the soil or are eaten by animals and converted into animal proteins. In animal metabolism the latter are eventually broken into nitrogenous wastes and excreted as urea, $(NH_2)_2CO$, uric acid, $C_5H_4N_4O_3$, or ammonia, NH_3. By action of other bacteria in soil or water, such wastes are converted to ammonia and to nitrites: by further bacterial action, either the nitrogen is returned to the air or the nitrites are converted into nitrates.

Certain **minerals,** or inorganic chemical substances, are essential for both plants and animals (Secs. 2-13, 5-13) in small but definite amounts that differ in various species. Plants obtain mineral constituents from the soil solution around their roots, and these return to the soil only by decay or burning of the plants. The supply for animals is taken partly from their food and partly from water, and in some cases directly from the soil. Minerals from animals return to the soil or water in excretions and feces and upon decay of their bodies (Fig. 12-3).

Phosphorus is needed in small amounts for all basic syntheses and energy transfers (Sec. 2-23). Phosphates are made available to plants through erosion from the great reservoirs of past ages in the rocks. Phosphorus is normally utilized in food of plants and animals, then returned to the soil through

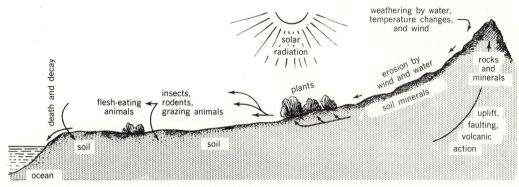

Figure 12-3 The mineral cycle. Materials uplifted into mountains become disintegrated into soil minerals. By erosion and water or wind transport, these move gradually toward the lowlands and the sea; meanwhile, some pass once or many times through plants and animals.

death and decay. Some is carried to the sea, where a part is lost in the deep sediments (until raised again by uplift of land that can be eroded). Much phosphorus enters ocean food chains and is recaptured by fishes and marine birds or mammals. Humans seek the vast deposits of bird droppings (guano), as on coastal islands of Peru and Nauru Island in the South Pacific, for phosphate fertilizer to bring it back into the cycle. They also obtain phosphorus by mining phosphate rock to fertilize crops and by eating fishes.

The biologic environment and interrelations of animals

During some 3.4 billion years of evolution, interdependence among the earth's organisms has become extensive and complex, as species have evolved together in varied environments which they themselves have helped create. Competition, engendered by high rates of reproduction and limited resources, has produced many modes of life in which the struggle for food, living space, shelter, and mates is minimized. The resulting diversification of organisms has been furthered by the increase in environmental complexity that has accompanied growth in numbers of species. Interdependence can be seen in both broad and detailed relationships, from the dependence of animals on plants for food to complicated interactions among members of a species.

12-6 Food Most plants compete with their neighbors for the same things—sunlight, soil min-

erals, and water—but animals are more diverse in their needs. Animal food, of whatever kind, is derived ultimately from plants. Each species of animal requires a certain amount of food of the right kinds. Human beings, rats, and houseflies can subsist on a variety of foods and change from one to another as necessary. Many species, however, are more specialized and can exist only where and when their particular foods are available. The beaver eats only the inner bark of willows and poplars; the larva of the cabbage butterfly requires leaves of cruciferous plants; some leafhoppers subsist only on the juices of particular species of plants; and horseflies require mammalian blood for reproduction. Some food supplies are seasonal, and the species depending upon them must, at other times of the year, shift to other foods, become dormant, migrate elsewhere, or perish.

The ocean, unlike the land, has few conspicuous plants; a sea hare (*Aplysia*), a mollusk, eats seaweed, and some snails and limpets rasp off the dense short algae covering shore rocks. The foundation of ocean pasturage, however, is the **plankton,** composed mainly of microscopic plants (diatoms and other algae) and animals (crustacea, larvae) that float and drift freely in the water. It varies in amount and species composition through the season (like the wild flowers in a field). Plankton is the food of innumerable small crustaceans, chaetognaths, and other animals and of the larvae of mollusks, annelids, and echinoderms.

Seaweeds are commonest in bays, estuaries, and shore waters, many of them being annuals (living only one year or season). When they die and sink, they are reduced by a rich flora of bottom-dwelling bacteria. The resulting microscopic **detritus** (plant material plus bacteria) forms a bottom scum. This is the food of various worms and other invertebrates that secrete mucous nets or have special ways to select fine particles from the water or mud. Sea cucumbers and some sea urchins ingest bottom mud in quantity to extract detritus for food. There is also a supply of larger organic particles, or **debris,** derived mainly from animals that sink after death. Debris is the food of scavengers—certain shrimps, crabs, sea urchins, and others. In the ocean, however, above the microscopic level in size, animals are the principal food supply. Minute creatures that feed on plankton are eaten by other swimming invertebrates; these in turn are the food of larger crustaceans and small fishes, which then are captured by larger fishes, and so on.

Plant-eating animals are the primary consumers in any animal community. They in turn serve as food for other animals (secondary consumers), which are eaten by still others. Several **trophic** (feeding) **levels** can thus be recognized within a living system: the producers, the green plants, occupy the first trophic level; the herbivores, or primary consumers, the second; and so on. The energy originally derived from the sun by plants thus passes through a **food web**—all the feeding relationships among and between the species of a biotic community. Food webs are usually composed of many interlocking **food chains** which represent single pathways up the web.

Any food web or chain is essentially a system of energy transfers. Solar energy trapped by plants passes through the successive animal consumer levels. Each gives off some as heat (of chemical transformation) to the environment, and the total declines progressively through the chain (about 90 percent of

Figure 12-4 Some interrelations among inorganic substances, plants, and animals and protistans in nature. Arrows indicate the path of materials from the primary (inorganic) sources through various organisms and back to the soil, water, or air. Any one circuit is a food chain.

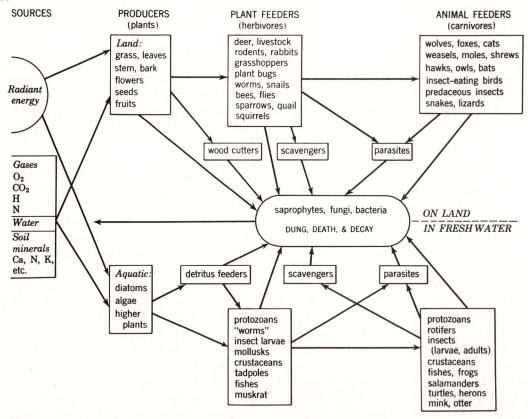

the energy is lost at each level). Decay of plants yields much heat, and the decay of animals results in a lesser amount.

Food webs are very complex (Fig. 12-4), even in a small community, but may be illustrated by two simplified examples. In a pond, photosynthetic bacteria and diatoms synthesize materials, and then in sequence small organisms are eaten by larger ones, thus:

Bacteria and diatoms → small protozoans → larger protozoans → rotifers, small crustaceans → aquatic insects → fishes

The large fishes, or any intermediate organisms, by death and decay become food for saprophytic bacteria, or the fishes may be taken by herons, minks, or humans and their residues end up elsewhere. On land a food web may include the following:

Plants → plant-eating insects, rodents, or grazing mammals → predaceous insects or small carnivores → larger carnivores

This ends with death and decay as in the water. In any predator chain the successive members are larger in size but fewer in total numbers. Chains are not strictly linear but have many branches and alternative links. Any animal above the smallest takes food (prey) in relation to its size and the means of consuming it, neither so small that it cannot profitably gather enough nor too large for it to overcome successfully. Most food chains on land are shorter (fewer trophic levels) than those in water, because land plants are generally larger and many large land animals feed directly on them. A food chain seldom has more than four or five species because of the energy lost with each energy transfer in moving up the chain. The many chains and their interconnections aid prey and predator populations to adjust to environmental changes, and they thus confer a measure of stability on the ecosystem.

12-7 Pyramids of numbers, biomass, and energy

The food web in any community can be depicted as a **pyramid of numbers** (Charles Elton)—the density of individuals present at each trophic level at a given interval of time. Typically animals at the base are small and abundant, whereas those at the apex are

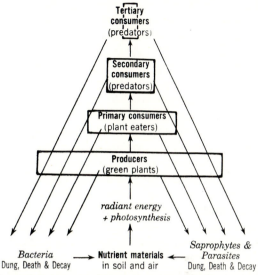

Figure 12-5 The food cycle and a pyramid of energy. Going upward, increasingly less energy is available for the next trophic level, as suggested by changing size of boxes.

few but large; of those between, there is a progressive increase in individual size but decrease in number. In a deciduous woods, for example, aphids and other minute plant-feeding insects may be enormously abundant, spiders and carnivorous beetles are fairly common, insectivorous birds are fewer, and the hawks and weasels preying on the birds are numerically scarce. The smallest kinds, because of their size, can grow and multiply rapidly, whereas larger members of the chain reproduce more slowly. The largest predators at the apex are relatively so scarce that in turn they cannot serve profitably as prey for another species. A similar **pyramid of biomass** can be used to show the total bulk of organisms (by weight) at any one time at each trophic level. Biomass is the total amount of living material in a given area. The pyramids may be inverted in some communities, the consumers outnumbering or outweighing the producers, as in some aquatic ecosystems in which the turnover in phytoplankton is much faster than that of the grazing zooplankton.

The **pyramid of energy** reflects progressive loss of energy (as heat) between one trophic level and the next (Fig. 12-5). Energy transfers are not entirely efficient. Producers use part of the energy derived from the sun in respiration. Each consumer in the chain expends or loses energy in obtaining food and digesting and metabolizing it and in the maintenance of bodily activities; some food is also passed unassimilated. It is estimated that only approximately 10 to 20 percent of the energy at a given trophic level is available for the next one. This explains why most food chains contain no more than four or five links. Not enough energy remains above the top predators to maintain another trophic level. In a Minnesota bog, respiration of producers consumed 21 percent of their energy intake, that of primary consumers 30 percent, and that of predators 60 percent. The great loss for predators results from their large size and activity.

A detailed account of energy transfer in a natural community is provided in an elaborate study by Howard T. Odum at Silver Springs, Fla. The huge artesian freshwater spring has been stable for many years, discharging upward of 1135 million L (300 million gal)/day at about 23°C (73°F). Primary production is by submerged plants, mainly *Sagittaria loricata* and the encrusting layer (*Aufwuchs*) of green algae. Dominant herbivores are turtles, mullet, sunfishes, other fishes, snails, some insects, etc. The smaller carnivores include sunfishes, catfishes, predaceous beetles, and small invertebrates. Top carnivores are large-mouthed black bass, gars, and alligators. **Decomposers** are bacteria and crayfishes. The turnover of biomass averages about eight times per year for a total primary production of 6.39 kg/m² (57,100 lb/acre) of organic matter. The pyramid, in grams per square meter of biomass, comprises producers, 809; herbivores, 37; carnivores, 11; top carnivores, 1.5; decomposers, 5. Differences between the output of one level and the input of the next (Table 12-1) result from decay products and materials washed downstream. The net plant production is supplemented by bread fed to fishes by tourists (486 kcal/m²/year).

Table 12-1
ENERGY BUDGET FOR SILVER SPRINGS, FLORIDA

| | Energy: Kilocalories/ square meter/year | | Efficiency for | |
	Totals	Used in respiration	Growth,* percent	Use, † percent
Sunlight				
Total	1,700,000			
Not absorbed	1,290,000			
Plant producers				
Used in photosynthesis	410,000	389,190	5	
Production:				
Gross	20,810			
Net	8,833	11,977	42	2.15
Herbivores				
Input	3,368			
Output	1,478	1,890	44	16.7
Carnivores				
Input	383			
Output	67	316	17	4.5
Top carnivores				
Input	21			
Output	6	13	29	9.0
Decomposers				
Input	5,060			
Output	460	4,600		
Downstream export	2,500			
Community respiration, total 18,796				

* Input versus output.
† Output versus output of prior level.
(After H. T. Odum. 1957, Ecol. Monographs, vol. 27, pp. 55–112.)

12-8 Shelter and breeding places Animals that live on plains or in large bodies of water can avoid capture through superior ability in locomotion. Many species of the smaller waters and on the land live in or near various types of **cover** and use a retreat, or **shelter,** to avoid predators and for other purposes. A rabbit, surprised in the open by a fox, darts for the first briar patch. A white-footed mouse, when come upon, seeks the safety of a hole—in the ground, in a log, or under a rock; the same retreat may serve as shelter from rain or excessive cold, as a storehouse for food, as sleeping quarters, and as a place to rear young.

Different mammals, birds, lizards, and insects live in cover such as grassland, shrubbery, trees, or rocks; various marine fishes and invertebrates dwell amid seaweeds, rocks, or coral in shore waters, and some freshwater fishes inhabit stands of aquatic plants. Moles, pocket gophers, some snakes, certain insects, earthworms, and other invertebrates live more or less continually within the soil. In all these cases the animals find their food in the surroundings and are able to escape enemies and avoid extremes of weather.

Most animals also have special requirements for **breeding places** where the eggs or young are produced. For some the shelter serves, but others build special **nests,** as is done by many birds, some fishes, and various insects. Small animals breed wherever conditions are suitable. In many kinds of birds and some mammals, however, each pair or group establishes a **territory** to supply the food requirements of the parents and offspring for the breeding season; the territory is defended against invasion by others of the same species.

The availability of suitable cover, shelter, and breeding places is important in regulating the number of a species that can live in an area. If "housing" is in short supply, few can be accommodated and the others are exposed to weather and predators. Human beings increase populations of game mammals, birds, and fishes by providing artificial cover or modifying natural conditions to provide better environment for the species of their choice. Contrariwise, they may reduce objectionable species by destroying their cover: removal of brush in Africa limits breeding and dispersal of tsetse flies. Such manipulations, however, disturb the natural community structure.

12-9 The niche and competition   The **ecological niche** is the way a species population is specialized within a natural community to gain the resources needed for its existence—how it actually uses its environment. Specialization reduces or eliminates competition and permits two or more species to coexist as members of the same community, for each is assured needed resources.

In a broad sense a species' niche includes its total spectrum of adaptations to its particular environment. Thus a full description of the niche would include an infinite number of biological and physical characteristics. In practice, attention is often centered on differences in spatial location, timing of activities, resources used, interactions with other species, and population control mechanisms. Niches are not fixed; they shift with time as the organismal unit, through natural selection, tracks a changing environment.

The **habitat** of a species refers to the place or environment(s) where it lives. The habitat of the eastern gray squirrel (*Sciurus carolinensis*) is the eastern hardwood forest; its niche is defined in terms of its food, time of feeding, nest site, etc.

Species vary in size and character of their niches. The house mouse and Norway rat have successfully followed humans over the world partly because of broad niches for food and shelter. The food niche for some species is narrow, centering on one or a few prey species: the thorny devil (*Moloch horridus*), a spiny lizard of Australia, feeds heavily on a single species of ant. In comparative studies species with broad niches are often called **generalists,** those with narrow niches **specialists.** The latter may be more efficient in use of a resource but are more vulnerable because they cannot change to another. Today species that are specialists are decreasing in many places because they cannot adapt to changes in the environment brought about by human beings.

Two or more species in the same habitat that have the same requirements for resources limited in supply are said to be in **competition.** If the niches of two or more species overlap, they compete with each other. Where two species have niches that overlap broadly, the result is usually the survival of but one. This concept is called **Gause's principle,** or the **principle of competitive exclusion,** which states that no two species occupying the same niche can coexist in the same place at the same time. Gause was able to test this principle experimentally by confining *Paramecium caudatum and P. aurelia* in a closed container with a fixed amount of food. The result was that one species always died out.

Whenever a number of similar species coexist in the same area (such as the species of antelope on the African plains), they are able to do so because they occupy different niches or because no necessity of life is in short supply.

Members of the same species compete with one another, a situation termed **intraspecific competition.** This competition is important in controlling population size and is notable in territorial animals, which establish and defend areas of the environment against encroachment by others of the same species (competition for space). It ensures a sufficient food supply and a population level that does not exceed the resource.

12-10 Predation Every animal that consumes another animal is called a **predator,** and the animal eaten is its **prey.** Any that consumes members of its own species is termed **cannibalistic,** and one that eats dead animals is a **scavenger.** Any food chain, after the first plant-eating animal, is a succession of predations. Predation differs from parasitism in that a predator destroys its prey outright whereas a parasite usually continues to feed on its living host. As Elton says, predators live on capital and parasites on income. In a food chain each predator is usually larger than its prey, whereas the parasite is always much smaller than its host.

There is a kind of coevolutionary race between predator and prey in the development of methods of offense and defense. An evolutionary advance in one causes a selective response in the other. Over the long term such reciprocal evolutionary escalations have resulted in some elaborate adaptations. The predator is commonly believed to regulate the numbers of its prey. Actually there is an **interaction** between predator and prey related to changes in their population densities. If a prey population increases, it will support more predators; if the latter then also increase, a heavier toll will be exacted of the prey, whereupon the latter will be reduced. In either type of change the numbers of the predator tend to lag behind those of the prey because the predator, being larger, usually has a slower rate of increase, although it lives longer. The amount of cover available in relation to the prey population also is a factor; more prey will escape the predators if there is much cover. When the prey population falls below a certain level, it becomes unprofitable for the predator to pursue that particular kind; it must shift to some other source or suffer decline. The red fox preys variously on woodchucks, cottontail rabbits, squirrels, mice, birds, and insects but also eats fruits and berries; its diet changes with season and place according to the foods most readily available. Such a predator is likely to vary less in numbers than does the arctic fox, which lives mainly on lemmings or rabbits in the far North, where the population of both those animals fluctuates widely from year to year.

Predation is the major mechanism by which excess animal productivity is redistributed by conversion to other animal tissue at higher trophic levels. It helps to maintain populations within the carrying capacity of their habitats and lessens sudden explosions and population crashes in prey species. It helps maintain the health of prey populations by the "culling effect" on sick and old individuals. It is one of the most important mechanisms of natural selection, for it generally removes less fit individuals and thus eliminates inferior genes from the population. Predators usually act as executioners for weak animals destined to die anyway.

12-11 Parasites and disease Virtually every animal species is subject to diseases produced by various types of organisms—viruses, rickettsias, bacteria, protozoans, parasitic worms, and arthropods. The disease organisms themselves must be considered as populations that in turn are affected by various factors in their respective environments, and they in turn influence the numbers and well-being of the animals on which they live. Attention here will be confined to parasites, although the role of disease in controlling animal abundance is probably far greater than is generally recognized.

A **parasite** is an organism that lives on or in another species, the **host,** obtaining food and shelter at the latter's expense. The host can live without the parasite, but the latter normally cannot exist without its host. The parasitic organisms include many PROTOZOA, such as those causing malaria, amoebic dysentery, and other diseases; flukes, tapeworms, and other PLATYHELMINTHES; hookworms, root nematodes, and many other NEMATODA; all hairworms (NEMATOMORPHA) and spiny-headed worms (ACANTHOCEPHALA); various leeches (ANNELIDA); and many ARTHROPODA, including fish lice, some barnacles and other CRUSTACEA, the mosquitoes, flies, fleas, and various other INSECTA, and the ticks and many mites (ARACHNIDA).

A species of parasite inhabits one host species or a group of similar hosts or alternates between two or more host species; each kind of parasite is usually restricted to a certain site in its host. **Ectoparasites** such as leeches and lice live on the skin, and **endoparasites** dwell within the body, in the gut cavity (many worms) or other organs, in muscles (trichina) or other tissues, in the blood (some worms and protozoans), or even in blood cells (malarial parasite). Some parasitic insects and ticks in turn are **intermediate hosts** for parasitic protozoans or other organisms that they transmit to other or **definitive hosts** (Chaps. 15, 18 to 20, 23 to 25).

Different parasites evidently arose separately in various phyla from free-living ancestors and have become variously specialized or degenerate for the parasitic mode of existence. Many have hooks or suckers for holding to their hosts; the gut is simplified (absent in tapeworms), because their fluid food is obtained directly by pumping or absorption from the host; and the reproductive organs are usually elaborated to produce enormous numbers of eggs or larvae to overcome successfully the hazards of reaching new hosts.

Some parasites have scant effect on their hosts, others injure them temporarily or permanently by the destruction of tissues or production of toxic secretions, and some kill their hosts. Parasitoids (Sec. 25-27), whose actions resemble a predator, are an example of the latter. In general, a "good" parasite does not kill its host, at least not until it and its host have reproduced. Parasites that cause disease are termed **pathogenic.** A host that recovers from the initial attack or damage often becomes a **carrier,** retaining some of the parasites, which continue to pass out eggs or larvae that may infect other hosts. Parasitism is the mode of life for many species. It, along with some diseases, is one factor in the regulation of populations of host animals. Some of the most notable instances of such control occur among insects, where the matter is further complicated by secondary parasites, or **hyperparasites,** that parasitize the primary parasites (Sec. 25-27).

12-12 Symbiosis Parasitism and some other kinds of special interrelations between two organisms of different species are termed **symbiosis** (Gr., living together). When one gains benefit by living with, on, or in another and without harm or benefit to the second, the case is called **commensalism.** The commensal "lives aboard his neighbor's vessel but does not eat his provisions"; some, however, do get scraps from the host's "table." In some cases the two are continuously associated with each other, as the special types of barnacles that attach only to whales or sea turtles, the crabs that live regularly in the tubes of some annelid and echiurid worms, and the crabs that dwell in the mantle cavities of sea mussels. With others the association is not continuous—like the remora (Fig. 30-30), which attaches by its dorsal sucker to some other fish for transport, the fish (*Fierasfer*) that shelters in the cloaca of a sea cucumber, and the elf owl of the southwestern American desert that roosts and nests in holes made by the gilded flicker, another bird, in the giant saguaro cactus.

More intimate is the relation of **mutualism,** where both parties are benefited. An example of "on and off" mutualism is seen in the birds that alight on large grazing mammals (cowbird or oxbird, on cattle, rhinoceros, etc.) to pick off ticks—the birds obtain food, and the big beasts are relieved of parasites; also the birds, by their behavior, warn the mammals of approaching enemies. Some marine fishes are similarly served by smaller fishes and crustaceans. Ants and aphids may feed separately, but certain ants keep aphids in their nests and put them on roots or stems of corn and other plants to feed. The aphids are protected, and the ants stroke them to obtain food, a sweet fluid from the hindgut of the aphids.

In other cases two types of organism are continuously associated. A striking example is that of certain termites and flagellate protozoans. The termites eat wood but cannot digest cellulose, but the flagellates in the termite gut can, so food is available for both. Termites experimentally rid of the flagellates soon die of starvation, and the protozoans cannot live a free existence. Some species of sponges, hydroids, and sea anemones live regularly on shells of hermit and other crabs. The riders are carried to new feeding areas and are not stranded at low tide, while the crab is somewhat camouflaged and the riders are protected in a degree from predators. Animal-plant mutualism is seen in the green hydra (*Chlorohydra*) that has green algae (zoochlorellae) in its cells, the one producing carbon dioxide and the other oxygen as a by-product; sealed in a tube of water, both live for some time by mutual aid. Various protozoans, sponges, sea anemones, corals, and flatworms have yellow or brown zooxanthellae (flagellates), and a similar O_2-CO_2 exchange occurs. Certain species of ants, beetles, and termites grow and tend "gardens" of peculiar types of fungi that live only under such care, while the fungi serve those insects as their sole food. Even humans have a mutualistic relation with the crops of wheat and maize and domestic animals that grow only with human aid.

The **cross-fertilization,** or **pollination,** of plant blossoms by insects (occasionally by birds or bats) is a mutual relation of wide occurrence and great importance, because many plants are self-sterile. Often both flower and insect are structurally and functionally specialized to achieve successful transfer of pollen from flower stamen to insect body to (another) flower pistil. Hive bees visit blossoms to obtain both nectar and pollen for their own needs (Sec. 25-15) and in so doing carry pollen between the flowers they visit. Many human staple food and forage crops depend on these flower-insect relationships. The great adaptive radiation of the flowering plants (angiosperms) following their origin in the Cretaceous appears to have involved mutualistic relationships with evolving insect pollinators.

12-13 Colonies and societies Vertebrates, most arthropods, and many other invertebrates are **free-living** in that each individual gets about by its own efforts. By contrast, sponges, many hydroids, corals, bryozoans, tunicates, and others are **sessile,** being fixed to some substratum of rocks, plants, or the shells of other animals. Among both categories, many species are **solitary** in that each individual is more or less independent, whereas others live in groups, or **colonies.** The numerous "individuals" of a colony among sponges, bryozoans, and tunicates are bound together structurally. In other cases the individuals in colonies of insects, schools of fish, flocks of birds, and herds of hoofed mammals are structurally separate but often integrated by behavior.

Many carnivorous mammals (mountain lions, minks, etc.), hawks, flycatchers, snakes, and predaceous insects forage best independently and pair up only for reproduction. Members of some other species assemble without reference to age or sex. In the winter flocks of robins, ducks, or starlings, many

eyes and ears, afford better protection when foraging or sleeping; groups of bats, snakes, or lady beetles gather for hibernation with some advantage. Among active groups of mammals or birds there is often a "leader" and a gradation in dominance between individuals, known as the **peck order**; this may be seen among fowls of one sex in a barnyard flock. Sexual aggregations brings males and females together for mating, as with frogs, toads, gulls, and fur seals.

Social organization occurs where many individuals of a species live together in an integrated manner so that each contributes in some specialized way to the welfare of all. The social habit has arisen independently in several orders of insects; it is highly developed among the termites and in many ants, bees, and wasps. The transition from solitary to social life apparently is correlated with lengthening of adult life and increasing parental care. Solitary bees provision their nests, lay eggs, and depart, never to see their offspring. Social bees, on the other hand, feed their young regularly during development. Distinct **castes** have developed to perform such tasks as feeding, guarding the colony, etc. Worker and soldier castes differ in structure and physiology and cannot live independently. Success with them is measured in terms of the colony and not of the individual (Sec. 25-32). Human society comprises integrated groups of like individuals that specialize in different trades or professions with benefit to both the individuals and the group.

Social integration and communication among individual animals is achieved by visual, acoustic, tactile, and chemical means. **Pheromones,** chemical signals which act between members of the same species, also are commonly employed and are usually detected by the senses of smell or taste. They may slowly (or sometimes rapidly) bring an animal into a condition of responsiveness for mating or other behavior. Various ungulates, carnivores, reptiles, and other animals mark their territories with distinctive scents. Females of some species of moth release pheromones that attract males from several miles or more. Male newts evidently identify females by their odor. Certain ants, minnows, and tadpoles, when threatened, release an alarm substance that quickly alerts others of their species in the vicinity.

Some kinds of ants respond with attack behavior. A pheromone secreted onto the ground guides foraging ants on routes between the nest and food.

Four "pinnacles" of social evolution may be recognized, as set forth by Wilson.[1] (1) **Colonial invertebrates** such as corals, siphonophores, and bryozoans have evolved nearly perfect societies. Beginning with clusters of self-sufficient individuals (zooids), a graded evolutionary series leads to aggregations that resemble a multicellular organism. In the latter the physically united individuals are specialized for several functions and are closely integrated in structure. Cooperation and "altruism" among members for the good of the colony are virtually complete. (2) In the **higher social insects**—ants, termites, and certain wasps and bees—social structure is less perfect, but some similarity with the lower vertebrates exists. There are specialized sterile castes that "altruistically" feed nestmates and defend the colony and queen, often with suicidal stings. The colony, although made up of unattached individuals, is held together by intricate forms of communication (pheromones, etc.). However, success of insect colonies depends to a considerable extent on independence of individuals. Workers make sorties from the nest to procure food for the colony. There is often some antagonism. Queens are not always the exclusive egg layers; workers occasionally appear to vie with the queen in laying eggs that produce the haploid males. In some wasp colonies egg laying becomes the right of the dominant female, and among some colonial insects, upon loss of a queen, workers fight to take her place. (3) In **vertebrate societies,** including mammals, independent behavior and antagonism are carried much further. Individual behavior is primarily selfish. Cooperation and association with the group are mainly expediencies that improve chances of individual survival and reproduction. "Altruism" is based chiefly on kinship. It is directed primarily toward personal offspring but may extend to other relatives and associates. It is best developed among mammals and some birds. How-

ever, such cooperating subgroups tend to pursue their own ends, thereby limiting the extent to which the society can function as a unit. In contrast with the invertebrates, communication is generally more elaborate: groupmates respond to one another as individuals, bonds may form among them, and learning and tradition assume prominence. (4) **Human social structure** is sufficiently distinct to be recognized as a fourth "pinnacle" of social evolution. It has arisen only once, whereas the others have repeatedly evolved independently. Its level of complexity is unique. Divergence from other vertebrate societies has occurred through the evolution of intelligence and the ability to learn from the past and to plan ahead. Selfishness persists, but kinship ties have developed far beyond those of any other social species: reciprocal altruism often spans generations. Cooperation approaches that of insect societies and, through language, far exceeds them in communication.

The first and second pinnacles are considered to be primarily the result of group selection, the third and fourth the result of individual selection. From siphonophore through social insect to "selfish" mammal, there is reduction in cooperation among individuals within the society. Social evolution exhibits a downward trend. Humans seem to have reversed the trend; they have been capable of a remarkable degree of intergroup cooperation with little or no loss to personal survival and reproduction. Under high population densities, however, their capacity for such cooperation is strained.

Some comments are in order concerning "altruism." The term need not imply conscious self-sacrifice. Altruism reduces personal fitness. How can it be a product of natural selection? When the genes causing the altruism are shared by two individuals because of common descent, and an altruistic act by one individual increases the joint contribution of these genes to the next generation, then the propensity to altruism will spread through the gene pool (Wilson). It is thus based on kinship.

---

[1] *Mostly paraphrased from E. O. Wilson, 1975, Sociobiology: The new synthesis. Cambridge, Mass., The Belknap Press of Harvard University Press.*

12-14 Populations   All the animals of one species that occupy a given area constitute a **population.**

Sometimes the limits of the area are set by the observer; in other cases populations have evident spatial boundaries. Beyond the activities of its constituent members, a population has definite structure and organization. It grows and declines and has a certain composition as to ratio of the sexes and age groups that may change with circumstances. Population is stated in terms of density, the number per unit of area. Rate of change—increase or decrease—is determined by the number of new individuals added (birth rate, or natality, and immigrants) versus the losses from all causes (death rate, or mortality, and emigrants). When additions exceed losses, the population increases, and vice versa. The course of a population with time and changes in its age structure can be expressed in graphs (Figs. 12-6, 12-7).

Few people realize the huge populations of animals and plants that exist and the rapid rate of turnover of many of them. Diatoms and protozoans may exceed 1 million per liter of seawater. Censuses of the upper $\frac{1}{2}$ in. of topsoil near Washington, D.C., revealed small organisms at the rate of 1,200,000 animals and 2,100,000 seeds or fruits per acre of forest soil; meadow soil contained 13,600,000 animal items and 33,800,000 of plant materials. On croplands, grasshoppers sometimes number 20,000 to 200,000 per acre, and heavy infestations of eggs and larvae of the alfalfa weevil may total 8,000,000 to 22,000,000 per acre. The numbers of larger organisms are naturally much smaller.

Some winter flocks of ducks have included more than 200,000 birds; the bison on the Great Plains originally numbered several million, and of these about 3 million were killed from 1872 to 1874 alone; in eastern North America, flocks of the now extinct passenger pigeon once darkened the skies for days during migration, and nearly 12 million were killed and sold in one Michigan town in 40 days.

The populations of some organisms that constitute stable natural communities are rather constant through time. Others fluctuate around a fairly constant mean. Yet each species has a large reproductive or **biotic potential.** This potential is geometric for either large or small organisms, as Malthus observed over 180 years ago. A female housefly (*Musca domestica*) can produce 120 eggs. If all lived, grew, and reproduced through four generations, there would be more than 25 million flies. Since this does not happen, there obviously is another force acting in opposition, which is termed **environmental resist-**

Figure 12-6 Ideal curves of population growth.

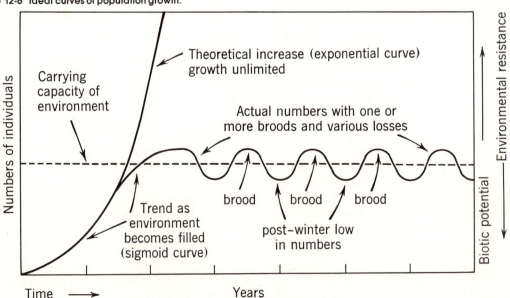

ance. It is manifested as an increase in mortality or a decrease in natality (birth rate) or both.

The biotic potential, or intrinsic rate of increase of a species, is symbolized by r. It depends on the maximum number of offspring an individual can produce and the total number of reproducing individuals in the original population, symbolized by N. The mathematical expression for geometric growth of a population of biotic potential r and population size N is the differential equation $dN/dt = rN$, where dN/dt stands for the rate of growth of the population with time (t). In other words, the change in population size with each unit increase in time is equal to rN. For example, if the initial population size (N) is 100 and r is 2, then at the end of the first breeding season (t_1) the population will have increased by 200 (i.e., $rN = 2 \times 100 = 200$). This figure then becomes the population size for the next breeding cycle, at the termination of which (t_2) the population size will be 400 (i.e., $rN = 2 \times 200 = 400$). Subsequent population sizes following each successive breeding season (t_3, t_4, t_5, etc.) will be 800, 1600, 3200 etc. Thus the change in population size is geometric with time. Population increase, however, is checked by environmental resistance. The latter is usually designated by K, the carrying capacity of the environment for the population, or the maximum population number that can be sustained under existing conditions. Adding K to the growth equation yields the logistic equation (Volterra-Gause equation) that indicates the rate of change with changing population density:

$$\frac{dN}{dt} = rN\left(\frac{K - N}{K}\right)$$

As N approaches K, the expression in parentheses approaches zero and so does the whole equation. Thus, as a population approaches the carrying capacity of its environment, the rate of growth (dN/dt) approaches zero. This equation describes only simple or theoretical population curves. It applies to human as well as other animal populations.

The characteristic growth curve for a population when introduced into a new area (or a population of fruit flies or flour beetles started in a laboratory cul-

ture) is sigmoid or ∫-shaped[1] (Fig. 12-6). Growth may be slow at first because the few individuals do not readily meet for mating. Even with free interbreeding it is relatively slow because of the nature of exponential growth. After establishment, growth becomes very rapid, increasing in exponential fashion (like compound interest) but eventually levels off as (1) the population reaches the limit of its food supply, or (2) all suitable habitats are occupied, or (3) parasites and predators multiply greatly. Also various **feedback mechanisms** operate to limit population size. Flour beetles in a culture bottle, for example, supplied with excess food and protected from enemies, nevertheless reach a peak and then decrease because of accumulation of waste products (feces). Water in which some fish or tadpoles have grown under crowded conditions contains material that inhibits growth of others of their own kind. An **equilibrium** is reached and the population fluctuates around this level, depending on variations in climatic and biotic factors. The equilibrium is also known as the **saturation level,** or **carrying capacity,** for a given locality, assuming that environmental conditions remain relatively constant; actually conditions change from season to season, year to year, and over periods of years. The "balance of nature," therefore, is a dynamic one, ever changing with shifts in the many environmental factors, some of which are also subject to cyclic changes.

Following the season for production of young in any species there is often a surplus of individuals, resulting in **population pressure.** The reaction is a reduction in numbers by several means: dispersal to other areas and losses by predation, disease, or starvation. The effects are **density-dependent,** that is, related to the numbers per unit of area. These factors of loss usually operate more severely when numbers are high, their effects lessening as the population density declines. Sometimes only one density-dependent factor need operate to effect population regulation.

---

[1] *The curve may be J-shaped rather than sigmoid when growth is rapid, resulting in sudden overexploitation of resources. Decline is then precipitous and occurs before self-limiting processes within the population become important. The population is said to crash.*

Severe unseasonable weather (hurricanes, droughts, floods, excessive heat or cold) is **density-independent** in its effect on decline in a species. The kill is in proportion to the number of animals present, but there is no special effect related to the density of numbers when the calamity occurs unless crowding deprives some individuals of shelter.

Some populations remain relatively stable from year to year with only slight fluctuations; on an annual basis the numbers added balance those lost. There is, of course, seasonal change through the year, numbers being largest just after the production of young and then declining until the next breeding season (Fig. 12-7). Such populations are usually of long-lived species in constant or predictable environments. However, fluctuation and extinction are commonplace events in local populations of most species. Quail populations near Davis, Calif., underwent a twofold fluctuation in November densities over a period of only 4 years (Fig. 12-7). Some species show conspicuous cyclic fluctuations: Eurasian locusts, meadow mice, lemmings, ruffed grouse, snowshoe hares, and arctic lynx are a few examples. The numbers rise phenomenally in some years, then decline abruptly in others. The varying hare (*Lepus americanus*) in Canada has fluctuated

over a 1:1000 ratio but more commonly in a 1:10 ratio. The cycle approximates 10 years. Lemmings have a cycle of about 4 years.

Cyclic population changes are still not well understood despite extensive research. Predator cycles can be explained as a response to cyclic behavior of prey, but the factors governing prey cycles are less clear. In small rodents such as meadow mice and lemmings, there is some evidence that their rhythmicity may result in part from periodic overexploitation of forage vegetation. Populations then crash and require about 4 years to reach density levels that again deplete the food supply. Density-dependent factors of predation, disease, shifting selection pressures, and competitive stress mount as high densities are approached and contribute to precipitous population decline. However, random fluctuations of population numbers can be easily mistaken for cycles of short period (3 or 4 years).

12-15 The biotic community The populations of plants and animals living together in a given environment and interacting with one another to form a distinctive living system constitute a **biotic community.** Communities are often named for some

Figure 12-7 **Age composition and changes in numbers of a valley quail population at Davis, Calif., through several years, as determined from banded birds.** Despite the large crop of young in 1936 the total population declined. (*After Emlen, 1940.*)

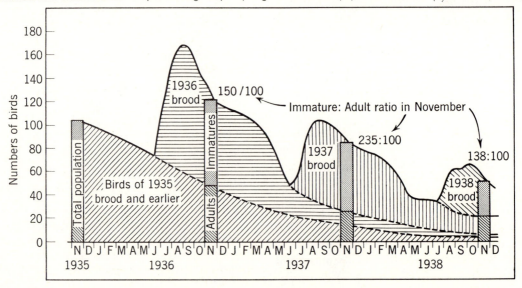

dominant feature, biotic or physical; e.g., *redwood* forest, *sagebrush* desert, *mudflat* community. They may range in size from a rotting log to major forest formations and the planktonic community of the world's oceans. In part, limits are set by the observer. The species composition of a community depends on climate and historical (evolutionary) factors. In terrestrial communities soil conditions and, in some areas, fire are further contributing influences. In widely separated areas plant communities have evolved similar growth form, stature, and other adaptive characteristics as a result of similarities in climate and soil, but species composition differs. Associated animals follow suit.

Some communities have rather sharp boundaries where ranges of some of the more conspicuous species stop, but other communities grade into one another in varying degrees. Sudden changes often occur when environmental gradients (in temperature, moisture, soil conditions) are steep or change abruptly. Whereas the ranges of some species are contained within the community, others extend beyond. Some species are distributed independently on environmental gradients.

The extent of interaction among community members varies. In mature communities in long-enduring stable environments, linkages among species may be great. Symbiosis is common, and there may be many interdependent relationships. In changing environments, species associations may be loose and species composition in a state of flux. In general, diversity tends to increase in communities with time by the addition of species differing in niche and habitat.

A natural community has been compared by analogy to an organism; but the integration of the component parts (the species) in a community is far less than that of the cells of an organism, and there is no central control. Each species is a separate entity with its own hereditary mechanism responding to natural selection in its own way, although influenced by its community context. Undisturbed natural communities are able to undergo changes in species composition and population densities without major disruption. Such changes trigger responses in other species, causing them to expand or contract. Thus as communities evolve, the functional complexity of the living system as a whole is maintained or changes more or less harmoniously.

In certain communities the members may have characteristic form and size: those in a fallen log are small and sometimes flattened; those in flowing water are streamlined. The larger communities of both land and water show **stratification.** A forest has characteristic animals of the treetops, the lower branches, the bark, the leaf litter, and soil. A lake has surface dwellers (water striders, ducks), others in intermediate depths (plankton, fishes), and bottom inhabitants (snails, worms, insect larvae, etc.). In any community one or a few species are **dominant** over the others in numbers or physical characteristics or both. In a pine forest the pine tree is dominant, exerting a controlling influence by shading other plants and the ground and by producing a carpet of needles (of acid reaction). Because of these conditions, the kinds and numbers of plants and animals that can live there are largely determined by the pine forest.

12-16 The ecosystem   A community of organisms and their nonliving environment at any one place together constitute an ecologic system, or **ecosystem.** There is a flow of energy and materials throughout a balanced ecosystem and feedback mechanisms that contribute to its stability. For example, if a predator multiplies, its numbers later will be reduced by scarcity of prey; then, as the predator population declines, the prey may revert to former numbers. This pattern of checks and balances resembles the physiologic homeostasis, or steady state, within an organism. An ecosystem may or may not be balanced, as can be determined by studying the total input and outgo of energy and materials. Humans degrade ecosystems by destroying their feedback mechanisms as well as their living components. When Guam was devastated by a typhoon in 1962, the overall balance was not disturbed and recovery to former conditions was rapid. By contrast, the introduction of domestic rabbits on Laysan Island in 1902 led to near extinction of the plant life by 1923, when the last rabbits were killed. Later a gradual repopulation of the plants restored a functioning ecosystem. Many supposedly "balanced" natural

ecosystems are undergoing long-term, perhaps irreversible, changes due to human activities.

An ecosystem may be closed in that materials (nutrients, water) for the most part flow within its boundaries, or open with flow both in and out. However, all ecosystems are "leaky" to some extent. Ecosystems vary in size from those in a small volume of soil or water to the entire earth as a single living system. The limits are arbitrary, as set by the observer. A self-sustaining ecosystem usually must include the following components: producers (mostly green plants), consumers (chiefly animals), decomposers (decay organisms), and the physical environment that provides minerals, water, and sunlight (Secs. 12-6 and 12-15).

The actions of the components of an ecosystem are seen clearly on a small island, where there is a certain potential for primary production and a well-defined carrying capacity for each kind of consumer level.

12-17 Ecologic succession   (Fig. 12-8) No ecosystem is permanent; some change abruptly or slowly, others persist for years or centuries. In a bare area, such as a region of exposed rock in the mountains, there will be a sequence, or **primary succession,** of communities: first a pioneer stage involving soil formation then gradual change, and ultimately a relatively stable phase, the **climax.** When the sequence occurs in an area formerly occupied by a community and where soil is already present, it is called a **secondary succession.** The sequence of communities that replace one another is called the **sere,** and each seral stage influences the development of the next, culminating in the climax. An example of a primary succession is the lake-pond-swamp-meadow-forest sequence to be seen in many areas once glaciated, then gradually filled and covered with soil by inflowing streams. The glacial lake is low in dissolved nutrients and hence has poor plank-ton supply and few fish; the shallower pond has higher mineral content, more marginal plants, better plankton, and a variety of small crustaceans and insect larvae and more fishes; the swamp has profuse rooted vegetation, few fish, and many aquatic invertebrates; the meadow has grasses and herbs and insects that forage on them, earthworms, toads and frogs, various birds, meadow mice, and shrews; finally, the forest is a drier habitat with other kinds of animals, including species that use trees for food or shelter.

The principle of ecologic succession is of practical importance in human affairs. Any field that is plowed and then left fallow has a changing sequence of plant growth. On range land the vegetational cover will be altered in a succession that depends on both the seasonal pattern and the intensity of grazing by livestock. In the Eastern states when forest is removed, trees soon grow, but the species composition of such a secondary succession may not resemble the original stand. In the arid West when a climax forest is cleared off—by lumbering or fire—the resulting increased sunlight, erosion, and competition by herbs, shrubs, and rodents make for a long succession before the original pattern is reestablished. There is also succession on a geologic scale, with major changes in climax communities. Cylindrical cores cut from the bottom of a Connecticut lake by Deevey showed a vertical succession of pollen types extending back 11,000 years (dated by radioactive carbon; Sec.12-29). The climate, indicated by vegetational types, had varied from cold to warm (pine and spruce to deciduous hardwoods). Transition from deep, clear waters to a shallow pond was indicated by remains of insect larvae: a freshwater midge (*Tanytarsus*) at the lower level and a stagnant-water type (*Chironomus*) higher up.

12-18 Ecology and conservation   Any change in the physical or biological characteristics of an en-

---

Figure 12-8   Succession in soil making and plant cover at edge of a large lake, providing new habitats for animals. Tree zones are indicated by dominant species. Over several centuries the shoreline moves out into the original lake area, and new types of plants progressively occupy the new land. Through time, replacement generations of animals keep to their respective habitats—fish and ducks to the water and the killdeer (a shorebird) to the advancing beach. As new habitats develop, they also become occupied—earthworms and other invertebrates in the humus, arboreal squirrels in the trees, and so on. The rock on the original beach is a "reference." (*Adapted from R. and M. Buchsbaum, Basic ecology, by permission of Boxwood Press, Pittsburgh, Pa.*)

vironment affects member species of plants and animals in different ways. To the natural forces that act upon wild populations, human influences have been added in many places and with increasing intensity in recent centuries and decades. Civilization is essentially a human effort to manipulate and use the environment for the advantage, usually immediate, of human kind. In early stages human beings "lived off the land"; bison, elk, deer, waterfowl, and land game together with beaver and other fur bearers have been greatly reduced or completely eliminated in the zeal to obtain flesh and pelts. Later, competitors such as the wolf, coyote, mountain lion, and others that prey on livestock or game and the rodents and insects that eat crops have been subjected to extensive continuing control. More direct environmental manipulation has involved removing forests, draining swamps, irrigating arid lands, and planting huge acreages of crops. Each ecologic alteration has affected animals. Forest removal takes the food and shelter of tree-using species, but planting shrubs and trees in gardens and parks provides habitats for some animals that previously could not occupy those sites. Draining swamps and lakes destroys living places for muskrats, beavers, ducks, some fishes, and others, whereas irrigation of once arid lands enables some aquatic and marsh-dwelling species to replace the rodents, reptiles, and other original denizens of those dry localities.

Agriculture, forestry, reclamation and irrigation, public health activities, and aid to game or fur species are all direct or indirect ecologic manipulations. The perfection of heavy bulldozers, gang plows, and power saws and of poisons that can be distributed in quantity by airplane or from the ground to control insects or weeds has greatly multiplied human ability to alter the surroundings. Much of this, however, is not constructive: accelerated erosion by wind and water after plowing up grass sod, stripping forests faster than they will regenerate, building up poisonous spray residues that enter food chains, and various other effects are evident in many places. The disappearance of former large human civilizations that once flourished in the Near East and the decline in human "carrying capacity" of many other areas demonstrate that much human manipulation has been too hasty and ill-advised; it has been largely exploitation with little regard for the future. A human civilization, like any plant or animal community, cannot long continue if its environment is damaged beyond recovery. Some current efforts at conservation of renewable (biologic) natural resources are attempting to correct the evils of environmental manipulations.

The earth's present natural ecosystems are the result of over 3 billion years of evolution. Their structure and functioning have been integrated and refined over vast stretches of time. Without them the cycles of basic elements, nutrients, and water and the processes of decomposition, which free elements for return to the biotic system, would be greatly impaired, if not eliminated. They provide animal pollinators and natural pest-control agencies (insects, birds, etc.) essential to agriculture. Without the natural systems the stability of plant and animal populations would be jeopardized. The great species diversity they contain results in a variety of feedback mechanisms that regulate population sizes. Humans have eliminated or disturbed natural ecosystems nearly everywhere. How much further they can go without causing irreversible and widespread deterioration remains to be seen. A prudent course would minimize further damage and would place high priority on the conservation of natural environments everywhere, from small urban remnants to large sections of major biotic communities. This would require stabilization and, ultimately, reduction of the human population. Such a course calls for worldwide development of a land ethic (Sec. 36-13) that views the human being as a member of the natural world, with a moral obligation to protect wildlife and natural ecosystems wherever possible. Besides guarding against the foregoing major threats, such conservation efforts would provide other benefits: (1) a continuing source of wild species to serve our needs—development of new pharmaceuticals, crop plants, biological controls of pests, etc.; (2) a reservoir of genetic resources for the protection and improvement of cultivated species through introduction of genes from wild relatives; (3) control areas (base lines) for assessing, understanding, and managing disturbances in manipulated ecosystems (crops, forests, etc.); (4) an early warning system signaling pollution effects that might endanger humans, since wild species may act as sensitive indicators of contamination (DDT and reproductive failure in predatory birds); (5) opportunities for advancement of science, human knowledge, and understanding through study of natural biological processes, the historical properties of which will never again be repeated; and (6) recreational and esthetic enjoyment; our psychologic need for natural, unmanaged surroundings may be far greater than we know, for it was in such environments that we evolved.

Distribution

No animal species occurs uniformly over the whole world; each is restricted to a definite **range,** or area of **distribution.** The entire extent of land or water over which a species may occur is termed its **geographic range,** and the kind of environment in which it lives is its **ecologic range;** both fluctuate over time. The **geologic range** of a species is its occurrence in the past (Figs. 12-9, 12-15). The study of animal distribution and of the factors controlling it is known as **zoogeography.** All the animals living in a particular area, large or small, are collectively termed the **fauna** (the

Figure 12-9 Relations among the three types of animal distribution: geographic, ecologic, and geologic.

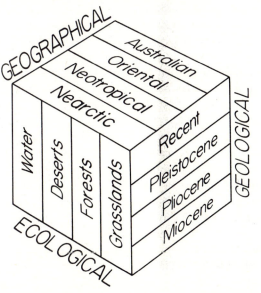

equivalent term for plants is *flora;* the plants and animals together are the *biota*).

The beaver, for example, has a geographic range that embraces much of North America and Europe; its ecologic range comprises freshwater lakes and streams bordered by aspens, poplars, or willows that may be cut to serve for food and for building dams; and its geologic range includes, besides the present or Recent geologic time, the Pleistocene and Pliocene epochs, when it had a wider geographic range. Some animals have large ranges: the harbor seal lives along most northern seacoasts, and the sperm whale formerly occurred in most oceans between 56°N and 50°S. Other species are local in occurrence, like the small Devils Hole pupfish that lives in a single cave spring in southwestern Nevada; some insects are known only from patches of plants covering but a few acres. The mallard duck inhabits freshwater marshes through much of the Northern Hemisphere, but the Laysan duck occurs only on a remote Pacific island of 5 km² (2 mi²).

12-19 Factors regulating distribution Since every species produces offspring in excess of the numbers that can survive within its normal range, there is a population pressure by which individuals tend to expand the boundaries of their range. Other factors such as competition, predators, disease, shortage of food, adverse seasonal weather, and decrease in available shelter act to reduce the population and its range boundaries. The distribution of all organisms, from protozoans to humans, is consequently dynamic rather than static and always subject to change. This is equally true of plants, on which so many animals depend. Most plants, being rooted to the ground, cannot extend their range as individuals but only by dispersal of seeds.

The external factors that limit distribution are termed **barriers.** A barrier is an area ecologically unsuitable for the species in question. An area may be unsuitable because of (1) **physical barriers** such as land for aquatic species and water for most terrestrial forms; (2) **climatic barriers** such as temperature (average, seasonal, or extreme), moisture (as rain, snow, air humidity, or soil moisture), or amount of sunlight; and (3) **biologic barriers** such as absence of appropriate food or presence of effective competitors, predators, or diseases. Many kinds of insects are limited to particular species of plants for their food, shelter, or breeding places, so that their distribution is controlled by factors that regulate these plants.

Every species of plant and animal has a limit of **tolerance**—maximum and minimum—to each factor in its environment. Tolerance to a poison in the soil or food may be narrow, whereas that to different wavelengths of light is wide. Changes in a factor beyond the tolerance limits result in avoidance or death, or survival of only those individuals better suited (more tolerant) to the altered conditions. A species is limited in distribution by the sum total of external influences, many of which are interdependent. However, the range and size level of a population may be limited by only a few factors or even a single factor. In the latter case the population is subject to the **law of the minimum,** (originally formulated by Liebig 1840). It is limited by the essential factor present in least amount or by some critical stage or condition for which the species has a narrow range of adaptability. Oysters, for example, can live in various saline waters but breed only if the temperature exceeds a certain minimum. The law of the minimum is applicable only under steady-state conditions, when inflows balance outflows of energy and when alternatives to a critical factor are not available. Under such conditions, deficiency in a single factor can cause a population to decline. On the other hand, in a fluctuating population many factors may rapidly replace one another as limiting and the single-factor hypothesis is unsatisfactory. Similar arguments apply to maximal levels of tolerance (to heat, light, etc.).

Population densities are highest in the more favorable parts of the range of species and are reduced in peripheral areas where the limits of tolerance for one or more factors are approached. Thus species ranges ordinarily do not abruptly stop at some point but rather trail off from a central dense area.

12-20 Methods of distribution Many free-living animals become distributed by their own efforts. Birds, fishes, and others that migrate are quick to settle in any suitable new situation. A species able to disperse only 3 km (2 mi) in a year or generation could extend its range over 300 km (200 mi) in a century, or across the entire North American continent in 1500 years, if conditions for spread remained favorable. Small aquatic animals and larvae and occasional large forms are carried about passively by water currents. Many small insects are wafted about in the air, mostly within a few hundred feet of the earth but some at higher levels, and thus may be deposited in new places. Rafts of trees, soil, and debris that carry animals pass down large rivers and occasionally are seen far at sea; oceanic islands may have been populated in part by such means. Violent hurricanes sometimes transport small living animals; some showers of organic matter in the United States have included earthworms and small fishes, the latter found after a heavy local rain in a previously dry cornfield in South Carolina. Parasites and commensal animals are transported by their hosts into new localities.

A species does not necessarily occur in all places suitable for it, only in those to which it has access, and this depends on its own history or that of its ancestors. Animal distribution of today is the joint result of barriers and environmental conditions in the past. The continents have been separated probably at least since the Cretaceous or early Tertiary period and have undergone many local alterations by elevation and erosion of mountain ranges, changes in the existence of lakes and streams, and the draining or flooding of lowlands. Some continents have been connected at times by **land bridges** and separated by seas at other periods. Warm climates extended to the polar regions in some periods, whereas glaciers blanketed much of the Northern Hemisphere several times during the Pleistocene epoch preceding the present. All such changes have altered the distribution of plants and animals. Older areas of land or water have been reduced or eliminated, and new ones have become available. Living organisms have been forced to move about, many species have been exterminated, and new species or groups have evolved to take advantage of new areas or environments.

Many North American animals are more closely related to species of eastern Asia than to those in South America. From this we infer that the shallow

Bering Strait (now about 90 m, or 300 ft, deep) between Alaska and Siberia was the site of a wide land bridge for terrestrial organisms in the late geologic past (there are fossil remains of redwoods, *Metasequoia*, on the Alaskan peninsula). The Isthmus of Panama is another bridge which has been interrupted at various times in the past, thus separating the Americas for considerable periods. Terrestrial species also need appropriate "ecologic bridges" of suitable environment through which to migrate. The bison could not pass from North to South America, because no open grasslands existed in the areas between.

Geographic distribution

12-21 Zoogeographic regions Alfred Russel Wallace (1823–1913) first recognized that several major regions of the earth have distinctive faunas composed of taxonomic groups differing from those in adjacent regions. He named the largest of these zoogeographic realms, defined their geographic limits, and listed the characteristic land animals in each (see front end papers). The regions are as follows:

1 **Australian** Australia, Tasmania, New Guinea, New Zealand, and oceanic islands of the Pacific. All monotremes, most marsupials (no placental mammals but bats and rodents), emu, cassowaries, brush turkeys, lyre birds, birds of paradise, most cockatoos, Australian lungfish. New Zealand has sphenodon and kiwi.

2 **Oriental** Asia south of the Himalayas: India, Ceylon, Malay peninsula, Sumatra, Borneo, Java, Celebes, and the Philippines. Tarsiers, macaques, gibbons, orangutan, Indian elephant and rhinoceros, jungle fowl, peacock.

3 **Ethiopian** Africa including the Sahara Desert, Madagascar, and adjacent islands. Gorilla, chimpanzee, African elephant, rhinoceros and lion, hippopotamus, zebras, giraffes, many horned antelopes, ostrich, guinea fowls, secretary bird. Many lemurs in Madagascar.

4 **Neotropical** South and Central America, Mexican lowlands, and West Indies. Llama, alpaca, peccaries, arboreal sloths, armadillos, anteater, guinea pig, vampire bats, rheas, toucans, curassows and guans, most hummingbirds.

5 **Nearctic** North America from the Mexican highlands to the Arctic islands and Greenland. Mountain goat, prong-horned antelope, caribou, muskrat.

6 **Palearctic** Eurasia south to the Himalayas, Afghanistan, Persia, and Africa north of the Sahara. Hedgehog, wild boar, fallow and roe deer.

The limits of each region and its fauna reflect the past history of animal groups and also of changes in the earth's surface that either permitted or prevented animal migrations. The Australian region has evidently been isolated longest and has many unique animals and plants. Its mammals include the egg-laying monotremes and many marsupials, the latter having radiated into a great variety of forms from huge jumping kangaroos to small burrowing marsupial moles (Chap. 34). There is much evidence to indicate that the great Asiatic land mass was for long a center where various animal stocks originated and whence they migrated to other regions. Thus the great flightless birds are now in southern regions, the emu and cassowaries in the Australian, the ostrich in the Ethiopian, and the rheas in the Neotropical. Such **discontinuous distribution** occurs also with the tapirs in Malaysia and Central and South America, the limbless amphibians (caecilians) in the tropics of the New and Old Worlds, and others. The Palearctic and Nearctic regions are least separated, and their faunas have much in common, so they are often combined as the **Holarctic region.** This is characterized by the elk (red deer), moose, bison, beaver, marmots, most bears and sheep, mallard duck, golden eagle, trouts, and salmons. Few or no species or groups occur throughout any single region, but some range in parts of two. The tiger occurs from India to northern China, the opossum from South America into the United States, and the mountain lion and rattlesnakes through both the Americas. Various subdivisions of each region can be distinguished, each with a more or less distinct fauna.

12-22 Insular faunas **Continental islands** stand in shallow waters on the continental shelf close to the continents from which they were probably separated in the recent geologic past by changing sea levels. The fauna of each often resembles that of the nearby mainland, having identical species or closely related subspecies. It often includes various small mammals, reptiles, and amphibians probably resident in the area at the time of its separation from the continent, since they cannot travel through salt water.

Oceanic islands arise by volcanic activity from great depths in the sea. "High islands" have weathered rocks well above sea level. The fauna usually lacks amphibians and small mammals save for bats and occasional rodents, the latter possibly transported in native boats. Wide-ranging fishes, seabirds, and marine mammals visit their shores. The land birds and insects are often different from continental species and often include wingless forms, which are less likely to be swept away by storms. Long-distance dispersal is the only way such a biota can be established, which explains why the fauna is restricted and why certain forms represented on mainland areas are almost invariably absent. The biota results from a balance between extinction and immigration. The Galápagos Islands at the equator off Ecuador have bats and some land birds, reptiles, and other organisms related to mainland species. The principal land birds (many belonging to the family GEOSPIZIDAE; Fig. 13-21) comprise about 40 local forms, mostly finchlike but with some resembling warblers and one species behaving somewhat like a woodpecker. The separate islands have a total of 15 races of giant land tortoises, relatives of which occur today only on certain islands in the Indian Ocean. The Hawaiian Islands, also oceanic, have one native bat. The bird family DREPANIDAE has radiated into various forms resembling finches, warblers, creepers, and other ecologic types. New Zealand is a continent with the character of an oceanic island. It lacked native mammals except one bat but had flightless birds, the now extinct moas and the living kiwi. Its most distinctive reptile, *Sphenodon*, is the sole living species of an otherwise extinct order (Sec. 32-18), and its one frog genus, *Leiopelma*, likewise belongs to an ancient group. On "low oceanic islands" the original volcanic rock has weathered away or sunk below the sea surface, and fringing coral reefs have formed an atoll enclosing a lagoon (Fig.

17-16). Coral atolls are 6 m (20 ft) or less above sea level and have a very limited biota. The only fresh water available is that from tropical rains. Coconut palms are the dominant trees, and the ground cover consists of low shrubs, sedges, etc. The only mammals are rats, and the birds are mostly immigrants. In addition to the thousands of atolls in the Pacific and Indian Oceans, there are many sea mounts (guyots) below the surface. Some of these are the tops of high submarine mountains that once rose above the sea and probably supported a biota like that of existing high islands. Most atolls and guyots are arranged in arcs; they presumably served as steppingstones for the dispersal of plants and animals.

A few generalizations can be made concerning island biogeography. (1) Large islands tend to support more species than small ones. (2) Islands support fewer species (especially at higher trophic levels) than mainland areas of comparable habitat and equal size. (3) In stable environments, a balance tends to be reached between the increase in number of species and the rate of extinction; species density reaches an equilibrium. (4) Although islands have fewer species than comparable mainland areas, it appears that in general they are supporting about all the species they can.

Islands are of great interest to ecologists and evolutionary biologists because of their isolation and their less complex and circumscribed biota, which facilitates study. Of interest also are "terrestrial islands" of isolated habitat such as sand dunes, mountain tops, or springs in deserts.

12-23 Introduced animals Many kinds of animals have been moved by humans into regions where they were not native, some deliberately and others by accident. Many such **aliens** soon disappear and some remain scarce, but others become widespread and abundant. In each case the result depends on the suitability of the new environment, the degree of competition with native species, and the extent to which the alien is affected by predators and diseases in its new home. Fleas, lice, tapeworms, and other parasites of human and domestic animals have been accidentally spread with their hosts to new lands; some have then shifted to native hosts, as the sheep tapeworm to American deer. The European corn borer, Japanese beetle, cotton boll weevil, codling moth, Argentine ant, and garden nematode are conspicuous alien pests now important on crops in the United States. The domestic rats and mice, housefly, and bedbug are aliens that are common nuisances. Quarantine laws and inspections attempt to restrict the further spread of many such pests.

Some deliberate but harmful introductions include the European gray rabbit in Australia and New Zealand, where it competes for sheep pasturage; the Indian mongoose, which was taken to Jamaica and Hawaii to control rats but which destroyed native birds instead; and the "English sparrow," which was brought into the United States in hope of controlling the introduced gypsy and brown-tail moths but which has merely become a nuisance.

Spread of a pest may be rapid. The starling (*Sturnus vulgaris*) was introduced in New York in 1890. By 1935 it had dispersed halfway across the United States and by 1955 had reached the Pacific Coast, ranging north to central Canada and southeastern Alaska and south into northern Mexico. In 65 years in the United States alone its population grew from less than 200 to an estimated 100 million birds.

Some deliberate introductions have been useful, such as the lady beetles imported for the biologic control of various scale insects harmful to fruit trees; the ring-necked pheasant established in many states as an additional game bird; the trout transplanted into various waters, and the striped bass, shad, and other fishes placed in lowland waters of the Pacific Coast to supplement the few native food and game fishes originally there.

In general, the practice of transplanting alien domesticated animals and plants has benefited the human race. In North America nearly all domesticated animals (the turkey is an exception) have been imported from other places, and most of our crop plants are likewise aliens. The practice, however, is risky. Parasites and diseases often are more destructive in their new homes. Plants and animals harmless in their native lands may become pests when transplanted; both the prickly pear cactus and the European rabbit, when introduced into Australia, ran wild, covering millions of acres. The prickly pear was finally controlled by introducing the insect ene-

mies which kept it in check in its native home on the American desert, and the rabbit has been reduced by bringing in a disease, myxomatosis. The introduction of wild species is even more hazardous and probably has done more harm than good.

Ecologic distribution

That part of the earth containing living organisms is known as the **biosphere.** Within this relatively thin layer are many places, large and small, suitable for plants and animals (Fig. 12-10). The most obvious major divisions of environments used by animals are salt water, fresh water, and land, but even these grade into one another (Fig. 12-10). Water contains 7 ml or less of oxygen per liter, its temperature changes slowly, and for animals the hazard of drying occurs only in temporary waters. The displacement support of water permits some active marine animals to be of huge size. On land, oxygen abounds (210 ml per liter of air), temperatures may vary widely and change abruptly, and plant food usually is abundant. There is often a dearth of moisture so that land animals must adjust their habits to avoid desiccation or must have a drought-resistant body covering.

12-24 Salt water Oceans, seas, and bays cover about 71 percent of the earth, providing habitats that are extensive and stable. Their physical features include (1) temperatures from 32°C (89°F) in the tropics to −2.2°C (28°F) in some polar regions, but rarely with an annual variation more than 5°C (9°F) at any one place; (2) dissolved gases varying with temperature and depth; (3) salt content averaging 3.5 percent ($NaCl$, 2.35; $MgCl_2$, 0.5; Na_2SO_4, 0.4; $CaCl_2$, 0.11; KCl, 0.07; $NaHCO_3$, 0.02; and others); (4) average depth of oceans about 3800 m (12,500 ft), but 11,500 m (11.5 km, or 7 mi) in the greatest "deep"; (5) pressure increasing about 1 atmosphere for each 10 m (33 ft) of depth, so that animals of deep waters live under enormous pressures, which, however, are equalized throughout their bodies; and (6) light penetrating decreasingly down to about 180 m (600 ft), a trace to 1200 m (4000 ft), with complete permanent darkness below.

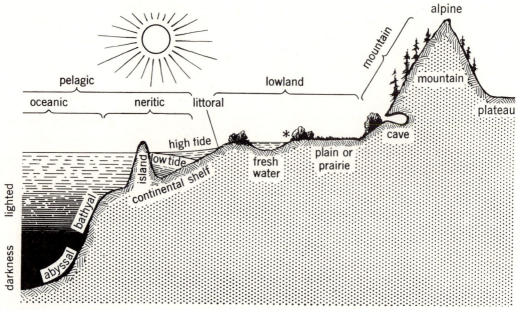

Figure 12-10 An ideal section at the margin of a continent with indication of some common ecologic environments available to animals; the region marked with asterisk (*) is enlarged in Fig. 12-11.

Marine animals include representatives of all phyla and of all classes except the centipedes, millipedes, ONYCHOPHORA, and amphibians. The ctenophores, brachiopods, chaetognaths, echinoderms, and lower chordates are exclusively marine. Species and individuals are most abundant near the surface and decrease with depth, but some occur down to the greatest depths.

Marine habitats are classified ecologically as follows:

1 **Pelagic** Open waters of the oceans; subdivided into smaller units both horizontally and vertically

   a **Neritic** Open water above the continental shelves

   b **Oceanic** Remainder of the open water overlying the ocean basins; divided vertically into the following zones:

   1 **Epipelagic** Uppermost layer of the ocean; well lighted and populated with numerous plants and animals; the only area where plant production occurs and sometimes called the photic zone in contrast to the black, or aphotic, zones below it; depth to 200 m

   2 **Mesopelagic** First aphotic zone; depth from 200 to about 700 m; limited below by the 10°C isotherm in the tropics; animals moderate in number, no plant; animals black or red; luminescent organs common; deep scattering layers

   3 **Bathypelagic** Depth about 700 to 4000 m; limited below by the 4°C isotherm; animals few; luminescent organs few; eyes small

   4 **Abyssal pelagic** Depth about 4000 to 6000 m; animals few and pale; eyes small or absent

   5 **Hadal pelagic** Depth from 6000 to 10,000 m; the area of the trenches; fauna often endemic to a trench

2 **Benthic** The bottom; subdivided as follows:

   a **Littoral (intertidal)** From the line of highest tide to that of lowest tide; animal and plant life abundant

   b **Sublittoral** From the low-tide mark to the edge of the continental shelf; abundant animal life; kelp beds; major fishing grounds

   c **Bathyal** The continental slope; underlying the bathypelagic oceanic zone

   d **Abyssal** Bottom at and below the average depth of the ocean down to the upper level of the trenches

   e **Hadal** Bottom of the trenches

Marine animals are also distinguished ecologically as follows:

1 **Plankton** Organisms that float and are moved passively by winds, waves, or currents; mostly of minute or microscopic size with surface large in relation to bulk, often with elongate body parts or ciliated; includes many protozoans and crustaceans, some mollusks, a few worms, and a host of larvae (sponges to tunicates) and microplants (diatoms, dinoflagellates).

2 **Nekton** Animals that swim freely by their own efforts; includes squids, fishes, sea snakes and turtles, seabirds, seals, whales, etc. Plankton and nekton animals of the open sea are termed **pelagic.**

3 **Benthos** Animals which crawl on, are attached to, or burrow into the bottom substrate.

12-25 Fresh water Unlike the sea, fresh waters are scattered, smaller and shallower, and more variable in temperature, content of gases and salts, light penetration, turbidity, movement, and plant growth. ''Pure'' waters contain mere traces of salts, but some saline and alkaline waters have large amounts. Carbonates (especially $CaCO_3$) are usually commoner than other salts. Some fresh waters are nearly constant in volume, but those of arid regions often fluctuate from flood stages to small volume or dry up completely during a single season.

Freshwater organisms include many protozoans, a few sponges, cnidarians and bryozoans, many worms, rotifers and snails, various bivalve mollusks, crustaceans, larval and adult insects, and vertebrates from fishes to mammals. Many of the invertebrates produce eggs or other ''resting'' stages resistant to drying or freezing that may be blown about by winds or carried accidentally on the feet of waterfowl. They rarely have floating larvae, since these might be carried by streams to the sea and be lost. In any one place the kinds and numbers usually change markedly through the year.

The principal freshwater habitats are as follows:

1 **Running water** The cold mountain streams, spring-fed brooks, slower brooks and creeks, and rivers of various sizes contain mobile animals that are segregated according to the rate of water movement, temperature, oxygen content, and character of the bottom. Thus trout live only in cool, well-oxygenated waters, whereas carp thrive in warm and even foul waters. Inhabitants of rapid waters often are flattened or have means for holding to the bottom.

2 **Standing water** This comprises lakes, swamps, marshes, and bogs (Fig. 12-11), either permanent or temporary, and occurs in all regions from polar and alpine areas to the tropics. Lakes in cold climates may be frozen over for a long time in winter and those of temperate regions for shorter periods, but lakes of hot regions are always open. Large lakes afford more stable environments than flowing waters; they have littoral, benthic, and pelagic divisions and a plankton fauna. Water in large temperate lakes undergoes circulation because of temperature differences and wind action. In spring and autumn this involves all depths, but in summer only the surface portion circulates, over a definite plane, or **thermocline,** separating it from the cold and often poorly oxygenated water below, where there are but few animals.

12-26 Land The interaction of many physical, climatic, and biologic factors produces a wide variety of ecologic conditions on the continents and islands.

Lands differ in (1) the **chemical nature** and **physical texture** of the soil, sand, and rocks exposed on the surface; (2) the **topography,** which includes plains, rolling lands, hills, valleys, and mountains; and (3) the **altitude,** which varies from basins below sea level (Death Valley) to peaks exceeding 8500 m (28,-000 ft; Mount Everest). Some of the climatic variants are as follows: (1) **Air temperatures** at different places range from far below freezing (where soil also freezes) to 60°C (140°F) on some deserts; in many localities the temperature fluctuates widely, by day or through a season, but there is only slight change in many tropical areas. (2) **Moisture** as rainfall amounts to 1270 cm (500 in) annually in a few tropical places but only mere traces on some deserts; the moisture content of air and soil varies from complete saturation to slight amounts according to place and

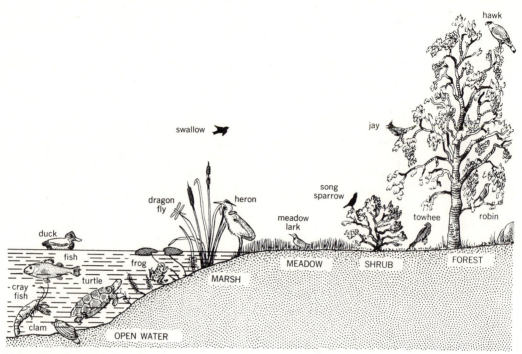

Figure 12-11 Ecologic distribution of some common animals of fresh water and land. In general, each keeps to a particular subdivision of the environment.

time of year. (3) **Winds** and **sunlight** affect both the temperature and the moisture content of the air and ground. The physical and climatic conditions influence the plant cover that may grow on any land area, and the plants in turn affect the animal populations, particularly where the latter depend directly on plants for food and cover.

The main land animals are mammals, birds, reptiles, insects, and arachnoids, with lesser numbers of amphibians, crustaceans, mollusks, and worms, in addition to protozoans. They are all mobile save for some parasites and live on the land surface, on plants, or at shallow depths in the ground. The subterranean habitat affords more uniform conditions than the surface. All animals that fly or "live" in the air return to the ground or to trees or rocks.

Ecologic classifications of land habitats are based either on climate (temperature, rainfall, relative humidity, etc.) or on the various associations or communities (biomes) of terrestrial plants and animals that live in more or less interdependent relations.

12-27 Biomes These are large and easily recognizable terrestrial communities resulting from interaction among the climate, biota, and substrate of a region. Plants adapted to a particular climate often have similar structure and growth form (physiognomy). The physiognomy of vegetation repeats itself in regions of similar climates, regardless of species composition, often in widely separated areas over the earth's surface. Thus broad, thick leaves and heavy canopy are characteristic of the tropical rain-forest biome, whether in Panama, New Guinea, or equatorial Africa; the grassland biome is similar in growth form though different in species composition on each continent; and the several great deserts support a type of vegetation (the desert biome) characterized by thorny plants having adaptations for conserving water. Areas where two biomes merge or blend are called **ecotones** (see back end papers).

Biomes are primarily determined by climate. The freezing temperatures and short growing season of the Arctic are favorable to the growth of sphagnum and dwarf vegetation of the tundra, whereas the dry conditions of deserts, caused by adjacent mountain ranges that deplete winds of their moisture, are habitable only by desert-type vegetation and associated animals. The biomes of North and Central America are as follows (Figs. 12-13A and 12-13B).

1 **Tundra** Treeless Arctic regions; only the surface soil thaws out in the short 60-day summer; below is permafrost; drainage poor, many ponds, marshes, and bogs; principal plants are bog (peat) mosses, lichens, sedges, grasses, and low herbs; mammals include musk ox, Barren Ground caribou, wolf, arctic fox, weasels, lemmings, arctic hare; conspicuous resident birds are snowy owl and ptarmigan (snow grouse); many migratory waterfowl and shorebirds nest there in summer but go south to winter; no reptiles.

2 **Coniferous forest** (evergreen) South of tundra to northern United States, with extensions southward along mountain systems (Sierra-Cascade, Rocky, and Appalachian) to Central America; winters bleak, summers cool, precipitation moderate; spruces, firs, pines, cedars; often includes shrubs and patches of grassland; mammals include moose and woodland caribou in the north and deer and elk toward the south, also the furbearers—red fox, Canada lynx, marten, fisher, wolverine—with black bear and mountain lion, snowshoe rabbit

(hare), and some small rodents; birds are various grouse, warblers, chickadees, jays, etc.; few reptiles and amphibians; trout, grayling, etc., in waters.

3 **Broad-leaved deciduous forest** (summergreen) Especially from Mississippi Valley eastward; winters cold, summers warm and humid with rain; oaks, maples, beeches, elms, walnuts, many shrubs and herbs; white-tailed deer, gray fox, bobcat, raccoon, and flying squirrels; many warblers, vireos, and other small songbirds; snakes and amphibians common.

4 **Grassland** (prairies and Great Plains) Mississippi Valley westerly, Texas to Canada; winters of severe continental cold, summers hot, thunder showers. Large areas of hardy grasses (buffalo, bluestem, grama), trees locally along streams; formerly bison, prong-horned antelope, and wolf; now, as then, coyote, badger, skunk, jackrabbit (prairie hare), cottontails in the thickets, ground squirrels; prairie chicken, burrowing owl, soaring hawks, meadowlarks; some snakes.

5 **Sagebrush** Great Basin plateau between Rocky Mountains and Sierra-Cascade system. Dry, winters cold with limited rain or snow, summers hot. Sagebrush (*Artemisia tridentata*) and other bitter-flavored shrubs, bunch grasses, piñon (nut) pine, juniper. Prong-horned antelope, jackrabbit, ground squirrels, other burrowing rodents, coy-

Figure 12-12 Comparison of the latitudinal and altitudinal zones or associations of plants that provide the appropriate environments for various kinds of animals.

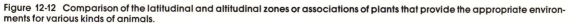

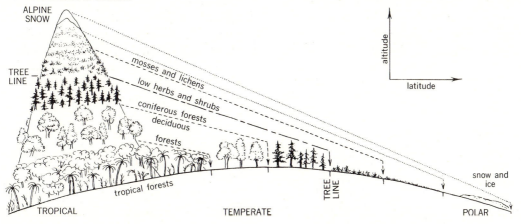

Arctic tundra   Coniferous forest

Broad-leaved deciduous forest   Grassland

Figure 12-13 A. **Biomes of North America.**

ote, badger; sage hen (largest American grouse), water birds in local ponds and marshes; many reptiles.

6 **Chaparral** California hills and parts of mountains. Rainy (snowy) winters, warm to hot dry summers. Chamise, manzanita, and other shrubs with thickened evergreen leaves. Mule deer, woodrat, chipmunks, brush rabbit, California thrasher, wrentit, lizards.

7 **Desert** Southeastern California to western Texas and southward. Soil rocky or sandy. High summer temperatures, scant water at any time. Vegetation scattered, many herbs grow and flower soon after the occasional rains: creosote bush and other shrubs, cacti, yuccas, trees small. Smaller carnivorous mammals, many rodents active by

night, few birds, many lizards and snakes, few amphibians.

8 **Tropical rain forest** Panama and parts of Central America. Sustained moderately high temperatures, abundant rainfall, and high humidity. Forest of evergreen broad-leaved trees, many vines, orchids, etc. Marmosas, opossums, sloths, anteaters, monkeys, great variety of birds, bats, reptiles, and amphibians.

Geologic distribution

12-28 Fossils The animals living today are only part of a vast and continuing population that has inhabited the earth through millions of years (Figs.

Sagebrush   Chaparral

Desert   Tropical rain forest

Figure 12-13 *B*. **Biomes of North America.**

Figure 12-14 Ecologic segregation during the nesting season of seven species of birds, all members of one family (PARULIDAE, wood warblers). Habitats are named below (fern glade, etc.) and the birds above (Tolmie warbler, etc.). Each has a separate forage niche and does not compete with the others. (*After Grinnell and Storer, 1924.*)

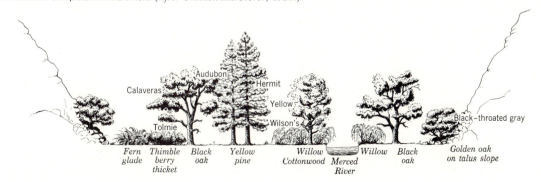

12-15, 12-16; Secs. 13-10 to 13-15). Evidence of former animals and plants is provided by **fossils.** A fossil (literally, something dug up) is an organic relic from a previous geologic period preserved by natural means in rocks or softer sediments that affords information about the character of the original organism.

Dead animals are usually destroyed by scavengers or by decay, but if soon covered by silt beneath water, by wind-blown soil, or by volcanic dust, decomposition will be slow and the hard parts may persist. If the surrounding material later becomes rock and is neither excessively crushed nor heated, the remains will survive for long periods. A fossil may be (1) an **unaltered hard part** such as a skeleton, a tooth, or a shell; (2) a **mold** where hard parts once present have been dissolved away by per-

colating waters to leave a cavity showing the original form; (3) a **petrifaction,** in which the original has been infiltrated with mineral matter, sometimes preserving fine detail; or (4) a **cast** of mineral which fills a mold to show only exterior features. Even soft parts may leave impressions in fine sediments. Some fossil records of animal activities survive as tracks, burrows, tubes, and droppings. Fossil plants are common as impressions, or casts. Special types of fossils are the carcasses (with flesh, hair, etc.) of the mammoth and woolly rhinoceros frozen in tundra soils of Siberia and Alaska; skin, hair, and dung of ground sloths dried in Nevada caves; skeletons of giant elk and other animals in peat bogs of Ireland and elsewhere; entire remains of insects (Fig. 25-16) and other small animals embedded in amber (*fossil resin*),

Figure 12-15   Distribution of the major groups of animals through geologic time (compare Fig. 12-18).   Changes in width of the black areas suggest relative numbers of species in each group during different eras and periods; broken lines suggest possible sources and time of origin for certain groups. (*After a chart by W. D. Matthew, University of California Museum of Paleontology.*)

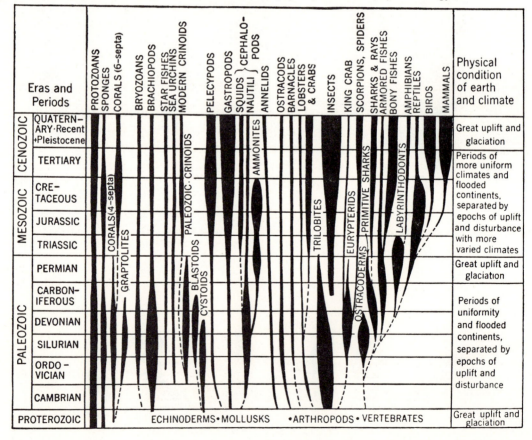

as along the Baltic Sea, in Arctic Alaska, and elsewhere; and skeletons of many species trapped in asphalt pits (former tar pools) of southern California.

Besides providing evidence of past life, fossils are used for identifying rock strata and for indicating a chronological or time sequence of such strata. They afford some information about ancient environments and climates and the interrelations of bygone animals; and they furnish important data on the organic evolution of animals and plants down through time (Chap. 13). Many present-day species of animals occur also as fossils, showing that they lived in earlier geologic epochs. Fossils show that in the past various groups such as the dinosaurs came to great prominence and later perished completely, whereas others like the coelacanth fishes (Sec. 30-26) have a few living survivors. The fossil record is fragmentary at best. Remains are more complete and numerous in rocks of later geologic periods; in the older formations they are scarcer and less perfect because the rocks have been disturbed and crushed by earth movements or changed by heat.

12-29 Geologic time Estimates of the age of fossils are derived mainly from study of radioactive minerals in fossil-bearing rocks.[1] The oldest recog-

nizable rocks (Archeozoic) are considered to be about 3.8 billion years old, and the oldest rocks (Cambrian) with numerous fossils with hard parts 600 to 700 million years old. The time since the Archeozoic may be visualized by comparing it to the distance from New York to San Francisco (about 5300 km, or 3300 mi). One year is represented by 3.5 mm (0.14 in), a human lifetime by 25 cm (9.7 in), the Christian era by 7 m (23 ft), and the time since early Pleistocene, when human life appeared, by about 3.5 km (2.2 mi)—but a short distance down the road of time.

12-30 Continental drift Alfred Wegener in 1912 offered a theory that all land masses originally formed one large supercontinent, Pangaea, before the Mesozoic. Later, by continental drift, they reached their present positions. For many years the theory had scant support, but recent evidence is convincing and bears importantly on the distribution and evolution of life. Two land masses have been envisioned: Gondwanaland in the Southern Hemisphere and Laurasia in the Northern Hemisphere, separated by the Tethys Sea. The southern continent may have been in the south-polar area during the Paleozoic, according to data from paleomagnetism of

[1] *Uranium 235 has a half-life of about 250,000 years and yields lead of atomic weight 206 (ordinary lead, atomic weight 207). Analyses of the ratio of uranium to lead 206 in undisturbed rocks provides an approximate time scale for various strata containing fossils in the "geologic column" and hence of the time when the animals were living. More recently the potassium-40/argon-40 method, applicable to volcanic deposits, has served to confirm and to extend time estimates by the uranium/lead method. The half-life of the potassium isotope is 1.3 billion years.*

More recent time scales are the chronology established by studying the number and spacing of tree rings (covering about 6000 years) and the percentage of disintegration of radioactive carbon in plant and some animal tissues (bone). As solar radiation bombards the atmosphere, some of the nitrogen is converted to a radioisotope, ^{14}C, which is then oxidized to CO_2. This radioactive CO_2 is mixed with the ordinary form (containing ^{12}C) present in the atmosphere. Since radioisotopes disintegrate (decay) at a constant rate by emitting particles (electrons, neutrons) and the solar production of radioactive CO_2 is considered to be relatively constant, an equilibrium exists in the atmosphere in the proportion of the two forms of CO_2. This

proportion appears to have been fairly constant since ancient times. Photosynthetic plants take up the two forms of CO_2, maintaining the equilibrium ratio, and incorporate the two kinds of carbon into their tissues in the proportion they exist in the atmosphere. When a plant dies, however, the proportion begins to change. The ^{14}C continues to disintegrate, whereas the amount of ^{12}C remains constant. The half-life of ^{14}C (time required for one-half of the radioisotope to disintegrate) is about 5600 years. Thus an estimate can be made as to the time of death by comparing the proportion of ^{14}C and ^{12}C in the plant remains (charcoal, pollen, wood) with the atmospheric proportions. A fossil with half the ^{14}C content of a modern plant would be about 5600 years old. Carbon-14 dating can be used on fossils up to around 70,000 years old.

Recent studies of growth rings in bristlecone pines, which presently give a ring sequence to 6270 B.C., reveal a discrepancy in age determinations by the ^{14}C method. Beyond 1000 B.C., ages estimated by ^{14}C fall increasingly short of the actual ages determined by counting rings. The proportion of ^{14}C in the atmosphere has evidently not been as constant as previously thought (Sec. 2-6).

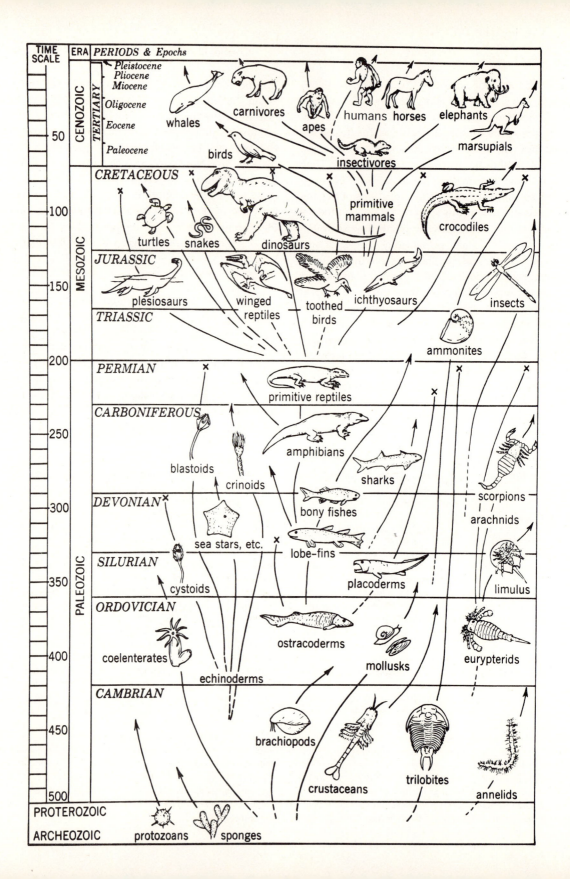

TIME SCALE

ERA | PERIODS & Epochs

CENOZOIC — TERTIARY
Pleistocene
Pliocene
Miocene
Oligocene
Eocene
Paleocene

whales · carnivores · apes · humans · horses · elephants · marsupials · insectivores · birds

50

MESOZOIC

CRETACEOUS
turtles · snakes · dinosaurs · primitive mammals · crocodiles

100

JURASSIC
plesiosaurs · winged reptiles · toothed birds · ichthyosaurs · insects

150

TRIASSIC
ammonites

200

PALEOZOIC

PERMIAN
primitive reptiles

CARBONIFEROUS
amphibians · blastoids · crinoids · sharks · scorpions

250

DEVONIAN
bony fishes · arachnids · sea stars, etc. · lobe-fins · limulus

300

SILURIAN
cystoids · placoderms

350

ORDOVICIAN
ostracoderms · mollusks · eurypterids · coelenterates · echinoderms

400

CAMBRIAN
brachiopods · crustaceans · trilobites · annelids

450

500

PROTEROZOIC
ARCHEOZOIC
protozoans · sponges

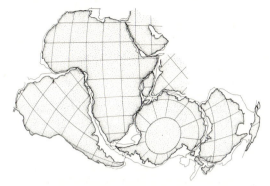

Figure 12-17 The theory of continental drift. Fit of the southern continents at the 500-fathom contour (Antarctica at 1000-m contour). Except for Ceylon, which has been fitted by inspection, the map is a tracing based on computer output. (*After A. G. Smith and A. Hallant, 1970.*)

iron-bearing rocks and glaciation. Gondwanaland yielded South America, Africa, India, Australia, and Antarctica. Laurasia gave rise to Eurasia, Greenland, and North America.

Evidence for former connection of the continents includes (1) geometric fit of continental contours below sea level on the continental shelf (Fig. 12-17); (2) age, structure, and presumed movements of rocks; (3) paleomagnetism records; (4) data on convection currents in the earth's mantle and seafloor spreading; (5) widespread deposits of a continental glacier in late Paleozoic of Africa, South America, India, and Australia; and (6) past and present distribution of life.

The theory of continental movement is as follows: The continents are plates of lightweight siliceous–granitic rocks that float on denser rock (chiefly basalt), which is thinned out on the ocean floors. Continental movement appears to depend in large part on convection currents in the earth's mantle. New ocean bottom is continuously being extruded along the crest of a worldwide system of oceanic ridges by upwelling of material from deep in the mantle. As the extruded rock moves outward on either side of the ridge, the continents are carried along. With seafloor expansion, older floor material returns to deeper layers. This occurs at oceanic trenches. A major oceanic ridge extends from the Gulf of Aden across the Indian Ocean, then east between Australia and Antarctica and north through the eastern Pacific to the west coast of North America. A mid-Atlantic ridge extends from Iceland south the length of the Atlantic. Major trenches or "subduction" zones occur along the west side of Central and South America and bound most of the western Pacific and eastern Indian Oceans. Except for the Pacific Ocean (which appears to be shrinking), the oceans are expanding. As molten rock emerges from an oceanic ridge or by vulcanism on land, particles of iron respond to the earth's magnetic field and behave like minute compass needles orienting toward the magnetic pole. When the rock solidifies they become fixed, providing a permanent record of the rock's position in relation to the pole at the time it was formed. Such paleomagnetic information has been of great value in charting movements of the seafloor and continents. Paleomagnetic events can often be pinpointed in time by radioisotope dating. In the breakup of Laurasia, Eurasia and North America separated along the Atlantic ridge, as did Africa and South America of Gondwanaland. India separated from Africa and Antarctica and made a landfall with Eurasia; Antarctica and Australia parted, and Antarctica separated from Africa. The time for the breakup of Gondwanaland is thought to have been from Permian to Cretaceous (by different authorities), with much rifting in late Jurassic and lower Cretaceous and dispersal of the continents continuing into the Tertiary. Resulting shifts in latitude probably subjected organisms to major climatic changes.

Facts pertaining to plant and animal distribution must now be scrutinized in the framework of continental drift. Dating of landmass separations and

Figure 12-16 Time chart of animal life—the fossil record of some conspicuous groups. Heavy curving lines extend up from the time (period) when each group first appeared; those topped with arrows are of stocks surviving to the present (Recent time); an **X** indicates time at which other groups became extinct; broken lines show presumed earlier origins of some types. Height of space for each period in the "geologic column" is somewhat proportional to its duration in millions of years (e.g., Tertiary, 63; Pleistocene, 1; Proterozoic and Archeozoic not to scale). Recent time omitted. Compare Fig. 12-15. (*Adapted in part from C. D. Dunbar, 1949, Historical geology, by permission of John Wiley & Sons, Inc.*)

landfalls becomes crucial in biogeographic studies. The evidence for continental drift is already providing explanations for some distributional enigmas— for example, archaic and strictly aquatic freshwater frogs (PIPIDAE) in Africa and South America, nearly identical Lower Triassic reptile faunas in South Africa and Antarctica, and the evolution of unique groups of mammals in South America, Africa, and Australia.

References

Articles of ecologic and conservation interest appear in the *Annual Review of Ecology and Systematics*, Palo Alto, Calif., Annual Reviews, Inc.

Allee, W. C., and others. 1949. Principles of animal ecology. Philadelphia, W. B. Saunders Company. xii + 837 pp., 236 figs. *Comprehensive; many examples of ecologic principles.*

Andrewartha, H. G. 1970. Introduction to the study of animal populations. 2d ed. London, Chapman and Hall. xiv + 283 pp., illus. *Theoretical aspects of animal numbers.*

_____ and **L. C. Birch.** 1954. The distribution and abundance of animals. Chicago, University of Chicago Press. xv + 782 pp.

Bolin, B. 1970. The carbon cycle. *Sci. Am.*, vol. 223, no. 3, pp. 124–132. *The main cycle of carbon is from carbon dioxide to living matter and back to carbon dioxide; some, however, is stored in large amounts in sedimentary rocks.*

Bormann, F. H., and **G. E. Likens.** 1970. The nutrient cycles of an ecosystem. *Sci. Am.*, vol. 223, no. 4, pp. 92–101. *Measurement of nutrient cycles in the Hubbard Brook Experimental Forest, New Hampshire.*

Carson, R. 1962. Silent spring. Boston, Houghton Mifflin Company. 368 pp. *The first comprehensive and well-documented warning on the dangers of the chemical pesticides.*

Clapham, W. B., Jr. 1973. Natural ecosystems. New York, The Macmillan Company. viii + 248 pp. *A synthetic approach to the principles by which natural ecosystems operate as dynamic, integrated systems.*

Cloud, P., and **A. Gibor.** 1970. The oxygen cycle. *Sci. Am.*, vol. 223, no. 3, pp. 110–123. *Atmospheric oxygen was put there by green plants; it made possible the evolution of the higher plants and animals.*

Cole, L. C. 1958. The ecosphere. *Sci. Am.*, vol. 198, no. 4, pp. 83–92. *A calculation is made as to how much living matter the earth is capable of supporting.*

Cott, H. B. 1940. Adaptive coloration in animals. London, Oxford University Press. xxxii + 508 pp., 48 pls., 84 figs.

Darlington, P. J. 1957. Zoogeography: The geographical distribution of animals. New York, John Wiley & Sons, Inc. xi + 675 pp., 80 figs. *Mainly land and freshwater vertebrates.*

Deevey, E. S., Jr. 1970. Mineral cycles. *Sci. Am.*, vol. 223, no. 3, pp. 148–158. *Living matter is mostly carbon, hydrogen, oxygen, and nitrogen, but other elements are essential constituents, notably phosphorus and sulfur.*

Delwiche, C. C. 1970. The nitrogen cycle. *Sci. Am.*, vol. 223, no. 3, pp. 136–146. *Nitrogen, abundant in the atmosphere, cannot be used by organisms until fixed by certain specialized forms of life or by industrial processes.*

Elton, Charles. 1935. Animal ecology. Rev. ed. London, Sidgwick & Jackson, Ltd. xx + 207 pp., 8 pls., 13 figs. *Pioneer book on modern ecologic concepts.*

_____. 1966. The pattern of animal communities. London, Methuen & Co., Ltd. 432 pp., 87 pls., 28 figs. *Based mainly on Wytham Woods near Oxford, England.*

Frey, D. G. (editor). 1963. Limnology in North America. Madison, The University of Wisconsin Press. xviii + 734 pp., 216 figs. *Chapters (26) by specialists on various regions; many references.*

Gates, D. M. 1971. The flow of energy in the biosphere. *Sci. Am.*, vol. 225, no. 3, pp. 88–100. *Solar energy warms the earth's surface and is radiated back into space; a very small fraction taken up by plants maintains all living matter.*

Hazen, W. E. 1970. Readings in population and community ecology. 2d ed. Philadelphia, W. B. Saunders Company. ix + 421 pp.

Hedgpeth, Joel (editor). 1957. Treatise on marine ecology and paleoecology, vol. I, Ecology Memoir 67. Geological Society of America. viii + 1296 pp.

Hesse, R., W. C. Allee, and **K. P. Schmidt.** 1951. Ecological animal geography. 2d ed. New York, John Wiley & Sons, Inc. xiii + 715 pp., 142 figs.

Kormondy, E. J. 1969. Concepts of ecology. Concepts of Modern Biology Series. Englewood Cliffs, N.J., Prentice-Hall, Inc. xiii + 209 pp.

Lack, David. 1954. The natural regulation of animal numbers. Fair Lawn, N.J., Oxford University Press. viii + 343 pp., 52 figs.

Landsberg, H. E. 1953. The origin of the atmosphere. *Sci. Am.*, vol. 189, no. 2, pp. 82–86. *The present atmosphere has lasted about a billion years and seems good for a few more billion if humans do not destroy it.*

Margalef, Ramon. 1968. Perspectives in ecological theory. Chicago Series in Biology, vol. 1. Chicago, The University of Chicago Press. viii + 111 pp.

Moore, H. B. 1958. Marine ecology. New York, John Wiley & Sons, Inc. xi + 493 pp., illus. *Organisms discussed by habitats, abyssal to estuarine.*

McKenzie, D. P., and **F. Richter.** 1976 Convection currents in the earth's mantle. *Sci. Am.,* vol. 235, no. 5, pp. 72–89. *Both large- and small-scale currents appear to play a role in plate tectonics.*

Myers, J. H., and **C. J. Krebs.** 1974. Population cycles in rodents. *Sci. Am.,* vol. 230, no. 6, pp. 38–46. *The 3- to 4-year population cycles in small rodents seem to result from changes in a population's genetic makeup.*

Nybakken, J. W. 1971. Readings in marine ecology. New York, Harper & Row, Publishers, Incorporated. xiii + 544 pp., illus.

Odum, E. P. 1971. Fundamentals of ecology. 3d ed. Philadelphia, W. B. Saunders Company. xiv + 574 pp., illus. *General text with emphasis on energy flow through ecosystems.*

————. 1966. Ecology. Modern Biology Series. New York, Holt, Rinehart and Winston, Inc. vii + 152 pp., illus. Paper. *General introduction to ecosystems and energy flow.*

Penman, H. L. 1970. The water cycle. *Sci. Am.,* vol. 223, no. 3, pp. 98–108. *Water is the medium of life processes and the source of their hydrogen.*

Reid, G. K. 1961. Ecology of inland waters and estuaries. New York, Reinhold Publishing Corporation. xvi + 375 pp., illus. *Introductory.*

Shelford, V. E. 1963. The ecology of North America. Urbana, University of Illinois Press. xxii + 610 pp., illus. *Discussed by biomes and subdivisions; physical and climatic features, principal plants and animals.*

Sverdrup, H. U., M. W. Johnson, and **R. H. Fleming.** 1942. The oceans: Their physics, chemistry and general biology. Englewood Cliffs, N.J., Prentice-Hall, Inc. x + 1087 pp., illus.

Tait, R. V. 1968. Elements of marine ecology. New York, Plenum Press. vii + 272 pp.

Tinbergen, N. 1953. Social behavior in animals. London, Methuen & Co., Ltd. xi + 150 pp., 8 pls., 67 figs.

Trivers, R. L. 1971. The evolution of reciprocal altruism. *Quart. Rev. Biol.,* vol. 46, pp. 35–57. *How natural selection can operate against the cheater in altruistic behavior. Cleaning symbiosis, warning cries of birds, and human reciprocal altruism are discussed.*

Vernberg, W. B., and **F. J. Vernberg.** 1972. Environmental physiology of marine animals. New York, Springer-Verlag New York, Inc. ix + 346 pp.

Wallace, A. R. 1876. The geographical distribution of animals. New York, Harper & Brothers. 2 vols., 1108 pp., illus. and maps. Reprint. 1962. New York, Stechert-Hafner, Inc. 2 vols.

Whittaker, R. H. 1975. Communities and ecosystems. 2d ed. New York, The Macmillan Company. xviii + 385 pp.

Wilson, E. O. 1975. Sociobiology: The new synthesis. Cambridge, Mass., The Belknap Press of Harvard University Press. ix + 697 pp. *Evolutionary theory and modern population biology applied to the study of animal societies.*

Wynne-Edwards, V. C. 1962. Animal dispersion in relation to social behavior. New York, Stechert-Hafner, Inc. xi + 653 pp., 11 pls.

————. 1964. Population control in animals. *Sci. Am.,* vol. 211, no. 2, pp. 68–74. *Social behavior limits reproduction in animals to avoid overexploitation of food resources.*

13

ORGANIC EVOLUTION

Human beings have long sought to learn how, when and where life originated and the ways in which the many kinds of animals and plants came into being. This chapter considers some theories on these subjects and the evidence on which they are based.

We have no knowledge of life except on the earth, but it is probable that life is present elsewhere in the universe. It has been estimated that in our galaxy alone there may be a billion planets with conditions suitable for life resembling that found on earth, and within the 100 million galaxies now within the range of the most powerful telescopes, as many as 10^{17} such planets. It appears, however, that within our solar system only extraterrestrial life of a very primitive or different sort could exist. Mars, long regarded as a likely habitat, has no liquid water, an atmosphere mainly of carbon dioxide, and night temperatures

that drop at least to $-86°C$ ($-123°F$). Perhaps in the depths of Jupiter, where temperatures rise and enormous gravitational pressure turns hydrogen into a solid, some form of life exists. Life on earth is based on carbon, hydrogen, nitrogen, oxygen, and liquid water, but other elemental systems may be capable of supporting life. Even so, it appears that we will not find organisms resembling earth's larger plants and animals on any of our sister planets.

13-1 Environments for life According to current theory, the sun and planets originated together from a vast cloud of dust and gas. The cloud was initially compressed by starlight, and the process was accelerated by gravitational forces. Free hydrogen was the most abundant element. It gravitated toward the center, where pressure and temperature became great enough to start the thermonuclear reaction of a star. The sun formed, and the wheeling mass of dust around it broke up into eddies that condensed to form the planets. While the primordial earth was still in the gaseous stage, the heavier elements (iron, nickel, aluminum, and silicon) gravitated toward its center and the lighter ones (hydrogen, oxygen, nitrogen, and carbon) remained toward the outside. As gravitational forces increased, gases such as water vapor, ammonia, methane, helium, and hydrogen were retained to form the primordial atmosphere, which was a reducing one. The common light elements on the earth's surface—carbon, hydrogen, oxygen, and nitrogen—probably formed many of the first molecules, and, as noted above, these elements are the primary constituents of living things. As the earth cooled, water could be retained on its surface. It filled depressions to form the oceans, which originally may have been very hot, as was the land. Life could not have existed until the waters and lands cooled.

13-2 Theories on origin of life   There are several principal theories on the origin of life:

1 **Spontaneous generation**   Earlier it was believed that life originated repeatedly from nonliving materials by spontaneous generation, even of worms and flies. This idea was discredited by experiments in the seventeenth and nineteenth centuries (Chap. 10). Modern theory, however, is that the first free-living, self-replicating molecules of DNA were formed from nonliving matter.

2 **Special creation**   Until the middle of the nineteenth century life was generally presumed to have been created by some supernatural power either once or at successive intervals, or each species was presumed to have been created separately. This idea is outside the realm of science and not subject to experimental study.

3 **Cosmozoic theory**   Resistant spores of living forms might have reached the earth accidentally from some other source in the universe. The extreme cold and dryness and the lethal radiations of interstellar space would not permit life as we know

it to survive unprotected. And this theory provides no explanation as to the actual origin of life.

4 **Naturalistic theory**   At some time more than a billion years ago temperature and moisture conditions became suitable for life. There was no free oxygen, but the atmosphere contained methane, ammonia, hydrogen, and water vapor. It is known from recent experiments that amino acids such as glycine and alanine are produced when the above-mentioned gases are exposed to ultraviolet light or electric discharges such as lightning. Also, the nucleic acid adenine has been produced in the laboratory by irradiation of a mixture of methane, ammonia, and water. Aggregations of such organic molecules would have accumulated, probably in shallow, confined pockets of seawater, because there were then no bacteria to cause decomposition. In the long course of chemical evolution such aggregates competed (natural selection) for the limited store of raw materials, and only the "fittest" survived. Some of them were able to act as catalysts and eventually became autocatalytic— able to catalyze the synthesis of molecules like themselves. These probably derived their energy from fermentation of simple sugars, like some present-day bacteria. Still later, autotrophs developed that could use light-absorbing pigments such as chlorophyll to synthesize complex carbon molecules and give off free oxygen (photosynthesis). This resulted in the earth's present store of oxygen, all of which is estimated to pass through living organisms about every 2,000 years. Using solar energy, the one-celled green algae developed and became the food for the first animal-like organisms, the one-celled protozoans. Once this stage was reached, cells could begin to form into groups of like units and later differentiate to form tissues with division of labor, as seen in higher organisms.

13-3 Where life originated   Since many simpler animals are aquatic and marine and since the cells and body fluids of all animals contain salts (NaCl and others), it is inferred that life began in the oceans. The earliest animal remains are all in rocks of marine origin. Various organisms later invaded the fresh waters and land. Some groups from the latter habitats have secondarily become marine, such as the early sharks and bony fishes, the plesiosaurs and other ancient reptiles, and the whales, seals, and sirenians among living mammals.

13-4 When life originated It is estimated that the earth was formed some 4.5 to 7 billion years ago. The oldest recognizable surface rocks (Archeozoic) are estimated to be about 3.8 billion years old, and the first (Cambrian) that contain numerous animal remains with hard parts were probably formed 600 to 700 million years ago (Sec. 12-29). Many groups of animals had by then already evolved. Unicellular fossil organisms (blue-green and green algae, bacteria, and possibly fungi) are found in rocks estimated to be 1 billion years old, and organisms resembling bacteria (*Eobacterium*) and blue-green algae in rocks over 3 billion years old. *Eobacterium* is the oldest known organism. There is no conclusive evidence of any complete break in the record of life, so conditions suitable for the existence of life must have prevailed somewhere on the earth throughout an enormous period of time.

13-5 Evolution The data of astronomers indicate that the stars and the solar and other systems of the universe have undergone change, or **cosmic evolution** (L. *evolve,* unroll). On the earth there is much evidence of gradual **geologic evolution** in the breakup of land masses and the formation of continents (Sec. 12-30) and the elevation and erosion of land, the transport of particles in water to form sediments, and the long-time changes in climates.

The animals and protistans now living and the many species of past times represented by fossils comprise a variety of forms progressively more complex, from the one-celled protozoans to the higher invertebrates and vertebrates. Biologists interpret the history of animals and plants on the earth as a continuing process of **organic evolution** which has produced the existing species. Existing organisms are considered to be the modified but lineal descendants of other species that lived in former geologic times. This is "descent with modification," the process termed by Charles Darwin "the origin of species." The processes of evolution are still in operation and therefore capable of experimental study. Existing knowledge of the pattern of evolution is summarized in the natural classification in Fig. 14-1, a "genealogic tree" of the Animal Kingdom. Some similarities and differences between phyla and classes are summarized in Tables 14-3 and 27-1. Such characteristics are used in the classification of various groups at the ends of Chaps. 15 to 34.

Evidence for organic evolution is derived from many sources: from comparative morphology, physiology, embryology, geographic distribution, study of fossils (paleontology); from animals and plants under domestication; and from experimentation. In recent years new evidence has derived from biochemistry, molecular biology, and genetics, including cytogenetics, and from other fields of biology. There is abundant evidence that evolution has occurred, but there is some difference of opinion about some of the mechanisms.

Evidence of evolution

13-6 Comparative morphology Most organisms are alike in being composed of cells with many common features. Among these are the four molecules composing the genetic code and mitochondria, the structures which transfer energy within cells. Cilia are widespread, occurring from protists to higher plants and animals, including humans. Their nearly omnipresent and characteristic pattern of filaments (Sec. 3-16) reveals that such divergent structures as auditory hair cells and retina photoreceptors have been derived from them. In recent years comparative study of chromosome number and morphology among groups of species and subspecies has provided evidence of evolutionary origins and relationships. Chromosome features particularly useful in such studies are distinctive loops or other configurations resulting from inversions, deletions, etc. (Sec. 11-19), and surface patterning as found in *Drosophila* salivary chromosomes (Fig. 11-20). Information on the **karyotype,** the array of chromosomes of a species, combined when possible with breeding tests, often permits an investigator to trace cytogenic events in evolution in considerable detail. The process is analogous to matching fingerprints or piecing together a message from scattered fragments. Such studies have confirmed that new polyploid species of plants have arisen naturally within historic times.

The larger groups of animals, although variously unlike in appearance, have similar organ systems for digestion, excretion, and other necessary functions. The members of any one group show greater structural resemblance; thus the insects have one pair of antennae, six legs, and many other features in common. The members of a species comprise animals of similar structure throughout.

In examining animals for structural evidences of evolution it is necessary to distinguish characters that are of common origin (homology), and hence indicative of common ancestry in descent, from purely adaptive features that are of similar function (analogy) but of unlike origin. Thus the skeletal elements in the wings of bats, birds, and pterodactyls (extinct flying reptiles) are homologous in that all are modifications of the common pattern of the forelimb in land vertebrates. The wings of insects, however, are only analogous to those of vertebrates; although used for flight, they are derived, not from limbs, but presumably from extensions of the body wall (Figs. 13-1, 13-2).

Studies in comparative morphology, embryology, and paleontology make it possible to trace the derivation of appendages of vertebrates from lateral folds on the bodies of lower chordates (Figs. 27-8, 29-3) to the fins of sharks and bony fishes. The fins of some fossil fishes (CROSSOPTERYGII) furthermore contain skeletal elements that may be homologized with bones in the limbs of land vertebrates (Fig. 13-3). Limbs of the latter show a wide range in adaptive modifications for special uses by changes in length or by fusion or reduction of parts (Fig. 13-2). Yet all are homologous, being derived from the pentadactyl limb (Gr. *penta*, five + *dactyl*, finger).

Homologies are present in every organ system of the vertebrates, from lowest to highest and including humans. The comparative account of organ systems in Chaps. 4 to 9 provides some of the most striking evidence for evolution (compare also Chaps. 27 to 35). In all vertebrates (1) the nervous system includes an anterior brain with comparable divisions, paired cranial nerves, a single dorsal nerve cord, and paired spinal nerves to each body somite; (2) the brain case is followed by a jointed spinal column of separate vertebrae that support the body and enclose the nerve cord; (3) the digestive tract is ventral to the vertebrae and always includes a liver and pancreas as the major digestive glands; (4) the ventrally placed heart connects to a closed system of vessels containing blood with both white and red corpuscles; and (5) the excretory and reproductive systems show many homologous features. In each system and organ there is agreement as to position in the body

Figure 13-1 *Analogy* **between wings of insects (no bony skeleton) and of vertebrates (with bony skeleton) — of like function but different origins;** *homology* **in the wing bones of vertebrates, all derived from the common pattern of the forelimb in land vertebrates but variously modified.** Pterodactyl (extinct reptile) with very long fourth finger; bird with first and fifth lacking, third and fourth partly fused; bat with second to fifth fingers greatly elongated.

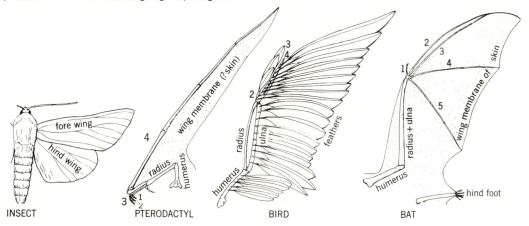

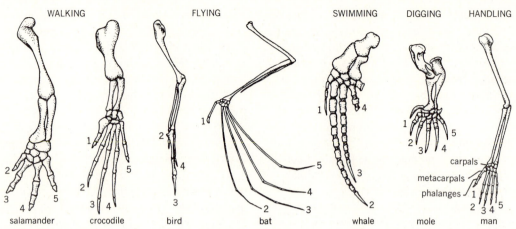

WALKING   FLYING   SWIMMING   DIGGING   HANDLING

salamander   crocodile   bird   bat   whale   mole   man

Figure 13-2 Homology and adaptation in bones of the left forelimb in land vertebrates. The limbs are *homologous* in being composed of comparable bones (humerus, carpals, etc.), which in each kind of animal are *adapted* for special uses by differences in the length, shape, and bulk of the various bones; one to five digits, or "fingers."

and general form and even in the microscopic structure of tissues. Consequently an amphibian, reptile, or mammal serves equally well for obtaining a basic knowledge of vertebrate anatomy.

The organ systems, however, are not exactly alike but show progressive changes from fishes to mammals. In the brain the trend (Fig. 9-6) is toward enlargement of the cerebral hemispheres, which are centers of higher mental activities, and also of the cerebellum, or center of coordination. The heart has two pumping chambers in fishes, is three-chambered in amphibians and most reptiles, and four-chambered in birds and mammals, eventually separating completely the venous and arterial blood (Fig.

13-4). In the excretory organs, drainage of wastes is first from the coelom and later only from the blood (Fig. 7-8).

In like manner there are many homologies among invertebrates. All arthropods have segmented bodies with chitinous covering, a paired series of jointed appendages, a double ventral nerve cord, and many other features in common. A double ventral nerve cord is present also in annelid worms, primitive mollusks, and some other invertebrates.

13-7 Comparative physiology Many basic similarities in physiologic and chemical properties parallel the morphologic features of organisms:

Figure 13-3 Probable evolution of the vertebrate forelimb. *A.* Lobed fin of a Devonian crossopterygian fish. *B.* Limb of an early amphibian. H, humerus; R, radius; U, ulna; S, scapula; C, clavicle; Cl, cleithrum. (*After Romer, 1949, The vertebrate body, Philadelphia, W. B. Saunders Company.*)

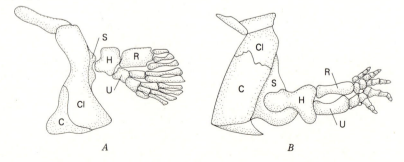

A   *B*

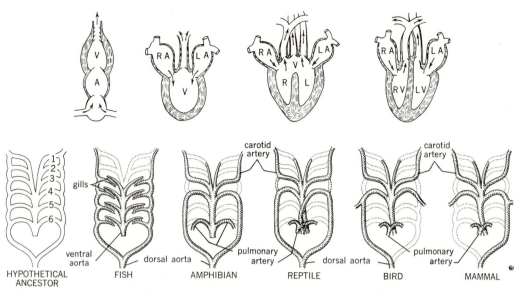

Figure 13-4 Homology and embryonic sequence in the aortic arches and heart chambers of vertebrates. *Below.* Six pairs of arches develop in the embryos of all vertebrates, but parts indicated by dotted lines later disappear. In land vertebrates the third pair always forms the carotid arteries; the fourth becomes the systemic arches to the dorsal aorta, but only the right persists in birds and the left in mammals; the sixth arch always forms the pulmonary arteries. *Above.* The embryonic heart always begins with one atrium (A) and one ventricle (V); it remains thus in fishes. The atrium becomes divided (RA, LA) in amphibians; the ventricle becomes partly divided in reptiles and completely so (RV, LV) in birds and mammals. In embryos of higher forms the arches and chambers develop progressively through the succession of stages shown. Arrows indicate paths of blood flow. (Compare Fig. 6-6.)

1 A classification based on the structure of oxyhemoglobin crystals from vertebrate blood parallels the classification based on body structure. Crystals from each species are distinct, but all from a genus have some common characteristics. Furthermore, those of all birds have certain resemblances but differ from crystals obtained from blood of mammals or reptiles.

2 Precipitin tests are reactions of the blood serum. In such tests human serum is least distinct from that of anthropoid apes (gorilla, chimpanzee, etc.), more so from other primates (monkeys), and still more distinct from that of other mammals. Sera of mammals, in turn, are more sharply distinguished from those of other vertebrates.

3 Some hormones derived from endocrine glands show like reactions when injected into widely different animals. The thyroid gland in cattle controls their rate of metabolism; extracts of that gland may be fed to human beings deficient in their own thyroid secretion to speed up bodily

metabolism. If beef or sheep thyroid hormone is given to frog tadpoles from which the thyroid gland has been removed, the tadpoles will grow normally and later metamorphose into frogs.

4 Many individual digestive enzymes present in different animals are essentially alike in physiologic action. Trypsin, which acts upon proteins, occurs in many animals and protistans, from protozoans to humans, and amylase, which acts on starches, is present from sponges to mammals.

13-8 Comparative embryology Except for a few specialized types of reproduction, every multicellular animal originates as a zygote, or fertilized egg (Chap. 10). The egg of each species has the distinctive ability to produce an individual of that species, but there are many features of embryonic development common to members of any animal group. Fertilized eggs segment, pass through a blastula stage and a two-layered gastrula stage, then become variously differentiated. In many kinds of invertebrates the egg yields a trochophore larva (Fig. 14-8). Eggs of vertebrates differ somewhat in mode of cleavage according to the amount of yolk present (Fig. 10-10), but the early embryos of all are much alike; later, those of each class become recognizable, and still later family and species characters become evident (Fig. 13-5). The beginning embryo in a hen's egg (Fig. 10-12) has at first the vertebrate essentials of a notochord, dorsal nervous system, and somites; later it acquires bird features such as a beak and wings; and much later there appear the characteristics of a chicken instead of a pigeon or duck.

A fish embryo develops paired gill slits, gills, aortic arches, and a two-chambered heart; these all persist in the adult to serve in aquatic respiration. Comparable structures appear in a frog embryo and are necessary during the fishlike life of the frog larva in water. When the larva transforms into an air-breath-

ing frog, however, the gills and gill slits disappear, lungs become functional for respiration in air, the aortic arches are altered to serve the adult structure, and the heart is three-chambered for circulation of the blood to both the body and lungs. The amphibian thus begins with certain fishlike features necessary for an aquatic larva, and later these are altered for terrestrial life. The early embryos of reptiles, birds, and mammals also develop a fishlike pattern of gill slits or pouches, aortic arches, and two-chambered heart (Fig. 13-6), although none of them has an aquatic larva and all respire only by lungs after birth. The embryonic gill slits soon close; the multiple aortic arches become the carotids and other arteries (Fig. 13-4); and the heart soon becomes three-chambered, later having four chambers in birds and mammals.

Gill pouches and multiple aortic arches in embryos of reptiles, birds, and mammals are ancestral relics. The fossil record indicates that aquatic, gill-breathing vertebrates preceded the air-breathing land forms. Their sequence of appearance was fishes, amphibians, reptiles, birds, and mammals (Fig. 12-16). The amphibians represent a transitional phase through which each frog or salamander still passes from aquatic respiration to air breathing (Fig. 31-16).

Many other features in embryonic development and adult structure have a similar significance. In most land vertebrates the first embryonic gill pouch

Figure 13-5 Series of vertebrate embryos in three successive and comparable stages of development. *Top.* All are much alike in the earliest stage. *Center.* Differentiation is evident. *Bottom.* Later the distinctive characteristics of each become evident. (*After Haeckel, 1891.*)

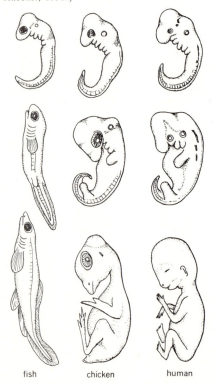

fish   chicken   human

Figure 13-6 Embryologic evidence of evolution. The human embryo (3 mm long) at right resembles adult shark at left in having a two-chambered heart, multiple aortic arches, and cardinal veins. (Compare Figs. 6-4, 6-6, 13-4, 29-5.)

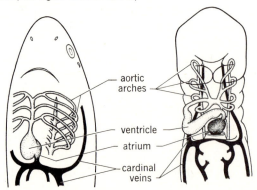

on each side is converted into an eardrum and Eustachian tube connecting the pharynx and cavity of the middle ear. In a fish the gill region is supported by a series of visceral (branchial) arches (Sec. 30-5); and in higher vertebrates, including humans, certain cartilages and bones of the jaws, middle ear, and larynx are derived from the embryonic visceral arches (Fig. 13-7).

The zoologic position of some animals that are of degenerate form in the adult stages has been established only by study of their embryonic and larval stages. The larvae of barnacles (Fig. 24-13) show that these animals belong among the crustaceans; the peculiar parasitic barnacle *Sacculina* (Fig. 24-21) can be recognized as a crustacean only during its larval existence. Likewise the tunicates (Fig. 27-3) were found to be chordates only by a study of their larval characteristics; the degenerate form of the adults gives no clue to their real position among animals.

These and many other facts illustrate the "laws," or basic principles, of embryonic development stated by Von Baer (1792–1876). (1) General characters appear before special characters. (2) From the more general the less general and finally the special characters develop. (3) An animal during development departs progressively from the form of other animals. (4) The young stages of an animal are like the young (or embryonic) stages of other animals

lower in the scale but not like the adults of those animals. The theory of recapitulation, or biogenetic "law," of Haeckel (1834–1919) stated that an individual organism in its development (ontogeny) tends to recapitulate, or repeat, the stages passed through by its ancestors (phylogeny). Von Baer's "laws" provide a more accurate statement. The pattern of embryonic development in a group of related animals may contain features reflecting their past, but many innovations are superimposed that often obscure the ancestral pattern. Thus there is no precise recapitulation. To the ancient (palingenetic) characters of embryos have been added other modern (cenogenetic) characters. Some of the latter appear early in the development of an individual, as, for example, the embryonic membranes of reptiles, birds, and mammals. These are "new" features, not present in the lower vertebrates but essential for protecting embryos of land vertebrates (Fig. 10-14). Another complication is the omission or telescoping of developmental features in relation to special environmental conditions, such as the absence of floating larvae in freshwater crustaceans and the omission of free-living larval stages of some frogs (e.g., *Eleutherodactylus*) and plethodontid salamanders. Larvae may sometimes reach sexual maturity and may reproduce (paedogenesis, Sec. 31-24). Larval ascidians with their gill slits and notochord could have started the chordate line, and the six-legged larva of a millipede, by maturing sexually yet retaining its larval features, has been postulated as the ancestor of the insects.

13-9 Vestigial organs   Structures seemingly without use and of reduced size are termed **vestigial organs.** They were at one time functional and necessary but appear to be in the process of disappearing. Various subterranean reptiles and cave-dwelling fishes, amphibians, crayfishes, and insects have the eyes reduced or absent, although their relatives that live out in the open have eyes. Traces of the pelvic girdle and hind limbs occur in boas and a few other snakes (Fig. 32-11) and in whales. Whalebone whales lack teeth as adults, but tooth buds occur in their embryos. Bird embryos have transient tooth buds, and certain fossil birds (Fig. 33-19) had teeth as adults. The flightless kiwi (Fig. 33-20) of New Zea-

Figure 13-7 Derivatives of human embryonic visceral arches compared with the branchial arches (I–VI) supporting the jaws and gills in an adult shark. In the human these become: I, lower jaw (only Meckel's cartilage) and malleus and incus of middle ear; II, styloid process of skull, stylohyoid ligament, lesser horn of hyoid, and stapes of middle ear; III, greater horn of hyoid; IV, thyroid cartilage of larynx and part of epiglottis; V, cricoid and arytenoid cartilages of larynx. Auditory ossicles shown in heavy black. The human maxilla (upper jaw) is a membrane bone of later origin.

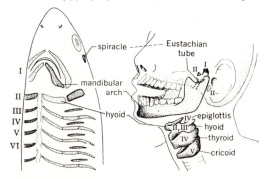

land has degenerate wings with only vestigial bones, and the great moas of that land lacked wings completely. The living horses have splint bones, which are vestiges of toes present in ancestral horses (Fig. 13-13).

Many vestigial features have been listed for the human body; examples are shown in Fig. 13-8. The horse, rodents, and some other mammals have a large caecum or appendix as an accessory digestive chamber. In humans the appendix is a slender vestige about 6.5 cm (2½ in) long that seems to have little function and sometimes is a site of infection requiring surgical removal. The external ears of mammals are moved by special muscles; in humans, lacking need for such movement, the muscles are usually reduced and nonfunctional. In the inner angle of the human eye is a pinkish membrane (plica semilunaris) representing the transparent nictitating membrane, or third eyelid, to be seen in the cat, bird, frog, and other land vertebrates. The human "wisdom teeth," or posterior molars, are often smaller and more variable than the other molars and irregular as to time or manner of eruption; this suggests

that they are becoming useless and may eventually disappear. A word of caution is necessary. An organ classed as vestigial and nonfunctional may, in fact, have an unknown, reduced, or changed function. The term, therefore, should be applied with reservation.

13-10 The fossil record Important evidence for evolution comes from the study of **fossils.**

Leonardo da Vinci (1452–1519) was among the first to recognize that fossils were evidence for animal life in the past.[1] The first comprehensive studies of fossils (vertebrates) were made by the French comparative anatomist, Georges Cuvier (1769–1832). In 1800 he published an account of fossil elephants, relating them to living forms; later , in a classification of fishes, he included both living and fossil species. He was the first to demonstrate a sequence of extinct faunas and to show the value of the comparative anatomic method. Cuvier, however, believed in special creation, and it was Darwin who first considered fossils as evidence of the evolution of organisms. **Paleontology** (Gr. *palaios*, ancient), the study of fossils, is now an important science that links zoology and geology and provides many facts of evolution.

The geologic record of past life is variously imperfect. It is like the remnants of a book that lacks all the beginning chapters, contains only scattered pages or parts of pages in the central portion, and retains an increasing number of intact pages or parts of chapters toward the end. Records of past life result from a succession of accidental events. (1) The remains of a dead animal escape destruction and (2) become buried in sediment or ash that (3) survives undue heating, crushing, or folding that would destroy the fossil. (4) The sediment or rock becomes elevated as a part of the land and (5) escapes destructive erosion by water or wind. Finally (6) the fossil becomes exposed or is dug out and comes to the attention of a paleontologist. Some fossil remains are complete, but many are fragmentary, and all the

Figure 13-8 Some vestigial structures in the human body. *(Partly after Kahn.)*

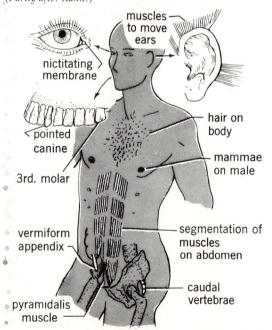

muscles to move ears

nictitating membrane

hair on body

pointed canine

mammae on male

3rd. molar

vermiform appendix

segmentation of muscles on abdomen

caudal vertebrae

pyramidalis muscle

---

[1] *The nature of fossils, means for estimating their age, and their distribution in time are discussed in Chap. 12. Various fossils are discussed in Chaps. 15 to 34, and the fossil history of humans is outlined in Chap. 35.*

Figure 13-9 A sample of the geologic column: succession of strata in north wall of the Grand Canyon, Ariz. (compare Fig. 12-18). Eras and periods are marked (names in parentheses are of local rock formations). One fault line is shown. Four unconformities (*Un*) mark long gaps in the record; below each the rocks were deformed, elevated, and finally far eroded before the layer next above was deposited. This site has no rocks of the Ordovician, Silurian, and Devonian periods or of the Mesozoic and Cenozoic eras. (*Photo by U.S. Geological Survey.*)

known fossils represent only a fraction of the many plant and animal species that lived in the past. Some species or groups may never have become fossils because they were soft-bodied or because they lived where fossilization could not occur. Many fossils have been destroyed by alteration of the rocks or by erosion, and any now in rocks deep in the earth or under the sea are inaccessible.

13-11 Invertebrates Rocks formed prior to Cambrian time have since been so folded and distorted that they now reveal few organic relics. Yet animals must then have been in existence for a long time, because the Cambrian rock strata, the oldest with

many fossils, contain a variety of invertebrate fossils —protozoans, sponges, jellyfishes, worms, brachiopods, echinoderms (sea cucumbers, crinoids), mollusks (gastropods, cephalopods), and arthropods (crustaceans, trilobites). The beginnings of most invertebrate phyla and of some classes cannot be traced, but the rise, continuance, and decline or extinction of others are well recorded. The trilobites (Fig. 23-2) were dominant when the record opens in the Cambrian; they increased in numbers and variety, then disappeared entirely in the Permian, when there were continental elevation, greater aridity, and glaciation. The lamp shells (BRACHIOPODA, Fig. 20-8) were abundant throughout the Paleozoic era (1300 genera) and less numerous in the Mesozoic (285 gen-

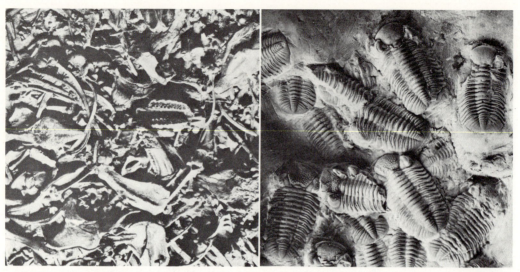

Figure 13-10 Representative fossils; covering matrix of rock removed. *Left.* Fossil bone bed from the Miocene of Nebraska containing remains of rhinoceroses and other mammals. *Right.* Accumulation of Devonian trilobites. (*Left, from American Museum of Natural History; right, from U.S. National Museum.*)

era) but persist today. About 30,000 extinct species have been described, but only some 63 genera and 260 species are still living. The living *Lingula* is much as it was in the Ordovician, 440 to 500 million years ago, and is perhaps the oldest living genus of animals. The ammonites (class CEPHALOPODA) began in the late Silurian. Like the pearly nautilus (Sec. 21-24), their shells comprise a succession of chambers, with the animal in the largest and terminal chamber. The gross form of the shell changed through time (Figs. 13-11, 13-12); sutures between successive chambers were simpler in the older stocks and more complex in those of later periods. The first chambers formed by some of the later ammonites were simple, like those of their ancestors. Then, surprisingly, some of the last of the ammonites had simpler sutures again, and their shells were coiled in the young but straight in the adult parts. Finally the entire line of ammonites ended in the Cretaceous. The great phylum ARTHROPODA was represented by aquatic crustaceans and trilobites in the Cambrian period and eurypterids in the Ordovician. Scorpions, the first air-breathing land animals, originated in the Silurian. The winged insects appear suddenly in the Carboniferous, as several differentiated orders, and give no ready clue as to which arthropod type they stem from.

13-12 Vertebrates The origins of vertebrates are shrouded by great gaps in the geologic record. The earliest remains are from freshwater deposits, and both structure and function of the vertebrate kidney favor such an origin. Yet echinoderms, from which the chordates may have arisen, are all marine, and many paleontologists favor a marine origin.

Figure 13-11 Trends in evolution of the fossil ammonites (phylum MOLLUSCA, class CEPHALOPODA), showing the coiling and later uncoiling of the shell and the changes in form of the sutures (shown by fine lines; black in *E*). All reduced, but not to same scale. (*From University of California Museum of Paleontology.*)

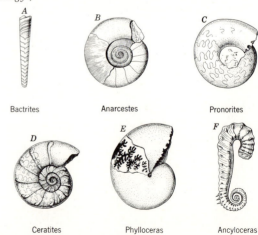

A Bactrites · B Anarcestes · C Pronorites · D Ceratites · E Phylloceras · F Ancyloceras

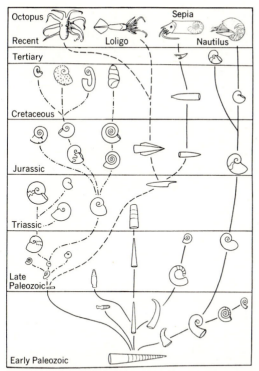

Figure 13-12 An interpretation of evolution in the cephalopods. NAUTILOIDEA (solid lines), COLEOIDEA (broken lines), and †AMMONOIDEA (dots and dashes). (Compare Figs. 13-11, 21-26.) *(Modified from Raymond, Prehistoric life, by permission of the President and Fellows of Harvard College.)*

Octopus
Recent   Loligo   Sepia   Nautilus
Tertiary
Cretaceous
Jurassic
Triassic
Late Paleozoic
Early Paleozoic

only four orders that persist today. The first small reptilelike mammals appeared in the Triassic, and the first known birds in Jurassic times. Early in the Tertiary (Paleocene) the mammals blossomed into a great variety, including many existing orders and others since vanished. The early forms were replaced by more modern types, and the mammals reached a peak of diversity in the Miocene. Since then they have declined, a considerable number having become extinct at the end of the Pleistocene, just preceding the present or Recent period in which we live. Thus, despite the fragmentary nature of the early record, ascending the geologic column the vertebrate types appear in an orderly time sequence that corresponds to the increasing structural complexity of the groups living today.

No vertebrate remains have been found in Cambrian rocks. Ordovician strata contain fragments possibly of ostracoderms, which were ancestral to cyclostomes, the lowest living vertebrates, without jaws or paired appendages. Silurian deposits have many ostracoderms and also spines and plates probably of placoderms, the earliest jawed vertebrates. By early Devonian time placoderms were abundant but ostracoderms less numerous. Further along in Devonian, however, both sharks and bony fishes appeared and then became abundant. Amphibians, with paired limbs, also are in the record of late Devonian. The reptiles undoubtedly began during the Carboniferous, because by the end of that period there were already several specialized types. Thenceforth, from the Permian to the Cretaceous, they were the dominant animals of land, sea, and fresh water. Many became large in size, such as the brontosaurs, tyrannosaurs, and plesiosaurs. All the great reptiles disappeared at the end of the Cretaceous, leaving

13-13 Horses The family EQUIDAE provides one of the most complete records of evolution in an animal series (Fig. 13-13), leading to the existing horses, asses, onagers, and zebras of the Old World. Much of their ancestral development occurred in North America, but horses died out there late in the Pleistocene (or early in Recent time), for reasons unknown. The wild horses of the Western states in the last five centuries derived from stocks that were brought in and escaped from early explorers and settlers.

The principal changes in the horses down through time were as follows: (1) increase in size from that of a small dog to some larger than existing horses; (2) enlargement and lengthening of the head anterior to the eyes; (3) increase in size and convolutions of the cerebrum; (4) increased length and mobility of the neck; (5) changes of the premolar and molar teeth from a structure suited for browsing to one adapted for grazing (small teeth with low rounded cusps were transformed into tall prismatic columns with complex enamel ridges); (6) elongation of the limbs for speedy running and fusion of bones in the foreleg to provide better hinge joints and support of the weight on a sturdy radius and tibia, although at the expense of rotational movement; and (7) reduction of the toes from five to one long toe (third) on each foot, which is covered by a hoof (claw); the lateral toes dwindle as "dew claws," and finally only small bones of the second and fourth toes persist as splints. By these changes the horse became

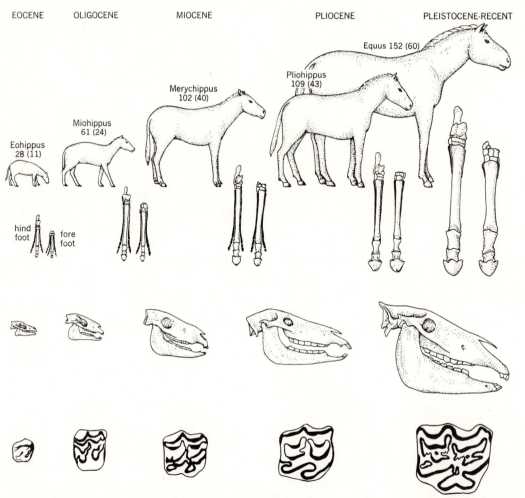

Figure 13-13 Evolution of the horse. *Top row.* Progressive change in size and conformation from the little forest-dwelling †*Hyraco-therium* (†*Eohippus*) of the Eocene epoch to the large, modern, plains-inhabiting *Equus* (numbers are height at shoulder in centimeters, followed by inches in parentheses). *Second row.* Bones of hind feet and forefeet, showing reduction in lateral toes (solid black), from †*Hyracotherium* with three hind and four front toes to *Equus* with only the third toe functional on each foot, the second and fourth represented by splints. *Third row.* Skulls, showing changes in size and outline and closing of orbit by postorbital process. *Bottom row.* Grinding surfaces of second upper molar, showing increasing complexity of enamel pattern (black). The structural changes shown are representative of stages through which the modern horse (genus *Equus*) must have passed. Neither *Merychippus* nor *Pliohip-pus* (in the strict sense), however, are in the direct line leading to *Equus*. (*Top row adapted from R. S. Lull, Fossils, courtesy of the University Society, New York; others from W. D. Matthew, 1913, and R. A. Stirton, 1940.*)

a long-legged, swift-running mammal suited to live and feed on open grasslands, with tall teeth having many enamel edges to grind harsh grassy vegetation through a long life.

The origin of horses is unknown. The record begins with †*Hyracotherium* (†*Eohippus*) in the Lower Eocene of North America and Europe. It was a browsing forest dweller, about 28 cm (11 in) tall, with

a short neck and head and a full set of 44 small short-crowned and rooted teeth that lacked cement. The front foot had four functional toes, but the hind foot only three, the first and fifth toes being represented by tiny splints. †*Miohippus* of the Oligocene was the size of a sheep, with taller but rooted molars and three functional toes on each foot; the lateral toes were smaller, and only one splint (fifth) persisted on the forefoot. By Miocene time, several lines (†*Parahippus*, †*Merychippus*) had developed. Both are stages in the evolution of the grazing line. *Anchitherium* was a persisting member of the browsing horses. During the Pliocene there were several distinct groups (†*Pliohippus*, etc.) grazing on the plains of North America. Some spread to Eurasia, and †*Hippidion* to South America, the latter giving rise to some short-limbed genera that did not survive the Pleistocene. The lateral toes were reduced to dew claws that did not touch the ground. The cheek teeth were longer, with short roots, more folding of the enamel, and cement between the folds. Finally the earliest one-toed horses developed, during the Pliocene in North America, and later they spread to all the continents except Australia. In the Pleistocene, there were ten or more forms (*Equus*) of various sizes in North America, all of which disappeared in prehistoric time. Evolution of the horses followed known changes in Tertiary landscapes from moist forest to dryish grasslands.

13-14 Elephants   The stout bones and teeth of elephants and their allies have left a well-documented story of evolution in the order PROBOSCIDEA (Fig. 13-14). †*Palaeomastodon* of the Oligocene in Egypt was about 2.4 m (8 ft) high. It had short tusks, but other incisors and the canines had disappeared to leave 26 teeth; all cheek teeth were functional at the same time. There was probably a long snout or proboscis. The descendants of this and other forms spread to India and later to Eurasia. †*Gomphotherium* of the Miocene and Pliocene reached North America. It had three molars in each jaw, long upper tusks, and a long lower jaw with short, broad tusks. Other collateral lines developed among the proboscidians but did not survive to later epochs. †*Stegodon*, with long tusks and trunk, appeared in the Pliocene but appears to have been a parallel to the elephants rather than ancestral.

No proboscidians are now native to the Americas, but four types inhabited North America during Pleistocene time, and mastodons lived in South America during the late Pliocene and Pleistocene. The mastodon (†*Mammut americanum*) was about 2.9 m (9½ ft) tall and had long coarse hair. It

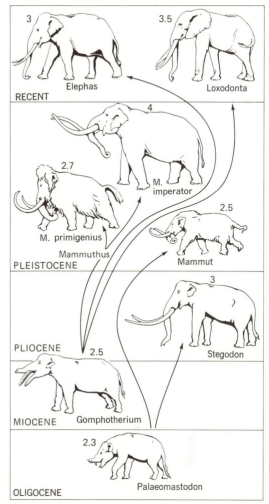

Figure 13-14 Phylogeny of the elephants, showing increase in size and in relative length of tusks and proboscis since Oligocene time; numbers are height at shoulder in meters (compare Fig. 13-15, showing changes in teeth). Other lines of proboscidians that became extinct are omitted. †*Mammut*, mastodon; †*Mammuthus primigenius*, woolly mammoth; †*M. imperator*, imperial mammoth; *Elephas*, Indian elephant; *Loxodonta*, African elephant. (*Adapted from H.F. Osborn, 1936–1942. Proboscidea, American Museum of Natural History.*)

inhabited forests of Canada and the United States and probably was hunted by the earliest native peoples. The mammoth (†*Mammuthus primigenius*) that ranged across the Northern Hemisphere was about 2.7 m (9 ft) high and covered with dense wool. It was hunted by Paleolithic people in Europe who pictured it in cave drawings; fleshy remains of mammoths have been recovered in the frozen soil of Siberia and Alaska. The Columbian mammoth (†*Mammuthus columbi*) was 3.4 m (11 ft) high and had incurved tusks; in early Pleistocene it roamed the warmer parts of North America south to Florida and the Mexican tablelands. The huge imperial mammoth (†*M. imperator*) grew to 4 m (13½ ft) high and was probably a plains dweller from Pliocene to mid-Pleistocene. The living elephants of India and Africa (*Elephas, Loxodonta*) have two large tusks in the upper jaw and but one molar functional in each jaw at a time. The molars have many transverse ridges of enamel, between which the softer dentine and cement are worn down by the coarse foods (Fig. 13-15).

13-15 Human evolution Many kinds of evidence (Chap. 35) indicate that humans are a product of evolution. In both gross and microscopic **structure** the human body resembles closely that of the anthropoid apes, is like that of other primates, and has much in common with mammals generally (compare Figs. 4-5, 4-6, 5-3). Homologies with other vertebrates are present in every organ system; some vestigial human organs have already been mentioned (Fig. 13-8). Strictly human characteristics such as the upright posture, opposable thumbs, flattened vertical face, scant body hair, and greatly increased brain are differences in degree and not in kind from other mammals. Many relationships in **function** (physiology) parallel those of structure; in both humans and anthropoids there are comparable blood groups, human blood can be distinguished by immunologic tests from that of all others except anthropoids, and some mouth protozoans are common to humans and other primates. The earlier **embryonic development** of humans is essentially the same as for other vertebrates (Fig. 13-5) and includes temporary gill pouches

Figure 13-15 Evolution of teeth in elephants (third lower left molar) through geologic time. × ⅓. The progressive increase in size was accompanied by change of grinding surface from few rows of low conical cusps to many cross ridges of enamel, together with deepening of the enamel folds. The five lower figures show exposed grinding surface; the upper two are longitudinal sections. The teeth of stegodonts were probably convergent on those of elephants.

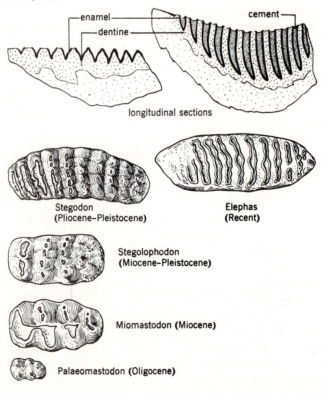

longitudinal sections

Stegodon
(Pliocene–Pleistocene)

Elephas
(Recent)

Stegolophodon
(Miocene–Pleistocene)

Miomastodon (Miocene)

Palaeomastodon (Oligocene)

and multiple aortic arches (Fig. 13-6). The human **fossil record** includes a series of types (Fig. 35-10) that gradually approach the form of existing humans. The present human dominance over other organisms results primarily from habits, behavior, and mental characteristics; these include social organization, modification of the environment to human advantage, development of tools and of language, and the ability to transmit learning by teaching.

The fossil record indicates a common pattern in evolution. A group of related, generalized species appears; their numbers increase and they radiate into a variety of habitats or niches, becoming increasingly specialized. A plateau of diversification is reached and the species extinction rate approximates that of speciation. Extreme specialization becomes typical. The group then begins to decline and finally is reduced to a few highly specialized forms or becomes extinct. This pattern has repeated itself many times in different groups of organisms. Examples are trilobites, jawless fishes, dinosaurs, and certain mammals.

Theories of evolution

Any effort to account for existing organisms and fossils should explain their origins, their likenesses and differences, their adaptations to various environments, and their distribution on the earth. Theories of organic evolution postulate that since life began on the earth it has been continuous and that later organisms have been derived from earlier forms by the inheritance of variations, either large or small, induced either by the environment or by processes within the organisms. Most evolutionary processes are slow and therefore difficult to test experimentally.

Various myths ascribe the origin of humans and other animals to creative acts of supernatural powers. Until the last century most persons, including such scientists as Linnaeus, Cuvier, Agassiz, and Owen, believed that species had been created separately. Cuvier thought the disappearance of fossil species had resulted from a series of catastrophes, the last being the biblical flood, and that after each of

these the earth had been repopulated by new creations of higher types. The belief in catastrophes was dispelled by the Scottish geologist Charles Lyell (1797–1875), who showed in his *Principles of Geology* (1830–1833) that the geologic processes of sedimentation, uplift, and erosion are essentially continuous or evolutionary.

Some early Greek philosophers had vague notions of an evolutionary process, but Aristotle (384–322 B.C.), the first notable zoologist, thought that organisms were molded by a "perfecting principle," and his ideas prevailed for centuries. Buffon (1707–1788) was the first modern biologist to discard the concept of special creation. He believed that animals were plastic, that small variations produced by the environment were accumulated to make larger differences, and that each animal in the ascending series of types was transformed from some simpler ancestor. Erasmus Darwin (1731–1802), grandfather of Charles Darwin, added the further idea that functional responses to external stimuli were inherited.

13-16 Lamarck and the inheritance of acquired characteristics   The first general theory of evolution was proposed by Jean Baptiste de Lamarck (1744–1829), an anatomist and student of classification. His theory was outlined in 1801 and set forth fully in his *Philosophie zoologique* (1809). Lamarck recognized a fundamental continuity in the diverse kinds of animals and believed that there had been progressive development in form and structure. His theory, briefed from his own words, is as follows: The environment affects the shape and organization of animals; frequent or continuous use develops and enlarges any organ, while by permanent disuse it weakens until it finally disappears; all acquisitions or losses wrought through influence of the environment, and hence through **use and disuse,** are preserved by reproduction.

The theory may be illustrated by two of his examples. Birds, he assumed correctly, were originally terrestrial. A land bird going to seek food in water would spread its toes to strike the water in moving about. The skin at the bases of the toes would be continually stretched, and muscular movements of the legs would promote an extra flow of blood to the feet. In consequence, the skin would become enlarged as webs between the toes, like those of ducks, pelicans, and other water birds. Disuse Lamarck illustrated by the structure of a snake. In crawling through grass its body would be stretched repeatedly to pass through narrow spaces, and the legs would not be used. Long legs would interfere with crawling, and four short legs could not move the body. Legs are characteristic of reptiles, and yet the snakes lost theirs. The eyes became lateral the better to see when on the ground, and the tongue developed as a protrusible sensory organ to detect objects in front of the snake.

There is no reliable evidence for Lamarck's theory, and it has little support today. The muscles of an athlete increase in strength and bulk with extensive use but become reduced in bulk if exercise is discontinued; children never inherit such acquired characteristics of a father. The docking of tails in horses, sheep, and bulldogs for many generations has not made these mutilations hereditary. Pavlov trained mice to come for food at the sound of a bell and claimed that fewer and fewer trials were needed to teach mice of succeeding generations, and MacDougall has claimed similar results in the training of rats, but neither of these experiments nor many others designed to test this theory have produced convincing results. This conclusion is not surprising when it is recalled that a new organism develops from the germ cells of its parents, not from their somatic cells. The germ cells are set aside early in the growth of an individual and are subjected to little or no effect from the body cells or environment (Chap. 10). This was demonstrated by Castle and Phillips, who replaced the ovaries of a white guinea pig with those from a black female. The former was then bred twice to a black male. All the six young produced were black and homozygous.

13-17 Darwin and the theory of natural selection
Charles Darwin (1809–1882) was a methodical, painstaking English naturalist of broad vision. As a young man he served (1831–1835) as naturalist on the *Beagle*, a vessel that explored South America, the Galápagos Islands, and other regions. From his detailed notes and studies he wrote excellent works on

barnacles, mammalian fossils, geology, and coral reefs. He began taking notes on the origin of species in 1837, writing in his notebook that he "had been greatly struck from about the month of March on [the] character of South American fossils, and species on [the] Galápagos Archipelago. These facts, especially latter, [are the] origin of all my views. . . ." The next year Darwin read Malthus's *Essay on Population*, wherein that author showed how populations increase in geometric ratio until checked by the limited food supply. On this Darwin wrote: "Being well prepared to appreciate the struggle for existence which everywhere goes on, from long continued observation of the habits of animals and plants, it at once struck me that under these circumstances favorable variations would tend to be preserved, and unfavorable ones to be destroyed. The result of this would be the origin of new species. Here then I had at last got a theory by which to work." In 1844 Darwin wrote a summary of his theory but continued to gather data from original researches and observations by himself and many other persons. Meanwhile Alfred Russel Wallace (1823–1913), another English naturalist, while studying the rich fauna and flora of the Malay archipelago, independently and quickly arrived at similar conclusions. In 1858 he sent an essay on the subject to Darwin. Some idea of Darwin's reaction is contained in a letter written to his friend, Sir Charles Lyell, on June 18, 1858: "I never saw a more striking coincidence; if Wallace had my MS. sketch written out in 1842, he could not have made a better short abstract! Even his terms now stand as heads of my chapters. . . . So all my originality, whatever it may amount to, will be smashed, though my book, if it will ever have any value, will not be deteriorated; as all the labour consists in the application of the theory." Darwin's first impulse was to withdraw in favor of Wallace, but geologist Charles Lyell and botanist Joseph Hooker persuaded him to prepare a brief of his conclusions which was read, together with Wallace's essay, at the meeting of the Linnean Society of London on July 1, 1858. In 1859 Darwin published his theory in a volume entitled *On the Origin of Species by Means of Natural Selection or the Preservation of Favoured Races in the Struggle for Life*.

This was the most important book of the nineteenth century. It contains (1) overwhelming evidence of the fact of evolution and (2) arguments for natural selection as the process. The doctrine of evolution was not original with Darwin, but his convincing presentation quickly won the support of scientists and of many laymen. Unscientific attacks on "Darwinism," as the theory was called, continued until after his death, and resistance to the teaching of evolution still persists. Meanwhile a great scientific search began for additional facts bearing on the theory, and there was much speculation on natural selection.

The essence of Darwin's theory is as follows:

1 **Variations** of all degrees are present among individuals and species in nature.
2 By the **geometric ratio of increase** the numbers of every species tend to become enormously large; yet the population of each remains approximately constant, because many individuals are eliminated by enemies, disease, competition, climate, etc.
3 This involves a **struggle for existence;** individuals having variations unsuited to the particular conditions in nature are eliminated, whereas those whose variations are favorable will continue to exist and reproduce.
4 A process of **natural selection** therefore is operative, which results in:
5 The **survival of the fittest,** or "the preservation of favored races."

13-18 Variation Among animals that reproduce sexually, offspring are rarely exactly alike (exceptions being identical twins and the quadruplets of armadillos). The individuals of every species vary in size, proportions, coloration, external and internal structure, physiology, and habits. Darwin recognized the widespread occurrence of variations; his theory assumes these but does not explain their origins. In his day the laws of inheritance (Chap. 11) were unknown, and often he could not distinguish the heritable variations, which alone are important in evolution, from nonheritable variations produced by differences in food, temperature, or other environmental factors. Darwin saw that domesticated animals and plants are more variable in many ways than wild species. He knew that humans have produced many domestic races by **artificial selection,** or breeding of individuals having heritable variations (characteristics) useful for human needs; also that practical breeders had established and improved the many breeds of livestock and varieties of cultivated plants by gradually accumulating small but useful hereditary differences through many successive generations. He rightly believed that, in most cases, all the domestic breeds of a species had been derived from one wild ancestral species—all breeds of rabbits from the European gray rabbit, and all pouters, fantails, racers, tumblers, and other widely differing breeds of pigeons from the rock dove. Many of these domestic breeds now differ so greatly from one another in appearance that if they occurred in the wild, any zoologist would classify them as distinct species and some as different genera! The domestic breeds of a species, however, all can mate with one another and produce fertile offspring. Having shown the wide diversity of domestic types produced from ancestral stocks by selection of small variations, Darwin assumed that small heritable variations in wild species were the materials of the evolutionary process in nature.

13-19 Geometric ratio of increase All forms of life have the potentiality of rapid increase. The protozoan *Paramecium* (0.25 mm long) can divide by fission about 600 times per year. If all offspring survived and continued to divide, their total bulk after some months would exceed that of the earth. The fruit fly *Drosophila* completes its life cycle from egg to egg in 10 to 14 days, and each female may lay 200 or more eggs. In 40 to 50 days, if all survived and bred, they would number 200 million; during one summer their numbers would become astronomical.

The brown rat, house sparrow, and European corn borer in the United States and the European rabbit in Australia are examples of pests that have multiplied somewhat in keeping with their theoretical possibilities when introduced into new and favorable environments. Plagues of native insects and of meadow mice result at times when abundant food supplies are suddenly available, predators are scarce, or other factors enable individuals to reproduce and mature rapidly.

13-20 Struggle for existence   Under ordinary conditions, however, animals never increase to such numbers as just indicated. The populations of most species tend to remain relatively stable because of various checks (Secs. 12-6 to 12-11). There are limitations in food supply, shelter, and breeding places; individuals of a species compete with one another for these necessities and also with other species having similar requirements; an enlarged population of any species soon is levied upon by its predators and is a fertile field for parasites and diseases. The "struggle for existence" is not always a spectacular battle, as of a rabbit trying to escape from a fox, but it is a continuing process in nature involving many factors, each of which eliminates some individuals. It acts at any stage in the life cycle of a species, from the egg, which may fail to be fertilized, through embryonic development, larval stages, and adult life. Any individual animal is "successful" in the struggle for existence if it survives long enough to reproduce.

13-21 Natural selection   Darwin assumed that in the struggle for existence individuals with slightly favorable variations enabling them to meet the conditions of life more successfully would survive and propagate their kind—a process that Herbert Spencer termed "the survival of the fittest." Under this sort of **natural selection** those lacking such variations would perish or fail to breed, so that the characters which they possessed would be eliminated from the population. In succeeding generations the process would continue and result in gradually adapting animals more perfectly to their environments. With a change in environmental conditions there would be a change in the sorts of characters surviving under natural selection. A species in a changing environment or one that had migrated to some new environment would be gradually altered to suit the new conditions. Animals failing to develop suitable new variations under any particular environmental conditions would soon be eliminated. In this manner Darwin conceived the development of adaptations of whatever sort, the "origin of species" in changed or

new environments, and also the disappearance of species in past geologic time. Two portions of a species population having to meet slightly differing conditions would tend to diverge from one another and in time would be separated, first by small differences as varieties or subspecies, later, when isolated from each other, as species that could not interbreed. A continuation of such divergence would lead in time to the production of still other species and in turn to wider differences (at the level of genus, family, etc.). In this manner he conceived the great number of species and larger categories of the Animal Kingdom to have been established through the long duration of geologic time.

Most biologists accept Darwin's theory as the best general explanation of evolution. They differ mainly in their later and better understanding of some of the essential biologic processes involved which were unknown in his day but which have been learned by later research. Modern interpretations, based on newer knowledge, are termed **Neo-Darwinism.**

Evolution: the modern synthesis

13-22 Origin of heritable variations   Darwin realized clearly that heritable variations occur in both wild and domesticated animals, but he had no knowledge of how they are produced or of the exact manner in which they are inherited. (Mendel's precise laws, although published in 1866, were not generally known until 1900.) Starting about 1875, however, biologists began to study the processes in germ cells and their relation to reproduction; a little later careful attention was given to experimental breeding. Soon there was a wealth of new knowledge that afforded a clear understanding of the manner of origin of heritable variations and of the ways they are passed from generation to generation. In recent years experimental breeding has been accompanied by study of the related changes in germ cells (these fields are combined as **cytogenetics**). The details of chromosome behavior and of genetic processes are all-important for understanding certain evolutionary processes. The essential points are as follows:

1 Chromosomes carry in linear arrangement the genes responsible for development of characteristics in an individual.

2 Meiosis segregates members of homologous chromosome pairs and halves the total number for each gamete (Chap 10.) Exchange of paternal and maternal genes may occur through crossing over (Sec. 10-5).

3 Fertilization, the random union of two gametes of unlike sex, brings together assortments of chromosomes (and therefore of genes) from two parents, resulting in production of individuals with different gene combinations (Chaps. 10, 11).

4 Mutations occur in genes, and chromosome rearrangements take place (Sec. 11-19); both result in altering the assortment of genes (hence characteristics) passed on to succeeding generations.

Many mutations first detected in laboratory stocks of *Drosophila* now are known to occur in wild populations. Conversely, the black and the silver mutations of the red fox, first known in nature, are now present in captive foxes on fur farms, where still other mutations have appeared and some have been preserved. From these and many other records it is now evident that new mutations are constantly appearing in nature, and studies of protein polymorphism reveal that many species populations are highly heterozygous (Sec. 13-31). Mutations are not as rare as formerly thought. Although they are uncommon at the level of the individual gene, at the species level they must be regarded as common when one considers the great number of genes and individuals that comprise most species. On the average there may be 2000 new mutations at each gene locus every generation in some insects. It has been estimated that current mutation rates range overall from about 1 in 10,000 to 1 in 50,000 to 200,000 (higher vertebrates) meiotic events. There is thus a wide range of hereditary variations in wild species. Whether any particular variation will become a persistent characteristic depends on the size of the population, the degree of isolation or segregation of small groups of individuals, and other factors.

Different mutations may be beneficial, neutral, or harmful. The huge mass of experimental data (im-portantly from *Drosophila*, microorganisms, and some plants) indicate that most of those found are harmful or neutral. They tend to disturb the well-adjusted adaptive characteristics of the organism. A high mutation rate is often severely disrupting. A slow rate, however, confers a great advantage by providing a constant source of new variability essential to the process of adaptation. Mutations useful to humans are well known among domestic animals and crop plants. In wild species mutations with superior adaptive value seem scarce, but this is to be expected, since any that appear are probably soon incorporated in the gene complex of the species to its advantage. Many harmful genes are damaging, however, only when homozygous (such as those of lethal characters); in the heterozygous condition, paired with their normal alleles, the majority have no unfavorable effect. A characteristic that is harmful by itself or under one set of environmental conditions may be beneficial in combination with others or under different conditions. Indeed, some experimental data show that two characters, each harmful when alone, prove beneficial when combined. Causes of mutations are discussed in Sec. 11-19.

13-23 Gene change and natural selection The primary agent in the change of gene frequencies in natural populations is natural selection, but gene flow (see below) and perhaps genetic drift (Sec. 13-29) may also play a role. The action of natural selection depends on heritable variation in populations combined with abundant multiplication of individuals and a variety of physical and biologic restraints to unlimited population growth (geometric ratio of increase). Under such constraints not all can survive. Births must ultimately balance deaths. Which individuals are favored by natural selection? It is those who have inherited traits that contribute to their reproductive success. An individual's fitness is measured by the proportion of its genes left in the population gene pool. Those which leave more successful offspring than their rivals will increase in the population and their rivals will eventually die out. It should be noted that the selective action of the environment is often highly sensitive to small changes in fitness of phenotypes.

Most natural populations contain much genetic diversity, often far more than realized. The total genetic information contained in a reproducing unit (population or species) is called the **gene pool.** The source of genetic variation in asexually reproducing organisms is mutation (Sec. 11-19). Bisexual populations have several important additional sources. (1) At any gene locus there are usually a number of alleles, in some cases over 40. The mating combinations possible for 10 alleles at each of 10 loci is 10 billion. Many species have thousands of alleles; a great many combinations are thus possible. (2) Shuffling of genes results from recombination and crossing-over. (3) Through independent assortment and recombination, a gene may be associated with any of its alleles and with many different groups of genes. The particular association in which a gene is found may influence how it acts, and it in turn may affect the behavior of other genes. Many different associations are possible. (4) New genetic information may be provided by immigration and, occasionally, by hybridization (Sec. 13-28). The movement of genes between and within populations is called **gene flow.** These mechanisms provide the many different phenotypes upon which natural selection can act.

Selection may take several forms. (1) In stable environments average phenotypes tend to be favored and the less fit extremes tend to be eliminated. Selection in favor of such intermediates is called **stabilizing selection.** (2) In a changing environment the average phenotypes may not be the most fit and **directional selection** occurs. There is a shift in the population mean toward a new phenotype. (3) **Disruptive selection** favors two or more different genotypes, and intermediates are at a disadvantage. It tends to occur in patchy environments where conditions differ from place to place. Polymorphisms (Sec. 13-26) are fostered by this kind of selection, and it is the type that occurs when two populations diverge on the two sides of a barrier.

Natural selection is not confined to action on individuals. Any entity having variation, reproduction, and heritability may evolve. Units of selection include (or have included) prebiotic molecules, cell organelles, genes, individuals, populations, species, and ecosystems. To illustrate: just as the individual may increase its proportional representation within a species through natural selection, in the same way certain species may be favored and others winnowed out in the evolution of a natural community.

13-24 Adaptations Organisms are fitted for existence in the environments they inhabit. They are **adapted** to the conditions of life. Adaptations are modifications in structure and function shared by the members of a group that aid survival. From the standpoint of evolution we are interested in those which are of hereditary origin and are the consequence of natural selection. They are present at all levels of biological organization—molecular, cellular, individual, and species. Modifications that took place early in evolution that are common properties of organisms—basic DNA structure and certain aspects of cellular metabolism—are usually not considered in discussions of adaptation, yet they were produced by the same selective process as the later ones. The same is true of many features at higher levels of biological organization shared with little variation by all members of a group—the vertebrate backbone, for example. In discussing adaptation, there is a tendency to focus on the more recent and divergent modifications that relate to obvious needs of survival. Such preoccupation must not cause us to overlook the many other aspects of adaptation that make perpetuation of the individual and species possible. Study of adaptations and their possible contribution to individual and group survival yields important information on evolutionary mechanisms and trends. Such study seeks the why of many aspects of the biology of organisms.

In view of the nature of the selective process, which acts mechanistically on the individual and its complex of characteristics, adaptations are rarely completely ideal solutions to problems of environmental adjustment. There is nothing inherent in the adaptation mechanism, for example, that ensures long-term reproductive success. The results of selection may be neutral, a precarious balance between useful and harmful, or even deleterious. Although short-term adaptive changes usually benefit the species (sometimes by maintaining the status quo), long-term trends such as increasing specialization may lead to species decline and even extinction. Indeed, extinction has been the fate of most species. Intraspecific selection can also lead to difficulties. Sexual selection for secondary sexual characteristics may reach such grotesque proportions as to threaten individual survival (display feathers of some male birds and the horns and antlers of some ungulates).

Natural selection cannot anticipate needs in changing environments. Usually at least several generations are required to adjust a population to new conditions. Thus at any given time the population will be made up of phenotypes varying in degree of fitness for existing conditions. How well the population as a whole is adapted will depend on its reproductive rate and inherited variability and on the rate and complexity of environmental changes.

The degree of adaptation differs in various groups, ranging from species that are narrowly or closely adapted (specialized) to those that are generalized. Every species, regardless of degree of specialization, is a constellation of adaptive characteristics involving structure, physiology, behavior, and mode of life. The honeybee (Chap. 25) has sucking mouth parts used for obtaining nectar, the ability to subsist on sugars, hairs and brushes used to gather pollen, wax produced and molded into shelters for food and young, and an intricate pattern of habits of three castes in a social colony. The human being, the brown rat, and other generalized species are able to do many things in various ways and to live in diverse environments. The mole, by contrast, is a specialized subterranean insectivore with slender teeth used in grasping worms, eyes covered and ears reduced, short forelimbs with huge palms bearing heavy claws employed in digging and "swimming" through soil, and short, reversible fur which is not disarranged by moving forward or backward. Other highly specialized forms are the many parasites that can live in only a single host species and some, like the malarial parasite and liver fluke, that must alternate between two particular hosts to complete their life cycles (Chaps. 15, 18).

Adaptive radiation is the relatively rapid diversification of a species or species population into a variety of different adaptive types (species), made possible by the rather abrupt availability of new habitats or niches. It is a common evolutionary phenomenon (Sec. 13-15). New environmental opportunities may result from colonization, environmental change, or genetic breakthroughs which allow entry into new or little-exploited habitats. A classic example is the radiation that occurred in the Australian marsupials. In the absence of competing placental mammals, these animals have radiated into diverse forms fitted to run, jump, climb, burrow, or glide. Adaptive radiation is an example of divergent evolution, or **cladogenesis. Adaptive convergence** often occurs when animals of different groups come to live in a common habitat. Large vertebrates of the ocean (Fig. 13-16) all have streamlined bodies and paddlelike fins or limbs, which enable them to swim more effectively. Such adaptive characters are superimposed on the fundamental ones that make the shark a cartilagi-

Figure 13-16 Oceanic vertebrates, sharks to mammals, showing adaptive convergence for swimming — bodies streamlined and fins or limbs paddlelike.

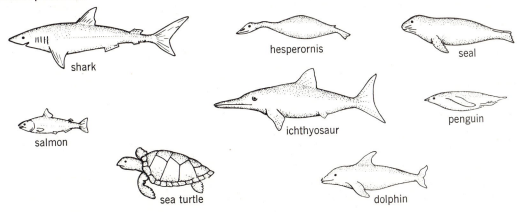

shark

hesperornis

seal

salmon

ichthyosaur

penguin

sea turtle

dolphin

nous fish and the seal a mammal. **Parallel evolution** is adaptive convergence of closely related forms.

Many adaptive features of animals are protective in a variety of ways—by structure, function, or coloration. The shells of armadillos, turtles, and most mollusks and the quill covering of porcupines are obvious structural adaptations that protect their owners. The sting of bees and wasps and the venom released by the skin glands of toads and other amphibians are examples of functional adaptations serving a similar purpose.

The concealing colors of animals are another sort of protective adaptation. Many animals more or less resemble the backgrounds against which they live, which renders them less likely to be attacked. Predators too may blend with their surroundings, gaining concealment from their prey. Hares, weasels, and ptarmigan of the far north molt into white coats when snow covers their surroundings; the ruffed grouse has a variegated pattern that blends with the leafy backgrounds of its woods habitat. Many planktonic invertebrates and fish larvae are transparent. Bark-inhabiting insects are commonly flecked with colors of the backgrounds on which they live.

Concealment may entail more than a simple color match. Even though an animal precisely matches the color of its background, this alone does not ensure concealment. A uniformly colored rounded or cylindrical object when illuminated from above is conspicuous, even against a background of the same color, because of the highlight on its upper surface and the shadow on its lower surface. In many animals this effect of overhead lighting is overcome by **countershading.** Surfaces of the animal that are habitually turned toward the light are dark and those on the underside are pale. The dark pigment dulls the intensity of the highlight and the pale color lightens the shadow, thereby reducing the appearance of three-dimensional form when the animal is viewed from the side. Examples of reversed countershading in animals that move or rest in an inverted position (e.g., heavy-bodied caterpillars) lend support to this interpretation. In aerial or aquatic environments, countershading reduces conspicuousness in another way: the pale lower surfaces tend to blend with the sky or water surface and the dark upper parts with the ground or bottom. Color patterns may be su-

perimposed on countershading. They break up relief and delay recognition of shape by blurring the outline of tails, fins, wings, limbs, eyes, and other characteristic structures. Disruptive markings often are bold and contrasting and intersect the outline. They are particularly effective against variegated backgrounds.

Flash markings are conspicuous markings suddenly revealed, then promptly concealed. They may help protect camouflaged animals when they move quickly, or they may serve as signals in intraspecific communication. The marks focus attention, but on an image that soon disappears. A predator, especially if inexperienced, may seek the shape and color briefly seen. Examples are the white outer tail feathers or conspicuous wing patches of some birds and the brightly colored flight wings of grasshoppers. Flash markings on some lizards are on an expendable part, the tail. Some schooling silver-sided fish produce distracting flashes when attacked, as individuals momentarily swim on their sides.

Warning coloration is ascribed to some butterflies and other insects considered to be distasteful to predators—they "advertise" their unpalatability. Bees and wasps with painful stings are often marked conspicuously with black and yellow. Certain nudibranch mollusks are brightly colored and usually unpalatable to predators. Flash markings may serve as a warning. They may be displayed when camouflage fails. When molested, the eyed hawk moth (*Smerinthus ocellatus*) spreads its forewings, uncovering conspicuous eyelike marks on bright red hind wings which frighten some species of birds. The ringneck snake (*Diadophis*), distasteful to some predators, abruptly coils its tail when disturbed, revealing bright red on the undersurface. The pyrgomorphid grasshopper (*Phymateus*) spreads its brilliantly colored hind wings as a repelling secretion exudes from glands on its back. The swayback posture of firebellied toads (*Bombina*) and newts (*Taricha*) momentarily reveals the bright ventral colors of these toxic animals.

In many instances harmless or palatable species resemble other stinging or unpalatable species. The viceroy butterfly **mimics** the often unpalatable (and distantly related) monarch butterfly (Fig. 13-17), and certain moths, beetles, and flies appear to have

Figure 13-17 Mimicry and protective resemblance. Viceroy butterfly (*Limenitis archippus*), above, which mimics the monarch butterfly (*Danaus plexippus*), below.

Figure 13-18 Protective resemblance. *Left.* Larva of geometrid moth positioned upright, head end touching curved twig. *Right.* Leaf insect (*Phyllium*) of Sri Lanka, appearing like yellow tree leaves—"veins" on wings and expanded forelegs, frayed edges, and "discolored" fungus spots.

Recognition marks and other signals are often important in intraspecific communication—between the sexes, parents and offspring, members of a flock, rivals, etc. They range from boldly conspicuous, such as the markings of certain insects, reef fishes, and male songbirds, to remarkably subtle, such as the differences in color of the eye ring in some seagulls.

Many adaptations evolve over long periods of time, but there are some recent instances of rapid adjustment. For many years hydrocyanic acid (HCN) gas proved successful in control of the red scale (family COCCIDAE) on citrus trees in southern California, but in 1914 the standard fumigation dose had become inadequate in one district and later in adjacent localities. Investigation by Quayle and others showed that there were two races of the insect, one cyanide-resistant and the other not. At a certain concentration of the gas, 45 percent of the former but only 4 percent of the latter survived. Crossing the two races showed them to differ in one sex-linked gene for HCN resistance. Similarly resistant stocks have appeared after years of spray control in other citrus pests and in larvae of the codling moth in apples. Use of DDT for pest control began in the mid-1940s, and resistance soon appeared—among houseflies in 1946 and shortly among mosquitoes

"copied" different species of wasps. Some long-horned beetles (CERAMBYCIDAE) resemble wasps in form and color and hover about flowers in wasp fashion. One African butterfly (*Papilio dardanus*) corresponds in color and form to three distasteful species of butterflies with which it occurs, three distinct types of females being produced by a single pair of parents.

Certain insects and other animals show **protective resemblance** to inanimate objects in their surroundings (Fig. 13-18). Some geometrid caterpillars in coloration, form, and position at rest appear much like the twigs of trees on which they live, and some walking sticks (ORTHOPTERA) look like dead or green twigs and others like leaves.

and other insects. By 1968 it was known in 127 agricultural pests and 97 of public health or veterinary importance. Behavioral avoidance evolved in some malarial mosquitoes that normally rest in homes. Those living indoors were killed off by DDT, but new outdoor-resting strains developed in Java, Mexico, and Panama. Resistance is commonly by selection of genes that produce enzymes able to detoxify the poison.

Pest control programs usually are ideal for rapid evolution of resistance, often within a few generations. Large areas are heavily dosed, killing off most susceptible individuals. If the material is short-lived, repeated applications are frequent. Susceptible insects are thereby removed from a large area long enough for the few surviving resistant individuals to multiply and establish a new population. Species having more than one generation per year and high reproductive rates adapt most quickly. A salt-marsh sand fly evolved resistance to dieldrin after three applications of 1 lb/acre. Some fishes and amphibians have developed insecticide resistance, but with only small annual broods a longer period is required. In these cases human beings have applied a new selective factor to the store of variability that included genes for both resistance and nonresistance.

Another rather quick adjustment is that of industrial melanism (or darkening) among moths, first observed in factory districts of England, Germany, and other European countries. Industrial melanism is now widespread throughout the Palearctic, wherever air pollution is taking place. The moths rest by day on trunks and branches of trees where originally their colors blended with the pale background of bark and light-colored lichens. Lichens are sensitive to air pollutants and die off as the fumes and soot increase. Darkening of the trees results from their death and accumulations of soot. Melanic (dark-colored) variants of over 100 species in Britain have increased in numbers and, in some areas, have all but replaced the original paler stocks. The dark form (*carbonaria*) of the peppered moth (*Biston betularia*) formed only 1 percent of the population near Manchester, England, in 1848 but had increased to 99 percent by 1898. Now most moths are dark in the eastern counties sooted by factory smoke, whereas only an occasional dark mutant is seen in the unpolluted woodlands of southwestern England, the Scottish highlands, or Ireland. Field experiments by Kettlewell and Tinbergen showed that selective feeding by spotted flycatchers, nuthatches, robins, thrushes, and other birds is responsible for this striking change in the population. The birds took six times as many dark and therefore conspicuous individuals as protectively colored pale specimens released in a clean forest in Dorset. The opposite occurred in a blackened wood near Birmingham, where three times as many light-colored individuals were taken as black forms. In areas of Britain where air pollution has been reduced (smokeless zones), the frequency of *carbonaria* has declined.

13-25 Preadaptations A **preadaptation** is a heritable characteristic or complex of traits already present which enables a species or subspecies to adapt to an environmental change or to adopt a new mode of life without going through the usual process of gradual natural selection. Preadaptation is passive and opportunistic. A preadaptation may or may not be functional prior to its use in the new evolutionary context. The adhesive toe pads of certain frogs, useful in locomotion on the ground, were perhaps a preadaptation that aided animals of this type to become arboreal. Gradual selection thereafter resulted in further enlargement of the pads, thereby enhancing their adhesive properties. An inherent capacity for cold tolerance, unexpressed in species in a warm climate, preadapts them to survive a cooling trend. For example, a species with the potential to complete its life cycle during a shorter growing season will be able to survive or to invade areas beyond its original range. A species is preadapted if able to survive under conditions to which it has not previously been exposed.

Larval forms and embryonic developmental stages may at times have been preadaptive steps and may have provided shortcuts in evolution. Neotenous amphibians are an example (Sec. 31-24) in which larval forms have given rise to species. The webbed, spatulate feet of certain neotropical arboreal plethodontid salamanders closely resemble those of their embryos and perhaps represent retention of a developmental trait in the adult animal—an example of **paedomorphosis.**

13-26 Polymorphism Genetic polymorphism is the occurrence in one habitat of more than two forms (or morphs) of a species in such proportions that the rarest cannot be accounted for by recurrent mutation (Ford). When only two forms are present, the species is said to be **dimorphic.** Polymorphism is widespread but tends to be reduced in small populations where homozygosity prevails. The number of morphs, or groups of like individuals, in a species may vary from two to many. Polymorphism aids the species in adjusting to environmental change. It may also represent a balance in the proportions of different kinds of individuals that have proved effective in maintaining the species in a stable environment. Such balances sometimes last for many generations. Polymorphic differences may exist in structure, physiology, and/or behavior. The more obvious ones are in color or structure. Color morphs are very common, being found in many insects and other invertebrates and in all vertebrate groups. Familiar examples in North America are the brown and black phases of the black bear and the gray and red morphs of the screech owl. Polymorphic variations sometimes have a fairly simple genetic inheritance and in many cases seem to be maintained by selection in favor of the heterozygote. The sickle-cell anemia allele (Ss) in humans is an example. The homozygous morphs differ markedly in the appearance of their red blood cells—normal or sickle-shaped. In malarial areas of Africa, the heterozygotes (even though mildly anemic) have a selective advantage because of their resistance to falciparum malaria, not present in the homozygous individuals (SS) with normal hemoglobin. The homozygous recessives usually die of the disease before reaching sexual maturity.

A variety of genetic and selective forces may bring about and maintain polymorphism in a species. Often **supergenes** are involved. These are coadapted, closely linked groups of genes that segregate together. A balanced polymorphism in some prey species may result from the tendency for predators to form search images of the more abundant morphs in the population. High population density seems to favor this form of selection. The predator looks for the more familiar prey. Concealing color is apparently not involved. There is thus selective maintenance of the rarer morphs in the population. In other instances, selection for background matching may maintain two or more color morphs in an environment where two or more predominant backgrounds alternate spatially or seasonally. A parallel case is found in examples of mimicry in which a palatable insect species contains several morphs which resemble several different species of distasteful models. The mimics must not become excessively abundant, otherwise predators will tend to lose their association of repugnance with a particular color pattern. The pervasiveness of genetic polymorphism is now evident from biochemical studies (Sec. 13-31), and it may have great durability over time in stable environments. The eastern African land snail (*Limicolaria martensiana*) has maintained five color-pattern morphs in comparable proportions for some 8000 to 10,000 years.

13-27 Isolation and speciation Darwin's title for his book *The Origin of Species* indicates a topic of primary concern to the evolutionary biologist and a problem that is not yet solved. Changes (mutations, chromosome rearrangements, etc.) occur in individual animals, and then by sexual reproduction are either preserved and spread through a population or are eliminated (natural selection). Thus the species is at the basis of the evolutionary process. A species is a natural interbreeding population that differs and is isolated (with some exceptions) reproductively from other populations. In asexual or parthenogenetic animals the question of interbreeding does not apply, and any distinctive type or strain may be called a species. **Speciation** is the process by which new species are formed, and **isolation** often is an essential step in their formation (Sec. 14-4). Isolation is often temporal or spatial. In temporal isolation a species gradually evolves into another through an unbroken evolutionary line over a long period. Spatial isolation involves the physical separation of portions of an original population. However, there are other modes of speciation.

The number of individuals in many species is

enormous—hundreds of thousands in common birds and mammals and millions or billions in widespread insects. The population of any one species, however, is not uniform in either distribution or characteristics. First, the individuals are not distributed evenly throughout the entire geographic range but are subdivided into smaller groups more or less isolated from one another. Each occupies a part of the range, and groups often do not intermingle except along their boundaries because of limited powers of locomotion or various barriers. Second, the groups differ from one another quantitatively (size, color, etc.); these differences blend where representatives of different groups can interbreed (hybridize) along group boundaries. The term **race,** or **subspecies** (Fig. 13-19), is applied to such local populations, and the total combined population is called a polytypic species. In some species the population is not broken into distinguishable groups but shows gradual continuous change of characteristics along a gradient—north to south, lowlands to highlands, or dry to moist climate. Such cases, known as **clines,** are exemplified by species of birds and mammals that are larger in cooler climates (Bergmann's rule) or darker in warm humid regions. Evidence from several fields—taxonomy, migration, experimental breeding, cytology, serology—shows that many a "species" in the ordinary sense is actually composed of numerous biologic strains, stocks, or races.

Races of some species may evolve rapidly. The house sparrow (*Passer domesticus*), introduced into North America from Europe in the period 1852 to

Figure 13-19 Geographic distribution of the subspecies of the tiger salamander (*Ambystoma tigrinum;* **Mexican forms omitted).**   Each differs in pattern and occupies a distinct range as shown. Adjacent subspecies intergrade along their boundaries. Unshaded portions indicate areas of unsuitable habitat. The dorsal patterns of several subspecies are shown. (*Patterns mostly from Stebbins, 1951, Amphibians of western North America, University of California Press.*)

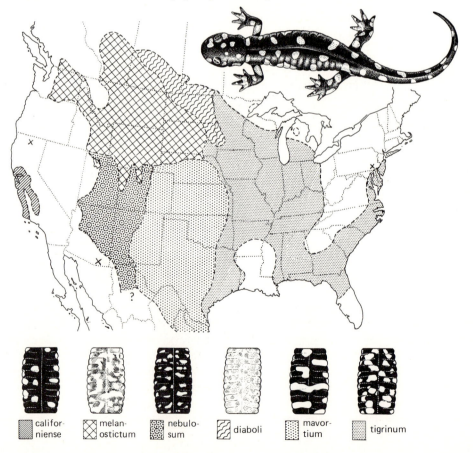

califor-niense   melan-ostictum   nebulo-sum   diaboli   mavor-tium   tigrinum

1860, spread rapidly across the continent. It has undergone changes in weight, wing and bill length, proportions of skeletal elements, and coloration in geographically separated areas in North America in some 50 to 115 generations. The variation in these birds is adaptively organized and follows ecologic expectations. The birds are larger where winter temperatures are low (Bergmann's rule) and have relatively smaller appendages (Allen's rule). They are dark in the Vancouver district (Gloger's rule) and around Mexico City. The equivalents of true geographic races have evolved. About 5000 years has been given as the minimum time for the evolution of an avian subspecies (Moreau), but it has happened in the house sparrow around Mexico City in about 30 years.

Populations or subdivisions thereof, separated geographically (Fig. 13-19), are termed **allopatric** (Gr. *allos,* other + L *patria,* country). Two or more different populations occupying the same or over-

Figure 13-20 **The four species of** *Zonotrichia* **sparrows and their nesting ranges.** Being distinct species, they do not interbreed, even where the ranges overlap. Winter ranges are mainly in the United States, where three occur together over wide areas and all four locally in the Pacific Northwest. (*Based on maps from U.S. Fish and Wildlife Service.*)

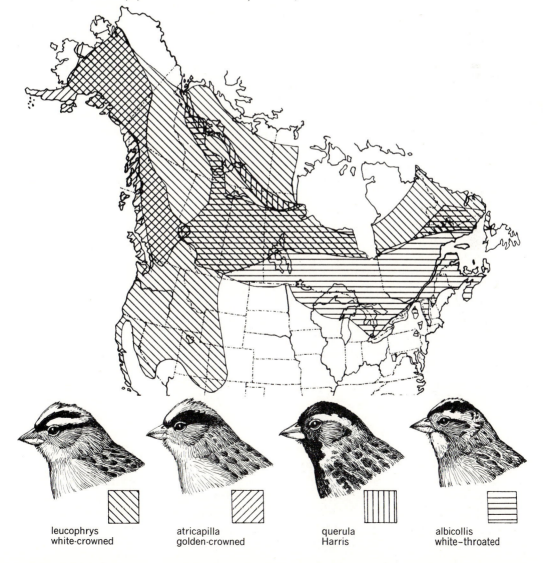

leucophrys
white-crowned

atricapilla
golden-crowned

querula
Harris

albicollis
white–throated

lapping ranges that maintain their distinctness (Fig. 13-20) are **sympatric** (Gr. *syn*, together). Populations are **parapatric** when they meet in a narrow geographic zone and retain their distinctness.

Isolation, the segregation of stocks into smaller units, may be of several kinds: (1) **geographic,** by physical separation in distance; (2) **ecologic,** in different types of environments although in the same general region; (3) **seasonal,** where two populations breed at different times of year; (4) **physiologic,** where there is functional incompatibility in mating or in the production, fertilization, and survival of gametes; and (5) **behavioral,** where animals of two different groups will not mate with one another.

The classic view of speciation has been that divergence usually begins with geographic isolation; then types 2 to 5 arise secondarily. There are two forms of allopatric speciation. (1) A widely distributed species becomes subdivided into two or more populations by external barriers. Genetic differences accumulate by small steps (microevolution) in the isolates as each responds to the somewhat different selective forces in the environment it occupies. Premating reproductive isolating mechanisms arise fortuitously and appear to be perfected first in the hybrid zone between the diverging forms. The process is often relatively long. (2) A new colony is established by a relatively small number of founders or in some cases perhaps even by a single gravid female (certain insects). This may be by accidental dispersal from mainland localities to islands or to new terrestrial or aquatic habitats. Early stages of reproductive isolation arise fortuitously as in the previous case. Groups predisposed to this type of speciation are often those coming from small, semi-isolated, peripheral species populations adapted to some degree of inbreeding and are often products of a period of rapid population increase. An important element of chance enters into the formation of the genotype of the founding colonies. The founders carry only a small fraction of the total genetic variation of the parental population, and the colony they establish will contain only the relatively few genes they brought with them except as replaced or augmented by mutation and contributions from additional immigrants. This phenomenon can result in testing of many different genotypes in a variety of environ-

ments. Such founder populations, however, are often vulnerable to extinction because of their small size and limited variability, which makes them less adaptable to environmental changes.

Parapatric speciation occurs when populations diverge to the species level within a continuous cline without spatial isolation. Such speciation is most common in animals that are sedentary. The development of reproductive isolating mechanisms accompanies entry into and use of the new habitat by the genetically divergent individuals.

In sympatric speciation, premating reproductive isolation arises in individuals within the area occupied by an interbreeding population and its dispersing offspring before the diverging population shifts to a new niche. Divergence of populations may occur well within the species range rather than at the periphery, and the first stage may be a stable polymorphism. Speciation starts with initially nonadaptive events such as polyploidy, chromosome rearrangements, or even single gene substitutions if they interfere sufficiently and permanently with gene flow between diverging populations. In such cases premating reproductive isolation arises before the population shifts to a new niche. Sympatric speciation has evidently occurred in many parasitic and parasitoid insects through a shift in their plant or animal host selection or in pheromone signals. Such insects are the most abundant of all plant and animal species. Polyploidy (Sec. 11-19) has apparently often been a mode of sympatric speciation in plants. A plant species was created artificially in this way when Karpechenko produced a synthetic plant genus, *Raphanobrassica*, by crossing a radish (*Raphanus sativus*) and a cabbage (*Brassica oleracea*). He obtained a tetraploid which bred true and was reproductively isolated from its parents. The haploid chromosome number in each parental species is 9, and the hybrid contained 18 chromosomes (9R + 9B). In most cases the chromosomes failed to synapse, and most gametes were inviable. However, a few fertile hybrids were obtained which contained 18 chromosomes that synapsed at meiosis.

Profound genetic changes may sometimes be required for speciation. In many organisms much of the species genotype is conservative and forms a closed system of established and integrated function. It contains **palaeogenes** or entrenched gene complexes that are important to the stability of long-tested genotypes. In some species the genetic unity extends over wide geographic areas and maintains uniformity in the frequency of even rare alleles. The closed system usually cannot be disrupted without endangering species survival. On the other hand, many species also contain genes in an open system involved in recombination and selection and in most of the adaptive responses of populations. Many of these genes (**neogenes**) are relatively recent and not so involved in entrenched species-maintenance complexes. The problem in speciation of such entrenched species is to break out of the restraints of the closed system and achieve the rather large amount of reorganization in the genome that seems to be required for speciation. The founder population has been proposed as one answer. Situated at the periphery of the range it exposes the species' cohesive genotype to strong new selective forces that allow for major reorganization—a kind of genetic revolution. This might come about through a population crash that would reduce the population to a few individuals (or even a single gravid female). This model recalls the "hopeful monster" of Goldschmidt: an individual (a gravid female) which in one mutational step is able to enter a new niche and is reproductively isolated from the parental stock. If several sexually compatible "hopeful monsters" are simultaneously produced, so much the better.

In bacteria, significant and abrupt genetic changes occur through transfer of blocks of DNA by virus particles from cell to cell. Bacterial antiobiotic resistance can be transmitted in this way. Restriction enzymes in the recipient cell screen the foreign genes and eliminate them if they are incompatible. Such gene transfer has not yet been observed in eukaryotes.

13-28 Hybridization and introgression Species differ in their adaptive gene complexes and are separated by reproductive isolating mechanisms that vary in potency, preventing or minimizing hybridization. In general, interbreeding in divergent forms results in wasted reproductive effort or the intrusion of untested foreign genes, which disturb the species' adapted gene complex. Sometimes, however, the new genetic input provides a beneficial increase in variability and the capacity to adapt to new conditions (Sec. 13-23). In stable environments, hybrids are usually at a disadvantage, for most are less well adapted than the parental stocks to the niches occupied by the latter, and other niches are filled with well-adapted organisms. Under conditions of rapid environmental change, however, when new niches become available and established species are subjected to new selection pressures, hybrids may take hold. Furthermore, the changing conditions may bring together species that were formerly ecologically separated, increasing the chances of hybridization. Both glaciation and modern human activities have been accompanied by extensive hybridization among natural populations.

Hybrids are more likely to mate with members of one or the other of their parental stocks than with each other, because the former are usually more abundant. Progeny of a parental backcross are more likely to be well adapted to the habitat in which they are found, which may be a modified form of one of the parental habitats, than progeny from hybrid matings. Frequent breeding may therefore occur with one of the parental species, and selection favors genotypes having mostly genes from that parental species but also a few genes introduced from the other parental form. This process, involving hybridization and frequent mating with one of the parental stocks and the stabilization of such backcross genotypes by selection, is known as **introgression.** The genes of one species intrude into the genetic complex of the other. Introgression is often an important source of variability in evolving species.

13-29 Genetic drift In the study of population genetics (Sec. 11-23), the findings of Mendelian heredity are applied to population phenomena under the Hardy-Weinberg law. This applies to large and freely interbreeding populations where there is genetic equilibrium and where variability usually remains constant. Under such conditions evolution does not occur. It is the deviations from this norm, the mutations, selective mating, and survival of desirable traits that result in change. Changes are more rapid in small, isolated populations such as occur in times of stress or when a few individuals reach a new habitat or remote island. Under these conditions variability is reduced rapidly or abruptly and the genotype tends to become homozygous. This change in gene frequency results from sampling error and depends on the size of the breeding population. Since the two genes of an allelic pair segregate independently of other genes (Mendel's laws), allelic behavior at meiosis can be compared to flipping a coin (Balootian and Stiles). Let genes H and h represent heads and tails, and the number of trials the individuals in the population. With many trials (a large population), the frequency with which H and h appear will be essentially the same, but with few trials (a small population), variance in their frequency will be large and alleles may even be lost, as when all tosses are H and no gametes receive the gene h. Some small populations may contain only a few breeding adults. Thus it appears that sometimes purely by chance small populations may come to differ from the original population in slight ways and become distinct stocks or races. In house mice (*Mus musculus*) it may occur where no physical barriers to interbreeding exist; tribal inbreeding, polygamy, and territorial behavior reduce intertribal breeding, with the consequence that random genetic events have an important role in determining the genetic composition of local demes.

A **deme** is a small, freely breeding population. Demes serve as testing grounds for evolutionary innovations. Homozygosity permits recessive traits to be expressed in the phenotype and tested by natural selection. In a large population they may be lost. The random genetic events described above and different selection pressures acting on demes provide diversity and benefit the species as a whole by providing flexibility in adapting to environmental changes.

In practice it is difficult to distinguish selective forces from drift, and more field studies of natural populations will be required before it can be determined how important drift may be in evolution.

13-30 Darwin's finches   A classic example of the effects of isolation on speciation is shown in the birds of the subfamily GEOSPIZINAE on the Galápagos Islands 965 km (600 mi) off Ecuador (Fig. 13-21). This group of oceanic islands, the largest 129 km (80 m) long, were formed by volcanic eruptions. The plant cover is diverse, thorn and cactus in lowlands where the ground is of jagged lava changing to moist forest in rich black soil on the mountains (610 to 1220 m, or 2000 to 4000 ft). Giant land tortoises and iguanas are the only large native land animals. Most distinctive of the 26 naturally occurring land birds are the finches, which have developed into 13 species. In turn, on separate islands, some have diverged further into subspecies or less distinct populations. In the Darwinian view, small differences so produced may lead to "incipient species," and speciation is regarded as complete when two such forms later meet and remain distinct. Thus on most of the islands there are three closely related species of ground finches (*Geospiza magnirostris, G. fortis,* and *G. fuliginosa*) that feed on seeds of relatively different hardness and that do not interbreed; presumably they evolved on different islands and later came together.

These finches illustrate another aspect of evolution—**adaptive radiation.** Ancestors of the Galápagos terrestrial plants and animals must have been transported there by accidental means (Sec. 12-20). When the ancestral finches arrived (colonization is thought to have been by a single mainland finch species), there were probably few or no competitors. Following establishment on one or more islands, competition increased and dispersal occurred to other islands in the archipelago. Since the islands differ in size, altitude, food conditions, and other factors and the ocean is more or less a barrier to interisland movements of these birds, adaptive radiation occurred and distinct island ecotypes evolved. Genetic divergence led to reproductive isolation. Subsequent movements caused intermingling of species on different islands. Coexistence increased competition for food and for other requirements, and selection accentuated differences evolved originally in isolation. Three species coexist on each of the 16 main islands of the archipelago, and three of the larger islands each have 10 species. They are able to live together because of the diversity in their structure and habits. For example, marked difference in beak size (which influences the size of

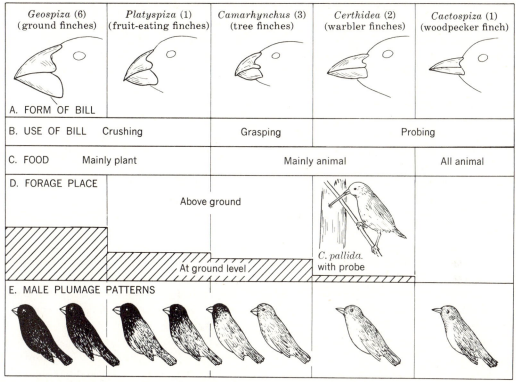

Figure 13-21 Genera of Darwin's Galápagos finches (number of species in parentheses). *A*. Form of bill. *B*. Use of bill. *C*. Types of food. *D*. Forage areas utilized. *E*. Plumages of adult males. Inset in *D* shows *Cactospiza pallida* with probe. (*Adapted from Bowman, 1961, and Lack, 1947.*)

seeds eaten) evidently reduces competition between the widespread seed-eating finches *G. fuliginosa* (small beak) and *G. fortis* (large beak), which coexist on many islands of the archipelago. However, where they occur separately, on the tiny islets of Daphne and Crossman, their beaks are nearly the same size. Within this insular subfamily of birds have evolved the biologic equivalents of seven continental familial groups. Some of the types among the "finches" are (1) seed eaters (*Geospiza*), with stout conical beak; (2) a bud, leaf, and fruit eater (*Platyspiza*), with beak suggestive of a parrot; (3) insect eaters (*Certhidea*), with slender bill and habits of a warbler; and (4) a woodpecker type (*Camarhynchus*), with short, stout beak for digging in tree trunks. The related *Cactospiza pallida*, lacking the long tongue of a true woodpecker, uses a twig or cactus spine to herd or poke insects out of crevices. On one of the smallest and most remote islands (Wenman) the sharp-

beaked ground finch (*G. difficilis septentrionalis*) feeds on the blood of boobies, which it obtains by piercing the skin.

These inconspicuous finches, on a group of remote tropical islands, have had an important part in evolutionary thought ever since 1839, when the young Darwin (*Voyage of the Beagle*) wrote: "By far the most remarkable feature in the natural history of this archipelago . . . is, that the different islands to a considerable extent are inhabited by a different set of beings. . . . I never dreamed that islands about fifty or sixty miles apart, and most of them in sight of each other, formed of precisely the same rocks, placed under a quite similar climate, rising to a nearly equal height, would have been differently tenanted. . . . Hence, we seem to be brought somewhat near to that great fact—that mystery of mysteries—the first appearance of new beings on this earth.''

13-31 Molecular evolution   Molecular biology seeks insight into evolution at the level of the gene. Although nucleic acids, components of the genes themselves, are studied, it is often technically more feasible to investigate enzymes and other proteins one step removed from the genes. By comparing the amino acid sequence in homologous (closely related) proteins among species, it is possible to determine the extent to which mutations have occurred and therefore the amount of genetic differentiation between species. Such studies of gene evolution provide a basis for determining more precisely than has been possible in the past (1) the nature of genetic variation in natural populations, (2) degrees of relationship among populations and species, and (3) rates of evolutionary change.

Studies of the extent of genetic variation in natural populations is of great importance in understanding the process of speciation (Secs. 13-23, 13-27). A variety of techniques for assessing such variation at the molecular level has been developed. They include nucleotide and amino acid sequencing, DNA and RNA hybridization, immunological techniques (microcomplement fixation, etc.), and electrophoresis. Proteins are the most frequently studied molecules—especially transferrins, esterases, dehydrogenases, albumins, etc., from blood or other tissues, or whole organisms.

In **electrophoresis,** a popular technique, molecules are allowed to move in a starch or acrylamide gel under the influence of an electric field. Differences in electrostatic charge, and sometimes size and configuration of the molecules, are reflected in different mobilities. A mobility difference between polypeptides can usually be considered as evidence for at least one amino acid difference. Specific histochemical stains are applied, revealing the positions of the proteins on the gel. The bands of stain so produced mark regions of protein concentration. By comparing the proteins encoded by many structural genes in two populations it is possible to determine the degree of electrophoretic similarity in their proteins and the extent of divergence between the populations.

Using such procedures, estimates have been made of the number of different alleles that exist at a given gene locus and their proportion in individuals and populations. In some organisms allelic diversity has been determined for many loci. Since the proteins, the products of genes, are a step removed from the genes themselves, it is appropriate to refer to the variation described as protein polymorphism. However, it is believed to reflect the genetic heterozygosity that exists. In studies of morphologic variation in color, size, and proportions, the amount of loci polymorphism can only be inferred.

Molecular investigations have revealed widespread protein polymorphism in natural populations, estimated to range from 30 to 80 (often 25 to 50) percent of the structural gene loci in most organisms studied. The proportion of heterozygous loci in such genes in the average individual ranges from 2 to 20 percent in large continental populations. There are often great differences in amount of heterozygosity between species and their smaller subdivisions, even to the level of the deme (Sec. 13-29). These findings stand in contrast to the classical view of much genetic uniformity in natural populations.

Study of biochemical loci can be used to assess the amount of gene flow (including introgression, Sec. 13-28, etc.) between populations and the extent of their genetic isolation. **Genetic distance**—the average number of electrophoretically detectable substitutions per locus which have accumulated since two populations diverged from a common ancestor—can be estimated. It is noteworthy that the degree of molecular change is frequently not correlated with the degree of morphological change. Major differences in biochemical polymorphism may appear between morphologically indistinguishable species, and may be present even in small intraspecific populations where no physical barriers to interbreeding exist (Sec. 13-29).

Molecular studies are also contributing to the solution of problems in phylogeny, i.e., the degree of divergence and time of separation of species from common ancestors. Species are compared as to the immunological properties and amino acid sequences of selected homologous proteins.

Let us assume that we wish to determine the degree of relationship among a group of salamanders. In immunological studies, a protein, such as serum albumin, from species A is purified and injected into a rabbit. Antibodies are formed in the rabbit's blood against the salamander albumin, a foreign protein.

Thereafter these antibodies will react not only against salamander A's albumin but against related proteins of the other salamander species as well. The strength of the immunological reaction increases with increasing similarity between the protein of species A and that of the other species tested. Since the degree of similarity between homologous proteins increases with closeness of taxonomic relationship, immunological reactions are stronger between closely related species than distantly related ones. This reflects the differences that have accumulated in the amino acid sequences of their proteins and the mutations in the DNA templates that produced them. Degrees of dissimilarity between the homologous proteins of species are expressed in units of "immunological distance."

Immunological distances have been determined for a variety of organisms—mammals, birds, amphibians, and others. On the basis of differences in their albumin proteins, humans, chimpanzee, and gorilla appear to be more closely related to each other than any one of them is to the orangutan and all are quite different from Old World monkeys. The correlation between albumin immunological distance and paleontological estimates of divergence time of taxonomic groups is often good, as in carnivores and ungulates.

Another method for comparing proteins is **amino acid sequencing.** A comparison is made between the kinds and order of the individual amino acids that make up homologous proteins of two species. The degree of relationship and extent of divergence from a common ancestor can be inferred from the degree of differentiation of their proteins. Proteins vary in their utility in providing information on branching of phylogenetic trees. The enzyme cytochrome c, crucial in cell respiration in all higher plants and animals, has changed slowly over geologic time and is useful in determining relationships of distantly related organisms. A phylogeny based upon it alone is remarkably close to that established on traditional criteria. More rapidly evolving proteins such as carbonic anhydrases and fibrinopeptides (in blood) are needed for study of more closely related species.

Some molecular biologists believe that the majority of gene differences are selectively neutral, unrelated to organismal evolution, and that their accumulation is uniform, reflecting inherent long-term mutation rates for the genes involved, and is proportional to time. This has led to the concept of an **evolutionary clock.** Recent studies, however, cast doubt on the constancy of protein changes. Rates of molecular evolution are evidently not constant. Protein changes may be used as an approximate evolutionary clock when they are averaged over many proteins and organisms.

Although the molecular approach shows great promise in further elucidating the process of evolution, it currently has some limitations. Probably less than 10 percent of the DNA of eukaryotic genomes produces functional enzymes or other proteins; some 30 percent of DNA base changes result in no modification in the amino acid sequences of proteins; most substitutions do not alter the protein's electric charge; much effort is involved, and often insufficient numbers of loci are investigated to provide an adequate determination of average heterozygosity; and little information has been obtained on the levels of genetic variability at loci that affect the kinds of morphology usually associated with the evolutionary process. Future investigations will no doubt resolve some of these difficulties. One approach that has particular promise is direct comparison of the sequences of nucleotides in DNA of different species. Methods have been developed to compare short segments of DNA and more extensive comparisons may soon be possible.

The total picture that emerges from the study of evolution is an inspiring one: starting with simple compounds of carbon and progressing to replicating molecules and cells, life has increased in diversity and perfection of adaptations through the ages, as revealed by the wide variety of both living and extinct forms. What are the potentialities for the future? Much depends on the behavior of one species —human beings.

References

The following are a few books and articles selected from the enormous literature on evolution:

Ayala, F. J. (editor). 1976. Molecular evolution. Sunderland, Mass., Sinauer Associates, Inc. x + 277 pp.

Bodmer, W. F., and **L. L. Cavalli-Sforza.** 1976. Genetics, evolution, and man. San Francisco, W. H. Freeman and

Co. xv + 782 pp. *Human genetics and evolution and their relevance to social problems.*

Bowman, R. I. 1961. Morphological differentiation and adaptation in the Galápagos finches. *Univ. Calif. Publ. Zool.,* vol. 58. Berkeley, University of California Press. vii + 326 pp.

Bush, G. L. 1975. Modes of animal speciation. *Ann. Rev. Ecol. and Syst.,* vol. 6, pp. 339–364.

Clarke, B. 1975. The causes of biological diversity. *Sci. Am.,* vol. 223, no. 2, pp. 50–60.

Colbert, E. H. 1969. Evolution of the vertebrates: A history of the backboned animals through time. 2d ed. New York, John Wiley & Sons, Inc. xvi + 535 pp., 144 figs.

———. 1971. Tetrapods and continents. *Quart. Rev. Biol.,* vol. 46, pp. 250–269.

Crow, J. F. 1959. Ionizing radiation and evolution. *Sci. Am.,* vol. 201, no. 3, pp. 138–160. *Living things evolve as a result of random mutations. What role does ionizing radiation play in the evolutionary process?*

Darwin, Charles. 1839. Journal of researches into the geology and natural history of the various countries visited during the voyage of H.M.S. "Beagle" under the command of Captain FitzRoy, R.N. from 1832–1836. London, H. Colburn. xiv + 615 pp., illus., maps. 1845. 2d ed. London, John Murray (Publishers), Ltd. viii + 519 pp., illus. Reprinted many times under the title *Voyage of the Beagle.* 1958. New York, Bantam Books, Inc. 439 pp.

———. 1859. On the origin of species by means of natural selection, or the preservation of favoured races in the struggle for life. London, John Murray (Publishers), Ltd. ix + 502 pp. Five subsequent editions by author; often reprinted.

De Vries, Hugo. 1909. The mutation theory: Experiments and observations on the origin of species in the vegetable kingdom. La Salle, Ill., The Open Court Publishing Company. 2 vols., xv + 582 pp., 119 figs.; vii + 683 pp., 6 pls., 147 figs.

Dobzhansky, T. G. 1970. Genetics of the evolutionary process. New York, Columbia University Press. ix + 505 pp. *Population and evolutionary genetics.*

———, **F. J. Ayala, G. L. Stebbins,** and **J. W. Valentine.** 1977. Evolution. San Francisco, W. H. Freeman and Co. xiv + 572 pp.

Ehrlich, P. R., R. W. Holm, and **D. R. Parnell.** 1974. The process of evolution. New York, McGraw-Hill Book Company. xv + 378 pp.

Emmel, T. C. 1976. Population biology. New York, Harper & Row, Publishers, Incorporated. xii + 371 pp.

Fisher, R. A. 1930. The genetical theory of natural selection. Oxford, Clarendon Press. Revised 2d ed. 1958. New York, Dover Publications, Inc. xiv + 291 pp., illus.

Hadži, Jovan. 1963. The evolution of the Metazoa. New York, Pergamon Press. xiii + 499 pp., 62 figs.

Huxley, J. 1942. Evolution: The modern synthesis. 3d ed., 1974. New York, Hafner Press. lxxvii + 705 pp. *One of the great evolutionary classics; the third edition contains a section on recent advances in evolutionary biology.*

Johnston, R. F., and **R. K. Selander.** 1971. Evolution in the house sparrow: II. Adaptive differentiation in North American populations. *Evol.,* vol. 25, no. 1, pp. 1–28.

Lamarck, J. B. 1809. La philosophie zoologique. Translated by Hugh Elliot, 1914, as Zoological philosophy. London, Macmillan & Co., Ltd. xcii + 410 pp. Reprint 1963. New York, Stechert-Hafner, Inc.

Lewontin, R. C. 1970. The units of selection. *Ann. Rev. Ecol. Syst.,* vol. 1, pp. 1–18.

———. 1974. The genetic basis of evolutionary change. New York, Columbia University Press. *Electrophoretic studies of protein variation in relation to population and evolutionary genetics.*

MacArthur, R. H., and **J. H. Connell.** 1966. The biology of populations. New York, John Wiley & Sons, Inc. xv + 200 pp. *Factors involved in changes in population density and in the genetic composition of populations.*

Mayr, E. 1942. Systematics and the origin of species. New York, Columbia University Press. xiv + 334 pp. *One of the great classics in evolutionary biology.*

———. 1970. Populations, species, and evolution. (An abridgement of animal species and evolution.) Cambridge, Mass., The Belknap Press, Harvard University Press. xv + 453 pp.

Moore, R. C., C. G. Lalicker, and **A. G. Fischer.** 1952. Invertebrate fossils. New York, McGraw-Hill Book Company. 765 pp.

Olson, E. C. 1971. Vertebrate paleozoology. New York, Interscience Publishers, a division of John Wiley & Sons, Inc. xv + 839 pp.

Pianka, E. R. 1974. Evolutionary ecology. New York, Harper & Row, Publishers, Incorporated. viii + 356 pp.

Raup, D. M., and **S. M. Stanley.** 1971. Principles of paleontology. San Francisco, W. H. Freeman and Company. x + 388 pp.

Romer, A. S. 1959. The vertebrate story. Chicago, The University of Chicago Press. vii + 437 pp. paper.

———. 1966. Vertebrate paleontology. 3d ed. Chicago, The University of Chicago Press. viii + 468 pp., 443 figs.

Salthe, S. N. 1972. Evolutionary biology. New York, Holt, Rinehart and Winston, Inc. viii + 437 pp.

Savage, J. M. 1969. Evolution. (2d ed.) New York, Holt, Rinehart and Winston, Inc. vii + 152 pp. *Emphasis on the mode of development of isolating mechanisms in the origin of species and the processes responsible for the origin of major evolutionary changes above the species level.*

Scientific American. 1978. Evolution. *Sci. Am.,* vol.

239, no. 3, pp. 46–230. *A collection of articles by leading evolutionary biologists.*

Selander, R. K., and **W. E. Johnson.** 1973. Genetic variation among vertebrate species. *Ann. Rev. Ecol. and Syst.,* vol. 4, pp. 75–91. *An excellent review of the subject of heterozygosity in the vertebrates.*

Simpson, G. G. 1953. The major features of evolution. New York, Columbia University Press. xx + 434 pp., 52 figs.

———. 1967. The meaning of evolution: A study of the history of life and of its significance for man. New Haven, Yale University Press. xvii + 368 pp., 38 figs. *The course of evolution, its interpretation and implications for humans.*

Stebbins, G. L. 1971. Process of organic evolution. 2d ed. Englewood Cliffs, N.J., Prentice-Hall, Inc. xiii + 193 pp.

Stebbins, R. C., and **B. Allen.** 1975. Simulating evolution. *Amer. Biol. Teacher,* vol. 37, no. 4, pp. 206–211. *Describes a method for demonstrating the mechanism of natural selection using colored chips (the animals) on a multicolored fabric background (the habitat).*

Wald, G. 1964. The origins of life. *Proc. Nat. Acad. Sci.,* vol. 52, no. 2, pp. 595–611.

PROTOZOANS

AND THE ANIMAL KINGDOM

14

CLASSIFICATION AND NOMENCLATURE

To facilitate the study of animals and to communicate about them it is necessary to name and describe them and to recognize their similarities and differences. Hence some means of grouping animals according to their relationships is necessary; this is one objective of the science of **taxonomy,** or **systematic zoology.** Taxonomy has two important divisions, **classification,** the arrangement of the kinds of animals in a hierarchy of smaller and larger groups, and **nomenclature,** the procedure of assigning names to the kinds and groups of animals to be classified. Without a recognized system of naming and classifying animals, all studies of animals would be fraught with uncertainty as to just what kind or group of animals is being discussed.

14-1 Numbers of animals Many kinds of animals now inhabit the earth, and many more have lived here during past geologic time. The more conspicuous kinds of land and sea animals are now fairly well known, but for many groups our knowledge is still imperfect.

The first widely accepted enumeration of all known animals was made by Linnaeus in 1758 and included 4236 kinds. In 1859 Agassiz and Bronn counted 129,370, and by 1911 Pratt estimated that 522,400 had been named. The total now is well over one million (Fig. 14-2). New forms are still being named, and there may be two million kinds of living animals in the world today. For certain groups in particular regions there are more precise figures.

Thus in North America there are over 850 species of mammals from Panama northward; north of Mexico the species of birds number 775, the reptiles 240, and the amphibians 160. California contains 97 species of mammals, 461 birds (286 breeding), 78 reptiles, and 45 amphibians. The numbers for animal groups at the ends of chapters in Part II are mostly estimates by specialists.

Classification

14-2 Methods and purposes Various degrees of resemblance and difference are easily seen in any mixed assemblage of animals. Of the domestic animals on a farm, the cow and sheep both have horns and cloven hoofs but differ in size, shape, color, and body covering. A horse resembles the cow and sheep in having long ears and teeth of the grinding type but lacks horns and has solid hoofs. A dog differs from all three in having nails and pads on its separate toes and having teeth of the stabbing and shearing types; it agrees in being covered with hair. The cat resembles the dog more closely than the hoofed animals. All these animals have hair and teeth, they produce living young which they suckle, and they show many other features in common. As a group they all differ from the chickens and ducks, which are covered with feathers, lack teeth, and lay eggs—but these and all other birds have eyes, lungs, four limbs, and other characters like the four-footed animals named. So it is, by likenesses and differences, that animals may be divided into minor and major groups.

Zoologists concerned with classification of animals attempt to separate out significant similarities and differences among organisms and to employ these features as a framework for a classification scheme which will represent, within the limits of current knowledge, the phylogenetic relationships among various groups. Since our knowledge of zoology is constantly increasing, and since this new information may well affect any system of classification, no such system can be regarded as static. Furthermore, since our knowledge is incomplete and because many facts or pieces of data are difficult to

interpret or may be interpreted in more than one way, there is also some difference of opinion among zoologists as to how well any scheme of classification represents phylogeny.

The inherent peculiarities, or **characters,** of animals are the basis of classification. These include structural features, size, proportions, and coloration as well as countable or **meristic,** features such as number of teeth, number of fin rays, etc. A character is usually more significant if constantly associated with others; thus every bird has, besides feathers, a beak, wings, clawed feet, a four-chambered heart, and warm blood.

A first purpose of classification is convenience, but a more important one is to show relationships based on **phylogeny.** Phylogeny is the evolutionary history of a group. Animals may be classified in various ways, as by grouping together all with shells, all of wormlike form, etc.; this was done by early zoologists. Increasing knowledge has shown that such arrangements bring together animals otherwise greatly different. The modern "natural system" of classification uses all available data as to structure, physiology, embryology, distribution, biochemistry, and other features; each group is distinguished by several to many characteristics. The natural classification is explained by the theory of evolution; it seeks to show relationship by descent from more simple organisms (Fig. 14-1).

For purposes of classification, characters that show **homology,** or similarity of origin (and hence relationship), must be distinguished from those that exhibit **analogy,** or similarity of use (but not necessarily of origin). The arms of man, forelegs of mammals and frogs, and wings of birds are homologous, being essentially similar as to structure of the bones, muscles, blood vessels, and nerves, although used for different purposes. By contrast, the wings of birds and butterflies are analogous, both serving for flight, but are unlike in embryonic development and adult structure (Fig. 13-1).

14-3 Kingdoms of organisms Since the time of Linnaeus until quite recently, all organisms were usually classified under one of two kingdoms: Plant or Animal. Plants have photosynthetic pigments and

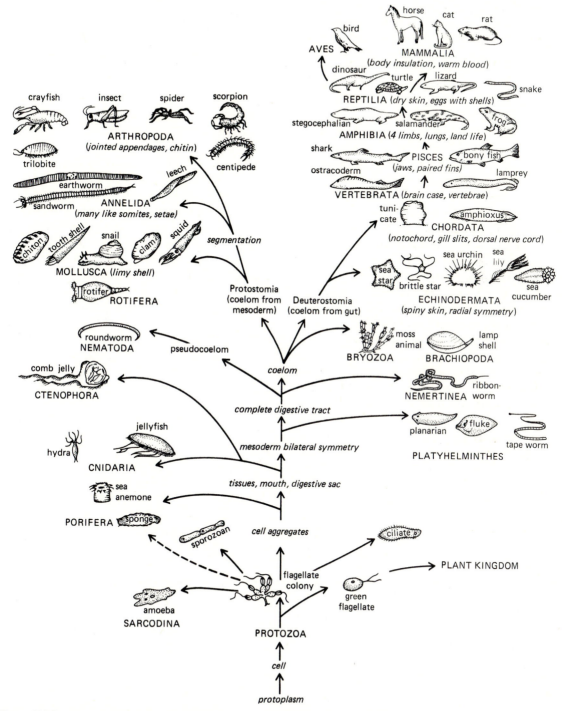

Figure 14-1 Kingdom Animalia and Phylum Protozoa. A "genealogic tree" to indicate the probable relationships and relative position of the major groups (named in capital letters). All groups above a given characteristic (named in italics) possess that character. Figures not to same scale.

are generally nonmotile, while animals have to obtain food by feeding on plants or other animals and generally move about. Certainly if one concentrates on larger (macroscopic) plants and animals, assignment seems easy, save, perhaps, for the mushrooms and their relatives (fungi), which are nonphotosynthetic. However, they still retain the plant form and seemingly similar growth pattern. Hence they were assigned to the Plant Kingdom. Major problems with the two-kingdom sequence arose when the microscopic organisms were encountered and closely studied. Here, for example, we have organisms with all the characteristics of animals and plants, such as *Euglena* (Chap. 15). Furthermore, these microscopic organisms are unicellular, yet the individual cells are more complex than the cells of the bodies of multicellular animals. Some microscopic organisms may associate in colonies but do not form tissues as in multicellular animals. The two-kingdom system could not accommodate these organisms without straining the definitions. The result was that botanists claimed motile animals with photosynthetic pigments as plants in their considerations and the zoologists claimed them as animals because they move. In the 1950s and 1960s, biologists began to learn a great deal more about the subcellular structure and biochemistry of cells. Because of this new knowledge, it became increasingly difficult to classify some organisms as either plants or animals. For example, bacteria and blue-green algae have genetic material but no nuclei. Furthermore, these organisms lack the subcellular organelles of the single-celled animals, such as mitochondria, vacuoles, and Golgi apparatus. Certain metabolic processes are also different. What should be done with

Table 14-1

Five Kingdoms	Four Kingdoms	Three Kingdoms
MONERA ⟶	MONERA ⎫	
PROTISTA ⟶	PROTISTA ⎬ ⟶	PROTOCTISTA
FUNGI ⎫		
PLANTAE ⎭ ⟶	PLANTAE ⟶	PLANTAE
ANIMALIA ⟶	ANIMALIA ⟶	ANIMALIA

them? They are not animals, and yet they do not have the defined characters of plants.

Another group that does not fit into a two-kingdom system is the fungi. Traditionally grouped with the plants, fungi show some very unplantlike characteristics. For example, plants obtain nourishment through photosynthesis, thereby manufacturing inorganic molecules out of carbon dioxide and water. Fungi do not carry out photosynthesis. Instead, they secrete enzymes into the substrate on which they grow. The enzymes break down complex organic molecules found in animal wastes and dead organisms, and the simpler molecules are then absorbed by the fungus. Another characteristic of fungi is the lack of definite boundaries between cells. As a result, large organic molecules and even nuclei can move from cell to cell. Plants, then, are as different from fungi as they are from one-celled amoebas and bacteria.

As a result of these seemingly fundamental differences, several scientists have proposed dividing the living world into from three to five kingdoms (Table 14-1). The minimum, three, would give the kingdoms PROTOCTISTA, PLANTAE (also called METAPHYTA), and ANIMALIA (also called METAZOA). In

Table 14-2
CHARACTERISTICS OF FIVE KINGDOMS OF ORGANISMS

Kingdom	Type of Cell	Cell Organelles	Nutrition Type
Monera	Prokaryotic	No membrane around organelles; no plastids; no mitochondria	Absorptive or photosynthetic
Protista	Eukaryotic	All cell organelles	Absorptive, ingestive, photosynthetic
Plantae	Eukaryotic with walls	Present but cells simpler	Photosynthetic mainly
Fungi	Eukaryotic	Lack plastids and photosynthetic pigments	Absorptive
Animalia	Eukaryotic without walls	Lack plastids and photosynthetic pigments	Ingestive

the three-kingdom system the fungi remain with the plants, while all single-celled organisms are grouped under the Protoctista. The four-kingdom arrangement would be Monera, Protista, Plantae (Metaphyta), and Animalia (Metazoa). The four-kingdom format divides one-celled organisms into Monera, which are without nuclei (prokaryotic) and Protista, with nuclei (eukaryotic). Finally, the five-kingdom system retains the above divisions of the four-kingdom system but makes Fungi a separate kingdom. Characteristics of the five kingdoms are given in Table 14-2.

In this text the five-kingdom system is used. With one exception, the following chapters are restricted to the description of animals. The exception is the Protozoa, a group of motile protistans that, until recently, were considered animals. The protozoans are included because they have traditionally been of concern to zoologists and because a zoology student should have some familiarity with one-celled forms of life. The reader should realize that the protozoan cell is far more complex and contains many organelles not found in the cells of metazoans. Furthermore, all protistans lack a type of tissue differentiation characteristic of metazoans.

14-4 Species The basic unit, or building stone, in biological classification is the **species** (not specie; the plural is also species). A species is a group of individuals which is naturally reproductively isolated from other such groups. That is, the individuals of a species are all derived from a common ancestry and can breed with one another to produce fertile offspring that resemble the parents. In nature separate species do not usually interbreed, though hybrids between species do occur occasionally. Examples of common species are the house fly, yellow perch, bullfrog, and house sparrow. Often the total of individuals constituting a species can be subdivided into smaller groups known as **subspecies** that differ from one another primarily in that each subspecies occupies a separate geographic range, and specimens from the boundary ranges of two adjacent subspecies are usually intermediate in their characteristics (Fig. 13-19).

Species of animals are always given a scientific name which is a dual name, much as human beings have a first name and a surname. This principle of usage of two words to designate a species is termed **binomial nomenclature** and has been employed since 1758, when Linnaeus, the great Swedish systematist and father of systematic biology, first used the system. The first name is always capitalized and is the generic name. The second is the trivial (or specific) name and is never capitalized. Each scientific name is unique in the Animal Kingdom; if two species are discovered to have the same name, the one having it the longest retains it while the other must be changed. An example of a scientific name is *Rana pipiens*, the common leopard frog.

14-5 Higher groups Species of animals are considered to be natural groups of actually or potentially interbreeding populations, but the other categories of the taxonomic hierarchy are arbitrary in that they are human constructions. They are intended to reflect differing levels of relationship among animals based on evolution, but because our knowledge of

Table 14-2 (Continued)
CHARACTERISTICS OF FIVE KINGDOMS OF ORGANISMS

Cellular Organization	Reproduction	Motility	Representative
Unicellular and/or colonial	Asexual by fission	Motile using flagella	Blue-green algae, bacteria
Unicellular and/or colonial	Asexual and sexual	Nonmotile or use cilia or flagella	Protozoa
Multicellular with tissues	Sexual and asexual	Nonmotile	Higher plants
Syncytial (lacks definite boundaries between cells)	Sexual and asexual	Nonmotile	Mushrooms, molds
Multicellular with tissues	Usually sexual	Motile using contractile fibers	Any animal

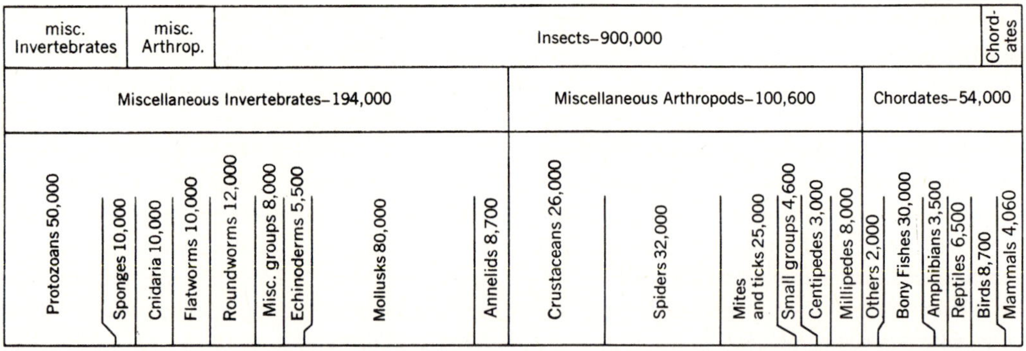

misc. Invertebrates	misc. Arthrop.	Insects–900,000																		Chordates
Miscellaneous Invertebrates–194,000									Miscellaneous Arthropods–100,600							Chordates–54,000				
Protozoans 50,000	Sponges 10,000	Cnidaria 10,000	Flatworms 10,000	Roundworms 12,000	Misc. groups 8,000	Echinoderms 5,500	Mollusks 80,000	Annelids 8,700	Crustaceans 26,000	Spiders 32,000	Mites and ticks 25,000	Small groups 4,600	Centipedes 3,000	Millipedes 8,000	Others 2,000	Bony Fishes 30,000	Amphibians 3,500	Reptiles 6,500	Birds 8,700	Mammals 4,060

Figure 14-2   Approximate numbers of living species in major divisions of the Kingdom Animalia and Phylum Protozoa.

evolution is yet incomplete, there are many interpretations of relationships and hence a certain amount of disagreement among biologists as to which higher group certain species should belong to. The taxonomic hierarchy consists of six major units: phylum, class, order, family, genus, and species.

Two or more species with certain characters in common form a **genus** (pl. **genera**). In turn, genera having common characters constitute a **family,** the families are combined into **orders,** the orders into **classes,** and the classes into **phyla** (sing. **phylum**). All the phyla together make up the **Animal Kingdom,** which is comparable to the Plant Kingdom. The scheme of classification is like a tree having many leaves (species), with one to many on a fine stem (genus), several stems on a larger twig (family), two or more of these on a little branch (order), a number of these on a larger branch (class), and the latter borne on the main framework (phyla), the whole forming a tree (kingdom). Intermediate categories (subfamilies, superclasses, subphyla, etc.) sometimes are needed to indicate properly the degree of relationship. There are cases where a group—

genus or higher—contains only one representative because it is distinct from all others; such a group is termed **monotypic.**

14-6 Embryologic features in classification
Many of the basic characters which are used in animal classification to differentiate the higher taxa, particularly phyla, and to assess the relationships among phyla which allow them to be placed in phylogenetic sequences (see Fig. 14-1) are embryologic. The strongest links between phyla and the arrangement of phyla together into lines of descent depend on fundamental processes which occur in the egg or embryo. Some of these were covered in Chap. 10; others which were not are reviewed here prior to discussing such phylogenetic arrangements.

Among animals, several types of eggs can be classified with respect to yolk distribution (Fig. 14-3). **Isolecithal,** or **homolecithal,** eggs are generally small, with the yolk equally distributed throughout. Such eggs, after fertilization, generally undergo a complete, or **holoblastic,** cleavage during develop-

Figure 14-3   Types of eggs. *A.* Isolecithal or homolecithal egg, with little yolk equally distributed. *B.* Telolecithal egg with yolk concentrated near vegetal pole but with holoblastic cleavage. *C.* Telolecithal egg with large amount of yolk and protoplasm limited to clear layer on top, meroblastic cleavage. *D.* Centrolecithal egg with protoplasm around inner yolk, meroblastic cleavage. 1, yolk. (*After Hyman.*)

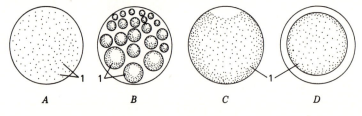

A   B   C   D

ment, and the resulting blastomeres are nearly equal in size. **Telolecithal** eggs have the yolk concentrated near the vegetal pole of the egg. Such eggs may undergo one of two types of cleavage, depending on the amount of yolk present at the vegetal pole. Where the amount of yolk is not great, the cleavage planes will extend through the whole egg (holoblastic), but these planes usually are retarded so that they produce two different-sized blastomeres: a set of large, yolk-filled ones at the vegetal pole called **macromeres** and a set of smaller ones at the animal pole called **micromeres.** In subsequent development the macromeres usually form the endoderm and the micromeres the ectoderm. Where the amount of yolk is very large, the cleavage planes do not penetrate through the yolky area and cleavage is restricted to the superficial layer of protoplasm at the animal pole (the germinal disc, or blastoderm). This is **meroblastic** cleavage. The final egg type is the **centrolecithal,** in which the yolk is concentrated in the center, with most of the living protoplasm surrounding it at the outside. In such eggs cleavage is again meroblastic.

Another fundamental embryologic feature used to group phyla has to do with the pattern of cleavage

in the fertilized egg. Two basic cleavage patterns are known. In **radial cleavage** the cleavage planes producing the successive sets of blastomeres are at right angles to each other and parallel or perpendicular to the polar axis of the fertilized egg (Fig. 14-4). This produces tiers of blastomeres. In this pattern of cleavage it is not possible to follow individual blastomeres to see which parts of the adult animal they will produce. In other words, the cleavage is **indeterminate.** In indeterminate cleavage the ultimate fate of each blastomere remains undecided until late in development. This means that animals with this form of development may have their blastomeres separated from each other after several cleavages, and each will form a complete embryo and adult. In **spiral cleavage** the cleavage planes tend to be oblique or diagonal to the polar axis of the egg. Successive cleavages in this pattern produce blastomeres arranged spirally around the polar axis. Here each successive tier of blastomeres rests above the grooves between the blastomeres in the tier below (Fig. 14-4). Spiral cleavage is **determinate** because the fate of the blastomeres is known. Each blastomere can be followed through to the particular organ or tissue it gives rise to in the adult.

Figure 14-4 Radial and spiral cleavage. *A–C.* Radial cleavage as seen from the side, animal pole uppermost. Successive blastomeres in tiers. *D.* Radial cleavage, view of blastomeres from top of animal pole. *E.* Spiral cleavage, early, seen from the top. *F.* Spiral cleavage as seen from the side, successive blastomeres offset from ones below. *G.* Spiral cleavage, view of blastomeres from top of animal pole. (*A, B, C, E, F after Berrill and Karp; D after Gardiner; G after Hyman.*)

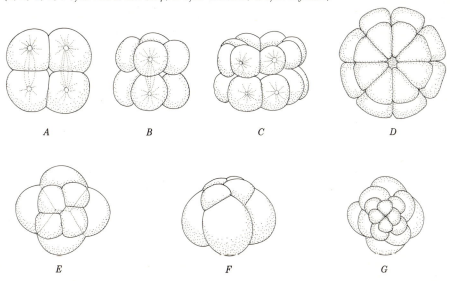

Following the appearance of the gastrula stage in development (Sec. 10-14), those phyla which develop a body cavity in the adult (all phyla except PROTOZOA, CNIDARIA, CTENOPHORA, NEMERTINEA, GNATHOSTOMULIDA, and PLATYHELMINTHES) begin development of this cavity. Two basic cavities occur in the Animal Kingdom. A **pseudocoelom** is the persistence into the adult stage of the embryonic blastocoel cavity found in the gastrula stage. The pseudocoelom is a cavity which is not lined by mesoderm. A true **coelom** is a body cavity which is lined with mesoderm (called *peritoneum* in the adult). Both the pseudocoelom and the coelom in the adult contain the internal or visceral organs. A true coelom may arise in one of two ways, either **enterocoelous** or

schizocoelous. In the enterocoelous method, the coelom arises as pouches which bud off the archenteron of the gastrula and subsequently fuse (Fig. 14-5). In the schizocoelous formation the coelom arises as a split in the mesoderm which is forming in bands near the blastopore.

Among coelomate animals there is a strong tendency for spiral, determinate cleavage and schizocoelous coelom formation to be linked together, while radial, indeterminate cleavage is found in those animals showing enterocoelous coelom formation. As a result of these fundamental differences in embryology, it is possible to divide the coelomate animal phyla into two fundamental divisions or lines of evolution. The **deuterostome** division of the Ani-

Figure 14-5 Relationships among the various body cavities in three phylogenetic lines. Gray indicates endoderm, hatched ectoderm, and clear mesoderm. In deuterostome and protostome lines mesoderm and coelom formation obliterates the original blastocoel, and mesoderm lines the body cavity in both cases. Pseudocoelomate animals retain the original blastocoel, and the internal organs are not lined with mesoderm.

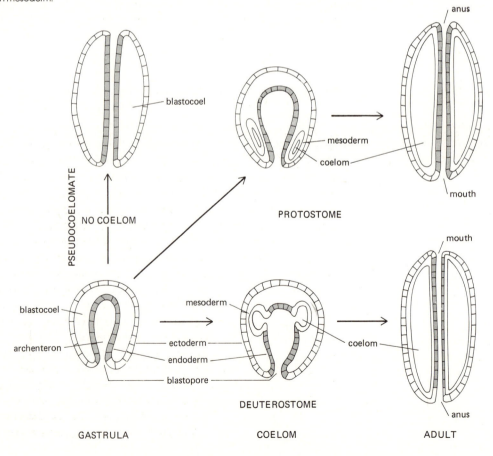

mal Kingdom includes those phyla with radial indeterminate cleavage and enterocoelous coelom formation, while the **protostome** division includes those phyla with spiral determinate cleavage and schizocoelous coelom formation. In deuterostomes the mouth arises away from the blastopore, while in protostomes it arises at or near the blastopore.

14-7 General characteristics   The wide occurrence of some characters makes possible the recognition of groups larger than phyla (see classification outline farther on in this chapter). Within the Animal Kingdom, the sponges (PORIFERA) form the branch PARAZOA, with no digestive cavity and with body wall pierced by pores, in contrast to the branch EUMETAZOA (all higher animals), with such a cavity and with body wall unperforated. The EUMETAZOA are divided into two great lines of evolution, one termed the **Protostomia,** which culminate in the phyla ARTHROPODA, ANNELIDA, and MOLLUSCA, and the other termed **Deuterostomia,** of which the ECHINODERMATA and CHORDATA are the most important phyla. These two lines are distinguished on the basis of previously discussed embryologic characteristics and also on larval types. Protostomes, if they have a larva, usually have a top-shaped type known as a *trochophore*. Deuterostomes, by contrast, if they have a larva, do not have a trochophore type. The EUMETAZOA are further divided according to the number of germ layers laid down in the embryo: two, or **diploblastic** (CNIDARIA, CTENOPHORA); three, or **triploblastic** (all others). The higher phyla

(BRYOZOA to CHORDATA) make up the EUCOELOMATA, having a body cavity or **coelom** lined by peritoneum whence excretory and reproductive ducts lead to the exterior. In MOLLUSCA and ARTHROPODA the coelom is greatly reduced and the blood circulates in spaces between the internal organs called a **hemocoel.** Body spaces in the ENTOPROCTA, NEMATODA, ROTIFERA, and allies are unlined and termed a **pseudocoel.** Two lower phyla lacking body spaces, the PLATYHELMINTHES and NEMERTINEA, are termed the ACOELOMATA. Invertebrates include all animals that lack a backbone of vertebrae, in contrast to the vertebrates (phylum CHORDATA: cyclostomes to mammals), which have a segmented vertebral column.

Besides the features just mentioned, some other characteristics are useful in classification (Fig. 14-7).

1 Symmetry   Many protozoans are **asymmetrical** because they are not divisible into equivalent halves; a few show **spherical symmetry.** The cnidarians and adult echinoderms are usually **radially symmetrical** around a median axis through the mouth; planes through this axis will divide the animal into radial sectors (antimeres) (Fig. 14-6). Members of most other phyla are **bilaterally symmetrical;** a lengthwise vertical (sagittal) plane divides the animal into equal and opposite halves. In such animals the part that moves forward (and usually contains the mouth) is termed **anterior** and the opposite end **posterior;** the back or upper surface is termed **dorsal,** and the undersurface (usually toward the ground) is termed **ventral** (L.

Figure 14-6. Types of symmetry, and the axes, planes, and regions in animal bodies.

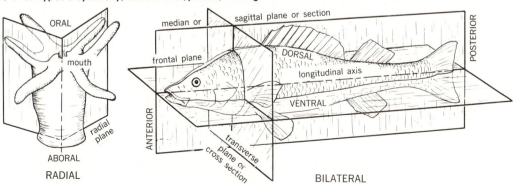

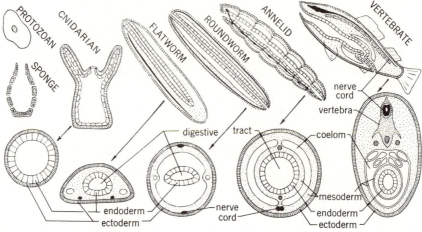

Figure 14-7 Body structure in various types of animals and protistans (diagrammatic). *Above.* In median section. *Below.* In cross section, with indication of embryonic germ layers.

venter, belly). Structures on or toward the central longitudinal axis are termed **medial,** and those toward the sides are said to be **lateral** (Fig. 14-6).

2 **Segmentation** (Fig. 27-1) In the annelids, arthropods, and chordates there is a linear repetition of body parts known as segmentation (metamerism); each repeated unit is a **somite** (metamere). In earthworms the successive somites are essentially alike, but they are unlike in different body regions of a crayfish or insect. Metamerism is conspicuous both externally and internally in annelids, is mostly external with arthropods, and is mainly internal in humans and other chordates (vertebrae, body muscles, some blood vessels, and nerves). There is a tendency among metameric animals for the more advanced forms to have certain of the segments grouped together. These are often morphologically different from other such sets and specialized to perform certain functions in the animal. Such a regionalization of segments is called **tagmatization,** and each set of segments a **tagma** (pl. tagmata). A good example is the division of an insect body into three tagmata: head, thorax, and abdomen.

3 **Appendages** Protruding parts that serve in locomotion, feeding, and other ways are termed **appendages;** examples are the tentacles of sea anemones, minute setae of earthworms, antennae and legs of arthropods, and the fins, legs, and wings of vertebrates.

4 **Skeleton** Most land dwellers and many aquatic animals have a skeleton for support or protection; it may be **internal** (frog, human, starfish, etc.) or **external** (coral, crab, insect) and may be of either inorganic or organic material.

5 **Sex** An animal having both **female** and **male** sex organs in one individual is termed monoecious (hermaphroditic); members of most higher phyla are **dioecious** (gonochoristic), each individual being either male or female.

6 **Larvae** The young stages known as **larvae** (Sec. 13-8) often provide important information on relationships that are not evident in the adult animals. Many have features that are obviously adapted to particular environments (such as cilia for swimming), but their basic structure is usually characteristic for each group (phylum, class). Barnacles and tunicates, for example, were first placed properly in their phyla by study of their larvae.

A common larval pattern in many aquatic invertebrates is the minute, transparent, and free-swimming organism called a **trochophore,** or trochosphere (Fig. 14-8). It is often pear-shaped, with the large end uppermost, and encircled by one or two lines of cilia that beat so as to suggest a rotating wheel—hence the name *trochophore* (Gr. *trochos,* wheel + *phoros,* bear). The upper end bears an apical plate with a tuft of cilia and a sense organ (eyespot). With various modifications, this is the early larva of many marine flatworms, nemer-

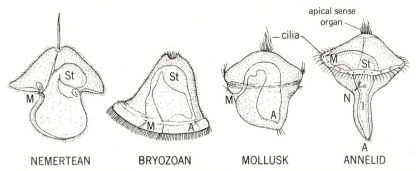

Figure 14-8 Early larvae of some invertebrates: nemertean pilidium, bryozoan cyphonautes, molluscan trochophore (*Patella*), annelid trochophore (*Polygordius*). M, mouth; St, stomach; I, intestine; A, anus; N, nephridium. (*After various authors.*)

teans, brachiopods, bryozoans, mollusks, and some annelids.

14-8 History of classification Aristotle (384–322 B.C.), the "father of zoology," indicated how animals might be grouped according to their characteristics, and from his writings other persons outlined the following classification that served for about two thousand years:

1 Enaima (vertebrates), with red blood
  a Viviparous
    1 Humans
    2 Whales
    3 Other mammals
  b Oviparous
    1 Birds
    2 Amphibians and most reptiles
    3 Snakes
    4 Fishes
2 Anaima (invertebrates), no red blood
  a Cephalopods
  b Crustaceans
  c Insects, spiders, etc.
  d Other mollusks, echinoderms, etc.
  e Sponges, cnidarians, etc.

John Ray (1627–1705) was the first biologist to have a modern concept of species and to make some efforts to classify a few groups. It was Carolus Linnaeus (1707–1778) who laid the real basis for modern classification and nomenclature. He divided and subdivided the Animal Kingdom down to species on the basis of structural characters and gave each spe-

cies a distinctive name. In his *Systema naturae* (tenth edition, 1758) he recognized six "classes," as follows: Mammalia, Aves (birds), Amphibia (reptiles and amphibians), Pisces (fishes), Insecta, and Vermes (all other invertebrates). Linnaeus thus did not divide the invertebrates with as much discrimination as Aristotle had done long before him. Because of the limited knowledge then available, a Linnaean "genus" might include animals now divided among several orders or families or placed in different phyla. The principle of his system is, however, the basis for our present method of classification.

In 1829 Cuvier (1769–1832) divided animals into four branches: Vertebrata (mammals to fishes), Mollusca (mollusks and barnacles), Articulata (annelids, crustaceans, insects, spiders), and Radiata (echinoderms, nematodes, cnidarians, rotifers, etc.). Anatomy and classification were of great interest through the nineteenth century, and many systems were proposed, a few of the important ones being by Lamarck (1801, 1812), Leuckart (1848), Owen (1855), Milne-Edwards (1855), and Agassiz (1859). Ernst Haeckel in 1864 and E. Ray Lankester in 1877 outlined the principal features of the zoologic classification that is used today.

14-9 Difficulties in classification All divisions used in classification above the level of species from genera to phyla, are human constructions designed to represent phylogeny. Only individuals and populations exist in nature. Whether a particular group shall be called a subspecies, species, genus, etc., may ultimately be a matter of judgment based on study of specimens. The species has an objective definition as a group of interbreeding individuals reproductively isolated from other such populations, but there are some populations and taxonomic situations in which it is difficult to apply the term. For example, all animals dredged from the very deep sea have only been seen alive for brief periods of time, hardly enough to assess reproductive isolation or lack of it. Hence the criteria for separating species are usually based on morphologic differences, and it is not surprising that biologists differ in appraising the relative importance of various characters and the amount of variation to be recognized in any group before subdividing it. Taxonomic "splitters" make many subdivisions, whereas conservative "lumpers" recognize fewer categories in the same material. Thus different recent authorities recognize 20 to 40 orders of insects, and so on. As new facts are discovered, there is a tendency toward further changes. Zoologists are in fair agreement on much of animal classification, particularly at the level of phyla and among the vertebrates, but no two may have exactly the same opinion on all details; as a result, different books seldom contain identical schemes of classification.

Because animal groups differ in their evolutionary history and in their success in Recent environments, there is no uniformity in the content of groups above species. A certain genus may contain one or a few species and another a hundred or more; different phyla have one to several classes. Attempts to indicate lines of descent by constructing "genealogic trees," especially in the higher units of classification (family, order, etc.), are difficult and may be misleading because they are based on so little actual data.

Because of the subjective nature of much of the work in taxonomy, which has led to the many discrepancies described above and a lack of a rigorous quantitative approach that has characterized advances in many other fields of biology, taxonomy has been shunted aside by some biologists as not worthy of study, as an art rather than a science. More recently, however, many taxonomists have broadened their approach and have greatly increased the number and kinds of characteristics employed in determining relationships. In addition to classic morpho-

logic features of form, structure, and coloration, many other traits may be considered, from blood proteins and chromosomes (Sec. 13-6, 13-7) to behavioral and ecologic characteristics. Taxonomists of the new school are called **biosystematists.** A further development is the field of **numerical taxonomy,** which seeks to evaluate the similarities and differences between taxonomic units objectively through numerical methods based on multiple unweighted characters. This involves the measurement of a large number (sometimes up to 100 or more) characters in specimens followed by computer analysis, which establishes the numerical degree of affinity among individuals. The affinity values are then used to erect a hierarchical order of taxonomic categories in which for a given numerical value a certain taxonomic category is assigned. In this way a specimen, for example, is placed in a certain species when it has a numerical value which falls into a certain range and into a different species for another range of values. Numerical taxonomy has challenged the traditional approach and has demonstrated the usefulness to taxonomy of the quantitative approach and of a powerful new tool, the computer. It has been severely criticized, however, by many traditional taxonomists; it is yet too early to estimate the value that this approach will have to the future of taxonomy. Its effectiveness in elucidating relationships still depends on the judgment of the taxonomist in selecting traits for evaluation.

The sequence of phyla reflects in general the trend of evolution, from simple to complex—one cell to many, two germ layers to three, etc. It is impossible to represent the three-dimensional evolutionary tree on the successive pages of a book. Certain mollusks, insects, and vertebrates are recognized as the most evolved, or complex, members of their respective lines, but it is a matter of opinion which of these phyla has evolved most recently. In this book all phyla have been grouped so that those which appear to be most closely related by evolution are discussed together in a single chapter or two. The chapters further progress from the more simple phyla to the more complex. The synopses of classification at the ends of other chapters of Part II are analytic, dividing the Animal Kingdom into phyla, classes, and orders; some important families are listed in a few chapters.

Nomenclature

14-10 Common, or vernacular, names Each country has its own names for well-known animals. Thus a common sparrow of Western Eruope that is now established throughout the United States is known in different countries as follows:

United States: English sparrow

England: House sparrow

France: *Moineau domestique*

Spain: *Gorrion*

Portugal: *Pardal*

Italy: *Passera oltramontana*

Germany: *Haussperling*

Holland: *Musch*

Denmark and Norway: *Graaspurv*

Sweden: *Hussparf*

Within one country a species may have different local names; for example, in the United States the mallard duck is also called wild duck, greenhead, gray duck, English duck, stock duck, and some thirty other names. Still other names apply to the male (drake), female (duck, hen mallard), and young (ducklings). Thus confusion is likely between peoples of different nationalities or even within one country.

14-11 Scientific names Until about the eighteenth century, manuscripts and most books were written or printed in Latin, the language of scholars. Few ordinary people then could read. When printed books began to appear in the languages of various countries, Latin was retained for the technical descriptions and names of animals, the names being long descriptive polynomials that were useful but cumbersome. Thus in the first real natural history book in North America, Mark Catesby's *Natural History of Carolina, Florida . . .* (1731–1743), which had the text in parallel columns of English and French, the mockingbird was designated as

Turdus minor cinereo-albus maculatus

meaning

Thrush small grayish-white spotted

When Linnaeus began his *Systema naturae*, which described and named all known animals (plants and minerals) with terse Latin descriptions of each, he started with Latin polynomials but later contracted these. In the tenth edition (1758) he consistently used but two names for each, a generic name and a trivial name. This **binomial nomenclature** soon became a universal means in all countries for naming animals and plants scientifically. Linnaeus designated the mockingbird of America as *Mimus polyglottos*, the English sparrow (to him, *hussparf*) as *Passer domesticus*, and the mallard as *Anas platyrhynchos*. When a species is divided into subspecies, the latter are designated by **trinomials;** thus *Passer domesticus domesticus* is the subspecies of continental Europe, and *Passer domesticus niloticus* (*P. d. niloticus*) is the slightly differing form of the Nile Valley.

14-12 Rules of scientific nomenclature With so many kinds of animals and numerous taxonomists at work in naming and describing them in different countries, some confusion in nomenclature has arisen. Sometimes the same name (homonym) has been given to different animals or different names (synonyms) to the same animal. Some rules of nomenclature were used by Linnaeus, and others were proposed later. The International Congress of Zoology created a permanent Commission to prepare the International Rules of Zoological Nomenclature and to render decisions on difficult cases. The Rules, or Code, were adopted in 1901 (revised 1961) and, with modifications, have helped to stabilize the names of animals. The Rules deal with all scientific names and provide in essence as follows:

(1) Zoologic and botanic names are distinct (the same genus and species name may be used, but is not recommended, for both an animal and a plant). (2) No two genera in the Animal Kingdom may bear the same name, and the same applies to two species in a genus. (3) No names are recognized prior to those included by Linnaeus in the *Systema naturae*,

tenth edition, 1758. (4) Scientific names must be either Latin or latinized and preferably are printed in italics. (5) The genus name should be a single word (nominative singular) and begin with a capital letter. (6) The species name should be a single or compound word beginning with a small letter (usually an adjective agreeing grammatically with the genus name). (7) The author of a scientific name is the person who first publishes it in a generally accessible book or periodical with a recognizable description of the animal. (8) When a new genus is proposed, the type species must be indicated. (9) A family name is formed by adding -IDAE to the stem of the name of the type genus and a subfamily name by adding -INAE.

In publishing the description of a new species, it is common practice to designate a particular, or type, specimen, to describe it and to indicate the collection in which it is placed. Early writers often indicated no types and gave imperfect descriptions, so that confusion has arisen as to which of their species are actually distinct and who were their authors. If a genus or species has been described more than once, the earliest name and author, according to the above rules, is recognized under the **law of priority,** although this law may be set aside under certain conditions to preserve a long-standing name. In taxonomic works the name of the author, sometimes abbreviated, is written after the genus or species, thus: *Passer domesticus* Linnaeus (or Linn., or L.).

14-13 Origin and evolution of the Metazoa Zoologists have speculated for years on the origin and subsequent evolution of the animal (or metazoan) phyla. Several theories have been proposed, but the only area of agreement is that animals arose from the protistans. Two major theories may be mentioned here. One theory has animals evolving from colonies of primitive protistan flagellates in which the metazoan condition arose through differentiation of cells into tissues, beginning with the reproductive cells. In this scheme the first animal group derived from the colonial flagellates is the CNIDARIA. This theory is supported by the similarity of uninucleate flagellate cells to metazoan body cells, by the presence of flagella in metazoan body cells, and by the resemblance of the cnidarian planula larvae to a flagellate colony.

A second theory has the metazoans arising from ciliate protistans. This theory is based on the fact that the protozoan cell performs all life functions and is equivalent to a metazoan animal. The theory holds that the metazoa arose by each of the several nuclei in a ciliate organism taking control of a local differentiated area of cytoplasm and then becoming metazoan by developing cell membranes between the nuclei. According to this scenario, the first metazoan was a turbellarian flatworm (Fig. 14-9).

The above two theories are the major ones but not the only ones, and others may be expected as our knowledge increases. While any and all such

Figure 14-9 Diagrammatic representation of two theories of the origin of the lower metazoan phyla.

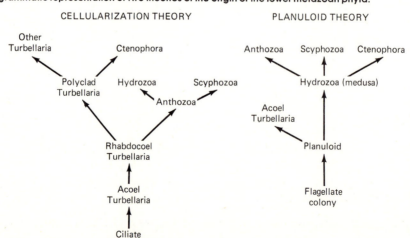

Table 14-3.
Some characteristics of the principal phyla of animals

Cells	Germ layers	Symmetry	Digestive tract	Excretory organs	Coelom	Circulatory system	Respiratory organs	Segmentation	Phylum	Distinctive features (exceptions omitted)
Incipient tissues	2, diploblastic	v	0	0	0	0	0	0	PORIFERA	Body perforated by pores and canals
		Radial	Incomplete	0	0	0	0	0	CNIDARIA	Nematocysts; digestive tract sac-like
		Biradial		0	0	0	0	0	CTENOPHORA	Comb plates for locomotion
Cells many, arranged in layers or tissues (METAZOA)	3, triploblastic	Bilateral		+	0	0	0	0	PLATYHELMINTHES	Flat, soft; digestive tract branched or none
			Complete (with anus)	+	0	+	0	0	NEMERTINEA	Slender, soft, ciliated; soft proboscis
				+	ps	0	0	0	ROTIFERA	Microscopic, cilia on oral disc
				+	ps	0	0	0	NEMATODA	Only longitudinal muscles, tough cuticle; no cilia
				0	+	0	0	0	BRYOZOA	Grow as moss-like or encrusting colonies
				+	+	+	0	0	BRACHIOPODA	Dorsal and ventral limy shell; a fleshy stalk
		v		+	+h	+	+	0	MOLLUSCA	External limy shell of 1, 2, or 8 parts, or none; segmentation rare; fleshy lobe, the mantle, covering body
				+	+	+	+0	+	ANNELIDA	Slender, of many like segments; fine setae as appendages
		Bilateral		+	+h	+	+	+	ARTHROPODA	Segmented, with jointed appendages; exoskeleton containing chitin
				+	+	0	0	0	CHAETOGNATHA	Small; arrow-shaped, transparent; lateral fins
		Radial		0	+	+	+	0	ECHINODERMATA	Adult symmetry 5-part radial; tube feet; spiny endoskeleton; larvae bilateral
		Bilateral		+	+	+	+	+	CHORDATA	Notochord, dorsal tubular nerve cord, gill slits; usually fins or limbs

+, present; 0, absent; ps, pseudocoel, not lined by peritoneum: h, coelom reduced, body spaces a hemocoel; v, symmetry various, or none.

theories are useful and intellectually stimulating, it should be remembered that they are speculative. We do not have, and probably never will have, the fossil evidence necessary to reconstruct the origin of the metazoans, which took place more than 1 billion years ago. Theories must then be based on other, indirect evidence, as above, hence we will never know for certain. It is also probable that the metazoa are not monophyletic (all derived from a single common ancestor) but rather polyphyletic in origin, with different groups evolving independently from different ancestors. Table 14-3 lists some characteristics of the principal animal phyla.

Synopsis of Protozoa and Kingdom Animalia

The following outline of the Kingdom Protista and Kingdom Animalia will serve for identification of most specimens to phylum and class. Other characters for these groups and for the orders are given in the more detailed classifications ending Chaps. 15 to 34. A few minor groups are omitted here. Some alternative names used in other books are included in parentheses. Groups marked † are extinct. Names of a few representative genera are given in italic, followed by common names when available.

KINGDOM PROTISTA

Unicellular organisms

Phylum 1. Protozoa.

Protozoans. Each individual one-celled or in colonies of similar cells; no tissues; size usually microscopic. (Chap. 15)

Subphylum A. Sarcomastigophora.

Organelles for locomotion are flagella, pseudopodia, or none; nuclei of one kind.

Class 1. Mastigophora.

Flagellates. One to many flagella for locomotion. *Ceratium, Euglena, Trypanosoma, Volvox.*

Class 2. Opalinata (Protociliata).

Many cilialike organelles in oblique rows; no cytostome; all parasitic. *Opalina.*

Class 3. Sarcodina (Rhizopoda).

Pseudopodia for locomotion. *Amoeba, Actinophrys, Globigerina, Badhamia.*

Subphylum B. Sporozoa.

No locomotor organelles or contractile vacuoles; all parasitic.

Class 1. Telosporea.

Elongate sporozoites, no polar capsules. *Monocystis, Gregarina.*

Class 2. Piroplasmea.

No spores; in blood cells of vertebrates. *Babesia* in cattle.

Subphylum C. Cnidospora.

Spore with 1 to 4 filaments.

Class 1. Myxosporea.

Spores originate from several nuclei. *Myxidium, Heliosporidium.*

Class 2. Microsporea.

Spores small. *Nosema.*

Subphylum D. Ciliophora.

Cilia or sucking tentacles in at least one stage of life history; nuclei of two kinds.

Class 1. Ciliata (Infusoria).

Ciliates. *Paramecium, Oxytricha, Vorticella, Podophrya.*

KINGDOM ANIMALIA

Multicellular or tissue animals. Body of many cells, usually arranged in layers or tissues.

Sarcodina   Mastigophora   Sporozoa   Ciliata   Opalinata

Figure 14-10 Phylum PROTOZOA.   Representatives of five classes or superclasses.

Branch 1. Mesozoa.

Digestive cells few, external, ciliated.

Phylum 1. Mesozoa.

Wormlike, small; symmetry bilateral; an external layer of ciliated digestive cells surrounding one or several reproductive cells; parasitic in cephalopods and other invertebrates. *Dicyema.* (Chap. 20)

Branch 2. Parazoa.

Cells loosely organized; no organs; no digestive cavity.

Phylum 2. Porifera.

Sponges. Symmetry radial; body cylindrical, globose, branching, or irregular; skeleton internal, of minute spicules or of fibers (spongin); surface with many pores, connecting to canals and chambers lined by flagellated collar cells (choanocytes) and one or more large exits (oscula); aquatic, sessile. (Chap. 16)

Class 1. Calcarea.

Calcareous sponges. Spicules limy ($CaCo_3$), 1-, 3-, or 4-rayed; body surface bristly; marine, in shallow waters. *Leucosolenia, Scypha.*

Class 2. Hexactinellida.

Glass sponges. Spicules siliceous, 6-rayed, and in definite arrangement; marine, at 90 m (300 ft) or deeper. *Hyalonema; Euplectella,* Venus's flower basket.

Class 3. Demospongiae.

Skeleton siliceous, of spongin, of both, or none; canals complex; some large and brilliant; mostly marine. *Halisarca; Spongilla,* freshwater sponge; *Euspongia (Spongia),* bath sponge; *Cliona,* boring sponge.

Class 4. Sclerospongiae.

Coralline sponges. Massive skeleton of calcium carbonate; living tissue a thin veneer on the surface of the skeleton. *Ceratoporella, Merlia.*

Branch 3. Eumetazoa.

True tissues present; a digestive cavity present (unless secondarily lost).

Division A. Radiata.

Symmetry radial or biradial, no organs; two germ layers (diploblastic) only (lacking mesoderm).

Phylum 3. Cnidaria. (Coelenterata)

Symmetry radial; the individual a sessile cylindrical polyp, often in colonies, or a bell-like free-floating medusa with much gelatinous mesoglea; stinging capsules (nematocysts) present; digestive cavity saclike, sometimes branched; mouth surrounded by

Figure 14-11 Phyla PORIFERA, CNIDARIA (three classes), and CTENOPHORA.

Porifera   Hydrozoa   Scyphozoa   Anthozoa   Ctenophora

soft tentacles; no anus, head, or other organ systems; all aquatic, chiefly marine, attached or floating. (Chap. 17)

Class 1. Hydrozoa.

Hydroids (and some medusae). Mouth opens directly into a digestive cavity that lacks partitions; hydroid stage usually in colonies; medusa with velum. *Hydra*, freshwater polyp; *Obelia*, colonial hydroid, marine; *Millepora*, "stinging coral"; *Physalia*, Portuguese man-of-war.

Class 2. Scyphozoa (Scyphomedusae).

Jellyfishes. Small to large medusae, chiefly of gelatinous mesoglea, and of bell or umbrella shape, margined with tentacles; digestive cavity with branched canals; polyp stage minute or none; medusae sexual, dioecious; all marine. *Aurelia*, common jellyfish.

Class 3. Anthozoa (Actinozoa).

Sea anemones, corals, etc. All polyps (no medusae); a flat oral disk with tentacles; mouth connects to stomodeum (gullet); digestive cavity divided by radial partitions; corals with limy exoskeleton; all marine and attached, some gregarious. *Metridium*, sea anemone; *Gorgonia*, sea fan (horny); *Astrangia Orbicella*, stony corals; *Tubipora*, organ-pipe coral; *Stylatula*, sea pen.

Phylum 4. Ctenophora.

Comb jellies. Symmetry biradial; body subspherical with much mesoglea, or flat; 8 external rows of ciliated comb plates for locomotion; no nematocysts; digestive cavity with branched canals; marine; solitary. *Pleurobrachia*; *Cestum*, Venus's girdle; *Beroë*. (Chap. 17)

Division B. Bilateria.

Symmetry bilateral (secondarily radial in ECHINODERMATA); with organ systems and mostly with spaces between body wall and internal organs; digestive tract usually complete, with anus; mesoderm present.

Subdivision 1. Protostomia.

Mouth from blastopore; cleavage spiral and determinate.

Section *a*. Acoelomata.

No body cavity; space between body wall and internal organs filled with parenchyma.

Phylum 5. Platyhelminthes.

Flatworms. Body depressed, thin and soft, leaf-or ribbonlike, digestive tract straight or branched and without anus, or absent; parenchyma fills spaces between organs; a pair of anterior ganglia or a nerve ring and 1 to 3 pairs of longitudinal nerve cords; usually hermaphroditic. (Chap. 18)

Class 1. Turbellaria.

Free-living flatworms. Body ribbonlike to disklike, epidermis ciliated, with many mucous glands, no hooks or suckers; often pigmented, some with brilliant markings; usually a digestive tract with ventral mouth; mostly free-living; marine, freshwater, or terrestrial. *Dugesia*, *Leptoplana*.

Class 2. Trematoda.

Flukes. Body often leaflike, with thick cuticle and no cilia; ventral suckers or hooks or both; mouth usually anterior, digestive tract 2-branched; all parasitic. *Fasciola*, liver fluke.

Class 3. Cestoda.

Tapeworms. Body narrow, flat, elongate, comprising a scolex, with suckers or hooks or both (for attachment), and a chain of few to many proglottids (pseudosegments), each with complete reproductive organs; cuticle thick, no cilia; no mouth or digestive tract; all parasitic. *Taenia*, *Dipylidium*.

Phylum 6. Gnathostomulida.

Body cylindrical, semitransparent, ranging in size up to 1 mm, covered with cilia; no segmentation; mouth ventral with complex jaws bearing teeth; no

Figure 14-12 Phyla PLATYHELMINTHES (three classes), NEMERTINEA, and NEMATODA.

Turbellaria   Trematoda   Cestoda   Nemertinea   Nematoda

anus; no circulatory system; no coelom; nervous system with anterior sensorium; hermaphroditic; marine in anaerobic sand and mud. (Chap. 20)

Phylum 7. Nemertinea (Rhynchocoela).

Ribbon worms. Body slender, soft, very elastic, and covered with cilia; no segmentation; mouth anterior and with a long eversible proboscis in a separate cavity (rhynchocoel); digestive tract complete, with anus; a blood vascular system; body spaces filled with parenchyma; no coelom; nervous system with anterior ganglia (brain) and 3 longitudinal nerve trunks; sexes usually separate; free-living, mostly marine, few freshwater and terrestrial. *Cerebratulus, Paranemertes.* (Chap. 18)

Section *b*. Pseudocoelomata.

Spaces between body wall and internal organs a persistent blastocoel; anus present.

Phylum 8. Entoprocta.

Individuals minute, solitary or colonial, each on a contractile stalk, with calyx bearing single circle of many ciliated tentacles; digestive tract U-shaped, both mouth and anus within circle of tentacles; body spaces a pseudocoel filled with parenchyma; monoecious or dioecious; attached to objects or animals in salt water or fresh water. *Pedicellina, Urnatella.* (Chap. 20)

Phylum 9. Rotifera (Rotatoria).

Wheel animalcules. Body of trunk and tapering "tail," the latter often jointed and with "feet" having adhesive glands for attachment; anterior end with trochal disk bearing cilia, used for locomotion and feeding, that move so as to produce a wheellike appearance; males minute or absent. *Hydatina.* (Chap. 19)

Phylum 10. Gastrotricha.

To 0.54 mm long; slender, flexible; ventral surface flat, with 2 lengthwise rows of cilia for locomotion;

mouth anterior, surrounded by bristles; aquatic. *Chaetonotus.* (Chap. 19)

Phylum 11. Kinorhyncha (Echinoderes).

To 1 mm long; cylindrical, head of 2 rings, encircled by spines; mouth with spiny retractile proboscis; body of 11 (or 12) zonites (rings) covered by stout cuticle and bearing spines; sexes separate; marine. *Echinodera.* (Chap. 19)

Phylum 12. Nematoda (Nemathelminthes).

Roundworms. Body rounded, slender, often tapered at ends; covered with tough cuticle; longitudinal muscles only, producing flexing motions but no elongation or contraction; an anterior nerve ring and 6 lengthwise nerve cords; sexes separate; free living in soil or water or parasitic. *Ascaris*, roundworm; *Necator*, hookworm. (Chap. 19)

Phylum 13. Nematomorpha (Gordiacea).

Hair snakes, or horsehair worms. Body threadlike, not tapered, anterior end blunt; cuticle rough, opaque; body spaces lined with cells; sexes separate; larvae parasitic in insects, adults free-living in water. *Gordius.* (Chap. 19)

Phylum 14. Acanthocephala.

Spiny-headed worms. Body flat and rough in life, cylindrical and smooth when preserved; cuticle thin; anterior end with retractile proboscis bearing rows of recurved spines; no digestive tract; sexes separate; parasitic, larvae in arthropods, adults in vertebrates. *Echinorhynchus, Gigantorhynchus.* (Chap. 19)

Section *c*. Eucoelomata (Schizocoela).

With true coelom, which arises as a split in mesoderm.

Figure 14-13 Miscellaneous phyla.

| Nematomorpha | Acanthocephala | Rotifera | Bryozoa | Brachiopoda |

Phylum 15. Bryozoa (Ectoprocta).

Moss animals. Colonies branched and plantlike, or as low incrustations on rocks or shells, or as gelatinous masses; individuals many, minute, each in separate housing (zooecium); ciliated tentacles around mouth; digestive tract U-shaped, complete; coelom developed and lined; monoecious; free-living in salt water and fresh water. *Bugula, Pectinatella.* (Chap. 20)

Phylum 16. Phoronida.

Body wormlike, cylindrical, unsegmented; live in self-secreted membranous tube; anterior end bearing ciliated tentacles on a horseshoe-shaped lophophore; digestive tract complete, U-shaped, mouth and anus outside lophophore; coelom of 6 compartments, with lining; blood vessels present, contractile; monoecious; marine. *Phoronis.* (Chap. 20)

Phylum 17. Brachiopoda.

Lampshells. External limy shell of dorsal and ventral valves, attached to substrate by a fleshy stalk; interior with 2 spiral arms (lophophores) bearing ciliated tentacles; digestive tract with or without anus; coelom well developed; heart small; sexes separate; marine. *Lingula, Terebratulina.* (Chap. 20)

Phylum 18. Mollusca.

Mollusks. Symmetry bilateral or asymmetrical (viscera and shell coiled in some); usually no segmentation; body soft, covered by a mantle that usually secretes a limy shell of 1, 2, or 8 parts; usually an anterior head and a ventral muscular foot for locomotion; digestive tract complete; a heart and blood vessels; respiration usually by special gills called ctenidia; coelom reduced; nervous system of a few paired ganglia and connectives; sexes usually separate; in salt water and fresh water, some on land. (Chap. 21)

Class 1. Monoplacophora.

Body and foot oval, shell single, 5 or 6 pairs of gills; segmentally arranged kidneys; marine. *Neopilina.*

Class 2. Aplacophora.

Solenogastres. Body wormlike, no shell; foot reduced or absent; mantle with spicules. *Chaetoderma.*

Class 3. Polyplacophora.

Chitons. Body usually elliptical, shell of 8 dorsal plates; head reduced; marine. *Chiton.*

Class 4. Scaphopoda.

Tooth shells. Shell and mantle tubular, slenderly tapered; foot conical; no gills. *Dentalium.*

Class 5. Gastropoda.

Univalve mollusks. Viscera usually asymmetrical in spirally coiled shell (shell in some reduced or absent); head distinct, with 1 or 2 pairs of tentacles; foot large, flat. *Helix,* snail; *Limax,* slug (shell internal); *Acmaea,* limpet (shell conical); *Haliotis,* abalone.

Class 6. Bivalvia (Pelecypoda).

Bivalve mollusks. Body enclosed in shell of 2 lateral valves (right and left), with dorsal hinge; no head or jaws: foot often hatchet-shaped, extends between valves when moving. *Ostrea,* oyster; *Mya,* clam; *Teredo,* shipworm.

Class 7. Cephalopoda.

Squids, octopuses, etc. Shell internal or external, or none; head large, with conspicuous complex eyes; mouth with horny jaws and surrounded by 8, 10, or many arms or tentacles; marine. *Loligo,* squid; *Octopus,* devilfish or octopus; *Nautilus,* pearly nautilus.

Phylum 19. Annelida.

Segmented worms. Body elongated, usually of many like segments with fine bristlelike setae for locomotion; cuticle thin; digestive tract complete, tubular; coelom usually large; blood vascular system closed; nervous system of dorsal "brain" and a ventral nerve cord having ganglia and lateral nerves in each somite; mostly free-living. (Chap. 22)

Figure 14-14 Phylum MOLLUSCA. Five of the classes.

Polyplacophora   Scaphopoda   Bivalvia   Gastropoda   Cephalopoda

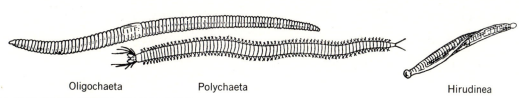

Oligochaeta   Polychaeta   Hirudinea

Figure 14-15 Phylum ANNELIDA. The three classes.

Class 1. Polychaeta.

Sandworms, tube worms, etc. Segmentation conspicuous, with many somites having lateral projections (parapodia) that bear numerous setae; head region evident, with tentacles; sexes usually separate; chiefly marine. *Neanthes* (*Nereis*), clam worm, in sand; *Chaetopterus*, in tube; *Polygordius*, in sand.

Class 2. Oligochaeta.

Earthworms, etc. Segmentation conspicuous; no head or parapodia; setae usually few per somite; monoecious; chiefly in fresh water and moist soil. *Enchytraeus; Lumbricus*, common large earthworm.

Class 3. Hirudinea.

Leeches. Body flattish; somites inconspicuous, each divided externally into several annuli; no setae or parapodia; a large posterior sucker, and often a smaller one at anterior end; coelom reduced; monoecious; in salt and fresh water or on land. *Hirudo, Macrobdella*.

Phylum 20. Sipuncula.

Peanut worms. Body slenderly gourd-shaped, highly contractile; slender anterior end (introvert) retractile, with short hollow tentacles around mouth; no segmentation or setae; digestive tract slender, spiraled, anus dorsal at base of introvert; coelom large, undivided, ciliated, containing blood with corpuscles; anterior dorsal nerve ganglion and ventral nerve cord; sexes separate; marine. *Sipunculus, Phascolosoma*. (Chap. 20)

Phylum 21. Echiura.

Body sausagelike; anterior end with trough-shaped elastic proboscis (nonretractile) leading to mouth; intestine spiraled, anus at posterior end, joined by 2 anal pouches; circulatory system of dorsal and ventral vessels; 1 to 3 pairs of nephridia anteriorly; one pair of large ventral setae below mouth; adults unsegmented, larvae with 15 vestigial somites; dioecious; marine. *Echiurus, Urechis*. (Chap. 20)

Phylum 22. Priapulida.

Sausage-shaped; anterior end with swollen introvert bearing lengthwise rows of spines; body narrower, with transverse striations (not segmented); digestive tract straight, anus in posterior end surrounded by gill lobes; body cavity large; adult lacks blood system, nephridia, and sense organs; sexes separate; marine. *Priapulus*. (Chap. 20)

Phylum 23. Pogonophora.

Beard worms. Cylindrical, diameter 0.5 to 2.5 mm, length 50 to 350 mm, in chitinous tubes; tentacles one to many, ciliated; body of 3 parts, the last with rings or adhesive papillae; no digestive tract; marine. *Siboglinum, Polybrachia*. (Chap. 20)

Phylum 24. Onychophora.

Elongate; no head, but anterior end with paired short "antennae" and oral papillae; body somewhat cylindrical, unsegmented; 15 to 44 pairs of stumpy, unjointed legs; terrestrial. *Peripatus*. (Chap. 20)

Phylum 25. Arthropoda.

Joint-footed animals. Body typically of head, thorax, and abdomen formed of segments and with body segments (somites) like or unlike, variously separate or fused; 4 or more pairs of jointed appendages; chitinous exoskeleton covering all parts, molted at intervals; digestive tract complete, straight; coelom reduced, body spaces a hemocoel; heart dorsal; respiration by gills, tracheae, or book lungs; "brain" dorsal, nerve cord ventral and paired, with ganglia in each somite or concentrated anteriorly; sexes usually separate; terrestrial or aquatic, free-living, commensal, or parasitic. (Chaps. 23–25)

Subphylum A. †Trilobita.

Trilobites. Body divided by 2 lengthwise furrows into 3 lobes; head distinct, abdomen of 2 to 29 somites and a fused caudal plate; all somites except last with biramous appendages; marine. Cambrian to Permian. †*Triarthrus*. (Chap. 23)

Subphylum B. Chelicerata.

No antennae; body of cephalothorax, with 6 pairs of appendages (chelicerae, pedipalpi, and 4 pairs of legs), and abdomen; chiefly terrestrial. (Chap. 23)

Class 1. Merostomata.

Cephalothorax broadly joined to abdomen on which are 5 or 6 pairs of appendages; with compound lateral eyes; aquatic. *Limulus,* horseshoe crab; †*Eurypterus,* eurypterid. (Chap. 23)

Class 2. Pycnogonida (Pantopoda).

Sea spiders. Mostly small to minute; body short, thin; mouth suctorial, on long proboscis; legs extremely elongated; marine. *Pycnogonum.* (Chap. 23)

Class 3. Arachnida.

Spiders, scorpions, mites, ticks, etc. Abdomen lacks locomotor appendages; eyes all simple; no gills; terrestrial. *Epeira,* spider; *Sarcoptes,* itch mite; *Ornithodorus,* tick (Chap. 23)

Subphylum C. Mandibulata (Antennata).

Body of 2 parts (head and trunk) or 3 parts (head, thorax with walking legs, and abdomen); 1 or 2 pairs of antennae; 1 pair of jaws (mandibles), 1 or more pairs of maxillae, and 3 or more pairs of walking legs.

Class 1. Crustacea.

Lobsters, crabs, water fleas, barnacles, etc. Two pairs of antennae, 1 pair of jaws, and 2 pairs of maxillae; some appendages biramous; respire mainly by gills; mostly aquatic. *Daphnia,* water fleas; *Balanus,* barnacle; *Astacus, Cambarus,* crayfishes. (Chap 24)

Class 2. Insecta (Hexapoda).

Insects. One pair of antennae; head, thorax, and abdomen distinct; thorax typically with 3 pairs of legs and 2 pairs of wings; mainly terrestrial. *Melanoplus,* grasshopper; *Musca,* fly; *Apis,* bee. (Chap. 25)

Class 3. Chilopoda.

Centipedes. Body long, flattened, of 15 to 181 somites, each with a pair of legs; terrestrial. *Lithobius, Scolopendra.* (Chap. 23)

Class 4. Diplopoda.

Millipedes. Body long, usually cylindrical; thorax of 4 somites, with 1 pair of legs on each except the first, which is legless; abdomen of 9 to more than 100 double somites, each with 2 pairs of legs; terrestrial. *Julus.* (Chap. 23)

Class 5. Pauropoda.

Minute; no eyes; antennae 3-branched; body cylindrical, of 11 (12) somites and 9 (10) pairs of legs; sex opening midventral on 3d somite; terrestrial. *Pauropus.* (Chap. 23)

Class 6. Symphyla.

To 6 mm long; no eyes; adult with 12 pairs of legs; sex opening midventral between 4th pair of legs; terrestrial. *Scutigerella,* garden centipede. (Chap. 23)

Phylum 26. Pentastomida (Linguatulida).

Wormlike, soft, unsegmented, but abdomen ringed; 2 pairs of ventral hooks beside mouth; parasitic in vertebrates. *Linguatula.* (Chap. 20)

Phylum 27. Tardigrada.

Water bears, or bear animalcules. To 1 mm long; body cylindrical, unsegmented; 4 pairs of stumpy unjointed legs with claws; in moss or marine waters or fresh waters. *Echiniscus.* (Chap. 20)

Subdivision II. Deuterostomia.

Mouth arising remote from blastopore. Mesoderm develops from archenteron.

Figure 14-16 Phylum ARTHROPODA.   Representatives of the five principal classes.

Crustacea   Insecta   Arachnida   Chilopoda   Diplopoda

Asteroidea   Ophiuroidea   Echinoidea   Crinoidea   Holothuroidea

Figure 14-17 Phylum ECHINODERMATA. The living classes and subclasses.

Phylum 28. Chaetognatha.

Arrow worms. Small, slender, transparent; body of head, trunk, and tail; bristles or hooks about mouth; paired fins on trunk and a terminal tail fin; digestive tract complete; coelom of 3 paired cavities; monoecious; free-living, marine. *Sagitta.* (Chap. 20)

Phylum 29. Echinodermata.

Echinoderms. Adult symmetry radial, usually 5-parted, around an oral-aboral axis; no segmentation; body wall with calcareous plates, usually forming a rigid or flexible endoskeleton with external spines; digestive tract with (or without) anus; coelom includes water vascular system; with external tube feet for locomotion; sexes usually separate; all marine. (Chap. 26)

Class 1. Crinoidea.

Sea lilies, feather stars. Body flowerlike, a boxlike calyx of many plates bearing slender branched arms; some species on aboral stalk. *Antedon.*

Class 2. Stelleroidea (Asteroidea and Ophiuroidra).

Starfishes and brittlestars. Free-living echinoderms with a body of central disc and radially arranged arms. *Asterias; Pisaster; Ophiura,* brittlestar; *Gorgonocephalus,* basket star.

Class 3. Echinoidea.

Sea urchins, sand dollars. Body hemispherical, disk- or egg-shaped, in shell (test) of fused plates that bear movable spines and pedicellariae; diges-

tive tract long, coiled. *Arbacia, Strongylocentrotus,* sea urchins; *Dendraster,* sand dollar.

Class 4. Holothuroidea.

Sea cucumbers. Body wormlike, wall soft, fleshy; no arms, spines, or pedicellariae; mouth anterior, surrounded by retractile tentacles; digestive tract S-shaped, anus posterior. *Holothuria, Thyone.*

Phylum 30. Hemichordata.

Tongue worms, pterobranchs. Symmetry bilateral, unsegmented, body of 3 divisions; either slender, wormlike, or vaselike in secreted tube; gill slits many, 2, or none; digestive tract complete, straight or U-shaped; marine. *Balanoglossus, Cephalodiscus.* (Chap. 20)

Phylum 31. Chordata.

Chordates. Having at some stage or throughout life, the following: an axial rodlike notochord for support of the body, a single dorsal tubular nerve cord, and paired gill slits between the pharynx and exterior; segmentation usually evident; tail behind anus. (Chap. 27)

Subphylum A. Urochordata (Tunicata)

Tunicates. Larvae minute and tadpolelike with gill slits and with both notochord and nerve cord in tail; adults tubular, globose, or irregular in form, covered with tunic or test, often transparent; with many gill slits, but notochord usually absent and nervous system reduced; marine. (Chap. 27)

Figure 14-18 Phyla HEMICHORDATA, CHAETOGNATHA, **and** POGONOPHORA.

Hemichordata   Chaetognatha   Pogonophora

Class 1. Larvacea (Appendicularia).

Individuals are minute neotenous larvae, separate and free-swimming, with notochord, "brain" and nerve cord, and 2 gill slits, test not permanent. *Appendicularia.*

Class 2. Ascidiacea.

Ascidians. Size and form various; solitary, colonial, or compound; usually sessile after metamorphosis when tail, nerve cord, and notochord are lost and "brain" is reduced to a ganglion, but many gill slits persist; test well developed, permanent. *Ciona, Molgula,* simple ascidians; *Botryllus,* compound ascidian.

Class 3. Thaliacea.

Chain tunicates. Size various; adults free-living, with no tail or notochord; test permanent, with circular muscle bands. *Salpa, Doliolum.*

Subphylum B. Cephalochordata (Leptocardii).

Lancelets. Body small, slender, elongate, fishlike, distinctly segmented; notochord and nerve cord extending entire length of body, and many gill slits enclosed in an outer atrium, all permanent. *Branchiostoma,* amphioxus or lancelet. (Chap. 27)

Subphylum C. Vertebrata

Vertebrates. With cranium (skull), visceral arches, and "spinal column" of segmented vertebrae, all cartilaginous in lower forms but bony in higher ones; notochord extends from tail to base of cranium; anterior end of nerve cord with enlarged brain of specialized parts; head region with paired special sense organs (smell, sight, hearing); paired semicircular canals for equilibration; circulatory system closed, with heart of 2 to 4 chambers, and red blood cells.

Class 1. †Ostracodermi.

Extinct armored fishes. Head and body armored with large scales, often fused into a cephalothoracic shield. †*Cephalaspis,* †*Pteraspis.*

Class 2. (Cyclostomata) Agnatha

Lampreys and hagfishes. Body cylindrical, slender, with median fins only; skin smooth, no scales; aquatic; no true jaws; one nasal opening; 5 to 16 baglike pairs of gill pouches opening on sides of body; heart 2-chambered. *Petromyzon,* lamprey; *Myxine,* hagfish. (Chap. 28)

Superclass a. Pisces.

Fishes. One pair of visceral arches modified as jaws; with median fins supported by fin rays; paired fins usually present; skin usually with scales containing limy material; nasal capsules not connected to mouth cavity; heart with one atrium; respiration by gills; aquatic.

Class 1. †Placodermi.

Ancient fishes. Jaws primitive; hyoid unspecialized, followed by a complete gill slit, no spiracle; paired fins variable; body covering of bony scales or plates; skeleton bony. †*Dinichthys,* arthrodire; †*Pterichthyodes,* antiarch. (Chap. 29)

Figure 14-19 Phylum CHORDATA. Two lower subphyla and classes of living vertebrates.

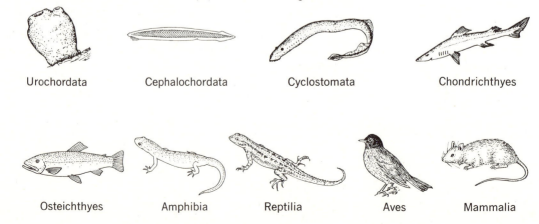

Urochordata   Cephalochordata   Cyclostomata   Chondrichthyes

Osteichthyes   Amphibia   Reptilia   Aves   Mammalia

Class 2. Chondrichthyes.

Cartilaginous fishes. Skeleton cartilaginous, notochord persistent; skin covered with placoid scales; chiefly marine. *Squalus,* shark; *Raja,* ray; *Chimaera,* chimaera. (Chap. 29)

Class 3. Osteichthyes.

Bony fishes. Skeleton more or less bony; mouth usually terminal; gills covered by opercula; scales usually cycloid. *Acipenser,* sturgeon; *Salmo,* trout, salmon; *Perca,* perch; *Neoceratodus,* lungfish. (Chap. 30)

Superclass b. Tetrapoda.

Four-legged land vertebrates. Typically with two pairs of 5-toed limbs, variously modified, reduced, or absent; skeleton of bone; nasal capsules connect to mouth cavity; some with external auditory canals; heart with 2 atria and double circulation of blood.

Class 1. Amphibia.

Amphibians. Living forms covered by a soft, moist, glandular skin; heart 3-chambered; larvae typically aquatic, respiring by gills; adults commonly terrestrial, usually respiring by lungs. *Ichthyophis,* caecilian; *Ambystoma,* salamander; *Bufo,* toad; *Rana,* frog. (Chap. 31)

Class 2. Reptilia.

Reptiles. Body covered with dry, cornified skin, usually with scales, or scutes; toes usually ending in claws; feet and limbs reduced or absent in some; respire by lungs; heart imperfectly 4-chambered; mostly terrestrial, some aquatic. *Chrysemys,* turtle; *Sphenodon,* tuatara; *Sceloporus,* lizard; *Thamnophis,* garter snake; *Alligator,* alligator. (Chap. 32)

Class 3. Aves.

Birds. Body covered with feathers; forelimbs modified as wings for flight; respire by lungs; heart completely 4-chambered; "warm-blooded" (endothermal); terrestrial or aquatic. *Struthio,* ostrich; *Larus,* gull; *Anas,* duck; *Corvus,* crow; *Passer,* sparrow. (Chap. 33)

Class 4. Mammalia.

Mammals. Body usually covered with hair; respire by lungs; heart as in birds; endothermal; females with mammary glands providing milk for nourishment of young. *Ornithorhynchus,* duckbill; *Macropus,* kangaroo; *Didelphis,* opossum; *Homo,* man; *Rattus,* rat; *Canis,* dog, wolf; *Equus,* horse. (Chaps. 34, 35)

References

(See also the larger books and general works listed in Chap. 1.)

Aristotle. The works of Aristotle. IV. Historia animalium. Translated by D'A. W. Thompson. 1910. New York, Oxford University Press. xv + about 300 pp.

Blackwelder, R. E. 1967. Taxonomy: A text and reference book. New York, John Wiley & Sons, Inc. xiv + 698 pp.

Clark, R. B. 1964. Dynamics in metazoan evolution. Oxford, Clarendon Press. 313 pp.

Hadzi, Jovan. 1963. The evolution of the metazoa. New York, The Macmillan Company. xii + 499 pp.

International code of zoological nomenclature. Rev. ed. 1964. London, International Trust for Zoological Nomenclature. xvii + 176 pp. In French and English with glossary.

Linnaeus, Carolus. 1758. Systema naturae per regne tria naturae, secundum classes, ordines, genera, species cum characteribus, differenties, synonymis, locis. Editio decima, reformata. Tomus I. Holmiae [The system of nature of three natural kingdoms, classes, orders, genera, species with characters, differences, synonyms, distribution. 10th ed., rev. Vol. 1. Stockholm], Laurentii Salvii 8 + 5–824 pp. *The starting point for scientific nomenclature.*

Mayr, Ernst. 1969. Principles of systematic zoology. New York, McGraw-Hill Book Company. xi + 428 pp., illus.

Simpson, G. C. 1961. Principles of animal taxonomy. New York, Columbia University Press. 247 pp., 30 figs.

Sneath, Peter H. A., and **Robert R. Sokal.** 1973. Numerical taxonomy. San Francisco, W. H. Freeman and Company. xv + 573 pp.

Whittaker, R. H. 1969. New concepts of kingdoms of organisms. *Science,* vol. 163, pp. 150–160.

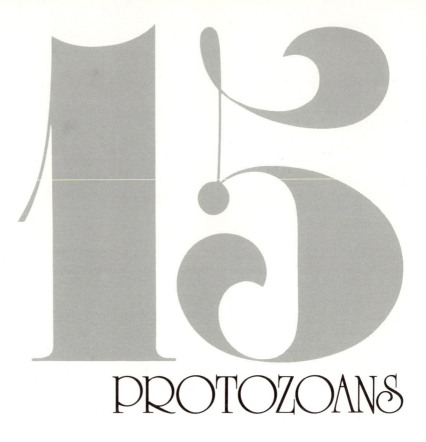

15

PROTOZOANS

Phylum Protozoa

The phylum PROTOZOA (Gr. *protos*, first + *zoön*, animal) are mostly one-celled animallike protistans of microscopic size. Structurally and functionally the single cell of a protozoan is more complex than the cell of a metazoan animal, and for that reason these organisms are classified in the Kingdom PROTISTA. Some protozoans are very simple in structure, and others are complex, with **organelles** (cell organs) which serve particular vital processes and which are functionally analogous to the organ systems of multicellular animals. About 50,000 kinds of PROTOZOA are known, and as individuals they far exceed animals in numbers. Each species lives in some particular moist habitat—in seawater or on the ocean bottom; in

fresh, brackish, or foul waters inland; in soil or decaying organic matter. Many are free-living and free-swimming, whereas others are sessile, and some in both categories form colonies. Still others live on or in protistans, some plants, and all sorts of animals, including humans. In different cases the interrelationships vary from casual occurrence to strict parasitism. In turn, some kinds of bacteria live on or in certain protozoans as casuals, symbionts, or parasites. Many protozoans serve as food of other minute organisms. Some are helpful in the purification of filter and sewage beds, but disease-producing species such as those causing amoebic dysentery, malaria, and African sleeping sickness are a scourge of humankind.

Classification of the PROTOZOA is quite complex, with four **subphyla and eight classes or superclasses** now recognized by most zoologists. In this chapter only the major classes are discussed separately (Fig. 15-1). There is a classification of the phylum down to orders at the end.

15-1 Characteristics

1 Small, usually one-celled, some in colonies of few to many similar individuals; symmetry none, bilateral, radial, or spherical.

2 Cell form usually constant, oval, elongate, spherical, or otherwise, varied in some species and changing with environment or age in many.

3 Nucleus distinct, single or multiple; other structural parts as organelles; no organs or tissues.

4 Locomotion by flagella, cilia, pseudopodia, or movements of the cell itself.

5 Some species with protective housings, or tests; many species produce resistant cysts or spores to survive unfavorable conditions and for dispersal.

6 Mode of life free-living, commensal, mutualistic, or parasitic.

7 Nutrition various: (*a*) **holozoic,** subsisting on other organisms (bacteria, yeasts, algae, various protozoans, etc.); (*b*) **saprophytic,** living on dissolved substances in their surroundings; (*c*) **saprozoic,** subsisting on dead animal matter; (*d*) **holophytic,** or autotrophic, producing food by photosynthesis as in plants. Some combine two methods.

8 Asexual reproduction by binary fission, multiple fission, or budding; some with sexual reproduction by fusion of gametes or by conjugation (in Ciliata).

The extreme age of the phylum Protozoa is proved by finding the hard remains of Radiolaria and Foraminiferida in Precambrian rocks. Many of the flagellate subclass Phytomastigina have the cells embedded in a common gelatinous matrix and show physiologic coordination between the individuals. Some are connected to one another by protoplasmic threads, and in *Volvox* there is differentiation into vegetative and reproductive cells. These conditions parallel the formation of tissues and the segregation of somatic and germ cells in the Animal Kingdom. Certain protozoans, ordinarily free-living, are occasionally found living within the bodies of other animals, thus affording a hint as to how strictly parasitic species may have been derived from free-living forms. Some chlorophyll-bearing flagellates resemble the green algae in structure and physiology and suggest a common origin for plants and animals. The Mastigophora are probably the most primitive, the Ciliata the most structurally advanced, and the Sporozoa probably morphologically simplified but also greatly specialized as a result of their strictly parasitic manner of life.

15-2 Size Protozoans are mostly so small that they are measured in micrometers (one micrometer, μm, $= \frac{1}{1000}$ mm). Some are only 2 or 3 μm in length. A dozen *Babesia* (piroplasmid) may inhabit one red blood cell, or several hundred *Leishmania* (flagellate) a single tissue cell. Most species are less than 250 μm long; but *Spirostomum* (ciliate) grows to 3 mm, and *Porospora gigantea* (sporozoan) to 16 mm.

Subphylum Sarcomastigophora
Class Sarcodina (Amoebas, etc.)

The common amoeba (*Amoeba proteus*) of clean fresh waters that contain green vegetation serves as an introduction to the Protozoa and to the Sarcodina (Gr. *sarcodes,* fleshy). The amoeba (Fig. 15-2) appears

Figure 15-1 Common representatives of the phylum Protozoa.

| *Amoeba* | *Euglena* | *Gregarina* | *Paramecium* | *Opalina* |
| Sarcodina | Mastigophora | Sporozoa | Ciliata | Opalinata |

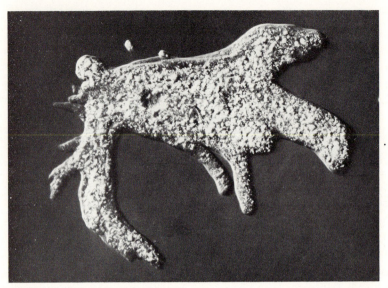

Figure 15-2 *Amoeba proteus* **as seen under oblique illumination.** (*Courtesy Carolina Biological Supply Co.*)

to be about the simplest possible living protistan, an independent cell with nucleus and cytoplasm but few permanent organelles. It has been studied intensively in the hope of discovering some of the fundamental features of life. Despite its seeming simplicity it can move, capture, digest, and assimilate complex food, egest indigestible residues, respire, produce secretions and excretions, respond to changes (stimuli) of various kinds in both its internal and external environment, grow, and reproduce itself in kind. It thus performs all essential living activities and shows physiologic specialization in various processes without having many structurally differentiated parts to perform these functions.

15-3 Structure The living amoeba is a mass of clear, colorless, jellylike protoplasm, up to 0.60 mm long, that is flexible, of irregular shape, and undergoes frequent change of form. It consists of (1) a very thin elastic external cell membrane, or **plasmalemma,** and beneath this (2) a narrow zone of clear nongranular **ectoplasm** surrounding (3) the main body mass of granular **endoplasm.** The latter consists of (*a*) an outer, stiffer **plasmagel** and (*b*) an inner **plasmasol** in which streaming movements (cyclosis) are visible. Within the endoplasm are (4) a disklike

nucleus, not easily seen in the living animal; (5) a spherical fluid-filled **contractile vacuole** (water expulsion vesicle), which at intervals moves to the surface, contracts, discharges its contents into the surrounding water, and then re-forms; (6) one or more **food vacuoles** of various sizes, containing bits of food under digestion; and (7) various other **vacuoles, crystals, oil globules,** and other **cell inclusions** ranging downward in size to or below the limit of the light microscope.

Briefly, the functions of these parts are as follows: (1) The cell membrane retains protoplasm within the cell, but permits the passage of water, oxygen, and carbon dioxide. (2) The ectoplasm lends form to the cell body. (3) The endoplasm contains the other structures and serves in locomotion. (4) The nucleus controls vital processes of the organism. (5) The contractile vacuole regulates water content. (6) The food vacuoles contain food undergoing digestion. (7) The other cell inclusions are reserve foods or other materials essential to metabolism. If the amoeba is cut in two, the cell membrane soon surrounds each piece and prevents loss of protoplasm; the part without a nucleus can still move and ingest food but is unable to digest or assimilate the food and soon dies, whereas that with the nucleus will continue to grow and reproduce. An isolated nucleus, however, cannot survive. Thus the nucleus and cytoplasm are interdependent.

15-4 Locomotion   The amoeba moves by forming and extending temporary fingerlike extensions, or **pseudopodia** (Gr. *pseudos,* false + *podos,* foot) at any place on its cell body (Fig. 15-3). This sort of irregular flowing is termed **amoeboid movement;** it occurs in many PROTOZOA and also in the amoebocytes of sponges and the white blood cells of vertebrates. Amoeboid movement is probably a basic characteristic of unspecialized protoplasm and has never been completely satisfactorily explained, though various theories have been offered. It is probable that amoeboid movement is not a single type but differs among various organisms employing it, much as pseudopodia also vary. S. O. Mast stated that movement in the amoeba is the result of changes within the colloidal protoplasm, from the fluid sol to the more solid gel condition, and vice versa. In this theory a stimulus causes the ectoplasm at some point on the surface to become liquid and turn to endoplasm (change from gel to sol). Internal pressure then creates a flow at the point of liquefaction, causing a pseudopodium to form. Recent work suggests that the sol and gel conditions are due to the relaxation and contraction of long-chain proteins, making the amoeboid movement more like muscular contraction. Three outstanding features of locomotion are (1) attachment to the substratum, possibly by a secretion; (2) transformation of plasmagel to plasmasol at the hinder end and the opposite process at the forward end of the animal; and (3) an increase in the elastic strength of the plasmagel as it passes backward. Contraction of the plasma membrane and gel may be part of locomotion. Attachment is best on rough surfaces but depends on the nature of the fluid about the animal and on the physiologic condition of the amoeba.

15-5 Feeding   The amoeba eats other protozoans, algae, rotifers, and dead organisms, preferring small live flagellates and ciliates. It may eat several paramecia or several hundred small flagellates daily and exhibit choice in selecting food. The amoeba is attracted by movements of the intended prey or by substances diffusing from it; unwanted or indigestible materials are usually avoided, as are organisms that show intense activity. Food may be taken in at any part of the cell surface. The amoeba extends pseudopodia that encircle the food (Fig. 15-4), which with some water is taken into the endoplasm as a

Figure 15-3 Locomotion in *Amoeba.*   The fluid inner plasmasol flows forward and is converted into plasmagel at the forward-moving pseudopod; the reverse process occurs in the opposite end and in pseudopodia that are being withdrawn. Large arrow indicates direction of movement of entire animal; small arrows, that of endoplasm. (*After Mast, 1926.*)

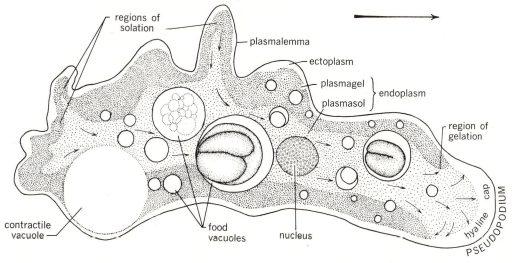

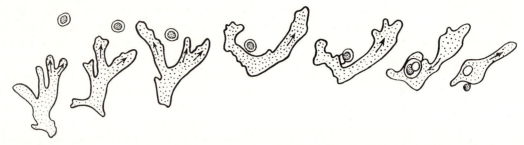

Figure 15-4 *Amoeba.* Stages in ingestion of food (small rounded particle); total time 8 min. At the end, digestible part is in food vacuole, remainder discarded. Arrows indicate movement of protoplasm in pseudopodia. (*After Schaeffer*, 1917.)

food vacuole. More water is included with an active item than with one that is quiet. The vacuoles are moved about by streaming movements in the endoplasm and thus come in contact with various parts of it. A recently formed vacuole gives an acid reaction (to litmus or neutral red), probably because of a secretion that kills the prey quickly. Later the reaction becomes alkaline and the action of enzymes secreted by the endoplasm is evident. Food particles lose their sharp outlines, swell, become more transparent, and then lessen in amount as the products of digestion are absorbed by the surrounding protoplasm. Evidence is clear as to the digestion of proteins but less certain as to fats, starches, and sugars. The absorbed materials serve for growth and reproduction as well as providing the energy for locomotion. The vacuoles decrease in size as digestion proceeds, and undigested residues are passed to the outside (Fig. 15-5) at any place on the cell surface.

15-6 Respiration and excretion The water in which the amoeba lives contains dissolved oxygen. This diffuses through the cell membrane, as in the internal respiration of cells in higher animals that derive their oxygen from the blood or other body fluids. Metabolism results in the production of waste products such as carbon dioxide and urea. For the well-being of the organism these must be eliminated; their disposal, mainly by diffusion through the cell membrane, is the process of excretion.

The contractile vacuole probably serves in part for excretion, but its principal function is to regulate the water content of the cell body. Water enters in food vacuoles, is a by-product of food oxidation, and also probably passes into the cell by osmosis, since the protoplasm contains a higher concentration of salts than the surrounding water. An amoeba placed in water with a higher salt content than usual forms a smaller vacuole that discharges less often than before, and some marine amoebas have no vacuole. The contractile vacuole forms gradually by fusion of smaller vacuoles that start from fine droplets. When filled, it is surrounded by a temporary "condensation membrane" that vanishes as the vacuole discharges through the cell membrane into the sur-

Figure 15-5 *Amoeba verrucosa.* Stages in passing out a waste pellet; *xy*, ingrowth of new cell wall to prevent loss of endoplasm. (*After Howland*, 1924.)

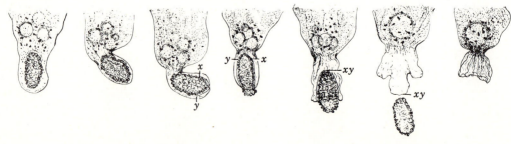

rounding water. In some amoebas, however, the vacuole may be dissected out under water with the membrane intact. Marine sarcodines lack contractile vacuoles, suggesting that maintenance of water content is the major role of the vacuole.

15-7 Behavior The responses of an amoeba to stimuli or changes in its environment, either internal or external, constitute its **behavior** (Fig. 15-6). Hunger is an internal stimulus to which the amoeba responds by searching for food. Contact with a food item is an external stimulus to which it responds by sending out pseudopodia to capture the food. The response to **touch,** or **contact** (thigmotaxis), is varied. A floating amoeba, with spread pseudopodia, responds positively to contact with a solid object by fastening to it; but a creeping amoeba, touched lightly with a needle, responds negatively by drawing back and moving away. Since the amoeba is exposed to **light** of varying intensity in nature, it does not respond to a gradual change in illumination. But if a strong beam is suddenly flashed on an individual in weak light, it responds negatively; the protoplasm streams away from the stimulus, and pseudopodia

soon form to aid the withdrawal. Furthermore, if an amoeba comes repeatedly into a field of intense light (Fig. 15-6E), the number of attempts to continue in the original direction decreases as the number of trials increases—a change in behavior similar to the learning process of higher animals. Response to **temperature** is shown by changes in both locomotion and rate of feeding. These both lessen as the water temperature approaches freezing. Locomotion is speeded in higher temperatures but ceases above 30°C (86°F). The amoeba responds negatively to strong salt solutions (or other chemicals) (Fig. 15-6D) but positively to substances diffusing from food. Responses by the amoeba to stimuli are such as to benefit the individual and the species by avoiding unfavorable conditions and seeking those that are useful to it.

15-8 Reproduction When the amoeba attains a certain size, it reproduces asexually by **binary fission.** The cell body becomes spherical and covered by short pseudopodia, then lengthens, and finally constricts into two parts; meanwhile, the nucleus has divided by mitosis. Under ordinary laboratory

Figure 15-6 *Amoeba.* *A–D.* Reactions to stimuli. Arrows indicate direction of movements. *E.* Successive stages at about 30-s intervals in reaction to strong light focused continuously on a microscope slide. (*A–D, after Jennings,* 1904; *E, after Mast,* 1910.)

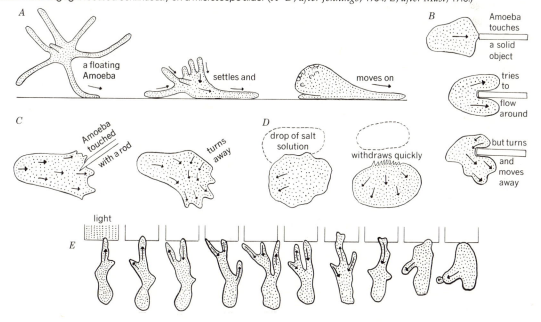

conditions the amoeba divides every few days, and mitosis requires about 33 min at 24°C (75°F).

15-9 Other Sarcodina The genus *Amoeba* includes many species that differ from *A. proteus* in size, shape of pseudopodia, and other features. These and members of other genera of amoeba-like protozoans inhabit fresh water, brackish water, and salt water. Still others produce a shell, or test, to enclose the cell body, examples being *Arcella,* which secretes a thick test, and *Difflugia* (Fig. 15-8), which cements together a test of sand grains or other foreign particles. Besides these free-living forms the order AMOEBIDA includes many commensal or parasitic amoebas. Species of the genus *Endamoeba* inhabit the gut in cockroaches and termites, and those of the genus *Entamoeba* live chiefly in the digestive tract of land vertebrates. Each is usually restricted to certain organs in a particular species of host animal. These amoebas may form resistant cysts by which they pass from an infected host to other individuals.

Six kinds of amoebas may occur in humans; of these, *Entamoeba gingivalis* lives in the mouth and *E. coli* in the intestine. Another intestinal species, *E. histolytica,* is pathogenic and may produce the disease known as *amoebic dysentery* (Fig. 15-9). When food or water containing cysts of *E. histolytica* enters the digestive tract, the amoebas are soon liberated and may invade glands of the intestinal wall to feed on the blood and tissues and to multiply. They cause the formation of abscesses that rupture and discharge blood and mucus into the intestine, resulting in thin feces and diarrhea. After such an acute condition the patient may recover somewhat and become a carrier, with less severe symptoms but continuing to discharge thousands of encysted amoebas and so able to spread the disease. In some persons an intestinal infection never produces any severe symptoms, but in others the amoebas may invade the liver, spleen, or brain with fatal results. Microscopic examination of fecal smears for cysts of *E. histolytica* is the basis for diagnosis, and medical treatment with drugs is necessary for a cure. The intestinal abscesses heal very slowly. Fecal contamination of drinking water and raw vegetables or contamination by infected and careless food handlers is the means of spread. About 10 percent of people in the United States are infected, and the proportion is much higher in tropical regions.

The order FORAMINIFERIDA (Figs. 15-7, 15-10) includes about 18,000 species, mostly marine, that have a covering shell (or test) through which the long threadlike reticulate pseudopodia stream out and in. Some fix to plants and hydroids or the sea bottom, others creep, and still others are pelagic, but all have some free-moving stage. The shells of various species are 0.01 to 190 mm in diameter, chambered, and of spherical, tubular, spiral, or other shapes. They are composed of gelatinous, chitinous, or limy material or of selected bits of sand, sponge spicules, or other debris. The organism inhabits all the chambers and extends thin pseudopods called reticulopods out through perforations in the shell for feeding. FORAMINIFERIDA have inhabited the seas since Precambrian time, and their shells have accumulated as bottom deposits that became rock strata. About 124,320,000 km² (48,000,000 mi²; 35 percent) of the present ocean bottom is covered with ooze formed of shells of certain pelagic forms, particularly *Globigerina*. The great pyramids near Cairo, Egypt, were carved from limestone deposits made of shells of an Early Tertiary foraminiferan, †*Nummulites*. Petroleum geologists study the foraminiferans obtained in drilling exploratory wells to identify oil-bearing strata.

Members of the order HELIOZOA are spherical; the cell body may be naked, in a gelatinous matrix, or in a latticelike shell. The many fine radiating pseudopodia suggest the name *sun animalcules*. Most of the species live in fresh waters. Protozoans of the order RADIOLARIA somewhat resemble heliozoans, but the cytoplasm is divided into outer and inner portions by a central capsule, and the skeleton is either of silica or of strontium sulfate (Fig. 15-7). They are all marine and pelagic, from the ocean surface to depths of 4575 m (15,000 ft), and very abundant as to species and individuals. About 5,180,000 km² (2,000,000 mi²) of ocean bottom, at depths of 3600 to 7300 m (12,000 to 24,000 ft) are covered with radiolarian ooze derived from empty skeletons of these animals. Many rock formations contain fossil radiolarians.

The order MYCETOZOA (MYXOMYCETES of botanists) includes the slime molds. The species form

Figure 15-7 *A*. Scanning electron micrograph of skeletons of RADIOLARIA, ×100. *B*. Scanning electron micrograph of FORAMINI-FERIDA showing characteristic perforations of the shell. (*S.T.E.M. Laboratories Inc. and Fisher Scientific Co.*)

masses of granular protoplasm (plasmodia), up to several inches in diameter, that contain thousands of nuclei and contractile vacuoles. They live on wood or other decaying vegetation in damp places on land.

Reproduction among the SARCODINA is varied.

Besides binary fission, the parasitic amoebas undergo multiple division of the nucleus while in cysts. FORAMINIFERIDA are dimorphic, each species having microspheric individuals or schizonts that reproduce asexually, giving rise to megalospheric indi-

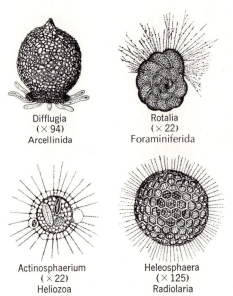

Figure 15-8 **Class SARCODINA.** Representatives of four orders. (*After Wolcott, Animal biology.*)

viduals or gamonts which develop gametes which then fuse to regenerate the microspheric individuals. The HELIOZOA reproduce sexually or multiply by fission. The RADIOLARIA have flagellated swarmers and may undergo fission. The MYCETOZOA produce multinucleate cysts to survive dry conditions.

Class Mastigophora (Flagellates)

The presence of one or more long slender **flagella** (sing. *flagellum*) at some or all stages in the life cycle is characteristic of the MASTIGOPHORA (Gr. *mastix*, whip + *phoros*, bearing). The flagella serve for loco-

motion and food capture and may be sense receptors. The cell body is usually of definite form—oval, long, or spherical—covered by a firm pellicle, and armored in certain groups of flagellates. Many species contain plastids with colored pigments; those with chlorophyll can synthesize food by the aid of sunlight and are often classified as plants. Many flagellates are free-living and solitary, others are sessile, and some form colonies of a few to thousands of individuals. They abound in fresh and salt water, where, with diatoms, they are a major part of the food supply for minute aquatic animals. A number of species inhabit the soil. Many others are parasites of humans and other animals, and some cause diseases of major importance. Reproduction usually is by longitudinal fission, but some undergo multiple fission, and there is sexual reproduction in at least two groups. Free-living flagellates may encyst to avoid unfavorable conditions. Some SARCODINA also possess flagella at times, and certain MASTIGOPHORA have amoeboid stages; so these two classes are rather closely related, hence their placement in the subphylum SARCOMASTIGOPHORA.

15-10 Structure *Euglena* is a common, solitary, free-living flagellate that contains chlorophyll; it may be cultured and studied easily in the laboratory. The slender cell body is up to 0.1 mm in length (Fig. 15-11), of constant shape, with a blunt **anterior end** that habitually travels forward; the opposite or **posterior end** is pointed. The body shape is maintained by a thin flexible covering membrane, or **pellicle,** that is marked spirally by parallel striations or thickenings. Within is a thin layer of clear **ectoplasm** around the main mass of granular nonflowing **endo-**

Figure 15-9 *Entamoeba histolytica,* **the parasitic amoeba of humans that causes amoebic dysentery.** (*After Cleveland and Sanders, 1933.*)

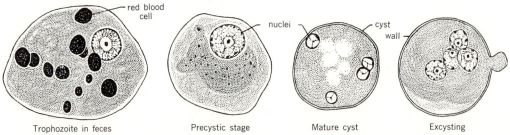

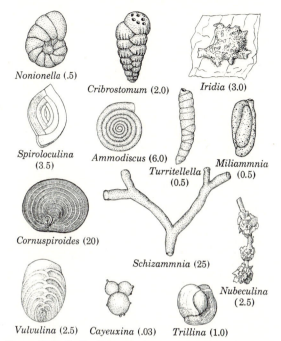

Nonionella (.5)

Cribrostomum (2.0)

Iridia (3.0)

Spiroloculina (3.5)

Ammodiscus (6.0)

Turritellella (0.5)

Miliammnia (0.5)

Cornuspiroides (20)

Schizammnia (25)

Vulvulina (2.5)

Cayeuxina (.03)

Trillina (1.0)

Nubeculina (2.5)

Figure 15-10 Shells of FORAMINIFERIDA. Numbers in parentheses indicate size in millimeters. (*After Galloway, A manual of Foraminifera.*)

plasm. The anterior end contains a funnellike **cytostome** that leads into a short tubular **cytopharynx.** A long **flagellum** extends out through the cytostome; it consists of a filament surrounded by a delicate sheath and arises from a **basal granule** (one or two), the **blepharoplast,** within the anterior cytoplasm. Behind the cytopharynx is a permanent spherical **reservoir,** and nearby a vacuole into which several minute **contractile vacuoles** empty. Fluid collected from the cytoplasm by the vacuoles passes into the

reservoir and out the cytopharynx. Near the reservoir is a red **stigma** (eyespot) that is perhaps sensitive to light. The round **nucleus** is near the center of the cell. *Euglena* is of green color owing to the **chloroplasts** (chromatophores) containing chlorophyll that often crowd its cytoplasm. Other cell inclusions are the **paramylum bodies,** which consist of a carbohydrate allied to starch.

15-11 Movement   The flagellum beats back and forth to draw *Euglena* through the water with a spiral rotation, so that it follows a straight course (see *Paramecium*, Sec. 15-20). The animal may also crawl by spiral movements of the cell body. At times it performs wormlike "euglenoid movements" by local expansions and contractions that suggest the peristalsis in a vertebrate's intestine. *Euglena* reacts positively to light by swimming toward a source of favorable intensity, as a green plant turns to the light, but it performs an avoiding reaction to direct sunlight.

15-12 Nutrition   Some free-living flagellates capture small organisms, which are taken into the cytopharynx and digested in food vacuoles in the cytoplasm, but such **holozoic nutrition** is rare or lacking in *Euglena*. The latter utilizes **holophytic nutrition,** whereby some food is synthesized within the body. This is done by photosynthesis, as in green plants, through the action of chlorophyll in the presence of light. *Euglena* also commonly subsists by **saprophytic nutrition,** which is the absorption of nutrient materials dissolved in the water where it lives. In rich nutrient solutions, cultures of some euglenas

Figure 15-11 Structure of *Euglena,* a free-living flagellate (class MASTIGOPHORA).

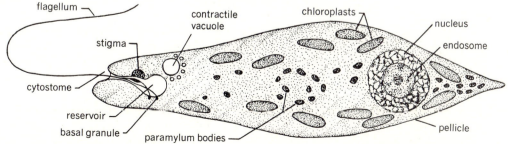

flagellum

contractile
vacuole

chloroplasts

nucleus

stigma

endosome

cytostome

reservoir

basal granule

paramylum bodies

pellicle

will persist and multiply rapidly in weak light or darkness.

15-13 Reproduction In active cultures, *Euglena* reproduces frequently by longitudinal **binary fission** (Fig. 15-12). The nucleus divides in two by mitosis, then the anterior organelles—flagellum, blepharoplast, cytopharynx, reservoir, and stigma—are duplicated, and the organism splits in two lengthwise. *Euglena* also has inactive stages when it becomes nonmotile and encysts. In hot weather *E. gracilis* may do this temporarily as a protective measure, without undergoing any change while encysted. *Euglena* may also lose the flagellum, encyst, and then divide by longitudinal fission. There may be further multiplication by longitudinal fission, resulting in cysts that contain 16 or 32 small daughter euglenas. Encystment is stimulated by lack of food or by the presence of chlorophyll (as in strongly illuminated cultures).

15-14 Other Mastigophora Many members of subclass PHYTOMASTIGOPHOREA, to which the genus *Euglena* belongs, are free-living, colored by chromatophores, and holophytic in nutrition. Species of the order DINOFLAGELLATA are mostly marine and

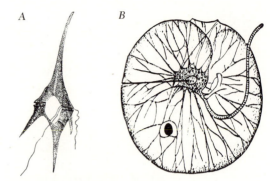

Figure 15-13 Class MASTIGOPHORA. Two common marine flagellates. *A. Ceratium,* ×105. *B. Noctiluca,* ×27.

generally possess two flagella in grooves, chromoplasts, and a celluloselike armor of two to many plates (Fig. 15-13). *Ceratium* and *Peridinium,* along with diatoms, are important members of the microscopic plankton, or "ocean meadows," on which the minute larvae of many marine animals feed. Certain dinoflagellates sometimes become excessively abundant (20 to 40 million per liter) in shallow marine waters, causing "red tides" on the sea by day and luminescence at night. A "flowering" of *Gymnodinium brevis* in 1947 on the west Florida coast killed billions of fishes and some sea turtles by the toxins it produced. *Gonyaulax catenella* on the California coast is fed on regularly by bivalves including sea mussels (*Mytilus*). It produces a substance which is harmless to the mussels but which causes mussel poisoning in humans if these mollusks are eaten during periods of high dinoflagellate abundance (usually the summer), when the mollusks subsist largely on this protozoan.

The order VOLVOCIDA includes several green freshwater flagellates (*Pandorina, Pleodorina, Volvox,* etc.) that form floating colonies (Fig. 15-14). *Volvox globator* is a hollow sphere (0.5 mm in diameter), filled with watery jelly. Embedded in the gelatinous exterior wall are 8000 to 17,000 minute (4 by 8 μm) individual cells, each with nucleus, contractile vacuole, red stigma, green chloroplasts, and two flagella. Protoplasmic threads connect adjacent cells and provide physiologic continuity between them. The flagella beat collectively to roll the colony through the water. Beginning in the spring, the first colonies develop from zygotes formed the previous season. Sev-

Figure 15-12 *Euglena viridis.* Stages in longitudinal fission. (*Modified from Tannreuther, 1923.*)

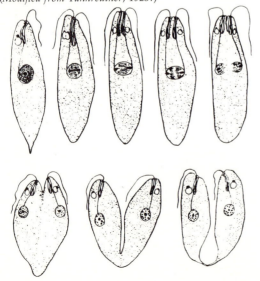

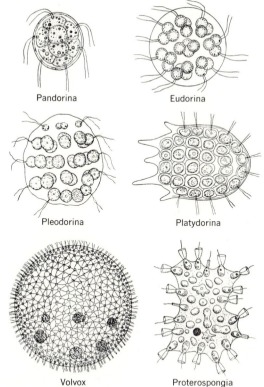

Pandorina

Eudorina

Pleodorina

Platydorina

Volvox

Proterospongia

Figure 15-14 Class MASTIGOPHORA. Colonial flagellates. (*After Hyman, The invertebrates.*)

eral asexual generations follow. In each colony a few cells (gonidia) lose flagella, enlarge, and divide to form hollow masses of small cells within the parent colony; the latter disintegrates, and each "embryo" mass grows asexually into a new colony. Then a sexual generation ensues. From 20 to 30 nonflagellated cells within a colony enlarge, each to become a **macrogamete** ("female sex cell," or "ovum"); a larger number of others divide repeatedly to form flat packets of 128 or more slender **microgametes** ("male sex cells," or "sperm"). Both types of gametes are discharged into the fluid interior of the colony, each "ovum" is penetrated by a "sperm," and the two fuse as a **zygote.** The latter secretes a cyst about itself and is released into the water when the parent colony disintegrates. It remains in a dormant stage until the following year, then escapes from the cyst to start a new colony by asexual fission.

The volvox colony thus bears some resemblance to a multicellular animal. It is composed of cells of two kinds, unlike in structure and function. The **somatic cells** of the colony wall can synthesize food and cooperate in swimming but cannot reproduce and eventually die, as does the body of a higher animal. The **germ cells** are unable to feed, are of two distinct kinds (sexes), and serve only to maintain the species by sexual reproduction.

The subclass ZOOMASTIGOPHOREA comprises flagellates, lacking chloroplasts, that are simple to complex in structure. The free-living forms are holozoic in nutrition, and the many parasitic species are saprozoic. The order KINETOPLASTIDA includes both types. *Bodo* (Fig. 15-15) and *Oikomonas* are common small solitary types in fresh waters and soil. The order CHOANOFLAGELLIDA contains freshwater flagellates such as *Codosiga, Proterospongia,* and some other colonial forms which have a flagellum surrounded by a collar, like the choanocytes of sponges.

All the family TRYPANOSOMATIDAE are parasitic. *Phytomonas* lives in the latex cells of milkweeds and is carried from one plant to another by a bug (*Oncopeltus*). *Crithidia* and *Herpetomonas* are gut parasites of insects and spread by cysts or fecal droplets. The genus *Trypanosoma* (Fig. 15-16) contains slender leaflike flagellates that inhabit the blood of verte-

Figure 15-15 Class MASTIGOPHORA. Representative free-living flagellates of fresh water; all enlarged to about same scale.

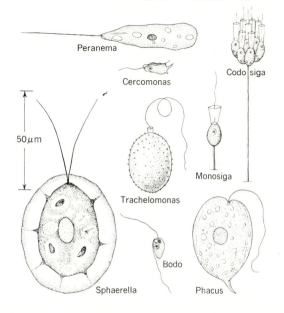

Peranema

Cercomonas

Codosiga

50 μm

Monosiga

Trachelomonas

Bodo

Sphaerella

Phacus

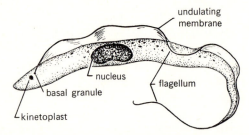

Figure 15-16 Class MASTIGOPHORA. Structure of *Trypanosoma.* Greatly enlarged.

brates, form no cysts, and are carried from one host to another by bloodsucking invertebrates. Species in fishes, salamanders, frogs, and reptiles are transmitted by leeches. *T. lewisi* of the rat, spread by rat fleas (Fig 15-17), usually does no harm to its host, but many other species are disease producers. *T. evansi* causes surra in domestic animals of North Africa, the Middle East, and the Orient and is carried by biting flies (*Tabanus*). In Africa *T. brucei, T. congolense,* and *T. vivax* occur in antelopes and other game animals. When transmitted to domestic livestock by bloodsucking tsetse flies (*Glossina*), they cause the lethal disease nagana (trypanosomiasis). The native animals have evolved an immunity to the disease. *T. gambiense* and *T. rhodesiense* are the causative agents of two types of human sleeping sickness[1] in Africa,

[1] *Encephalitis, or sleeping sickness, in the United States is caused by a virus.*

but Gambian sleeping sickness results in a slow death whereas Rhodesian sleeping sickness kills within weeks. Both are transmitted by tsetse flies, Gambian directly from human to human, Rhodesian via a third host in the native antelope. *T. cruzi* causes Chagas' disease in Central and South America, being transmitted by bugs (*Triatoma, Rhodnius*). The organism occurs in these insects and in wood rats (*Neotoma*) and other mammals in the Southwestern United States, where only a few human cases have been recorded. Flagellates of the genus *Leishmania* attack the endothelial cells in human blood vessels. *L. donovani* causes kala azar, and *L. tropica* the skin disease called *Oriental sore*; both occur in Asia, the Mediterranean region, and parts of South America. The related *L. brasiliensis* of Mexico and Central and South America also produces a skin disease which may cause disfigurement and is more serious than the disease produced by *L. tropica*. The spread of leishmanias is variously from rodents by bloodsucking flies (*Phlebotomus*), by direct contact between people, and by dogs.

The order HYPERMASTIGIDA comprises species with many flagella and a complex internal structure. They abound in the gut of cockroaches (*Cryptocercus*) and woodroaches and in about one-fourth of the species of termites, where they constitute an outstanding example of mutualism (Fig. 15-18). Termites (Chap. 25) eat wood, but many of them are unable to

Figure 15-17 Class MASTIGOPHORA. Life cycle of the rat trypanosome, *Trypanosoma lewisi;* natural size, 25 μm long. (*Adapted from Minchin and Thompson, 1915.*)

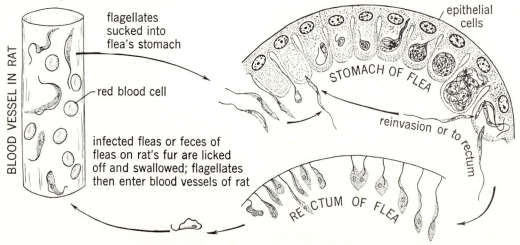

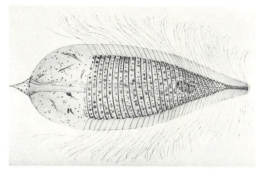

Figure 15-18 Class MASTIGOPHORA. *Spirotrichonympha bispira,* a complex flagellate that lives only in the gut of termites. (*After Cleveland,* 1928.)

digest the cellulose that it contains. The flagellates that receive "lodging and board" are able to digest this material for themselves and their hosts. Termites experimentally defaunated (by exposure to a high oxygen atmosphere or by being kept at 36°C (96.8°F) for 24 h to kill the flagellates) will starve and die in 10 or more days, even when wood is eaten. If, however, they are reinfected with the flagellates, they can then utilize the wood and survive. The flagellates form pseudopodia to ingest the wood fragments in the termite's gut, the cellulose is digested, and the soluble products become available for absorption by the termites. Other nonsymbiotic protozoans occur in termites.

nucleus. They resemble flagellates in that their asexual reproduction division plane is similar.

Subphylum Sporozoa[1]

The SPOROZOA (Gr. *spora,* seed + *zoön,* animal) are all parasites. The simple cell body is rounded or elongate, with one nucleus and no locomotor organelles (except in coccidian and malarial microgametes) or contractile vacuole. Some move by change in shape of the cell body. Food is absorbed directly from the host (saprozoic nutrition), and respiration and excretion are by simple diffusion. Most sporozoans have complex life cycles with alternating sexual and asexual generations. In the asexual phase they increase rapidly by multiple fission, or schizogony; the cell becomes multinucleate by repeated mitoses, and then the cytoplasm divides. The resulting **merozoites** continue to spread the sporozoan through the host tissues. Following a maturation phase, merozoites, through a process of gametogony, produce sexual **macro-** and **microgametes;** these join in pairs of opposite kind to form **zygotes.** The latter, in many species, produce **sporozoites** by **sporogony,** and in this stage the organisms are spread from one host individual to another. The typical coccidian life cycle is thus:

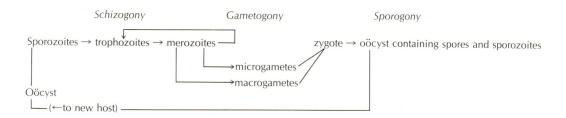

Class Opalinata (Opalines)

The OPALINATA (*Opalina,* etc.) mostly inhabit the intestines of frogs and toads (Fig. 15-1). They have two to hundreds of nuclei, all alike, and no cytostome; in "sexual reproduction" their gametes fuse permanently. The opalines resemble ciliates in having numerous cilialike organelles in oblique rows over the body surface and in having more than one

SPOROZOA are possibly the most widely occurring parasites in living organisms. Distinct species occur in various organisms from protozoans to mammals. Some live within the host's cells and others in its

[1] *All* SPOROZOA *were once placed together in a single class, but now that the life histories of many are better known, they are proving to be so diverse that it is advisable to erect two or three subphyla and several classes.*

body fluids and cavities. They inhabit variously the digestive tract, muscles, blood, kidneys, or other organs. Malaria in humans, coccidiosis in fowls and rabbits, certain fevers in cattle, and pébrine disease of silkworms are examples of serious diseases caused by sporozoans.

Class Telosporea (Gregarines, Coccidia, Haemosporiina)

15-15 Subclass Gregarinia. These organisms, some of which are the largest of the sporozoa, are extracellular or intracellular parasites of invertebrates. Usually the body is divided into an anterior **protomerite** and a posterior **deuteromite** (Fig. 15-19). A common example is *Monocystis,* which lives in the seminal vesicles of earthworms. The minute **spore** is a spindle-shaped case containing eight **sporozoites.** Upon escaping from the spore, each sporozoite enters a clump of immature sperm cells and transforms into a **trophozoite** that consumes the cells. The trophozoites then become grouped as pairs of **gametocytes,** each pair being surrounded by a **cyst.** Sporulation follows, every nucleus dividing many times with later partitioning of the cytoplasm so that many **gametes** of about equal size result. A pair of gametes, each from a different parent cell, then unite as a **zygote,** which secretes a thin, hard spore case, or **oocyst,** around itself while still within the cyst. The zygote nucleus inside divides successively into two, four, and eight daughter nuclei, and each combines with part of the cytoplasm to become a sporozoite.

Figure 15-19   Class TELOSPOREA. A gregarine from the mealworm. Greatly enlarged. (*From Hyman.*)

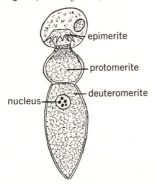

epimerite

protomerite

deuteromerite

nucleus

The manner by which spores are spread from one worm to another is unknown.

15-16 Subclass Coccidia   These parasites live in epithelial cells of many vertebrates, some myriapods, and a few other invertebrates. They occur chiefly in the lining of the intestine, but also in the liver, bile duct, kidneys, testes, blood vessels, and coelom. The life cycles involve sequential schizogony, gametogony, and sporogony, with change of host by some species. Coccidia produce the disease coccidiosis, which may be serious and even fatal. Species of economic importance occur in poultry, domestic mammals, and many kinds of wild animals. *Eimeria stiedae* (Fig. 15-20) is common in both domestic and wild rabbits. When food contaminated with oocysts is swallowed, the host's digestive juices dissolve the cysts and release the sporozoites. These enter epithelial cells in the bile duct and enlarge as feeding trophozoites that undergo rapid schizogony to produce many merozoites. In turn, the merozoites invade other cells to repeat the cycle. Epithelial tissue in the bile duct and liver may be destroyed, with consequent harm to the host. Later some merozoites develop into gametocytes, also within epithelial cells; some become enlarged macrogametes, and others divide to yield many microgametes. Union of a macrogamete and a microgamete forms a zygote that secretes an elliptical covering and becomes an oocyst. In this stage the coccidia pass out of the rabbit in the feces. Later the zygote within divides into four spores, each enclosing two sporozoites, and this is the infective stage for a new host. Among domestic rabbits this parasite often causes severe epidemics with many deaths, particularly in young animals. Rearing rabbits in separate hutches, keeping their food off the floor, and cleaning the hutches frequently are practices essential to keep down losses from coccidiosis.

Toxoplasma is medically the most important genus of the suborder EIMERIINA. The organisms reside in the tissue cells of many birds and mammals and attack a variety of tissues in humans, resulting in the disease **toxoplasmosis;** the severity of the disease varies among humans according to the tissues infected. The incidence of human infection by this parasite is thought to be high.

Invertebrates

1. A sea anemone of the genus *Tealia*.
2. The scyphozoan jellyfish *Pelagia panopyra*.
3. The nemetean worm *Baseodiscus punnetti* on the bryozoan *Hippodiplosia insculpta*.
4. *Spirobranchus*, a sedentary polychaete worm.

Photos 1 to 4 courtesy of Don Wobber

5

Invertebrates (continued)

7

6

5. *Hopkinsia rosacea*, a nudibranch mollusk.
6. *Orchestoidea*, an amphipod crustacean.
7. Pyrgomorphid grasshopper, *Pymateus morbillosus*.
8. The sea star *Pycnopodia helianthoides*.

Photos 5, 6 courtesy of Don Wobber; Photos 7 courtesy of Edward S. Ross

8

Vertebrates

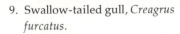

9. Swallow-tailed gull, *Creagrus furcatus.*
10. Lioness, *Panthera leo.*
11. Yarrow's spiny lizard, *Sceloporus jarrovi.*
12. San Francisco garter snake, *Thamnophis sirtalis tetrataenia.*

Photos 11 and 12 courtesy of Nathan W. Cohen

13

14

Vertebrates (continued)

15

13. Striped seaperch, *Embiotoca lateralis*, among white sea anemones, *Metridium senile*.
14. California treefrog, *Hyla cadaverina*.
15. Large-blotched salamander, *Ensatina eschscholtzi klauberi*.
16. Pacific pond turtle, *Celmmys marmorata*.

Photo 13 courtesy of John H. Tashjian; Photos 15 and 16 courtesy of Nathan W. Cohen

16

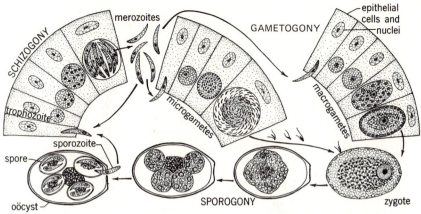

Figure 15-20 Class TELOSPOREA. Life cycle of *Eimeria stiedae,* a coccidium from the liver of the rabbit. (*Adapted from Wasielewski,* 1924.)

15-17 Suborder Haemosporiina These TELO-SPOREA are parasites of blood cells and tissues in vertebrates; they form no resistant spores but are transferred by bloodsucking arthropods as intermediate hosts. The most familiar example is *Plasmodium,* which causes malaria (Fig. 15-21). This disease, a scourge of humans since ancient times, has caused an enormous amount of illness and innumerable deaths, especially in tropical and subtropical regions. Possibly it is the most important of all diseases. The parasites are transmitted to humans by females of certain species of mosquitoes of the genus *Anopheles* (Fig. 25–40). When the mosquito's mouth parts pierce the skin to obtain blood, the infective sporozoites pass from its salivary glands into the wound. They enter parenchyma cells in the liver and multiply. Infection later spreads into red blood cells, where amoebalike **trophozoites** develop. Each trophozoite in an individual blood cell grows to be a schizont, which by multiple fission (schizogony) divides into 6 to 26 daughter **merozoites,** according to the species. By rupture of the corpuscle these escape into the blood plasma, to invade other red cells and repeat the cycle. After about 10 days the parasites are so numerous that the shock of their nearly simultaneous release produces a chill, followed by a violent fever in response to toxins from the liberated parasites. The chill-and-fever cycle thenceforth depends on the species of parasite, occurring every 48 h in benign tertian malaria caused by *P. vivax* and every

72 h in quartan (fourth-day) malaria caused by *P. malariae*. In aestiveo-autumnal (malignant tertian) malaria caused by *P. falciparum,* the cycle is lacking or irregular. *P. ovale,* with a 48-h cycle, produces a mild infection ending in about 15 days.

After a period of schizogony, some merozoites become **gametocytes** but do not change further in the human host. If taken with blood into the gut of an appropriate female *Anopheles,* the female gametocyte soon becomes a **macrogamete** and the male gametocyte divides into six to eight spermlike **microgametes.** Two gametes of opposite sex fuse as a zygote. This becomes a wormlike **ookinete** that penetrates the gut wall to lie under the membrane surrounding the gut. It then absorbs nutrients from the insect and enlarges as a rounded **oocyst.** A single mosquito may contain 20 to 30 or more oocysts. In approximately 1 week the contents of each cyst divide (sporogony) into thousands of slender sporozoites. The cysts burst, the sporozoites migrate through the body spaces, and many enter the salivary glands to await transfer to a human host. The sexual cycle in the mosquito requires about 1 to 3 weeks before the insect becomes infective.

Acute symptoms of malaria in humans usually continue for some days or weeks and then often subside as the body develops an **immunity** to the disease, but relapses may occur at irregular intervals. In different persons the infection disappears in time, or lingers and causes damage to other organs, or may

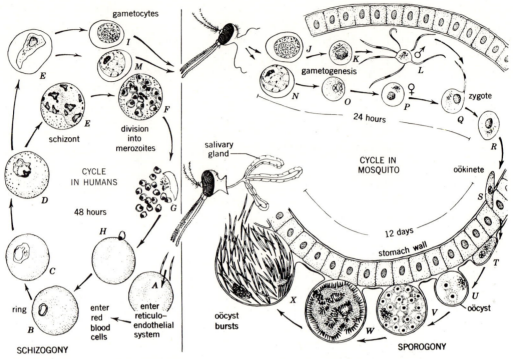

Figure 15-21 Class TELOSPOREA, suborder HAEMOSPORIINA. Life cycle of *Plasmodium vivax*, the tertian malarial parasite of humans and anopheline mosquitoes. Sporozoites injected during the biting of the mosquito invade parenchyma cells (reticuloendothelial system) of the liver (not shown in figure) and multiply asexually (schizogony); then the resulting parasites (*A*) enter red blood cells and multiply further (*B–H*). Later gametocytes (*I, M*) are produced; if drawn up by a biting mosquito, they transform into gametes in the insect's stomach (*J–L, N–P*). The zygotes encyst externally on the gut wall and form large oocysts. By asexual multiplication (sporogony; *U–W*) many sporozoites are produced (*X*) that invade the salivary glands.

result in death. Malaria patients may be treated with chloroquine. Healthy persons living in malarial regions often take small doses of pyrimethamine or chloroquine as a prophylactic. General reduction or control of malaria in a region requires (1) treating the human victims, (2) effectively screening dwellings to exclude mosquitoes, and (3) draining off, oiling, or poisoning the water to reduce mosquitoes by killing the larvae. Insecticides have been highly effective in killing adult mosquitoes when used judiciously to leave a toxic residue on walls and ceilings where mosquitoes rest. Mosquito fish (*Gambusia affinis*) are planted in waters to eat mosquito larvae and pupae.

Other species of *Plasmodium* occur in mammals, birds, and reptiles; several in primates closely resemble the species in humans.

15-18 Other Sporozoa The subphylum CNIDO-SPORA is composed of sporozoans that are largely

parasitic in fishes (MYXOSPORIDIA) or fishes and arthropods (MICROSPORIDIA). Myxosporidians are often found in the cavities of internal organs, but microsporidians are intracellular parasites. The microsporidian parasites of the genus *Nosema* are of considerable economic importance, being responsible for serious diseases among honeybees and silkworms.

The PIROPLASMEA include a group of small parasites currently included with the SPOROZOA. *Babesia bigemina,* a minute parasite in the red blood cells of cattle, causes red water, or Texas cattle fever. The intermediate host is a tick, *Boöphilus annulatus* (Fig. 23-10*A*). Following asexual stages in the blood of cattle, the merozoites are withdrawn when the tick feeds. The sexual stages occur in the blind gut of the tick, and the zygotes then penetrate to the ovary and eggs. Later they multiply in the developing salivary glands of the embryo tick. When the latter hatches and feeds on a bovine animal, the sporozoites are injected into the new host individual. Control of this disease, which once ravaged herds in many Southern states, has been accomplished by dipping the cattle to remove ticks and rotating their pasturing, so that clean herds have no chance to acquire infected ticks.

Subphylum Ciliophora
Class Ciliata (Ciliates, or Infusorians)

The CILIATA (L. *cilium,* eyelash) possess cilia during some part of their life cycle which serve for locomotion and food capture. Universally present below the outer pellicle is a complex system of basal granules and fibrils for operation of the surface cilia. Each species is of constant and characteristic form and has one or more large macronuclei concerned with the routine or vegetative functions and one or more small micronuclei that are important in reproduction. Ciliates are the most specialized of protozoans in having various organelles to perform particular vital processes. This results in a division of labor

between parts of the organism, analogous to that between organ systems in a multicellular animal. On the whole, in fact, the ciliates seem the most animal-like of all the protozoans. Ciliates constitute the largest protozoan group, with nearly 6000 species, and they abound in fresh and salt waters. Many are free-living, some are commensal or parasitic in other animals, and some are sessile. They are much used in experimental studies because of their relatively large size and the ease with which some may be reared. *Tetrahymena,* for example, can even be grown axenically (i.e., in the absence of any other living organism) and has been the subject of extensive biochemical research.

Paramecium (Fig. 15-22) is a ciliate common in fresh waters that contain some decaying vegetation. It multiplies rapidly in the laboratory in an infusion made by boiling a little hay or some wheat grains in water. As many as 13 species differing in size, shape, structure, and biochemical and mating reactions, have been described. *P. aurelia* is 0.1 to 0.2 mm long, and *P. caudatum* measures 0.15 to 0.3 mm in length. The following account deals mainly with *P. caudatum.*

Figure 15-22 Structure of *Paramecium caudatum,* a ciliate of fresh waters (class CILIATA). *A.* The entire animal. Lines of dots indicate rows of cilia over the surface. Arrows show path of food vacuoles in endoplasm. *B.* Enlarged sketch of a few cilia showing how they beat in coordinated waves to drive the paramecium forward (to the left). *C.* Outline of body form (cross sections) at different positions along its length.

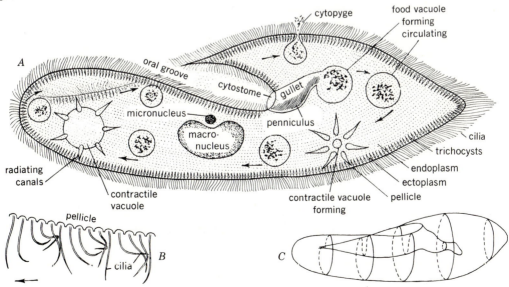

15-19 Structure The long cell body is blunt at the forward-moving or **anterior end,** widest behind the middle, and tapered at the **posterior end.** The exterior surface is covered by a distinct elastic membrane, the **pellicle,** with fine **cilia** arranged in lengthwise rows and of uniform length save for a posterior **caudal tuft** of longer cilia. Within the pellicle, the cell contents, as in *Amoeba,* consist of a thin, clear external layer of dense **ectoplasm** around the larger mass of more granular and fluid **endoplasm.** The ectoplasm contains many spindle-shaped **trichocysts,** alternating between the bases of the cilia, that may be discharged as long threads to serve perhaps in attachment or defense. From the anterior end a shallow furrow, or **oral groove,** extends diagonally back about halfway along the lower or **oral surface** and has the **cytostome** (cell mouth) at its posterior end. The cytostome opens into a short tubular **cytopharynx** (gullet), ending in the endoplasm. In the gullet cilia are fused to form two lengthwise dense bands (the **penniculus**). On one side, just behind the cytopharynx, is the **cytoproct** (cell anus), which can be seen only when particles are discharging through it. In the endoplasm are **food vacuoles** of various sizes that contain material undergoing digestion, and toward each end of the cell body is a large, clear **contractile vacuole** (water expulsion vacuole). The small rounded **micronucleus** is partly surrounded by the larger **macronucleus.** *P. aurelia* has two micronuclei.

Paramecium and similar ciliates have a complex system of organelles associated with the pellicle and cilia. The cilia arise individually or in groups from cuplike depressions which are surrounded by inflated areas of the pellicle termed **alveoli.** Below the alveoli, and alternating with them, are the trichocysts, which, in *Paramecium,* may be discharged in defense (Fig. 15-23). Also, each cilium connects beneath the alveolar layer to a basal granule, or **kinetosome.** All the kinetosomes in a given row are interconnected via longitudinal fibrils termed **kinetodesma.** For a given row the kinetosomes and the kinetodesma linking them together constitute a **kinety.** The kinety system appears to be common to all ciliates, even those which lack cilia during some stage in their life history. The kinety system is thought by some to coordinate ciliary action, but this has not been demonstrated in a convincing fashion.

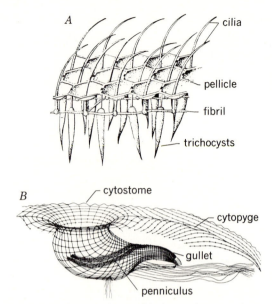

Figure 15-23 *Paramecium multimicronucleatum.* *A*. Structure of the pellicle and associated organelles. *B*. The fibrillar system in part of the cell body. (*After Lund,* 1935.)

Contractile fibrils (myonemes) occur in ciliates such as *Stentor* and *Vorticella* but not in *Paramecium.*

15-20 Locomotion The cilia beat backward to carry the paramecium forward in the water, and, as their stroke is oblique, it rotates on its longitudinal axis. The cilia in the oral groove beat more vigorously than the others, so that the anterior end swerves aborally. The combined effect is to move the organism forward in a spiral course, counterclockwise as viewed from behind (Fig. 15-24). Thus the asymmetrical organism may travel in a direct path. To swim backward the ciliary beat is reversed, as is the path of rotation. If, when swimming forward, the paramecium meets some unfavorable chemical stimulus, it executes an **avoiding reaction** (Fig. 15-25): the ciliary beat reverses, the paramecium moves backward a short distance, and then it rotates in a conical path by swerving the anterior end aborally while pivoting on the posterior tip. While doing this, cilia in the oral groove bring "samples" from the water immediately ahead; when these no longer contain the undesirable stimulus, the paramecium moves forward again. The reaction is similar upon

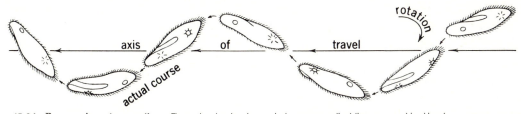

Figure 15-24 *Paramecium:* **locomotion.** The animal swims in a spiral manner so that the asymmetrical body progresses on a direct course. (*Adapted from Jennings, Behavior of the lower organisms, Columbia University Press.*)

Figure 15-25 Reactions of *Paramecium.* CONTACT. *A*. Negative, or avoiding, reaction. Successive positions (1–6) in avoiding a solid object. *B*. Weak reaction. Anterior end swings in small circle. *C*. Strong reaction. Swings in large circle. Positive reaction: *D*. Resting at food mass; arrows indicate water current produced by cilia. *E*. Resting on a cotton fiber.

CHEMICALS. Negative reaction: *F*. Drop of 0.5% NaCl solution introduced into a paramecium culture. *G*. Drop still empty 4 min later. Positive reaction: *H*. Path of an individual within a drop of acid. *I*. Individuals gather about a drop of 0.02% acetic acid. *J*. Culture 2 min after bubble of air and another of CO_2 are introduced; individuals gather densely about CO_2 (weak acid). *K*. Twenty minutes later; paramecia spread as CO_2 diffuses outward.

TEMPERATURE. *L, M, N*. The paramecia avoid extremes of cold and heat. (*After Jennings, Behavior of the lower organisms, Columbia University Press.*)

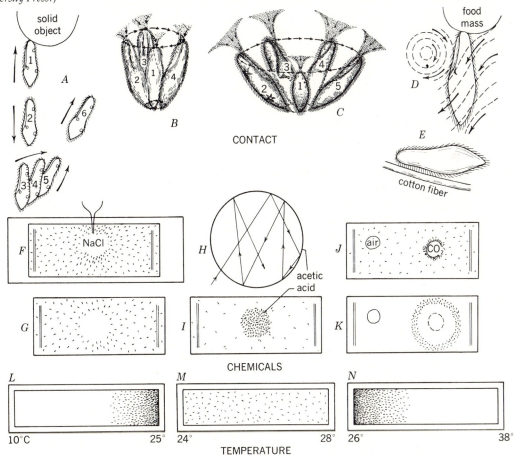

encountering a solid object: it reverses, rotates, and goes forward, repeating the process if necessary until a clear path is found.

15-21 Feeding and digestion *Paramecium* feeds on bacteria, small protozoans, algae, and yeasts. The constant beating of cilia in the oral groove sweeps a current of water containing food toward the cytostome, and movements of the penniculus gather the food at the posterior end of the cytopharynx in a watery vacuole. The vacuole reaches a certain size, constricts off, and begins to circulate in the endoplasm as a **food vacuole;** another then begins to form in its place. Streaming movements (cyclosis) in the endoplasm carry the vacuoles in a definite route, first posteriorly, then forward and aborally and again posteriorly to near the oral groove. The contents of the vacuoles are acid at first and gradually become alkaline, as shown by use of Congo red and other indicator dyes. As in the amoeba, the food is digested by the action of enzymes secreted by the endoplasm. This process continues until digested material is absorbed by the surrounding protoplasm and either stored or used for vital activity and growth. The vacuoles gradually become smaller, and any indigestible residues are egested at the cell anus.

P. caudatum will live and multiply in a bacteria-free suspension of liver extract, dead yeast, and kidney tissue that provides the necessary organic materials. *P. aurelia* requires certain salts and bacteria, but certain strains can be grown axenically. *P. bursaria,* which contains spherical green algae (zoochlorellae) in its endoplasm, will multiply in a solution of salts alone if kept in the light—this is a case of mutualism, where the organisms are interdependent. Carbon dioxide produced by the protozoan in respiration serves the algae with their chlorophyll to synthesize organic materials and produce oxygen, both of which are necessary for the paramecium.

15-22 Respiration and excretion As with amoebas, respiration in paramecia corresponds to the internal respiration of cells in multicellular animals. Oxygen dissolved in the surrounding water diffuses through the pellicle and then throughout the organism; carbon dioxide and organic wastes resulting from metabolism are probably excreted by diffusing outward in the reverse direction.

The contractile vacuoles (water expulsion vacuoles) regulate the water content of the body and may serve also in excretion of nitrogenous wastes such as urea and ammonia. Liquid within the cytoplasm is gathered by a series of 6 to 11 **radiating canals** that converge toward and discharge into each vacuole. The canals are most conspicuous as a vacuole is forming. When each vacuole has swelled to a certain size, it contracts and discharges to the exterior, probably through a pore. The vacuoles discharge alternately, at intervals of 10 to 20 s. If water containing a paramecium is densely filled with particles of carbon or carmine, the discharged contents from a vacuole will be evident momentarily as a clear spot in the surrounding clouded fluid before dispersed by action of the cilia.

The rate of vacuole discharge varies with temperature, is higher in an inactive paramecium than in one that is swimming about, and is higher in water with a scant supply of dissolved salts than with stronger concentrations. Since the body of a paramecium contains dissolved substances, water tends to pass through the pellicle, which functions as a semipermeable membrane. The contractile vacuoles offset this tendency and maintain an optimum concentration of water in the body protoplasm by disposing of the excess.

15-23 Behavior The responses of a paramecium to various kinds of stimuli are learned by study of its reactions and of the grouping or scattering of individuals in a culture. The response is **positive** if the paramecium moves toward a stimulus and **negative** when it moves away. To an adverse stimulus it continues to give the avoiding reaction until it escapes. All adjustments are made by **trial and error.** The intensity of the reaction may differ according to the kind and intensity of the stimulus. Experiments indicate that the anterior end of the animal is more sensitive than other parts.

To **contact,** the response is varied; if the anterior end is lightly touched with a fine point, a strong avoiding reaction occurs, as when a swimming paramecium collides with some object in the water, but if touched elsewhere there may be no response. A slow-moving individual often responds positively to contact with an object by coming to rest upon it; since food organisms are common about masses of algae or plant stems, this response is advantageous, and individuals often concentrate about such materials. Paramecia seek an optimum **temperature** of 24 to 28°C (75 to 84°F); in a temperature gradient they congregate in spots between these limits (Fig. 15-25). Greater heat stimulates rapid movement and avoiding reactions until they escape or are killed; in ice water they also seek to escape but may be benumbed and sink. To **gravity** there is generally a negative response, as seen in a deep culture in which many individuals gather close under the surface film with their anterior ends uppermost. In a gentle **water current** the paramecia will mostly align with the flow, their anterior ends upstream, and in a constant, or direct, **electric current,** they head toward the negative pole (cathode) from which electrons stream toward the positive pole (anode). To most **chemicals** the response is negative. If a drop of weak salt solution (0.5%) is introduced in a paramecium population on a microslide, they respond with the avoiding reaction and none enters the drop. To acids, however, the response is positive, even when the concentration is of sufficient strength to kill them. This latter phenomenon probably is related to the fact that although paramecia often live in alkaline waters, local concentrations of them produce sufficient carbon dioxide to render the immediate environment acid, and it is to this condition that they are most adapted. The trichocysts of paramecia (Fig. 15-26) and certain other ciliates have been thought to be organelles of defense against attack, but even when discharged they have seldom been observed to save the organism from its predator.

Although *Paramecium* is of almost universal occurrence in fresh waters, its means of dispersal is unknown. It has been reported to form cysts, but this observation lacks general confirmation.

15-24 Reproduction Paramecia reproduce by fission and also undergo several types of nuclear reorganization—conjugation, autogamy, and others. In **binary fission** (Fig. 10-1) the micronucleus divides by mitosis into two micronuclei that move toward opposite ends of the cell, and the macronucleus divides transversely by amitosis; a second gullet forms, two new contractile vacuoles appear, and then a transverse furrow divides the cytoplasm into two parts. The resulting two daughter paramecia are of equal size, each containing a set of cell organelles. They grow to full size before another division occurs. Fission requires about 2 h to complete and may occur one to four times per day, yielding 2 to 16 individuals. A single paramecium thus gives rise to 2, 4, 8, 16 . . . 2^n individuals; all those that result by fission (uniparental reproduction) from a single individual are known collectively as a **clone.** Upward of 600 "generations" per year may be produced. The rate of multiplication depends on external conditions of food, temperature, age of the culture, and population density; also on internal factors of heredity and physiology. If all descendants of one individual were to survive, they soon would equal the earth in volume. Stocks of *P. aurelia* have been maintained artificially for 25 years through 15,300 generations of fission without conjugation, by continually isolating daughter individuals in a fresh culture medium. In ordinary laboratory cultures with many individuals, the rate of fission gradually declines and some type of nuclear reorganization ensues.

15-25 Conjugation In paramecia and other ciliates there is, at intervals, a temporary union of individuals in pairs with mutual exchange of micronuclear materials that is known as **conjugation** (Fig. 15-27). The organisms become "sticky," adhering to one another by their oral surfaces, and a protoplasmic bridge forms between them. The pairs continue to swim about during this process. A sequence of nuclear changes then occurs in each. The macronucleus disintegrates and eventually disappears into the cytoplasm. The micronucleus enlarges, its chromatin forms long strands, a division spindle appears, and the micronucleus divides by meiosis; a second meiotic division of each follows at once, yielding four micronuclei. Three of the four micronuclei degenerate, and the remaining one divides by mitosis to yield two pronuclei in each organism. One from each then migrates through the protoplasmic

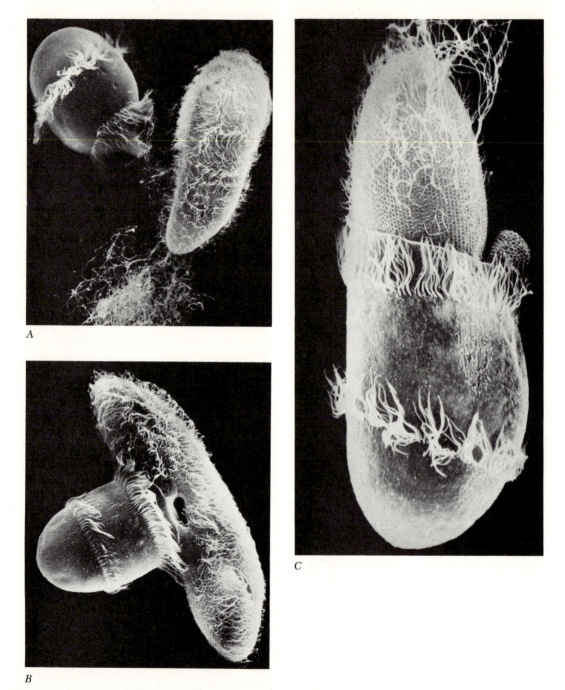

Figure 15-26 *Paramecium* **attacked and consumed by** *Didinium.* *A. Paramecium* discharging trichocysts, ×450; *B. Didinium* grasping *Paramecium,* ×350; *C. Didinium* engulfing *Paramecium,* ×850. Scanning electron micrographs. (*Courtesy of H. S. Wessenberg and G. A. Antipa. A, unpublished; B and C, 1970, J. Protozool., vol. 16.*)

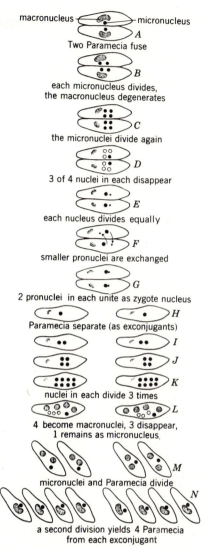

macronucleus — micronucleus

A

Two Paramecia fuse

B

each micronucleus divides,
the macronucleus degenerates

C

the micronuclei divide again

D

3 of 4 nuclei in each disappear

E

each nucleus divides equally

F

smaller pronuclei are exchanged

G

2 pronuclei in each unite as zygote nucleus

H

Paramecia separate (as exconjugants)

I

J

K

nuclei in each divide 3 times

L

4 become macronuclei, 3 disappear,
1 remains as micronucleus.

M

micronuclei and Paramecia divide

N

a second division yields 4 Paramecia
from each exconjugant

Figure 15-27 *Paramecium caudatum.* Diagram of conjugation.

bridge into the opposite individual, uniting there with the other pronucleus, which has remained in place. Soon thereafter the two paramecia separate as exconjugants. In each the fused micronucleus, or zygote, divides successively into two, four, and eight micronuclei. Four of these enlarge to become macronuclei, and three of the others degenerate. Then the micronuclei and organisms divide twice, so that from each exconjugant there are four paramecia, each with one macronucleus and one micronucleus.

Thenceforth the organisms multiply by binary fission as already described.

Conjugation differs from the sexual union of gametes in other PROTOZOA and in METAZOA, because progeny are not a direct product of the fusion; after conjugation each individual continues asexual fission. The net result, however, is like the fusion of gametes (syngamy) in animals. Conjugation is a process that provides for hereditary transfer, as the two exconjugants are genetically altered by their exchange of micronuclear materials (chromosomes). The nuclear changes bear comparison with those of meiosis and fertilization in METAZOA, with divisions (reduction) of the micronucleus, the joining (fertilization) of one micronucleus (egg pronucleus) by a migrant micronucleus (sperm pronucleus) to form a fusion micronucleus (zygote). The macronucleus which directs the vegetative processes in a paramecium has been compared to the body (soma) of the multicellular animal, and its disintegration to the death of the body. The fusion micronucleus (zygote) then produces another macronucleus (soma) and continues as the reproductive component (germ plasm) of the exconjugant.

15-26 Mating types For years no differences were detected between the two paramecia joining in conjugation. In 1937 Sonneborn reported stocks of *P. aurelia* containing two kinds ("sexes") of individuals; only members of opposite kinds would conjugate. These two categories were termed **mating types** I and II. By separating progeny of exconjugants that multiplied by binary fission, it was shown that some clones produced only type I, others only type II, and still others both types I and II. Sonneborn further showed that mating types were determined by a single Mendelian factor. Single-type clones were aa, and double types Aa or AA. Recent work has shown that mating-type determination in ciliates is not just a simple matter of Mendelian genetics. Cytoplasmic and environmental factors may also play a role.

15-27 Autogamy Nuclear reorganization resembling conjugation but occurring within one individual (hence one cell generation) is termed **autogamy**

(Diller, 1936). In *Paramecium aurelia* (Fig. 15-28), which has two micronuclei, these divide twice (prezygotic) to yield eight. The macronucleus becomes skeinlike and later breaks up. One of the eight micronuclei, as a gamete nucleus, enters a protoplasmic cone bulging near the cell mouth (the other seven micronuclei disintegrate), where it undergoes yet another division yielding two gamete nuclei. The two fuse as a synkaryon (zygote), which then divides twice (postzygotic). Of the four resulting nuclei, two continue as micronuclei and two become macronuclei. The cell and micronuclei then divide to yield two daughter paramecia, each with a (new) macronucleus and two micronuclei. Thereafter binary fission resumes.

Autogamy is a rhythmic process and one in which hereditary changes occur; lines homozygous as to mating type can be produced. There is decreased survival only when, as the result of a cross, lethal genes are received.

Other kinds of nuclear reorganization are **hemixis,** in which only the macronucleus divides, and **cytogamy,** which is like conjugation but without the mutual exchange of pronuclei.

15-28 Variation and heredity A wild stock of *Paramecium* in nature will contain individuals that differ as to total length and other features. Such variations are of three kinds: (1) age variation, (2) random variations due to environmental factors, and (3) inherited variations due to genetic factors. From a wild stock, Jennings isolated eight "races," or biotypes, that differed in average size. Within each race the individual and random variations produced organisms of various lengths, but the average of these was the same in successive generations by asexual fission because of their hereditary makeup.

DeGaris obtained conjugation between individuals of races that differed greatly in size, one (198 μm long) being twenty times the bulk of the other (73 μm long). After the exconjugants had separated, each produced a series of generations by fission. Those from the larger individual became successively smaller, and the opposite occurred with the smaller individual; after 24 days the offspring of both lines were alike. Before conjugation each race continued at its characteristic size; after conjugation the two individuals had like nuclear content but differed as to cytoplasm. The subsequent divisions show that for a period of days the size of descendants was affected by *both cytoplasm and nucleus,* but finally only by nuclear constitution. In other crosses the final size from both conjugants was not strictly intermediate but might be more nearly like either one or the other. Thus in these PROTOZOA, as in METAZOA, recombining the chromosomal materials produces different combinations in different cases.

In summary, paramecia show (1) structural specialization in organelles, (2) physiologic division of labor between parts to perform special functions, (3) a stereotyped avoiding reaction, (4) reproduction by binary fission, (5) nuclear reorganization by autogamy and other means, (6) conjugation as a mecha-

Figure 15-28 *Paramecium aurelia.* Diagram of autogamy. (*Adapted from Diller,* 1936.)

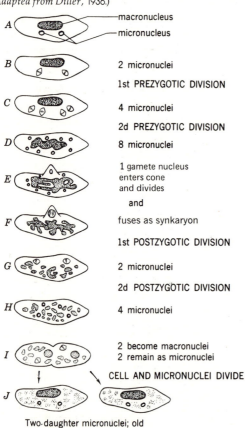

A — macronucleus — micronucleus

B — 2 micronuclei — 1st PREZYGOTIC DIVISION

C — 4 micronuclei — 2d PREZYGOTIC DIVISION

D — 8 micronuclei

E — 1 gamete nucleus enters cone and divides — and

F — fuses as synkaryon — 1st POSTZYGOTIC DIVISION

G — 2 micronuclei — 2d POSTZYGOTIC DIVISION

H — 4 micronuclei

I — 2 become macronuclei — 2 remain as micronuclei — CELL AND MICRONUCLEI DIVIDE

J — Two-daughter micronuclei; old macronucleus disintegrates

nism providing for exchange of hereditary factors, (7) segregation into many mating types that are mutually exclusive in conjugation, (8) pure lines that are genetically distinct in a wild population, and (9) the relative influence of nuclear and cytoplasmic materials in inheritance.

15-29 Other Ciliata Examination of a drop of water from any quiet freshwater, brackish water, or saltwater source will usually reveal one or more species of ciliates and testify to the wide occurrence of these protozoans (Fig. 15-29). Free-swimming species are the fastest moving of the protozoa and are commonly ellipsoidal to spherical in shape, whereas creeping forms are often flattened. Anteroposterior and oral-aboral relations in structure and locomotion are usually evident. The cilia on some types are reduced in numbers and grouped to form membranelles, or cirri, as in *Uronychia,* by which the ciliates move about. Some ciliates of both solitary (*Vorticella, Stentor*) and colonial habit (*Zoothamnium*) are attached by stalks or other means. The ciliates as a group exhibit tremendous variety in configuration, placement of cilia, form of the macronucleus, and other features. Most are preeminently free-living organisms.

The class Ciliata is divided into four subclasses. These are recognized primarily on the basis of ciliary structure and location on the body.

Members of the subclass Holotrichia, which includes *Paramecium,* are uniformly ciliated and have no adoral membranelles. They commonly encyst. This is the largest subclass, with 3500 described species. The species are diverse in form and either live in various kinds of water or are parasitic. *Balantidium coli* is a common ciliate parasite of the intestine in pigs that occurs rarely in humans. Human infection probably is acquired from cysts carried into the mouth by the hands or with food. It sometimes invades the intestinal wall and produces ulcers. This is the only pathogenic ciliate of humans.

The subclass Peritrichia is distinguished by a disklike oral region that is conspicuously ciliated, with few or no cilia elsewhere; the opposite end of the cell has a structure for temporary or permanent

Figure 15-29 Class Ciliata. Common freshwater ciliates; all enlarged to about same scale. Macronucleus, heavy stipple; contractile vacuole, clear.

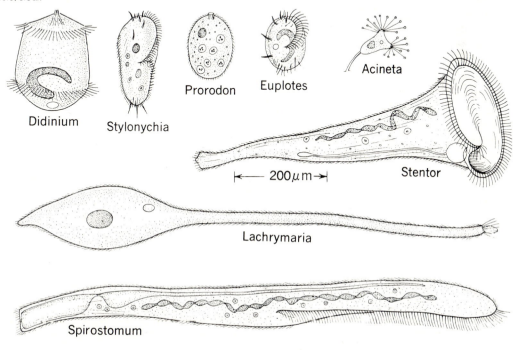

Didinium Stylonychia Prorodon Euplotes Acineta Stentor

|← 200 μm →|

Lachrymaria

Spirostomum

attachment to objects in the water. Some species are colonial, and some secrete a case, or lorica, into which the organism may withdraw. The common freshwater genus *Vorticella* is a representative.

Adults of the subclass SUCTORIA are sessile protozoans that have delicate protoplasmic "tentacles" but neither cilia nor a cytostome (Fig. 15-29, *Acineta*). Their young are ciliated and free-swimming. The cell body of suctorians is spherical, slender, or branched in different species, usually attached by a stalk or disk to some object and covered by a pellicle. Some of the tentacles are tipped with rounded knobs that act as suckers to catch and hold the small ciliates used for food. Other tentacles are pointed and pierce the prey to take up its soft parts, which stream through these tentacles into the cell body. Reproduction is by fission or budding; the young swim for a time and then settle to lose their cilia and transform into the adult stage. Various species inhabit fresh water, brackish water, and salt water and attach to inanimate objects, plants, and small aquatic animals; some are parasitic. *Podophrya* is a common, free-living, freshwater representative; *Dendrosoma* is branched and up to 2.5 mm tall; *Sphaerophrya* is spherical and parasitic in *Paramecium* and the ciliate *Stentor*; *Trichophrya micropteri* lives on the gills of bass; and *Allantosoma* attaches to ciliates in the colon of horses.

The subclass SPIROTRICHIA is characterized by highly developed cilia around the mouth. In addition to the many free-living species, some of peculiar structure inhabit parts of the digestive tract in herbivorous mammals—the caecum and colon of horses

and the rumen and reticulum (Sec. 34-20) of sheep, cattle, deer, and other ruminants (Fig. 15-30). They may number up to 500,000 or 1,000,000/cm³ of gut contents, so that a mature ox may contain 10 billion to 50 billion. These ciliates digest starches, fats, and proteins in the host's food. They probably are only messmates (commensals), without either benefit or harm to their hosts.

Ciliates are of considerable utility to human society in sewage processing, where they function to clarify water by flocculation. They also remove particles and bacteria from effluent, thus improving its quality.

Classification

Phylum Protozoa.

Protozoans. Single-celled or in colonies of like cells (no tissues), with more organelles than METAZOA; symmetry spherical, bilateral, or none; size usually microscopic; Precambrian to Recent, 50,000 species.

Subphylum A. Sarcomastigophora.

Locomotor organs are pseudopodia or flagella; nuclei of one kind.

Class 1. Mastigophora (Flagellata).

Flagellates. With one to many flagella for locomotion; fission binary; with or without pseudopodia.

Subclass 1. Phytomastigophorea. Usually with chromatophores; mostly free-living.

Figure 15-30 Class CILIATA. *Epidinium (Diplodinium) ecaudatum,* a complex ciliate from the rumen of cattle, as seen in median section. (*After Sharp,* 1914.)

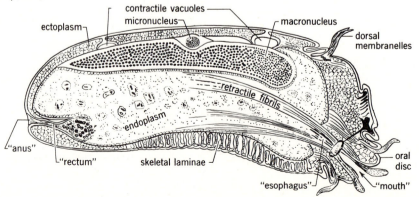

Order 1. Chrysomonadida. Small, solitary or colonial, often amoeboid; flagella 1 or 2 (or 3); chromatophores yellow, brown, or none; no gullet; nutrition holophytic or holozoic; produce endogenous cysts of silica. *Chromulina; Synura uvella* and *Uroglena americana,* both colonial, in fresh waters, causing bad flavor in water supplies.

Order 2. Silicoflagellida. Tiny; internal skeleton of siliceous material; flagella single or absent; marine. *Dictyocha, Distephanus.*

Order 3. Coccolithophorida. Small marine flagellates with an external covering of calcareous plates; 2 flagella. *Coccolithus, Rhabdosphaera.*

Order 4. Heterochlorida. Siliceous cyst walls; two unequal flagella. *Heterochloris, Myxochloris.*

Order 5. Cryptomonadida. Small, oval, usually flattened, not amoeboid; flagella 2; chromatophores 1 or 2 yellow-brown or none; usually with gullet; mostly holophytic; some live as symbionts in other Protozoa and in Metazoa. *Chilomonas,* in foul waters, saprophytic.

Order 6. Volvocida (Phytomonadida). Small, solitary or in swimming colonies; some enclosed in cellulose membrane; flagella 2 or 4 (or 8); no cytostome or gullet; chromatophores green or none; nutrition holophytic or saprophytic; reserve food is starch; mostly in fresh waters. Solitary forms: *Chlamydomonas nivalis,* on snow; *Haematococcus pluvialis,* appears suddenly, causing "red rain" in pools. Colonial forms (Volvocidae), in fresh waters: *Gonium,* as squares of 4 or 16 cells; *Pandorina,* globular, 16 cells; *Eudorina,* 32 cells near surface of gelatinous sphere; *Pleodorina,* 32, 64, or 128 cells in solid gelatinous ball; *Volvox,* to 15,000 cells on hollow gelatinous sphere, to 2 mm in diameter.

Order 7. Ebriida. Skeleton internal, siliceous; 2 flagella; chromoplasts absent; mostly fossil. *Ebria.*

Order 8. Euglenida. Form slender, usually definite; pellicle rigid or soft; flagella usually 1 or 2, in anterior gullet, near contractile vacuole; chromatophores green or none; nutrition holophytic, holozoic, or saprophytic; reserve food is paramylum; mostly in fresh waters. *Euglena,* some to 0.5 mm long; *Astasia,* colorless; *Trachelomonas,* in shell; *Copromonas,* in feces of toads and frogs.

Order 9. Chloromonadida. Flagella 2; chromatophores many if present; no stigma; reserve food stored as fat. *Gonyostomum.*

Order 10. Dinoflagellida. Flagella usually 2, in grooves, one in girdle around body, one trailing; body often of fixed form, in celluloselike armor of 2 to many plates; chromatophores many, brown, yellow, or green; many species important in marine plankton, some in fresh water, some parasitic. *Gymnodinium,* unarmored; *Peridinium* and *Ceratium,* both "armored," in salt or fresh waters; *Noctiluca,* spherical, to 2 mm in diameter, marine, luminescent; *Gonyaulax,* marine, producing "red tides"; *Blastodinium,* in intestine and eggs of copepods.

Subclass 2. Zoomastigophorea. No chromatophores; mostly parasitic or symbiotic.

Order 1. Choanoflagellida. One anterior flagellum surrounded basally by delicate collar; free-living. *Codosiga,* on stalks, and *Proterospongia* in gelatinous matrix, possibly related to sponges.

Order 2. Bicosoecida. Two flagella, one attaching posterior part of animal to a shell and the other free. *Salpingoeca, Poteriodendron.*

Order 3. Rhizomastigida. Flagella one or more; pseudopodia varying in number. *Mastigamoeba hylae,* in gut of tadpoles and frogs; *Histomonas meleagris,* cause of blackhead in turkeys and other fowls, transmitted by contact or by nematode (*Heterakis gallinae*).

Order 4. Kinetoplastida (Protomonadida). Flagella 1 or 2; body plastic but without pseudopodia; conspicuous DNA containing mitochondrial kinetoplast situated near flagellar base; in fresh or foul waters, free-living or parasitic, holozoic: *Bodo,* solitary. Parasitic or saprozoic: *Herpetomonas,* in gut of flies; *Proteromonas,* in gut of amphibians and reptiles; *Phytomonas,* in milky juices of milkweeds, euphorbias, etc., transmitted by plant-sucking bugs; *Trypanoplasma,* in blood of fishes, transmitted by leeches; *Trypanosoma,* narrowly leaf-like, in blood of vertebrates, mostly transmitted by arthropods (African sleeping sickness by tsetse flies; Chagas' disease by triatomine bugs in South America); *Crithidia,* in gut of mosquitoes.

Order 5. Retortamonadida. Flagella 1 to 4, one near ventral cytostomal area, turned backward; parasitic. *Chilomastix,* mainly in gut of vertebrates.

Order 6. Diplomonadida. Bilaterally symmetrical, 4 pairs of flagella; mostly parasitic. *Giardia,* in intestines of vertebrates, common in humans, transmitted by cysts in feces.

Order 7. Oxymonadida.   Flagella 4 or more, typically in 2 pairs, 1 or more turned posteriorly; all parasitic. *Oxymonas, Pyrsonympha,* commensal in gut of termites.

Order 8. Trichomonadida.   Flagella typically 4 to 6, 1 bent backward and often associated with undulating membrane; all parasitic. *Trichomonas,* in digestive and reproductive tracts of humans and other animals. *T. foetus,* causing abortion in cows; *Devescovina,* in gut of termites.

Order 9. Hypermastigida.   Structure complex, many flagella and parabasal bodies; holozoic, mostly ingest wood; symbiotic in gut of termites and cockroaches and necessary for digestion of cellulose by these insects. *Trichonympha,* in termites and wood roaches; *Lophomonas,* in cockroaches; *Staurojoenina,* in termites.

Class 2. Opalinata (Protociliata).

Order Opalinida.   Many cilialike organelles in oblique rows over body surface; cytostome absent; nuclei two to many, sexual reproduction via flagellated gametes; parasitic. *Opalina, Cepedia,* in intestine of amphibians.

Class 3. Sarcodina (Rhizopoda).

With pseudopodia for locomotion and food capture; often with external or internal skeletons; mostly free-living.

Subclass 1. Rhizopoda.   Protoplasmic processes various but no central filament.

Order 1. Amoebida (Lobosa).   Pseudopodia short, lobose, changing; naked, in thin membrane; fresh water and other waters or parasitic. *Amoeba; Chaos (Pelomyxa),* to 5 mm long; *Naegleria,* with flagellate stage; *Hartmannella,* in foul waters, feeds on bacteria; *Entamoeba histolytica,* parasitic in human intestine and tissues, causing amoebic dysentery; *E. gingivalis,* in human mouth, and *E. coli,* in intestine of humans, apes, and monkeys, both harmless; others in frogs, cockroaches, etc.

Order 2. Arcellinida (Testacea).   Pseudopodia lobose, extruded through definite aperture in one-chambered shell of silicon or $CaCO_3$; mostly in fresh waters. *Arcella,* with bowllike shell of siliceous prisms set in tectin; *Difflugia,* shell of sand grains over organic base.

Order 3. Aconchulinida.   Tapering, branched filipodia; naked. *Penardia.*

Order 4. Gromiida.   Protoplasmic processes tapering and branching; shell with a distinct aperture. *Euglypha,* shell of siliceous scales and spines.

Order 5. Athalamida.   Reticulopoda arising from any position, delicate. *Biomyxa.*

Order 6. Foraminiferida.   Protoplasmic processes slender, branched, sticky; with simple or chambered shell (test) of calcareous, chitinous, or foreign materials with one or many openings; shell 0.01 to 5 mm (some 190 mm) in diameter; reproduce asexually by multiple fission and sexually; mostly marine and bottom dwellers, some pelagic. Cambrian to Recent, 18,000 species living and fossil. *Allogromia,* no shell, in fresh waters; *Elphidium; Globigerina,* pelagic, its empty shells (in globigerina ooze) cover about one-third of the ocean floor at depths of 2500 to 4500 m (8000 to 15,000 ft); †*Camerina* (†*Nummulites*), early Tertiary, European limestone.

Order 7. Mycetozoa.   Slime molds. Adult phase a sheet of multinucleate protoplasm with streaming movements; feed on decaying wood or leaves or live fungi. *Ceratiomyxa,* on wood, masses to several feet long; *Badhamia,* on fungi.

Order 8. Labyrinthulida.   Pseudopodia not obvious; locomotion by gliding on mucous track which forms a network; body spindle-shaped. *Labyrinthula,* on marine plants.

Subclass 2. Actinopoda.   Pseudopodia delicate and radiating, each with central filament.

Order 1. Radiolaria.   Radiolarians. Protoplasm divided into inner and outer parts by porous spherical capsule of chitin perforated by 1, 3, or many pores; skeleton or spicules of silica; diameter microscopic to 6 cm (colonies); marine, pelagic, surface to depths of 4500 m (15,000 ft). Precambrian to Recent. *Aulacantha,* solitary; *Sphaerozoum,* colonial.

Order 2. Acantharia.   Central capsule with thin membrane without pores; skeleton of radiating spines of strontium sulfate. *Acanthometra.*

Order 3. Heliozoa.   Sun animalcules. No central capsule; skeleton of siliceous scales and spines; chiefly in fresh waters. *Actinophrys; Clathrulina,* in latticed sphere of silica.

Order 4. Proteomyxa.   Without shell; protoplasmic processes radiating, branched, tending to fuse; primarily parasites of aquatic plants; *Vampyrella,* punctures cells of algae and sucks the contents.

Subphylum B. Sporozoa.

Sporozoans. No locomotor organs or contractile vacuoles; reproduction by multiple asexual fission and sexual phases, usually producing spores which lack polar filaments; all internal parasites, usually with intracellular stages.

Class 1. Telosporea.

Sporozoites elongate; no polar capsules in spores.

Subclass 1. Gregarinia. Mature trophozoite, large and wormlike, 10 μm to 16 mm long, extracellular; zygote producing one-walled spore containing eight sporozoites; in digestive, coelomic, and other cavities of invertebrates. *Ophryocystis,* in Malpighian tubules of beetles; *Monocystis,* in sperm balls of earthworms; *Gregarina,* in grasshoppers, mealworms, etc.

Subclass 2. Coccidia. Zygote nonmotile; spores with one to many walls; with alternate asexual schizogony followed by sporogony; small and intracellular, chiefly in epithelial tissues of mollusks, annelids, arthropods, and vertebrates.

Suborder 1. Eimeriina. Zygote nonmotile, sporozoites enclosed in sporocyst, life cycle complete in a single host in which they inhabit tissue cells. *Toxoplasma,* in mammals including humans; *Eimeria (Coccidium),* chiefly in digestive epithelium of arthropods and vertebrates, especially domestic birds and mammals; *E. stiedae,* cause of coccidiosis in domestic rabbits.

Suborder 2. Haemosporiina. Zygote motile, producing naked sporozoites; reproduction alternately by schizogony, within red blood cells and blood system in vertebrates and by sporogony in bloodsucking intermediate arthropod host. *Plasmodium,* in mosquitoes (*Anopheles, Culex*), cause of malaria in birds, mammals, and humans; *Haemoproteus,* in bloodsucking flies (HIPPOBOSCIDAE), transmitted to birds and reptiles; *Leucocytozoon,* in black flies (*Simulium*), causes disease in ducks; *Theileria parva,* infects cattle in Africa, causing East Coast fever.

Class 2. Piroplasmea.

Order Piroplasmida. No flagella or cilia; locomotion by body flexion or gliding. *Babesia bigemina,* parasitic in red blood cells of cattle, transmitted by ticks (*Boöphilus annulatus*), causes Texas cattle fever.

Subphylum C. Cnidospora.

Spore with 1 to 4 polar filaments for attaching to host.

Class 1. Myxosporea.

Spores originate from several nuclei.

Order 1. Myxosporida. Spore large, bivalve; 1 to 4 polar filaments; parasitic in cavities and tissues of lower vertebrates, especially fishes, where heavy infections cause severe losses. *Sphaeromyxa, Myxidium.*

Order 2. Actinomyxida. Spore with 3 valves, 3 polar filaments. In gut or coelom of aquatic annelids. *Triactinomyxon.*

Order 3. Helicosporida. Spore barrel-shaped, one coiled filament. *Heliosporidium,* in fly and mite larvae.

Class 2. Microsporea.

Order 1. Microsporida. Spore small, 1 or 2 polar filaments. Intracellular in arthropods and fishes. *Nosema bombycis,* causes pébrine disease in silkworms; *N. apis,* causes nosema disease in honeybees.

Order 2. Haplosporida. Spores few in cyst, small. *Haplosporidium,* chiefly in annelids.

Subphylum D. Ciliophora.

Class Ciliata.

Ciliates. With cilia or ciliary organelles in at least one stage of life cycle; nuclei of two kinds.

Subclass 1. Holotrichia. Cilia simple, over all or part of cell body; adoral cilia usually absent or inconspicuous.

Order 1. Gymnostomatida. No adoral cilia. *Didinium,* attacks other ciliates, especially *Paramecium.*

Order 2. Trichostomatida. Vestibular but no buccal cilia. *Balantidium,* in intestines of various animals; *Isotricha,* commensal in digestive tract of cattle.

Order 3. Chonotrichida. No cilia on surface of mature forms; body vase-shaped; marine, attached to crustaceans by stalk. *Spirochona,* on gills of amphipods.

Order 4. Apostomatida. A rosette near small cytostome; cilia spirally arranged on body. *Chromidina,* with polymorphic life cycle in marine invertebrates.

Order 5. Astomatida. Cilia uniform over body; no mouth; mostly parasitic in annelids; *Anoplophrya*, in invertebrates.

Order 6. Hymenostomatida. Body cilia uniform; mouth ventral, mouth cilia in the form of membranelles. *Paramecium*, free-living; *Tetrahymena.*

Order 7. Thigmotrichida. With anterior tuft of cilia. In mantle cavity of bivalve mollusks. *Ancistrum.*

Subclass 2. Suctoria.

Order Suctorida.
Cilia only in young stages; adults attached by stalk; tentacles for feeding; no cytostome; mostly sessile; in salt or fresh waters. *Acineta,* free-living; *Sphaerophrya,* spherical, parasitic in *Paramecium* and *Stentor; Trichophrya micropteri,* on gills of bass.

Subclass 3. Peritrichia.

Order Peritrichida.
Adoral row of cilia beginning to left of peristome and passing around to right; few or no other cilia; body typically bell- or vaselike; mostly sessile, often colonial. *Vorticella,* solitary on stalk; *Epistylis, Zoothamnium, Opercularia,* in colonies; *Trichodina,* on aquatic animals.

Subclass 4. Spirotrichia. Cilia generally sparse and compound; adoral cilia conspicuous, of many membranelles, beginning at right of peristome and passing around to left.

Order 1. Heterotrichida. Body cilia, when present, short, uniform. *Stentor,* trumpet-shaped, feeds on small ciliates; *Spirostomum,* slender, to 3 mm long.

Order 2. Oligotrichida. Body cilia sparse or absent; mouth membranelles prominent. *Strobilidium,* marine.

Order 3. Tintinnida. Membranelles feathery; body with vase-shaped case (lorica). *Tintinnidium,* marine and pelagic.

Order 4. Entodiniomorphida. No simple body cilia; oral membranelles restricted. *Epidinium (Diplodinium), Ophryoscolex,* of complex structure, commensal in digestive tract of ruminants.

Order 5. Odontostomatida. Oral cilia of 8 membranelles; body cilia sparse. *Saprodinium,* polysaprobic.

Order 6. Hypotrichida. Body cilia compound, confined to tufts on the ventral side of body; body flattened dorsoventrally; freshwater and marine forms. *Stylonychia* and *Oxytricha,* with cirri restricted to a few ventral groups and a marginal row; *Euplotes,* with marginal row absent.

References

Adam, K. M. G., J. Paul, and **V. Zaman.** 1971. Medical and veterinary protozoology. Edinburgh, Churchill Livingstone. viii + 200 pp., 186 figs.

Baker, J. R. 1969. Parasitic Protozoa. London, Hutchinson & Co. (Publishers), Ltd. 176 pp., illus.

Chandler, A. C., and **C. P. Read.** 1961. Introduction to parasitology. 10th ed. New York, John Wiley & Sons, Inc. PROTOZOA, *pp. 37–212, figs. 1–45, especially human parasites.*

Chen, T. (editor). 1967–1972. Research in protozoology. New York, Pergamon Press. 1967. Vol. 1, vii + 428 pp., illus. 1967. Vol. 2, vii + 399 pp., illus. 1969. Vol. 3, vii + 744 pp., illus. Vol. 4, 413 pp., illus.

Corliss, J. O. 1978. The ciliated Protozoa: Characterization, classification, and guide to the literature. 2d ed. New York, Pergamon Press. 500 pp., 22 pls.

Curtis, H. 1968. The marvelous animals: An introduction to the Protozoa. Garden City, N.Y., Natural History Press. xvi + 189 pp., 57 figs.

Cushman, J. A. 1948. Foraminifera, their classification and economic use. 4th ed. Cambridge, Mass., Harvard University Press. viii + 605 pp., 55 pls.

Doflein, F., and **E. Reichenow.** 1949–1953. Lehrbuch der Protozoenkunde. 6th ed. Jena, Germany, Gustav Fischer Verlagsbuchhandlung. viii + 1213 pp., 1151 figs.

Dogiel, V. A. 1965. General protozoology. New York, Oxford University Press. xiv + 747 pp., 316 figs.

Edmondson, W. T. (editor). 1959. Fresh water biology. 2d ed. New York, John Wiley & Sons. xiii + 1248 pp. *Keys to free-living protozoans.*

Faust, E. C., P. F. Russell, and **R. C. Jung.** 1970. Craig and Faust's Clinical parasitology. 8th ed. Philadelphia, Lea & Febiger. PROTOZOA, pp. 75–311, figs. 2–87.

Grassé, Pierre-P., and others. 1952–1953. Traité de zoologie. Paris, Masson & Cie. Vol. 1, Protozoaires (except ciliates), pt. 1, xii + 1071 pp., 829 figs.; pt. 2, 4 + 1060 pp., 833 figs.

Grell, K. G. 1973. Protozoology. New York, Springer-Verlag. viii + 554 pp., 437 figs.

Hall, R. P. 1953. Protozoology. Englewood Cliffs, N.J., Prentice-Hall, Inc. 6 + 682 pp., illus.

————. 1964. Protozoa. New York, Holt, Rinehart and Winston, Inc. 124 pp., illus.

Hammond, D. M., and **P. L. Long** (editors). 1973. The Coccidia. Baltimore, University Park Press. x + 482 pp., illus.

Honigberg, B. M., and others. 1964. A revised classification of the Phylum Protozoa. *Journal of Protozoology,* Vol. 11, pp. 7–20. *By a committee of the Society of Protozoologists.*

Hutner, S. H., and **Andre Lwoff.** 1951–1964. Biochemistry and physiology of Protozoa. New York, Academic Press, Inc. 3 vols.

Hyman, Libbie H. 1940. The invertebrates. New York, McGraw-Hill Book Company. Vol. 1, Protozoa through Ctenophora. pp. 44–232, figs. 6–67.

Jahn, T. L., and **F. F. Jahn.** 1949. How to know the Protozoa. Dubuque, Iowa, William C. Brown Company Publishers. 6 + 234 pp., 394 figs.

Jennings, H. S. 1906. Behavior of the lower organisms. New York, Columbia University Press. xiv + 366 pp., 144 figs.

Jeon, K. W. (editor). 1973. The biology of Amoeba. New York, Academic Press, Inc. xviii + 628 pp., illus.

Jones, A. R. 1974. The ciliates. New York, St. Martin's Press, Inc. 207 pp., illus.

Jurand, A., and **G. G. Selman.** 1969. The anatomy of *Paramecium aurelia.* London, Macmillan & Co., Ltd. xiv + 218 pp., illus.

Kudo. R. R. 1966. Protozoology. 5th ed. Springfield, Ill., Charles C Thomas, Publisher. xi + 1174 pp., 388 figs. *General account and classification.*

Leedale, G. F. 1967. Euglenoid flagellates. Englewood Cliffs, N.J. Prentice-Hall, Inc. xiii + 242 pp., 176 figs.

Levine, N. D. 1973. Protozoan parasites of domestic animals and of man. 2d ed. Minneapolis, Burgess Publishing Co. x + 406 pp., illus.

MacKinnon, D. L., and **R. S. J. Hawes.** 1961. An introduction to the study of Protozoa. New York, Oxford University Press. xix + 506 pp., 180 figs. *Systematic, bibliography.*

Manwell, R. D. 1961. Introduction to Protozoology. New York, St Martin's Press, Inc. xii + 642 pp., illus. *General and by groups, glossary.*

Moore, R. C. (editor). 1954–1964. Treatise on invertebrate paleontology. Lawrence, University of Kansas Press. Protista 2, 1964. Vol. 1, xxxi + 510 pp., 399 figs. Vol. 2, ii + 390 pp., 254 figs. Protista 3, 1954. xii + 195 pp., 92 figs.

Murray, J. W. 1973. Distribution and ecology of living benthic foraminiferids. New York, Crane, Russak & Co. 274 pp.

Pitelka, D. R. 1963. Electron-microscopic structure of Protozoa. New York, Pergamon Press. 269 pp., illus.

Sleigh, M. A. 1973. The biology of Protozoa. New York, American Elsevier Publishing Company, Inc. viii + 315 pp., illus.

Tartar, V. 1961. The biology of *Stentor.* New York, Pergamon Press. 413 pp., illus.

Westphal, A. 1976. Protozoa. Glasgow, Blackie & Son, Ltd. 336 pp., illus.

Wichterman, Ralph. 1953. The biology of Paramecium. New York, McGraw-Hill Book Company. xvi + 527 pp., 141 figs.

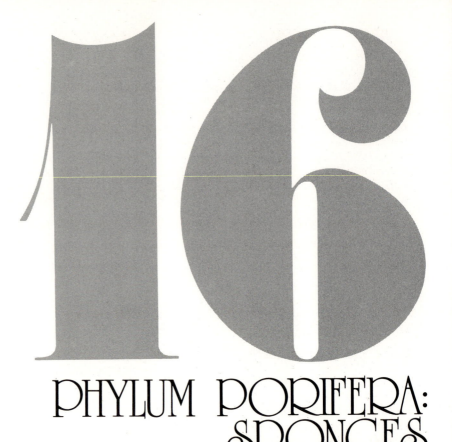

16

PHYLUM PORIFERA: SPONGES

The sponges are lowly multicellular animals, incapable of movement, that resemble various plants in appearance (Figs. 16-1, 16-2). Different species are thin flat crusts, vaselike, branched, globular, or varied in form, and from 1 mm to 2 m in diameter. Many are gray or drab-colored, others are brilliant red, orange, yellow, blue, violet, or black. Most sponges are marine, from Arctic to tropical seas, living from the low-tide line to depths of 5.5 km (3½ mi); two families are widespread in fresh waters. All are sessile, attached to rocks, shells, and other solid objects in the water. The name PORIFERA (L. *porus*, pore + *ferre*, to bear) refers to the porous structure of the body, with many surface openings.

16-1 Characteristics

1 Symmetry radial or none; multicellular; cells imperfectly arranged as tissues with mesenchyme between.

2 Body with many pores, canals, or chambers through which water flows.

3 Some or all interior surfaces lined with choanocytes (flagellated collar cells).

4 No organs, movable parts, or appendages; digestion intracellular.

5 Usually with an internal skeleton of separate crystalline spicules or of irregular organic fibers, or both.

6   Reproduction asexual by buds or gemmules; also sexual by eggs and sperm; larva ciliated, free-swimming.

7   Development of larvae unique, with external flagellated cells moving inside and internal cells outward.

Sponges are an enigmatic group in that although they are multicellular animals they lack true tissues or organ systems. The cells further show a certain amount of independence, with no overall coordination via a functioning nervous system; and in no other metazoan is the principal opening exhalent, as the osculum is in the sponge. For these and other reasons, sponges are usually set apart from the main line of invertebrate evolution in the Subkingdom or Branch PARAZOA.

Sponges were once thought to be plants. Their animal nature was learned in 1765, but their place in the Animal Kingdom was uncertain until about 1857. They resemble some colonial flagellate protozoans (*Proterospongia*) in having groups of flagellated collar cells and intracellular digestion but differ in having greater cell arrangement, limited cellular division of labor, and a body with many pores.

16-2 Structure of a simple sponge One of the simple sponges is *Leucosolenia* (Fig. 16-3), which attaches to seashore rocks just below the low-tide line. It consists of a small group of vaselike, slender upright tubes united at their bases by irregular horizontal tubes. Each upright portion is a thin-walled sac, enclosing a central cavity, the **spongocoel,** with one large opening, or **osculum,** at the summit. The wall is made up of (1) an outer **pinacoderm** of thin, flat cells and (2) a continuous inner lining of flagellated collar cells, or **choanocytes,** in loose contact with one another; between these two layers is (3) a gelatinous **mesenchyme,** which contains (4) free cells, or **amoebocytes,** of several kinds and (5) many minute crystallike **spicules** of calcium carbonate supporting the soft wall. Some spicules are slender rods (monaxon), and others are three- or four-rayed (triaxon or tetraxon). The wall is pierced by many minute incurrent openings, or **pores** (ostia), that extend from the external surface to the central cavity, each pore being a canal through a tubular cell, or **porocyte,** of the pinacoderm.

The sponge cannot move, and contraction is slight at best, but the porocytes can open or close; in some sponges the osculum can close slowly. In life

Figure 16-1 Phylum Porifera. Class CALCAREA: *Scypha* (formerly called *Sycon*). Class HEXACTINELLIDA: *Regadrella*, glass sponge. Class DEMOSPONGIAE: *Poterion*, Neptune's goblet; *Spongia*, bath sponge; *Microciona*; *Haliclona*, encrusting sponge. (*Regadrella and Scypha after Lankester, Treatise on zoology, A. & C. Black. Ltd.*)

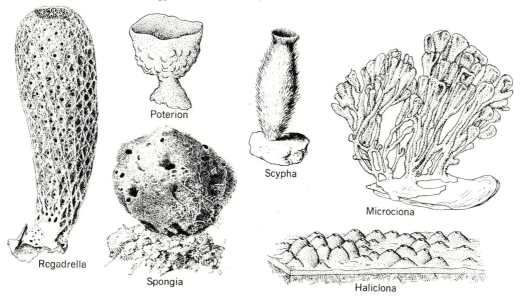

Poterion

Scypha

Microciona

Regadrella

Spongia

Haliclona

Figure 16-2 Sponges. *A. Leucilla nuttingi,* Class CALCAREA. *B. Adocia gellindra,* Class DEMOSPONGIAE.

the amoebocytes move freely within the mesenchyme, which is soft, colloidal, or even watery in consistency. Each choanocyte is a rounded or oval cell, resting against the mesenchyme, its free end with a transparent contractile collar encircling the base of a single whiplike flagellum. The flagella beat back and forth, producing a continuous flow of water into the many pores, through the central cavity, and out the single osculum. The water brings oxygen and food and removes wastes. A sponge only 10 cm (4 in) tall is estimated to pass 95 liter (25 gal) of water daily through its body. The food is plankton—micro-

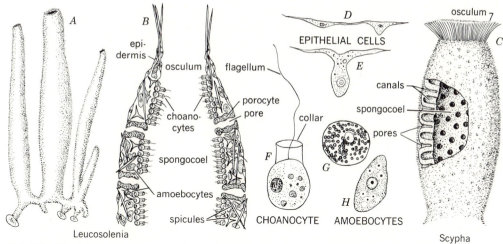

Figure 16-3 Structure of simple sponges. *A. Leucosolenia,* a small colony. *B. Leucosolenia,* enlarged section at top of body. *C. Scypha,* entire individual with part of body wall cut away (compare Fig. 16-4). *D–H.* Cells of sponges. (*Adapted from Hyman.*)

scopic animals and plants—and bits of organic matter. The choanocytes are responsible for food capture. Electron microscope studies of the collar cells have shown how food is captured. The "collar" is not a solid structure but rather an upright set of closely spaced rods, much like a picket fence. When the flagellum creates a water current, water passes through the "fence," but any small particles which are larger than the space between the "pickets" are caught and fall down to the base, where they are ingested by the collar-cell body. The collar cell may pass food on to amoebocytes for transport to other areas and for final digestion. Indigestible material is passed into the water current by choanocytes. There are no obvious sensory or nerve cells, but sponges do react slightly to touch, especially around the osculum, and stimuli are conducted slowly, presumably from cell to cell.

16-3 Other sponges A somewhat more complex type is *Scypha* (formerly called *Sycon* or *Grantia*), a slender form rarely over 2.5 cm (1 in) tall, with one tubular central cavity and a single osculum (Figs. 16-3, 16-4). The thick body wall is folded to form many short horizontal canals. The **incurrent canals** open from the exterior through small pores but end blindly toward the inner surface, whereas the **radial canals** begin blindly near the outer surface and open

by minute **apopyles** into the spongocoel. Other smaller canals, or **prosopyles,** connect the incurrent and radial canals. The exterior surface is covered by thin **dermal pinacoderm,** the spongocoel is lined with a thin **gastral epithelium,** and the radial canals are lined with **choanocytes.** The canal systems greatly increase the surface exposed to the water. The body substance between these layers and outside the canals is of gelatinous **mesenchyme** (the mesohyal layer) containing amoeboid cells as in *Leucosolenia.* The spicules supporting the body are chiefly of three kinds: long straight (monaxon) about the osculum, short straight (monaxon) surrounding the ostia, and Y-shaped (triaxon) lining the spongocoel and in the body wall. Those about the osculum and incurrent pores protrude beyond the body surface, producing a bristly appearance.

16-4 Canal Systems In most sponges the structure is complex, with various kinds of canal systems (Fig. 16-5). (1) The simple **ascon** type (*Leucosolenia*) has a thin body wall perforated by short straight pores leading through **porocytes** directly to the spongocoel, which is lined with choanocytes. (2) The **sycon** type (*Scypha*) contains two types of canals, but only the radial canals are lined with choanocytes. (3) The **leucon** type has a body of thick, dense mesen-

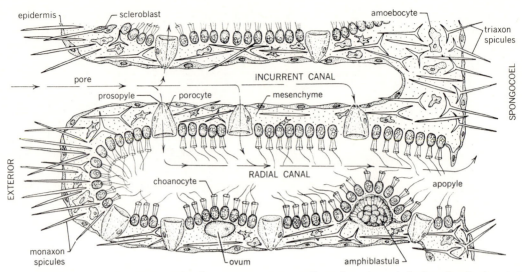

Figure 16-4 Structure of the body wall of a *Scypha* sponge; diagrammatic section. Arrows indicate paths of water movement.

chyme penetrated by complexly branched canal systems, with choanocytes restricted to small, spherical flagellated chambers.

16-5 Skeleton The soft bodies of sponges are supported by many minute crystalline spicules or by organic fibers, forming a "skeleton" (Fig. 16-6). Calcareous sponges such as *Leucosolenia* and *Scypha* have spicules of calcium carbonate ($CaCO_3$); those in "glass" sponges are of siliceous material (chiefly $H_2Si_3O_7$). Spicules are of many kinds, sizes, and shapes; some simple forms have been mentioned,

but others are of complex types. Spicules of some deep-sea glass sponges become fused together in a framework. The bath sponges and some other horny sponges contain fine interjoined and irregular fibers of spongin. Like hair, nails, and feathers, this is a sulfur-containing scleroprotein, insoluble, inert chemically, and resistant to protein-digesting enzymes. Spicules are secreted by special mesenchyme cells (scleroblasts) and spongin by others (spongioblasts). A monaxon spicule begins within a binucleate scleroblast as an organic axial thread around which calcium carbonate is deposited, the one cell dividing into two as the process continues (Fig.

Figure 16-5 Canal systems of sponges, diagrammatic sections. Epithelium, light lines; mesenchyme, stippled; areas with collar cells, heavy black; arrows indicate water currents. *Upper right,* portion of an encrusting sponge; *lower right,* freshwater sponge.

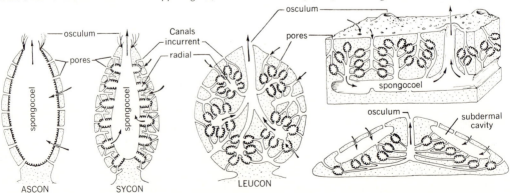

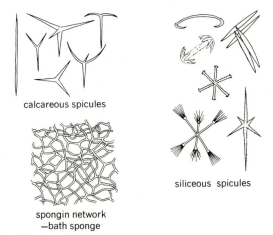

calcareous spicules

siliceous spicules

spongin network
—bath sponge

Figure 16-16 Spicules and fibers of sponges. (*After Hyman, The invertebrates.*)

of sponges is based importantly on the kinds and arrangement of these skeletal materials.

16-6 Histology The outer pinacoderm and the lining of the central cavity in complex sponges are both of thin, flat cells called **pinacocytes.** The choanocytes form a loose layer of cells wherever they occur. These are the organized cell layers of sponges. Amoeboid cells (amoebocytes) in the mesenchyme have many functions. They serve to store and transport food, remove excretory materials, produce spicules and/or spongin fibers, and give rise to the reproductive cells (ova, sperm, gemmules). They also have the ability to transform into pinacocytes (Fig. 16-8). With only choanocytes and amoebocytes it is thus possible to produce all the other cell types found in a sponge. Recognizable types of amoeboid cells in the mesenchyme include (1) **scleroblasts;** (2) **spongioblasts;** (3) **collencytes,** or connective tissue cells of stellate form with threadlike pseudopodia; (4) **myocytes,** or long contractile cells that form sphincters around pores and the osculum; and (5) **archeocytes** of amoeboid

16-7). One cell is called the **founder cell** and moves along, determining the length of the spicule; the other, called the **thickener,** follows the founder, depositing calcium carbonate. Both wander off when the spicule is complete. A complex spicule is formed by the cooperation of several cells. The classification

Figure 16-7 Spicule formation in sponges. *A–C.* Early stages in secretion of a monaxon spicule. *D.* Later stage of secretion of a Triaxon spicule. (*After Hyman, 1940.*)

A

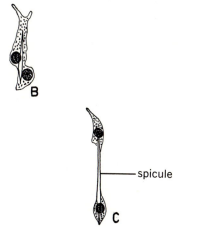

B

spicule

C

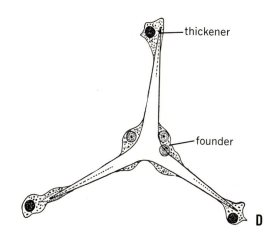

thickener

founder

D

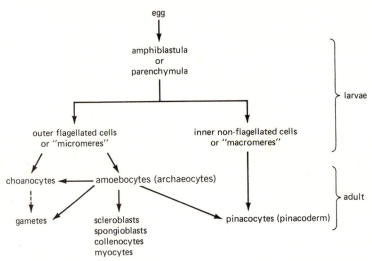

Figure 16-8 Derivation of the main cell types in sponges. Note the central position of the amoebocytes.

character and various functions. These last have large nuclei with nucleoli. Recent work on cell differentiation in sponges suggests that archeocytes are also the precursors of all other amoeboid cells in sponges, including scleroblasts, spongioblasts, collencytes, and myocytes (Fig. 16-8). Sponges thus seem to have retained a plasticity of cell organization not found in most other metazoan organisms after their developmental period.

16-7 Individuality in sponges Some cells in sponges are arranged in layers but much of the vital activity is carried out by separate cells. Sponges are thus cell aggregates without organs, and it is difficult or impossible to decide what is an "individual." Examples of *Leucosolenia* or *Scypha* each with a single osculum may be so designated, but in complex sponges with multiple canal systems and many oscula it is impossible to define the limits of an "individual."

16-8 Physiology The functioning of a sponge depends largely on the constant flow of water through the body. The water furnishes a constant source of oxygen for respiration, and metabolic wastes are removed from the sponge by the same system. Food is removed from the water and gametes

and larvae dispersed out of the sponge via the water current. The flow of water is created by the flagella of the choanocytes but is not coordinated by a nervous system. The flow is apparently regulated only by a limited reduction or enlargement of the osculum by the myocytes surrounding it.

Because of their dependence on water movement, sponges in general are intolerant of stagnant water. Oscula in sponges are often raised above the level of the rest of the sponge body in order to direct the excurrent water away from the incurrent ostia on the body surface. This is an adaptation to prevent recirculating the water which has already had the food and oxygen removed and metabolic wastes added in a previous passage through the sponge.

16-9 Reproduction Sponges multiply both asexually and sexually. Parts of a sponge lost by injury will be replaced by **regeneration.** If certain species of sponges (*Microciona prolifera, Ophlitospongia seriata, Suberites ficus*) are squeezed through fine meshed silk so that the cells are separated, they will come together by amoeboid movements, unite, and reaggregate into a sponge like the original. Many kinds of sponges increase commonly by **budding,** the buds either separating from the original sponge as growth proceeds or else remaining attached and thus increasing the number of parts or the bulk of the mass.

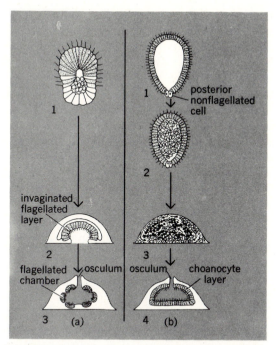

Figure 16-9 Larval structure and metamorphosis in calcareous sponges. (*a*) Developmental stages characteristic of *Sycon*: 1, amphiblastula larva; 2, newly settled larva showing metamorphosis; 3, young syconoid-stage. (*b*) Developmental stages characteristic of *Esprella*: 1, coeloblastula larvae; 2, parenchymula larvae; 3, newly settled larva showing gastrulation; 4, young sponge of asconoid grade. (*From McGraw-Hill Encyclopedia of Science & Technology.*)

Asexual reproduction in sponges occurs via buds, which may be of two types. External buds arise in certain marine sponges (DEMOSPONGIAE and HEXACTINELLIDA) and are aggregations of amoebocytes which migrate from the mesenchyme to the surface of the sponge, where they develop into small sponges and then drop off to the bottom to take up an independent existence. Internal buds, or **gemmules,** are more common and are produced especially by freshwater sponges to carry the species through unfavorable conditions such as cold or drought (Fig. 16-10). Groups of archeocytes (statocytes) enriched with food materials gather in the mesenchyme and are surrounded by a resistant covering, sometimes containing spicules. As the sponge dies and disintegrates, the minute gemmules drop out and survive. When conditions again become suitable, as in spring, the cell mass escapes from within the covering and starts growth as a new sponge. In marine sponges that form gemmules, the gemmules do not develop the resistant covering and often develop into free-swimming **parenchymula** larvae. (See below.)

In **sexual reproduction** some sponges are dioecious (gonochoristic), but most (e.g., *Scypha*) are monoecious (hermaphroditic). No true gonads exist in sponges; gametes may develop only in certain areas or throughout the body. Sperm and ova arise from either choanocytes or archeocytes. Spermatogenesis usually occurs in special spermatic cysts. The first part of oogenesis often takes place in the flagellated chambers, after which the developing ova migrate to positions under the choanocytes. There they complete development and undergo rapid growth by absorbing nutrients from special nurse cells or by actually engulfing the nurse cells. Fertilization is internal in most sponges; the shed, free-swimming sperm enter the flagellated chambers of the sponges, penetrate into a choanocyte, and are transported to an ovum by the choanocyte or are transferred to an amoebocyte which carries the sperm to the ovum.

The first three cleavages are vertical and produce a disk of eight pyramidal cells. A horizontal cleavage then yields eight large cells (further epidermis) and eight small cells (future choanocytes). The latter increase rapidly and elongate, forming a blastula, and each acquires a flagellum on its inner end facing the blastocoel. The few large cells become rounded and granular, and in their middle an opening forms that functions as a mouth to ingest adjacent maternal cells. The embryo then turns inside out (like a stage in the protozoan *Volvox*) through its mouth, bringing the flagellated cells outside. In this **amphiblastula** stage the oval larva escapes through the osculum of the parent to swim for some hours with the flagellated cells foremost (Fig. 16-9). It then settles down and attaches. The flagellated half of the amphiblastula then invaginates into, or is overgrown by, the larger nonflagellated cells (*gastrulation*). This gastrulation process in which the micromeres become internal and the macromeres external is exactly the opposite of the development seen in all metazoan animals and has supported the separation of sponges in the PARAZOA. After gastrulation the larva begins growth as a young sponge (olynthus stage in CALCAREA and rhagon stage in the DEMOSPONGIAE).

Mesenchyme cells arise from both layers. The amphiblastula larva is characteristic of only a few calcareous sponges. Most sponges of the DEMOSPONGIAE and HEXACTINELLIDA have a **parenchymula** larva. This larva differs from the amphiblastula in having the entire outer surface flagellated around an inner solid mass of amoebocytes. The parenchymula larva also develops in the mesenchyme of the parent, is set free, swims for a while, and settles down. It gastrulates by having the outer flagellated cells move inward to form flagellated chambers (Fig. 16-9).

16-10 Freshwater sponges   The family SPONGILLIDAE includes about 15 genera and 150 species, of worldwide occurrence but restricted to freshwater streams or ponds and lakes. They grow as tufts or irregular masses, to 30 cm (1 ft) or so in width, on sticks, stones, or plants (Fig. 16-10). Some are yellow or brown; others are green when exposed to sunlight because of zoochlorellae (freshwater algae) in the mesenchyme. *Spongilla lacustris* is common in sunlit running waters, whereas *S. fragilis* avoids the light.

16-11 Coralline sponges   A new class in PORIFERA, SCLEROSPONGIAE, has been proposed to accommodate some recently discovered sponges from the deeper slopes of Jamaican coral reefs. These sponges differ from all other living sponges in that they pro-

Figure 16-10   Freshwater sponge (family SPONGILLIDAE). Gemmules about 1 mm in diameter. (*After Jammes; Wolcott; and Hyman.*)

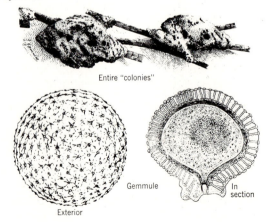

Entire "colonies"

Gemmule

In section

Exterior

duce a massive calcareous skeleton of aragonite, making them superficially resemble corals. All the species so far discovered have a thin covering of living tissue, with a structure similar to the DEMOSPONGIAE, overlying the massive calcareous base. The surface of the calcareous mass is often pitted and has raised prominences which are the sites of the oscula of the animal. Tissue extends into the pits, which are slowly filled with calcium carbonate as the sponge grows upward. The calcareous skeleton is laid down by the sponge on a network of organic fibers which serve as centers of calcification. Also secreted are siliceous spicules in which the base is embedded in the organic fiber matrix. The spicules are eventually trapped in the growing calcareous matrix.

Ostia lead into inhalent or subdermal, cavities which, in turn, communicate with small flagellated chambers. The excurrent channels converge on the raised oscula and often leave patterns visible on the underlying calcareous skeleton.

These sponges seem most closely related to the DEMOSPONGIAE. It has been suggested that a strange group of fossils known as stromatoporids, the taxonomic position of which has never been resolved, are in fact the calcareous skeletons of coralline sponges.

These sponges are abundant on coral reefs and may be of considerable importance in the building of coral reef frames.

16-12 Relations with other animals   Sponges are seldom attacked or eaten by other animals, probably because of their skeletons and the unpleasant secretions and odors that they produce. The main predators on marine sponges are the opisthobranch mollusks, primarily the nudibranchs (Fig. 21-17). Insect larvae of the order NEUROPTERA (spongilla flies) live in and feed on freshwater sponges. The interior cavities of many marine sponges are inhabited by crustaceans, worms, mollusks, and other animals. In a sponge the size of a washtub, A. S. Pearse counted 17,128 animals, mostly one kind of shrimp but also a considerable variety of others, including some slender fishes up to 13 cm (5 in) long. Certain kinds of crabs tear off bits of living sponges, which are held against the shell until they attach; continued growth provides the crab with a protective covering of sponge having a disagreeable odor and containing spiny spicules. The sponge *Suberites* commonly grows on snail shells inhabited by some hermit crabs, soon absorbing the shell so that the crab actually inhabits a cavity in the sponge. Another sponge, *Cliona,* grows on mollusk shells and other calcium carbonate substrates, boring into and eventually destroying the animal and its shell. Some sponges that grow on bivalve mollusks become competitors of the latter, since both sponge and mollusk feed on plankton; these sponges may become a nuisance on oyster beds.

16-13 Phylogeny of sponges Sponges are no doubt an ancient group, and fossils are known from as far back as Cambrian time. The origin of sponges, however, and their relation to other organisms remain somewhat obscure, partly because no metazoan animal has such a low level of organization, with a body constructed around a flow-through canal system, or evidences such unique embryologic features. Such an array of unique features has led biologists to place the sponges in the separate Branch Parazoa (=Subkingdom PARAZOA) off the main line of metazoan evolution. At least two ideas are proposed about the origin of sponges. In one, sponges are thought to have arisen from the protistan choanoflagellate group, which have collar cells. The objections to this theory stem mainly from the fact that in sponge embryology the flagellated cells are not choanocytes and only become such after gastrulation. A second theory has sponges arising from some colonial flagellate, probably one similar to those which might have given rise to the rest of the METAZOA. This theory has support in that the sponge larvae are similar to the colonial flagellate colonies. The inversion of the larvae when flagellated cells become internal is also found in certain colonial flagellated colonies such as *Volvox.* Perhaps, as certain biologists believe, there was a common ancestor to both sponges and metazoans which was a free-swimming ciliated ball of cells in which an opening appeared in the central mass. This became the spongocoel in

the line leading to the PORIFERA and the digestive cavity in the METAZOA.

The fact that sponges have so many unique features not shared by metazoan animals leaves little doubt that they are not in the main line of metazoan evolution, that they diverged early, and that they have not given rise to more advanced groups. It would seem that they are well placed in a separate branch or subkingdom.

Classification

Phylum Porifera.

Sponges. Form flat, vaselike, rounded, or branched; symmetry radial or none; color various; no organs; "body" with many pores; some or all inner cavities lined with choanocytes; an internal supporting skeleton of crystalline spicules, irregular spongin fibers, or both, rarely none; sessile and marine, two families in fresh water; about 10,000 species. Precambrian to Recent.

Class 1. Calcarea (Calcispongiae).

Calcareous sponges. Spicules calcareous, monaxon or 3- or 4-branched; surface of body bristly; dull-colored; mostly under 15 cm (6 in) long.

Order 1. Homocoela. Body of asconoid type; body wall thin, interior not folded, lined continuously with choanocytes. *Leucosolenia, Clathrina.*

Order 2. Heterocoela. Body type synconoid and leuconoid; body wall thickened, folded internally; lining of choanocytes in radial canals not continuous. *Scypha (Grantia).*

Class 2. Hexactinellida (Hyalospongiae).

Glass sponges. Spicules siliceous and 6-rayed (hexactine), some always fused to form networks, some skeletons resembling spun glass; surface lining a syncytium; choanocytes only in finger-shaped chambers (synconoid); body mass commonly cylindrical to funnel-shaped, some curved or flat or with basal stalk; length to 90 cm (3 ft); exclusively marine, at depths of 90 m (300 ft) to over 5 km (3 mi).

Order 1. Hexasterophora. Small spicules 6-rayed, no amphidisks. *Euplectella aspergillum,* Venus's flower basket.

Order 2. Amphidiscophora. Small spicules (amphidisks) with hooks at both ends, no 6-rayed spicules. *Hyalonema.*

Class 3. Demospongiae.

Skeleton of siliceous spicules, of spongin, of both, or none; spicules when present not 6-rayed, mostly monaxon and tetraxon; leuconoid only.

Subclass 1. Tetractinellida. Spicules 4-rayed or none; no spongin; body mostly rounded or flattened; no branches or projections; in shallow waters.

Order 1. Myxospongida (Dendroceratida). No spicules; structure simple, without skeleton. *Oscarella; Halisarca.*

Order 2. Carnosa (Homosclerophora or Microsclerophora). Spicules present, all of about the same size. *Plakina, Plakortis.*

Order 3. Choristida. Spicules present, both large and small. *Thenea; Geodia.*

Subclass 2. Monaxonida. Spicules monaxon; spongin in some; body form various; mostly in shore waters to 45 m (150 ft), but some to depths of 5.5 km ($3\frac{1}{2}$ mi); abundant; most common of all sponges.

Order 1. Hadromerida (Astromonaxonellida). Large spicules separate, mostly in radial bundles; no spongin. *Suberites; Cliona,* boring sponge.

Order 2. Halichondrida. Large spicules 2-rayed or 1-rayed, or both, mixed, not in bundles; spongin present. *Halichondria.*

Order 3. Poecilosclerida. Large spicules united by spongin into a regular network. *Microciona.*

Order 4. Haplosclerida. Spicules 2-rayed, no special arrangement, usually no small spicules. *Spongilla,* freshwater sponges; *Haliclona,* marine.

Subclass 3. Keratosa. *Order Dictyoceratida.* Horny sponges. Skeleton a network of spongin fibers, no spicules; form usually rounded, often of considerable size; surface leathery, coloration dark, chiefly black. *Phyllospongia,* leaf-shaped; *spongia (Euspongia),* bath sponge; *Hippospongia,* horse sponge.

Class 4. Sclerospongiae.

Coralline sponges. Thick skeleton of calcium carbonate crystals (aragonite) on an organic fiber network; spicules siliceous, monaxon styles or rods with whorls of spines; living tissue a thin veneer on surface of skeleton; size to 1 m in diameter; marine, at depths of 8 to 100 m (25 to 325 ft); Devonian to Recent. *Ceratoporella, Merlia, Stromatospongia, Goreauiella, †Stromatopora.*

References

DeLaubenfels, M. W. 1933. The marine and freshwater sponges of California. U.S. National Museum, Proceedings, vol. 81, art. 4, pp. 1–140, illus.

———. 1936. The sponge fauna of the Dry Tortugas . . . and the West Indies . . . [and] a revision of the Proifera. Washington, D.C., Carnegie Institution of Washington, Publ. 467. 225 pp., 22 pls.

———. 1953. A guide to the sponges of Eastern North America. University of Miami, Marine Laboratory, Special Publication. 32 pp.

———. 1954. The sponges of the west-central Pacific. Corvallis, Ore., Oregon State College. x + 320 pp., illus., maps.

Florkin, M., and **B. T. Scheer** (editors). 1968. Chemical Zoology. Vol. II, Porifera, Coelenterata and Platyhelminthes. New York, Academic Press, Inc. xviii + 639 pp., illus.

Fry, W. G. (editor). 1970. The biology of the Porifera. New York, Academic Press, Inc. xxviii + 512 pp., illus.

Giese, A., and **J. S. Pearse** (editors). 1974. Reproduction of marine invertebrates. Vol. I, Acoelomate and pseudocoelomate metazoans. New York, Academic Press, Inc., xi + 546 pp., illus.

Harrison, F. W. and **R. R. Cowden** (editors). 1976. Aspects of sponge biology. New York, Academic Press, Inc., xiii + 354 pp., illus.

Hyman, Libbie H. 1940. The Invertebrates. Vol. 1, Protozoa through Ctenophora. New York, McGraw-Hill Book Company. Sponges, pp. 284–364, figs. 77–105.

Jewell, Minna. 1959. Porifera. In W. T. Edmondson (editor), Ward and Whipple, Fresh-water biology, 2d ed. New York, John Wiley & Sons, Inc. pp. 298–312, 30 figs.

Moore, J. C. (editor). 1955. Archaeocyatha, Porifera. Part E, Treatise on invertebrate paleontology. New York, Geological Society of America. xviii + 122 pp., 728 figs.

Pennak, R. W. 1953. Fresh-water invertebrates of the United States. New York, The Ronald Press Company. Sponges, pp. 77–97, illus.

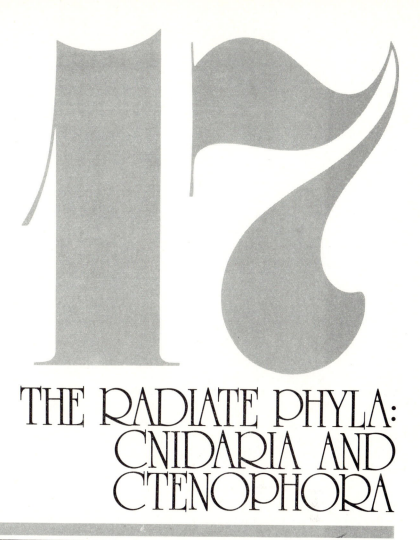

17

THE RADIATE PHYLA: CNIDARIA AND CTENOPHORA

The phyla CNIDARIA and CTENOPHORA are closely related and were long considered as a single phylum, the COELENTERATA , a name which is still often used in connection with the CNIDARIA. These two phyla share certain common fundamental features not found in other phyla, hence their consideration together here. Both phyla are diploblastic animals of the tissue grade of organization without organs or organ systems, and they each have a pattern of symmetry which is radial or biradial. No other invertebrate phyla have such a combination. The two tissue layers are separated to a greater or lesser degree by a jellylike nonliving mesoglea. Another similar feature is the central gastrovascular cavity, the only internal space.

These two phyla are distinguished from sponges in that both have true tissue layers. Hence these are the first METAZOA. In addition, their main internal cavity is a digestive cavity.

Although the two phyla are superficially similar, share a few fundamental morphologic similarities, and were once grouped in a single phylum, there are enough differences between them to warrant separate phylum status. These differences may be summarized as follows. Cnidarians have nematocysts, whereas ctenophores (except one species) lack them; ctenophores have ctene rows, anal pores, and a complex sense organ, all of which are lacking in CNIDARIA. Cnidarians are polymorphic, having various polyp and medusa stages, whereas ctenophores are exclusively monomorphic. Finally, ctenophores have a unique form of development of the embryo not seen in cnidarians.

Phylum Cnidaria (Cnidarians)

The lowest animals with definite tissues are the CNIDARIA (Gr.*knide,* nettle). The individuals are either solitary or in colonies and of two basic body types: (1) the **polyp,** or hydranth, with a tubular body having one end closed and attached and the other with a central mouth usually surrounded by soft tentacles, and (2) the usually free-swimming **medusa,** with a gelatinous body of umbrella shape, margined with tentacles, and having the mouth on a central projection of the concave surface. Both are variously modified, and both appear in the life cycle of many species. Each individual has a digestive cavity, some muscle fibers, and many minute

stinging capsules (nematocysts). All are aquatic, and nearly all are marine. The phylum includes the hydroids and hydromedusae, etc. (class HYDROZOA), the jellyfishes (class SCYPHOZOA), and the sea anemones and corals (class ANTHOZOA) (Fig. 17-1). Many hydroids grow as small, dense plantlike colonies, the jellyfishes and some hydroid medusae swim feebly, the flowerlike anemones abound on rocky ocean coasts, and the corals with their limy skeletons form reefs on tropical shores. Anemones and corals are often brightly colored, as are some medusae, and many cnidarians are luminescent. Cnidarians are of slight economic importance; some coral is used for jewelry and decorative art, and the stings of certain jellyfishes and siphonophores occasionally injure bathers (a few are highly dangerous).

17-1 Characteristics

1 Symmetry radial or biradial about an oral-aboral axis; no head or segmentation.

2 Body of two layers of cells, an external epidermis and an inner gastrodermis, with a varying amount of mesoglea between; nematocysts in either or both layers.

Figure 17-1 Phylum CNIDARIA. Some marine representatives in characteristic habitats, all reduced but not to same scale. Class HYDROZOA, order HYDROIDA: *Tubularia, Plumularia, Gonionemus;* order SIPHONOPHORA: *Physalia,* Portuguese man-of-war. Class SCYPHOZOA, order STAUROMEDUSAE: *Haliclystus;* order CORONATAE: *Periphylla;* order SEMAEOSTOMAE: *Aurelia,* common jellyfish. Class ANTHOZOA, order GORGONACEA: *Gorgonia,* sea fan; order PENNATULACEA: *Stylatula,* sea pen; order ACTINIARIA: *Edwardsia, Epiactis,* sea anemones; order MADREPORARIA: *Astrangia,* stony coral; order CERIANTHARIA: *Cerianthus.*

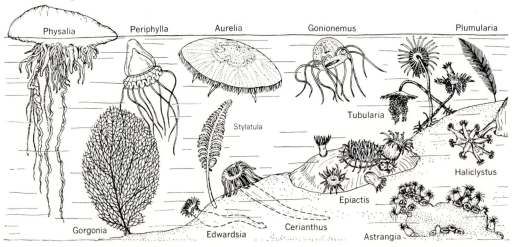

3 Skeleton limy, horny, or none; muscle fibers in epithelia.

4 Mouth surrounded by soft tentacles and connecting to a saclike digestive (gastrovascular) cavity that may be branched or divided by septa; no anus.

5 No circulatory, respiratory, or excretory organs.

6 A diffuse network of unpolarized nerve cells in body wall, but no central nervous system; some with eyespots or statocysts.

7 Reproduction commonly with asexual budding in the attached (polyp) stage and with sexual reproduction by gametes in the medusa stage; monoecious or dioecious; some with simple gonads but no sex ducts; cleavage holoblastic; a ciliated planula larva; mouth forms from blastopore.

8 Cnidocyte cells containing nematocysts (stinging organelles) present in all and employed extensively for food capture and defense.

The stinging qualities of cnidarians were known to Aristotle, who considered these organisms intermediate between plants and animals. They were long included in Zoophyta (Gr., animal-plants) together with various forms from sponges to ascidians. When their animal nature was established in the eighteenth century, Linnaeus and others classed them with the echinoderms as Radiata because of their symmetry. Leuckart in 1847 grouped the sponges, cnidarians, and ctenophores as the COELENTERATA, and only in 1888 did Hatschek separate these as distinct phyla, the PORIFERA, CNIDARIA, and CTENOPHORA. The CTENOPHORA and CNIDARIA are distinguished from sponges in being *tissue animals* (METAZOA) that have a distinct digestive cavity.

17-2 Size Individual hydroid polyps are usually microscopic, but colonies of various species are a few millimeters to 2 m (6½ ft) in length. The jellyfishes are from 12 mm to over 2 m (½ in to over 6½ ft) in diameter, the largest (*Cyanea arctica*) having tentacles up to 10 m (33 ft) long. Anemones range from a few millimeters to 1 m (39 in) in diameter. The individual polyps on corals are minute, but their skeletons form massive growths in warm seas.

Class Hydrozoa (Hydroids)

17-3 Hydra The small, solitary polyp known as hydra serves well as an introduction to both the METAZOA and the CNIDARIA. It is slender and flexible, 10 to 30 mm (½ to 1¼ in) long, with several delicate tentacles at one end (Fig. 17-5). It lives in the cool, clean, and usually permanent fresh waters of lakes, ponds, and streams, attaching to stones, sticks, or aquatic vegetation. Of the nine species in the United States, the "white" hydra, *Hydra americana*, is gray or tan, with tentacles shorter than the body and no stalk; the brown hydra, *H. oligactis* (formerly *H. fusca*), has a slender base or stalk to the body and tentacles three or four times as long as the latter; and the green hydra, *Chlorohydra viridissima* (formerly *H. viridis*), has symbiotic algae (zoochlorellae) in its inner cells that make the body grass-green. The following account applies to any common species.

17-4 General features The body (Fig. 17-2) is a cylindrical tube with the lower end closed to form a **basal disk,** or "foot," used for attaching to objects and for locomotion. The opposite and free oral end contains the **mouth** as a small opening on a conical **hypostome,** encircled by 6 to 10 slender **tentacles.** The mouth leads into the digestive cavity, or **gastrovascular cavity,** which occupies the interior of the body and connects to the slender cavities in the tentacles. The number of tentacles differs among species and increases with the age and size of an individual. The entire animal is very flexible. The body may extend as a slender tube, may bend in any direction, or may contract to a short spherical form. The tentacles move independently or together and may extend as long delicate threads or contract down to slight knobs. The side of the body may bear lateral **buds** that give rise to new individuals by asexual reproduction; at times it bears other rounded projections, the **ovaries** or **testes,** concerned with sexual reproduction.

17-5 Cellular structure and function The wall of the body and tentacles consists of but two cell layers (Fig. 17-2), a thin external **epidermis,** of cuboidal

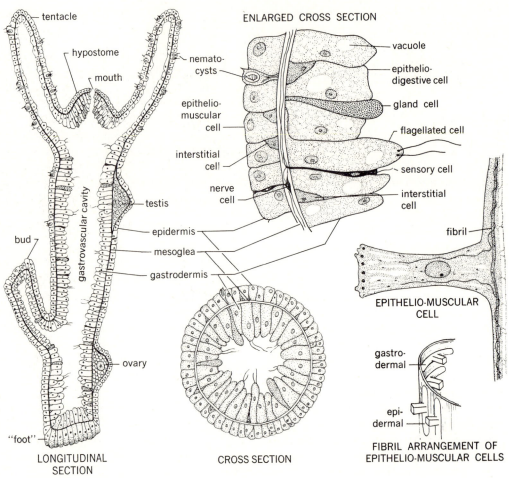

Figure 17-2 *Hydra.* Structure as seen in microscopic sections. (*Epitheliomuscular cell after Hyman, The invertebrates.*)

cells, chiefly protective and sensory in function, and inside, a thicker **gastrodermis,** of tall cells, serving mainly in digestion. Between the two is a thin noncellular **mesoglea,** or supporting lamella, secreted by and attached to both layers, that provides an elastic framework for both the body and the tentacles.

Both layers are composed of several cell types: (1) epitheliomuscular, (2) glandular, (3) interstitial, (4) cnidocytic, and (5) sensory. A network of nerve cells lies in the epidermis and also in the gastrodermis. The cells of each type are specialized in both structure and function to perform particular life processes; collectively, they are responsible for all the activities of the living hydras, as in all multicellular animals. Many of the cells in the gastrodermis

bear one or two whiplike flagella, each arising from a blepharoplast within the cell.

The **epitheliomuscular cells** are ⊥-shaped, having a bulbous outer portion and an elongate base containing a contractile fibril placed against the mesoglea. In the epidermis these cells are closely placed and collectively form the exterior surface of the body; their fibrils attach lengthwise to the mesoglea and act as longitudinal muscles that contract to shorten the body stalk and tentacles. The epitheliodigestive cells of the gastrodermis are the principal part of the lining of the gastrovascular cavity and function actively in the digestion of food. Their contractile fibrils attach transversely inside the mesoglea so that they act as circular muscles to reduce the diameter and extend the length of the body. Circular contractile fibrils around the mouth and the bases of the tentacles act as sphincters to close off these openings.

The basal disk is covered exclusively by tall **gland cells** that secrete a sticky mucus by which hydras attach to objects in the water; these cells can also produce a gas bubble. Gland cells are uncommon elsewhere in the epidermis, but large ones occur about the mouth, and many others are scattered in the gastrodermis, where they secrete the enzymes which digest the food.

The **interstitial cells** are small, round undifferentiated cells with large nuclei, found between the bases of the epidermal cells. They have the potential to produce all other cell types, such as cnidocytes and gametes, and hence are the basis for regeneration and repair of all parts of the body. **Cnidocytes** are specialized cells which contain the unique cnidarian stinging apparatus, the **nematocyst.** Each nematocyst is a minute oval capsule filled with fluid and containing a coiled **thread tube** that may be everted to aid in the capture of prey, in defense, or in locomotion (Fig. 17-3). On the exterior of the cnidocyte is a triggerlike **cnidocil.** Some nematocysts occur singly, but often one large and several small ones are grouped in a single large epithelial cell to form a small surface tubercle, or "battery." Nematocysts are most abundant on the tentacles, but some occur throughout the epidermis, except on the basal disk.

In hydras these cells are not found in the gastrodermis.

Hydras (Fig. 17-4) have four kinds of nematocysts. (1) The large spherical **penetrant** (16 μm in diameter) has a long thread tube that is coiled and bears at its base three long spines and three rows of small barbs; when discharged the thread tube shoots out to pierce the skin of small animals and "inject" a fluid (hypnotoxin) that paralyzes the prey. (2) The pear-shaped **volvent** (9 μm long) contains a short, thick thread in a single loop; upon discharge this coils tightly around bristles or hairs of the prey. (3) The oval **streptoline glutinant** (9 μm) has a long thread, in three or four transverse coils, that bears minute thorns and may coil upon discharge. (4) The small **stereoline glutinant** (7 μm) discharges a straight, unarmed thread. The first two types are of particular value in capturing prey; the others produce a sticky secretion possibly used in locomotion as well as food getting.

The cnidocyte is an independent effector in that discharge of the nematocyst is due, not to a nerve impulse, but to some stimulus affecting the cnidocil directly. It was formerly thought to act like a trigger, but direct mechanical stimulus, as by touching with a glass rod or by the protozoans that live and move about on the surface of hydras, is not ordinarily effective. Substances diffusing in the water from the small crustaceans, worms, and larvae on which the hydra feeds will usually provoke discharge, as will

Figure 17-3 Discharged nematocyst of *Corynactis californica.* **×1000.** (*Courtesy of Dr. R. Mariscal, Florida State University.*)

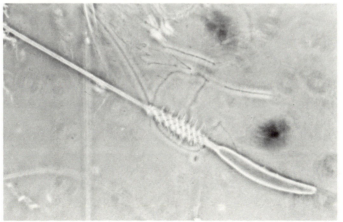

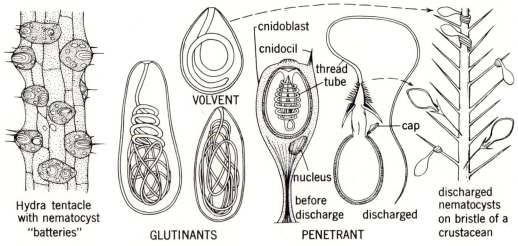

Figure 17-4   The nematocysts, or stinging capsules, of hydra. (*After Schulze.*)

acetic acid added to the water. Eversion of the thread evidently is caused by increased osmotic pressure within the capsule.

The cnidocyte containing a nematocyst develops in the body epidermis, from which it migrates into and through the fluid in the gastrovascular cavity and again passes through the body or tentacle wall to its final location in the epidermis; none originate in the tentacles. The thread tube when discharged cannot be withdrawn, and the cnidocyte cannot form another nematocyst; such parts return to the gastrovascular cavity and are digested, to be replaced by new cells and capsules arriving in the manner just described.

The **sensory and nerve cells** require special staining methods for their demonstration. Scattered through the epidermis are many slender **sensory cells** of several kinds, with bristles or flagellated tips. They are most numerous on the tentacles, about the mouth, and around the basal disk; a small number occur in the gastrodermis. The bases of the sensory cells connect to the **nerve cells** that form a network in the epidermis and adjacent to the mesoglea (Fig. 17-6). The nerve cells have slender processes of various kinds, which often conduct impulses in both directions, not in a single direction as in higher animals. These processes (dendrites and axons) join via synapses to other nerve cells and to contractile fibers of epitheliomuscular cells. The combination provides a **sensory-neuromotor mechanism;** the sensory cells **receive** stimuli, the nerve cells **conduct** impulses, and the contractile fibers **react** to the latter. Excluding the neuromotor organelles of some protozoans, this is the first and simplest neural mechanism to be seen. It provides for coordination in movement of the body and tentacles. There is no central ganglion or brain such as occurs in flatworms and higher metazoan animals.

17-6 Locomotion   Hydra lives attached by its basal disk to objects in the water but is able to twist about, to perform movements for the capture of prey, and to change its location. It can move in several ways. Commonly it bends over, attaches the tentacles to the substratum by use of the glutinant nematocysts, releases and moves the basal disk to a new site, disengages the tentacles, and again assumes an upright position (Fig. 17-5). This "walking" resembles the looping action of a leech or measuring worm. Sometimes it travels inverted, using the tentacles as legs. Again, it may glide along the substratum by pseudopodialike action of cells on the basal disk. Occasionally it uses a gas bubble, secreted in mucus by the disk, to rise in the water and float at the surface. *H. oligactis* can "climb" by attaching its long tentacles to some object, releasing the basal disk, and then contracting the tentacles. The white and brown

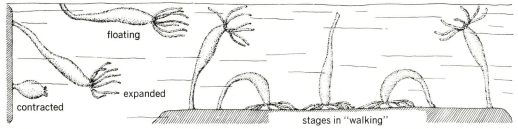

Figure 17-5 Phylum CNIDARIA. *Activities of hydra, a freshwater polyp; natural size to 30 mm (1¼ in) long. ("Walking," after Wagner, 1905.)*

hydras often remain fixed for considerable periods, whereas green hydras move about often, especially when seeking food.

17-7 Feeding and digestion Hydra feeds mainly on minute crustaceans, insect larvae, and similar animals; at times it may swallow prey larger than itself. A hungry individual usually remains attached by its base with the body extended and the tentacles stretched to wave about in search of prey. When a tentacle touches any small animal, nematocysts are discharged into it at once. The penetrants puncture the victim and give off the paralyzing hypnotoxin, the volvents wrap about appendages or other parts, and the glutinants may fasten to its surface. Other tentacles may perform coordinated movements and discharge nematocysts to aid the capture. The tentacles bend inward and carry the food toward the mouth. The latter opens and moves around the food, which is swallowed and passed into the upper part of the gastrovascular cavity by contractions of the hy-

Figure 17-6 Nerve cells in a young hydra. *(After Hadzi, 1909.)*

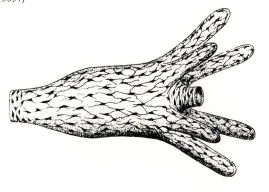

postome and body wall. Glutathione, which is released when nematocysts penetrate the prey, stimulates the hydra's mouth to open.

Gland cells in the gastrodermis become active and discharge secretions of protein-digesting enzymes that act upon the food. Expansion and contraction of the body wall and whipping movements of flagella on the digestive cells bring these secretions against all parts of the food. Soft parts of the latter soon become separated and liquefied, whereas harder portions (such as chitin) are unaffected. Some digestion is completed extracellularly in the gastrovascular cavity, and the partially digested materials are absorbed by cells of the gastrodermis, where digestion is completed intracellularly. In intracellular digestion the free ends of epitheliomuscular cells in the gastrodermis send out pseudopodia that draw food particles into vacuoles within the cells, a process similar to that in protozoans and sponges. Hydra thus combines the digestive procedures of forms both lower and higher than itself. Indigestible residues pass out through the mouth, which thus functions also as an anus. Absorbed food (especially glycogen) is stored locally in cells of the gastrodermis, from which the needs of the epidermis are probably supplied by diffusion. Stored materials tend to be concentrated where metabolism is active and where buds or gonads are forming.

17-8 Respiration and excretion The oxygen necessary for respiration diffuses from the surrounding water directly into the cells of hydra, and metabolic wastes such as carbon dioxide and nitrogenous compounds are lost by diffusion, mainly from the epidermis.

17-9 Behavior All movements result from action of the opposed sets of contractile fibers in the body wall and tentacles and show an obvious coordination between the different parts of the organism that results from transmission of impulses through the nerve net (Fig. 17-6). The response of any individual hydra to stimuli from its environment is conditioned by its physiologic state at the time. A hungry individual usually responds more actively than one that has recently fed, and there are also differences in response between the species of hydras.

A slight touch, as with a needle will cause the part touched to turn away, and a stronger stimulus such as jarring the dish containing the animal will usually result in sudden and complete contraction of both tentacles and body. Besides thus responding to external stimuli, the hydra reacts also to internal stimuli; an undisturbed specimen will, from time to time, suddenly contract and then expand slowly. Such spontaneous movements probably are related to food getting, being more frequent with hungry individuals. Each species responds to a particular optimum of light intensity, the green hydra seeking a stronger intensity than other species. In the main, they avoid either very strong or very weak illumination and usually move about by trial and error until the most favorable condition is found. This probably also is related to feeding, since most of hydra's prey seek well-lighted areas. Hydra prefers cool waters and in a temperature gradient will seek the colder portion. It endeavors to avoid injurious chemical substances. Under rich feeding, high temperature, foul water, and certain other conditions, hydra undergoes a lowered metabolic state known as *depression.* There is a gradual shortening and loss of the tentacles and column, beginning at the distal end; this may proceed to disintegration or may be followed by recovery.

Hydras occur in enormous numbers in some lakes, to depths of 55m (180 ft), but commonly disappear from surface waters at 21°C (70°F) or higher.

17-10 Budding Hydra produces new individuals either by asexual budding or by sexual means involving eggs and sperm. A **bud** forms as an evagination on the body wall containing an outpocketing of the gastrovascular cavity (Fig. 17-2). This lengthens, acquires blunt tentacles and a mouth distally, and later is constricted and detached at the base, to become an independent hydra. Occasionally, several buds form on a single "parent" and these in turn may produce secondary buds, to form temporarily a group somewhat resembling a colonial hydroid. Prior to bud formation the gastroderm cells at the site become well supplied with food for the initial development; later, the bud receives food through the connection of its enteron with that of the parent individual. Budding may occur at almost any season but is most common in summer.

17-11 Regeneration and grafting The ability of certain animals to restore, or **regenerate,** lost parts was first reported in 1744 by Trembley, an English naturalist, from studies on hydra. If a living specimen is cut across into two or more pieces, each will regenerate into a complete but smaller hydra. The hypostome and tentacles alone will form a new individual, and a hydra split part way through the mouth will form a "two-headed" polyp, if the portions are held apart. Even minute fragments will regenerate completely, and pieces too small to grow independently may fuse and regenerate; but the germ layers will not mix. Epidermis joins only with epidermis and gastrodermis with gastrodermis, cells of the latter sending out protoplasmic processes that interweave with one another.

Parts of two hydras may be brought together and **grafted** in various arrangements. The parts usually retain their original polarity, that originally toward the oral end producing tentacles and the opposite end a basal disk, but an oral region grafted to an aboral end may **induce** an oral growth with tentacles in the latter. The active growth region lies just below the tentacles. From this point cells move gradually downward, being sloughed off at the stalk, or "foot," after about 4 weeks. Thus hydra undergoes continuous replacement of parts and therefore might be regarded as immortal.

17-12 Sexual reproduction Most species of hydra are dioecious, any one individual producing only male or female sex organs and cells; a few species, however, are monoecious. The formation of gonads and sexual reproduction ordinarily occur in the autumn but may be induced at other seasons by reducing the water temperature. The **gonads** are temporary structures on the sides of the body, the **ovaries** producing eggs and the **testes,** or spermaries, producing sperm. Both the male and female gonads arise from interstitial cells in the epidermis (Fig. 17-7).

Each of the several testes is a conical outgrowth containing a number of elongate cysts (Fig. 17-7A,B). The interstitial cell (spermatogonium) at the base divides repeatedly to produce many spermatocytes (C). These migrate outward, and the cysts become rounded. As each cell goes through two maturation divisions (D), four nuclei are produced, but the cytoplasm does not separate until later. The spermatids (E) transform into sperm, which escape in numbers to the distal end of the cyst and emerge through a pore in the testis to swim in the water (F); there they may remain active for a day or more.

The future egg forms from an interstitial cell under the epidermis (G); it enlarges and is joined by other interstitial cells, which serve as food in the egg's formation. Two maturation divisions ensue, with production of two polar bodies and reduction of the egg chromosomes from 12 to 6. The mature ovum (H) squeezes through a small opening in the epidermis covering the ovary to lie free on the surface of the latter (I). Fertilization then occurs with entrance of a sperm head, bringing six chromosomes.

The egg soon begins to divide (J), cleavage being total and equal. The blastula is a sphere formed of a single layer of cells enclosing a small cavity. These cells shortly secrete a chitinous shell, or cyst, around the blastula (K). The outer ends of the blastula cells become the ectoderm, destined to form the epidermis, and their inner ends divide off to form the endoderm, which will produce the gastrodermis; the endoderm is solid (L) at first but later hollows out to make the enteron. Mesoglea is produced later between the two cell layers. The cyst containing the embryo soon hardens and drops into the water. After some time (10 to 70 days) the cyst softens, and a young hydra with short tentacles hatches out. There is no larval stage.

17-13 Colonial hydroids   Unlike hydra, most members of the class HYDROZOA are marine and colonial. They include the hydroids, the stinging corals, some jellyfishes, and the free-floating siphonophores. *Obelia* (Fig. 17-8) is a typical hydroid, of mossy or hairy form, found on rocks or shells or piling in the shallow waters of seacoasts. The small whitish or brownish **colony** is fastened by a rootlike base (hydrorhiza) bearing slender branched stems (hydrocauli) on which grow hundreds of micro-

Figure 17-7   Sexual reproduction of hydra. (*Adapted from Tannreuther, 1908, 1909.*)

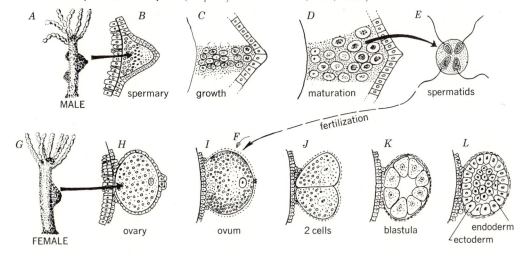

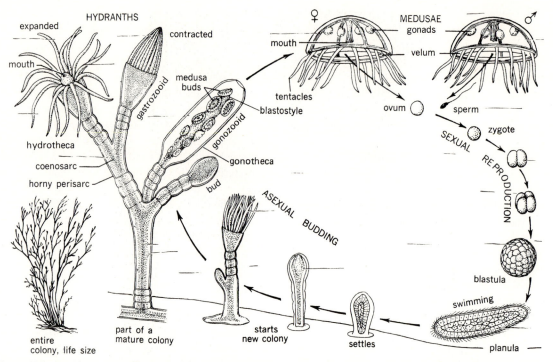

Figure 17-8 Class HYDROZOA. Structure and life cycle of a colonial marine hydroid, *Obelia*. The colony comprises polyps of two types, the feeding gastrozooids and the reproductive gonozooids, both formed by asexual budding on branched stems attached to the substratum by a rootlike hydrorhiza. Free-swimming medusae of separate sexes bud off from the gonozooids and later produce ova and sperm. The zygote develops into a ciliated swimming planula larva; this soon attaches and forms a new colony by budding. The three kinds of individuals illustrate polymorphism, and the alternation of asexual and sexual generations is termed *metagenesis*. *(Modified from Wolcott, Animal biology, McGraw-Hill Book Company.)*

scopic polyps of two kinds. The feeding polyp, or **gastrozooid,** is hydralike, with 20 or more solid tentacles, and is set in a transparent, chitinous, vase-shaped **hydrotheca** that affords it protection. These polyps capture minute animals by use of their nematocysts and tentacles. The reproductive polyp or **gonozooid** (gonangium) is of cylindrical form, covered by a transparent **gonotheca,** and contains a central axis or **blastostyle** on which lateral buds form that develop into medusae. The common stem supporting both kinds of polyps comprises an external transparent and noncellular **perisarc** that is continuous with the hydrotheca and gonotheca and an internal hollow **coenosarc** (common gastrovascular cavity) of cellular structure connecting the gastrovascular cavities of various polyps. Digested food circulates through the coenosarc. Both types of polyp are produced asexually by buds on the stem.

The **medusa** is a minute jellyfish, shaped like an umbrella and rimmed with tentacles (Figs. 17-8, 17-9); on its concave side is a central projecting **manubrium.** This contains the mouth, which leads to a gastrovascular cavity in the middle of the bell, whence four **radial canals** extend to a **ring canal** in the bell margin. A gelatinous mesoglea fills the space between the epidermis over the bell, tentacles, and manubrium and the gastrodermis that lines the digestive tract and its branches. Although the feeding polyps and the medusae differ markedly in appearance, their basic structure is essentially the same.

The medusae escape from the gonozooids to float and feed in the sea. They are of separate sexes, and their gonads develop from interstitial cells clustered in the epidermis from which eggs and sperm are released into the water. There each zygote develops into a minute ciliated **planula** larva. This larva is characteristic of the cnidarians. It is elongate or oval

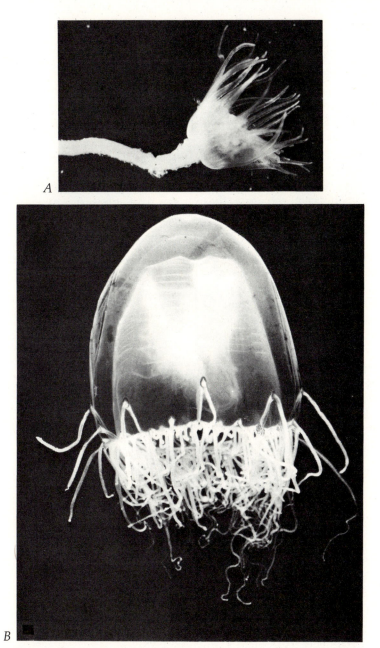

Figure 17-9 Class HYDROZOA. *A. Tubularia,* a large solitary polyp. *B. Scrippsia,* a hydrozoan medusa.

in shape, with a ciliated outer layer of cells and an inner nonciliated cell mass. There is no internal cavity. The planula has distinct anterior and posterior ends. The larva soon settles and attaches with its blastopore uppermost, then grows to be a small polyp, which by asexual budding begins a new colony.

Hundreds of species of hydroids occur at various depths, a majority being found in shallow coastal waters. Although many show an alternation of generations, others have no medusa stage (*Tubularia,* Fig. 17-9) and the gonozooids produce a planula directly. In HYDROZOA of the order TRACHYLINA the polyp generation is reduced or lacking and the medusa stage is a jellyfish of some size with a thin, narrow velum around the inner margin of the umbrella. Here the planula larva gives rise to the medusa stage directly or via a second larva, the **actinula.** *Gonionemus* is a common representative in some localities along the Atlantic and Pacific coasts (Fig. 17-1). The order MILLEPORINA contains hydroids which produce erect skeletons of calcium carbonate, thus superficially resembling true corals.

17-14 Polymorphism A wide range in the complexity of colony structure and polyp function is to be seen among various hydroids. In *Obelia* the gastrozooids perform all functions except that of reproduction and the gonozooids are solely reproductive, forming the medusae for dispersal. *Hydractinia,* which lives on rocks or the shells of hermit crabs, has feeding, reproductive, skeletal, and fighting polyps, the last being of slender form, with no mouth, and bearing many nematocysts. The greatest diversity is seen in members of the order SIPHONOPHORA, which form floating or swimming colonies of specialized individuals. Thus in *Physalia pelagica,* the Portu-

guese man-of-war, each colony includes at least four types of polyps, as follows: (1) the pneumatophore, or float, into which gas is secreted to render the colony buoyant; (2) feeding polyps; (3) defensive or fighting polyps with nematocysts; and (4) reproductive polyps. Other siphonophores have swimming polyps (nectophores) and leaflike protective polyps (bracts). Such diversity in the form and function of the individuals in a colony when more than two body forms are present at the same stage in the life cycle is known as **polymorphism.** All hydroid colonies have at least two different types of polyp (dimorphic).

Class Scyphozoa (Jellyfishes)

In the SCYPHOZOA (Gr. *skyphos,* cup + *zoön,* animal) the medusa stage dominates the life cycle. It may be from 2.5 cm to 2 m (1 in. to 7 ft) in diameter and consists largely of gelatinous mesoglea, or "jelly." The polyp stage of the life cycle is minute or lacking. *Aurelia aurita* is a common jellyfish of coastal waters, often seen singly or in companies.

17-15 Structure The body (Fig. 17-10) is shallowly convex above and concave below and is

Figure 17-10 Class SCYPHOZOA Structure of a jellyfish, *Aurelia;* one-fourth of the body cut away to show internal structure.

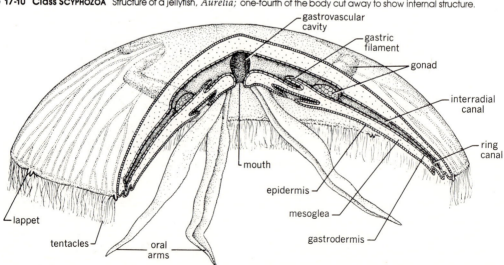

fringed by a row of closely spaced delicate **marginal tentacles.** These are interrupted by eight equally spaced indentations, each with a **sense organ** (rhopalium) between two small **lappets.** There is no velum except in one order. Circular muscle fibers are numerous in the bell margin. Central in the concave oral surface is the **mouth,** on a short **manubrium,** between four tapering **oral arms,** which are grooved and bear nematocysts along their edges. A short **gullet** through the manubrium connects to the gastrovascular cavity. Off the latter are four **gastric pouches** containing slender, tentaclelike **gastric filaments** with nematocysts. Many **radial canals** extend through the mesoglea from the pouches to a **ring canal** in the bell margin. There are four U-shaped **gonads,** one in the floor of each gastric pouch. The **nerve net** is best developed about the bell margin. Each marginal sense organ comprises (1) a pigmented **eyespot** sensitive to light; (2) a hollow **statocyst,** off the circular canal, that contains minute calcareous granules and gives direction for swimming movements; and (3) two **sensory pits,** one lateral and one medial, probably chemoreceptors having to do with food recognition. The exterior surface is covered by epidermis; and the lining of the digestive system and canals from the mouth inward, the gastric tentacles, and the gonads are of gastrodermis.

17-16 Natural history *Aurelia* may occur singly or in great schools. They float quietly and can swim feebly by rhythmic contractions of the bell but are largely at the mercy of currents and waves. Great numbers are sometimes cast on shore during storms. The food of *Aurelia* is mainly small particles trapped in mucus secreted on the lower surface and conveyed to the mouth via flagella on the oral arms. Other scyphozoans are predaceous on small invertebrates and fishes captured by nematocysts on the tentacles. Some of the food digested in the gastrovascular cavity passes to the radial and circular canals for absorption, and undigested parts are cast out the mouth. Respiration and excretion are presumably performed by the whole body surface. The nerve net serves to coordinate the pumping contractions of the bell and action of the oral lobes.

The sexes are alike but separate (Fig. 17-11). Sperm from the gonads (testes) of a male pass out of its mouth and into the gastrovascular cavity of a female to fertilize the eggs produced in her gonads (ovaries). The zygotes emerge to lodge on her oral

Figure 17-11 Class SCYPHOZOA. Life cycle of the jellyfish, *Aurelia*. Adults of separate sexes produce eggs and sperm, and the zygotes develop on the oral arms of the female. The ciliated planula larva swims and later attaches to become a small scyphistoma; this, by transverse fission (strobilation), yields several ephyrae which grow to be adult jellyfishes. Medusae reduced, other stages enlarged. (*After Agassiz; and Wolcott, Animal biology, McGraw-Hill Book Company.*)

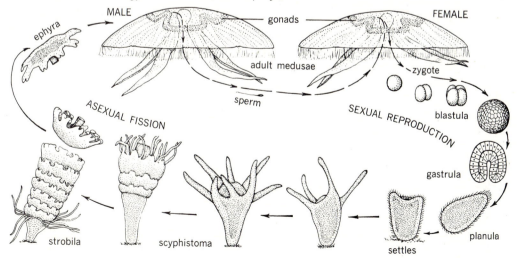

MALE — gonads — FEMALE

ephyra

adult medusae — zygote

sperm

ASEXUAL FISSION

SEXUAL REPRODUCTION

blastula

gastrula

strobila   scyphistoma

planula

settles

arms, where each develops into a ciliated planula larva. This escapes to swim for a while, then settles and attaches to the sea bottom. Losing the cilia, it becomes a minute trumpet-shaped polyp (**scyphistoma**) with basal disk, mouth, and tentacles. It feeds and grows to about 12 mm ($\frac{1}{2}$ in) long and may produce lateral buds that become separate polyps, as with hydras. In autumn and winter a type of transverse fission (strobilation) ensues; horizontal constrictions form around the body and deepen so that the organism somewhat resembles a pile of minute saucers with fluted borders, the edge of each being formed into eight double lobes. These flat, eight-lobed **ephyrae** later separate, invert, and swim about, and each grows to be an adult jellyfish.

17-17 Other Scyphozoa *Aurelia* is an example of the most common of the scyphozoans, as is *Pelagia* (color plate). Other scyphozoans include those of the order RHIZOSTOMAE such as *Cassiopeia*. These jellyfish have the oral arms greatly enlarged and branched, obliterating the original mouth. They have numerous small openings to the gastrovascular cavity and are usually plankton feeders. *Cassiopeia* is unique in possessing symbiotic algae like the corals and in living upside down on the bottom in shallow tropical waters, where the algae furnish it with much of its food. *Chironex fleckeri* in the order CUBOMEDUSAE, occurring over sand bottoms in the Australian region, may be the most dangerously venomous animal in the sea: contact with its virulent nematocysts has been known to kill a human being in 3 min. *Haliclystus* (Fig. 17-12) and other members of the order STAUROMEDUSAE are unique among jellyfish in that they are attached to the substrate by a stalk.

Class Anthozoa (Sea anemones, corals)

The ANTHOZOA are marine polyps of flowerlike form (Gr. *anthos*, flower + *zoön*, animal), of small to large size and rather firm texture, with a tendency toward biradial symmetry in arrangement of the gullet and internal septa. Besides the familiar sea anemones and stony corals, this class includes the soft, horny, and black corals, the colonial sea pens and sea pansies, and others, all of which lack a medusa stage.

Figure 17-12 *Haliclystus,* an attached scyphozoan of the order STAUROMEDUSAE.

17-18 Structure A common sea anemone such as *Metridium senile* has a short cylindrical body (Fig. 17-13). On the upper and flat **oral disk** are many short hollow **tentacles** around a slitlike **mouth;** the base, or **pedal disk,** serves for attachment to solid objects in the sea. The **gullet,** or stomodeum, is a flat tube connecting the mouth and gastrovascular cavity. Along the sides of the gullet is a smooth ciliated furrow, the **siphonoglyph,** in which water passes to the gastrovascular cavity. Internally the body is divided into radial compartments by six pairs of complete **septa,** or mesenteries, that extend vertically from the body wall to the gullet; between these are other, incomplete septa (secondary, tertiary) attached to the body wall but not reaching the gullet. In the septa, beneath the oral disk, are openings, or **ostia,** through which water can pass between the internal compartments. The free inner margin of each septum is a three-lobed **septal filament,** continued below as a threadlike **acontium;** both parts bear nematocysts, and the septal filaments also bear gland cells. Other nematocysts occur on the tentacles. The acontia can be protruded through pores in the body wall or through the mouth, to aid in protection. The **gonads** form along margins of the septa.

The exterior surface is covered completely by tough epidermis, with cilia on the oral disk, tentacles, and gullet, and the gastrovascular cavity is en-

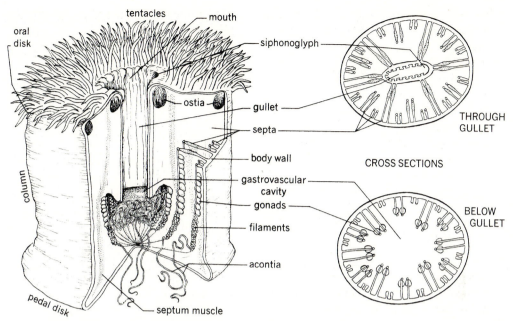

Figure 17-13 Class ANTHOZOA. Structure of a sea anemone, *Metridium*. Part of the body has been cut away to show internal features. Cross sections through and below the gullet show the arrangement of the septa.

tirely lined by gastrodermis. The contractile or muscular bundles in anemones are much more specialized than in other classes of CNIDARIA. The muscular fibers are mainly in the gastrodermis, where bundles of muscles exist for shortening and lengthening the column. Between the epidermis and gastrodermis of the body wall and disk, and within the septa, there is mesoglea composed of cellular corrective tissue. The epidermis contains a nerve net, and nerves occur in the septa, but no localized sense organs are present.

17-19 Natural history Most sea anemones live attached to some firm surface but can creep slowly on their pedal disk. When undisturbed and covered by water, the body and tentacles are widely extended. If irritated or if exposed by a receding tide, the oral disk may be completely inturned and the body closely contracted. Cilia on the tentacles and disk beat so as to keep these surfaces free of debris. A constant current of water moves down the siphonoglyphs to circulate in the gastrovascular cavity for respiratory purposes and to keep the body turgid, and an outward current passes up elsewhere in the gullet.

The food is mollusks, crustaceans, other invertebrates, and fishes, paralyzed by nematocysts and carried by the tentacles to the mouth; some prey is gripped directly by the mouth and gullet, both of which can gape widely. The food passes to the gastrovascular cavity, is digested by enzymes secreted from the filaments, and is absorbed by the gastrodermis. Undigested wastes are cast out the mouth. Certain anemones (*Anthopleura*) contain symbiotic algae in their tissues which provide them with some of their nutritive needs.

Anemones may have separate sexes or are hermaphroditic. The gonads are located in the gastrodermal layer of the septa. Fertilization may occur in the gastrovascular cavity or externally in sea water. The zygote develops into a solid blastula and then into a typical planula larva. While still free-swimming the planula develops a gastrovascular cavity and septa. After settlement the larva forms tentacles and becomes a miniature anemone. Certain anemones (*Epiactis*) brood their young in external pouches on the column.

Anemones also reproduce asexually. They may divide by longitudinal or transverse fission, and such continued fission produces clones of genetically

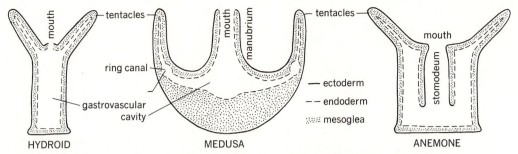

Figure 17-14 Phylum CNIDARIA. Comparison of the hydroid polyp, the medusa (inverted), and the anthozoan polyp.

similar anemones. Another means of asexual reproduction is pedal laceration. Here parts of the disk are left behind as the anemone moves and grow into new anemones.

Most kinds of anemones are sedentary, attached to rocks, shells, or other surfaces, but most can move if necessary; a few minute forms (*Boloceroides*) can swim by lashing their tentacles, and some slender types (*Edwardsia, Cerianthus*) burrow in the bottom, leaving only the tentacles and oral disk exposed (Fig. 17-1). Anemones are eaten by fishes, crabs, and other crustaceans, starfishes, and nudibranch mollusks; cod, flounders, and other fishes use them in some quantity. Some nudibranchs are immune to the toxins of anemones and even incorporate the undischarged nematocysts into their own bodies. Several instances of longevity in captive anemones are recorded, one of *Cereus pedunculatus* to over 65 years.

Despite the simple organization of the anemones, recent work has discovered that they have definite behavior patterns. Adjacent clones of *Anthopleura elegantissima* often actively repel each other along their common borders, and *Metridium senile* has been found to have special tentacles called **catch tentacles** which are employed in intraspecific combat. Certain crabs have developed symbiotic relationships with anemones, placing anemones on their carapaces or shells or carrying them in their claws to fend off predators.

17-20 Corals Members of the order MADREPORARIA are the principle reef builders, growing as rigid masses able to resist the constant pounding of waves. Other invertebrates and marine algae also contribute to reef formation. Many kinds of animals live among or in corals; some commensal crabs become imprisoned in galls on corals, and in the cells of all shallow-water reef-building corals are symbiotic algae (zooxanthellae).

The coral organism is a small anemonelike polyp with short tentacles, scant musculature, and no pedal disk, that lives in a stony cup with radial ridges in the bottom, which is secreted by the epidermis. Generations of such polyps in dense colonies produce the calcareous coral skeletons (Fig. 17-15) and coral reefs.

Reef-building (hermatypic) corals require water of 18°C (64.5°F) or warmer throughout the year to grow well. Moreover, they cannot live deeper than

Figure 17-15 Types of corals (class ANTHOZOA). *A.* Solitary, and *B–E*, parts of colonial stony corals (order MADREPORARIA). *F.* Horny coral (order STOLONIFERA). (*After Wolcott, Animal biology, McGraw-Hill Book Company.*)

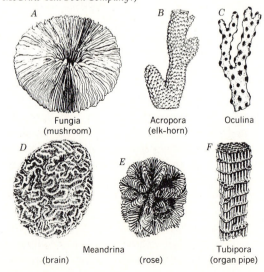

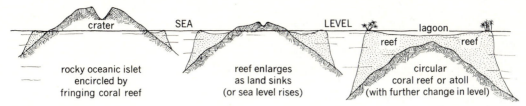

Figure 17-16 Formation of a coral atoll according to Darwin.

about 36 m (120 ft), since they must have enough light for their algal symbionts. Most are intolerant of sediment-laden water and large changes in salinity. Within these restrictions coral reefs are distributed throughout tropical seas, but reef development is greatest in the Indo-Pacific region.

Although corals are carnivores, the presence of the symbiotic algae (zooxanthellae) in the tissues seems to be obligatory for the health of the coral colonies. Recent work has shown that the zooxanthellae are important to the coral in laying down the calcium carbonate skeleton, and they may also contribute to the nutrition of the polyps.

Most of the corals are colonial, but a few (*Fungia*) are solitary. Nonhermatypic corals (without algae) are found throughout the world's oceans, and some occur to depths of 7600 m (25,000 ft).

A coral reef may be classified as (1) a **fringing reef** extending out from shore without a lagoon between it and shore; (2) a **barrier reef** separated by a lagoon of some kilometers in width and a depth to 55 m (180 ft), and (3) an **atoll,** or circular reef, encircling a lagoon and not enclosing an island. Most famous is the Great Barrier Reef along northeast Australia, about 1900 km (1200 mi) long and from a few to 145 km (90 mi) off shore.

There are various theories as to the origin of coral reefs. Darwin inferred that growth began on a sloping shore and that, as the shore sank or was weathered away, this became a barrier reef and later an atoll when the land was completely submerged (Fig. 17-16). Another theory by Daly assumes that withdrawal of water to form the great icecaps of the last glacial period lowered the sea about 60 m (200 ft) below its present level. Various terraces were then cut or islands leveled by wave action. Later, with rising temperatures, corals began growing and built up with the gradual rise in sea level as the ice melted. Recent borings on Bikini showed sand containing FORAMINIFERIDA of the Oligocene (about 30 million years ago) at a depth of 760 m (2500 ft). Deeper borings to 1400 m (4630 ft) on Eniwetok Atoll ended in hard volcanic rock, thus confirming Darwin's theory at least for Pacific Ocean coral reefs.

17-21 Other Anthozoa Other anthozoans belonging to the subclass OCTOCORALLIA are conspicuous inhabitants of marine waters. All OCTOCORALLIA are characterized by their eight pinnate tentacles and internal skeleton (Fig. 17-17). They are always colonial. The sea fans and sea whips (order GORGONACEA) have the soft mesoglea reinforced with spicules and are abundant on coral reefs. Precious red coral (*Corallium*) used in making jewelry is the fused spicular skeleton of a member of this order. Sea pens and sea pansies (order PENNATULACEA), on the other hand, inhabit soft sand and mud bottoms and are bioluminescent.

17-22 Fossil Cnidarians Relics of the HYDROZOA and ANTHOZOA occur from Cambrian time onward, and scant impressions of the SCYPHOZOA appear in Cambrian, Permian, and Jurassic rocks. Fossil reefs of stony corals occur in Iowa, Kansas, and Kentucky and in Europe, and a great reef surrounds the "Michigan basin," from Wisconsin to Ohio. The †TABULATA of Lower Ordovician to Permian times had colonies of many tubes with cross partitions, some pierced by pores; they were probably a subclass of ANTHOZOA. The †STROMATOPORIDEA (†*Stromatopora, etc.*) of Cambrian to Cretaceous times left masses of wavy calcareous plates that formed reefs in ancient oceans, they are considered to be related to the PORIFERA.

Figure 17-17 OCTOCORALLIA. Pinnate tentacles of *Telesto riisei*. (*Courtesy of J. L. Larson.*)

Phylum Cnidaria.

Cnidarians, coelenterates. Symmetry radial or biradial; the individual either a sessile cylindrical polyp, often in colonies, or a free-floating bell-like medusa with much mesoglea; with stinging nematocysts; gastrovascular cavity the only internal cavity; unbranched; soft tentacles about mouth; no anus, no head, no other organ systems; nervous system diffuse; some with eyespots or statocysts; reproduction usually asexual in polyps and sexual in medusae; dioecious or monoecious; no sex ducts; all aquatic, chiefly marine, attached or floating; Lower Cambrian to Recent, 10,000 species.

Class 1. Hydrozoa.

Hydroids (and some medusae). No stomodeum; gastrovascular cavity lacks partitions and nematocysts; mesoglea noncellular; medusa usually small and with velum (craspedote); chiefly in shallow salt waters; colonial or solitary; 3700 species.

Order 1. Hydroida. Polyp generation well developed, solitary or colonial, usually budding off small free medusae that bear ocelli and ectodermal statocysts.

Table 17-1
TYPES OF REPRODUCTION IN THE CNIDARIANS

Hydrozoa:		
Hydra	Small solitary polyp	→ bud → polyp → gametes → zygote → polyp
Obelia	Dimorphic colony of minute polyps	→ bud → gastrozooid → bud → gonozooid → bud → minute medusa → gametes → zygote → ciliated planula → polyp → buds → polyp colony
Tubularia	Monomorphic colony of large polyps	polyp → gonophore → sessile medusae → planula larvae → actinula larvae → polyp colony
Aglaura	Small medusa	medusa → gametes → planula → actinula → medusa
Scyphozoa: *Aurelia*	Large medusa	→ gametes → zygote → ciliated planula → minute polyp → bud → polyp → transverse fission → ephyra → medusa
Anthozoa: *Metridium*	Large solitary polyp	→ fission → polyp → fragmentation → polyp → gametes → zygote → ciliated larva → polyp

Suborder 1. Anthomedusae (Gymnoblastea). Hydranths lack hydrothecae; gonophores naked; medusae tall; gonads on manubrium; no statocysts. *Cordylophora,* colonial, in rivers and brackish water; *Corymorpha,* solitary, marine; *Tubularia, Bougainvillia, Eudendrium* sessile, colonial, and marine.

Suborder 2. Leptomedusae (Calyptoblastea). Hydrothecae present; gonophores in gonothecae; medusae flattish; gonads on radial canals; usually with statocysts. *Obelia, Sertularia, Plumularia,* sessile, colonial, marine; *Polyorchis,* Pacific Coast, hydroid stage unknown.

Suborder 3. Limnomedusae. Polyps small and sessile, with or without tentacles; medusae with or without statocysts; gonads borne on manubrium or radial canals. *Proboscidactyla,* commensal on tubes of sabellid polychaete worms; *Craspedacusta sowerbyi,* freshwater jellyfish, diameter to 20 mm, polyp stage 2 mm tall; *Hydra, Chlorohydra,* all in fresh water, solitary, no medusoid stage; *Gonionemus.*

Order 2. Milleporina. Polyps minute and dimorphic (short, plump gastrozooids and slender dactylozooids) protruding through pores in massive calcareous skeleton. Tentacles capitate. *Millepora,* stinging coral, with powerful nematocysts, on Florida coast and on tropical reefs down to 30 m (100 ft). Triassic to Recent.

Order 3. Stylasterina. Large calcareous skeleton in which the gastrozooid cup has a spine (style) at the base; Dactylozooids solid, no tentacles. *Stylantheca,* California.

Order 4. Trachylina. Polyp generation reduced or none; medusa of some size, with velum beneath bell margin and tentacles arising above margin; with statocysts and endodermal tentaculocysts; sexes separate.

Suborder 1. Trachymedusae. Bell margin smooth, gonads on radial canals, manubrium very long. *Liriope, Aglantha,* all marine, warm seas, surface to depths of 3000 m (10,000 ft).

Suborder 2. Narcomedusae. Bell margin scalloped by tentacle bases; gonads in floor of gastrovascular cavity, no radial canals, and manubrium lacking or very short. *Cunina, Aegina, Solmaris.*

Order 5. Siphonophora. Siphonophores. Swimming colonies comprising several kinds of polyps and medusae; no oral tentacles; upper end of colony usually a supporting float; nematocysts many, large, and powerful; medusae incomplete, attached to stem or disk, rarely free; marine, pelagic, especially in warm seas. *Physalia pelagica,* Portuguese man-of-war, float inflated; *Agalma, Muggiaea, Rhizophysa.*

Order 6. Chondrophora (Physophorida). Colonial with a chitinous float, the underside of the float with a large single central gastrozooid surrounded by concentric rings of gonozooids and marginally by dactylozooids. *Velella, Porpita.*

Order 7. Actinulida. Minute, ciliated, solitary hydrozoans; monomorphic polyps; development direct; actinulalike in appearance; interstitial in marine sand. *Halammohydra, Otohydra.*

Class 2. Scyphozoa (Scyphomedusae).

Jellyfishes. Chiefly free-swimming medusae of bell or umbrella form, with strong 4-part radial symmetry and much gelatinous mesoglea containing amoebocytes; no true velum; no stomodeum; gastric tentacles about mouth; central gastrovascular "stomach," usually divided into 4 gastric pouches; notches in bell margin with sense organs (rhopalia) having endodermal statoliths; medusae sexual, dioecious, with gonads in gastral cavity; polyp generation reduced (scyphistoma) or lacking, producing medusae directly or by transverse fission; all marine; 200 species.

Order 1. Stauromedusae (Lucernariida). Goblet-shaped; marginal sense organs lacking or as modified tentacles; sessile, attached by oral stalk to seaweeds; bays and coastal waters of colder regions. *Haliclystus, Lucernaria.*

Order 2. Cubomedusae (Carybdeida). Bell cubical and margin bent inward; tentacles: 4 or in 4 groups; tropical and subtropical waters, shores, and open seas; feed mostly on fish. *Tamoya,* Atlantic Coast; *Chironex,* sea wasp, most dangerous jellyfish found in Australia; *Chiropsalmus,* Pacific.

Order 3. Coronatae. Bell surrounded by circular furrow, above scalloped margin; chiefly in deep waters. *Periphylla, Nausithoë, Linuche.*

Order 4. Semaeostomae. Mouth central, corners prolonged as 4 frilly lobes; tentacles present; usually with scyphistoma stage; tentacles often on bell margin. *urelia,* common jellyfish; *Cyanea,* to 2 m (7 ft) in diameter, from Arctic waters south along both coasts of North America; *Pelagia,* in open ocean, no scyphistoma stage.

Order 5. Rhizostomae.   Oral arms fused and each doubled (8 in all); no central mouth, many small mouths in the expanded, fused oral arms; no tentacles on bell margin. *Cassiopeia,* lives upside down on bottom in shallow water, Florida; *Rhizostoma,* arms with slender tips.

Class 3. Anthozoa.

Corals, sea anemones, etc. All polyps and attached (no medusae); oral disk flat, with hollow tentacles; mouth leading into stomodeum (gullet), usually with siphonoglyph; gastrovascular cavity divided by vertical septa bearing nematocysts on inner margins; mesoglea a connective tissue; with or without skeleton; gonads (endodermal) in septa; all marine; solitary or colonial; 6100 species.

Subclass 1. Octocorallia (Alcyonaria).   With 8 pinnately branched tentacles and 8 single complete septa; one ventral siphonoglyph; endoskeleton; colonial.

Order 1. Stolonifera.   Polyps arising separately from common stolon or mat; skeleton of separate spicules, sometimes fused as tubes. *Clavularia,* California coast; *Tubipora musica,* organ-pipe coral, in warm waters on coral reefs.

Order 2. Telestacea.   Colonies of stems, each an axial polyp with lateral polyps, on a slender base. *Telesto* (Fig. 17-17).

Order 3. Alcyonacea.   Soft corals. Polyps with lower parts fused in a fleshy mass and only oral ends protruding; skeleton of separate limy spicules, not axial; mostly in warm shore waters. *Xenia, Alcyonium, Anthomastus.*

Order 4. Coenothecalia.   Skeleton massive, of crystalline calcareous fibers penetrated on the surface by tubes of two sizes. *Heliopora,* blue coral of Indo-Pacific region.

Order 5. Gorgonacea.   Horny corals. Colony usually of plantlike form; axial skeleton of calcareous spicules, of hornlike gorgonin, or both; polyps short; 1000 species. *Corallium,* red coral, used for jewelry; *Gorgonia,* sea fan.

Order 6. Pennatulacea.   Colony fleshy, of one long axial polyp, with many secondary dimophic polyps along sides, above bare stalk; skeleton of limy

spicules; 300 species. *Stylatula,* sea pen, featherlike; *Renilla,* sea pansy, disk-shaped.

Subclass 2. Hexacorallia (Zoantharia).   Tentacles few to many (never 8), simple, never pinnate; siphonoglyphs 2, 1, or none; skeleton not of spicules, if present.

Order 1. Actiniaria.   Sea anemones. No skeleton; polyp of some size, columnar, with muscular wall and usually a pedal disk; stomodeum usually with siphonoglyphs; septa paired, often in multiples of 6; on rocks, on sand, or on invertebrates; sedentary but not fixed; essentially solitary, some closely grouped; 1000 species.

Suborder 1. Actiniaria.   Filaments with ciliated areas. *Metridium, Gonactinia; Anthopleura,* Pacific Coast, to 30 cm (12 in) tall; *Adamsia, Actininia,* on hermit crab shells; *Edwardsia,* in "burrows."

Suborder 2. Ptychodactiaria.   No ciliated areas on filaments; primitive septal arrangement; no basilar muscles; no capitate tentacles. *Ptychodactis, Dactylanthus,* Arctic and Antarctic waters.

Order 2. Corallimorpharia.   No ciliated areas on filaments; with capitate tentacles, usually in radial series; similar to true corals but without skeleton. *Corynactis,* central California.

Order 3. Madreporaria. (Scleractinia).   Stony corals. Exoskeleton compact, calcareous; polyps small or minute, in cups on skeleton; tentacles in multiples of 6 usually; no siphonoglyph; muscles feeble; mostly colonial in warm seas; Precambrian to Recent; 2500 living, 5000 extinct species. *Fungia,* solitary; *Balanophyllia,* California coast; *Astrangia danae,* Atlantic Coast, *A. insignifica,* southern California. Reef-building corals: *Orbicella,* in Florida and West Indies; *Acropora, Montipora, Meandra, Isophyllia, Siderastrea,* in the Indo-Pacific (Hawaii to Australia) and Africa.

Order 4. Zoanthidea.   No skeleton or pedal disk; polyps usually united by basal stolons, some solitary with stalked base; single siphonoglyph; many species on exterior of various invertebrates. *Epizoanthus,* on hermit crab; *Parazoanthus.*

Order 5. Antipatharia.   Black corals. Skeleton plantlike, of stems (some branched) composed of horny material and bearing small polyps; tentacles, 6; in deeper tropical waters. *Antipathes,* West Indies, etc.

Order 6. **Ceriantharia.** Slender, elongate, anemonelike; many tentacles in 2 circles; no pedal disk, one siphonoglyph; solitary, enclosed in mucous tube encrusted with sand; many septa. *Cerianthus,* inhabits slimelined vertical tubes in sea bottom.

References

Darwin, C. 1842. The structure and distribution of coral reefs. 3d ed. 1896. New York, D. Appleton-Century Company, Inc. xx + 344 pp., illus. Paperback ed. with foreword by H. W. Menard. 1962. Berkeley, University of California Press. xii + 214 pp., 3 pls.

Florkin, M. and **B. T. Scheer** (editors). 1968. Chemical zoology. Vol.2, Porifera, Coelenterata and Platyhelminthes. New York, Academic Press, Inc. xviii + 639 pp., illus.

Fraser, C. McL. 1937. Hydroids of the Pacific Coast of Canada and the United States. Toronto, University of Toronto Press. 207 pp., 44 pls.

————. 1944. Hydroids of the Atlantic Coast of North America. Toronto, University of Toronto Press. 451 pp., 94 pls.

Giese, A. and **J. S. Pearse** (editors). 1974. Reproduction of marine invertebrates. Vol 1. Acoelomate and pseudocoelomate metazoans. New York, Academic Press, Inc. xi + 546 pp., illus.

Hyman, L. H. 1940. Cnidaria. In The invertebrates. New York, McGraw-Hill Book Company. Vol. 1, pp. 365–661, Figs. 106–208. 1959. Vol. 5, pp. 718–729.

Lenhoff, H. M., and **W. F. Loomis** (editors). 1961. The biology of Hydra and of some other Coelenterates. Coral Gables, Fla., University of Miami Press.

Lenhoff, H. M. (editor). 1971. Experimental coelenterate biology. Honolulu, University of Hawaii Press. 288 pp., illus.

Mackie, G. O. (editor). 1976. Coelenterate Ecology and behavior. New York, Plenum Press. xiii + 744 pp. illus.

Mayer, A.G. 1910. Medusae of the world. Carnegie Institution of Washington Pub. No. 109, 3 vols., 735 pp., 76 pls.

Moore, R. C. (editor). 1956. Treatise on invertebrate paleontology. Lawrence, University of Kansas Press. Part F. Coelenterata, pp. 1–498, 358 figs. *General introduction to living and fossil forms, with chapters by specialists.*

Muscatine, L., and **H. M. Lenhoff** (editors). 1974. Coelenterate biology, reviews and perspectives. New York, Academic Press, Inc. ix + 501 pp.

Rees, W. H. (editor). 1966. The Cnidaria and their evolution. New York, Academic Press, Inc. xviii + 449 pp., illus.

Russell, F. S. 1953-1970. The medusae of the British Isles. New York, Cambridge University Press. Vol. I, 544 pp., 319 figs., 54 pls. Vol.II, 284 pp., 128 figs., 16 pls.

Totton, A. K. 1965. A synopsis of the Siphonophora. London, The British Museum. vii + 230 pp., 153 figs., 40 pls.

Phylum Ctenophora

17-23 *The ctenophores* The phylum CTENOPHORA (Gr. *ktenos,* comb + *phoros,* bearing) comprises about 90 species of free-swimming marine animals with transparent gelatinous bodies. They are often called *comb jellies* because of the comblike plates on the body; those with spherical bodies are known as *sea gooseberries.* Ctenophores show some resemblance to cnidarian medusae and formerly were classified with them, but they are distinct in structure and biology (Fig. 17-18).

17-24 *Characteristics*

1 Symmetry biradial (radial + bilateral), on an oral-aboral axis; three germ layers, with much mesoglea; no segmentation.

2 Body usually with eight external rows of comb plates (ctenes); no nematocysts (except in *Euchlora*); tentacles with adhesive colloblast cells.

3 Cellular epidermis and gastrodermis; mesoglea

Figure 17-18 Phylum CTENOPHORA. Representative forms; not to scale. (*After Hyman, The invertebrates.*)

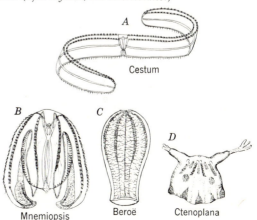

A

Cestum

B

Mnemiopsis

C

Beroë

D

Ctenoplana

with amoebocytes and muscle cells (may be considered cellular, making animals triploblastic).

4 Digestive system with mouth, "pharynx" of infolded epidermis, and stomach which branches into an elaborate canal system.

5 Nervous system diffuse, with a characteristic aboral sense organ (statocyst).

6 Monomorphic.

7 Hermaphroditic; reproductive cells formed from endoderm in digestive canals; development with a distinct larval type, the cydippid; no asexual development.

8 Anal pores terminating digestive tract.

Ctenophores resemble cnidarians in having (1) basic radial symmetry, (2) parts arranged around an oral-aboral axis, (3) a gastrovascular cavity with branches, (4) gelatinous mesoglea, (5) no internal spaces except the digestive system, and (6) no other organ systems. They differ in having (1) eight rows of comb plates, (2) mesenchymal or mesodermal muscles, (3) more advanced organization of the digestive system with anal pores, (4) no nematocysts (except *Euchlora*), (5) mosaic development, and (6) an aboral sensory region. The eight-part system of comb plates (with evidence for eight-part distribution of nervous elements) is, to some biologists, a foreshadowing of the eight nerve strands in certain flatworms.

17-25 Structure *Pleurobrachia* is a common ctenophore along North American coasts (Fig. 17-19). Its transparent body is about 2 cm ($\frac{3}{4}$ in) in diameter and nearly spherical. The **mouth** is at the larger, or oral, end, and a **sense organ** at the opposite, or aboral, end; these establish the radial axis. Eight equally spaced **comb plates** extend as meridians from pole to pole. Each plate is a slight ridge, bearing a succession of small transverse paddles, or combs, formed of fused cilia. In life, these combs beat with quick strokes toward the aboral end, and a wave of beats progresses along each row from aboral to oral end. This propels the ctenophore with the mouth end forward. Near the aboral end, on opposite sides, are two blind sacs from each of which a long, flexible **tentacle** protrudes. The tentacle is solid and contains muscle fibers; it may be trailed out to about 15 cm (6 in) in length or contracted, but does not assist in locomotion. Its surface is covered with **colloblasts,** or glue cells, that secrete an adhesive material to entangle small animals, which are then conveyed to the mouth. The position of the tentacles and some features of the digestive tract establish biradial symmetry.

The **mouth** leads into the **pharynx** (*stomodeum*), where extracellular digestion begins. Beyond is the **stomach,** from which a biradial system of canals branches in definite pattern to beneath the comb plates, tentacle sacs, and pharynx. The system is gastrovascular, serving for both the digestion and the distribution of food. From the stomach a canal runs aborally and gives off four branches, two of which open by anal pores on opposite sides of the sense organ. Undigested wastes pass out through either the mouth or these pores.

Figure 17-19 Structure of a ctenophore. *A. Pleurobrachia,* entire specimen. *B.* Sense organ. (*Both after Hyman, The invertebrates.*) *C.* Reproductive cells in digestive canals under comb plates. (*From Bourne in Lankester, Treatise on zoology, A. & C., Black, Ltd.*)

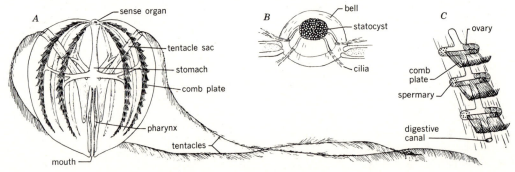

The exterior surface, mouth, pharynx, and tentacles are covered with thin ciliated **epidermis,** and the stomach and canals are lined by ciliated **gastrodermis.** Gelatinous **mesenchyme** fills all spaces between body structures and contains scattered muscle fibers, connective tissue cells, and amoebocytes of mesodermal origin. The animal is thus triploblastic.

The **apical sense organ** contains four elongated tufts of cilia (balancers) that support a small rounded **statolith** of calcareous material, the whole in a covering like a bell jar. Each of the four ciliary tufts supporting the statolith branches as it leaves the sense organ, giving rise to two ciliated furrows, each running beneath a comb row. The apical sense organ functions to control the position of the animal with respect to gravity. When the ctenophore is tilted, a differential pressure is exerted on one cilia tuft. This produces a signal via the ciliated furrows to the appropriate comb rows, which then alter their beat to bring the ctenophore back into proper orientation.

17-26 Natural history   Most ctenophores are planktonic animals spending their lives freely drifting in the open waters of the world oceans. Most striking is the genus *Cestum* (Venus's girdle), a laterally flattened ctenophore up to a meter (40 in) long which swims by graceful undulations (Fig. 17-18*A*). *Beröe* is representative of a different ctenophore group, those without tentacles. It has an enormous mouth and literally scoops up prey as a whale does krill (Fig. 17-18*C*). Most unusual of all are the ctenophores such as *Ctenoplana* (Fig. 17-18*D*), which have been greatly flattened on an oral-aboral axis and have taken up a benthic crawling existence.

Ctenophores abound in warm seas, and some occur in temperate or Arctic regions. They are mostly in surface water, but a few live at various depths, even to 3000 m (10,000 ft). They rest vertically in the water and can swim only feebly, so that currents or tides may concentrate them in large numbers. In the dark they emit a luminescence (light) from beneath the comb plates. Their food is small planktonic animals, including copepods, larvae of mollusks and crustaceans, fish eggs, and small fishes; digestion is rapid, beginning extracellularly in the pharynx and completed intracellularly by cells lining the diges-

tive canals. Over oyster beds at spawning time they may consume numbers of oyster larvae.

Ctenophores are monoecious, both eggs and sperm being produced by the endodermal lining in digestive canals beneath the comb plates. Eggs are extruded through the mouth, and fertilization occurs in the water. Cleavage is determinate, the fate of parts of the blastula being determined, and is reminiscent of the development seen in advanced protostome invertebrates. Ctenophore development is unusual in two respects. First, the cleavage pattern, after the initial two cleavages, is unique, a pattern called biradial in which the embryo develops as a saucerlike structure. Secondly, the larvae which result from development often reproduce as larvae, then regress and becomes sexually mature and reproduces again as an adult, a phenomenon called **dissogeny.** All ctenophores pass through a larval stage called a **cydippid,** which metamorphoses or develops directly into an adult.

Classification

Phylum Ctenophora

(see Characteristics).

Class 1. Tentaculata.

With tentacles.

Order 1. Cydippida.   Body spherical; tentacles branched and retractile into sheaths. *Pleurobrachia, Hormiphora.*

Order 2. Lobata.   Body laterally compressed; two large oral lobes; tentacles short and without sheaths. *Bolinopsis; Mnemiopsis leidyi,* Atlantic Coast.

Order 3. Cestida.   Body elongate, ribbonlike, compressed in plane of tentacles. *Cestum veneris,* Venus's girdle, to over 90 cm (3 ft) long and 5 cm (2 in) wide, in tropical seas, travels by sinuous movements.

Order 4. Platyctenea.   Body compressed on oral-aboral axis, to flattened creeping form, greatly modified. *Gastra* (*Gastrodes*), early larva parasitic in tunicates (*Salpa*); *Coeloplana, Ctenoplana.*

Class 2. Nuda.

Order. Beroidea. No tentacles. Body saclike; mouth wide, pharynx very large. *Beroë,* to 20 cm (8 in) tall, often pink-colored, in colder waters.

References

Horridge, G. A. 1974. Recent studies on the Ctenophora. In L. Muscaline and H. L. Lenhoff (editors), Coelenterate biology reviews and new perspectives, pp. 439–468. New York, Academic Press, Inc.

Hyman, L. H. 1940. Ctenophora. In The invertebrates. New York, McGraw-Hill Book Company. Vol. 1, pp. 662-696, figs. 209–221. 1959. Vol. 5, pp. 730–731.

Mayer, A. G. 1912. Ctenophores of the Atlantic Coast of North America. Carnegie Institution Publication 162, 58 pp., 17 pls., 12 figs.

Pianka, Helen D. 1974. Ctenophora. In. A. C. Giese and J. S. Pearse (editors), Reproduction of marine invertebrates. Vol. I. Acoelomate and pseudocoelomate metazoans, New York, Academic Press, Inc., pp. 201–265.

18

ACOELOMATES: THE PHYLA PLATYHELMINTHES AND NEMERTINEA

With the two phyla which constitute the acoelomates, PLATYHELMINTHES and NEMERTINEA, we arrive at the organ level of construction in the Animal Kingdom. The acoelomates are also the first animals to show the development of a third germ layer, the **mesoderm.** They are thus **triploblastic** animals. It is the presence of this third layer in the embryo which makes possible the development of most of the organ systems seen for the first time here and in most subsequent groups.

Concomitant with the development of the mesoderm and organ systems is the development of **bilateral symmetry,** so that we can now speak of anterior and posterior regions of the body and even of a head and a tail region. The acoelomate phyla are the first to show such bilateral symmetry as opposed to the radial or biradial symmetry in the previous phyla. Bilateral symmetry means that animals have their external features and many internal features arranged symmetrically on either side of a median or sagital plane. All subsequent phyla show bilateral symmetry as a primary symmetry and may collectively be termed the Bilateria. The few exceptions are generally sessile or sedentary animals, hence we associate bilateral symmetry with an active mode of life and radial or biradial symmetry with a more sedentary life.

Additional advances of these phyla over the radiate phyla include development of excretory organs, concentration of the nervous system into an anterior "brain" and a variable number of posteriorly directed nerve cords, and formed reproductive organs.

Both phyla show a determinate, spiral cleavage pattern in the fertilized egg. They are thus the first **protostome** phyla, the dominant line of invertebrate evolution. Of the two phyla, the PLATYHELMINTHES are generally considered to be the more primitive.

Phylum Platyhelminthes (Flatworms)

Lowest of the worms are the PLATYHELMINTHES (Gr. *platy*, flat), which have thin soft bodies. This phylum includes three classes: the TURBELLARIA, or free-living flatworms, most of which inhabit fresh water, salt water, or moist places on land; the TREMATODA, or flukes, which are either external or internal parasites; and the CESTODA or tapeworms, the adults of which are intestinal parasites of vertebrates. Flukes and tapeworms are important parasites of humans, livestock, and wild animals, some causing serious illness or death to these hosts.

18-1 Characteristics

1   Symmetry bilateral; three germ layers; body usually flattened dorsoventrally; no true segmentation.

2   Epidermis soft and ciliated (TURBELLARIA) or covered by cuticle and with external suckers or hooks or both for attachment to host (TREMATODA, CESTODA).

3   Digestive system incomplete, a mouth but no anus, and usually much branched; none in ACOELA or CESTODA.

4   Muscle layers well developed; no body cavity; spaces between internal organs filled by loose parenchyma.

5   No skeletal, circulatory, or respiratory systems; excretory system with many flame cells connected to excretory ducts (protonephridia).

6   Nervous system a pair of anterior ganglia or a nerve ring connecting to one to three pairs of longitudinal nerve cords with transverse commissures.

7   Sexes usually united (monoecious); reproductive system of each sex with gonads, ducts, and accessory organs; fertilization internal; eggs microscopic, each enclosed with several yolk cells in a shell; development either direct (some TURBELLARIA, monogenetic TREMATODA) or with one or more larval stages (digenetic TREMATODA, some TURBELLARIA, CESTODA); asexual reproduction in some forms.

The PLATYHELMINTHES show many advances over the PORIFERA, CNIDARIA, and CTENOPHORA in having (1) bilateral symmetry, with anteroposterior and dorsoventral relations; (2) a nervous system of enlarged anterior ganglia and nerve cords extending along the body; (3) mesoderm as a third germ layer (in place of mesoglea), producing muscles and other organs between the ectoderm and endoderm; (4) layers and bundles of muscles making various movements possible; and (5) internal gonads with permanent reproductive ducts and copulatory organs. Flatworms differ from most higher animals in having (1) no body cavity; (2) the gut branched to various parts of the body; and (3) usually no anus.

The only free-living platyheminthes are found among the turbellarians. Flukes are either external or internal parasites, and all tapeworms are internal parasites. Structural and physiologic specializations parallel this trend in habits. The turbellarians have a delicate ciliated epidermis, whereas both flukes and tapeworms are covered with cuticle resistant to digestion and have suckers and hooks for attaching to their hosts. Those which live internally lack sensory organs, and the tapeworms have no digestive tract. The species that are internal parasites produce enormous numbers of eggs, as is often necessary for organisms that have complex life histories.

18-2 Size
Turbellarians are mostly under 50 mm (2 in) long, and some ACOELA are microscopic, but a few land planarians become 500 mm (20 in) long.

Various flukes are 0.5 to 75 mm (3 in) in length, and different species of tapeworms are 3 mm to 30 m (100 ft) or more in length.

Class Turbellaria (Free-living flatworms)

The common freshwater planarians of North America, *Dugesia* (*Euplanaria*) *tigrina,* etc., are all small free-living animals. They inhabit cool, clear, and permanent waters, including streams, ponds, marshes, and springs, where they cling to the undersurfaces of submerged plants, rocks, and logs and avoid the light.

18-3 Structure *Dugesia* is a thin, slender, soft worm about 15 (5 to 25) mm ($\frac{1}{2}$ in) long, with a bluntly triangular anterior end, or "head," and a tapered body patterned with dark pigment (Fig. 18-1). The head region bears two black **eyespots** middorsally. The **mouth** is on the ventral surface near the middle of the body. Through it a tubular **pharynx,** or **proboscis,** with muscular walls, can be extended to capture food. Minute **excretory openings** are present laterally on the dorsal surface but are difficult to see. Sexually mature worms have a small **genital pore** on the ventral surface behind the mouth.

The body (Fig. 18-3) is covered by **epidermis** of a single layer of cuboidal to columnar cells, resting on an elastic basement membrane. The epidermal cells contain small bodies called **rhabdites,** of uncertain purpose, and many deep-lying **unicellular glands** that open on the surface. Gland cells produce mucus which may be used for adhesion, prey capture, or as a roadway for movement. The ventral epidermis is largely covered with **cilia** that serve for locomotion. Beneath the basement membrane are layers of **muscle fibers,** circular, longitudinal, and diagonal, also dorsoventral fibers. Spaces between the muscles and internal organs are filled with **parenchyma** (mesenchyme), a loose meshwork or cells lacking definite walls (syncytium); there is no body cavity. In the parenchyma are scattered free **formative cells,** which by mitosis produce new parts in regeneration.

The digestive system (Fig. 18-1) consists of the **mouth,** the **pharynx,** and the **intestine** of three main branches, one anterior and two posterior, with many smaller lateral subdivisions. It is composed of columnar epithelium, derived from endoderm.

There is no skeleton other than the elastic basement membrane, and no respiratory system. Oxygen–carbon dioxide exchange occurs through the epidermis. The **excretory system** (Fig. 18-2) is a **protonephridial system** which comprises two longitudinal ducts connecting to a network of tubules that branch throughout the body and end in many large **flame cells;** the latter are between various body cells, from which they collect excess water or fluid wastes. The central cavity within a flame cell contains a group of flickering cilia that drive the collected fluid into the tubules and to the canals that open on the body surface. Flame cells can be seen in unpigmented worms or in preparations made by crushing living worms between two glass slides.

Figure 18-1 Class TURBELLARIA: Planaria, a free-living flatworm. *A*. External features. *B*. General structure (somewhat diagrammatic). On the right (upper) side the testes, ductus deferens, and parts of the digestive tract are omitted; on the left (lower) side the nerve cord, yolk glands, and oviduct are omitted; only a bit of the excretory system is shown on the left anteriorly; pharynx withdrawn into mouth cavity.

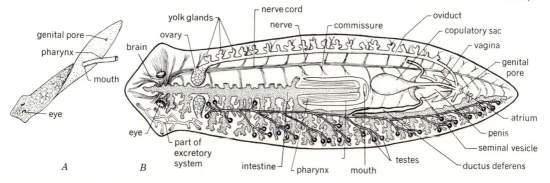

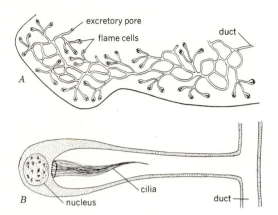

Figure 18-2 Planarian. *A.* Part of the excretory system. *B.* One flame cell. Both enlarged. (*After Hyman.*)

The **nervous system** of planarians is more highly organized than the diffuse nerve net of cnidarians. In the head region, beneath the eyes, are two **cerebral ganglia** joined to form a "brain," from which short nerves extend to the anterior end and the eyes, and two longitudinal **nerve cords** pass back, one along either side, with many **transverse connectives** and **peripheral nerves.** The **eyespots** (Fig. 18-3) are sensitive to light from certain directions but form no image. Auricular sense organs on the sides of the head region may be chemoreceptors of "taste" or "smell."

A sexually mature worm has both male and female reproductive systems and hence is monoecious (hermaphroditic). Both the testes and ovaries develop from formative cells of the parenchyma. The **male reproductive system** includes (1) several hundred small spherical **testes** along both sides of the body, each connected by (2) a minute **ductus efferens** to (3) a larger **ductus deferens,** which extends along each side; the two ducts enter (4) a median **seminal vesicle** for sperm storage, which connects to (5) the muscular **penis** opening into (6) the **genital atrium,** just within the genital pore. The **female reproductive sytem** consists of (1) two rounded **ovaries** near the anterior end of the body, connecting to (2) two **oviducts** paralleling the nerve cords. Along each duct are (3) many **yolk** or **vitelline glands,** which supply yolk cells when eggs are produced; the two oviducts join (4) the median **vagina** opening into (5) the **genital atrium;** to the vagina is connected (6) a bulbous **copulatory sac** that receives sperm at mating. Soon the sperm move to (7) the **seminal receptacles,** which are slight enlargements between the ovaries and oviducts.

18-4 Natural history (Fig. 18-4) Planarians avoid strong light and by day rest quietly on the undersurfaces of objects in water, often in groups of about 6 to 20. After dark, they crawl actively in laboratory aquaria and presumably do so in nature. They adhere to objects or surfaces by a sticky mucus secreted by the epidermal glands and do not swim independently in water. Locomotion is usually by **gliding,** with the anterior end forward and slightly raised. This is accomplished by backward strokes of

Figure 18-3 *Planaria.* *A.* Section through eye. *B.* Cross section of the body (excretory structures omitted).

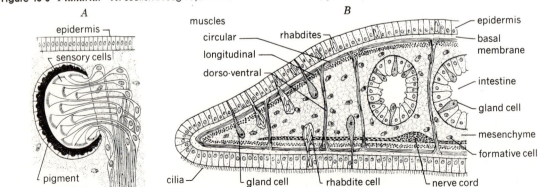

SECTION OF EYE

CROSS SECTION OF BODY

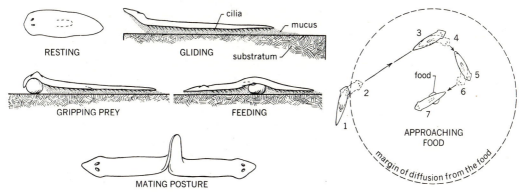

Figure 18-4 Some activities of planarians. (*After Pearl, 1903.*)

the cilia on the ventral surface, over a slime track produced by the glands. Less often a worm travels by **crawling.** This results from muscular movements; contraction of the circular and dorsoventral muscles elongates the body, the anterior end is then affixed by mucus, and the posterior part is drawn up by contraction of the longitudinal muscles. Differential action of local muscle groups produces turning or twisting movements. A moving worm tests its environment by turning the head region from side to side.

When weak stimuli are applied to the head, the worm turns toward the stimulus, but strong stimuli result in a negative reaction. The same is true in lesser degree when the middle of the body is stimulated; a strong stimulus at the posterior end causes the worm to move ahead. Compared with a cnidarian, the planarian shows much more coordination in the action of its parts. This results from the higher development of the nervous system, aided by sensory structures scattered in the epidermis, particularly in the head region. The complexity of muscular fibers under direction of the nervous system makes a greater variety of movements possible.

Studies reveal a form of learning and memory. If a planarian is exposed repeatedly to bright light at intervals of a minute or so, it will react only to the first few bursts. When exposed repeatedly to mild electric shocks, however, it continues to react (contract). If these two experiments are combined (repeated bursts of light, then after a few seconds a shock), the planarian will subsequently react if exposed to light

alone (in "anticipation" of the shock). This acquired reactivity to light is not centered in the brain, because when a trained worm is cut in half and each half is permitted to regenerate, both the "anterior half" worm and the "tail" worm react to light. Further, if an untrained worm eats a trained one, the cannibal shows a significantly greater response to light. Thus memory is somehow stored throughout the body, presumably in dispersed parts of the nervous system, and can be transmitted through cells and tissues.

Planarians feed on other small animals living or dead. When hungry, they travel about actively, gathering quickly on any edible material. They evidently feed on small crustaceans trapped in mucus that the worms secrete. When one planarian begins feeding, others are soon attracted, either by substances diffusing from the food or by digestive juices from the worm. The common method of collecting planarians is to place bits of meat in shallow water; juices diffusing from the meat will soon attract them in numbers. A small food item may first be "gripped" by the head region (Fig. 18-4). Then the muscular pharynx is protruded on and into the food, enzymes are secreted, and bits of food are drawn into the mouth and enteron by suction of the pharynx. Digestion is completed intracellularly within vacuoles of cells lining the enteron. Products of digestion gradually pass to other body tissues, probably in fluid in the mesenchyme. Any undigested material must be disposed of through the mouth, since there is no anus. A planarian starved for several weeks will exhaust its stored food; then certain organs degenerate and are utilized, and the worm decreases markedly in size.

18-5 Reproduction Asexual multiplication occurs by **transverse fission;** a worm constricts in two, usually behind the pharynx, pulls apart, and the missing parts on each piece then grow and differentiate. Planarians possess great powers of regeneration; when they are injured either naturally or experimentally, any part of the body can be replaced, and entire small worms will result from artificial cutting of a larger individual into pieces (Fig. 10-2).

In sexual reproduction two planarians bring their posterior ventral surfaces together, and copulation is mutual, the penis of each being inserted into the genital atrium of the other; sperm from the seminal vesicle in the male system of each passes to the female seminal receptacle of the other. Such exchange of sex products between separate individuals is called **cross-fertilization;** the direct transfer of sperm from male to female organs is termed **internal fertilization.** The worms separate after mating, and sperm migrate up the oviducts to fertilize the eggs. Several zygotes and many yolk cells are later combined in a separate capsule, or eggshell; this is derived from yolk cells. Development is direct, without a larval stage.

18-6 Other Turbellaria Most turbellarians are marine. The members of the order ACOELA are the most primitive living flatworms. These animals have a mouth but no digestive cavity. Food is taken in through the mouth and passed directly to the mass of digestive cells. Acoels lack excretory organs and have no distinct gonads, gametes being produced directly from parenchyma cells. The nervous system in some is an epidermal net system much like that of cnidarians, but in others it is formed into cords. Many are free-living, and others live in the intestines of sea urchins and sea cucumbers. *Convoluta,* an acoel of sand beaches in Europe, has symbiotic algae in the tissues and migrates up out of the sand at low tide to illuminate the symbionts.

Some turbellarians have a simple unbranched digestive tract. Most of them are free-living, some are commensals on the interior and exterior of marine invertebrates, and a few are true endoparasites in mollusks, echinoderms, and crustaceans. The free-living *Microstomum* feeds on hydras, the nematocysts of which pass from the worm's gut to its own epidermis and serve for defense against other animals. This worm reproduces asexually like a planarian, but the parts may remain attached and form a chain of 8 or 16 individuals (Fig. 10-1C).

Except for the orders TRICLADIDA and POLYCLADIDA, most of the other orders of turbellarians consist mainly of small marine and freshwater forms with simple intestines. Many are common interstially in marine sand beaches and in surface layers of mud but are overlooked. Another large group are commensal or parasitic in or on many larger marine organisms.

The order TRICLADIDA, to which *Dugesia* belongs, has other members in fresh waters, a few that are marine, and many land planarians in the humid tropics and subtropics. Some of the latter are large and brilliantly colored. They travel in damp places on slime paths and can descend from leaves or branches by hanging on slime threads. *Bipalium kewense* is a common land planarian in greenhouses around the world, having been transported widely with potted plants. Two species of *Geoplana* from South America are in California gardens.

The POLYCLADIDA are exclusively marine, living among rocks, sessile animals, or plants of the seashore. They are leaflike, some quite broad, some with numerous eyespots, some with tentacles; the gut is many-branched. Some have a free-swimming larval stage. Many, especially in the tropics, are strikingly colored and rival the nudibranch mollusks as the most colorful invertebrates. Most are predators on other marine invertebrates (Fig. 18-5).

Class Trematoda (Flukes)

These flatworms are all parasitic, mostly in vertebrates. Whereas turbellarians have a ciliated epidermis, trematodes are covered externally by a nonciliated syncytium (tegument), representing extensions of cells lying deeper in the mesenchyme (Fig. 18-6). Beneath the syncytium are consecutive layers

Figure 18-5 Order POLYCLADIDA. A free-living marine polyclad, *Eurylepta Californica*. *(Courtesy of G. McDonald.)*

of circular, longitudinal, and diagonal muscle. There is usually a sucker about the mouth and one or more on the ventral surface. The mouth is anterior, and the digestive tract is usually somewhat λ-shaped, with two main trunks having smaller branches (Fig. 18-8). The food consists of the tissues or body fluids of the host, which are sucked in by action of the muscular pharynx. Structurally the trematodes are more like the turbellarians than are the cestodes; complex muscle layers, excretory organs, and a nervous system are present as in the TURBELLARIA. Most differences relate to their parasitic mode of life, such as development of a resistant cuticle, suckers, and hooks and general lack of sense organs. Sensory organs such as

dorsal eyespots occur in some larve and in some flukes that are ectoparasites.

18-7 Subclass Monogenea Members of this group inhabit only one host. They are chiefly ecto-parasites of fishes, amphibians, and reptiles, but some inhabit the mouth cavities or urinary bladders. A monogenetic trematode has at the posterior end a well-developed adhesive organ with one or more suckers, and often with chitinous hooks or anchors as well. The reproductive organs resemble those of TURBELLARIA, and cross-fertilization is usual. Only one or a few eggs are produced at a time, either laid in water or attached to the host. A ciliated larva

Figure 18-6 Histologic structure of the trematode *Fasciolaria hepatica*. *(After Hyman, 1951.)*

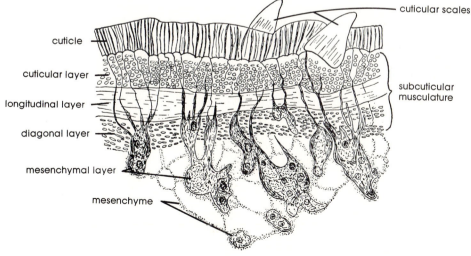

cuticle

cuticular layer

longitudinal layer

diagonal layer

mesenchymal layer

mesenchyme

cuticular scales

subcuticular musculature

hatches from the egg to swim about and find a host or die. Development is direct to the adult form, so that each egg, if successful, yields one fluke. A common example is the minute *Sphyranura* (Fig. 18-10), which lives on the fins, skin, and gills of carp, trout, and other freshwater fishes. It sometimes becomes abundant on fishes in hatcheries and kills many of them.

18-8 Subclass Aspidobothrea The flukes of this group are internal parasites of lower vertebrates and mollusks and have but a single host. They are characterized by the presence of a huge adhesive structure (sucker) covering the whole ventral part of the body. They show features of both MONOGENEA and DIGENEA and may have one or two hosts (Fig. 18-7).

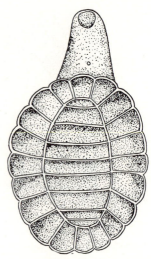

Figure 18-7 Subclass ASPIDOBOTHRIA, genus *Cotylaspis*, ventral aspect. (*After Hyman, 1951.*)

18-9 Subclass Digenea This is by far the largest group of trematodes. These flukes are all internal parasites. Each species has a succession of stages that must live in certain organs of two or more host species to complete the life cycle, the larvae in a certain snail or other invertebrate (intermediate host) and the adult in some vertebrate (primary host). Various species live in parts of the digestive tract, lungs, urinary bladder, blood vessels, or other organs. Some have about the most complex life cycles of any animals. Although the life history pattern, number of larval forms, and intermediate hosts vary, a general pattern is as follows:

18-10 Sheep liver fluke The common liver fluke of sheep, *Fasciola hepatica*, inhabits the bile ducts and sometimes invades other organs. It is commonest in sheep and cattle but is sometimes found in other mammals and occasionally in humans, producing the disease known as *liver rot*. Moderate infections cause sheep to be unhealthy and subject to other diseases, and heavy infestations result in many deaths. Fluke disease thrives only on ranges or pastures with marshy areas where the approriate snails occur but is common in many parts of the United States, Europe, and other areas of the world (Figs. 18-8, 18-9).

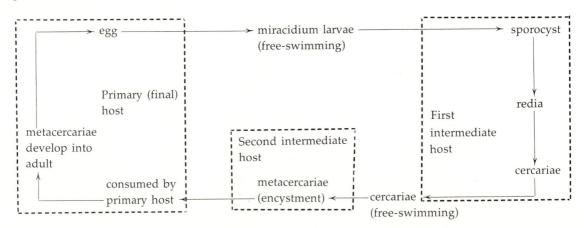

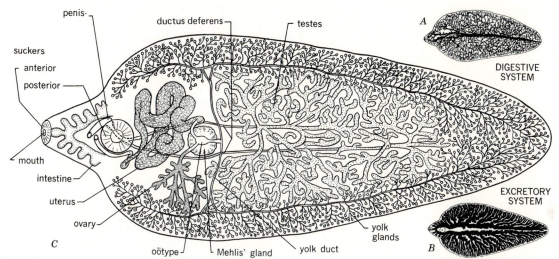

Figure 18-8 Class TREMATODA: **liver fluke of sheep,** *Fasciola hepatica.*   *A.* Excretory system. *B.* Digestive system; both natural size. *C.* Reproductive system in ventral view and enlarged; digestive system shown only at anterior end. (*Modified from Sommer and Landois, 1880.*)

The liver fluke has a leaf-shaped body up to 30 mm (1¼ in) long, rounded anteriorly and bluntly pointed behind (Fig. 18-8). The **anterior sucker** is terminal, surrounding the mouth; and close behind is the **ventral (posterior) sucker,** which serves for attachment in the host. Between them is the **genital opening.** The **digestive system** comprises the mouth, muscular pharynx, short esophagus, and two-branched enteron with many subdivisions throughout the body. The **muscles** are complex, and

Figure 18-9 Life history of the liver fluke of sheep, *Fasciola hepatica;* **larval stages, about × 90; snail about natural size.** (*Details from Thomas, 1883.*)

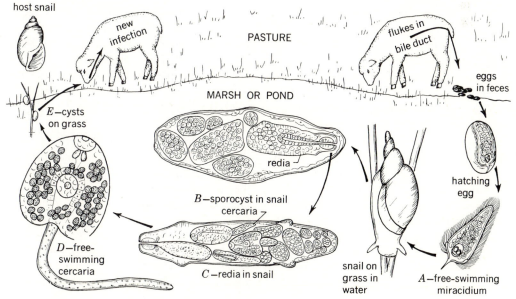

parenchyma fills spaces between the internal organs. The **excretory system** has many flame cells joined to one main canal opening in a single posterior pore. The **nervous system** (visible when stained with methylene blue) includes a double ganglion near the esophagus, two lengthwise nerve cords, and various nerves. Both sex systems are highly developed in a mature worm. The **male reproductive system** has two much-branched testes along the middle of the body, each connected by a ductus deferens to the single seminal vesicle, to which is also joined a prostate gland. The vesicle connects to the penis just within the genital opening. The **female reproductive system** comprises a branched ovary on the right side connected by an oviduct to the median ootype, which is surrounded by the Mehlis ("shell") glands. Off the oviduct is a seminal receptacle. Laurer's canal, of unknown function, extends dorsally from the receptacle. Along each side of the body are many yolk glands, joined to two lengthwise yolk ducts with a common entry into the ootype. From the latter an irregularly coiled uterus extends to the genital opening.

Each fertilized egg is combined with several yolk cells and surrounded by an eggshell in the ootype. Egg production is practically continuous; the uterus is crowded with eggs that are beginning to segment. The eggs (Fig. 18-8) emerge from the genital pore, pass through the bile duct and intestine of the sheep, and are voided with its feces (Fig. 18-9). In warm moist surroundings they develop in 9 days or more, development being retarded at lower temperatures. When it is too cold for any development, eggs are able to survive for several years. From each egg there emerges a larva that is barely visible, multicellular and ciliated, with a pointed anterior rostrum, two eyespots, a nerve ganglion, and two nephridia. If this **miracidium** (Fig. 18-9*A*) hatches in water, it swims about for not over 24 h, and will die unless it finds a certain kind of snail, *Stagnicola bulimoides*, or *Fossaria modicella* in the United States. In the latter event it burrows into the soft tissues and enters usually the pulmonary chamber or lymph vessels. The larva loses its outer cilia, enlarges, and becomes a saclike **sporocyst** (*B*). Within the sporocyst are germ cells, each of which can develop by

parthenogenesis (without fertilization) into another larval stage termed the **redia** (*C*). Each sporocyst produces 3 to 8 elongate saclike rediae, with a mouth and short gut. Within about 8 days they burst from the sporocyst and migrate to another organ, usually the liver. There the rediae may produce other rediae parthenogenetically for one or two generations. Finally, each original or daughter redia yields several larvae of another type, known as a **cercaria** (*D*). This has a slender tail and disk-shaped body, with both oral and ventral suckers and a forked gut. The cercaria burrows out of the snail to swim through the water by use of its tail. After a few hours it settles on a grass blade or other vegetation near the water surface, loses the tail, and becomes a **metacercaria** (*E*) in a tough enclosing cyst. The encysted larvae remain viable for some weeks or months on grass or even on damp hay, if not subjected to high temperatures. When infested vegetation bearing metacercariae is eaten by a sheep or other suitable host, the cysts are digested off and the larvae burrow through the intestinal wall to the body cavity to reach the liver. They burrow for several weeks and damage liver tissue before entering the bile duct to mature and live for years.

Many chances of failure beset the path of a species that must complete such a life cycle. These include the dropping of eggs in dry places, failure of miracidia to find quickly a snail of the right kind, death of infected snails, drying of ponds, or encystment on vegetation where sheep do not happen to graze. As an offset of these hazards, each adult fluke may produce up to a half million eggs, and the parthenogenetic multiplication in the snail may yield up to 300 larvae from a single egg.

18-11 Other Trematoda   About 6000 species of digenetic flukes (Fig. 18-10) have been described from various hosts—fishes to mammals—but the life histories of very few are known. Important parasitic flukes of humans are:

Intestinal fluke, *Fasciolopsis buski:* China to India and adjacent islands. Larval stages in snails. Cercariae encyst on water plants (water nuts, caltrop) used as raw vegetables in the Orient. In parts of

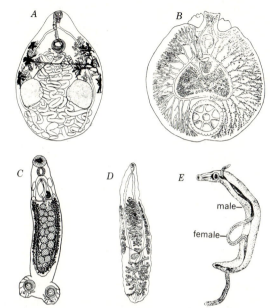

Figure 18-10   Class TREMATODA: representative flukes, variously enlarged. *A. Prosthogonimus macrorchis. B. Tristoma coccineum. C. Sphyranura osteri. D. Opisthorchis sinensis. E. Schistosoma haematobium. (A, after Macy; B, after Lameere; C, after Furhmann; D, E, after H. B. Ward.)*

China, over half the human population is infected. Also in pigs and dogs.

Liver fluke, *Opisthorchis (Clonorchis) sinensis:* Japan, China and adjacent islands, to Vietnam. Also in dogs and cats. Earlier larval stages in freshwater snails and later ones (metacercariae) in muscles of many freshwater minnows and carp. Human infections result from eating raw fish. Adult flukes inhabit bile ducts, and infections may last from 5 to 20 years (Fig. 18-10*D*).

Lung fluke, *Paragonimus westermani:* In Japan, China, the Philippine Islands, New Guinea, India, Africa, Yucatán, and Peru; occasional in Oriental immigrants in the United States. Also in cats, dogs, pigs, goats, rats, and various wild mammals of regions where human infections are present, the flukes showing little host specificity. Eggs discharged in host's sputum, larvae occur in snails, and cercariae encyst in freshwater crabs and crayfish. By eating these crustaceans uncooked, the final hosts become

infected; young flukes occur in various organs, and mature flukes encyst near the surface of the lungs.

Blood flukes: *Schistosoma haematobium* about the southern and eastern Mediterranean, in Africa, Madagascar, and southwestern Asia; *S. mansoni* in Egypt, tropical Africa, South Africa, South America, and the West Indies; and *S. japonicum* from Japan and China to the Philippines and Celebes; the last also parasitizes cats, dogs, rodents, pigs, and cattle. Adult flukes inhabit veins of the digestive or urinary tract. Sexes separate and unlike (Fig. 18-10*E*), male 9 to 22 mm ($\frac{1}{3}$ to $\frac{4}{5}$ in) long, with widened and infolded body enclosing the slender female (14 to 26 mm; $\frac{1}{2}$ to 1 in). Eggs of *S. haematobium* are usually deposited in venules of bladder; those of *S. mansoni* and *S. japonicum* in venules of intestine, where they break through the walls of these organs and leave in urine or feces. Larvae in various freshwater snails which live in slow-moving waters and cercariae with forked tails burrow through human skin or are taken in drinking water. The disease **schistosomiasis** is one of the most important diseases over most of the tropics, and its incidence is rising with the increase in numbers of artificially created lakes in these areas.

Cercariae of some nonhuman schistosome flukes burrow into the skin and produce "swimmer's itch" in persons frequenting freshwater lakes of the North Central states, Pacific Coast, Canada, and Europe.

Class Cestoda (Tapeworms)

The cestodes (Gr. *cestus,* girdle + *oid,* like) are mostly slender and elongate, with a flat body usually of many short similar sections; hence the name *tapeworm.* Cestodes are set off from the other two classes by the complete lack of a digestive system. They lack cilia, are covered with a tegument like the trematodes, and have complex muscle layers, parenchyma, paired excretory ducts with flame cells, and a nerve ring with three pairs of nerve cords. Food is absorbed directly through the body wall. All are endoparasites, the adult worms in the intestines of vertebrates and the larvae (with one exception) in tissues of some alternate host.

18-12 Subclass Cestodaria This group includes a few species of small worms lacking a scolex, each with a sucker for attachment, the body undivided, and one set of reproductive organs. They are similar in outward appearance to trematodes, but the larva is like that of tapeworms and they lack a digestive tract. The few species are all parasites of cartilaginous and primitive bony fish (Fig. 18-11). *Gyrocotyle*, found in the chimaeroid fishes, is an example.

18-13 Subclass Eucestoda All other tapeworms belong to this group and have a scolex. A common example is *Taenia solium*, the pork tapeworm of humans, known since ancient times, and from 2 to

Figure 18-11 Class CESTODA, Subclass Cestodaria. The chimaeroid pasasite *Gyrocotyle*. (*After Hyman, 1951.*)

7.5 m (6 to 25 ft) long when mature (Fig. 18-12). It has a minute knoblike "head," or **scolex,** with four muscular **suckers** on the sides and a circle of **hooks** on the elevated tip, or **rostellum.** A short "neck," or budding zone, joins the scolex to the body, or **strobila,** which consists of a series of up to 1000 **proglottids.** The suckers and hooks serve to fasten the scolex to the intestinal wall of the host, and the chain of proglottids lies free in the intestinal cavity. New proglottids are constantly forming by transverse budding in the neck and remain connected to be forced backward by growth of still others. As they move backward, the proglottids increase in size, mature, and finally become detached.

The scolex contains a nerve ring, with nerves to the suckers and rostellum and joined to three pairs of longitudinal nerves extending backward in the proglottids. In it also is an excretory structure connecting to a pair of excretory canals through the proglottids. Each proglottid contains muscles, parenchyma, sections of the excretory canals connected by a cross canal, many flame cells, and the nerves. A complete set of both male and female sex organs develops in every proglottid when some distance beyond the scolex. Those in any proglottid are comparable with the entire reproductive system in a turbellarian or trematode. The male system matures first. Self-fertilization between the systems in one proglottid or in separate proglottids or cross-fertilization between parts in two worms in one host are all possible. Eggs then develop in numbers, each including a fertilized ovum and several yolk cells enclosed in a resistant shell formed in the ootype. These pass into the uterus, which gradually becomes a branched sac crowded with thousands of eggs; other organs degenerate. Development of the eggs begins at once and continues as ripe proglottids are cast loose, pass out in the host's feces, and disintegrate (Fig. 18-13). In eggs thus scattered on the ground is a six-hooked embryo (oncosphere).

If such eggs are eaten by a pig, the shells are digested off in the pig's intestine and the six-hooked larvae burrow into blood or lymph vessels, to be carried finally to voluntary muscles, where they encyst. The cyst enlarges, becomes filled with fluid, and is then called a *bladder worm,* or **cysticercus.** One side of the inner wall thickens to form a hollow pa-

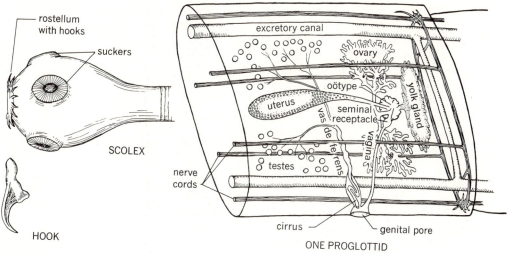

Figure 18-12 **Class CESTODA.** Scolex and one proglottid of the pork tapeworm, *Taenia solium.*

pilla projecting into the sac. In the cavity of the papilla a scolex develops, with suckers and hooks, but there is no further growth. When raw or imperfectly cooked pork containing such a cyst is eaten by humans, the outer cyst wall is digested off, the papilla everts to form a scolex and "neck," and the bladder disappears. The scolex attaches to the host's intestine, and a new tapeworm begins to form. Larvae of this tapeworm occur rarely in monkeys, dogs, cats, and sheep. In addition to harboring adult worms, humans may become infected with larvae of this parasite, often with serious consequences if the cysticercus lodges in a vital organ.

The budding, gradual maturing, and freeing of proglottids by a tapeworm resemble somewhat the production of ephyrae by the strobila of a scyphozoan cnidarian (Sec. 17-16). The proglottids, however, degenerate to free the eggs, whereas each ephyra becomes a complete, free-living, and mature jellyfish that later reproduces sexually.

Figure 18-13 **Life cycle of the pork tapeworm,** *Taenia solium.* (*Adapted from Buchsbaum, Animals without backbones, The University of Chicago Press.*)

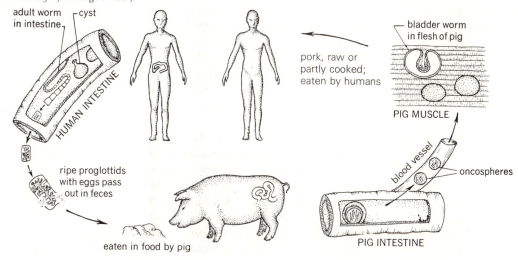

18-14 Adjustments to parasitic life The tapeworm, as an internal parasite, has many special physiologic adjustments in comparison with a free-living animal. (1) The integument protects against digestion by host alkaline digestive juices but is freely permeable to water and nutrients. The latter are mostly amino acids and simple sugars broken down from complex molecules by the host. Adults are resistant to alkaline pancreatic juice. (2) The internal osmotic pressure is lower than that of the surrounding host fluid or tissue. (3) The pH tolerance is high, 4 to 11 (see Sec. 2-8). (4) The tissues have a high glycogen content (to 60 percent of dry weight) and much lipid but far less protein. Fatty acids are derived from anaerobic metabolism of glycogen. Carbohydrate in the host diet induces growth, longevity, and egg production, while elimination of protein has little effect. (5) Oxygen is used in respiration if available, but anaerobic respiration predominates. (6) Eggs eaten by the intermediate host must experience both acid and alkali and hence are released only in the intestine. A cysticercus swallowed by the final host meets both acid and alkaline digestive juices, and its bladder is digested away, but the inverted scolex (exposed only to alkaline juice) is not; bile stimulates the scolex to evert.

In addition to the above adaptations seen in cestodes, a few adaptations are common to both trematodes and cestodes and to internal parasites in general: (1) sense organs are usually much reduced or absent, (2) some mechanism of attachment to the host tissues is developed (hooks, suckers), and (3) usually there is increased egg production coupled with a larval stage to facilitate survival and infection of new hosts.

18-15 Other cestodes More than 1500 species of tapeworms live in various vertebrates from fishes to mammals. The adult stage of each is usually in a final host that preys upon the intermediate host, consuming the flesh of the latter, which contains the infective stages. Larvae of many groups are in arthropods; those of most CYCLOPHYLLIDEA are in mammals. Some tapeworms are harmless; others, particularly in the larval stage, may produce severe symptoms and occasionally death. Human infections are decreasing in industrialized countries because of a

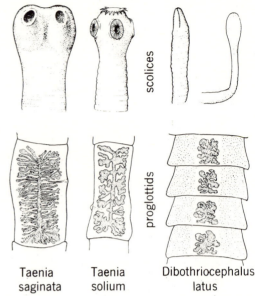

scolices

proglottids

Taenia saginata Taenia solium Dibothriocephalus latus

Figure 18-14 Scolices and mature ("ripe") proglottids of three human tapeworms (beef, pork, and fish).

wider knowledge of how infections may be avoided, because of the detection and treatment of persons harboring tapeworms, and because of the inspection and refrigeration of meat. A few common tapeworms are:

Beef tapeworm, *Taenia saginata:* Larvae in flesh of cattle, adult worms in man; 4 to 12 m (13 to 39 ft), even 25 m (82 ft) long, proglottids up to 2000 (Fig. 18-14). Preference for rare beef may cause an increase in human infections.

Dog and cat tapeworm, *Taenia pisiformis:* Larvae in liver and mesenteries of rabbits, adults in dogs and cats.

Dog tapeworm, *Dipylidium caninum:* Larvae in biting dog louse and fleas of dogs, cats, and humans; adults in dogs and cats, occasional in humans; 15 to 40 cm (6 to 15½ in) long, not over 200 proglottids, each with 2 sets of reproductive organs.

Rat tapeworm, *Hymenolepis diminuta:* Larvae in earwigs, flour beetles, fleas, other insects, and myriapods; adults common in rats and mice, occasional in humans; 20 to 60 cm (8 to 23 in) long, 800 to 1000 proglottids.

Dwarf tapeworm, *Hymenolepis nana:* No interme-

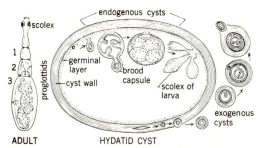

Figure 18-15 Life cycle of the hydatid worm, *Echinococcus granulosus.* The adult, 3 to 6 mm (⅛ to ¼ in) long, of 3(or 4) proglottids, lives in carnivorous mammals; the larval stage is in various domestic and wild herbivorous mammals and humans, forming multiple cysts.

diate host required, but flour beetles and fleas may serve as such; larvae live in intestinal villi and adults in intestine of humans and rodents; 10 to 45 mm (½ to 1¾ in) long, to 200 proglottids.

Sheep tapeworm, *Moniezia expansa:* Larvae in free-living mite (*Galumna*); adults in sheep, goats, etc.; up to 6 m (19½ ft) long.

Hydatid worm, *Echinococcus granulosus* (Fig. 18-15): Adults in intestine of dog, wolf, jackal, only 3 to 6 mm (⅛ to ¼ in) long, 3 (or 4) proglottids; larvae in humans, cattle, sheep, pigs, and other domestic and wild animals; larval cysts up to football size after 12 to 20 years in humans, the size depending on the organ invaded; delicate inner

lining of cyst forms brood capsules, each with numerous scolices; daughter or "exogenous" cysts also form and may scatter to secondary centers in body; dangerous and sometimes fatal to humans; almost cosmopolitan. Another hydatid worm, *E. multiocularis,* may infect humans. Adults are usually in foxes, larvae in various rodents.

Fish tapeworms, *Dibothriocephalus latus* (Fig. 18-16): First larval stage (0.5 mm long) in copepod crustaceans (*Diaptomus, Cyclops*), second (1 to 20 mm) in perch, turbot, trout, and other freshwater fishes; adult worms in humans, pigs, cats, dogs, foxes or other fish-eating mammals; to 18 m (60 ft) long, with up to 4000 proglottids; human infections from eating fish raw or insufficiently cooked; common in Europe and Japan; also present in northern United States.

Classification

Phylum Platyhelminthes.

Flatworms. Body soft, flattened dorsoventrally, often elongate, covered by ciliated epidermis or by cuticle; some with external suckers or hooks, or both; digestive tract incomplete or none; excretion by flame-bulb protonephridia with ducts; no coelom; body spaces filled with parenchyma; ner-

Figure 18-16 Life cycle of the broad tapeworm of fish, *Dibothriocephalus latus.* Adult worm and fish host much reduced, larval stages variously enlarged. (*Adapted partly from Kükenthal.*)

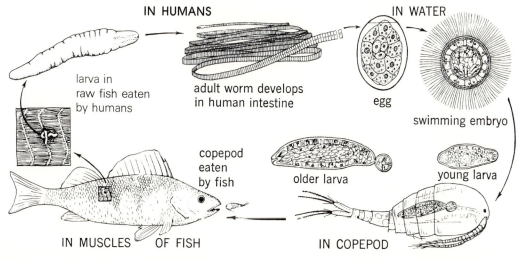

vous system with anterior ganglia or nerve ring and 1 to 3 pairs of longitudinal nerve cords; no skeletal, circulatory, or respiratory organs or true segmentation; usually monoecious; fertilization internal; eggs microscopic, with shell and containing yolk cells; development direct or with larval stages; free-living, commensal, or parasitic; more than 10,000 species.

Class 1. Turbellaria.

Free-living flatworms. Body undivided; epidermis with cilia, rodlike rhabdites, and many mucous glands; usually pigmented, some species brilliantly marked; usually with ventral mouth and intestine (except Acoela); development usually direct; asexual reproduction in some.

Order 1. Acoela. Length 1 to 4 mm; with mouth and occasionally a pharynx but no intestine, protonephridia, oviducts, or definite gonads; marine. *Convoluta; Amphiscolops,* on sargassum weed; *Ectocotyla,* commensal on hermit crab.

Order 2. Catenulida. Length 0.9 to 9 mm ($\frac{1}{2}$ in); single or in chains up to 32 zooids; slender, threadlike, pale; pharynx simple; intestine without diverticula; 1 median protonephridium; in stagnant fresh water. *Catenula, Stenostomum.*

Order 3. Macrostomida. Length 0.8 to 11 mm ($\frac{1}{2}$ in), single or in chains up to 18 zooids; flat or cylindrical; color pale; epidermis of *Microstomum* often with nematocysts obtained from feeding on *Hydra;* pharynx simple; intestine with small lateral diverticula or a preoral blind sac; 2 protonephridia; in fresh water. *Macrostomum, Microstomum.*

Order 4. Neorhabdocoela. Length 0.3 to 15 mm ($\frac{1}{16}$ to $\frac{3}{5}$ in); colorless or pale, some green with zoochlorellae; pharynx bulbous; intestine variable; 2 protonephridia; in salt water, brackish water, or fresh water, some commensal or parasitic. *Mesostoma, Castrella, Castrada, Phaenocora.*

Order 5. Temnocephalida. Length to 14 mm (to $\frac{3}{5}$ in); flat, mostly colorless; 2 to 12 fingerlike tentacles anteriorly; a posterior adhesive disk; pharynx cask-shaped, intestine saclike; 2 protonephridia; commensal mainly on freshwater crustaceans, also on turtles and snails. *Temnocephala.*

Order 6. Archoophora (Proplicastomata). To 1 mm ($\frac{1}{25}$ in) long; pharynx plicate; no female ducts;

frontal gland. *Proporoplana* is the only genus, marine.

Order 7. Lecithoepitheliata. To 6 mm ($\frac{1}{4}$ in) long; marine or freshwater forms without yolk glands and usually a penis stylet; pharynx simple. *Prorhynchus, Geocentophora.*

Order 8. Prolecithophora (Holocoela, Cumulata). Marine and freshwater worms with distinct yolk glands, intestine simple, and unarmed penis. *Plagiostomum, Hydrolimax.*

Order 9. Seriata. Marine and freshwater worms with a lobulated intestine; pharynx plicate; most with statocyst; with numerous yolk glands. *Monocelis, Otoplana, Bothrioplana.*

Order 10. Tricladida. Usually small—2 or 3 to 500 mm (20 in); mouth midventral, with proboscis; intestine 3-branched; 1 to 4 protonephridial tubules on each side. *Dugesia* (= *Euplanaria*), pigmented, and *Procotyla* and *Dendrocoelum,* milky white, all in fresh waters; *Bdelloura,* on gills of horseshoe crab; *Procerodes, Spalloplana,* white, eyeless, in Mammoth Cave, Ky.; *Bipalium,* common in greenhouses; *Geoplana,* terrestrial.

Order 11. Polycladida. Small to 150 mm (6 in) long; usually thin and oval; eyes many; intestine with many irregular branches; marine, a few pelagic. *Notoplana (Leptoplana), Planocera; Stylochus,* often feeds on oysters.

Class 2. Trematoda.

Flukes. Body undivided, covered by cuticle (no epidermis or cilia); one or more suckers for attachment; mouth usually anterior and digestive tract 2-branched; 1 ovary; all parasitic.

Subclass 1. Monogenea. Oral sucker weak or none; posterior end with adhesive disk, usually with hooks; excretory pores 2, anterior, dorsal; eggs few; larva ciliated; no intermediate host; chiefly ectoparasites of cold-blooded vertebrates, a few on cephalopods and crustaceans.

Order 1. Monopisthocotylea. Trematodes with an opisthaptor, a simple posterior organ of attachment, and a weakly developed (or absent) oral sucker. *Gyrodactylus,* on gills of freshwater fishes.

Order 2. Polyopisthocotylea. Oral sucker present; opisthaptor with several distinct suckers or other modifications for attachment. *Polystoma,* larva in gills of tadpole, adult in bladder of frog.

Subclass 2. Aspidobothrea. No oral sucker or anterior adhesive organs; ventral surface with 1 big sucker or row of suckers; excretory pore 1, posterior; endoparasitic in 1 host. *Aspidogaster,* in pericardial and renal cavities of freshwater bivalves (UNIONIDAE); *Stichocotyle,* slender, in spiral valve and bile duct of skates.

Subclass 3. Digenea. Suckers usually 2, 1 around mouth, 1 ventral; no hooks; excretory pore 1, posterior; uterus long, eggs many; 1 or more larval stages reproducing in intermediate host(s) before metamorphosis to adult form; chiefly endoparasites, larvae in mollusks (also crustaceans, fishes), adults in vertebrates.

Order 1. Strigeatoidea. Cercaria larvae with forked tails; miracidia with 2 protonephridia (primitive); *Schistosoma,* the human blood fluke.

Order 2. Echinostomida. Cercaria larvae with unforked tails; miracidia with 2 protonephridia; *Fasciola,* sheep liver fluke.

Order 3. Opisthorchida. No stylet; excretory ducts posterior in the tail; *Opisthorchis,* Chinese liver fluke.

Order 4. Plagiorchiida. Stylets present, tail without excretory ducts; *Prosthogonimus.*

Class 3. Cestoda.

Tapeworms. Body covered with cuticle, no epidermis or external cilia; unpigmented; no digestive tract or sense organs in adult; usually an anterior scolex for attachment, with adhesive grooves (bothria), or leaflike outgrowths (bothridia), or suckers, or hooks; body usually of one to many proglottids (segments), each containing 1 or 2 complete monoecious reproductive systems; embryo hooked; all endoparasitic, usually with alternate hosts, adults in intestines of vertebrates.

Subclass 1. Cestodaria. Body undivided; no scolex; larva 10-hooked.

Order 1. Amphilinidea. No suckers or bothria; with proboscis, or boring muscle. *Amphilina,* in coelom of sturgeon.

Order 2. Gyrocotylidea. An anterior cuplike sucker and a posterior rosettelike organ of attachment. *Gyrocotyle,* in intestine of chimaeroid fishes.

Subclass 2. Eucestoda. Body long, ribbonlike, with 4 to 4000 proglottids; scolex with adhesive organs; embryo 6-hooked.

Order 1. Proteocephaloidea. Scolex with 4 lateral suckers and terminal sucker or glandular organ. *Proteocephalus,* in freshwater fishes, amphibians, and reptiles.

Order 2. Tetraphyllidea (Phyllobothrioidea). Scolex with 4 bothridia, often with hooks. *Phyllobothrium,* in elasmobranchs.

Order 3. Disculicepitidea. Scolex a large cushionlike pad without suckers or hooks; proglottids square. *Disculiceps,* in gray shark, *Carcharias,* Massachusetts.

Order 4. Lecanicephaloidea. Scolex of 2 parts, upper with disk or branches, lower with 4 suckers. *Polypocephalus,* in elasmobranchs.

Order 5. Trypanorhyncha (Tetrarhynchoidea). Scolex with 4 bothridia and 4 slender, protrusible spiny proboscides. *Tentacularia (Tetrarhynchus),* larvae in marine invertebrates or bony fishes, adults in elasmobranchs.

Order 6. Cyclophyllidea (Taenioidea). Scolex with 4 deep suckers and often with hooks at tip; sex openings lateral; proglottids free or joined when mature; includes most tapeworms of higher animals and humans. *Dipylidium, Echinococcus, Hymenolepis, Moniezia, Taenia.*

Order 7. Aporidea. Scolex with 4 suckers, rostellum armed; no yolk glands, sex ducts, or pores. *Gastrotaenia,* in swans.

Order 8. Nippotaeniidea. Small; no scolex; 1 terminal sucker. *Nippotaenia,* in freshwater fishes of Japan.

Order 9. Caryophyllidea. Scolex ill-defined, no true suckers or bothria; only 1 set of reproductive organs. *Caryophyllaeides,* in freshwater fishes; *Archigetes,* matures in freshwater oligochaetes.

Order 10. Spathebothridea. Scolex without true bothria or suckers; body lacks proglottids; testes in two lateral bands. *Spathebothrium,* in marine fishes.

Order 11. Pseudophyllidea. Scolex not always

distinct, bothria 2 to 6, some lack adhesive organs. *Triaenophorus,* larvae in copepods, adults in freshwater fishes; *Dibothriocephalus latus,* fish tapeworm of humans.

References

Brumpt, E. 1949. Précis de parasitologie. 6th ed. Paris, Masson et Cie. xii + 2138 pp., 4 pls., 1308 figs.

Bychowsky, B. E. 1961. Monogenetic trematodes, their systematics and phylogeny. J. W. Hargis, Jr. (editor), translated by P. C. Oustinoff. Washington, D.C., American Institute of Biological Sciences. xx + 627 pp., 315 figs.

Caullery, Maurice. 1952. Parasitism and symbiosis. Translated by A. M. Lysaught. London, Sidgwick & Jackson, Ltd. xii + 340 pp., 80 figs.

Chandler, A. C. 1961. Introduction to parasitology. 10th ed. by E. C. P. Read. New York, John Wiley & Sons, Inc. xii + 822 pp., 342 figs.

Dawes, Ben. 1946 The Trematoda with special reference to British and other European forms. New York, Cambridge University Press. xvi + 644 pp., 81 figs.

Erasmus, D. A. 1973. The biology of trematodes. New York, Crane Russak and Co. 312 pp., illus.

Faust, E. C., and **P. F. Russell.** 1964. Craig and Faust's Clinical parasitology. 7th ed. Philadelphia, Lea & Febiger, 1099 pp., 8 col. pls., 352 figs.

Florkin, M., and **B. T. Scheer.** 1968. Chemical zoology. Vol. II Porifera, Coelenterata and Platyhelminthes. New York, Academic Press, Inc. xviii + 639 pp.

Grassé, Pierre-P. 1961. Traité de zoologie. Paris, Masson et Cie. Vol. 4, fasc. 1. Platyhelminthes, 692 pp., 587 figs. Classification by J. G. Baer and others.

Hyman, Libbie H. 1951. Platyhelminthes. In The invertebrates. New York, McGraw-Hill Book Company. Vol. 2, Platyhelminthes and Rhynchocoela. pp. 52–458, figs. 13–173, 1959. Vol. 5, Smaller coelomate groups, pp. 731–738.

————, and **E. R. Jones,** 1959. Turbellaria. In W. T. Edmondson (editor), Ward and Whipple, Freshwater biology. 2d ed. New York, John Wiley & Sons, Inc. Pp. 323–365, 72 figs.

Noble, E. R., and **G. A. Noble.** 1972. Parasitology: The biology of animal parasites. 3d ed. Philadelphia, Lea & Febiger. 617 pp.

Riser, N. W., and **M. P. Morse.** 1974. Biology of the Turbellaria. New York, McGraw-Hill Book Company. xxv + 530 pp., illus.

Wardle, R. A., and **J. A. McLeod.** 1952. The zoology of tapeworms. Minneapolis, The University of Minnesota Press. xxiv + 780 pp., 419 figs. Reprinted 1968. New York, Stechert-Hafner, Inc.

Yamaguti, S. 1958. Systema helminthum. Vol. 1. The digenetic trematodes of vertebrates. New York, Interscience Publishers, Inc. Parts 1 and 2, x + 1575 pp., 106 pls. 1959. Vol. 2. The cestodes of vertebrates, 860 pp., 70 pls.

Phylum Nemertinea (Ribbon worms)

18-16 The nemerteans. These are slender worms with soft, flat, unsegmented bodies, capable of great elongation and contraction. A few are only 5 mm long, many are a few centimeters in length, and one species may stretch to 30 m (100 ft). The 650 species are variously red, brown, yellow, green, or white, some solidly colored and others striped or crossbanded. Most nemerteans are marine, living closely coiled beneath stones, among algae, or in burrows, usually in shallow water, but some are inhabitants of deep water and a few are pelagic. The genus *Prostoma* lives in fresh waters, the tropical *Geonemertes* in moist earth, *Malacobdella* in the mantle cavities of bivalve mollusks, and *Carcinonemertes* among the gills or egg masses of certain crabs. There are no true parasites.

18-17 Characteristics

1   Symmetry bilateral, three germ layers.

2   Body slender, soft, highly contractile, and unsegmented.

3   An eversible proboscis separate from the digestive tract in a fluid-filled cavity, the rhynchocoel.

4   Digestive tract straight and complete.

5   No coelom or respiratory organs; body spaces filled with mesenchyme.

6   Circulatory system with two major pulsating longitudinal vessels, often additional vessels; no heart; two lateral excretory canals, branched, with flame cells.

7   Nervous system with paired anterior ganglia and

a pair of lateral longitudinal nerves, some species also with middorsal and midventral nerve trunks.

8 Sexes usually separate, with multiple pairs of gonads, and development direct or through a pilidium larva; asexual reproduction by fragmentation.

9 Body wall with circular and longitudinal muscle bands in two or three layers.

The NEMERTINEA appear to be closely related to the PLATYHELMINTHES and may have evolved as an offshoot from the free-living TURBELLARIA, with which they share a number of features. Similar features include the acoelomate body plan, a complex ciliated epidermis containing rhabdites and gland cells, presence of a protonephridial excretory system with flame cells, similar nervous system and sense organs, absence of respiratory system, and lack of external segmentation. Nemerteans have, however, become more highly organized than turbellarians and represent the culmination of acoelomate development and evolution. They differ from turbellarians in having a complete digestive tract with mouth and anus, a closed circulatory system, a unique proboscis, and a more simplified reproductive system without accessory organs and copulatory apparatus and in being generally dioecious as opposed to the hermaphroditic turbellarians. Larval development, where it occurs, is also strikingly different.

18-18. Structure and natural history (Fig. 18-17) The soft **epidermis** is of columnar cells interspersed with gland cells and sensory cells and backed by a connective tissue dermis. Beneath this are muscle layers, both circular and longitudinal. **Mesenchyme** tissue fills all the internal spaces between the muscle layers and the proboscis sheath and digestive tract. Locomotion is by action of the external cilia and the muscles. At the anterior end, above the mouth, is a pore through which the soft **proboscis** may be extended far out or completely withdrawn, like inverting the finger of a glove; the proboscis is contained in a cavity **(rhynchocoel)** inside the body (Fig. 18-18). It is thought to serve for offense and defense, being armed with stylets or spines in some species. The rhynchocoel sheath is muscular, and the muscles acting on the fluid-filled cavity function as a hydraulic system to push out the proboscis. Retraction is via a muscle. The **digestive tract** is a ciliated tube, extending the entire length of the body, having both a mouth and an anus, and sometimes also paired lateral caeca; the food is other animals, both living and dead. The **circulatory system** includes a pair of lengthwise lateral vessels and usually a middorsal vessel, with cross connections at the head and about the digestive tract (Fig. 6-1). The blood fluid is colorless and circulates back and forth as a result of movements of the body and pulsations of the vessels themselves. The path is irregular. The blood contains corpuscles bearing pigments of various colors which impart a red, orange, green, or yellow color to the whole blood.

Figure 18-17 A ribbon worm (phylum NEMERTINEA). *A*. External form with proboscis extended. *B*. Internal structure, diagrammatic. Left side removed except for part at middle of body; proboscis retracted into sheath.

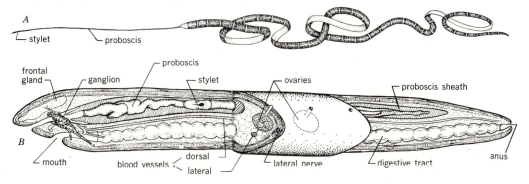

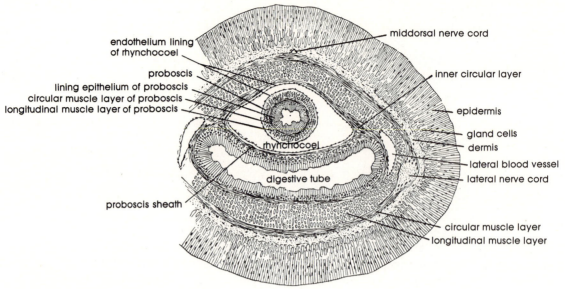

Figure 18-18 A ribbon worm (phylum NEMERTINEA). Cross section showing proboscis in rhynchocoel sheath, digestive tract, and body-wall layers. (*After Hyman, 1951.*)

Respiration is by diffusion through the body wall. The **excretory system** comprises a pair of lateral canals with numerous branches and flame cells that lie close to blood vessels. Waste is picked up from the blood by the flame bulbs and transported via the canals to exit through the nephridiopores. The **nervous system** includes two large cerebral ganglia in the head region, each of which is bilobed, one lobe located dorsally and one ventrally to the rhynchocoel on each side. These lobes are connected by commissures to encircle the rhynchocoel. From these ganglia two large lateral nerves extend posteriorly along the side of the body, with ganglion cells and cross connectives; some species have also a middorsal and a midventral nerve trunk with nerves under the epidermis or among the muscles. The **sense organs** include few to many pigment-cup ocelli in the head region, variously arranged cephalic grooves in the head (probably chemosensory), and the unique cerebral or cephalic sense organs which connect directly to the brain but whose function is not understood. Tactile sensory cilia occur over much of the body.

The sexes are usually separate (a few species are monoecious, some protandric). The **reproductive system** consists of pairs of simple sacs or multiple pairs of gonads between the intestinal pouches, developed within the parenchyma and opening directly on the body surface when the eggs or sperm are mature. There are no copulatory organs, and eggs and sperm are usually discharged into seawater, where fertilization occurs. Cleavage is spiral and determinate, giving rise to a hollow blastula followed by a solid gastrula. Certain nemerteans (LINEIDAE) have a free-swimming **pilidium larvae;** development in others is direct, and some are viviparous. The pilidium larvae undergo a strange development to form the adult. Sections of the pilidium ectoderm sink into the larva and rearrange themselves to form a young worm, which subsequently "hatches" from the pilidium. The adults have great powers of regeneration, and some reproduce regularly during warm weather by fragmentation of the body, each piece then growing into a complete worm.

Classification

Phylum Nemertinea

(Nemertea, or Rhynchocoela; see Characteristics).

Class 1. Anopla.

Mouth posterior to brain; proboscis unarmed; nervous system under epidermis or among muscles.

Order 1. Palaeonemertini. Muscles of body wall in 2 or 3 layers (innermost circular); dermis gelatinous. *Tubulanus.*

Order 2. Heteronemertini. Muscles of body wall in 3 layers (innermost longitudinal); dermis fibrous. *Lineus, Cerebratulus.*

Class 2. Enopla.

Mouth anterior to brain; proboscis may be armed; nervous system inside muscles of body wall.

Order 1. Hoplonemertini. Proboscis with 1 or more stylets; intestine straight with paired diverticula. *Paranemertes; Carcinonemertes,* commensal on gills and eggs of crabs; *Prostoma,* in fresh waters; *Geonemertes,* terrestrial on shores; *Pelagonemertes,* in North Atlantic at depths of 1000 m (3300 ft) or more.

Order 2. Bdellonemertini. Proboscis unarmed; intestine sinuous, no diverticula. *Malacobdella,* in mantle cavity of marine clams.

References

Bayer, F. M., and **H. B. Owre.** 1968. The free living lower invertebrates. New York, The Macmillan Company. viii + 229 pp., 271 figs.

Coe, W. R. 1905. Nemerteans of the west and northwest coasts of North America. *Bull. Museum Comp. Zool.* vol. 47, pp. 1–318, pls. 1–25, figs. 1–62.

———. 1943. Biology of the nemerteans of the Atlantic coast of North America. *Conn. Acad. Arts Sci. Trans.,* vol. 35, pp. 129–328, 4 pls., 1 col.

Gibson, R. 1972. Nemerteans. London, Hutchinson & Co. (Publishers), Ltd. 224 pp., 33 figs. 2 pls.

Gontcharoff, Marie. 1961. Nemertiens. In P.-P. Grassé, Traité de zoologie. Paris, Masson et Cie. Vol. 4, fasc. 1, pp. 783–886, figs. 686–788, 3 col. pls.

Hyman, L. H. 1951. Rhynchocoela. In The invertebrates. New York, McGraw-Hill Book Company. Vol. 2, pp. 459–531, figs. 174–208. 1959. Vol. 5, pp. 738–739.

19

THE PSEUDOCOELOMATE PHYLA

The PLATYHELMINTHES and NEMERTINEA represent the highest development of metazoan animals which lack any internal body cavity. The great majority of animals, and all considered subsequently, possess some sort of internal fluid-filled body cavity as adults, though in some phyla this cavity is much reduced. That such a cavity should be developed in all successful phyla and a majority of all animal species suggests that it has contributed some useful survival value to the organisms. Scientists have speculated and continue to speculate on what this value (or values) may be, and although there is no universally accepted single reason, suggested reasons include increased freedom of movement for the internal organs, ability to use the fluid in hydraulic systems and skeletons, provision of more space for internal

organs, aid in the circulation of materials through the body and in locomotion (particularly for worm-like animals).

One such body cavity is the pseudocoelom, and the pseudocoelomate animals include the following phyla: ROTIFERA, GASTROTRICHA, NEMATODA, NEMATOMORPHA, KINORHYNCHA, ACANTHOCEPHALA, and ENTOPROCTA. This is a very diverse group in which relationships are often not clear. Phylogentically certain of the phyla may not be closely related or stem from a single ancestral line. Others of the phyla seem more closely related, and in fact, the ROTIFERA, GASTROTRICHA, NEMATODA, NEMATOMORPHA, and KINORHYNCHA were formerly included in a single phylum, the ASCHELMINTHES.

With the exception of the NEMATODA, the pseudocoelomate phyla are all small in terms of numbers of species, and most are also small in size, with rotifers and gastrotrichs, for example, being of the same size as ciliate protozoans.

The groups formerly considered as Aschelminthes share a few similarities, such as possession of a pseudocoelom, absence of a well-formed head, body covered with a cuticle, restriction or absence of epidermal body cilia, absence of respiratory and circulatory systems, a specialized pharynx, and a complete digestive tract. Presently biologists disagree as to whether these features are important enough to retain the phylum Aschelminthes or whether the considerable differences (discussed below) among these groups warrant separate phylum status for each. We have taken the latter route here because it is consistent with the majority of current texts in invertebrate zoology.

Phylum Nematoda (Nematodes)

19-1 Nematodes The phylum Nematoda (Gr. *nematos,* thread) comprises the unsegmented roundworms with slender cylindrical bodies, resistant cuticle, and a triradial arrangement of mouth structures. Among metazoan animals they are probably second only to insects in numbers of individuals. Some marine bottom muds have been estimated to have more than 4 million nematodes per square meter, while good soil may have billions per acre. They are found everywhere; perhaps no other taxonomic group is so pervasive in habitat. Many are free-living in soil and water, and many others are parasites in the tissues or fluids of animals and plants. Still others are restricted to peculiar habitats: the roots of plants, seeds of wheat, gum of tree wounds, and the intestines, blood, and other organs in animal bodies. They are mostly small or minute, but a few grow to a meter (3 ft) in length. Their eggs are microscopic and resistant to adverse environmental conditions.

19-2 Characteristics

1 Symmetry bilateral; three germ layers; no true segmentation, appendages, or proboscis.

2 Body slenderly cylindrical, tapered at both ends, covered with tough resistant cuticle; epidermis unique, with cell nuclei restricted to four cords arranged middorsally, ventrally, and laterally.

3 Digestive tract complete, a straight tube with mouth and anus at opposite ends of body.

4 Body wall with longitudinal muscle fibers only; space within body an unlined pseudocoel.

5 No circulatory or respiratory organs; excretory organs simple, two, one, or none.

6 Nerve ring around esophagus connecting to six anterior nerves and to several posterior nerves or cords.

7 Sexes usually separate, male smaller than female; gonads continuous with reproductive ducts, single (or double) in male, paired (or single) in female; fertilization internal; eggs microscopic, each covered by a chitinous shell; development determinant and direct, "larva" with several molts; no asexual reproduction or regeneration.

8 Internal organs usually with a definite number of cells (cell constancy).

The nematodes differ from flatworms in shape, in the absence of cilia and suckers, in the presence of a complete digestive tract and a body cavity, and in having separate sexes. Their muscles are all longitudinal, aligned with the body axis, and restrict movement to only dorsoventral and lateral bending. They are thus less complex than those in flatworms, nemerteans, and annelids.

Zoologic opinion has varied as to the proper place of nematodes in classification. They have often been considered a class in the phylum Aschelminthes but are now generally considered to be a separate phylum.

19-3 Structure of Ascaris The common intestinal roundworm of humans and of the pig (*Ascaris lumbricoides*) shows the general features of a nematode (Fig.

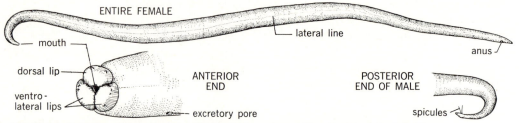

Figure 19-1 Phylum NEMATODA. External features of the intestinal roundworm of the pig, *Ascaris lumbricoides*.

19-1). The female is 20 to 40 cm (8 to 16 in) long and about 0.6 cm ($\frac{1}{4}$ in) in diameter; the male is of lesser size, 15 to 25 cm (6 to 10 in) long. Fresh specimens are yellow to pink in color. The **body** is slender and round, tapering toward either end. It is covered with smooth, tough, elastic **cuticle,** bearing minute striations. Four whitish **longitudinal lines** extend along the body, one dorsal, one ventral, and two lateral. The **mouth** opens at the anterior end, between three rounded **lips;** the dorsal lip bears two double papillae, and each ventrolateral lip has one double and two single papillae. The **anus** is a transverse slit close to the posterior end of the ventral surface. The **male** has a sharply curved posterior end, and two penial spicules project from the male genital pore just within the anus. The **female** is straighter, and the genital pore, or vulva, is midventral on the body at about one-third the distance from the anterior end.

The **body wall** (Fig. 19-2) is thin, consisting of (1) the **cuticle,** which is noncellular and secreted by the epidermis; (2) the **epidermis,** a thin protoplasmic layer that contains nuclei but no cell walls, hence is a syncytium; and (3) the **muscle layer,** of longitudinal fibers. Each muscle cell consists of a spindle-shaped fiber extending lengthwise beneath the epidermis and a club-shaped process that contains the nucleus and projects medially. The muscle layer is divided into four lengthwise parts by the riblike inner extensions of the four lateral lines. These lines and the muscle cells are the irregular outer boundary of the body space, or **pseudocoel,** in which the other internal organs lie free.

The straight **digestive tract** (Fig. 19-3) extends the length of the body and consists of (1) the **mouth;** (2) a small **buccal cavity;** (3) a muscular sucking **pharynx,** or esophagus, about 1 cm ($\frac{1}{2}$ in) long, that acts to draw in food; (4) a long narrow **intestine,** which is nonmu-

Figure 19-2 *Ascaris.* *A.* Cross section of a female. *B.* One muscle cell. Both enlarged.

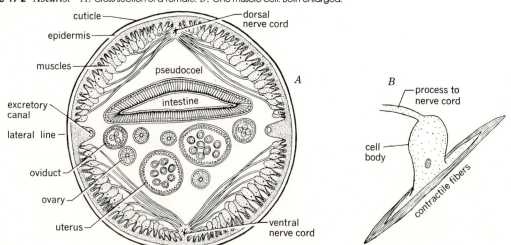

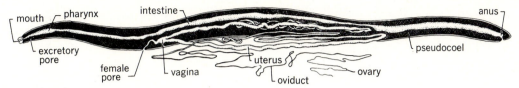

Figure 19-3 *Ascaris.* Internal structure of a female.

scular, composed of a single layer of tall endoderm cells that absorb the digested food; and (5) a short **rectum** discharging through (6) the **anus.** No circulatory or respiratory organs are present. The inner part of each lateral line contains an **excretory canal,** both emptying through a minute midventral pore just behind the mouth. A **nerve ring** surrounds the esophagus; it connects to six short anterior nerves and to six posterior **nerve cords,** a larger dorsal and ventral cord each in the corresponding longitudinal line and others in the body wall, with various lateral branches and cross connectives. The principal sense organs are on the body surface and include various papillae. Free-living nematodes have chemosensory **amphids,** which are cuticular pits containing highly modified cilia. A few have eyespots.

Each reproductive organ is a slender tube, closed at its inner end, of gradually increasing diameter, coiled back and forth in the body space and unattached save where connected to the genital pore. The gonad and reproductive duct are continuous. The **male reproductive system** is single; its parts are (1) the **testis** for sperm production; (2) the **ductus deferens** for conduction; (3) the **seminal vesicle** for storing mature sperm; (4) the **ejaculatory duct** for ejecting sperm; and (5) a sac containing two **penial spicules,** which are inserted in the female's vulva to join the male and female during copulation. The **female reproductive system** is ≺-shaped; each branch is up to 125 cm (4 ft) long and consists of (1) an **ovary,** (2) an **oviduct,** and (3) a **uterus.** The two uteri are joined in (4) a single short **vagina** opening at (5) the **vulva.**

19-4 Natural history The adult ascarid lives as a parasite within the intestine of its host and has problems of existence somewhat different from those of a free-living animal. Locomotion and maintenance of position are accomplished chiefly by dorsoventral or lateral bendings of the body. The cuticle protects the living worm against the digestive juices of its host (antienzymes also may be a factor). Food is obtained from the semifluid materials in the host's intestine, being pumped in by the worm's muscular esophagus, and after digestion passes through its intestinal wall to be distributed by fluid in the body space to other tissues. Respiration is primarily anaerobic, since the intestinal contents of its host contain little free oxygen. The sensory papillae probably are receptors for chemical and tactile stimuli.

19-5 Reproduction Male and female worms copulate within the host's intestine. The eggs are fertilized in the oviducts of the female, and each becomes covered by a tough shell. A large female may contain 27 million eggs at one time and lay 200,000 or more per day. The eggs pass out of the female, into the host's intestine, and leave with the feces. A period of development is necessary before they become infective for another host. In drought or cold they may lie dormant for many months, but in a warm, moist, shady site development requires 3 or 4 weeks. If such "embryonated" eggs (containing embryo worms) are swallowed by the proper host with food or water, they pass to the intestine. There the larvae (0.2 to 0.3 mm long) latch, burrow into the veins or lymph vessels in the intestinal wall, and travel through the heart and to the pulmonary capillaries in the lungs, meanwhile growing in size. In a few days they break into the air passages and move via the trachea, esophagus, and stomach to the intestine again, where they grow to maturity.

No intermediate host is necessary to complete the life cycle (Fig. 19-4). Young pigs usually acquire larvae from infected soil in pig yards or in dirt on the sow's udder when nursing. Adult pigs are relatively

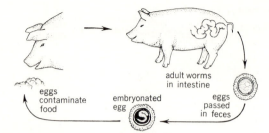

Figure 19-4 *Ascaris.* Life cycle in the pig.

resistant to infection; therefore any worms they contain were probably acquired when young. Human infection with *Ascaris* is very common; a high percentage of persons in some rural localities are infected. Worms found in humans and pigs are identical structurally but differ physiologically in that infective eggs of the human *Ascaris* do not ordinarily produce mature worms in pigs, and vice versa. Passage of many larvae through the lungs may cause local inflammation and symptoms like pneumonia. Adult ascarids in the intestine may produce secretions toxic to the host, and the worms when numerous may obstruct the intestine. Worms sometimes migrate up to the mouth or nose or even penetrate the intestinal wall and invade other organs, causing serious illness or death of the host. Some animals and people become sensitized or allergic to the secretions of ascarids.

19-6 Other nematodes Many species, both parasitic and free-living, resemble *Ascaris* in general structure but differ in various details. Some parasitic forms have teeth, hooks, or cutting plates in the mouth for attachment and feeding; others that live on plant roots have a sharp, hollow "spear" to puncture the cells and a muscular pharynx to withdraw cell sap. The cuticle on many free-living species bears minute bristles, spines, or transverse rows of scales that aid in crawling through soil. Some predatory nematodes have enlarged mouths margined with teeth or other projections (Fig. 19-5). The orders of nematodes are diverse in structure of mouth and esophagus, with lesser differences in muscles, excretory organs, and reproductive structures. Males may have two, one, or no penial spicules, and in some the posterior end has an expanded caudal bursa to hold the female while mating. The vulva is variously located, and females of some species have a single ovary, oviduct, and uterus.

There is a great host of free-living nematodes whose life histories are virtually unknown. They occur in moist soil and in some deserts, in the sand of ocean beaches, in fresh water, stagnant water, and salt water, on the shores and bottoms of lakes and rivers, on the filter beds of waterworks, on mountain tops, in hot springs, and in polar seas and ice. The upper 8 cm (3 in) of an acre of alluvial soil may contain 3 billion nematodes and the top 2.5 cm (1 in)

Figure 19-5 Some free-living nematodes; all much enlarged. (*After Cobb, 1915.*)

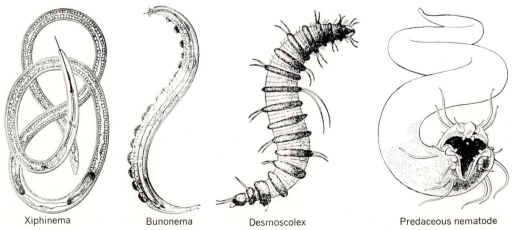

Xiphinema   Bunonema   Desmoscolex   Predaceous nematode

of beach sand a third as many. Other nematodes live in various parts of plants: roots, seeds, fruits, gums, leaf axils, bark crevices, and galls; the eggs or larvae of some are transported by insects. They are truly ubiquitous animals. It has been estimated that only one-fiftieth of the nematode species have as yet been described.

19-7 Physiology and ecology Despite the fact that the best-known nematodes are parasitic, there are many free-living forms, and all types of feeding and nutrition are found. Nematodes may feed on algae, on the fluids of large plants, on detritus, and on dead organic matter. Many are predaceous on other organisms. In predaceous forms the mouth parts may be modified into jaws for seizing or a stout piercing spine for puncturing the prey organisms. Enzymes are secreted into the esophagus and intestine, where digestion takes place extracellularly in the lumen. Absorption is from the intestine also.

The nematode excretory system is peculiar in that it consists of a very few large cells called **renette cells** in the primitive forms and an H-shaped canal system in more advanced forms. Since ammonia is the major nitrogenous waste and is excreted through the body wall, the importance of the simplified excretory system is unknown. It may, however, function in osmoregulation.

Because nematodes have only longitudinal muscles, they are more restricted in types and patterns of movement than other wormlike animals. Locomotion is usually the result of the interaction of the longitudinal muscles and the elastic cuticle and the action of these two on the fluid-filled pseudocoelom. The presence of an aquatic medium appears to be a necessary requisite for locomotion in most nematodes. Most directional movement in nematodes results from undulatory motions produced by the longitudinal muscles acting against particles in the substrate or against a water film. Some nematodes employ a crawling locomotion using bristles on the cuticle and/or adhesive glands.

The dominant sense organs in nematodes are various setae and **amphids** and **phasmids.** Amphids are pits or depressions in the cuticle, commonly near the anterior end, and appear to be chemoreceptors.

Phasmids are glandular structures at the posterior end which are also chemosensory. Eyespots are uncommon in nematodes, being found in a few species.

One of the unique features of the reproductive physiology of nematodes is that the sperm lack flagella and must move by some amoeboid method. The majority of nematodes are dioecious, and internal fertilization is the rule. Development of the eggs often begins in the female reproductive tract prior to egg deposition. The embryology of nematodes is characterized by spiral cleavage and early separation of future gonads from the cells which will constitute the rest of the body. Cell constancy is the rule in nematode internal organs, and most organs attain their total cell count at about the time the eggs hatch. Eggs hatch into juvenile worms (sometimes called larvae) and attain adult size after several molts of the cuticle. Molting does not occur in adults.

19-8 Relation to humans The many species of nematodes parasitic in humans, domesticated animals, and cultivated plants are of great practical importance. Some do little or no damage, but others cause impaired efficiency, unhealthiness, illness, or death of the host. The effects depend upon the species of worm, the number present, and other factors. More than 50 species are parasites of humans. Others parasitize farm livestock and poultry, all kinds of wild vertebrates, and various invertebrates. Almost any organ of a vertebrate may be invaded by some kind of nematode. Each species is usually confined to one or a few related hosts, occasionally invades others, and ordinarily occupies a particular organ. Most parasitic nematodes have a free-living egg or larval stage in soil or water, and the filarial worms require an intermediate host. Occasionally a host contains enormous numbers, such as a pint of stomach worms in a human being or 40,000 in a 25 kg (50 lb) mammal. Infections with intestinal worms can be diagnosed by searching for the microscopic eggs in the feces, those of most species being of characteristic size, shape, and structure. Treatment to eliminate nematodes requires use of a drug, not too toxic for the host, that will cause the worms to sicken and be carried out in the feces.

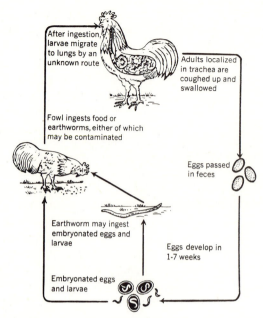

Figure 19-6. After ingestion, larvae migrate to lungs by an unknown route — Adults localized in trachea are coughed up and swallowed — Fowl ingests food or earthworms, either of which may be contaminated — Eggs passed in feces — Earthworm may ingest embryonated eggs and larvae — Eggs develop in 1-7 weeks — Embryonated eggs and larvae

Figure 19-6   Life cycle of the gapeworm, *Syngamus trachea.* (*From Koutz and Rebrassier, Ohio State University Press, 1951.*)

19-9 Root nematodes   Many species of minute to microscopic nematodes live in or about the roots of plants. Some, such as the sugar-beet nematode *Heterodera schachtii*, are restricted to one or a few kinds of plants, but others are less specialized. The common root-knot nematodes *Meloidogyne* (*Heterodera*) *radicicola* and *M. marioni* have been reported in over 1000 varieties of plants and commonly infest over 75 garden and field crops, fruit and shade trees, shrubs and weeds (Fig. 19-7). Their eggs are deposited in roots or soil, and the young upon hatching penetrate rootlets to feed on the tissues within. The roots react by forming small galls of scar tissue—the root knots —about the worms. The adult male is slender (1.2 to 1.5 mm long), and the female becomes a whitish teardroplike object (0.8 to 0.5 mm) with a swollen body producing 500 to 1000 eggs. Fertilization may occur but is unnecessary. The eggs are laid in a yellow gelatinous secretion. The life cycle requires nearly 3 months at 14°C (58°F) but less than a month in soil at 27°C (81°F). Several generations per year are thus possible in the warm Southern and Western states. The larvae within galls can withstand some drying and survive 0°C (32°F) but die at lower temperatures.

Root knot causes weakening or death of the plant, and soil infested with these worms often fails to yield profitable crops. The nematodes are carried into clean soils with plants, soil, manure, farm implements, or even irrigation water from infested lands. Once established, they are difficult to eradicate. Plant breeders have developed selected strains of some economic plants that are resistant to root nematodes. Certain chemicals (methyl bromide, etc.) are used to reduce the worms in infested soils.

19-10 Hookworm   This scourge of humankind lives in moist tropical and subtropical areas. In populations that go barefoot and use no sanitary toilets from 50 to 95 percent may be infected, and workers in mines, tunnels, and brickyards often harbor hookworms. *Ancylostoma duodenale* and *Necator americanus* are both common in humans, the latter especially in the United States.

The adult worm is 8 to 13 mm ($\frac{1}{3}$ to $\frac{1}{2}$ in) long and has cutting plates, or teeth, in its mouth (Fig. 19-8). It

Figure 19-7.   The root-knot nematode, *Meloidogyne* (*Heterodera*). Egg, larva, and adults, all greatly enlarged. Root galls produced by the worms, about natural size. (*After Tyler, 1933.*)

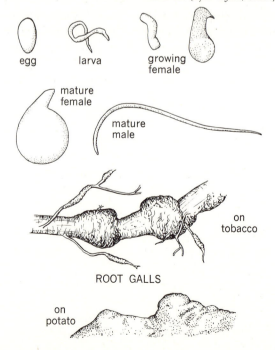

egg   larva   growing female   mature female   mature male   on tobacco   ROOT GALLS   on potato

Figure 19-8 Hookworms. *A.* Mouth of *Ancylostoma duodenale* in anterior view, showing the teeth. *B. A. caninum* attached to intestinal wall in dog. Both enlarged. (*After Stiles; and Loess.*)

attaches to the inner wall of the small intestine, where blood, lymph, and bits of mucous membrane are drawn into the worm's gut by action of the pumping pharynx. Feeding is aided by production of a secretion that prevents coagulation of the host's blood. The worm may pump much more blood than it digests, and the wound it makes may bleed after the worm has moved elsewhere; the host, in consequence, loses blood into the intestine and becomes anemic.

Hookworms mate in the host's intestine, and each female produces several thousand fertile eggs daily, which pass out with the feces. In moist, warm, shady places these hatch in 24 to 48 h (Fig. 19-9); the larvae feed on excrement or other organic debris in the soil, later to undergo two molts. When about 0.5 mm long, they become infective for humans. Entry is usually by burrowing through the soft skin between the toes and on the sides of the foot, causing "ground itch." The larvae travel in blood and lymph vessels, passing through the heart and to the lungs, from capillaries to air cavities, and thence up the trachea into the esophagus and to the intestine. With two further molts they soon mature and may live some months or even several years. Continued reinvasion of the host maintains the infection.

Children harboring a hundred or more hookworms may be retarded in physical and mental growth, and persons of all ages with many worms become anemic; their energy is lessened, and their susceptibility to other diseases is increased. Treatment under medical supervision will rid a person of these parasites. Wearing shoes and sanitary disposal of human feces prevent infection. Other species of hookworm occur in dogs, foxes, cats, cattle, sheep, and pigs.

19-11 Trichina worm *Trichinella spiralis* occurs as a minute living larva encysted in the striated muscles (Fig. 19-10) of pigs, house rats, and humans, also in cats, dogs, black bears, and polar bears. If one of these eats the flesh of another containing such larvae, the cysts are dissolved by the digestive juices and the larvae are liberated in the intestine of the new host. In about 2 days they become sexually mature and mate; male worms are then about 1.5 mm long and females 3 to 4 mm. The females burrow into

Figure 19-9 Development of the hookworm, *Ancylostoma duodenale.* (*After Stiles, 1902.*)

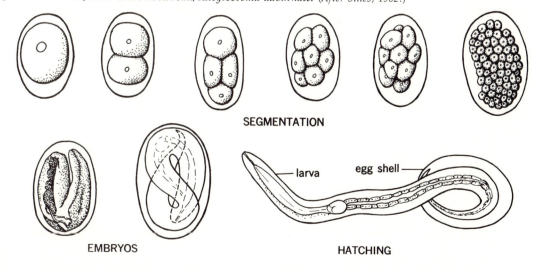

SEGMENTATION

larva   egg shell

EMBRYOS   HATCHING

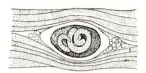

larva encysted in muscle

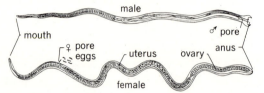

Figure 19-10 **The trichina worm,** *Trichinella spiralis;* **enlarged.** Adults below.

the intestinal wall to live for some time, each producing up to 1500 living larvae about 0.1 mm long. These enter lymph spaces, are carried in the bloodstream, and burrow into skeletal muscles. They grow to about 1.0 mm long, then coil up and become enclosed in cysts, which later may become calcified. Such larvae cannot mature unless the flesh containing them is eaten by another susceptible mammal, but they may live for months or years in humans.

Mild infections cause no particular symptoms, and some encysted trichinae occur in about 16 percent of people in the United States. Heavy infections cause the disease **trichinosis,** which may be severe and even fatal. Intestinal disturbances and abdominal pain occur first; with spread of the larvae there is fever, the muscles become swollen, hard, and painful, and there may be difficulty in swallowing and breathing. No specific treatment is known, and recovery is slow. Hogs, rats, and the other mammals become infected by eating slaughterhouse scraps, garbage, or animal carcasses that contain encysted larvae. Humans acquire the infection by eating pork (occasionally bear meat) containing the microscopic cysts. Examination of pork in slaughterhouses for trichinae is economically impractical, and some infections may be missed. The only protection is to *cook pork thoroughly,* by heating all parts to at least 58°C (137°F). Pink color in freshly roasted pork is evidence of inadequate cooking. Salami, headcheese, and other pork products may be dangerous unless thoroughly cooked. Refrigeration at −23°C (−10°F) for 3 days will usually kill all larvae, but cooking is safer.

19-12 Pinworms One of the commonest parasitic nematodes in humans is the pinworm *Enterobius vermicularis* (Fig. 19-11). Pinworms are especially common in children and are more common in tropical countries than in temperate ones. Adult pin-

Figure 19-11 **The pinworm,** *Enterobius vermicularis.* *A.* Female. *B.* Male. (*From Hyman, 1951.*)

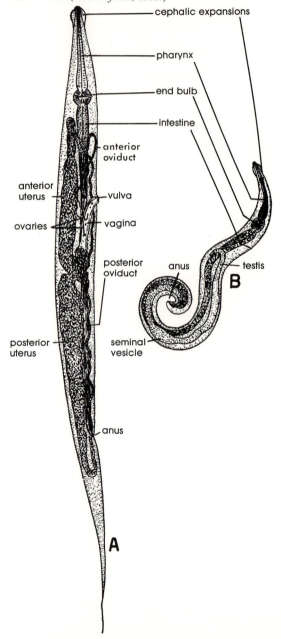

worms inhabit the gut, from whence the females migrate to the anal region of the host to lay eggs. Deposition of eggs and movement of the female worms cause itching in the host. Scratching by the host to relieve itching usually causes the eggs to be picked up. If they are subsequently ingested they will hatch and thus reinfect the host child or another child. The incidence of infection of children in tropical countries may reach 100 percent. Although pinworms cause no serious or fatal disease, heavy infestations cause intestinal upset and lowering of resistance and thus contribute to a poor state of health.

19-13 Filarial worms

The filariae of humans and various vertebrates are nematodes that require an intermediate anthropod host to complete the life cycle. One of these is *Wuchereia bancrofti* (Fig. 19-12), a human parasite in all warm regions of the world. The minute larvae, or microfilariae, are about 0.2 mm long. They live by day chiefly in large blood vessels, passing at night to small vessels in the skin, but cause no trouble even if numerous. If taken up by certain kinds of nocturnal mosquitoes, the larvae pass to the insect's stomach, then to the thoracic muscles, undergo metamorphosis, and when about 1.4 mm long move into the proboscis. At the mosquito's next meal, the larvae crawl onto and enter the human skin, pass into lymph vessels, coil up in lymph glands, and later mature. The adult female grows to 80 mm (3 in) long, the male 40 mm (1½ in), but neither is over 0.3 mm in diameter. If both sexes are present, they mate, and each female produces large numbers of microfilariae. Adults obstruct the lymph circulation, and if present in numbers may, after some years, cause the disease **elephantiasis,** in which the limbs or other parts grow to enormous size. An amazing biologic adjustment occurs in many localities, where the microfilariae increase in the superifical vessels at night when mosquitoes are active.

19-14 Guinea worm

Dracunculus medinensis is a human parasite in India, Arabia, and Africa. It occurs in several Old World mammals and has been found in dogs, foxes, raccoons, and mink in the United States but not in humans. The larvae occur in water fleas (*Cyclops*), which may be swallowed by the final host in drinking water. Upon reaching the host's stomach the larvae escape and burrow to subcutaneous tissues, occasionally to the heart or brain meninges. There a year or less of growth yields a female up to 1200 mm (48 in) long by 1 or 2 mm in diameter (males are virtually unknown). The mature gravid female migrates, often to the host's ankle or foot, where an ulcer forms. Upon immersion of the limb in water, larvae escape to seek the intermediate

Figure 19-12 Stages in the life cycle of the filarial worm, *Wuchereria bancrofti.* (*After Francis; Fulleborn; and Sambon.*)

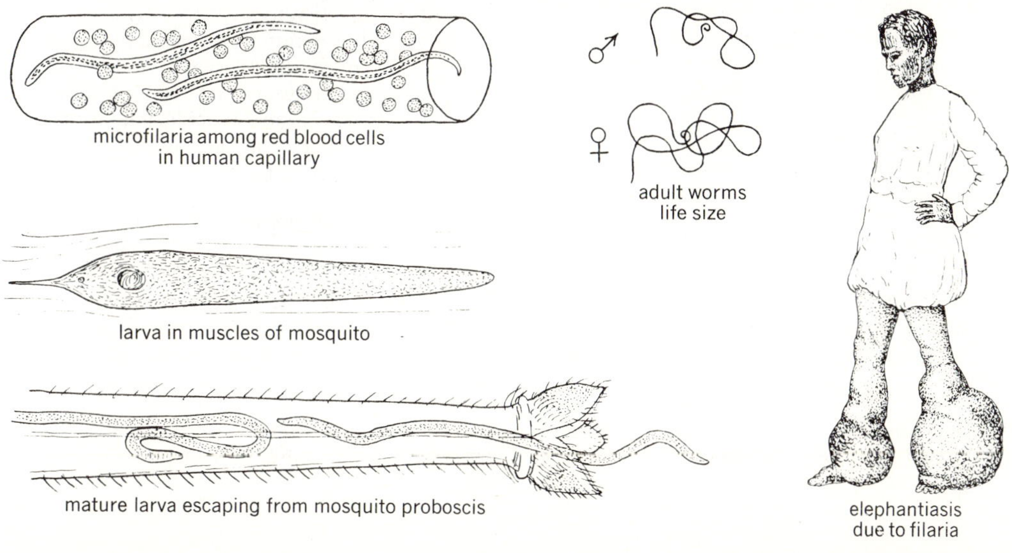

microfilaria among red blood cells in human capillary

adult worms life size

larva in muscles of mosquito

mature larva escaping from mosquito proboscis

elephantiasis due to filaria

host. Native peoples commonly remove a worm by winding it on a stick, a few turns per day; secondary infections are common. Another species (*Ophiodracunculus ophidensis*) found in Michigan, has the larval stage in *Cyclops viridis*, may occur in frog tadpoles, and becomes adult in the garter snake, *Thamnophis sirtalis*, females growing to 250 mm (10 in) long.

Classification

Phylum Nematoda (Nemathelminthes).

Roundworms. Body slender, cylindrical, often tapered toward ends; covered with cuticle; no segmentation or cilia; digestive tract complete, straight; no proboscis; no circular muscles; body space a pseudocoel; sexes usually separate; male with cloca and usually 1 or 2 copulatory spicules; no cloaca in female; 12,000 species.

Class 1. Adenophorea (Aphasmidia).

No phasmids (caudal sensory organs); amphids (anterior sensory organs) large, various, rarely porelike; rudimentary or no execretory canals; coelomocytes and mesenterial tissue well developed; males usually lack caudal alae; usually free-living.

Order 1. Chromadorida. Esophagus of 3 regions and not elongate; esophageal glands never duplicate, multinucleate, or opening anteriorly in esophagus.

Suborder 1. Monhysterina. Esophagointestinal valve not triradiate or vertically flattened; teeth none, 1 or 3 small, or 6 outward-acting; fresh or salt waters. *Monhystera, Plectus, Cylindrolaimus.*

Suborder 2. Chromadorina. Esophagointestinal valve triradiate or vertically flattened; teeth 1 large, or 6 inward-acting, or 3 jaws; amphids spiral, circular, etc.; in salt and fresh waters. *Chromadora, Cyatholaimus, Desmoscolex.*

Order 2. Enoplida. Esophagus of 2 parts (anterior muscular, posterior glandular), often long, cylindrical or conoidal; amphids pocketlike, elongate, or porelike.

Suborder 1. Enoplina. Intestine lacks muscles; no caudal sucker or stylet; male with 2 spicules; free-living in salt water, brackish water, or fresh water or in soil. *Enoplus, Trilobus.*

Suborder 2. Dorylaimina (Trichurata). Intestine lacks muscles; no caudal sucker or glands; male with 2, 1, or no spicules. *Dorylaimus,* in soil and fresh water or brackish water. *Trichinella spiralis; Trichuris* (*Trichocephalus*) *trichiura,* whipworm; *Capillaria,* hairlike in liver, gut, and elsewhere in vertebrates; *Mermis,* larvae in insects, adults free-living but do not feed.

Suborder 3. Dioctophymatina. Intestine with 4 rows of muscles; male with caudal sucker and 1 spicule. *Dioctophyma renale,* giant nematode, male to 450 mm (18 in) long, female to 900 by 15 mm (36 by $\frac{1}{2}$ in), in body cavity of mammals; *Hystrichis,* in intestine of birds.

Class 2. Secernentea (Phasmidia).

Phasmids usually present; amphids porelike; excretory system with paired lateral canals; coelomocytes 4 to 6, mesenterial tissue scant; no caudal or lateral hypodermal glands.

Order 1. Rhabditida. Esophagus of 3 regions, especially in larvae.

Suborder 1. Rhabditina. Lips 6, 3, 2, or 0; vagina short, transverse; caudal alae, if present, with papillae; no stylet; free-living and saprophagous in soil or parasitic in plants or animals. *Rhabditis,* some free-living in soil, others parasitic; *Strongyloides, Rhabdias,* alternate generations of parasitic females and of free-living males and females.

Suborder 2. Tylenchina. Lips 8, 6, or 0; stylet present; lateral excretory canals only on 1 side; vagina short, usually transverse; feed on fluids of living cells or in body spaces of insects. *Tylenchus,* meadow nematode; *Heterodera schachtii,* sugarbeet nematode, and *Meloidogyne,* root-knot nematodes, in roots; *Anguina,* wheat eelworm; *Chondronema,* in insects.

Order 2. Strongylida. Lips 6, 3, or 0, small; uterus heavily muscular, with ovijector; bursa with rays; parasitic in land vertebrates. *Oesophagostomum; Syngamus trachea,* gapeworm; *Ancylostoma, Necator,* hookworms; *Trichostrongylus,* hairworm; *Haemonchus,* stomach worm; *Metastrongylus,* lungworm.

Order 3. Ascaridida. Lips usually 3 large, or 6 small, or none; some with 2 lateral jaws; vagina and posterior uterus muscular; caudal alae with papillae (no muscles or rays); parasitic in land snails, insects, and vertebrates. *Ascaris; Enterobius,* pinworm; *Toxocara, Heterakis, Thelasoma.*

Order 4. Spirurida. Esophagus of 2 regions, anterior muscular and posterior glandular; adults in gut or tissues of vertebrates.

Suborder 1. Spirurina. Esophageal glands multinucleate; larval phasmids porelike; intermediate stages usually in insects *Gonglyonema, Thelazia, Tetrameres.*

Suborder 2. Camallanina. Esophageal glands uninucleate; phasmids in larvae large, pocketlike; copepods are intermediate hosts. *Dracunculus medinensis,* Guinea worm; *Camallanus,* in turtles and fish.

Suborder 3. Filariina. Mouth lacking lips; amphids small or reduced; adults parasitic in vertebrates, often in blood vessels or coelomic spaces. *Wuchereria, Filaria; Loa,* eyeworm.

References

(See works by Brumpt; Chandler; Faust and Russell; and Noble and Noble cited in references Chap. 18.)

Baylis, H. A., and **R. Daubney.** 1926. A synopsis of the families and genera of Nematoda. London, British Museum. xxxvi + 277 pp.

Chitwood, G. G., and **M. B. Chitwood.** 1974. Introduction to nematology. Baltimore, University Park Press. 334 pp., 200 figs.

Filipjev, I. N., and **J. H. Schuurmans Stekhoven, Jr.** 1941. A manual of agricultural helminthology (nematology). Leiden, Netherlands, E. J. Brill, NV. xv + 878 pp., 460 figs.

Florkin, M. and **B. T. Scheer** (editors). 1969. Chemical zoology. Vol. 3, Echinodermata, nematoda, and acanthocephala. New York, Academic Press, Inc. xx + 687 pp., illus.

Goodey, T. 1933. Plant parasitic nematodes. New York, E. P. Dutton & Co., Inc. xx + 306 pp., illus.

Goodey, J. B. 1963. Soil and freshwater nematodes. New York, John Wiley & Sons, Inc. xvi + 544 pp., illus.

Human, L. H. 1951. Nematoda. In The invertebrates. New York, McGraw-Hill Book Company. Vol. 3, pp. 197–455, figs. 93–200. 1959. Vol. 5, pp. 742–747.

Kükenthal, W. and others. 1928–1934. Handbuch der Zoologie. Berlin, Walter de Gruyter & Co. Nematoda, Vol. 2, pt. 1, sec. 4, pp. 249–402, figs. 267–426.

Lee, D. L. 1965. The physiology of the nematodes. San Francisco, W. H. Freeman and Company. x + 154 pp., 46 illus.

Rogers, W. P. 1962. The nature of parasitism. New York, Academic Press, Inc. ix + 287 pp. 40 figs. *Emphasis on nematodes.*

Sasser, J. N., and **W. R. Jenkins** (editors). 1960. Nematology: Fundamentals and recent advances with emphasis on plant parasitic and soil forms. Chapel Hill, The University of North Carolina Press, pp. xv + 480, illus.

Throne, Gerald. 1961. Principles of nematology. New York, McGraw-Hill Book Company. xiv + 553 pp., illus. *Plant parasitic nematodes.*

Yamaguti, S. 1961. Systema helminthum. Vol. 3. The nematodes of vertebrates. New York, Interscience Publishers, Inc. Part 1, 679 pp. Part 2, pp. 681–1261, 102 pls.

Phylum Nematomorpha (Gordiacea) (Horsehair worms)

19-15 The Nematomorpha The gordian worms (Fig. 19-13) are long, slender parasites of insects and crustaceans but free-living as adults. Different species are from 10 to 700 mm ($\frac{1}{2}$ to 28 in) long but only 0.3 to 2.5 mm in diameter, females being longer than males. The exterior is opaque yellow, gray, brown, or black. Adult worms are often seen wriggling in the water of ponds, quiet streams, rain puddles, or drinking troughs. Their presence in such places, sometimes in numbers, is responsible for an old belief that they are long horsehairs that have "come to life" in water. About 80 species are known. With the exception of *Nectonema,* which is marine, all are freshwater or terrestrial.

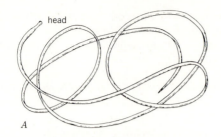

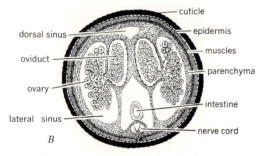

Figure 19-13 Phylum NEMATOMORPHA: the horsehair worm, *Gordius.* *A.* Entire specimen. *B.* Cross section of female, enlarged. *(After Kükenthal.)*

19-16 Characteristics

1 Body cylindrical, bluntly rounded at ends; symmetry bilateral; no segmentation.

2 Surface of cuticle bearing complex plates or papillae.

3 Layer of longitudinal muscles, incomplete at ends of body.

4 Pseudocoel open (*Nectonema*) or nearly filled with mesenchyme (GORDIOIDEA).

5 Digestive tract degenerate, young worms absorb food from hosts, adults nonfeeding.

6 No circulatory, respiratory, or excretory organs.

7 Nerve ring around esophagus, connecting to midventral nerve cord (also a dorsal cord in *Nectonema*).

8 Sexes separate, gonads one or two in body cavity; reproductive ducts paired, emptying into cloaca; no copulatory spicules; eggs of GORDIOIDEA minute, in long swollen gelatinous strings (to 230 cm, or 91 in!), on aquatic vegetation.

19-17 Natural history

In *Gordius* the larva (to 0.25 mm long) swims for a day in water, then is thought to encyst on damp vegetation. If taken with food by a cricket, grasshopper, or certain beetles, it burrows into the insect's hemocoel, feeds on surrounding tissues, absorbing the material through the body wall, and in several weeks or months, following a number of molts, becomes an adult; if the host then comes near the water, the worm emerges. Mating is by spiral coiling of the posterior ends of the two sexes and migration of sperm from the male into the seminal receptacle of the female. *Nectonema* is a marine pelagic form up to 20 cm (8 in) long and covered with swimming bristles. Specimens have been taken on both sides of the North Atlantic, in the Gulf of Naples, and at a few other places. The larvae are parasitic in crabs.

These animals have often been grouped with the nematodes because of their similar body form, presence of a thick cuticle, absence of segmentation, and only longitudinal musculature; they differ in having the body space filled with mesenchyme and one nerve cord (or two).

Classification

Phylum Nematomorpha.

Order 1. Cordioidea. No swimming bristles; pseudocoel filled with mesenchyme; 2 gonads. *Paragordius,* to 300 mm (12 in) long, common; *Gordius robustus,* to 890 mm (35 in) long, 0.5 to 1 mm in diameter.

Order 2. Nectonematoidea. Two rows of swimming bristles; pseudocoel unfilled; 1 gonad. One genus, *Nectonema,* to 200 mm (8 in) long, marine, pelagic near coasts, larvae parasitic in crustaceans (*Palaemonetes*).

References

Hyman, L. H. 1951. Nematomorpha. In The invertebrates. New York, McGraw-Hill Book Company. Vol. 3, pp. 455–472, figs. 201–208.

Pennak, R. W. 1953. Nematomorpha. In Fresh-water invertebrates of the United States. New York, The Ronald Press Company. Chap. 10, pp. 232–239, 6 figs.

Rauther, Max. 1929. Nematomorpha. In W. Kükenthal and others, Handbuch der Zoologie. Berlin, Walter de Gruyter & Co. Vol. 2, pt. 1, sec. 4, pp. 403–448, figs. 427–480.

Phylum Rotifera (Rotifers, or "wheel animalcules")

19-18 The rotifers These are minute to microscopic (mostly under 1 mm long), attractive in form and color, active in movements, and favorites of amateur microscopists. They abound in many freshwater environments, and a few occur even in the leaf axils of mosses. Most of the 1700 species are free-living, but some are fixed in protective tubes, a few are external or internal parasites, and a few inhabit salt water. The phylum and common names refer to the beating cilia on the anterior end of the body, which suggest the rotation of microscopic wheels. This ciliated area, the absence of external cilia elsewhere, and movements of the chewing pharynx (mastax) distinguish rotifers from all other minute aquatic animals.

19-19 Characteristics

1 Symmetry bilateral, no true segmentation, three germ layers; body usually composed of a fixed number of cells.
2 Body somewhat cylindrical, usually with ciliated coronal disk anteriorly and forked "foot" posteriorly.
3 Body wall a syncytium covered with hardened cuticle (lorica); body cavity without special lining.

4 Digestive tract with a complex triturating mill, the mastax.
5 Two protonephridia, coiled and branched, with flame cells.
6 A dorsal nerve ganglion and various nerves (no cord), sense organs tuftlike or as eyespots.
7 Sexes separate; males usually minute and degenerate or lacking; female with ovary, yolk gland, and oviduct; reproduction both parthenogenetic and sexual; usually oviparous; no larva.

The ROTIFERA resemble the PLATYHELMINTHES in having excretory organs with flame cells (protonephridia) and the NEMATODA in being composed of relatively few cells and having a body cavity without special lining. Some adult rotifers resemble a trochophore larva. As a group, the rotifers display an amazing variety in structure, suggesting that they have had a long geologic history. Unfortunately they are unknown as fossils, probably because of their minute size and lack of a hard skeleton.

19-20 Structure and function The typical rotifer, such as *Philodina* (Fig. 19-14), is composed of an anterior **head** region, an expanded **trunk,** and a narrow taillike posterior **foot,** usually movable and often ending in two slender **toes.** Each toe contains a **cement gland** providing a sticky secretion by which the animal may attach temporarily to some object. The **body wall** is a thin syncytium with a constant number of nuclei, covered by a thin, glassy, chitinlike cuticle (lorica). On the anterior end is a retractile disk, or **corona** (often double), rimmed with cilia. These cilia beat with a whirling motion that draws water containing oxygen and food toward the head end, carries off wastes, and serves for locomotion. The digestive tract is lined by cilia, except in the pharynx, and includes (1) the **mouth** below the corona; (2) a rounded muscular pharynx, or **mastax,** that has a "dental mill" of chitin-like **jaws** (trophi) with **teeth** used to grasp, cut, and grind the food, (3) a short **esophagus;** (4) a sizable **stomach,** of large cells and with a pair of digestive glands; (5) a short **intestine;** (6) the oval **cloaca;** and (7) the **anus,** which

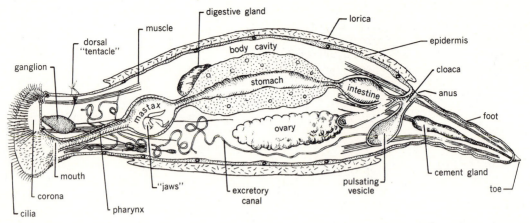

Figure 19-14 Phylum ROTIFERA. Structure of a female rotifer, in section from the left side; much enlarged. (*After Delage and Herouard.*)

is dorsal at the end of the trunk. Some species lack digestive organs beyond the stomach. Between the body wall and digestive tract is a fluid-filled **pseudocoelom** with branching syncytial cells. Body muscles lie beneath the syncytial epidermis. Both longitudinal and circular muscles occur, but not in distinct layers as in the flatworms. Other muscles serve to withdraw the corona and to move or contract the foot. Excretion is accomplished by two slender **protonephridia** connected to an enlarged **pulsating vesicle,** or **bladder,** that discharges relatively large amounts of water into the cloaca. This movement through the kidney of such large amounts of water suggests that the bladder has a primary function of regulating water and salts (osmoregulation). Each protonephridium is irregularly coiled and has several lateral branches beginning in flame cells. A large **nerve ganglion,** dorsal to the mouth, gives off nerves to various organs. There is usually a pair of short sensory tufts (*lateral antennae*) on the sides of the body posteriorly; some rotifers have also a tuft (or two) above the brain, and one to three dorsal eyespots. Ectoderm provides the external covering, the nervous structures, and the lining at the ends of the digestive tract; the midgut is from endoderm; and the other structures grow from mesoderm.

The preceding account applies to a typical female (BDELLOIDEA). Females dominate the populations of most rotifers and in some species males are unknown. Females have a single ovary combined with a yolk gland in a single unit. Eggs move via an oviduct to the cloaca. The male, in species where known, is much smaller than the female, lacks mouth and anus, and is short-lived. Its body cavity contains one large **testis;** this opens by a duct, either in a **penis** that can be protruded dorsally or in the foot.

In various rotifers (Fig. 19-15) the body is slender (*Rotaria neptunius*), broad (*Macrochaetus*), saclike (*Asplanchna*), flattened (*Ascomorpha*), or even spherical (*Trochosphaera*). The lorica may bear ridges, rings, or spines. The foot may be long, short, or absent; it serves for attachment and as a rudder in swimming, also in some as an organ for leaping. The corona may have one or two concentric wreaths of cilia or be modified into a funnel trap with bristles but no cilia (*Collotheca*); the mastax, jaws, and teeth are of varied form and may be employed by carnivorous rotifers to capture or grasp prey. Many of these differences are used in classification.

19-21 Natural history   Rotifers are cosmopolitan; the same species may occur in America, Eurasia, and Australia, wherever conditions are appropriate. Two lakes of like physical and biologic characteristics but hundreds of miles apart may contain similar assemblages of rotifers; conversely, two others un-

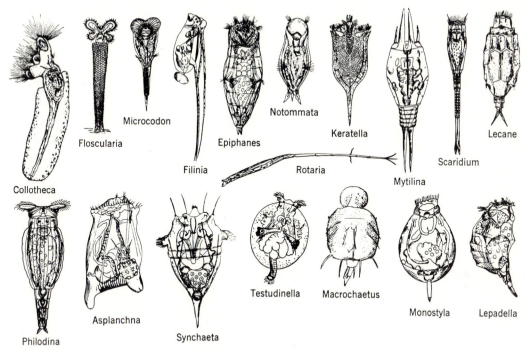

Figure 19-15 Phylum ROTIFERA. Various rotifers, all enlarged. (*After Jennings, 1901.*)

like but close together may have quite different kinds. The rotifer fauna of any body of water undergoes some change as to the constituent species during one summer season. The great majority are free-living and solitary. They are commonest in quiet fresh waters with much aquatic vegetation. Less than a hundred species are strictly marine, but certain freshwater species are also found in the sea. Over a million individuals have been found per liter of damp shore sand. Some species swim in open water, but more live about or on freshwater plants. Several species live attached, each in a self-constructed protective case either of its own secretions or of foreign particles; thus *Collotheca* inhabits a transparent tube, and *Floscularia* constructs a tube of microscopic "spherical bricks." Certain species are colonial, some free-swimming, and others fixed. The BDELLOIDEA include the most common and best-known freshwater forms. They are typically bottom forms which can creep like leeches but also can

swim. After years of drying, some species have been revived by wetting for only 10 min. The SEISONACEA consist of one genus and a few species which are found on marine crustaceans. *Seison nebaliae* lives as a commensal on a crustacean (*Nebalia*). True parasites are rare in rotifers. *Pleurotrocha parasitica* is an external parasite and *Albertia naidus* an internal parasite of freshwater annelids, *Ascomorpha* lives within colonies of the protozoan *Volvox,* and *Proales wernecki* forms microscopic galls in *Vaucheria,* an alga.

Rotifers that feed on unicellular algae have short broad teeth; those subsisting on the juices of larger plants have pointed teeth to pierce the plant cells and a muscular pharynx that acts like a piston to suck out the juices; and those of predatory habit have slender jaws armed with teeth that can be projected through the mouth to grasp the protozoans, rotifers, and other animals used as food. Rotifers aid in keeping the water clean by feeding on bits of organic debris as well as on other organisms. In turn, various

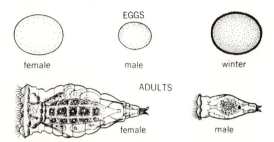

EGGS

female   male   winter

ADULTS

female   male

Figure 19-16   Stages in the life cycle of a rotifer, *Epiphanes senta;* **all enlarged.** (*After Whitney, 1916.*)

rotifers serve as food for small worms and crustaceans and are an important part of the food chain in fresh waters.

19-22 Reproduction   (Figs. 19-16, 19-17) The females, at different seasons, usually produce two kinds of eggs, "summer" (amictic) and "winter" (mictic). The first are thin-shelled, develop at once by parthenogenesis, and yield only females (amictic); in favorable waters a series of such generations may occur. Then, with some change in environmental conditions not clearly understood, a generation is produced in which the females (mictic) lay eggs which if unfertilized develop into males. If fertilized by these males the eggs form "winter eggs" with thick resistant shells like protozoan cysts. These require a rest period and may remain dormant for long periods, can survive drought or freezing, and may be blown about by winds or carried on the feet or feathers of birds. Later, in favorable waters, they develop into females that resume the "summer"

phase. Such means for surviving adverse conditions and for dispersal explain the universal distribution of rotifers.

Classification

Phylum Rotifera.

(Rotatoria; see Characteristics.)

Class 1. Seisonacea.

Body elongate; corona reduced; 2 ovaries; males fully developed; only 2 species. *Seison,* commensal on a marine crustacean (*Nebalia*).

Class 2. Bdelloidea.

Corona of two disks; retractile; toes none to 4, foot glands numerous; 2 ovaries; males unknown; obligatory parthenogenetic reproduction. *Habrotrocha, Philodina, Rotaria (Rotifer).*

Class 3. Monogononta.

Toes none to 2; 1 ovary; males usually present but degenerate, *Proales, Mytilina, Epiphanes (Hydatina).*

References

Donner, J. 1966. Rotifers. New York, Frederick Warne & Co., Inc. xi + 80 pp., illus.

Edmondson, W. T. 1959. Rotifera. In W. T. Edmondson (editor), Ward and Whipple, Fresh-water biology. 2d ed. New York, John Wiley & Sons, Inc. Pp. 420–494, 125 figs.

Figure 19-17 Diagram of the life cycle of the monogonatid rotifer, *Epiphanes senta.*

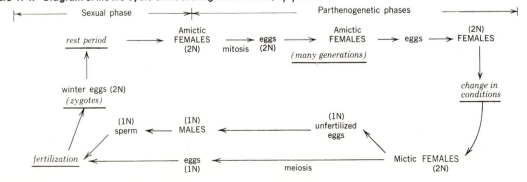

Harring, H. K. 1913. Synopsis of the Rotatoria. *U.S. Natl. Museum Bull.* Vol. 81, pp. 7–226, illus.

Hyman, L. H. 1951. Class Rotifera. In The invertebrates. New York, McGraw-Hill Book Company. Vol. 3, pp. 52–151, figs. 26–69. 1959. Vol. 5, pp. 739–740.

Pennak, R. W. 1953. Rotatoria. In Fresh-water invertebrates of the United States. New York, The Ronald Press Company. Pp. 159–213, figs. 96–132.

Thane, Anne. 1974. Rotifera. In A. C. Giese and J. S. Pearse (editors), Reproduction of marine invertebrates. New York, Academic Press, Inc. Vol. I, pp. 471–484.

Wesenberg-Lund, C. 1929. In W. Kükenthal and others, Handbuch der Zoologie. Berlin, Walter De Gruyter & Co. Vol. 2, pt. 1, sec. 4, pp. 8–120, figs. 2–131.

Phylum Gastrotricha

19-23 The gastrotrichs This group includes about 400 species of microscopic free-living forms (0.07 to 2.0 mm long) common between the sediment grains and among bottom debris in marine and fresh waters. The number of species is about equally split between fresh and marine waters.

19-24 Characteristics

1 Small, wormlike, with cilia confined to the flat ventral surface; symmetry bilateral, no segmentation.

2 Outer surface with a cuticle usually elaborated into spines, scales, or bristles; epidermis syncytial.

3 Pseudocoelom reduced to a small space between body wall and digestive tract.

4 Digestive tract with muscular pharynx and straight stomach-intestine, anus posterior.

5 Excretory system of two protonephridia with a single flame bulb each, or one or more ventral glands.

6 Adhesive tubes of varying form usually present.

7 Sexes united or only females, some with two kinds of eggs; development direct.

19-25 A typical gastrotrich (*Macrosdasys*, Fig. 19-18) is slender and flexible; the posterior end is pointed and contains cement glands for temporary attachment. The flat ventral surface has two lengthwise bands of cilia used for gliding locomotion, and on the arched dorsal surface are many slender scales or spines. The mouth is anterior, surrounded by bristles and fine sensory hairs. Gastrotrichs are omnivorous, with bacteria probably constituting the bulk of the diet. They ingest food by the pumping action of the muscular pharynx. The body wall is underlaid by delicate longitudinal and circular muscle bands. Two unbranched protonephridia with single flame cells opening midventrally occur in Chaetonoidea but are absent in Macrodasyoidea, where ventral glands serve a similar function. A saddle-shaped nerve ganglion anteriorly connects to two longitudinal lateral nerves and, by fibers, to the sensory hairs. In freshwater species only females are known and reproduction is parthenogenetic, as in many rotifers. The marine forms are monoecious. The simple ovary fills much of the body cavity and produces one to five large eggs, each in a tough shell. Sperm produced in the testes are transferred to the female system either directly via a penislike struc-

Figure 19-18 Phylum Gastrotricha: *Macrodasys*, enlarged, external appearance, dorsal view.

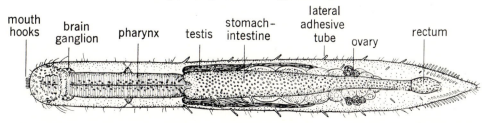

ture or as spermatophores attached to the cuticle. Eggs exit via an *X* organ, a separate opening, or a body-wall rupture. Some eggs have hooks that fasten to materials in the water. Development is direct, and hatchlings reach maturity in a few days.

Gastrotrichs resemble rotifers in size, presence of a cuticle, syncytial epidermis, and protonephridia. The digestive system, however, resembles that of nematodes, as do the bristles and elaborations of the cuticle and adhesive tubes.

Classification

Phylum Gastrotricha

Class 1. Macrodasyoidea.

Adhesive tubes anterior, lateral, and posterior; no protonephridia; monoecious; marine shores, interstitial in sand. *Cephalodasys, Macrodasys.*

Class 2. Chaetonotoidea.

Adhesive tubes usually only on tail; 2 protonephridia; reproduction parthenogenetic (lacks male system); mainly in fresh waters on vegetation. *Chaetonotus, Lepidodermella; Neodasys,* marine.

References

Beauchamp, P. 1965. Classe des gastrotriches. In P. P. Grassé (editor), Traité de zoologie. Paris. Masson et Cie. Vol. 4, pp. 1381–1430.

Brunson, R. B. 1959. Gastrotricha. In W. T. Edmondson (editor), Ward and Whipple, Fresh-water biology. 2d ed. New York, John Wiley & Sons, Inc. Pp. 406–419, 34 figs.

Hummon, William D. 1974. Gastrotricha. In A. C. Giese and John S. Pearse (editors). Reproduction of marine invertebrates. New York, Academic Press, Inc. Vol. I, pp. 485–506.

Hyman, L. H. 1951. Class Gastrotricha. In The invertebrates. New York, McGraw-Hill Book Company. Vol. 3, pp. 151–170, figs. 70–81. 1959. Vol. 5, pp. 740–741.

Pennak, R. W. 1953. Gastrotricha. In Fresh-water invertebrates of the United States. New York, The Ronald Press Company. Pp. 148–158, 6 figs.

Remane, Adolph, 1963. Gastrotricha. In Bronn, Klassen und Ordnungen des Tierreichs. Leipzig, Akademische Verlagsgesellschaft Geest & Portig KG. Vol. 4, div. 2, book 1, pt. 2, pp. 1–242, illus.

Phylum Kinorhyncha (Echinodera)

19-26 The kinorhynchs   This group consists of about 70 species of marine worms up to 1 mm long (Fig. 19-19), found in the surface layers of bottom mud and sand of shallow to deep seas. They feed on algae and detritus.

19-27 Characteristics

1   Bilaterally symmetrical, body of 13 (or 14) overlapping joints (zonites), the first anterior segment forming a head and the second a neck, both retractile into the body.

2   Cuticle chitinous, with dorsal and lateral spines but no cilia.

3   Body space a pseudocoel.

4   Mouth at anterior end of head, pharynx and esophagus lined with cuticle, stomach-intestine unlined, anus terminal in retractile ring.

Figure 19-19   Phylum KINORHYNCHA: *Echinoderes,* **dorsal view, enlarged.**

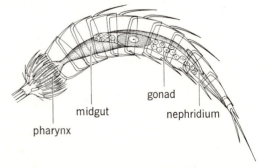

gonad

midgut

nephridium

pharynx

5   Excretory tubes two, each with flagellated flame cell, with duct opening dorsally on trunk ring 9.

6   Nervous system in epidermis, a nerve ring around the pharynx connecting to midventral nerve cord; eyespots in a few.

7   Sexes separate, each with two gonads opening ventrally beside the anus; male system with copulatory spines.

8   Development includes a series of juvenile stages.

19-28 Natural history   The KINORHYNCHA are rather poorly known. Most collections are from European coasts, but recent collections of species from Africa, Japan, the Antarctic, the Arctic, and coasts of the Americas indicate that the group is worldwide. Kinorhynchs burrow in the mud at the mud-water interface, where they are deposit feeders; a few feed on diatoms. Their body structure is similar to that of gastrotrichs and rotifers and is distinguished from these two groups primarily by the lack of cilia, the markedly jointed body, and the retractile head. They resemble nematodes but differ in having paired flagellated excretory organs and paired sex organs. Kinorhynchs have separate but virtually indistinguishable sexes. The young differ from the adults and attain adulthood after several molts. Adults do not molt.

Classification

Phylum Kinorhyncha (Echinodera)

Order 1. Homalorhagida.   Both head and neck retractile; first trunk ring with 3 ventral plates. *Pycnophyes, Trachydemus.*

Order 2. Cyclorhagida.   Only first head ring retractile. *Echinoderes, Campyloderes, Centroderes, Caterla, Semnoderes.*

References

Higgins, R. P. 1965. The homalorhagid Kinorhyncha of northeastern U.S. coastal waters. *Trans. Am. Microscop. Soc.,* vol. 84, pp. 65–72.

———1968. Taxonomy and postembryonic development of the Cryptorhage, a new suborder for the mesopsammic kinorhynch genus *Cateria. Trans. Am. Microscop. Soc.,* vol. 87, pp. 21–39.

———1974. Kinorhyncha. In A. C. Giese and J. S. Pearse (editors). Reproduction of marine invertebrates. New York, Academic Press, Inc. Vol. 1, pp. 507–518. 14 figs.

Hyman, L. H. 1951. Class Kinorhyncha. In The invertebrates. New York, McGraw-Hill Book Company. Vol. 3, pp. 170–183, figs. 82–86.

Remane, Adolph. 1936. Kinoryncha, In Bronn, Klassen und Ordnungen des Tierreichs. Leipzig, Akademische Verlagsgesellschaft Geest & Portig KB. Vol. 4, div. 2, book 1, pt. 1, pp. 243–372, figs. 210–297.

Zelinka, Karl. 1928. Monographie der Echinodera. iv + 396 pp., 27 col. pls., 73 figs.

Phylum Acanthocephala (Spiny-headed worms)

19-29 The acanthocephalans   These parasites of peculiar structure and function live as adults in the intestines of vertebrates and as larvae in anthropods. The distinctive feature for which the group is named (Gr. *acanthos,* spine + *kephale,* head) is the anterior cylindrical eversible **proboscis** that bears rows of recurved spines serving to attach a worm to the host's gut (Figs. 19-20, 19-21).

19-30 Characteristics

1   Anterior end with retractile proboscis, armed with hooks; worm-shaped, female usually larger than male; symmetry bilateral.

2   Cuticle thin, secreted by a syncytial epidermis containing canals and large vesicles.

3   Muscle layers, outer circular, inner longitudinal.

Figure 19-20 Phylum ACANTHOCEPHALA: two acanthocephalan worms from the gut of sea otters. (*Courtesy Gary McDonald.*)

4 Body space a pseudocoel.

5 Neck region with two internal slender sacs (lemnisci), serving as fluid reservoirs when proboscis is retracted.

6 No digestive, circulatory, or respiratory organs.

7 Protonephridia two or none.

8 Nerve ganglion in proboscis sheath, nerves to proboscis and body.

9 Sexes separate, gonads in ligament between

proboscis sheath and posterior end; male with two testes; ovary of female not persistent.

10 Nuclear constancy in the body.

19-31 Natural history Adult acanthocephalans attach to the intestinal wall of their vertebrate hosts, where food is absorbed directly from the host's intestine. In reproduction sperm pass through a duct to the genital pore in the bell-like posterior bursa and are transferred via a protrusible penis to the female. Ova develop on the ligament, are set free in the body space and fertilized there, and are then surrounded usually by three membranes. Development proceeds in the female until a larval stage with hooks is formed, at which time the larvae are encased in a shell. They are then released to escape the host in the feces. When eggs are eaten by the intermediate anthropod host, the larvae emerge and enter the hemocoel. At this stage the larva is called an **acanthor.** Completion of the life cycle requires that the larval host in turn be eaten by the host of the adult worm.

The 500 or more species of acanthocephalans are 1.5 to 650 mm (26 in) in length and parasitize different vertebrates from fishes to mammals. Larvae of

Figure 19-21 Phylum ACANTHOCEPHALA: spiny-headed worm, enlarged; left side of body removed to show internal structure. *A.* Male. *B.* Female (reduced). *C.* Anterior end with proboscis withdrawn. (*Adapted from Lynch, 1936.*)

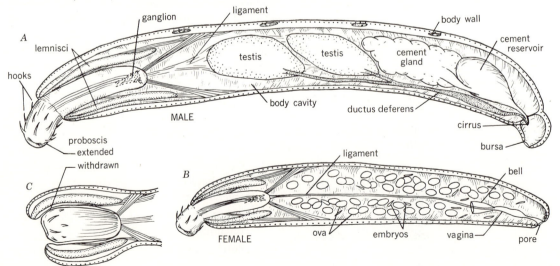

those in terrestrial hosts usually develop in insects and those of aquatic vertebrates in crustaceans. When present in numbers, acanthocephalans can do considerable damage to the intestinal wall of the host. *Macracanthorhynchus hirudinaceus* is common in pigs and occasional in humans; the male is 50 to 100 by 3 to 5 mm (2 to 4 by 0.1 to 0.2 in), and the female 100 to 650 by 4 to 10 mm (4 to 26 by 0.16 to 0.4 in). Its intermediate host is the white grub or larva of the june beetle (Scarabaeidae), which pigs often eat when in pastures.

The Acanthocephala resemble the aschelminth phyla in possessing a syncytial epidermis, a cuticle, and nuclear constancy. Relationship to the Turbellaria has been claimed, on the basis of structure of the nervous system and the presence of protonephridia, and it is difficult to determine true relationships.

Classification

Phylum Acanthocephala

(see Characteristics above).

Order 1. Archiacanthocephala. Proboscis spines concentric; protonephridia present; 2 persistent ligament sacs in females; 8 cement glands in males; in terrestrial hosts. *Gigantorhynchus,* in anteaters; *Macracanthorhynchus,* in beetles and pigs; *Moniliformis,* in beetles or cockroaches and rodents.

Order 2. Palaeacanthocephala. Proboscis spines in alternate radial rows; no protonephridia; ligament sacs transient in females; usually 6 cement glands in males; mostly in aquatic hosts. *Leptorhynchoides,* in amphipods and fishes; *Polymorphus,* in crustaceans and water birds; *Acanthocephalus,* in isopods, fishes, and amphibians.

Order 3. Eocanthocephala. Proboscis spines arranged radially; no protonephridia; persistent ligament sacs in females; cement gland syncytial with reservoir; in aquatic hosts. *Neoechinorhynchus,* in aquatic arthropdos and fishes.

References

Baer, J. C. 1961. Acanthocéphales. In P.-P. Grassé, Traité de zoologie. Paris, Masson et Cie. Vol. 4, fasc. 1 pp. 733–782, figs. 616–685.

Florkin, M. and **B. T. Scheer** (editors). 1969. Chemical zoology, Vol. 3. Echinodermata, Nematoda and Acanthocephala. New York, Academic Press, Inc., xx + 687 pp., illus.

Hyman, L. H. 1951. Acanthocephala. In The invertebrates. New York, McGraw-Hill Book Company. Vol. 3, pp. 1–52, figs. 1–25. 1959. Vol. 5, p. 739.

Meyer, A. 1933. Acanthocephala. In Bronn, Klassen und Ordnungen des Tierreichs. Leipzig, Akademische Verlagsgesselschaft Geest & Portig KG. Vol. 4, pt. 2, book 2, sec. 2, pp. 333–582, illus.

Ward, H. L. 1951; 1952. The species of Acanthocephala described since 1933. *Tenn. Acad. Sci. J.,* vol. 26, pp. 131–149, vol. 27, pp. 282–311.

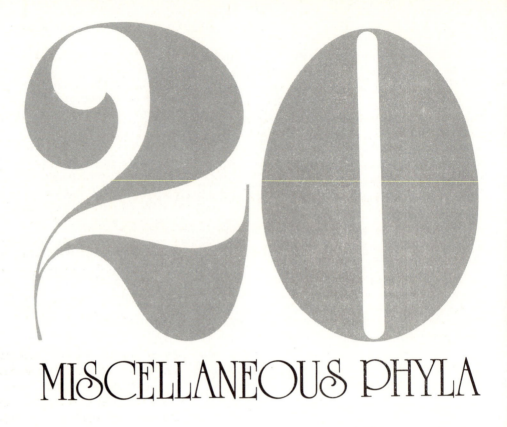

MISCELLANEOUS PHYLA

Besides the major phyla described in other chapters, there are other phyla of multicellular animals, not often studied by beginners in zoology, which are minor in terms of numbers of species or individuals and ecologic importance. They are described here to complete the account of the Animal Kingdom. Most of these animals are found in marine waters, a few in freshwater and terrestrial habitats, and one group is parasitic. Some are abundant and of wide occurrence, others uncommon.

The majority of the phyla treated in this chapter are coelomate animals which fall into one of the two major coelomate lines. One group may be termed the lesser protostomes, because they belong to the protostome line of invertebrate evolution. This includes the Sipuncula, Echiura, Pentastomida, Tardigrada, Onychophora, and Pogonophora. The second group may be termed lesser deuterostomes, as they show the traits of this line. Included in this are the phyla Hemichordata and Chaetognatha. Finally there is a group which evidences a mixture of protostome and deuterostome features but tends to have more protostome characteristics. The three phyla in this group are all related in possessing a ciliated tentacular crown called a lophophore and are collectively termed lophophorates. Included here are the phyla BRYOZOA, PHORONIDA, and BRACHIOPODA. The ENTOPROCTA are also placed near this latter group for reasons discussed later. The PRIAPULIDA is an enigmatic group of uncertain affinity. The remaining two phyla, MESOZOA and GNATHOSTOMULIDA, are acoelomate and difficult to place in any phylogenetic scheme. These are discussed first.

The existence of small phyla is a logical consequence of evolution—various groups have originated and flourished in past geologic time. Some died out completely, while others persist with smaller representation in Recent time. The bryozoans and brachiopods abounded in Paleozoic seas over 200 million years ago and now are less abundant. Soft-bodied forms such as hemichordates, chaetognaths, and phoronids have left no fossil record to suggest their relation to other existing phyla, and such relations have been inferred on the basis of comparative morphology and embryology of living forms.

Phylum Mesozoa

20-1 The mesozoans The MESOZOA (Gr. *mesos,* middle + *zoön,* animal) are minute wormlike solid organisms with the simplest structure of any multicellular animals but with very complicated life cycles. all are endoparasites of marine invertebrates. The class DICYEMA (*Dicyema*) (Fig. 20-1) are all parasites in the nephridia of octopuses and squids. Adults, called **nematogens,** are up to 7 mm

long and composed of an outer layer of about 25 ciliated somatic cells (number constant for each species) surrounding one or more long axial reproductive cells. The anterior eight or nine cells form a polar cap. Repeated fission of the axial cell yields cells (agametes, with no observed maturation) that develop asexually into vermiform larvae which then escape from the adult and develop into new individuals. This process builds up the population quickly in young cephalopods. When the cephalopod becomes adult the nematogen axial cells begin to produce a different larva called the **infusorigen.** The mesozoans which produce infusorigen larvae are now called **rhombagens.** Infusorigen larvae escape the rhombagens and the host cephalopod, but the remainder of the life history is unknown.

The ORTHONECTIDA are rare parasites in tissues and cavities of flatworms, nemerteans, brittle stars, annelids, and one species of clam. About 43 species of mesozoans are known.

20-2 Characteristics

1 All parasitic; symmetry bilateral or none.

2 Cells in two layers, number constant in a species, inner layer reproductive, not digestive.

3 No segmentation, no organ systems.

4 Life cycle complex, alternate sexual and asexual generations, possibly in different hosts.

The MESOZOA resemble some colonial protozoans in having external cilia through much of the life cycle, in having intracellular digestion by the exter-

Figure 20-1 Phylum MESOZOA. *Dicyema. A.* Outline of fullgrown individual, 3 mm long. *B.* Young nematogen stage, 0.4 mm. (*Adapted from McConnaughey, 1949.*)

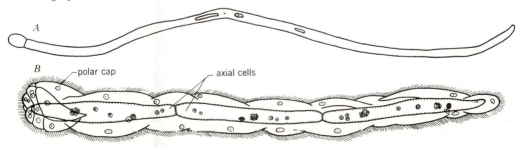

nal cells, and in having special internal reproductive cells (as in *Volvox*). Their two cell layers are not comparable with the ectoderm and endoderm of META-ZOA, and they have no internal digestive cavity. Either they are intermediate between protistans and multicellular animals or else are degenerate forms, possibly related to the PLATYHELMINTHES. They have been likened to the planula larvae of cnidarians. The taxonomic position of this odd group remains an enigma.

Classification

Phylum Mesozoa

(see above).

Class 1. Dicyema.

Usual form (nematogen) to 7 mm long; body elongated, bilaterally symmetrical; one internal cell.

Order 1. Dicyemida. Adults ciliated over all the surface. *Dicyema.*

Order 2. Heterocyemida. Adults without cilia. *Conocyema, Microcyema.*

Class 2. Orthonectida.

Sexual forms under 1 mm long; adult body a multinucleate plasmodium with inner reproductive cells. *Rhopalura.*

References

Grassé, P.-P. 1961. Traité de zoologie. Vol. 4, fasc. 1, pp. 693–729, figs. 588–615.

Hyman, L. H. 1940. Mesozoa. In The invertebrates. Vol. 1, pp. 223–247, figs. 68–71. 1959. Vol. 5, pp. 713–715.

McConnaughey, B. H. 1949. Mesozoa of the family Dicyemidae from California. Univ. *Calif. Berkeley Publ. Zool.,* Vol. 55, pp. 1–34, 7 pls.

————. 1951. The life cycle of the dicyemid Mesozoa. Ibid., pp. 295–335, 13 figs.

Stunkard, H. W. 1954. The life history and systematic relations of the Mesozoa. *Quart. Rev. Biol.,* vol. 29, pp. 230–244.

————. 1972. Clarification of taxonomy in the Mesozoa. *Syst. Sys. Zool.,* vol. 21, pp. 210–214.

20-3 The gnathostomulids The GNATHOSTOMU-LIDA (Gr. *gnath*, jaw + *stoma*, mouth) is the most recently described animal phylum. These animals are acoelomate wormlike organisms, often semitransparent, ranging in size from a few hundred micrometers to about 1 mm (Fig. 20-2). About 80 species are known.

20-4 Characteristics

1 Completely ciliated externally, each cell with a single cilium.
2 Pharynx muscular with cuticularized single basal plate and paired jaws.
3 No anus.
4 Hermaphroditic.
5 Bilaterally symmetrical.

20-5 Natural history Gnathostomulids have a completely ciliated body in which each epidermal cell bears a single cilium, a sensory system of long, stiff cilia anteriorly, a ventral mouth hardened by cuticular plates bearing teeth, and the presence in the mouth cavity of jaws which range from simple pincer types to complex types with many teeth. The animals have a digestive system consisting of large cells surrounding a cavity but have no anus. Gnathostomulids are hermaphroditic, with a complex reproductive system. The nervous system is simple and lies beneath the epidermis; nervous tissue is concentrated near the anterior end. There is no circulatory system, but an excretory system has recently been found. The muscular system consists of paired

Figure 20-2 Phylum GNATHOSTOMULIDA. *Gnathostomula.* Outline of an adult individual, about 1 mm. (*Adapted from Kersteuer, 1969.*)

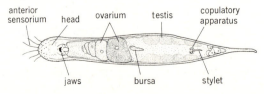

groups of longitudinal fibers beneath a thin layer of circular muscles. Eggs are extruded through the body wall, development proceeds in the external environment, and the animals hatch resembling the adults but without certain adult structures, such as jaws, which are gained with further development.

Gnathostomulids are almost unique among metazoan animals in that they are primarily inhabitants of marine anaerobic mud or mud-sand environments, where they presumably feed on fungi and bacteria. They appear to be very abundant in these areas. The geographic distribution of the phylum is from the tropics to the polar regions.

Gnathostomulids appear most closely related to platyhelminths, with which they share features such as lack of an anus, ciliated epidermis, ventral mouth, and complex hermaphroditic reproductive system. They have, however, also been considered to be closely allied to the rotifers and gastrotrichs, with which they share the features of pharyngeal structures and sensory bristles. Because of their mixture of features, they are best considered a separate phylum at present.

Classification

Phylum Gnathostomulida.

Order 1. Filospermoidea. Reproductive system lacking vagina, bursa, and penis; body much elongated. *Hapagnatha, Pterognathia.*

Order 2. Bursovaginoidea. Reproductive system with a bursa and penis, usually a vagina; body shorter and stouter. *Agnathiella, Gnathostomula, Austrognathia.*

References

Ax, P. 1956. Die Gnathostomulida, eine rätselhafte Wurmgruppe aus dem Meeressand. *Abhandl. Akad. Wiss. Lit. Mainz Math.-Naturw. Kl.,* no. 8, pp. 530–562.

_____. 1960. Die Entdeckung neuer Organisationstypen im Tierreich: Die neue Brehm-Bücherei. Wittenberg, A. Ziemsen Verleg. 116 pp.

Riedl, R. J. 1969. Gnathostomulida from America. *Science,* vol. 163, pp. 445–452.

Sterrer, W. 1971. On the biology of the Gnathostomulida. *Vie et Milieu,* Suppl., vol. 22, pp. 493–508.

_____. 1971. Gnathostomulida. In N. C. Hulings and J. S. Gray (editors). A manual for the study of meiofauna. Smithsonian Cont. Zoology, No. 76.

_____. 1974. Gnathostomulida. In A. C. Giese and J. S. Pearse (editors), Reproduction of marine invertebrates. New York, Academic Press, Inc.

_____. 1972. Systematics and evolution within the Gnathostomulida. *Syst. Zool.,* vol. 21, pp. 151–173.

Phylum Priapulida

20-6 The priapulids   These are cylindrical wormlike creatures (Fig. 20-3), yellow or brown and up to 15 cm (6 in) long, that inhabit mud or sand under shallow waters from Boston and the Belgian coast northward and from Patagonia south to Antarctica. A few live down to 460 m (1500 ft). Some have been dredged from mud on the bottom of Tomales Bay, California, in 10-fathom water. They live in mud but can burrow to leave piles of castings on the bottom. Only eight species are known.

20-7 Characteristics

1   Symmetry bilateral, three germ layers, no segmentation.

Figure 20-3 Phylum PRIAPULIDA. *Priapulus,* external form, × 1½. *(From Wolcott, Animal biology.)*

2   Body cylindrical and fleshy, comprising an anterior retractile proboscis with longitudinal rows of spines and papillae, a trunk with superficial rings and folds, and usually one or two posterior processes with soft, hollow, gill-like outgrowths.

3   Body wall with outer cuticle, epidermis, circular and longitudinal muscles.

4   Digestive tract straight, with muscular pharynx, thin-walled intestine, and terminal anus.

5   Body space large, questionably eucoelomate, undivided, connecting to posterior outgrowths, lined by thin membrane, filled with fluid containing rounded cells.

6   Two excretory organs (protonephridia) connecting to sex ducts.

7   Nervous system subepithelial, with ring around mouth (no dorsal ganglion) and ventral ganglionated nerve cord.

8   Sexes separate, one pair of gonads.

9   Larval stage in a case.

Little is known about the biology of priapulids other than that they appear to be slow-moving predators upon benthic polychaete worms, which they seize with the spines of the everted proboscis (introvert). Although the animals are known to have a larval form which closely resembles a rotifer, the complete sequence of events in embryologic development has never been reported. As a result the critical question of how the body cavity is formed has not been answered, and the group has been placed with various phyla at one time or another. The priapulids appear to have most features in common with the aschelminth group of pseudocoelomates (nematodes, kinorhynchs, rotifers, and gastrotrichs), including a cuticle, an eversible proboscis like the kinorhynchs, and a larval form like the rotifers. They lack the cell constancy found in most aschelminths and have cellular tissues rather than the syncytial arrangement common to most aschelminths. If the recent evidence that their body cavity is a true coelom is sustained, it would suggest a more distant relation to the aschelminths than the present morphologic features suggest. The group is tentatively placed here because of the current questionable situation regarding its body cavity.

References

Baltzer, F. 1931. Priapulida. In W. Kükenthal, Handbuch der Zoologie. Vol. 2, pt. 2, sec. 9, pp. 1–14, 14 figs.

Hyman, L. H. 1951. Priapulida. In The invertebrates. Vol. 3, pp. 183–197, figs. 87–92.

Shapeero, W. L. 1961. Phylogeny of the Priapulida. Science, vol. 113, pp. 879–880.

Van der Land, J. 1970. Systematics, zoogeography and ecology of the Priapulida. Zool. Verhandelingen. Rijksmus. Nat. Hist. Leiden, No. 112, 118 pp., 4 pls., 89 figs.

Lophophorates

The three phyla which constitute the lophophorates are morphologically quite dissimilar as adults but share one common anatomic feature, the lophophore. A **lophophore** is an anterior fold of the body wall surrounding the mouth and bearing a series of hollow, ciliated tentacles. The hollow portions of the tentacles are extensions of the coelom. In addition to the characteristic lophophore, these animals all lack a well-defined head and are sessile. Lophophorates are an enigma because in their embryology they tend to show a mixture of protostome and deuterostome features. The protostome features tend to dominate; the larvae are modified trochophores in phoronids and bryozoans, and the mouth is derived from the blastopore in phoronids and brachiopods, but in all three phyla the cleavage pattern tends to be radial rather than spiral, a deuterostome feature. Brachiopods also show enterocoelic mesoderm formation. Because of this mixture of features, these phyla have been difficult to place in any phylogeny. They have often been considered to be near the base stock from which the deuterostome and protostome lines evolved.

Phylum Bryozoa (Moss animals)

20-8 The bryozoans   Many bryozoans (Gr. *bryon,* moss + *zöon,* animal) are small tufted colonies attached to objects in shallow seawater. This is the largest and most abundant of the lophophorate phyla. All are colonial, and the individuals are extremely small. Some appear like hydroids or corals, but their internal structure is more advanced. Their form suggested the names *moss animals* and *zoophytes* (plantlike animals), and the many individuals in a colony led to the name POLYZOA (Gr. *poly*, many). Some zoologists use the term ECTOPROCTA for the BRYOZOA, since the ENTOPROCTA (Sec. 20-18) were separated from the latter. Most bryozoans are matlike, forming thin encrustations on rocks, shells, or kelp. All members of the phylum are aquatic, and most of them are marine, but one class is restricted to fresh waters (Fig. 20-4). Bryozoans are often included among the "seaweeds" pressed and dried by amateurs at the shore. They are widely distributed as fouling organism. Some marine bryozoans are unique among the lophophorates in that they show polymorphism. Each colony consists of several different "morphs" or individuals specialized for different tasks.

Most bryozoan orders appear in Ordovician rocks. About 4000 living species are known, and a larger number from the Paleozoic onward. Their skeletons aided in forming lime-bearing rocks of many strata. Many of the species had a short geologic (time) range but wide geographic distribution; they are thus useful in correlating geologic strata and are of economic importance in studying the cores brought up in drilling test wells for petroleum.

Figure 20-4 Phylum BRYOZOA. Freshwater bryozoans. *Plumatella. A.* Colony, natural size, on a twig. *B.* Part of colony much enlarged, showing both expanded and contracted individuals. *C.* Statoblast. *D. Cristatella,* entire colony is free-moving. (*After Allman, 1856.*)

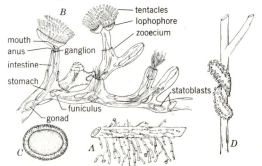

20-9 Characteristics

1 Symmetry bilateral, no segmentation, three germ layers.

2 Colonial, individuals minute, each in separate housing (zooecium); polymorphism in some.

3 Digestive tract complete, U-shaped; mouth surrounded by a retractile lophophore bearing ciliated tentacles; anus opening outside lophophores.

4 Coelom well developed in two parts (marine forms) or three parts (freshwater forms); no circulatory or respiratory organs.

5 No nephridia.

6 A nerve ganglion between mouth and anus.

7 Sexes usually united and gonads developed from peritoneum; eggs fertilized in coelom or externally and usually brooded in a modified zooecium called an ooecium, among the tentacles, in the coelom, or in a portion of the reproductive individual; larva a trochophore; colonies formed by asexual budding, also by statoblasts in freshwater species.

20-10 Structure and natural history *Bugula* is a common bryozoan that grows as branched brown or purple tufts, 5 to 8 cm (2 or 3 in) long, attached to objects in shallow seawater (Fig. 20-5). Each "stem" includes numerous individuals less than 1 mm long, closely united in four longitudinal rows. An individual, or **zooid,** consists of a tubular chitinous housing, or **zooecium,** with the soft living parts, or **polypide,** inside. Many marine bryozoans are polymorphic. In *Bugula* the zooecia of the typical zooids with lophophores bear externally smaller and highly modified zooids, called **avicularia.** An avicularium is shaped like a bird's head, with jaws that open and snap shut to keep minute animals from settling on the surface. It does not feed (Fig. 20-5). The anterior end, or **introvert,** of a typical living polypide is a raised ridge, or **lophophore,** that bears a circle of slender, hollow, flexible, and ciliated **tentacles.** The lophophore is U-shaped in the freshwater bryozoans and circular in the marine bryozoans. The cilia of the tentacles create a water current which

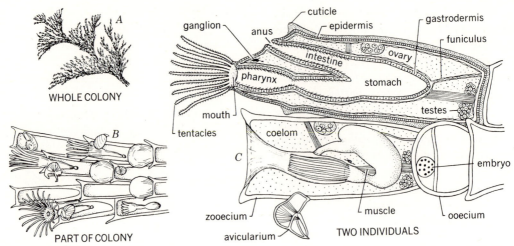

Figure 20-5 Phylum BRYOZOA. Structure of *Bugula*. *A*. Whole colony, natural size. *B*. Part of colony, enlarged. *C*. Two individuals in longitudinal section, the upper expanded and the lower contracted.

enters at the top of the circle of tentacles and passes out between the tentacles. The current and the cilia both sweep microscopic food organisms toward the mouth, where they are ingested. The lophophore may also aid in respiration. The lophophore and associated parts of the polypide can be quickly withdrawn into the zooecium by action of the large **retractor muscle** (Fig. 20-5). To push the introvert out again the bryozoan contracts other muscles, called **parietal muscles,** which act to increase the pressure on the coelomic fluid, thus everting the introvert hydraulically.

The **digestive tract** is complete, with a partially ciliated lining and muscle fibers which allow food to be moved by peristalsis. It includes the **mouth** between the tentacles, a wide **pharynx,** a slender **esophagus,** an enlarged **stomach** of U or V shape, and a slender **intestine** leading to the **anus,** which opens just outside the lophophore. The dorsal region is reduced so that the mouth and anus are close together. There are no circulatory, respiratory, or excretory organs. The body cavity is a true coelom divided by septa and filled with fluid containing coelomocytes. Respiration occurs over the general body surface, and the coelomic fluid operates as a circulatory system. Coelomocytes in the coelomic fluid engulf waste material. Some bryozoans have a **nerve ganglion** just below the lophophore, with

fibers to the tentacles and elsewhere. The epidermis of the body wall secretes the **zooecium.** Under certain circumstances the lophophore and digestive tract may degenerate in an individual. In that case, the body wall produces a new polypide.

Bugula is monoecious; both ovaries and spermaries develop from the coelomic lining, and eggs and sperm are shed into the coelom. Most marine bryozoans brood their eggs. In *Bugula*, brooding occurs in a special external chamber called an **ooecium.** Eggs escape the coelom via a coelomopore, while sperm leave via pores at the tips of the tentacles. Each egg becomes a ciliated trochophorelike larva which is liberated in the sea, soon to settle, apical end downward, as an **ancestrula** and found a new colony by asexual budding. Individuals are short-lived, degenerating in a few weeks and then regenerating fully.

20-11 Other bryozoans The largest zooids are not over 3 mm long by 1 mm in diameter, and many are minute. In different species the zooecium is tubular, conical, or urnlike and is usually calcareous or chitinous. The gelatinous cylinders of *Alcyonidium* are 60 to 90 cm (2 to 3 ft) long. Some bryozoans (*Penetrantia*) burrow into shells of mollusks and barnacles. The freshwater forms (PHYLACTOLAEMATA) produce large gelatinous housings bearing zooids on the surface, and they have a special mode of asexual reproduction to survive unfavorable conditions. Internal buds called **statoblasts** form in the funiculus and are enclosed in a chitinous shell. Upon the death and decay of the parent colony these are set free in the water, either to float or to sink; they can survive freezing or drought and later produce new colonies. Freshwater bryozoans (e.g., *Paludicella*) may grow in water pipes, occasionally in such quantity as to obstruct the flow.

Classification

Phylum Bryozoa

(Polyzoa, or Ectoprocta; see Characteristics).

Class 1. Phylactolaemata.

Lophophore U-shaped; a lip (epistome) overhangs mouth; body wall muscular; colonies monomorphic; overwintering egg a statoblast with chitinous shell. Recent, in fresh waters. *Plumatella*, colony much branched or creeping; *Pectinatella*, zooids exposed on surface of gelatinous mass of zooecia; *Cristatella mucedo*, creeping gelatinous mass to 25 cm (10 in) long, in ponds.

Class 2. Gymnolaemata.

Lophophore circular; no lip overhanging mouth; body wall without muscles; colonies often polymorphic; zooecia complex; mostly marine; includes many fossil and most living forms.

Order 1. Ctenostomata. Zooecia chitinous or gelatinous, not limy, orifice same diameter as zooecium and closed by a flexible fold of body wall bearing pleats; colonies low, encrusting on rocks or shells. Ordovician to Recent. *Alcyonidium, Penetrantia*, marine; *Paludicella*, in fresh waters.

Order 2. Cheilostomata. Zooecia chitinous or limy, boxlike, with avicularia; operculum usually present. Jurassic to Recent. *Bugula, Menipea*, colonies tufted; *Membranipora*, colonies encrusting on kelp and sargassum weed.

Class 3. Stenolaemata.

Zooecia tubular, limy; openings terminal; no operculum; embryos in large ovicells. Upper Cam-

brian to Recent. †*Archaeotrypa*, Upper Cambrian; *Crisia, Tubulipora*, marine.

References

Bassler, R. S. 1953. Bryozoa. In R. C. Moore (editor), Treatise on invertebrate paleontology. Lawrence, Kansas, Geological Society of America. Pt. G, 253 pp., 175 figs.

Hyman, L. H. 1959. Phylum Ectoprocta. In The invertebrates. Vol. 5, pp. 275–515, figs. 98–182.

Larwood, G. P. (editor). 1973. Living and fossil Bryozoa. New York, Academic Press, Inc. xvii + 634 pp., illus.

Osburn, R. C. 1950. Bryozoa of the Pacific coast of America, pt. 1, Cheilostomata-Anasca. Allan Hancock Pacific Expeditions, vol. 14, pp. 1–269, 29 pls.; 1952, pt. 2, Cheilostomata-Ascophora, vol. 14. pp. 270–611, 64 pls; 1953, pt. 3, Cyclostomata, Ctenostomata, Entoprocta, and addenda, vol. 14, pp. 612–841, 82 pls.

Rogick, M. D. 1959. Bryozoa. In W. T. Edmondson (editor), Fresh-water biology. 2d ed. Ward and Whipple. Pp. 495–507, 17 figs.

Ryland, J. S. 1970. Bryozoans. London, Hutchinson University Library. 199 pp., illus.

Phylum Phoronida

20-12 The phoronids These are slender wormlike sedentary creatures that inhabit the bottom of shallow seas. They may occur singly or in clusters but are not colonial. Each individual is separately housed in a self-secreted chitinous tube of leathery or membranous texture from which its tentacles are extended to feed (Fig. 20-6). Only two genera are known, *Phoronis* and *Phoronopsis*. Some of the 15 or more species are less than 1 mm long, but *Phoronopsis viridis* on the Pacific Coast grows to 20 cm (8 in) or more, with tubes of 20 to 45 cm (8 to 18 in).

20-13 Characteristics

1   Symmetry bilateral, three germ layers, no segmentation.
2   Body cylindrical, crowned by a double spiral lophophore bearing 20 to 500 or more hollow ciliated tentacles in two rows which are fused at their bases to a ridge.

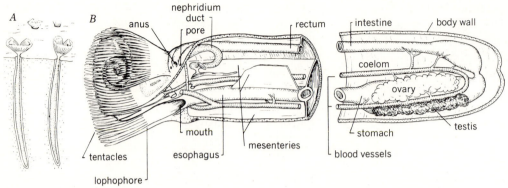

Figure 20-6 Phylum PHORONIDA. *Phoronis. A.* Group of individuals as in life, in bottom sand or mud under shallow water, each in a self-excavated tube. *B.* Internal structure, enlarged and partly diagrammatic; left side of lophophore and body wall removed; only part of left lateral mesentery shown. (*After Benham, 1889.*)

3 Body wall of cuticle, epidermis, circular muscles (layer), and longitudinal muscles (bundles).

4 Digestive tract U-shaped, mouth inside and anus outside lophophores.

5 Coelom throughout body and into tentacles, fluid-filled.

6 Circulatory system closed, of two lengthwise contracting trunks with lateral branches in body and vessels into bases of tentacles; blood with corpuscles containing hemoglobin.

7 No respiratory organs.

8 Two ciliated nephridia opening near anus.

9 Nervous tissue below epidermis, a ring surrounding mouth and fibers (some giant) to parts of body.

10 Sexes united, gonads (testis, ovary) develop from peritoneum beside blood vessels, gametes released into the coelom and escape via nephridia; larva (actinotroch) somewhat like trochophore.

20-14 Natural history Phoronids feed like other lophophorates by extending the mucus-coated tentacles into the water column, where the cilia of the tentacles create a current down between the tentacle rows. Plankton organisms in the water are then caught in the mucus of the tentacles or in the groove between the tentacle rows and transported to the

mouth. When uncovered by the receding tide or when disturbed, they withdraw to the shelter of their tubes. The lophophore, if torn off, can be regenerated. The bodies and tentacles of different species are red, orange, or green and when numerous may give brilliant color to the sea floor. At Naples colonies of *Phoronis hippocrepia* form a network of their interwined tubes 5 to 8 cm (2 or 3 in) thick on submerged pilings.

Phoronids show an advance over bryozoans in that they have a circulatory system and an excretory system, a necessary set of organs in an animal of this size. Like bryozoans, phoronids shed gametes into the coelom, where they escape via the nephridia. Fertilization is external, but eggs and embryos may be brooded in the lophophore. The hatching form is a distinctive larva called an **actinotroch,** which later settles and develops into an adult.

References

Benham, W. B. 1889. The anatomy of *Phoronis australis. Quart. J. Microscop. Sci., n.s.,* vol. 30, pp. 125–158, pls. 10–13.

Cori, C. J. 1937. Phoronidea. In W. Kükenthal and others, Handbuch der Zoologie. Vol. 3, pt. 2, sec. 5, pp. 5–138, figs. 82–151.

Emig, C. C. 1977. Embryology of Phoronida. *Am. Zool.,* vol. 17, pp. 21–37.

Dawydoff, C., and **P.-P. Grassé.** 1959. Phoronidiens. In P.-P. Grassé, Traite de zoologie. Vol. 5, fasc. 1, pp. 1008–1053, figs. 835–875, 1 col. pl.

Hyman, L. H. 1959. Phoronida. In The invertebrates. Vol. 5, pp. 228–274, figs. 82–97.

Phylum Brachiopoda (Lamp shells)

20-15 The brachiopods These bear a superficial resemblance to bivalve mollusks in having an external shell of two valves but are dorsal and ventral, rather than lateral as in the mollusks. Brachiopods have inhabited the sea bottom since animal life was first common in Cambrian time. Their shells are widespread and abundant fossils in rock strata of marine origin and are useful for correlating such deposits. Some 456 genera are known from Paleozoic rocks and 177 in the Mesozoic; the 68 genera and 260 living species are but a remnant of those now extinct. About 30,000 fossil species have been described. The living genus *Lingula* is much as it was in Ordovician time—over 400 million years ago—possibly the oldest living genus of animals. All living brachiopods are marine, solitary, and usually attached; most of them are in temperate waters. Shells range in size from 5 mm to 7.5 cm ($\frac{1}{5}$ to 3 in), but some fossils were 38 cm (15 in) wide. The name of the phylum (L. *brachio,* arm + Gr. *pod,* foot) refers to the lophophore within the shell, and the term *lamp shell* indicates the resemblance to an old Roman oil lamp.

20-16 Characteristics

1 Symmetry bilateral, no segmentation, three germ layers.

2 Shell external, dorsal and ventral valves unlike, usually with fleshy peduncle for attachment.

3 Mouth preceded by an extensive two-armed lophophore; digestive tract with or without anus.

4 Coelom well developed, fluid-filled; circulatory system open; blood colorless and with coelomocytes.

5 Excretion by one or two pairs of nephridia serving also as reproductive ducts.

6 A nerve ring about gullet.

7 Sexes usually separate, each with paired gonads; eggs and sperm discharged into seawater; a free-swimming ciliated larva; no asexual reproduction.

BRACHIOPODA are similar to phoronids and bryozoans in that they possess a lophophore for feeding. In brachiopods the lophophore is enclosed within a shell and much coiled and convoluted to increase the surface area for feeding. The shells of brachiopods open only a short distance when feeding, and water containing food is drawn into the shell by the cilia of the lophophore. The food particles are trapped in mucus on the tentacles and transported via cilia to the mouth.

20-17 Structure and natural history (Magellania) The soft parts are contained between two stout, scoop-shaped calcareous **valves,** or shells, which are covered externally with organic periostracum (Fig. 20-7). The larger ventral valve has a posterior projecting **beak** perforated for passage of the fleshy stalk, or **peduncle,** by which the animal attaches permanently to the sea bottom (Fig. 20-8). The actual **body** occupies only the posterior part of the space between the shells. The body wall consists of an external epidermis, stout connective tissue, and a ciliated coelomic lining. Two double folds of this wall extend anteriorly as the dorsal and ventral **mantle lobes,** which line the interior surface of the respective shells and have fine papillae penetrating the shell substance. A thin limy loop supports a large W-shaped **lophophore** that attaches to the anterior surface of the body and lies in the space between the mantle lobes (Fig. 20-8). The mouth opens from the lophophore base into a short **gullet,** followed by the larger **stomach** (with paired **digestive glands** or "liver") and a blind **intestine.** Three pairs of **muscles** serve to close and open the valves, and two other pairs attaching to the peduncle and shell permit the animal to turn about. The large fluid-filled **coelom** contains the internal organs, which are supported on

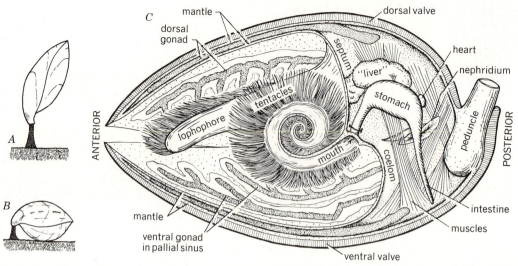

Figure 20-7 Phylum BRACHIOPODA. *A,B.* Typical positions of Recent lamp shells in life. *C. Magellania,* internal structure from left side; shells cut to midline, mantle and lophophore of left side removed. (*A,B, after Shrock and Twenhofel, Invertebrate paleontology.*)

mesenteries; branches of the coelom extend into the mantle lobes and lophophore. A small contractile **heart** and middorsal blood channel are present. The circulatory system is open, the blood returning through tissue spaces. At each side of the intestine is a large **nephridium,** with a fringed nephrostome draining from the coelom and a small exit into the mantle cavity. A **nerve ring** surrounds the gullet, with nerves to various organs, but special sense organs are absent. The sexes are usually separate; in each are two **gonads,** dorsal and ventral, respectively, and the nephridia serve as reproductive ducts, discharging eggs and sperm to the exterior. The fertilized egg grows into a free-swimming larva of unique form, which later metamorphoses to the adult.

Order 1. Atremata. Peduncle emerging from parts of both valves; shell with calcium phosphate. Lower Cambrian to Recent. *Lingula,* Pacific and Indian Oceans, and *Glottidia,* both coasts of the United States, lives between tide lines, burrows in mud or sand with long contractile peduncle.

Order 2. Neotremata. Peduncle emerging from the notch in ventral valve only. Cambrian to Recent. *Crania,* ventral valve flat, fastened to rock, dorsal valve conical.

Class 2 Articulata.

Dorsal and ventral valves unlike, densely formed of oblique limy prisms ($CaCO_3$); with hinge, and usually with beak for peduncle; calcareous support for lophophore; digestive tract blind; no anus. Cambrian to Recent.

Classification

Phylum Brachiopoda

(see Characteristics).

Class 1. Inarticulata.

Two valves nearly alike, of chitinlike material containing calcium phosphate; no hinge, beak, or calcareous support for the lophophore; digestive tract complete with anus.

References

Beauchamp, P. de, and **J. Roger.** 1960. Brachiopodes. In P.-P. Grassé, Traité de zoologie. Vol. 5, fasc. 2, pp. 1380–1499, figs. 1270–1370.

Hyman, L. H. 1959. Brachiopoda. In The invertebrates. Vol. 5, pp. 516–609, figs. 183–212.

Moore, R. C. (editor). 1965. Treatise on invertebrate paleontology, pt. H, Brachiopoda. Vol. 1, xxxii + 521 pp., 397 figs.; vol. 2, pp. 523–927, figs. 378–746. Lawrence, Kansas, Geological Society of America.

A.   *B.*

Figure 20-8 Phylum BRACHIOPODA. *A. Laqueus californicus* attached to a rock. *B. Laqueus californicus* opened showing large lophophore. (*Courtesy of G. McDonald.*)

Muir-Wood, H. M. 1955. History of the classification of the phylum Brachiopoda. London, British Museum of Natural History. 124 pp., 12 figs.

Rudwick, M. J. S. 1970. Living and fossil brachiopods. London, Hutchinson University Library. 199 pp., 99 figs.

Schrock, R. R., and **W. H. Twenhofel.** 1953. Principles of invertebrate paleontology. Pp. 260–349, 60 figs.

Phylum Entoprocta

20-18 The entoprocts The ENTOPROCTA (Kamptozoa) are small creatures—microscopic to 5 mm tall—solitary or colonial, that live attached to aquatic organisms or other objects in shallow coastal waters (Fig. 20-9). The body is a vaselike **calyx,** slightly flattened laterally, rimmed by a circle of ciliated **tentacles,** which are extensions of the body wall. Within the circle of tentacles the mouth opens at one side and the anus at the other, the latter often on a cone. The base of the calyx joins a stalk serving for attachment. Organisms may be solitary, with a single stalk attaching the calyx to the substrate, or colonial, with several stalks connected to a basal plate or with stalks arising some distance apart from a basal stolon. Some species have swellings on the stalk which contain muscles and are responsible for the peculiar nodding motion of these animals. The body has a thin cuticle overlying the epidermis. Within the calyx is a large U-shaped digestive tract surrounded by a gelatinous-filled pseudocoelom. These animals

formerly were included among the BRYOZOA, which they resemble superficially, but they appear to possess a pseudocoelom rather than the true coelom of the BRYOZOA and hence have been given separate phylum status for many years. Recent work with the larvae, however, suggests that they are indeed closely related to the BRYOZOA, and some workers would reunite them. For this reason, and because any relationship to other pseudocoelomate phylum is difficult to discern, they are discussed here.

Figure 20-9 Phylum ENTOPROCTA. *Pedicellina. A.* Part of colony. *B.* Calyx, median section, enlarged. (*Modified from Becker, 1937.*)

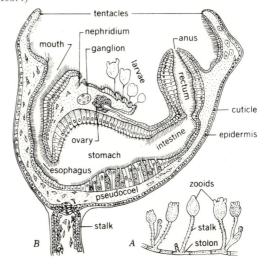

20-19 Characteristics

1 Symmetry bilateral, no segmentation, three germ layers.

2 Individual zooids minute, each with calyx, stalk, and enlarged base.

3 Edge of retractile circular calyx surrounded by one row of many slender, active, ciliated tentacles.

4 Both mouth and anus open within circle of tentacles, which are extensions of the body wall.

5 Digestive tract U-shaped, complete, ciliated, with esophagus, stomach, and rectum.

6 Body space a pseudocoelom filled with gelatinous parenchyma.

7 Muscles limited to tentacles and stalk.

8 No circulatory or respiratory organs.

9 One pair of protonephridia for excretion.

10 Nerve ganglia present.

11 Monoecious, or sexes separate; two gonads, with ducts.

12 Eggs and larvae develop in ovary, then trochophore larvae swim freely in water before attaching to grow.

13 Calyx often lost during unfavorable periods and later regenerated from stalk.

20-20 *Natural history* Of the 60 or more species, *Urnatella gracilis*, usually found in pairs, is the only freshwater form, living on the undersides of stones in running waters of the Eastern and Central states and in Lake Erie. *Loxosomella* and *Loxosoma* grow from buds that separate so that each individual is distinct; large numbers live as commensals on marine annelid worms and less often on sponges or ascidians. Common colonial types are *Pedicellina* and *Barentsia*, which grow on shells, rocks, or algae in shallow seawater. Asexual reproduction is by budding. Buds may be produced by the calyx in solitary species and from the stolon in colonial forms. Sexual reproduction is by eggs fertilized in the ovary and then brooded in a **vestibule** in the center of the calyx. Eggs hatch into ciliated trochophore larvae not unlike those of annelids

and mollusks. After a short swimming period the larva settles down and metamorphoses into an adult. Entoprocts are filter feeders, removing small organisms from the water current passing through the tentacles. Cilia on the tentacles create the current and capture the organisms. When disturbed, entoprocts fold the tentacles over the top of the calyx.

Classification

Phylum Entoprocta

(see Characteristics).

LOXOSOMATIDAE. Solitary, stalk attached by simple adhesive disk; on sponges, polychaetes, ascidians, and other animals. *Loxosoma, Loxocalyx.*

PEDICELLINIDAE. Colonial, with basal stolons. *Pedicellina, Barentsia.*

URNATELLIDAE. Small colony from a basal plate. One genus, *Urnatella*, fresh water.

References

Brien, Paul. 1959. Endoproctes. In P.-P. Grassé, Traité de zoologie. Vol. 5, fasc. 1, pp. 927–1007, figs. 736–834.

Hyman, L. H. 1951. Entoprocta. In The invertebrates. Vol. 3, pp. 521–554, figs. 209–223.

Mariscal, R. N. 1965. The adult and larval morphology and life history of the entoproct *Barentsia gracilis* (M. Sars, 1835). *J. Morphol.*, vol. 116, pp. 311–338.

Nielsen, C. 1964. Studies on Danish Entoprocta. *Ophelia*, vol. 1, pp. 1–76.

_____. 1966. On the life cycle of some LOXOSOMATIDAE (Entoprocta). Ibid., vol. 3, pp. 221–247.

_____. 1966. Some LOXOSOMATIDAE (Entoprocta) from the Atlantic coast of the United States. Ibid., vol. 3, pp. 249–275.

_____. 1971. Entoproct life cycles and the entoproct/ectoproct relationship. Ibid., vol. 9, pp. 209–341.

The Lesser Protostomes

This is a group of small phyla which all share the common features of protostome development. They are not necessarily closely related, however, and several show close relationships to certain of the larger protostome phyla. They are grouped here for convenience.

Phylum Sipuncula

20-21 The sipunculids This is a small phylum of about 330 species of unsegmented marine worms living shallowly in the sand or mud, nestling in empty shells, the holdfasts of kelps, about rocks or boring into rocks. When disturbed, the animals retract the anterior fringed end to assume the shape of a short club or even of a peanut kernel (Fig. 20-10). The sipunculids are sometimes called *peanut worms*. Most are relatively large animals, ranging in size from 2 mm to 72 cm (28 in).

20-22 Characteristics

1 Symmetry bilateral, three germ layers, no somites.

2 Body usually bluntly cylindrical posteriorly, with narrow anterior retractile portion (introvert) bearing fine chitinous papillae and ending in a circle of short, hollow, ciliated tentacles.

3 Body wall of slightly roughened cuticle, epidermis, dermis with glands and sense organs, and three layers of muscle: circular, oblique, and longitudinal.

4 Mouth centered in tentacles, digestive tract slender, with ciliated groove, extending to posterior end and spiraled back on itself, anus dorsal near base of introvert.

5 Coelom large, undivided, peritoneum ciliated, coelomic fluid reddish (hemerythrin), with several kinds of corpuscles.

6 No true circulatory organs but a tentacular system of dorsal and ventral vessels with ring sinus below tentacles.

7 Nephridia (brown tubes) two or one, opening near anus.

8 Nervous system of dorsal bilobed ganglion near tentacles with connectives to single ventral unsegmented nerve cord, many lateral nerves.

9 Sexes separate but alike, no permanent gonads, sex cells develop in tissues on retractor muscles of introvert, and gametes escape through nephridia.

10 Larva usually a trochophore (no trace of somites).

20-23 Natural history Sipunculids are common shallow-water marine worms but are often not observed because of their habit of living in holes, cracks, or crevices. Some are borers into soft rock, but the mechanism is unknown. Sipunculids are deposit feeders, ingesting organic particles from the substrate. When feeding, the tentacles are extended on the sea bottom and the beating cilia entrap micro-

Figure 20-10 Phylum SIPUNCULA. *Sipunculus.* A. Internal structure. B. Introvert retracted into body. C. External appearance, introvert and tentacles extended.

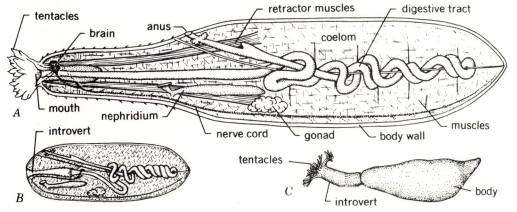

organisms in mucus, which is swallowed. Some sipunculids ingest sand as they burrow, then digest the adhering microorgansims. When a worm is disturbed, the introvert is drawn entirely into the posterior part of the body by retractor muscles, but the tentacles are not inverted. Then the introvert can be extended by contraction of muscles in the wall of the fluid-filled body. The tentacles are hollow, connecting to a system of contractile canals and sacs. The coelomic fluid functions as a circulatory system. It contains a respiratory pigment, hemerythrin. The nervous system is annelidlike, with a dorsal brain and ventral nerve cord. Sense organs are found primarily on the end of the introvert and include nuchal organs, which may be chemosensory, as well as eyespots in a few species. Sipnuculids are dioecious, and gametes are produced from the epithelial lining of the retractor muscles. Eggs and sperm exit via the nephridia. Fertilization is external in seawater, and the larva is a trochophore. Common widely distributed sipunuclids include the genera *Sipunculus*, *Dendrostomum*, and *Golfingia* (*Phascolosoma*). *Phascolion*, which lives at depths to 1000 fathoms from the West Indies to the Arctic, inhabits a snail shell cemented full of sand to form a tube, which the worm carries about. These worms are considered to be close to the ANNELIDA, with which they share the following features: (1) large coelom, (2) trochophorelike larva, (3) nervous system, and (4) structure of body wall muscles.

References

Fisher, W. K. 1952. The sipunculid worms of California and Baja California. *U.S. Natl. Museum Proc.*, vol. 102, pp. 371–450, pls. 18–39.

Hyman, L. H. 1959. Sipunculida. In The invertebrates. Vol. 5, pp. 610–696, figs. 213–241.

Peebles, Florence, and **D. L. Fox.** 1933. The structure, functions, and general reactions of the marine sipunculid worm *Dendrostoma zostericola*. Scripps Inst. Oceanog. Univ. Calif. Bull. Tech. Ser., vol. 3, pp. 201–224, 11 figs.

Rice, Mary. 1967. A comparative study of the development of *Phascolosoma agassizii*, *Golfingia pugettensis*, and *Themiste pyroides* with a discussion of developmental patterns in the Sipuncula. *Ophelia*, vol. 4, pp. 143–171.

Stephen, A. C., and **S. J. Edmonds.** 1972. The phyla Sipuncula and Echiura. London, British Museum, 528 pp., illus.

Tétry, A. Sipunculiens. In P.-P. Grassé. Traité de zoologie. Vol. 5, fasc. 1, pp. 785–854, figs. 575–673, 1 col. pl.

Phylum Echiura

20-24 The echiurids These comprise about 100 species of marine worms, similar to sipunculans in size, that inhabit the mud, sand, or cracks and crevices in rocks under shallow coastal waters of all warm and temperate seas (Fig. 20-11). The commoner forms live in burrows in soft bottoms or in shells or other cavities out of which they extend their long proboscises that collect detrital food from the bottom. The name *spoon worm* derives from the proboscis shape when contracted. A few species dwell in empty shells. *Thalassema mellita* lives in the tests of sand dollars off the southeastern coast of North America. *Urechis caupo* of California grows to 25 cm (10 in), but others are smaller, down to 7 mm or less. *Urechis* is unusual in that it forms a U-shaped burrow and secretes a mucous net in the burrow to capture food (Fig. 20-11). The Japanese *Ikeda taenioides* is 40 cm (16 in) long, with a 145-cm (58-in) proboscis.

20-25 Characteristics

1 Symmetry bilateral, three germ layers, no segmentation in adult.

2 Body commonly sausage-shaped, soft, with anterior proboscis, either spatulate or threadlike and long, that is flexible and contractile but cannot be withdrawn into body.

3 Body wall thick, muscular, epidermis covered with thin cuticle bearing rings of fine papillae.

4 One pair of ventral anterior setae, some (*Echiurus*) also with one or two rings of setae around anus.

5 Digestive tract complete, mouth at base of proboscis, pharynx muscular, intestine long and much coiled, rectum with two long vesicles, anus posterior and terminal.

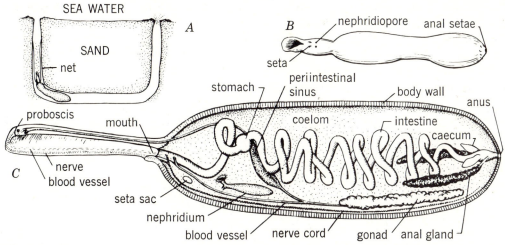

Figure 20-11 Phylum ECHIURA. *Urechis caupo. A.* The echiurid worm in its burrow with mucous net spread for food capture. *B.* Ventral view, proboscis partly retracted. (*Both after Fisher and MacGinitie, 1928.*) *C.* Internal structure with left side of proboscis and body wall removed.

6   Circulatory system usually of dorsal and ventral vessels joined anteriorly, but no lateral vessels.

7   Coelom large, undivided, crossed by many muscle strands supporting intestine.

8   Nephridia usually one to four pairs anteriorly or one large, all serving as gonoducts; commonly a pair of excretory sacs.

9   A midventral nerve cord in body with paired lateral nerves and a nerve trunk around edge of proboscis but no anterior or body ganglia and no sense organs.

10   Sexes separate, dimorphic in some, gonad single, median and posterior.

11   Larva a trochophore.

20-26 Natural history The widely distributed *Echiurus* inhabits a burrow and feeds by extending the mucus-covered proboscis on the surface of the mud to trap debris. If the animal is disturbed, the proboscis is cast off and then regenerated. Echiurids have a closed circulatory system (except *Urechis*) similar in general form to that of annelids. The system is also aided by the coelomic fluid, which contains some cells with the respiratory pigment hemoglobin. Excretory organs include the pairs of nephridia and also the peculiar outpocketings of the

rectum called anal sacs, which have a structure similar to primitive kidneys. Unlike sipunculids, echiurids have no special sense organs, but the nervous system is generally similar to that of sipunculids. The sexes are normally separate, and the gonads are in the ventral coelomic mesentery. Gametes are released to the coelom and exit via the nephridia; fertilization occurs in seawater. The larva is a trochophore.

Urechis caupo lives permanently in a tunnel dug by use of its setae and flushed with water expelled from the rectum. To feed, a band of epidermal cells behind the setae secrete a mucous cylinder (with pores only 0.004 μm in size) attaching to the burrow walls and forepart of the body. Then peristaltic contractions in the body wall draw water through the burrow. Microorganisms are sieved by the mucous net, which later is swallowed. As commensals, a goby, a polychaete worm, and a small crab live also in the burrow. *Bonellia,* which uses deserted burrows on the sea bottom, is noted for sexual dimorphism. The female has a small, green ovoid body and a threadlike proboscis, up to 1 m long, forked at the tip. The male is minute and ciliated, like a turbellarian worm, but has no proboscis, mouth, or anus. Early in life it enters the female's gut, later to inhabit the nephridium as a "parasite." What is unusual about *Bonellia* is the mechanism of sex determination. Larvae have the potential to be of either sex. If a larva settles on the bottom it becomes a female. If it lands on a female, it becomes a male.

Echiurids, like sipunculids, appear to be most closely related to the ANNELIDA, with which they share similarities in the construction of the nervous system, circulatory system, excretory system, and larva and larval development. They also have a large but undivided coelom. The major difference from the annelids in both sipunculids and echiurids is the complete absence of metamerism or segmentation of the body, a fundamental difference which suggests early divergence from the annelid line of evolution. Hence their classification as separate phyla.

Classification

Phylum Echiura

(see Characteristics).

Class 1. Echiurida.

Proboscis and anterior setae usually present; anal sacs 2.

Order 1. Echiuroinea. Blood-vascular system closed; nephridia usually one to four pairs. *Echiurus, Bonellia.*

Order 2. Xenopneusta. No vascular system; coelomic fluid with large corpuscles containing hemoglobin; nephridia usually one to four pairs. *Urechis.*

Order 3. Heteromyota. Blood-vascular system closed; nephridia 200 to 400, unpaired. *Ikeda,* Japan.

References

Baltzer, F. 1931. Echiurida. In W. Kükenthal and others, Handbuch der Zoologie. Vol. 2, pt. 2, sec. 9, pp. 63–168, figs. 49–132.

Dawydoff, C. 1959. Echiuriens. In P.-P. Grassé, Traité de zoologie. Vol. 5, fasc. 1, pp. 853–904, figs. 674–717.

Fisher, W. K. 1946. Echiuroid worms of the north Pacific Ocean. U.S. Natl. Museum Proc., vol. 96, pp. 215–292, figs. 1–19.

Fisher, W. K., and **G. E. MacGinitie.** 1928. The natural history of an echiuroid worm. *Ann. Mag. Natl. Hist.,* ser. 10, vol. 1, pp. 204–213.

Newby, W. W. 1940. The embryology of the echiuroid worm. *Urechis caupo. Am. Phil. Soc. Mem.,* vol. 16, xx + 219 pp., 85 figs.

Stephen, A. C., and **S. J. Edmonds.** 1972. The phyla Sipuncula and Echiura. London, British Museum. viii + 528 pp., illus.

Phylum Pentastomida

20-27 The pentastomids These are wormlike parasites that live in the lungs or nasal cavities in vertebrates, primarily tropical reptiles (Fig. 20-12). About 70 species are known. They have five anterior projections, four of which bear hooks while the fifth contains the mouth. There are no special circulatory, respiratory, or excretory organs. The nervous system is arthropodlike, and the sexes are separate, with a well-developed reproductive system. Completion of the life cycle requires an intermediate host, usually another vertebrate. Although highly specialized for parasitic life, these animals show affinities to the arthropods, having a chitinous cuticle, which is molted in the larvae, striated muscles, and a segmentally organized nerve cord. The two pairs of "legs" in one group are like those of tardigrades and *Peripatus.*

20-28 Characteristics

1 Form wormlike, soft, unsegmented.
2 Cephalothorax short, two pairs of ventral retractile hooks beside mouth.
3 Abdomen elongate, ringed.
4 No circulatory, respiratory, or excretory organs.
5 Sexes separate.
6 Parasitic.

Figure 20-12 A linguatulid "worm" (*Porocephalus,* phylum PENTASTOMIDA).

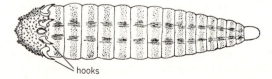

hooks

Classification

Phylum Pentastomida

(see Characteristics).

Order 1. Cephalobaenida. Mouth hooks without a basal arm (fulcrum); sex opening anterior; development direct, no intermediate host. *Cephalobaena tetrapoda,* in South American snakes; *Reighardia sternae,* in gulls and terns.

Order 2. Porocephalida. Mouth hooks with a basal arm; sex opening anterior in male, posterior in female, development indirect, larvae in a different host from adult. *Linguatula serrata,* larva 5 mm long, in liver, lung, etc., of rabbit, horse, goat, among others; adult female to 130 mm (5 in) long, male to 20 mm (1 in), in nasal cavities of fox, wolf, and dog, occasional in horse and goat, rare in humans. *Sibekia,* in crocodiles; *Kiricephalus,* in water snakes; *Porocephalus crotali,* larva in rodents, adult in rattlesnakes; other species in boas and pythons (Fig. 20-12).

References

Cuenot, L. 1949. Les Onychophores, les Tardigrades, et les Pentastomides. In P.-P. Grassé, Traité de zoologie. Vol. 6, pp. 3–75, figs.

Heymons, R. 1935. Pentastomida. In H. G. Bronn, Klassen und Ordnurgen des Tierreichs. Vol. 5, pt. 4, book 1, pp. 1–268, 148 figs.

Nicoli, R. M. 1964. Phylogenese et systematique le phylum Pentastomida. *Ann. Parasit. Hum. Comp.,* vol. 38, pp. 483–516.

Phylum Tardigrada

20-29 The tardigrades   The water bears, or bear animalcules (Fig. 20-13), are creatures up to 1 mm long with soft bodies, a slightly constricted anterior region, two retractible stylets as mouth parts, and four pairs of short unjointed legs, each with four or more claws or sticky pads. The body is covered with a nonchitinous cuticle secreted by an underlying epidermis which has a constant number of cells. The cuticle is molted periodically. Internally the body wall lacks muscle layers and muscles are all separate,

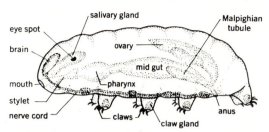

Figure 20-13   A water bear (phylum TARDIGRADA); length to 1 mm.

constituting single cells. Although true coelomate animals, the coelom is confined to the gonads and the large internal cavity is a hemocoel. The digestive tract is complete, with a complex piercing apparatus of two stylets and muscles in the buccal region. The stylets are used by the animals to pierce plant cells, thus allowing the tardigrade to suck out the contents. Three glands near the end of the intestine, called Malpighian tubules, serve as excretory organs. Sexes are separate, and fertilization, if it occurs, is internal. Eggs are laid and development is direct. In many tardigrades males are rare or unknown. Tardigrades usually live in damp moss or lichens, but a few are interstitial in marine beach sands. They can withstand extreme temperatures and desiccation, and during periods of drought or dry weather they lose water, contract, and enter a state of **cryptobiosis** (anabiosis) similar to that seen in rotifers. Placed in water they revive quickly even after years in a cryptobiotic state.

20-30 Characteristics

1   To 1 mm long.

2   Body cylindrical, rounded at ends, no visible somites.

3   Four pairs of stumpy, unjointed legs ending in four or more claws or sticky pads, last pair at hind end of body.

4   Anterior end with retractile snout.

5   Mouth with stylet.

6   No circulatory or respiratory organs.

7   Three Malpighian tubules (rectal glands) for excretion.

8   Sexes separate.

9 One gonal.

10 Direct development.

11 About 350 species.

The position of the tardigrades in any phylogenetic scheme is unclear. They show, in cell constancy, cryptobiosis, and lack of males, a relation to the rotifers and the gastrotrichs, but their development indicates that they are coelmate. This, plus the cuticle and molt, suggests a relation to the arthropods, hence their position as a separate phylum.

Classification

Phylum Tardigrada

(see Characteristics).

Order 1. Heterotardigradida. Separated claws or sticky pads; no cuticular thickenings in pharynx; gonoduct preconal; lateral cirri present. *Tetrakentron,* in buccal tentacles of a sea cucumber.

Order 2. Eutardigradida. Claws in 2 pairs with a large and a small claw united; 3 excretory glands; gonopore and anus opening together. *Batillipes,* in marine sandy beaches, *Macrobiotus,* 0.7 mm, in fresh waters; others marine and terrestrial.

References

Cuenot, L. 1932. Tardigrades. Fauna de France, vol. 24, pp. 1–96.

———. 1949. Les Onychophores, les tardigrades, et les pentastomides. In Grassé, P. (editor). Traité de Zoologie. Vol. 6, pp. 3–75.

Marcus, E. 1929. Tardigrada. In H. G. Bronn, Klassen und Ordnungen des Tierreichs. Vol. 5, pt. 4, book 3, viii + 608 pp., 1 pl., 398 figs.

———. 1959. Tardigrada. In W. T. Edmondson (editor), Fresh-water biology, 2d ed. Ward and Whipple. Pp. 508–521.

Phylum Pogonophora

20-31 The pogonophorans Also called *beard worms;* these are very long, slender tube dwellers that inhabit the depths of the ocean. They were unknown to science until the first species was dredged from Indonesian waters in 1900. Since that time increased oceanographic research has turned up about 80 species classified in 11 genera, mostly from deep water in the northwestern Pacific, but they have also been found in the North Atlantic, the Indian Ocean, and off the coasts of Central and South America. Specimens have been taken at depths from 18 m to 9 km (60 to 30,000 ft). The worms are cylindrical, ranging from 50 to 350 mm (2 to 14 in). They live in chitinous tubes up to 1.5 m (5 ft) long, presumably erect in the bottom ooze. The tubes are smooth or of rings or funnel-shaped pieces. The most remarkable feature of the worms is that they are the only free-living animals without any evidence of a digestive system. All have anterior tentacles characterized by ciliated tracts and pinnules which form a cylindrical chamber where feeding and external digestion presumably take place.

20-32 Characteristics

1 Symmetry bilateral, three germ layers.

2 Body cylindrical, of an anterior protosome with one to many tentacles, a short mesosome, and a long metasome or trunk with rings or adhesive papillae and a terminal segmented opisthoma with setae.

3 Tentacles hollow, covered with short pinnules, each a slender extension of a single epidermal cell, the tentacle with two longitudinal rows of cilia; gland cells on either side of pinnules.

4 Body wall of cuticle with a middorsal ciliated strip and a midventral trough anteriorly; epidermis of columnar cells with many glands; a thin layer of circular muscles and a thick layer of longitudinal muscles.

5 No mouth, anus, or digestive tract in adults.

6 Coelom unlined, that of protosome extending into tentacles, the mesosome and trunk each with a pair of compartments.

7 Circulatory system closed, with a middorsal (blood flowing posteriorly) and midventral

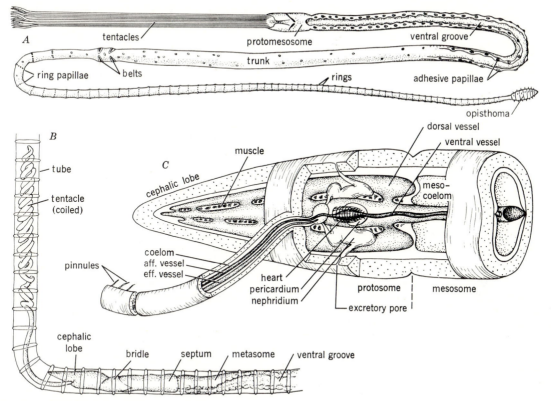

Figure 20-14 Phylum POGONOPHORA. *Lamellisabella. A.* Entire animal, enlarged. *Siboglinum. B.* Anterior part of worm in tube, one tentacle. *C.* Interior structure of anterior end, part of body wall cut away. *(After Ivanov, 1960.)*

(flowing anteriorly) vessel; ventral vessel enlarged in protosome as a heart; two blood vessels in each tentacle with a loop in each pinnule.

8 Two coelomoducts (nephridia) in protosome, opening to exterior.

9 Nervous system in epidermis enlarged in cephalic lobe, from which arise tentacular nerves, and a middorsal unpaired nerve cord.

10 Sexes separate, gonads paired, testes long, in posterior half of trunk, sperm ducts opening behind mesosome-metasome septum; ovaries in anterior half of trunk, oviducts open at midregion; eggs large, rich in yolk.

20-33 Natural history Because of their inaccessible habitat, details of pogonophoran biology are little known. Presumably they move up and

down in their long tubes and feed by forming the tentacles into a cylinder. In the space thus formed the pinnules intermesh to form a food-catching net. Rows of cilia bring food to the pinnules, where it is digested externally and then absorbed. Several species have been taken in the same dredge in the Kurile-Kamchatka Trench at a depth of over 7.5 km (25,000 ft). POGONOPHORA are dioecious, with gonads in the trunk coelom. The mechanism of fertilization is unknown, but some species brood the young. Development is unusual in that mesoderm formation occurs in two stages, the latter of which is characteristically protostome. It is not known whether a larva is formed.

The POGONOPHORA were formerly considered to be deuterostomes, but in 1970 the most posterior segment of the body, the opisthoma, was discovered; it was segmented and contained setae similar to those of annelids. This circumstance and the mechanism of mesoderm formation suggest that these animals are protostomes closely related to annelids.

Classification

Phylum Pogonophora

(see Characteristics).

Order 1. Athecanephria. Hind part of trunk without ventral adhesive papillae; nephridiopores lateral; pericardial sac present. *Siboglinum,* 1 tentacle, widely distributed in shallow or deep waters; *Oligobrachia,* tentacles 6 to 12, separate.

Order 2 Thecanephria. Hind part of trunk with ventral transverse rows of adhesive papillae; tentacles many; nephridiopores medial; no pericardial sac. *Polybrachia,* tentacles separate; *Lamellisabella,* tentacles fused into a cylinder.

References

Hyman, L. H. 1959. Pogonophora. In The invertebrates. Vol. 5, pp. 208–277, figs. 75–81.

Ivanov, A. V. 1960. Pogonophores. In P.-P. Grassé, Traité de zoologie. Vol. 5, fasc. 2, pp. 1521–1622, figs. 1384–1468, 3 col. pls.

_____. 1963. Pogonophora. Translated from Russian by D. B. Carlisle. New York, Consultants Bureau. xvi + 479 pp., 176 figs. *Much detail.*

Nörrevang, Arne (editor). 1975. The phylogeny and systematic position of the Pogonophora. Berlin, Paul Parey. 143 pp., 104 figs.

Southward, E. C. 1971. Recent researches on the Pogonophora. *Ann. Rev. Oceanogr. Mar. Biol.,* vol. 9, pp. 193–220.

Phylum Onychophora

20-34 Onychophora (from Gr. *onychus,* claw + *phorus,* bearing). Members of this group are small "walking worms" distributed discontinuously from the Himalayas to New Zealand, parts of Africa and South America, Central America, Mexico, and the West Indies. The 70 species live in dark moist places, in rock crevices, under stones, or beneath bark on rotting logs.

Peripatus capensis, a representative species, is about 50 mm (2 in) long, with a cylindrical body but no distinct head (Figs. 20-15, 20-16). The anterior end bears (1) two short **antennae;** (2) a pair of small **eyes** dorsally; (3) a midventral **mouth** rimmed by a fleshy fold and containing two small horny **jaws** used to tear or grind food; and (4) two blunt **oral papillae,** one at each side of the mouth. The **skin** is thin and lightly chitinized, with many transverse rings of fine papillae, each with a spine. The only features of segmentation are 14 to 44 pairs of short stumpy **legs,** each ending in two claws. At the inner base of each leg is a **nephridial opening.** The **anus** opens at the bluntly conical posterior end, preceded by the single ventral **genital opening.**

The **body wall** includes a flexible cuticle, an epidermis, and a thin but complex series of muscle layers, circular, longitudinal, and transverse, all unstriated. Special muscles operate the legs and parts at the anterior end. Within the body is an undivided cavity, or **hemocoel,** containing the internal organs. At either side is a large and much-branched **slime gland,** one opening through each oral papilla. These produce slimy mucus that can be squirted several centimeters and adheres to any object; it serves for defense and to capture insects and other small arthropods used as food. The **digestive system** is a straight tube, from mouth to anus, with a buccal cavity, two large salivary glands, muscular pharynx, short esophagus, long stomach-intestine, and rectum. The **circulatory system** consists of a single mid-

Figure 20-15 Phylum Onychophora. *Peripatus.* A. Entire animal, slightly enlarged. B,C. Anterior end in lateral and ventral views. (*After Snodgrass, 1938.*)

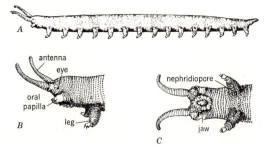

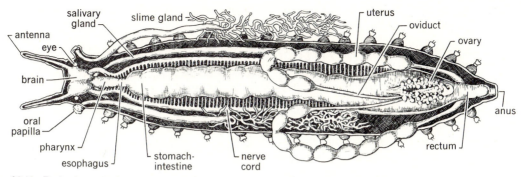

Figure 20-16 *Peripatus.* Internal structure in dorsal view. *(After Snodgrass, 1938.)*

dorsal vessel having muscle fibers in its walls and a pair of openings, or ostia, in each segment. Many delicate tubular ingrowths of the body wall, termed **tracheae,** presumably serve for respiration. The **excretory organs** are pairs of nephridiumlike structures; each has a sac closed at its inner end, a ciliated funnel and duct, a slightly expanded bladder, and a nephridiopore. The **nervous system** comprises a pair of oval cerebral ganglia (the brain) above the mouth, several anterior nerves, two circumpharyngeal connectives, and a pair of separate ventral nerve cords without true ganglia that extend the length of the body and have many transverse connections. The eyes resemble those of annelids.

The sexes are separate, and the reproductive organs are paired, but with a single external opening. The **male** system includes paired testes, seminal vesicles, ductus deferentia, accessory glands, and an ejaculatory duct. In the **female** the paired ovaries and oviducts join to a common auricle at the genital opening. Most species of Onychophora are viviparous, and part of each oviduct is specialized as a

uterus where the embryos develop. In some forms a placentalike trophoblast develops by which the embryo obtains nourishment from the uterine wall. A large female may produce 30 to 40 young per year, which at birth resemble adults except in size and color.

The Onychophora resemble annelids in the structure of the eyes, the segmental nephridia, the ciliated reproductive ducts, and the simple gut. Some of their arthropod characteristics are the jaws that are derived from appendages, the hemocoel body cavity, the dorsal "heart" with ostia, the coelom reduced to the cavities of the nephridia and reproductive ducts, and the general structure of the reproductive organs. Their tracheae somewhat resemble those of insects. Features peculiar to the Onychophora are the single pair of jaws, scant metamerism, arrangement of the tracheal apertures, nature of the skin, and separate nerve cords with no true ganglia. *Peripatus* is evidently of an ancient stock, as indicated by Cambrian fossils and by its discontinuous distribution at present.

Figure 20-17 Living *Peripatus.* *(Courtesy of California Academy of Sciences.)*

The Lesser Deuterostomes

The deuterostome line of evolution culminates in two phyla, ECHINODERMATA and CHORDATA, considered in subsequent chapters. The only other deuterostome phyla are the CHAETOGNATHA and the HEMICHORDATA, both small in terms of numbers of species and perhaps ecologically as well, hence the term *lesser* here. As deuterostomes they show the typical embryologic features of mouth arising away from the blastopore, mesoderm formation by outpocketing from the archenteron, and absence of a trochophore type of larva.

Phylum Chaetognatha (Arrow worms)

20-35 The chaetognaths   These are little torpedo-shaped marine animals, from 5 to 140 mm (0.2 to 5.5 in) long, of which 9 genera and about 65 species are known. With the exception of the genus *Spadella*, all are planktonic. A fossil species of Cambrian age has been described. The planktonic species are abundant and are important components of the ocean plankton. They are eaten by animals that feed on plankton. Arrow worms are voracious carnivores feeding on small crustaceans, their larvae, fish larvae, and other small marine life, darting about like arrows—hence the common name and that of the principal genus, *Sagitta* (L., arrow). Being transparent, they are also called *glass worms.* The genus *Spadella* lives in shallow water, clinging to rocks and algae by adhesive papillae near the tail. The phylum name refers to the bristles about the mouth (Gr. *chaeton*, bristle + *gnathos*, jaw).

20-36 *Characteristics*

1   Mouth region margined by grasping bristles which may be concealed under a hood.
2   Symmetry bilateral, three germ layers, no segmentation.
3   Body slender, with lateral fins.
4   Digestive tract complete, anus ventral.
5   Coelom well developed as three pairs of cavities.
6   No circulatory, respiratory, or excretory organs.
7   Nervous system with dorsal and ventral ganglia and sensory organs.
8   Monoecious, development direct.
9   Muscles longitudinal only.

The chaetognaths show a mixture of features which make them difficult to place in a phylogenetic sequence. On the one hand they have a cuticle, lack a peritoneal lining to the coelom, and have only longitudinal muscles, all characteristics of pseudocoelomates. But they are coelomate, with enterocoelous coelom formation. Furthermore they are not segmented, they lack cilia, and the tail is postanal as in no other phylum except the chordates. The adult structure is therefore somewhat like that of the pseudocoelomate phyla, but the embryonic development resembles that of chordates. There is no larva, so no comparisons of larvae are possible.

20-37 Structure and natural history   The body is cylindrical, with **head, trunk,** and **tail** regions (Figs. 20-18, 20-19), and covered with thin cuticle. There are one or two pairs of **lateral fins** on the trunk and a **tail fin** across the end. The **mouth** is a ventral slit on the broadened head, followed by the muscular **pharynx,** slender straight **intestine,** and **anus** at the end of the trunk. Each side of the mouth has rows of small

Figure 20-18 Phylum CHAETOGNATHA. Arrow worm, *Sagitta,* ventral view; natural size 20 to 70 mm (1 to 3 in).

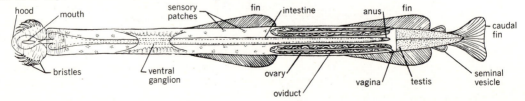

hood
mouth
sensory patches
fin
intestine
anus
fin
caudal fin
bristles
ventral ganglion
ovary
oviduct
vagina
testis
seminal vesicle

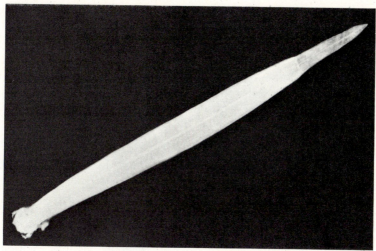

Figure 20-19 Phylum CHAETOGNATHA. A specimen of *Sagitta*. *(Courtesy of G. McDonald.)*

teeth and a lobe bearing several sickle-shaped grasp-ing **spines** of chitin (Fig. 20-20); these are worked by muscles to serve in capturing food, which is swal-lowed whole. A **hood,** attached at the base of the head, can be drawn forward over the sides and top. The spacious **coelom** consists of three pairs of cav-ities separated by median **mesenteries** above and below the intestine and between the head, trunk, and tail. The **body wall** is made up of an **epidermis** with several layers of large vacuolated cells over a basement membrane and a thin layer of striated **muscles.** Four longitudinal bands of muscles, two dorsolateral and two ventrolateral, provide for loco-motion. The **intestinal wall** has digestive epithe-lium internally and coelomic lining outside, with a

Figure 20-20 Phylum CHAETOGNATHA. The head of *Sagitta*. Note the bristles. *(Courtesy of G. McDonald.)*

thin layer of muscle between. The **nervous system** is of ectodermal origin and includes a pair of cere-bral ganglia dorsal to the pharynx, a pair of con-nectives around the latter, and a ventral ganglion midway on the trunk, with nerves to various parts. Dorsally on the head are two **eyes.** Small **tactile papillae** with stiff bristles occur over the body. Extending back from the head is an odd **ciliary loop,** the function of which is unknown but may be sensory.

20-38 Reproduction Chaetognaths are monoe-cious, and the sex cells are derived from the coelomic epithelium. In each trunk coelom is a long solid **ovary,** with a slender **oviduct** which is unique in that it consists of two tubes, one inside the other. The inner tube opens to the exterior, but the outer tube, which connects to the ovary, is blind. Each caudal coelom contains a narrow solid **testis** from which immature cells are released to mature as free sperm in the coelom, then are collected by the ciliated funnel of a **sperm duct** opening laterally on the tail. Sperm are gathered in a seminal vesicle and formed into a spermatophore. Reproduction occurs through much or all of the year. Fertilization is internal, with the help of nurse cells surrounding the ova, which pierce the wall of the blind oviduct, allowing sperm to enter. Ova leave the body through rupture of the body wall or a special duct. A typical gastrula is formed, with the coelom produced as lateral out-growths of the gastrocoel. The young at hatching resemble the adult.

References

Alvarino, A. 1965. Chaetognaths. *Ann. Rev. Oceanogr. Mar. Biol.*, vol. 3, pp. 115–194.

Beauchamp, P. de. 1960. Chétognathes. In P.-P. Grassé, Traité de zoologie. Vol. 5, fasc. 2, pp. 1500–1520, figs. 1371–1384.

Ghirardelli, E. 1968. Some aspects of the biology of chaetognaths. *Adv. Mar. Biol.*, vol. 6, pp. 271–375.

Hyman, L. H. 1959. Chaetognatha. In The invertebrates. Vol. 5, pp. 1–71, figs. 1–19.

Phylum Hemichordata
(Tongue worms, acorn worms, etc)

20-39 The hemichordates   These are small soft-bodied animals that live singly or in colonies on sandy or muddy sea bottoms. The body and coelom are divided into three regions, there are usually paired gill slits, and the nervous tissue is both dorsal and ventral in the epidermis. Certain tissue in a short anterior projection from the mouth cavity formerly was interpreted as a notochord, thus placing this group in the phylum CHORDATA. Later studies have shown that this is not a notochord but an extension from the mouth, here termed a **buccal pouch** (stomochord). Between hemichordates and chordates there are resemblances in the pharyngeal gill slits and collar cord of the nervous system—its origin from dorsal epidermis, sometimes hollow form, and dorsal location. The embryo and early larvae of hemichordates and asteroid echinoderms are much alike, in ciliated bands, digestive tract form, derivation of the anus from the blastopore (deuterostome) and especially in the enterocoelous origin, and subdivision of the coelom into three parts. These features strongly suggest a common origin.

20-40 Characteristics

1   Symmetry bilateral, unsegmented, three germ layers; body of three divisions (proto-, meso-,

metasome), either wormlike (enteropneusts) or vaselike in a case (pterobranchs).

2   Paired gill slits many, two, or none.

3   Body wall of epidermis, nervous tissue, muscle fibers, and peritoneum lining coelom.

4   Buccal pouch anterior to mouth cavity.

5   Digestive tract straight and anus terminal (enteropneusts) or U-shaped and anus near mouth (pterobranchs).

6   Coelom in three parts, paired in mesosome and metasome.

7   Circulatory system open, dorsal and ventral vessels or channels in mesosome, pulsating vessel in protosome.

8   Excretion possibly by glomerulus (nephridium?) near buccal pouch.

9   Nervous system diffuse, in epidermis, middorsal and midventral.

10   Sexes separate, alike; enteropneusts with many paired gonads, development with tornaria larva or direct; pterobranchs with two gonads and a larval stage; adults reproduce by asexual buds.

20-41 Class Enteropneusta   The acorn or tongue worms are slender and 25 to 2500 mm (1 to 100 in) long. Most of them live in shallow water, but a few go deeper. They burrow shallowly by means of a soft proboscis, and a sticky mucus secreted by glands in the skin causes formation of a tubular case of sand and other debris in which each animal lives. The burrows are U-shaped, with two openings, and coils of fecal casts are put out of the rear opening. About 12 genera and 70 species are known. *Balanoglossus* is worldwide in distribution, and *Ptychodera* occurs in tropical waters. *Saccoglossus* (= *Dolichoglossus*) inhabits both coasts of North America; *S. kowalevskii* occurs from Beaufort, N.C., to Massachusetts Bay and grows to about 175 mm (7 in) long. *S. pusillus* of southern California becomes 75 mm (3 in) in length and is brilliantly orange; in *S. kowalevskii* the proboscis is yellowish white, the collar red-orange, and the trunk orange-yellow to green-yellow posteriorly. Some tongue worms have persistent and often unpleasant odors.

The body (Fig. 20-21) comprises a **proboscis,** a **collar,** and a long **trunk.** Behind the collar are many

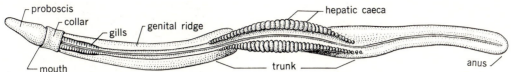

Figure 20-21 **The tongue worm,** *Saccoglossus* **(phylum HEMICHORDATA), dorsal view.** (*After Spengel.*)

gill slits on either side. Ventral to these is a lateral ridge marking the presence of the gonads. Some species have paired transverse ridges, dorsally behind the gills, that indicate the hepatic caeca. The **mouth** opens widely at the anterior ventral margin of the collar, behind the proboscis; a dilated **buccal cavity** follows, then the **pharynx,** with U-shaped gill openings high on either side that connect to the exterior. The straight **intestine,** with dorsal **hepatic caeca** (liver sacs), leads to the terminal **anus.**

In life the cavities in the proboscis and collar (Fig. 20-22) are believed to fill with water through pores on the dorsal surface; when these parts become turgid, the animal burrows through the sand or mud, aided by muscular movements of the trunk. The mouth remains open, so that a mixture of water and sand containing organic debris is forced into the buccal cavity. The water passes through the gill slits for respiration, the organic material serves as food, and the sand passes out the anus.

The **circulatory system** includes a middorsal vessel, in which colorless blood flows anteriorly (as in annelids), and a midventral vessel. The two are joined in a central sinus dorsal to the buccal pouch, and there are other branches near the gill slits. Above the central sinus is the "heart vesicle," which contains muscle fibers and whose pulsations help to drive the blood around, first anteriorly to the glo-merulus and then to the ventral vessel to supply the body. Contractions of the larger vessels probably also cause the blood to circulate. The small unpaired glomerulus (proboscis gland) is thought to be the **excretory organ.**

In the **body wall** (Fig. 20-23), the thick ciliated epidermis contains many mucus-secreting cells; it rests on a basement membrane beneath which are muscle layers. The **nervous system** consists of cells and fibers in the base of the epidermis. Concentrations of these provide a middorsal and a midventral nerve "cord" of small size, with ringlike connective tissue between the two anteriorly in the collar. A thickened cord, hollow in some species, lies in the collar dorsal to the mouth cavity and has many nerve fibers in the epidermis of the proboscis.

The buccal pouch is a short anterior out-pocketing of the digestive tract in the hind part of the proboscis. There is also a small proboscis skeleton of firm material. The **coelom** has the typical three-part division of deuterostomes, with one in the proboscis and one each in the collar and the trunk.

The sexes are separate, with multiple **gonads** in two dorsolateral rows from behind the collar to the hepatic caeca; each, when mature, releases its contents immediately to the exterior through a separate pore, and fertilization is external. In some species the egg produces a small ovoid tornaria larva, quite

Figure 20-22 **Phylum HEMICHORDATA.** The tongue worm, *Saccoglossus.* Median section of anterior portion; diagrammatic.

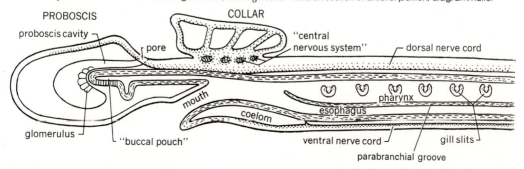

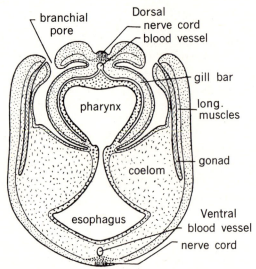

Figure 20-23 Phylum HEMICHORDATA. Generalized cross section through pharynx. *(Modified from Sedgwick, Textbook of zoology, George Allen & Unwin, Ltd.)*

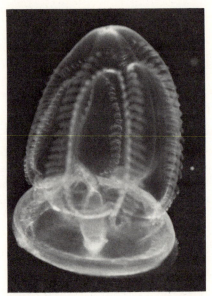

Figure 20-25 Phylum HEMICHORDATA. A living tornaria larva. *(Courtesy of R. L. Larson.)*

transparent, with surface bands of cilia (Figs. 20-24, 20-25); at metamorphosis the proboscis and collar become evident. The American species of *Saccoglossus* develop directly without a larval stage. Tongue worms can regenerate the trunk region, proboscis, and collar.

20-42 Class Pterobranchia These small forms of both shallow and deep seas are very unlike the ENTEROPNEUSTA in basic features and resemble the BRYOZOA in appearance. About 20 species are known in three genera, primarily from the Southern Hemisphere. Individuals of *Rhabdopleura* are about 1 mm long, living in ringed tubes branching from a com-

Figure 20-24 Phylum HEMICHORDATA. Tornaria larva. *(After Spengel.)*

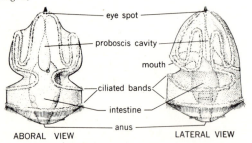

eye spot
proboscis cavity
mouth
ciliated bands
intestine
anus
ABORAL VIEW   LATERAL VIEW

mon base attached to rocks or shells on the bottom, a colony being up to 10 cm (4 in) wide (Fig. 20-26). The colony develops by budding from one individual, and all are connected by a cord. There is one pair of arms bearing ciliated, hollow tentacles and one gonad, but no gill openings. By contrast, *Cephalodiscus* forms a bushy growth with many aggregated individuals, usually under 5 mm long. Each individual has a pair of gill openings, many tentacle-bearing arms, a U-shaped digestive tract, and two gonads. *Atubaria* lives solitarily and is not in a tube. Pterobranchs reproduce both sexually and by budding. Fossil forms occur in the Ordovician period and from the Cretaceous onward.

A related group (class PLANCTOSPHAEROIDEA) is known only from transparent pelagic larvae (diameter 10 mm) with U-shaped digestive tract, some coelomic sacs, and much-branched ciliated bands on the surface.

The HEMICHORDATA appear to be most closely related to the echinoderms, with which they share numerous embryologic features and similar larvae. A close relationship to the CHORDATA is also indicated by the presence of gill slits in the hemichordates, the only other phylum outside the chordates to have such gill slits.

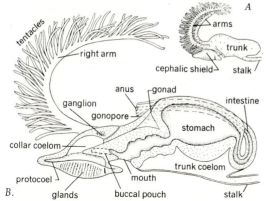

Figure 20-26 Phylum HEMICHORDATA, class PTEROBRANCHIA: *Rhabdopleura.* *A.* External form. *B.* Median section. Glands in cephalic shield (proboscis) secrete ringed tube. Coelomic cavities of collar and trunk paired, separated by mesentery. (*Adapted from L. H. Hyman, 1959, The invertebrates.*)

tive tract U-shaped; gill slits 2 or none; collar with hollow branched ciliated arms bearing tentacles; reproduce sexually or by buds.

Order 1. Rhabdopleurida. Two-armed; no gill slits; gonad 1; colonial, individuals connected by a stolon, each enclosed in a secreted tube. *Rhabdopleura.*

Order 2. Cephalodiscida. Several arms; 2 gill slits; gonads 2; individuals separate, free or grouped in colonies housed in a common secreted case. *Cephalodiscus, Atubaria.*

Class 3. Planctosphaeroidea.

Known only from a peculiar larval form.

Classification

Phylum Hemichordata

(Stomochorda; see Characteristics).

Class 1 Enteropneusta.

Tongue, or acorn, worms. Wormlike, solitary, body 25 to 2500 mm (1 to 100 in) long, fleshy, contractile, comprising proboscis, collar, and trunk; paired gill slits 10 to 80 or more; digestive tract straight; coelom distinct, 3-parted; no tentacles. About 63 Recent species. *Ptychodera*, tropical coral reefs; *Balanoglossus, Saccoglossus*, worldwide in sand or mud.

Class 2. Pterobranchia.

To 7 mm (3 in) long, body vaselike, usually in groups or colonies enclosed in a secreted case; diges-

References

Barrington, E. 1965. The biology of Hemichordata and Protochordata. San Francisco, W. H. Freeman and Company. vi + 176 pp., illus.

_____. and **R. Jefferies** (editors). 1975. Protochordates. Symposia of the Zoological Society of London No. 36. New York, Academic Press, Inc. xiii + 361 pp.

Dawydoff, C. 1948. Stomocordes. In P.-P. Grassé, Traité de zoologie. Vol. 11, pp. 367–532, 134 figs.

Hyman, L. H. 1959. Hemichordata. In The invertebrates. Vol. 5, pp. 72–207, figs. 20–74.

Morgan, T. H. 1894. The development of Balanoglossus. *J. Morphol.*, vol. 9, pp. 1–86, illus.

Van der Horst, C. J. 1927–1939. Hemichordata. In H. G. Bronn, Klassen und Ordnungen des Tierreichs. Vol. 4, pt. 4, book 2, pp. 1–737, 732 figs.

21
PHYLUM MOLLUSCA: MOLLUSKS

Mollusks (L. *mollis*, soft) have soft bodies, consisting typically of an anterior head, a ventral foot, and a dorsal visceral mass. The body is more or less surrounded by a thin, fleshy mantle, which is an outgrowth of the body wall, and is commonly sheltered in an external limy shell. The phylum comprises seven classes of diverse appearance and habits: *Neopilina* (class MONOPLACOPHORA), the chitons (POLYPLACOPHORA), solenogasters (APLACOPHORA), tooth shells (SCAPHOPODA), snails and slugs (GASTROPODA), clams, oysters, and other bivalves (BIVALVIA), and the nautili, squids, and octopuses (CEPHALOPODA).

The phylum MOLLUSCA is the second largest animal phylum, with over 80,000 living species, and is a major group in the protostome line of invertebrate evolution. Mollusks are of wide distribution in both

time and space, having a continuous record since Cambrian time; many are abundant as individuals and are important ecologically. Although most mollusks are marine, various snails and slugs have invaded freshwater and terrestrial environments, and bivalves are common in fresh water. Along with the arthropods, the mollusks show adaptation to the greatest number of habitat types of any invertebrate phylum. Most mollusks are free-living but slow-moving creatures that show a close association with the substrate. Some attach to rocks, shells, or wood; some burrow; others float; and the squids and octopuses can swim freely. Of major economic importance are the clams, scallops, oysters, squids, and others that serve as human food and a few bivalves that produce pearls. Some snails and slugs are agricultural pests and feed on cultivated plants, certain freshwater snails are intermediate hosts for parasitic worms, one group of freshwater clams has larvae parasitic on fishes, and the shipworms damage wooden ships and wharves. Malacology is the science that deals with mollusks; conchology is the study of shells, especially those of mollusks. Shell collecting is a popular hobby and has added significantly to the knowledge of mollusks. Fossil shells are often well preserved and serve as useful indicators of early geologic epochs.

21-1 Characteristics

1 Body usually short and partially or wholly enclosed by a fleshy outgrowth of the body wall called the **mantle,** which may be variously modified; between the mantle and the visceral mass is a **mantle cavity** containing components of several systems (secondarily lost in a few groups).

2 A shell, if present, secreted by the mantle and of one, two, or eight parts; head and ventral muscular foot closely allied, with foot variously modified for crawling, burrowing, swimming, or food capture.

3 Digestive tract complete, complex, with ciliary tracts for sorting small particles; mouth with **radula** bearing transverse rows of minute chitinous teeth to rasp food (except BIVALVIA); anus opening in mantle cavity; a large digestive gland and often salivary glands.

4 Circulatory system open (except for the CEPHALOPODA) and typically includes a dorsal heart with one or two atria and one ventricle, usually in a pericardial cavity, an anterior aorta, and other vessels and many blood spaces (hemocoels) in the tissues.

5 Respiration by one to many uniquely structured **ctenidia** (gills) in the mantle cavity (secondarily lost in some), by the mantle cavity, or by the mantle.

6 Excretion by kidneys (nephridia), one, two, or six pairs or a single one, usually connecting to the pericardial cavity and exiting to the mantle cavity; coelom reduced to cavities of the nephridia, gonads, and pericardium.

7 Nervous system typically a circumesophageal nerve ring with various pairs of ganglia and two pairs of nerve cords, one pair innervating the foot and the other the visceral mass; many with organs for touch, smell, or taste, eyespots or complex eyes, and statocysts for equilibration.

8 Sexes usually separate (some monoecious, a few protandric); gonads four, two, or one, with ducts; fertilization external or internal; mostly oviparous; egg cleavage determinate, spiral, unequal, and total (meroblastic in CEPHALOPODA); **trochophore** and **veliger larvae,** or parasitic stage (UNIONIDAE), or development direct (PULMONATA, CEPHALOPODA).

9 Unsegmented (except MONOPLACOPHORA); symmetry bilateral or asymmetrical.

21-2 Phylogenetic relationships

Although early zoologists confused brachiopods and barnacles with the mollusks, the phylum is one of the more clearly defined and all mollusks are usually easily recognized as belonging to it. They have three features not found in any other phylum: mantle, radula, and ctenidia.

Because of their shells, the mollusks have left a good fossil record, and all the major classes are known from Paleozoic times. However, the various molluscan classes are often very different from one another, and the fossil record does not indicate any intermediate forms which would either link together the various classes or indicate which is the ancestral group. A similar situation exists with respect to the relation of the mollusks to the other invertebrate phyla. Lacking such fossils, biologists have been forced to use features of living mollusks to attempt to construct phylogenetic relationships and the features of a primitive, generalized mollusk.

The mollusks are undoubtedly protostome invertebrates, for they show in development the characteristic spiral cleavage, mesoderm formation, coelom formation, and trochophore larvae of this line. The other dominant coelomate protostome phyla are the ARTHROPODA and the ANNELIDA, and a diagnostic basic feature of their body construction is the presence of segmentation. This feature has been a stumbling block in efforts to relate the mollusks to these phyla. In 1959, however, a new class of mollusks was erected to receive the genus *Neopilina*, a mollusk with some segmentally arranged organs which was considered to be the missing link. Since the advent of *Neopilina* it has become easier to relate the mollusks to the annelid-arthropod line of evolution (Fig. 21-1).

The mollusks and the annelids share a number of developmental features which suggest a close relationship. These include a strikingly similar pattern of spiral cleavage and a similar trochophore larva. The segmental arrangement of *Neopilina* adds further evidence suggesting a close relationship and/or origin from the annelids.

Mollusks, however, lack a large coelom and move generally slowly by ciliary-mucous methods, and many have a ladderlike nervous system. These features have suggested to other zoologists that perhaps the mollusks originated from the turbellarian flatworms, which share certain of the above features as well as those of spiral, determinate cleavage.

Even accepting a close relationship to the annelids, there is a question as to the arrangement of the molluscan classes, and several phylogenetic schemes can be suggested. One case assumes that the segmented arrangement of *Neopilina* is the primitive condition of the phylum and makes this group the link to the annelids, with segmentation being lost in the evolution of the other classes. A second scenario derives all the mollusk classes except the MONOPLACOPHORA from the protostome line before the advent of segmentation and separately derives the MONOPLACOPHORA after segmentation occurred, thus making the mollusks diphyletic in origin. A final one would derive mollusks from the annelid-arthropod line of evolution before the advent of segmentation and then have the MONOPLACOPHORA arising later

from the mollusks. In this scheme, segmentation is thought to have arisen twice in evolutionary history: in the annelid-arthropod line and in the MOLLUSCA in the line leading to the MONOPLACOPHORA.

Given the lack of intermediate forms, either living or fossil, and the great morphologic differences among the various living molluscan groups, the true origin and relationships must remain obscure.

21-3 General features   Mollusks are a diverse group both in body form and in habitats occupied (Fig. 21-1). Although the various classes differ among themselves, most of the mollusks share a number of anatomic and physiologic features which are either unique to or characteristic of the phylum. Those features are reviewed here before considering each of the classes in detail. There have been a number of attempts by zoologists to group these basic features together to construct a generalized or hypothetical ancestral mollusk which could then be considered the common ancestor of the whole phylum. Although this construction is complicated because of the disagreements noted above as to the position of the MONOPLACOPHORA and segmentation, it is still possible, without actually constructing the animal, to discuss basic molluscan features for which there is broad agreement among zoologists.

The most primitive living mollusks—chitons, solenogasters, and monoplacophorans—and indeed most living mollusks, and presumably also the ancestral forms, are slow-moving animals living upon the substrate in aquatic environments. Thus it is assumed that the ancestral mollusk lived in a similar place and possessed a broad, flat **foot** with the head at the anterior end. The rest of the body containing the major visceral organs was projected dorsally

Figure 21-1 Phylum MOLLUSCA. Relations in six classes of the shell (heavy lines), foot (stippled), digestive tract (shaded), mouth (*M*), and anus (*A*).

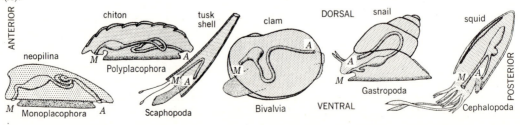

above the foot and covered with a **mantle** which secreted externally a single caplike shell. A **mantle cavity** is presumed to have existed in the rear between the mantle and the visceral mass. The mantle cavity probably contained a pair of **ctenidia** (gills), and into it opened the ducts from the nephridia as well as the anus. The digestive tract was complete, with a rasping organ, the **radula,** in the buccal region. The animals were probably feeders on algae scraped from the substrate by the radula, and the stomach was modified to deal with small particles. A pair of kidneys (nephridia) probably connected the pericardial cavity to the mantle cavity. The kidneys removed metabolic wastes and also served to transport the gametes to the outside from the gonads located adjacent to and discharging into the pericardium. The coelom was confined mainly to the pericardium, and the heart had one ventricle and two atria. The circulatory system was open, with many large blood spaces in the tissues. The nervous system consisted of a circumesophageal ring and longitudinal cords.

As a characteristic feature of the mollusks, the mantle forms the shell, the siphons, and also the mantle cavity. In land snails the mantle cavity is vascularized and serves as a lung. The molluscan **shell** is composed of an outer layer of proteinaceous material called the **periostracum** and two layers of calcium carbonate, a middle **prismatic** layer, and an inner **nacerous** layer (Fig. 21-19). The outer edge of the mantle has two grooves, the outer of which, the periostracal groove, secretes the periostracum. The cells near the edge of the mantle secrete the prismatic layer, while the whole outer surface of the mantle secretes the nacreous layer. The shell thus grows in length at the edge but in thickness all over. The innermost mantle groove is margined by two lobes, the inner one being muscular and responsible for forming siphons when they are present.

Respiration in primitive mollusks, as in most living mollusks, is by the unique molluscan gills called **ctenidia** (Fig. 21-3). Each ctenidium consists of a central axis containing blood vessels. From each side of the axis extends a series of flattened filaments, those on one side alternating with those on the other. Each filament has lateral cilia on the face, with frontal and abfrontal cilia on the edges. Each is strengthened by a skeletal rod. The lateral cilia create a water cur-

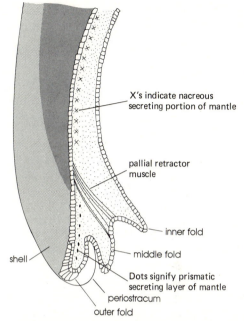

Figure 21-2 Generalized cross section of the mantle edge of a mollusk, showing the shell layers and secretory regions of the mantle. (*After Russel-Hunter, 1968.*)

rent passing over the ctenidia, while the frontal and abfrontal cilia remove particles to keep the gill clean. The vessels in the axis distribute blood to the filaments where gas exchange takes place and return blood to the heart. Primitively a pair of these ctenidia were located in the mantle cavity.

The mantle cavity also contains the primary molluscan chemical sense organ, the **osphradium,** which is situated to test the incoming water before it passes the ctenidia. Water leaving the gills in the mantle cavity then encounters the openings from the kidneys and the anus. On the roof of the mantle cavity near these openings is a **hypobranchial gland** which secretes mucus and probably has as its primitive function to tie up fecal particles from the anus into pellets and prevent them from fouling the ctenidia.

Although they are coelomate animals, the coelom has been greatly reduced; the only real manifestation of it is the pericardium surrounding the heart. In this respect the mollusks are similar to the arthropods. The heart primitively consists of two atria which receive blood from the ctenidia and a single ventricle which pumps the blood anteriorly in a large aorta. The blood, however, soon leaves the vessel and per-

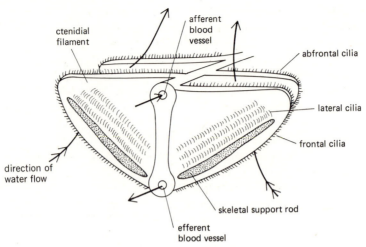

Figure 21-3   Diagrammatic illustration of a molluscan ctenidium, showing the structures. (*After Russell-Hunter, 1968.*)

colates through the tissues, collecting in sinuses. The system is thus open. Blood is returned to the ctenidia from sinuses, where it is again oxygenated. Such a system is inefficient in delivering oxygen, and blood movement is sluggish, but it is suitable for animals which move slowly as the mollusks do. Because of its limitations, however, this system cannot be used in the fast-moving cephalopods (see Sec. 21-22).

The gonads primitively discharged the ova and sperm into the pericardial cavity, from which they passed to the outside via the kidney ducts. Only in more advanced mollusks have the gonads established separate ducts. Because of this, external fertilization was the primitive condition.

Characteristic of the molluscan digestive tract and unique to the phylum is the feeding organ called the **radula.** The radula lies in a ventral outpocketing of the mouth cavity called the **radula sac.** It consists of a long chitinous ribbon set with numerous rows of chitinous teeth (Fig. 21-4). The ribbon is stretched over a tough cartilaginous structure called the **odontophore** (Fig. 21-5). A complex series of muscles at-

Figure 21-4   Scanning electron micrographs of molluscan radulas. *A.* Radula from a squid. *B.* Radula from a heteropod (*Carinaria*). (*Courtesy of G. McDonald.*)

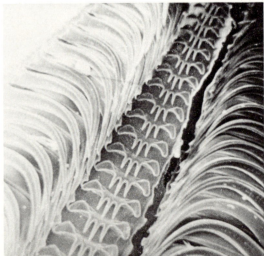

A.   *B.*

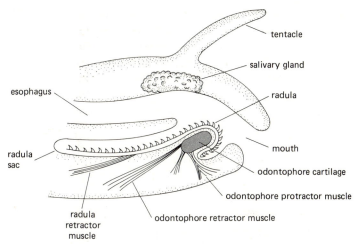

Figure 21-5 Diagrammatic representation of a longitudinal section of a mollusk head, showing radular apparatus. (*After several authors.*)

tached to the odontophore serves to push the whole apparatus out of the mouth of the animal and press it against the substrate. The animal then either drags the chitinous teeth over the substrate while moving forward or else moves the chitinous ribbon independently. Either way the result is to scrape off particles of plant material, which are then carried back into the mouth when the radula is retracted. A radula is found in all molluscan classes except BIVALVIA, and it is assumed that the primitive feeding mechanism employed the radula in the same fashion.

The remainder of the molluscan digestive system is primitively designed to deal with such small particles. The stomach is often complex, with many ciliary sorting tracts to separate out particles before sending them to the digestive gland or liver, where digestion proceeds intracellularly.

Locomotion in primitive mollusks, and indeed in many advanced mollusks, is via the large flat **foot**, another characteristic mollusk feature. The foot is muscular and has numerous mucous glands and cilia on its sole. Locomotion is either by ciliary-mucus movement, much as seen in turbellarian flatworms, or by muscular action. A mucus trail is laid down and the cilia act upon it to propel the animal slowly forward. Muscular movement is accomplished via one or more continuous waves of muscular contraction and expansion moving over the sole of the foot, each wave serving to move the animal forward a small distance. In bivalves, where the animals dig

into the substrate, the foot is extended into the substrate by pumping blood into it. After it is anchored, the animal is pulled up to the foot by powerful retractor muscles.

A **trochophore larva** is usually the product of embryologic development (except in CEPHALOPODA and terrestrial forms) of the zygote, and it is similar to the annelid trochophore. In the dominant molluscan classes, however, the trochophore develops into a second larval form, the **veliger,** which is unique to the phylum (Fig. 21-6). This larva swims by means of two enlarged, ciliated lobes and has a shell. At meta-

Figure 21-6 Photomicrograph of the veliger larva of the prosobranch genus *Conus*.

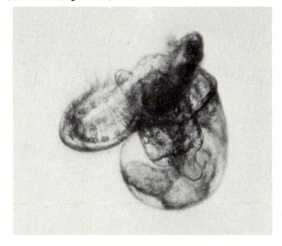

morphosis it settles down to the substrate and casts off its swimming lobes to take up a crawling existence.

21-4 Size Various chitons are 1 to 30 cm ($\frac{1}{2}$ to 12 in) long, *Cryptchitonn* (*Amicula*) of the Pacific Coast being the largest. The tooth shells are mostly under 6.5 cm ($2\frac{1}{2}$ in) long, but some grow to 15 cm (6 in.) The gastropods range from minute snails under 1 mm in diameter to the north Australian spindle shell (*Syrinx aruanus*) 70 cm (28 in) long; most species are less than 5 cm (2 in) in diameter or height. The shells of bivalves are from 1 cm ($\frac{1}{2}$ in) long up to 1.5 m ($4\frac{1}{2}$ ft) in the giant clam (*Tridacna gigas*) of the tropical Pacific, which may grow to weigh 250 kg (550 lb). Some squids and octopuses are but 2.5 cm (1 in) long, but the giant squid (*Architeuthis*) is recorded with a 6-m (20 ft) body and 10.5-m (35 ft) tentacles, the largest invertebrate known (Fig. 1-2).

Class Monoplacophora

21-5 Structure *Neopilina* (Figs. 21-7, 21-8) is oval in outline with a single, bilaterally symmetrical, cap-shaped **shell** up to 40 mm ($1\frac{1}{2}$ in) long, its apex raised

and bent anteriorly. The undersurface consists of a small head region and a flat, almost circular **foot** surrounded by a **mantle cavity** (pallial groove). The foot is ciliated and has many gland cells. The **mouth** is anterior, with a broad ciliated flap (velum) on each side and branching tentacles behind. Inside the mouth is a well-developed **radula** with 16 rows of teeth, followed by the **pharynx** with two large diverticula and a single salivary gland. The straight esophagus extends to the roughly triangular **stomach,** which has a large **digestive gland** on each side. The **intestine** is coiled and the anus posterior. There are five or six pairs of segmentally placed **ctenidia** (gills), each having five to eight main filaments on one side of the central axis and vestigial branches on the other. Opening at the bases of the gills are five pairs of **nephridia** and a sixth pair anteriorly. There are five pairs of major foot-retractor muscles and a group of three additional pairs anteriorly. The nervous system consists of a circumoral nerve ring joining two main pairs of longitudinal cords, one pair in the foot and the other pair laterally in the pallial fold. Nerves extend from these cords to the gills. There is a cross commissure anteriorly in the foot, and the longitudinal cords join posteriorly. On each side there are 10 connectives between the foot

Figure 21-7 Class MONOPLACOPHORA: *Neopilina.* *A.* Left side of shell. *B.* Dorsal surface of shell. *C.* Ventral view. *D.* Internal structure from left side. (*After Lemche and Wingstrand, 1959.*)

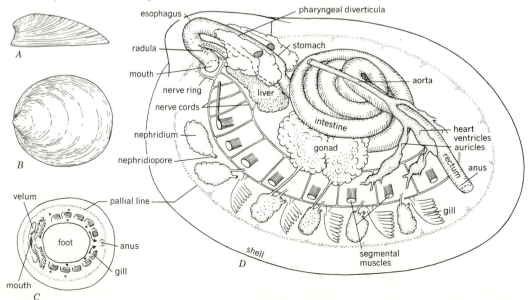

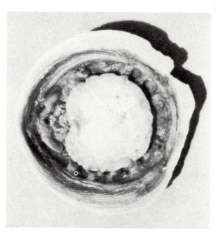

Figure 21-8   Ventral view of a specimen of *Neopilina*, class MONOPLACOPHORA. (*Courtesy of Shane Anderson.*)

and pallial cords. The circulatory system has a **dorsal heart;** it receives arterial blood from efferent gill vessels through two pairs of atria. The paired ventricles join anteriorly in a dorsal aorta extending forward to open into anterior blood sinuses. Anterior to the heart lie a pair of large, fluid-filled sacs which have been termed *coelomic sacs.* If these sacs are coelomic, they would be the largest coelom in the mollusks and suggest a closer relationship to the annelids. The **gonads** are on either side of the intestine and open through the two middle pairs of nephridia. The sexes are separate.

21-6 Natural history All species of *Neopilina* are abyssal, so little is known of their life history or habits. The first 10 specimens (of *N. galatheae*) were taken in 1952 from dark muddy clay at 3350 m (11,000 ft) off the coast of Costa Rica. Subsequently five additional species were described from the Indo-Pacific and South Atlantic oceans. All have come from abyssal depths. The stomach contents of *N. galatheae* included many radiolarians, suggesting that *Neopilina* is a detritus feeder, possibly filtering small organisms by means of the posterior tentacles. At present, since the animals come up dead and observation in their natural habitat is not feasible, virtually nothing is known of their ecology. Reproduction and development are also unknown.

With respect to other mollusks, *Neopilina* has a number of features in common with the chitons, such as a similar radula, certain similarities in shell

construction, and the presence of a special sense organ, the **subradula organ.** The nervous system also seems to be similar.

Fossil relatives of *Neopilina* are classified in two orders, six families, and 21 genera. They lived from the Lower Cambrian to the Devonian. The shells were recognized by their paired muscle scars, but the significance of the group was not fully realized until living specimens were dredged by the Danish Galathea Expedition in 1952.

Class Polyplacophora (Chitons)

21-7 Structure A chiton (class POLYPLACO-PHORA, Figs. 21-9, 21-10) has an elliptical **body** with a convex dorsal surface bearing eight overlapping limy **plates** (valves); these are jointed to one another and covered at the sides or entirely by a thick fleshy **girdle** (part of the mantle) containing bristles or spines. The flat **foot** occupies most of the ventral surface. Between the foot and mantle is a **mantle cavity** (pallial groove). Under the anterior margin of the girdle is the small **head,** which contains the **mouth** but has no eyes or tentacles. The floor of the mouth cavity has a long **radula** with many crossrows of fine **teeth** that may contain the iron compound magnetite. Also present is a chemosensory structure, the **subradula organ.** A short **pharynx** leads to the rounded **stomach,** to which the **digestive gland** (liver) connects. The long coiled **intestine** ends at the **anus** posteriorly in the mantle cavity. The **heart** lies posterodorsally in a pericardial cavity; it comprises two atria and a ventricle connecting to an anterior aorta. Two slender **nephridia** drain wastes from the pericardial cavity into the mantle cavity. In the mantle cavity are the **gills,** 6 to 80 in different species. The **nervous system** comprises a ring about the mouth that joins two pairs of ventral longitudinal nerve cords, the pedal in the foot and pallial in the girdle; the cords have many cross-connectives but no ganglia. The system is very similar to that of turbellarian flatworms. Chitons are unique among mollusks in that they have in their shells channels opening at the surface and containing sensory structures called **esthetes.** These esthetes appear to be sensitive to changes in light intensity.

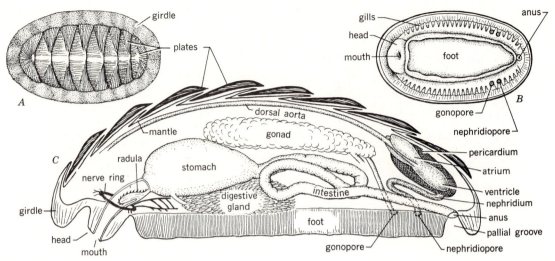

Figure 21-9 Class POLYPLACOPHORA: a chiton. *A*. Dorsal view. *B*. Ventral view. *C*. Internal structure from the left side, with the shell, mantle, and foot shown in median section.

The sexes are separate, each having one fused gonad from which a duct extends on either side to a gonopore posterior in the mantle cavity.

21-8 Natural history All chitons are marine, living most commonly and abundantly on rocks in shallow water, though a few are found at great depths. Only about 600 species are known, and all are similar in both morphology and ecology. Although an ancient group, they have not undergone much adap-

Figure 21-10 Dorsal view of the chiton *Mopalia ciliata.* (*Courtesy of the California Academy of Sciences.*)

tive radiation. Chitons are slow-moving microphagous feeders, scraping algae and other small invertebrates from the substrate with their massive radula. Before the radula is employed, the chemosensory subradula organ is protruded out of the mouth to test the substrate. Chitons are well adapted to live in wave-swept areas, as their muscular foot allows them to cling to rocks tightly and the flexible eight-piece shell permits close attachment to irregular surfaces. Respiration is effected by raising a portion of the girdle (mantle) at both anterior and posterior ends, allowing the lateral cilia of the ctenidia to create a current through the U-shaped mantle cavity. Chitons are dioecious, and fertilization occurs either externally in seawater or in the mantle cavity of the female. There are no copulatory organs; sperm are shed into the sea and brought into the female mantle cavity with respiratory currents. Eggs are either brooded or shed into the sea, where they develop into trochophore larvae. The trochophore larva, however, metamorphoses into an adult without the intermediate veliger stage.

Chitons are primitive mollusks that have changed little since the Paleozoic. They retain many of the features proposed for the ancestral mollusk, and those which are different, such as the shell in eight plates, heavy mantle, and multiple gills, seem special adaptations to their special habitat.

Class Aplacophora (Solenogasters)

21-9 The aplacophorans The class APLACO-
PHORA includes about 130 species of small wormlike
forms in which the mantle covers the entire body and
contains fine limy spicules as the only evidence of a
shell (Fig. 21-11). The foot has been reduced to a
ciliated ridge in a ventral groove or else is lacking.
The digestive tract is simple, with an anterior radula.
One group is hemaphroditic, the other dioecious.
Little is known of their reproduction except that in
some a trochophore larva is produced which develops into the adult stage without passing through a
veliger stage. Some live on corals or hydroids at considerable depths in the ocean; others are burrowers
in mud in deep water, and a few are found in shallow
waters. They appear to be primitive mollusks, and
their development, wormlike body, lack of a foot and
shell, and burrowing locomotion all suggest a close
relation to annelids. The ladderlike nervous system
and thick mantle with spicules link them to the chitons, with which they were once grouped.

Class Scaphopoda (Tooth shells, or tusk shells)

21-10 Structure and natural history In the SCA-
PHOPODA (Gr. *skaphe*, trough + *podos*, foot) the body
is slender dorsoventrally, surrounded by the mantle,
which secretes an elongated conical shell, open at
both ends and slightly curved and tapered, giving it
the form of an elephant tusk (Fig. 21-12). The 200

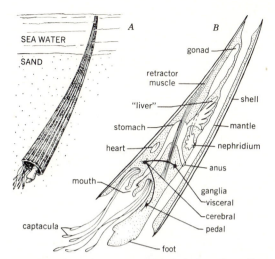

Figure 21-12 Class SCAPHOPODA: The tooth shell, *Dentalium*.
A. Position in life. *B*. Internal structure from the left side; diagrammatic.

species of scaphopods are all marine, in shallow
waters and to depths of 4570 m (15,000 ft), living
partly buried obliquely in the mud or sand. The
pointed foot protrudes from the larger ventral end of
the shell for use in burrowing. About the mouth are
several delicate ciliated and contractile tentacles
(captacula) with expanded tips; these are sensory
and prehensile, serving to capture the protistans
used as food. The head has been reduced to little
more than a mouth, but the radula is retained. Ctenidia are absent, and the large mantle cavity serves for
respiration. Cilia on the mantle cavity create a current of water into it. Periodically this water is vio-
lently ejected by contraction of the foot and the process begins again. The circulatory system consists
only of blood sinuses, without heart or vessels. Scaphopods are dioecious; eggs are discharged individually and develop into first trochophore and then
veliger larvae. After the larval stages the young animals sink to the bottom, where a gradual metamorphosis takes place. Scaphopods appear to be closely
related to bivalve mollusks, with which they share
features such as similar foot and burrowing habit,
reduced head, embryonic mantle, and symmetry
and orientation of the body in the shell (Fig. 21-13).
Scaphopod (*Dentalium*) shells kept on strings were
the money of Indians on the Pacific Coast from California to Alaska.

Figure 21-11 Class APLACOPHORA: the solenogaster *Neomenia carinata.* (*After Hyman, 1967.*)

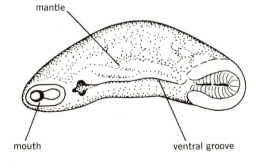

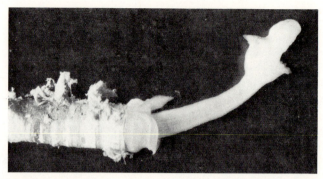

Figure 21-13 Anterior portion of the scaphopod *Dentalium rectius,* showing the shape of the foot and mantle edge. Compare Fig. 21-12. (*Courtesy of G. McDonald.*)

Class Gastropoda
(Snails, whelks, conchs, slugs, etc.)

21-11 General features and evolution The class GASTROPODA is by far the largest, most successful, and most diverse group of the MOLLUSCA. They are found in all major habitats, terrestrial, freshwater, and marine, but are most numerous and diverse in marine waters. More than 15,000 fossil species are known and 45,000 or more living species. Although most gastropods are slow-moving animals crawling over or burrowing into the bottom substrate, adaptive radiation has produced successful swimming forms as well. Gastropods are abundant in number of species and as individuals, as anyone who has observed mud snails in an exposed mud flat, limpets or periwinkles on a rocky shore, or slugs in the garden can attest.

Gastropods are so diverse in structure and form that it is difficult to give a defining feature for the class. Although several features are characteristic of the group, one or another of these seems to be missing in one or more subgroups of the class. The single feature which can define the class is that all living gastropods have undergone the unique process of **torsion** (Fig. 21-14). The process of torsion occurs late in larval life, when an asymmetrical muscle pulls the whole visceral mass 180° counterclockwise relative to the head-foot, thus bringing the mantle cavity, which was in the rear, around to lie just behind the head. As a consequence of this, the digestive tract became more or less U-shaped and the nerve cords

are crossed in a figure eight. With the mantle cavity just behind the head, the anus and nephridiopores, which discharged into the mantle cavity, now discharged virtually on the head of the animal. The adaptive significance of torsion has been argued among zoologists for years, but no universally acceptable explanation has been forthcoming. What does seem apparent is that placing the mantle cavity in front created a fouling problem relative to the situation in the ancestral mollusk with the mantle cavity in the rear. This fouling occurs because the water now passes directly into the cavity from the front, goes over the ctenidia, picks up the wastes from the anus and nephridia, and essentially exits on top of the animal's head, where it may be deposited. In the ancestral mollusk this material would have been deposited posteriorly, away from the animal. Many of the major evolutionary trends seen in the gastropods and the definition of *primitive* have to do with attempts to solve the fouling problem. In the most primitive gastropods the problem is solved by having slits or holes in the shell over the mantle cavity so that water with wastes exits farther back on the shell. More advanced gastropods have solved the problem by extending the anal and nephridial openings to the far edge of the mantle cavity and developing a siphon to channel an oblique water flow through the mantle cavity. In this the water enters on one side and exits on the other side, away from the head region. Finally, in one group, OPISTHOBRANCHIA, the animals have undergone detorsion and the whole mantle cavity is displaced back toward the rear or is lost.

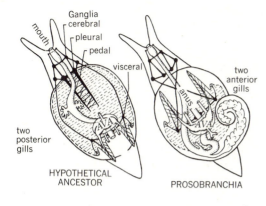

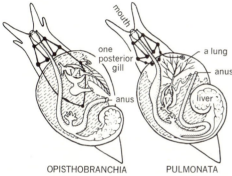

Figure 21-14 Class GASTROPODA. Plan of organization in the three subclasses, showing rotation of the viscera and crossing of nerve commissures, with loss of the left ventricle, gill, and kidney in OPISTHOBRANCHIA and a lung replacing the gill in PULMONATA. Shell and body outlined, mantle area shaded, nervous system black. *A.* atrium, *V,* ventricle; sex organs omitted. (*Modified from Stempell.*)

Most gastropods are slow-moving sedentary animals closely associated with the substrate but within that restriction show a remarkable adaptive radiation of shell types, feeding types, and specializations to various microhabitats and niches. The few swimming or pelagic forms are not fast swimmers.

Most gastropods possess a **radula** (Fig. 21-4), which in this class also shows the greatest diversity of shape and form. The radula is the main food-capturing organ for the gastropods and retains its primitive structure and use in the herbivorous gastropods. It has, however, been modified in many advanced gastropods to be used not to scrape off small particles but to capture prey or to bore into prey organisms. In a few the ctenidia are employed in feeding. Hence this class also shows the greatest diversity of feeding types of any mollusk group.

Internally and externally gastropods resemble the previously described primitive or hypothetical ancestral mollusk more closely than another class, except that the shell, if present, is usually spirally coiled and hence the visceral mass is also coiled. This makes them asymmetrical rather than bilaterally symmetrical.

Most gastropods are dioecious (except PULMONATA and OPISTHOBRANCHIA), with the single gonad primitively discharging gametes to the exterior by way of the kidneys. In advanced groups, however, separate genital ducts are developed to carry gametes to the mantle cavity. With this development also comes the potential of internal fertilization. Except for terrestrial slugs and snails, most gastropods form a **trochophore** and then a **veliger larva.** These larvae are often free-swimming in the plankton and represent the means by which these usually sedentary forms are dispersed to new habitats.

21-12 Classification   Despite the large size of this class, only three major subclasses are recognized. PROSOBRANCHIA is the largest subclass, with 25,000 or more species, and is characterized by the fact that none of the species have undergone any detorsion and all have the nerve cords in a figure eight (streptoneurous condition). Almost all species are marine. This group is the most diverse of the gastropods in terms of size, structure, food habits, and habitats occupied. The most primitive gastropods belong in this subclass, as do many of the advanced forms. Both of the remaining subclasses are considered to be derived from this group.

The most primitive prosobranchs are in the order ARCHAEOGASTROPODA, which have two atria, two nephridia, and one or two ctenidia. Many have holes or slits in the shell over the mantle cavity to allow water to exit and prevent fouling. The radula is characterized by the large number of teeth per row (except limpets), and most are herbivorous. Included here are such common mollusks as limpets (*Acmaea, Patella*), abalone (*Haliotis*) (Fig. 21-15), and turban and top shells (*Tegula, Calliostoma, Turbo*). The other two orders in the subclass are MESOGASTROPODA and NEOGASTROPODA, both of which are characterized by having but a single ctenidium in the mantle cavity

A.

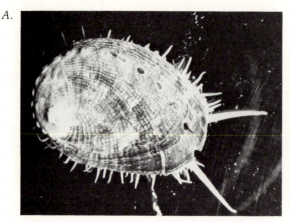

B.

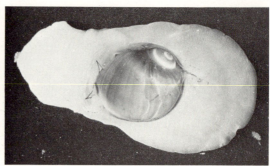

Figure 21-15 Gastropods of the subclass PROSOBRANCHIA. *A.* Order ARCHAEOGASTROPODA, *Haliotis rufescens. B.* Order MESO-GASTROPODA, *Polinices lewisii. (Courtesy of G. McDonald.)*

and but one atrium and one nephridium. The ME-SOGASTROPODA (Fig. 21-15) have a radula with seven teeth per row. In this diverse group are found such common snails as periwinkles (*Littorina*), slipper shells (*Crepidula*), conchs (*Strombus*), and cowries (*Cypraea*). The NEOGASTROPODA are characterized by having a very large osphradium and a radula of only three teeth per row or less. Most neogastropods are carnivores and include common forms such as oyster drills (*Urosalpinx*), moon snails (*Polinices*; Fig. 21-15), whelks (*Busycon, Buccinum*), mud snails (*Nassarius*), and cone snails (*Conus*).

The subclass OPISTHOBRANCHIA is a small but very diverse group of mostly marine gastropods in which some degree of detorsion has taken place, so that the mantle is displaced to the right or lost. In most the shell has become internal or is completely lost. All are hermaphroditic. About eight orders are recognized; the most interesting are the SACOGLOSSA and NUDIBRANCHIA (Fig. 21-29). Both have lost the shell and mantle cavity. The SACOGLOSSA are sea slugs which feed on algae, and some have developed the remarkable ability to transfer the individual chloroplasts from the algae to the dorsal surface of their own bodies, where they are cultured to produce food for the slug. Nudibranchs are brightly colored naked mollusks which feed upon such unlikely animals as sponges and hydroids. Many that feed on hydroids have developed the ability to transfer, undischarged, the nematocysts from their hydroid prey to their own dorsal processes, where they are employed by the nudibranch for defense.

The last subclass, PULMONATA, includes all terrestrial and most freshwater gastropods. They are characterized by the complete loss of ctenidia from the mantle cavity, which has been vascularized and is used as a lung. All are hermaphroditic. Two main groups are recognized, those with one pair of tentacles (BASOMMATOPHORA) and another with two pairs of tentacles (STYLOMMATOPHORA). The structure of one of the latter, *Helix,* is given below.

21-13 Structure A common gastropod such as the garden snail (*Helix aspersa,* Fig. 21-16) has a fleshy **head** bearing two pairs of retractile **tentacles,** a pair of **eyes** (on tentacles), and a **mouth.** The head joins directly to a muscular **foot,** on top of which is the **shell;** the latter is of calcium carbonate, covered externally by a horny periostracum. A mucous epithelium covers all exposed fleshy parts. On the right side, the **genital pore** opens beside the head, and the small **anus** and larger **respiratory pore** are in the soft mantle margin at the edge of the shell. The **mantle** is a thin membrane that secretes and lines the shell and surrounds the viscera within. All soft parts can be drawn entirely into the shell by the **columella muscle,** which extends internally to the spire at the top of the shell.

The **digestive system** includes (1) the mouth; (2) a muscular pharynx with a dorsal horny "jaw" and ventral radula; (3) a slender esophagus; (4) the large thin-walled crop; (5) a rounded stomach; (6) a long twisted intestine; and (7) the anus. Two flat salivary

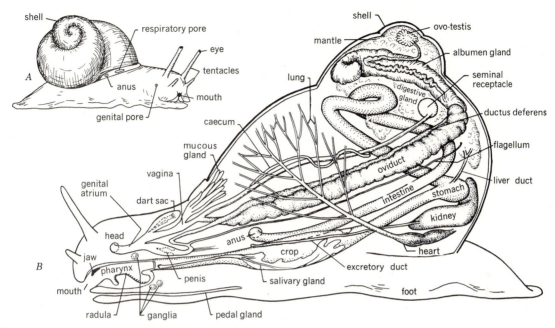

Figure 21-16 Class GASTROPODA: brown garden snail, *Helix aspersa.* *A*. External features from the right side. *B*. Internal structure from the left side. The lung is indicated by the branching blood vessels of the mantle cavity that connect to the heart.

glands beside the crop send ducts to the pharynx, and a two-lobed digestive gland, high in the shell, connects to the stomach. Replacing the ctenidia of other gastropods and mollusks, the land snail has a **lung.** The lung is a network of blood vessels on the outer wall of the mantle in the large **mantle cavity** within the shell; air enters and leaves by the respiratory pore. Blood collected from the body is aerated in the lung and then pumped by the heart through arteries to the head, foot, and viscera. The **heart** has one atrium and a ventricle. A single **kidney** drains from the pericardial cavity around the heart and discharges into the mantle cavity. The **nerve ganglia** are condensed, the cerebral pair dorsal to the pharynx and the buccal, pedal, and visceral pairs close beneath it; nerves from the ganglia extend to all organs. The tip of each posterior tentacle has an **eye,** with cornea, lens, and retina, and also probably an olfactory organ. Below the pedal ganglia lie a pair of **statocysts,** the organs of equilibrium, each containing calcareous bodies, cilia, and sensory cells. There are other sensory structures in the epidermis of the head and foot.

Each individual has a combined male and female **reproductive system.** High in the shell is the **ovo-**

testis, which produces both eggs and sperm. A **hermaphroditic duct** connects to the **albumen gland** of the female system. A slender **ductus deferens** conducts sperm to the **penis,** which lies in a sac off the common **genital atrium;** the long **flagellum,** joined to the penis sac, presumably forms the spermatophore (sperm packet) in which sperm are transferred at mating. The **oviduct** (larger than the ductus deferens) leads from the albumen gland to the **vagina,** which in turn connects to the genital auricle. Joined to the vagina are the slender duct of the **seminal receptacle,** the **mucous gland,** and the **dart sac.**

21-14 Natural history *Helix* is most active at night and in damp weather. It glides slowly and smoothly by waves of muscular action on the ventral side of the foot, over a slime track of mucus secreted by the large pedal gland below the mouth. The food is green vegetation, moistened by salivary secretions, held by the jaw, and rasped off in small bits by movements of the radula. Land snails, when present in numbers, may do severe damage to garden field plants and even to trees. By day the snails hide beneath objects on the ground, in crevices, or in bur-rows; the soft head and foot are then drawn into the shell. In dry weather a temporary covering (epiphragm) of mucus and lime is secreted to cover the shell aperture and avoid desiccation.

Reproduction is preceded by a mating perform-ance between two snails, during which a dart from each is discharged into the body of the other. Copu-lation is reciprocal, the penis of each being inserted into the vagina of the other for transfer of a spermato-phore; the snails then separate. Each later deposits one or more batches of eggs in damp places or shallow slanting burrows. Development requires many days and is direct, the young emerging as minute snails.

21-15 Other Gastropoda There is great vari-ety in the shell and soft parts of gastropods. The shell may be tall or short, conical, turbanlike, spindle-shaped, or cylindrical, white or variously colored, and plain or ornamented with ridges, spines, etc. Land slugs have the shell reduced and internal or lacking, and the nudibranchs have none as adults. Most gastropods can withdraw completely into their shells, and many have a permanent plate (opercu-lum) to cover the aperture. In most species the shell is right-handed (dextral), being coiled clockwise as seen from the spire, but some are left-handed (sinis-

tral). In the HETEROPODA the foot is a vertically com-pressed fin, and the PTEROPODA have a pair of lateral fins near the head; both of these groups swim in the open sea (Fig. 21-17).

The more primitive gastropods (ARCHAEOGAS-TROPODA) are chiefly marine and possess two gills; others (MESOGASTROPODA, NEOGASTROPODA) have but one gill, nephridium, and auricle, the left of each having disappeared and the true right organ being on the left side (Fig. 21-14). They evidently devel-oped in the sea, whence some forms in time mi-grated to fresh water; some types invaded the land to become the PULMONATA. Certain of the latter have returned to fresh water, where, having a lung instead of gills, they must come to the surface at intervals to respire. Some gastropods (limpets, nudibranchs) have regained an external bilateral symmetry, but the internal viscera are coiled, implying their deriva-tion from coiled types.

Gastropods abound in salt water and fresh water and on land, occurring from the tropics to subpolar regions, to depths of 5200 m (17,000 ft) in the sea and up to 5500 m (18,000 ft) in the Himalayas. Land snails are found from moist tropical regions to deserts. A number of them and some slugs have been spread with cultivated plants to places where they were not native, and some marine snails have been trans-planted with oysters.

Figure 21-17 Class GASTROPODA: some marine forms in life and some empty shells. Subclass PROSOBRANCHIA: *Fissurella*, limpet; *Haliotis*, abalone; *Crepidula*, slipper shell; *Tegula*, turban shell; *Carinaria*, heteropod; *Murex*, rock shell; *Buccinum*, edible whelk. Subclass OPISTHOBRANCHIA: *Aplysia*, sea hare; *Creseis*, pteropod; *Doris*, nudibranch.

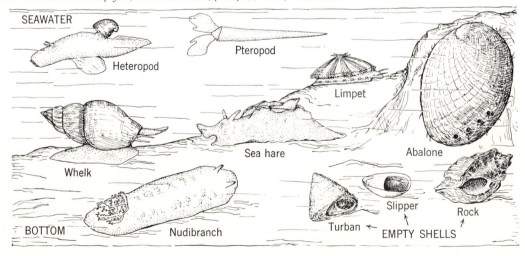

Many snails are the necessary intermediate hosts for trematode flatworms (Sec. 18-11). Gastropods are eaten by other invertebrates and by many kinds of vertebrates. Univalves have long served as human food. *Helix pomatia* is still eaten commonly in Europe, abalone *(Haliotis)* is a delicacy in Japan and in the western United States, and many other snails are consumed in Japan. Shells of gastropods provided the money of various native races, including the wampum of American Indians.

Class Bivalvia
(Bivalves: clams, mussels, oysters)

Members of this class are bilaterally symmetrical and laterally compressed, with the soft body enclosed in a rigid shell of two parts—hence called *bivalves;* there is no head. The foot usually is wedge-shaped (hence Pelecypoda, an earlier name for the class), and the gills are thin and platelike (hence Lamellibranchiata, another earlier name for the class). Bivalves inhabit both salt water and fresh water. Some creep on the bottom, others attach to solid objects, and many burrow in sand or mud.

21-16 General features and evolution Bivalves are the second largest molluscan class, with perhaps 20,000 species, widely distributed in both fresh and salt waters. They do not, however, display the great adaptive radiation found in the gastropods, and most are sessile or sedentary burrowing animals, filtering small food particles from the water or sediment. The great similarity in life-style is reflected in their morphology, which is much more uniform than that of gastropods. The major anatomic changes from the primitive mollusks are those concerned with the change from a crawling habit to one of burrowing into soft substrates and the elaboration of the mantle cavity and ctenidia for the filtration mechanism. Thus the foot has changed from a flat crawling structure to a hatchet shape, which is used for digging into the sand or mud. In this process blood is pumped into the foot to extend it into the sand or mud, then retractor muscles pull the shell down to the foot. The mantle is greatly enlarged and envelops the whole body, forming a large mantle cavity.

Within the mantle cavity the ctenidia have become greatly enlarged. Furthermore, the ctenidial filaments have been elongated and folded back on themselves to increase the surface area for filtering. The original molluscan ctenidia were V-shaped in cross section, but the bivalves have expanded the ctenidia to a W shape. The mantle in most bivalves has evolved into siphons to maintain contact with the water column while burrowing, so that filtering can continue. The mantle also secretes the two valves of the shell and the dorsal ligament which binds the two valves together. The ligament is elastic and under compression when the shell valves are closed. One or two large adductor muscles between the valves serve to hold the valves shut when contracted. If the adductors are relaxed the shell valves open because of the release of the compression pressure on the hinge ligament. Since the entire body is now within a shell and feeding is by filtering, the head and its associated sense organs, which were prominent in primitive mollusks, have all but disappeared, leaving only the mouth and labial palps. The radula has also vanished.

Feeding in bivalves is based on ciliary mucous systems. The lateral cilia on the ctenidial filaments cause the water to enter the mantle cavity and pass through the ctenidia. Particles in the water are trapped in mucus on the ctenidia and transported in mucous strands by cilia to the base of the ctenidia and thence via labial palps to the mouth. The labial palps sort some of the particles, rejecting some, which fall to the floor of the mantle cavity and are later ejected as pseudofeces. The remaining particles are taken into the stomach, where an elaborate set of ciliary sorting tracts further segregates them prior to digestion. A characteristic feature of the stomach is the **crystalline style,** a cylindrical rod passing into the stomach from a ciliated style sac. The crystalline style rotates because of ciliary action in the style sac. As it rotates against one wall of the stomach it abrades off enzymes for digestion as well as helping to pull into the stomach the mucous thread with the particles constantly being gathered by the ctenidia. Digestion is ultimately intracellular.

Most of the other organ systems in bivalves are similar to that described for other mollusks.

Most bivalves have a structure and function as

described above and are grouped in the subclass LAMELLIBRANCHIA, which contains more than 90 percent of the bivalve species. The most primitive bivalves belong in the subclass PROTOBRANCHIA, in which the ctenidia are not enlarged and folded back, the foot retains its flat surface, and long palp proboscides extend from the mouth area. These bivalves feed on particles in the sediment, which they obtain via cilia on the proboscides. The final subclass is the SEPTIBRANCHIA, a group of bivalves which live in deep water and lack ctenidia altogether. They are carnivorous or omnivorous, sucking in small organisms through the incurrent siphon.

Most bivalves are burrowers in soft sediments, but others are attached to or burrow into hard substrates.

The typical anatomy of a bivalve is illustrated in the sections that follow by the freshwater mussels *Anodonta* or *Unio* (Fig. 21-18).

21-17 Structure The somewhat oval **shell** is a firm exoskeleton that protects the body and provides for muscle attachments. It is of symmetrical right and left **valves,** the thinner margins of which are ventral and the thicker dorsal. In the dorsal region are (1) the **hinge teeth** (none in *Anodonta*), which align the valves with each other and serve as a pivot when the shell opens or closes; (2) an elastic **hinge ligament** between the valves that tends to hold them together

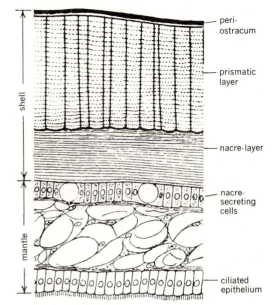

Figure 21-19 Freshwater clam. Enlarged cross section of shell and mantle.

dorsally; and (3) an anterior swollen **umbo** on each valve, representing its oldest portion. Around the umbo are many concentric **lines of growth** indicating intervals between successive growth stages, the annual lines being more conspicuous.

The shell (Fig. 21-19) is of three layers: (1) the external **periostracum,** a thin, colored, horny covering that protects parts beneath from being dissolved

Figure 21-18 Class BIVALVIA: freshwater clam, *Anodonta.* Shell and external features. Lines radiating from the umbo show the path of the muscle attachments as the shell grows in size.

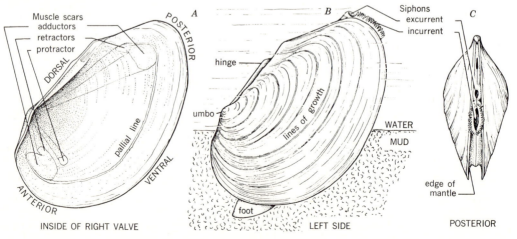

by carbonic acid in the water; the umbo on old shells is often corroded where this layer wears off; (2) a middle **prismatic layer** of crystalline calcium carbonate; and (3) the inner **nacre,** or mother-of-pearl, lining, formed of many thin layers of calcium carbonate and having a slight iridescence. The outer two layers are produced by the edge of the mantle and the nacre by the entire surface of that membrane. The shell grows in area by addition around the margin and in thickness by successive deposits or nacre, being thickest under the umbo.

The soft **body** within the shell (Fig. 21-20) consists of (1) a plump median **visceral mass,** attached dorsally and containing various organs; its anteroventral part forms (2) the muscular **foot.** On each side of these hangs (3) a thin W-shaped **ctenidium** (gill), outside of which is (4) the **mantle lobe,** a thin sheet of tissue adhering to the inner surface of a valve. The free margins of the mantle are muscular and can be brought together to close the **mantle cavity** within. Posteriorly the mantle margins form two short tubes: a ventral or **incurrent siphon** and a dorsal or **excurrent siphon.** Water moves in and out these openings by the action of cilia covering surfaces of the ctenidia.

Scars on the inner surface of each valve indicate the attachments of **muscles** (Fig. 21-18A). These are the large **anterior** and **posterior adductors,** both transverse, which draw the valves together; the **anterior** and **posterior retractors,** which draw the foot into the shell, and the **anterior protractor,** which helps to extend the foot.

The **digestive system** includes (1) the small **mouth,** just behind the anterior adductor and between two thin fleshy **labial palps;** (2) a short **esophagus;** (3) a rounded **stomach** dorsal in the visceral mass and joined by ducts from (4) the paired **digestive gland** (liver); (5) the slender **intestine** coiled in the visceral mass above the foot; (6) the dorsal **rectum** surrounded by the heart; and (7) the **anus** opening into the excurrent siphon. Off the stomach is a pouch (*pyloric caecum*), usually containing a transparent flexible rod, the **crystalline style,** that provides a starch reducing enzyme useful in digesting plankton. The rectum has a longitudinal fold, or **typhlosole,** that increases its inner surface (as in earthworms). The mouth has no jaws or radula.

The **circulatory system** comprises a dorsal **heart** of two atria joining a muscular ventricle; the latter is around the rectum and within a **pericardial cavity.** The ventricle pumps blood forward in an **anterior aorta** supplying the foot and viscera (except the kidneys and gills) and backward in a **posterior aorta** delivering to the rectum and mantle. Blood oxygen-

Figure 21-20 Freshwater clam, *Anodonta.* *A.* Internal structure as seen with shell, mantle, and gills of the left side removed. *B.* Cross section through the heart region. Both diagrammatic. (*B, after Stempell.*)

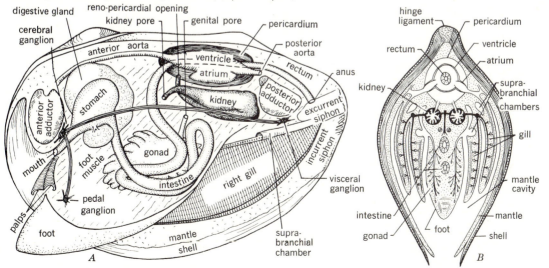

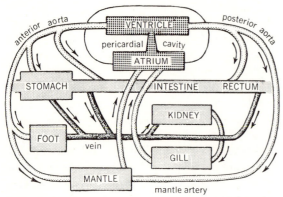

Figure 21-21 Freshwater clam. Diagram of the circulatory system and blood flow.

ated in the mantle returns directly to the atria, but that which circulates through other organs is collected in a **vein** to the kidneys and thence passes to the gills for oxygenation before returning to the heart (Fig. 21-21). Some arterial blood enters vessels lined by epithelium, and some discharges into blood sinuses without cellular lining, as in the foot; some blood also diffuses to intercellular spaces. The blood carries oxygen and dissolved nutrients to all parts of the body; also it disposes of carbon dioxide in the

gills and mantle and of organic wastes in the kidneys.

The function of respiration is performed jointly by the mantle and two double ctenidia. The **ctenidium** is a W-shaped structure in section, each half of two thin, platelike **lamellae** completely joined at the ventral margin, where there is a **food groove** (Fig. 21-22). Each lamella is made up of many vertical **gill bars,** strengthened by chitinous rods and connected to one another by horizontal bars, with small pores (ostia) between. Cross partitions (interlamellar junctions) between the two lamellae divide the interior of a gill into many vertical **water tubes.** Dorsally the water tubes of each gill join a common **suprabranchial chamber** that extends posteriorly to the excurrent siphon. There is a slit-like passage from the mantle cavity to the suprabranchial chamber of each inner gill. Blood from veins in the kidneys passes through fine afferent and efferent vessels in the interlamellar junctions for aeration before returning to the heart.

Just below the pericardium are two **kidneys,** which remove organic wastes from the blood and pericardial fluid. Each is a U-shaped tube with a ciliated aperture draining from the pericardial cav-

Figure 21-22 Freshwater clam: diagram of gill structure. *A*. Outer half of left gill partly cut away to show internal structure. ←---, path of water; ←——, path of food particles caught in mucus and carried to mouth; ←--·-, path of rejected particles. *B*. Portion of gill enlarged. ·······, path of blood flow, down in afferent and up in efferent vessels. *C*. Cross section of two gill bars much enlarged to show ciliated surfaces and blood cells.

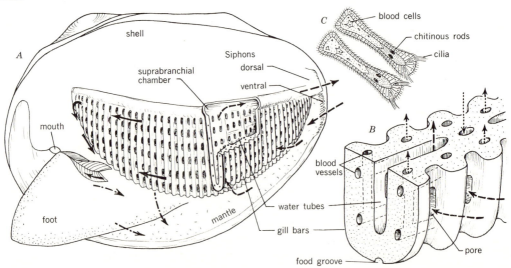

ity, a glandular region adjacent to the vein, and a bladder discharging through a ciliated opening into the suprabranchial chamber of the gill.

The **nervous system** includes three pairs of **ganglia,** the cerebral beside the esophagus, the pedal in the foot, and the visceral below the posterior adductor muscle. Each pair is joined by a commissure, and there are also cerebropedal and cerebrovisceral commissures besides nerves to various organs. The **sensory structures** include light-responsive devices in the siphon margins, tactile organs in the mantle edge, a pair of **statocysts** (for equilibration) in the foot, and an **osphradium** in the incurrent siphon over each visceral ganglion. The osphradia evidently test incoming water for silt, stimulating a reduction of intake when the silt content is high.

The sexes are separate but usually alike externally. In each are two much-branched **gonads** about the intestinal coils in the visceral mass, each discharging by a short **duct** near the kidney aperture.

21-18 Natural history Freshwater mussels live in ponds, lakes, and streams, some in quiet and others in flowing waters. They may migrate to shallows by night and retire to deeper places by day and may change their habitat with the seasons. They are usually embedded partly in sand or mud or are wedged between rocks, with the valves slightly spread, the mantle margins closed, and the siphons exposed (Fig. 21-18). They can travel slowly by extending the foot between the valves to expand or hook it into bottom materials and then draw the body along by contraction of the foot muscles. This action is effected by the filling and emptying of blood sinuses in the foot. Dragging the shell along leaves parallel furrows in the bottom. The siphons open and close in response to light, touch, and other stimuli. By the action of cilia covering the mantle and ctenidia, water is drawn through the incurrent siphon into the mantle cavity. Organic particles and microorganisms (e.g., diatoms, protozoans) suspended in the water constitute the food; this is trapped in mucus on the gills and carried ventrally by the beating cilia to the food groove along the ventral edge of the gill and to the labial palps and mouth. Food is digested in the stomach by the aid

of the digestive gland secretions and absorbed in the intestine; any residues pass out the anus. The water provides oxygen for aeration of blood in the gills; it passes through the pores, up the water tubes to the suprabranchial chambers, and out the exhalent siphon carrying away carbon dioxide and feces and sex products.

21-19 Reproduction Most bivalves set their eggs and sperm free, but the freshwater bivalves of the family UNIONIDAE have a peculiar mode of reproduction. In a female the ripe eggs pass from the ovaries to the suprabranchial chambers and there are fertilized by sperm discharged from a male and brought in water entering the female. The eggs attach (by mucus) in the water tubes of her gills, which enlarge as **brood chambers** (marsupia); some species use all gill lamellae, others only the outer two (Fig. 21-23). Each egg, by total but unequal cleavage, develops into a minute larval **glochidium,** 0.1 to 0.4 mm wide, with two valves closed by an adductor muscle and a long larval thread; in some (e.g., *Anodonta*) the valves have ventral hooks, but others (*Unio, Quadrula, Lampsilis*) lack hooks. The larvae are shed into the water through the female's excurrent siphon, then sink to the bottom or are scattered by water currents. They can open or close their shells but cannot move independently. Hooked glochidia attach to soft exterior parts of freshwater fishes resting on the bottom; hookless forms clamp to the gill filaments of fishes, being carried in the respiratory water of the latter. In a few hours either type is covered with a capsule formed by migration and mitosis of cells of the host's epithelium. The parasitic larvae feed and grow by absorbing nutrients from the host's body fluids. Later the cyst weakens, the young mollusk opens and closes its valves, extends its foot, and escapes to the bottom to become free-living. In some clams (*Quadrula, Unio*), breeding occurs in summer (May to August) and the glochidial stages last 10 to 70 days, depending on the species and water temperature. Others (*Anodonta, Lampsilis*) produce eggs in late summer and the glochidia are retained in the female until the following spring. Individual fishes in nature may carry up to 20 glochidia, but a fish 7.5 to 10 cm (3 to 4 in) long may be artificially infected

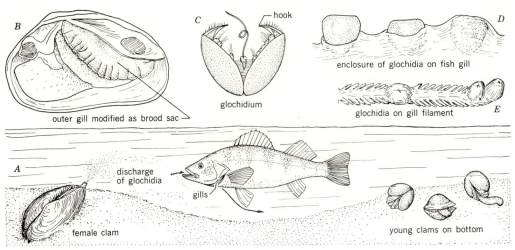

Figure 21-23 Freshwater clam. *A*. Diagram of the life cycle. *B*. The female's outer gill modified as a brood sac. *C*. One glochidium. *D*. Enclosure of glochidia by epithelium on gill of a fish. *E*. Glochidia on a gill filament. *C–E*, much enlarged. (*B–D, after Lefevre and Curtis, 1910.*)

with several hundred that will grow to metamorphosis. The glochidial stage serves to disperse the young over a wide area. At times glochidia in hatcheries cause losses among trout.

21-20 Other Bivalvia The bivalves are all aquatic, mostly marine, and chiefly sedentary bottom dwellers. They are commonest from the tide lines into shallow waters, but some occur down to 5000 m (17,000 ft). A few creep slowly on the bottom, but most marine species, including various edible clams, burrow in the sand or mud and extend their siphons up to the water (Fig. 21-24). The sea mussel (*Mytilus*) attaches to solid objects by a **byssus** of threads secreted by glands in the reduced foot. Edible oysters *(Ostrea)* become attached permanently to rocks or shells by cementing one valve to the substrate. Some bivalves *(Pholas)* can burrow in hard clay or soft rocks by abrading away the rock or clay, using the shell valves. The shipworms, or teredos *(Teredo, Bankia),* are highly specialized, with very slender bodies and small anterior shells; using the latter like a rasp, they burrow in the wood of ships or wharves immersed in seawater. They also have the ability to digest the cellulose wood fibers. The scal-

Figure 21-24 Class BIVALVIA. Positions of some marine forms in life; reduced but not to same scale. Order NUCULACEA: *Nucula; Yoldia*. Order ANISOMYARIA: *Ostrea*, edible oyster; *Mytilus*, sea mussel with byssus; *Pecten*, scallop. Order HETERODONTA: *Mya*, mud clam; *Venus*, quahog; *Tagelus*, jackknife clam; *Ensis*, razor clam. Order ADAPEDONTA: *Pholas*, rock borer; *Teredo*, pileworm, or shipworm.

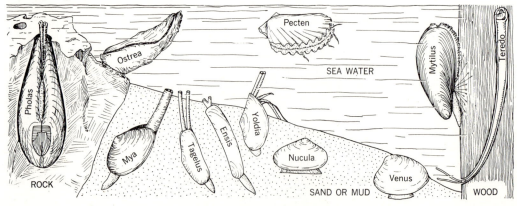

lop (*Pecten*) attaches by a byssus when young but later frees itself and swims by clapping its shells together. In both *Pecten* and *Lima* the mantle edge is prolonged as filaments beyond the shell. Still other bivalves are commensals on echinoderms, some live in the burrows of worms and crustaceans, others are embedded in sponges and the tests of ascidians, and *Entovalva* is parasitic in holothurians.

Some bivalves are enormously abundant. An estimated $4\frac{1}{2}$ billion clams live on 1800 km² (700 mi²) of the Dogger Bank, east of England. Shells of the fossil †*Exogyra* of Cretaceous time occur in rocks from New Jersey to Mexico, over 4000 km (2500 mi). Oysters, mussels, and other bivalves inhabit various coastal waters in large numbers.

In the great majority of BIVALVIA, both ova and sperm are discharged into the water, where fertilization occurs, and the zygote develops first into a ciliated **trochophore** larva somewhat resembling an annelid trochophore (Fig. 22-12) and then into a **veliger** larva. After a free-swimming period, this larva sinks to the bottom and becomes a miniature bivalve. Since the hazards in such reproduction are great, these mollusks produce vast numbers of eggs—from 16 million to 60 million may issue from a female oyster in one season.

In the common American oyster (*Crassostrea virginica*) of the Atlantic Coast the sexes are separate. Eggs and sperm unite in the water, and within 48 h the larva has a bivalved shell. It swims freely for about 2 weeks and grows, then goes to the bottom and moves around, using the foot. Finally it settles on the left valve and quickly becomes fixed to some solid object by a secretion from the mantle. The European oyster (*Ostrea edulis*) and the Pacific Coast oyster (*O. lurida*) are both protandric, the gonad of an individual first producing sperm and then eggs, in rhythmic alternation. The eggs of *O. lurida* are fertilized and grow to bivalved larvae within the mantle cavity of the parent "female" and then have a brief free existence before attaching. Attached young oysters are called *spat*.

21-21 Relation to humans Since humans first visited the seashore, they have used bivalves as food. On many coasts are shell mounds where ancient peoples, for generations, ate mollusks and discarded the shells. An Indian shell mound on San Francisco Bay contained over 28,000 m³ (1 million ft³) of debris accumulated over an estimated 3500 years. Good natural oyster beds in shallow water are highly valued, and spat is planted to replace the adult oysters that are harvested by tongs or dredges. Both the Atlantic oyster and the Japanese oyster (*C. gigas*) have been introduced on the Pacific Coast. Humans eat all of the body of an oyster or clam but only the single large adductor muscle of a scallop (*Pecten*). Oyster shell is used on roads and to supply lime for poultry and other needs. Various kinds of clams are dug by hand, and in many places a limit has been placed on size, number, and the season of take to conserve the stocks.

Pearls are formed about foreign objects between the mantle and shell of some bivalves. A bit of the mantle encloses the object and secretes successive layers of nacre about it, in the manner that the shell lining is produced. The most valuable kinds come from marine pearl oysters (*Pinctada margaritifera*) of the Persian Gulf and Indian Ocean. The Japanese artificially introduce small particles in the mantle and then retain the bivalves, *Pinctada* (*Meleagrina*) *mertensii*, in cages for several years until "culture" pearls are produced.

The burrowing of teredos in wooden marine structures sometimes causes serious damage, since the piles supporting wharves may be weakened and collapse. Wharves are now constructed of creosoted wood, concrete, or metal to avoid such troubles. An introduced Asiatic freshwater clam (*Corbicula fluminea*) is now so abundant in some California canals that it impedes water flow.

Class Cephalopoda
(Nautili, squids, and octopuses)

21-22 General features and evolution The CEPHALOPODA (Gr. *kephale,* head + *podos,* foot) are the most highly evolved and structurally different group of mollusks. In many respects they may even be considered the most advanced group of living invertebrates. The class is represented by only about 650 living species and appears to be a declining group, since about 10,000 fossil species are known, stretching back to the Cambrian. It includes the largest living invertebrate in the giant squid *Architeuthis,* which reaches a length of 16 m (52 ft). Cephalopods have diverged most from the ancestral mollusk type, as evidenced in many aspects of their anatomy and physiology. In contrast to the stereotyped picture of the mollusks as slow-moving animals functioning with ciliary-mucous mechanisms, the cephalopods are fast-moving and mainly pelagic predators. In order to become such swift carnivores, basic changes were necessary in the original molluscan body design and functioning. One of the requirements for fast movement is an increase in metabolic rate, which in turn requires delivery of more oxygen and removal of more carbon dioxide. The original open molluscan circulatory system is clearly inadequate to transport the oxygen at the needed rate, hence the cephalopods have evolved a closed circulatory system. This closed system is further enhanced by the addition of a second set of hearts at the base of the ctenidia whose function is to increase the rate of blood flow through them. Similarly, the use of lateral cilia on the ctenidia to create the respiratory current across them could not pass enough water across the ctenidia to permit extraction of adequate amounts of oxygen to sustain the cephalopod. This problem is overcome in octopuses and squids by eliminating cilia on the ctenidia and employing the mantle, now highly muscularized, to pump water over the gills. By combining this pumping of water with a siphon developed from the molluscan foot, the cephalopods also obtained the necessary increased speed and mobility via jet propulsion. It would not have been possible for them to become swift pelagic predators using the primitive molluscan locomotory mechanisms of crawling, or even swimming, on the broad foot. In order to employ the mantle as a muscular pump irrigating the ctenidia and in movement, it was necessary to free it from the shell. This has been accomplished in all living cephalopods except *Nautilus.* Predators such as squids and octopuses must have considerable nervous control and sensory input to seek out and capture food. To facilitate this the cephalopods have centralized the originally diffuse nervous system of mollusks into a single "brain" and have evolved eyes which are in most respects the equivalent of those of vertebrates.

The eyes are supplemented by an impressive array of tactile and chemical sense organs.

As animals with a highly evolved centralized nervous system and sense organs, the cephalopods also show considerably more awareness of their environment plus complex behavior patterns. Many cephalopods show definite courtship behavior prior to mating, and laboratory tests on octopuses have indicated considerable ability to learn. The highly evolved image-forming eyes coupled with complex chromatophores under direct nervous control have made these animals masters of camouflage in their ability to match their backgrounds.

Prey capture is effected by the array of eight or ten sucker-bearing arms which surround the mouth and are derived from the original molluscan foot. Unique to the cephalopods is the "beak" which surrounds the mouth. This parrotlike chitinous jaw apparatus is employed to bite off chunks of food, which are then transported to the pharynx by the radula. Movement of food through the digestive system is primarily by muscular action (peristalsis) as in the vertebrates and not by cilia as in most mollusks.

The cephalopods are dioecious, fertilization is internal, and large, yolky eggs are laid. Again, they differ from most mollusks in that cleavage is meroblastic and there are no larval stages, the young hatching as juveniles.

Cephalopods depend on their swift swimming to escape most potential predators, but many also have an ink sac attached to the rectum which can secrete a cloud of black material ejected through the siphon which may distract or obscure the vision of a potential predator.

Many deep-dwelling squids display a remarkable development of complex light-producing organs, the function of which is not altogether understood.

The following account details the anatomy of a common squid, *Loligo pealii* of the Atlantic Coast or *Loligo opalescens* of the Pacific Coast, as an example of this class.

21-23 Structure and natural history   (Fig. 21-25)
The large **head** has two conspicuous **eyes** and a central **mouth,** which is surrounded by 10 fleshy **arms** bearing cuplike **suckers;** the fourth pair of arms are

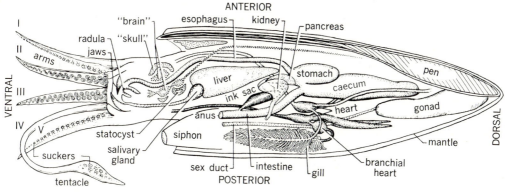

Figure 21-25 Class CEPHALOPODA: the squid, *Loligo.* Internal structure as seen with body wall and arms removed on left side.

long retractile **tentacles.** The slender conical **body** has a fleshy triangular **fin** along either side of the tapered end. The head and body join by a **neck,** around which the free edge of the mantle forms a loose **collar.** Below the neck is a muscular funnel, or **siphon.** The skin contains many **chromatophores,** each with yellow or brown pigment in an elastic capsule surrounded by muscle cells. These expand and contract rhythmically, causing the animal to be alternately light and dark. The squid is longest dorsoventrally, the head is morphologically ventral, and the arms and siphon represent the foot of other mollusks.

In the upper (anterior) wall of the body is a horny **pen,** the shell, stiffening the body; a cartilage-like case surrounds the "brain"; there is a nuchal cartilage over the neck and similar support for the siphon and fins. The mantle, fins, siphon, and arms are all muscular. The **mantle** is a conical envelope surrounding the internal organs. By its alternate expansion and contraction, water is drawn into and expelled from the mantle cavity. For respiration alone the water passes into and out of the space between the neck and collar, but for "jet" locomotion the mantle closes around the base of the siphon and water is forcibly ejected from the latter (Fig. 21-26). To swim "tail first," the siphon is directed toward the arms; to move head foremost the tip of the siphon is bent around to force the water tailward. The fins aid in steering and also serve in swimming.

Figure 21-26 Class CEPHALOPODA. *Loligo,* squid; *Sepia,* cuttlefish; *Nautilus,* pearly nautilus; *Argonauta,* paper nautilus; *Octopus,* octopus. Variously reduced.

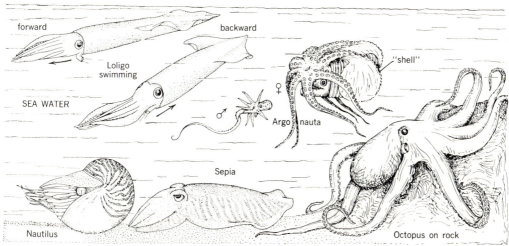

The digestive system includes (1) the **mouth;** (2) a muscular **pharynx** (buccal mass) with a pair of horny **jaws,** like a parrot beak inverted, and a **radula;** (3) a slender **esophagus;** (4) the saclike muscular **stomach,** joined by (5) a thin-walled **caecum** with a complex valve between; (6) the slender **intestine** and (7) **rectum** extending to (8) the **anus,** which opens into the mantle cavity. Two pairs of **salivary glands** connect to the pharynx, and the long **liver** and a small **pancreas** join by ducts to the stomach. The squid eats crustaceans, mollusks, and fishes. It swims forward with the arms together, then darts at the prey by suddenly ejecting water from the siphon; the arms are spread, the prey grasped and drawn to the mouth, bitten and cut into pieces by the jaws, and swallowed. Small bottom animals are quietly covered by the spread arms and then gathered into the mouth.

Above the rectum is the glandular **ink sac,** with a duct opening near the anus; the ink is a dark pigment that can be forced out the siphon to produce an aquatic "smoke screen," under cover of which the squid can escape from a predator. In each side of the mantle cavity is an elongate **gill.** Blood in veins from body tissues is pumped by a **branchial heart** through capillaries in the gill filaments and then is collected in an atrium on each side; these two join the **systemic heart,** which sends the blood through arteries and capillaries to the gut, head, arms, and other organs. A pair of **kidneys** drain from the pericardial cavity around the heart into the mantle cavity. The pairs of **nerve ganglia** are concentrated in the "brain box" around the pharynx. The **eyes** are unique among invertebrates; each has a cornea, lens, anterior and posterior chambers, and retina with rods and can form a real image (Fig. 9-17D). They are structurally like those of vertebrates. Below the brain are two **statocysts** serving for equilibration.

The sexes are separate, each with one **gonad** near the tip of the mantle cavity and a **duct** opening toward the funnel. At mating, a spermatophore (sperm packet) is transferred by the specialized tip (*hectocotylus*) of the bottom left (fifth) arm on the male to the mantle cavity of the female, where the tip becomes detached and remains. The eggs are large, with much yolk, and are laid in long gelatinous capsules. Cleavage is superficial, somewhat as in birds, and there is no larval stage; the young hatch as miniature adults able to swim and feed at once.

21-24 Other Cephalopoda   This class has swarmed the seas since Cambrian time and has an abundant fossil record of 10,000 species (Figs. 13-11, 13-12). Nautiloids were dominant through Paleozoic times and the ammonites in the Mesozoic; the coleoides (such as *Loligo*) abound today. Both the earlier groups had external limy shells that varied in size, shape, and ornamentation but were always divided into chambers by transverse septa. The animal probably lived in the outermost compartment, as does the living pearly nautilus. Early forms had slenderly conical shells, 2.5 cm to several meters in length. †*Endoceras* of Ordovician time was 4.5 m (15 ft) long, the largest shelled invertebrate known; an ammonite, †*Pachydiscus,* of the Cretaceous had a coiled shell 2.5 m ($8\frac{1}{2}$ ft) in diameter. The only living relic of this great assemblage is the pearly nautilus *(Nautilus)* that first appeared in the Jurassic. Three species are generally recognized, living near the bottom at depths to 550 m (1800 ft) in the eastern Pacific and Indian Oceans (Fig. 21-27). The flat, coiled shell is up to 25 cm (10 in) in diameter, closely coiled in one plane, and divided by septa. The animal lives in the outermost chamber, having previously occupied each of the others in turn as it and the shell grew. Contact with previous chambers is kept via a fleshy structure, the **siphuncle,** which runs back through each chamber. The siphuncle and chambers probably serve as a hydrostatic device to regulate buoyancy. The head bears 60 to 90 tentacles without suckers but with sticky surfaces. The eye is a simple pinhole type without lens. Four ctenidia are present, and the siphon is folded but unfused and pumps water itself in the absence of a free mantle.

The living cephalopods are all predaceous and mostly free-swimming (Figs. 21-26, 21-28). Squids range from species 2.5 cm long to the giant *Architeuthis,* of cold depths off Newfoundland and elsewhere, which is eaten by sperm whales. Small squids like *Loligo* sometimes occur in enormous schools; they are eaten by some sharks, marine birds, many fishes, and marine mammals and are used as bait by fishermen. Squids are also an article

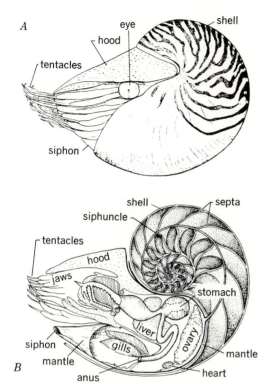

A

eye
hood
shell
tentacles
siphon

shell
siphuncle
septa
tentacles
hood
jaws
stomach
liver
ovary
siphon
gills
mantle
mantle
heart
B
anus

Figure 21-27 Class Cephalopoda: **the pearly nautilus,** *Nautilus pompilius.* *A.* External features. *B.* Internal structure; shell and mantle (except at siphon) cut away to midline; jaws and radula shown in median section; two left gills removed.

of human food in many countries. The cuttlefish (*Sepia*) of European Atlantic waters has a short oval body bordered by fins; its long tentacles are completely retractile. Its internal calcareous shell is used to regulate buoyancy by changing the volume of gas in the narrow chambers of the shell. For centuries its ink has provided the sepia pigment of artists. The devilfishes, or octopuses, vary from 5 cm to 8.5 m (2 in to 28 ft) in spread. The flexible bulbous body lacks a shell and has eight long sucker-bearing tentacles. Octopuses crawl about rocks and tide pools along the seacoasts but can swim by use of the siphon. They feed mainly on crabs, paralyzing them with secretions from their salivary glands. Large squid and octopuses are feared more rightly for their dangerously powerful beaks than for their supposed ability to grip a person under water by their tentacles. In the paper nautilus (*Argonauta*) of temperate oceans, the female is up

to 20 cm (8 in) long and secretes a shell only as an egg case; the male is only about 2.5 cm (1 in) long.

Classification

Phylum Mollusca.

Mollusks. Symmetry bilateral or none; body soft, covered by thin mantle that commonly secretes limy shell of 1, 2, or 8 parts; usually an anterior head and ventral muscular foot; mouth with radula (except bivalves); heart dorsal with 1 or 2 atria and one ventricle (except Monoplacophora); respiration by gill-like ctenidia, secondary gills, or "lung"; nephridia 6, 2, or 1 pairs or single; nervous system with a circumesophageal ring and usually with paired ganglia, connectives and nerves; sexes separate or united; gonads 4, 2, or 1, with ducts; mostly oviparous; with larval stages (trochophore, veliger, or glochidium) or development direct; mostly in salt water or fresh water, some Gastropoda on land; 80,000 living, 40,000 fossil species.

Class 1. Monoplacophora.

Body oval, shell single, bilaterally symmetrical in living forms; 2 ventricles; 5 or 6 pairs of gills; 6 pairs of nephridia; Lower Cambrian to Devonian, 6 Recent species (order Tryblidiacea): *Neopilina galatheae,* 5 pairs of gills; *N. (Vema) ewingi,* 6 pairs of gills.

Class 2. Aplacophora

Solenogastres. Form wormlike; no shell or foot; integument thick, with minute limy spicules; foot vestigial or absent; 250 species.

Order 1. Neomenioidea. Foot reduced to a midventral groove; monoecious; *Neomenia, Proneomenia, Lepidomenia.*

Order 2. Chaetodermatoidea. Foot completely absent; dioecious; *Chaetoderma.*

Class 3. Polyplacophora.

Chitons. Body elliptical; shell a middorsal row of 8 broad plates, surrounded by fleshy girdle; foot large, flat; ctenidia, 6 to 80 pairs, in groove around foot; sexes separate; 1 gonad; a trochophore larva; on rocks, chiefly in shallow coastal waters; 600 living, 150 fossil species; Ordovician to Recent.

Order 1. Lepidopleurida. Valves of shell either lacking insertion plates or, if these are present, lack-

Figure 21-28   Class CEPHALOPODA: *Histioteuthis,* **a pelagic, deepwater squid.** (*Courtesy of Gary McDonald.*)

ing insertion teeth. *Lepidopleurus,* without insertion plates; *Hanleya,* with insertion plates.

Order 2.   Chitonida.   Valves of shell with insertion plates and teeth. *Cryptochiton* the largest living chiton; *Katherina,* the gum boot; *Placiphorella,* a carnivorous chiton; *Mopalia.*

Class 4. Scaphopoda.

Tooth shells or tusk shells. Shell and mantle slenderly tubular, slightly curved, open at both ends; foot conical; delicate captacula around mouth; no ctenidia; sexes separate; 1 gonad; a veliger larva; marine, fresh water, or terrestrial. Upper Cambian to (15,000 ft); 350 living, 300 fossil species, Devonian to Recent. *Dentalium.*

Class 5. Gastropoda.

Univalves: limpets, whelks, snails, slugs, etc. Shell usually spiral (uncoiled, reduced, or absent in some); head distinct, with scraping radula, commonly with tentacles and eyes; foot large, flat, for holdfast or creeping; visceral mass typically turned 180° counterclockwise (torsion) on head and foot, and coiled in shell; ctenidia 2 or 1 or replaced by secondary gills or lung; nephridia 2 or 1; sexes separate or united; 1 gonad; mostly oviparous; larva with trochophore and veliger stages (except land forms);

marine, fresh water, or terrestrial. Upper Cambian to Recent. About 40,000 living species.

Subclass 1. Prosobranchia (Streptoneura).   Visceral torsion 180°; nervous system a figure eight; tentacles 2; ctenidia anterior to heart; mantle cavity opens anteriorly; sexes usually separate, 1 sex opening. Upper Cambrian to Recent, mostly marine; 25,000 living, 10,000 fossil species.

Order 1.   Archaeogastropoda (Aspidobranchia, Diotocardia).   Nervous system little concentrated; ctenidia 1 or 2 (or none), plumelike with 2 rows of filaments; 2 atria in heart; 2 nephridia. Upper Cambrian to Recent. *Acmaea, Patella, Lottia,* limpets, shell flatly concial; *Fissurella,* keyhole limpet; *Haliotis,* abalones, shell ear-shaped, aperture large.

Order 2.   Mesogastropods (Pectinobranchia, Monotocardia).   Nervous system not concentrated; 1 ctenidium with filaments in 1 row; 1 atrium in heart; 1 nephridium; edge of shell opening usually lacks a siphonal notch or canal; osphradium monopectinate and radula with 7 teeth per row. Ordovician to Recent. *Cypraea,* cowries; *Crepidula,* slipper or boat shells; *Littorina,* periwinkles; *Viviparus (Paludina), Goniobasis,* in fresh waters; *Pleuroplaca gigantea,* giant conch, southern Florida and West Indies, shell 250 mm (10 in) or more, largest univalve in United States.

Order 3. **Neogastropoda (*Stenoglossa*).** Nervous system concentrated; 1 ctenidium with filaments in 1 row; 1 atrium in heart; 1 nephridium; osphardium bipectinate; radula with 3 or fewer teeth per row; edge of shell opening with a siphonal notch or canal. Ordovician to Recent. *Murex,* rock shells; *Urosalpinx,* oyster drill; *Buccinum,* edible whelks; *Conus,* cone shells, with poison gland.

Subclass 2. Opisthobranchia. Shell reduced, internal or none; viscera and nervous system secondarily "unwound"; opening of mantle cavity lateral or posterior; usually 1 nephridium, 1 atrium; ctenidium 1, posterior, or replaced by secondary gills; monoecious. Carboniferous to Recent. 1500 living, 300 fossil species (Fig. 21-29).

Order 1. **Cephalaspidea.** Bubble shells. Shell present, large, either internal or external; head with large cephalic shield; parapodia usually present, large. *Bulla,* bubble shell; *Gastropteron; Navanax,* preying on other opisthobranchs; *Acteon, Aplustrum* (Fig. 21-29).

Order 2. **Anaspidea.** Sea hares. Shell internal, reduced; no cephalic shield, parapodial lobes large, concealing a small mantle cavity on the right side. *Aplysia,* the sea hare, to 30 cm (12 in) long; *Phyllaplysia, Akera.*

Order 3. **Thecosomata.** Pteropods. Shell present; planktonic; large mantle cavity; foot modified into winglike parapodia for swimming; filter feeding. *Cavolina,* shell uncoiled, bilaterally symmetrical; *Peraclis, Cymbulia, Creseis.*

Order 4. **Gymnosomata.** Naked pteropods. Shell absent; no mantle cavity; parapodia small; buccal area highly developed with buccal cones; arms bearing suckers. *Cliona, Cliopsis, Pneumoderma.*

Figure 21-29 Subclass OPISTHOBRANCHIA: **representatives of the eight orders, not to the same scale.** Order CEPHALASPIDEA: *Aplustrum.* Order ANASPIDEA: *Aplysia,* the sea hare. Order THECOSOMATA: *Creseis.* Order GYMNOSOMATA: *Pneumoderma.* Order NOTASPIDEA: *Umbraculum.* Order ACOCHILIDIACEA: *Hedylopsis.* Order SACOGLOSSA: *Oxynöe.* Order NUDIBRANCHIA: *Acanthodoris.* (*After Grassé, Traité de Zoologie.*)

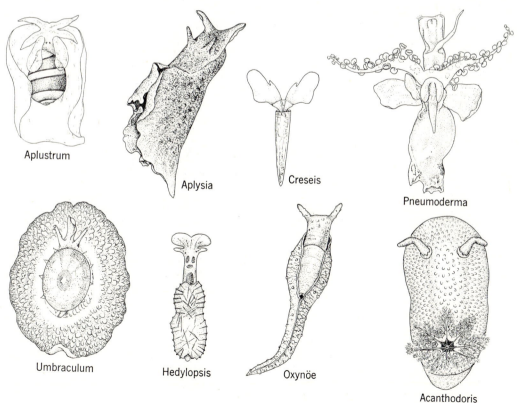

Aplustrum

Aplysia

Creseis

Pneumoderma

Umbraculum

Hedylopsis

Oxynöe

Acanthodoris

Order 5. Notaspidea. Shell internal, external, or absent; mantle cavity absent; a single plicate gill on right side. *Umbraculum, Pleurobranchus, Tylodina.*

Order 6. Acochilidiacea. Minute; no shell, gill, or jaws; visceral mass larger than and sharply demarcated from foot; often with spicules; interstitial in sand. *Hedylopsis, Microhedyle.*

Order 7. Sacoglossa. Radula and buccal area modified for piercing and sucking algae; shell present or absent; *Berthelinia,* a "bivalved gastropod" from tropical seas; *Lobiger,* 4 large winglike parapodia; *Oxynöe, Stiliger, Elysia.*

Order 8. Nudibranchia. Sea slugs. Shell-less, externally bilateral opisthobranchs; no true gills, but often with secondary gills around anus; mantle cavity absent; dorsal surface often with extensive projections *(cerata)* which contain extensions of the digestive glands. *Aeolidia, Dendronotus, Dendrodoris, Tritonia, Acanthodoris* (Fig. 21-29).

Subclass 3. Pulmonata. Freshwater and land snails and slugs. Mostly small; shell a simple spiral (or none); head with 1 or 2 pairs of tentacles, 1 pair of eyes; mantle cavity anterior, its vascular lining an air-breathing lung, opens to a contractile pore on right side; no gills; monoecious; gonad single; mostly oviparous; development direct; chiefly terrestrial. Upper Carboniferous to Recent; more than 16,000 living, 1000 fossil species.

Order 1. Stylommatophora. Land snails. Tentacles 2 pairs, eyes at tip of hind pair. *Polygyra, Zonites, Helminthoglypta,* mostly in moist places; *Helix,* European garden snails, some species now in the United States; *Testacella haliotidea,* greenhouse slug, with small terminal shell, in greenhouses, preys on earthworms; *Limax, Deroceras (Agriolimax), Ariolimax,* slugs with vestigial shell in mantle; *Arion,* slugs with no shell; *Achatina fulica,* giant African land snail, native to eastern Africa, widely introduced in islands of Pacific, causing much agricultural damage; *Gonaxis,* predaceous snail.

Order 2. Basommatophora. Freshwater snails. One pair of tentacles; eyes near tentacle base. *Lymnaea, Helisoma (Planorbis), Physa; Ferrissia,* freshwater limpet.

Class 6. Bivalvia (Lamellibranchiata, Pelecypoda).

Bivalved mollusks. Shell of 2 lateral valves, usually symmetrical, with dorsal hinge and ligament, and closed by 1 or 2 adductor muscles; mantle of flattened right and left lobes, the posterior margin commonly forming siphons to control flow of water in and out of mantle cavity; fleshy labial palps beside mouth; 1 (or 2) pairs of ctenidia (gills or branchia), commonly platelike; no head, jaws, or radula; usually dioecious, some protandric; gonad opening into mantle cavity; a veliger larva or glochidial stage; mostly marine, some in fresh waters. Ordovician to Recent, 20,000 living, 15,000 fossil species.

Subclass 1. Protobranchia. Gill with central axis bearing 2 divergent rows of short flat filaments, unreflected; foot flat ventrally; 2 adductor muscles.

Order 1. Nuculacea. Nut clams. Without siphons; feed by proboscides which are outgrowths of enlarged labial palps. *Nucula, Yoldia.*

Order 2. Solenomyacea. Siphons present; feeding via gills. *Solenomya,* awning clam.

Subclass 2. Lamellibranchia. Gills enlarged, turned back to give a W shape; either ciliary connections between adjacent filaments (filibranch) or fleshy connections (eulamellibranch).

Order 1. Taxodonta. Numerous similar teeth on the hinge; gills without interlamellar junctions. *Arca, Glycymeris.*

Order 2. Anisomyaria. Anterior adductor muscle small or none, posterior large; no siphons; gill lamellae flat, similar filaments, filibranch condition; often with a byssus. *Mytilus,* sea mussel; *Ostrea,* oyster; *Pecten,* scallop.

Order 3. Heterodonta. Hinge teeth varied; equal or subequal adductor muscles, gills eulamellibranch; mantle often produced into siphons. *Tellina; Tresus,* horseneck clam; *Cardium,* cockle; *Venus,* quahog, or hard-shelled clam; *Macoma; Panope,* "goeduck"; *Saxidomus,* Washington clam; *Tagelus,* jackknife clam; *Ensis,* razor clam; *Mya,* mud clam; *Mactra* (Fig. 21-24).

Order 4. Schizodonta. Characteristic schizodont hinge; gills eulamellibranch. *Unio,* freshwater mussel.

Order 5. Adapedonta. Shell gapes, ligament weak or absent; mantle margins fused save for pedal gape; siphons large, long, and united; burrowers in hard substrates. *Pholas,* burrowing in clay or rock; *Barnea,* "angel wings"; *Teredo, Bankia,* shipworms, wormlike, boring in wood (Fig. 21-24); *Saxicava.*

Order 6. Anomalodesmata. Gills eulamellibranch, outer demibranch (half gill) reduced and upturned; fused mantle edges; foot small. *Brechites,* watering-pot shell; *Pandora, Clavagella.*

Subclass 3. Septibranchia. Gill reduced to horizontal muscular position dividing mantle cavity. *Cuspidaria, Poromya.*

Class 7. Cephalopoda.

Nautili, squids, and octopuses. Shell external, internal, or none; head large, eyes conspicuous and complex; mouth with horny jaws and radula and surrounded by 8 or 10 arms or many tentacles; a siphon; nerve ganglia grouped in head as "brain," in cartilagelike covering; dioecious; development direct; all marine; 650 living, 10,000 fossil species.

Subclass 1. Nautiloidea. Nautili. Shell external, coiled in one plane, divided by internal septa; 2 pairs of gills and nephridia; siphon 2-lobed; no ink sac; tentacles many, no suckers; eyes without lens; 1 living genus: *Nautilus,* western Pacific and Indian Oceans at depths to 550 m (1800 ft); about 2500 fossil species; Upper Cambrian to Recent.

Subclass 2. †Ammonoidea. Ammonites. Shell external, chambered, with complex, irregular suture lines; Silurian to Cretaceous; †*Ammonites.*

Subclass 3. Coleoidea. (Dibranchia). Octopuses, squid. Shell internal and reduced, or none; body cylindrical or globose, often with fins; arms 8 or 10, with suckers; one pair of gills and nephridia; siphon tubular; an ink sac; eyes with lens.

Order 1. Decapoda. Ten arms. Triassic to Recent. (Belemnites of Carboniferous to Cretaceous with internal chambered shell. †*Belemnites.*) Living forms: *Spirula,* shell spiral, internal, deep tropical seas, shells occasional on Atlantic Coast; *Sepia,* cuttlefish, Atlantic waters; *Rossia,* body stubby, to 75 mm (3 in); *Loligo,* squid; *Architeuthis,* giant squid.

Order 2. Octopoda. Eight arms. Cretaceous to Recent. *Argonauta,* paper nautilus; *Octopus,* octopus.

Order 3. Vampyromorpha. Vampire squids. Eight arms united by an arm web and 2 coiled, tendril-like arms. *Vampyroteuthis,* in the deep sea.

References

Abbott, R. T. 1974. American seashells. 2d ed. New York, D. Van Nostrand Company, Inc. 663 pp., 24 color pls., illus. *Semipopular; describes 2000 and lists 4500 North American marine species.*

Emerson, W. K., and **M. K. Jacobson.** 1976. The American Museum of Natural History guide to shells, land, freshwater and marine from Nova Scotia To Florida. New York, Alfred A. Knopf, Inc. viii + 482 pp., illus.

Fretter, V. (editor). 1968. Studies in the structure, physiology and ecology of molluscs. New York, Academic Press, Inc. xvii + 377 pp., illus.

_____ and A. Graham. 1962. British prosobranch Mollusca. London, The Ray Society. xvi + 755 pp., 317 figs.

_____ and **J. Peake** (editors). 1975. Pulmonates. Vol. I. Functional anatomy and physiology. New York, Academic Press, Inc. xxxii + 418 pp., illus.

Grassé, P.-P. (editor). 1960. Traité de zoologie. Paris Masson & Cie. Mollusques (except Gastropods and Cephalopods). Vol. 5, pt. 2, pp. 1623–2164, figs. 1469–1830, 2 col. pls. Vol. 5, pt. 3, Gastropods and Scaphopods, 1083 pp., 517 figs.

Hyman, L. 1967. Mollusca I. In The invertebrates. New York, McGraw-Hill Book Company. Vol. 6, vii + 792 pp., 1250 illus.

Keen, A. M. 1971. Seashells of tropical west America. 2d ed. Stanford, Calif., Stanford University Press. xiv + 1064 pp., illus. 22 color pls.

_____ and **E. Coan.** 1974. Marine molluscan genera of western North America: An illustrated key. 2d ed. Stanford, Calif., Stanford University Press. vi + 208 pp., illus. *Keys and systematic lists.*

Lemche, Henning, and **K. G. Wingstrand.** 1959. The anatomy of *Neopilina galatheae* Lemche, 1957 (Mollusca Tryblidiacea). Copenhagen. Galathea Report, vol. 3, pp. 9–71, 56 pls., 169 figs.

Mead, A. R. 1961. The giant African snail: A problem in economic malacology. Chicago, The University of Chicago Press. xvii + 257 pp., 15 figs.

Moore, R. C. (editor). 1957–1964. Treatise on invertebrate paleontology. Lawrence, University of Kansas Press. Mollusca 1, 1960. xxiii + 351 pp., 216 figs.; Mollusca 3, 1964. xxviii + 519 pp., 361 figs. Mollusca 4, 1957. xxii + 490 pp., 558 figs.

Morris, P. A. 1947. A field guide to the shells of our Atlantic Coast. Boston, Houghton Mifflin Company. xvii + 190 pp., 40 pls. (8 in color).

_____. 1966. A field guide to shells of the Pacific Coast and Hawaii. Boston, Houghton Mifflin Company. xxxiii + 297 pp., 72 pls.

Morton, J. E. 1967. Molluscs. 4th ed. London, Hutchinson

& Co. (Publishers), Ltd. 244 pp., 41 figs. *General account and outline of classification.*

Pilsbry, H. A. 1939–1948. Land Mollusca of North America (north of Mexico). Philadelphia Academy of Natural Sciences, Monograph 3, vol. 1, pts. 1 and 2, 994 pp., 580 figs.; vol. 2, pts. 1 and 2, 1113 pp., 584 figs., 1 col. pl. *Complete monograph.*

Purchon, R. D. 1977. The biology of the Mollusca. 2d ed. New York, Pergamon Press. xxv + 560 pp., 185 figs.

Raven, C. P. 1966. Morphogenesis. The analysis of molluscan development. 2d ed. London, Pergamon Press. xiii + 365 pp., illus.

Robson, G. C. 1929, 1932. A monograph of the recent Cephalopoda. London, British Museum. Pt. 1, xi + 236 pp., 89 figs., 7 pls.; pt. 2, xi + 359 pp., 79 figs., 6 pls.

Runham, W. W., and **P. J. Hunter.** 1971 Terrestrial slugs. London, Hutchinson University Library. 184 pp.

Solem, A. 1974. The shell makers: Introducing mollusks. New York, John Wiley & sons, Inc. xii + 289 pp., illus.

Theile, J. 1929–1935. Handbuch der systematischen Weichtierkunde. Jena, Gustav Fischer. Vol. 1, 778 pp., 783 figs. Vol. 2, 376 pp., 114 figs.

Thompson, T. E. 1976. Biology of the opisthobranch molluscs, Vol. I. London, The Ray Society. 207 pp., 106 figs.

Wells, M. J. 1962. Brain and behavior in cephalopods. Stanford, Calif., Stanford University Press. 171 pp.

Wilbur, Karl M., and **C. M. Yonge** (editors). 1964–1966. Physiology of the Mollusca. New York, Academic Press, Inc. Vol. 1, xiii + 473 pp. Vol. 2, xiii + 645 pp.

Yonge, C. M. 1960. Oysters. The New Naturalist. London, Collins Clear-type Press. xiv + 209 pp., 71 figs., 17 pls.

———— and **T. E. Thompson.** 1976. Living marine molluscs. London, Collins Clear-type Press. 288 pp., 162 figs.

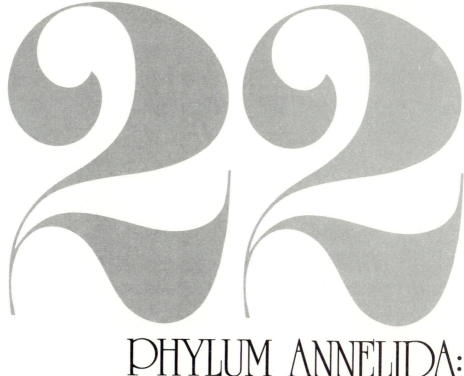

PHYLUM ANNELIDA: SEGMENTED WORMS

In contrast to the various kinds of worms previously discussed, those of the phylum ANNELIDA (L. *annelus,* little ring) have bodies composed of many essentially similar and ringlike segments, or somites. This segmentation usually shows in both external and internal features, including muscles, nerves, and circulatory, excretory, and reproductive organs. Most earthworms and their allies (class OLIGOCHAETA) are inhabitants of damp soil and fresh waters; the marine worms (class POLYCHAETA) are found chiefly along the seashore, and the leeches (class HIRUDINEA) are mainly in fresh waters or on moist ground. Some annelids are free-living, many inhabit burrows or dwell in tubes, some are commensals on other aquatic animals, a few are ecto- or endoparasites, and many of the leeches attach to vertebrates.

22-1 Characteristics

1 Symmetry bilateral; three germ layers; body elongate and usually conspicuously segmented both externally and internally.

2 Appendages are minute rodlike chitinous setae, few to many per somite; POLYCHAETA have fleshy tentacles on the head and the setae borne on lateral fleshy parapodia; most HIRUDINEA lack setae.

3 Body covered by thin moist cuticle over columnar epithelium containing unicellular gland cells and sensory cells.

4 Body wall and digestive tract both with layers of circular and longitudinal muscles; coelom well developed (except in HIRUDINEA), and divided by septa in OLIGOCHAETA and POLYCHAETA.

5 Digestive tract complete, tubular, extending length of body.

6 A closed circulatory system of longitudinal blood vessels (or coelomic sinuses) with lateral branches in each somite; blood plasma usually contains dissolved hemoglobin and free amoebocytes.

7 Respiration by the epidermis, or by gills in some tube dwellers.

8 Excretory system typically of one pair of nephridia per somite, each removing wastes from the coelom and bloodstream directly to the exterior.

9 Nervous system with pair of cerebral ganglia (brain) and connectives to a solid (double) midventral nerve cord extending length of body, with a ganglion and pairs of lateral nerves in each segment; sensory cells and organs for touch, taste, and light perception.

10 Sexes united and development direct (OLIGO-CHAETA, HIRUDINEA), or sexes separate and with trochophore larval stage (POLYCHAETA); some OLIGOCHAETA and POLYCHAETA reproduce asexually by budding.

22-2 General features and evolution The annelids, together with the MOLLUSCA and ARTHROPODA, constitute the three major protostome phyla. Of the three, they are the smallest in terms of species, only about 10,000 species being known. Annelids are the most highly advanced and most successful of the various "worm" phyla and are abundant in marine waters, fresh waters, and terrestrial environments, with the largest number of species and greatest adaptive radiation occurring in marine waters.

The dominant characteristics of the annelids are the organization of the body in a linear series of similar segments, an arrangement termed **metamerism,** and the presence of a large, compartmentalized coelom. The mollusks lack both these features, and the arthropods, while metameric, lack the large coelom. The success of the annelids also rests on these two key features. Annelids are soft-bodied animals in which the large fluid-filled coelom serves as a hydrostatic skeleton; that is, the contraction of the body-wall muscles acts against the internal coelomic fluid rather than against a rigid skeletal part as in the arthropods or chordates, and the animals operate like a hydraulic lift or other hydraulic system. Thus annelids need no rigid skeletal parts in order to effect movement. Compartmentalization of the coelom by transverse septa in each of the segments means that differential pressure may be exerted on the coelomic fluid in different parts of the body, thus increasing the precision of the movements. In annelids, contraction of the longitudinal muscles of the body wall forces coelomic fluid to move laterally and causes the worm or the segment to shorten. Contraction of the circular muscle layer forces coelomic fluid both anteriorly and posteriorly, and the worm or segment elongates. Compartmentalization of the coelom into segments allows this contraction or elongation to be confined to one or a few segments while the reverse is happening in other segments. Through nervous control, waves of muscular contraction can thus pass down the body, alternately elongating and contracting groups of segments and exerting a powerful force permitting the worm to move forward. Such a motion is particularly effective in moving an elongated animal through a burrow or soft substrate, and indeed, most annelids are burrowers. Metamerism also has another advantage in that it gives the potential of specialization of segments or groups of segments for different functions. Such specialization or **tagmatization,** is not well developed in the annelids but is highly evolved in the arthropods.

Most biologists feel that metamerism has probably arisen twice in evolution, once in the protostome line, giving rise to annelids and arthropods, and once in the deuterostome line, which gave rise to the phylum CHORDATA.

In addition to the metameric arrangement of the body-wall muscles, annelids have metameric arrangement of the nephridia, and the nerve cord has ganglia in each segment.

The metameric animal may have arisen from an unsegmented stock by forming chains of individuals through asexual fission as occurs in some PLATYHELMINTHES, but with the products of fission remaining united and acquiring both structural and physiologic unity. Another theory presumes the segmental division of muscles, nerves, coelom, nephridia, etc., within a single individual as seen in the formation of somites by both larvae and adults of some annelids.

The closest relatives of ANNELIDA are ECHIURA, which have setae but are not segmented as adults, and the SIPUNCULA, which lack setae and are unsegmented. In addition to metamerism, the ANNELIDA resemble the ARTHROPODA in having a body covered by cuticle that is secreted by the epidermis, in the structure of the nervous system, and in the forming of mesoderm from special embryonic cells. They differ in having a large coelom, only simple unjointed appendages, little or no specialization of somites in different parts of the body, and no succession of larval stages, or molts. The phylum ONYCHOPHORA is in some respects intermediate between annelids and arthropods (Sec. 20-33). The ANNELIDA resemble MOLLUSCA in embryonic features such as spiral cleavage and mesoderm formation. Also marine annelids have a trochophore larva (Fig. 22-12), like mollusks.

Annelids appeared early in the fossil record. Fossil worm tracks and tubes have been reported from Precambrian rocks, but the earliest undoubted worms are from the Middle Cambrian in British Columbia.

22-3 Size Some of the smallest OLIGOCHAETA (*Aeolosoma, Chaetogaster*) are under 1 mm long, but the giant earthworms (*Rhinodrilus fafner* of Brazil and *Megascolides australis* of Australia) grow to over 2 m (7 ft) long and 2.5 cm (1 in) in diameter. Most earthworms are only a few centimeters in length. The smallest POLYCHAETA are minute, but *Neanthes brandti* of the California coast grows to 1.5 m (5½ ft) and *Eunice gigantea* to nearly 3 m (10 ft). The leeches range from 10 to 200 mm (½ to 8 in), most of them being small.

Class Oligochaeta (Earthworms)

This class includes annelids with few setae per segment (Gr. *oligos,* few + *chaete,* spine). The large earthworm of Europe and eastern North America, *Lumbricus terrestris,* which may grow to 30 cm (12 in) long and 1 cm (⅜ in) in diameter, is the basis for the following account.

22-4 External features The body (Fig. 22-1) is long and cylindrical, bluntly tapered at each end, and somewhat depressed posteriorly; the ventral side is flattened and paler than the dark dorsal surface. There is no distinct head. A mature worm is divided into 115 to 200 ringlike segments, or **somites,** separated by transverse grooves. The **mouth** is in somite I, overhung by a fleshy lobe, the **prostomium,** and the vertically oval **anus** is in the last somite. The **clitellum** (L., pack saddle) is a conspicuous glandular swelling over somites XXXII to XXXVII; it secretes material forming cocoons to contain eggs. On each somite except the first and last are four pairs of minute bristlelike **setae** that project slightly on the ventral and lateral surfaces. Each seta is a fine chitinous rod, contained in an epidermal seta sac within the body wall, and is secreted by a large cell in the sac. A seta can be moved in any direction and extended or withdrawn by protractor and retractor muscles attached to the sac. The setae serve as holdfast devices when a worm is in its burrow or moving over the ground.

Besides the mouth and anus there are many minute external openings on the body, including (1) a **dorsal pore,** connecting the body cavity and exterior, middorsally in the furrow at the anterior border of each somite from VIII or IX to the anal end; (2) a **nephridiopore,** or excretory opening, lateroventrally

Figure 22-1 Class OLIGOCHAETA: **the earthworm,** *Lumbricus terrestris,* **external features.** I, V, etc., somites.

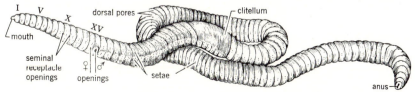

on either side of each somite except I to III and the last; (3) the four openings of the **seminal receptacles** laterally in the furrows between IX and X and X and XI; (4) the openings of the two **oviducts** ventrally on XIV; and (5) the openings of the paired **sperm ducts,** with swollen lips, ventrally on XV, from which two glandular ridges extend to the clitellum.

22-5 Body wall The exterior is covered (Fig. 22-2) by a thin transparent **cuticle,** marked with fine cross-striations that produce a slight iridescence. This is secreted by the **epidermis** beneath, a single layer of columnar epithelium that includes many **unicellular glands** producing mucus to pass through pores in the cuticle and keep the latter soft. The cuticle is also perforated by pores over many **sensory cells** in the epidermis. Both kinds of pores are seen easily in pieces of cuticle stripped off and spread to dry on a glass slide. The epidermis rests on a basement membrane, beneath which is a thin layer of **circular muscles** and a thicker layer of **longitudinal muscles.** The inner surface of the body wall is covered by a thin smooth epithelium, the **peritoneum.** Pigment cells, connective tissue, and blood capillaries occur among the circular muscle fibers. There is no skeleton. The form of the animal is maintained by the elasticity and contraction of the body wall on the coelomic fluid within. Contraction of the circular muscles elongates the body, contraction of the longi-

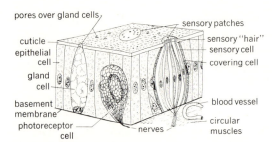

Figure 22-2 Earthworm. Diagram of cuticle and epidermis in perspective. (*Adapted from Hess, 1925; and Langdom, 1895.*)

tudinal muscles shortens it, and local or differential action of these same muscles produces the bending movements so characteristic of earthworms.

22-6 Internal structure (Fig. 22-3) If the body wall is slit middorsally and spread, the earthworm is seen to consist essentially of two concentric tubes, the outer **body wall** and the straight **digestive tract** within; the space between them is the body cavity, or **coelom.** It is divided into a succession of compartments by the **septa,** which are thin transverse partitions of connective tissue aligned with the grooves on the exterior surface; septa are absent between somites I and II and incomplete between III and IV and between XVII and XVIII. The coelom and all organs within it are covered by peritoneum. The cavity contains watery coelomic fluid with free colorless amoe-

Figure 22-3 Earthworm. Internal structure of anterior portion from the left side; body wall and digestive tract cut in median section. Two hearts shown in place; nephridia omitted; reproductive organs of right side included. I, X, XX, somites.

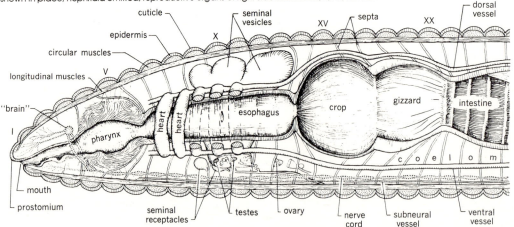

bocytes. Small pores in the septa permit this fluid to pass from somite to somite in keeping with movements of the body, and some of it may pass out the dorsal pores to moisten the exterior surface (Fig. 22-4).

22-7 Digestive system This consists of (1) the **mouth,** followed by a short **buccal cavity** (somites I to III); (2) the **pharynx** (IV to V), with glands to lubricate the food and muscle fibers in its external walls; (3) the slender, straight **esophagus** (VI to XIV) joined on either side by three pairs of **calciferous glands** (X to XIV), which serve to excrete excess calcium from the body; (4) a thin-walled and enlarged **crop** (XV to XVI), for storage; (5) the **gizzard** (XVII to XVIII), with thick, firm muscular walls and lined with cuticle; and (6) the **intestine,** which continues to (7) the **anus.** The intestine is thin-walled and bulges laterally in each somite, and its dorsal wall carries an

infolded **typhlosole;** the bulges and typhlosole afford increased surface for digestion and absorption of food. The internal lining is of columnar ciliated epithelium, containing many club-shaped gland cells; next are blood vessels, and then longitudinal and circular muscles; and the exterior of the intestine is covered by rounded, yellowish **chloragog cells,** which are modified peritoneum. These cells function much as a liver and are the location for glycogen and fat synthesis and storage, urea formation, and deamination of amino acids. Waste materials may also accumulate in these cells.

The food is chiefly dead leaves, grasses, and other vegetation; it is moistened by salivalike secretions from the mouth region and then drawn in by muscular action of the prostomium, "lips," and pharynx. Any organic acids present in the food are neutralized by calcium carbonate secreted by the calciferous glands of the esophagus, so that food undergoing digestion is alkaline in reaction. Food is stored tem-

Figure 22-4 Earthworm: diagrammatic cross section. Left half shows an entire nephridium and a dorsal pore but omits setae; right half includes setae but no nephridium.

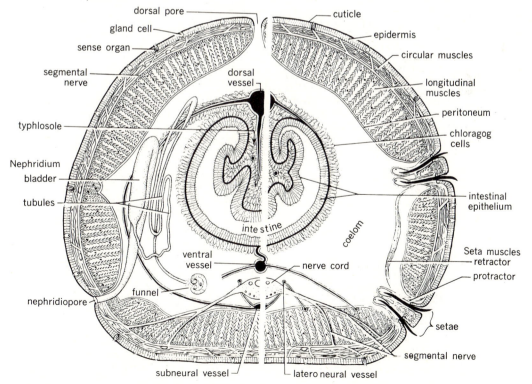

porarily in the crop and later ground up in the muscular gizzard with the aid of sand grains present there. The digestive tract secretes enzymes, including pepsin and trypsin acting on proteins, lipase on fats, cellulase on cellulose, and amylase on carbohydrates. Digested material is absorbed in the intestine, and residues are passed out the anus. Small animals in the ground may be eaten, and captive worms will eat bits of fat or meat and dead worms. Earthworms utilize any organic food contained in the earth taken in while burrowing.

22-8 Circulatory system The **blood** consists of a fluid **plasma** that contains free colorless **corpuscles** (amoebocytes). The plasma is colored red by **hemoglobin,** a respiratory pigment dissolved in it. Blood circulates to and from all parts of the body in a system of closed **blood vessels** with capillaries. There are five principal vessels extending lengthwise of the body and five pairs of aortic arches or "hearts" in somites VII to XI. Each somite from XII posteriorly has paired segmental vessels connecting the longitudinal vessels and various organs, and special vessels serve structures in the anterior somites (Fig. 22-5).

22-9 Respiration There is no organized respiratory system. Blood circulating in capillaries close to the moist cuticle of the body wall receives oxygen there and gives up carbon dioxide. The oxygen combines with the hemoglobin in the plasma and is carried to various tissues. In an emergency an earthworm can survive for several hours without a fresh supply of oxygen.

22-10 Excretory system Every somite except the first three and the last has a pair of nephridia (Fig 22-6) for excretion. Each **nephridium** begins as a ciliated funnel, or nephrostome, at the anterior base of the septum beside the nerve cord. It connects through the septum by a fine ciliated tubule to the main part of the nephridium in the somite behind. The tubule there is of several loops of increasing size and a larger bladder that discharges to the exterior through a **nephridiopore** opening near the ventral pair of setae. Cilia of the funnel and tubule beat to draw wastes from the coelom, and other waste is received from blood vessels surrounding the nephridium. Excretory products (ammonia, urea, creatine) are formed in the body wall and gut wall and enter both the blood and coelomic fluid. The nephridia act much like tubules in a human kidney—by filtration, resorption, and tubular secretion they yield a protein-free urine and maintain the steady state in the body.

22-11 Nervous system Above the pharynx (in somite III) is a pair of **suprapharyngeal (cerebral)**

Figure 22-5 Earthworm: circulatory system. *A.* One pair of hearts and other vessels as in somite x. *B.* Relations of blood vessels in any somite behind the clitellum. Arrows indicate paths of blood flow. (*Adapted from Johnston and Johnston, 1902; and Bell, 1947.*)

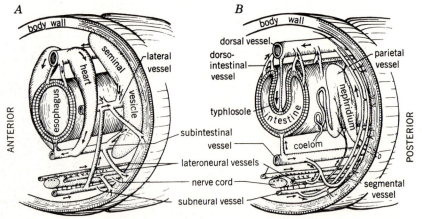

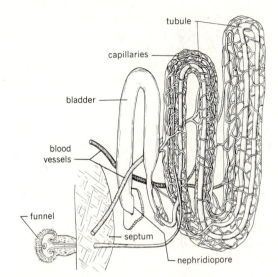

Figure 22-6 Earthworm: an entire nephridium.

between the two muscle layers of the body wall and includes both **sensory fibers,** carrying impulses in from the epidermis to the nerve cord, and **motor fibers,** carrying directive impulses outward from the cord to muscles and eipdermal cells. These nerves also include fibers of both types connecting to the nephridia, seta sacs, and other organs.

The epidermis has many **sense organs,** each a group of sensory cells surrounded by special supporting cells (Fig. 22-2). These abound on the anterior and posterior ends, on the swollen part of each somite, and in the buccal cavity—the parts of the body most likely to receive stimuli from the environment. Each sense cell has a hairlike tip projecting through a pore in the cuticle, and a sensory nerve fiber connects to the base of the cell. Free endings of nerve fibers also occur between cells of the epithelium. Special **photoreceptor cells** are present in the epidermis and on nerves at both ends of the body. These cells are most numerous on the prostomium, anterior somites, and anal somite, the parts most sensitive to light. Each cell contains a clear transparent organelle ("lens") that focuses light from any direction on a network of neurofibrils ("retina") in the periphery of the cell. The neurofibrils are probably the actual light receptors, and they connect to sensory nerves. These photoreceptors resemble the visual cells of leeches in structure and function. Stimuli on the light receptors enable the worm to distinguish daylight and dark.

ganglia, the "brain" (Fig. 22-7A). Two **circumpharyngeal connectives** extend around the pharynx to the bilobed **subpharyngeal ganglia** (base of somite IV), whence the **ventral nerve cord** extends along the floor of the coelom to the anal somite. Several nerves connect to the prostomium and mouth region. The ventral nerve cord in each somite has an enlarged **ganglion** and gives off three pairs of **lateral nerves** (Fig. 22-7B). The cord and ganglia appear to be single but develop as paired structures. Each lateral nerve extends around half a somite

Figure 22-7 Earthworm. *A.* Ganglia and larger nerves of the anterior end, in lateral view. (*After Hess, 1925.*) *B.* Stereogram of the ventral nerve cord and body wall to show a simple reflex arc. Sensory cells (receptors) in the epidermis connect to sensory fibers that pass in a lateral nerve to the nerve cord; the sensory axon joins through a synapse to a motor nerve that leads to the body muscles (effector). Arrows indicate the direction of nerve impulses.

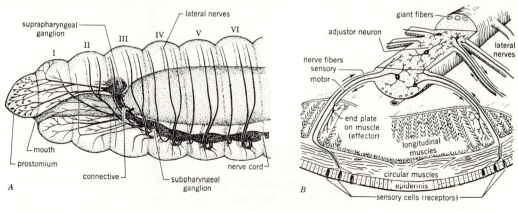

22-12 Reproductive system   The earthworm is monoecious; both male and female sex organs are ventral and anterior (Figs. 22-3, 22-8). The **male reproductive system** includes (1) two pairs of minute **testes** (somites X, XI) and back of each (2) a ciliated **sperm funnel** connecting to (3) a short duct, the **ductus efferens.** The two ducts on each side connect (in XII) to (4) a **ductus deferens** that leads to (5) the **male pore** (XV). The testes and funnels are contained in (6) pairs of large saclike **seminal vesicles** (in IX to XIII) that extend dorsally around the esophagus. Immature sperm cells separate from the testes to complete their differentiation in the vesicles, and mature sperm are discharged through the funnels and ducts during copulation. The **female reproductive system** includes (1) a pair of **ovaries** (anterior in XIII) that discharge mature ova into the coelom, whence they are collected by (2) the two **oviducal funnels** (posterior in XIII), with egg sacs, connecting to (3) the **oviducts** opening on somite XIV. The female system also includes (4) two pairs of **seminal receptacles** (in IX, X) where sperm received in copulation are stored until needed to fertilize eggs in cocoons.

22-13 Natural history   Various species of earthworms live in most lands of the world, including oceanic islands and subarctic regions. They are numerous in good soils with much humus and abundant moisture but scarce in poor, acid, sandy, or dry situations. They inhabit burrows for protection against predators and unfavorable climatic conditions. The burrows are nearly vertical at the top, then wind about to various depths, some extending to 2 m (6 ft) or more below the surface. In soft topsoil, a worm makes the burrow by contracting the circular muscles of the body wall hydraulically, forcing its slender anterior end as a wedge to enter any crevice, and then contracts the longitudinal muscles, forcing the soil particles outward by swelling its pharynx region. In heavier or deeper soil the worm excavates by actually eating its way. The earth passes through its digestive tract and is deposited on the ground surface as small mounds of feces, or castings, to be seen wherever worms are plentiful. The upper parts of a regular burrow may be smoothly lined with slime, earth from castings, leaves, or fine pebbles. Earthworms in cold climates retire below the frost line in winter and remain inactive; in arid regions they do likewise in summer to avoid the drought and heat, each rolling up in a close ball.

When the soil is moist and the temperature moderate, each worm lies by day in the upper part of its burrow, anterior end foremost. The entrance may be plugged with bits of leafy material. After dark this is pushed aside and the worm extends its forward end out over the ground surface, to explore, forage, or mate; the posterior end remains in the burrow, holding by the setae, so that, if frightened, the entire animal can withdraw to safety by a quick contraction of its body. At times a worm will leave its burrow and travel on the surface, and a sudden flooding of the ground, as by a heavy rain, will cause many worms to emerge.

Earthworms avoid all but the weakest of light, remaining hidden by day and at night withdrawing quickly if a light is flashed on them; they tolerate a red light but avoid a blue one. They are sensitive to mechanical vibrations such as heavy footfalls but evidently do not "hear" mere sound vibrations in air. Various vibratory devices reportedly are effective in bringing worms to the surface when they are being collected for fish bait. They react favorably to contact, being used to that of the ground surface or the surrounding burrow, and if turned up, as in the spading of a garden, will seek contact with the earth, thus also avoiding daylight. Tactile stimuli are all-important in retaining contact with the ground and

Figure 22-8   Reproductive system of earthworm in dorsal view, other organs omitted; seminal vesicles cut away on right side.

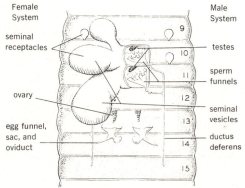

Female System

seminal receptacles

ovary

egg funnel, sac, and oviduct

Male System

testes

sperm funnels

seminal vesicles

ductus deferens

9
10
11
12
13
14
15

burrow. Since respiration depends upon the cuticle being kept moist, they respond positively to a moist environment rather than to dryness. The chemical responses, comparable to taste and smell, are more difficult to study, but it is evident that they serve in the choice of food. Unpleasant or irritating chemical vapors in the air will cause withdrawal into the burrow.

22-14 Reproduction   Earthworms reproduce through much of the year but most actively in warm moist weather. Mating occurs at night and requires 2 or 3 h. Two worms stretch out from their burrows and bring their ventral anterior surfaces together, with the anterior ends pointing in opposite directions (Fig. 22-9). the clitellum (XXXII to XXXVII) on each grips somites VII to XII of the other, and a lesser contact of somite XXVI on each is made with XV of the other. Special ventral setae (VI to X and on clitel-

Figure 22-9 *A.* Two earthworms in copulation (X = tenth somite). *B.* Secretion of slime tube and cocoon. *C.* Slime tube and cocoon slip forward. *D.* Free tube containing cocoon. *E.* Cocoon. (*A, after Grove, 1925; B–D, after Foot and Strobell, 1902.*)

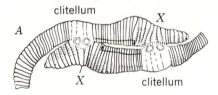

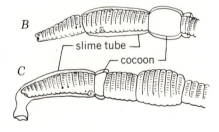

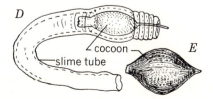

lum) of each actually penetrate the body of the other to aid in holding the worms together. Each worm secretes a slime tube about itself, covering somites IX to XXXVI. On each worm a pair of seminal grooves forms (XV to clitellum), along which masses of sperm pass to enter the seminal receptacles of the other (between IX and X and X and XI); after this the worms separate. There is thus reciprocal fertilization and cross-fertilization.

Each worm later produces cocoons containing eggs. In *Eisenia foetida,* each worm forms 2 to 10 cocoons at intervals of 3 to 5 days, containing one to eight eggs, of which an average of two develop. The cocoons of *Lumbricus terrestris* measure about 7 by 5 mm, and each has several eggs of which only one or two develop. For each cocoon a slime tube is secreted around the clitellum and anterior somites (IX to XXXVI), and within this the cocoon forms as a separate secretion over the clitellum; eggs and albumen enter the cocoon while it is still on the clitellum. The tube and cocoon then slip forward, the sperm fertilizes the eggs, entering while they pass over the seminal receptacles, and, as the worm withdraws from the tube, the cocoon closes into a lemon-shaped case that is deposited in damp earth.

Cleavage is holoblastic but unequal, making a hollow blastula of small ectoderm cells above and larger endoderm cells below (Fig. 22-10). Two large mesoblast cells develop on the future posterior surface and later pass inside to lie between the ectoderm and endoderm. The spherical blastula flattens and elongates as an oval plate, the endoderm cells become columnar, and the ectoderm grows down marginally. In gastrulation the ventral surface becomes concave and the edges curl down, meeting first posteriorly. This closure proceeds forward, the endoderm thus invaginating as a tube (future gut) ending with the blastopore (future mouth) at the anterior end. The anus forms much later. Meanwhile two cords of mesoblast cells multiply from behind forward, between ectoderm and gut; they spread as two flat plates giving rise to a series of pockets (the somites) and the body muscles. After gastrulation albuminous material in the cocoon enters the primitive gut (gastrocoel) to nourish the embryo. Development requires several weeks, since young hatch in 2 to 3 weeks from coccons kept in a laboratory.

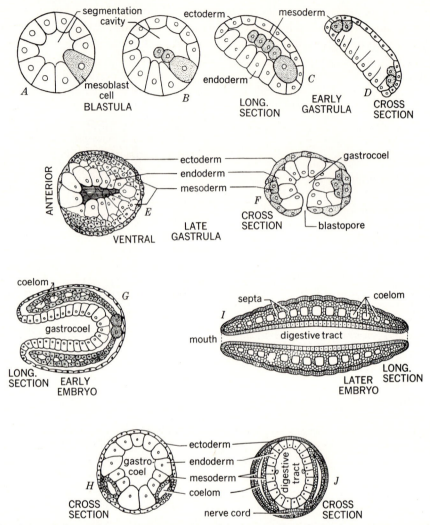

Figure 22-10 Development of an earthworm: diagrammatic sections (except *E*). BLASTULA: *A*. Early, with mesoblast cell. *B*. Growth of mesoblast (mesoderm). GASTRULA: *C*. Early longitudinal section. *D*. Early cross section. *E*. Later entire gastrula in ventral view. *F*. Cross section of later gastrula. *G*,*H*. Early embryo; two layers of mesoderm separate to yield coelom in somites. *I*,*J*. Later embryo; mouth derived from blastopore. (*Adapted from Wilson, 1889; and Sedgwick and Wilson, Introduction to general biology, Henry Holt and Company, Inc.*)

Adult earthworms can regenerate somites removed at the ends of the body by accident or experiment. At the anterior end up to six (or nine), if removed, may regenerate, including a "head" region. At the opposite end regeneration follows removal of more somites. Experiments in grafting have produced worms with two tail ends, short worms from two terminal portions, or extra long worms by joining parts of three worms.

22-15 Relation to humans Earthworms are used as bait for fishing, and many "farms" grow worms for this market. Polychaete worms are important as bait for saltwater fishes. On lawns and golf greens earthworm castings can be a nuisance, and control by poison has been used. Ancient medical writers mention uses of earthworms in human medicine, and some of these are still followed in parts of Japan and China.

In nature the long-time effects of earthworms have certain practical values. In many soils thousands are present per acre, and their burrowing during most of the year turns over much surface soil. According to Charles Darwin, in favorable locations they may bring up 18 tons of soil per acre in a year, and a layer of cinders or gravel may be completely covered by the castings over a period of years. In extreme instances the burrows may cause water seepage through irrigation ditches or speed soil erosion on sloping lands, but generally worm cultivation helps by turning over the top portion and allowing air and water to penetrate. Indeed, the depth of arable topsoil in less fertile areas may be gradually increased by the worms.

22-16 Other Oligochaeta This class includes over 3100 species, most of which are smaller than *Lumbricus terrestris*. Most species are terrestrial or occur in fresh water. Only a few are estuarine or marine. The most common over North America is the small *Allolobophora caliginosa*. The brandling, or "stinking earthworm," *Eisenia foetida*, ringed with maroon and yellow, is a favorite fishworm that lives in piles of manure or compost. Earthworms are passive carriers of the gapeworm of fowls (*Syngamus*) and intermediate hosts for a cestode of fowls (*Amoebotaenia sphenoides*) and a lungworm of pigs (*Metastrongylus elongatus*); the last-named carries a virus that, in combination with a bacterium, causes swine influenza. Other oligochaetes inhabit damp banks of lakes and streams and either fresh or polluted waters. Some live deep in lakes, some beneath stones or decaying seaweeds of the ocean shore, and a few are marine at shallow depths. Others occur in mountains to 3800 m (12,500 ft) (Kashmir), and some inhabit the soil under snow fields from Oregon to Alaska. Most oligochaetes feed on vegetation, but a few are probably carnivorous (*Chaetogaster*). The reddish *Tubifex* lives in tubes on the bottom of polluted waters, feeding on the bottom muck, and is capable of extracting oxygen from water which has too low an oxygen concentration to sustain most animals. Some oligochaetes are important in the diet of moles, various birds, amphibians, fishes, and predatory invertebrates. No parasitic oligochaetes are known, but some of the smaller species are found living in association with other invertebrates.

The BRANCHIOBDELLIDAE are intermediate between oligochaetes and leeches, having a caudal sucker, 11 trunk segments, two pairs of nephridia, and no setae. Their length is from 1 to 12 mm, and they are commensals on crayfish, living on the gills or generally over the body surface.

Aquatic oligochaetes in some genera reproduce asexually by transverse fission or fragmentation, with several to many "generations" per year. In *Enchytraeus* some species reproduce sexually but one lacks sex organs. It multiplies by fragmenting into pieces (of about five somites), each regenerating into a complete worm in about 10 days. Individuals of *Chaetogaster* and *Aeolosoma* form chains of zooids.

Class Polychaeta
(Sandworms, tube worms)

The polychaetes (Gr. *poly*, many + *chaete*, spine or bristle) constitute the largest class of annelids, with over 8000 species. They also show the greatest adaptive radiation in terms of body form, food habits, and habitats occupied. Virtually all species are marine, ranging from intertidal down to abyssal depths. Many species are abundant as individuals, particularly in sand and mud bottoms, and they are an ecologically important group in marine ecosystems. A few are pelagic.

The polychaetes include some species that are presently considered to be the most primitive annelids. Ancestral polychaetes were probably characterized by having a body composed of an array of similar segments, without tagmatization, and they burrowed in or moved over the substrate to feed on small particles. Such primitive features are best displayed today in certain of the free-living errant polychaete species. Tubicolous sedentarian polychaetes tend to be more advanced.

Discerning the evolution and phylogenetic arrangement of the polychaetes is difficult, first because the presently living polychaete families are distinct from one another and second because, on account of the lack of skeletal materials in the body, there is virtually no fossil record for the group (except tubes and tracks). The class is often divided into two subgroups, Errantia and Sedentaria, but these are admittedly arbitrary, and there is no universal agreement as to the higher classification within the class. The errant polychaetes are those which are free-living and crawl or swim. They usually have a body of similar segments with well-developed parapodia, a muscular eversible proboscis with teeth or jaws, and a distinct head with sense organs. Sedentary polychaetes are those confined permanently to tubes, often with the body divided into two or more regions or tagma, without a protrusible proboscis with jaws, and with a reduced head.

Polychaetes differ from the Oligochaeta in having (1) a differentiated head with sensory appendages, (2) many setae in bundles in each body somite, often borne on a pair of lateral parapodia, (3) the sexes usually separate and without permanent gonads, and (4) a free-swimming trochophore larval stage (Fig. 22-12). Some taxonomists have recently grouped the Oligochaeta and the Hirudinea together in a single class, Clitellata, emphasizing the closer relationship of earthworms and leeches as opposed to the polychaetes. Both leeches and oligochaetes share features of hemaphroditism, complex reproductive organs, lack of parapodia, and absence of setae in bundles. The older system is adhered to here.

The basic polychaete structure is best exemplified in an errant polychaete. The clamworm, *Neanthes (Nereis) virens* (Fig. 22-11), is such a representative polychaete, living near the low-tide line. It hides by day beneath stones or in a temporary burrow, with only the head protruding; at night it reaches out, crawls over the sand, or swims by lateral wriggling of the body. Large individuals may be 450 mm (18 in) long.

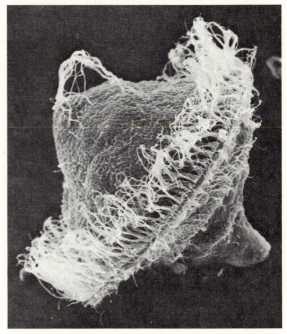

Figure 22-12 Scanning electron micrograph of a trochophore larva of the polychaete, *Serpula vermicularis.* (*Courtesy of F. Shiang Chia, University of Alberta.*)

22-17 Structure and natural history The long, slender, greenish **body** is rounded above and flattened ventrally and composed of 200 or more similar **somites.** A distinct **head** is formed by the prostomium and peristomium (somites I + II). The prostomium bears two short **prostomial tentacles** medially, a pair of stubby conical **palps** laterally, and two pairs of small **eyes** dorsally. The peristomium surrounds the ventral **mouth** and carries four pairs of **peristomial tentacles** dorsally (Fig. 22-13). These specialized sensory organs of touch, smell, and sight serve in finding food, avoiding predators, and other activities. On either side of each somite there

Figure 22-11 Clamworm, *Neanthes virens* **(class POLYCHAETA).** External features.

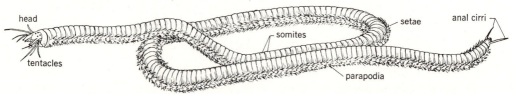

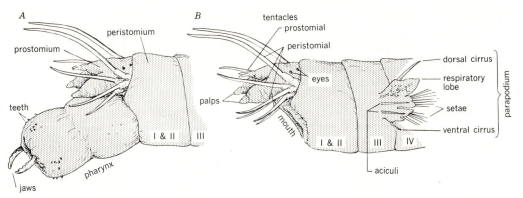

Figure 22-13 Clamworm, *Neanthes*: head region from the left side. *A.* Pharynx extended. *B.* Pharynx retracted.

is a flat **parapodium** (Gr. *para*, beside + *podos*, foot) of two lobes (dorsal notopodium, ventral neuropodium), each bearing a **cirrus** and a bundle of bristlelike **setae.** Within are two thick chitinous **acicula,** which extend into the body and serve to support and move the parapodium (Fig. 22-14). The parapodia serve for locomotion, in both creeping and swimming. The **anus** is in the last somite, on which are two soft sensory **anal cirri.**

The body is covered by **cuticle** over an **epidermis,** and beneath the latter are **muscles,** a thin layer of circular muscles and four large bundles of longitudi-nal muscles (two dorsal, two ventral). The parapodia are moved by oblique muscles in each somite, and there are special muscles in the head. Within the body wall is the **coelomic cavity,** lined by periton-eum and divided by **septa** between most of the so-mites, as in the earthworm. There are also median dorsal and ventral **mesenteries** between the body wall and gut that divide the coelom of each somite into right and left compartments.

The **digestive tract** is a straight tube and includes (1) the **mouth;** (2) a protrusible **pharynx** with two horny, serrated jaws and groups of chitinous "teeth"

Figure 22-14 Clamworm: diagrammatic cross section. Left half with ova free in coelom, right half through the parapodium.

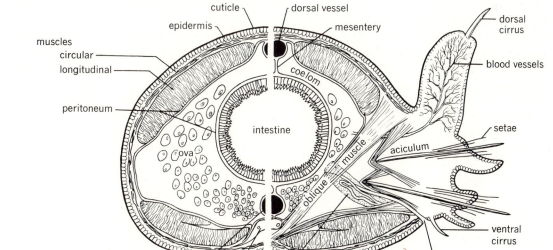

(paragnaths); (3) a short **esophagus** joined by two large digestive glands; and (4) the **stomach-intestine,** which extends from about somite XII to (5) the **anus.** Most clam worms are omnivorous, taking a varied diet of plants and animals.

The **circulatory system** comprises a dorsal vessel and a ventral vessel, both longitudinal, with transverse branches in each somite to the gut, nephridia, body wall, and parapodia. The blood plasma is red, containing dissolved hemoglobin and colorless corpuscles; circulation results from peristaltic contractions of the dorsal vessel. **Respiration** is effected by capillaries in the parapodia and body wall. **Excretion** is performed by paired nephridia that differ from those of the earthworm in being more compact and having much of the ducts ciliated. The **nervous system** includes a brain, nerves to the head and tentacles, connectives to the midventral nerve cord, and a pair of ganglia and lateral nerves in each somite.

The sexes are separate, and **gonads** are present only in the breeding season. Eggs and sperm form from cells of the coelomic peritoneum and, when mature, pass out through the nephridia or burst through the body wall. Fertilization occurs in the sea, and the zygote develops into a ciliated trochophore larva (Fig. 22-12), which later transforms into a young worm.

22-18 Other Polychaeta The polychaetes are such a diverse group that the few species mentioned here can give only a sampling of their diversity. Most of the species live from the tide line down to about 36 m (120 ft), but a few go down to 5500 m (18,000 ft). Some are brilliantly colored. Various species crawl on the bottom, hide beneath rocks or plants, dwell in temporary burrows, or live in permanent tubes. A few swim in the open sea, and a few others inhabit fresh water. The ARCHIANNELIDA are primarily dwellers between the sand grains on marine beaches, and the small oval MYZOSTOMIDAE are parasites or commensals on echinoderms. Many polychaetes feed on small particles of living or dead material, some burrowing forms feed on bottom muck, and tube dwellers subsist mainly on plankton. Others are carnivorous, preying on other polychaetes and invertebrates.

The sea mouse, *Aphrodite,* has an oval body with pairs of large plates (elytra) dorsally and is covered by long hairlike setae (Fig. 22-20). The lugworm, *Arenicola,* is a soft-bodied worm that burrows by ingesting sand or mud and produces castings like an earthworm. Other species plaster the sandy walls of their burrows with a mucous secretion that binds the sand grains against collapse and produces a smooth lining. Among the SEDENTARIA, *Terebella* builds a tube of selected sand grains, bits of shell, or other materials set in mucus (Fig. 22-20). *Pectinaria* cements a tapering tube of fine sand grains, open at both ends, which the worm may carry about. The minute *Spirorbis* secretes a spiral calcareous tube that is fastened to the surface of rocks or seaweed, and *Protula* forms a larger tube of the same sort (Fig. 22-15). Some tube dwellers can burrow in rocks or shells. *Chaetopterus* is a specialized worm living permanently in a parchmentlike U-shaped tube, up to 2 cm ($\frac{3}{4}$ in) in diameter, that is set with both openings just above the mud or sand (Fig. 22-16), at depths to 15 fathoms. Three pairs of parapodia on the middle of its body are modified as fans to draw in water containing oxygen and microscopic organisms. The mouth is a wide funnel. Other animals live as commensals in the tube. *Serpula* and other tube dwellers have a crown of many ciliated tentacles, derived from the prostomial palps, that collect microscopic food (Fig. 22-20); the tentacles contain blood vessels and so serve also as gills for respiration. The tip of one filament is enlarged as an operculum to close the tube when the worm draws within. Some tube worms, such as *Sabellaria,* often occur in large, closely set colonies. Other polychaetes have gills on the parapodia or, exceptionally, at the posterior end.

Errant polychaetes include many carnivores such as *Glycera* which have a large, protrusible pharynx armed with jaws. *Glycera* burrows through sand or mud in search of prey. The primitive family AMPHINOMIDAE have large bundles of brittle, hollow setae containing a toxin. If these worms are picked up, the setae break off in the flesh, producing an inflammation. They are called *fireworms.* The family SYLLIDAE includes many small, abundant polychaetes which are of interest because of their often bizarre ways of reproducing by budding off reproductive individuals from the parapodia and/or the

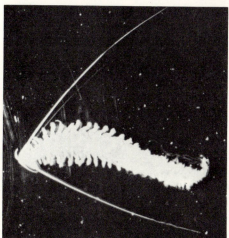

Figure 22-15 Polychaetes. *A. Protula,* a sedentary tube worm of the family SERPULIDAE. *B. Tomopteris,* a pelagic polychaete.

posterior end. Two families, TOMOPTERIDAE and ALCIOPIDAE, are pelagic (Fig. 22-15). The alciopids are of some interest because they have large, vertebratelike eyes, the only polychaetes to have such well-developed eyes.

In most polychaetes the sexes are alike at maturity. The germ cells develop on the peritoneum or the blood vessels, usually in the posterior somites, which may become different in appearance from the anterior, or nonsexual, part. This production of a

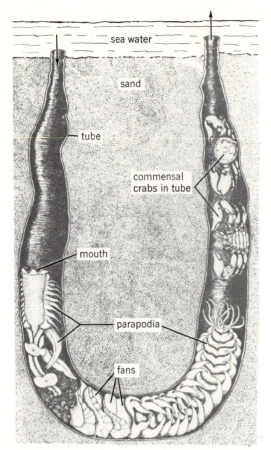

Figure 22-16 Class POLYCHAETA: *Chaetopterus,* **a specialized worm dwelling in a secreted tube in the sea bottom.** The "fans" are modified parapodia that draw water through the tube. Commensal crabs (*Polyonyx*) also inhabit the tube. (*After Pearse, 1913.*)

morphologically different sexual individual, or **epitoke,** is common among errant polychaetes (Fig. 22-17). In *Syllis,* the hindmost sexual portion is budded off and develops a head and the nonsexual part forms a new tail end. In *Autolytus,* regeneration of segments begins before such separation, and the sexual part may divide several times to produce a chain of zooids. In *Syllis racemosa* there is lateral budding that yields branched chains of zooids.

The palolo worm, *Eunice viridis* (Fig. 22-17), of Samoa and Fiji, has its burrows in coral reefs and produces many posterior segments that become crowded with eggs or sperm. These are cast off as a unit by each worm, to rise and swarm at the surface

of the sea, when the water becomes milky with the millions of discharged gametes. Swarming occurs regularly on the first day of the last quarter of the October-November moon, continuing for 2 or 3 days. People gather large numbers of swarming worms, which they consider a delicacy.

The worm *Odontosyllis enopla* of Bermuda and the Florida coasts swarms on the second, third, and fourth days after full moon, appearing at the surface an hour after sunset. The female glows steadily and thus attracts the males, which flash intermittently while converging on her. During mating they rotate together, scattering eggs and sperm, after which the luminescence ceases abruptly.

The ARCHIANNELIDA (Fig. 22-20) are very small

Figure 22-17 *Eunice viridis,* **the palolo worm, illustrating the epitoke condition.** (*After Woodworth.*)

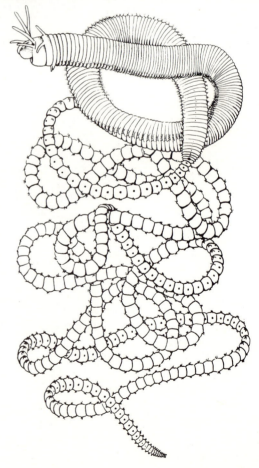

worms which are probably specially adapted for the interstitial environment. This is not a natural division of polychaetes but includes a few small families of uncertain relationships.

Class Hirudinea (Leeches)

The leeches are aquatic or terrestrial worms, of predatory or parasitic habits, that have enlarged terminal suckers for locomotion and attachment. Their annelid characteristics include a ventral nerve cord with segmental ganglia and the segmentally arranged nephridia and gonads in the reduced coelom. They resemble most closely the Oligochaeta in being monoecious with formed gonads in a few segments, in producing cocoons, and in lacking tentacles, parapodia, and a larval stage. They differ from all other Annelida in the lack of setae (except *Acanthobdella*) and in the presence of copulatory organs and of genital openings on the midventral line. Most are fluid-feeding animals, pumping blood or tissue fluids from various animals.

22-19 Structure The **body** of a leech at rest is long or oval in outline and usually flattened dorsoventrally; it is very flexible and may be greatly stretched, contracted, or dilated. It consists fundamentally of 34 **somites,** as shown by the nerve ganglia, but the exterior is marked by transverse furrows into many **an-**

nuli (one to five per somite). At the posterior end is a rounded **sucker,** formed from seven somites; another sucker (of two somites) surrounds the mouth at the anterior end in many species.

The body is covered by **cuticle** secreted by a single-layered **epidermis,** and many unicellular mucous glands open on the surface. Beneath is a **dermis** with pigment cells and blood capillaries. The **muscular system** is elaborate and powerful, with circular, longitudinal, oblique, and dorsoventral bands of fibers. Mesenchyme between the muscles and internal organs reduces the **coelom** to a system of canals and sinuses, from which the nephridia drain and in which the gonads and major blood vessels are located.

The **digestive tract** includes (1) the mouth, (2) a muscular pharynx with unicellular salivary glands and usually either a proboscis or three horny-toothed jaws, (3) a short esophagus, (4) a long crop with up to 20 pairs of lateral pouches or caeca, (5) a slender intestine, (6) a short rectum, and (7) the anus opening dorsally before the posterior sucker. The **circulatory system** is of longitudinal sinuses, dorsal, ventral, and lateral, with many cross-connections; pulsations in some of these cause the blood to circulate. **Respiration** is by a network of capillaries beneath the epidermis. **Excretion** is by up to 17 pairs of peculiar nephridia that sometimes are branched and sometimes have closed nephrostomes. The **nervous system** resembles that of other annelids, with a pair of dorsal ganglia and paired connectives to the ven-

Figure 22-18 The medicinal leech, *Hirudo medicinalis* **(class Hirudinea).** Internal structure as seen in dissection from the ventral surface; I to XXV, somites.

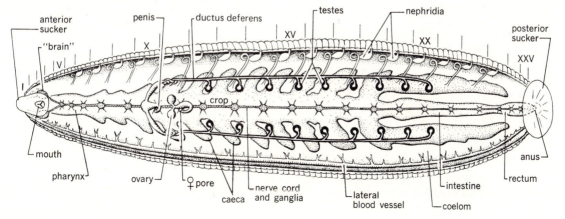

tral nerve cord, which has segmental ganglia. Four of the anterior ganglia and seven at the posterior end are fused. Each ganglion gives off several pairs of nerves. The sensory structures include taste cells in the mouth, tactile organs on the lips and body, one to five pairs of eyes anteriorly, and other eyelike organs (sensillae) on several annuli.

The sexes are united. The **male reproductive system** includes 4 to 12 pairs of testes beneath the crop; those on each side join a ductus deferens running anteriorly, and the two ducts enter a median penis to which accessory glands connect. The penis is within the male genital pore on the midventral surface and toward the anterior end of the body. The **female reproductive system** comprises two ovaries and oviducts joining a single albumen gland and a median vagina that opens just behind the male pore.

22-20 Natural history Most leeches inhabit fresh waters, a few occur on marine fishes, and some live in moist places on land (Fig. 22-19). They are largely nocturnal but may be attracted by food during the day. They travel by looping movements of the body—like a measuring worm—using the suckers for attachment, or they move like oligochaetes, using the body-wall muscles without employing the suckers; some also swim by graceful undulations of the body. Leeches are scavenging, predatory, or parasitic; some feed on dead animals, and others prey on small worms, insects, larvae, mollusks, etc. The group is best known for the blood-sucking habits of some species. These attach to the exterior or mouth cavity of various vertebrates, from fish to humans, some being restricted to one or a few host species. Such a leech fastens on by its suckers and pierces the skin, usually without being noticed, since there is no pain; then blood is sucked out by action of the muscular pharynx and stored in the distensible crop. A salivary enzyme (hirudin) prevents coagulation of the blood. At one feeding a leech may ingest several times its own weight of blood. Much of the liquid is soon excreted, but the concentrated part of such a "full meal" remains fluid

Figure 22-19 Class HIRUDINEA. *Left.* Leeches in fresh water. (*After Schmeil.*) *Right.* Successive stages *A–I* in locomotion of the medicinal leech. (*After Uexkull, 1905.*)

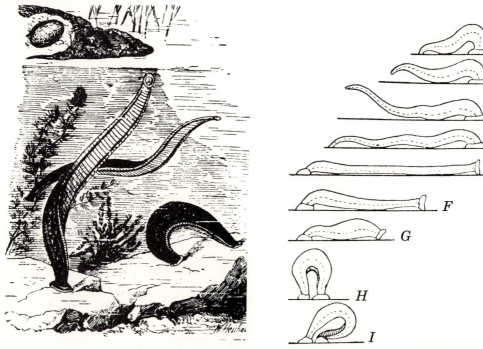

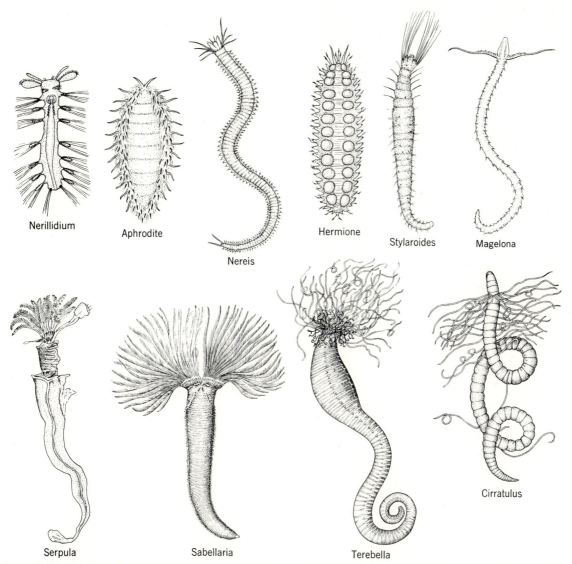

Nerillidium

Aphrodite

Nereis

Hermione

Stylaroides

Magelona

Serpula

Sabellaria

Terebella

Cirratulus

Figure 22-20 Class POLYCHAETA: some of the various marine polychaetes, not to the same scale. Subclass ARCHIANNELIDA: *Nerillidium;* subclass ERRANTIA: *Aphrodite, Nereis,* and *Hermione;* subclass SEDENTARIA: *Stylaroides, Magelona, Serpula, Sabellaria, Terebella,* and *Cirratulus. (After Grassé, Traité de Zoologie.)*

and is absorbed slowly over a period of several months. Digestion is by proteolytic enzymes and by slow decomposition due to action of a bacterium (*Pseudomonas hirudinis*). Because leeches consume large amounts of food when feeding and because digestion is so slow, they can exist for periods of a year or more between feedings.

Reproduction takes place in the warmer months.

Copulation and reciprocal fertilization occur in some (family HIRUDINIDAE); in others, packets of sperm (spermatophores) are deposited by one leech on the exterior body surface of another, whereupon the sperm penetrate to the ovaries and effect fertilization. Most leeches produce cocoons containing fertilized eggs, and these are placed either in water or in earth. In the GLOSSIPHONIIDAE the eggs are attached ventrally to the parent, where they develop and the young remain for a time. Leeches do not reproduce asexually, and they do not regenerate lost parts as do other annelids.

Leeches are eaten by various aquatic vertebrates and sometimes are used as fish bait. In North Temperate regions they are a minor nuisance to persons wading or bathing in some waters, but they seldom produce serious effects on humans or other large vertebrates. In some localities they may attach to the buccal cavity of humans or other mammals who are drinking. The land leeches of southeastern Asia and neighboring islands are extremely abundant and sometimes cause severe injury to human beings. Most famous is the medicinal leech (*Hirudo medicinalis*) of Europe. It grows to 10 cm (4 in) in length, can stretch 20 to 30 cm (8 to 12 in), and can ingest a large quantity of blood. Since ancient times it has been used for bloodletting. During the early nineteenth century this was a common, though erroneous, method of medical treatment. The medicinal leech was collected or reared in ponds in many parts of Europe.

Classification

Phylum Annelida.

Segmented worms. Body long, usually segmented and with paired setae; cuticle thin, nonchitinous; digestive tract complete, usually tubular; coelom usually large; blood system closed; paired nephridia; a pair of dorsal cerebral ganglia connected to a solid midventral nerve cord, with ganglia and lateral nerves in each somite; worldwide, marine, fresh water, and on or in soil; more than 10,000 species.

Class 1. Polychaeta.[1]

Sandworms, tube worms. Body segmented externally and internally, somites numerous, with lateral parapodia that bear many setae; a head region with tentacles; no clitellum; sexes usually separate; no permanent gonads; fertilization commonly external;

[1] *Most zoologists consider the polychaete families to be such a heterogeneous group that they cannot be satisfactorily grouped into orders, and most do not consider the* ERRANTIA *and* SEDENTARIA *to be good natural or phylogenetic divisions. The* ARCHIANNELIDA *are now acknowledged to be an unrelated group of puzzling polychaetes.*

a trochophore larval stage; asexual budding in some species; predominantly marine; more than 8000 species.

Subclass 1. Errantia. Somites alike except in head and anal regions; parapodia alike along entire body; pharynx usually protrusible, often with jaws; head distinct with sensory structures; free-living or in free or attached tubes; often predatory, some pelagic, *Neanthes* (*Nereis*), clamworm; *Eunice viridis*, palolo worm; *Aphrodite,* sea mouse; *Syllis, Autolytus,* with asexual budding; *Eurythöe,* fireworm; *Myzostomum,* with much parenchyma and no internal segmentation, parasitic or commensal on echinoderms. Freshwater forms: *Manayunkia,* New Jersey and Great Lakes; *Neanthes limnicola,* in Lake Merced, San Francisco, and northward.

Subclass 2. Sedentaria. Body of 2 or more regions with unlike somites; parapodia reduced; prostomium indistinct, head appendages modified or none; pharynx without jaws, usually nonprotrusible; gills anterior or none; in burrows or tubes, feed on detritus or plankton. *Chaetopterus,* no gills; *Cirratulus,* with long gills; *Arenicola,* with branched segmental gills and vestigial parapodia; *Hydroides, Spirorbis, Sabellaria, Serpula,* in tubes, with anterior gills.

Subclass 3. Archiannelida. Size small; segmentation chiefly internal; parapodia and setae usually absent; nervous system in epidermis; usually dioecious; gonads numerous; usually a trochophore larva; marine. *Polygordius, Dinophilus, Chaetogordius.*

Class 2. Oligochaeta.

Earthworms. Segmentation conspicuous externally and internally; no head or parapodia; setae usually few per somite; seldom with gills; monoecious, gonads few and anterior; with clitellum secreting cocoon for eggs; no larva, development direct; chiefly in fresh water and moist soil, 3100 species.

Order 1. Lumbriculida. Openings from the testes in the same segment as the testes; clitellum encompassing both male and female gonopores; 4 setae per segment. *Lumbriculus* in fresh water.

Order 2. Monilogastrida. Openings from the testes in the segment behind the one bearing the testes, but gonoducts do not pass through a septum; clitellum encompassing both male and female gonopores; 4 setae per segment. *Monilogaster,* terrestrial in Asia.

Order 3. Haplotaxiada. Male gonoducts pass through 1 or 2 septa.

Suborder 1. Haplotaxina. Two pairs of testes; 4 setae or 4 bundles of 2 sete each per segment. *Haplotaxis.*

Suborder 2. Tubificina. Numerous setae per bundle; male gonopores open on ovarian segment; single pair of testes. *Tubifex,* often in delicate tubes in the mud; *Enchytraeus,* in soil and sea beaches.

Suborder 3. Lumbricina. Male gonoducts pass through 2 septa; 8 setae per segment. *Lumbricus,* in soil; *Megascolides,* giant Australian earthworm.

Class 3. Hirudinea.

Leeches. Body pigmented, usually depressed; a large posterior sucker and often a smaller one at anterior end; no tentacles, parapodia, or setae (one exception); somites 34, subdivided externally into many annuli; coelom filled by connective tissue and muscles; monoecious; eggs usually in cocoons; no larva; in fresh water or salt water or on land; 500 species.

Order 1. Acanthobdellida. No anterior sucker, proboscis, or jaws; 2 pairs of setae each on somites II-IV, coelom segmented. *Acanthobdella,* one species on salmonid fishes in Lake Baikal, Siberia, and elsewhere in Russia and Finland.

Order 2. Rhynchobdellida. A protrusible proboscis; no jaws; blood colorless. *Branchellion,* with gills, marine, on rays and skates; *Piscicola,* on bony fishes; *Placobdella,* abundant on turtles and in fresh waters; *Glossiphonia,* on mollusks in fresh waters.

Order 3. Gnathobdellida. No proboscis; usually 3 chitinous jaws; blood red; includes most leeches. *Hirudo medicinalis,* European medicinal leech, introduced in eastern United States; *Haemopis marmorata,* horseleech, in mud, usually feeds on invertebrates; *Macrobdella decora,* to 300 mm (12 in) long, in fresh waters of Northern states, attacks humans, cattle, frogs, and fish, eats invertebrates; *Herpobdella; Haemadipsa,* tropical land leech, Southeastern Asia to Tasmania.

Order 4. Pharyngobdellida. No protrusible proboscis; 1 or 2 stylets but no jaws; blood red. *Erpobdella,* worm leeches and jawless land leeches.

References

Brinkhurst, R. O., and **B. G. Jamieson.** 1972. Aquatic Oligochaeta of the world. Toronto, University of Toronto Press. 860 pp., illus.

Clark, R. B. 1969. Systematics and phylogeny: Annelida, Echiura and Sipuncula. In M. Florkin and B. T. Scheer (editors), Chemical zoology, vol. 4, pp. 1–68. New York, Academic Press, Inc.

Dales, R. P. 1963. Annelids. London. Hutchinson & Co. (Publishers), Ltd. 200 pp., 19 figs. *Compact, modern, general account with references.*

Darwin, C. 1881. The formation of vegetable mould through the action of worms, with observations on their habits. London, John Murray. vi + 326 pp., 13 figs.

Day, J. H. 1967. The Polychaeta of southern Africa. London, The British Museum. Pt. 1, Errantia, xxix + 458 pp., illus. Pt. 2, Sedentaria, xvii + 421 pp., illus.

Edmondson, W. T. (editor). 1959. Ward and Whipple, Fresh-water biology, 2d ed. New York, John Wiley & Sons, Inc. Oligochaeta, by C. J. Goodnight, chap. 21, pp. 522–537, 16 figs.; Polychaeta, by Olga Hartman, chap. 22, pp. 538–541, 4 figs.; Hirudinea, by J. P. Moore, chap. 23, pp. 542–557, 13 figs.

Grassé, P.P. (editor). 1959. Traité de zoologie, Vol. 5, Paris, Masson et Cie. 784 pp., 574 figs. Polychètes, by P. Fauvel, pp. 13–196, figs. 4–163, 2 col. pls.; Archiannélides, by P. de Beauchamp, pp. 197–223, figs. 164–191; Oligochètes, by M. Avel, pp. 224–470, figs. 192–318; Hirudinées, by H. Harant and P.-P. Grassé, pp. 471–593, figs. 319–427, 1 col. pl.

Hartman, O. 1968. Atlas of the errantiate polychaetous annelids from California. 828 pp., illus. 1969. Atlas of the sedentariate polychaete annelids from California. 812 pp., illus. Los Angeles, Allan Hancock Foundation.

Laverack, M. S. 1963. The physiology of earthworms. New York, The Macmillan Company. ix + 206 pp., 58 figs.

Mann, K. H. 1962. Leeches (Hirudinea), their structure, physiology, ecology and embryology. New York, Pergamon Press. vii + 201 pp., 2 pls. (1 col.), 90 + 23 figs. International series of monographs on pure and applied biology, vol. 11.

Michaelsen, W. 1900. Oligochaeta. In Das Tierreich. Berlin, Friedlander & Sohn. 10th Lieferung. xxiv + 575 pp., 13 figs.

Pennak, R. W. 1953. Fresh-water invertebrates of the United States. New York, The Ronald Press Company. ix + 769 pp., 470 figs.

Stephenson, J. 1930. The Oligochaeta. Oxford, Clarendon Press. xv + 978 pp., 242 figs.

23

PHYLUM ARTHROPODA: GENERAL FEATURES, CHELICE-RATES, AND MINOR GROUPS

The phylum ARTHROPODA (Gr. *arthros,* joint + *podos,* foot) contains the great majority of the known animals, about 1 million species, and many of them are enormously abundant as individuals. It includes such common and well-known forms as the crabs, shrimps, barnacles, and other crustaceans (class CRUSTACEA), the insects (class INSECTA), the spiders, scorpions, ticks, and their allies (class ARACHNIDA), the centipedes (class CHILOPODA), and the millipedes (class DIPLOPODA), as well as a host of other less familiar and fossil forms. Ecologically the phylum is one of the most important, in that it dominates all terrestrial and aquatic ecosystems in numbers of species or individuals or both and in that a major share of the energy moving through these systems

passes through arthropod bodies. The body is segmented externally in varying degree, and the appendages are jointed, being differentiated in form and function to serve special purposes (Fig. 23-1); Table 23-1). All exterior surfaces are covered by an organic exoskeleton containing chitin. The nervous system, eyes, and other sense organs are proportionately large and well developed, making for quick response to stimuli. This is the only major invertebrate phylum with many members adapted for life in all areas on land, and the insects are the only invertebrates capable of flight.

Arthropods occur at altitudes of over 6000 m (20,000 ft) on mountains and crustaceans to depths of more than 9750 m (32,000 ft) in the sea. Different species are adapted for life in the air, on land, in soil, and in fresh water, brackish water, or salt water. Others are parasites of plants and on or in the bodies of other animals. Some are gregarious, and several kinds of insects have evolved social organizations with division of labor among members of different castes. Many arthropods are important economically. The larger crabs, lobsters, and shrimps are eaten by humans, small crustaceans are the major herbivores in the sea and the basis of most food webs, and insects and spiders are eaten by many land vertebrates. Although many insects are beneficial, others are the chief competitors of humans, eating their crops, stored foods, household goods, or clothing. Some insects and ticks injure or carry diseases to humans, domestic animals, and crops.

23-1 Characteristics

1 Symmetry bilateral; three germ layers; body usually segmented and jointed externally; head, thorax, and abdomen variously distinct or fused (head somites always fused).

2 Appendages one pair per somite or less, each with few to many hinge joints[1] and containing opposed sets of muscles; variously differentiated, sometimes reduced in number or parts, rarely lacking.

3 A hardened exoskeleton containing chitin, secreted by the epidermis and molted at intervals.

4 Muscles striated, often complex, usually capable of rapid action.

5 Digestive tract complete; mouth parts with lateral jaws adapted for chewing or for sucking; anus terminal.

6 Circulatory system open (lacunar); heart dorsal, distributing blood by arteries to organs and tissues, whence it returns through body spaces (hemocoel) to the heart; coelom reduced.

7 Respiration by gills, tracheae (air ducts), book lungs, or body surface.

---

[1] For brevity the divisions or segments of appendages are referred to as "joints"—three-jointed, etc.

8 Excretion by coxal or green glands or by two to many Malpighian tubules joined to gut.

9 Nervous system with paired dorsal ganglia over mouth and connectives to a pair of ventral nerve cords, with a ganglion in each somite or ganglia concentrated; sensory organs include antennae and sensory hairs (tactile and chemoreceptive), simple and compound eyes, auditory organs (INSECTA), and statocysts (CRUSTACEA).

10 Sexes usually separate, male and female often unlike; fertilization mostly internal; eggs with much yolk, in shells; either oviparous or ovoviviparous, cleavage usually superficial; usually with one to several larval stages and gradual or abrupt metamorphosis to adult form; parthenogenesis in some crustaceans and insects.

23-2 Phylogeny and relationships

The arthropods stand at the pinnacle of the protostome line of evolution and are dominant in the present ecologic systems of the world. Furthermore, the fossil record of the group is extensive, extending back to Cambrian times. Unfortunately this fossil record lacks forms which would link together or show relationships among the various arthropod groups; hence there has been disagreement among zoologists as to certain aspects of the origin and evolution of the phylum.

Most zoologists agree that the arthropods probably arose from some primitive polychaete group or at least from some segmented ancestral form common to both stocks. The ARTHROPODA and ANNELIDA still share a number of characteristics which reflect their common origin. They are the only two phyla with conspicuous segmentation; both groups, at least primitively, also had paired appendages on each segment; and the nervous system is similar, with dorsal brain and ganglionated ventral nerve cord. Additional similar features include an epidermally produced cuticle, a tubular digestive tract running from an anterior mouth to a posterior anus, and the concentration of sense organs in the head. The major differences between annelids and arthropods reflect the different structural and physiologic features associated with the soft, hydraulically operated annelid body and the hard, lever-operated arthropod body. Major features present in arthropods but not in annelids include (1) great reduction of the coelom, (2) presence of a hard, inflexible exoskeleton, (3) an open lacunar circulatory system, (4) absence of cilia in all organ systems, (5) jointed appendages, and (6) compound eyes.

The phylum ONYCHOPHORA (Sec. 20-34) shows a mixture of arthropod and annelid features. It has segmental nephridia, soft body, and uniramous appendages like a polychaete but lacks internal septa and has antennae and jaws like an arthropod. With this single exception, however, no living or fossil forms show relationships between the two groups or, in the case of the arthropods, even among the major divisions within the phylum.

Traditionally three major subphyla of the arthropods have been recognized: CHELICERATA, †TRILOBITA, and MANDIBULATA. In the past these three have been assumed to be natural groupings, and the whole phylum has been assumed to have a monophyletic origin. Under this scheme the †TRILOBITA would be the most primitive, followed by the CHELICERATA, leaving the MANDIBULATA as the most highly evolved. Recently this concept has been challenged by some zoologists because there is little evidence relating the three major groups, even in the fossil record, and there appear to be major differences between the two dominant groups in the

MANDIBULATA, namely, the CRUSTACEA and the INSECTA. This has led some zoologists to suggest that the ARTHROPODA are not monophyletic but polyphyletic in origin. In such a scheme, the trilobites and chelicerates seem to form one natural phyletic grouping, the CRUSTACEA another, and the INSECTA and two small classes, DIPLOPODA and CHILOPODA, a third (Table 23-1). Such an arrangement resolves particularly the great disparity between the two mandibulate groups. While retaining the three subphyla for convenience, our coverage here of the biology of the ARTHROPODA will follow the polyphyletic subdivision, because it gives a more natural arrangement and combines forms with similar structure and physiology. The sequence of groups is mainly one of convenience, although the trilobites and crustaceans are probably the most ancient. The trilobites must have originated in Precambrian time, because fossils are abundant and diverse in Cambrian rocks.

23-3 General features The dominant arthropod feature, and the one most responsible for the structural and physiologic features unique to the group as well as for the success of the group, is the presence of a hard chitinous exoskeleton. In contrast to the thin, flexible annelid cuticle, the arthropods have a multilayered thick cuticle which gives rigidity to the

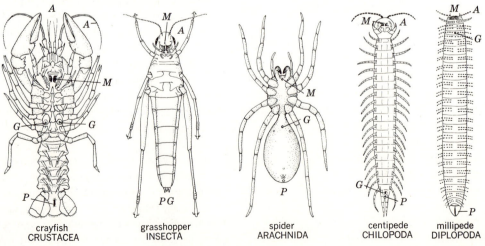

Figure 23-1 Phylum ARTHROPODA: examples of five major groups in ventral view, showing body divisions, somites, appendages with their divisions, and the body openings. *A*, antennae; *M*, mouth; *P*, anus; *G*, genital opening(s).

| crayfish CRUSTACEA | grasshopper INSECTA | spider ARACHNIDA | centipede CHILOPODA | millipede DIPLOPODA |

Table 23-1

PHYLUM ARTHROPODA: GENERAL CHARACTERISTICS OF THE PRINCIPAL CLASSES

	Crustacea (crustaceans)	Insecta (insects)	Arachnida (spiders, etc.)	Chilopoda (centipedes)	Diplopoda (millipedes)
Body divisions	Usually cephalo-thorax and abdomen	Head, thorax, abdomen	Cephalothorax and abdomen	Head with body of similar segments	Head, short thorax, long abdomen
Paired appendages:					
Antennae	2 pairs	1 pair	None	1 pair	1 pair
Mouth parts	Mandibles Maxillae, 2 pairs Maxillipeds	Mandibles Maxillae, 1 pair Labium	Chelicerae Pedipalpi	Mandibles Maxillae, 2 pairs	Mandibles Maxillae, 1 pair
Legs	1 pair per somite or less	3 pairs on thorax (+wings)	4 pairs on cephalo-thorax	1 pair per somite	2 pairs (or 1) per somite
Respire by	Gills or body surface	Tracheae	Book lungs or tracheae	Tracheae	Tracheae
Sex openings	2, hind part of thorax	1, end of abdomen	1, second somite of abdomen	1, end of abdomen	1, third somite near head
Development	Usually with larval stages	Usually with larval stages	Direct, except mites and ticks	Direct	Direct
Principal habitat	Salt water or fresh water, few on land	Mainly terrestrial	Mainly terrestrial	All terrestrial	All terrestrial

body. The cuticle itself consists of an outermost layer of nearly impervious proteins and waxes called the **epicuticle** and a thicker inner layer, or layers, called collectively the **procuticle,** composed of layers of chitin, a nitrogenous sugar, and often containing pigments and calcium carbonate. The cuticle is noncellular and is secreted by the underlying epidermis. The cuticle may vary in thickness in different species or in different body regions of the same species. In order to allow movement, the cuticle is thin and flexible between joints in the limbs and between body parts such as adjacent segments. Throughout the body the cuticle bears various sensory setae which connect to underlying cells, thus permitting the animal to obtain information. The cuticle also often bears pores opening from underlying gland cells, allowing passage of secretions to the outside.

Because the cuticle is rigid and also virtually impermeable to gases and liquids, it serves both as a means of support and protection for the arthropod and also as a restriction to exchange of gases, nutri-ents, and liquids. It is the impermeability of the cuticle which has contributed greatly to the success of the arthropods on land, because the cuticle has prevented water loss, a major problem for animals in terrestrial environments.

Another advantage of the rigid exoskeleton is in locomotion. The presence of a series of rigid skeletal units in each limb joined together by flexible articulating membranes has made possible very efficient locomotion. Where annelid movement depends on body-wall muscles acting indirectly through coelomic fluid, the arthropods move by using muscles directly to move adjacent rigid skeletal units. The exoskeleton provides a solid place for muscle attachment. Muscles are attached across an articulating membrane to two rigid limb or body pieces in antagonistic pairs (extensor and flexor). Contraction of the flexor muscle draws the distal unit of the limb inward, and contraction of the extensor moves the limb outward. Contraction of these muscles acting across many such joints in limbs leads to extremely precise and accurate movement when coordinated by the more highly evolved nervous system. The locomotion is thus very similar to that of vertebrates in which sets of muscles operate limbs across joints, except that here the rigid skeletal pieces are external, not internal (Fig. 23-1). Such a system of locomotion is suitable not only in water but also on land, because the rigid skeletal pieces serve to support the body as well—another reason for the success of the arthropods on land.

Since the exoskeleton is inflexible and nonliving and covers the entire body, it presents one serious problem in that it does not allow for growth. In order for an arthropod to grow, the old skeleton must be shed at certain intervals and be replaced by a newer, larger one. The process of shedding an old exoskeleton and forming a new one is called **molting,** or **ecdysis.** In molting, the underlying epidermis first detaches from the old procuticle, then secretes enzymes which partially digest away some of the old inner procuticle without affecting muscle and nerve connections. A new cuticle is then produced but remains soft. The old skeleton then splits along certain lines and the animal slips out. While the new cuticle is still soft and expandable, the arthropod expands its body by pumping it up with air or water. After hardening of the new cuticle, the animal replaces the air or water with real tissue growth.

The entire molt cycle is controlled by the endocrine system. At least two major hormones are involved, **ecdysone,** which initiates the cycle, and another hormone which inhibits molting (Sec. 8-2).

In contrast to the annelids, the arthropods are all characterized by a greatly reduced coelom. The only remaining manifestations of the coelom are the cavities in the gonads and excretory organs. This reduction in size and importance of the coelom may be related to the abandonment of hydraulically assisted locomotion. As with the mollusks, the only fluid-filled body spaces are hemocoels or blood spaces associated with the characteristic open circulatory system. The heart is now an elongated tube lying not in a pericardium but in a hemocoel space on the dorsal side of the body. Its many openings, or ostia, allow blood to flow in and to be pumped out to other hemocoels. The blood itself is similar to annelid blood, with respiratory pigments, if present, dissolved in the plasma.

Most arthropods show a high degree of concentration and development of the central nervous system. Correlated with this is the appearance of well-developed sense organs, including large eyes, antennae, and sensory bristles and hair. Concomitant with the large brain, the arthropods have developed complex behavioral patterns, including societal organizations.

Arthropods are usually dioecious animals. The terrestrial forms have internal fertilization, but the aquatic species may have either internal or external fertilization. Arthropod eggs usually have large amounts of yolk in them, which precludes typical protostome holoblastic determinant cleavage. Instead, a special superficial cleavage occurs which results in the surrounding of the inner nondeveloping yolk cells by the developing tissue layers. Development is thus more akin to that seen in birds (Sec. 10-17, Fig. 25-11). Most arthropods have a larval stage, and the adult stage is reached through metamorphosis.

23-4 Size Because of the limiting weight of the exoskeleton, no arthropod is of great size. Some fossil eurypterids grew to be 3 m (9 ft) long, and a living Japanese crab (*Macrocheira kaempferi*) spreads to 3.5 m (12 ft) with its slender legs. The Atlantic lobster is recorded to 60 cm (24 in) in length (weight to 15 kg, or 34 lb), and a living isopod (*Bathynomus giganteus*) to nearly 35 cm (14 in). No living insect exceeds 28 cm (11 in) in length or spread. The smallest crustaceans, insects, and mites are under 1 mm long.

23-5 Classification Because there are so many arthropods, it is useful to outline initially the larger taxonomic subdivisions so that one can have a framework in which to place the various forms. The phylum is divided into three major subphyla: †Trilobita, Chelicerata, and Mandibulata. The trilobites are all extinct but probably represented the most primitive arthropods. They lacked specialized head appendages. The chelicerates differ from the mandibulates in two primary ways: chelicerates do not have antennae, and their first appendages on the anterior end are pincerlike chelicerae. Mandibulates all have antennae as the first appendages on the anterior end. The Chelicerata include two small classes of marine organisms, PYCNOGONIDA and MEROSTO-MATA, but the class ARACHNIDA (spiders, ticks, and mites) constitutes the bulk of living chelicerates. The MANDIBULATA is by far the largest subphylum and is dominated by two classes: CRUSTACEA and INSECTA. CRUSTACEA can be distinguished from INSECTA by their two pairs of antennae on the head; the insects have only one pair. The class CHILOPODA is characterized by one pair of legs per segment while DIPLOPODA have two pair per segment. The classes PAUROPODA and SYMPHYLA are small terrestrial arthropods of uncertain affinities. A detailed higher classification of the ARTHROPODA is given at the end of this chapter.

Subphylum †Trilobita (Trilobites)

23-6 Structure Trilobites are the most primitive of all the arthropod groups and therefore an appropriate taxon with which to begin a description of the phylum. These primitive marine arthropods, quite abundant during Paleozoic times, had a three-lobed body divided by two lengthwise furrows. They were covered by a hard segmented shell that could be rolled up. On the distinct head, of four fused segments, was a pair of slender antennules (antennae), four pairs of biramous appendages, and often a pair

of compound eyes. The thorax consisted of 2 to 29 short separate somites, and the abdominal somites were fused in a caudal plate, or pygidium. All somites but the last bore similar biramous limbs consisting of an inner walking limb and an outer limb bearing filaments which may have been respiratory. The fossil record for these animals is so good that a series of developmental stages is also known. They appear to have gone through three larval periods. The larva hatched as a pronauplius (protaspis) that gained somites in successive molts. After several molts a merapsis stage was reached, in which the body regions became more defined. Finally a holaspis larva emerged which appeared like the adult. Details of structure and growth have been learned from many entire and sectioned specimens, especially of †*Triarthrus becki* from the Ordovician shale near Rome, N.Y. (Figs. 13-10, 23-2).

Trilobites were differentiated in Precambrian time, were at a peak in the Cambrian and Ordovician, and died out in the Carboniferous except for one Permian genus. More than 4000 species have been described. The adults were usually 50 to 75 mm (2 to 3 in) long, but different species were from 10 to 675 mm (0.4 to 27 in) in length. Many evidently burrowed or pushed through the bottom mud and sand of ancient seas; the larvae and some adults probably swam. Some species were local in occurrence and others were cosmopolitan.

Subphylum Chelicerata (Spiders, horseshoe crabs, and sea spiders)

The CHELICERATA (Gr. *chele*, claw + *keros*, horn) are a varied assemblage comprising the spiders, mites, ticks, harvestmen, scorpions, whip scorpions, pseudoscorpions, horseshoe crabs, and others (Figs. 23-3, 23-4). Members of this group differ from mandibulates in body form and the nature of their appendages. They are mostly free-living terrestrial animals of small size, and the great majority are more numerous in warm dry regions than elsewhere. Many have poison glands and poison jaws or fangs by which they kill insects and other small animals, whose fluids and soft tissues they suck as food. The spiders and some other arachnids also have special glands that secrete fine threads of silk. This material

Figure 23-2 Subphylum †TRILOBITA: a trilobite (†*Triarthrus becki*) of Ordovician time. (*After Beecher.*)

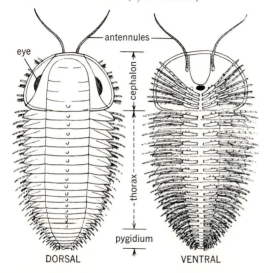

eye
antennules
cephalon
thorax
pygidium
DORSAL   VENTRAL

is used to build nests, shelters, and egg cases and for other purposes.

Some members of this subphylum, because of their peculiar appearance, are feared by many persons, although the great majority are quite harmless. Spiders and some others are actually beneficial to humans because they feed on various kinds of insects. A few spiders and scorpions, however, have bites or stings that may cause serious human illness and even death. Some mites injure plants, and certain mites and ticks which parasitize humans and other animals produce injury, sickness, and death. Several kinds of ticks are the intermediate hosts for protozoans and viruses that cause specific diseases (Table 25-1).

23-7 Characteristics[1]

1 Body usually of distinct cephalothorax (prosoma) and abdomen (copisthosoma) (except ACARINA); typically with six pairs of jointed appendages: chelicerae, pedipalpi, and four pairs of legs, all on cephalothorax; no antennae or mandibles.

[1] Compare ARTHROPODA, Table 23-1.

2 Mouth parts and digestive tract mainly suited for sucking; some with poison glands.

3 Respiration by book lungs, tracheae, or book gills.

4 Excretion by paired Malpighian tubules or coxal glands, or both.

5 Nervous system with dorsal ganglia (brain) and ventral nerve cord having paired ganglia or else concentrated anteriorly; eyes usually simple and in pairs; tactile hairs or bristles on body.

6 Sexes mostly separate; sex opening one (or two), anterior on abdomen; fertilization usually internal; mainly oviparous; development direct or through a larval stage.

7 Chiefly terrestrial and solitary; either free-living and predaceous or parasitic.

The CHELICERATA are of ancient origin, some stocks having been developed in Cambrian time. The lack of antennae and mandibles is a distinctive characteristic. The TRILOBITA of the Paleozoic differed in having biramous appendages and one pair of antennae. Among the MANDIBULATA (with mandibles), the CRUSTACEA have biramous appendages and two pairs of antennae; the INSECTA have wings (absent in some) and three pairs of legs; and the

Figure 23-3 Class ARACHNIDA: some common members in their characteristic habitats; not to scale. Order ARANEAE, spiders: *Salticus*, jumping spider; *Latrodectus*, black widow spider; *Bothriocyrtum*, trap-door spider; *Dugesiella (Aphonopelma)*, American "tarantula"; *Lycosa*, hunting or wolf spider; *Argiope*, orb-weaving spider with egg case. Order SCORPIONIDA: *Vejovis*, scorpion. Order UROPYGI: *Mastigoproctus*, whip scorpion (vinegarone). Order SOLPUGIDA: *Eremobates*. Order PSEUDOSCORPIONIDA: *Apocheiridium*, false scorpion. Order OPILIONES: *Phalangium*, harvestman. Order ACARINA: *Dermacentor*, tick.

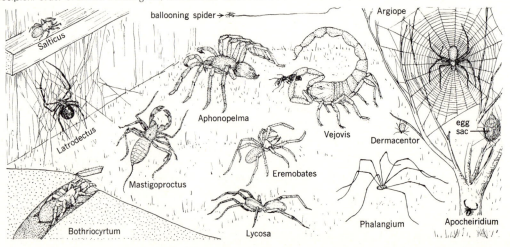

remaining classes are slender, with one or two pairs of jointed legs on each segment. None of the other small arthropod groups is likely to be confused with the CHELICERATA.

23-8 Size The horseshoe crab (*Limulus*) grows to 500 mm (20 in) in length. Most spiders are under 25 mm (1 in) long, the extremes being *Microlinypheus,* which is under 1 mm and *Theraphosa leblondi,* which grows to 90 mm (3½ in) long. The smallest scorpion (*Microbuthus pusillus*) is 13 mm (½ in) long and the largest centipede (*Scolopendra gigantea*) 312 mm (12½ in). The SOLPUGIDA are 8 to 70 mm (0.3 to 3 in) long, *Galeodes caspius* being the largest. The smallest mites are under 0.5 mm long, and the largest tick (*Amblyomma*) is only 30 mm (1 in) in length.

Class Arachnida (Spiders and their allies)

The arachnids (Gr. *arachne,* spider) include spiders, scorpions, pseudoscorpions, whip scorpions, sun spiders, ticks, mites, and a few obscure groups. In general they are terrestrial, with two pairs of mouth parts (chelicerae and pedipalps) and four pairs of legs, thus differing from the other chelicerate classes —the MEROSTOMATA and PYCNOGONIDA.

23-9 General features and evolution The arachnids are by far the largest and most successful group of chelicerates. With few exceptions they are terrestrial, but they are believed to have evolved from aquatic forms similar to eurypterids. The fossil record includes water scorpions dating back to the Silurian period. The transition to a terrestrial existence for arachnids probably came early in geologic history, and they may have been among the first of the phylum to inhabit dry land. As a result of this transition to a terrestrial existence, certain changes occurred in the anatomy and physiology of the group. One such change is in the reproductive system, where, to prevent water loss, fertilization of the eggs is internal and eggs are protected against desiccation by being deposited in humid burrows, retained in the female (viviparity), or protected by an outer covering. Similarly, free-swimming larval forms are not possible, and larval stages are either passed in the egg, necessitating large amounts of yolk, or are parasitic on other animals. Other adaptations included development of a more impermeable exoskeleton to reduce water loss and the conversion of the original book gills to air-breathing book lungs or a tracheal system.

In arachnids the body is in two parts, an anterior **cephalothorax,** or **prosoma,** and a posterior **abdomen,** or **opisthoma.** Only the cephalothorax bears appendages, of which there are six pairs. The most anterior pair are the fanglike **chelicerae,** which are employed in feeding. The second pair are **pedipalps,** which may be variously modified and have different uses. The remaining four are walking legs. The greater number of arachnids are predaceous, feeding on other arthropods, which are seized and held with the chelicerae and/or pedipalps. Many have developed venom glands associated with chelicerae or other parts of the body, which are used to immobilize the prey while enzymes are secreted into it to

Figure 23-4 Subphylum CHELICERATA. Representatives of eight groups in ventral view (not to scale), showing general form, division between cephalothorax and abdomen (broken line), chelicerae (black), pedipalpi (outlines), and legs (stippled), with joints in each. (Compare Fig. 23-1.)

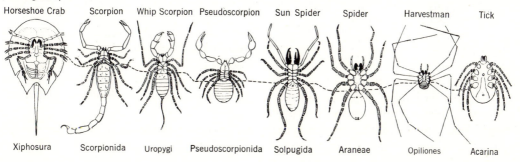

| Horseshoe Crab | Scorpion | Whip Scorpion | Pseudoscorpion | Sun Spider | Spider | Harvestman | Tick |

| Xiphosura | Scorpionida | Uropygi | Pseudoscorpionida | Solpugida | Araneae | Opiliones | Acarina |

begin the digestive process and reduce it to a semidigested liquid or slurry. Arachnids have no mandibles and ingest only liquid material or small particles. The digestive system is modified to deal with liquid food, and digestion is completed internally.

The chief metabolic waste produced in arachnids is guanine, which is removed via the coxal glands, thin-walled sacs opening into the coxae of the appendages, or the Malpighian tubules, which are slender tubes absorbing material from the hemocoels and discharging into the intestine.

The nervous system is concentrated, with a large dorsal brain. This large brain is correlated with the often complex behavior patterns seen among arachnids, including nest and web building, courtship behavior, and parental care of young. Sense organs include sensory hairs, eyes, and sensory slits. Eyes of arachnids are simple ocelli comprising a combined cornea and retina. Eyes may be constructed as direct, with the light-sensory cells facing toward the light source, or indirect, wih the sensory cells facing the membrane behind the eye. Sensory hairs are abundant and function to provide the animal with information about surrounding objects, gravity, air movement, surface features, etc. Slit organs appear to function in detecting sound vibrations and proprioception.

Respiration in arachnids is by tracheae or book lungs or a combination of both. Book lungs are more primitive and consist of a pleated invagination of the ventral abdominal wall which forms a series of lamellae separated by bars. Gas exchange occurs across the lamellae. The tracheae are similar to those of insects (Sec. 25-8), as is the circulatory system.

All arachnids are dioecious animals, often with sexual dimorphism. Sperm are often packaged as spermatophores and transferred to the female after a more or less elaborate courtship. Parental care of the young is developed in many arachnids.

23-10 Order Araneae, the spiders   More than 30,000 species of spiders are known. They live in various habitats, from sea level to the highest mountains, from the seashore and freshwater swamps to the driest of deserts, among rocks, in or on sandy or other soils, in forests, bushes, and grasses, and about buildings. The following general account applies to most common species.

23-11 External features[1]   The body consists of a distinct **cephalothorax** and an **abdomen**, both rounded and usually unsegmented and joined by a slender "waist," or **peduncle**. The cephalothorax commonly has eight simple **eyes** anteriorly, and its ventral surface bears six pairs of appendages. Each of the two **chelicerae** has a basal segment and a terminal clawlike fang with a duct near its tip, connecting to a poison gland within the cephalothorax. The pair of **pedipalpi** are short, six-jointed, and leglike, with enlarged bases that form "maxillae" used to squeeze and chew the food; in mature males the tip becomes a specialized container for the transfer of sperm. There are four pairs of **walking legs,** each of seven joints (coxa, trochanter, femur, patella, tibia, metatarsus, and tarsus) and ending in two or three toothed claws; on some spiders the tip bears a pad of "hairs" (scopula) by which the animal may cling to a wall or similar surface. All external parts are covered by cuticle bearing many bristles and hairs, some of a sensory nature.

The external openings are as follows: (1) The minute **mouth,** between the maxillae; all others are ventral on the abdomen and include (2) the **genital opening** anteriorly in the midline and covered by a plate (epigynum) in females; (3) the slitlike entrances to the two or four **book lungs,** or respiratory organs, at each side of the epigynum; (4) a **spiracle** anterior to the anus and connecting to short tracheae; (5) the two or three pairs of bluntly conical **spinnerets,** sometimes jointed, each with many fine tubes through which the "silk" for webs and other purposes is secreted from glands within the abdomen; and (6) the terminal **anus.**

23-12 Internal structure   (Fig. 23-5) The **digestive system** comprises (1) the mouth; (2) a slender esophagus connecting to (3) a sucking stomach operated by muscles extending from its dorsal surface to the cephalothorax; (4) the main stomach in the

---

[1] *Details of structure may be examined closely on a live spider enclosed in a petri dish under a binocular microscope.*

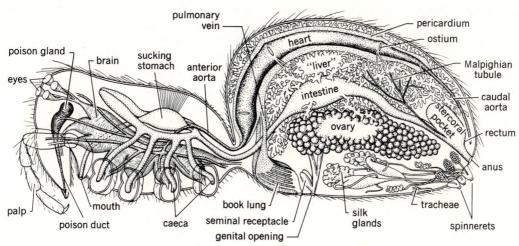

Figure 23-5 Class ARACHNIDA: structure of a spider as seen with left side of body removed. (*Modified from Leuckart.*)

cephalothorax, with five pairs of caeca, or pouches, one dorsal and one toward each leg; (5) the straight intestine in the abdomen, which is joined by ducts of (6) the much-branched digestive gland ("liver") and extends to (7) the rectum, where (8) an enlarged stercoral pocket connects just before (9) the anus.

The **circulatory system** is somewhat like that of insects; the heart is a slender, muscular, contractile tube dorsally placed in the abdomen, with three pairs of openings, or ostia, and surrounded by a tubular pericardium. From the heart a caudal aorta extends backward, and an anterior aorta sends paired arteries to the stomach, legs, eyes, and poison glands. The colorless blood contains amoeboid corpuscles and dissolved hemocyanin as a respiratory pigment. The heart pumps blood through the aortae and into sinuses among the tissues; thence it passes to the book lungs for aeration and returns by "pulmonary veins" to the pericardial cavity, to reenter the heart through the ostia.

Respiration is accomplished mainly by the **book lungs,** which are peculiar to arachnids; each consists of 15 to 20 leaflike horizontal plates containing fine blood vessels. Air entering the external slit on the abdomen circulates between the plates, where the O_2—CO_2 exchange occurs. The **tracheae,** if present, are like those of insects but restricted to the abdomen. **Excretion** is performed by paired **Malpighian tubules** connected posteriorly to the intestine and by a pair or two of **coxal glands** in the floor of the cepha-

lothorax that empty by ducts between the legs. These glands are considered homologous with the green glands in crustaceans.

The **nervous system** is concentrated, comprising a bilobed ganglion above the esophagus that joins by two thick connectives to a large ventral ganglionic mass, whence nerves radiate to all organs. Paired ganglia occur in the abdomen of young spiders but not in adults. The **eyes** are simple, with a chitinous lens, epithelial layer, optic rods, and retinal cells; vision is keener in some spiders than in others. The pedipalpi and many of the external hairs are evidently sensitive to touch. The response to sound is uncertain, although some spiders have sound-producing mechanisms. The sense of smell seems well-developed and may reside in minute "lyriform organs" on the body and appendages.

The sexes are separate and often unlike in markings and size, females being larger. The **male** has two testes below the intestine that join by coiled ductus efferentia to a single seminal vesicle leading to the genital opening. In the **female** the two ovaries are large and hollow, each with an oviduct joining the single vagina; two lateral seminal receptacles attach to the vagina.

23-13 Natural history Spiders are free-living, commonly solitary, and predaceous, feeding mainly on insects. Hunting spiders either wait for or wander in search of food, run it down (*Lycosa*), or jump on it (*Salticus*). Other spiders entrap their prey in webs. Small prey is grasped, killed by the poison-bearing fangs, and "eaten"—actually, while being held, it is partly digested by enzymes poured out from the gut. Larger items of food may first be bound in silk shroud or fastened to the web and then killed. The poison kills invertebrates quickly, and that of some large spiders (*Eurypelma*) will overcome small vertebrates. When food is available spiders feed often, but in captivity some may fast for many months. In the majority of species individuals live only about a year, but some large "tarantulas" have lived in captivity for 25 years. Spiders are attacked mainly by birds, lizards, and members of two orders of insects (HYMENOPTERA and DIPTERA). Some wasps are external parasitoids of spiders, while others are solitary egg parasitoids. One family of flies (ACROCERIDAE) are solitary internal parasitoids of immature spiders.

Spider silk is a proteid secretion of special abdominal glands that passes out the many microscopic tubes in the spinnerets and solidifies into a thread upon contact with the air (Fig. 23-6); it serves many purposes. Terrestrial hunting spiders pay out a "dragline" as they travel. Some spiderlings are dispersed by climbing to some height and spinning a long thread on which they are carried by the wind; trees sometimes become covered by such gossamer,

and numbers of "ballooning" young spiders are occasionally seen by ships at sea. The most primitive webs are scarcely more than many draglines radiating from a spider's retreat (*Segestria*). Some (*Amaurobius*) add a sheet of fine silk "carded" by a plate (cribellum) anterior to the spinnerets, and others form a hammocklike sheet over a definite framework (*Tegenaria*). Some webs are irregular networks of silk (*Latroductus*). The orb weavers (*Argiope, Epeira, Miranda*) outline a flat rectangular framework within which radial threads are neatly spaced; then a spiral of silk, bearing sticky droplets, is attached for the capture of prey. The webs are repaired or renewed at frequent intervals. Snares, shelters, nests for hibernating, molting, or mating, and egg cocoons are spun with various kinds and colors of silk by different species.

23-14 Reproduction   The male, when mature, may spin a small web on which a droplet of fluid containing sperm is deposited and then taken into cavities in his pedipalps (Fig. 23-6). He then seeks a female and may go through a nuptial performance with her before transferring the sperm by inserting his pedipalps in her genital opening. The female may kill and eat the male after mating, but this is unusual. Later she spins a padded cocoon in which the eggs are laid. Females of some species fasten the cocoon to or near the web, and others carry it about; the female *Lycosa* carries the young on her abdomen for some days after they hatch. The black widow (*Latrodectus mactans*), for example, lays 25 to over 900 eggs in a cocoon and produces one to nine cocoons per season. Her eggs hatch in 10 to 14 days, and the young remain in the sac for 2 to 6 weeks, leaving after the first molt. Males undergo about five molts before becoming sexually mature, and females seven or eight. At successive molts the spiders increase in size and change in form, proportions, and color pattern.

Figure 23-6 Structural details of spiders. *A*. Front of head. *B*. Enlarged pedipalpi of male. *C*. End of leg with claws hooked on web. *D*. Spinnerets on abdomen. *A, D*. Ctenid spider. *B, C*. Black widow spider.

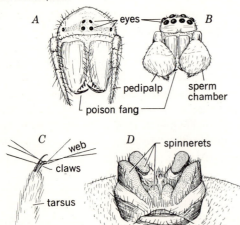

23-15 Other Arachnida   (Figs. 23-3, 23-4, 23-7) The scorpions (SCORPIONIDA) are elongate, with big pincer-ended pedipalpi, small chelate chelicerae, and a slender 12-segmented abdomen bearing a terminal sharp poison sting. The prosoma is short and covered with a carapace which has a pair of eyes in

Figure 23-7 Female scorpion with young. *(Courtesy of California Academy of Sciences.)*

the middle of the dorsal surface. Other smaller eyes are found along the anterior border of the carapace. The four walking legs attached to the prosoma terminate in two pairs of claws. The second abdominal segment bears a pair of comblike structures called **pectines,** unique to scorpions and probably sensory in function. Scorpions are most abundant in tropical and subtropical areas, hidden under stones or in shallow burrows by day and running actively at night to catch the insects, spiders, and scorpions used for food. Prey is grasped by the pedipalps and torn slowly apart by the chelicerae, larger animals first being paralyzed by the sting. Mating is preceded by a courtship dance, during which the male fixes a spermatophore to the ground and then maneuvers the female over it to receive the ejected sperm. The female produces living young, which ride for some days on her abdomen.

Whip scorpions (Uropygi) somewhat resemble scorpions but have a slender abdominal whip and lack a poison sting. They have large glands in the abdomen which open near the anus. These glands produce a secretion which is mainly acetic acid (84 percent) combined with caprylic acid (15 percent). When disturbed the animals spray the attacker with the fluid, which can produce acid burns on humans. The odor is strongly of vinegar, hence they are called *vinegarones.* The large pedipalpi have pincers, and the first legs are tactile. The whip scorpions live in

warm countries, are nocturnal, and prey on insects. The pseudoscorpions (Pseudoscorpionida) are like miniature scorpions, but they lack the long abdomen, the sting, and the pectines and have tracheae. The pedipalps resemble those of scorpions except they have poison glands in them. Silk from glands issuing on the chelicerae is spun into nests for molting and hibernation. The pseudoscorpions live under stones, moss, or bark, and some in books or furniture; their food is minute insects. Sometimes they are abundant—150,000 per acre—but are rarely seen because of their small size. The sun spiders (Solpugida) are of spiderlike form but with the prosoma divided into two sections and the pedipalpi leglike. They are characterized by enormously enlarged chelicerae in the form of pincers, making them look formidable. However, they lack both poison and silk glands. Their habitat is warm dry regions. Harvestmen, or daddy longlegs (Opiliones), have compact ovoid bodies with no constriction between the prosoma and opisthosoma and extremely long legs by which they run readily. They lack poison but have "stink glands" for protection. They are common in temperate regions, and they appear to be more omnivorous than other arachnids.

The mites and ticks (order Acarina) are small to microscopic, with the head, thorax, and abdomen fused closely and unsegmented. Over 25,000 species are known, and the group is virtually ubiquitous on

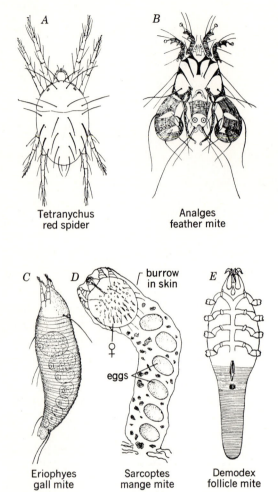

Figure 23-8 Order ACARINA: some representative mites; all enlarged, not to same scale. (*After Banks, 1905.*)

land. Many are parasites, often on humans or their livestock and crops, hence this group is the most important of the arachnids from the human point of view. Most are 1 mm or less in size. The group is so diverse that it is difficult to generalize about any feature. Perhaps the most universal and striking features are the lack of body segments and of any recognizable division between the prosoma and opisthosoma and the presence of a buccal cone. The body covering is membranous or leathery, sometimes with hard plates or shields. A slender anterior region, often jointed to the body, bears the mouth parts. The eight legs are laterally placed and often bear bristles. The sexes are separate. In most species there hatches from the **egg** a six-legged **larva** that feeds and molts into an eight-legged **nymph** having no genital opening. With further feeding and often three molts, the **adult** sexual stage is reached.

Mites (Fig. 23-8) abound as to species (20,000) and individuals in soil, humus, stored foods, fresh water and salt water, on plants, and as parasites of both plants and animals. Some feed on fresh or decaying plant and animal materials, others suck plant juices, and still others subsist on the skin, blood, or other tissues of land vertebrates.

The ticks (Figs. 23-9, 23-10) feed on the blood of reptiles, birds, and mammals. Upon finding a host, the tick's mouth parts pierce the skin and the sucking pharynx draws in blood, which is kept fluid by a salivary anticoagulant; its stomach expands and its body becomes greatly enlarged (Fig. 23-10). In spring the female tick deposits her eggs under shelter on the ground, and hatching occurs a month or so later. The larvae climb on bushes to wait for appro-

Figure 23-9 Order ACARINA: female Rocky Mountain spotted fever tick (*Dermacentor andersoni*); internal structure, from the left side. (*After J. R. Douglas.*)

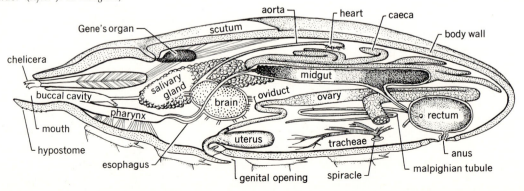

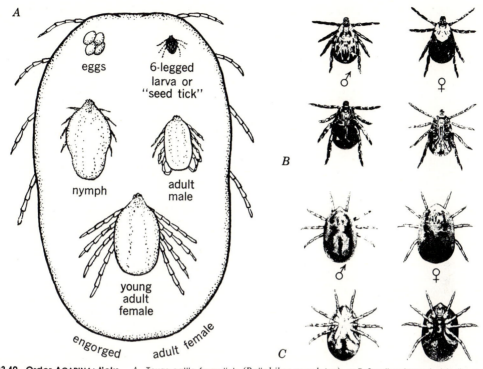

Figure 23-10 Order ACARINA: ticks. *A.* Texas cattle fever tick (*Boöphilus annulatus*), × 5. Smaller stages inside the outline of an engorged female. (*After U.S. Department of Agriculture.*) *B.* Rocky Mountain spotted fever tick (*Dermacentor andersoni*, family IXODIDAE). *C.* Relapsing fever tick (*Ornithodoros hermsi*, family ARGASIDAE). Upper row, dorsal; lower row, ventral. (*From W. B. Herms.*)

priate hosts, to which they fasten; then they drop off after feeding and molt. In turn, the nymphs and then the adults do likewise. Ticks can survive long periods without feeding, some for a year or more if they fail to find a host animal. The spotted fever tick normally requires 2 years to complete its life cycle. Adults of the soft, or argasine, ticks hide by day and emerge at night to feed briefly, whereas all stages of ixodine ticks remain in place on their hosts for some time. In the latter the males are smaller, with a scutum covering the whole dorsal surface; the scutum on females is smaller and anterior.

23-16 Relation to humans The bite of most spiders is harmless to humans. Even the large American "tarantulas" (*Dugesiella*) cause no more injury than a wasp. Bites of small brown spiders (*Loxosceles*) in Southern and Midwestern states and South America, however, produce pain, ulcers, and sometimes other more serious disturbances and death. Females of the genus *Latrodectus* are distinctly dangerous. The black widow spider

(*L. mactans*) occurs from southern Canada to South America and the Hawaiian Islands, living in cliffs, piles of rock and lumber, and outdoor privies and other buildings. The adult female is shiny black, with an hourglass mark of red beneath the abdomen. Severe pain, muscle spasms in the abdomen, and restlessness are common symptoms which occur after the bite of the black widow. Serum from victims of previous poisonings and antivenin are used in treatment, with magnesium chloride injections to relieve the muscle spasms.

Some scorpions have powerful neurotoxic venom; that of *Androctonus* of the Sahara is as toxic as cobra venom. Species of *Centruroides* in Mexico sting thousands of people annually and have caused hundreds of deaths, mostly in children. Those in the southwestern United States seem to be less dangerous. Treatment by antivenin has decreased the danger in recent years. The hairy scorpions (*Hadrurus*) of the southwestern United States often sting people, causing severe but transient pain.

Of the mites that suck plant juices, the red spider

idiosoma   gnathosoma

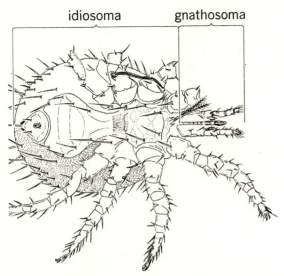

Figure 23-11 Ventral view of Laelaps echidnina, the spiny rat mite (Parasitiformes), showing major body divisions. *(From McGraw-Hill Encyclopedia.)*

(*Tetranychus telarius*) causes blisters and injures leaves of orchard trees; the blister mite (*Eriophyes*) injures buds and fruits of pears, apples, grapes, and other crops; and the bulb mite (*Rhizoglyphus echinopus*) tunnels in bulbs and roots so that fungi and bacteria are able to enter. The itch mite (*Sarcoptes scabiei*) burrows in and irritates human skin; related species cause mange in hogs, dogs, and other mammals; the scab mite (*Psoroptes*) produces serious damage to sheep; and the chicken mite (*Dermanyssus gallinae*) causes unthriftiness in fowls. Chiggers in the larval stage attack humans, causing severe itching. In the Orient they transmit the disease scrub typhus.

Ticks, when numerous, may make the host anemic and liable to various diseases or may cause its death. Common species are the fowl tick (*Argas persicus*) on poultry and the spinose ear tick (*Ornithodoros megnine*) and dog or wood ticks (*Dermacentor variabilis* and *D. occidentalis*), which infest both wild and domestic mammals and humans. The salivary secretions injected in the biting of some ticks produce severe, slow-healing wounds; certain ticks are intermediate hosts of disease. *D. andersoni* transmits the rickettsia organism causing Rocky Mountain spotted fever, a disease of native rodents and larger mammals in the arid West (Montana to eastern Oregon) that produces serious human illness. Spotted

fever has also been recognized in various eastern states from Illinois to New Jersey and North Carolina. Vaccine grown on incubated hen's eggs is now made to protect persons exposed to tick bite. Texas cattle fever of the Southern states is due to a protozoan (*Babesia bigemina*) carried by a cattle tick (*Boophilus annulatus*). In both these diseases the infective agent passes from the female tick to her eggs. Texas cattle fever was the first disease in which an arthropod vector was proved to be the essential agent transmitting the causative organism from infected to healthy hosts.

Class Merostomata (Horseshoe crabs and eurypterids)

These large chelicerates are characterized by their abdominal appendages modified into gills and by the presence of a long, pointed terminal structure, the telson. Two distinct subclasses are recognized, the extinct †EURYPTERIDA and the living XIPHOSURA, represented by the horseshoe crabs.

23-17 Horseshoe crabs The subclass XIPHOSURA includes many fossil forms and five living species of horseshoe crab. Of these, *Limulus polyphemus* (Fig. 23-12) inhabits shallow quiet waters of the Atlantic Coast from Nova Scotia to Yucatán. The carapace covering the prosoma is horseshoe-shaped, convex, dark brown, and unsegmented. The abdominal shield is broadly hexagonal, with pairs of short lateral movable spines and a hinged bayonetlike telson posteriorly. Besides the six pairs of appendages on the cephalothorax as in other arachnids, the abdomen bears six pairs of broad, thin appendages joined medially. On the hind surface of the last five pairs are exposed book gills, each of 150 to 200 leaves, containing blood vessels for respiration. A pair of four-lobed coxal glands in the cephalothorax serves for excretion. The nerve ring about the mouth connects to a ventral nerve cord with ganglia. On the carapace are two lateral compound eyes, of different structure from those of other arthropods, and two medial simple eyes.

The horseshoe crab burrows by shoving its carapace into the sand and pushing with its telson and appendages. It can walk on the bottom, can swim by flapping the abdominal plates, and may hop by aid

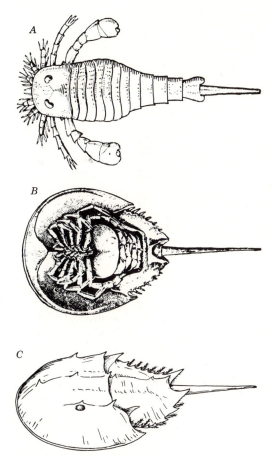

Figure 23-12   Class MEROSTOMATA.   *A.* Extinct eurypterid (†*Eurypterus*, subclass †EURYPTERIDA), in dorsal view. *B, C.* The living horseshoe crab (*Limulus polyphemus*, subclass XIPHOSURA). *B*, ventral view; *C*, laterodorsal view. Length to 45 cm (18 in).

of the telson. It is active by night and is omnivorous, feeding on nereid worms, algae, and soft mollusks, which are grasped by the chelicerae, aided by the pincers on other appendages, and chewed by the bases of the legs (*gnathobases*), which serve as jaws. In early summer the animals enter intertidal sandy areas, where the smaller male mounts the female and numbers of large eggs, fertilized externally, are deposited in shallow depressions scooped out by the female. Eggs are abandoned by the adults after laying. The larva at hatching resembles a trilobite in having a segmented abdomen without appendages or telson, but these features gradually change toward the adult condition with each of the many molts. The XIPHOSURA date from the Ordovician and the genus *Limulus* from Triassic time.

23-18 Eurypterids   Subclass †EURYPTERIDA (Fig. 23-12) was a dominant group in seas of Paleozoic times. In form they were somewhat scorpionlike and were the largest arthropods that have ever lived, some being up to 3 m (9 ft) long. They probably were much like horseshoe crabs in body morphology except that the cephalothorax was smaller, the abdomen larger and segmented, and the fourth pair of walking legs paddlelike. They probably swam or crawled on the bottom of bodies of salt water and brackish waters, feeding on worms and small fishes.

Class Pycnogonida (Sea spiders)

23-19 Sea spiders   (Figs. 23-13, 23-14) This is a small (600 species) group of marine arthropods of rather aberrant spiderlike appearance. Because of their unique mixture of features not found in any other arthropod group, there has been considerable controversy as to their taxonomic position. They are here placed among the chelicerates because they have chelicerae, but the presence of a segmented prosoma, special ovigerous legs, additional segments on the legs, multiple gonopores, and often additional somites and legs is not observed in any other chelicerate group. The body is small, with four (rarely five or six) pairs of legs; the mouth is at the end of a rounded sucking proboscis; and typically the head, which is fused to the first trunk segment, bears four pairs of appendages. The first pair are chelicerate; the second are sensory palpi; the third, ventral in origin, are ovigers carrying eggs in the

Figure 23-13 A sea spider (*Nymphon*, class PYCNOGONIDA).

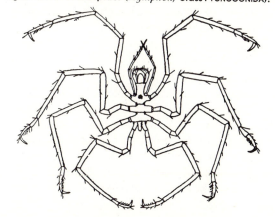

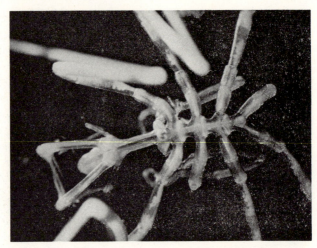

Figure 23-14 The pycnogonid *Nymphon.* *(Courtesy of The California Academy of Sciences.)*

male and used for grooming in the large deep-water species; the fourth, on the trunk segments, are walking legs. The prosoma typically has three segments, each bearing a pair of walking legs. The head bears a protuberance with four eyes. Internally the sea spiders lack respiratory and excretory systems, and the circulatory system is simple. The digestive system extends into the bases of the legs, as do the gonads. The nervous system is crustacean-like. In size (spread of legs) they measure 3 to 500 mm (20 in). Sea spiders occur from the shore to depths of 3650 m (12,000 ft) and are most abundant in colder waters. They feed on the soft parts of cnidarians and sponges. Sexes are separate. Males carry the eggs during early stages of development. The pycnogonids have a free larval stage, the **protonymphon,** found otherwise in the arthropods only in the CRUSTACEA. Like the crustacean nauplius larva, this larva has three pairs of appendages, but they are not homologous in the two larvae.

Subphylum Mandibulata

This is by far the largest subdivision of the arthropods, including the CRUSTACEA (Chap. 24) and INSECTA (Chap. 25), together with centipedes, millipedes, and their allies. All are alike in having the first

pair of jaws as mandibles, which are variously modified. The four minor groups described below are all similar in that they have an elongated trunk of similar segments bearing legs. This feature has been used to join them all in a single class, MYRIAPODA, presently not recognized by most zoologists, which, however, indicates their close relationship.

Class Chilopoda (Centipedes)

23-20 Structure and habits The centipedes (Fig. 23-15) are of slender segmented form and are flattened dorsoventrally. The head bears a pair of long antennae with 12 or more joints, a pair of mandibles, and two pairs of maxillae. In different species, the body is of 15 to 181 somites. The first trunk somite bears a pair of four-jointed poison claws, and on each of the other somites except the last two is a pair of small seven-jointed walking legs. The digestive tract is straight, with two or three pairs of salivary glands at the mouth and two long Malpighian tubules pos-

Figure 23-15 A centipede (class CHILOPODA).

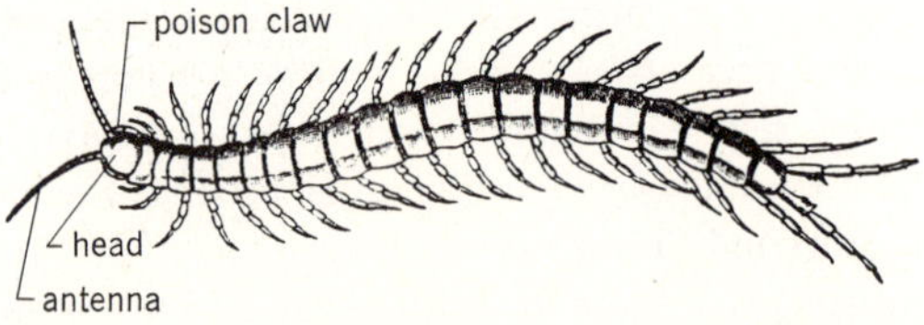

poison claw
head
antenna

teriorly for excretion. The heart extends the length of the body, surrounded by pericardium, with a pair of ostia and of lateral arteries in each somite. In some somites there are spiracles joining a fused system of tracheal tubes that serve for respiration. The sexes are separate, each having one dorsal gonad and paired accessory glands connected to a ventral genital opening near the posterior end.

Centipedes are found throughout temperate and tropical regions of the world, hiding by day under stones or logs and running swiftly about at night to prey on other arthropods; large species may capture small vertebrates. Prey is killed quickly by poison from a duct in the poison claw and is chewed by the mandibles. Some centipedes lay eggs and others are viviparous. The young resemble adults, with the same or a lesser number of somites. Some tropical species are 15 to 25 cm (6 to 10 in) long, and their bites are painful to humans. The small, agile house centipedes (*Scutigera*) have 15 pairs of extremely long, fragile legs. They eat insects and are harmless to humans.

Class Diplopoda (Millipedes)

23-21 Structure and habits The millipedes have cylindrical bodies of many somites, and the body wall includes deposits of lime salts (Fig. 23-16). Many are brightly colored. The head bears two clumps of many simple eyes and a pair each of short antennae and mandibles. A platelike structure (*gnathochilarium*) may represent the fused maxillae. The thorax is short, of four single somites, all but the first with one pair of legs. The long abdomen has 9 to over 100 double somites, each containing two pairs of spiracles, ostia, and nerve ganglia and bearing two pairs of seven-jointed legs. The digestive, circulatory, and excretory systems are somewhat as in centipedes. There are many separate tracheae, each a

tufted pouch connected to a spiracle in front of a leg. The gonad is single and ventral, with a duct opening on the third somite. The male has modified appendages (gonopods) on the seventh somite. Fertilization is internal, and eggs are deposited in clusters in soil or humus, in a covering of regurgitated food, or in a nest. The young at hatching have but seven somites and three pairs of legs, others being added in front of the anal somite at the successive molts (up to 10) during growth.

Millipedes live in humid dark places under stones or beneath or within rotting logs and shun the light. They travel slowly, with the body extended, and test the route of travel with the antennae. The many legs seemingly move in a series of waves from behind forward. Their food is primarily dead plant materials, but they also will eat animal matter. The pill millipedes (Oniscomorpha) roll into a ball when disturbed; long-bodied forms coil into a protective spiral. Members of some orders have "stink glands" that spray a fluid containing cyanide and iodine to repel enemies.

23-22 Other Mandibulata Two other classes are commonly recognized. The Pauropoda are minute creatures, resembling millipedes but with only 11 or 12 trunk segments and 9 or 10 pairs of legs. They are characterized by branched antennae. They live in moist soil or among decaying leaves and are thought to feed on fungi. The Symphyla (Fig. 23-17) include the garden centipedes (*Scutigerella*), which have small, slender, pale-colored bodies with one pair of antennae, a pair of unjointed posterior appendages, and 12 pairs of jointed legs ending with double claws. The sex opening is on the third body somite. They live on vegetable material in the soil, occasionally damaging field crops and greenhouse plants. Symphylids resemble primitive wingless insects (Diplura) in head structure and the eversible sacs

Figure 23-16 A millipede (*Spirobolus*, class DIPLOPODA). (*From Haupt.*)

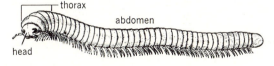

Figure 23-17 The garden centipede (*Scutigerella*, class SYMPHYLA), to 6 mm long. (*After Michelbacher.*)

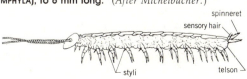

and styli on abdominal somites, hence are considered possible ancestors of the insects.

Classification

Phylum Arthropoda.

Joint-footed animals. Body with head, thorax, and abdomen, of like or unlike somites, variously fused, each typically with a pair of jointed appendages; exoskeleton containing chitin, molted entire at intervals; digestive tract complete, of fore-, mid-, and hindgut; coelom reduced, body spaces a hemocoel; circulatory system lacunar, heart dorsal; respiration by gills, tracheae, book lungs, or body surface; excretion by Malpighian tubules or coxal glands (nephridia rare); brain dorsal, nerve cord ventral, and paired ganglia in each somite or concentrated anteriorly; with simple and compound eyes, antennae, sensory hairs (and statocysts); sexes usually separate, gonads paired, and fertilization internal; cleavage usually superficial; commonly with larval stages and metamorphosis; some parthenogenesis; terrestrial or aquatic; free-living, sessile, commensal, or parasitic; fully 1 million living species.

Subphylum A. †Trilobita.

Body 10 to 675 mm ($\frac{1}{2}$ to 27 in) long, divided by 2 lengthwise furrows into 3 lobes; head distinct; abdomen of 2 to 29 somites and a fused caudal plate; all somites except last with biramous appendages fringed by setae; development with larval stages; marine. Cambrian to Permian; over 4000 species. †*Triarthrus.*

Subphylum B. Chelicerata.

Body usually of 2 parts (except ACARINA); cephalothorax unsegmented (except SOLPUGIDA), with 2 chelicerae bearing chelae or claws, 2 pedipalpi variously modified, and 4 pairs of legs; no antennae or true jaws; abdomen either of 6 to 14 somites or undivided and with no locomotor appendages (except MEROSTOMATA); respiration by gills, book lungs, or tracheae; excretion by Malpighian tubules or coxal glands; sexes usually separate, male often smaller; sex opening anterior on abdomen; mostly oviparous; some with larval stages; mostly predaceous; chiefly terrestrial, some in fresh water or salt water.

Class 1. Merostomata.

Cephalothorax with compound lateral eyes and broadly joined to abdomen, on which are 5 or 6 pairs of appendages bearing exposed gills and a terminal bayonetlike telson; aquatic.

Subclass 1. Xiphosura. Horseshoe crabs. Cephalothorax an arched, horseshoe-shaped carapace; abdomen wide, unsegmented; chelicerae 3-jointed, pedipalpi and legs 6-jointed; marine, in shallow waters; Ordovician to Recent. *Limulus polyphemus,* Nova Scotia to Yucatán; *Tachypleus,* Japan to Malaysia and East Indies.

Subclass 2. †Eurypterida (Gigantostraca). Eurypterids. Carapace not expanded; abdomen of 12 somites, narrowed behind; Ordovician to Permian. †*Eurypterus,* †*Pterygotus.*

Class 2. Pycnogonida (Pantopoda).

Sea spiders. Mostly small; body short, thin, of 1 cephalic (fused with first trunk somite) and 3 or 4 trunk somites, abdomen vestigial; 4 eyes; mouth suctorial, on long proboscis; usually 4 pairs of head appendages, 1 pair being 10-jointed ovigerous (egg-bearing) legs; 4 (5 or 6) pairs of long thin walking legs, 8- or 9-jointed; sexes usually separate, eggs carried by male; development direct, or a 3-legged larva with metamorphosis; marine, on hydroids, anemones, or algae or under stones, from tide lines down to more than 3650 m (12,000 ft); Devonian, Recent; 600 species. *Pycnogonum; Nymphon; Colossendeis,* spreads to 60 cm (24 in); *Achelia.*

Class 3. Arachnida.

Abdomen lacks locomotor appendages; eyes all simple; cuticle often with sensory hairs or scales; pedipalpi usually sensory, modified in male for transfer of sperm to female in the ARANEAE; no gills; mostly oviparous; no metamorphosis (except ACARINA); chiefly terrestrial.

Order 1. Scorpionida. Scorpions. Body elongate; cephalothorax (prosoma) compact, broadly joined to abdomen; chelicerae 3-jointed, small; pedipalpi large, 6-jointed, with stout pincers; abdomen long, of 12 ringlike somites, second with pair of comblike and tactile pectines; last 6 somites (metasomal) narrow, with a sharp poison sting at end; 4 pairs of book lungs; viviparous; nocturnal, in warm regions; Silurian to Recent; 800 species. *Hadrurus hirsutus,* hairy scorpion, and *Vejovis spinigerus,* both to 100 mm (4 in) long, Texas to California;

V. boreus, small, Nebraska to Nevada; *Centruroides,* several species with toxic venom in southwestern United States.

Order 2. Uropygi (Pedipalpi). Whip scorpions. Cephalothorax narrowly joined to flat abdomen of 12 somites; abdomen with slender, terminal whip-like flagellum; odoriferous glands; 8 eyes; chelicerae small, 2-jointed, chelate; pedipalpi strong, 6-jointed, tarsus of first leg a long, many-jointed feeler; nocturnal; Carboniferous to Recent; 130 species. *Mastigoproctus giganteus,* to 130 mm (5 in) long, Florida to Arizona.

Order 3. Amblypygi. Abdomen without a whip no odoriferous glands. *Tarantula* (not a tarantula!), 20 mm (1 in) long, Florida, Texas, and California.

Order 4. Palpigrada. Size minute; chelicerae stout, 3-jointed, pedipalpi like walking legs; no eyes; abdomen oval, of 11 somites, with slender "tail" of 15 segments; live under stones. *Prokoenenia,* 2.5 mm long, Texas, California.

Order 5. Araneae (Araneida). Spiders. Cephalothorax and abdomen both unsegmented, joined by narrow "waist"; chelicerae small, 2-jointed, a poison duct in terminal fang; pedipalpi short, 6-jointed, leglike, the basal joint on each an enlarged "maxilla" for chewing; legs 7-jointed; eyes 8 to none; book lungs 2 to 4, some with tracheae; tubercle-like spinnerets on abdomen extrude silk; oviparous, eggs commonly in cocoons; chiefly terrestrial; Carboniferous to Recent; 61 families, 32,000 species. Some common families and members (Fig. 23-3) are:

Suborder 1. Orthognatha (Mygalomorphae). Chelicerae horizontal, 2 pairs book lungs.
CTENIZIDAE. Trap-door spiders. Tarsal claws 3. *Pachylomerus* (southeastern United States) and *Bothriocyrtum* (California), nest hole in ground closed by hinged door of earth and silk.
THERAPHOSIDAE. The "tarantulas." Tarsal claws 2; body size to 10 cm (4 in); live to 20 years. *Dugesiella (Aphonopelma),* American "tarantula," to 48 mm (2 in) long, Arkansas and Texas to California; on ground or in trees; bite painful but not dangerous to humans; *Eurypelma,* South America.

Suborder 2. Labidognatha (Araneomorphae). Chelicerae vertical, 1 pair book lungs.
THERIDIIDAE. Tarsal claws 3; legs without spines. *Latrodectus mactans,* black widow spider, web irregular, bite dangerous and sometimes fatal; *Theridion tepidariorum,* house spider, web flat, in corners.
AGELENIDAE. Funnel-web spiders. Tarsal claws 3; legs with spines; eyes almost equal in size. *Agelenopsis (Agelena) naevia,* grass spider, web flat with funnel at side, in grass and in houses; *Tegenaria agilis,* house spider, naturalized from Europe, webs in corners.
LYCOSIDAE. Hunting spiders. Tarsal claws 3; legs with spines; eyes unequal in size. *Lycosa,* wolf spider, to 25 mm (1 in) long; active on ground, hunts prey or builds turrets to watch for prey; female carries egg cocoon and young.
ARGIOPIDAE. Orb weavers. Tarsal claws 3; webs of geometric design. *Miranda aurantia,* orange garden spider, web to 60 cm (2 ft) across, with zigzag band, in grass or bushes; *Epeira marmorea,* with nest of leaves at side of web, where spider sits holding a thread to web.
THOMISIDAE. Crab spiders. Tarsal claws 2; front pair of eyes not larger than others; body short, wide, appearance and gait crablike, no web. *Misumena,* bright- or white-colored, lives in flowers and mimics colors; *Philodromus,* mottled; *Xysticus,* under bark or leaves.
SALTICIDAE (Attidae). Jumping spiders. Tarsal claws 2; front pair of eyes much larger than others; run and jump in all directions, no web. *Salticus scenicus,* on fences and houses; *Phidippus,* on plants, nest baglike, among leaves.

Order 6. Solpugida (Solifugae). Sun spiders. Cephalothorax of 6 somites, I to III a "head," eyes 4 or 6; chelicerae swollen, 2-jointed, with chelae, and joined to carapace; pedipalpi leglike, 6-jointed; first legs tactile, not locomotory; no "waist"; abdomen of 10 somites; no spinning organs; in warm dry regions; Carboniferous to Recent; 800 species. *Eremobates,* southwestern United States; *Ammotreca; Galeodes,* "tarantula" of Egypt.

Order 7. Pseudoscorpionida (Chelonethida). False scorpions. Length 1 to 7.5 mm; unsegmented cephalothorax broadly joined to abdomen of 11 somites; no sting; eyes 2, 4, or none; chelicerae 2-jointed, with comblike serrula handling silk for nests from 2 glands in cephalothorax; pedipalpi large, 6-jointed, chelate; under moss or stones, in trees, or about buildings; Tertiary to Recent; 1000 species. *Garypus,* California seacoast, to 7.5 mm long; *Chthonius; Chelifer cancroides,* house or book scorpion, 3.5 mm long, about buildings.

Order 8. Ricinulei (Podogona). Small; movable hood on cephalothorax over small chelicerae; no eyes; abdomen of 6 somites, I and II forming a "waist"; tropical; Carboniferous to Recent. *Cryptocellus*, few specimens known.

Order 9. Opiliones (Phalangida). Harvestmen "daddy longlegs." Body short, ovoid, cephalothorax broadly fused to faintly segmented abdomen; chelicerae 3-jointed, slender; pedipalpi 6-jointed, not chelate; legs 7-jointed, very long and delicate, tarsi many-jointed; stink glands on cephalothorax; no book lungs or silk glands; sexes much alike; in fields, woods, and buildings; Carboniferous to Recent; 3200 species. *Phalangium, Liobunum*.

Order 10. Acarina (Acari). Mites and ticks. Small to microscopic; body compact, ovoid, cephalothorax and abdomen fused; no segmentation; chelicerae and pedipalpi various; legs widely separated, 6- or 7-jointed; respire by tracheae or body surface; with larval and nymph stages; free-living or parasitic; worldwide; Devonian to Recent; 60 families, 25,000 species (Figs. 23-8 to 23-10).

Suborder 1. Onychopalpida (Notostigmata, Tetrastigmata). Two or four pairs of lateral stigmata; pedipalps with reduced claws; labrum with radula-like organ; free-living, predaceous. *Neocarus*, Texas.

Suborder 2. Mesostigmata. One pair of stigmata with a slender tube lateral to legs; free-living, predaceous, or commensal. *Parasitus*, on insects; *Laelaps*, on rodents; *Spinturnix*, on bats; *Halarachne*, in bronchi of seals; *Pneumonyssus*, in lungs of Old World monkeys; *Ornithonyssus bacoti*, tropical rat mite, common on alien rats (*Rattus*); *Dermanyssus gallinae*, chicken mite, attacks poultry and poultrymen; *Allodermanyssus sanguineus*, on house mouse (*Mus*), transmits *Rickettsia akari*, cause of rickettsialpox in humans.

Suborder 3. Ixodides (Metastigmata). Ticks. One pair of stigmata, with stigmal plate, posterior or lateral to legs.

ARGASINE TICKS. No dorsal shield. *Argas persicus*, fowl tick, 8 to 10 mm, on poultry, quail, pigeons; *Ornithodoros megnini*, spinose ear tick, in ears of horses, cattle, sheep, and humans; *O. turicata*, 6 to 7 mm, on hogs, cattle, and humans; *O. moubata*, carries African tick fever; *O. hermsi*, on wild rodents, transmits *Spirochaeta recurrentis*, cause of relapsing fever in humans.

IXODINE TICKS. A dorsal shield (Fig. 23-10). *Ixodes ricinus*, castor-bean tick, 3 to 10 mm, on various mammals; *Dermacentor variabilis* and *D. occidentalis*, dog ticks, to 15 mm ($\frac{1}{2}$ in) long, on

domestic mammals and humans, common in woods and brush; *D. andersoni*, on domestic and wild mammals and humans—this and other species transmit *Dermacentroxenus rickettsii*, causing spotted fever in humans; *Boöphilus* (*Margaropus*) *annulatus*, on cattle, transmits *Babesia bigemina* (PROTOZOA), causing Texas cattle fever (Table 25-1).

Suborder 4. Trombidiformes (Prostigmata). One pair of stigmata on or near mouth or stigmata absent; chelicerae usually modified for piercing; palpi as pincerlike clasping organs or as sensory organs.

RED SPIDER MITES. *Tetranychus*, 0.4 mm, red or yellow, suck leaves of plants making pinpoint lesions; female spins silken web.

CHIGGERS. *Trombicula alfreddugesi* (*irritans*), larvae puncture and irritate the skin. Oriental species transmit a serious rickettsial disease, scrub typhus.

WATER MITES. Length 0.5 to 8 mm, colored, in fresh water or salt water to 1050 m (3500 ft) deep in the ocean. *Hydrachna, Halacarus*.

GALL MITES. Body slender, legs 4, anterior; suck juices from plants, stimulating formation of open-ended galls. *Eriophyes*, blister mite, damages buds, leaves, and fruits of trees and shrubs.

FOLLICLE MITES. Slender, ringed, 8 legs, 3-jointed. *Demodex folliculorum*, 0.4 mm, in hair follicles and sebaceous glands of humans and domestic mammals.

Suborder 5. Sarcoptiformes (Astigmata, Cryptostigmata). No stigmata or with tracheae opening on various parts of body; chelicerae modified for chewing; palpi simple. *Galumna*, horny or beetle mite, cuticle hard, in moss or tree bark, eats decaying matter; *Acarus siro*, cheese mite; *Rhizoglyphus echinopus*, bulb mite, in bulbs and roots; *Analgopsis megninia*, feather mite, in bird feathers and quills, not harmful.

ITCH AND MANGE MITES. Body ovoid, legs 8, with suckers, claws, or bristles. *Sarcoptes scabiei*, female burrows and oviposits in skin, causing itch in humans and mange in domestic animals; *Psoroptes*, scab mites of domestic animals; *Knemidokoptes*, itch and scaly-leg mites of poultry.

Subphylum C. Mandibulata (Antennata).

Body of 2 parts (head and trunk) or 3 parts (head, thorax with walking legs, and abdomen); 1 or 2 pairs of antennae, 1 pair of jaws (mandibles), 1 or more pairs of maxillae, and 3 or more pairs of walking legs; respiration by gills or tracheae; excretion by Malpighian tubules or antennal glands; sexes usually separate; oviparous or ovoviviparous; usually with larval stages; terrestrial or aquatic, in fresh water or salt water.

Class 1. Crustacea.

Crustaceans. (See Chap. 24.)

Class 2. Insecta (Hexapoda).

Insects. (See Chap. 25.)

Class 3. Chilopoda.

Centipedes. Body long, flattened dorsoventrally; head with 1 pair of jointed antennae, 1 pair of jaws, and 2 pairs of maxillae; body somites, 15 to 181, each with 1 pair of legs; first pair of body appendages 4-jointed, hooklike, with poison duct opening in terminal claws; sex opening midventral on next to last somite; respiration by tracheae; excretion by Malpighian tubules; development direct; terrestrial; nocturnal; Tertiary to Recent; 3000 species (Fig. 23-15).

Order 1. Scutigeromorpha.   Legs long, 15 pairs; spiracles middorsal. *Cermatia (Scutigera) forceps,* house centipede, to 25 mm (1 in) long, in buildings, swift-moving, eats insects.

Order 2. Lithobiomorpha.   Legs short, 15 pairs; spiracles lateral; includes smallest centipedes, from 3 mm (0.1 in) long. *Lithobius,* to 30 mm (1 in).

Order 3. Scolopendromorpha.   Legs, 21 or 23 pairs; some species to 265 mm (10 in) long. *Scolopendra,* to 250 mm (10 in) long, 21 pairs of legs, widely distributed.

Order 4. Geophilomorpha.   Legs 31 to 181 pairs. *Geophilus,* legs 31 to 93 pairs, body thin.

Class 4. Diplopoda.

Millipedes. Body long, usually cylindrical; head with 1 pair each of short, 7-segmented antennae, jaws, and maxillae (usually fused as gnathochilarium); thorax of 4 somites, I legless, II to IV each with 1 pair of legs; abdomen of 9 to over 100 segments (double somites) closely spaced, each with 2 pairs of legs; sex opening midventral on IIId somite; respiration by tracheae; excretion by 2 to 4 Malpighian tubules; development direct; terrestrial in moist places; Silurian to Recent; 7500 species (Fig. 23-16).

Subclass 1. Pselaphognatha. Body somites 13; back and sides with large tufts of serrated setae; integu-

ment soft; no stink glands; 2 leglike maxillae; worldwide. *Polyxenus,* 2.5 to 3 mm (0.1 in) long.

Subclass 2. Chilignatha.   Body with hardened integument; setae simple; stink glands present or absent.

Order 1. Glomeridesmida (Limacomorpha). Body somites 22; male gonopods (clasping organs) are 1 or 2 pairs of legs on last somite; body unable to be rolled into a ball; Old and New World tropics. *Glomeridesmus,* 10 mm (0.4 in).

Order 2. Oniscomorpha.   Body somites 14 to 16; male gonopods, 1 or 2 pairs on last somite; body able to roll into a tight ball; Old World. *Glomeris,* 10 to 15 mm ($\frac{1}{2}$ in).

Order 3. Polydesmida.   Body somites 19 to 22; male gonopods, 1 or 2 pairs on VIIth somite; no spinning glands; repugnatorial glands; no eyes. *Polydesmus,* flattened, to 28 mm (1 in), Texas.

Order 4. Chordeumida (Nematomorpha).   Body somites 26 to 60; male gonopods, 1 or 2 pairs on VIIth somite; end of body with 2 or 3 pairs of spinning glands and bristles; eyes present. *Striaria,* 30 body somites.

Order 5. Juliformia (Opisthospermophora). Body somites 40 or more; male gonopods, 1 or 2 pairs on VIIth somite; no spinning glands. *Julus virgatus,* 12 mm ($\frac{1}{2}$ in) long, body somites 30 to 35; *Narceus (Spirobolus),* to 100 mm (4 in) long, to 100 pairs of legs (50 somites), North America, China.

Order 6. Colobognatha.   Body somites 30 to 60; male gonopods are second pair of legs on VIIth somite and first pair on VIIIth somite; head conical. *Platydesmus,* Southeastern states and California.

Class 5. Pauropoda.

Length 0.5 to 1.8 mm; resemble millipedes but antennae 3-branched; no eyes; body cylindrical, of 11 (12) somites, with 6 dorsal plates and 9 (10) pairs of legs; sex opening midventral on IIId somite; 4 larval instars; in dark damp places under logs, stones, or leaves and in soil; about 360 species. *Pauropus huxleyi,* Eastern states and Europe; others in California.

Class 6. Symphyla.

Garden centipede. To 6 mm long; white; no eyes; with antennae, jaws, and 2 pairs of maxillae; adult with 12 pairs of legs; sex opening midventral between fourth pair of legs; terrestrial, in damp places with humus. *Scutigerella immaculata,* garden centipede (Fig. 23-17), often injures seeds and young shoots of sugar beets, asparagus, and other crops.

References

Baker, E. W., and **G. W. Wharton.** 1952. An introduction to acarology. New York, The Macmillan Company. xiii + 465 pp., 377 figs.

Cloudsley-Thompson, J. L. 1958. Spiders, scorpions, centipedes and mites. New York, Pergamon Press. xiv + 228 pp., 40 figs., 17 pls. *Natural history and ecology.*

Comstock, J. H., and **W. J. Gertsch.** 1940. The spider book. 2d ed. Garden City, N.Y., Doubleday & Company, Inc. xi + 729 pp., 766 + figs. *Other* Arachnida *briefly described.*

Edney, E. B. 1957. The water relations of terrestrial arthropods. New York, Cambridge University Press. 109 pp., illus. *Water uptake and loss through the cuticle and excretion and osmotic regulation.*

Evans, G. Owen, J. G. Sheals, and **D. Macfarlane.** 1961. The terrestrial Acari of the British Isles. Vol. 1, Introduction and biology. London, British Museum (Natural History). v + 219 pp., 216 figs. *Includes a general classification.*

Fabre, J. H. 1919. The life of the spider. New York, Dodd, Mead & Company, Inc. 404 pp., illus.

Gertsch, W. J. 1949. American spiders. Princeton, N.J., D. Van Nostrand Company, Inc. xiii + 285 pp., 32 pls., some colored.

Grassé, P.-P. 1949. Traité de zoologie. Paris. Masson et Cie. Vol. 4, Onychophores, Tardigrades. Arthropodes, Trilobitomorphes, Chelicerates. pp. 1–979, 719 figs.

Hedgpeth, J. W. 1954. On the phylogeny of the Pycnogonida. Acta Zoologica, vol. 35, pp. 193–213, 9 figs.

Helfer, H., and **E. Schlottke.** 1935. Pantopoda. In Bronn, Klassen und Ordnungen des Tierreichs. Leipzig Akademische-Verlagsgesellschaft Geest & Portig, KG. Vol. 5, pt. 4, book 2, pp. 1–314, 223 figs.

Herms, W. B. 1961. Medical entomology. 5th ed. Rev. by M. T. James. New York, The Macmillan Company. xi + 616 pp., 185 figs.

Kaston, B. J., and **E. Kaston.** 1953. How to know the spiders. Dubuque, Iowa, William C. Brown Company Publishers. vi + 220 pp., 552 figs.

King, P. E. 1973. Pycnogonids. New York, St. Martin's Press, Inc. 144 pp., 40 figs.

Kükenthal, W., and others. 1926– . Handbuch der Zoologie. Berlin, Walter de Gruyter & Co. Vol. 3, pt. 2, Arachnoidea, Onychophora; vol. 4 pt. 1, Diplopoda, Chilopoda, etc. About 1400 pp., 1800 figs.

Levi, H. W., and **L. R. Levi.** 1969. A guide to spiders and their kin. New York, Golden Press, 160 pp., illus. in color.

Petrunkevitch. A. 1933. The natural classification of spiders based on a study of their internal anatomy. *Connecticut Academy of Sciences Transactions,* vol. 31, pp. 299–389.

_____. 1939. Catalogue of American Spiders. Ibid., Vol. 33, pp. 133–338.

Pratt, H. S. 1935. Manual of the common invertebrate animals exclusive of insects. New York, McGraw-Hill Book Company. Arachnoidea, etc., pp. 468–555, figs. 639–739.

Savory, T. H. 1964. Arachnida. New York, Academic Press, Inc. xii + 291 pp., illus.

Sharov, A. G. 1966. Basic arthropodan stock, with special reference to insects. New York, Pergamon Press, xii + 271 pp., 89 figs. *Presents a diverse approach to phylogeny and classification.*

Verhoeff, K. W. 1934. Symphyla and Pauropoda. In Bronn, Klassen und Ordnungen des Tierreichs. Leipzig, Akademische Verlagsgesellschaft Geest & Portig, KG. Vol. 5, pt. 2, book 3, pp. 1–200, 136 figs.

CLASS CRUSTACEA: CRUSTACEANS

The CRUSTACEA (L. *crusta*, hard shell) include the fairy shrimps, water fleas, barnacles, crayfishes, crabs, and their kin. Most of them are marine, occurring in all oceanic habitats, but many live in fresh water or brackish water, and a few, like the sow bugs, are in moist places on land. The great majority are free-living, and some planktonic species occur in vast schools. The barnacles are either sessile or parasitic; certain other crustaceans are commensal or parasitic on various aquatic animals, from hydroids to whales. Some parasitic species are so greatly modified that their status as crustaceans is shown only in their larval stages. Despite their inclusion with the insects and myriapods in the MANDIBULATA, the crustaceans, most zoologists agree, are quite different from insects and myriapods and probably arose independently. Hence the MANDIBULATA is more a taxon of convenience than one representing a natural assemblage.

The CRUSTACEA is the dominant aquatic arthropod class. Although small in number of species when compared to insects, they are more varied in morphology and in habitats occupied. They are also a very old group, with fossils back to Cambrian times, and they retain many primitive living forms. This great diversity of form, the presence of many primitive stocks, and the great adaptive radiation are reflected in the more complex higher taxonomy of the group.

24-1 Characteristics

1   Head of five fused somites with two pairs of antennae, one pair of lateral mandibles for chewing,

and two pairs of maxillae; thorax of 2 to 60 somites, distinct or variously fused; abdominal
somites, usually distinct, with telson at end;
often with a carapace over head and parts of thorax, as a dorsal shield or as two lateral valves;
appendages variously modified, some usually
biramous.

2 Respiration by gills (rarely by body surface);
pseudotracheae on pleopods of some land isopods.

3 Excretion by antennal or maxillary glands; no
Malpighian tubules.

4 Sexes separate (except in Cirripedia and a few
others); sex openings paired; eggs often carried
by female; parthenogenesis in Cladocera and
some other Branchiopoda; usually one or more
larval stages.

24-2 General features and evolution Although
crustaceans are a very diverse group, they are readily
separated from other mandibulates by their two
pairs of antennae. The body is usually conspicuously
segmented, with primitive forms having large numbers of segments (to 60), while more advanced forms
have fewer, usually 19 (Malacostraca). Primitive
crustaceans show little specialization of the segments, the body being a linear array of similar units.
More advanced crustaceans have the segments differentiated into groups, or **tagmata.** Usually three
tagmata are present, head, thorax and abdomen, but
among different crustacean groups these regions
may be obscured because of fusion. Thus in many
advanced crustaceans the head and thoracic segments are fused into a single **cephalothorax.** In terms
of the number of segments making up each of the
tagmata, the head is the most uniform throughout
the class. It consists of five segments: the most anterior two bear the **antennae,** the third the **mandibles,**
and the last two are food-manipulating appendages
called **maxillae.** The thorax and abdomen are more
variable as to the number of segments in each, but
the number tends to be consistent within any of the
taxonomic groups. Thus all Malacostraca have a
thorax of eight segments.

Crustacean trunk appendages are usually **biramous,** composed of two jointed branches, an inner
endopodite and an outer **exopodite,** attached to a

basal **protopodite.** In the primitive condition these
appendages were used for locomotion, food capture,
and respiration, a situation now observed only in a
few forms such as brine shrimp (*Artemia*). Most
modern crustaceans have limbs which are variously
specialized for different functions. In the crayfish,
for example, certain thoracic limbs are specialized to
handle food (maxillipeds), others for capturing and
holding food (chelipeds), and still others for locomotion (walking legs, pereiopods). In addition, many
crustaceans have abandoned, partially or wholly, the
biramous condition, so that a single ramus, either
endopodite or exopodite, is the only functional unit.

In contrast to chelicerates and insects, the crustaceans have the cuticle hardened and strengthened by
the deposition of calcium carbonate.

The circulatory system is similar to that of chelicerates, with gills the usual means of obtaining oxygen in this primarily aquatic group. Gills are usually
flattened, thin processes extending from the body
wall or the bases of the limbs. In higher forms they
are often restricted to certain limbs, while in very
small crustaceans they may be absent.

The excretory system consists of glands in either
antennae or maxillae segments. The structure of such
glands is similar to that of the coxal glands of chelicerates. The main excretory product is ammonia.

As noted for arthropods in general, the nervous
system tends to be centralized and the animals have a
large number of sense organs such as eyes, statocysts, proprioceptors, chemoreceptors, and sensory
setae. As opposed to the chelicerates, the eyes are
large and well developed. Compound eyes and simple eyes are present, but compound eyes are found in
most species. Some evidence of color vision has been
reported. Chemosensory functions seem to be located on the antennae.

With certain exceptions, crustaceans tend to be
dioecious, and internal fertilization is the rule, as the
sperm lack a locomotory flagellum. Eggs are often
brooded, and the hatching stage is typically a free-
swimming **nauplius larva** (Sec. 24-21). In most
crustaceans the nauplius larva gives rise to one or
more further larval stages before the adult is reached.
Adult form may be reached gradually through successive larval stages or suddenly through metamorphoses of a larval stage.

Crustaceans are known from the Cambrian

period, but the origin of the group and its phylogenetic relation to other mandibulates is not clear. It is assumed that the ancestral crustaceans were small, pelagic animals which had a body consisting of a linear array of similar segments, each bearing a pair of biramous appendages which served the functions of food capture, locomotion, and respiration. Only the ANOSTRACA among living crustaceans approximate these conditions. The major evolutionary trends in crustaceans have thus been toward increased tagmatization, fusion of segments, specialization of the limbs, and, with few exceptions, a benthic existence.

Subclass Malacostraca (Crayfish, lobsters, crabs, etc.)

The crustaceans are conveniently divided into two groups, a miscellaneous group of higher taxa, generally small in size, collectively called *entomostraca*, and the subclass MALACOSTRACA, which comprises all the well-known, larger crustaceans. More than two-thirds of all crustaceans are in the MALACOSTRACA, hence it is convenient to use a member of that

subclass to detail the anatomy and physiology of the crustaceans.

24-3 The crayfish This animal (*Cambarus, Astacus*, etc.) serves well for an introduction to the CRUSTACEA (Fig. 24-1). It is up to 15 cm (6 in) long and has appendages differentiated to serve special purposes. The common English names *crayfish* and *crawfish* (and even the American *crawdad*) probably derive from the French word *écrevisse* (Old German *krébiz*, crab). Different crayfishes inhabit freshwater streams, ponds, and lakes over much of the world. The lobster (*Homarus americanus*) of salt waters along the Atlantic Coast is much larger but resembles the crayfish closely in structure.

24-4 External features An exoskeleton containing chitin covers the entire body and is hardened except at joints, where it is thin and soft to permit movements. The body comprises an anterior rigid **cephalothorax** (head + thorax) and a posterior jointed **abdomen.** It is composed of **somites** (head 5, thorax 8, abdomen 6), each with a pair of jointed **appendages** ventrally. The skeletal elements of a single somite, as in the abdomen, include a transverse

Figure 24-1 Class CRUSTACEA: external features of a crayfish.

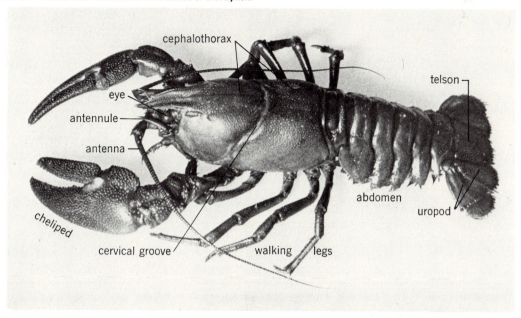

dorsal plate, or **tergum,** and a ventral crossbar, or **sternum,** joined together at either side by a lateral **pleuron.** Somites of the cephalothorax are covered by a continuous **carapace** shielding the dorsal surface and extending as the **brachiostegite** on the lateral surfaces. On this a transverse **cervical groove** marks the division between the head and thorax, and at the anterior end is a median pointed **rostrum.**

Beneath the rostrum on either side is a **compound eye,** stalked and movable. The **mouth** is between the mandibles and the **anus** opens ventrally in the broad median **telson** (not a somite) at the end of the abdo-

Table 24-1
APPENDAGES OF A CRAYFISH

Somite* and name of appendage	Number and structure of parts			Function
	Protopodite	Endopodite	Exopodite	
Head:				
I Antennule	3, statocyst in dorsal base	Short jointed feeler	Short jointed feeler	Equilibrium, touch, taste
II Antenna	2, excretory pore in ventral base	Long feeler of many joints	Thin pointed blade	Touch, taste
III Mandible	2, heavy jaw + base of palp	2, distal parts of palp	0	Biting food
IV 1st maxilla	2, thin medial plates	1, small, unjointed	0	Food handling
V 2d maxilla	2, bilobed plates	1, narrow exopodite and epipodite form scoop to draw water over gills		
Thorax:				
VI 1st maxilliped	2, broad medial plates + epipodite	2, slender, small	2, slender, minute	Touch, taste, food handling
VII 2d maxilliped	2, short, with gill	5, short, stout	2, slender	
VIII 3d maxilliped	2, with gill	5, larger	2, slender	
IX 1st walking leg	2, with gill	5, stout; heavy pincer at tip	0	Pincer for offense and defense
X 2d walking leg	2, with gill	5, slender, small pincer	0	Walking, grasping
XI 3d walking leg	2, with gill; sex opening in ♀	5, slender, small pincer	0	
XII 4th walking leg	2, with gill	5, slender, no pincer	0	Walking
XIII 5th walking leg	2, no gill; sex opening in ♂	5, slender, no pincer	0	
Abdomen:				
XIV 1st swimmeret	In ♀, reduced or none			Transfer sperm from ♂ to ♀
XV 2d swimmeret	In ♂: protopodite + endopodite fused, tubular			
	In ♂: 2 joints	Rolled, conical	Filamentous	
	In ♀: same as XVI			
XVI 3d swimmeret	2, short	Jointed filament	Jointed filament	Water circulation in both sexes; carrying eggs and young in ♀
XVII 4th swimmeret	2, short	Jointed filament	Jointed filament	
XVIII 5th swimmeret	2, short	Jointed filament	Jointed filament	
XIX Uropod	1, short broad	1, flat oval plate	1, flat oval plate, with hinge	Swimming; egg protection in ♀

(Left margin brackets indicate: head = somites I–V; thorax = somites VI–XIII; abdomen = somites XIV–XIX.)

* Some authorities consider that a somite without appendages precedes that bearing the antennule, which would make 6 somites in the head and 20 in all.

men. The **gills** are beneath the brachiostegite on either side of the carapace. The paired **female sex openings** are at the base of the third pair of walking legs on the thorax and those of the **male** at the fifth pair.

24-5 Paired appendages

24-5 Paired appendages (Table 24-1; Figs. 24-1, 24-2). Unlike the earthworm with its many simple and solid setae, the crayfish has one pair of jointed appendages on each somite; these contain opposed sets of muscles for their movement. The joints on the thoracic appendages are not all in the same plane, so these members may be moved in various directions. The appendages on different somites are of parts that differ in number, structure, and function.

The short **antennules** and long **antennae** are mobile sensory structures that receive and test stimuli from the environment. The stout **mandibles** crush the food, which is manipulated by the **maxillae** and **maxillipeds.** The large **chelae** (pincers) serve in offense and defense, and the other **walking legs** (*pereiopods*) are used for locomotion, handling food, and cleaning the body. The abdominal **swimmerets** (*pleopods*) aid in respiration and carry the eggs on females, and the wide **uropods,** with the telson, form a broad terminal paddle for swimming and for egg protection.

24-6 Muscles

24-6 Muscles In crayfish and other arthropods the muscles are complex, and all are contained within the exoskeleton, instead of being part of the body wall as in annelids or external to the skeleton as in vertebrates. The muscles attaching to many parts are in opposed pairs; a **flexor** to draw the part toward the body or point of articulation and an **extensor** to straighten it out. The largest muscles are the flexors which bend the abdomen forward under the body when the crayfish swims backward.

24-7 Digestive system

24-7 Digestive system (Fig. 24-3) This includes (1) the **mouth,** opening above the mandibles; (2) a short tubular **esophagus;** (3) the large thin-walled **stomach** in the thorax that is divided into a swollen anterior **cardiac chamber** and smaller **pyloric chamber** behind; (4) a short **midgut;** (5) the slender tubular **intestine** extending dorsally in the abdomen to (6) the **anus;** and (7) two large **digestive glands** (hepatopancreas, or "liver"), beneath the stomach. Each gland is three-lobed and contains many fine tubules. Food is brought to the mouth by the chelate second and third pairs of legs, the soft parts being torn and crushed to small size by the mandibles, then passed through the esophagus to the cardiac chamber. Projecting into this are strongly calcified **teeth,** one median and two lateral, that are moved by muscles attached outside the stomach, forming a **gastric mill** for further grinding of the food. The entrance to the pyloric chamber bears many hairlike setae that permit only fine particles to enter. The hepatopancreas secretes digestive enzymes, stores glycogen, fat, and

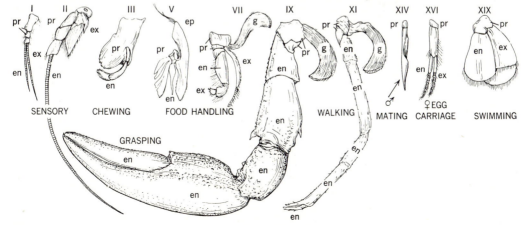

Figure 24-2 Representative appendages of a crayfish (right side, ventroposterior view), showing structural differentiation to serve various functions. *pr*, protopodite; *en*, endopodite; *ex*, exopodite; *ep*, epipodite; *g*, gill.

SENSORY   CHEWING   FOOD HANDLING   WALKING   MATING   CARRIAGE   SWIMMING

GRASPING

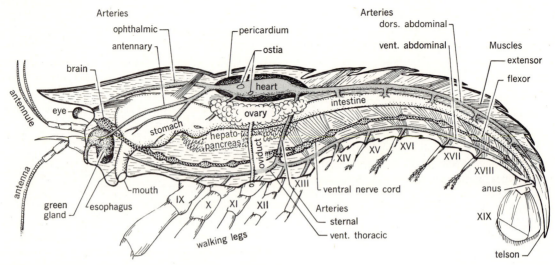

Figure 24-3 Crayfish: internal structure of a female. IX–XIX, somites.

calcium, and, together with the midgut, absorbs digested food. Hard particles are rejected by the mouth or cardiac mill, and undigested materials are formed as feces in the intestine to pass out the anus. The digestive tract, except the midgut, is lined by delicate chitin, continuous at the mouth and anus with the external cuticle; the entire lining is shed at each molt. Two gastroliths form on walls of the cardiac chamber prior to molt. Presumably they store lime (CaCO₃), withdrawn from the old exoskeleton, that later will be deposited in the new body covering.

24-8 Circulatory system The short muscular **heart** is of irregular shape and suspended in a large middorsal **pericardial sinus** of the thorax by six ligaments that attach to the sinus walls. Blood in the sinus enters the heart through three pairs of **valves** (ostia) and is pumped, by contraction of the heart, into six arteries that distribute to all parts of the body (Table 24-2). The arteries contain valves that prevent a return flow. From the finer arteries the blood flows into open spaces, or **sinuses,** between body organs, whence it collects in a large **sternal sinus** in the floor

Table 24-2
PRINCIPAL ARTERIES OF A CRAYFISH

Artery	Origin	Direction of blood flow	Supplies:
Ophthalmic (1) anterior aorta)	Heart	Anterior to head	Eyes, antennules, brain
Antennary (2)	Heart	Anteroventral	Antennae, green glands, anterior muscles, stomach
Hepatic (2)	Heart	Ventral	"Liver," pyloric stomach, midgut
Dorsal abdominal (1) (posterior aorta)	Heart	Posterior, top of abdomen	Intestine, telson, abdominal muscles
Sternal (1)	Base of posterior aorta	Ventral (passes through nerve cord)	Ventral thoracic and abdominal arteries, gonads
Ventral thoracic (1)	Sternal artery	Anterior, floor of thorax	Mouth, esophagus, appendages III–XI*
Ventral abdominal (1)	Sternal artery	Posterior, floor of abdomen	Abdominal muscles, appendages XII–XIX*

* See Table 24-1.

of the thorax and passes in afferent channels to the gills. There the CO_2–O_2 exchange of respiration occurs, and the blood returns in efferent channels to the **branchiocardiac sinuses** that lead up the inner sides of the thoracic wall to the pericardial sinus. This **lacunar** or **open system,** distributing between the tissues and without veins, is characteristic of the arthropods and is in striking contrast to the closed circulatory systems of annelids and vertebrates. The nearly colorless **blood plasma** has a dissolved respiratory pigment (hemocyanin) that transports oxygen to the tissues; there are free **amoeboid corpuscles** in the plasma.

24-9 Respiratory system The **gills** (Figs. 24-4, 24-5) are delicate plumelike projections of the body wall, containing blood channels and located along either side of the thorax in a **gill chamber.** The latter is covered by a lateral part (*branchiostegite*) of the carapace but is open ventrally and at both ends. A paddlelike projection (*scaphognathite*, or gill bailer) of the second maxilla moves back and forth in a channel under each side of the carapace to draw water containing dissolved oxygen over the gill filaments. The gills are arranged in lengthwise rows, the **podobranchiae** being attached to the coxopodites of appendages VII to XII, and the double row of **arthrobranchiae** (anterior VII to XII, posterior VIII to XII) to the membranes joining these appendages to the thorax, making 17 on each side. *Pacifastacus* has one thoracic **pleurobranchia,** derived from the body wall, above appendage XIII.

Figure 24-4 Crayfish: cross section of the body through the heart. Arrows indicate the course of blood flow. Dark, unoxygenated; light, oxygenated blood.

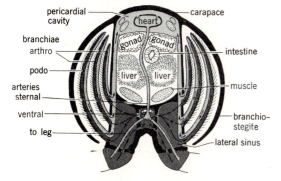

Figure 24-5 Crayfish: the gills (podobranchiae) exposed by removal of the branchiostegite. (*After Huxley*.)

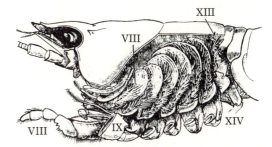

24-10 Excretory system Two large **green glands** (antennal glands), ventral in the head and anterior to the esophagus, serve to remove organic wastes from the blood and body fluids. Each consists of a glandular region, an expanded "bladder" with thin walls, and a duct opening ventrally on the basal segment of the antenna. The cavities in the excretory and genital organs are the only relics of the coelom in arthropods; the spaces within the body form a **hemocoel,** not comparable with the large coelom of annelids.

24-11 Nervous system In a crayfish, this system resembles that of an earthworm but is proportionately larger (Fig. 24-6). The **supraesophageal ganglia** ("brain") in the head send nerves to the eyes, antennules, and antennae, and a pair of **circumesophageal connectives** join it around the esophagus to the **subesophageal ganglia,** behind the mouth, at the anterior end of the double **ventral nerve cord.** The fused subesophageal ganglia of the adult represent six (or five) pairs of ganglia that are separate in the embryo; nerves from these pass to the mouth appendages, esophagus, green glands, and anterior muscles. Along the nerve cord each somite from VIII to XIX contains a pair of joined ganglia sending nerves to the appendages, muscles, and other organs. Giant nerve fibers arise from cell bodies in the brain and aid in the rapid escape reflex of the crayfish.

24-12 Sense organs As with most higher crustaceans, crayfish have a highly evolved set of sense organs. These enable a crayfish to test its environment continually, to find shelter, food, and mates, and to avoid unfavorable surroundings and preda-

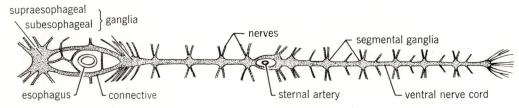

Figure 24-6 Crayfish: nervous system, dorsal view.

tors. Much of the body is sensitive to **touch,** especially the chelae, chelipeds, mouth parts, underside of the abdomen, and edge of the telson. On these regions are many **tactile hairs,** which are delicate plumelike projections of the cuticle containing fibers connected to sensory nerves lying beneath the cuticle. The **chemical sense** (taste + smell) resides in hairs on the antennules, tips of the antennae, mouth parts, and ends of the chelae. If meat juice is brought in a fine pipette to any of these, the animal vibrates its mouth parts, becomes excited, and turns towards the stimulus. Meat placed slightly away from a crayfish is "discovered" when juices diffusing from it through the water reach these sensory organs. The sense of **equilibrium,** or orientation to gravity, is served by a small chitin-lined sac, the **statocyst,** that opens dorsally under fine hairs on the basal segment of each antennule. It contains a ridge with many fine upright **sensory hairs** to which sand grains become attached by mucus and serve as **statoliths.** The action of gravity on the statoliths produces stimuli through the hairs, which are attached to an "auditory" nerve leading to the brain. Tilting or inverting the body alters the direction of gravitational pull on the statoliths and sets up a changed stimulus to the brain, which leads the animal to right itself. At a molt, the statolith lining and the sand grains are lost, but new grains are soon acquired from bottom debris. In proof of their function, if a freshly molted crayfish is placed in filtered water containing no foreign particles, so that no new statoliths can be acquired, its equilibrium will be disturbed. Should iron filings be placed in such water, some will enter the statocysts; if then a magnet is placed above or at the side of the crayfish, the latter will react by turning its ventral surface toward the magnet.

Each **compound eye** (Fig. 24-7) has an outer rounded surface covered by transparent cuticle, the **cornea,** which is divided into about 2500 microsco-

pic square **facets.** The facet is the outer end of a slender tapering visual unit, or **ommatidium.** Each of these consists of (1) the corneal facet, or **lens;** (2) two **corneagen cells,** which secrete the lens; (3) a **crystalline cone** of four cone cells; (4) **distal pigment cells** about the cone; (5) a long tapering **retinula** of eight cells, which form a central **rhabdom** where they meet; (6) **basal pigment cells** around the retinula and separating the rhabdom from adjacent rhabdoms; and (7) a **tapetum cell** between the inner bases of the retinular cells. The inner ends of the retinular cells penetrate the **basement membrane** on which all the ommatidia rest, and each connects by a **nerve fiber** through four **optic ganglia** to the **optic nerve,** which joins the brain.

The eye forms images in two ways (Fig. 24-7D, E). In strong light the pigment is extended to isolate each ommatidium from those adjacent. Light from any small area on an object that passes axially through a lens will reach the light-sensitive retinula and stimulate its nerve fibers, but any rays entering at an angle are refracted into the pigment and absorbed. Of any object, the eye will form a **mosaic image,** the light from different parts registering on separate ommatidia. Any slight shift of the object will stimulate other ommatidia; hence the compound eye is especially efficient in registering movement. In weak light the pigment recedes toward the distal and basal parts of the ommatidia, the light rays spread to adjacent ommatidia, and there forms on the retinulae a continuous or **superposition image,** which is probably less distinct than the apposition image but provides a proportionately stronger stimulus in relation to the amount of light received.

24-13 Reproductive system The sexes are separate in crayfish, and the female has a proportionately broader abdomen than the male. The first two pairs

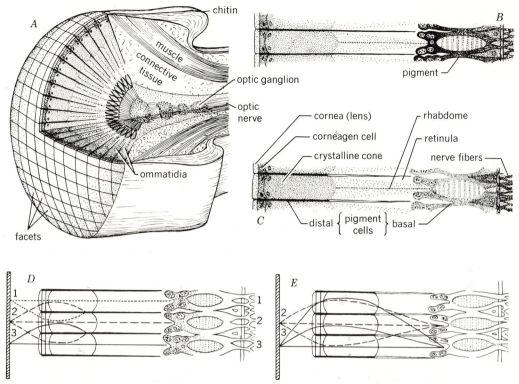

Figure 24-7 Crayfish: the compound eye; diagrammatic. *A*. Entire eye sectioned to show its general structure. *B*. One ommatidium in the light, pigment extended. *C*. Ommatidium in the dark, pigment contracted. *D*. Apposition image formed of separate images on retinulae from points 1, 2, 3 on object. *E*. Superposition image; each retinula receives oblique and direct rays from one point. (*Adapted in part from Imms, Textbook of entomology, E. P. Dutton & Co., Inc.*)

of appendages on the male's abdomen are tubular, for sperm transfer. The gonads are hollow and three-lobed in both sexes. In the **male** the two soft, white **testes** are fused and lie beneath the pericardial sinus. On each side a slender coiled **ductus deferens** extends ventrally to open at the base of the fifth walking leg. In the female the two **ovaries** are of comparable form and location with the testes of the male. Eggs develop in follicles on the inner surface and are discharged through an **oviduct** opening near each third walking leg.

24-14 Distribution and ecology Various genera and species inhabit fresh waters of Eurasia, Australia, New Zealand, North America from Canada to Guatemala, and southern South America. None are found in Africa or the Orient. In the United States there are five Pacific Coast species and six genera, with about 175 species from the Mississippi Valley

eastward. Common species are *Cambarus diogenes* and *bartoni* in the Eastern states, *Onconectes virilis* and others in the Midwest, and *Pacifastacus leniusculus* on the Pacific Coast. *Procambarus clarkii* of the Gulf states has become acclimatized in much of California.

Species are segregated ecologically. Many of them dig burrows 30 to 90 cm (1 to 3 ft) deep, straight or branched, that are topped by mud chimneys under the water. Some live habitually in burrows, others wander out during rainy periods, and some burrow in wet places only during drought. Nonburrowing kinds shelter under rocks or bottom debris, variously in ponds, lakes, or ditches or in sluggish or flowing streams. A few blind species are confined to underground waters in caves.

24-15 Natural history The crayfish is a solitary bottom dweller, hiding by day under stones or in crevices or burrows, where it keeps as much of its body as possible in contact with surrounding objects. The animal faces the entrance of its retreat, with the chelae extended, the antennae waving about and the gill bailers and swimmerets performing respiratory movements; other appendages are moved at times to receive stimuli or detect food. It grasps any food that passes within reach and will emerge to seize prey and then go back into its retreat. It resists attack by use of the strong chelae and tries to avoid being drawn out of its shelter; if the latter is opened or removed, the animal darts off to a new hiding place. The crayfish can walk forward, sideways obliquely, or backward with the fourth pair of legs bearing much of the weight. Through extension of the abdomen, uropods, and telson and sudden flexion of them under the body, the resistance offered these broad parts by the water enables the animal to "swim" or dart backward, and quick repetition of this action often enables it to escape danger. Crayfish are most in evidence when feeding actively in spring and early summer; in cold weather they retire to burrows or other safe retreats under water. During drought, the burrows are retreats for fishes and other aquatic animals that later emerge to reoccupy transient ponds.

The food includes live insect larvae, worms, crustaceans, small snails, fishes, and tadpoles, besides some dead animal matter. The burrowing species subsist extensively on stems and roots of plants, and other species use some such materials. The predators of crayfish besides humans, include certain fishes; large salamanders, turtles, and water snakes; herons, kingfishers, and other birds; and some aquatic mammals.

24-16 Molt Since the exoskeleton is rigid, it must be molted at intervals to permit increase in body size. The young molt about every 2 weeks and adults usually twice a year, in late spring and again in summer. Much experimental evidence shows that the X-organ hormone inhibits molt but that its action is suspended periodically; thereupon the Y-organ hormone induces molt (Sec. 8-2).

The entire process involves four stages: (1) premolt (proecdysis), including thinning of the old cuticle by removal and storage of lime in gastroliths or the hepatopancreas and laying down of new soft cuticle; (2) molt (ecdysis), shedding of old skeleton and swelling by water uptake; (3) postmolt (metecdysis), rapid deposition of chitin and salts in the cuticle; and (4) intermolt, a period when reserves are stored for the next molt and tissues grow. Prior to molt, the old inner layers of cuticle are digested away, calcium is resorbed and a new soft exoskeleton grows beneath and separates from the older, and the muscles and other structures within the appendages soften and are reduced in bulk. The old cuticle then opens dorsally between the carapace and abdomen, and the animal slowly withdraws itself, leaving its former covering complete, even to such minor features as the gut lining, eye facets, and setae. At this time much water is taken into the gut, increasing the body volume and thereby stretching the new cuticle. The animal hides from predators while the cuticle is hardening.

24-17 Regeneration A crayfish can replace lost parts, chiefly the appendages and eyes, but to a lesser extent than lower animals such as cnidarians or earthworms. This ability of **regeneration** is greater in the young animals than in older ones. Upon loss or removal of an appendage, the new one is partly formed at the next molt and increases in size with successive molts until it may be fully restored. Removal of the tip of an eye may be followed by normal regeneration, but if the entire structure is cut away, an antennalike replacement may result. Such regeneration of a part different from that which was removed is called **heteromorphosis.**

24-18 Autotomy The crayfishes and other crustaceans, particularly crabs, have the power of self-amputation, or **autotomy,** of the thoracic legs. If a cheliped or walking leg is broken or severely handled, the terminal five segments are cast off—sacrificed to the predator. The fracture is at a definite breaking plane, marked by a fine encircling line on the basal segment of the endopodite on a cheliped and at the third joint of the others. This procedure is a precaution to avoid undue loss of blood. Across the inside of the appendage, on the proximal side of the breaking plane, there is a diaphragm with a small opening through which the

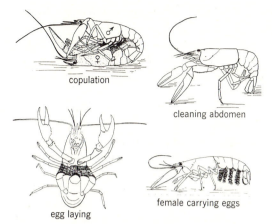

Figure 24-8 Crayfish: activities in reproduction. *(After Andrews, 1904.)*

copulation

cleaning abdomen

egg laying

female carrying eggs

nerves and arteries pass. The joint is fortified by interlocking external spines, and breaking is effected by a special muscle distal to the diaphragm. When the leg is cast off; the opening is quickly stopped by a blood clot and a small fold or valve and soon closes by cell growth. A miniature replacement of the limb begins to grow, and at subsequent molts this increases toward normal size and complete replacement.

24-19 Reproduction *Cambarus limosus (affinis)* of the Eastern states mates in March. The male grasps and inverts a female, stands over her, seizes her walking legs with his two chelae, and flexes his telson tightly over the end of her abdomen, so that she is held motionless (Fig. 24-8). He uses one of his fifth pair of walking legs to press the tips of the two modified swimmerets on his somite XIV against the sperm receptacles (annuli) between somites XII and XIII on her thorax. Sperm then pass in mucus to lodge in her receptacles, after which the animals separate. Some days or weeks later the female cleans her abdomen and swimmerets thoroughly, lies upside down with her abdomen sharply flexed, and a slimy secretion issues from glands on the swimmerets. Soon the 200 to 400 eggs (2 mm in diameter) are extruded from the oviducts, are fertilized by sperm from the seminal receptacles, and become affixed by the secretion to her swimmerets. She later resumes normal position and backs into a shelter, whereupon the eggs hang like berries and are aerated by movements of the swimmerets. Five or more weeks ensue until hatching. Each young is a miniature crayfish (Fig. 24-9) about 4 mm long, transparent, and attached to cuticle molted within the shell before hatching. Additional

Figure 24-9 Development of a crayfish; figures variously enlarged. *B.* Eggs in ovary surrounded by follicle cells; when laid they become attached to swimmerets on female (Fig. 25-9). *D, E.* Cleavage is superficial. *F.* The shallow gastrula forms beneath the yolk. *F–H.* Ventral views. *I.* Young removed from egg just before hatching. (*A, B,* from G. B. Howes, *Atlas of zootomy,* The Macmillan Company; *C–E,* after Zehnder, 1934; *F–I,* after Huxley.)

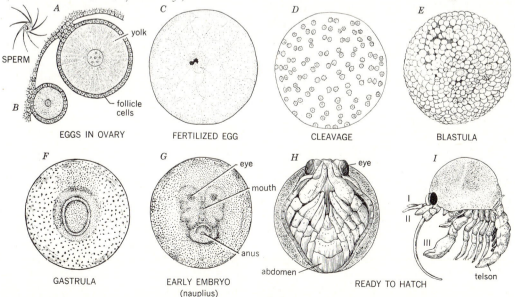

SPERM

yolk

follicle cells

EGGS IN OVARY

FERTILIZED EGG

CLEAVAGE

BLASTULA

GASTRULA

eye

mouth

anus

EARLY EMBRYO (nauplius)

eye

abdomen

READY TO HATCH

I

II

III

telson

molts occur frequently during the first months of life, the young enlarging at each. They remain attached to the female until the second stage, stay near her a few days more, and then become independent. By late autumn they are 35 to 50 mm (1½ to 2 in) long and are colored and shaped like the adult. *Procambarus clarkii* in Louisiana mates in late spring and produces eggs late in summer that hatch in about 15 days. The young grow through the winter, and some are of market size (85 mm, or 3½ in) by February. In the cold waters of Europe, *Astacus fluviatilus* produces eggs in October or November that do not hatch until the following June.

24-20 Other Malacostraca. The subclass MALACOSTRACA constitutes more than two-thirds of all the crustaceans and includes all the larger and better known forms. The structure of this subclass is con-

siderably more uniform than that of the various entomostracan groups. Hence the previously described anatomy of the crayfish serves well to summarize these animals. Other malacostracans bear mentioning here, however, to give a feeling for the diversity of the subclass.

The most primitive malacostracans are 10 species of the order LEPTOSTRACA which are unique among malacostracans in that they have eight rather than six abdominal segments. They are also the first of the malacostracans to appear in the fossil record in Silurian time.

The mantis shrimp of the order STOMATOPODA are striking animals which appear much like an aquatic praying mantis (Fig. 24-12). They have a small thorax and head region and a large abdomen. The second thoracic appendages are enlarged, with the terminal digit fitted with sharp spines or sharpened like a knife blade and folding into a groove in the immedi-

Figure 24-10 Class CRUSTACEA: some marine forms in their characteristic habitats; mostly reduced, but not to same scale. Subclass CIRRIPEDIA, order THORACICA: *Pollicipes,* goose barnacle; *Balanus,* acorn barnacle. Subclass MALACOSTRACA, order LEPTOSTRACA: *Nebalia.* Order MYSIDACEA: *Mysis,* opossum shrimp. Order ISOPODA: *Ligia.* Order AMPHIPODA: *Orchestia,* sand hopper; *Caprella,* skeleton shrimp. Order STOMATOPODA: *Squilla.* Order DECAPODA: *Penaeus,* shrimp; *Panulirus,* spiny rock lobster; *Pagurus,* hermit crab; *Cancer,* edible crab; *Uca,* fiddler crab.

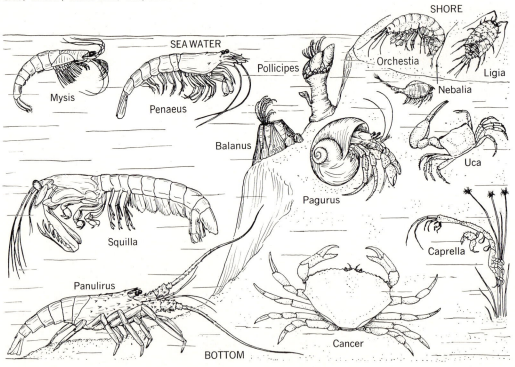

ate proximal segment of the appendage, much like a pocket-knife blade. These appendages are employed to capture prey by quickly snapping the blade shut. Some of the larger forms are called "thumb splitters" because of this ability.

The MALACOSTRACA are dominated by two superorders: PERACARIDA and EUCARIDA. These two contain more than 90 percent of all malacostracan species. They represent the two major lines of evolution within this subclass and as such are very successful. The PERACARIDA are distinguished from the EUCARIDA by the following features: 1) no larval stages, young hatch as juveniles; 2) a ventral brood pouch in which eggs are carried; and 3) a carapace which, if present, is not attached to all thoracic segments. EUCARIDA, on the other hand, have a number of larval forms, no ventral brood pouch, and a carapace attached to all thoracic segments. In both superorders the same evolutionary tendencies can be observed. Thus the primitive forms are all filter feeders, whereas the advanced forms are omnivorous or predaceous. Most peracarideans are small organisms. Members of the order ISOPODA are generally dorsoventrally flattened inhabitants of hard substrates, where they scrape or bite off food (Figs. 24-10, 24-11, 24-13). One group is of interest in that it eats its way through wood and destroys pilings. Many isopods are also parasitic, some becoming aberrant morphologically. The AMPHIPODA include the common sand or beach hoppers found among

Figure 24-12 The stomatopod *Lysiosquilla*. (*Courtesy of R. Larson, U. S. National Museum.*)

rotting seaweed on beaches (Figs. 24-10, 24-11, 24-13) and also such odd forms as whale lice (cyamids), which live on whales, and caprellids, which are much like miniature praying mantises. In contrast to isopods, which have been successful on land, amphipods are almost exclusively aquatic.

The dominant eucaridean order is the DECAPODA,

Figure 24-11 Class CRUSTACEA: some freshwater inhabitants in their characteristic habitats. Subclass BRANCHIOPODA, order ANOSTRACA: *Branchinecta*, fairy shrimp. Order NOTOSTRACA: *Lepidurus*. Order DIPLOSTRACA: *Daphnia*, water flea. Subclass OSTRACODA, order PODOCOPA: *Eucypris*. Subclass COPEPODA, order CYCLOPOIDA: *Cyclops*. Subclass MALACOSTRACA, order ISOPODA: *Porcellio*, sow bug (on land). Order AMPHIPODA: *Hyalella*. Order DECAPODA: *Cambarus*, crayfish. Some enlarged (*Daphnia, Eucypris, Cyclops*) and others reduced, but not to same scale.

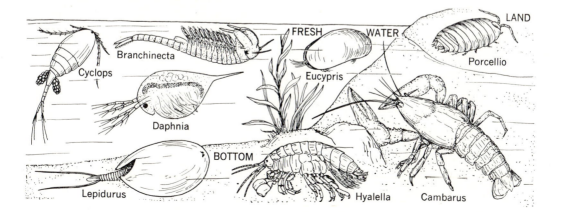

A.

B.

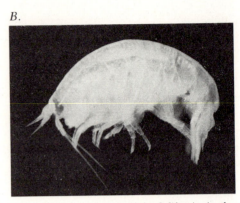

Figure 24-13 *A.* **The isopod** *Cirolana harfordi.* *B.* **The amphipod** *Orchomene obtusa.* (*Courtesy of the California Academy of Sciences and G. McDonald.*)

of which the crayfish is a member (Fig. 24-14). This is a large and extremely diverse group found in all habitats. Since this group is the most highly evolved, it also shows the greatest amount of concentration of the central nervous system and consequently behavioral patterns. Many of the adaptations to habitats and survival in general are truly remarkable and include complex commensal relationships with other invertebrates such as sponges, cnidarians, and mollusks, as well as the "cleaning behavior" exhibited by many shrimp in removing parasites from free-swimming fishes. Still others, such as hermit crabs, select and employ mollusk shells as refuges,

while others display complex reproductive behavior.

Crabs may become covered with marine growths —algae, sponges, or barnacles—and some pick off and attach bits of sponge to the shell. In one kind of hermit crab, living in a snail shell, a sponge (*Suberites*) grows over and dissolves the shell, leaving the crab encased in the sponge as a shelter.

24-21 *Malacostracan larvae*   Like some other invertebrates of fresh waters, the young crayfish at hatching resembles the adult except for minor

Figure 24-14 Decapoda. *A.* The ghost shrimp *Callianassa californiensis.* *B.* The crab *Cancer gracilis.* (*Courtesy of G. McDonald.*)

B.

A.

details. Most crustaceans, however, have several larval stages, the younger of which are quite unlike the parent animals. These minute larvae represent the means of dispersal, especially of sedentary forms. The first three larval stages of the lobster have exopodites on the thoracic legs, as in an adult mysid, so are termed *mysid larvae*.

Shrimps and prawns (*Penaeus*) begin as a minute **nauplius larva** (Fig. 24-15) with an unsegmented body, a simple median eye, and three pairs of appendages (the later antennules, antennae, and mandibles). With molt the **metanauplius** and **protozoea** stages appear, the latter with seven pairs of appendages and beginning somites. Then follows the **zoea,** with distinct cephalothorax and abdomen, eight pairs of appendages, and six more beginning. This molts into the **mysis** (schizopod) larva, having 13 pairs of appendages on the cephalothorax and those of the thorax bearing exopodites that serve in swimming. A further molt yields the adult, with 19 pairs of appendages. In a crab the first, or **zoea,** larva has a helmetlike carapace with long dorsal and anterior

Figure 24-16   A megalops larva of a decapod crustacean. (*Courtesy of G. McDonald.*)

spines and sessile eyes; the thorax bears two pairs of biramous swimming legs (maxillipeds), but the slender mobile abdomen lacks swimmerets. In the last, or **megalops,** larva the carapace is broad but lacks spines, the eyes are large and stalked, the thorax has five pairs of walking legs, and there are functional swimmerets on the abdomen (Fig. 24-16). Both these minute stages swim in surface waters. Later the megalops sinks to the bottom and molts into typical crab form, with a still broader carapace, the abdomen folded beneath the carapace, and its swimmerets useless for locomotion.

Lower Crustacea

24-22 *Lower Crustacea*   As the most morphologically diverse mandibulates, the thousands of crustacean species present a wide variety in body structure, coloration, habitat, and mode of life (Figs. 24-10, 24-11). The smaller and more primitive crustaceans collectively termed *Entomostraca* include a number of taxonomically different groups. Perhaps the most primitive living crustaceans are the few tiny species of the subclass CEPHALOCARIDA, first discovered in the 1950s living in the mud of Long Island Sound. These shrimplike animals have all trunk appendages as well as the maxillae similar and triramous. Unfortunately, little is known of their biology. The subclass BRANCHIOPODA are mainly freshwater crustaceans inhabiting ponds and lakes or are present in stressful environments otherwise unsuitable for most other life, such as saline lakes

Figure 24-15   Larval stages and adult of a prawn or shrimp, *Penaeus,* **showing the succession of changes in body form and in the appendages (1–19).** (*After F. Müller, and Huxley.*)

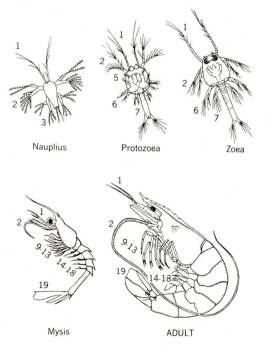

Nauplius   Protozoea   Zoea

Mysis   ADULT

and salt ponds. They are an important constituent of freshwater plankton, and most common forms are pelagic, employing their biramous leaflike appendages in locomotion, respiration, and filter feeding. This is considered the primitive condition for the whole class. Among some species males are uncommon or absent, and successive generations consist only of females produced parthenogenetically, a condition similar to that encountered in rotifers (Sec. 19-22). Eggs are often extremely resistant to desiccation and temperature and are able to hatch after months or years of drying and/or cold storage. In contrast to most crustaceans, branchipods usually have direct development (a nauplius is produced in the brine shrimp, *Artemia*). Some branchiopods, such as the water fleas (*Daphnia*), also show a curious change in external morphology in successive generations over the seasons, a phenomenon termed **cyclomorphosis.**

The subclass OSTRACODA includes a large number of usually small crustaceans that are completely enclosed in a chitinous bivalve shell. They have the appearance of small bivalve mollusks. Most species scuttle around on the bottom of fresh and marine waters, employing mainly their large antennae to move them over the substrate. One of the most striking features of the ostracod body is that it consists mainly of the segments and appendages of the head, the trunk being drastically reduced in size and number of segments.

The subclass COPEPODA has the largest number of species of any of the entomostracan groups, 7500, and is widely distributed in fresh and salt waters (Fig. 24-17). Copepods are enormously abundant as individuals and constitute perhaps the most important group in the plankton of the world's oceans. They are ecologically perhaps the most important of the crustacea in that they are the major herbivores, grazing on the phytoplankton and thus forming the basis of most food chains in the sea. Most free-living copepods are very small, with a short body of head plus 10 trunk segments. Compound eyes are absent, but a single simple median eye is present. The first antennae are long and conspicuous. The copepods also include a large number of parasitic forms, most of which are ectoparasites on fishes (Fig. 24-20), but they may be found on or within certain inverte-

Figure 24-17 The copepod *Gaussia princeps*. (*Courtesy of G. McDonald.*)

brates as well. The parasites are often so highly modified that they may be recognized as copepods only in their larval stages. Copepods are dioecious animals. Eggs hatch as nauplius larvae and go through several copepodite larval stages before becoming adults.

The subclasses BRANCHIURA and MYSTACOCARIDA are small, little-known groups. The BRANCHIURA (Fig. 24-20) are all specialized ectoparasites of fishes, whereas the MYSTACOCARIDA are tiny free-living animals found between the sand grains on marine beaches.

The subclass CIRRIPEDIA (barnacles) are highly modified crustaceans in which the adults are hermaphroditic, enclosed in a calcareous shell, and of sessile habit. The common goose barnacles (*Lepas, Pollicipes*) and rock or ship barnacles (*Balanus*) attach to fixed objects (Figs. 24-10, 24-18, 24-19); other species fasten only to gorgonians, crabs, sharks, sea turtles, or whales, and still others are parasitic (Fig. 24-21). The egg develops in the parent's mantle cavity and hatches out a microscopic nauplius larva, which swims, feeds, and molts one to three times in a week or so, with slight changes in form. Another molt produces the very different **cypris larva,** with a bivalve shell and additional appendages and containing fat globules for buoyancy. This stage lasts for 4 days to 10 to 12 weeks in different species; the larva then settles to the bottom and hunts a place of attach-

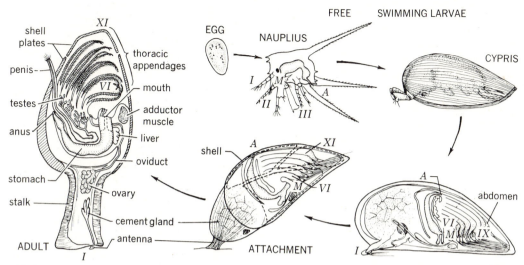

Figure 24-18 Subclass CIRRIPEDIA: goose barnacle, *Lepas.* Adult with right side of mantle and shell removed. From the egg a free-swimming nauplius larva hatches that feeds and molts to become a cypris larva. The latter attaches by its antennules and cement gland, then transforms to the sessile adult stage. Egg and larvae much enlarged. *M*, mouth; *A*, anus; I, antennule; II, antenna; III, mandible; VI–XI, other appendages.

Figure 24-19 The goose barnacle *Lepas fasicularis.* (*Courtesy of G. McDonald.*)

ment, to which it adheres by the antennules, aided by secretion from a cement gland. In a rather complete metamorphosis the bivalve shell is lost, the body mass alters in form, and the valves of the adult shell appear. The barnacle thenceforth remains fixed, literally "standing on its head." When feeding, the long delicate thoracic appendages repeatedly fan out well above the shell opening, then curve and quickly withdraw, carrying food to the mouth. The parasitic barnacles attack echinoderms, corals, and other crustaceans (Fig. 24-21). Still other barnacles bore into calcareous substrates such as mollusk shells.

24-23 Commensal and parasitic crustaceans

Most species of crustaceans are free-living, but others are associated with various animals in relations ranging from casual attachment to full parasitism. Some crabs dwell in the tubes inhabited by annelid worms (Fig. 22-16); certain copepods, amphipods, and porcelain crabs are regular tenants of the snail shells used by hermit crabs; and other small crabs (*Pinnotheres*) live within the shells of oysters and mussels. Besides the many free-living isopods and copepods, others are ectoparasites on various fishes (Fig. 24-20); their anterior limbs are modified as hooks or suckers, and some (*Aega*) have mouth parts specialized to penetrate and suck blood from their hosts. Certain forms (such as *Cymnothoa*) are free-swimming as larvae but permanently attached as adults, and others are parasitic also as larvae. The copepod *Lernaea* is so far modified as to be of wormlike form. Isopods of the suborder EPICARIDA are all parasitic on prawns and crabs. The most extreme of parasitic crustaceans are barnacles of the order RHIZOCEPHALA, of which *Sacculina* is a common example (Fig. 24-21). Its

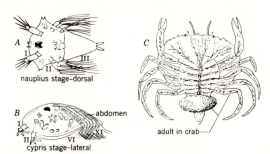

Figure 24-21 Subclass CIRRIPEDIA: *Sacculina*, **a barnacle parasitic in crabs.** I–XI, appendages. Compare Fig. 24-18. (*After Stempell.*)

earliest (nauplius) larva resembles that of other crustaceans and barnacles but lacks a digestive tract. The later (cypris) larva fastens to a gill on a crab, then discards its shell, and burrows into the host. It becomes a mere mass of cells that pass into the crab's bloodstream, attach to the intestine, and grow. When next the crab molts, part of the parasite protrudes on the crab's abdomen as an ovary, later to be packed with eggs. The remaining internal parts become branched and rootlike, penetrating all parts of the crab's body and absorbing nourishment from its tissues. The host thenceforth neither grows nor molts, and its sex organs degenerate so that it cannot breed.

24-24 Relation to humans

Flesh of certain crustaceans is much esteemed as human food. In the United States, some crayfish are eaten in the Northwest and enormous quantities in the Southern states. These and other crustaceans are also eaten in other parts of the world. Shrimps are captured with seines, but crabs, lobsters, and crayfishes are lured into baited traps (lobster pots, etc.) of wire, wood, or net. The leg muscles of crabs and the abdominal muscles of the others are either freshly boiled or canned. The blue crab (*Callinectes*) of the Atlantic Coast is caught and held captive until it molts, then sold in the soft-shelled condition; after removal of the viscera, the whole animal is cooked and eaten. The small crustaceans that abound in salt water and fresh water are important links in the food cycle for many fishes and for other aquatic animals. *Artemia* and *Daphnia* are collected as adults or eggs and

Figure 24-20 Parasitic CRUSTACEA that live on fishes; enlarged. A. *Caligus* (subclass COPEPODA), second antennae hooked. B. *Argulus* (subclass BRANCHIURA), second maxillae become sucking disks.

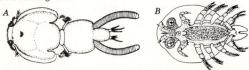

sold by aquarium dealers as fish food. Some cope-pods are the intermediate hosts for worm parasites of humans and various vertebrates. Crayfishes in the Gulf states often damage fields of cotton and corn by eating the young plants, and occasionally their bur-rows injure levees. Sow bugs sometimes eat plants in gardens and greenhouses, and the gribble, a wood-boring isopod, burrows into and damages wharves in salt water. Barnacles are among the most impor-tant fouling organisms on the hulls of ships.

Classification

Phylum Arthropoda.

Joint-footed animals. (See Chap. 23.)

Class 1. Crustacea.

Crustaceans. Head (of 5 fused somites) with 2 pairs of antennae, a pair of jaws, and 2 pairs of maxil-lae; body usually with a dorsal carapace and ending in a telson containing the anus; exoskeleton with lime deposits; appendages often biramous; respira-tion by gills (or body surface); excretion by antennal or maxillary glands; sex openings paired, anterior; eggs usually carried by female; development usually with larval stages; mainly aquatic; 26,000 species.

Subclass 1. Cephalocarida. Small, slender; anten-nules and antennae small, distinct; second maxillae well developed, and 9 pairs of postcephalic append-ages, all triramous; no eyes or carapace; monoecious; most primitive living Crustacea, resemble †Lepido-caris of Middle Devonian; marine, in mud at bottom of bays of Atlantic Coast and San Francisco; 4 spe-cies. Hutchinsoniella.

Subclass 2. Branchiopoda. Phyllopods. Antennules small, uniramous; second maxillae reduced or ab-sent; thoracic appendages 4 pairs or more, leaflike margined by gills; sexes separate, parthenogenesis common; mostly in fresh water; Cambrian to Recent, 800 species.

Order 1. Anostraca. Fairy shrimps. Slender, no carapace; eyes stalked; trunk appendages 11 to 19 pairs. *Artemia salina,* brine shrimp, 10 mm (0.4 in) long, in salty pools and lakes, often cultured for fish food; *Branchinecta; Eubranchipus vernalis,* to 23 mm (0.9 in) long, in temporary spring pools and ponds.

Order 2. Notostraca. Tadpole shrimps. Cara-pace low, oval; trunk appendages 35 to 70 pairs; eyes sessile. *Lepidurus; Triops (Apus),* to 30 mm (1.2 in) long.

Order 3. Diplostraca. Clam shrimps and water fleas. Laterally compressed carapace enclosing trunk and appendages; second antennae large.

Suborder 1. Conchostraca. Clam shrimps. Cara-pace bivalved, enclosing compressed body; ap-pendages 10 to 32 pairs; eyes sessile. Devonian to Recent. *Leptestheria,* Kansas to California; *Lyn-ceus,* Eastern states and southward.

Suborder 2. Cladocera. Water fleas. Minute; cara-pace, usually bivalved, not enclosing head; paired eyes fused, median, and sessile; trunk ap-pendages 4 to 6 pairs; second antennae enlarged for swimming, movements jerky. *Daphnia pulex,* water flea, 2 mm long; *Alona; Leptodora,* 12 to 18 mm ($\frac{1}{2}$ to $\frac{3}{4}$ in) long.

Subclass 3. Ostracoda. Seed shrimps. Minute; carapace bivalved, compressed, enclosing body made up mainly of head segments; only 2 pairs of trunk appendages; fresh water or salt water, mostly on or near bottom; 2000 species. Ordovician to Recent.

Order 1. Myodocopa. Carapace notched; second antennae enlarged at base, alone used in swimming. *Cypridina, Concoecia,* northern oceans.

Order 2. Cladocopa. Carapace unnotched; both pairs of antennae used in swimming, second pair with 2 branches; marine. *Polycope.*

Order 3. Podocopa. Carapace unnotched; sec-ond antennae leglike, clawed at tips. *Eucypris, Dar-winula,* both in fresh waters; *Entocythere,* on crayfish gills; *Cythereis,* marine.

Order 4. Platycopa. Carapace unnotched; both pairs of antennae large but not used for swimming, second pair flattened, biramous; marine. *Cytherella.*

Subclass 4. Mystacocarida. Size microscopic; body cylindrical; antennules and antennae long, promi-nent; 4 eyespots, no compound eyes; thorax of 4 seg-ments, each with 1 pair of simple appendages; excre-tion by both antennal and maxillary glands; ventral nerve cord paired; sexes separate; 3 species. *Dero-cheilocarus typicus,* damp intertidal sand near Woods Hole, Mass.

Subclass 5. Copepoda. Mostly small to microscopic; form various; typically 10 free trunk somites, with 4 lacking appendages, but reduced in parasitic species; appendages mostly biramous; 3 ocelli often fused as median eye, no compound eyes; excretion by maxillary glands; eggs on abdomen of female in 1 or 2 egg sacs; fresh water and salt water, free-living, commensal or parasitic; 4500 species.

Order 1. Calanoida. Body constricted behind somite of fifth leg; first antennae very long, second antennae biramous; generally planktonic. *Calanus finmarchicus,* "brit," 4 mm long, northern oceans, important food of herring, mackerel, and whales; *Diaptomus,* in freshwater lakes, ponds, and some inland saline water.

Order 2. Harpacticoida. Body scarcely constricted between somites of fourth and fifth legs; first antennae short, second antennae biramous; generally benthic; salt water, brackish water, and fresh water. *Harpacticus.*

Order 3. Cyclopoida. Body constricted between somites of fourth and fifth legs; antennae uniramous; in freshwater lakes and ponds and salt water, some parasitic. *Cyclops.*

Order 4. Notodelphyoida. Body articulation between IVth and Vth thoracic somites, (male) and between Ist and IId abdominal somites (female). *Doropygus,* commensal in tunicates.

Order 5. Monstrilloida. No antennae or mouth parts; larvae parasitic in marine worms. *Monstrilla,* with intermediate larvae in polychaete worms.

Order 6. Caligoida. Body articulation between IIId and IVth thoracic somites, may be absent in female; ectoparasitic in gill chambers of freshwater and marine fishes, attached by antennae. *Caligus.*

Order 7. Lernaeopodoida. Segmentation and appendages reduced or none; ectoparasites of freshwater and marine fishes, attached by second maxillae. *Lernaea,* larva on gills of flatfish, wormlike female on gills of codfish; *Salmincola,* dwarf male permanently attached to female that parasitizes gills of trout.

Subclass 6. Branchiura. Fish lice. Body flat; carapace large, disklike, covering head and thorax; abdomen small, bilobed; first maxillae modified as suckers; compound eyes present; sessile; 75 species. *Argulus,* parasitic on fishes of fresh and salt waters.

Subclass 7. Cirripedia. Barnacles. Adults sessile, attached or parasitic; attach by cement gland on first antenna; carapace becomes mantle surrounding body, usually with limy plates; 6 or fewer pairs of slender and bristly biramous appendages behind mouth used in food gathering; abdomen vestigial; usually monoecious; larvae free-swimming, marine; 900 species. Ordovician to Recent.

Order 1. Thoracica. Mantle and 6 pairs of trunk appendages. *Lepas, Pollicipes,* goose barnacles, body on fleshy stalk; *Balanus,* acorn shell or rock barnacle, body in irregular conical shell, no stalk.

Order 2. Acrothoracica. Boring barnacles. Mantle present; chitinous disk for attachment; fewer appendages. *Trypetesa.*

Order 3. Ascothoracica. Parasitizing echinoderms and corals; digestive tract with branches into mantle; 6 pairs of trunk appendages. *Laura,* parasitic in black coral.

Order 4. Rhizocephala. A mantle but no shell, appendages, or gut; body saclike, with absorptive "roots" penetrating host. *Sacculina,* parasitic on crabs.

Subclass 8. Malacostraca. Lobsters, crayfishes, crabs, pill bugs, beach hoppers, mantis shrimp. Body typically of 19 somites (5 head, 8 thorax, 6 abdomen); head fused to one or more thoracic somites; mandibles and maxillae with palps; commonly with carapace; abdomen with appendages; 18,000 species.

Series 1. Phyllocarida. Abdomen of 8 somites.

Order 1. Leptostraca. (Nebaliacea). Carapace bivalved; marine. Cambrian to Recent. *Nebalia,* to 12 mm ($\frac{1}{2}$ in) long.

Series 2. Eumalacostraca. Abdomen of 6 (or fewer) somites.

Superorder 1. Syncarida. No carapace.

Order 1. Anaspidacea. *Anaspides,* fresh waters of Australia.

Superorder 2. Peracarida. Carapace, when present, leaving 4 or more thoracic somites distinct; females with thoracic brood pouch where young develop; no larvae, development direct.

Order 1. Mysidacea. Opossum shrimps. Carapace over much of thorax, but fused to only 4 segments; uropods forming tail fan; mostly marine; 450 species. Carboniferous to Recent. *Mysis,* Great Lakes and Eurasia.

Order 2. Cumacea. Small; carapace with 2 anterior extensions often joined over head; abdomen slender, mobile; uropods slender; mostly marine, burrowing in surface sand or mud. *Diastylis,* 10 mm (0.4 in) long.

Order 3. Tanaidacea. Mostly minute; carapace small; second thoracic appendage with chela; telson unjointed; marine, to depths of 3,650 m (12,000 ft) in mud or in tubes. *Apseudes, Tanais.*

Order 4. Isopoda. Pill bugs, wood lice. Thoracic appendages except first all similar; body usually depressed dorsoventrally; no carapace; abdomen short, partly or all fused; in salt water or fresh water among plants or under stones, some terrestrial, many parasitic on fish and crustaceans; 4000 species; Pennsylvanian to Recent. Marine: *Cirolana; Aega psora,* salve bug of fishermen; *Limnoria lignorum,* gribble, 3 mm long, burrows in marine timbers, damaging wharves; *Idothea* and *Ligia,* free-living. Fresh water: *Asellus communis,* to 15 mm (0.6 in) long. Land: *Oniscus asellus* and *Porcellio scaber,* sow bugs; *Armadillium vulgare,* pill bug, rolls up. Parasitic: *Hemioniscus,* on barnacles; *Hemiarthrus,* on decapod crustaceans.

Order 5. Amphipoda. Sand hoppers. Thoracic and abdominal appendages each in at least two different functional groups. Body often laterally compressed; no carapace; abdomen flexed ventrally between IIId and IVth somites; telson usually distinct; mostly marine; 4600 species. Tertiary to Recent. *Orchestia,* beach flea, on sand or under seaweed; *Hyalella,* in fresh water; *Gammarus,* in both fresh and salt waters; *Caprella,* body cylindrical, abdomen vestigial, on seaweeds; *Cyamus,* whale louse, on skin of whales, body narrow, legs with hooked claws, abdomen vestigial.

Superorder 3. Hoplocarida. Head with two movable anterior somites bearing eyes and antennules.

Order 1. Stomatopoda. Mantis shrimps. Antennal scale enlarged; gills on abdominal appendages; second thoracic appendage modified for raptorial feeding; carapace small; marine on bottom in sand or crevices; 200 species. Carboniferous to Recent. *Squilla; Chloridella empusa,* Florida to Cape Cod, to 250 mm (10 in) long, edible.

Superorder 4. Eucarida. Carapace large, fused to and covering all of thorax; eyes stalked; gills thoracic; no brood pouch; larval stages present.

Order 1. Euphausiacea. Thoracic appendages all biramous and similar, none modified as maxillipeds; marine. *Euphausia,* "krill," to 25 mm (1 in) long, often abundant, an important whale food. 90 species.

Order 2. Decapoda. Thoracic appendages mostly uniramous; first three thoracic appendages maxillipeds; walking legs 10 (5 pairs); mostly marine, some in fresh water, few terrestrial; many edible; Triassic to Recent, 8500 species. Abdomen long, *Penaeus,* shrimp, and *Palaemonetes,* prawns; *Panulirus,* spiny rock lobster; *Homarus,* lobster, on Atlantic shores; *Astacus, Cambarus,* crayfishes, in fresh waters. Abdomen short (REPTANTIA, crabs): *Pagurus,* hermit crab, lives in snail shell; *Emerita (Hippa),* sand crab; *Libinia,* spider crab, legs long; *Birgus latro,* coconut crab, to 30 mm (12 in) long, terrestrial in tropics; *Cancer,* large edible rock crab; *Callinectes,* edible blue, or soft-shelled, crab of Atlantic Coast; *Paralithodes,* Alaska king crab; *Panopeus,* mud crab; *Uca,* fiddler crab; *Pinnotheres,* oyster or mussel crab, in mantle cavity of oysters and mussels.

References

Adiyodi, K. G. and **R. G. Adiyodi.** 1970. Endocrine control of reproduction in decapod Crustacea. *Biol. Rev.,* vol. 45, pp. 121–165.

Borradaile, L. A., and **F. A. Potts.** 1958. The Invertebrata. 3d ed. New York, Cambridge University Press. Rev. by G. A. Kerkut, Arthropoda, pp. 317–577, figs. 218–391.

Bousfield, E. L. 1973. Shallow water gammaridean Amphipoda of New England. Ithaca, Cornell University Press. xii + 312 pp., illus.

Calman, W. T. 1909. Crustacea. In E. R. Lankester. Treatise on zoology. London, A. & C. Black, Ltd. Pt. 7, fasc. 3,viii + 346 pp., 194 figs.

———. 1911. The life of Crustacea. London, Methuen & Co. xvi + 289 pp., 32 pls., 85 figs.

Carlisle, D. B., and **Francis Knowles.** 1959. Endocrine control in crustaceans. New York, Cambridge University Press. 119 pp., 18 figs., 5 plates, some in color.

Carthy, J. D. 1965. The behaviour of arthropods. San Francisco, W. H. Freeman and Company. 148 pp., 41 illus.

Darwin, C. 1851–54. A monograph on the subclass Cirripedia. 2 vols. London, The Ray Society.

Edmondson, W. T. (editor). 1959. Ward and Whipple, Fresh-water biology. 2d ed. New York, John Wiley & Sons, Inc. Crustacea, pp. 558–901, illus.

Green, J. 1961. A biology of crustacea. London, H. F. & G. Witherby Ltd. xv + 180 pp., 58 figs.

Gurney, Robert. 1942. Larvae of the decapod crustacea. London, The Ray Society. vi + 306 pp., illus. Reprinted 1960. Codicote, England, Wheldon and Wesley.

Kaestner, A. 1970. Invertebrate zoology. vol. 3. Crustacea. New York, John Wiley & Sons, Inc., 523 pp., illus.

Kükenthal, W., and others. 1927. Handbuch der Zoologie. Berlin. Walter de Gruyter & Co. vol. 3, pt. 1, ARTHROPODA, pp. 211–276, figs. 182–215; CRUSTACEA, pp. 277–1078, figs. 216–1171.

Lockwood, A. P. M. 1967. Aspects of the physiology of Crustacea. San Francisco, W. H. Freeman and Company. x + 328 pp., 68 figs.

Marshall, S. M., and A. P. Orr. 1972. The biology of a marine copepod. New York, Springer-Verlag. 195 pp., 63 figs. (Reprint of the 1955 edition.)

Moore, R. C. (editor). 1955–1969. Treatise on invertebrate paleontology. Lawrence, University of Kansas Press. Arthropoda 1, 1959. xix + 560 pp., 415 figs. Arthropoda 2, 1955. xvii + 181 pp., 565 figs. Arthropoda 3, 1961. xxiii + 442 pp., 334 figs. Arthropoda 4, 1969. xxxvi + 398 pp., 216 figs.

Neale, John (editor). 1969. The taxonomy, morphology and ecology of recent Ostracoda. Edinburgh, Oliver & Boyd Ltd. ix + 553 pp., illus.

Pennak, R. W. 1953. Fresh-water invertebrates of the United States. New York, The Ronald Press Company. CRUSTACEA, pp. 321–469, figs. 203–292.

Rathbun, M. J. 1918. The grapsoid crabs of America. *Bull. U.S. Natl. Museum,* vol. 97, xxii + 461 pp., 161 pls., 172 figs.

———. 1925. The spider crabs of America. *Bull. U.S. Natl. Museum,* vol. 129, xx + 613 pp., 283 pls., 153 figs.

———. 1930. The cancroid crabs of America. *Bull. U.S. Natl. Museum,* vol. 152, xvi + 609 pp., 230 pls., 85 figs.

Richardson, Harriet. 1905. A monograph of the isopods of North America. *Bull. U.S. Natl. Museum,* vol. 54, liii + 727 pp., 740 figs.

Schmitt, W. L. 1965. Crustaceans. Ann Arbor, University of Michigan Press. 204 pp., 75 figs.

Shrock, R. R., and W. H. Twenhofel. 1953. Principles of invertebrate paleontology. New York, McGraw-Hill Book Company. ARTHROPODA, pp. 536–641, figs. 13-1 to 13-58.

Snodgrass, R. E. 1938. Evolution of the Annelida, Onychophora, and Arthropoda. *Smithsonian Institution, Miscellaneous Collections,* vol. 97, no. 6, pp. 1–159, 54 figs.

———. 1952. A textbook of arthropod anatomy. Ithaca, N.Y. Comstock Publishing Associates. viii + 363 pp., 88 figs.

Waterman, T. H. (editor). 1960. The physiology of crustacea. New York, Academic Press, Inc. Vol. 1. Metabolism and growth, xvii + 670 pp., illus. Vol. 2. Sense organs, integration, and behavior, xiv + 681 pp., illus.

25

CLASS INSECTA: INSECTS

The grassphoppers, flies, lice, butterflies, beetles, bees, and a host of similar small arthropods that comprise the class INSECTA (L. incised, into distinct parts) number perhaps 900,000 species. They are the most abundant and widespread of all land animals, being the principal invertebrates that can live in dry environments and the only ones able to fly. These habits are made possible by the chitinous body covering that protects the internal organs against injury and loss of moisture, by the extensions of this covering that form the wings, and by the system of tracheal tubes that enable insects to breathe air. The ability to fly helps them to find food and mates and to escape predators. Because their life cycles usually are short, they can multiply rapidly under favorable conditions. Insects abound in all habitats except the sea; various kinds live in fresh waters and brackish water, in soil, on and about plants of all kinds, and on or in other animals. Different species eat all sorts and parts of plants—roots, stems, leaves, sap, blossoms, seeds, or fruits; many flower-visiting insects aid in pollination. Others utilize the tissues, fluids, and excretions of animals, and the scavenger insects consume dead animals and plants. Parasitic insects live on or within other animals and plants at the latter's expense but often do not kill their hosts. Parasitoids deposit their eggs in the eggs, larvae, pupae, or adults of other insects, and their larvae consume the host, then emerge as free-living individuals. Some insects transmit diseases—virus, bacterial, protozoan, or other—to plants, animals, and humans. Insects in turn are eaten by other insects, spiders, scorpions, and many vertebrates from fishes to mammals. The predaceous, parasitoid, and parasitic species serve importantly to regulate the numbers of other insects.

Entomology (Gr. *entomon*, insect) is the science dealing with insects. Because of the numbers and many biologic relations of insects, they are of great economic significance; some are useful and others are harmful to humans.

25-1 Characteristics[1]

1 Head, thorax, and abdomen distinct; head with one pair of antennae (except PROTURA); mouth parts for chewing, sucking, or lapping, consisting of mandibles, maxillae, and a labium (fused second maxillae); thorax (of three somites) with three pairs of jointed legs and usually two (or one or no) pairs of wings; abdomen of 11 or fewer somites, terminal parts modified as genitalia.

2 Digestive tract of fore-, mid-, and hindgut; mouth with salivary glands.

3 Heart slender, with lateral ostia and an anterior aorta; no capillaries or veins; body spaces a hemocoel (coelom reduced).

4 Respiration by branched cuticle-lined tracheae that carry oxygen from paired spiracles on sides of thorax and abdomen directly to the tissues (except some PROTURA and COLLEMBOLA); some aquatic forms with tracheal or blood gills.

5 Excretion by two to many fine Malpighian tubules attached to anterior end of hindgut (except COLLEMBOLA).

6 Nervous system of supra- and subesophageal ganglia connecting to double ventral nerve cord, with one pair or fewer ganglia per somite; sense organs usually include simple and compound eyes, chemoreceptors for smell on antennae and for taste about mouth, and various tactile hairs; some with means for sound production and reception; no statocysts.

7 Sexes separate; gonads of multiple tubules with one median duct posteriorly; fertilization inter-

nal; ova with much yolk and protecting shells; cleavage superficial (except COLLEMBOLA); development with several molts and direct, or with several nymphal stages and gradual metamorphosis, or with several larval stages, a pupa, and complete metamorphosis to adult form; parthenogenesis in aphids, thrips, gall wasps, etc.

25-2 General features and evolution

The insects are the most successful terrestrial animals living today as measured by the enormous number of species and the tremendous adaptive radiation they have undergone. Insects have occupied essentially all niches available on land as well as a significant number in fresh water. They have, however, been notably unsuccessful in penetrating the oceans. No single factor can be pointed to as the one responsible for this tremendous success, but certainly one of the unique attributes of insects among all invertebrates is their power of flight. This has undoubtedly contributed to their success in allowing access to more habitats, permitting maximum dispersal, and making possible escape from potential predators.

Insects are distinguished from other mandibulates by having a single pair of antennae on the head, a body in three parts (head, thorax, and abdomen), and three pairs of legs on the thorax. Wings, usually two pairs, are borne on the thorax. Although superficially insects resemble the crustaceans and hence have been grouped with them in the MANDIBULATA, they are really quite different. Insects lack the second pair of antennae found in crustaceans, and their limbs are never biramous even in embryonic stages. Respiration is by a tracheal system, which is not present in any crustacean, or by gills which are not homologous to those in crustaceans. For these and other reasons, many zoologists feel that the two groups are not closely related and have evolved from separate ancestral forms.

Insects are known from as far back as Devonian time, and current theory is that they may have evolved from an ancestral form which resembled the current living forms of the class SYMPHYLA.

The sudden appearance of winged insects in Carboniferous rocks is a spectacular feature of the fossil

---

[1] *Compare phylum* ARTHROPODA, *Chap 23, Table 23-1.*

record. Several theories have been proposed to account for the origin of insect wings. The paranotal theory is based on the presence in many Carboniferous insects (†PALAEODICTYOPTERA) of lateral expansions of the prothoracic terga, suggesting that the functional wings of meso- and metathorax may have arisen from similar broad flaps. Presumably these could have been used for gliding or planing even before they became movable as true wings. The paranotal theory would derive winged insects from forms like the present day DIPLURA, and these, in turn, are derivable from myriapodlike ancestors that resembled present-day SYMPHYLA. The tracheal-gill hypothesis is based on the resemblance between wings and the movable gills found on present-day mayfly nymphs. Primitive mayflies were present in Paleozoic times, but this theory would require an unknown aquatic ancestor for winged insects on a line separate from the terrestrial APTERYGOTA.

25-3 Size   Some insects are smaller than large protozoans, and others exceed the smallest of the vertebrates. Some beetles (TRICHOPTERYGIDAE) are but 0.25 mm (0.01 in) long, and a few egg parasites (MYMARIDAE) are even smaller. Most insects are 2 to 40 mm (0.8 to 1½ in) long. The longest include *Pharnacia serratipes* (ORTHOPTERA), 260 mm (10 in); a Venezuelan beetle, *Dynastes hercules*, 155 mm (6 in); and a bug, *Lethocerus grandis*, 115 mm (4½ in). The wingspread is greatest in some tropical moths, *Thysania (Erebus) agrippina*, 280 mm (11 in), and *Attacus atlas*, 240 mm (9½ in); some fossil insects (†*Meganeura*) exceeded 700 mm (28 in) in wingspread.

Grasshopper

To indicate some of the variety in the structure and function of insects, this chapter describes a grasshopper and includes a comparative account of the honeybee. A grasshopper is generalized as to anatomy, has chewing mouth parts, undergoes a gradual (paurometabolous) or incomplete metamorphosis from the young, or nymph, stages to the adult, and lives independently for a single season.

Grasshoppers[1] occur over the world, mainly in open grasslands, where they eat leafy vegetation. The following description will serve for any common species such as the Carolina grasshopper (*Dissosteira carolina*), the American grasshopper (*Schistocerca americana*), or the eastern lubber grasshopper (*Romalea microptera*) (Fig. 25-1).

25-4 External features   The body comprises a head of six fused somites, a thorax of three somites with legs and wings, and a long segmented abdomen ending with reproductive organs (Fig. 25-2). It is covered by a cuticular exoskeleton containing chitin, secreted by the epidermis beneath and molted periodically in the nymphs to permit increase in size; adults do not molt. The exoskeleton is formed into hard plates, or **sclerites,** separated by soft cuticle that permits movement of the body segments and appendages. Pigment in and under the cuticle provides

[1] *Order* ORTHOPTERA, *Family Acrididae. A grasshopper is an essentially solitary and resident species, often abundant in individuals, but may occasionally migrate. The term "locust" applies properly to gregarious and migratory Old World forms* (Locusta); *the North American periodic cicada* (Magicicada, *order* HOMOPTERA), *however, is also called "locust."*

Figure 25-1 Class INSECTA. *Above.* Long-winged grasshopper (*Dissosteira longipennis*), ×1¼. *Below.* Short-winged, or lubber, grasshopper. (*Brachystola magna*), ×¾. *(After Walton, 1916.)*

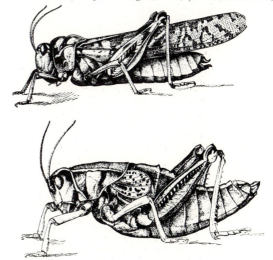

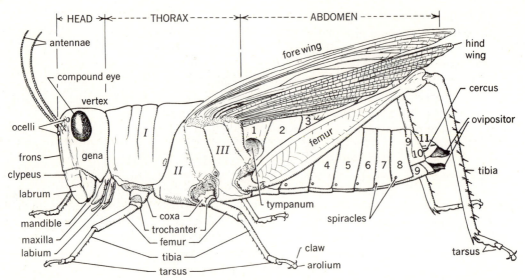

Figure 25-2 External features of a grasshopper, a generalized insect; female. I to III, somites of thorax; 1 to 11, somites of abdomen.

protective coloration by which grasshoppers resemble their environments.

The **head** (Fig. 25-3) bears one pair of slender, jointed **antennae** with fine sensory bristles, two lateral **compound eyes** that are unstalked but constructed like those of the crayfish, and three simple

eyes, or **ocelli.** Much of the head is enclosed in a fused case, or head capsule, with a dorsal **vertex,** lateral cheeks or **genae,** and the anterior **frons.** Below the latter is a broad plate, the **clypeus.** The mouth parts are of the chewing, or mandibulate, type, ventral on the head, and include (1) a broad upper lip, or

Figure 25-3 A grasshopper. *Left.* Front view of the head. *Right.* Mouth parts in anterior view. Both enlarged.

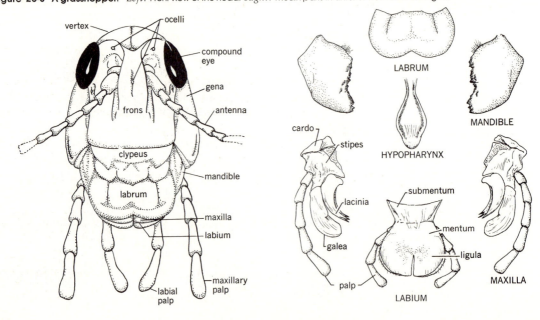

labrum, hinged to the clypeus; (2) a median tongue-like **hypopharynx;** (3) two heavy blackish lateral jaws, or **mandibles,** each with teeth along the inner margin for chewing food; (4) a pair of **maxillae** of several parts and with slender sensory palps at the sides; and (5) a broad median lower lip, or **labium,** with two short palps.

The **thorax** consists of the large anterior **prothorax** (with a dorsal saddlelike pronotum), the **mesothorax,** and the posterior **metathorax;** each bears a pair of jointed legs, and the meso- and metathorax each a pair of wings. The sclerites on each somite form a dorsal **tergum** of four fused plates, a **pleuron** of two plates on each side, and a single ventral **sternum.** Each **leg** is a linear series of segments: (1) the short **coxa,** which articulates to the body; (2) a small **trochanter** fused to (3) the stout **femur;** (4) a slender spiny **tibia;** and (5) the **tarsus** of three parts, the proximal bearing four pairs of ventral **pads** and the distal having a fleshy **arolium** between two **claws.** The arolia enable the grasshopper to hold on to smooth surfaces, and the claws serve in rough places. All the legs are used in walking and climbing. Each metathoracic leg has a large femur containing muscles and a long tibia, which serve for leaping. The narrow **forewings** (tegmina) are colored and slightly stiff. The **hind wings** are broad and membranous, with many veins, and fold under the forewings at rest. Each wing develops as a saclike projection of the

body covering and flattens to a thin double membrane that encloses tracheae, nerves, and blood sinuses. The cuticle thickens along the sinuses to form strengthening **veins** (Fig. 25-4). When of full size the wings become hard and dry, but blood flow continues in some veins. The wing veins are of such constant pattern in species and higher groups of insects as to be useful in classification.

The slender cylindrical **abdomen** consists of 11 somites, the terminal ones being modified for copulation or egg laying. Along the lower sides of the thorax and abdomen are 10 pairs of small openings, the **spiracles,** connecting to the respiratory system.

The first abdominal somite is divided about the insertions of the hind legs, with the sternum firmly united to the mesothorax; its tergum contains on either side an oval **tympanum** over an organ of hearing. On somites II to VII the tergum is ∩-shaped and joins the ∪-shaped sternum by lateral membranes that permit the abdomen to pulsate in breathing. In a male, somite VIII resembles those preceding, the terga of IX and X are fused, and the tergum of XI forms the suranal plate over the anus; a small spine or **cercus** (relict appendage) projects behind X on either side, and the long ventral sternum of IX encloses the male copulatory organ. In a female, the terga of VIII to XI and cerci are as in the male, the sternum of VIII is large and has a posterior median egg guide; that of IX is long; a lateral plate of XI is

Figure 25-4 The wings and body covering of insects; diagrammatic. *A.* Generalized wing showing the principal veins. *B.* Cross section of wing and veins. *C.* Section of body covering at junction of two somites. *D.* Structure of the body wall or exoskeleton. (*A, B, after Metcalf and Flint; C, D, adapted from Snodgrass.*)

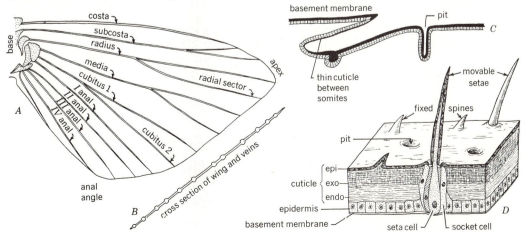

present; and the end of the abdomen bears three paired lobes forming the ovipositor (from somites VIII and IX).

25-5 Muscles The head contains complex small muscles that move the antennae and mouth parts. In the thorax are large muscles that manipulate the wings and legs. Segmental muscles are most conspicuous in the abdomen but are small as compared with those of a crayfish; some perform the respiratory movements, and others control the reproductive structures.

25-6 Digestive system (Fig. 25-5) The mouth parts surround (1) the **mouth cavity** which opens into (2) a slender muscular **pharynx** that widens in (3) a short **esophagus** joining to (4) the large thin-walled **crop.** Below the crop are small branched salivary glands that discharge through ducts opening at the labium. Beyond the crop is (5) a small **proventriculus,** or gizzard, lined by plates. The preceding parts constitute the foregut. Next is (6) the midgut, **stomach** (ventriculus) joined by (7) a series of six double finger-shaped glandular **gastric caeca.** The hindgut, or (8) **intestine,** consists of a tapered anterior part, slender middle portion, and enlarged **rectum** that opens at (9) the **anus.** Food is held by the forelegs, labrum and labium, lubricated

by the salivary secretion (which contains some enzymes), and chewed by the mandibles and maxillae; the palps bear organs of taste. Chewed food is stored in the crop, further reduced in the proventriculus, and strained into the stomach. There it is digested by enzymes secreted by the gastric caeca and absorbed. In the rectum, excess water is withdrawn from the undigested material, which is formed into slender fecal pellets and passed out the anus.

25-7 Circulatory system The slender tubular **heart** lies against the dorsal wall of the abdomen in a shallow **pericardial cavity** formed by a delicate transverse **diaphragm.** Blood enters the heart through pairs of minute lateral openings, or **ostia,** with valves, and is pumped forward by contractions of the heart into a dorsal **aorta** extending to the head. There it emerges into the body spaces, or **hemocoel,** between the internal organs and moves slowly backward around these organs, finally returning to the pericardial sinus. Some blood circulates in the appendages and wing veins. The system is open, or lacunar, as in other arthropods, with no capillaries or veins. The clear plasma contains colorless blood cells that act as phagocytes to remove foreign organisms. The blood serves mainly to transport food and wastes, as there is a separate respiratory system. The **fat body** is a loose network of tissue in

Figure 25-5 A grasshopper; female. Internal structure as seen with left side of body wall removed; tracheae omitted. Compare Figs. 23-5, 24-3.

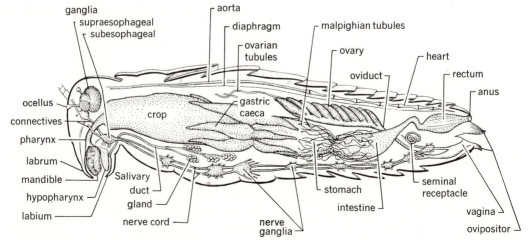

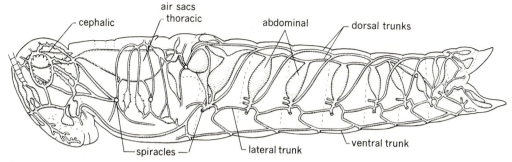

Figure 25-6 A grasshopper: respiratory system. Tracheal trunks, air sacs, and spiracles of the left side. (*After Albrecht, 1953.*)

spaces between the organs that stores food reserves, especially in young insects before metamorphosis.

25-8 Respiratory system   (Figs 25-6, 25-7) The paired **spiracles** connect to a system of elastic ectodermal air tubes, or **tracheae,** that branch to all parts of the body. The finest branches, or **tracheoles,** carry oxygen to and remove carbon dioxide from the tissue cells. The tracheal wall is a single layer of thin cells that secretes a lining of cuticle (cast off at molting), and the larger tubes are reinforced by a spiral thread to prevent their collapse; longitudinal air trunks connect to the spiracles. The grasshopper, unlike some insects, has several large, thin-walled **air sacs** in the abdomen where alternate contraction and relaxation of the body wall serve to pump air in and out of the tracheal system. At inspiration the first four pairs of spiracles are open and the posterior six closed, and at expiration the arrangement is reversed, so that there is a definite circulation of air in the tracheae. The finest tracheoles contain fluid in which the oxygen dissolves before actually reaching the tissue cells.

Some harmful insects can be killed by the use of oil films, emulsions, or dusts that cover or clog the spiracles and stop respiration.

25-9 Excretory system   To the anterior end of the hindgut are joined a number of threadlike **Malpighian tubules;** these lie in the hemocoel and have their free ends closed. The tubule wall is composed of a single layer of large cells that remove urea, urates, and salts from the blood and discharge into the intestine (Fig. 7-7).

25-10 Nervous system   (Figs. 25-5, 25-8) The brain, or **supraesophageal ganglion,** in the head comprises three pairs of fused ganglia (proto-, deuto-, tritocerebrum) with nerves to the eyes, antennae, and other head organs. It joins by two **connectives** around the esophagus to the **subesophageal ganglion,** which is also of three pairs (mandibular, maxillary, and labial). From the latter, the ventral **nerve cord** extends posteriorly as a series of paired

Figure 25-7 The tracheae of insects. *A.* Large trunks and branches. *B.* Cellular wall of a tracheal tube and its internal spiral thread. *C.* Terminal branches around muscle fibers. *D.* Fine tracheoles distributed over muscle fibers. (*After Snodgrass.*)

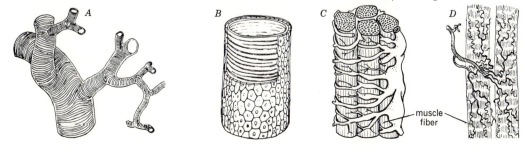

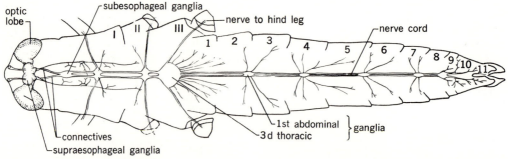

Figure 25-8   A grasshopper: nervous system in dorsal view.   I–III thoracic somites; 1–11, abdominal somites. (*After Riley, 1878.*)

ganglia and longitudinal connectives. Each thoracic somite contains a pair of ganglia with nerves to the legs, wings, and internal organs. There are only five pairs of abdominal ganglia; some originally separate have become fused. These send nerves to various posterior organs. There is also a visceral or **sympathetic nervous system,** made up of an esophageal portion with ganglia and nerves connecting to the brain, foregut, midgut, and heart, and a posterior portion from the last abdominal ganglion to the hindgut and reproductive system. A fine pattern of peripheral nerves lies beneath the epidermis of the body wall.

25-11 Sense organs   The sensory receptors of a grasshopper are adapted for receiving stimuli from the air and the land. They include (1) **tactile hairs** on various body parts, especially the antennae, mouth palps, abdominal cerci, and distal leg segments; (2) **olfactory organs** on the antennae; (3) **organs of taste** on the palps and other mouth parts; (4) the **ocelli,** which are sensitive to light and shade but do not

form images; (5) the **compound eyes,** which function essentially like those of a crayfish, and (6) the **organ of hearing.** The last (Fig. 25-14) consists essentially of a stretched tympanic membrane that is set into movement by sound vibrations in the air; this affects a slender point beneath the membrane connecting to sensory nerve fibers. A grasshopper produces sound by rubbing the hind tibia, which has a row of minute pegs along the medial surface, against a wing vein to set the latter into vibration.

25-12 Reproductive system (Fig. 25-9).   The sexes are separate and show secondary sexual characters in the terminal parts of the abdomen (Sec. 25-4). In a **male,** each of the two **testes** comprises a series of slender tubules or **follicles,** above the intestine, that are joined to a lengthwise **ductus deferens.** The two ducts unite as a common median **ejaculatory duct,** which is joined by **accessory glands** and opens at the end of the large ventral male **copulatory organ** (penis). In a **female,** each **ovary** is composed of several tapering egg tubes, or **ovarioles,** in which

Figure 25-9   Reproductive systems of insects; diagrammatic. (*After Snodgrass.*)

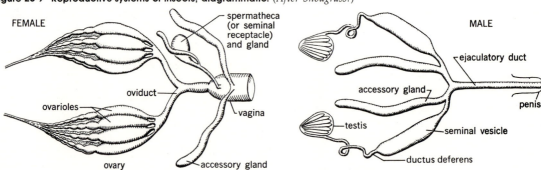

the ova are produced, and is joined to an **oviduct.** The two oviducts unite as a median **vagina** that leads posteriorly and is joined by a duct from the **seminal receptacle** (spermatheca), where sperm received at copulation are stored. The female tract opens close to the ventral **egg guide.**

25-13 Natural history In the warm days of early spring the nymphs hatch from eggs laid in the soil the previous autumn. They resemble adults but differ in proportions and have no wings or reproductive organs (Fig. 25-11). They feed on tender vegetation and hide under plants or in crevices to avoid predators and desiccation. After a few days the cuticle softens and is molted; the emerging nymph increases in volume and its fresh cuticle hardens and darkens. Each individual has five (or six) nymphal stages, and its entire growth period requires 30 to 50 or more days. The wings first appear as small pads, which become larger at successive molts and unfold to full size after the last molt into the adult stage.

Both nymphs and adults eat many kinds of vegetation, especially succulent types; they often migrate into new feeding grounds and may damage or ruin farm and garden plantings. Feeding is most active in the midmorning hours of quiet sunny days. When food is scarce, these insects will eat cotton or woolen fabrics, wood, and disabled grasshoppers. Adults of some species under conditions of crowding, sometimes undertake long migrations.

Grasshopper eggs are eaten by some beetles, bee flies, moles, skunks, and mice, the nymphs by robber flies and digger wasps, and both nymphs and adults by large predatory insects and by frogs, reptiles, birds, and mammals. One-tenth of all insects found in bird stomachs examined by the U.S. Biological Survey were grasshoppers and their close allies. Eggs of grasshoppers are parasitized by certain insects. Flesh flies (*Sarcophaga*) lay living maggots on adults, and tachinid flies deposit their eggs on grasshoppers in flight; the larvae of both burrow into their hosts and consume the fat tissues. Parasitized grasshoppers become lethargic and fail to reproduce, or die. The parasitic insects thus constitute a factor in grasshopper control. Both fungal and bacterial diseases also destroy numbers of grasshoppers at times.

Humans practice control by using chemical sprays and poisoned baits on fields where nymphs and adults feed and also by plowing weed or stubble fields to expose the egg masses.

25-14 Reproduction Some days after the adult stage is reached in late summer, the grasshoppers begin to mate and may do so several times. The male clings to the female's back, inserts his genitalia into her vagina, and transfers spermatozoa. After a further interval, egg laying (Fig. 25-10) begins and continues into the autumn. The female uses her ovipositor to form a short tunnel in the ground in which the eggs are placed, surrounded by a sticky secretion that fastens them together as an egg pod. The eggs are 3 to 5 mm (0.1 to 0.2 in) long. About 20 are laid at a time, and one female may lay up to 10 lots. The adults die some days after breeding is ended.

In the ovary, each egg is enclosed by a delicate inner **vitelline layer,** or **membrane,** and a brownish flexible shell, or **chorion,** that contains a minute pore, or **micropyle,** through which sperm enter dur-

Figure 25-10 A female grasshopper, with greatly extended abdomen, depositing eggs in ground; below, a complete pod, or packet eggs. (*After Walton, 1916.*)

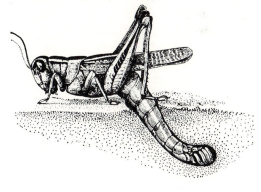

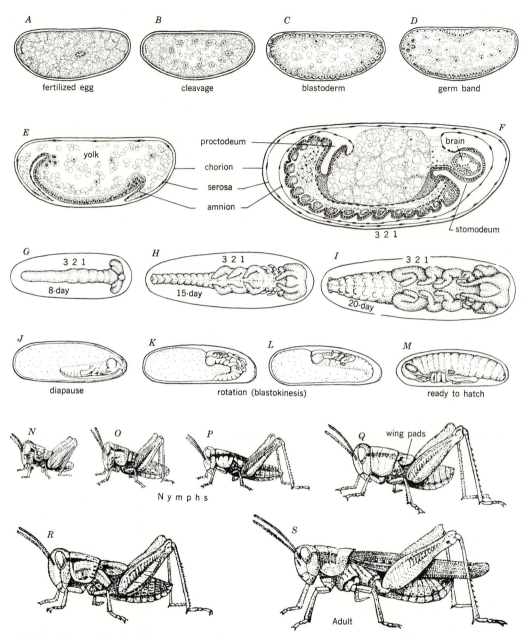

Figure 25-11 Development of a grasshopper. *A*. Fertilized egg surrounded by chorion. *B*. Cleavage, nuclei scattered. *C*. Blastoderm cells surround yolk. *D*. Germ band formed. *E*. Early embryo, serosa, and amnion forming. *F*. Later embryo with somites. *G*. 8-day embryo. *H*. 15-day embryo. *I*. 20-day embryo, ready for diapause; 1, 2, 3, thoracic somites. *J*. Embryo in diapause. *K,L*. Blastokinesis, or rotation of the embryo. *M*. Embryo before hatching. *N–R*. The five nymphal stages with gradual increase in size and development of wing pads. *S*. Adult with wings. (*A–F, generalized sagittal sections, after Johannsen and Butt, 1941; G–I, ventral views, after Slifer, 1932; J–M, lateral views, after Burkholder, 1934; N–S, modified from Emerton in Riley, 1878.*)

ing laying. Development (Fig. 25-11) begins at once and continues for about 3 weeks until the embryo is well formed. Then a rest, or **diapause,** ensues until spring, when growth is resumed and the young hatch and crawl to the ground surface. The diapause is a means of surviving the adverse conditions of cold and lack of food in winter.

The fertilization nucleus within the egg yolk divides into scattered cleavage nuclei. These migrate to the periphery of the yolk, where each is surrounded by cytoplasm and a cell membrane, and the cells form an epithelium, or blastoderm, around the yolk. Those of a limited ventral area thicken as a **germ band** that will produce the **embryo,** and the lateral and dorsal cells become the embryonic envelope, or **serosa.** At the ends and sides of the germ band, folds then form; their outer layers, inside the serosa, become the **amnion,** which encloses the embryo in an amniotic cavity. A lengthwise ventral furrow along the germ band folds up to form a layer (mesoderm + ?endoderm) above the germ band. The latter then divides by cross furrows, from before backward, into a linear series of somites that give rise to the head and its appendages, the thorax and its legs, and the segmented abdomen. The future foregut (*stomodeum*) forms as a pit at the anterior end and the hindgut (*proctodeum*) similarly at the posterior end. Later the midgut forms from endoderm cells and the gut becomes a continuous tube. The tracheae develop as paired lateral invaginations of ectoderm. The nervous system arises as an infolding of ventral ectoderm into two lengthwise strands of cells that later produce the nerve cords, ganglia, and brain.

The embryo shows six primitive head somites: (1) preoral, with compound eyes; (2) antennal, and (3) intercalary, respectively homologous with the antennule and antennal somites of the crayfish; (4) mandibular, behind the mouth; (5) maxillary; and (6) labial, with a pair of embryonic appendages that provide the fused labium of the adult. There are six pairs of nerve ganglia in the head, three preoral pairs forming the supraesophageal ganglia and three postoral forming the subesophageal ganglia. Rudiments of abdominal appendages appear but later disappear except posteriorly, where they form parts of the external genitalia.

The Honeybee

25-15 A social insect   The honeybee, *Apis mellifera* (order HYMENOPTERA; Gr. *hymen,* membrane + *pteron,* wing), resembles a grasshopper in general structure but is specialized in many features. It has mouth parts suited for both sucking and chewing, undergoes complete metamorphosis (holometabolous) from the wormlike larva through a pupal stage to the flying adult, feeds on nectar and pollen, and lives socially in a permanent colony comprising many individuals of three **castes** (Fig. 25-12). The **queen** lays the eggs; the males or **drones,** serve only to fertilize new queens; and the thousands of sterile females, or **workers,** build and guard the hive, provide the food for all castes, attend the queen, and rear the young. Wild honeybees live in natural cavities of trees or rocks, but humans have partly domesticated this species and house it in hives of wood. Each colony lives amid vertical **combs** of wax that contain small lateral cells used to store honey or pollen and rear the young. Workers collect fluid **nectar** from flowers; this is chemically altered and stored as the sirupy carbohydrate solution that we call **honey.**

Figure 25-12 The honeybee, *Apis mellifera* (order HYMENOPTERA); growth stages and three adult castes slightly enlarged; portion of comb reduced. (*After Phillips, 1911.*)

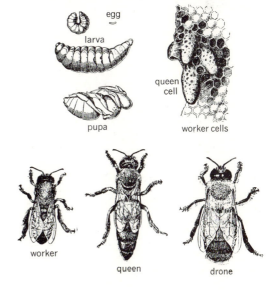

They collect **pollen** (bee bread) to provide proteins for growth of the larvae. Bees also gather resins from plant buds which, as **propolis** (bee glue), serve to cement and varnish crevices in the hive against wind and water. The workers have **stings** to protect the colony and its honey against robbery by other animals.

25-16 Structure and function   The body of a bee (Fig. 25-13) is densely covered by hairs having short lateral barbs on which pollen grains lodge easily. Unbranched hairs occur on the compound eyes and legs. On the foreleg, the tibia is margined by an **eye brush** of stiff hairs for cleaning the compound eyes, and its distal end bears a flat movable spine, or **fibula.** The latter closes over a bristle-lined notch on the proximal end of the tarsus to form an **antenna comb** through which the antenna is drawn to remove pollen or other foreign material. Long hairs on the large first segment of the tarsus form a cylindrical **pollen brush** to gather pollen from the foreparts of the body. On the middle leg, the flat tarsus also has a pollen

brush to remove pollen from the forelegs and body; and the inner distal end of the tibia bears a **spur** used to pick up wax. On the hind leg, the wide tibia is slightly concave externally and margined by incurving hairs to form a **pollen basket** (corbicula). This has a comb of stiff hairs, the **pecten,** at its distal end and just below a flat plate, or **auricle.** The outer surface has a **pollen brush** for cleaning the body posteriorly, and its inner surface carries about 10 rows of stiff downward-pointing spines forming a **pollen comb.**

The thin delicate **wings** lie flat over the back at rest. In flight the two on each side are locked together by fine hooks bordering the hind wing that catch to a groove along the rear margin of the forewing (Fig. 25-13C). The wings may vibrate up to 400 times per second, with the tips moving in a figure-eight path. Workers are capable of long flights, even up to 13 km (8 mi).

The smooth **mandibles** of workers serve to gather pollen and also to mold wax in making combs. The maxillae and labial palps form a tube around the slender **tongue,** or labium; by movements of the

Figure 25-13 The honeybee worker.   *A.* Mouth parts, pollen-collecting structures, and sting. *B.* The hairs. *C.* The wing-locking mechanism. Leg segments: *cx,* coxa, *tr,* trachanter; *f,* femur; *ti,* tibia; *ts,* tarsus. (*Adapted in part from Casteel, 1912.*)

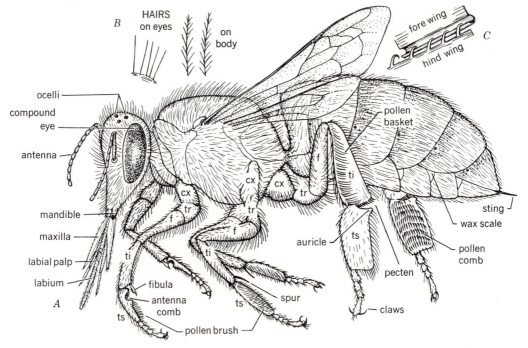

tongue and a pumping action of the pharynx, fluid nectar is drawn into the large crop, or **honey stomach.** Behind the latter are four triangular lips that form a **valve** (honey stopper) to prevent nectar or honey from entering the stomach except when wanted for food. The slender **intestine** is joined by about 100 **Malpighian tubules,** and the large rectum serves to accumulate feces for discharge through the anus after a bee leaves the hive.

The **sting** is a modified ovipositor, present only in workers and queens. It comprises (1) a hollow dorsal **sheath** and (2) two **darts,** grooved along their inner surfaces so that they may slide along each other by the action of muscles at their inner bases, (3) at either side a sensory **sting palpus,** and (4) a large median **poison sac** supplied by two acid glands and a slender alkaline gland. The fluid is pumped into the wound made by the darts. A worker dies about 2 days after using its sting, as all the poison apparatus and some adjacent parts pull off in the process. The queen's slender sting serves to combat rival queens and can be used more than once.

Among the many adaptive features, each short **antenna** has many olfactory pits that provide a keen sense of smell (1600 on a queen, 2400 on a worker, 18,900 on a drone). Each **compound eye** has many ommatidia (4900 on a queen, 6300 on a worker, 13,-000 on a drone). The brain is proportionately large. Bees evidently find their way about and seek food by both scent and sight. They can be trained to visit a food supply having a particular scent or associated with certain colors, except red, which they cannot distinguish from black. Glands on the abdomen produce a scent when bees are disturbed and may serve to "mark" new sources of food afield. Honeybees have a good sense of orientation, and each returns to its own hive. If a hive is moved the absent workers return to the old site, but if confined within during the moving, they take account of the new location upon leaving and will return to it.

25-17 Food A worker bee, upon discovering a food supply in the field, fills her nectar stomach, returns to the hive, and either deposits the gathered nectar or feeds young bees. When the source is less than 100 m (300 ft) away, she performs a "round dance," turning right and left in quick succession. If the source is more distant, she executes a different dance, one which informs other bees of its direction and distance. Beginning at a given spot, she makes a semicircular run, then a straight "wagging run" back to the point of origin, next a semicircle in the opposite direction, and again a run to the starting place. The total pattern is like a compressed figure eight, repeated several times.

Experiments by Wenner and associates have demonstrated, however, that despite the "dance" and presumed "language" of the bees, it is possible to explain nearly all bee ability to find food material on the basis of scent alone. The nature of the food source is communicated by odor from the plant source on the bee's body or in the nectar brought. Other bees keep their antennae touching the scout during her dance in the darkness of the hive and then can find the food source by following airborne scent trails.

When pollen is gathered from a flower, (1) it is taken by the mandibles, moistened with honey, and (2) mixed with that gathered on the pollen brushes of the forelegs; (3) it is then moved to the brushes on the middle legs, which in turn are (4) drawn between the pollen combs of the hind tarsi; (5) each of the latter is then scraped over the opposite leg to deposit pollen on the pecten, or outer surface of the auricle, and (6) by flexing the tarsus on the tibia, the pollen is pushed upward and patted into the pollen basket. The bee thus accumulates a bulging load of sticky pollen in both baskets, returns to the hive, and pushes her load into a cell, to be tamped down by the heads of young workers. Propolis is carried similarly but is removed by other workers.

Nectar held in the honey stomach is acted upon by salivary enzymes, cane sugar being inverted to dextrose and levulose. Upon returning to its hive, a worker regurgitates this fluid into a cell of the comb, where the young "house bees" work it over in their mouths, causing further chemical changes; they evaporate the excess water by fanning with their wings and then seal the cell with wax. Honey averages 17 percent water and 77 percent sugars, with small amounts of minerals, enzymes, and pollen; its color is water-white to dark, and the flavor varies according to the nectar source.

25-18 Reproduction The reproductive system is vestigial in workers but highly developed in queens. About 7 days after emerging, a young queen mates with a drone high in the air; his copulatory organs then are torn away, to remain in her genital bursa until removed by workers after her return to the hive. The spermatozoa thus received into her spermatheca must serve for all the fertilized eggs she will ever lay. Her ovaries enlarge to fill the long abdomen, and in a day or two she begins to lay. She can control the process of fertilization. Unfertilized eggs produce drones, or males (genetically haploid, 16 chromosomes), and fertilized eggs yield females (diploid, 32 chromosomes). In the season of nectar flow, a queen lays up to 1000 eggs per day, gluing each to the bottom of a cell. The tiny wormlike **larva** has no legs or eyes. For 2 days all larvae are fed on *royal jelly* produced by pharyngeal glands of young workers. Thereafter, drone and worker larvae receive mainly honey and pollen, but queen larvae continue chiefly on royal jelly, which causes them to develop differently and to become larger. Each larva has five molts and grows; then its cell is capped with wax, and the larva within spins a thin **cocoon.** There, as a **pupa,** it undergoes complete metamorphosis and finally cuts the cell cap with its mandibles to emerge as a young bee. The time of development for each caste is standardized because of the temperature regulation in the hive:

Queen: egg, 3; larva, $5\frac{1}{2}$; pupa, $7\frac{1}{2}$ = 16 days
Worker: egg, 3; larva, 6; pupa, 12 = 21 days
Drone: egg, 3; larva, $6\frac{1}{2}$; pupa, $14\frac{1}{2}$ = 24 days

To humans the honeybee is a symbol of industry and cooperation, gathering food in time of plenty against the needs of winter. When the warmth of spring brings early flowers, the workers gather nectar and pollen, the queen lays rapidly, and new workers soon swell the colony population. Overcrowding leads to **swarming;** the queen and several thousand workers emerge as a dense swarm and fly to a temporary site. Then some workers scout for a new home; upon returning to the swarm, they communicate its direction and distance by a dance similar to that indicating a new food source (Sec. 25-17). If different sites are reported, the conflict is finally resolved when all scouts agree on a single location, to which the swarm moves. Prior to this, some queen larvae are started in the old colony. One of the queens emerges, usually stings the other queen larvae, is fertilized in a mating flight, and returns to serve the old hive in egg laying. A queen may live for three to five seasons and lay a million eggs. Drones are produced during active nectar flow; but when brood production ceases, they are mostly driven out to starve and die. Many workers hatched in autumn survive until spring, but those born earlier in the year age more rapidly and live only 6 to 8 weeks.

25-19 The hive Each comb (Fig. 25-12) in a hive is a vertical sheet of wax, fastened to the top (and sides) of a cavity and covered with hexagonal cells. The **worker cells,** where workers are reared and honey or pollen is stored, are about 5 mm across, and the **drone cells,** 6 mm across, serve to rear drones and for storage. Large vertical peanutlike **queen cells,** open below, are built along the lower comb margins for queen rearing. The wax is secreted as small flakes by glands in pockets under the abdomen on workers. Once formed, the combs are used for years, the cells being cleaned and polished for reuse.

Honeybees have achieved "air conditioning." In summer they fan their wings vigorously to ventilate the hive, to keep the temperature inside at about 33°C (91°F) for brood rearing, and to evaporate excess water from honey in open cells. In hot dry weather they carry in water to humidify the colony and to dilute the honey if neccessary. During winter, when the stored honey is used as food, they form a compact cluster and produce heat by active body and wing movements. Clusters form at 14°C (57°F) or below and can raise the hive temperature to 24°C (75°F) or 30°C (86°F), even when the outside air is at or below freezing.

Bee colonies are reduced by dearth of nectar or pollen and by exhaustion of honey stores in winter. Adult bees are eaten by toads, skunks, and bears, the latter having a proverbial liking for honey as well. Two serious forms of foulbrood (bacterial disease) cause heavy losses in colonies if unchecked.

Other Insects

25-20 Form and function Except for a few degenerate or specialized forms, the adults of all insects are alike in having one pair each of antennae and compound eyes, a fused head, a thorax of three somites with six legs (hence sometimes call Hexapoda), and a distinct abdomen. Within these limits members of the various orders, families, and species show great diversity in details of structure and habits. Many features are well suited to particular modes of life.

The **coloration** is produced by pigments (chemical), or by surface structure (physical), or by a combination of these. Some pigments are deposited in the cuticle and others in the epidermis or deeper. The iridescent colors of certain beetles, butterflies, and others are produced by differential interference of light falling upon microscopically fine surface ridges or parallel plates of cuticle.

The **cuticle** is often waxy and not easily wetted; this is of advantage to the insects but causes some difficulty in attempts to control harmful species by sprays. The **body** is adaptively streamlined in many aquatic bugs and beetles, depressed in cockroaches and others that live in crevices, and laterally compressed in fleas that move between hairs and feathers of their hosts. The hairy covering on some nocturnal moths may help to insulate them against chilling, and the hairs on many flies and bees serve in the gathering of pollen.

All insects have **eyes** except some larvae that live concealed from the light, some adults that inhabit caves or nests of termites or ants, some biting lice, and the nonsexual castes of some ants and most termites. The **mouth parts** are of two main types–chewing or sucking–as in a grasshopper and bee, respectively. Many species with sucking mouth parts have means for piercing tissues, such as the mosquitoes and fleas that "bite" other animals and the aphids that puncture plants.

The **legs** are modified for running in tiger beetles, jumping in fleas, swimming in water bugs and beetles, skating in water striders, burrowing in mole crickets, and spinning webs in embiids. The **wings** are commonly thin and membranous. The forewings are hardened as elytra in beetles; they are firm in the ORTHOPTERA, and the forward half is thickened in the HEMIPTERA. In flies (DIPTERA) and male coccids the hind pair is represented by minute knobbed halteres (balancing organs), and the fore pair is reduced on male stylopids. Primitive insects (PROTURA to THYSANURA) and lice, fleas, and some other parasitic forms lack wings; ants and termites are also wingless except for the sexual castes. The embryo in many insects has abdominal appendages, but there are rarely any on adults except for the cerci and copulatory organs. The **muscles** are numerous and complex; about 2000 are present in a caterpillar. Many insects are disproportionately powerful as compared with larger animals; a honeybee, for example, can pull 20 times its own weight and can lift a load equal to four-fifths of its weight in flight.

The **digestive system** shows modifications in relation to food habits. The salivary glands of bloodsucking insects produce anticoagulants that keep blood fluid during ingestion and digestion. **Respiration** is performed by a tracheal system in most insects but through the thin body covering in most COLLEMBOLA, some PROTURA, and some endoparasitic larvae. Adult aquatic insects that dive carry along a film of air on the exterior of their nonwettable bodies or beneath their wings to serve for respiration. Some aquatic larvae, such as those of mosquitoes, must extend their spiracles above the water to breathe, but the larvae of caddis flies and others have thin **gills**–tracheal, anal, or blood–that take up oxygen dissolved in water (Fig. 7-2). None of these is homologous with the gills of CRUSTACEA.

The **nervous system** is annelidlike in the lowest insects and in many larvae of higher forms, with paired ganglia in each body somite, but in the adults of some flies and other insects the (posterior) ganglia are concentrated forward in the body. **Sound production** is common in many ORTHOPTERA (Fig. 25-14), HEMIPTERA, cicadas, certain moths, some mosquitoes and other flies, and some beetles and bees. The mechanism of production varies and in some may be merely incidental to their manner of flight. **Light** is produced by glowworms, fireflies (LAMPYRIDAE), and a few others. The light is cold (80 to 90 percent efficient as compared with 3 per-

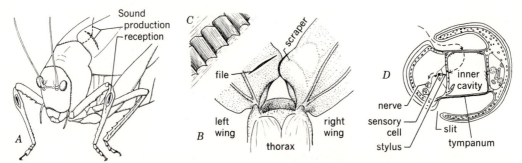

Figure 25-14 *A*. Foreparts of a katydid showing the location of structures for sound production and reception. *B*. Undersurface of the fore wings with file and scraper that are rubbed together to produce sounds. *C*. Enlarged detail of the file. *D*. Cross section of foreleg showing the sound receptor, or "ear," with paths of sound waves (broken line) and nerve impulses (solid arrow). (*Partly after Dahlgren, 1925.*)

cent in an electric lamp), and the radiation is all within a range visible to the human eye.

25-21 Distribution Some insects are of wide occurrence, over a continent or more, whereas other species are limited to a few acres. Insects occur from sea level to well above 6000 m (20,000 ft) on the highest mountains. Surveys by airplanes show that many inhabit the air, especially by day in summer; they are mostly below 300 m (1000 ft) but some have been taken up to 4200 m (14,000 ft). By such means, species are constantly being carried to new localities. Some beetles and other insects live at the seashore between tide lines, and a few water striders (*Halobates*) on the ocean surface; there are few submerged marine insects.

25-22 Seasonal occurrence Life on land exposes insects to greater extremes of environmental conditions than are experienced by most inhabitants of the sea and fresh waters. They have to withstand changes in temperature and food supply between summer and winter and in the abundance or scarcity of plants. Different species are adapted to these problems in various ways. Many are abundant in warm seasons and much reduced at other times. Some, such as yellow jackets, overwinter as adults that hide away to hibernate in shelters, where their bodily metabolism is greatly reduced. Others survive as pupae or larvae. In many species all individuals die at the close of the warm season and then are

represented only by eggs that will develop and hatch the following spring.

25-23 Sensory perception and behavior Insects respond to many of the stimuli that cause human sensations, including light, chemical stimuli (smell, taste), touch, and sound, but their perceptions differ in both kind and magnitude. They detect chemical stimuli far too delicate for the human nose or tongue, and some react to ultraviolet rays but not to red or infrared.

The simplest response is a **reflex** such as the action of a bee's sting, which operates if touched, even when severed from the body. The invariable type of response by which an animal orients itself to or away from a stimulus is known as a **taxis.** *Drosophila* is guided by its olfactory organs to overripe fruit that bears the yeast upon which the fly feeds by a **positive chemotaxis** to certain alcohols and organic acids in the fermenting fruit. Many insects find mates by a similar taxis to the delicate scents (*pheromones*) emitted by the opposite sex. The moth that flies directly to (and even into) a flame shows **positive phototaxis;** and the cockroach that runs for cover when suddenly exposed to light, a **negative phototaxis.** Aquatic larvae of caddis flies often show **positive rheotaxis** by aligning themselves head foremost in a current of water. The term **kinesis** is used to describe undirected locomotor reactions in which the speed of movement or frequency of turning depends on the intensity of stimulation. The cockroach, besides avoiding light, seeks shelter in crevices where its body is in contact with the shelter–this is thigmokinesis. Taxes and kineses, and some other types of reaction enable insects and many other animals to find and inhabit the small environmental microclimate in which each kind is most successful.

Many details of behavior in insects are based on **instincts** (Sec. 9-10)—series, or chains, of coordinated reflexes by the whole organism. The complex sequence of events by which a solitary wasp constructs a nest, deposits an egg, provisions it with paralyzed insects, and skillfully closes the cell is an example of chain instincts. Other features of insect behavior are plastic, or modifiable, and bring to bear some features of experiences of an individual that are registered as memory. An example here is the hive bee, which can be trained to associate color with a food supply.

25-24 Flight Insects, birds, and bats are the only animals capable of true flight. The wings of insects (Figs. 25-4, 25-15) are unique structures in being derived as extensions of the body integument, quite unlike the limb-wings of the vertebrates (Fig. 13-1). The ability to fly enables insects to extend their feeding ranges and to disperse and occupy new territory.

Insect wings are attached to the dorsal tergum (Fig. 25-15), but a short distance distally they rest on a vertical pleural process which is a fulcrum on which wings move up and down. Up-and-down movement of the wings is, then, produced by muscles, the tergosternal muscles producing an upward stroke by depressing the tergum and the longitudinal muscles producing a downward stroke by raising the tergum.

Simple up-and-down movement of the wings is insufficient in and of itself to produce forward motion in flight. There must also be forward and backward motion of the wings, hence the wings of insects describe a figure eight or an ellipse during a complete up-and-down cycle. The rate at which wings move, and hence the speed of insect flight, varies greatly among species. Slow-moving insects, such as butterflies, have wing beats of 4 to 20 per second, while gnats may have as many as 1000 beats per second. Since insects are ectothermic (poikilothermic), low environmental temperatures affect their metabolism and their flying ability. When environmental temperatures are low, it may take some warm-up time before an insect is able to generate internal temperatures sufficient to enable it to fly.

25-25 Water conservation When the ancestors of insects left the water and moved to the land and air, they experienced changes analogous to those of amphibians and reptiles as compared with fishes. The sense organs became adapted to function in air, the cuticle served to resist the loss of body fluids by evaporation and the tracheae provided a means for breathing air. All insects except those in humid environments have a problem in the conservation of body water that does not beset aquatic animals, since moisture may be lost in respiration and in evacuating food residues. The valves in the spiracles may limit the amount of air (and moisture) exhaled under some conditions, but the more vigorous respiratory movements during flight serve incidentally to deplete the supply of fluids. Insects in dry situations extract water from food residues in the rectum, and

Figure 25-15. Wing movements in the flight of an insect. *A. Upstroke:* Contractions of the tergosternal muscles depress the tergum and carry the wing bases downward, each acting over a wing process as a fulcrum. Rotation of the wings is produced by still other muscles and movement of some sclerites. *C. Downstroke:* Contraction of the longitudinal (and other) muscles causes the tergum to bulge upward and forces the wing tips downward. (*After Snodgrass.*)

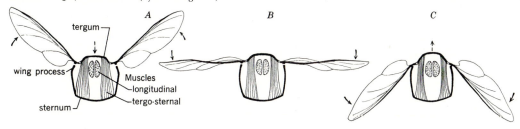

they gain some metabolic water as a by-product of the oxidation of food materials within the body.

25-26 Food Fully half the known species of insects are **phytophagous,** feeding on the tissues or juices of plants. Grasshoppers eat a great variety of plants, the potato beetle uses only those of the family SOLANACEAE, the larva of the monarch butterfly eats only the milkweeds (genus *Asclepias*), and the food of a common copper butterfly is restricted to a single species of sorrel (*Rumex acetosella*). Most termites and certain beetles subsist on wood, but some of them and certain ants eat fungi exclusively; among such insects are some that produce their own food by planting, fertilizing, and tending "fungus gardens". The **saprophagous** insects include beetles, fly larvae, and others which eat dead animals, and the **carnivorous** species are those which capture and devour insects or other living animals.

25-27 Predators and parasites Because of their abundance, insects are preyed upon by a great variety of other insects and vertebrates of the land and fresh waters. They are also subject to the attacks of parasites and parasitoids and are affected by many diseases that reduce their numbers.

A predator, to reach adult size, consumes many prey animals, but the immature stage of a parasite often develops at the expense of one host individual, which usually is not killed. A parasitoid is parasitic as a larva and destroys the host—thus resembling a predator–but is free-living as an adult. Examples of parasitoids on other insects are the larvae of tachinid flies (DIPTERA) and of the ichneumonid, braconid, and chalcid wasps and eggs of scelionid wasps (HYMENOPTERA). These invaders kill their hosts or destroy the eggs, serving thus to reduce or control populations of the latter. The number of parasitoids varies in keeping with that of the hosts. By contrast, parasites usually only weaken their hosts. The situation is further complicated by **hyperparasitism,** in which parasites are themselves parasitized, or by hyperparasitoidism, in which parasitoids are killed by other species. A tussock moth caterpillar (*Hemerocampa leucostigma*) studied by L. O. Howard was found to have 23 primary parasites (6 DIPTERA, 17 HYMENOPTERA), which in turn had 13 secondary parasites, or hyperparasites; the latter were subject to 2 (or 5) tertiary parasites. Such are the complexities in the balance of nature.

25-28 Reproduction Fertilization is always internal and either directly into the female reproductive tract or by secondary transfer of sperm (ODONATA) or by spermatophores. Most species are **oviparous** and lay their eggs singly or in clusters, either on or in the ground, on the plants or animals where their larvae feed, or in plant tissues. Those with aquatic larvae oviposit in or near water. The eggs of some hatch in a few hours, but those of some ORTHOPTERA and LEPIDOPTERA require many months to hatch. The aphids and several DIPTERA are **viviparous,** producing living young. Eggs of tachinid flies, which are deposited on other insects, will hatch almost as soon as laid. The living young of tsetse and hippoboscid flies develop within the "uterus" of the female parent, where they are nourished by special secretions. **Parthenogenesis,** or reproduction from eggs without fertilization, occurs in aphids, thrips, gall wasps, sawflies, and others. The generations of aphids (Fig. 25-32) in spring and summer consist only of females that reproduce by parthenogenesis, but later both sexes are produced in the same manner. These mate, and the fertilized females then lay eggs that remain quiescent over winter, to produce females for the next spring. A special type of parthenogenesis known as **paedogenesis** occurs in the fly *Miastor* (CECIDOMYIDAE) and a few other insects. Each larva produces 7 to 30 larvae, and these in turn yield others. Some later larvae pupate to become male and female flies. The chalcid wasps that parasitize eggs of LEPIDOPTERA exhibit **polyembryony;** a chalcid egg begins devlopment, then divides into 100 or more masses, each of which grows to be a larva that transforms into a wasp.

25-29 Gall insects The cynipids, or gall wasps (HYMENOPTERA), and the cecidomyids, or gallflies (DIPTERA), are small insects that oviposit in plant tissues. Some substance then injected or resulting from growth of the larvae causes the plant to produce a characteristic swelling, or **gall.** These are of distinctive form and location (stem or leaf) according to the host plant and the kind of insect involved. Galls are also produced by some aphids and psyllids (Ho-MOPTERA) and by certain gall mites (ACARINA).

25-30 Number of offspring This varies in different insects from the single larva hatched at a time by some viviparous flies to the million or so eggs laid by a queen bee. The actual number from any one female is less important than the rate of increase; in some species with short life cycles this is very rapid. The pomace fly (*Drosophila*) lays up to 200 eggs per female, and the entire cycle requires only 10 days at about 27°C (80°F). The housefly may complete its cycle in 8 to 10 days during hot weather. Parthenogenesis in aphids and others likewise leads to extremely rapid multiplication of the population under optimum conditions of temperature, moisture, and food supply. The successive generations from a single aphid could cover the earth in one season if all survived!

25-31 Growth and metamorphosis Since an insect lives in an armorlike exoskeleton, it can change form or increase in size only after a molt, and none molts after attaining the adult stage. The increase in linear dimensions at each successive molt is about 1.4 but varies, depending on the duration of each stage. Also parts may grow at different rates, the size of a particular part being an exponential function of the whole (allometric growth). The primitive orders PROTURA to THYSANURA hatch as young which are similar to the adult but smaller and attain adult form and size by slightly graded changes. There is no metamorphosis, hence these are called the AMETABOLA (Gr. *a*, not + *metabola*, change; Table 25-2). The HE-MIMETABOLA (ODONATA to THYSANOPTERA) undergo a gradual or incomplete metamorphosis, as with the grasshopper. The young hatches as a small **nymph,** crudely resembling the adult, with compound eyes. In successive stages, or **instars,** the wings appear externally as small wing pads that enlarge at successive molts to become functional in the **adult,** or **imago.** In the HOLOMETABOLA (MECOPTERA to HY-MENOPTERA), the young emerges as a small wormlike segmented **larva** having the head, thorax, and abdomen much alike and often with short legs but no wings or compound eyes. The successive larval instars increase in size through several molts. Each then enters a "resting" stage as a **pupa,** within the last larval skin, in a special puparium or in a cocoon. Many larval organs then break down and are resorbed, while new structures for the adult arise concurrently. When these profound changes are completed, the adult emerges. This is complete metamorphosis as seen in the honeybee.

Experimental evidence (Sec. 8-2) indicates that both molt and metamorphosis are induced by **ecdysiotropin** secreted by the brain. This hormone stimulates the ecdysial glands to produce **ecdysone,** which permits molting to occur. At the same time the corpora allata glands produce **juvenile hormone.** In the presence of juvenile hormone a larval form will continue to undergo simple molt to produce successively larger larvae. When the corpora allata cease to produce juvenile hormone, the ecdysone stimulates molting to the adult stage.

25-32 Social insects Most insects are **solitary,** each individual living unto itself; the sexes associate only to mate, and the female deserts her eggs or dies after laying. **Gregarious** species assemble in large numbers, as in swarms of locusts and hibernating ladybird beetles. Among such insects the parents usually never see or live coincidentally with their offspring. About 6000 species of insects, however, exhibit social instincts, the female or both parents living cooperatively with their offspring in a common shelter. These conditions begin with **subsocial relations,** as in the female earwig (order DERMAPTERA), which guards her eggs and later the young. Cockroaches and crickets, some beetles and bugs, the EMBIOPTERA, and the ZORAPTERA do likewise. A solitary wasp provisions individual egg cells with insects as food for her larvae, which later hatch out and grow independently.

True social organization occurs with all termites, all ants, and certain wasps and bees. In this arrangement many individuals of both sexes live together in a complex organization usually involving a definite division of labor among themselves. Often but a single reproductive individual, the **queen,** is present in the colony, and the other individuals of both sexes serve various functions to ensure survival of the group. The complexity of the organization and the division of labor vary with the species of insect involved. Queens of some termites and ants may live for several years. In the simpler colonies the female merely remains with successive broods or feeds them daily. From this condition there is a graded series to the complex colonial life of termites, ants, and hive bees, which have a division of labor among several often morphologically different groups of individuals called *castes*. The larger populations with social life require an enhanced food supply. The ants show a progression in food habits such as probably occurred in human history. The lowest kinds hunt insects or flesh. Pastoral ants attend and shelter aphids (*ant cows*) from which they obtain honeydew as food, and harvester ants gather and store seeds in summer to tide them through the winter. Finally, the fungus ants (*Atta*) grow their own pure crops of certain fungi in underground gardens fertilized with organic debris; each young queen upon setting out to found a new colony carries a seed stock of fungal hyphae in a pouch below the mouth. Communication in the colonies is enhanced by the secretion of **pheromones** (communication chemicals) which serve variously to identify individuals of the colony, to mark trails which others can follow, to maintain the castes, to give alarms regarding attack, and to enhance group activity.

25-33 Relation to humans Virtually every person is affected by some insects, from apartment dwellers who eat honey, wear silk, and swat flies to aboriginal people who are plagued by lice, fleas, and flies. Economic entomology deals with the several thousand species of insects of importance to agriculture, forestry, and the food industries and medical entomology with those affecting the health of humans and domestic animals.

There are many **beneficial insects.** The bees and others that go from flower to flower to gather pollen are essential in cross-fertilizing blossoms of apples, cherries, blackberries, clover, and other crops which otherwise will not set fruit or seed. Hives of bees are placed in orchards or fields to ensure such fertiliza-

tion. The Smyrna fig, grown in California, produces only female flowers; to produce good fruit, they need "caprification" (pollination) by a small wasp (*Blastophaga*) that develops only in the small nonedible caprifig, whence it brings pollen. Throughout the world honeybees produce honey, which serves as human food; they also yield beeswax, used in polishes, candles, modeling and to wax thread. Raw silk is obtained in the Orient and Europe from the silkworm (*Bombyx mori*). The larvae are reared in domestication on a diet of white mulberry leaves, and each spins from its salivary secretions a cocoon of silk, which is the silk of commerce. The shellac of commerce is obtained from waxy secretions of certain lac or scale insects (Coccidae) of India, and the dyes known as *cochineal* and *crimson lake* are derived from the dried bodies of some tropical scale insects of cactus.

Many harmful plant-eating insects are devoured by a host of **predaceous insects**–ground beetles, syrphid flies, and wasps. Scale insects that feed on and damage citrus and other trees are eaten by larvae of lady beetles (Coccinellidae); certain species of these beetles have been imported, reared, and liberated in orchards in attempts to control scale insects. Parasitoids that oviposit in eggs or young of plant-feeding insects and whose larvae destroy the latter are another useful group; some are reared artificially and liberated to serve in biological control of harmful species. Such insects are in turn subject to hyperparasitoids which help to regulate parasitoid populations.

Other useful insects are the scavenger beetles and flies that clean up the dung and dead bodies of animals. Flesh flies lay quantities of eggs in animal carcasses, which their voracious larvae soon reduce to skin and bone. Termites, some ants, and beetles slowly reduce the remains of dead trees and other plants, but termites also do much damage to buildings and other wooden structures. Finally, many insects are indirectly useful as the food for fishes, game birds, fur mammals, and other wild vertebrates and at times for poultry.

Many species of **harmful insects** injure agricultural crops, stored foods, and property; and others affect the comfort and health of wild and domestic animals and humans. Every cultivated plant has insect pests, and important crops such as corn, cotton, wheat, and tobacco have a hundred or more each. These levy a steady toll in damage or loss of crops and in expenditures for control by poison sprays, dusts, and parasites. Some major **native pests** are the potato beetle, chinch bug, and grasshoppers; among the many **introduced pests** are the Hessian fly of wheat, European corn borer, cotton boll weevil, and codling moth of apples. Federal and state quarantines are maintained to limit the spread of such insects.

Human foods are eaten or ruined by ants, cockroaches, and weevils, and are dirtied by houseflies; stored cereals are damaged by grain weevils and moths; woolen clothing, carpets, furs, and feathers are riddled by clothes moths and carpet beetles; and books are damaged by silverfish, beetle larvae, and termites. Bedbugs, stable flies, mosquitoes, and gnats bite humans and their animals; attacks of biting lice cause poultry and livestock to become unhealthy; bloodsucking tabanid flies annoy horses, and horn flies do the same to cattle; larvae of botflies are a source of irritation in the stomachs of horses; and larvae of ox warble flies burrow in the backs of cattle, causing them to lose flesh and damaging the hides for leather.

Many insects and some ticks act as intermediate hosts for various diseases of man and the larger animals and plants; a few important examples are given in Table 25-1.

Development of modern insecticides (DDT and others) has been viewed as a technological triumph. Pest problems were dramatically alleviated over vast areas, with immediate reduction in human deaths and suffering and in economic losses by agriculture. These benefits, however, were obtained by considerable damage to the environment (Sec. 36-8).

Insect control in the future needs to be based more on ecologic principles, with a greater reliance on biological and nonchemical methods. It may be necessary to accept some pest damage. An integrated approach will be required that considers the many aspects of control in advance and attempts to anticipate side effects and overall impact on the ecosystem.

Already there are some alternatives to "hard" control chemicals. Some examples are (1) introduction of harmless insects to control harmful ones; (2) use of pathogens—pest-specific viruses or bacteria; (3) release of sterilized male insects into natural populations, causing infertile matings and population decline (e.g., screwworm botfly control in southern United States); (4) use of pheromones–sex attractants that will draw insects to sites where they can be sterilized or killed; (5) repellents; (6) use of electro-

Table 25-1
EXAMPLES OF DISEASES TRANSMITTED BY INSECTS AND TICKS

Disease	Causative organism	Intermediate host	Order
Dutch elm disease	*Ceratostomella* (fungus)	Bark beetle (*Scolytus*)	COLEOPTERA
Cucumber wilt	*Erwinia tracheiphila*	Cucumber beetle (*Diabrotica*)	COLEOPTERA
Curly top of sugar beets	Virus	Beet leafhopper (*Circulifer*)	HOMOPTERA
Human yellow fever	Virus	Mosquito (*Aëdes aegypti*)	DIPTERA
Bubonic plague of rats and humans	*Pasteurella pestis*	Flea (*Xenopsylla* and others)	SIPHONAPTERA
Tularemia	*Pasteurella tularensis*	Deer fly (*Chrysops*) and others	DIPTERA
Human typhus fever	*Rickettsia*	Body louse (*Pediculus*)	ANOPLURA
Human malaria	*Plasmodium* (protozoan)	Mosquito (*Anopheles*)	DIPTERA
Chagas' disease	*Trypanosoma cruzi*	Bug (*Triatoma*)	HEMIPTERA
Filariasis	*Wuchereria bancrofti* (nematode)	Mosquito (*Culex* and others)	DIPTERA
Dog tapeworm	*Dipylidium caninum*	Louse (*Trichodectes*) and flea	MALLOPHAGA SIPHONAPTERA
Rocky Mountain spotted fever of humans	*Dermacentroxenus rickettsii*	Tick (*Dermacentor andersoni*)	ACARINA
Texas cattle fever	*Babesia bigemina*	Tick (*Boöphilus annulatus*)	ACARINA

magnetic radiation–radio waves, infrared, ultraviolet, or ionizing radiation–to destroy insects in wood, grains, etc.; (7) application of hormones to inhibit or accelerate development and cause emergence at unfavorable seasons; (8) use of biodegradable control products based on chemical compounds derived from plants and animals (e.g., pyrethrum); and (9) breeding of crops resistant to plant pests. Chemistry is important in many of these approaches but is directed more toward understanding and simulating natural chemical controls than toward the development of new nonbiological compounds. The latter may still be useful if made rapidly biodegradable and applied selectively in an ecologic context.

Biological control is not a panacea for all pest problems. Introductions of some species of insects for control might cause more problems than they solve.

25-34 Fossil insects   Despite their fragile nature, remains of insects as fossils have been found in Australia, China, Russia, Europe, and the United States, and over 10,000 species have been named. The oldest (COLLEMBOLA) are in Middle Devonian rocks, about 375 million years old. In the coal-forming forests of Carboniferous times winged insects flourished, including the primitive †PALEODICTYOPTERA, which lasted into Permian time, and the order ORTHOPTERA, which includes the living cockroaches. The †PROTODONATA resembled dragonflies, and some (†*Meganeura*) had a wingspread up to 74 cm (29 in)! Some Paleozoic orders died out before the end of that era, but about 13 Recent orders were established by the end of the Permian, 230 million years ago. The rising land, colder climates, and appearance of seasonal seed plants in those times evidently favored the evolution of pupae to withstand adverse conditions, since the earliest scorpion flies (MECOPTERA) are so much like existing species as to suggest that metamorphosis was an early achievement.

The transparent amber (fossil resin) of the Oligocene, found along the Baltic coast of Europe, contains many insects with all external details beautifully preserved and easily seen (Fig. 25-16). Present-day families and genera are common in the amber, and some species are very close to those now existing.

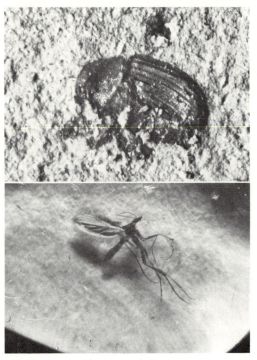

Figure 25-16   Examples of fossil insects.   *Above*. Weevil (*Anthonomus*, family CURCULIONIDAE), ×7, in shale rock at Florissant, Colo. *Below*. Fungus gnat (*Bradysia*, family SCIARIDAE), × 3½, in amber (fossil resin) from Simojovel, Chiapas, Mexico. Both Oligocene.

The ants even show polymorphism, and some of them had learned to attend plant lice. Ants in amber of Upper Cretaceous age found in New Jersey indicate the existence of social life among insects approximately 100 million years ago. Amber deposits are also known from Canada and northern Alaska. Arctic amber is of Cretaceous age also, and the insects found thus far differ from modern genera but belong to present-day families. No ants have been found in the Canadian or arctic amber, in contrast to the Baltic amber fauna. One of the richest known fossil insect faunas is preserved at Florissant, Colo., near Pikes Peak, in Oligocene lake deposits of volcanic sand and ash that later became shale. Most of the genera of insects of Tertiary time still survive; others have disappeared, and some are now less widely distributed. The tsetse fly, *Glossina*, is known as a fossil in Florissant shales but is restricted to parts of Africa at the present time.

Table 25-2
CHARACTERISTICS OF RECENT ORDERS OF THE CLASS INSECTA
(Adults; exceptions as to wings or other features omitted)

Subclass	Metamorphosis	Order and common names	Mouth parts — Chewing	Mouth parts — Sucking	Wings — Fore	Wings — Hind	Distinctive features
APTERYGOTA	AMETABOLA: no metamorphosis	PROTURA	C		None		No antennae
		COLLEMBOLA: springtails	C		None		Spring (furcula) on abdomen
		DIPLURA: japygids	C		None		2 pincers or cerci on abdomen
		THYSANURA: bristletails	C		None		Body fine-scaled; 3 "tails" on abdomen
PTERYGOTA: typically winged; wings reduced or absent in some	HEMIMETABOLA: young are nymphs with compound eyes, and wings grow externally; metamorphosis gradual (incomplete)	ODONATA: dragonflies	C		Filmy, not folded, nearly alike		Large; eyes big; no cerci
		EPHEMEROPTERA: mayflies	C		Filmy, not folded — Larger	Smaller	Mouth parts vestigial, "tails" 2 or 3
		ORTHOPTERA: roaches, grasshoppers	C		4 or none — Leathery	Thin	Usually with cerci
		DERMAPTERA: earwigs	C		Hard, short	Thin, fanlike	Forceps at end of abdomen
		PLECOPTERA: stone flies	C		Filmy, narrower	Pleated, broader	2 long cerci
		ISOPTERA: termites	C		Sexual forms with like wings; others wingless		Sexual forms pigmented; others pale, uncolored
		EMBIOPTERA: embiids	C		♂ winged; ♀ wingless		Tarsi of forelegs enlarged for spinning
		MALLOPHAGA: chewing lice	C		None		Minute, flat; head wide
		ANOPLURA: sucking lice		S	None		Minute, flat; mouth parts retractile; head narrow
		PSOCOPTERA: book lice	C		4, folded, roofed, or none		A maxillary "pick"
		HEMIPTERA: true bugs		S	Half leathery	Filmy	A triangular scutellum; base of beak far forward on head
		HOMOPTERA: aphids, scale insects		S	Texture uniform; 4, 2, or none		Base of beak close to thorax
		THYSANOPTERA: thrips		S	Fringed with hairs		Tarsi bladderlike
	HOLOMETABOLA: young are larvae with no compound eyes; wings grow internally; metamorphosis complex (complete)	MECOPTERA: scorpion flies	C		Filmy, roofed, nearly alike		Head elongate as a beak; cerci short
		NEUROPTERA: ant lions, dobson flies	C		Filmy, roofed, nearly alike		No cerci
		TRICHOPTERA: caddis flies	C		Filmy, roofed		Wings hairy-coated
		LEPIDOPTERA: moths, butterflies		S	Covered by fine over-lapped scales		Maxillae as coiled proboscis for feeding
		DIPTERA: true flies		S	No hind wings		Halteres ("balancers") replace hind wings
		SIPHONAPTERA: fleas		S	None		Small; laterally compressed
		COLEOPTERA: beetles, weevils	C		Hard, veinless	Filmy, folded	Prothorax large, mesothorax reduced
		STREPSIPTERA: stylops	C		Hind wings only in ♂; none in ♀		♀ maggotlike, having head and thorax fused
		HYMENOPTERA: ants, wasps, bees	C	S	Filmy; 2 pairs or none		Base of abdomen usually constricted

♂ = male; ♀ = female; filmy = membranous; roofed = wings at rest over abdomen thus ⋀.

Classification

(Lengths are given for body, not including wings; fossil orders and many families omitted)

Class 2. Insecta (Hexapoda).

Insects. Head, thorax, and abdomen distinct; 1 pair of antennae (except PROTURA); mouth parts for chewing, sucking, or lapping; thorax typically with 3 pairs of jointed legs and 2 pairs of wings, variously modified, reduced, or absent; a slender middorsal heart and aorta; respiration usually by branched tracheal tubes conveying air from spiracles directly to tissues; excretion by Malpighian tubules joined to hindgut (except COLLEMBOLA); brain of fused ganglia, double ventral nerve core with segmental ganglia, often concentrated anteriorly; eyes both simple and compound; sexes separate; sex opening usually single, at end of abdomen; usually oviparous; development after hatching with gradual or abrupt metamorphosis; in all habitats, few in ocean; about 900,000 species.[1]

Subclass 1. Apterygota (Ametabola). Primitively wingless; little or no metamorphosis; abdomen with ventral appendages (styli) and usually with cerci.

Order 1. Protura. Minute (0.6 to 1.5 mm), primitive, mouth parts folded inward, piercing; no antennae, true eyes, or wings; abdomen of 11 somites, each of first 3 with pair of minute appendages; first nymphal stage with 9 abdominal somites, 1 added at

---

[1] *The accounts that follow use mostly estimates made in 1948 on numbers of species in the various orders, totaling about 686,000 (Sabrosky in Bishopp,* Insects: The yearbook of agriculture, *1952). Annually 6000 to 7000 new species are named, giving a gross figure for 1978 of about 900,000.*

each molt by division of terminal somite; no metamorphosis; in damp places between decaying leaves, under bark on logs or twigs, or in moss; 4 families, 90 species. *Acerentulus* (Fig. 25-17A).

Order 2. Collembola. Springtails. Minute to 5 mm, colored or white; antennae 4- to 6- jointed; chewing mouth parts folded inward; no wings, compound eyes, or Malpighian tubules, usually no tracheae; abdomen of 6 partly fused somites; most species leap by action of ventral springing organ (furcula) on IVth abdominal somite, when released by hook (hamula) on IIId somite; a ventral tube on Ist somite receives sticky secretion from gland behind labium by which animal adheres to smooth surfaces; no metamorphosis; in damp places under leaves, moss, bark, or stones and on water and snow; food of decaying matter; 10 families, 2000 species; Devonian to Recent (Fig. 25-17B). *Anurida maritima*, intertidal; *Smithurus hortensis*, garden flea, damages young vegetables; *Achorutes armatum*, in manure, damages mushrooms; *A. nivicola*, snow flea, sometimes gets into sap at maple-sugar camps.

Order 3. Diplura (Entotrophi). Japygids. Size to 50 mm (2 in); antennae long; no eyes; chewing mouth parts sunk within head; no wings; Malpighian tubules reduced or absent; abdomen of 11 somites, with 2 slender jointed cerci or a pair of pincers; no metamorphosis; 3 families, 500 species. *Campodea*, in rotten wood or under leaves or stones; *Japyx* (Fig. 25-17C); cerci pincerlike, under stones.

Order 4. Thysanura. Bristletails. Minute to 30 mm (1 in); antennae long; chewing mouth parts exserted; no wings; body usually scaly; abdomen of 11 somites, with 2 slender jointed cerci at end and a median jointed filament; run swiftly or jump; no metamorphosis; 9 families, 500 species.

Suborder 1. Archaeognatha (Microcoryphia). Mandibles joined to head by one articulation like

Figure 25-17 *A.* Order PROTURA. *Acerentulus barberi*, 1 mm long. *B.* Order COLLEMBOLA. Springtail (*Entomobrya laguna*), 2 mm. *C.* Order DIPLURA. Japygid (*Japyx*), to 25 mm (1 in) long. *D.* Order THYSANURA. Silverfish (*Lepisma*), 10 mm. *E.* Order THYSANURA. Machilid (*Pedetontus*), to 20 mm. Life cycle (egg, young, adult)—no metamorphosis. *(After various authors.)*

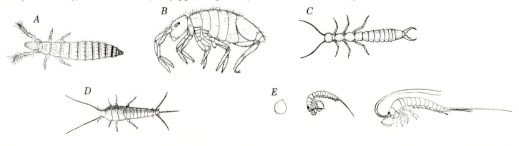

Figure 25-18 Order ODONATA: dragonflies. *Aeshna: A.* Nymph capturing prey, with labium extended. *B.* Nymphal skin. *Libellula: C.* Nymph. *D.* Nymphal skin. *E.* Adult at rest with wings extended. (*After Brehm.*)

other APTERYGOTA. Upper Carboniferous and Permian to Recent. †*Dasyleptus; Machilis, Pedetontus* (Fig. 25-17E), in leaf litter and rocky places.

Suborder 2. Zygentoma. Mandibles joined to head by 2 articulations as in PTERYGOTA. Triassic to Recent. *Lepisma saccharina,* silverfish (Fig. 25-17D), in buildings, eats starch in books, glazed paper, and clothing; *Thermobia domestica,* firebrat, about fireplaces and bake ovens.

Subclass 2. Pterygota. Winged insects. Wings usually present, held outward or erect in ODONATA and EPHEMEROPTERA (PALEOPTERA), usually folded backward over body (NEOPTERA) in all others; no abdominal appendages except cerci and genitalia.

Division 1. *Hemimetabola* (Exopterygota). Young stages are nymphs with compound eyes; wings develop externally; metamorphosis gradual to adult form.

Order 1. Odonata. Dragonflies, damselflies. Large; body often brightly colored; chewing mouth parts; eyes huge, prominent, with up to 30,000 om-matidia; 2 pairs of transparent membranous wings with complex cross veins, a strong cross vein and notch, or nodus, in anterior margin of each; abdomen slender, elongate; adults strong-flying, predaceous; legs used to capture other insects in flight but not for locomotion; mate in air; eggs dropped in water or inserted in aquatic plants; nymphs (naiads) aquatic, capture prey with extensible labium having hooklike "jaws" at end; 11 to 15 molts, 3 months to 5 years in water with gradual metamorphosis, climbing out for final molt to winged adult stage; 23 families, 5000 species; Permian to Recent.

Suborder 1. Anisoptera. Dragonflies. Hind wings wider at base than forewings; wings held laterally in repose; nymphs with gills in rectum, water for aeration drawn in and expelled, aiding also in locomotion. *Gomphus, Anax, Aeshna, Libellula, Sympetrum* (Fig. 25-18)

Suborder 2. Zygoptera. Damselflies. Wings alike, held vertically over back in repose; nymphs with 3 external leaflike caudal gills. *Hetaerina,* ruby spot; *Lestes, Argia* (Fig. 25-19).

Figure 25-19 Order ODONATA: damselflies. *Argia: A.* Nymph with gills. *B.* Adult.

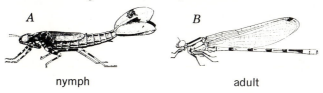

nymph   adult

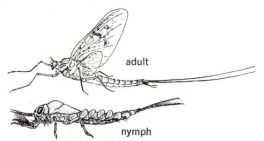

Figure 25-20 Order EPHEMEROPTERA: **mayflies.** (*From General Biological Supply House.*)

Order 2. Ephemeroptera. Mayflies. To 25 mm. (1 in) long; body soft; mouth parts for chewing, but vestigial; antennae short; wings 4, membranous, forewings much larger, all held vertically above body at rest; end of abdomen with 2 or 3 long, many-jointed "tails"; nymphs aquatic, with lateral tracheal gills; feed on plant materials; up to 21 molts, few months to 3 years in water; finally float at surface, and flying subimago emerges and soon molts to imago that lives but a few hours or days, reproduces and dies without feeding; 11 families, 1500 species; Permian to Recent. *Ephemera; Leptophlebia; Callibaetis* (Fig. 25-20).

Order 3. Orthoptera. Grasshoppers (locusts), crickets, etc. Size medium to large; chewing mouth parts; forewings (tegmina) narrow, parchmentlike, and veined; hind wings membranous, broad, many-veined, folding fanlike beneath forewings; some wingless; abdomen with cerci; mostly plant feeders; 50 families, 23,000 species; Carboniferous to Recent (Fig. 25-21).

BLATTIDAE. Cockroaches. Body depressed, head partly under prothorax; antennae long; tarsi 5-jointed; eggs in hard, purselike case; dark-colored, foul-smelling; hide in crevices, run rapidly in dark; omnivorous, spoil foods, clothing, leather, and books. *Periplaneta americana,* American cockroach; *Blatta orientalis,* oriental cockroach; *Blatella germanica,* croton bug or European cockroach; all in United States.

MANTIDAE. Praying mantis. Head free, prothorax long; forelegs enlarged to grasp insects used as food; tarsi 5-jointed, eggs laid in masses attached to twigs, etc.; movements stealthy; mostly tropical. *Stagmomantis carolina,* praying mantis; north to Ohio and New Jersey, 60 mm (2½ in) long.

PHASMIDAE. (Phasmatidae). Walkingsticks. Form sticklike, mimicking twigs on which the insect rests; wings small or none; meso and metathorax and legs long; tarsi 5-jointed; eggs dropped singly; slow-moving; feed on foliage; mostly tropical, some to 330 mm (13 in) long, some mimic leaves. *Anisomorpha (Diapheromera) femorata,* north to Canada, wingless.

Figure 25-21 Order ORTHOPTERA *A.* Walkingstick (*Anisomorpha*), 75 to 100 mm (3 to 4 in) long. *B.* Praying mantis (*Paratenodera*), 70 mm (3 in). *C.* American cockroach. (*Periplaneta americana*), 45 mm (2 in). *D.* Katydid (*Microcentrum*), 70 mm (3 in). *E.* Field and house cricket (*Gryllus*), 30 mm (1 in). *F.* Tree cricket (*Oecanthus*), 19 mm (¾ in). *G.* Grylloblatta (*Grylloblatta*), 25 mm (1 in). (*After various authors.*) See also Fig. 25-1.

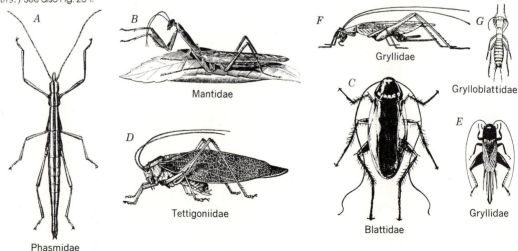

ACRIDIDAE. (Locustidae). Grasshoppers, locusts. Body deeper than wide, head and eyes large, antennae short; tarsi 3-jointed; hind legs elongate for jumping; eggs laid in ground; males produce sounds; diurnal; feed on green plants, often damage crops. *Melanoplus femur-rubrum,* red-legged grasshopper; M. *spretus,* Rocky Mountain grasshopper; *Dissosteira carolina,* Carolina locust, and *Schistocerca americana,* American locust, both of wide distribution; *Romalea microptera,* lubber grasshopper, almost wingless, in Southern states; *Locusta migratoria,* winged migratory locust of Old World, cause of biblical locust plagues.

TETTIGONIIDAE. (Locustidae). Long-horned grasshoppers, katydids, etc. Form delicate; often green; antennae often longer than body; tarsi 4-jointed; hind legs long for jumping; nocturnal; males produce sounds. *Microcentrum,* katydid; *Anabrus simplex,* Mormon cricket, wingless, a crop pest.

GRYLLIDAE. Crickets. Short, dark-colored antennae long, tarsi 3-jointed; some wingless; often leap on long hind legs; males produce chirps. *Gryllus,* field and house crickets; *Gryllotalpa,* mole cricket, burrows in ground; *Oecanthus,* tree cricket, small, greenish-white, puncture twigs and berry canes to lay eggs.

GRYLLOBLATTIDAE. Body narrow, to 2.5 cm (1 in) long; wingless; eyes reduced or absent, no ocelli; legs similar, 5-jointed; cerci long, 8-jointed. *Grylloblatta,* in mountains of western North America to 3650 m (12,000 ft) under stones near glaciers and snowbanks.

Order 4. Dermaptera. Earwigs. Elongate; chewing mouth parts; forewings short, leathery, no veins; hind wings large, semicircular, membranous, veins radial, fold beneath forewings at rest; some wingless; tarsi 3-jointed; cerci form stout horny forceps at end of abdomen; metamorphosis gradual; feed on green plants and on other insects; hide in crevices, sometimes in buildings; 32 families, 1100 species; Jurassic to Recent. *Labia minor,* 6 mm long, common; *Forficula auricularia* (Fig. 25-22), European earwig, now a local garden nuisance in Eastern states and on Pacific Coast; *Anisolabis maritima,* wingless, acclimatized from Europe on beaches of Atlantic Coast.

Order 5. Plecoptera. Stone flies. Size moderate to large; body soft; mouth parts for chewing but often absent in adults; antennae long, with setae;

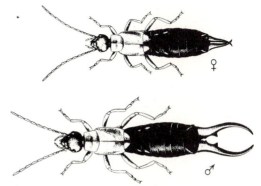

Figure 25-22 Order DERMAPTERA: European earwig (*Forficula auricularia*), 11 to 15 mm long. (*From W. G. Herms.*)

wings 4, membranous, posterior pair held pleated but flat on back at rest, hind pair larger; tarsi 3-jointed; end of abdomen usually with 2 long many-jointed cerci; nymphs aquatic, often with a tufted tracheal gill behind each leg; live under stones, eat small insects; used as fish bait; adults of weak flight; metamorphosis gradual; 16 families, 1500 species; Permian to Recent. *Pteronarcys; Brachyptera pacifica,* salmonfly, occasionally damages fruit-tree buds in Washington (Fig. 25-23).

Order 6. Isoptera. Termites. Body soft; thorax joins broadly to abdomen; chewing mouth parts; workers and soldiers wingless, sterile, blind, colorless (hence "white ants"); sexual males and females pigmented, with 2 pairs of similar, narrow, membranous wings, carried flat on back at rest and detached after nuptial flight; tarsi usually 4-jointed; cerci short; metamorphosis gradual; 6 families, over 2000 species, mostly in tropics; Eocene to Recent (Figs. 25-24, 25-25). *Reticulitermes,* subterranean termites, in earth and into dry wood, Mexico to southern Canada; *Kalotermes,* dry-wood termites, in wood, north to Virginia and California; *Zootermopsis,* damp-wood termites, western Canada to Virginia and southward; *Amitermes,* in earth of deserts, Texas to California; *Macrotermes,* in tropics, constructs firm

Figure 25-23 Order PLECOPTERA: stone fly (*Brachyptera pacifica*). *A.* Nymph. *B.* Adult, 13 mm long. (*B, After Newcomer, 1918.*)

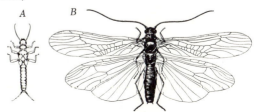

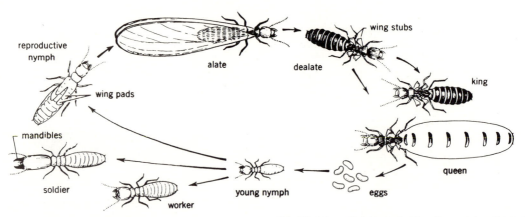

Figure 25-24   Order ISOPTERA: castes and life cycle of a termite. (*Modified from Kofoid et al., Termites and termite control, University of California Press.*)

above-ground nests (termitaria) to 9 m (30 ft) tall, of earth particles cemented by salivary secretions; *Nasutitermes*, soldiers with jaws reduced, secrete a repellent fluid on pointed head; in tropics, build above-ground nests of "carton" (paperlike wood debris); some with "fungus gardens" for food.

Primitive, social, live in closed tunnels within earth, wood, or nests, in colonies of few to thousands of individuals comprising 3 or more castes: (1) **sexuals** (kings and queens), swarm out in flights for dispersal, then settle in pairs, remove wings at pre-

Figure 25-25   Activities of the subterranean termite (*Reticulitermes***).** Necessary contact with earth is by dirt "chimneys" over concrete foundation. Adults fly after rains. (*Adapted from Light, 1929.*)

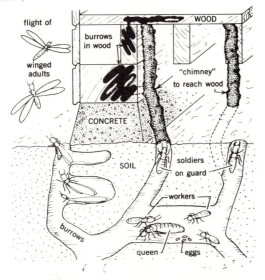

formed breaking joint near base, excavate new nest, mate, and produce colorless, wingless, and sterile young; (2) **soldiers,** heads and mandibles large, used to defend colony (some produce poisonous spray); (3) **workers** excavate tunnels and build nests, collect food and feed other castes and rear young with aid of older nymphs. **Supplementary reproductives** derived from certain nymphs may become sexually mature and replace king and queen. Feed on cellulose, which in some termites is digested by symbiotic flagellates (Sec. 15-14); termites help to reduce stumps, downed logs, etc. and form humus but do serious economic damage, especially in tropics, by tunneling in wooden buildings, bridges, posts, and lumber, also in furniture, books, and papers; some attack living trees, and some distribute wood-destroying fungi.

Order 7. Embioptera. Embiids. Small; body long, straight-sided, and soft; chewing mouth parts; forelegs with first segment of tarsi inflated, containing spinning organs; cerci 2-jointed; tarsi 3-jointed; males usually winged, females all wingless; in grass or under objects on ground, construct silk-lined tunnels joined to underground nests; colonial; eat decayed plant matter; tropical and subtropical; 7 families, 200 species; Oligocene to Recent. *Anisembia texana,* Texas; *Haploembia solieri,* California; *Oligotoma,* Africa, eastern Asia, etc. introduced in United States (Fig. 25-26).

Order 8. Mallophaga. Chewing lice. Small, to 6 mm long; body depressed, flat; wingless; head broad; mouth parts highly modified for chewing (biting); antennae short, 3- to 5-jointed; eyes reduced or none; thorax short, more or less fused; legs short; tarsi 1- to 2-jointed, clawed for clinging to

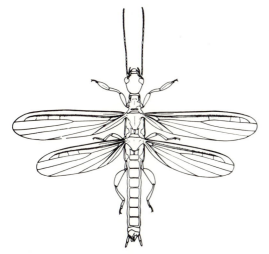

Figure 25-26 Order EMBIOPTERA: *Oligotoma saundersi,* **9 to 12 mm long.** (*E. S. Ross.*)

host; no cerci; metamorphosis slight; exclusively ectoparasitic on birds and a few mammals; feed on fragments of feathers, hair, or epidermis; eggs glued to host; 12 families, 2700 species, none fossil. *Menopon, Philopterus, Colpocephalum,* various species on different wild birds; *Menopon gallinae* and *Menacanthus stramineus,* common hen lice of poultry (Fig. 25-27A,B); *Goniodes, Lipeurus,* on various domestic fowls; *Trichodectes,* various species on wild and domestic mammals; *T. bovis,* on cattle; *T. canis,* on dogs, intermediate host of tapeworm (*Dipylidium caninum*).

Order 9. Anoplura. Sucking lice. Small, to 6 mm long; body depressed, flat; wingless; head narrow; mouth parts modified for puncturing skin and sucking blood, can be withdrawn into head when not in use; eyes reduced or none; antennae short, 3- to 5-jointed; thorax fused; no cerci; tarsi 1-jointed, each with single claw; metamorphosis slight; exclusively ectoparasitic on mammals; feed on blood and transmit various diseases; 7 families, 250 species, none fossil (Fig. 25-27C–E). *Pediculus humanus,* human louse, or cootie, on head and body, eggs, or nits, on hair or clothing; adults transmit typhus fever, trench fever, European relapsing fever, and other human diseases; related species on apes and monkeys; *Haematopinus suis,* hog louse, on domestic swine; *Polyplax spinulosa,* spiny rat louse; *Echinophthirius horridus,* on harbor seals; others on sea lions, walrus, and fur seals.

Order 10. Psocoptera (Corrodentia). Book lice, or psocids. Small to minute; antennae slender; chewing mouth parts; prothorax small; wings 4 (or none), veined, membranous, roofed over abdomen at rest; no ovipositor; metamorphosis gradual; 13 families, 1100 species; Permian to Recent.

PSOCIDAE. Antennae 12- to 50-jointed; a maxillary "pick"; tarsi 2- or 3-jointed; no cerci; feed on starches, cereals, paste, and decaying materials; under bark, in bird nests among fungi, and carry spores. *Psocus,* bark lice, winged, some under silken tents; *Liposcelis* (*Troctes*) *divinatoria,* book louse, wingless (Fig. 25-28), others winged on trees, etc.

Order 11. Zoraptera. Antennae 9-jointed; tarsi 2-jointed; cerci short; wings shed by basal fracture. *Zorotypus,* 16 species, under 3 mm long; Texas, Florida, other states, under bark, in humus, near termite nests.

Order 12. Hemiptera. True bugs. Mostly large; mouth parts piercing-sucking; jointed beak attached far forward on head, composed of labium with piercing mandibles and maxillae; pronotum large; trian-

Figure 25-27 Order MALLOPHAGA. Hen louse (*Menacanthus stramineus*) 2 to 2.5 mm long. *A. Male. B. Female.* Order ANOPLURA. *C.* Hog louse (*Haematopinus suis*), 5 to 6 mm. *D.* Human body louse (*Pediculus humanus*), 3 mm. *E.* Crab, or pubic, louse (*Phthirus pubis*), 2 mm. (*After U.S. Department of Agriculture.*)

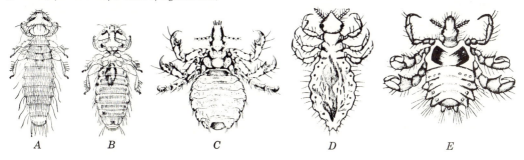

A   *B*   *C*   *D*   *E*

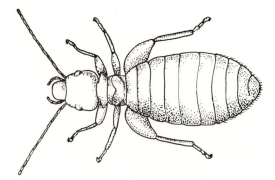

Figure 25-28   Order PSOCOPTERA (CORRODENTIA: book louse (*Liposcelis divinatora*), 1.0 mm long. (*After U.S. Dept. Agr. Farmers' Bull. 1104.*)

gular scutellum between wing bases; wings 4 (or none), forewings thick and horny at bases, membranous behind, crossed at rest; hind wings membranous, fold under forewings; aquatic or terrestrial; food of plant sap or animal body fluids; many important economically; 56 families, 40,000 species; Triassic to Recent (Fig. 25-29).

Suborder 1. Cryptocerata   Short-horned bugs. Antennae short, under head; aquatic; mostly predaceous.

CORIXIDAE.   Water boatmen. Swim upright by fringed oarlike hind legs; eat microplants and animals. *Sigara*, *Cenocorixa*, freshwater ponds and streams; *Trichocorixa*, brackish waters.

NOTONECTIDAE.   Backswimmers. Swim by long hind legs, with convex dorsal surface downward; bloodsucking; may bite humans. *Notonecta*; *Buenoa*, with reddish hemoglobin.

NEPIDAE.   Water scorpions. Usually slender, sticklike, 25 to 50 mm (1 to 2 in) long; long anal breathing tube; crawl in shallow waters. *Ranatra*, slender; *Nepa*, broad.

BELOSTOMATIDAE.   Giant water bugs. To 115 mm (4½ in) long; beak short but stout, bite severely if handled; abdomen with 2 short, straplike tails; often under lights. *Lethocerus* electric-light bug, or toe-biter; *Abedus*, *Belostoma*, females glue eggs to back of male.

Suborder 2. Gymnocerata.   Long-horned bugs. Antennae long, visible from above, usually 4-jointed; mostly terrestrial or on water surface.

GERRIDAE.   Water striders. Delicate, long-legged, walk on surface film of water; preda-

Figure 25-29   Order HEMIPTERA.   *A.* Water scorpion (*Ranatra*), 50 mm (2 in) long. *B.* Water strider (*Gerris*), 20 mm (¾ in). *C.* Lace bug (*Corythucha*), 3 mm. *D.* Bedbug (*Cimex*), 5 mm. *E.* Water boatman (*Cenocorixa*), 10 mm long. *F.* Corsair bug (*Rasahus*), 18 mm. *G.* Giant water bug (*Lethocerus*), 60 mm (2½ in). *H.* Squash bug (*Anasa*), 16 mm. *I.* Plant bug (*Lygus*), 6 mm. *J.* Milkweed bug (*Lygaeus*), 12 mm. *K.* Backswimmer (*Notonecta*), 12 mm (½ in). (*After various authors.*)

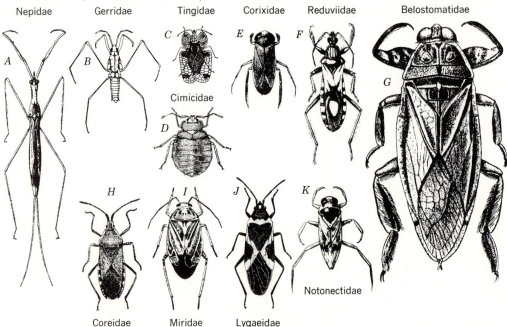

ceous. *Gerris,* on fresh waters; *Halobates,* on warm oceans even far from land.

REDUVIIDAE. Assassin bugs. Head narrow, beak stout, curved under head into stridulatory groove of prosternum; predaceous. *Rasahus,* corsair bug, predatory on insects, also bites humans severely; *Rhodnius* and *Triatoma,* cone noses, suck blood from mammals; *T. protracta,* to 24 mm (1 in) long, Mexico and Texas to California, in nests of wood rats (*Neotoma*) and human dwellings, often infected with *Trypanosoma cruzi,* which causes Chagas' disease in South America (Sec. 15-14).

CIMICIDAE. Bedbugs. Flat, oval; wing pads only; nocturnal, bad-smelling, bloodsucking. *Cimex lectularius,* common bedbug of humans; other species on humans, bats, poultry, and swallows.

MIRIDAE.   Plant bugs. To 14 mm ($\frac{1}{2}$ in) long; no ocelli; forewings with fracture. *Lygus lineolaris,* tarnished plant bug, damages many crop plants.

TINGIDAE.   Lace bugs. No ocelli; pronotum and forewings expanded and lacy, of many cells. *Corythucha,* on leaves of many plants.

LYGAEIDAE.   Chinch bugs. With ocelli; antennae below center of eye. *Blissus leucopterus,* seriously harmful to wheat and other cereals.

COREIDAE.   Squash bugs. With ocelli; antennae high on head. *Anasa tristis,* squash bug, harmful to melons, etc.

PENTATOMIDAE.   Stink bugs. Body oval or shield-shaped; antennae 5-jointed; scutellum over part or all of abdomen; bad-smelling; some predaceous, many plant feeders. *Murgantia histrionica,* harlequin bug (Fig. 25-30), on cruciferous and other plants; *Perillus bioculatus,* eats caterpillars and beetles.

Order 13. Homoptera.   Cicadas, aphids, scale insects. Mostly small; mouth parts piercing-sucking; base of beak close to thorax; wings 4 (or none), membranous, of like texture throughout, roofed over abdomen at rest; puncture plants and feed on sap, many destructive, some transmit diseases; 63 families, 32,000 species; Permian to Recent (Fig. 25-31).

Suborder 1. Auchenorhyncha.   Rostrum (free) on head; antennae short, a bristle at end; tarsi 3-jointed; active.

CICADIDAE.   Cicadas, or "locusts." Usually large; femur of foreleg enlarged; males "sing" loudly, by 2 sets of vibrating plates on forepart of abdomen, or click with wings. *Magicicada septendecim,* periodic cicada, or seventeen-year locust, nymphs grows 13 or 17 years in ground, sucks sap from the roots, then emerges as flying adult; eggs laid by chisellike ovipositor, damaging fruit trees.

CERCOPIDAE.   Spittlebugs, froghoppers. To 12 mm ($\frac{1}{2}$ in) long; ocelli 2 or none; tibiae usually spined; brown; nymphs hide in whipped-up white froth. *Aphrophora.*

CICADELLIDAE.   (Jassidae). Leafhoppers. Small; antennae bristlelike, between eyes; can run sidewise. *Circulifer* (*Eutettix*) *tenellus,* beet leafhopper, transmits virus causing curly top disease in sugar beets.

MEMBRACIDAE.   Treehoppers. To 10 mm ($\frac{1}{2}$ in) long; prothorax variously enlarged and projecting in grotesque shapes. *Ceresa bubalus,* buffalo treehopper, on various trees.

Figure 35-30 **Order HEMIPTERA: harlequin cabbage bug (*Murgantia histrionica*), adult 10 mm long.** (*After U.S. Dept. Agr. Farmers' Bull. 1061.*)

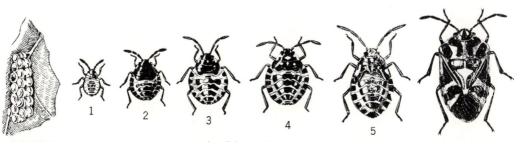

egg mass   nymphs (5 instars)   adult

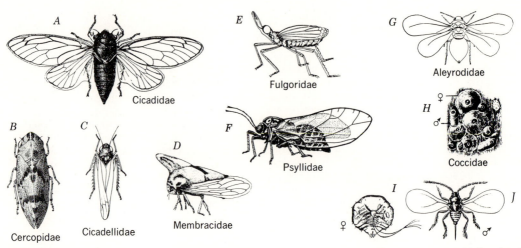

Figure 25-31 Order HOMOPTERA. *A.* Periodic cicada (*Magicicada septendecim*), 25 mm (1 in) long. *B.* Spittle bug (*Aphrophora*), 10 mm. *C.* Leafhopper (*Empoasca*), 3.5 mm. *D.* Treehopper (*Platycotis*), 10 mm. *E.* Planthopper (*Scolops*), 6 mm. *F.* Psyllid (*Paratrioza*), 1.4 mm. *G.* Whitefly (*Dialeurodes*), 2 mm. *H–J.* San José scale (*Aspidiotus perniciosus*): *H.* Scales on tree bark; *I.* Female, 1.0 mm. *J.* Winged male. (*After various authors.*)

FULGORIDAE.   Planthoppers. To 40 mm (1½ in) long; head variously enlarged and prolonged. *Scolops,* 6 to 8 mm long, in grasses and shrubs.

Suborder 2. Sternorhyncha.   Small; rostrum apparently arising on thorax; antennae longer, no bristle; tarsi 1- or 2-jointed.

PSYLLIDAE.   Jumping plant lice. To 6 mm long; leap by enlarged hind legs. *Psylla pyricola,* on pears; *P. mali,* on apples.

APHIDAE.   Aphids, or plant lice. Small to minute, delicate; usually with 2 honeydew tubes (cornicles) dorsal on abdomen; mostly wingless, but males and some females winged; tarsi 2-jointed; parthenogenesis common (Fig. 25-32); suck sap from plants, much of which is secreted as honeydew and often eaten by ants. *Phylloxera vitifoliae,* grape phylloxera, damages roots; *Anuraphis maidiradicis,* corn root aphid, harmful to roots; *Aphis gossypii,* cotton or melon aphid, on various vegetables.

Figure 25-32 Order HOMOPTERA: family APHIDAE. Stages in the life cycle of the spring grain aphid (*Toxoptera graminum* adult male 1.5 mm long), an insect having both parthenogenetic and normal sexual development. Also shown: parasitization of aphid by wasp. (*After U.S. Bur. Entomol. Bull. 110.*)

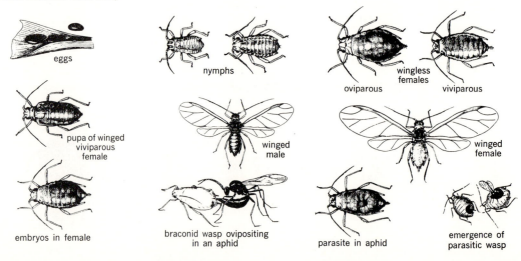

ALEYRODIDAE. Whiteflies. 1 to 3 mm long, body and wings covered by white powder; nymphs flat, scalelike, sessile; all adults winged. *Trialeurodes vaporariorum,* greenhouse whitefly.

COCCIDAE. Scale insects and mealybugs. Small to minute; males with one pair of wings; most females lack eyes, legs, and wings, covered by secreted wax or shell-like scale; tarsi 1-jointed; many harmful to crop and garden plants. *Saissetia oleae,* black scale, on many host plants, causes great damage; *Aspidiotus perniciosus,* San Jose scale, quite harmful (Fig. 25-31*H-I*); *Pseudococcus citri,* citrus mealybug, damages various plants; *Laccifer lacca,* lac insect of Southeast Asia, secretion used to make shellac.

Order 14. Thysanoptera. Thrips. Mostly minute (0.5 to 8 mm), slender; mouth parts conical, rasping-sucking; antennae 6- to 10- jointed; wings 4, alike, narrow, few veins, fringed by many long hairs, and laid flat on body at rest; some wingless; tarsi 1- or 2-jointed, ending in protrusible "bladder", no cerci; parthenogenesis common, males unknown in some species; larvae often gregarious; metamorphosis gradual to complete; 1 to 10 generations per year in different species; both larvae and adults scrape epidermis on flowers, leaves, and fruits to suck out juices of plants and leave whitened scars; many species injurious to crops, and some transmit virus diseases of plants; 19 families, 4000 species; Jurassic to Recent.

Suborder 1. Terebrantia. Females with sawlike ovipositor to deposit eggs in epidermis of plants. *Thrips tabaci,* onion thrips, worldwide on hundreds of plant hosts; *Hercothrips fasciatus,* bean thrips (Fig. 25-33), on many cultivated crops; *Heliothrips haemorrhoidalis,* greenhouse thrips, on roses, tomatoes; *Scirtothrips citri,* citrus thrips, 0.75 mm long, damage oranges and lemons.

Suborder 2. Tubulifera. Females lack ovipositor, eggs laid on surface of plants; less harmful. *Leptothrips.*

Division 2. *Holometabola* (Endopterygota). Young stages are larvae without compound eyes; wings grow internally; metamorphosis complete (complex).

Order 15. Mecoptera. Scorpion flies. Small to large; antennae and legs long; chewing mouth parts on downward-projecting beak; wings 4, alike, slender, membranous, many-veined; cerci small; some males carry end of abdomen upcurved like scorpion; inhabit dense herbage; larvae and adults carnivorous; larvae caterpillarlike with 3 pairs of thoracic legs and 8 of prolegs on abdomen, live in burrows and feed on ground surface; 5 families, 400 species; Permian to Recent. *Panorpa,* scorpion fly (Fig. 25-35*H*); *Bittacus,* resembles crane fly; *Boreus,* on winter snow in Northern states.

Order 16. Neuroptera. Dobson flies, ant lions. Large to small; antennae long; chewing mouth parts; wings 4, alike, large, membranous, many cross veins, roofed over abdomen at rest; tarsi 5-jointed; no cerci; larvae spindle-shaped with grooved suctorial mandibles, carnivorous; metamorphosis complete; 23 families, 4700 species; Permian to Recent. (Fig. 25-34).

Suborder 1. Megaloptera. Veins not forked near wing margins; prothorax short, no long ovipositor; larvae (hellgrammites) with chewing mouth parts, aquatic, with long lateral gills. *Corydalis cornuta,* dobson fly, male with huge mandibles but harmless; *Sialis,* alderfly.

Suborder 2. Raphidiodea. Snakeflies. Veins usually not forked near wing margins; prothorax long; ovipositor long and slender; larvae with chewing mouth parts, terrestrial. *Agulla, Inocel-*

Figure 25-33 Order THYSANOPTERA: bean thrips (*Hercothrips fasciatus*), adult 1 mm long. (*From Stanley F. Bailey.*)

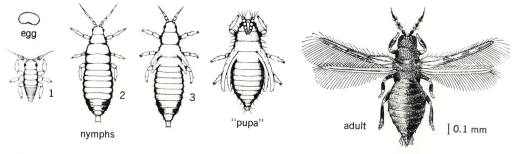

egg

1 2 3 "pupa" adult] 0.1 mm

nymphs

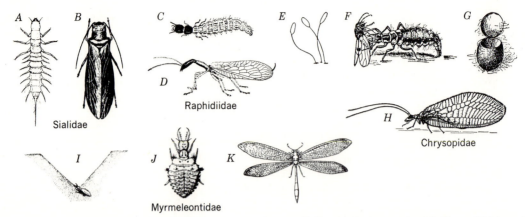

Figure 25-34 Order NEUROPTERA. *A,B.* Suborder MEGALOPTERA. SIALIDAE. *Sialis,* alderfly; larva and adult, 20 mm (¾ in) long. *C, D.* Suborder RAPHIDIODEA. RAPHIDIIDAE. *Agulla,* snakefly: larva and adult, 15 mm (½ in) long. *E–H.* Suborder PLANIPENNIA. CHRYSOPIDAE. *Chrysopa,* green lacewing: eggs, larva eating an insect, cocoon, adult, 12 mm (½ in) long. *I–K.* MYRMELEONTIDAE. *Myrmeleon,* ant lion, 20 mm (¾ in) long: larva in pit, larva, and adult. (*After various authors.*)

lia, prey on small insects under bark and on foliage. (Fig. 25-34*C,D*).

Suborder 3. Planipennia. Veins forked near wing margins; prothorax usually short; no long ovipositor; larvae mostly terrestrial, with sucking mouth parts.

CHRYSOPIDAE. Lacewings or goldeneyes. Small; wings delicate, green; eggs on stalks attached to plants; larvae (aphid lions) predaceous, beneficial, feed on aphids. *Chrysopa.*

SISYRIDAE. Spongillaflies. Wings brown, anterior veins few, rarely forked; eggs in clusters on plants over water; larvae aquatic, with long bristlelike mouth parts to penetrate tissue of freshwater sponges (Sec. 16-10). *Sisyra, Climacia.*

MYRMELEONTIDAE. Ant lions. Adults resemble damselflies but have knobbed antennae; eggs

laid in sand or loose earth; larva, or "doodlebug," makes a conical crater to trap ants, aphids, etc., which fall in and from which it sucks the body fluids. *Myrmeleon, Hesperoleon.*

Order 17. Trichoptera. Caddis flies. Adults 3 to 25 mm (1 in) long, soft-bodied; mouth parts rudimentary; antennae and legs long; wings 4, membranous, many longitudinal and few cross veins, roofed over abdomen at rest; body and wings clothed with hairs, scalelike on some; larvae carnivorous or feed on algae, aquatic, each species living in distinctive movable case of leaves, plant stems, sand, or gravel, bound by lining of silk secreted by glands opening near mouth; some in rapid water construct nets and funnels to capture prey; often used as fish bait; 19 families, 4500 species; Jurassic to Recent. *Limnophilus, Hydropsyche* (Fig. 25-35*A-G*).

Order 18. Lepidoptera. Moths and butterflies. Size various, spread 3 to 250 mm (10 in), mouth parts

Figure 25-35 Order TRICHOPTERA. Caddis fly: *A.* Larva, 15mm (½ in) long. *B.* Adult (*Phryganea*), 10 mm. *C–G.* Cases of various larvae. *H.* Order MECOPTERA. Scorpionfly (*Panorpa*). (*After various authors.*)

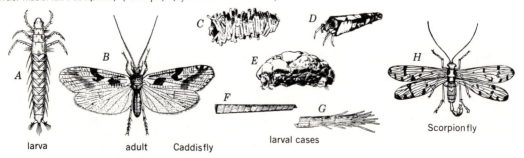

for chewing in larvae, for sucking in adult; usually no mandibles; maxillae joined as coiled tube (proboscis) for sucking fluids; antennae long; eyes large; wings, 4, membranous, usually broad with few cross veins and covered by microscopic overlapping scales; body scaly or hairy; coloration brilliant to obscure; larvae (caterpillar, cutworm, etc) wormlike with 3 pairs of legs plus prolegs on abdomen; 2 silk glands on labium used to spin cocoon to contain pupa; 118 families, 112,000 species; Jurassic to Recent. (Fig. 25-36, 25-37, 25-38).

Suborder 1. Zeugloptera.   Very small diurnal metallic moths; mandibles functional; venation of fore and hind wings similar; maxillary palpi 5-segmented, folded; midtibia without spurs; 1 family. MICROPTERYGIDAE. *Micropteryx.*

Suborder 2. Monotrysia.   Small to very large moths; female with 1 or 2 genital openings on sternites 9-10; no separate copulatory aperture on sternite 8 (genital opening serves in copulation); 9 families. HEPIALIDAE. Mouth parts vestigial. *Hepialus.*

Suborder 3. Ditrysia.   Small to very large moths and all butterflies; female with 2 genital openings, copulatory aperture on sternite 8 and genital aperture on sternites 9 to 10; 89 families of moths, 12 familes of butterflies.

AEGERIIDAE.   (Sesiidae). Clearwinged moths. Wings mostly scaleless, transparent; adults resemble wasps but sluggish; larvae bore in wood of trees and shrubs. *Sanninoidea (Aegeria) exitiosa,* peach-tree borer; *Ramosia tipuliformis,* currant borer, from Europe.

GELECHIIDAE.   Small; larvae burrow in seeds, tubers, twigs, or leaves. *Sitotroga cerealella,* angumois grain moth, 12 mm ($\frac{1}{2}$ in) wide, larvae in grains and corn; *Pectinophora gossypiella,* pink bollworm, larvae in cotton bolls, 4 to 6 generations per year.

TINEIDAE.   Clothes moths. Small, native to Europe, widely naturalized. *Tinea pellionella,* case-making clothes moth, larvae in "silk"-lined cases, covered by food debris; destroys woolens, furs, and feathers.

COSSIDAE.   Carpenter, or goat moths. Adults large, some spread to 180 mm (7 in); larvae bur-

Figure 25-36 Order LEPIDOPTERA: moths; size is total spread. *A.* Tiger moth (*Estigmene acraea*), 35 mm (1½ in). *B.* Clear-wing moth (*Aegeria*), 30 mm (1¼ in). *C.* Plume moth (*Adaina*), 14 to 20 mm (½ to ¾ in). *D.* Gypsy moth (*Porthetria dispar*), 50 to 60 mm (2 to 2½ in). *E.* Tussock moth (*Hemerocampa vetusta*), female, wings reduced, length 12 to 15 mm. *F.* Codling moth (*Carpocapsa pomonella*), 15 to 22 mm. *G.* Leopard moth (*Zeuzera pyrina*), 60 mm (2½ in). *H.* Fall armyworm moth (*Laphygma frugiperda*), 25 to 30 mm. *I.* Aquatic moth (*Parargyractis*), 20 mm. *J.* Tobacco hornworm moth (*Protoparce sexta*), 100 to 125 mm (4 to 5 in). (*After various authors.*)

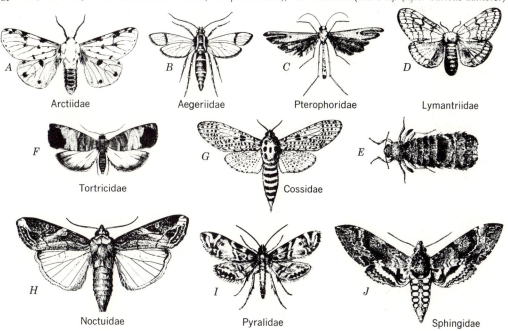

Arctiidae   Aegeriidae   Pterophoridae   Lymantriidae

Tortricidae   Cossidae

Noctuidae   Pyralidae   Sphingidae

row in solid wood of deciduous trees, live 2 or 3 years. *Prionoxystus robiniae*, carpenter or goat moth; *Zeuzera pyrina*, leopard moth.

TORTRICIDAE. Leaf rollers. Larvae often roll up and feed in leaves. *Carpocapsa pomonella*, codling moth, from southeastern Europe, larvae in fruits of apples and pears, 2 or 3 generations in warm regions, very harmful.

PYRALIDAE. Snout moths. Spread to 25 mm (1 in); larvae often concealed, roll in leaves or bore in stems, fruits, or stored materials. *Pyrausta nubilalis*, European corn borer, in Eastern states, bores in stems and kills plants; *Plodia interpunctella*, Indian-meal moth, damages dried foods; *Galleria mellonella*, bee moth, larva (wax worm) eats combs of hive bees; *Nymphula, Parargyractis*, larvae aquatic.

PTEROPHORIDAE. Plume moths. Spread to 25 mm (1 in); wings deep-notched and feathery. *Adaina ambrosiae*, ragweed plume moth, larva pale green.

PSYCHIDAE. Bagworm moths. Larvae in cases of silk and foliage scraps, females wingless, remain in cases. *Thyridopteryx ephemeraeformis*.

LASIOCAMPIDAE. Tent caterpillars. Adults with proboscis atrophied, cannot feed; larvae spin community web among branches of trees and forage outside. *Malacosoma americanum*, tent caterpillar (Fig. 25-37), may defoliate trees.

SPHINGIDAE. Sphinx, or hummingbird, moths. Forewings narrow, flight rapid, hover over flowers at dusk to feed with long proboscis. *Sphinx chersis*, great ash sphinx; *Protoparce sexta*, tobacco hornworm, and *P. quinquemaculata*, tomato hornworm, both with naked larvae to 100 mm (4 in) long, adults spread to 125 mm (5 in).

GEOMETRIDAE. Measuring worm moths and cankerworm moths. Many resemble butterflies, some females wingless; larvae naked, feed on exposed foliage, and walk by looping body that has legs only near ends, hence called loopers, or measuring worms. *Paleacrita vernata*, spring cankerworm, on trees.

SATURNIIDAE. Giant silkworm moths. Some yield silk from cocoons. *Hyalophora* (*Simia* or *Platysamia*) *cecropia*, cecropia moth, North America, wingspread to 165 mm (6½ in); *Antheraea polyphemus*, polyphemus moth, spread to 120 mm (5 in); *Attacus*, in tropics, females spread to 240 mm (9½ in).

BOMBYCIDAE. *Bombyx mori*, silkworm moth, native to China, entirely domesticated; larvae feed on mulberry leaves and spin dense cocoons that yield the genuine white and yellow silk of commerce—a most valuable insect; larvae subject to pébrine disease (caused by *Nosema bombycis*, a sporozoan), studied by Louis Pasteur.

Figure 25-37 Order LEPIDOPTERA: life history of the tent caterpillar (*Malacosoma americanum*); adult 20 mm (⅘ in) long. Stages in transformation or metamorphosis take place inside the cocoon. Not to scale. (*After Snodgrass, Insects, Smithsonian Institution Series, Inc.*)

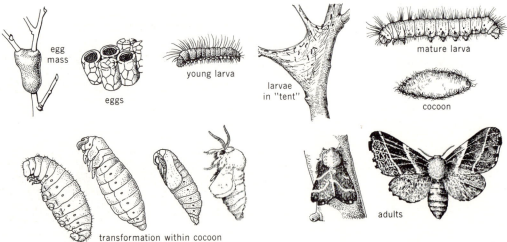

egg mass

eggs

young larva

larvae in "tent"

mature larva

cocoon

transformation within cocoon

adults

ARCTIIDAE. Tiger moths. Adults bright-colored; larvae densely hairy, the "woolly bears." *Hyphantria cunea,* fall webworm; *Estigmene acraea,* acraea moth.

NOCTUIDAE. Owlet moths. Colors somber, some adults spread to 150 mm (6 in); larvae feed variously, some in ground as cutworms. *Heliothis zea,* corn earworm; *Pseudaletia unipuncta,* armyworm of North America, larvae sometimes "march" across country in great numbers; *Thysania (Erebus) agrippina,* largest moth.

LIPARIDAE. (Lymantriidae) Tussock moths. Larvae thickly covered with hairs. *Porthetria dispar,* gypsy moth, and *Nygmia phaeorrhoea,* brown-tail moth, both of Europe and now in Northeastern states, defoliate and kill shade trees. *Hemerocampa vetusta,* tussock moth, female with reduced wings.

HESPERIIDAE. Skippers. Antennae with hooked tips, abdomen heavy; larvae with constriction behind head. *Atalopedes campestris,* field skipper; *Calpodes ethlius,* canna leaf roller.

PAPILIONIDAE. Swallowtails. Spread to 125 mm (5 in); bright-colored; usually a taillike projection on each hind wing; larvae mostly naked; pupae angular, not in cocoons. *Papilio* (Fig. 25-38C–F).

PIERIDAE. White and sulphur butterflies. Abdomen slender. *Pieris rapae,* cabbage butterfly, wings white, on cabbage, etc. *Colias eurytheme,* alfalfa butterfly.

NYMPHALIDAE. Brush-footed or four-footed butterflies. Forelegs reduced, not functional, folded under body; tibiae hairy. *Danaus plexip-pus,* milkweed or monarch butterfly, tawny and black; *Nymphalis (Aglais) antiopa,* mourning cloak butterfly, blackish; *N. californica,* tortoiseshell butterfly, occasionally irrupts and migrates in great numbers; *Vanessa cardui,* painted lady; *Limenitis (Basilarchia) archippus,* viceroy, coloration mimics that of monarch (Fig. 13-17).

LYCAENIDAE. Blues, hairstreaks, and gossamers. Mostly small, delicate, often brown, eyes white-rimmed; front tarsi reduced on males; larvae sluglike. *Lycaena hypophlaeas,* the copper; *Feniseca tarquinius,* larvae prey on woolly aphids; *Strymon melinus,* wings gray.

Order 19. Diptera. True flies. Forewings transparent, few veins; hind wings represented by short knobbed halteres; some wingless; mouth parts piercing-sucking or "sponging," often forming a proboscis; body divisions distinct, abdomen of 4 to 9 visible somites; larvae usually footless; metamorphosis complete; mostly diurnal; 138 families, over 85,000 species; Upper Permian to Recent (Fig. 25-39).

Suborder 1. Nematocera. Antennae long, 6- to 39-jointed.

TIPULIDAE. Crane flies. Slender, like enlarged mosquitoes, legs long and fragile, wings narrow; a V-shaped mark on back; larvae are "leather worms," living about grass roots or semiaquatic. *Tipula.*

PSYCHODIDAE. Moth flies and sand flies. To 4 mm long, woolly, wings velvety; larvae and adults in sewers or damp organic wastes, some larvae in water. *Psychoda; Phlebotomus,* sucks blood of vertebrates, carries human "three-day fever" in Mediterranean region and verruga peruana in Peruvian Andes.

Figure 25-38 Order LEPIDOPTERA: butterflies. *A.* Skipper (*Calpodes ethlius*), spread 25 to 30 mm (1 to 1¼ in). *B.* Alfalfa butterfly (*Colias eurytheme*), spread 40 to 60 mm (1½ to 2½ in). *C–F.* Black swallowtail (*Papilio polyxenes*), spread 80 to 100 mm (3¼ to 4 in). *C.* Larva. *D.* Pupa, *E.* Resting adult. *F.* Spread adult. (*After various authors.*)

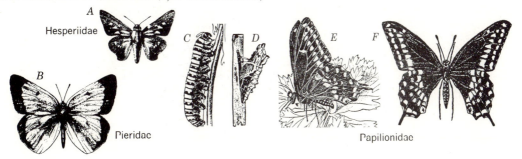

Hesperiidae

Pieridae

Papilionidae

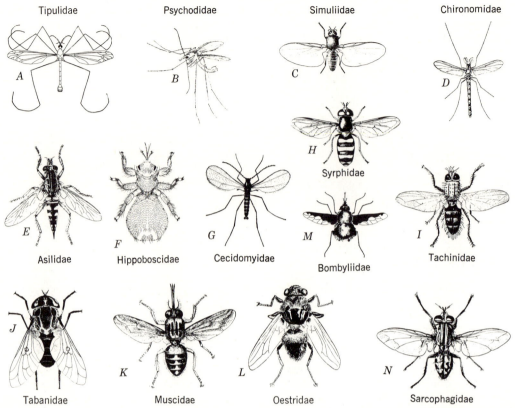

Tipulidae

Psychodidae

Simuliidae

Chironomidae

Syrphidae

Asilidae

Hippoboscidae

Cecidomyidae

Bombyliidae

Tachinidae

Tabanidae

Muscidae

Oestridae

Sarcophagidae

Figure 25-39 Order DIPTERA: flies. *A.* Crane fly (*Tipula*), 10 to 25 mm (½ to 1 in) long. *B.* Sand fly (*Phlebotomus*), 1.5 mm. *C.* Black fly (*Simulium*), 2 mm. *D.* Midge (*Tendipes = Chironomus*), 3 to 10 mm. *E.* Robber fly (*Erax*), 25 mm (1 in). *F.* Sheepked (*Melophagus ovinus*), 6 mm. *G.* Hessian fly (*Phytophaga destructor*), 2.5 mm. *H.* Hover fly (*Syrphus*), 10 mm. *I.* Tachinid fly (*Tachina*), 8 mm. *J.* Horsefly (*Tabanus*), 25 mm. *K.* Tsetse fly (*Glossina*), 13 mm. *L.* Botfly (*Hypoderma bovis*), 12 mm. *M.* Bee fly (*Bombylius*), 10 mm. *N.* Flesh fly (*Sarcophaga*), 7 mm. (*After various authors.*)

CULICIDAE. Mosquitoes. Slender, delicate; proboscis long and piercing in females; body humped; wings fringed by scales; larvae ("wrigglers") with large head, long abdomen, and breathing siphon, in various waters; adult males suck plant juices; adult females chiefly bloodsuckers, on birds, mammals, and humans; abundant, annoying, some transmit serious diseases (Fig. 25-40); 1500 species. *Culex,* many species; *C. pipiens,* common house mosquito, transmits bird malaria; *Anopheles,* some species transmit *Plasmodium* (SPOROZOA) of human malaria (Fig. 15-21); *Aëdes,* different species in salt marshes, swamps, tree holes, and snow waters; *Aëdes, (Stegomyia) aegypti,* tropics and subtropics, transmitter of human dengue fever and yellow fever.

CHIRONOMIDAE. Midges, gnats, and punkies. Mostly minute, delicate, mosquitolike, mouth parts short, wings scaleless; larvae in water or damp places; adults often "dance" as swarms in air at end of day; few marine; a few reproduce parthenogenetically. *Chironomus* (= *Tendipes*), larvae are bloodworms, important as fish food.

CECIDOMYIDAE. Gall gnats. Minute, mosquitolike, antennae with whorled hairs; many produce galls on plants, in which larvae are reared; others are predaceous or parasitic or feed on decaying matter or on plants. *Phytophaga destructor,* Hessian fly, feeds in stems of wheat, very harmful; *Diarthronomyia chrysanthemi,* chrysanthemum gall midge, produces galls; *Miastor,* reproduces by paedogenesis.

MYCETOPHILIDAE. Fungus gnats. Small, delicate, mosquitolike; antennae and coxae long; in damp woods, basements, etc; larvae feed on

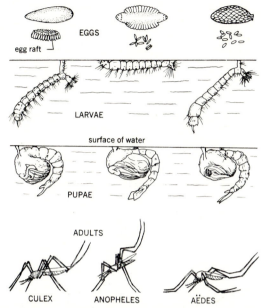

Figure 25-40 Order DIPTERA: family CULICIDAE. Stages in the life cycles of three important types of mosquito. Larvae and pupae in typical postures in the water, adults posed as when resting or biting.

fungus, some gregarious. *Mycetophila mutica,* in mushrooms and toadstools.

SIMULIIDAE. Black flies. Short, chunky, wings broad; first tarsal joint dilated; larvae attach by anal end to rocks in running water. *Simulium,* females suck blood of domestic birds and mammals and of humans; *S. pecuarum,* buffalo gnat.

Suborder 2. Brachycera. Antennae short, usually 5 or fewer true segments.

TABANIDAE. Horseflies. Stout; eyes large, lateral, often banded; proboscis projecting; eggs and larvae in water or damp places; male flies suck plant juices and honeydew, females blood-sucking. *Tabanus,* horsefly, buzzes about and bites horses and cattle; *Chrysops,* deer fly; *C. discalis,* transmits tularemia from rodents to humans.

ASILIDAE. Robber flies. Large, robust; body bristly; hairy "beard" about mouth; catch insects in air and suck out their juices. *Erax, Asilus.*

BOMBYLIIDAE. Bee flies. Densely haired, resembling bees; legs and proboscis long, slender; larvae parasitic or prey on other insects; adults hover and feed in flowers, harmless. *Bombylius.*

Suborder 3. Cyclorrhapha. Antennae 3-segmented, aristate; last larval skin a hard puparium with circular lid, opened by protrusion of a sac-like structure (ptilinum) from head of emerging fly; a suture may or may not form following retraction of ptilinum.

Tribe 1. Aschiza. No suture (often termed *frontal suture*) around antenna base.

SYRPHIDAE. Hover flies. Bright-colored, no bristles, mimic wasps or bees; larvae prey on aphids and other small insects or (rattail maggots) scavenge in filth; adults hover over and feed in flowers. *Syrphus,* hover fly; *Lampetia, Eumerus,* bulb flies, larvae damage narcissus, onions, etc.

Tribe 2. Schizophora. Muscoid flies. Suture present around antenna base.

TEPHRITIDAE. (Trypetidae). Fruit flies. Tip of abdomen narrow. *Ceratitis capitata,* Mediterranean fruit fly, and *Dacus dorsalis,* Oriental fruit fly, boring larvae of both injure fleshy fruits.

DROSOPHILIDAE. Pomace flies. Small brown flies with eyes usually red. *Drosophila melanogaster,* about decaying fruit; much used in genetics (Fig. 11-1).

ANTHOMYIIDAE. Root maggot flies. Larvae on roots or decaying materials, some carnivorous. *Hylemya brassicae,* cabbage maggot, and *H. antiqua,* onion maggot, both damage crops.

MUSCIDAE. House flies, etc. Arista usually plumed to tip; larvae grow in decomposing organic materials; adults about wild and domestic mammals and humans, obnoxious and harmful; *Musca domestica,* common house fly (Fig. 25-41),

Figure 25-41 Order DIPTERA: housefly (*Musca domestica*), 4 to 6 mm long. (*After W. B. Herms.*)

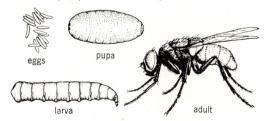

eggs   pupa

larva   adult

with sponging mouth parts, may contaminate food with filth and spread disease; *Calliphora vomitoria,* bluebottle, and *Phaenicia (Lucilia) sericata,* greenbottle, both "blow" (oviposit on) flesh; *Stomoxys calcitrans,* stable fly, and *Haematobia irritans,* horn fly, both bite domestic mammals; *Callitroga hominivorax,* screwworm, maggots in wounds, nose, or navel of domestic mammals, cause injury and death; *Glossina,* tsetse fly, carries trypanosomes of African sleeping sickness (Fig. 15-16).

SARCOPHAGIDAE. Flesh flies. Larvae consume dead animals, feces, garbage, etc., or live in insects. *Sarcophaga.*

TACHINIDAE. Tachinid flies. Eggs or larvae deposited on insects, destroying them; adults feed in flowers. *Gonia,* in cutworms.

OESTRIDAE. Botflies. Adults large, beelike, hairy, about domestic animals. *Gasterophilus intestinalis,* horse bot, eggs, or "nits," laid on hairs, removed by tongue; larvae burrow from mouth to stomach, pupae emerge in feces; *Hypoderma bovis,* larvae live as warbles under skin on back of cattle, hosts become unhealthy and hides are damaged for leather; *Oestrus ovis,* in sheep, living larvae deposited about nose, live as head maggots in nasal region; *Cuterebra,* under skin of rodents.

HIPPOBOSCIDAE. Louse flies. Body fat, louse-like; winged or wingless; abdomen rubbery, expanded; deposit pupae on hosts; ectoparasitic, blood sucking. *Melophagus ovinus,* sheep tick or "ked," 6 mm long, on domestic sheep, harmful; *Lynchia,* on domestic pigeons; *Ornithoica,* on wild birds.

BRAULIDAE. Bee louse. *Braula coeca,* minute, wingless, on queens and drones of honeybees.

Order 20. Siphonaptera. Fleas. Body laterally compressed, tough; no wings; mouth parts piercing-sucking; antennae short, in grooves; eyes simple or none; legs long, adapted for leaping, coxae enlarged, tarsi 5-jointed; eggs laid in habitat of or on host; larvae minute, legless, feed on organic debris; pupae in cocoons; adults periodically ectoparasitic and bloodsucking on birds and mammals, avoid light and seek warmth; metamorphosis complete; 7 families, 1100 species; Oligocene to Recent. *Pulex irritans,* human flea, also on rats, etc.; *Ctenocephalides canis,* dog flea, and *C. felis,* cat flea (Fig. 25-42), both on dogs, cats, and humans; *Xenopsylla cheopis,* Indian rat flea, transmits *Pasteurella pestis,* which causes bubonic plague; also transmits murine typhus in Southern states; *Ceratophyllus,* various species on wild birds and mammals; *Echidnophaga gallinacea,* sticktight flea of poultry, often harmful; *Tunga (Dermatophilus) penetrans,* chigoe, or jigger, on feet of humans and other mammals, female burrows in skin, abdomen distends to size of small pea; *Hystricopsylla gigas,* largest flea, 5 mm long, on some California rodents.

Order 21. Coleoptera. Beetles, weevils. Minute to large, cuticle heavy; chewing mouth parts, some snoutlike; fore wings (elytra) thick, leathery, veinless, meet along middorsal line; hind wings membranous, few veins, fold forward under fore wings at rest; some wingless; antennae usually 11-jointed; prothorax enlarged, movable; meso- and metathorax united to abdomen; larvae wormlike, usually with 3 pairs of legs; pupae rarely in cocoons; metamorphosis complete; 184 families, more than 280,000 species; late Permian to Recent (Fig. 25-43).

Suborder 1. Adephaga. First abdominal segment completely divided by hind coxae into 2 or 3 parts; larvae with legs of 5 segments and 2 claws.

CICINDELIDAE. Tiger beetles. Adults long-legged, bright-colored, run actively in open places, flight fast; larvae in deep burrows, catch insects at surface. *Cicindela,* in sandy places; *Omus,* black, wingless, nocturnal.

CARABIDAE. Ground beetles. Mostly blackish, long-legged, and terrestrial; tarsi 5-jointed; adults and larvae mainly carnivorous; some burrow, some blind in caves, some emit foul spray if disturbed. *Calosoma,* large, preys on caterpillars and cutworms.

DYTISCIDAE. Predaceous diving beetles. To 37 mm (1½ in) long; oval, smooth, shiny; hind legs flat, fringed with hairs, serve as oars in swimming; air stored under elytra when diving; adults and larvae aquatic and predaceous; food of larvae digested externally. *Dytiscus.*

Figure 25-42 Order SIPHONAPTERA: **cat flea** (*Ctenocephalides felis.*), **2.5 mm long.** (*After W. B. Herms.*)

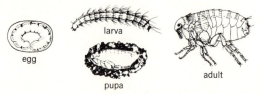

egg

larva

pupa

adult

Cicindelidae   Carabidae   Dytiscidae   Staphylinidae   Silphidae   Hydrophilidae   Lampyridae   Elateridae   Buprestidae

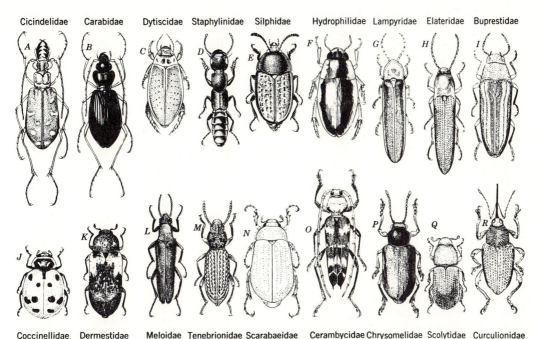

Coccinellidae   Dermestidae   Meloidae   Tenebrionidae   Scarabaeidae   Cerambycidae   Chrysomelidae   Scolytidae   Curculionidae

Figure 25-43 Order COLEOPTERA: beetles; not to scale. *A*. Tiger beetle (*Cicindela*), 14 to 16 mm long. *B*. Ground beetle (*Pterostichus*), 8 to 17 mm. *C*. Predaceous water beetle (*Rhantus*), 10 to 12 mm. *D*. Rove beetle (*Staphylinus*), 12 to 25 mm. *E*. Carrion beetle (*Silpha*), 11 to 18 mm. *F*. Water scavenger beetle (*Tropisternus*), 8 to 11 mm. *G*. Firefly (*Photinus*), 6 to 14 mm. *H*. Click beetle (*Limonius*), 5 to 12 mm. *I*. Flatheaded borer (*Buprestis*), 14 to 19 mm. *J*. Lady beetle (*Hippodamia*), 6 to 8 mm. *K*. Larder beetle (*Dermestes*), 6 to 7 mm. *L*. Blister beetle (*Epicauta*), 7 to 14 mm. *M*. Darkling ground beetle (*Nyctoporis*), 11 to 15 mm. *N*. June beetle (*Phyllophaga*), 13 to 24 mm. *O*. Long-horned beetle (*Megacyllene*), 13 to 23 mm. *P*. Leaf beetle (*Chrysochus*), 9 to 11 mm. *Q*. Bark beetle (*Dendroctonus*), 5 to 8 mm. *R*. Acorn weevil (*Balaninus*), 5 to 9 mm. (*Drawn by Frieda L. Abernathy.*)

GYRINIDAE. Whirligig beetles. Small, oval, lustrous, eyes divided horizontally; skate rapidly on water surface, gregarious; predaceous. *Gyrinus*.

Suborder 2. Polyphaga. First abdominal segment complete; legs of larvae with not more than 4 joints and 1 claw; some legless.

STAPHYLINIDAE. Rove beetles. Slender; elytra short, hind wings with simple straight veins; abdomen below with 7 or 8 segments; scavengers or predators. *Creophilus*, preys on carrion fly maggots; *Myrmedonia*, resembles ants and lives in ant nests; *Termitomimus*, in termite nests, ovoviviparous.

SILPHIDAE. Carrion beetles. Minute to 37 mm (1½ in) long; characters as above but abdomen with 6 visible ventral segments; many flattened; eat carrion, fungi, or decaying plant material. *Necrophorus*, burying beetle, buries small animals as food for larvae; *Silpha*, some feed on crop plants.

HYROPHILIDAE. Water scavengers. Resemble DYTISCIDAE; black, a lengthwise keel under body; antennae shorter than palps, with hairy club which breaks surface film to obtain air; eat decaying material in water or damp places. *Hydrous triangularis*, to 27 mm (1 in) long.

COCCINELLIDAE. Ladybird beetles. Many 3 to 6 mm long, rounded, convex, elytra cover abdomen; bright-colored, spotted; head turned down in notch of prothorax, all tarsi 3-jointed; larvae soft, spiny; most larvae and adults beneficial, preying on aphids and scale insects. *Coccinella* and *Hippodamia*, predaceous; *Epilachna varivestis*, Mexican bean beetle, eats beans and other crops, harmful.

DERMESTIDAE. Skin beetles. Small dark, hairy; all tarsi 5-jointed; both adults and larvae destructive, eating furs, woolens, meat, and hides. *Anthrenus scrophulariae*, carpet beetle, 3 mm long, very injurious.

LAMPYRIDAE. Glow-worms, fireflies. To 12 mm ($\frac{1}{2}$ in) long, straight-sided; prothorax semicircular over downturned head; all tarsi 5-jointed; nocturnal; with light-producing organs on abdomen. *Lampyris, Photinus.*

ANOBIIDAE. Powder-post beeltes; etc. Length 3 to 8 mm, cylindrical or globular, dark; all tarsi 5-jointed. *Stegobium paniceum,* drugstore beetle, damages stored plant materials and upholstery; *Xestobium rufovillosum,* deathwatch beetle, larva bores in solid wood, adult taps with head.

BUPRESTIDAE. Metallic wood borers. Body iridescent; antennae serrate; all tarsi 5-jointed; larvae with expanded flat prothorax, head minute. *Chrysobothris femorata,* flatheaded apple-tree borer, larva feeds under bark.

ELATERIDAE. Click beetles. Adults hard, narrow; can snap into air when placed on back, by releasing catch on ventral keel; all tarsi 5-jointed; larvae are tough-skinned wireworms, some harmful to seeds and roots of crops. *Alaus,* to 45 mm (2 in) long; *Melanotus.*

TENEBRIONIDAE. Darkling beetles. Some resemble CARABIDAE, but the hind tarsi are 4-jointed; body shape various. *Tenebrio molitor,* mealworm, larvae damage stored cereals, used to feed caged animals; *Tribolium confusum,* flour beetle, larvae in flour, etc.

MELOIDAE. Blister, or oil, beetles. Length 7 to 30 mm (1 in); prothorax narrow, necklike; hind tarsi 4-jointed; some lack hind wings, some with hypermetamorphosis (extra larval stages and prepupa); adults eat foliage and flowers. *Epicauta,* adults damage crops; *Lytta,* larvae eat grasshopper eggs, *L. vesicatoria,* Spanish fly, yields cantharidin, used in poultices; *Meloe, Sitaris,* larvae in nests of solitary bees.

LUCANIDAE. Stag beetles. Mandibles often enormous in males; front tibiae enlarged for digging in soil; antennae loosely clubbed and elbowed; abdomen with 5 visible segments; larvae in roots or wood under ground, some live 4 or more years. *Lucanus.*

SCARABAEIDAE. Scarabs, June beetles. Small to large; body broad, deep, convex; legs spiny; antennae clubbed or elbowed, of 3 to 7 plates; abdomen with 6 visible segments; nocturnal. *Phyllophaga,* larvae (white grubs) on grass roots, etc. adults on foliage; *Popillia japonica,* Japanese beetle, destructive, introduced in Eastern states; *Macrodactylus,* rose chafer, adults eat leaves of cultivated trees and shrubs; *Scarabaeus sacer,* Egyptian sacred scarab or dung beetle, adults roll up balls of dung for food of larvae.

CHRYSOMELIDAE. Leaf beetles. To 12 mm ($\frac{1}{2}$ in) long; color brilliant, often metallic; all tarsi apparently 4-jointed; adults eat leaves; larvae often in ground, feed on roots; many destructive; *Leptinotarsa decemlineata,* Colorado potato beetle, widespread; *Diabrotica,* cucumber beetles; *Donacia,* larva aquatic, obtains oxygen by piercing air chambers in stems of aquatic plants.

CERAMBYCIDAE. Long-horned beetles. Slender, to 75 mm (3 in) long; colors often bright; antennae may exceed body in length; all tarsi apparently 4-jointed; adults eat foliage or soft bark; larvae are roundheaded borers in solid but usually dead wood. *Saperda candida,* roundheaded apple borer; *Ergates,* pine sawer.

CURCULIONIDAE. Weevils, or snout beetles. Small to 50 mm (2 in); hardshelled, often rough-surfaced; snout prolonged; antennae usually elbowed; femora often swollen; adults feign death when disturbed; larvae legless, feed within seeds, fruits, stems, or roots; many very destructive to crops. *Anthonomus grandis,* cotton boll weevil; *Sitophilus granarius,* granary weevil; *Hypera postica,* alfalfa weevil (Fig. 25-44).

SCOLYTIDAE. Bark beetles. Short, cylindrical; antennae clubbed; adults and larvae feed under bark, "engraving" branched channels. *Scolytus rugulosus,* short-hole borer, in various trees; *Dendroctonus,* pine beetle, in various conifers; *Xyleborus,* ambrosia beetle, grows fungus ("ambrosia") for food.

Order 22. Strepsiptera. Stylops. Minute; metathorax enlarged; mouth parts for chewing but small (or none); males with fan-shaped hind wings but fore wings reduced to clubbed halteres; mobile, live 1 or 2 days; females larvalike, no antennae, eyes, wings, or legs; head and thorax fused; females and larvae permanently parasitic in body spaces of bees, wasps, HOMOPTERA; absorb nutriment from host and modify host structure ("stylopize" it); a hypermetamorphosis; 8 families, 300 species; Oligocene and Recent (Fig. 25-45). *Stylops,* in bee (*Andrena*); *Xenos,* in wasp (*Polistes*).

Order 23. Hymenoptera. Wasps, ants, bees, etc. Mouth parts chewing or chewing-lapping; wings 4

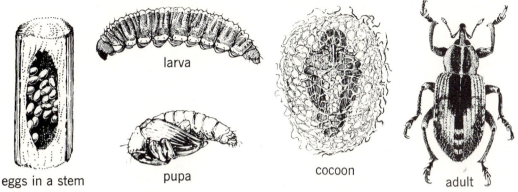

Figure 25-44 Order COLEOPTERA: the alfalfa weevil (*Hypnera postica*), adult 3 to 5 mm long. (*After U.S. Bur. Entomol. Circ. 137.*)

(or none), small, membranous, few veins, interlocked in flight; female with ovipositor for sawing, piercing, or stinging; larvae caterpillarlike or legless; pupae commonly in cocoons; metamorphosis complete; most species solitary, but some social in colonies, others parasitic as larvae; 109 families, 105,000 species, Jurassic to Recent (Fig. 25-46).

Suborder 1. Symphyta (Chalastrogastra). Abdomen broadly joined to thorax; larvae caterpillarlike, with legs, feed on plants.

SIRICIDAE. Horntails, wood wasps. Body straight-sided; abdomen with spine at end; female with projecting ovipositor for boring; larvae in wood of trees or shrubs. *Sirex; Tremex,* to 50 mm (2 in) long.

TENTHREDINIDAE. Sawflies. Body robust; female with ovipositor concealed and sawlike; parthenogenesis common; larvae eat foliage or bore in stems, etc. *Cimbex americanus,* American sawfly, wingspread to 50 mm (2 in); *Caliroa cerasi,*

larva—pear or cherry "slug"—eats epidermis of tree leaves.

Suborder 2. Apocrita (Clistogastra). Base of abdomen a narrow "waist" behind thorax; larvae legless.

ICHNEUMONIDAE. Ichneumon wasps. Length 4 to 38 mm ($1\frac{1}{2}$ in); ovipositor very long (to 150 mm in *Thalessa*); larvae parasitize caterpillars and other insects, live and pupate in host. *Therion, Ephialtes.*

BRACONIDAE. Braconid wasps. Mostly under 3 mm; like ichneumonids in form and habits but usually pupate outside of host. *Apanteles congregatus,* on larvae of sphingid moths; *Lysiphlebus testaceipes,* in aphids (Fig. 25-32).

CHALCIDIDAE. Chalcid wasps. Mostly minute; antennae elbowed; wing veins reduced; some larvae in seeds or stems, others parasitic, some in other parasites. *Bruchophagus,* larvae in clover seed; *Harmolita grandis,* larvae in wheat stems, harmful; *Trichogramma minutum,* millions reared and released for control of pest insects such as European corn borer. *Aphytis,* parasitic on scale insects.

AGAONTIDAE. Fig wasps. Length to 2.5 mm; antennae elbowed; males wingless. *Blastophaga psenes,* breeds in wild capri figs and serves to pollinate Smyrna figs. Introduced to California from Mediterranean region.

CYNIPIDAE. Gall wasps. Small to minute; abdomen laterally compressed; eggs laid in plant tissues, mainly on oaks, which form swollen galls around the developing larvae; males rare, par-

Figure 25-45 Order STREPSIPTERA: *Stylops,* male 2 mm long. (*After Bohart, 1941.*)

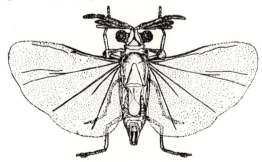

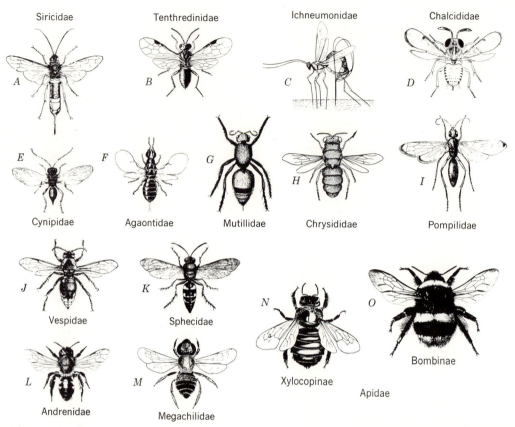

Siricidae

Tenthredinidae

Ichneumonidae

Chalcididae

A

B

C

D

E

F

G

H

I

Cynipidae

Agaontidae

Mutillidae

Chrysididae

Pompilidae

J

K

N

O

Vespidae

Sphecidae

Bombinae

L

M

Xylocopinae

Andrenidae

Megachilidae

Apidae

Figure 25-46 Order HYMENOPTERA. *A*. Horntail (*Urocerus* or *Sirex*), 30 mm (1.2 in). *B*. Sawfly (*Endelomyia*), 15 mm (0.6 in). *C*. Ichneumon wasp (*Megarhyssa*), to 38 mm (1.5 in). *D*. Chalcid wasp (*Aphytis*), 0.8 mm. *E*. Rose gall wasp (*Rhodites*), 3 mm (0.1 in). *F*. Blastophaga wasp (*Blastophaga psenes*), 2.5 mm. *G*. Velvet ant (*Sphaerophthalma*), 12 mm (0.5 in). *H*. Cuckoo wasp (*Parnopes*), 6 to 12 mm ($\frac{1}{4}$ to $\frac{1}{2}$ in). *I*. Spider wasp (*Pepsis*), 20 to 40 mm (0.8 to 1.6 in). *J*. White-faced hornet (*Vespula maculata*), 12 to 19 mm ($\frac{1}{2}$ to $\frac{3}{4}$ in). *K*. Mud dauber wasp (*Sphecius*), 12 to 20 mm ($\frac{1}{2}$ to $\frac{3}{4}$ in). *L*. Mining bee (*Andrena*), 6 to 18 mm ($\frac{1}{4}$ to $\frac{3}{4}$ in). *M*. Leaf-cutter bee (*Megachile*), 12 mm ($\frac{1}{2}$ in). *N*. Carpenter bee (*Xylocopa*), 12 to 20 mm ($\frac{1}{2}$ to $\frac{3}{4}$ in). *O*. Bumblebee (*Bombus*), 10 to 20 mm (0.4 to 0.8 in). (*After various authors.*)

thenogenesis common. *Andricus*, on oaks; *Diplolepis*, on rose roots.

FORMICIDAE. Ants. Length 2 to 18 mm ($\frac{3}{4}$ in); pronotum enlarged; abdominal pedicle narrow; polymorphic, with sexual females (queens), males, and wingless (female) workers; other secondary castes in some species; social, in colonies of few to 100,000; queen removes wings after nuptial flight, starts colony, later only lays eggs and is fed liquids from mouths of workers; most species are scavengers and useful in cleaning up dead materials; few predaceous; many eat seeds, often are harmful; some colonize aphids and cause much trouble to agriculture. *Ponera*, primitive, few in colony; *Eciton*, legionary ants of tropics and Southern states, no permanent nests, march in large companies, predaceous; *Mono-*

morium, small black ant, nests in soil or wood; *Solenopsis geminata*, fire ant, stings severely, damages crops; *Pogonomyrmex*, harvester ant, eats seeds and often clears ground about entrance to nest in ground; *Atta*, fungus ant, feeds on fungus cultured on organic debris in underground galleries; *Formica*, builds large anthills of debris over nest in ground; *F. sanguinea*, rears workers of *F. rufa* to serve as slaves; *Iridomyrmex humilis*, Argentine ant, a household and agricultural pest, colonizes and attends aphids and scale insects to obtain honeydew as food; *Camponotus*, carpenter ant (Fig. 25-47), large, in dead wood or ground. Many beetles and other insects inhabit ant nests, including (1) scavengers or predators; (2) indifferent guests; (3) true guests, which are fed and even reared by ants; and (4) parasites.

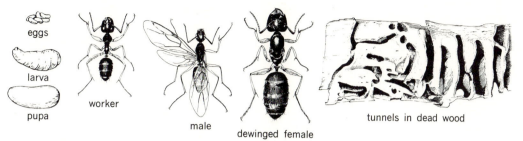

eggs

larva

pupa

worker

male

dewinged female

tunnels in dead wood

Figure 25-47 Order HYMENOPTERA: family FORMICIDAE. Carpenter ant (*Camponotus*): eggs, larva, pupa (cocoon); worker 6 to 10 mm (0.24 to 0.4 in) long, winged male (same length), dewinged female to 15 mm (0.6 in) long.

MUTILLIDAE. Velvet ants. Body black, densely covered with colored hair; males winged, females wingless; larvae parasitic in cocoons of solitary wasps and bees. *Mutilla.*

SCOLIIDAE. Vespoid wasps. Large, black, hairy, larvae parasitic in white grubs or larvae of scarabaeid beetles. *Scolia.*

CHRYSIDIDAE. Cuckoo wasps. Integument hard; colors metallic; abdomen flat beneath, curled under body when attacked; larvae in nest cells of solitary wasps and bees. *Chrysis.*

POMPILIDAE. Spider wasps. Abdomen not narrowed at base; sting paralyzes spiders; nest in ground; *Pepsis,* large, blue or black, some with orange wings.

VESPIDAE. Vespid wasps. Social and solitary. Yellowjackets and hornets, sting powerful; young females overwinter, and each starts colony that lasts one season; nest of paper (chewed wood) with horizontal combs of cells to contain larvae; many small females or workers; males produced in late summer from unfertilized eggs; adults feed on nectar, honeydew, or fruit juices; young carnivorous. *Vespa diabolica,* yellow jacket, nest in or above ground, combs enclosed by paper walls; *Vespula maculata,* bald-faced hornet, 12 to 19 mm ($\frac{1}{2}$ to $\frac{3}{4}$ in) long, nests large, enclosed, in trees; *Polistes,* paper wasp, nest flat and open below, one female; *Eumenes,* potter wasps, solitary, make nests of mud or in plant stems, each with one egg stocked with prey that is stung but not killed to feed larva.

SPHECIDAE. Sphecid wasps. Some are thread-waisted, others are stout bodied. *Sphex, Sceliphron,* mud daubers, thread-waisted, abdomen slender, long, enlarged at end, nests in ground or mud cells provisioned with stung prey for

food of larvae; *Bembix,* beelike, live in separate nests but close together; larvae in unsealed cells, fed daily by adults.

ANDRENIDAE. Mining bees. Each female prepares a nest, lays eggs, and feeds nectar and pollen to larvae; nests close together. *Andrena.*

MEGACHILIDAE. Leaf-cutting and mason bees. Resemble hive bees, head broad, tongue long, mandibles sharp; nests in plant stems, wood, or earth and provisioned with nectar and pollen. *Megachile,* leaf-cutter bee, nests lined with disks cut from leaves; *Anthidium,* mason bee, nests of mud cells lined with plant down, etc.

HALICTIDAE. Halictid bees. Small, dark-colored to brownish or metallic green; nest in burrows in ground, often in colonies, usually in areas of sparse vegetation; members of brood may share same passageway to outside; provision nests with pollen or nectar. *Halictus.*

APIDAE. Honeybees and others. First two segments of labial palps elongated and flattened; tongue long and slender. Honeybees, eyes hairy; workers with pollen baskets on hind legs; native to Old World, probably Asia. *Apis mellifera,* hive bee (Figs. 25-12, 25-13); *Xylocopa,* carpenter bees, large, black, resemble bumblebees but less hairy; powerful mandibles used to tunnel in solid wood for nest cells; *Bombus (Bremus),* bumblebees (humblebees), medium to large, densely haired; tongue long, hence useful in pollinating clover and many other crop plants; overwintering young queens establish new colonies that last one season, with queen, many worker females, and drones (males); colonies in mouse nests or rodent burrows surrounded by grass, with wax cells to contain larvae and open cells (honeypots) for food storage.

References

The literature of entomology comprises a great number and variety of books, pamphlets, and periodicals. Besides the many works on general entomology, others are devoted to individual orders, families, or species and to the insect faunas of particular regions, large or small. Only a small selection is given here.

Albrecht, F. O. 1953. The anatomy of the migratory locust. London, Athlone Press. xvi + 118 pp., 141 figs. *Complete anatomy and instructions for dissection.*

Bishopp, F. C., and others. 1952. Insects: The yearbook of agriculture. Washington, D.C., U.S. Government Printing Office. xviii + 780 pp., 72 col. pls., many figs. *Harmful and useful species of economic importance; popular.*

———, **D. M. Delong** and **C. A. Triplehorn.** 1976. An introduction to the study of insects. New York, Holt, Rinehart and Winston, Inc. viii + 852 pp., illus.

Borror, D. J., and **R. E. White.** 1970. A field guide to the insects of America north of Mexico. Boston, Houghton Mifflin Company. xi + 404 pp., 16 pls., many figs.

Brues, C. T., A. L. Melander, and **F. M. Carpenter.** 1954. Classification of insects. Rev. ed. Cambridge, Mass. *Museum of Comparative Zoology Bulletin*, vol. 108, v + 917 pp., 1219 figs. *Comprehensive keys to families and subfamilies; includes immature stages and fossils.*

Campbell, F. L. (editor). 1959. Physiology of insect development. Chicago, The University of Chicago Press. xiv + 167 pp., illus.

Chapman, R. F. 1969. The insects: Structure and function. New York, American Elsevier Publishing Company, Inc. xii + 819 pp., 509 figs.

Chu, H. F. 1949. How to know the immature insects. Dubuque, Iowa, William C. Brown Company Publishers. 234 pp., 631 figs. *illustrated keys to common types.*

Clark, L. R., P. W. Geier, R. D. Hughes, and **R. F. Morris.** 1967. The ecology of insect populations. London, Methuen & Co., Ltd. xiii + 232 pp., 43 figs.

DeBach, Paul (editor). 1964. Biological control of insect pests and weeds. New York, Reinhold Book Corporation. xxiv + 844 pp., 123 figs.

Elzinga, R. J. 1977. Fundamentals of entomology. New York, Prentice–Hall, Inc. 320 pp. illus.

Gilmour, Darcy. 1965. The metabolism of insects. San Francisco, W. H. Freeman and Company 195 pp., 32 illus.

Grassé, Pierre-P., and others. 1949. Traité de zoologie. Paris, Masson et Cie. Insects, vol. 9, 1117 pp., 752 figs; vol. 10 (2 pts.),1948 pp., 1648 figs. *Complete treatment but no keys for identification.*

Herms, W. B. 1969. Medical entomology. 6th ed. Rev. by M. T. James. New York, The Macmillan Company. xi + 484 pp.

Holland, W. J. 1951. The butterfly book. Rev. ed. Garden City, N. Y., Doubleday & Company, Inc. xii + 424 pp., 198 figs. 77 pls., mostly color. *Popular and scientific guides to biology and identification.*

———1968. The moth book; A guide to the moths of North America. Rev. by A. E. Brower. New York, Dover Publications, Inc. xxiv + 479 pp., 48 pls. in color, 262 figs.

Horn, D. J. 1976. Biology of insects. Philadelphia, W. B. Saunders Company. 439 pp., illus.

Imms, A. D. 1964. A general textbook of entomology. 9th ed. Rev. by O. W. Richards and R. G. Davies. London, Methuen & Company., Ltd. x + 886 pp., 609 figs.

Jeannel, R. 1960. Introduction to entomology. London, Hutchinson & Co. (publishers), Ltd. 344 pp., pls. in color, 150 figs. *General anatomy and classification, biology, paleontology, and geographic distribution of insects; excellent information on fossils.*

Jacobson, M. 1972. Insect sex pheromones. New York, Academic Press, Inc. 382 pp.

Jaques, H. E. 1947. How to know the insects. 2d ed. Dubuque, Iowa, William C. Brown Company Publishers. 205 pp., 411 figs. *Keys to common families.*

Johnson, C. G. 1969. Migration and dispersal of insects by flight. London, Methuen & Co., Ltd. xxii + 763 pp., 217 figs.

Kilgore, W. W. and **R. L. Doutt.** 1967. Pest control. New York, Academic Press, Inc. xii + 477 pp. *Biologic, physical, and selected chemical methods for the control of harmful insects and vertebrates.*

Klots, A. B., and **E. B. Klots.** 1959. Living insects of the world. Garden City, N.Y., Doubleday & Company., Inc. 304 pp., illus.

Lanham, Urless. 1964. The insects. New York, Columbia University Press. 240 pp., illus.

Metcalf, C. L., and **W. P. Flint.** 1962. Destructive and useful insects. 4th ed. Rev. by R. L. Metcalf. New York, McGraw-Hill Book Company. xii + 1099 pp., illus.

Michener, C. D., and **M. H. Michener.** 1951. American social insects. Princeton, N.J., D. Van Nostrand Company, Inc. xiv + 267 pp., 109 figs., some in color.

Novak, V. J. A. 1975. Hormones. 2d ed. New York, John Wiley & Sons, Inc. 600 pp.

Oldroyd, Harold. 1958. Collecting, preserving and studying insects. London, Hutchinson & Co. Publishers, Ltd. 327 pp., 15 pls., 136 figs.

Patton, R. L. 1963. Introductory insect physiology. Philadelphia, W. B. Saunders Company. vi + 245 pp., illus. *Brief general treatment with citations to recent literature.*

Pesson, Paul. 1959. The world of insects. Translated from French by R. B. Freeman. New York, McGraw-Hill Book Company. 204 pp., 225 photos, inc. 52 in color. *Semi-popular, pictorial.*

Pimentel, David (editor). 1975. Insects, science and society. New York, Academic Press, Inc. 284 pp.

Price, Peter W. 1975. Insect ecology. New York, John Wiley & Sons, Inc. 514 pp.

Pringle, J. W. S. 1957. Insect flight. New York, Cambridge University Press. Monographs in Experimental Biology, no. 9. viii + 132 pp., 52 figs.

Rockstein, Morris (editor). Physiology of the Insecta. 2d ed. New York, Academic Press, Inc. Vol. 1, 1973, 544 pp.; 1974. Vol. 2, 568 pp.; Vol. 3, 517 pp.; Vol. 4, 448 pp.; Vol. 5, 682 pp.; Vol. 6, 548 pp.

Roeder, K. D. (editor). 1953. Insect physiology. New York, John Wiley & Sons, Inc. xiv + 1100 pp., 257 figs. *Structure, physiology, behavior, and development.*

Romoser, W. S. (editor). 1973. The science of entomology. New York, The Macmillan Company. 449 pp., illus.

Snodgrass, R. E. 1952. A textbook of arthropod anatomy. Ithaca, N.Y., Comstock Publishing Associates. viii + 363 pp., 88 figs.

_____1956. Anatomy of the honey bee. Ithaca, N.Y., Comstock Publishing Associates. xiv + 334 pp., 107 figs.

Steinhaus, E. A. (editor). 1963. Insect pathology. New York, Academic Press, Inc. Vol. 1, xvii + 661 pp., illus. Vol. 2, xiv + 689 pp., illus. *An advanced treatise with chapters by leading authorities.*

Swain, R. B. 1948. The insect guide (to) orders and major families of North American insects. Garden City, N.Y., Doubleday & Company., Inc. xlvi + 261 pp., 454 figs., incl. 330 in color.

Usinger, R. L. (editor). 1956. Aquatic insects of California, with keys to North America genera and California species. Berkeley, University of California Press. 518 pp., illus.

von Frisch, Karl. 1967. The dance language and orientation of bees. Translated by Leigh E. Chadwick. Cambridge, Mass., The Belknap Press, Harvard University Press. xiv + 556 pp., illus.

Wigglesworth, V. B. 1964. The life of insects. Cleveland, Ohio, The World Publishing Company. xii + 360 pp., 164 figs. (8 in color). *General introduction to insect biology, references, and glossary.*

_____1972. The principles of insect physiology. 7th ed. New York, John Wiley & Sons, Inc. 827 pp., illus.

Williams. C. B. 1958. Insect migration. New York, William Collins Sons & Co., Ltd. xiii + 235 pp., 49 figs., 11 col. pls.

Wilson, E. O. 1971. Insect Societies. Cambridge, Harvard University Press.

PHYLUM ECHINODERMATA: ECHINODERMS

The ECHINODERMATA (Gr. *echinos*, hedgehog + *derma*, skin) is one of the most distinctive and readily recognizable phyla of the Animal Kingdom. It includes the well-known starfishes or sea stars and brittle stars (class STELLEROIDEA), sea urchins and sand dollars (ECHINOIDEA), sea lilies and feather stars (CRINOIDEA), and sea cucumbers (HOLOTHUROIDEA), in addition to a large number of extinct classes (Figs. 26-1, 26-2). All are large animals, and none is parasitic or colonial. Virtually all are benthic in habit and are either permanently attached to the ocean bottom or move slowly over the substrate. They are unique among animals in having no head, an internal skeleton, bilateral larvae which metamorphose into radially symmetrical adults, and an internal subdivision of the coelom which is used for locomotion and food capture. All echinoderms presently living are marine in distribution and are common and abundant animals in all the world's oceans. Together with the phylum CHORDATA they constitute the major deuterostome animals.

26-1 Characteristics

1   Symmetry usually radial in adults, bilateral in larvae; three germ layers; most organs ciliated; no segmentation in larva or adult.

2   Body surface of five symmetrical radiating areas, or ambulacra, whence the tube feet project, and alternating between these five interambulacra (interradii).

3   Body covered by delicate epidermis over a firm

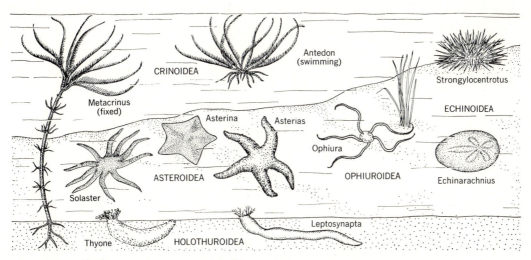

Figure 26-1 Representative echinoderms as they live in the sea; all reduced in size, but not to same scale. Class CRINOIDEA: crinoids (*Metacrinus*, attached; *Antedon*, free-swimming). Subclass ASTEROIDEA: starfishes (*Asterias, Asterina*), sun star (*Solaster*). Subclass OPHIUROIDEA: brittle star (*Ophiura*). Class ECHINOIDEA: sea urchin (*Strongylocentrotus*), sand dollar (*Echinarachnius*). Class HOLOTHUROIDEA: sea cucumbers (*Thyone, Leptosynapta*).

mesodermal endoskeleton of movable or fixed calcareous plates, usually in definite pattern; often with spines (skin leathery and plates usually microscopic in HOLOTHUROIDEA).

4 No head, body arranged on an oral-aboral axis.

5 Coelom enterocoelous, large and lined with ciliated peritoneum and subdivided during development to give rise to the unique water-vascular system, with tube feet serving for locomotion, food handling, and respiration; a complex hemal system.

6 Respiration by minute dermal branchiae (skin gills) or papulae protruding from the coelom, by tube feet, and in HOLOTHUROIDEA by cloacal respiratory trees.

7 Nervous system diffuse and composed typically of three rings centered on the mouth region with radiating branches.

8 Sexes separate (rare exceptions), alike externally; gonads large, with simple ducts; ova abundant, usually fertilized in the sea; larvae bilateral, microscopic, ciliated, transparent, and usually free-swimming, with conspicuous metamorphosis.

9 No excretory system.

26-2 Relationships and phylogeny The echinoderms are an ancient group of animals with an abundant fossil record dating back to early Cambrian times. However, as with other dominant phyla, such as MOLLUSCA and ARTHROPODA, the fossils do not indicate either origin or relationships of the phylum. Many of our inferences about the origin and relationships of this unique group are therefore drawn from study of the living members, particularly from embryonic development. All living echinoderms have a typical deuterostome embryology, with radial, indeterminate cleavage of the fertilized egg, mouth arising separate from the blastopore, and coelom forming from pouches budded off the archenteron (enterocoelous). In addition, the larval forms are bilaterally symmetrical and characterized by a series of complex ciliated bands coupled with a very involved metamorphosis. The fact that the larvae are bilaterally symmetrical suggests that the ancestral echinoderms were also bilateral animals and that the present radial condition is associated with their present sedentary or sessile mode of life, as with other sessile groups (CNIDARIA, PORIFERA). The only other major phylum which shares all the above deuterostome embryologic features is the CHORDATA. In addition, the chordates and the echinoderms have a well-de-

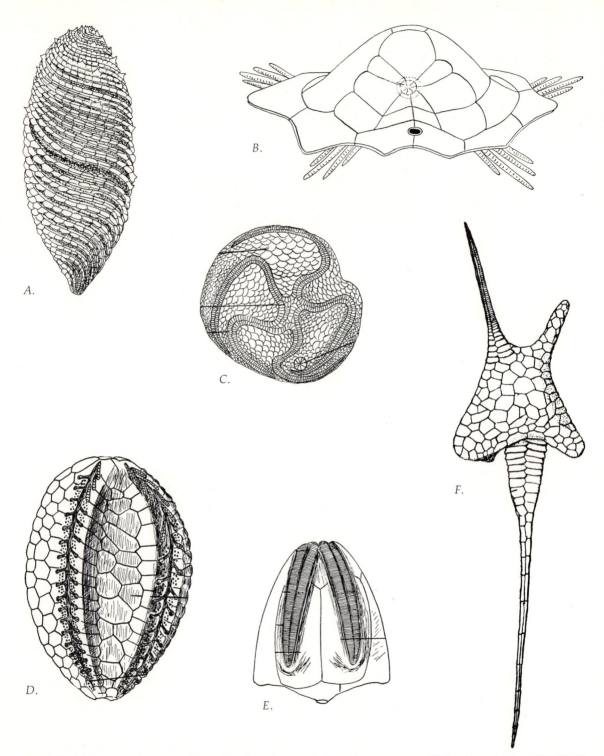

Figure 26-2 Representatives of six extinct echinoderm classes. *A.* Class †HELICOPLACOIDEA: †*Helicoplacus. B.* Class †OPHILOCISTIOI-DEA †*Volchovia. C.* Class †EDRIOASTEROIDEA: †*Lepidodiscus. D.* Class †CYSTOIDEA: †*Proteroblastus. E.* Class †BLASTOIDEA: †*Pentre-mites. F.* Class †HOMOIOSTELA: †*Dendrocystites. (A, from McGraw-Hill Encyclopedia of Science and Technology; B–F, after Hyman).*

veloped internal skeleton. For these reasons the echinoderms have long been considered to be most closely related to the chordates, and the two phyla may stem from a common ancestor. The present divergent anatomic features in adults of these two phyla and the long but distinct fossil record in both suggest, however, that they have evolved along separate lines for a long period of time. No intermediate fossil forms are known.

There is considerable controversy as to which groups within the phylum are the most primitive, some zoologists favoring the concept that the sessile, attached forms such as the living crinoids are the most primitive, whereas others suggest that the free-living forms or their ancestors are more primitive. The oldest fossil echinoderms represent both attached and free-living forms, so the fossil record cannot answer this (Fig. 26-2). While the crinoids are obviously separated from the other living classes in the structure of both adult and embryo, the relationships among the other living classes are obscured by the fact that adult and larval structures contradict each other. Thus ASTEROIDEA and OPHIUROIDEA appear more similar to each other as adults than to other classes or subclasses, but the larval forms are different, with the ophiuroid larvae resembling echinoids closely. Similarly the HOLOTHUROIDEA and ASTEROIDEA appear very different as adults, but the larvae are structurally similar. It is thus not possible to establish acceptable relationships among the living classes. It should also be noted that because of the conflicting evidence on relationships and the numerous fossil forms there is considerable controversy regarding the higher classification in the phylum. The classification system here employed follows that found in most newer invertebrate zoology texts.

26-3 General features Echinoderms are unique animals, and much of what makes them so different from other invertebrates resides in the elaborate divisions of the coelom and the use made of it. Echinoderms are the only radially symmetrical organisms which have a coelom. The coelom in echinoderms is divided during development into three parts: (1) a conspicuous system of fluid-filled tubes called the

water-vascular system, which manifests itself on the external body surface in the form of **tube feet,** or **podia;** (2) a second inconspicuous set of internal tubes enclosing strands of tissue called the **perihemal system;** and (3) the large, fluid-filled **perivisceral coelom** which surrounds the internal organs. The water-vascular system is a hydraulic system which functions variously in locomotion, feeding, and respiration, depending on the particular echinoderm class. In the most common forms, such as starfishes and sea urchins, the system operates primarily in locomotion (see Sec. 26-5), although some gas exchange probably takes place across the permeable walls of the tube feet. In the brittle stars and crinoids, the tube feet lack suction cups and are not used in locomotion. In these forms the tube feet function mainly to capture food. If crinoids can be considered representative of the more primitive condition of the phylum, then it appears that the ancestral use of the water-vascular system was primarily to capture food and perhaps secondarily for respiration. All water-vascular systems have a ring canal around the mouth, radial tubes leading out to the tube feet, and an axial component running from the ring canal aborally to a pressure-equalization plate, or **madreporite.** The madreporite allows fluid to enter the system to allow for pressure changes and losses due to breakage of the tube feet, etc. Since the whole system functions hydraulically (fluid under pressure), a positive pressure must be maintained.

The perihemal system is not well understood. The tubes of this system parallel those of the water-vascular system, but there is also an **axial organ** which appears as an enlargement of the tube system. The function of this whole system is not at all clear. It may be involved in the transport of some materials and may also be involved in producing amoebocytes to defend the animal against diseases and to remove wastes.

Only the CHORDATA and the ECHINODERMATA have an internal skeleton. This is formed in the dermis in echinoderms and consists of separate units called **ossicles,** which grow as the animal grows. In some forms, such as sea stars and sea urchins, these ossicles have projections which penetrate the dermis and appear on the surface. Usually these spines appear uncovered, giving rise to the belief that they are external and hence that the animals have an external skeleton. In all cases, however, a thin layer of epidermis covers them, and they are part of the internal skeleton. The function of the ossicles is to provide a rigid skeleton against which muscles may operate, furnish protection to internal systems, and give the rigidity necessary to allow the tube feet to operate in locomotion.

An unusual feature for such an advanced phylum is that the echinoderms have no centralized nervous system. The nervous system is poorly developed and consists of several strands which parallel the water-vascular system and usually join a ring nerve which surrounds the mouth area. No brain is present. Despite the lack of a brain and of ganglia in general, the echinoderms are somehow able to effect coordination in such things as movement, food capture, and righting themselves when turned over. With few exceptions, echinoderms also do not have special sensory organs. It appears that sensory reception of all types—touch, chemicals, light, etc.—is a function of the unspecialized sensory cells in the epidermis.

No special excretory, circulatory, or respiratory systems are usually present (although some HOLOTHUROIDEA have special respiratory organs), which is also unusual for such large animals. Excretion is probably effected by the amoebocytes in the perivisceral coelomic fluid. These amoebocytes have the ability to engulf material and to escape from the coelom to the exterior. Excretion may also be aided by the axial organ. Absence of any defined excretory organs also means that echinoderms have little if any osmoregulatory ability (Sec. 2-11), hence we would expect them to be a completely marine group, which in fact, they are. Respiration is often carried out by the tube feet of the water-vascular system or by some projection out of or into the large perivisceral coelom, such as **dermal branchiae** or **papulae** in starfishes and **respiratory trees** in sea cucumbers. Gases diffuse into coelomic fluid, and transport within the body is by coelomic fluid. In the absence of a separate circulatory system, the large coelom with its coelomic fluid serves to transport food and oxygen to the tissues and carry off the metabolic waste and carbon dioxide.

Orientation is to the surface which bears the mouth and the surface opposite it. These are, respectively, the **oral surface** and the **aboral surface.** Most echinoderms are oriented with the oral surface against the substrate. Crinoids are oriented with the aboral surface against the substrate, while sea cucumbers lie on their side (Fig. 26-3).

With few exceptions echinoderms have separate sexes. No copulatory organs exist, nor is there any external difference between sexes. Both eggs and sperm are released into seawater, where fertilization occurs. In order to ensure fertilization of the maximum number of eggs, spawning of one animal of either sex usually stimulates all others in the area to release gametes—an effect further enhanced by the gregarious or clumped distribution of many species. The fertilized egg usually develops into a bilaterally symmetrical swimming larvae (although a few echinoderms brood young which hatch as immature adults). These larvae are distinctive (Fig. 26-4), with many ciliated tracts. The two main types are the **pluteus** and **auricularia.** All are hypothesized to have originated from a **dipleurula** ancestral form (Fig. 26-4). They are unique among larvae in that they undergo perhaps the most complex metamorphoses,

Figure 26-3 Phylum ECHINODERMATA. Diagrammatic sections showing for the living classes and subclasses the relations of the mouth (M), anus (A), tube feet (T), and spines (S). The digestive tract is outlined.

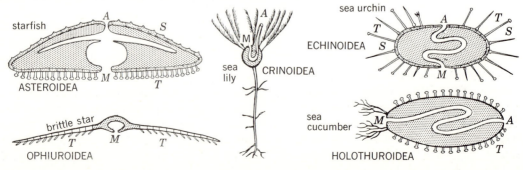

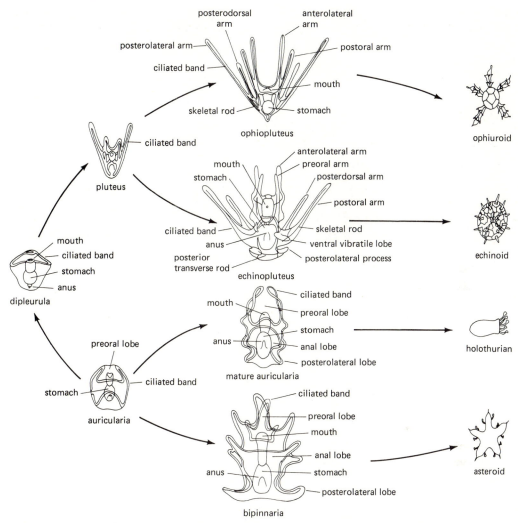

Figure 26-4 **General scheme indicating relationships among echinoderm larvae and the hypothetical dipleurula larva.** (*From McGraw-Hill Encyclopedia of Science and Technology.*)

to change into radially symmetrical adults (Fig. 26-12). Since the adult echinoderms are usually sessile or sedentary, these free-swimming larvae serve as the main dispersal mechanism.

26-4 Size Most echinoderms are of moderate size. The largest starfish (*Pycnopodia helianthoides*) spreads to about 80 cm (32 in). The largest deep-sea urchin (*Echinosoma hoplacantha*) has a shell about 30 cm (12 in) across, and some tropical urchins (*Diadema*) have 30-cm (12-in) spines. One sea cucumber (*Synapta maculata*) becomes 2 m (6 ft) long and 5 cm

(2 in) in diameter. Some fossil sand dollars measure only 0.5 cm ($\frac{1}{4}$ in), whereas the stem of a fossil crinoid was 21 m (70 ft) long.

Class Stelleroidea

Subclass Asteroidea (Starfishes)

Starfishes abound on most seacoasts, especially on rocky shores and about wharf piling. Various species live from the tide lines to considerable depths on sand and mud (Fig. 26-5). Common North American species are *Asterias forbesi* from the Gulf of Mexico to

A.

B.

Figure 26-5 *A*. The starfish *Leptasterias hexactis*. *B*. The starfish *Dermasterias imbricata*. (*Courtesy of G. McDonald.*)

Maine, *A. vulgaris* from Cape Hatteras to Labrador, and *Pisaster ochraceus* along the Pacific Coast. The description below is a general one for starfish structure.

26-5 Structure The body (Figs. 26-6, 26-7) consists of a central **disk** and five tapering rays, or **arms.** The axes of the arms are termed **radii,** and the spaces between them on the disk are **interradii.** On the upper or **aboral surface** are many blunt calcareous **spines,** which are parts of the skeleton. Small, soft **dermal branchiae** (papulae) project from the body

cavity and between the spines to function in respiration and excretion. Around the spines and among the papulae are many minute pincerlike **pedicellariae** (Fig. 26-8). Each pedicellaria has two jaws moved by muscles that open and snap shut when touched; they keep the body surface free of debris or small organisms and may help to capture food. The **anus** is a minute opening near the center of the aboral surface, and nearby is the rounded **madreporite.** The **mouth** is in the middle of the lower or **oral surface,** surrounded by a soft peristomial membrane. A median **ambulacral groove,** bordered by large spines, extends along the oral surface of each arm, and from

Figure 26-6 Subclass ASTEROIDEA. Starfish, *Asterias forbesi.*

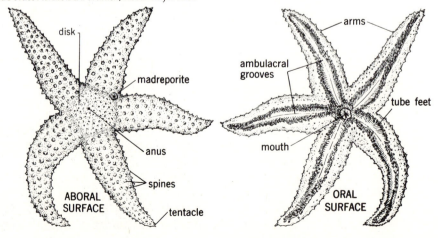

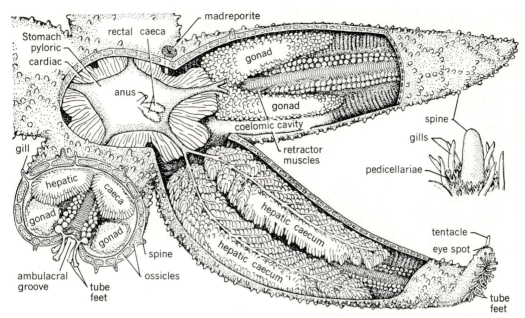

Figure 26-7 Starfish: general structure. Three arms are cut off, one is seen in cross section at the left, the disk and aboral surface of two arms are removed, and the hepatic caeca are removed from the upper right arm. Enlarged inset shows spine, gills, and pedicellariae.

it many slender **tube feet** protrude in four (or two) rows. On the tip of each arm is a small, soft tactile **tentacle** and a light-sensitive **eyespot.**

The entire body is covered by ciliated **epidermis.** Beneath is dermis that produces and contains the **endoskeleton,** a framework of many small calcareous **ossicles** of various but definite shapes arranged in a regular pattern (Fig. 26-9). The ossicles are bound together by connective tissue and joined by muscle fibers. Inside the skeleton is the large **coelom,** lined by ciliated epithelium, that contains the internal organs. It is filled with a lymphlike fluid containing free amoebocytes that participate in circulation, respiration, and excretion. Extensions of the coelom into the dermal branchiae bring the fluid near the surrounding seawater, separated only by the thin peritoneum and external epidermis, where respiratory exchanges are easily made. The amoebocytes gather wastes in the fluid and then escape from the branchiae to the exterior.

The water-vascular system (Fig. 26-10) is a specialized part of the coelom. Its parts are (1) the sieve-like **madreporite** through which seawater may enter; (2) the **stone canal** connecting to (3) the **ring canal**

Figure 26-8 Pedicellaria from a starfish showing structure. (*Modified from Hyman.*)

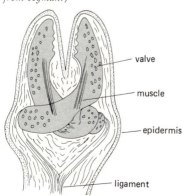

Figure 26-9 Starfish, *Pisaster ochraceus.* Part of cleaned skeleton showing the framework of ossicles.

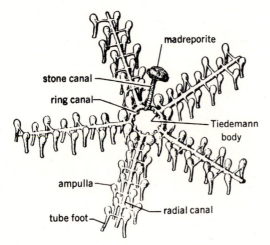

Figure 26-10 **Starfish: Diagram of the water vascular system.** *(From Coe, 1912.)*

around the mouth, whence (4) five **radial canals** extend, one in each arm above the ambulacral groove. Each of the latter gives off (5) many **lateral canals,** one to (6) each of the **tube feet,** all with valves. The margin of the ring canal bears **Tiedemann bodies,** which are thought to produce the free amoebocytes present in the fluid filling the system. Each tube foot is a closed cylinder with muscular walls, having a sucker at the outer, or free, end and a bulblike **ampulla** at its inner end within the body cavity. When an ampulla contracts, the fluid it contains is forced into the tube foot and extends the latter as a slender flexible process that can be twisted about by muscles in its walls. If the tip touches an object, contraction of a retractor muscle raises the center of the tip of the tube foot. This lessens the pressure within the tip and causes it to adhere to the substrate much like a suction cup because of the greater pressure of seawater or atmosphere outside. Adherence is aided by the secretion of an adhesive. Contraction of the tube-foot muscles then returns the fluid to the ampulla and the foot shortens, pulling the starfish ahead. The tube feet act either independently or in a coordinated manner. They serve to hold the starfish to the substratum, for locomotion, and in the capture and handling of food.

The **digestive system** comprises (1) the **mouth;** (2) the saclike **stomach,** which is of two parts, a large lower, or cardiac, portion with thin folded muscular walls, to which five pairs of retractor muscles attach, and a smaller aboral, or pyloric, portion joined by the ≺-shaped ducts of five pairs of **hepatic caeca** (digestive glands) located in the arms; and (3) a short, minute **intestine** joined by two rectal caeca and leading to (4) the **anus.**

The **perihemal system** is reduced and difficult to see; it includes vessels encircling the mouth and five radial vessels, one in each arm under the canal of the water vascular system.

The **nervous system** includes a circumoral nerve ring and nerve cords in the arms. In each arm there is (1) a cord in the epidermis within the ambulacral groove, (2) a pair of deep nerves inside this epidermis, and (3) a cord in the peritoneum on the aboral side. Minute nerves extend to the tube feet, epidermis, and internal structures.

The sexes are separate. A pair of **gonads** lie in the coelom of every arm, and a minute duct from each opens aborally on the central disk.

26-6 Natural history Starfishes are sedentary and spend most of their time quietly attached to some solid object. The body seems stiff and rigid, and indeed can be broken in parts by rough treatment; yet the animal voluntarily bends or twists both the disk and the arms when moving about or when its body is fitted into irregular spaces among rocks or other shelter. On a rough or upright surface the animal holds and travels by the tube feet. To move, the ray or rays pointing in the given direction are raised slightly and the tube feet on them are stretched out an inch or so; these grip the new surface and then contract, pulling the body forward. Some starfishes live on sand and mud. These have no suckers on the tube feet, which are pointed and inserted into the substrate for purchase in order to pull the animal forward. There is no "head" and no one arm that habitually moves forward, although some experiments suggest that the part near the madreporite may often be carried forward. A starfish can be trained experimentally to use a particular arm, and the habit will persist for some time. The animal can progress in any direction over a surface and once started, shows coordinated action of the arms and tube feet. If turned upside down, the arms are twisted until some of the tube feet touch and attach to the substratum, then the entire body slowly folds over so that the oral surface is again downward (Fig. 26-11).

Starfishes feed on mollusks, crustaceans, tube worms, and other invertebrates, including other echinoderms. Some species are suspension feeders, trapping particles in mucus on the body and then transporting it to the mouth. Small active animals, even fishes on occasion, may be caught by the tube feet or pedicellariae and passed to the mouth. Starfishes that feed on bivalves lie over the prey and wait until the shells gape, then insert the everted stomach into the shell. Also, a starfish can grip the opposite valves with its tube feet and gradually pull them apart; then the starfish everts its stomach over the soft body of the bivalve. The stomach secretes mucus, and the hepatic caeca give off enzymes to digest the food which is taken into the stomach; then the stomach and contents are withdrawn into the body. Other starfishes do not evert the stomach but take small organisms directly into the stomach whole. Any large bits of waste are cast out the mouth, since the small intestine and anus are nearly nonfunctional. Starfishes may feed voraciously but also can go for long periods without food. On commercial oyster beds, starfishes may cause serious losses by eating the oysters. Owners of such beds use a rope drag to capture the starfishes, which are then killed in hot water or carried ashore; or they sprinkle

lime on the beds to kill the starfishes. Recently the crown-of-thorns starfish (*Acanthaster planci*) has experienced a population explosion on many Pacific Ocean coral reefs. This starfish eats only living coral and has destroyed large areas of coral reefs which may take years to regenerate. It could be a serious threat to the whole coral-reef ecosystem in some areas.

26-7 Reproduction Eggs and sperm are shed into the seawater, where fertilization occurs. Cleavage (Fig. 26-12*B–D*) is rapid, total, equal, and indeterminate, yielding on the second day a spherical ciliated **blastula** (*E*; 0.2 mm in diameter) that swims. Invagination produces an oval **gastrula** (*G*), and the blastopore later becomes the anal end, as in chordates. Mesenchyme forms from cells budded off the inner blind end of the primitive gut, where also two outpocketings bud off to produce the coelom and its mesodermal lining. The mouth (*I*) results from an ectodermal inpocketing (stomodeum) that connects to the esophagus, stomach, and intestine derived from the gastrocoel. Ciliated bands (*I,J*) develop on the exterior that serve in locomotion and food capture. The bilaterally symmetrical larva swims anterior end foremost, with a clockwise rotation, feeding throughout its free existence. Later, three lobes form on each side of the body to produce the **bipinnaria** larval stage. Still later, these become modified as long ciliated processes that move and contract, at which stage the larva is termed a **brachiolaria** (2 to 3 mm long). After 6 or 7 weeks the larva settles to the bottom, the anterior end becomes a stalk by which it attaches, and the posterior end enlarges and bends to the left. Five lobes form on the right side of this posterior part, which becomes the aboral surface of the future starfish; the left side produces the oral surface. Skeletal elements appear as the first lobes (arms) become evident. Two pairs of outgrowths from the coelom (hydrocoele) in each lobe become the first tube feet and serve for attachment. A complex internal reorganization occurs during the metamorphosis, but no parts are cast off. Any bilateral symmetry in the adult is about a plane different from that in the larva. Some starfish (*Henricia, Leptasterias*;

Figure 26-11 Starfish. An inverted individual turning over (in direction of arrows); *a–e*, same arms in successive positions. (*After Cole, 1913.*)

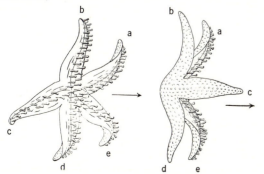

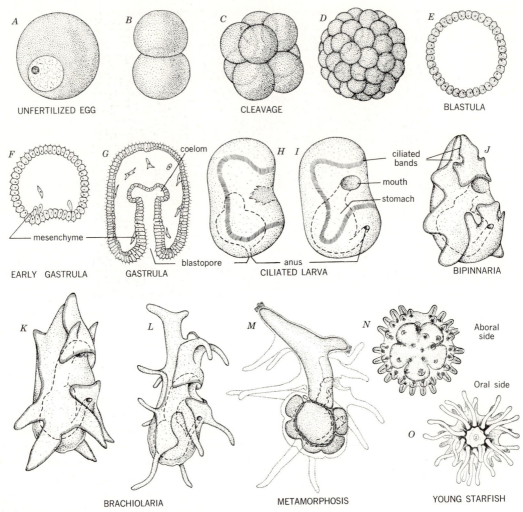

Figure 26-12 Development of a starfish, _Asterias vulgaris._ The blastula (_E_) and gastrula (_F, G_) are sectioned; the latter shows migration of mesenchyme cells and budding of coelomic cavities off the gastrocoel. The blastopore becomes the anus (_H_), and the stomodeum invaginates to form the mouth (_I_). The bilaterally symmetrical bipinnaria larva (_J_) produces three pairs of lateral lobes that lengthen in the brachiolaria larva (_K, L_), as do others on the ventral surface. The starfish forms on the lower right side of the brachiolaria (_M_), the upper parts of which are absorbed. (_Adapted from Field; Goto; and Brooks._)

Fig. 26-5_A_) produce few large-yolked eggs that are held beneath the disk by arching the rays; the larval stages are abbreviated, and the young emerge as miniature adults.

Starfishes suffer injury in nature and may break off an arm (autotomy) if severely handled. The arms regenerate readily, and individuals often are seen with one or more of them growing out. Experimental removal of four or, exceptionally, all arms may be followed by complete regeneration.

Subclass Ophiuroidea (Brittle stars and serpent stars)

26-8 Structure Brittle stars have a small rounded disk with five distinct arms that are long slender, jointed, and fragile (Figs. 26-1, 26-13). An arm consists of many similar segments, each comprising two central fused ossicles covered by four plates, the laterals with spinelets and the dorsal and ventral spineless. The interior of a segment is almost filled by solid

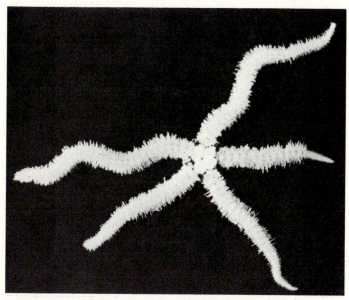

Figure 26-13 The brittle star *Ophiothrix spiculata.* *(Courtesy of G. McDonald.)*

cylindrical ossicles ("vertebrae") excavated on the proximal face and convex distally so that the adjacent vertebrae articulate with one another by a complex ball-and-socket joint. Four muscles between each two vertebrae enable the arm to be bent readily. In the arm is a small tubular coelom, nerve cord, hemal space, and branch of the water-vascular system. The small tube feet are ventrolateral, without suckers. They are sensory, they aid in respiration, and they may pass food to the mouth. There are no pedicellariae or dermal branchiae. All digestive and reproductive organs are in the disk. The mouth is centered orally and surrounded by five groups of movable plates that serve as jaws. There is a saclike stomach, but no caeca or anus; indigestible wastes are cast out the mouth. Five pairs of saclike bursae open by narrow slits about the mouth; they function in respiration and receive the gonad ducts. The madreporite is on the oral surface. In the basket stars (*Gorgonocephalus*) the arms are repeatedly branched, with tendrillike tips (Fig. 26-14).

26-9 Natural history Ophiurans live in shallow to deep water, sometimes in great numbers, hiding

beneath stones or seaweed or burying themselves in the mud or sand, to become active at night. They move by rapid snakelike movements of the arms, holding to objects by one or more arms and pushing with the others so as to jerk the body along. The podia are not used in locomotion. They can swim by use of their arms much as a person would do. Their food is small crustaceans, mollusks, and other animals and bottom debris; ophiurans in turn are eaten by fishes. The arms break or can be cast off easily, and some species can discard the disk except for the mouth framework, such parts being readily regenerated. The sexes are usually separate. They release their sex cells into the sea or else brood the eggs in the bursae, from whence young escape via the slits or by rupture of the oral disk. Nonbrooding forms develop a free-swimming microscopic **ophiopluteus larva** with characteristic long ciliated arms which undergoes a metamorphosis resembling that of starfishes. In a few, asexual reproduction by fission occurs. Ecologically, brittle stars are abundant and dominant members of many communities. Basket stars appear to be plankton feeders, crawling up on objects and spreading out their many tendrillike arms as a net to intercept small organisms.

Figure 26-14 Subclass OPHIUROIDEA: basket star *Gorgonocephalus.* (*Courtesy of the California Academy of Sciences.*)

Class Echinoidea (Sea urchins, sand dollars, and heart urchins)

26-10 Structure Members of this class have rounded bodies lacking free arms or rays but bristling with slender movable spines (Fig. 26-15). Sea urchins (*Arbacia, Strongylocentrotus*) are hemispherical in shape, heart urchins (*Spatangus*) are ovoid, and sand dollars, or "sea cakes" (*Echinarachnius, Dendraster*), are disklike.

In a common urchin the viscera are enclosed in a rigid **test,** of 10 double rows of plates, usually firmly sutured together. Five areas (*ambulacra*), corresponding to the arms of starfishes, are perforated to allow passage of a double series of long slender **tube feet;** the other alternating areas (*interambulacra*) are wider but lack tube feet. On the plates are series of low, rounded **tubercles** over which the spines articulate. Each spine (a single crystal of calcite, $CaCO_3$) has a cup-shaped base fitting over the tubercle and can be moved by muscles around the base. Among the spines are three-jawed **pedicellariae** on long flexible stalks; some echinoids have several kinds, and a few bear poison glands. The pedicellariae keep the body clean and capture small prey. In the ambulacral areas and in the membrane surrounding the mouth are minute spherical sensory organs called **sphaeridia,** each containing a statocyst with enclosed statolith. The anus is centered on the aboral surface in a membrane, the **periproct,** which also contains a varying number of small ossicles. Bordering the periproct are **genital plates** containing the **genital openings.** The large mouth is on the oral surface surrounded by a thickened membrane, the **peristome,** containing five pairs of gills and five pairs of buccal tube feet. Inside the mouth are five strong teeth; these are supported by a five-sided framework and complex muscle array inside the shell that is known as **Aristotle's lantern.**

The long **digestive tract** is looped around inside the test (Fig. 26-16). From the mouth a slender esophagus leads to a more expanded intestine, which

Figure 26-15 Class ECHINOIDEA: sea urchin, *Strongylocentrotus.* The tube feet and spines are in place on the left side of the figure but removed on the right half. Pedicellariae omitted.

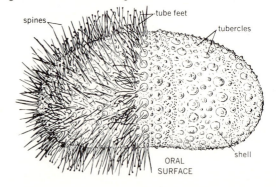

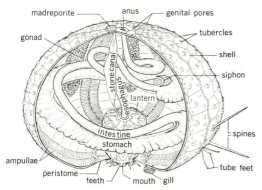

Figure 26-16 Sea urchin, *Arbacia,* **general structure.** Pedicellariae and most of spines and tube feet omitted.

loops around the oral side of the coelom and then ascends along the oral-aboral axis via the rectum to the anus. A slender tubelike **siphon,** lined with strong cilia, extends from the esophagus to the beginning of the large intestine; presumably it serves to remove water from the food. In sand dollars the mouth is central but the anus is near the edge of the disk; in heart urchins both the mouth and the anus are marginal, at opposite sides of the disk, providing a secondary bilateral symmetry. The madreporite is aboral; there is a ring canal around the esophagus, and five meridional radial canals (on the shell interior) connect to the tube feet. The water-vascular system is similar to that of starfishes. A **nerve ring** surrounds the mouth, and the five radial **nerves** accompany the radial canals. The **gonads** (five, four, or two) are attached by strong mesenteries to the inner aboral surface, and from each a fine duct leads to a genital opening. Respiration is effected by the five pairs of gills in the peristomial membrane. Eggs and sperm are discharged into the sea, and the minute fertilized egg becomes an **echinopluteus** larva that metamorphoses after 5 or 6 weeks. Some echinoids brood their eggs.

26-11 Natural history The urchins live on rocks or mud of the seashore and bottom. They move by joint use of the spines and tube feet, the feet serving to grip objects on the sea floor as well as to hold pieces of debris against the animal to camouflage it. Some shore dwellers shift to tide pools or hide under seaweeds at low tide. Others live permanently in self-excavated depressions in hard clay or soft rock under shore waters. The sand dollars and heart urchins bury themselves shallowly in sand. Neither have peristomial gills but have special tube feet modified as gills. All echinoids clean their bodies by movements of the spines and pedicellariae, and wastes from the anus are similarly moved off. Many regular urchins feed variously on seaweed, dead animal matter, and small organisms scraped up with the Aristotle's lantern. Irregular urchins (sand dollars and sea biscuits) feed on organic particles in the sand or mud, which they obtain either by direct ingestion of the substrate or by trapping in mucous nets. Echinoids have many commensal ciliates in the digestive tract, and various other commensals and parasites live on or in their bodies. Fishes, sea stars, crabs, and predaceous birds and mammals are their chief predators. Most urchins have little power of regeneration. Gonads of echinoids, raw or roasted in the half shell, are eaten in several cultures.

Class Holothuroidea (Sea cucumbers)

26-12 Structure Unlike other echinoderms, sea cucumbers have slender bodies elongated on an oral-aboral axis. The body wall is thick and leathery and contains only microscopic calcareous ossicles. The mouth is surrounded by 10 to 30 retractile tentacles, which are modified from the buccal tube feet found in other echinoderms. *Thyone* and *Cucumaria* are common examples (Figs. 26-17, 26-18). A holo-

Figure 26-17 Class HOLOTHUROIDEA: **sea cucumber,** *Thyone briareus;* **external features.** *(After Coe, 1912.)*

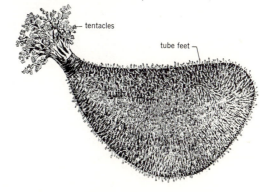

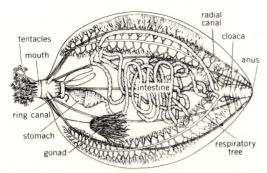

Figure 26-18 Class HOLOTHUROIDEA: sea cucumber, *Thyone*; internal structure. Body wall cut lengthwise and laid open. (*After Coe, 1912.*)

thurian lies usually with one side (the dorsal) uppermost. On some the dorsal side has two lengthwise zones of tube feet that are tactile and respiratory in function. The ventral side has typically three zones of **tube feet,** with suckers, that serve in locomotion. Other cucumbers have the tube feet restricted to the sole (*Psolus*, Fig. 26-19B) or distributed all over the body (*Eupentacta*; Fig. 26-19A), and still others lack tube feet (*Leptosynapta*). The body wall comprises a cuticle over a nonciliated epidermis, a dermis which contains ossicles, a layer of circular muscles, and then five double bands of powerful longitudinal muscles along the radii. The large, fluid-filled, and

Figure 26-19 Class HOLOTHUROIDEA. *A. Eupentacta quinquesemita; B. Psolus chitonoides.* (*Courtesy of G. McDonald.*)

A.

B.

undivided coelom contains various amoebocytes. Action of the muscles over the fluid-filled body enables the holothurian to extend or contract its body and to perform wormlike movements.

The long digestive tract is slender and looped within the coelom. A short esophagus leads from the mouth to the enlarged stomach; the long intestine is supported by mesenteries and connects to a muscular **cloaca** ending at the posterior anus. Two much-branched tubes, the **respiratory trees,** extend forward from the cloaca into the coelom. In most cucumbers water pumped in and out of these tubes by action of the muscular cloaca serves for respiration and excretion. The water-vascular system includes a madreporite within the coelom, a ring canal around the esophagus, short canals to each of the tentacles, and five radial canals connecting to tube feet along each muscle band. The hemal system is more evident than in other echinoderms because of the vessels along the intestine. A nerve ring circles the esophagus, and nerves extend along the radii. The sexes are separate (united in a few). The gonad is unique among echinoderms in that it is single. Morphologically it is brushlike, with fine tubules joining a single duct that opens middorsally near the tentacles. Many holothurians extrude their eggs and sperm into the sea, where fertilization occurs, and the resulting larva is an **auricularia,** which resembles the bipinnaria of asteroids. Later this larva develops into a **doliolaria** larva before metamorphosis. Other holothurians brood the eggs either on or in the body.

26-13 Natural history Sea cucumbers move sluggishly on the sea bottom or burrow in the surface mud or sand to leave only the ends of the body exposed; when disturbed they contract slowly. The food is organic material in the bottom debris that is pushed into the mouth or plankton trapped in mucus on the tentacles. They travel slowly by use of the tube feet or by muscular movements of the body. The free-swimming *Pelagothuria* is an exception, floating with the aid of a web and supporting papillae. Various commensals and parasites live on or in holothurians, including annelid scale worms, crabs, and, in the tropics, a fish, *Carapus (Fierasfer),* which dwells in the cloaca but emerges to feed. Some holothurians have Cuvierian organs associated with

the rectum. When the animal is irritated or attacked, these organs are expelled out the anus and immobilize the attacker in a mass of sticky, adhesive tubules. The holothurian crawls off and later regenerates the tubules. Still others (*Holothuria, Stichopus*) rupture the body wall and cast out various organs, which later regenerate. In the Orient the body wall of some sea cucumbers is boiled and then dried in the sun to produce trepang, or beche-de-mer, used for soup. Over 10,000 tons are produced annually for export. Holothurians are often the dominant invertebrates in the deepest parts of the oceans, and many taxa are restricted to deep water.

Class Crinoidea (Feather stars and sea lillies)

26-14 Structure and natural history These flowerlike echinoderms live from below the low-tide line to abyssal depths. The body is a small cupshaped **calyx** of limy plates to which are attached five flexible **arms** that fork at the calyx edge to form 10 or more narrow appendages, each bearing many slender lateral **pinnules,** arranged like barbs on a feather. Sea lilies (Fig. 26-20, 26-21) have a long jointed **stalk** from the lower or aboral surface of the calyx that attaches the crinoid to the sea bottom by rootlike outgrowths; feather stars (*Antedon*) lack a stalk but have flexible **cirri** for gripping objects in the water (Fig. 26-1). Both mouth and anus are on the upper, or oral, surface of the disk, the anus often on a raised cone. The oral surface of each arm and pinnule has an open ambulacral groove, lined with cilia and bordered by the tentaclelike tube feet. The food is of microscopic plankton and detritus, caught by the tentacles and conveyed by the cilia to the mouth. There is no madreporite. In contrast to other echinoderms, the nervous system is mainly aboral, located in a complex organ at the apex of the calyx. From this central location nerves are given off to each arm and to the cirri. The coelom is scant and the gonads are usually not distinct, the gametes being produced from epithelial tissue in the pinnules. Some crinoids shed the eggs directly into the water, but in other species they remain attached to the pinnules until hatched. Eggs are

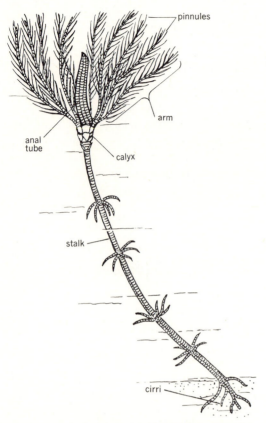

Figure 26-20 Class CRINOIDEA: a simple attached crinoid.

Figure 26-21 Class CRINOIDEA: a free-moving crinoid of the genus *Antedon*. (*Courtesy of G. McDonald.*)

brooded in some species. In most a vitellaria (doliolaria) larva is hatched from the egg. The larva subsists on yolk in the egg, having no mouth. After a few days of free life it attaches by the anterior end, and then the stem, disk, and arms are developed. Crinoids have great powers of regeneration, casting off the arms or much of the calyx and then renewing these parts. No animals are reported to feed regularly on crinoids. There are many commensals and parasites, notably bizarre polychaetes (*Polynöe, Myzostomum*) on the arms and disk. Some small gastropods bore into crinoids to eat out the soft parts.

Crinoids were especially abundant in Paleozoic times, and many thick limestone beds over the world are mainly of fossil crinoids. Of living species, about 80 are sea lilies that live attached to the bottom, on coral reefs, and elsewhere, forming extensive "gardens." The remainder are free-living feather stars like *Antedon* that can swim by using their long arms but commonly grip objects on the bottom with their cirri. Many living crinoids are brilliant yellow, red, white, green, or brown in life. Feather stars are particularly abundant on coral reefs in the Pacific.

Classification

Phylum Echinodermata.

Echinoderms. Symmetry radial, body usually of 5 ambulacra (radii) that bear tube feet alternating with 5 interambulacra (interradii), around an oral-aboral axis; no segmentation; body wall with epidermis over limy mesodermal plates that usually form a flexible or rigid boxlike endoskeleton with projecting spines or low bosses (none in sea cucumbers); commonly with minute jawed pedicellariae; coelom large, subdivided into perivisceral coelom, watervascular system, and perihemal system; sexes usually separate but alike; larvae bilaterally symmetrical and free-living before metamorphosis; all marine; Cambrian to Recent; about 5480 living species.

Subphylum A. Homalozoa.

Carpoids. Extinct forms which lack any manifestation of radial symmetry and have a flattened body with peculiar appendages. Cambrian to Devonian.

Class 1. †Homostela.

Feeding arm absent, stem or stalk present; body spoon-shaped.

Class 2. †Homoiostela.

Both feeding arm and stalk present; body asymmetrical.

Class 3. †Stylophora.

Feeding arm present but stalk absent; body asymmetrical.

Subphylum B. Crinozoa.

Both mouth and anus on upper surface; body in cup- or calyx-shaped skeleton; usually attached by aboral stalk or by aboral surface.

Class 4. †Cystoidea.

Calyx oval, attached directly or by a vertical stalk; usually not 5-parted. Ordovician to Devonian. †Echinosphaerites, Ordovician.

Class 5. †Blastoidea.

Calyx budlike, 5-parted, of 13 major plates; attached directly or by a short vertical stalk. Ordovician to Permian. †Pentremites, Carboniferous.

Class 6. †Eocrinoidea.

Calyx oval, attached by stalk; no thecal pores, no distinct pore rhombs; unbranched arms. Cambrian to Ordovician. †Macrocystella.

Class 7. †Paracrinoidea.

Calyx with numerous plates not arranged serially; lacking a tegmen. Ordovician.

Class 8. Crinoidea.

Sea lilies, feather stars. Calyx cuplike, symmetrical, 5-parted; arms branched, attached by a stalk or free-moving; mouth and anus on oral surface; Cambrian to Recent; about 630 living and 5000 fossil species. †Eocystites, pedunculate, free-living, Lower Cambrian; Antedon, feather star.

Subphylum C. Echinozoa.

Without arms, globoid and unattached.

Class 9. †Edrioasteroidea.

Calyx of discoid shape, without stalk or arms, of many small plates; lying free or attached by aboral

surface. Cambrian to Carboniferous. †Stromatocystites, flat, free-living, Middle Cambrian; †Edrioaster, probably fixed by basal sucker, Ordovician.

Class 10. †Helicoplacoidea.

Calyx spindle-shaped, spirally pleated, expansible, without stalk or arms; free-living; oral and aboral poles at opposite ends. Lower Cambrian. †Helicoplacus, found with †Eocystites, the earliest echinoderms.

Class 11. Holothuroidea.

Sea cucumbers. Body long, sausage- or worm-shaped, wall leathery to thin; no arms, spines, or pedicellariae; skeleton usually only of microscopic plates in body wall; tube feet (podia) usually present; mouth anterior, ringed by retractile tentacles; gut long, anus posterior; cloaca usually with respiratory trees. Ordovician to Recent; about 900 living species.

Order 1. Dactylochirota. Tentacles simple, body enclosed in a flexible test. Sphaerothuria.

Order 2. Aspidochirota. Tentacles usually 20 (15 to 30), each branched (peltate) from central stalk; tube feet many; respiratory trees present, Holothuria, Stichopus, Bathyplotes.

Order 3. Elasipoda. Tentacles 10 to 20, peltate; tube feet few; mouth usually ventral; respiratory trees absent; mainly deep sea at depths to 5000 m (15,000 ft). Pelagothuria, Psychropotes.

Order 4. Dendrochirota. Tentacles usually 10 (up to 30), with treelike branches; tube feet many; respiratory trees present. Cucumaria, Thyone, Psolus.

Order 5. Molpadida. Tentacles 15 (rarely 10), small, finger-shaped; tube feet only as anal papillae; respiratory trees present. Molpadia, Caudina.

Order 6. Apodida. Tentacles 10 to 20 or more, finger- or feather-shaped; no tube feet or respiratory trees. Leptosynapta, Chiridota.

Class 12. Echinoidea.

Sea urchins, sand dollars. Skeleton (test) usually rigid and globular, disklike or heart-shaped; with movable spines and 3-jawed pedicellariae; slender tube feet with suckers; gut long, slender, coiled; mouth and anus either central or lateral. Ordovician to Recent; about 860 living and 7200 fossil species.

Subclass 1. Perischoechinoidea. Interambulacra of 1 to 14 rows of plates, ambulacra of 2 to 20 columns; no sphaeridia.

Order 1. Cidaroida. Test rigid, globular; madreporite interambulacral; interambulacra of 2 or 4 and ambulacra of 2 rows of plates; each interambulacral plate with a large spine. Carboniferous to Recent. †*Archaeocidaris, Cidaris.*

Subclass 2. Euechinoidea. Interambulacra and ambulacra of 2 rows of plates each; sphaeridia present.

Superorder 1. Diadematacea. Test regular; lantern present, teeth grooved.

Order 2. Diadematoida. Spines usually finely toothed. Jurassic to Recent. *Diadema*, slender spines to 30 cm (12 in) long, Pacific, West Indies, and Florida.

Order 3. Echinothurioida. Spines smooth. Cretaceous to Recent. *Echinothuria.*

Order 4. Pedinoidea. Test rigid, spines solid, peristomial membrane with 10 buccal plates. *Caenopedina*, single living genus.

Superorder 2. Echinacea. Test rigid, regular; lantern present, teeth keeled.

Order 5. Phymosomatoida. Compound ambulacral plates 3, of full size; interambulacral plates usually with more than 1 tubercule; bars at top of lantern unjoined; primary tubercles not perforate. Triassic to Recent. †*Phymosoma*, Eocene; *Glyptocidaris*, Japan, the only living species.

Order 6. Arbacioida. Compound ambulacral plates composed of a median primary plate and a short plate on either side; bars at top of lantern unjoined; primary tubercles imperforate. Jurassic to Recent. *Arbacia*, common sea urchin.

Order 7. Temnopleuroida. Compound ambulacral plates of 2 primary plates with 1 narrow plate between outer ends; grooves for gills usually deep and sharp. Jurassic to Recent. *Temnopleurus, Toxopneustes.*

Order 8. Echinoida. Compound ambulacral plates of 2 primary plates with 1 narrow plate between their outer ends; grooves for gills shallow; bars at top of lantern joined. Cretaceous to Recent. *Echinus, Strongylocentrotus.*

Order 9. Salenioidea. Large anal plate causes anus to be located off center in periproct. Jurassic to Recent. *Acrosalenia.*

Order 10. †Hemicidaroidea. Suranal plates absent from periproct; primary tubercles of test plates perforate. Triassic to Cretaceous.

Superorder 3. Gnathostomata. Test irregular; lantern present, teeth keeled.

Order 11. Holectypoida. Ambulacra narrow on oral surface; lantern teeth with lateral flanges. Jurassic to Recent. †*Holectypus, Echinoneus.*

Order 12. Clypeasteroida. Sand dollars. Ambulacra wider than interambulacra on oral surface; no lateral flanges on teeth. Cretaceous to Recent. *Echinocyamus, Clypeaster,* many from Oligocene to Pliocene, only 12 Recent species.

Superorder 4. Atelostomata. Test rigid, irregular; no lantern.

Order 13. Nucleolitoida. Ambulacra of aboral surface petal-shaped; petals all similar, with unequal pores. Lower Jurassic to Recent. †*Nucleolites*, †*Galeropygus, Apatopygus.*

Order 14. Cassiduloida. Ambulacra of aboral surface petal-shaped; petals all similar with equal pores. †*Cassidulus; Echinolampas,* Jurassic to Recent in Red Sea, Antilles, and Africa.

Order 15. Holasteroida. Ambulacra of aboral surface petal-shaped; petals not all similar, not sunk. Jurassic to Recent. †*Holaster, Calymne.*

Order 16. Spatangoida. Heart urchins. Ambulacra of aboral surface petal-shaped; petals not all similar, sunk inward. Cretaceous to Recent. †*Toxaster, Echinocardium, Spatangus.*

Class 13. †Ophiocistioidea.

Discoid, enclosed in plates except at mouth; 6 pairs of giant tube feet in each ambulacrum. Ordovician to Devonian. †*Sollasina.*

Subphylum D. Asterozoa.

Free-moving, radially symmetrical echinoderms with strongly developed arms.

Class 14. Stelleroidea.

Pentamerous, free-living echinoderms with body of central disk and radially arranged arms. Cambrian to Recent. 3600 living species.

Subclass 1. Somasteroidea. Petaloid arms with pinnately arranged ossicles. No skeletal ring (mouth frame) around mouth.

Order 1. Platyasterida. Ambulacra widely open; arms with series of long interambulacral ossicles (virgalia). Lower Ordovician to Recent. †*Villebrunaster*, earliest known starfish; *Platasterias*, living fossil found off west coast of Mexico, all relatives extinct for 400 million years.

Subclass 2. Asteroidea. Starfishes, or sea stars. Body flattened, star-shaped or pentagonal; arms 5 to 50, usually not sharply set off from disk; endoskeleton flexible, of separate ossicles; spines and pedicellariae short; ambulacral grooves open, with 2 or 4 rows of tube feet; madreporite aboral; oral surface downward; stomach large; mostly predaceous. Cambrian to Recent; 1600 living and 300 fossil species.

Order 1. Phanerozonia. Arms with 2 rows of marginal plates; papulae all aboral; no crossed pedicellariae; tube feet in 2 rows. Ordovician to Recent. †*Petraster; Ctenodiscus*, mud star; *Luidia*, disk small, arms flexible; *Dermasterias*, leather star.

Order 2. Spinulosa. Marginal plates small; pedicellariae simple or lacking. *Henricia*, arms slender, cylindrical; *Solaster*, sun star, up to 14 arms; *Pteraster*, cushion star, with brood chamber for young.

Order 3. Forcipulata. Marginal plates inconspicuous; pedicellariae with crossed jaws. *Pycnopodia*, Pacific Coast, large, rays 18 to 24, disk soft and flat; *Asterias, Pisaster*, common starfishes; *Leptasterias*, broods young; *Stephanasterias*, up to 9 arms.

Subclass 2. Ophiuroidea. Brittle stars. Body a central disk and 5 slender, jointed flexible arms; tube feet in 2 rows, sensory, no suckers or pedicellariae; stomach saclike, no caeca or anus; madreporite aboral; free-living, active. Carboniferous to Recent; about 2000 living and 180 fossil species; 3 extinct orders and 2 Recent.

Order 4. Ophiurae. Arms unbranched, cannot twist or turn to mouth; disks and arms usually covered with plates. *Ophiothrix; Ophiura*, Cretaceous to Recent.

Order 5. Euryalae. Arms often branched, can twist and turn to grip objects; disk and arms with thick skin. *Gorgonocephalus*, basket star.

References

Binyon, John. 1972. Physiology of echinoderms. New York, Pergamon Press. x + 264 pp., 25 figs.

Boolootian, R. A. (editor). 1966. Physiology of Echinodermata. New York, Interscience Publishers, a division of John Wiley & Sons, Inc. xvii + 822 pp., illus.

Clark, A. H. 1921. Sea-lilies and feather stars. Smithsonian institution, Miscellaneous Collections, vol. 72, no. 7, pp. 1–43, 16 pls., 63 figs.

————. 1915–1950. A monograph of the existing crinoids. Vol. 1, *U.S. National Museum Bulletin* 82, 4 + pts. *Incomplete.*

Cuénot, L. 1948. Echinodermes. In P.-P. Grassé, Traité de zoologie. Paris, Masson et Cie. Vol. 11, pp. 1–363, 399 figs.

Durham, J. W., and **K. E. Caster,** 1963. Helicoplacoidea: A new class of echinoderms. *Science,* vol. 140, pp. 820–822, illus.

Fell, H. B. 1962. A surviving somasteroid from the Eastern Pacific Ocean. *Science,* vol. 136, pp. 633–636, 3 figs.

Fisher, W. K. 1911–1930. Asteroidea of the North Pacific and adjacent waters. *U.S. National Museum Bulletin,* 76, 3 pts., 1020 pp., 296 pls.

Florkin, M., and **B. Scheer** (editors). 1969. Chemical zoology. vol. 3. Echinodermata. New York, Academic Press, Inc. xx + 687 pp., illus.

Galtsoff, P. S., and **V. L. Loosanoff.** 1950. Natural history and methods of controlling the starfish (*Asterias forbesi* Desor). *U.S. Bureau of Fisheries Bulletin,* vol. 49, pp. 75–132, 32 figs.

Harvey, E. B. 1956. The American Arbacia and other sea urchins. Princeton, N.J., Princeton University Press. xiv + 10 + 298 pp., 16 pls., 12 figs. *Natural history, embryology, summary of experimental studies.*

Hyman, L. H. 1955. Echinodermata. In The invertebrates. New York, McGraw-Hill Book Company. Vol. 4, vii + 763 pp., 280 figs. *Comprehensive, with bibliography.*

Jennings, H. S. 1907. Behavior of the starfish *Asterias forreri* De Loriol. *University of California Publications in Zoology,* vol. 4, pp. 53–185, 19 figs.

MacBride, E. W. 1896. The development of *Asterina*

gibosa. Quarterly Journal of Microscopical Science, n.s., vol. 38. pp. 339–411, pls. 18–29.

Millott, N. (editor). 1967. Echinoderm biology. New York, Academic Press, Inc. xiv + 240 pp.

Moore, R. C. (editor). 1966–1967. Treatise on invertebrate paleontology. Lawrence, University of Kansas Press. Part S. *Echinodermata* 1, 1967. Vol. 1, xxix + 296 pp., 176 figs. 1967. Vol. 2, 297–650 pp., 224 figs. *Echinodermata* 3, 1966. Vol. 1, xxx + 366 pp., 272 figs. 1966. Vol. 2, 329 pp., 262 figs.

Mortensen, Theodore. 1927. Handbook of the echinoderms of the British Isles. London, Oxford University Press. ix + 471 pp., 269 figs.

_____. 1928–1951. Monograph of the Echinoidea. Copenhagen, C. A. Reitzel. "5" vols. (= 16), 4434 pp., 551 pls., 2468 figs.

Nichols, David. 1969. Echinoderms. 4th ed. London, Hutchinson & Co. (Publishers), Ltd. 200 pp., 26 figs. *Compact, modern, general account with references.*

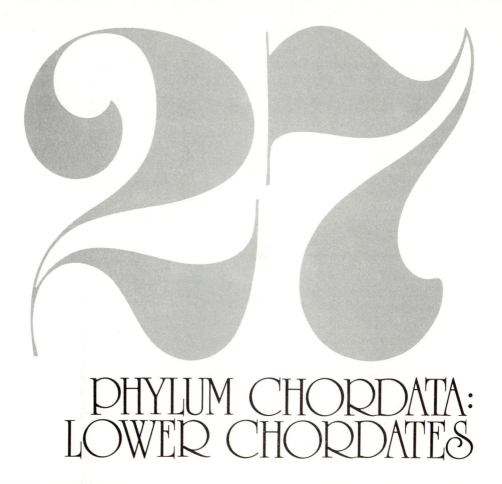

PHYLUM CHORDATA: LOWER CHORDATES

The phylum CHORDATA (Gr. *chorda,* string) is the largest and ecologically most significant phylum in the deuterostome line of evolution. It includes some invertebrate groups, as well as all the vertebrate animals. Like other large and successful phyla, such as MOLLUSCA and ARTHROPODA, it is distributed in all habitats—marine, fresh water, and terrestrial. The phylum includes all the large animals present on the earth today (with the possible exception of the cephalopods) and is therefore the one with which humans are often most familiar. Because of this, and because the phylum includes humans, it has received a disproportionate share of interest and attention from zoologists over the years. It has thus become probably the best known of all the phyla. The phylum really comprises two different groups of organisms. The *lower chordates* are all marine and small

in size and lack vertebrae; they include the tunicates, salps, and lancelets. All other chordates are *free-living vertebrates,* encompassing the fish, amphibians, reptiles, birds, and mammals. This is by far the largest group. This chapter describes the origin and evolution of the phylum, its common features (Fig. 27-1), the structure and natural history of the lower chordates, and the general characteristics of the vertebrates.

27-1 Characteristics

1 A rodlike dorsal supporting notochord present during at least part of the life cycle.

2 A dorsal hollow nerve cord present at some time in the life cycle.

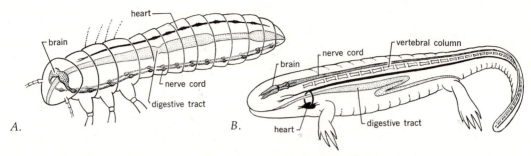

Figure 27-1 Basic differences between (*A*) a nonchordate (insect) and (*B*) a chordate (salamander) in relative positions of the nervous system, digestive tract, and heart; diagrammatic.

3 Gill slits present in the pharyngeal region during some stage of the life cycle.

4 Often with a tail projecting posterior to the anus.

5 Bilateral symmetry, with three germ layers and a segmented body.

6 Coelom enterocoelous in origin and well developed (except UROCHORDATA).

7 Skeleton, if present, an endoskeleton formed in mesoderm.

8 Closed circulatory system with a ventral heart (except UROCHORDATA).

9 Sexes usually separate (a few hermaphroditic or protandric); oviparous or viviparous.

Among the above list of characteristics, the first three taken together are definitive in that they set the CHORDATA apart from all other phyla (Fig. 27-2).

Figure 27-2 Phylum CHORDATA: basic features of three subphyla; diagrammatic.

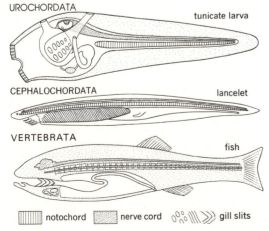

UROCHORDATA

tunicate larva

CEPHALOCHORDATA

lancelet

VERTEBRATA

fish

▦ notochord ▦ nerve cord ⌇ gill slits

These features all form in the early embryo of a chordate; they persist, are altered, or may disappear in the adult.

The **notochord** is the first supporting structure of the chordate body. In the early embryo it forms above the primitive gut as a slender rod of cells containing a gelatinous matrix and is sheathed in fibrous connective tissue. It forms a flexible, yet rigid, rod against which muscles may act to effect locomotion. In tunicates it is present in the tail and only during the larval stages. In the lancelets and higher forms it extends almost the length of the body. It persists throughout life as the main axial support in lancelets and lampreys, but in the fishes to mammals it is later replaced by the cartilaginous or bony vertebral column.

The **nerve cord** forms on the dorsal surface of the early embryo soon after the gastrula stage. Infolding of the ectoderm produces the hollow tubular cord that lies above the notochord. The anterior end becomes enlarged as a simple cerebral vesicle in tunicate larvae and in lancelets, but in all vertebrates it thickens and differentiates as the brain, to become progressively more complex in higher forms. In tunicates the cord and vesicle degenerate to a ganglion at metamorphosis. From the lampreys onward the nerve cord later becomes surrounded by neural arches of the vertebrae that protect it from injury, and the brain is enclosed by a brain box, or cranium.

Paired **gill pouches** develop on the sides of the embryonic pharynx (digestive tract). Each is formed by an outpocketing of endoderm in the pharynx and a corresponding inpocketing of ectoderm on the outside of the body; the intervening wall breaks through to form a gill slit. In higher vertebrates which respire by lungs, such as birds and mammals, the gill slits develop only in the embryo, do not break through, and disappear before birth or hatching. They are never functional. In fishes, where breathing is by gills, each slit is margined by many slender filaments containing blood vessels, forming a functional gill. Water containing dissolved oxygen passes into the mouth and pharynx and out over the filaments, where the blood gives up its carbon dioxide and acquires oxygen, so that the gill serves the process of external respiration. All aquatic chordates from tunicates to amphibians respire by gills. In amphibians which transform from aquatic larvae to air-breathing adults, the gills are lost at metamorphosis.

27-2 *Origin and evolution of chordates*   No remains of chordates have been found in the Cambrian rocks, where animal fossils first were common. The earliest were probably soft-bodied, without hard skeletal elements likely to be preserved. The first vertebrates were several groups of fishes in Silurian and

Table 27-1
MAJOR DIVISIONS OF THE PHYLUM CHORDATA
(For other details, see Classification in Chaps. 14 and 28 to 34)

Subphyla and Superclasses	Classes and their principal characteristics	See Chap.
UROCHORDATA: Notochord and nerve cord only in larva; adult contained in secreted tunic	LARVACEA: Minute, tadpolelike; tunic temporary; 2 gill slits ASCIDIACEA: Ascidians. Tunic with scattered muscles; many gill slits THALIACEA: Salps. Tunic with circular muscle bands	27
CEPHALOCHORDATA: Notochord and nerve cord along entire body and persistent, as are the gill slits	LANCELETS. Slender, fishlike, segmented; epidermis 1-layered, no scales; many gill slits	
VERTEBRATA: With cranium, visceral arches, vertebrae, and brain	†OSTRACODERMI: Ancient armored fishes. Scales large, often fused as cephalothoracic shield CYCLOSTOMATA: Cyclostomes. Skin without scales; mouth suctorial; gills 5 to 16 pairs	28
Superclass PISCES: Paired fins, gills, and skin with scales	†PLACODERMI: Ancient fishes. Jaws primitive; complete gill slit before hyoid	29
	CHONDRICHTHYES: Sharks and rays. Skin with placoid scales; skeleton of cartilage; 5 to 7 pairs of gills in separate clefts	
	OSTEICHTHYES: Bony fishes. Skin with cycloid or ctenoid scales; 4 pairs of gills in common cavity under opercula	30
Superclass TETRAPODA: Paired limbs, lungs, cornified skin, and bony skeleton	AMPHIBIA: Amphibians. Skin moist, soft, no external scales	31
	REPTILIA: Reptiles. Skin dry, with scales or scutes	32
	AVES: Birds. Skin with feathers; forelimbs are wings; warm-blooded	33
	MAMMALIA: Mammals. Skin with hair; warm-blooded; suckle young	34

Ordovician time (Fig. 12-17). Thenceforth vertebrates have been common and often dominant elements in the fossil record and show a progressive series of developments toward the living types. Fossils of the earliest sharks and bony fishes appear in freshwater deposits, and later both groups invaded the seas. The amphibians probably derived from crossopterygian fishes. They appear first in Devonian rocks and were abundant and varied by Carboniferous times. The first salamanders and frogs (Jurassic) are distinct, like their modern successors. Reptiles appeared in Permian time and expanded widely into a great variety of dominant types in the Mesozoic era; then most of the orders died out at the end of the Cretaceous. The first bird (†*Archaeopteryx*) appeared in the Upper Jurassic. The mammals arose from the reptiles, beginning in the Triassic, and differentiated widely in the early Tertiary period. The study of fossils has contributed importantly to the classification of vertebrates and has indicated the probable origins of most of the larger groups.

For more than 100 years now zoologists have speculated about the origin of the phylum CHORDATA, and more particularly the origin of the vertebrates as the dominant group. Perhaps more has been written about the origin and evolution of the vertebrates and chordates than all other groups combined. In this process virtually all the large and successful invertebrate phyla have been suggested as ancestors and several different theories proposed. Two major reasons for the existence of so many divergent theories on the origin of the vertebrates are the significant difference in morphology between vertebrates and invertebrate phyla and the complete lack of any intermediate forms in the fossil record. As a result speculation on their origin has centered primarily on embryologic features as observed in living forms and on the comparative morphology and embryology of the most primitive of the chordates, the tunicates and amphioxus. Of the many theories proposed, only two are mentioned here.

The **annelid theory** points out that annelids and chordates are both bilaterally symmetrical and segmented, with segmental excretory organs, a well-developed coelom, and a closed circulatory system with longitudinal blood vessels. Inverting an annelid would place its nerve cord dorsal to the digestive tract, and the path of blood flow would resemble that of chordates; however, the mouth would then be dorsal and the brain ventral, unlike a chordate, and other dorsoventral relations would be altered. Annelids, moreover, have no structure suggestive of the notochord or gill slits. Even more fundamentally, their embryologic pattern of development (protostome) is different from that of the chordates (deuterostome). Hence this theory has been largely discredited and discarded.

The **echinoderm theory** is based on the patterns of development. It was apparent by the twentieth century that the chordates were most similar to echinoderms and hemichordates, with which they shared the typical deuterostome embryologic features of radial indeterminant cleavage, blastopore becoming the anus, and enterocoelous coelom formation. The two invertebrate phyla are further linked via their larvae. The tornaria larva of tongue worms and the bilateral bipinnaria larva of echinoderms are both minute and transparent, with almost identical external ciliated bands and a like number of coelomic cavities; both have a dorsal pore (Fig. 20-25, Sec. 14-7). Indeed, the first known tongue worm (hemichordate) larva was identified as that of a starfish by a famous zoologist! If chordate ancestry is involved in such small, soft embryologic forms, the chance of finding any conclusive fossil record is remote. Hence we are left in the rather unsatisfactory position of being able to show a certain level of relationship at the developmental level but with no conclusive evidence in the form of fossils or intermediate forms. Perhaps this theory is correct, but it cannot be proved.

27-3 General features In the other dominant phyla, such as arthropods and mollusks, uniformity of the basic adult body plan and structure throughout the phylum gives the phylum many of its distinctive features. In the chordates as a whole this is less true, because the major defining features are embryologic rather than adult and because the phylum includes one group, the tunicates, which is very different in body plan from the rest of the chordates. Tunicates, as adults, show only one of the defining features of the phylum, namely, pharyngeal gill slits, and are otherwise unrecognizable as chordates. If they are excluded, however, the phylum shows a remarkable consistency of structure and function, especially in the largest and most successful subdivision, the subphylum VERTEBRATA. Indeed, it is entirely due to the vertebrates that the phylum is so successful, hence it is appropriate to consider briefly here some of the general features of vertebrates which may have led to their success.

One outstanding feature of vertebrate structure is the presence of an internal skeleton, or **endoskeleton.** In all vertebrates the endoskeleton is first manifested by the notochord, which remains as the only internal skeleton in lancelets but in other vertebrates is replaced by a more extensive framework of cartilage and/or bone. This endoskeleton, also present in echinoderms, serves for the protection of soft and/or vulnerable body parts such as the brain, as a rigid support for the body itself, as a surface for attachment of muscles, and as a rigid support which permits muscles to operate.

A second major feature contributing to the success of the vertebrates is the nervous system. It is in the vertebrates that the development of the nervous system reaches its zenith. Vertebrates show the greatest centralization of the nervous system, the largest brains, and the most elaborate and complex behavior patterns. Correlated with these they also show the most highly evolved sense organs, which give them much more complete information about the physical and biological environment around them. In many ways they are the antithesis of the insects, which are successful but mainly instinctive. The vertebrates are successful by virtue of the high development of the nervous system and attendant sense organs, which permit precise coordination of body movements and development of behavior patterns allowing adaptation to varied environmental conditions.

Finally, the respiratory and circulatory systems are more efficient than those of the invertebrate phyla. A closed circulatory system with an efficient oxygen carrier (hemoglobin) and a strong pump (heart) has allowed the vertebrates to have a rapid exchange of gases and other materials in the cells, permitting them to become quick-moving and also large in body size. The insect tracheal system is also efficient but is limited to animals of small size.

Subphylum Urochordata (Tunicates)

Tunicates are the most aberrant of the chordates in that the adults evince little or no resemblance to other chordates. All are marine. A few are free-living as adults, but most, after a short larval stage, are attached to rocks, shells, wharf pilings, or ship hulls. Tunicates may be solitary, colonial, or with many individuals grouped under a common covering. They vary in size from nearly microscopic forms to others 30 cm (1 ft) in diameter. About 2000 species are known, of which 100 are pelagic. Only the larval form has the chordate characteristics. Tunicates are usually hermaphroditic, but methods of reproduction are varied, some sexual, some asexual by budding. The group name refers to the self-secreted tunic, or saclike covering, over the body. Three classes are recognized: ASCIDIACEA, THALIACEA, and LARVACEA. The best known tunicates are the sea squirts, or ascidians, of the class ASCIDIACEA. The THALIACEA and LARVACEA are small groups, both pelagic.

Aristotle (384–322 B.C.) described a simple ascidian as Tethyum, but not until early in the nineteenth century were many species described in detail. Cuvier then gave the name Tunicata to the group and placed it between the Radiata and Vermes; later, its members were classified as mollusks. In 1866 Kowalewsky, by careful study of the developmental stages of a larval ascidian, brilliantly demonstrated the true position of the group to be among the chordates.

The tunicates are best understood by considering first the free-living larva of an ascidian and then the adult. The larva shows chordate characteristics, but some of these are absent in the adult and others are obscured by adaptations to the sessile mode of life.

27-4 The larval ascidian The small fertilized egg segments to form a blastula and then a gastrula; mesoderm forms from the archenteron, but there is no pouching, and a coelom never develops. With continued development the embryo elongates and soon hatches as a transparent free-swimming larva, 1 to 5 mm in size, with a long tail (Fig. 27-3A). Its tail contains a supporting **notochord** (whence the name UROCHORDATA), a dorsal tubular **nerve cord,** and se-

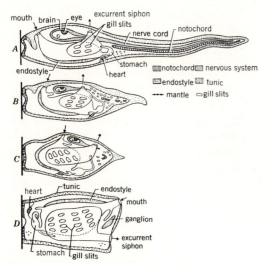

Figure 27-3 Subphylum UROCHORDATA. Stages in the metamorphosis of a simple ascidian from the free-swimming larva to the sessile adult. Arrows indicate entrance and exit paths of water currents. (*Adapted from Kowalewsky; and Herdman.*) *A*. Larva attaches to a solid object by anterior mucous suckers. *B*. Tail resorbed, notochord and nerve cord reduced. *C*. Notochord disappears, internal organs begin to rotate. *D*. Metamorphosis complete, with rotation (90–180°) of internal organs and external apertures; branchial sac enlarges, tunic (or test) is secreted, and nervous system reduced to a ganglion. (See Fig. 27-4.)

rial pairs of lateral, segmented muscles. The other organs are confined to the anterior and larger head + body region. The anterior end bears three mucous, or adhesive, glands. The **digestive tract** is complete, with mouth, perforate **gill slits** opening into an **atrium**, endostyle, intestine, and anus; there is a **circulatory system** with blood vessels. The **nervous system** (continuous with the nerve cord in the tail) and sensory structures include (1) a "brain" and posterior to it (2) a trunk ganglion, (3) a median eye with retina, lens, pigment, and cornea, and (4) a pigmented otolith, attached to delicate hair cells.

27-5 Metamorphosis After a few hours or days of free life, the small larva attaches vertically by its adhesive glands to a rock or wharf pile (Fig. 27-3). A rapid transformation (retrograde metamorphosis) follows in which most of the chordate features disappear. The tail is partly absorbed and partly cast off, its notochord, nerve cord, and muscles being withdrawn into the body and absorbed. Of the nervous system only the trunk ganglion persists. The bran-

chial, or gill, sac enlarges, develops many openings, and is invaded by blood vessels. The stomach and intestine grow. That portion of the body between the point of attachment and the mouth grows rapidly, causing the body within to rotate dorsoposteriorly nearly 180°, so that the mouth is at the upper nonattached end. Finally the gonads and ducts form in mesoderm between the stomach and intestine. The adhesive glands disappear, and the tunic grows upward to enclose the entire animal.

27-6 The adult ascidian A simple ascidian (e.g., *Ciona, Molgula*) is cylindrical or globose, attached by a **base** or stalk (Fig. 27-4). It is covered with a tough elastic layer, the test, or **tunic,** of cellulose (a material rare in animals), lined by a membranous **mantle** containing muscle fibers and blood vessels. There are two external openings, an **incurrent siphon** (branchial aperture) at the top and an **excurrent siphon** (atrial aperture) at one side. Water drawn into the incurrent siphon brings minute organisms, which serve for food, and oxygen for respiration; that

Figure 27-4 Subphylum UROCHORDATA. Structure of an adult simple ascidian; diagrammatic. Tunic, mantle, and upper half of branchial sac removed on left side. Arrows indicate path of water currents through the animal.

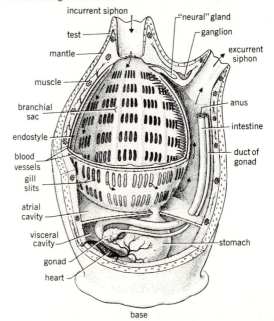

passed out the excurrent siphon removes wastes and the sex cells. Within the test and mantle is the **atrial cavity,** in which is a large **branchial sac** having many gill slits (stigmata) bordered by long cilia.

The digestive system begins with the incurrent siphon, followed by a circle of hairlike sensory tentacles at the entrance to the branchial sac. Along the midventral wall of the latter is the **endostyle,** a vertical groove lined with both flagellated and mucous cells, where food from the entering water is caught and moved downward. Feeding only on plankton, the ascidian needs no large appendages to handle food. From the base of the branchial sac a short **esophagus** leads to the dilated **stomach,** which connects to the **intestine;** the latter two organs are outside the branchial sac. The intestine curves to end at the **anus** below the excurrent siphon. The stomach of some ascidians has glandular outgrowths ("liver") discharging into its cavity.

The wall of the branchial sac has many gill slits margined by ciliated cells that beat to move water from within the sac to the atrial cavity, whence it flows to the excurrent siphon.

The circulatory system includes a tubular heart in the **visceral cavity** near the stomach. To each end is connected a large vessel, or **aorta,** one distributing to the stomach, tunic wall, and one side of the branchial sac, the other to the opposite side of the branchial sac. In the walls of the latter an intercommunicating series of small vessels surrounds the gill slits, to constitute the respiratory mechanism. These "vessels," however, are spaces within the tissues and lack an endothelial lining. The tunicate is unique in that the path of blood flow reverses at short intervals; the heart and vessels lack valves. The blood cells may be colorless or blue, red, etc.

Near the intestine is a structure, without a duct, considered to be excretory in function. The only relic of the nervous system is a slender **trunk ganglion** in the mantle between the two siphons with nerves to various parts. Nearby is a **neural gland,** possibly of endocrine nature and somewhat like a pituitary structure.

Ascidians are hermaphroditic. The **ovary** is a large hollow gland on the intestinal loop, with an **oviduct** paralleling the intestine and opening in the atrial cavity near the anus. The **testes** comprise many branched tubules, on the surface of the ovary and intestine, that discharge into a **ductus deferens** paralleling the oviduct. Some tunicates also reproduce asexually, by budding.

27-7 Class Larvacea   These small pelagic organisms (*Oikopleura, Appendicularia*) are up to 5 mm (0.2 in) long, larvalike in appearance, with a persistent notochord. The class name stems from the fact that the group appears to be neotonous, with a structure similar to regular ascidian tadpole larvae. They live in the upper levels of the sea and swim by contractions of the bent or curved tail. Each individual secretes a large gelatinous housing, or tunic, in which it lives and through which it filters water to capture food. Each "house" contains sets of fine filtering screens across the incurrent opening. There is also an excurrent opening. The animal, confined to the interior, beats its tail to create a current through the "house." During the passage, organisms are filtered out on the screens and ingested. Each "house" is temporary and continually shed and replaced. Only sexual reproduction is known.

27-8 Class Ascidiacea   In simple ascidians the individuals are solitary; that is, each has a distinct and separate test. They are common in shallow waters. The compound ascidians (Fig. 27-5) are those in which the "individuals" are buried in a common covering mass, or housing, and do not have separate tests. In many of this group, each "individual" has a separate incurrent siphon, but several or many (to

Figure 27-5 Subphylum UROCHORDATA. A compound tunicate (*Botryllus*); individuals with separate incurrent siphons but common excurrent siphon. (*After M. Edwards.*)

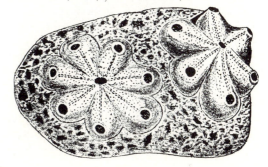

several hundred) have a common external aperture (cloaca) into which their respective excurrent siphons discharge. Compound ascidians reproduce asexually by budding and also produce eggs and sperm for sexual reproduction. With most species, development proceeds to the larval stage within the atrium, then the young emerge. Most ascidians (*Ciona, Botryllus*) are attached, but some (*Polycarpa*) are free-living in sand or mud.

27-9 Class Thaliacea   The salps are highly modified tunicates of peculiar form that live most commonly in the open tropical and subtropical seas and have specialized modes of reproduction. The body is cask-shaped, surrounded by bands of circular muscles, with the incurrent and excurrent openings at opposite ends (Fig. 27-6). Most salps are solitary and swim by contracting the ringlike muscles so as to force water out the posterior end. Others, such as *Pyrosoma*, are colonial, with a common large central cloaca and excurrent opening and many separate individual incurrent openings. They are wafted about by water currents. All salps have a life cycle involving asexual budding. In sexual reproduction some produce larvae while others brood eggs, and there is no larval stage (*Salpa, Pyrosoma*). In *Doliolum* (Fig. 27-7), the fertilized egg develops into a free-swimming tailed larva, somewhat like that of an ascidian. This metamorphoses into a barrel-shaped adult "nurse" stage that by asexual reproduction gives rise to many individuals. The latter eventually break free, swim away, and develop into sexually reproducing adults. *Salpa* is another type in which, by asexual means, a solitary adult produces long chains of individuals, sometimes several hundred connected to one another. Fragments of such a chain, comprising several animals, may break loose from the parent structure. Later these, while still connected or further separated, reproduce sexually. Each gives rise to one or several eggs that remain attached to the parent by a placentalike structure, analogous to that of higher

Figure 27-6   Subphylum Urochordata, **Class** Thaliacea: **the salp** *Thetys.*   (*Courtesy of G. McDonald.*)

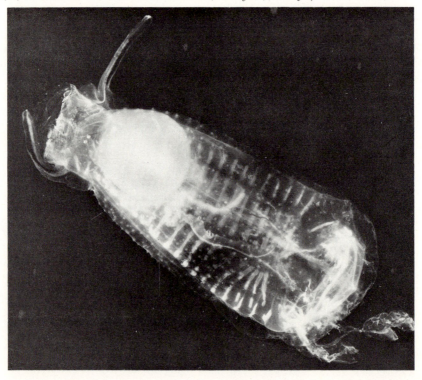

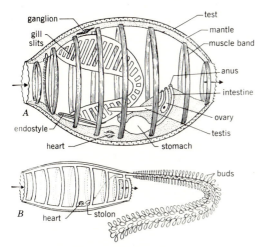

Figure 27-7 Subphylum UROCHORDATA: a salp (*Doliolum*).
A. Adult of sexual generation. *B.* Asexual adult ("nurse" stage)
with attached chain bearing many buds. Arrows indicate path of
water through the animal. (*Modified from Stempell.*)

chordates, where the maternal and embryonic blood
circulations are closely related during development.
The young that hatch out become the asexual form.

Subphylum Cephalochordata (Lancelets)

27-10 Amphioxus This group (subphylum
CEPHALOCHORDATA) comprises about 30 species of
fishlike animals (such as *Branchiostoma*), commonly
called amphioxus, that inhabit tropical and temperate seacoasts. *Branchiostoma virginiae* occurs
from Chesapeake Bay to Florida, and *B. californiense*
from San Diego Bay southward. The latter grows to
100 mm (4 in) long, but most species are smaller.
Amphioxus burrows in clean shifting sand of shallow

water, leaving only its anterior end protruding.
At times it emerges to swim by rapid lateral movements of the body. Amphioxus is of special zoologic
interest because it shows the three distinctive characteristics of the phylum CHORDATA in simple form
and is considered to resemble some ancient ancestor
of this phylum.

The **body** is slender, long, and laterally compressed, pointed at both ends, and has no distinct
head (Fig. 27-8). The low median **dorsal fin** along
most of the body and the **preanal fin** from atriopore
to anus are made of fin-ray chambers containing
short **fin rays** of connective tissue. The **tail** has a
membranous fin. Anterior to the ventral fin the body
is flattened ventrally, with a metapleural fold along
each side. The **mouth** is ventral at the anterior end,
the **anus** is on the left side near the base of the tail fin,
and the **atriopore** is an additional ventral opening
forward of the anus.

The body covering is a single layer of soft **epidermis** in which some cells bear sensory processes;
beneath this is a soft connective tissue. The **notochord** is the chief support of the body. It is a slender
rod of tall cells containing gelatinous material and is
surrounded by a continuous sheath of connective
tissue. Other supportive structures include a circular
cartilagelike reinforcement in the oral hood, the fin
rays, and delicate rods with cross connections in the
gill bars. Along each side of the body and tail, different species have 50 to 85 <-shaped muscles, or **myomeres,** each of lengthwise muscle fibers and separated from one another by thin septa of connective
tissue. Those on the two sides alternate in position,
whereas segmental muscles of other chordates are
symmetrically paired. These muscles contract to produce lateral bending for burrowing and swimming.

Figure 27-8 Subphylum CEPHALOCHORDATA: the lancelet, or amphioxus (*Branchiostoma*). Adult partly dissected from left side.
Natural size about 5 cm (2 in) long.

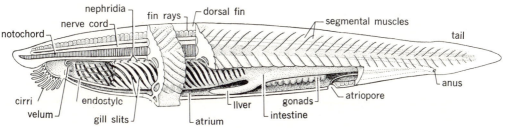

Transverse muscles in the floor of the atrial cavity, between the metapleural folds, serve to compress the cavity and force water to the exterior.

The straight and simple **digestive tract** begins with the anterior **oral hood** (vestibule), which is surrounded by about 22 delicate fleshy **buccal cirri;** behind the latter are several ciliated bars. The **mouth** proper is a circular opening in a membrane, or **velum,** posterior in the oral hood. It is guarded by 12 **velar tentacles,** which exclude large particles. Cilia in the hood, during life, produce a rotating effect and are called the *wheel organ.* On the cirri and tentacles and within the hood are sensory structures. Behind the mouth is the large compressed **pharynx,** with many diagonal **gill slits** at the sides. There follows the narrow straight **intestine,** which ends at the **anus.** A slender saclike **liver,** thought to secrete digestive fluid, attaches ventrally to the anterior part of the intestine.

The pharynx is suspended dorsally, beneath the notochord, but hangs free in a cavity, the **atrium,** within the muscles of the body wall. The atrium is an external cavity, lined by ectoderm (hence not a coelom), and connects to the atriopore. The pharynx contains a middorsal furrow, the **hyperbranchial groove,** lined with ciliated cells, and midventrally is a corresponding groove, the **endostyle,** having both ciliated and gland cells (Fig. 27-9). Water containing minute organisms is drawn into the mouth by action of the cilia; the food is trapped by mucus in the endostyle and carried posteriorly to the intestine, while the water passes between the gill bars to the atrium and thence to the exterior through the atriopore.

The **circulatory system** is somewhat on the plan of that in higher chordates but lacks a heart. Besides the definite blood vessels, there are open spaces where the colorless blood escapes into the tissues. Blood from the area of the digestive tract flows anteriorly in a **subintestinal vein** to the **hepatic portal vein** entering the liver; thence it collects in a **hepatic vein,** and with other blood from the posterior part of the body passes forward in a **ventral aorta** below the endostyle and into branches at each primary gill bar. Each branch has a small pulsating bulb, and together these function as a heart, forcing blood upward in

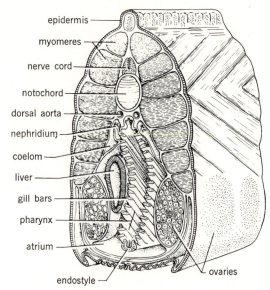

Figure 27-9 Subphylum CEPHALOCHORDATA: amphioxus. Enlarged section through pharynx. (*Adapted from Kükenthal.*)

the gill bars, where it is aerated and then collects in the paired dorsal aortas. The latter join behind the pharynx to form a single **dorsal aorta,** in which the blood moves posteriorly to supply the body and intestine and finally through capillaries to the venous side. Some oxygenated blood passes forward in the right dorsal aorta to the anterior end of the body. The general course of blood movement in amphioxus is thus like that in higher chordates and opposite to that in invertebrates such as annelids.

Respiration results from passage of water containing oxygen from the pharynx through 100 or more **gill slits** on each side, past the **gill bars** that contain blood vessels, and into the atrium. This water current is aided by cilia in cells on the gill bars. The gill system of amphioxus is like that of higher chordates during early larval life, its inner surface being of endoderm and the exterior of ectoderm. Later the exterior surface of the gill region of amphioxus is enclosed by growth of the covering wall that forms the atrial cavity outside the gills.

During development, the **coelom** is formed of five embryonic pouches (as in hemichordates); in the adult amphioxus it becomes reduced and complicated except around the intestine. The **excretory sys-**

tem comprises about 100 pairs of small ciliated **nephridia** in dorsal relics of the coelom above the pharynx; they connect the coelom to the atrial cavity and show some structural resemblance to the nephridia of polychaete worms. A pair of larger structures, the **brown bodies,** dorsal to the intestine, may also be excretory in function.

The **nervous system** lies above the notochord; it consists of a single dorsal **nerve cord** with a small central canal. The slightly large anterior end forms a median **cerebral vesicle,** with a mid-dorsal **olfactory pit,** a small nonsensory **eyespot** of black pigment, and two pairs of "cranial" **nerves.** The cord gives off to each myomere alternately a pair of **nerves,** the dorsal root being both sensory and motor, the ventral only motor in function. In the skin and the mouth region are ciliated cells thought to be sensory in function.

The sexes are separate; about 25 pairs of **gonads** (in two rows) bulge into the atrium. Eggs and sperm break into the atrial cavity to pass out through the atriopore, and fertilization is external. The egg is about 0.1 mm in diameter, with little yolk, and segmentation is holoblastic (Fig. 10-10A). During the breeding season, egg deposition usually occurs about sunset, and by morning a free-swimming ciliated larva is hatched; this feeds and grows for up to 3 months, gradually assuming the adult form, and then takes to burrowing in the sand.

Lancelets of the genus *Asymmetron* have gonads only on the right side.

Subphylum Vertebrata (Vertebrates)

27-11 Characteristics The classes CYCLOSTO-MATA to MAMMALIA constitute the major part of the phylum CHORDATA. The diagnostic features of the subphylum VERTEBRATA are an enlarged brain enclosed in a brain case, or cranium, and a segmental spinal column of vertebrae that becomes the axial support of the body. Typically the body comprises a head, neck, trunk, and tail. These classes show a progressive series of structural and functional advances in all organ systems besides the features of the notochord, nerve cord, and gill slits mentioned earlier in this chapter (Fig. 27-10; see also Chaps. 4 to 9).

1 The **body covering** is a stratified epithelium of epidermis and dermis with many mucous glands in aquatic species; most fishes are covered with protective scales; the exterior is cornified on land dwellers, with scales on reptiles, feathers on birds, and hair on mammals; feathers and hair form insulated body coverings.

2 The internal and jointed **skeleton** is of cartilage in lower vertebrates and of bone in higher groups; it supports and protects various organs; the cranium shelters the brain and has paired capsules to contain organs of special sense; a series of visceral arches supports the gill region,

Figure 27-10 Principal structures in a vertebrate (both lungs and gills indicated but rarely occur together); diagrammatic.

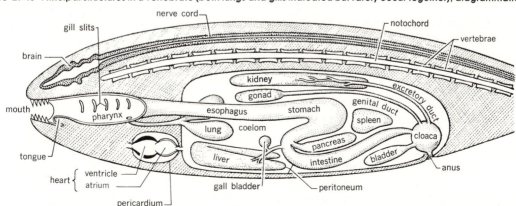

and certain arches become the jaws and other structures of the head region; the vertebral column extends from the base of the cranium to the end of the tail and has neural arches dorsally to house the nerve cord. Two pairs of appendages, the fins of fishes and limbs of tetrapods, with jointed skeletal supports, are articulated with the vertebral column through limb girdles.

3 On the skeleton are **muscles** that move its parts and provide for locomotion.

4 The complete **digestive tract** is ventral to the vertebral column; the mouth contains a tongue and usually teeth; the anus opens at the end of the trunk; the liver and pancreas are two large digestive glands that pour their secretions through ducts joined to the intestine.

5 The **circulatory system** includes a well-developed muscular heart of two, three, or four chambers, located ventral to the digestive tract; its contractions propel the blood through a closed system of arteries, capillaries, and veins, the flow being anteriorly on the ventral side and posteriorly in the dorsal arteries; the blood plasma contains both white and red corpuscles, the latter with hemoglobin as a respiratory pigment; a system of lymph vessels is present; paired aortic arches transport blood from the heart to the gills in lower vertebrates; progressive separation of the respiratory (pulmonary) and systemic blood paths through the heart contributes to regulated body temperature in the warm-blooded (endothermic) birds and mammals.

6 **Respiration** in the lower forms is by paired gills; terrestrial species have lungs developed from outpocketings of the digestive tract.

7 The paired **excretory organs,** or kidneys, discharge through ducts opening near or through the anus; in lower forms the organs are of segmental nature and drain wastes from both the coelom and the blood; in higher forms they are nonsegmental and drain only from the blood; a bladder for storage of urine occurs in many.

8 The **brain** becomes regionally differentiated as to structure and function; the cerebral hemispheres and cerebellum enlarge, especially in higher forms; there are 10 or 12 pairs of cranial nerves in the head that serve both motor and sensory function, including the paired organs of special sense (smell, sight, and hearing plus equilibration); from the **nerve cord** a pair of spinal nerves serves each primitive body somite; an autonomic nervous system regulates involuntary functions of internal organs.

9 A series of **endocrine glands** (thyroid, pituitary, etc.) provide hormones, transported by the bloodstream, that regulate bodily processes, growth, and reproduction.

10 With rare exceptions the **sexes** are separate, and each has a pair of **gonads** that discharge sex cells through ducts opening into or near the anus.

11 A well developed **perivisceral coelom** is present.

Classification

Phylum Chordata.

Chordates. At some stage or throughout life with notochord, dorsal nerve cord, and paired gill slits; segmentation and coelom usually evident; tail behind anus.

Subphylum A. Urochordata (Tunicata).

Tunicates. Larva free-living, minute, tadpolelike, with notochord and nerve cord in tail and with gill slits; adult tubular, globose, or irregular in form, covered in tunic (test), often transparent; usually no notochord, nerve cord reduced to ganglion; no coelom, segmentation, or nephridia.

Class 1. Larvacea (Appendicularia).

To 5 mm ($\frac{1}{5}$ in) long; larvalike in form and structure; notochord, "brain," and nerve cord persistent; 2 gill slits; tunic not persistent; free-swimming, pelagic, often inside secreted "houses." *Oikopleura, Appendicularia.*

Class 2. Ascidiacea.

Ascidians. Size and form various; solitary, colonial, or compound; usually sessile after metamorphosis when notochord, nerve cord, and tail are lost and brain reduced to a ganglion; gill slits many, persistent; tunic well developed, permanent; atrium opens dorsally; stolon simple or none.

Order 1. Enterogona. Body sometimes divided (thorax, abdomen); neural gland usually ventral to ganglion; gonad 1, in or behind intestinal loop; larva with 2 sense organs on "head." *Clavelina, Ciona, Ascidia*.

Order 2. Pleurogona. Body undivided; neural gland usually dorsal or lateral to ganglion; gonads 2 or more, in lateral walls of mantle; larva usually with 1 sense organ on "head." *Styela, Molgula*, simple; *Botryllus*, compound.

Class 3. Thaliacea.

Salps. Size various; adults lack notochord and tail; gill slits various; tunic permanent, with circular muscle bands; auricle opens posteriorly; reproduction includes asexual budding; stolon complex; adults free-living, pelagic.

Order 1. Pyrosomida. A compact, tubular colony closed at one end; common, large excurrent opening; individual incurrent openings; muscle bands at ends of body only; gill slits tall, numerous, to 50; phosphorescent in life; no larva. All oceans. One genus, *Pyrosoma*.

Order 2. Salpida. Cylindrical or prism-shaped; muscle bands incomplete below, convergent above; first gill slit in adult a single large opening; transparent in life; no larva. Widespread in coastal and deeper waters. *Salpa*.

Order 3. Doliolida. Barrel-shaped; muscle bands complete, with 8 regular rings; gill slits few or many, short; a tailed larva. In warm and temperate oceans, widely distributed. *Doliolum*.

Subphylum B. Cephalochordata (Leptocardii).

Lancelets, amphioxus. Slender, body fishlike, segmented; epidermis 1-layered, no scales; notochord and nerve cord along entire body and persistent, as are the many gill slits. Widely distributed in shallow waters. Only 2 genera: *Branchiostoma*, gonads paired; *Asymmetron*, gonads unpaired.

Subphylum C. Vertebrata.

Vertebrates. With cranium, visceral arches, vertebrae, and brain. See Classification in Chaps. 28 to 35.

References

Lower chordates

Barrington, E. J. W. 1965. The biology of Hemichordata and Protochordata. San Francisco, W. H. Freeman and Company. 176 pp., illus.

————, and **R. P. S. Jefferies.** (editors). 1975. Protochordates. Zoological Society of London Symposium Number 36. New York, Academic Press, Inc. xiii + 361 pp., illus.

Berrill, N. J. 1950. The Tunicata, with an account of the British species. London, The Ray Society. iii + 354 pp., 120 figs.

————. 1955. The origin of vertebrates. New York, Oxford University Press. viii + 257 pp., 31 figs. *Deals mainly with lower chordates.*

Bigelow, H. B., and **I. P. Farfante.** 1948. Lancelets. In Fishes of the western North Atlantic. New Haven, Yale University, Sears Foundation for Marine Research. Pt. I, pp. 1–28.

Bone, Q. 1972. The origin of chordates. Oxford Biological Readers. London, Oxford University Press.

Brien, P., and **P. Drach.** 1948. Procordés. In Grassé, Pierre-P. Traité de zoologie. Paris, Masson & Cie. Vol. 11, pp. 535–1040, 460 figs.

Huus, J., and others. 1933. Tunicata. In W. Kükenthal and others, Handbuch der Zoologie. Vol. 5, pt. 2, secs. 1–7, pp. 1–768, 581 figs.

Van Name, W. G. 1945. The North and South American ascidians. *American Museum of Natural History Bull.,* vol. 84, pp. 1–476, 31 pls., 327 figs. *A manual with keys to species.*

Vertebrates (see also Chaps. 1 and 28 to 34)

Blair, W. F., and others. 1968. Vertebrates of the United States, 2d ed. New York, McGraw-Hill Book Company, 701, pp., illus. *A manual with keys for identification, characteristics of groups and species, and distribution; omits saltwater fishes.*

Colbert, E. H. 1969. Evolution of the vertebrates. 2d ed. New York, John Wiley & Sons, Inc. 535 pp., illus. *Includes fossil and recent forms.*

DeBeer, G. R. 1951. Vertebrate zoology. New ed. London, Sidgwick & Jackson, Ltd. xv + 435, pp., 185 figs.

Goodrich, E. H. 1930. Studies on the structure and development of vertebrates. London, Macmillan & Co., Ltd. xxx + 837 pp., 754 figs. Reprinted 1958. New York, Dover Publications, Inc. *Noteworthy for the wealth of detail and comments on phylogeny.*

Jollie, Malcolm. 1962. Chordate morphology, New York, Reinhold Publishing Corporation. xiv + 478 pp., illus. *Emphasizes structure of head.*

Parker, T. J., and **W. A. Haswell.** 1962. A textbook of zoology. 7th ed. New York, The Macmillan Company. Vol. 2 on Chordates rev. by A. J. Marshall. xxiii + 952 pp., 660 figs. *A comprehensive work, first published in 1897.*

Romer, A. S. 1959. The vertebrate story. (Revision of Man and the vertebrates.) Chicago, The University of Chicago Press. vii + 437 pp., 122 pls., many figs.

————. 1966. Vertebrate paleontology. 3d ed. Chicago, The University of Chicago Press. ix + 496 pp., 443 figs. *Comprehensive, much detail.*

Torrey, T. W. 1962. Morphogenesis of the vertebrates. New York, John Wiley & Sons, Inc. x + 600 pp., illus. *Development and structure of organ systems, some histologic detail.*

Weichert, C. K. 1970. Anatomy of the chordates. 4th ed. New York, McGraw-Hill Book Company. vii + 402 pp., 261 figs.

Young, J. Z. 1962. The life of vertebrates. 2d ed. New York, Oxford University Press. xv + 820 pp., 514 figs. *Structure and biology.*

CLASS CYCLOSTOMATA: CYCLOSTOMES

The cyclostomes (Gr. *cyklos*, circular + *stoma*, mouth) include the lampreys and hagfishes (or slime eels) (Fig. 28-1). Lampreys are found in both salt and fresh waters but the hagfishes only in salt water. Cyclostomes are chiefly Temperate Zone forms which reach high latitudes and are found in cool waters of both hemispheres. These lowest of living vertebrates lack true jaws, and their nearest allies are the ancient ostracoderms of Silurian and Devonian times. The lamprey-hagfish lines of evolution diverged early and probably had separate origins among their ostracoderm ancestors (Stensiö). Some ichthyologists therefore place these fishes in separate classes.

28-1 Characteristics

1 Body long, slender, and cylindrical, tail region compressed; median fins supported by cartilaginous fin rays; skin soft and smooth, with many unicellular mucous glands; no scales, no true jaws, or paired fins.

2 Mouth ventroanterior, suctorial in lampreys, eversible and biting in hagfishes; olfactory organs paired but with single median opening on snout.

3 Skull and visceral arches (branchial basket) cartilaginous; notochord persistent; vertebrae repre-

sented by small imperfect neural arches (arcualia) over notochord.

4 Heart two-chambered, with atrium and ventricle; multiple aortic arches in gill region; blood with leukocytes and nucleated circular erythrocytes; multiple hearts in hagfishes (Sec. 28-13).

5 Gills in lateral saclike pouches off pharynx; 7 pouches in lampreys, 5 to 16 in hagfishes.

6 Brain differentiated, with 8 or 10 pairs of cranial nerves; each "ear" with two semicircular canals (or one, in hagfishes).

7 Mesonephric kidneys with ducts to urogenital papilla; pronephros persists in adult hagfishes; nitrogenous wastes chiefly ammonia.

8 Body temperature variable (ectothermal).

9 Gonad single, large, without duct; fertilization external; development direct (hagfishes and slime eels) or with long larval stage (lampreys).

The vertebrate features of cyclostomes include the differentiated brain and paired cranial nerves, eyes, and internal ears, beginnings of segmental vertebrae, characteristic organ systems, and presence of both red and white blood cells. Cyclostomes differ from other vertebrates in having no well-differentiated head, no true jaws, paired appendages, limb girdles, ribs, or reproductive ducts connecting to the gonads. They have but a single gonad and only one or two semicircular canals in each "ear." The fossil record is scant—a lamprey from the Carboniferous of Illinois.

28-2 Structure of a lamprey (Figs. 28-2, 28-3) The combined head and trunk, or **body,** is cylindrical, and the **tail** is laterally compressed. On the posterior dorsal region and tail are **median fins.** Ventrally on the head is a large cuplike **buccal funnel,** margined by soft papillae and lined within by conical yellow, horny **teeth.** The single **prenasal duct** (nostril) opens middorsally on the head, followed by thin skin over the pineal organ, a median eyelike structure. The two large **eyes** are lateral and covered by transparent skin but lack lids. Behind each are seven rounded **gill slits.** A row of small lateral **sense pits** extends segmentally along each side of the body and tail. The **anus** opens ventrally at the base of the tail, and close behind is a small **urogenital papilla,** pierced by a duct. The whole animal is covered by smooth epithelium containing many mucous glands but no scales.

28-3 Skeleton The **notochord** persists throughout life as the axial skeleton; it is a slender gelatinous rod sheathed in tough connective tissue. The other elements of the **skeleton,** all cartilaginous, include (1) the complex skull (cranium and sense capsules); (2) the stout lingual cartilage of the tongue, and a ring of annular cartilage surrounding the buccal funnel; (3) the elaborate paired visceral arches, or branchial basket, supporting the gill region; and (4) small rod-

Figure 28-1 Class CYCLOSTOMATA. *A*. California hagfish (*Eptatretus stouti*), with soft mouth, four pairs of tentacles, and 11 pairs of gill pouches. The enlarged posteriormost aperture provides for exit of water that may bypass the gills. *B*. Sea lamprey (*Petromyzon marinus*), with buccal funnel, eyes, and seven pairs of gill slits. (*A, after Wolcott, Animal biology; B, after Norman, Guide to fish gallery, British Museum.*)

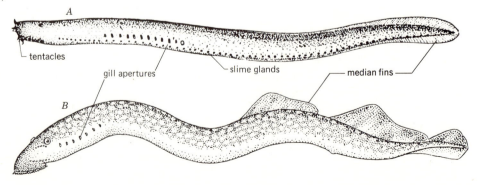

tentacles
gill apertures
slime glands
median fins

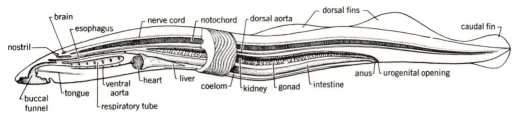

Figure 28-2 Class CYCLOSTOMATA. Structure of adult lamprey (*Entosphenus tridentatus*); left side of body mostly removed.

like pieces (arcualia) dorsal to the notochord, two pairs to most body segments, that represent the neural arches of higher vertebrates.

28-4 Muscular system   The trunk and tail regions are walled by short segmental muscles, that in side view are ≲-shaped, as in other fishes. Radial muscles operate the buccal funnel, and the tongue is moved by stout retractor and smaller protractor muscles.

28-5 Digestive system   The small **mouth** is centered in the buccal funnel and closed or opened by the fore-and-aft pistonlike movement of the **tongue.** On the tongue are horny teeth like those in the funnel (Fig. 28-5*A*). A short **pharynx** follows, then the tract divides into a dorsal **esophagus** and ventral **respiratory tube;** the entrance to the latter is guarded by a flexible transverse plate, the **velum.** There is no stomach. The posterior end of the esophagus opens, by a valve, into the straight **intestine.** Within the latter is a spiral longitudinal fold (typhlosole), or spiral

valve. The intestine ends at the small **anus.** The **liver** is usually without a bile duct; the pancreas consists of clusters of cells associated with the liver and in the typhlosole of the gut beneath the liver.

28-6 Circulatory system   The **heart** lies within the cup-shaped posterior end of the branchial basket in a pericardial sac that communicates with the coelom. It consists of a sinus venosus, atrium, and thick-walled ventricle (no conus arteriosus). The blood is pumped anteriorly into a **ventral aorta** that distributes in eight pairs of afferent branches to the gill filaments, then is collected in a median **dorsal aorta,** above the gills, that distributes both anteriorly and posteriorly. The venous system returns blood from the head and body and includes a hepatic system through the liver but no renal portal system; lymphatic vessels are also present.

28-7 Respiratory system   There are seven **gills** in pouches on either side, between the respiratory tube and the body wall, whence the name MARSIPO-

Figure 28-3 Adult lamprey (*Entosphenus*), anterior end, with left side of body removed to median plane, leaving parts of gill region and muscular body wall.

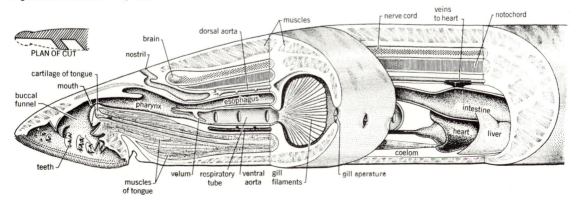

BRANCHII (Gr., pouch gills). Each gill contains many gill filaments with fine capillaries in which the blood is aerated by water in the pouches. In an adult lamprey the water currents serving for respiration pass both *in* and *out* of the gill pouches, a route unlike that in true fishes. This method is necessary because the lamprey often attaches by its buccal funnel to food or other objects, making passage of water through the mouth impossible. In a larval lamprey the respiratory current passes in through the mouth and out the gill openings, as in fishes.

28-8 Excretory system   The two **kidneys** (mesonephroi, Fig. 7-8) are dorsal in the body cavity; a tubular **ureter** extends from each to the **urogenital sinus,** which empties to the exterior through the **urogenital papilla.**

28-9 Nervous system   Two large paired **olfactory lobes,** followed by smaller **cerebral hemispheres,** attach to the **diencephalon;** under the latter is a broad **infundibulum** and, dorsally, a **pineal structure.** On the midbrain, dorsally, are two large **optic lobes.** The **cerebellum** is a small transverse dorsal band, and the ventral **medulla** is much larger. Large openings with choroid plexuses containing blood vessels occur between the optic lobes and over the medulla. Within the brain are four **ventricles,** as in other vertebrates. There are 10 pairs of **cranial nerves.** The **nerve cord** is flattened and bandlike. The dorsal and ventral roots of the paired **spinal nerves** emerge alternately from the cord and are not bound together as in other vertebrates. There is a poorly defined sympathetic system.

28-10 Sense organs   The single nasal opening leads into paired **olfactory sacs** receiving nerve endings from both olfactory lobes of the brain. Besides the paired **eyes,** an elaborate median **pineal eye** with clear lens and pigmented retina lies behind the nasal aperture. Each internal **ear** (organ of equilibrium) has two semicircular canals. There are taste buds in the pharynx and **lateral line** sense organs on the sides of the body and lower surface of the head.

28-11 Endocrine glands   Beneath the infundibulum is a **pituitary body;** a long sac extending posteriorly from the olfactory sac is termed the *pituitary pouch.* The endostyle of the larva resembles similar structures in tunicates and amphioxus and is considered to be the forerunner of the thyroid gland of the adult lamprey.

28-12 Reproductive system   Any one individual is either male or female; that is, the sexes are separate (dioecious) as in other vertebrates. At sexual maturity a single large long **gonad** fills much of the abdominal cavity. There is no genital duct; eggs or sperm discharge into the abdominal cavity and pass through paired **genital pores** into the urogenital sinus and thence to the outside.

28-13 Hagfishes and slime eels   As compared with lampreys, members of the order MYXINIFORMES differ in many ways, having (1) a small mouth with one large dorsal epidermal tooth and rows of small teeth; (2) several pairs of soft anterior tentacles around the mouth and nasal opening; (3) nostril terminal and median with canal (and pituitary pouch) under brain to roof of pharynx as a channel for water entering to aerate gills; (4) roof and sides of brain case membranous, branchial basket much reduced, arcualia few and posterior; (5) myomeres alternate; (6) gills in 5 to 16 pairs, far posterior; (7) segmental pronephros (with nephrostomes) in adult; (8) brain primitive, of four lobes, no recognizable cerebrum or cerebellum, dorsal and ventral roots of spinal nerves united; (9) eyes not visible; (10) pineal eye vestigial; (11) each ear with only one semicircular canal; (12) lateral line system far reduced; and (13) eggs large, with shells formed in ovary.

The pancreas consists of groups of cells along the liver and mesenteric blood vessels and in the intestinal epithelium. The endocrine (islet) tissue is around the common bile duct. The venous system of hagfishes has secondary "hearts" (contractile bulbs) that augment the flow of blood in their low-pressure circulatory system.

28-14 Natural history   Lampreys live in both fresh waters and salt waters; the hagfishes are marine, some descending to depths of over 1800 m (1000 fathoms). Some adult lampreys are nonparasitic (about half the species in North America). They are brook dwellers and have degenerate teeth and digestive tract. The parasitic species attach to fishes (Fig. 28-4) by suction of the funnel and use of the buccal teeth; a hole is rasped by the teeth on the tongue and an anticoagulant injected, maintaining the flow of blood from the host into the mouth of the lamprey. Healthy fishes are attacked and may be killed.

Hagfishes are intolerant of strong light and low salinities. Like marine invertebrates, their body fluids are isosmotic (in osmotic equilibrium) with seawater. They seek muddy parts of the seafloor, generally below depths of 24 m (80 ft), where they burrow. They eat worms and other invertebrates that live in or on bottom mud and dead and disabled fish and other animals that sink to the bottom. There is little evidence that they are predators or parasites on free-living fishes; however, they will eat their way into the bodies of netted fish. When feeding, rows of horny elongate teeth on each side of the floor of the mouth are moved outward over the lower lip, the rows on each side parting as they do so. When movement is reversed, they come together to bite. Food is apparently located by smell and usually is bitten off in chunks but not rasped or sucked into the mouth.

When irritated, hagfishes release copious quantities of mucus from their skin glands, hence the name *slime eel.*

Figure 28-4 Lamprey attaching and feeding; diagrams of the action of the buccal funnel and tongue. *A.* Buccal funnel is collapsed against an object and its interior volume is decreased by forcing water behind tongue into pharynx. *B.* Tongue closes passage between funnel and pharynx; funnel warped out to increase internal volume and degree of suction. *C.* Water in hydrosinus is forced behind velum; tongue continues to block oral passage. *D.* Velum prevents return of water from gill pouches; tongue is free to move back and forth in rasping flesh. (*After T. E. Reynolds, 1931.*)

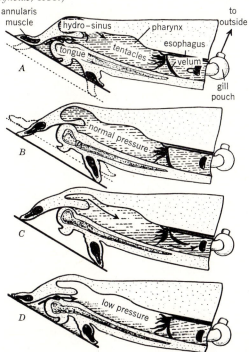

28-15 Reproduction   When lampreys, either marine or fresh water, become sexually mature, in the spring or early summer, the gonads swell and both sexes move into streams, sometimes "riding" on a passing fish or boat. They seek clear water and riffles in streams and by use of the buccal funnel move stones on the bottom and sweep away bottom debris by rapid body movements until a shallow rounded depression, free of silt, is prepared as a nest (Fig. 28-5*C*). The female then attaches to a stone on the upstream side, and the male fastens to the female, usually attaching to the top of her head, both using their buccal funnels. Partly entwined, they wriggle back and forth as eggs and sperm are discharged, fertilization being external. The eggs are small (1 mm in the brook lamprey) and adhesive; they quickly sink and are covered by silt and sand. A female brook lamprey may contain from 2000 to 65,000 eggs and the large sea lamprey up to 236,000 eggs. Several pairs usually spawn close together, at times in the same nest. Breeding may occur in autumn in some species in Europe. All adults die after spawning. The young emerge in a few weeks as minute larvae; when 12 to 15 mm (0.5 to 0.6 in) long, they leave the nest to seek quiet water. There each constructs and inhabits a U-shaped tunnel in the sand and silt where it removes microplankton from the water or emerges to feed on ooze covering the stream bottom (Fig. 28-5*D*). Water is drawn into the mouth by ciliary action and passes out through the gill aper-

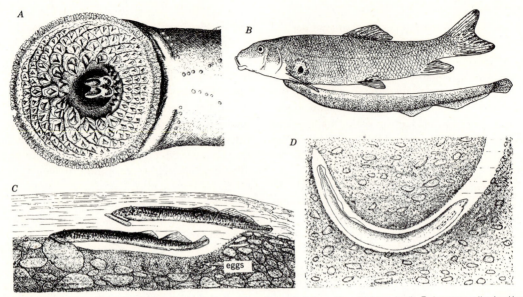

Figure 28-5 Structure and life of lake lampreys. *A*. Buccal funnel in ventral view with many horny teeth. *B*. Lamprey attached to a fish. Scar resulting from previous attack above pectoral fin. *C*. Nest with female attached to a stone, male carrying another stone; small eggs in rear of nest. *D*. Larval lamprey (ammocoete) in its burrow under the water. (*After S. H. Gage, 1893, 1929.*)

tures. The food is caught in mucus secreted by the endostyle on the floor of the pharynx, as in amphioxus, with which the larvae were formerly confused.

The larvae, known as ammocoetes, look much like amphioxus (Sec. 27-10). They are blind and toothless and have a long life, from at least 3 years (*Entosphenus tridentatus*) to perhaps 7 years (landlocked *Petromyzon*). They live in debris and mud in quiet parts of streams and filter minute organisms from the water. When metamorphosis occurs, two courses are pursued by different types of lamprey. In the more primitive group a functional digestive tract is retained and strong sharp teeth develop; such lampreys feed on fishes, continuing to live and grow in the sea or in large streams or lakes, according to the species. After having existed in this state for a year (perhaps longer in the largest species), they reascend small streams in the spring to spawn and then die. Members of the second group cease feeding and growth after metamorphosis in August to October; the alimentary tract and teeth partly degenerate, and after 4 to 11 months the animals breed and die. Such degenerate lampreys have developed

as offshoots of the normal type at several places, in both Eastern and Western states and other parts of the world.

Hagfishes are structurally hermaphroditic but functionally of separate sexes. They are first hermaphrodites, with the single gonad an ovary anteriorly and a testis posteriorly. When the ovarian portion develops and the testis is suppressed, a female results; the reverse yields a male. If both parts develop beyond the stage of normal sex differentiation, gonadal growth is retarded, degeneration occurs, and a sterile animal results. All stages between true males and true females may be found. A single animal may produce sperm one season and eggs the next. Hagfishes may be evolving from hermaphroditism toward bisexuality.

Eggs or sperm pass into the abdominal cavity, as in lampreys, and enter the coelomic ducts, well back in the abdomen, which connect to the cloaca. The eggs are few, with much yolk and a tough, horny shell; they are elongate oval in shape and may be large (10 × 30 mm). On the ends are threadlike holdfasts with anchor-shaped terminals that attach the eggs to seaweed or other bottom objects. Growth of the hatchlings to adult form apparently is direct, without a larval stage.

28-16 Relation to humans Cyclostomes, particularly hagfishes, feed upon crippled or diseased fishes, and hagfishes dispose of dead remains. In some marine environments they are among the most important vertebrate scavengers, an essential part of the ecosystem. But they have other effects on human concerns. Larval lampreys serve as bait for both commercial and sports fishing, and adults (especially sea lampreys) are used as food. Lampreys injure and destroy fishes by consuming their blood and flesh and also are responsible for secondary infections. The construction of the Welland Ship Canal during the nineteenth century, connecting Lake Ontario with Lake Erie, permitted sea lampreys (*Petromyzon marinus*) to invade the Great Lakes in large numbers, where they are now landlocked. They have caused great damage to commercial fisheries. Their attacks on lake trout at one stage were considered responsible for reducing the annual catch from about 7 million kg (15 million lb) to $\frac{1}{4}$ million kg ($\frac{1}{2}$ million lb). Attempts at control included electric weirs and selective poisons to kill the larvae. This was sufficiently successful to permit restocking of lake trout and a partial restoration of fishing.

Classification

Class 1. †Ostracodermi (Cephalaspida).

Small, mostly under 60 cm (2 ft) long, body often depressed and spade-shaped; mouth small and jawless; head often covered by bony shield and body by plates or spines; cranial skeleton ossified; 1 or 2 dorsal fins; scale-covered flaps resembling pectoral fins sometimes present; heterocercal tail (Fig. 30-11*A*), in some forms reversed, the enlarged portion projecting downward; 2 eyes and pineal eye; 2 semicircular canals; usually about 10 pairs of gill cavities; Silurian to Devonian, mainly on bottom in fresh waters where food was probably strained from mud by branchial apparatus; some, however, were probably active, surface-swimming types. †*Cephalaspis*; †*Birkenia*, probably surface-dwelling, with reversed heterocercal tail; †*Pteraspis*.

Class 2. **Cyclostomata (Agnatha, Marsipobranchii).**

Cyclostomes, hagfishes. Body eellike, long, cylindrical; skin smooth, no scales; no true jaws or paired fins; skeleton cartilaginous, gills in pouches, 5 to 16 pairs; heart 2-chambered; about 50 species.

Order 1. Petromyzontiformes. Lampreys. Mouth surrounded by large ventral suctorial funnel with many horny teeth; long protrusible tongue also armed with horny teeth; nasal sacs connected to single duct opening medially just in front of eyes, no internal opening to pharynx; 7 pairs of gill pouches, opening separately; branchial arches fused into "basket" enclosing gills; cartilaginous neural arches (arcualia) present; eggs small; a long larval stage; about 30 species (19 in North America); nearly worldwide along seacoasts and in streams and lakes. *Petromyzon marinus*, sea lamprey, Atlantic Coast from Chesapeake Bay northward and in Europe, grows to 1 m (3 ft); spawns in fresh water, landlocked in Finger Lakes, New York, also in Great Lakes; *Lampetra*, in Eurasian fresh waters and salt waters; *Entosphenus tridentatus*, Pacific lamprey of western shores, southern California to Alaska, spawns in fresh waters; *Lampetra lamottei*, American brook lamprey and others, nonparasitic, Eastern and Middle Western states; *Ichthyomyzon*, brook lamprey, Central United States and Canada. †*Mayomyzon*, Carboniferous (Pennsylvanian).

Order 2. Myxiniformes. Slime eels and hagfishes. Mouth nearly terminal, with marginal tentacles, no buccal funnel, teeth concealed when not feeding; nasal sacs open into duct that extends from near tip of snout to pharynx; eyes under skin; gill pouches 5 to 16 pairs; "branchial basket" and arcualia poorly developed; eggs 20 to 30, large, in horny cases, development probably direct; 21 species; marine, saltwater lakes. *Myxine glutinosa*, Atlantic hagfish, Cape Cod northward on Atlantic Coast, to 75 cm (30 in) long, also in Europe, but only half this size; *Bdellostoma*, Chile; *Eptatretus stouti*, Pacific hagfish, or "borer," lower California to Alaska.

References

(*See also works by Marshall, Nikol'skii, and others cited in references in Chap. 30.*)

Applegate, V. C., and **J. W. Moffett.** 1955. The sea lamprey. *Sci. Am.*, vol. 192, no. 4, pp. 36–41, illus. *The scourge of commercial fisheries in the Great Lakes; control measures are described.*

Brodal, A., and **R. Fänge.** 1963. The biology of *Myxine*. Oslo, Universitetsforlaget. xii + 588 pp.

Gage, S. H. 1893. The lake and brook lampreys of New York. . . . Ithaca, N.Y., Wilder Quarter-Century Book. Pp. 421–493, 8 pls. *Structure, function, and natural history; includes the landlocked form of the sea lamprey.*

Hardisty, M. W., and **I. C. Potter.** 1971. The biology of lampreys. New York, Academic Press, Inc. Vol. 1, xiv + 423 pp. 1972. Vol. 2, xiv + 466 pp.

Hubbs, C. L. 1924. The life-cycle and growth of lampreys. *Michigan Academy of Science, Arts and Letters, Papers*, vol. 4, pp. 587–603, figs. 16–22.

Jensen, David. 1966. The hagfish. *Sci. Am.*, vol. 214, no. 2, pp. 2–10, illus. *Aspects of its structure and biology.*

Lennon, R. E. 1954. Feeding mechanism of the sea lamprey and its effect on host fishes. *U.S. Fish & Wildlife Service, Fisheries Bull.*, vol. 56, pp. 245–293; 19 figs.

Parker, P. S., and **R. E. Lennon.** 1956. Biology of the sea lamprey in its parasitic phase. *U.S. Fish & Wildlife Service, Research Rept.* 44. 32 pp., 8 figs.

Reynolds, T. E. 1931. Hydrostatics of the suctorial mouth of the lamprey. *University of California Publications in Zoology*, vol. 37, pp. 15–34. 3 pls., 2 figs. *Anatomy and function of the head region.*

Stensiö, E. In T. Ørvig. 1968. The cyclostomes with special reference to the diphyletic origin of the Petromyzontida and Myxinoidea. In Nobel Symposium 4: Proceedings of the Fourth Nobel Symposium held in June 1967 at the Swedish Museum of Natural History (Naturhistoriska riksmuseet) in Stockholm. New York, Interscience Publishers, a division of John Wiley & Sons., Inc. 539 pp. *Evidence for independent origin of hagfishes and lampreys and summary of basic features uniting and separating groups of living and fossil agnathans.*

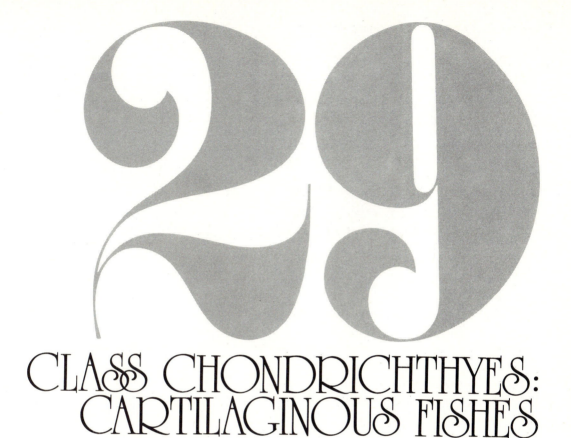

CLASS CHONDRICHTHYES: CARTILAGINOUS FISHES

The sharks and rays, and the chimaeras of the class CHONDRICHTHYES (Gr. *chondros*, cartilage + *ichthys*, fish; Figs. 29-1, 29-2) are the lowest living vertebrates with complete and separate vertebrae, movable jaws, and paired appendages. All are predaceous, and virtually all are ocean dwellers. The group is ancient and represented by many fossil remains, especially teeth, fin spines, and scales. Sharks are of major biologic interest, because some of their basic anatomic features are paralleled in early embryos of the higher vertebrates.

29-1 Characteristics

1 Skin tough, covered with minute placoid scales (Fig. 29-4) and having many mucous glands; both median and paired fins present, all supported by fin rays; pelvic fins with claspers in males; heterocercal tail.

2 Mouth ventral, with enamel-capped teeth; nostrils two (or one), not connected to mouth cavity; both lower and upper jaws present; intestine with spiral valve (Fig. 29-6).

3 Skeleton cartilaginous, no true bone; cranium joined by paired sense capsules; notochord persistent; vertebrae many, complete, and separate; pectoral and pelvic girdles present.

4 Heart two-chambered (one atrium, one ventricle), with sinus venosus and conus arteriosus, contains only venous blood; several pairs of aortic arches; some veins expanded as sinuses; red blood cells nucleated and oval.

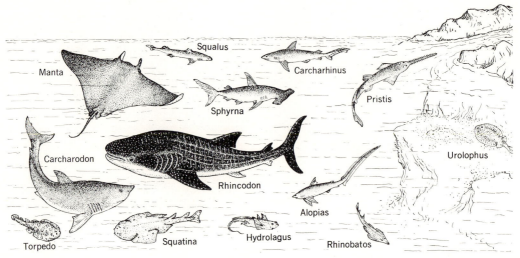

Figure 29-1 Class CHONDRICHTHYES: common cartilaginous fishes, all marine, variously reduced; bottom dwellers below. Order SQUALIFORMES: *Alopias*, thresher shark; *Carcharhinus*, gray shark; *Carcharodon*, white shark; *Rhincodon*, whale shark; *Sphyrna*, hammerhead; *Squalus*, spiny dogfish; *Squatina*, angel shark. Order RAJIFORMES: Rays. *Pristis*, sawfish; *Rhinobatos*, guitarfish; *Manta*, manta; *Torpedo*, torpedo (electric); *Urolophus*, stingray. Order CHIMAERIFORMES: *Hydrolagus*, ratfish.

5 Respiration by gills attached to opposing walls of five to seven pairs of gill pouches, each pouch with a separate slitlike opening; no swim bladder.

6 Ten pairs of cranial nerves; each "ear" with three semicircular canals.

7 Excretion by mesonephric kidneys; urea chief nitrogenous waste.

8 Body temperature variable (ectothermal).

9 Sexes separate; gonads typically paired; reproductive ducts discharge into cloaca; fertilization internal; oviparous or ovoviparous; eggs large,

Figure 29-2 Class CHONDRICHTHYES. *A.* Spiny dogfish, or dog shark (*Squalus acanthias*). B. Ray (*Raja*). C. Chimaera (*Chimaera colliei*). (*A. after Goode; B, after General Biological Supply House; C, after Dean, Fishes living and fossil, The Macmillan Co.*)

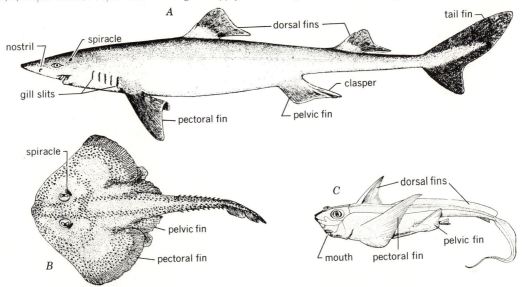

with much yolk, segmentation meroblastic; no embryonic membranes; development direct, no metamorphosis.

The cartilaginous fishes show changes from the cyclostomes in having (1) scales covering the body, (2) two pairs of lateral fins, (3) movable jaws articulated to the cranium, (4) enamel-covered teeth on the jaws, (5) three semicircular canals in each ear, and (6) paired reproductive organs and ducts. They differ from the bony fishes in having (1) the skeleton of cartilage, with no true bone, (2) placoid scales, (3) separate gill clefts, (4) a pair of spiracles connecting to the pharynx, and (5) no swim bladder.

Several theories have been proposed on the origin of paired fins in fishes. The theory of Balfour and others would derive the paired fins from lengthwise lateroventral fin folds supported by parallel fin rays (Fig. 29-3). As evidence, (1) amphioxus has such continuous (metapleural) folds joining the ventral caudal fin; (2) a small Devonian fish, †*Climatius*, had five accessory finlets between the pectoral and pelvic fins on each side; (3) †*Cladoselache*, a late Devonian shark, had paired fins with expanded bases supported by many parallel rays; and (4) embryos of some living sharks have lateral skin folds preceding the appearance of the paired fins. Other evidence from embryology, anatomy, and paleontology favors this theory.

29-2 Size Dogfishes (*Squalus*) grow to about 1 m (3 ft) in length, and most sharks are under 2.5 m (8 ft)

Figure 29-3 Possible origin of the fins on fishes from fin folds such as those on amphioxus. Compare Fig. 27-8. (*After Wiedersheim.*)

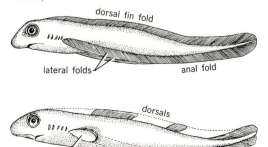

long, but the great white shark (*Carcharodon carcharias*) grows to 6 m (20 ft), the basking shark (*Cetorhinus maximus*) to 12 m (40 ft), and the whale shark (*Rhincodon typus*) to 18 m (60 ft). These are the biggest living vertebrates except whales. Most rays are 30 to 90 cm (1 to 3 ft) in length, but the greater devilfish (*Manta birostris*) grows to 5 m (17 ft) in length and 6 m (20 ft) across the pectoral fins. The chimaeras are often less than 1 m (3 ft) long.

The dogfish, or dog shark

29-3 External features The **head** is bluntly pointed, and the **trunk** is spindle-shaped, largest near the pectoral fins and tapering behind. There are two separate median **dorsal fins** (each preceded by a spine in the spiny dogfish, *Squalus*), a median **caudal fin,** and two pairs of lateral fins, **pectoral** and **pelvic.** In a mature male each pelvic fin bears a slender, grooved appendage, the **clasper,** used in copulation. The smooth dogfish (*Mustelus*) also has a median **anal fin.** The **tail** is heterocercal, the vertebral column turning upward and extending almost to the tip of the caudal fin (Fig. 30-11). Ventrally on the head are two **nostrils** and the wide **mouth;** the **eyes** are lateral and without lids. Five oval **gill slits** open anterior to each pectoral fin; a gill-like cleft, or **spiracle,** opens behind each eye. The **vent** is between the pelvic fins.

The caudal fin is bilobed, the upper part usually narrower than the ventral lobe, which is attached along the lower side of the vertebral column (Fig. 30-11). As the tail is moved from side to side in swimming, the ventral lobe provides lift. On each inward stroke the upper surface is tilted toward the midline and the ventral lobe trails behind. The resultant force has an upward component, and the tail tends to rise and pitch the foreparts downward. The pectoral fins, however, incline upward and cause the foreparts to rise , countering the tail action. The combined effect tends to raise the fish, a matter of importance for an animal that lacks the buoyancy of a swim bladder and is heavier than seawater.

29-4 Body covering The gray-colored **skin** is evenly covered with diagonal rows of minute **pla-**

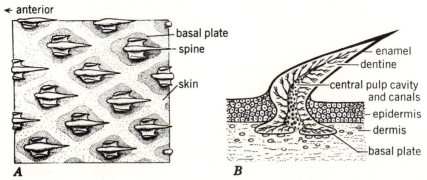

← anterior

basal plate
spine
skin

enamel
dentine
central pulp cavity and canals
epidermis
dermis
basal plate

A *B*

Figure 29-4 Placoid scales (enlarged). *A*. Skin with scales in surface view. *B*. Median section through a scale. (*After Klaatsch.*)

coid scales (Fig. 29-4), each with a backward-pointing spine covered by enamel and a basal plate of dentine in the dermis.

29-5 Skeleton The entire skelton is of cartilage (gristle) more or less reinforced with limy deposits; the axial parts are the **chondocranium, visceral skeleton,** and segmented **vertebral column.** Each **vertebra** has a spool-shaped centrum, concave on both ends (amphicoelous), and above this a neural arch to house the nerve cord; in the tail, each also bears ventrally a hemal arch shielding the caudal aorta and vein; the notochord persists. The chondocranium is made up of (1) the cranium housing the brain and (2) paired capsules for the olfactory, optic, and auditory organs. The visceral skeleton consists of the jaws, hyoid arch, and five pairs of branchial arches sup-

porting the gill region. The upper jaw is not united to the brain case. The **appendicular skeleton** includes (1) the U-shaped **pectoral girdle** supporting the pectoral fins, (2) the flatter **pelvic girdle** to which the pelvic fins attach, and (3) the many small jointed cartilages within and supporting each lateral fin. The median fins are supported by dermal **fin rays.**

29-6 Muscular system The body and tail **muscles** are of segmental character and serve to produce the lateral undulations of the trunk and tail necessary for swimming (Fig. 29-8). More specialized muscles serve the paired fins, gill region, and structures of the head.

29-7 Digestive system (Fig. 29-5) The broad **mouth** is margined with transverse rows of sharply

Figure 29-5 Spiny dogfish; internal structure.

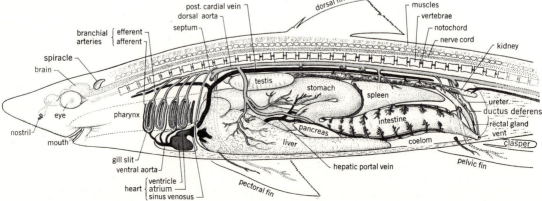

pointed **teeth.** The teeth have evolved from placoid scales (Fig. 29-4). They are embedded in flesh and not attached to the jaws as in bony fishes and higher vertebrates. Replacement teeth develop continuously in rows behind the functional teeth and move forward to replace those lost. The teeth serve to grasp prey such as small fishes, which are often swallowed entire. A flat **tonguelike structure** adheres to the floor of the mouth. On the sides of the wide **pharynx** are openings leading to the separate **gill slits** and **spiracles.** The short **esophagus** leads to the J-shaped **stomach** which ends at a circular sphincter muscle, the pyloric valve. The **intestine** follows and connects directly to the **cloaca** and **anus.** In the intestine is a spirally arranged partition, or **spiral valve,** covered with mucous membrane, that delays the passage of food and offers increased area for absorption (Fig. 29-6). The large **liver** is of two long lobes, attached at the anterior end of the body cavity. Bile from the liver collects in the greenish **gall bladder** and thence passes through the **bile duct** to the anterior part of the intestine. The bilobed **pancreas** lies between the stomach and intestine, its duct joining the latter just below the bile duct. In some sharks the liver oil averages about 20 percent of the body weight and affords considerable buoyancy. There is no swim bladder in sharks.

29-8 Coelom The stomach, intestine, and other internal organs lie in the large **body cavity,** or **coelom.** It is lined with a smooth glistening membrane, the **peritoneum,** which also covers the organs. The

Figure 29-6 **Intestine of dogfish opened to show spiral valve.** (*After Jammes.*)

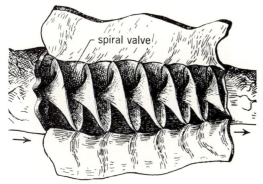

latter are supported from the middorsal wall of the coelom by thin **mesenteries,** also formed of peritoneum. A transverse **septum** separates the coelom from the cavity containing the heart.

29-9 Circulatory system The **heart** lies adjacent to the gill region in a sac, the pericardium; it consists of (1) a thin-walled dorsal sinus venosus that receives blood from various veins, followed by (2) the atrium, (3) the thick-walled ventricle, and (4) the conus arteriosus. From the latter, blood passes anteriorly into the **ventral aorta,** whence five pairs of **afferent branchial arteries** distribute to capillaries in the gills for aeration; four pairs of **efferent branchial arteries** then collect the blood into the **dorsal aorta,** which extends along the middorsal wall of the coelom. The principal **arteries** are (1) paired external and internal carotids to the head; (2) paired subclavians to the pectoral fins; (3) coeliac to the stomach, liver, and intestine; (4) anterior mesenteric to the spleen and hindpart of the intestine; (5) posterior mesenteric to the rectal gland; (6) several renal and genital (ovarian or spermatic) to the kidneys and reproductive organs; and (7) paired iliacs to the pelvic fins. Beyond the latter the caudal aorta continues in the tail.

In the **venous system,** blood in the caudal vein from the tail passes in (1) paired renal portal veins to the kidneys. Other blood from the posterior regions passes forward in (2) the paired postcardinal veins (sinuses) paralleling the kidneys and in (3) paired lateral abdominal veins on either side of the body cavity. Paired (4) jugular and (5) anterior cardinal veins (sinuses) return blood from the head region. All these veins enter large sinuses connected to the sinus venosus. Blood from the digestive tract flows in (6) the hepatic portal vein to be filtered through the liver, then is collected in (7) the hepatic veins connected to the sinus venosus. The blood passes through the heart but once in each circuit of the body, as in cyclostomes and most fishes, and the heart blood is all venous (unoxygenated; dark in Fig. 29-5; see also Fig. 6-6).

29-10 Respiratory system In benthic sharks water is drawn in through the mouth and over the gills by expansion of the **branchial pouches**. Upon contraction of the pouches, water is expelled through the gill slits. Pelagic species use a ram-jet method swimming with mouth and gill slits open, thereby maintaining a flow of water over the gills. (Fig. 30-6). The **gills** are composed of many parallel slender filaments that contain capillaries. Blood from the ventral aorta passes through these capillaries, discharges carbon dioxide and absorbs oxygen dissolved in the water, and then continues into the dorsal aorta.

29-11 Excretory system The two slender **mesonephric kidneys** lie immediately above the coelom on either side of the dorsal aorta. Urine from each is collected in a series of segmental tubules that join the **Wolffian duct** (ureter), leading posteriorly; the two ureters empty through a single **urogenital papilla** into the urogenital sinus at the cloaca. A slim **rectal salt gland,** attached at the junction of the intestine and cloaca, aids the kidneys in removing excess salts from the blood.

The large salt content of ocean waters presents a problem in osmotic regulation for many marine animals. The ancestors of both bony fishes and elasmobranchs lived in fresh waters. In such a hypotonic medium, the concentration of salts in their body fluids was lowered, thereby reducing osmotic uptake of water through gills and mucous membranes and avoiding expenditure of energy in eliminating the excess. Upon entering seawater they faced an opposite difficulty—loss of body water to the hyper-

tonic medium of the sea. To combat the osmotic imbalance, sharks and rays retain chloride and urea (a waste product of protein metabolism) in the blood at 2.0 to 2.5% (versus 0.01 to 0.03% in most other vertebrates), so that their blood and tissue fluids are slightly hypertonic to seawater. Little urea is lost through the gills or kidneys. They absorb some water through the gills and buccopharynx, thus they need not drink. Their output of urine is small. It is of interest that the freshwater stingrays (POTAMOTRYGONIDAE) have little urea in their blood.

29-12 Endocrine glands In elasmobranchs, the pituitary, below the base of the brain, has four subdivisions. The thyroid is compact, buried in the tongue or pharynx, and the islets of Langerhans are within the pancreas. Of the adrenals, that part comparable to the cortex in mammals is medial in the kidney region (interrenal bodies), while the suprarenals, producing epinephrine, are segmental. The latter are close to the sympathetic ganglia and project dorsally into the postcardinal veins.

29-13 Nervous system A shark's **brain** (Fig. 29-7) is of a more advanced type than that of the lamprey. From the two olfactory sacs in the snout, the large olfactory tracts extend to the **olfactory lobes,** which attach closely to the paired **cerebral hemispheres,** on the **diencephalon.** Dorsally the latter bears a pineal stalk and **pineal body** and ventrally the **infundibulum,** to which is attached the **hypophysis.** All these are part of the forebrain. Two rounded **optic lobes** lie

Figure 29-7 **Brain and nerves of the spiny dogfish from the left side.** I–X, cranial nerves; 1, 2, spinal nerves.

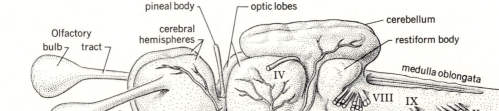

dorsally on the midbrain. The hindbrain comprises the large median dorsal **cerebellum** over the thin-roofed **medulla.** Ten pairs of **cranial nerves** serve structures, chiefly of the head, in approximately the same distribution as among other vertebrates (Table 9-1). The spinal **nerve cord** is protected by the neural arches of the vertebrae, an advance over the condition in cyclostomes. Paired **spinal nerves** to each body segment emerge between the neural arches of successive vertebrae. The **sympathetic nervous system** comprises a series of ganglia, roughly segmental, above the postcardinal veins.

29-14 Sense organs The two **nostrils** (olfactory sacs) ventral on the snout are sluiced by oncoming water and serve the sense of smell for dissolved materials. The pharynx contains scattered **taste buds.** The **eyes,** moved by three pairs of muscles as in other vertebrates, are without lids. As in most elasmobranch fishes, the retina contains only rods, and the eyes appear adapted for dim light. The **"ear"** seemingly is an organ of balance, with three semicircular canals at right angles to one another as in all higher vertebrates. A slender endolymphatic duct extends from each "ear" to a small pore dorsal on the head.

The **lateral line** is a fine groove along each side of the trunk and tail that contains a slender canal with many small openings to the surface. Within the canal are sensory hair cells connecting to a branch of the tenth cranial nerve. They evidently respond to low-frequency pressure stimuli in the surrounding water —a sort of distant "touch response." On the head are other **sensory canals** opening by pores; each pore leads to a small chamber (ampulla of Lorenzini) containing an electroreceptor (Sec. 30-15) with sensory hairs connecting to nerve fibers.

29-15 Reproductive system The sexes are separate. In the **male,** sperm develop in two long **testes** anterior in the body cavity; from each testis several **ductus efferentes** lead to the Wolffian duct or **ductus deferens,** which in sharks also passes urine. The duct empties into the urogenital sinus. At mating, the claspers are placed close together and the two organs are inserted into the cloaca of the female. Seminal fluid flows down the channel formed by adjacent clasper grooves.

The **female** reproductive system has two large **ovaries** (sometimes fused) supported by stout membranes. Two **Müllerian ducts,** or **oviducts,** extend most of the length of the body cavity; their anterior ends join in a single large funnel through which the eggs enter. The forward part of each duct is dilated as a **shell gland,** and in ovoviviparous species like the dogfish, the posterior part is enlarged as a **"uterus"** to contain the young during development. The oviducts open separately into the cloaca.

Other cartilagenous fishes

29-16 Structure and evolution Most other sharks resemble the dogfish in general anatomy. The skates and rays, however, have much depressed bodies with large pectoral fins broadly joined to the head and trunk, so that they usually are diamond- or disk-shaped in outline. They swim by flapping movements of the fins (Fig. 29-8).

The gill openings are on the flat ventral surface,

Figure 29-8 The swimming of cartilaginous fishes. *A.* Shark (dorsal view) swims by lateral undulations of the trunk and tail. *B.* Ray (lateral view) swims by flapping movements of the pectoral fins. (*After Marey.*)

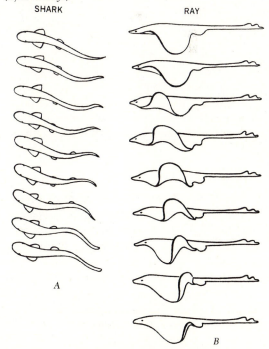

SHARK

RAY

A

B

and in the bottom-dwelling forms the spiracles serve for entry of the respiratory currents of water. This prevents clogging of the gills with bottom debris. The tail usually is slender and little used in locomotion.

Convergent evolution has occurred among some bottom-dwelling sharks and rays. Both groups have "sawfishes" with a long, flat bladelike rostrum margined on both sides by teeth—the saw sharks (Pristiophoridae) and the sawfish rays (Pristidae). The saw is used in defense and for disabling prey. Sawfish rays have a long body and tail, sharklike in both appearance and function. Again, in each group "sand sharks" have evolved that lurk partly buried in sand and forage chiefly on the bottom. The raylike angel sharks (Squatinidae), with flattened body and winglike pectorals, are paralleled by the guitarfishes (Rhinobatidae), rays with a robust sharklike tail used in sculling.

The chimaeras (or ratfishes) are taxonomically distinct from the sharks and rays. They are grotesque, large-eyed fishes of the deep-sea floor (low latitudes) or shallower waters (high latitudes). They are 0.5 to 2 m (2 to 6 ft) long. The skin is scaleless and the upper jaw completely fused with the cranium. Their jaws contain large flat plates. Chimaeras feed on seaweed, invertebrates, and fishes. The big pectoral fins serve in locomotion. Some fossil forms were large and diverse in dentition.

The teeth of elasmobranchs reflect their food habits. Those of sharks are awllike to triangular in shape—often with serrate edges used in cutting, tearing, or slashing. Guitarfishes and most other rays are bottom dwellers; their teeth usually are small, blunt, and pavementlike and used in crushing armored animals.

29-17 Natural history Most sharks and rays are marine, but a few live in tropical rivers and lakes in brackish water or fresh water. Sharks usually are in open waters and rays on the bottom, but the manta and other large rays swim near the surface. Some sharks lie on the bottom. Sharks are mostly predaceous, and their predators are chiefly other sharks. The sharks are active swimmers and usually feed among schools of fish. The smaller species also take

squid and crustaceans, while some of the largest forms may capture seals or sea lions. The big basking and whale sharks, however, feed on plankton. Rays eat various invertebrates (mollusks, crustaceans, etc.). Electric rays stun their prey with electric shocks (Sec. 30-15). The poison-covered spine(s) of stingrays, however, are used in defense.

29-18 Reproduction Elasmobranchs copulate, and fertilization is internal. Males are often aggressive, and some sharks induce females to mate by biting them or, in smaller species, even gripping them with the jaws. Mating is often cyclic, during a brief period, and females of some species usually produce young only in alternate years. Many sharks deposit their eggs or young in special nursery areas (where adults do not feed); some rays enter coastal bays to drop their young. The large eggs undergo meroblastic cleavage and develop slowly (16 to 25 months in the dogfish; 9 to 12 months in the chimaeras). The primitive heterodontid sharks (small bottom dwellers with a spine before each dorsal fin), some dogfishes and rays, and the chimaeras are oviparous, depositing each egg in a brown horny capsule (mermaid purse) of distinctive shape; some cases have tendrils that affix to seaweeds. Most sharks, dogfishes, and rays, however, are ovoviviparous, retaining the eggs for development internally and giving birth to living young. The walls of the "uterus" (oviduct) in the female produce numerous loops of blood vessels that lie against the yolk sac of the embryo and provide for the respiration of the latter.

In embryos of the stingray and electric ray, uterine villi secrete a milky nutritive fluid as yolk becomes depleted. In viviparous forms, such as the blue shark (*Prionace*), hammerheads, and smooth dogfishes (*Mustelus*), a yolk-sac placenta develops and interdigitates with uterine vascular villi that provide food and oxygen. When hatched or born, the young of all sharks and rays resemble their parents.

29-19 Relation to humans Sharks and rays are used as food in many countries, including the United States, but the product may be called grayfish or other names. In the Orient shark fins are dried, then boiled to yield a gelatinous material favored for soups. Shark livers contain much oil rich in vitamin A. Fisheries for this product have long existed in waters about Greenland, Iceland, and Norway and more recently along shores of the United States, but this industry has declined in the United States and elsewhere as shark oil has been replaced by synthetic vitamins. Formerly in the Old World sharkskin was tanned with the scales in place (then termed shagreen) to serve for casing fine books, jewel boxes, and sword handles and as an abrasive for polishing wood or ivory. With the scales crushed, the skins make acceptable leather for shoes and bags.

Sharks are a nuisance to fisherman because they tear nets and steal captured fish and take bait or fish from hooks. Rarely, large sharks may capsize small boats and injure or kill fishermen. Both small and large sharks menace swimmers and skin divers in shallow waters of Australia, East Africa, the Philippines, Japan, Florida to New York, and the California coast. The number of attacks may have decreased in recent years. In 1959, a peak year for attacks reported in modern times, there were 56 cases recorded world wide. The high count may have been partly due to improved surveillance. Some popular beaches are protected by net fences near the shore, and others have lookout guards to warn bathers when sharks approach. The growing popularity of scuba diving is bringing sharks and people together in increasing numbers. Sharks may be attracted to fish killed or injured by spearing. When excited they are reactive to human disturbances (swim fins of a swimmer breaking the surface, etc.) and may attack. Suprisingly, in Hawaii the number of attacks has been small despite many swimmers.

29-20 Fossil forms Earliest of the known fossil vertebrates with jaws are the †PLACODERMI of Devonian to Permian time. The jaws were primitive. Most placoderms had an operculum covering the gills. There were both median and paired fins, but the latter were more variable in number than in later fishes. The first CHONDRICHTHYES appear in late Devonian rocks (later than the first bony fishes) but already were well advanced in evolution of the jaws.

No true bone is recognized in any shark, fossil or living, and the cartilaginous skeleton seems more likely a degenerate than a primitive character. Some early types inhabited fresh waters, but the group has been predominantly marine; they were abundant and often well armored (placoderms) in the Paleozoic era but later decreased in numbers and had less armor.

Classification

Class 1. †Placodermi.

Primitive jawed fishes with transverse articulation between skull and shoulder girdle; full-sized functional gill arch preceded hyoid arch.

Order 1. †Rhenanida. Head covered with few separate plates or tubercles; Devonian.†*Gemuedina,* raylike; †*Jagorina,* sharklike.

Order 2. †Arthrodira. Includes arthrodires. Bony armor of 2 parts, head and gill region jointed to that of body; jaws bony, often massive; small to large; Devonian. †*Coccosteus,* in rivers; †*Dinichthys,* marine, to 6 m (20 ft) long.

Order 3. †Antiarchi. Head and forepart of body in bony depressed shield; hindpart of body tapered, scaly, or naked; pectoral fin long, with jointing of pectoral armor; on bottom in fresh waters; Devonain. †*Pterichthyodes;* †*Bothryolepis,* length to 1 m (3 ft).

Class 2. Chondrichthyes.

Cartilaginous fishes. Living forms with skeleton of cartilage; first gill slit a spiracle; scales minute, placoid; mouth and 2 nostrils ventral; males with claspers; ova few, large, much yolk, cleavage meroblastic; about 1000 species. (Fig. 29-1.)

Subclass 1. Elasmobranchii.

Order 1. †Cladoselachii (Pleuropterygii). Sharklike; pectoral fins broad at base; 2 dorsal fins; no claspers; marine; Upper Devonian to Upper Carboniferous. †*Cladoselache.*

Order 2. †Pleuracanthodii. Dorsal fin along

much of back, notched at base of tail; freshwater; Late Devonian to Triassic. †*Xenacanthus*.

Order 3. Chlamydoselachiformes CHLAMYDOSE-LACHIDAE. Frill sharks. Six pairs of gill slits; uncommon. *Chlamydoselachus anquineus*, frill shark.

Order 4. Hexanchiformes Hexanchidae. Cow sharks. Six or seven gill clefts; 1 dorsal fin; ovoviviparous. *Hexanchus griseus*, six-gill shark.

Order 5. Squaliformes. Two dorsal and one (or no) anal fin; dorsals with or without spines; body typically spindle-shaped; gill slits lateral, five pairs; pectoral fins moderate, constricted at base; swim by use of tail; Lower Carboniferous to Recent; about 250 living species.
HETERODONTIDAE. Horned sharks. Jaws with both crushing and piercing teeth; spines before dorsal fins; *Heterodontus francisci*, horned shark, to 1¼ m (4 ft).
ODONTASPIDIDAE. *Odontaspis taurus*, sand tiger; *O. ferox*, ragged-tooth shark.
LAMNIDAE. Mackerel sharks. *Cetorhinus maximus*, basking shark, to 9 m (30 ft) long, sluggish, feeds on plankton; *Lamna nasus*, porbeagle, to 3 m (10 ft) long, sometimes in schools; *Carcharodon carcharias*, white shark, or man-eater, to 6 m (20 ft), dangerous to swimmers.
ALOPIIDAE. Thresher sharks. Upper tail lobe long, slender. *Alopias vulpinus*, thresher shark, to 6 m (20 ft).
RHINCODONTIDAE. Whale sharks. *Rhincodon typus*, whale shark, to 18 m (60 ft) long, eats crustaceans, small fishes, and plankton.
CARCHARHINIDAE. Requiem sharks. *Carcharhinus*, ground shark; *Prionace glauca*, blue shark; *Mustelus*, smoothhound; *Galeocerdo cuvier*, tiger shark, to 5½ m (18 ft) long, teeth flattened, sickle-shaped, with fluted edge and triangular point projecting outward.
SPHYRNIDAE. Hammerhead sharks. A large eye-bearing lobe at each side of head. *Sphyrna zygaena*, hammerhead, to 4½ m (15 ft) long.
SQUALIDAE. Dogfish sharks. *Squalus acanthias*, spiny dogfish, to 1 to 1¼ m (3 or 4 ft) long, common in shore waters.
SQUATINIDAE. Angel sharks. Resemble rays but gill openings lateral and pectoral fins free from head, swim with tail; live on bottom, eat flatfishes, mollusks, and crustaceans. *Squatina*.
PRISTIOPHORIDAE. Saw sharks. *Pristiophorus* (five pairs of gill slits) and *Pliotrema* (six pairs), resemble *Pristis* but gill slits lateral and pectoral fins free from head; to about 1¼ m (4 ft) long.

Order 6. Rajiformes (Batoidei). Body depressed, pectoral fins enlarged, joined to sides of head and

body; gill slits ventral, five pairs; spiracle highly functional; swim chiefly by flapping or undulating pectoral fins. Upper Jurassic to Recent; chiefly on bottom in shore waters, some down to 2750 m (9000 ft); about 350 species.
PRISTIDAE. Sawfishes. Body sharklike, snout a long flat blade with toothlike scales at sides, used to forage along bottom or slashed from side to side, disabling fishes in a school. *Pristis pectinata*, small-tooth sawfish, to 6 m (20 ft) long.
RHINOBATIDAE. Guitarfishes. Somewhat sharklike. *Rhinobatos*.
TORPEDINIDAE. Electric rays. Pectoral fins rounded, with many branchial muscle cells modified as an electric organ producing severe shocks and used to stun prey (see Sec. 30-15). *Torpedo*.
RAJIDAE. Skates. Diverse; tail slender, swim with broad winglike pectoral fins; feed on mollusks, crustaceans, and fishes on the bottom; over 100 species. *Raja*.
DASYATIDAE. Stingrays. Round and diamond-shaped stingrays. *Dasyatis*.
MYLIOBATIDAE. Eagle rays. Small, tail slender and whiplike, with one to three saw-edged spines containing poison glands that produce ugly, slow-healing wounds, sometimes complicated by gangrene or tetanus, on persons bathing or fishing; live at bottom in shallows. *Aetobatus*.
MOBULIDAE. Mantas, or devilfishes. Width to 6 m (20 ft), swim at midwater levels or near surface, feed on plankton and small fishes caught by sievelike gill plates. *Manta*.
POTAMOTRYGONIDAE. Freshwater stingrays. Restricted to fresh water in South America and Southeastern Asia; little urea in blood. *Potamotrygon*.

Subclass 2. Holocephali.

Order 7. Chimaeriformes. Chimaeras (Fig. 29–2). Gill slits covered by operculum; no cloaca or spiracles; first dorsal fin far forward with strong spine; tail slender, whiplike; upper jaw united with skull; each jaw with large tooth plate; adults scaleless; oviparous. Upper Devonian to Recent. In cold ocean waters, down to 1800 m (6000 ft); 25 species. *Hydrolagus* (*Chimaera*) *colliei*, in North Pacific, from California northward, eats fish, invertebrates, and seaweed.

References

(See also works by Greenwood, Marshall, Nikol' skii, Norman, and others in references to Chap. 30).

Baldridge, H. D. 1974. Shark attack. New York, Berkeley Medallion Books. xi + 236 pp. Paper. *Attacks on humans, antishark measures, emergency action advice.*

Bigelow, H. B., and **W. C. Schroeder.** 1948–1953. Sharks, rays, and chimaeroids. In Fishes of the western North Atlantic. New Haven, Conn., Yale University, Sears Foundation for Marine Research. Pts. 1 and 2, 1164 pp., illus. *Keys, descriptions, distribution, and habits.*

Gilbert, P. W. 1962. The behavior of sharks. *Sci Am.,* vol. 207, no. 1, pp. 60–68. *The role of their senses in feeding behavior.*

_____(editor). 1963. Sharks and survival. Boston, D. C. Heath and Company. 792 pp. *Attacks on humans, anti-shark measures, field and experimental studies, sensory organs of sharks.*

_____, **R. F. Mathewson,** and **D. P. Rall** (editors). 1967. Sharks, skates, and rays. Baltimore, The Johns Hopkins Press. xv + 624 pp. *Collection of symposium papers on elasmobranch biology.*

Lineaweaver, T. H., III, and **R. H. Backus.** 1970. The natural history of sharks. Philadelphia, J. B. Lippincott Company. 256 pp.

McCormick, H. W., T. Allen, and **W. E. Young.** 1963. Shadows in the sea. Philadelphia, Chilton Company–Book Division. xii + 415 pp., illus. *Shark attacks on humans, protective measures, shark fishing, use for food and leather, accounts of principal species, and references.*

Moy-Thomas, J. A., and **R. S. Miles.** 1971. Paleozoic fishes. 2d ed. Philadelphia, W. B. Saunders Company. xi + 259 pp., illus.

Watson, D. M. S. 1937. The acanthodian fishes. *Phil. Trans. Roy. Soc. London Ser.* B, vol. 228, pp. 49-136, pls. 514, figs. 125.

30

CLASS OSTEICHTHYES: BONY FISHES

Many kinds of animals that live in water are called *fish*, from jellyfishes to starfishes, but the word applies properly to lower aquatic vertebrates. The Greeks knew the fishes as *ichthyes*, and *ichthyology* is the scientific study of fishes; the common name *fish* derives from the Latin *pisces*. The most typical or bony fishes have bony skeletons, are covered by dermal scales, usually have spindle-shaped bodies, swim by movements of body and fins, and breathe by means of gills (Fig. 30-1). Various species inhabit all sorts of waters: fresh water, brackish water, or salt water; warm or cold. Fishes have been a staple protein food of humans since antiquity, and many species provide recreation for sports fishermen.

30-1 Characteristics

1 Skin with many mucous glands, usually with embedded bony dermal scales of mesodermal origin (cycloid, ctenoid, sometimes ganoid; Fig. 30-8); some scaleless; a few with enamel-covered scales; both median and paired fins present (some exceptions), supported by fin rays of cartilage or bone.

2 Mouth usually terminal and with teeth; jaws well developed, articulated to skull; olfactory sacs two, dorsal, usually not connected to mouth cavity; eyes usually well developed, no lids.

3 Skeleton chiefly of bone (cartilage in sturgeons and some others); vertebrae many, distinct; tail usually homocercal in advanced forms (Fig. 30-11); relics of notochord often persist.

4 Heart with two pumping chambers (one atrium, one ventricle) with sinus venosus and conus arteriosus, containing only venous blood; four pairs of aortic arches; red blood cells nucleated and oval.

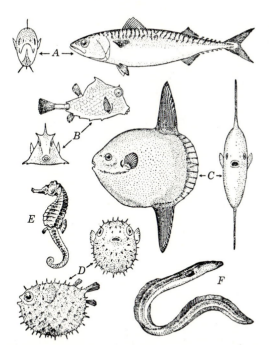

Figure 30-1 Representative bony fishes (class OSTEICHTHYES) of different bodily forms. *A.* Mackerel (*Scomber*), streamlined and fast-swimming. *B.* Trunkfish (*Ostracion*), body rigid, only fins movable. *C.* Marine sunfish (*Mola*), huge, thin, deep-bodied. *D.* Globefish (*Chilomycterus*), body spiny, swollen, fins small. *E.* Seahorse (*Hippocampus*), swims erect by small dorsal fin, tail prehensile. *F.* Common eel (*Anguilla*), long and highly flexible. (*After Norman, History of fishes, Ernest Benn, Ltd.*)

5 Respiration by pairs of gills on bony gill arches in a common chamber at each side of pharynx, covered by a bony operculum; usually with a swim bladder, sometimes with duct to the pharynx, and lunglike in DIPNOI and some others.

6 Brain with optic lobes and cerebellum well developed; ten pairs of cranial nerves.

7 Excretion by mesonephric kidneys; chief nitrogenous waste of larvae is ammonia, of most adults is urea.

8 Body temperature variable (ectothermal); may be metabolically elevated in large, active fishes.

9 Gonads typically paired; usually oviparous (some ovoviviparous or viviparous); fertilization external (some exceptions), eggs minute to 25 mm (1 in) (*Latimeria*), yolk various in amount; segmentation usually meroblastic; no embryonic membranes; early young (postlarvae) sometimes quite unlike adults.

Structurally these fishes are notable for their bony skeletons, thin, flexible scales, complex jaws (Fig. 30-13), and swim bladder and for advances in the structure of the brain. The tail usually is of homocercal or diphycercal form, occasionally heterocercal (Fig. 30-11). The caudal fin, apart from its skeletal support, varies greatly in shape according to habits. Most modern fishes have less bony armor in the scales than did ancient types, many of which were fully armored (with cosmoid or ganoid scales). The head is encased in a true skull made up of both cartilage or replacement bones and membrane or dermal bones. The fins in most bony fishes are supported by many parallel dermal rays, but in crossopterygians each paired fin has a single stout distinct base, supported by bony axial elements that articulate with the limb girdles. Bony parts of limbs of this type probably became the limb skeleton in land vertebrates (Fig. 13-3). The swim bladder is lunglike in some bony fishes, and certain species have accessory structures that enable them to breathe air in shallow mucky water.

30-2 Size The smallest fish appears to be a Philippine goby (*Pandaka*), only 10 mm (0.4 in) long. Most fishes are under 1 m (3 ft) in length. Some record large specimens are halibut at 2.7 m (9 ft), swordfish at 3.7 m (12 ft), white sturgeon at 3.8 m (12$\frac{1}{2}$ ft) and 583 kg (1285 lb), and ocean sunfish (*Mola*) at 907 kg (2000 lb).

Structure of a bony fish: the yellow perch

30-3 External features (Fig. 30-2) The body is spindle-shaped, higher than wide, and of oval cross section for easy passage through the water. The **head** extends from the tip of the snout to the hind edge of the operculum, the **trunk** from this point to the anus; the remainder is the **tail**. The large **mouth** is terminal, with distinct jaws that bear fine teeth. Dorsally on the snout are two double **nostrils** (olfactory sacs). The **eyes** are lateral without lids, and behind each is a thin bony gill cover, or **operculum,** with free edges below and posteriorly. Under each operculum are four comblike **gills.** The **anus** and **urogenital aper-**

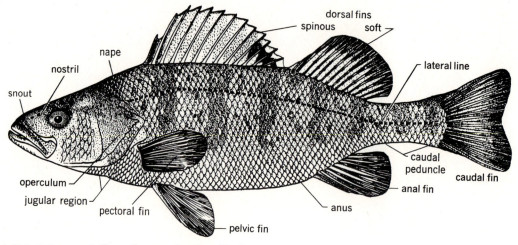

Figure 30-2 Yellow perch (*Perca flavescens*); **external features.**

ture precede the anal fin. Ichthyologists measure the standard length of a fish from the tip of the snout to the end of the last vertebra (*hypural plate*) to avoid error from wear of the tail fin, but fishermen include the fin.

On the back are two separate **dorsal fins,** on the end of the tail is the **caudal fin,** and ventrally behind the anus is the **anal fin;** all these are median. The lateral, or paired, fins are the **pectoral fins** behind the opercula and the ventral or **pelvic fins** close below.

The fins are membranous extensions of the body covering, supported by calcified **fin rays.** The spines are stiff and unjointed; the soft rays are flexible, have many joints, and usually are branched. The fins aid in maintaining equilibrium, in steering, and in locomotion.

30-4 Body covering (Fig. 30-4) A soft, mucus-producing **epidermis** covers the body and protects

Figure 30-3 Yellow perch: general structure. Operculum, pectoral fin, most of the skin and scales, and some trunk and tail muscles have been removed.

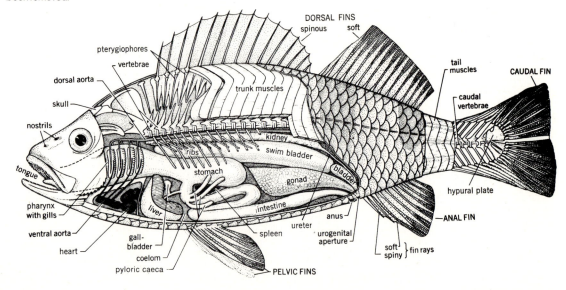

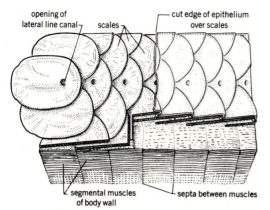

opening of
lateral line canal   scales   cut edge of epithelium
over scales

segmental muscles   septa between muscles
of body wall

Figure 30-4 Body wall of a bony fish (carp) near the lateral line, showing relations of the epidermis, scales, and muscles. (*Modified from Lankester, Treatise on zoology, A. & C. Black, Ltd.*)

against abrasion and entry of disease organisms. The trunk and tail bear thin, rounded dermal **scales,** in lengthwise and diagonal rows, their free posterior edges overlapping like shingles on a roof; each lies in a dermal pocket and grows throughout life. The free portion is covered with a thin layer of skin. The **lateral line,** along either side of the body, is a row of small pores connected to a lengthwise tubular canal mostly under the scales (see Sec. 30-12).

30-5 Skeleton The endoskeleton consists of the skull, vertebral column, ribs, pectoral girdle, and many small accessory bones (*pterygiophores*) supporting the fin rays.

The **skull** comprises the **cranium** housing the brain, the **capsules** for the paired organs of special sense (olfactory, optic, auditory), and the **visceral skeleton,** which provides the jaws and the supports for the tongue and gill mechanism. The skull has a double articulation with the vertebral column and is so firmly joined that a fish cannot turn its head. **Teeth** are usually present on the premaxillary, dentary, vomer, and palatine bones.

In the embryo and young, the cranium is of cartilage, later largely replaced by separate **cartilage bones.** To these are added many **membrane bones** that result from ossifications in the embryonic connective tissue.

During development the visceral skeleton comprises seven paired **arches,** first of cartilage and later ossified; these correspond to the arches of sharks and rays but not to the branchial basket of lampreys. The upper part (*palatoquadrate*) of the **mandibular arch** (no. 1) attaches to the cranium and becomes modified, and each half of the upper jaw is formed of two membrane bones (*premaxillary* and *maxillary*). The lower part of the arch, the primitive layer **lower jaw** (Meckel's cartilage), on each side is supplied with three bones, the dentary and angular (membrane bones) and the articular, the last hinging on the quadrate, which attaches to the cranium. The **hyoid arch** (no. 2) partly supports the tongue. The hyoid is involved in throat movements which enlarge or restrict the size of the buccal cavity. There follow five gill arches, four of which (nos. 3 to 6) bear a gill on the outer curvature and two rows of small nipple- or fingerlike gill rakers on the inner or pharyngeal border. The gill rakers contain many small teeth. When brought together, the rakers interlock, forming a sieve which protects the gills from injury by food and foreign matter. The small last arch (no. 7) has pharyngeal teeth but no gill.

The **vertebral column** is of many similar and separate **vertebrae;** each consists of (1) a cylindrical centrum, concave on both ends (amphicoelous); (2) a small dorsal neural arch over the spinal cord; above this (3) a slender elongate neural spine for muscular attachments; and (4) paired parapophyses, laterally on the centrum. Each caudal vertebra has also (5) a ventral hemal arch, around the caudal artery and vein, and (6) a slender hemal spine below (Fig. 4-4). The column ends in a fan of several small bones supporting the tail, the hypural plate. The vertebrae are bound together by ligaments, and between the centra are relics of the notochord.

Slender paired riblike bones attach to each trunk vertebra, and delicate **intramuscular bones** (*epipleurals*) extend lengthwise between some of the ribs. In the flesh between the spines of the vertebrae are interspinal bones (pterygiophores) that support and articulate the dorsal and anal fin rays.

Each half of the **pectoral girdle** is of several bones (scapular, coracoid, clavicle, etc.), articulated dorsally to the skull and providing attachment for muscles of the pectoral fins. Four short bones with expanded tips (actinosts) and several nodules of cartilage (distal pterygiophores) articulate between the pectoral girdle and rays of each pectoral fin. There is no pelvic girdle; the ventral (pelvic) fins attach to a pubic bone (basipterygium), having a tendon or ligament to the pectoral girdle but no connection to the vertebral column.

30-6 Muscular system The substance of the trunk and tail consists chiefly of **segmental muscles** (myomeres) that alternate with the vertebrae and produce the swimming and turning movements. Fish myomeres are broadly Σ-shaped, in four principal bands, and heaviest along the back. Between successive myomeres are delicate connective tissue septa; when a fish is cooked, these dissolve to leave the myomeres as individual "flakes." The muscles of the fins, gill region, and head are small.

30-7 Digestive system The jaws have many small conical **teeth** to grasp food, and farther back are pharyngeal and gill-raker teeth helpful in holding and crushing it. Mucous glands are numerous, but there are no salivary glands. The small **tongue** is attached to the floor of the mouth cavity and may aid in respiratory movements. The **pharynx** has gills on the sides and leads to a short **esophagus** followed by the recurved **stomach.** A pyloric valve separates the latter from the **intestine.** Three tubular **pyloric caeca,** probably absorptive in function, attach to the intestine. There is a large **liver** anteriorly in the body cavity, a **gallbladder,** and a bile duct to the intestine. The **pancreas** is usually diffuse.

30-8 Circulatory system The two-chambered **heart** lies below the pharynx in the pericardial cavity, an anterior portion of the coelom. Venous blood passes into the **sinus venosus,** to the thin-walled **atrium,** thence into the muscular **ventricle,** all separated by valves that prevent reverse flow. Rhythmic contractions of the ventricle force the blood through the **conus arteriosus** and short **ventral aorta** into four pairs of **afferent branchial arteries** distributing to capillaries in the gill filaments for oxygenation. It then collects in correspondingly paired **efferent branchial arteries** leading to the **dorsal aorta,** which

distributes branches to all parts of the head and body. The principal **veins** are the paired anterior cardinals and posterior cardinals and the unpaired hepatic portal circulation leading through the liver. The **blood** of fishes is pale and scanty compared with that of terrestrial vertebrates. The fluid plasma contains nucleated oval red cells (erythrocytes) and various types of white cells (leukocytes). The **spleen,** a part of the circulatory system, is a large red-colored organ near the stomach. A lymphatic system is also present.

30-9 Respiratory system (Figs. 30-5, 30-6) The perch respires by means of **gills,** of which there are four in a common gill chamber on each side of the pharnyx, beneath the operculum. A gill consists of a double row of slender **gill filaments;** every filament bears many minute transverse plates covered with thin epithelium and containing capillaries between the afferent and efferent branchial arteries. Each gill is supported on a cartilaginous **gill arch,** and its inner border has expanded **gill rakers,** which protect against hard particles and keep food from passing out the gill slits.

In "breathing," the opercula close and the mouth cavity enlarges, drawing water into the mouth. Simultaneously the gill chambers enlarge, producing a lower pressure there than in the mouth, and water flows over the gills. The mouth cavity then contracts, and the passive oral valves close, preventing outflow of water from the mouth; water is forced over the gills. The gill chambers contract, forcing water through the opercular openings. Backflow is prevented by membranes along the ventroposterior border of the operculum. Thus the mouth cavity and gill chamber act alternately as suction pump and pressure pump to maintain a steady flow of water over the gills. Each gill has a double row of filaments which, during respiration, curve outward away from one another and touch the filaments of the neighboring gill (Fig. 30-5D). The direction of blood flow in the gill lamellae is opposite to the flow of water across them, a **countercurrent mechanism** that ensures thorough oxygenation of the blood. Oxygen-poor blood entering the lamellae encounters water with an increasingly higher oxygen content

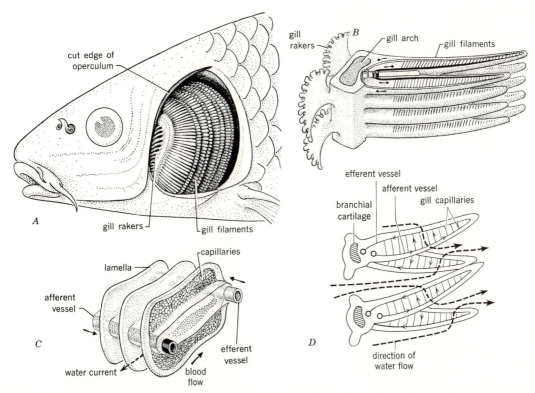

Figure 30-5 Gills of a bony fish (carp). *A*. Gills in the gill chamber with operculum cut away. *B*. Part of a gill showing gill rakers and filaments, with path of blood in the latter; afferent vessels dark, efferent vessels light (in many fishes the gill rakers are slender). *C*. Part of one filament, much enlarged; each lamella contains capillaries where blood is aerated. *D*. Position of gill filaments during respiration. Direction of blood and water flow shown, respectively, by solid and broken arrows. (*Partly after Goldschmidt, Ascaris, Prentice-Hall, Inc.*)

Figure 30-6 The respiratory mechanism of fishes; diagrammatic frontal sections (lobes of oral valve actually dorsal and ventral); arrows show paths of water currents. SHARK. *A*. Water enters ventrally placed mouth, which then closes, and floor of mouth region rises to force water over the gills and out the separate clefts. BONY FISH. *B*. *Inhalent*, opercula closes, oral valve opens, cavity dilates, and water enters. *C*. *Exhalent*, oral valve closes, buccal cavity contracts, water passes over gills in common cavities at sides of pharynx and out beneath opercula. (*Modified from Boas.*)

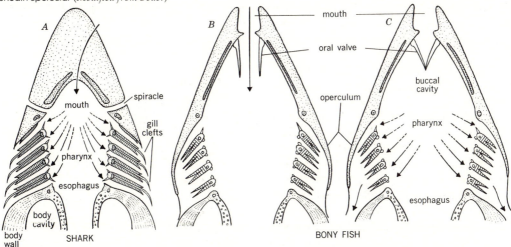

while flowing across to the opposite side. Oxygen thus continues to be picked up throughout the passage of the blood through the gills. The fish needs a constant supply of oxygen-bearing water and soon dies of asphyxiation if removed from water or if the water is depleted of oxygen. The phrase "to drink like a fish" is a misconception of the respiratory process, since most freshwater fishes take fluid into the stomach only with food. Marine forms, however, do drink.

A large thin-walled sac, the **swim bladder** (or air bladder), occupies the dorsal portion of the body cavity. It is connected to the pharynx by a pneumatic duct in some fishes, especially soft-rayed species, but not in the perch or spiny-rayed forms. The bladder is filled with gases (O_2, N_2, CO_2) and acts as a hydrostatic organ to adjust the specific gravity of the fish to that of the water at different depths. By secretion or absorption of the gases through blood vessels in the wall (Sec. 30-22), a fish makes this adjustment slowly as it moves from one depth to another; if a fish is suddenly hauled up from a considerable depth, the greater pressure within the swim bladder may, upon reaching the surface, force the stomach out of the mouth. In many fishes the bladder may aid respiration or serve as a sense organ or in sound production. The swim bladder is lunglike in the lung fishes (DIP-NOI) and a few others.

30-10 Excretory system The two slender dark **kidneys** lie dorsally between the swim bladder and vertebrae. Fluid nitrogenous waste (ammonia, urea) removed from the blood is carried posteriorly from each in a tubular **ureter,** both emptying into a **urinary bladder,** which in turn discharges through the **urogenital sinus** to the exterior. These substances are also diffused outward across the gills. The role of the excretory system in osmoregulation is discussed in Sec. 30-20.

30-11 Nervous system The perch **brain** is short, the olfactory lobes, cerebral hemispheres, and diencephalon being smaller than in a shark and the optic lobes and cerebellum larger. There are 10 pairs of **cranial nerves.** The **nerve cord** is covered by the neural arches and gives off a pair of lateral **spinal nerves** to each body segment.

30-12 Organs of special sense The dorsal **olfactory sacs** on the snout contain cells sensitive to substances dissolved in water. Each sac has an anterior and posterior opening. Water enters the front nostril and flows out the rear, passing over folds of sensory epithelium on the floor of each sac as it does so. **Taste buds** are present in and around the mouth. The large **eyes** probably focus clearly only on nearby objects but can detect moving objects at greater distances, including those above water, such as a person walking on the bank. Each internal **ear** contains three semicircular canals and **otoliths** serving the sense of equilibrium. The **lateral line system** (Fig. 30-7) includes the lateral line and various extensions on the

Figure 30-7 Body wall of carp in longitudinal section, showing the lateral line sensory system; compare Fig. 30-4. (*Modified from Lankester, Treatise on zoology, A. & C. Black Ltd.*)

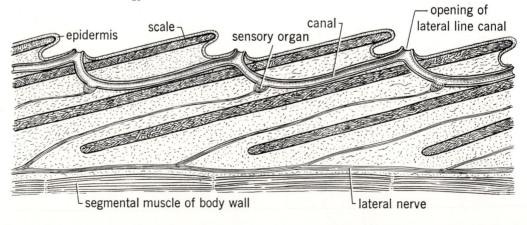

head and evidently detects slight changes in pressure or slow wave or current movements (not over six per second) such as might be experienced by a fish swimming close to some solid object from which water movements would be reflected (distant touch orientation). The presence of other fish, prey and predators and underwater flow patterns also are probably detected. The lateral line system contains mechanoreceptor hair cells (Sec. 9-11) in open canals or tunnels on the body surface. It is innervated by the lateralis branch of the Xth cranial nerve and is considered to be homologous in land vertebrates to the inner ear.

30-13 Reproductive system   In a **male** the two **testes** enlarge greatly in the breeding season, and at mating the milt, or sperm, passes in a **ductus deferens** from each to emerge from the urogenital aperture. In a **female** the eggs pass from the two united **ovaries** through the **oviducts.**

**Structure of other
bony fishes**

All bony fishes are enough alike in general form and structure to be recognized as members of the class OSTEICHTHYES, but they differ among themselves in many details. Many have the general shape of the perch; the flounders, soles, and some tropical reef fishes are thin-bodied, the eels are slender, and the porcupine fishes are globular.

30-14 Scales   Most bony fishes are covered with **scales** (Fig. 30-8). While usually thin and overlapping, the scales are separate and minute on eels, small and tuberclelike on some flounders, and slender spines on porcupinefishes. The overlapping scales of tarpon are up to 5 cm (2 in) in width; some fishes are scaleless or naked (some catfish). The scales are usually bony. On the perch and many other fishes the exposed hind part of each scale has many tiny spines, making a **ctenoid scale.** Scales that lack such spines are termed **cycloid** scales, and still other fishes have **ganoid** scales, which are bony and capped with ganoin, a hard, glassy, enamellike substance. Ganoid scales were characteristic of the early

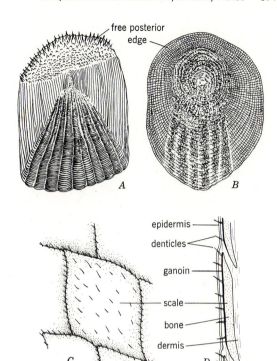

Figure 30-8 Scales of bony fishes, enlarged. *A.* Ctenoid (with fine teeth). *B.* Cycloid. *C, D.* Ganoid (*Lepisosteus*) in surface view and vertical section.

ray-finned fishes (palaeoniscids and holosteans) but in living teleosts are present only in the bichirs (*Polypterus*), reedfish (*Calamoichthys*), and garpikes (*Lepisosteus*). The heads and bodies of various fishes, living and fossil, are armor-plated with large stout scales, as in the trunkfish; living species so protected are usually small or sluggish.

The scales grow throughout life, increasing in size with the fish. There is no molt of the body covering, although occasional scales may be lost and replaced. In many species growth results in the formation of concentric rings of new scale substance along the scale margin. In spring and summer the rings are usually well separated, but as growth slows or stops in winter they become fewer and closer together, forming a definite "winter line" (Fig. 30-9). The rings are most noticeable in fishes from temperate regions. Physiologic demands of spawning may then cause resorption of parts of the scale margin. If growth then resumes, the junction between old and new scale substances is irregular, and a "spawning mark" is formed. These features make age determination pos-

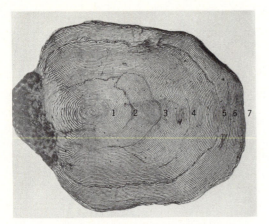

Figure 30-9 Scale of 3-kg (7-lb) female rainbow trout, showing concentration of winter growth rings (numbered), which indicate age in years. (*Enlarged photograph, courtesy of C. McC. Mottley.*)

sible in salmon, trout, bass, and others. (Otoliths also yield age data in fishes.) This provides important data on the rate of growth, age at spawning, and other features for some food and game fishes. The scale pattern is rather constant in a species; thus the structure, form, number, and arrangement of scales are of much value in classification.

30-15 Electric fishes There are about 250 species of electric fishes, including (1) the sea-bottom-dwelling skates and rays and (2) raylike torpedos, (3) stargazers (marine teleosts), (4) the South American knife fishes and the electric eel, (5) African snout fishes (mormyrids and gymnarchids), and (6) the African catfish. The torpedo rays, some stargazers, the electric eel, and the African catfish can produce violent shocks; discharges of the others are much weaker. Electric organs are modified muscle fibers, disklike multinucleate plates arranged in stacks like batteries in series and embedded in a gelatinous matrix. A large torpedo ray may have half a million plates. Muscles in different parts of the body may be modified—trunk and tail in the electric eel, hyobranchial apparatus in the electric ray, and extrinsic eye muscles in the stargazers.

Powerful electric organs emit pulses in brief bursts several times a second; the weaker ones discharge in an uninterrupted rapid series (300 per second in *Gymnarchus*). An oscillating electric field is

formed around the fish, and skin receptors inform the animal of disturbances in the field. In mormyrids the receptors are **mormyromasts,** modified lateral line organs, located at the bottom of jelly-filled pits. In elasmobranchs the ampullae of Lorenzini are electroreceptors. Some electric fishes are highly sensitive to disturbances of their electric fields, and others (*Gymnarchus*) can accurately locate disturbances at a great distance.

Strong electric organs serve to stun prey and ward off predators. A large torpedo ray (which may reach 2 m (6 ft) can produce a shock of more than 200 volts, capable of stunning a human. The weaker electric organs of mormyrids and knife fishes evidently are used to avoid obstacles or predators, find food, and detect others of their species. Electric fishes typically hold the body straight when swimming, which may be important in keeping the electricity-generating and -detecting systems aligned. Many have weak eyes and live in turbid water or other locations where visibility is poor.

These animals are an outstanding example of convergent evolution. The electric organs have evolved from muscles in different parts of the body among divergent groups of fishes.

30-16 Coloration Bony fishes are variously colored by pigment cells, or **chromatophores,** in the dermis, either external to or beneath the scales. Pigments of black, yellow, orange, and red are present in different species. Combinations of these result in green (black + yellow), brown (red or yellow + black), and other colors. In other cells (iridocytes), reflecting crystals of **guanine** and hypoxanthine provide "structural" colors resulting from interference phenomena. The crystals, often in stacks or platelets, reflect light of different wavelengths according to their width and that of the cytoplasmic layers which separate them. When these layers are about the same width as the crystals, light of long wavelength, such as red, is reflected; when they are less wide, reflection is of shorter wavelength such as yellow or green. The silvery sheen common in fishes is caused by the overlapping of layers of platelets that reflect different parts of the spectrum. A wave band not reflected by one layer may penetrate more deeply to be reflected by another. Together the layers may reflect light of full spectral range.

Many fishes are concealingly colored. Illumination under water is nearly constant in a horizontal plane and brightest overhead, a situation ideal for the functioning of countershading (Sec. 33-16). Orientation of the reflectors described above may contribute to this form of concealment. They are distributed from dorsum to belly of the fish, more or less parallel to one another in a vertical plane; thus in a fish in upright position the reflective surfaces are directed laterally. An observer approaching from the side receives light reflected mirrorlike from the fish's flanks. If this light equals in intensity that from the background—and it often does—the fish is very difficult to see. From above, only the edges of the reflectors are presented. Dark pigment on the back absorbs descending light, and the fish blends with the depths. From below, however, the fish is seen against a lighted surface, and even a silver-bellied fish may appear as a dark silhouette. Some fishes have specialized muscles (resembling optical fibers) that conduct the light striking their bodies from above to the ventral surface, apparently for the purpose of lightening the silhouette. The pony fish (*Leignathus equulus*) of the Indo-Pacific region has gone even further. Light from symbiotic luminescent bacteria living in a special chamber opening into its esophagus can be projected into the large swim bladder, where a silvery layer of guanin in the upper lining reflects the light downward through the fish's translucent belly. Light intensity is controlled by a movable flap at the esophageal opening.

Certain fishes change color by concentrating or spreading pigment in the chromatophores, slowly or rapidly. Some flounders can match the sea bottom on which they rest (Fig. 30-10). The stimulus to change is received through the eyes, as fish artificially blinded do not respond like normal ones. Some fishes, such as brook trout, undergo little or no change in pattern or color but can alter in shade. Fishes living in the perennial darkness of the deep sea are often black.

30-17 Fins The fins are diverse in form, size, and placement. The pectoral fins are usually near the gill apertures; the pelvics are on the abdomen in trout, near the gill openings on perch, on the throat (jugular) in blennies, and lacking in eels. The dorsal fin may be single, multiple, or continuous along the back; salmonids and catfishes have, besides the single dorsal, a small fleshy or adipose fin posteriorly. In the mosquito fish and other viviparous species, the anterior part of the anal fin is modified as a copulatory organ (Fig. 30-27).

The shape of the caudal fin varies greatly (Fig. 30-11). Fishes with sickle-shaped or forked tails

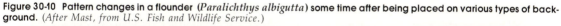

Figure 30-10 Pattern changes in a flounder (*Paralichthys albigutta*) some time after being placed on various types of background. (*After Mast, from U.S. Fish and Wildlife Service.*)

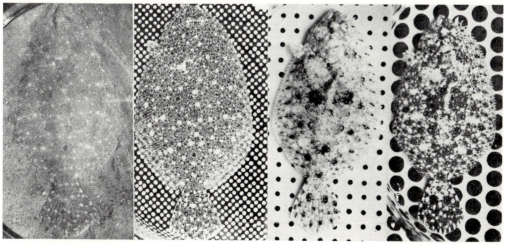

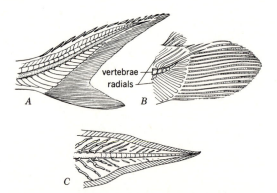

Figure 30-11 Types of tails on bony fishes, showing relation of the vertebrae (or notochord) and radials to the caudal fin. *A*. Heterocercal (sturgeon, *Acipenser*). *B*. Abbreviate heterocercal (bowfin, *Amia*). *C*. Diphycercal, equal lobes (lungfish). The homocercal tail, common to many fishes, is shown in Fig. 30-2.

usually are capable of sustained high speed, whereas those with squarish or rounded tails swim more slowly or in bursts and are more maneuverable.

30-18 Sounds Fishes produce sounds in several ways, depending on the species. They may move the fins against parts of the body, grind the pharyngeal teeth, release gas from the duct of the swim bladder, or vibrate its wall with special muscles. Sounds may serve to assemble individuals for breeding or feeding. In some species they are emitted during aggressive and defensive displays and may be important in territorial behavior.

In many species sounds seem to be transmitted through the body to the inner ear. In some teleosts, however, the inner ear is coupled to the swim bladder, the walls of which are vibrated by underwater sound waves somewhat as in a hydrophone. The coupling may be direct, with a stethescopelike connection to the ear capsule (tarpon, deep-sea cods, herrings), or indirect through several small bones, the Weberian ossicles (characins, carplike fishes, and catfishes).

Natural history

30-19 Distribution Fishes occur from the polar seas to the equator, from the surface to depths of more than 9000 m (30,000 ft) and in lakes at over 4500

m (15,000 ft) in the Andes. They live variously in open water, on sandy, rocky, or muddy bottoms, in crannies of reefs, in saline bays and estuaries, in fresh or alkaline rivers and lakes, in cave waters, and even in hot springs. Some live deep in lakes or the sea, others in polar waters [ice fishes, CHAENICHTHYIDAE of the Antarctic, lack red blood cells and live at −1°C (30 to 31°F)]. Others are restricted to shallow and warm tropical seas, and some pupfish live in springs up to 34°C (93°F) on the American deserts. Trout thrive in waters at 13 to 21°C (55 to 70°F). Most fishes are intolerant of temperature fluctuations beyond a span of about 9 to 11°C (12 to 15°F). The majority live either in fresh water or in salt water, but sticklebacks, sculpins, and others are at home in either sort. Many open-water marine species range widely in the oceans, but shore fishes are of restricted distribution, so along the coasts of the continents and in inland waters there are various definite fish faunas.

Many fishes, like birds, perform seasonal **migrations.** The barracuda and swordfish move north in spring and south in autumn, in a latitudinal migration. Tunas migrate across the Pacific from California to Japan. Various species migrate from salt water to fresh water for spawning, and many (eels, some gobies) do the reverse. The salmon, shad, striped bass, and some trout are examples of the former and are termed **anadromous;** the freshwater eel, which reverses this process, is said to be **catadromous.** Salmon make their way back to the stream where they were hatched by the "smell" of that stream, perhaps imprinted on them during embryonic growth. They appear to use celestial navigation in moving landward from the open sea. Cod and herring of the ocean perform an inshore migration to spawn on banks or shoals. Some deep-water fishes have a daily vertical migration.

30-20 Osmoregulation Fishes face special problems of osmoregulation, particularly species that move between salt and fresh water. In freshwater forms the salt content of blood and tissues is considerably higher than that of the water in which they dwell, hence they absorb water. They do so through their gills and the mucous membranes of the mouth and pharynx. Excess water is eliminated by the kidneys, and copious dilute urine is produced (commonly 10 times or more as much per kilogram of body weight per day than in saltwater species). Some salts are lost through the gills and in urine and feces but is replaced as needed by absorption from food and from the water bathing the gills and buccopharynx. Freshwater species do not drink.

In saltwater fishes, on the other hand, the body salts are at a lower concentration than seawater, and these fishes lose water through their gills and buccopharyngeal surfaces. This dehydration process is countered by drinking large quantities of seawater and eliminating the excess salts (sodium, potassium, chloride) acquired via the kidneys, gills, and perhaps other parts of the body. Saltwater fishes excrete very concentrated urine in small amounts, thereby conserving water. The fact that some marine fishes lack glomeruli or have vestigial glomeruli (deep-sea angler fish, various pipefishes, etc.) is evidently related to the reduced urinary output. Certain cells in the gills of both freshwater and saltwater fishes are specialized to take up or excrete chlorides. The "chloride" cells are especially well developed in **euryhaline** fishes (flounder, three-spined stickleback, etc.), which have a wide salinity tolerance. These fishes may be contrasted with **stenohaline** species that are restricted to either fresh or salt water. Some fishes (salmon, lampreys) that migrate between salt and fresh water are adapted in various stages of their life history to one or the other environment. The gills of both freshwater and marine fishes, in addition to their role in salt balance, also excrete the simpler products of nitrogen breakdown such as ammonia and urea. In some species (goldfish, carp) such excretion may be six to ten times the amount of the kidneys. See Sec. 29-11 for information on osmoregulation in elasmobranchs.

30-21 Habits   Skin divers, wearing air tanks (scuba gear) for prolonged submergence and equipped with cameras and powerful lights, are learning much about the habits of fishes. Some species are active at all hours, many are quiet all night, and a few are nocturnal. Some wrasses sleep lying on their sides. Fishes of freshwater streams habitually head against the current to maintain position, facilitate respiration, and catch food, as may be seen in trout in a brook or hatchery. Marine species are active at all seasons, but many freshwater types become inactive during the winter, descending to deeper water in lakes and rivers, and the carp engages in anaerobic metabolism. Some fishes are solitary, whereas others are gregarious and live in schools of various sizes, those of herring numbering a few thousand to many million individuals.

30-22 Locomotion   Most fishes swim by lateral undulations of the body and tail, which provide a backward thrust against the water. The movements are produced by alternate contractions of the muscles on the two sides. The other fins serve chiefly to maintain balance and change direction, but some fishes use the dorsal or anal and sometimes the pectorals for swimming. The pelvic fins and sometimes the pectorals may be used as brakes. The streamlined body of many fishes is efficient for rapid locomotion, and their slimy surface has been shown to greatly reduce water friction.

The center of gravity of a fish is usually in the swim bladder. The bladder functions as a hydrostatic organ which makes it possible for a fish to remain suspended in a state of weightlessness at any depth without swimming to do so. Energy is conserved, and fin action can be directed toward swimming and maneuvering. In ascending and descending, gas pressure in the swim bladder would be expected to decline or rise with changes in water pressure, the volume of the bladder increasing or decreasing. In fact, however, there are compensatory mechanisms. Many shallow-water fishes have a duct connecting the swim bladder to the gullet and adjust bladder pressure by releasing gas into the water or gulping air at the surface. In the many fishes (two-thirds of all teleosts, including many marine forms) that have a closed swim bladder, pressure is reduced through vascular **resorption** of gas by the "oval" in the upper rear part of the bladder and is increased through vascular **secretion** of gas by the "gas gland" in the lower front part of the bladder.

The gas gland has a countercurrent (Sec. 30-9) system of blood vessels (*rete mirabile*). Outgoing blood from the bladder has a high concentration of gases, incoming blood much less. Juxtaposition of vessels permits the incoming flow to pick up gas released (apparently by a lactic acid mechanism) from the outgoing blood and to build the high gas pressure needed to recharge the bladder. The swim bladder is reduced or absent in many bottom-dwelling fishes (flatfishes, clingfishes, blennies) and species in fast streams that engage in bottom-hugging dashes.

Trout, salmon, and other fishes may jump or leap from the water at times, when in pursuit of prey; they swim rapidly up to the surface where momentum alone carries them into the less dense air. The halfbeaks (HEMIRHAMPHIDAE) skitter along the surface, propelled by the tail, which remains submerged. Flying fishes (EXOCOETIDAE; Fig. 30-26) actually leave the water to glide. They swim rapidly, then "taxi" with the body above the surface while the submerged tail moves laterally up to 70 times per second (the pectoral fins may vibrate as a result of bodily movements); the speed may be up to 10 m/s (22 mi/h). Then the broad pectoral fins are instantaneously extended and the fishes rise and glide for up to 20 s, traveling 20 to several hundred meters, helped occasionally by tail action as the ventral lobe of the caudal fin touches the water. The small freshwater hatchetfishes (GASTEROPELICINAE) of Panama and tropical South America make flights of 1.5 to 3 m (5 to 10 ft) on vibrating pectoral fins.

Some fishes are able to move about on land and to respire for long periods (many hours to days) out of water. The climbing perch (*Anabas*), walking catfishes (*Clarias*), and the mudskippers (*Periophthalmus*) of tropical and subtropical shores are examples (Fig. 30-12). Mudskippers emerge from their burrows on mud flats at low tide and forage for small invertebrates in the exposed mud. They hitch forward by simultaneous movements of their pectoral fins, resembling a person using crutches, and keep their gills moist by gulping water from tide pools. Their terrestrial habit allows them to exploit a food source unavailable to strictly aquatic fishes. The walking catfish (*Clarius batrachus*) uses its head, pectoral fins, and slithering movements of its body to move on land, and when out of water breathes with modified lung-like gills. Individuals of this Asian

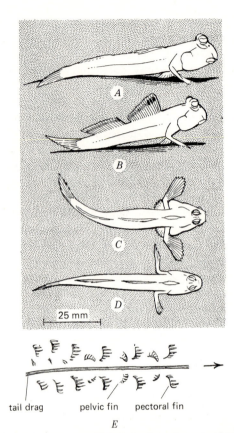

Figure 30-12 **An amphibious goby, the mudskipper (*Periophthalmus sobrinus*).** *A.* On land, poised to pounce on prey (small invertebrates). *B.* Dorsal fins elevated in aggressive display. *C.* Position of pectoral fins when walking, gill chambers distended with air and water; viewed from above. *D.* Water expelled from gill chambers. *E.* Track in mud.

species were released by a tropical fish dealer in Boca Raton, Florida, in 1968. The species expanded rapidly, displacing valuable native fishes and now ranges from Lake Okeechobee to Miami. Cold weather in recent years has fortunately reduced its populations.

30-23 Food A few freshwater fishes and some marine species eat aquatic vegetation, and the carp and suckers use bottom debris containing algae and minute invertebrates. Most fishes, however, even the young of herbivorous species, are predaceous, feeding on small aquatic animals or other fishes, and a few larger kinds may capture birds or mammals.

Methods of feeding are varied. Many seize or surround their prey with the jaws. Others suck food in by expanding the mouth and opercular chambers. Anchovies swim with mouth open and strain small organisms from the water with fine gill rakers. Angler fish engulf prey by suddenly opening the mouth. The jaws of fishes are the most complex among vertebrates. By movements of the premaxillary and maxillary bones, many species can protrude the mouth (Fig. 30-13). This action aids procurement of food by suction, encircling food with the jaws, or bottom foraging without steep tilting of the body axis, a position which would delay escape. Being ectothermal, most fishes use little or no energy to maintain the body temperature through internal heat production. Some can fast for long periods, using stored body fat and even proteins. A sunfish (*Lepomis*) takes in about 5 percent of its body weight in food per day in summer but less than 1 percent in winter.

30-24 Reproduction Most Temperate Zone fishes breed in spring or early summer, but some trout in autumn or winter. Tropical freshwater species breed mainly in the rainy season when streams rise or overflow, but a few "annual fishes" spawn in

Figure 30-13 Mouth protrusion in a sunfish (*Lepomis*), showing jack-in-the-box action of jaws.

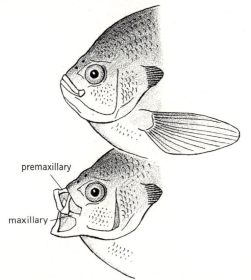

premaxillary

maxillary

drying ponds, and the eggs survive under the hard mud crust until the next rainy season. A majority of fishes have one annual brood, but mosquitofish (*Gambusia*) can produce up to six.

Fishes are mainly oviparous, but many individual species and all members of some groups are viviparous. Mosquitofish and the viviparous surfperches (Embiotocidae) bear but few young at a time, whereas most oviparous fishes produce many eggs. Brook trout lay 80 to more than 5600, according to size, Atlantic salmon up to 17,000, the cod to over 6 million, and the ocean sunfish to 300 million. The bodies of females are often swollen with the maturing eggs just before laying. Changes in coloration or other features, such as the hooked jaws of salmon, may occur in males during the mating season. Courting performances precede actual spawning among various kinds of fishes. Some freshwater fishes provide nests; the "redds" of trout and salmon are shallow depressions, cleared of fine debris and covered with gravel or sand after egg deposition. The male stickleback (*Gasterosteus*) makes a globular nest of fine vegetable fibers bound together with a sticky secretion from the kidneys; he later guards the eggs and small young. The viviparous surfperches (Fig. 30-31) and mosquitofish retain eggs to develop in the ovaries; a male seahorse (*Hippocampus*) carries the eggs in a brood pouch; and males of some marine catfishes and African freshwater cichlids carry the eggs in the mouth. Many bony fishes, marine and freshwater, provide some parental care.

The eggs usually are small, and the time required for development varies with the species and the water temperature. Eggs of some marine fishes hatch as small transparent larvae within a few hours, and those of most tropical freshwater species hatch in 20 to 48 h. By contrast, the eggs of brook trout need about 44 days at 10 to 11°C (50 to 52°F) and 90 days or more at below 4.5°C (40°F) (Fig. 30-14).

The young of some viviparous fishes resemble their parents when born; other young are larval on emergence and gradually assume the adult form. Flounders and soles (Heterosomata) when hatched are bilaterally symmetrical, with an eye on each side, but at an early age one eye begins to "migrate" to the opposite side, which thenceforth is the upper surface as the fish lies on the bottom. Young of the mos-

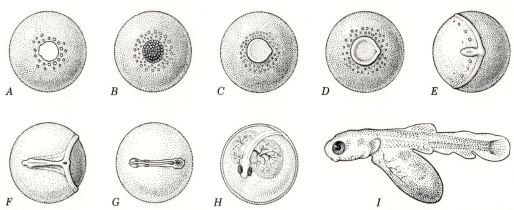

Figure 30-14 Early development of the trout, a bony fish. *A.* Germinal disk (white) concentrates after fertilization. *B.* Meroblastic segmentation. *C, D.* Gastrula forming, blastodisc elongates in axis of future embryo. *E.* Primitive streak begins, blastodisc spreads. *F.* Neural tube forms, blastoderm surrounds yolk. *G.* Embryo with eye and ear vesicles and 18+ somites. *H.* "Eyed egg" — embryo with large eyes; blood vessels spread over yolk. *I.* "Yolk sac" stage of hatched young. (*A–G, after Henneguy. 1888.*)

quitofish may mature and breed before 4 months of age, whereas the Chinook, or king salmon, of the Yukon River requires 5 or 6 years before spawning. Individuals of the Pacific salmon (*Oncorhynchus*) grow to sexual maturity, breed once, and then die.

30-25 Relation to humans Fishes have been important as human food from the earliest times, as evidenced by fishbones found in Paleolithic kitchen middens, to the present day. World fisheries take approximately 70 million metric tons annually of marine and freshwater fishes (figures for 1974 to 1975) (over 56 billion kg, or 124 billion lb, in 1968) and employ thousands of persons, but some fisheries are now declining because of overexploitation and environmental deterioration. The flesh of most fishes is white (or reddish) and flaky in texture. It contains 13 to 20 percent protein and has a food value of 660 to 3530 cal/kg (300 to 1600 cal/lb), depending on the oil content (to 17 percent in salmon). Fishes deteriorate rapidly after being caught and must either be consumed soon or preserved. In the fresh state they are iced or frozen. Methods of preservation include drying, smoking, and canning. Crude fish oils are used in paints and insecticidal sprays, and refined oils from livers of cod and other species provide a concentrated source of vitamin D. The scraps from canneries as well as entire fishes of some species not of interest to fishermen are ground and

dried into meal. In varying forms this is used as human food, for feeding pets (dogs and cats) and poultry, and for fertilizer. Liquid glues are rendered from heads and trimmings of fishes. Flesh of some tropical fishes is poisonous at times. The pufferfishes (TETRAODONTIDAE) contain a potent nerve poison (tetrodotoxin), evidently identical with the poison in newts (*Taricha*). The toxin occurs chiefly in the liver and roe.

Sportfishing is an outdoor recreation for thousands of persons and also a source of food. Much money is spent each year by anglers in pursuit of trout, salmon, perch, bass, and other game fishes. In the United States the federal and state governments rear millions of trout and other fishes in hatcheries and plant them in streams and lakes to replace some of those taken by fishermen. Pituitary implants (Sec. 8-12) are now used to induce full and prompt spawning in some hatchery fishes.

Captive fishes of many kinds in ponds and aquaria are kept and bred by fish fanciers and other persons, and many public institutions maintain large glass-fronted aquaria where both native and foreign fishes are displayed. Fish hobbyists have increased in recent years and spend much on their captives. The keeping of fishes in ponds is an ancient practice of the Oriental peoples, the Romans, and the Aztecs of Middle America. Artificial rearing was on record in Europe in the fourteenth century and is now a widespread practice. Intensive pond culture of fishes, especially carp in Central Europe and the Orient, supplies considerable protein to human populations in those regions. *Tilapia* is now widely cultured in the tropics and elsewhere and has spread through accidental introduction to the detriment of the native fish faunas. Catfish are cultured in rice fields of the United States. Experiments in the United States and elsewhere have shown that addition of natural or chemical fertilizers to fish ponds will increase the diatom-algae-invertebrate food chain upon which fishes depend, and 113 kg (250 lb) or more of fish may be produced in an acre of water annually.

Mosquitofishes have been propagated and distributed widely to aid in control of mosquitoes and malaria by devouring mosquito larvae. In some areas, however, they have decimated native fishes through competition and feeding on their young.

Of particular concern are the numerous introductions of nonnative fishes throughout the world, some between continents. Introductions have been made for food and sport, but many have been accidental—escapes from culture ponds or survival of fish used as bait. Many of these introductions are undocumented and were made before study of local fish faunas. It is now impossible to determine the evolution and natural distribution of many species.

30-26 Fossil fishes   Bony fishes probably derived from primitive Silurian ancestors with paleoniscoid ganoid or cosmoid scales. The oldest stocks, of predaceous habits, were dominant and widespread through Paleozoic time and continued through the Mesozoic. They lived in freshwater streams and lakes. Today they are represented only by the African bichir (*Polypterus*), the spoonbill (*Polyodon*) of the Mississippi Valley, and the sturgeons (ACIPENSERIDAE) of the Northern Hemisphere.

Early in the evolution of bony fishes (Devonian times), two major lines became established: fishes with fleshy fins, the SARCOPTERYGII, and those with ray fins, the ACTINOPTERYGII.

The earliest recognized modern fishes (†SEMIONOTIDAE, Triassic and Jurassic) were slow-swimming bottom dwellers with ganoid scales. The family gave rise to various offshoots, most of which in turn

became extinct during the Cretaceous except for the bowfins (*Amia*) and garpikes (*Lepisosteus*), and the mooneyes (e.g., *Hiodon*) still living in North America. Most modern families and a few existing genera probably date from the Eocene epoch (Fig. 30-15). Fossil fish remains of later periods often include traces or impressions of soft parts.

The order DIPNOI (lungfishes) flourished during the Paleozoic and dwindled in the Triassic, but in the latter period species resembling the living *Neoceratodus* of Australia were of almost worldwide occurrence. The five living species of lungfishes in Australia, Africa, and South America are the closest living relatives of the amphibians but are off the main line of amphibian descent (Fig. 30-17).

Of the crossopterygians, the suborder RHIPIDISTIA (†OSTEOLEPIDAE) of the Devonian and Carboniferous, with rounded lobelike fins, simple teeth, and bodies covered by ganoid scales, resemble the fish stock that probably gave rise to the amphibians. The suborder COELACANTHINI exhibits a most remarkable time range, occurring almost unchanged from Lower Carboniferous to Cretaceous. More remarkable, a 1.5-m (5-ft) living coelacanth (*Latimeria chalumnae*) was taken off East London, South Africa, on Dec. 22, 1938. By 1977, 88 specimens had been obtained. All but the East London animal have been taken near the shores of Grande Comore and Anjouan Comore islands in the Comoro Archipelago near Madagascar, where the fish has long been known to the native people. They have been caught with hand lines at moderate reef depths averaging about 100 m and as shallow as 35 m. Noteworthy

Figure 30-15 Fossil fish (†*Priscacara,* family POMACENTRIDAE) from Eocene Green River shale, Kemmer, Wyo. (*Photo by T. I. Storer.*)

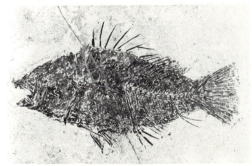

features of *Latimeria* are its cartilaginous skeleton, tubular vertebrae, small spiracles, obscure lateral line, vestigial swim bladder, absence of internal nostrils, single ventral kidney, rectal gland (shared with sharks and rays), and its ovoviviparity.

Classification[1]

Class Osteichthyes (Pisces).

Bony fishes. Skeleton more or less bony; skin usually contains embedded dermal (bony) scales, of cycloid, ctenoid, or ganoid form; both paired lateral and median fins usually present, supported by fin rays; no pelvic girdle; mouth usually terminal; gills always present, not in separate clefts; many with swim bladder; heart with 2 pumping chambers; no cloaca; ova usually small but to 21 mm (0.8 in) in marine mouth-breeding catfish, segmentation meroblastic; Devonian to Recent, in salt water, brackish water, and fresh water; probably about 30,000 species.

(Note: Of the 42 orders, most fossil and a few small living orders are omitted)

Subclass 1. †Acanthodii. Body armor of many small diamond-shaped bony (ganoid) scales; each fin except caudal preceded by a spine; small fishes, freshwater or marine; Devonian to Lower Permian. †*Climatius*; †*Acanthodius*. (Relationship uncertain.)

Subclass 2. Sarcopterygii. Nostrils usually connect to mouth cavity; each paired fin with large median lobe containing jointed skeleton and muscles, fringed with dermal rays (Fig. 30-16), caudal fin with epichordal lobe (a lobe situated above body axis); typically 2 dorsal fins (fused in modern lungfishes).

Order 1. Crossopterygii. Lobe-finned fishes. Premaxillae and maxillae present; teeth well developed, conical; predatory, fish-eating.

Suborder 1. Rhipidistia †Osteolepidae. Head normal, skull bony; teeth on both premaxillae and maxillae; a pineal eye; tail heterocercal. Devonian to Carboniferous. †*Osteolepis*, †*Eusthenopteron*. Suborder from which amphibians probably arose.

---

[1] *Divisions within the infraclass* Teleostei *are groupings recognized by Greenwood et al., 1966; other higher categories are mostly as treated by Romer, 1966.*

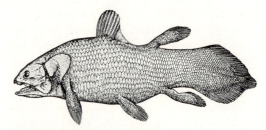

Figure 30-16 Coelacanth (*Latimeria chalumnae*, suborder Coelacanthini); length to 1.5 m (5 ft). (*After A. S. Romer, The vertebrate body, W. B. Saunders Company.*)

Suborder 2. Coelacanthini. Head short, deep; skull with much cartilage; teeth only at tips of premaxillae and dentary; choanae apparently absent; no pineal eye; vertebral column not upturned, tail 3-lobed; Devonian to Cretaceous. †*Bunoderma*, †*Macropoma*. Also Recent: *Latimeria chalumnae* (Fig. 30-16), Comoro Islands and off southeastern Africa; to 1.5 m, 82 kg (5 ft, 180 lb); preys on other fishes.

Order 2. Dipnoi. Lungfishes (Fig. 30-17). Body long, slender; no premaxillae or maxillae; 3 pairs of hard tooth plates on palate and edge of lower jaw; paired fins narrow; swim bladder lunglike; food mainly invertebrates; Devonian to Recent, 7 fossil families; lungfish burrows known from Permian.

Ceratodontidae. †*Ceratodus*, Triassic to Cretaceous, almost worldwide; *Neoceratodus forsteri*, barramunda, only living species; scales large, cycloid, overlapping; fins leaflike; lung not bilobed; length over 1.2 m (4 ft); in Burnett and Mary Rivers and transplanted to lakes and reservoirs, Queensland, Australia; inhabits quiet pools that become stagnant

Figure 30-17 Lungfishes (order Dipnoi). *A.* Australian (*Neoceratodus*), length to 1.8 m (6 ft), but most under 1.2 m (4 ft). *B.* South American (*Lepidosiren*), length to 80 cm (32 in). *C.* African (*Protopterus*), length to 1 m (3 ft). (*After Norman, Guide to fish gallery, British Museum.*)

during dry season, when the fish rises to surface and takes air into lung; cannot survive out of water.

LEPIDOSIRENIDAE. Body eellike, scales small, fins filamentous; lung bilobed; inhabit swamps; in dry season each retires to vertical burrow (nest) in mud, lined with mucus; eggs laid in nest, guarded by male. *Lepidosiren paradoxa*, plains of South America, length to 1 m (3 ft), makes burrow in mud under water; *Protopterus annectens* and 2 other species of central Africa retire to burrows in mud, where mucus dries to form a "cocoon" with lid and tube leading to mouth of fish for breathing; spawns after return of water.

Subclass 3. Actinopterygii. Ray-finned fishes. Nostrils not connected to mouth cavity except in stargazers; scant skeleton or muscle in paired fins; no epichordal lobe in caudal fin; 1 dorsal fin (often in 2 or 3 parts). Middle Devonian to Recent.

Infraclass 1. Chondrostei. Primitive forms, often with thick ganoid scales; dominant in Triassic; represented today by a few specialized and degenerate survivors; includes extinct palaeoniscoid fishes.

Order 3. Polypteriformes. Body slender; scales thick, ganoid, rhombus-shaped; 8 or more dorsal finlets, each preceded by spine; caudal fin arrow-shaped; vertebral centra biconcave, bony; swim bladder lunglike, used for respiration when water is stagnant; larvae with slender external gills; Paleozoic and Recent (none known between). *Polypterus*, bichir, tropical Africa; about 10 species (Fig. 30-18); *Calamoichthys calabaricus*, reedfish, eellike, Nigeria and Congo.

Order 4. Acipenseriformes. Snout long; tail heterocercal; no teeth; skeleton cartilaginous, vertebrae lack centra. Upper Cretaceous to Recent.

ACIPENSERIDAE. Sturgeons. Sharklike; mouth ventral, with barbels; body with 5 rows of bony (ganoid) scutes; eat worms, mollusks, small fishes, and aquatic plants; flesh palatable, eggs preserved as caviar, lining of swim bladder provides isinglass (pure form of gelatin) for clarifying jellies; about 20 species. *Acipenser oxyrhynchus*, Atlantic sturgeon (Fig. 30-19), on both sides of Atlantic, and *A. transmontanus*, great white sturgeon of Pacific Coast; both

Figure 30-19 Atlantic sturgeon (*Acipenser oxyrhynchus*, order ACIPENSERIFORMES); length to 3 m (10 ft). (*After Norman, Guide to fish guttery, British Museum.*)

ascend large rivers to spawn; *A. fulvescens*, lake sturgeon, and *Scaphirhynchus platorhynchus*, shovelnose sturgeon, both in Mississippi drainage, only in fresh waters; *Huso huso*, beluga, Caspian, Adriatic, and Black Seas, to 8.5 m (28 ft) and over 1130 kg (2500 lb).

POLYODONTIDAE. Paddlefishes. Body naked or few vestigial scales; 2 living species. *Polyodon spathula*, paddlefish or spoonbill, Mississippi drainage and south, to 2 m (6 ft) long, snout paddle-shaped; eats small invertebrates strained by fine-mesh gill rakers from bottom mud; *Psephurus gladius*, Yangtse River, China, up to 6 m (20 ft).

Infraclass 2. Holostei. Intermediate in structure between chondrosteans and teleosts; ganoid covering of scales reduced in many; 10 living species; characteristic ray-finned fishes of Mesozoic.

Order 5. Semionotiformes. LEPISOSTEIDAE. Garpikes and alligator gars. Snout and body long; scales ganoid in oblique rows; single dorsal fin set well to rear; tail fin short, heterocercal; vertebrae solid, concave behind; teeth stout, conical; prey commonly seized with sidewise movement of jaws; swim bladder lunglike, can be used to breathe air; Upper Cretaceous to Recent. *Lepisosteus oseus*, longnose gar (Fig. 30-20), Great Lakes and Mississippi Valley to Florida, length to 1.5 m (5 ft); *L. spatula*, alligator gar, lower Mississippi Valley, length to 3 m (10 ft); all voracious, prey on smaller fishes.

Order 6. Amiiformes. AMIIDAE. Snout normal; dorsal fin long; tail fin short, appearing homocercal but skeletal support heterocercal; scales thin, overlapping, cycloid; Upper Jurassic and Lower Tertiary; 1 living species. *Amia calva*, bowfin (Fig. 30-21),

Figure 30-18 Bichir (*Polypterus bichir*, order POLYPTERIFORMES); length to 1 m (3 ft). (*After Norman, Guide to fish gallery, British Museum.*)

Figure 30-20 Longnose gar (*Lepisosteus osseus*, order SEMIONOTIFORMES); length to 1.5 m (5 ft). (*After Norman, Guide to fish gallery, British Museum.*)

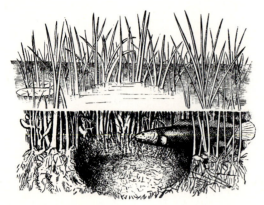

Figure 30-21 Bowfin and nest (*Amia calva*, **order AMII-FORMES); length to 1 m (3 ft)**. (*After Norman, Guide to fish gallery, British Museum.*)

Figure 30-22 American freshwater eel (*Anguilla rostrata*, **order ANGUILLIFORMES); length to 120 cm (4 ft)**. (*After Goode, 1884*).

Great Lakes to Texas and Florida, to 1 m (3 ft) long; swim bladder large, cellular, aids in respiration; eats fishes and invertebrates; eggs in crude nest, guarded by male.

Infraclass 3. Teleostei. Light skeleton, density of bones reduced and ganoid covering of scales reduced or absent; swim bladder a hydrostatic organ, rarely functioning as lung; homocercal tail fanning out from hypurals; premaxillae and maxillae mobile; tooth-bearing bones of upper jaw attached only at snout; most small, about 15 cm (6 in) long, but some reach over 3.5 m (12 ft) in length; over 20,000 species displaying great adaptive radiation; dominant fishes since Cretaceous.

Division Taeniopaedia.

Order 7. Elopiformes. ELOPIDAE. Tarpons. Medium to large, in salt water, prized for sport. *Megalops atlantica*, tarpon, common off Florida.

Order 8. Anguilliformes (Apodes). Diverse; some 25 families.

ANGUILLIDAE. Eels. Body long and slender; dorsal, caudal, and anal fins continuous, caudal rarely lacking; pelvic fins absent; gill openings small; scales minute or absent; swim bladder usually with duct; no oviducts; mostly marine; Cretaceous to Recent. *Anguilla*, freshwater eels (Fig. 30-22), grow to maturity in freshwater streams, adults migrate to sea in autumn, spawn in deep water, and die, the delicate transparent larvae (leptocephali) swim near the surface of the sea, feeding and growing, to enter the rivers perhaps a year later; *A. rostrata*, American eel, of Atlantic Coast, spawns near Bermuda islands.

MURAENIDAE. Morays. Marine, highly predaceous, many of striking coloration, abound in crevices of coral reefs. *Muraena*.

SACCOPHARYNGIDAE. Gulper eels. Body eel-shaped, naked; mouth very large; fin rays soft; tail long, slender; no ribs or caudal or pelvic fins; oceanic at considerable depths; *Saccopharynx*.

Order 9. Notacanthiformes (Lyopomi). Spiny eels. Body slender, tail long, no caudal fin; fins with spines; in deep seas. *Notacanthus*.

Division Archaeophylacea.

Superorder 1. Osteoglossomorpha. Primitive fishes; principal bite from bony tongue against roof of mouth; widespread in fresh waters, chiefly in tropics.

MORMYRIDAE. Snout fishes. Electric organs on each side of base of tail; mormyromasts in skin which may sense electric field; large brain; poor vision; swim bladder connects to ear; about 150 species; murky freshwater habitat; tropical Africa. *Gnathonemus*, elephant fishes.

Superorder 2. Clupeomorpha. Silvery compressed fishes, usually marine; primitive; fins in simple arrangement, lacking spines; pelvic fins abdominal; swim bladder connects with ear capsule, no auditory ossicles.

Order 10. Clupeiformes. Herrings. Fins without spiny rays; pelvic fins abdominal; swim bladder with open duct to pharynx; pyloric caeca usually numerous; Jurassic to Recent.

CLUPEIDAE. Small; teeth small; most feed on plankton; chiefly in salt water, often in schools numbering millions along both sides of North America, coming shoreward to spawn; about 300 species. *Clupea harengus*, common herring, in Atlantic and Pacific waters; marketed fresh, salt, smoked, or pickled, also used for bait; young canned as "sardines"; *Sardinops sagax*, pilchard, or Pacific sardine, formerly taken in abundance along Pacific Coast for canning and for reduction to fish oil and meal; *En-*

graulis ringens, anchoveta of Peru, basis of world's largest fishery; *Sardinia,* sardine of Spain and Italy; *Alosa sapidissima,* shad, native to Atlantic Coast and introduced on Pacific Coast, enters rivers to spawn, much esteemed for eating; *Brevoortia,* menhaden, Atlantic Coast, used for meal, oil, and fertilizer.

Division Euteleostei.

Superorder 1. Protacanthopterygii. Predominately slender, predatory fishes; no spiny rays; pelvic fins abdominal; pectorals low on body; adipose fin usually present; premaxillaries expanded posteriorly, tending to exclude maxillaries from gape.

Order 11. Salmoniformes.

SALMONIDAE. Includes salmons, trouts. Usually with a small adipose fin on back; in cool waters of Northern Hemisphere, much prized for game and food. *Oncorhynchus,* Pacific salmon (Fig. 30-23), 5 species, central California to Alaska, others in Asia; millions of kilograms taken in nets, used fresh or canned; spawn in fresh water, young migrate to sea and mature in 2 to 8 years, return to fresh water, spawn once, and die; *O. tschawytscha,* Chinook, king, or tyee salmon, averages 13 kg (30 lb), some over 45 kg (100 lb); other species smaller. *Salmo salar,* Atlantic salmon, eastern North America and Europe, spawns more than once, as do trouts; *S. gairdneri,* rainbow trout, resident in fresh waters of Pacific Coast, California to Alaska, has a sea-run race, the steelhead trout; *S. clarki,* cutthroat trout, Colorado and central California northward; *S. trutta,* brown, or Loch Leven, trout of Europe, widely introduced in United States; *Salvelinus fontinalis,* eastern brook, or speckled, trout, Iowa and Allegheny Mountains to Saskatchewan and Labrador; *Coregonus,* whitefish, and *Thymallus,* grayling, northern states to Canada and Alaska. Various trout and salmon have been transplanted to many new localities.

Figure 30-23 Chinook, or king, salmon and young (*Oncorhynchus tschawytscha,* **family SALMONIDAE); length 60 to 150 cm (2 to 5 ft).** (*After Goode, 1884.*)

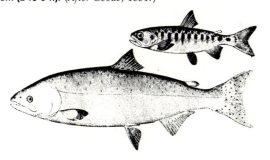

ESOCIDAE. Pikes. No pyloric caeca; slender-bodied with large mouths and conspicuous teeth, feed on other fishes. *Esox lucius,* northern pike or pickerel, Arkansas and Minnesota eastward, other species there and in Eurasia; *E. masquinongy,* muskellunge, upper Mississippi basin, Great Lakes, and northward, grows to 2.5 m (8 ft) and over 45 kg (100 lb), a choice game fish.

GONOSTOMATIDAE. Bristlemouths. Herringlike; one of several families of deep-sea luminous fishes (suborder STOMIATOIDEA); bristlemouths occur by billions at intermediate depths in all seas and are an important element of the "deep scattering layer" recorded by sonic depth-finding apparatus. *Cyclothone.*

Order 12. Myctophiformes. Lantern fishes. Premaxillaries exclude maxillaries from gape; light organs present or absent.

MYCTOPHIDAE. Light organs in definite pattern on sides; adipose fin present; swim bladder absent or filled with fat to withstand pressure differences; Cretaceous to Recent. *Myctophum,* lantern fish, deep seas to 2500 fathoms, with luminous spots for species and sex recognition.

SYNODONTIDAE. Lizard fishes. Shallow tropical waters. *Synodus.*

HARPADONTIDAE. *Harpadon nehereus,* Bombay duck, common food fish of India.

Superorder 2. Ostariophysi. Dominant group of freshwater fishes; anterior vertebrae fused, and lateral elements of 4 vertebrae detached to form small auditory ossicles (Weberian organ) between swim bladder and inner ear; swim bladder of 2 or 3 parts, usually with duct to pharynx; pelvic fins abdominal; fin spines often present; many with barbels around mouth. Eocene to Recent; over 5000 species.

Order 13. Cypriniformes. Includes characins, minnows, suckers, loaches, catfishes.

CHARACIDAE. Characins. Silver-sided, mostly small but some large food fishes. Over 800 species in Central and South America; *Serrasalmus,* piranha, strong, sharp teeth, occasionally dangerous to humans, 150 species in Africa; *Hydrocyanus,* tiger fish.

GYMNOTIDAE. Electric fishes. Slender; no dorsal, caudal, or ventral fins, anal fin long; tail with electric organs. *Electrophorus electricus,* electric eel, Orinoco and Amazon drainages in South America, grows to 2.5 m (8 ft), produces powerful electric shocks.

CYPRINIDAE. Includes carps, chubs, minnows. Jaws toothless, teeth in pharynx used to grind food; protrusible premaxillae; largest family of fishes, some 275 genera, about 2000 species; Europe, Africa, Asia, North America. *Cyprinus carpio,* common carp, native to fresh waters in eastern Asia, where reared for food and ornament for centuries; domestic races such as mirror carp (large scales) and leather carp (scaleless) introduced into Europe in thirteenth century and brought from Germany (hence "German" carp) to United States in 1872, where now widespread and abundant; *Carassius auratus*, goldfish, also native to Asia and long under domestication, with many artificial breeds or races of peculiar form; daces and minnows, many native to North America, of small size, often erroneously thought to be young fishes: *Notropis*, shiners or minnows, eastern North America; *Ptychocheilus lucius*, Colorado squawfish or "white salmon" of Colorado River basin, grows to 1.5 m (5 ft) and 36 kg (30 lb), the largest North American cyprinid, now scarce.

CATOSTOMIDAE. Suckers. With fleshy lips to suck up bottom mud containing small animals and plants used as food; *Castostomus, Carpiodes*, and others, more than 60 species in North America and 2 in Eastern Asia.

Order 14. Siluriformes (Nematognathi). Catfishes.

ICTALURIDAE. Freshwater catfishes. Mouth with teeth, nonprotractile, and with sensory barbels; body either naked (scaleless) or with bony plates; many species (2000+), mostly in tropics; highly diverse; over 30 species in fresh waters of United States, others in salt waters and southward; American species naked and with adipose fin (Fig. 30-24). *Ictalurus nebulosus*, brown bullhead, native to most quiet waters east of Rocky Mountains and introduced in the West; *Ictalurus punctatus*, channel or spotted catfish, in clear flowing waters from Great Lakes region to Gulf states.

ARIIDAE. Sea catfishes. Shallow coastal seas. *Arius; Galeichthys*.

Superorder 3. Paracanthopterygii. Mostly marine; stout, soft-bodied; carnivorous; upper jaw not protractile; pelvic fins usually well forward.

Order 15. Percopsiformes (Salmopercae).

AMBLYOPSIDAE. North American cave fishes. Five species, four in streams in limestone caves; cave species blind; females of *Amblyopsis* carry eggs in gill chambers. *Typhlichthys*, Indiana to Alabama.

PERCOPSIDAE. Trout-perches. Dorsal and anal fins preceded by 1 or 4 spines; North America, 2 species; *Percopsis*.

Order 16. Batrachoidiformes (Haplodoci). Toadfishes. Head depressed, mouth wide; a spinous dorsal fin and long soft-rayed dorsal and anal fins; bottom of tropical and subtropical seas; *Opsanus*; toadfish; *Porichthys*, singing fish, or "midshipman".

Order 17. Gobiescoiformes (Xenopterygii). Clingfishes. Naked; large subcircular adhesive disk on abdomen, by which the fish attaches to stones or shells; shore waters, tropic and temperate. *Gobiesox*, clingfish.

Order 18. Lophiiformes (Pediculati). Includes goosefishes (angler fishes), frogfishes. "Spinous" dorsal fin of a few flexible rays, the first located on head and with tip modified as a lure; pectoral fin with armlike base; luminescent organs in deepwater forms; bottom dwellers of shore waters or ocean depths, sargassum fish of open sea. *Lophius americanus*, goosefish, or angler fish (Fig. 30-25), broad, soft-bodied, with huge mouth, to 1.2 m (4 ft) in length, on both sides of North Atlantic down to 110 m (60 fathoms), captures fish, marine invertebrates, and birds; attracts prey with movable lure on modified first dorsal fin spine. *Ceratias*, deep-sea anglers, at 460 to 1830 m (250 to 1000 fathoms), mostly small, of blackish color, with a rod tipped with a luminescent winking lure moved by a basal bone that slides in a groove along the roof of the skull and the back; in some species male minute and attaches permanently to female, their bodies and bloodstreams growing together, such "parasitism" proba-

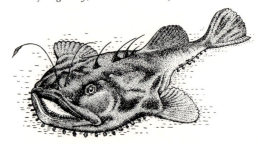

Figure 30-25 Goosefish, or angler fish (*Lophius americanus,***
order** LOPHIIFORMES**); length to 120 cm (4 ft).** (*After Norman,
Guide to fish gallery, British Museum.*)

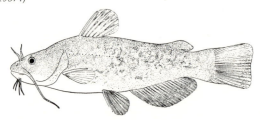

Figure 30-24 Brown bullhead (*Ictalurus nebulosus,* **order**
SILURIFORMES**); length 15 to 25 cm (6 to 10 in).** (*After Trautman,
1957.*)

bly an adaptation to ensure mating in the dark depths of the sea.

Order 19. Gadiformes (Anacanthini). Includes cods, pollacks, hakes, burbots. Fins soft-rayed; pelvics often many-rayed and thoracic or jugular in position; caudal fin, when present, formed mainly of dorsal and anal rays, without expanded hypural; swim bladder lacks duct; Northern Hemisphere; mostly marine bottom dwellers; of great commercial importance. *Gadus morhua,* Atlantic cod, both sides of North Atlantic, fished from ancient times; haddock (*Melanogrammus aeglefinus*), sold fresh or salted; *Merluccius,* hake; *Lota,* burbot (in fresh water).

Superorder 4. Atherinomorpha. Mostly small surface-feeding fishes, in marine waters, brackish water, and fresh water; with or without fin spines; upper jaw protractile in many; most freshwater species sexually dimorphic, males often with bony genitalia; many viviparous.

Order 20. Atheriniformes.

Suborder 1. Exocoetoidei (Synentognathi). Needlefishes, garfishes, halfbeaks, flying fishes. Dorsal fin above anal; pectorals high on body; lower pharyngeals united as one bone; swim bladder lacks duct; mostly marine.
    BELONIDAE. Needlefishes. *Belone, Strongylura.*
    SCOMBERESOCIDAE. Sauries, or skippers. *Scomberesox.*
    EXOCOETIDAE. Flying fishes and halfbeaks. Chiefly in warm seas, emerge to glide in air over the water (Fig. 30-26). *Hemiramphus, Cypselurus.*

Suborder 2. Cyprinodontoidei.
    CYPRINODONTIDAE. Killifishes, topminnows. Size small or diminutive; mouth protractile; lateral line weak; pelvic fins abdominal or absent; chiefly in tropical America, Africa, Asia, either fresh water or brackish water; feed chiefly on small animals or plants; some with conspicuous courtship performances. *Cyprinodon macularius* and others, desert pupfishes, in scattered spings, streams, and artesian waters, southern Nevada and southeastern California to Sonora; California forms threatened by drainage of springs and introduction of exotic fishes; *Fundulus heteroclitus,* common killifish in brackish coastal waters, Canada to Florida.
    POECILIIDAE. Live-bearers. Southern United States to South America. *Poecilia reticulata,* guppy; *P. latipinna* and others, mollie, used worldwide in aquaria and laboratories; *P. formosa,* Amazon mollie, all female species; *Xiphophorus,* swordtail; *Gambusia affinis,* mosquitofish (Fig. 30-27), in fresh waters, reared and planted widely to control mosquito larvae.
    ANABLEPIDAE. Four-eyed fish. *Anableps,* eyes divided horizontally, capable of vision above and below water as fish swims at surface; rivers of Central and South America.

Suborder 3. Atherinoidei.
    ATHERINIDAE. *Leuresthes tenuis,* California grunion. Spawns on Californian beaches at night following new or full moon at high tides; females, accompanied by one or two males, burrow into sand to lay eggs; young emerge after fortnight at next high tide.

Superorder 5. Acanthopterygii. Fishes of extremely variable form and habits, chiefly marine; upper jaw protractile in many; pelvic fins, when present, far forward, pectorals high on sides; fin spines and ctenoid scales common; a few viviparous.

Order 21. Beryciformes. Squirrelfish or soldierfish. Pelvic fins usually with 5 or more soft rays, caudal fin with 19 principal rays (17 branched); some spiny rays; Cretaceous to Recent; marine, mostly in deep water. *Holocentrus,* squirrelfish.

Figure 30-26 Flying fish (*Cypselurus,* family EXOCOETIDAE); length to 38 cm (15 in). (*After Norman, Guide to fish gallery, British Museum.*)

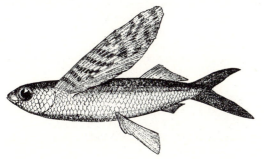

Figure 30-27 Mosquitofish (*Gambusia affinis,* family POECILIIDAE); about natural size. (*After Garman, 1895.*)

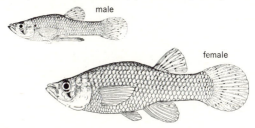

male

female

Order 22. Zeiformes. Anal fin preceded by 1 to 4 spines forming separate fins; marine. *Zenopsis ocellata,* American John Dory, Atlantic Coast.

Order 23. Lampridiformes. Fins with soft rays or dorsal with 1 or 2 spines, pelvic fin of 1 to 17 rays; marine, mostly oceanic; some large, basis of sea-serpent stories. *Lampris regius,* opah; *Trachypterus,* ribbonfish; *Regalecus glesne,* oarfish, slender, to 6 m (20 ft) long.

Order 24. Gasterosteiformes. Dorsal fin with spiny rays or preceded by 2 or more free spines; swim bladder closed.

GASTEROSTEIDAE. Sticklebacks. Free spines before dorsal fin; mouth normal. *Gasterosteus,* stickleback (Fig. 30-28), fresh water or brackish water, male builds nest of sticks and secretions, then guards eggs.

SYNGNATHIDAE. Pipefishes and seahorses. Snout tubular, suctorial; male usually with brood pouch under tail, shelters eggs until hatching; marine. *Syngnathus,* pipefish, slender, body covered with ringlike plates, fins minute; *Hippocampus,* seahorse (Fig. 30-29), head large, at right angle to body, swims vertically using dorsal fin; *Centriscus,* shrimpfish, transparent, swims head down.

Order 25. Synbranchiformes. Synbranchoid eels. Shaped like eels; dorsal and anal fins without fin rays and continuous with caudal; no pectoral fins; gill openings fused as transverse ventral slit; Africa, India, and tropical America in freshwater streams. *Synbranchus.*

Order 26. Scorpaeniformes (Scleroparei).

SCORPAENIDAE. Scorpionfishes. Mostly along seacoasts; used for food; fillets sold as "ocean perch." *Sebastes,* rosefish and rockfish (or rock cod).

COTTIDAE. Sculpins. Fins large, with slender flexible spines; on bottom in shallow seas and freshwater streams. *Cottus.*

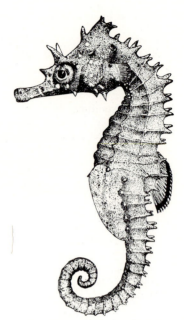

Figure 30-29 Seahorse (*Hippocampus,*** order GASTEROSTEIFORMES); length to 15 cm (6 in).** (*After Norman, Guide to fish gallery, British Museum.*)

Order 27. Pegasiformes (Hypostomides). Dragonfishes. Small; body enclosed in a broad bony box, tail in bony rings; tropical seas of Old World. *Pegasus.*

Order 28. Perciformes. Includes perches. Fin spines usually present; pectoral fins well up on sides; pelvic fins far forward, each usually with 1 spine and 5 soft rays; upper jaw bordered usually only by premaxilla, which is protrusible; swim bladder lacks duct; Upper Cretaceous to Recent; in all aquatic habitats, salt water and fresh water; largest order of vertebrates; 18 suborders, over 100 families, 8000 + species; diverse in form and habits.

CENTRARCHIDAE. Sunfishes, freshwater basses. Many species in Eastern states, prized for sport fishing; some transplanted into Western states, Europe, and South Africa. *Micropterus salmoides,* largemouth bass; *M. dolomieui,* smallmouth bass; *Lepomis,* sunfishes; *Pomoxis,* crappies.

PERCIDAE. Freshwater perches. *Perca flavescens,* yellow perch; *Etheostoma,* darters, eastern United States.

ECHENEIDAE. Remoras, or suckfishes. Upper surface of head with flat oval adhesive disk (modified spinous dorsal fin), transversely furrowed, by which the fishes attach to sharks, whales, porpoises, tur-

Figure 30-28 Stickleback (*Gasterosteus aculeatus,*** order GASTEROSTEIFORMES); length to 10 cm (4 in).** (*After Norman, Guide to fish gallery, British Museum.*)

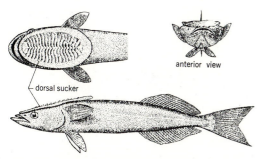

Figure 30-30 Remora, or suckfish (*Remora*), family ECHENEI-DAE); length to 53 cm (21 in). (*After Norman, Guide to fish gallery, British Museum.*)

tles, or other objects in the water; other fins normal and used for ordinary swimming; warm seas. *Echeneis, Remora* (Fig. 30-30). Remoras glean food from the surfaces of their hosts and feed on fragments of host's prey.

EMBIOTOCIDAE. Surfperches. California to Japan (Figs. 1-1, 30-31). *Crymatogaster.*

CICHLIDAE. Cichlids. *Tilapia, Cichlasoma,* important food and aquarium fishes. About 800 species in tropical America and Africa.

MUGILIDAE. Mullets. Head blunt, teeth small; in shallow estuaries, some in fresh waters; about 100 species. *Mugil.* Important food fishes in tropics along coasts, prized for flavor.

SPHYRAENIDAE. Barracudas. Slender, mouth and teeth large; some 2 m (6 ft) long; in warm seas. *Sphyraena,* excellent food fish; large *S. barracuda* often implicated as poisonous, however.

GOBIIDAE. Gobies. In shallow seashore and fresh waters, mostly tropical; largest family of marine fishes; about 1000 species. *Clevelandia; Periophthalmus* (mudskipper), amphibious.

SCOMBRIDAE. Mackerels and tunas. *Scomber*

Figure 30-31 Viviparous surfperch (*Crymatogaster agregata*, order PERCIFORMES, family EMBIOTOCIDAE), cut open to show fully formed young ready for birth; length 13 cm (5 in). (*After Jordon, 1925.*)

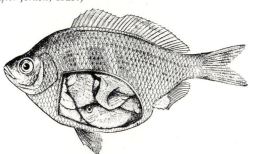

scombrus, Atlantic mackerel, commercially important; tunas (albacore, bluefin, skipjack, yellowfin), marine, open seas, some of large size, commercially important.

XIPHIIDAE. *Xiphias gladias,* swordfish. Big-game fishes, in all warm seas.

ISTIOPHORIDAE. *Makaira,* marlin. Big-game fishes, in all warm seas.

MASTACEMBELIDAE. Body eel-shaped; dorsal fin preceded by separate small spines; anterior nostril tubular on fleshy tentacle at end of snout; fresh waters of Africa and Southeastern Asia. *Mastacembelus.*

Order 29. Pleuronectiformes (Heterosomata). Flatfishes. Asymmetrical, with both eyes on one side of head and movable; body strongly compressed; dorsal and anal fins fringing the body (Fig. 30-32); bottom dwellers, lying on side, undersurface usually unpigmented; Upper Eocene to Recent; usually marine, on bottom in coastal waters; important food fishes, sought commercially; early young bilaterally symmetrical, but fish soon lies on one side and "lower" eye migrates around so that both eyes are on upper surface. *Hippoglossus,* halibut; *Paralichthys,* flounder; *Symphurus,* tonguefish; *Achirus,* lined sole.

Order 30. Tetraodontiformes (Plectognathi). Triggerfishes, trunkfishes, puffers, headfishes. Body form various, some globose, some inflate by swallowing water; jaws short, powerful; teeth usually strong incisors or fused into sharp-edged beak; scales generally spiny or bony; gill openings small; warm seas. *Balistes,* triggerfish; *Diodon,* porcupinefish; *Lactophrys,* trunkfish; *Mola mola,* ocean sunfish, body short, high, compressed (Fig. 30-1), anal and dorsal fins enlarged, skin leathery, specimens of 110 to 180 kg (250 to 400 lb) occasionally off New York, the largest nearly 3.5 m (11 ft) long and 910 kg (1 ton) in weight off California.

Figure 30-32 American plaice, or sand dab (*Hippoglossoides platessoides,* order PLEURONECTIFORMES); length to 61 cm (24 in). (*After Goode, 1884.*)

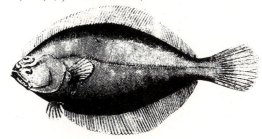

References

Many publications of the Bureau of Sport Fisheries and Wildlife, U.S. Department of the Interior, deal with the fishes and fisheries of the United States and dependencies. Various state game and conservation departments issue publications on the fishes of their states. *The Transactions of the American Fisheries Society*, the quarterly periodical *Copeia*, and other journals contain materials on fishes and fisheries.

Alexander, R. McN. 1967. Functional design in fishes. London, Hutchinson University Library. 160 pp., 16 figs. *Swimming, bouyancy, respiration, feeding, and sense organs of fishes.*

American Fisheries Society. 1970. A list of common and scientific names of fishes from the United States and Canada. 3d ed. Washington, D.C., American Fisheries Society Special Publ. No. 6, 114 pp. *Includes 2131 freshwater and marine species, fully indexed.*

Baerends, G. P., and **J. M. Baerends-Van Roon.** 1950. An introduction to the study of the ethology of cichlid fishes. Behaviour. Supplement 1. vii + 242 pp. *An excellent study of fish behavior.*

Breder, C. M., and **D. E. Rosen.** 1966. Modes of reproduction in fishes. New York, Natural History Press. xv + 941 pp. *Reproduction in fishes, lampreys to teleosts.*

Brett, J. R. 1965. The swimming energetics of salmon. *Sci. Am.*, vol. 213, no. 2, pp. 80–85. *Laboratory experiments help show how salmon can swim hundreds of miles upstream without food to spawn.*

Bullock, T. J. 1973. Seeing the world through a new sense: Electroreception in fish. *Am. Sci.*, vol. 61, pp. 316–325.

Carey, F. G. 1973. Fishes with warm bodies. *Sci. Am.*, vol. 228, no. 2, pp. 36–44. *A countercurrent system of arteries and veins conserves the heat of metabolism to increase the power of swimming muscles in tuna and mackerel shark.*

Fuhrman, F. A. 1967. Tetrodotoxin. *Sci. Am.*, vol. 217, no. 2, pp. 60–71. *A potent nerve poison found in puffer fish and newts.*

Gosline, W. A. 1971. Functional morphology and classification of teleostean fishes. Honolulu, University of Hawaii Press. ix + 208 pp. *Evolutionary changes in morphology and their relationship to the classification of teleosts.*

Grassé, P.-P. (editor). 1958. Traité de zoologie. Vol. 13, 3 parts. Agnathes et poissons. Paris, Masson et Cie. 2758 pp., 6 pls., 1889 figs. *Structure, physiology, habits and life histories, fossil record, classification; many references.*

Greenwood, P. H., and others. 1966. Phyletic studies of teleostean fishes, with a provisional classification of living forms. New York. *American Museum of Natural History Bulletin*, vol. 131, art. 4, pp. 341–455. 32 charts, 9 figs. *Main phyletic trends in teleostean fishes, based primarily on study of living forms.*

Grundfest, H. 1960. Electric fishes. *Sci. Am.*, vol. 203, no. 4, pp. 115–124, *A number of fishes generate current for purposes of attack, defense, and navigation.*

Hasler, A. D. 1966. Underwater guideposts, homing of salmon. Madison, The University of Wisconsin Press. 155 pp.

Herald, E. S. 1961. Living fishes of the world. Garden City, N.Y., Doubleday & Company, Inc. 304 pp., 300 photos, incl. 145 in color.

Hoar, W. S., and **D. J. Randall.** 1969–1971. Fish physiology. New York, Academic Press, Inc. 1969, Vol. 1, Excretion, ionic regulation, and metabolism, 465 pp.; 1969, Vol. 2, The endocrine system, 446 pp.; 1969, Vol. 3, Reproduction and growth: Bioluminescence, pigments and poisons, 485 pp.; 1970, Vol. 4, The nervous system, circulation, and respiration, 532 pp.; 1971, Vol. 5, Sensory systems and electric organs, 600 pp.; 1971, Vol. 6, Environmental relations and behavior, 559 pp.

Ivlev, V. S. 1961. Experimental ecology of the feeding of fishes. Translated from the Russian by Douglas Scott. New Haven, Yale University Press. viii + 302 pp. *A classic analysis of the factors determining what fishes eat.*

Johansen, K. 1968. Air-breathing fishes. *Sci. Am.*, vol. 219, no. 4, pp. 102–111.

Lagler, K. F., J. E. Bardach, and **R. R. Miller.** 1976. Ichthyology, 2d ed. New York, John Wiley & Sons, Inc. Illus. *Major fish groups, structure and physiology, reproduction, genetics and evolution, nomenclature, ecology and zoogeography.*

Leggett, W. C. 1973. The migrations of the shad. *Sci. Am.*, vol. 228, no. 3, pp. 92–98. *An important food fish migrates between the sea and its spawning grounds, guided apparently by temperature.*

Limbaugh, C. 1961. Cleaning symbiosis. *Sci. Am.*, vol. 205, no. 2, pp. 42–49. *Many marine organisms either live by cleaning other marine organisms or benefit by being cleaned.*

Lissmann, H. W. 1963. Electric location by fishes. *Sci. Am.*, vol. 208, no. 3, pp. 50–59. *Some fish produce a weak electric field in order to "see" objects in the water.*

Love, M. S., and **G. M. Cailliet.** 1978. The biology of fishes: Readings in ichthyology. Los Angeles, Goodyear Press. *A large selection of papers covering all aspects of ichthyology.*

Lüling, K. H. 1963. The archer fish. *Sci. Am.*, vol. 209, no. 1, pp. 100–108. *A Southeast Asian fish that downs insects with a jet of water spurted from its mouth.*

Marshall, N. B. 1966. The life of fishes. London, Weidenfeld and Nicolson. 402 pp., 43 illus., 86 figs. Republished in 1970 by Universe Books, New York. *Biology of fishes with emphasis on their adaptations to diverse habitats.*

————. 1971. Explorations in the life of fishes. Harvard Books in Biology, no. 7. Cambridge Mass., Harvard University Press. 204 pp. *Essays on the functional morphology and biology of fishes.*

Millot, J. 1955. The coelacanth. *Sci. Am.*, vol. 193, no. 6, pp. 34–39. *A discussion of its evolutionary relationships and biology.*

Nelson, J. S. 1976. Fishes of the world. New York, John Wiley & Sons, Inc. xiii + 416 pp. *A systematic treatment of all major fish groups.*

Nikol'skii, G. V. 1961. Special ichthyology. 2d ed. Translated from Russian. Jerusalem, Israel Program for Scientific Translations, Ltd. 11 + 538 pp., 312 figs. *Systematic accounts of cyclostome and cartilaginous and bony fishes, including fossil forms; characters for most families and higher groups and brief accounts of habits, food, etc.; references in Russian and other languages.*

———. 1963. The ecology of fishes. New York, Academic Press, Inc. xv + 352 pp., 140 figs.

Norman, J. R. 1975. A history of fishes. 3d ed. P. H. Greewood (editor). New York, John Wiley & Sons, Inc. xxiv + 467 pp. *Excellent general account of the biology of fishes.*

Rudd, J. T. 1965. The ice fish. *Sci. Am.*, vol. 213, no. 5, pp. 108–114. *A family of nearly transparent antarctic fishes lacks both red blood cells and hemoglobin.*

Scientific American offprint reader in ichthyology. Introduction by George W. Barlow. San Francisco, W. H. Freeman and Company. *A selection of articles on fishes from Scientific American.*

Shaw, E. 1962. The schooling of fishes. *Sci. Am.*, vol. 206, no. 6, pp. 128–138. *How fish maintain the constant parallel orientation that characterizes schools.*

Todd, J. H. 1971. The chemical languages of fishes. *Sci. Am.*, vol. 224, no. 5, pp. 98–108. *Catfishes use pheromones and a keen sense of smell in social interactions.*

Young, J. Z. 1962. The life of vertebrates. 2d ed. New York, Oxford University Press. xv + 820 pp.

CLASS AMPHIBIA: AMPHIBIANS

The AMPHIBIA include the toads and frogs (order ANURA), salamanders, (order CAUDATA), elongate, limbless tropical caecilians (order GYMNOPHIONA), and various fossil forms from Devonian time onward. The class name (Gr. *amphi,* dual + *bios,* life) appropriately indicates that most of the species live partly in fresh water and partly on land. In both structure and function, the amphibians stand between the fishes and reptiles, being the first group among the chordates to live out of water. Several "new" features adapt them for terrestrial life, such as legs, lungs, nostrils connecting to the mouth cavity, and sense organs that can function in both water and air. Amphibians are important predators on insects and other small invertebrates, and many are the chief vertebrate predators of moist small openings found in rotting logs, beneath stones and bark, in leaf litter,

and in the ground. They are important in nutrient cycling between fresh waters and upland terrestrial environments, as pond nutrients, incorporated into their larval structure, are transported to the land through dispersal and death of transformed individuals. They serve as food for various vertebrates, including humans, and many species are used for biological teaching and research.

31-1 Characteristics

1   Skin moist and glandular; no external scales.

2   Two pairs of limbs for walking or swimming (no paired fins); toes four to five or fewer (no limbs on caecilians, no hind limbs on SIRENIDAE); any median fins lack fin rays.

3 Nostrils two, connected to mouth cavity and with valves that exclude water and aid lung respiration; eyes often with movable lids; eardrums external on toads and frogs; mouth usually with small teeth; tongue often protrusible.

4 Skeleton largely bony; skull with two occipital condyles; ribs, if present, not attached to sternum.

5 Heart typically three-chambered, two atria and one ventricle, but atrial septum incomplete in salamanders, which lack or have reduced lung function; one (or three) pairs of aortic arches; red blood cells nucleated and oval.

6 Respiration by gills, lungs, skin, or the mouth lining, separately or in combination; gills present at some stage in life history; vocal cords in toads and frogs.

7 Brain with 10 pairs of cranial nerves.

8 Excretion by mesonephric kidneys; urea chief nitrogenous waste of transformed individuals.

9 Body temperature variable (ectothermal).

10 Fertilization external or internal; mostly oviparous; eggs with some yolk and enclosed in gelatinous coverings; cleavage holoblastic but unequal; no extraembryonic membranes; usually an aquatic larval stage with metamorphosis to adult form.

The AMPHIBIA are the earliest TETRAPODA (Gr. *tetra,* four + *podos,* foot), or land vertebrates. They undoubtedly derived from some fishlike ancestor, possibly in Devonian times. The transition from water to land involved (1) modification of the body for travel on land while retaining the ability to swim, (2) development of limbs in place of paired fins, (3) skin changes to facilitate respiration, (4) increased emphasis on lung respiration, usually with loss of gills in adults, (5) changes in the circulatory system to provide for respiration by the lungs and skin, (6) changes in metabolism and excretion to form less toxic nitrogenous wastes, and (7) acquisition of sense organs that function in both air and water. In the early larvae of all amphibians and in such salamanders as retain the gills throughout life, there are multiple aortic arches, as in fishes. After metamor-

phosis, the other salamanders, the toads and frogs, and the caecilians have but one pair of arches, as in reptiles. The amphibian skull is simpler, with fewer bones than in fishes, but the limb muscles are more complex than those of the lateral fins in fishes. Some primitive fossil labyrinthodont amphibians probably were ancestral to the oldest reptiles and so to all higher land vertebrates.

Salamanders have the head and neck distinct, the trunk long and either cylindrical or depressed dorsoventrally, and a long tail. Toads and frogs have the head and trunk joined in a broad depressed body, no tail, short forelimbs, long hind limbs, and the eardrums exposed. Caecilians are limbless and wormlike, and their skin may contain small recessed scales (Fig. 31-1).

31-2 Evolution The lungs and lobed fins of crossopterygian-type fish were key features in the vertebrate transition from water to land. Lung respiration became far more important as gills were lost in the adult with the advent of land life, and such respiration was augmented by the conversion of the scaly skin of the fish into the thin dermal respiratory membrane of the amphibian. This required loss of scales, an increase in the watery component of dermal secretion, and major changes in the circulatory system. There appeared for the first time the separation of pulmonary and systemic blood flow characteristic of all tetrapods. The aortic arches that passed between the gills were eliminated except for the last, used for passage of blood to the lungs, and the fourth, used to carry blood to the body. Partitioning of the heart into four chambers to accommodate the "double" flow of blood began but was not completed in amphibians.

Lobe fins, with their sturdy skeletal support, were modified into limbs. At this early stage in limb evolution, elbow and knee moved forward and backward in a horizontal arc and the feet were placed well out from the body, as in living salamanders and some limbed reptiles. The lateral undulatory movements characteristic of fish were retained (except in frogs), as were the segmented muscles of trunk and tail that powered such movements.

Emergence on land brought the threat of desiccation and greater risks of encountering thermal ex-

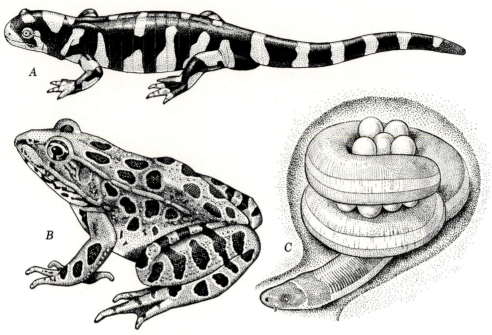

Figure 31-1 Representative amphibians (class AMPHIBIA). *A.* Tiger salamander (*Ambystoma tigrinum,* subspecies *mavortium,* order CAUDATA). *B.* Leopard frog (*Rana pipiens,* order ANURA). *C.* A tropical caecilian or limbless amphibian (*Ichthyophis glutinosus,* order GYMNOPHIONA).

tremes. But there was also increased opportunity for adaptation to higher temperatures with the competitive advantage of a higher metabolic rate, eventually to lead through the reptiles to the endothermic birds and mammals. Adaptation to dry, warm environments was limited, however, by the respiratory role of the skin and its function in maintaining body-water levels through cutaneous water absorption.

Soft-skinned, mostly small, and generally unprotected by weapons such as spines, claws, and sharp teeth, amphibians evolved an elaborate system of poison-secreting skin glands as protection against predators.

Land life required a change in nitrogenous waste excretion. Ancestral crossopterygians probably excreted chiefly ammonia, as do modern freshwater fishes and amphibian larvae. Ammonia is toxic, but when water is abundant it can be rapidly discharged and dissipated. On land, however, with water conservation at a premium, nitrogenous waste excretion shifted toward a less toxic substance, urea.

Evolution of a muscular, quick-acting, sticky, protrusible tongue used primarily to capture prey

was favored by life on land. Its skeleton was derived from the hyoid framework that supported the gills and throat of fish. Water density was an impediment to such development in water.

Faced with dryness and air-borne particulate matter, glands and movable eyelids evolved for lubrication, cleaning, and protection of the eyes, and a tear duct to carry off superfluous secretion. An eardrum was formed by extension of a membrane across the site of the fish spiracle, and a sound-transmitting bone, the columella, was constructed from the hyomandibular bone of the fish jaw suspension. This made possible amplification of feeble air-transmitted sounds. Freeing of this bone became possible with the tightening of connections between jaws and skull. Hearing improved as voice became an important method of vertebrate communication. The fish lateral line system (Sec. 30-12) was retained in aquatic amphibians. Ancestral amphibians were equipped with internal nostrils connecting the olfactory chambers with the mouth cavity, inherited from their lobe-finned ancestors. Aquatic ancestral forms, probably like their aquatic descendents, moved water carrying olfactory sense data across the olfactory epithelium by throat pulsations. On land the same movements transport air-borne sense data through the nose. Jacobson organs appear first in amphibians. They open inside the nasal passages and are innervated from a special area associated with the olfactory lobes. They respond to chemical sense data in the mouth or nose and appear to be accessory to the senses of smell and taste. In the aquatic environment they are bathed by water drawn into the nose by throat movements. In terrestrial salamanders of the family PLETHODONTIDAE, however, water-borne sense data are carried to the nose by the nasolabial grooves (unique to this family). The nasolabial groove is a hairline, nonciliated furrow extending from the edge of the upper lip to the nostril, and water is conveyed to the nose by capillary action as the animals tap their snouts against the ground.

Despite their many adaptations for life on land, amphibians as a group have been limited in their expansion into dry environments by their dependence on cutaneous respiration, inability to produce a concentrated urine, and failure to develop a desiccation-resistant land egg.

31-3 Size Most salamanders are 8 to 20 cm (3 to 8 in) long. The giant salamander *Andrias* (*Megalobatrachus*) *japonicus* grows to 1.5 m (over 5 ft), whereas a Mexican salamander (*Thorius pennatulus*) is only 4 cm (1½ in) in length. The giant frog (*Conraua goliath*) of the African Cameroons grows to 30 cm (12 in) in head-and-body length, and the smallest anuran is a Cuban tree toad (*Smithillus limbatus*), only 1 cm (⅜ in) when grown. Most toads and frogs are 5 to 13 cm (2 to 5 in) long, and caecilians 10 to 75 cm (4 to 30 in).

Structure of an amphibian: the frog[1]

31-4 External features The **skin** is soft, smooth, and moist. The head (Fig. 31-2) bears a wide **mouth,** small valvular **nostrils** (external nares), large spherical **eyes,** and behind each eye a flat **eardrum,** or **tympanic membrane.** Each eye has a fleshly, opaque upper **eyelid** and a lesser lower lid. Continuous with the latter and folded behind it when at rest is a transparent portion which functionally resembles the nictitating membrane of higher vertebrates. It moves upward over the eyeball to keep it moist in the air and protect the eye against injury. At the posterior end of the body is the **vent,** or **cloacal opening,** through which are discharged undigested food wastes, urine, and eggs or sperm. At midbody is the **sacral hump,** where the pelvic girdle is hinged to the sacrum.

The short **front leg** (arm) comprises an upper arm, forearm, wrist, and hand with four fingers (digits) and vestigial thumb (Fig. 31-5). The inner digit (2 in Fig. 31-5), usually called the "thumb," is thickened on males, especially in the breeding season. The **hind leg** includes a thigh, lower leg, ankle (tarsal region), and long foot with a narrow sole (metatarsus) and five slender toes connected by broad, thin webs. The palms and soles bear cornified

[1] *The following account applies to many common species.*

Figure 31-2 **Head and mouth of a male bullfrog.**

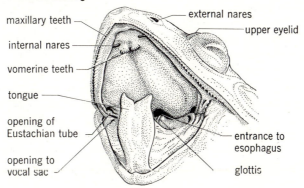

maxillary teeth

internal nares

vomerine teeth

tongue

opening of Eustachian tube

opening to vocal sac

external nares

upper eyelid

entrance to esophagus

glottis

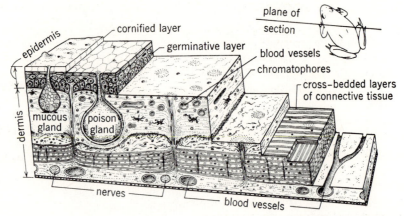

Figure 31-3 A frog's skin. Enlarged stereogram showing the component cell layers, fibers, glands, and other details (compare Fig. 4-1, human skin). The section is cut at a 45° angle to the main body axis, as indicated in the small outline figure.

tubercles that protect the feet against abrasion and provide traction on wet, slippery surfaces.

31-5 Body covering The skin is glandular and highly vascular (Fig. 31-3). The many small **mucous glands** secrete a colorless watery fluid that keeps the skin moist—essential to skin respiration—and slippery, a protection against predation. Scarcer but larger **poison glands** pour out a thick, whitish, granular, alkaloidal secretion that also protects the animal to some degree from predators. Each gland is ovoid, and its secretion accumulates in a central cavity, whence it can be forced out of the duct by the action of encircling muscle fibers. Unlike that of other vertebrates, the skin of the frog is attached to the body only along certain lines. Its flexibility and mucous-covered surface often enable a frog to slip from the grasp of a predator. Every month or so during the frog's active period a new layer forms beneath the old, and the old covering is molted, or sloughed off; it splits down the back, is worked off in one piece by the "hands," and usually is swallowed. Periodicity of molting is apparently governed by the level of circulating thyroid hormone.

The coloration of amphibians may be concealing or bright and colorful. The basic pattern is fixed on most species, but many frogs undergo a great change from dark to pale coloration under different physiologic and environmental conditions.

In the leopard frog and others, three kinds of

chromatophores are involved (Fig. 31-4). Just beneath the epidermis are xanthophores; alone they impart a yellowish hue. Next below are the iridophores, with light-reflecting organelles that produce light-scattering Tyndall blue, a structural color. This, filtered by the overlying yellow pigment, results in the green skin color. The black melanophores are extensively branched cells under the iridophores and surrounding them, with fingerlike processes that extend between them and the xanthophores. The skin darkens when melanin granules move outward into the branches and obscure the iridophores. The pituitary hormone intermedin causes dispersal of mela-

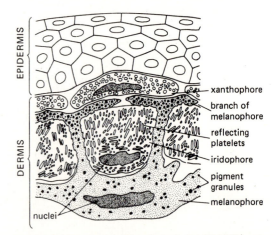

Figure 31-4 Transverse section through the skin of the green treefrog (*Hyla cinerea*), showing a dermal chromatophore unit in dark-background-adapted state. (*From Bagnara et al., 1968.*)

nin and contraction of the iridophores. A reverse movement of both produces a paler skin. Dark pigment in the epidermis usually is scant in amphibians that change color, since a heavy layer would conceal the underlying pigments.

31-6 Skeleton and muscles The skull is broad and flat, the brain case narrow, and the orbits large (Fig. 31-5). Teeth in the frog are present only on the upper jaw. The vertebral column consists of nine **vertebrae** and a slender rodlike **urostyle** of fused vertebrae. The urostyle helps stiffen the lower back. Such rigidity is required for effective transmission to the body mass of the force produced by the hind legs in jumping. The vertebral column is short, of few segments, and scarcely flexible (except at the sacrum), unlike the column in most vertebrates. There are no ribs.

The **pectoral girdle** is a U-shaped framework around the thorax that shelters organs within and supports the forelimbs; it is attached to the vertebrae by muscles (Fig. 31-5). The suprascapula and portions of the sternum are cartilaginous. Where the scapula and coracoid meet is a shallow depression (glenoid cavity) in which the head of the humerus articulates. The **pelvic girdle** is a rigid, narrow, V-shaped frame that connects the hind limbs to the vertebral column and transmits the power in locomotion from the hind limbs to the body. The head of the femur articulates in a cuplike socket, the **acetabulum.** Each ilium parallels the urostyle, and its anterior end attaches to a stout transverse process on the ninth (sacral) vertebra.

In an early frog larva the skeleton is entirely of **cartilage,** but many parts later become **bone** (Sec. 3-14). Cartilage persists on the ends of limb bones to form smooth joint surfaces and in parts of the skull and limb girdles.

The frog has far greater development of the muscles of the locomotor appendages and more differentiation of the muscles of the trunk than fish. The muscles are rather well defined for the first time and may be recognized individually. The conspicuous

Figure 31-5 Skeleton of the bullfrog.

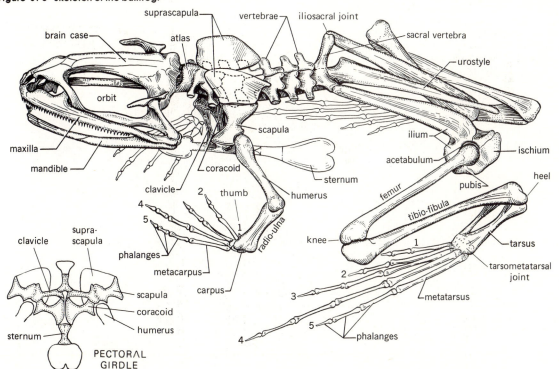

development of the nonsegmental muscles is necessary to enable the frog to move about on land. The muscles of the hind limbs are elongate and powerful, adapted for leaping and swimming. During the breeding season those of the forelimbs of males enlarge, aiding them in maintaining their grip on the female during amplexus.

31-7 Digestive system   The small animals taken as food are lubricated by mucus secreted in the mouth (a frog lacks salivary glands) and pass through the **pharynx** and **esophagus** to enter the stomach.

The digestive enzymes in the frog's stomach and intestine include proteases, which act on proteins; lipase, which acts on fats; and maltase and other enzymes, which act on starches. Hydrochloric acid secreted in the stomach activates certain of the digestive enzymes. Some absorption may occur in the stomach, but most of the mixed and finely divided contents enters the **small intestine.** The liver and pancreas supply secretions there, besides those from glands in the intestinal wall. Most digestion and absorption take place in the small intestine. Undigested residues are slowly moved by peristalsis into the **large intestine,** are formed into feces, and

are finally passed out through the cloacal opening. Much reserve food is stored in the liver as glycogen (animal starch), a carbohydrate that can be converted into glucose for use in the body as needed. Fat is stored mainly in the fat bodies, fingerlike growths anterior to the gonads (Fig. 31-9). Such food reserves are metabolized during hibernation and estivation (summer quiescence) and during the breeding season.

31-8 Circulatory system   The heart of the tadpole is fishlike, with one atrium and the ventricle. It receives only unoxygenated blood, which is pumped directly to the gills. Following metamorphosis the heart is three-chambered (Fig. 31-6) and consists of (1) a conical thick-walled **ventricle** posteriorly; (2) the left and right **atria** anteriorly, with thin muscular walls; (3) a thin-walled **sinus venosus** dorsally; and (4) a stout tubular **conus arteriosus** that leads forward from the anterior base of the ventricle. **Valves** between the chambers prevent backward flow of blood. The atria are separated by an interatrial septum. In the conus is a thin, flat, twisted **spiral valve.**

The course of circulation (Fig. 6-6) is as follows: Blood accumulates in the sinus venosus, which con-

Figure 31-6   Heart of bullfrog, enlarged.

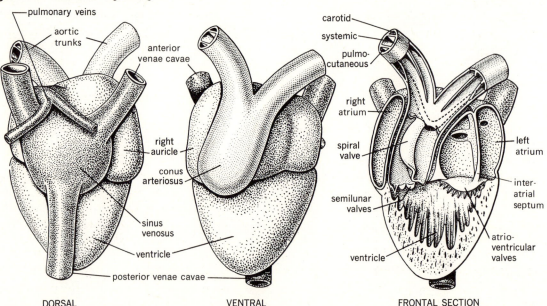

DORSAL   VENTRAL   FRONTAL SECTION

tracts to force it into the right atrium. That from the lungs passes into the left atrium. The two atria contract, forcing their contents into the ventricle. When the ventricle contracts, both oxygenated and unoxygenated blood are forced into the conus arteriosus, where mixing is partly prevented by the spiral valve. The left and right branches of the conus subdivide into three major vessels or "arches," the common **carotid** to the head, the **systemic** to the body and viscera, and the **pulmocutaneous** to the lungs and skin. Highly oxygenated blood goes into the carotid arches, that least oxygenated to the pulmocutaneous arches, and a mixture enters the systemic arches. Where the carotid divides, there is a **carotid sinus,** a spongy vascular enlargement that helps regulate blood pressure and perhaps respiration.

The two systemic arches curve around the esophagus to join as a median **dorsal aorta** extending posteriorly below the vertebrae. Each pulmocutaneous artery divides into a **pulmonary artery** to capillaries in the lungs and a **cutaneous artery** branching on the inner surface of the skin.

The venous system is somewhat more complex. Two **precaval veins** return blood to the sinus venosus from veins in the head, forelimbs, and skin. A median **postcaval vein** collects from the kidneys, gonads, and dorsal musculature. There are also three special venous paths: (1) Two **pulmonary veins** return blood that has been oxygenated in the lungs to the left atrium; all other veins deliver eventually to the sinus venosus and right atrium. (2) The **hepatic portal system** gathers blood from the capillary beds of the digestive tract (stomach and intestines) and carries it to the liver. There the veins break up into capillaries, and then the blood from the liver passes into the hepatic vein that enters the postcava. The hepatic portal circulation allows some materials in blood from the digestive tract to be either stored or filtered out during passage through the liver. In the frog the hepatic portal also receives an abdominal vein that collects from the hind limbs (femoral veins), bladder, and ventral body wall. (3) The **renal portal system** gathers blood from the hind limbs (sciatic and femoral veins) and posterior body wall and divides into capillaries within the kidneys. This blood is collected by the renal veins and returns to the heart in the postcaval vein.

The **lymphatic system** includes many delicate **lymph vessels** of varied diameter and shape that penetrate the organs and tissues but are difficult to see. Frogs and toads, unlike other vertebrates, have also several large **lymph sacs** or spaces between the skin and body. Behind the shoulder girdle and beside the vent are two pairs of **lymph hearts** that pulsate frequently. Dorsal to the coelom and peritoneum there is a large **subvertebral lymph sac** near the kidneys. The watery **lymph** in these structures contains leukocytes but lacks red cells and some proteins of blood plasma. It filters out from blood capillaries into the tissues, enters the lymph vessels, and later returns to the veins. Some small openings (stomata) in the peritoneum communicate with lymph vessels.

31-9 Respiratory system Amphibians have more means of respiration than any other animal group, reflecting the transition from aquatic to land habitats. In different species the gills, lungs, skin, and buccopharynx serve separately or in combination.

The **respiratory organs** are the **lungs, skin,** and **lining of the mouth and pharynx.** They all have moist surfaces (epithelium) overlying blood vessels. Oxygen from the air dissolves in the surface moisture and passes inward to the blood, while carbon dioxide passes in the opposite direction. The lungs (Fig. 31-7) are two thin elastic sacs with shallow internal folds that increase the inner surface. They are lined with capillaries of the pulmonary circulation. Each lung connects by a short **bronchus** to the voice box, or **larynx,** behind the glottis. A frog breathes with its mouth closed. Just before expiration the buccal floor is lowered, drawing air through the nasal passages into a depression in the floor of the buccopharyngeal cavity. At this time the glottis is closed. Upon expiration the glottis opens, the thoracic walls contract, and the elastic lungs recoil, expelling air at high velocity through the nostrils. This exhaled air mostly passes over that previously drawn into the buccopharynx; there is little mixing. Upon completion of exhalation, the nostrils close and the buccal floor is elevated, forcing air into the lungs. The glottis now closes and buccal pulsations occur, clearing the buccopharynx of any residual unexpired air

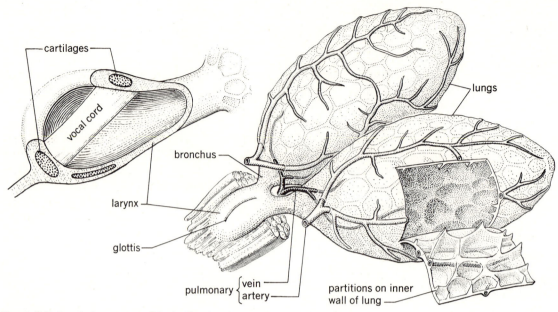

Figure 31-7 Respiratory organs of the bullfrog in laterodorsal view, with left lung cut open to show partitions on inner surface. Enlarged sketch at upper left shows larynx opened along midline with right vocal cord in place.

and filling the buccal chamber with outside air in preparation for the next breathing cycle. The frog's skin contains many blood vessels that serve for respiration either in air or water and especially during hibernation. Most oxygen uptake is by the lungs, and most carbon dioxide release is by the skin. The frog tadpole respires by **gills** (Fig. 31-10). These are slender extensions of the epithelium of the pharynx containing many blood capillaries; their function is comparable to that of the skin or lung of a frog (Fig. 31-7).

The larynx is reinforced by cartilages and contains two elastic bands, the **vocal cords** (Fig. 31-7).

When air is forced vigorously from the lungs, the cords vibrate and produce sound, the pitch of which is regulated by muscular tension on the cords.

31-10 Excretory system   The **kidneys** (Fig. 31-9) lie dorsal to the coelom and peritoneum. They are selective filters that remove soluble organic wastes (especially urea), excess mineral salts, and water gathered from the body cells and fluids by the blood (Chap. 7). When the frog is immersed or on a wet surface, they eliminate excess water absorbed by the permeable skin.

Figure 31-8 Nervous system of bullfrog in laterodorsal view, with the nerves and sympathetic trunk of the left side. I–X, cranial nerves; 1–10, spinal nerves (see Table 9-1).

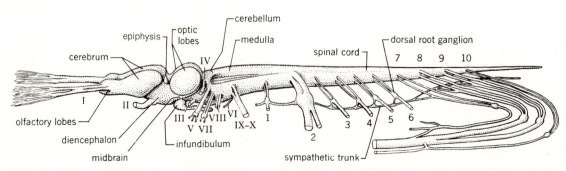

Each kidney is a compact mass of about 2000 microscopic **nephrons** bound together by connective tissue. A nephron comprises (1) a coiled knot, or **glomerulus,** of capillaries within (2) a double-walled cup, or **Bowman capsule,** which is connected to (3) a **uriniferous tubule** surrounded by capillaries (Fig. 7-10). The tubule joins through collecting tubules to the **ureter** (Wolffian duct), a fine white tube that extends along the outer margin of the kidney to the dorsal wall of the cloaca. Blood with both oxygen and wastes is brought to the kidneys in the renal arteries, which divide into capillaries in the glomeruli and continue about the tubules; the renal portal veins also join to these capillaries. Blood reduced in wastes leaves the kidneys by the renal veins. On the ventral surface of each kidney are many ciliated funnels (nephrostomes) that drain wastes and excess fluids from the coelom. They connect to uriniferous tubules in frog larvae and later to the renal veins, a structural arrangement peculiar to frogs and toads.

Urine collected in the kidneys passes down the ureters to the cloaca and may be voided at once through the cloacal opening or may be stored temporarily in the thin-walled **bladder** connected to the ventral side of the cloaca.

31-11 Nervous system and sense organs   The functions of the brain have been determined by studying the behavior of frogs after experimental injury or removal of parts of the brain and by stimulation of the brain with electric currents. The olfactory lobes serve the sense of smell (Fig. 31-8). The cerebral hemispheres are areas of memory, intelligence, and voluntary control in higher animals, but in the frog their function is less clear. Removal impairs memory, the frog is lethargic, and its movements are more machinelike. The diencephalon contains important regulatory centers and relay channels. The optic lobes integrate sensory information from the eyes and certain other senses. The medulla directs most bodily activities. If all the brain but the medulla is removed, the frog can leap, swim, capture and swallow food, recover normal position if inverted, and breathe normally. Death soon follows removal of the medulla.

The eyes are positioned and structured to see in nearly all directions at once. Vision is good both day and night. A frog is able to detect certain colors and is responsive to both small insect prey and distant predators, indicating a broad range in visual acuity.

The **eardrums** receive sound waves from air or water. Vibrations of the eardrum are transmitted across the middle ear cavity by the columella connecting the eardrum and inner ear. The latter lies within the prootic bone and contains a lymph-filled compartment with sensory endings of the auditory nerve (VIIIth cranial), where sound stimuli are received.

Large olfactory lobes indicate a well-developed sense of smell. The saclike **Jacobson organ** opens into the anterior medial portion of the nasal chamber. In salamanders, it is usually lateral in position.

Figure 31-9 Excretory and reproductive organs (urogenital system) in ventral view. Cloaca opened ventrally to show entrance of ducts (enlarged) and bladder turned aside. Aorta and arteries shown in male, postcaval vein in female; left ovary omitted. Posterior end of ureter in male enlarged as seminal vesicle in some species.

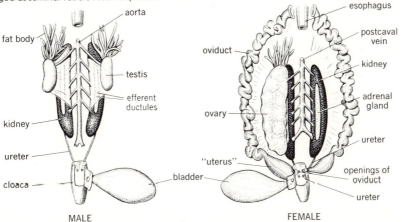

MALE   FEMALE

31-12 Endocrine system   The **anterior lobe** of the pituitary, in larvae and young, secretes a **growth-stimulating hormone** that controls growth, especially of long bones, and also affects the thyroid. Removal of the gland in larvae delays their growth and they do not change into frogs; replanting the gland restores these functions. Feeding or injecting an extract of the gland produces larvae of greater than normal size. In adult frogs the anterior lobe secretes a **gonad-stimulating hormone** responsible for release of ova or sperm from the reproductive organs. If anterior lobes are implanted on successive days in an adult but nonbreeding female, her eggs will mature and be laid. Similar implants into a male will hasten sexual maturity, the clasping reflex, and the discharge of spermatozoa. The **intermediate lobe** produces the melanophore-stimulating hormone, which causes darkening of the skin. The response is mediated chiefly by the eyes. On a dark surface the skin usually darkens, improving the frog's concealment. A secretion from the **posterior lobe,** probably arginine vasotocin, increases water intake by the skin and slows urinary water loss and increases resorption by the bladder. The thyroid gland regulates general metabolism and is important in amphibian metamorphosis (Sec. 8-4).

31-13 Reproductive system   The **ovaries** are attached dorsally in the coelom, near the kidneys, each supported by a mesentery (mesovarium). In early spring the ovaries of an adult contain hundreds of small black eggs that distend the abdomen, but a summer specimen has only a small mass of grayish ovarian substance. Each ovary is a hollow sac of four to seven lobes with thin double walls, and every egg is enclosed in a delicate follicle formed of cells between the two layers. The ovary is supplied with arteries that bring materials for growth of the ova. Along each side of the middorsal line of the coelom is the whitish twisted **oviduct;** its anterior end is an open ciliated funnel (ostium), and its posterior end joins dorsally to the cloaca.

When the eggs become mature, in the breeding season, each follicle ruptures (under stimulation of a pituitary hormone), and the eggs escape into the coelom. There they are moved anteriorly by the action of cilia covering the peritoneum and enter the oviduct funnels. They are moved down the ducts by cilia on lengthwise ridges lining the interior. Between these ridges are gland cells; these secrete albuminous material to form the jelly coatings that swell out around the eggs after they are laid (Fig. 31-15). The eggs may be accumulated in the enlarged posterior portion (uterus) before being laid.

The **testes** are attached near the kidneys by mesenteries (mesorchia). Each testis is a mass of coiled **seminiferous tubules** where spermatozoa are produced. The minute sperm, when mature, enter several fine ducts, the efferent ductules, that connect to uriniferous tubules in the anterior part of the kidney. The sperm then pass down the tubules and into the ureter (a joint urogenital canal). They may be stored in the posterior end, which is enlarged as a **seminal vesicle** in some species. At mating, the sperm are discharged through the cloacal opening to fertilize the eggs.

The **fat bodies** (Fig. 31-9) are largest just prior to hibernation and greatly reduced after breeding. They are of special importance during the breeding season for the males, which then take little or no food.

Structure of other amphibians

31-14 Body covering   Many amphibians, including most frogs (*Rana*), have a smooth skin; in others, however, it may be rough. Keratinized skin tubercles occur in newts (*Taricha*) when they are terrestrial, and warts (containing poison glands) are present in toads (*Bufo*). The former structures protect against abrasion and the latter against predators. In contrast to frogs and other anurans, the skin of salamanders and caecilians adheres closely to the body muscles.

31-15 Skeleton and muscles   Salamanders usually have at least twice as many neck and trunk vertebrae as frogs, and they have many caudal vertebrae. Elongate snakelike amphibians may have up to 100 (*Amphiuma*), and some caecilians have over 250. Ribs are present in salamanders, caecilians, and some primitive frogs. The pelvic girdle of salamanders is short, and caecilians lack both limbs and girdles. Most amphibians have fine teeth on the upper jaw and roof of the mouth; some salamanders have teeth on both jaws, but toads (BUFONIDAE) are toothless. The teeth are fastened to the surfaces of the bones and are continually and alternately replaced.

Segmental muscles are conspicuous on the trunk and tail of salamanders and trunk of caecilians. Gill-bearing species have special muscles to move the gills and to open or close the gill slits.

31-16 Respiratory structures   Three pairs of external **gills** occur in most embryos and larvae (Fig. 31-10) and persist in the adults of some strictly aquatic salamanders. In tadpoles water is drawn in through the mouth and nostrils, then forced over the gills and out the spiracle(s). Salamanders aid aeration by moving their gills, sometimes rhythmically. In aquatic species the lungs, in addition to their res-

piratory role, serve as hydrostatic organs, being inflated when the animals are floating. In some amphibians that inhabit swift mountain streams (*Rhyacotriton*) the lungs are reduced, and all salamanders of the essentially New World family PLETHODONTIDAE lack lungs. Lung buoyancy appears to have been an impediment to the control of locomotion in fast waters, and cutaneous respiration and an abundance of oxygen made their sacrifice possible. Ancestral plethodontids are thought to have originated in cold mountain streams of eastern North America.

The skin of all amphibians contains many blood vessels that aid aeration of the blood. Some aquatic species therefore can remain submerged for long periods and can hibernate in ponds.

Many species have **buccopharyngeal respiration;** pulsations of the throat move air in and out of the mouth cavity, and aeration of the blood occurs in vessels in the mucous membrane there. Although the proportion of capillaries is small (5 to 10 percent of those present in the skin), throat pulsations appear to contribute significantly to respiration, especially at higher temperatures. By moving air through the nose, these movements also aid olfaction.

Figure 31-10   Gills of amphibian larvae; enlarged insets show finer details of gill filaments.   *A*. Tiger salamander (dorsal view). *B*. Bullfrog (ventral view.)

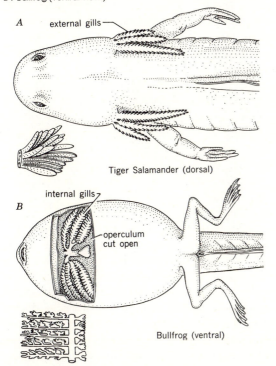

A   external gills

Tiger Salamander (dorsal)

internal gills

B

operculum cut open

Bullfrog (ventral)

Natural history

31-17 Distribution   Amphibians live mainly in fresh water or damp places; a few occur in brackish water, but none are marine. They are common in moist temperate regions, but most, including all caecilians, are tropical. Amphibians are the most abundant terrestrial vertebrates in many tropical forests. Several frogs range into the Arctic Circle; a frog and treefrog occur above 3650 m (12,000 ft) in the Sierra Nevada of California; and a toad (*Bufo*) ranges to over 4870 m (16,000 ft) in the Andes. Some toads, treefrogs, and other amphibians live on deserts, as in the American Southwest and Australia, where they hide in underground retreats during dry periods and at times are nocturnal.

The hellbender (*Cryptobranchus*), mud puppy (*Necturus*), congo eel (*Amphiuma*), mud eel (*Siren*), the pipid frogs (*Pipa, Xenopus*), and some caecilians are strictly aquatic. Bullfrogs live in or close to water, as do some other frogs, the wood frog on the moist floor of forests. Some frogs and treefrogs are partly or completely arboreal. Land salamanders hide commonly under stones or logs; some are arboreal, and the tropical caecilians burrow in moist earth or swim.

31-18 Locomotion The jumping locomotion of frogs may have evolved as a method of rapid retreat to water, primarily to escape terrestrial predators. Similar leg movements are employed in swimming. Structural modifications for jumping are evident in the compact body and elongate hind limbs, which, with the iliosacral and tarsometatarsal joints, constitute a lever system of five joints (Fig. 31-5). On land a frog rests with its short angular forelegs upright, its pelvic girdle flexed, and the long hind legs folded beside the body. When disturbed, it jumps by suddenly extending the girdle, hind legs and feet. In the water it swims by alternately flexing and extending the hind legs, the broad webs between the toes pushing against the water and carrying the animal forward. It may float with all legs extended and only the eyes and nostrils above water; from this position it can turn and dive to the bottom. On the hind foot in the spadefoot toads (*Scaphiopus*) and some other toads (*Bufo*) that live in arid habitats, the innermost tubercle becomes a horny cutting spade used to dig backward into the ground when seeking shelter. The treefrogs (HYLIDAE) and some other frogs (DENDRO- BATIDAE, RHACOPHORIDAE) have expanded adhesive disks on all toes by which they can climb, even on vertical surfaces. Some rhacophorid treefrogs of Southeast Asia parachute or glide from trees using their large, splayed-out, broadly webbed feet. Some Neotropical plethodontid salamanders have broad spatulate feet used in climbing. In general, webbing of the hind toes is least in strictly terrestrial species and extensive in the more aquatic forms.

31-19 Seasonal activity All amphibians must avoid temperature extremes and drought, because they have no internal regulation of body temperature and can lose water easily from the skin. In winter some frogs and aquatic salamanders hibernate deep in lakes and streams that do not freeze; toads burrow or go below the frost line, and terrestrial salamanders enter crevices or animal burrows. During hibernation all bodily processes are lessened, the heartbeat is slow, and the animal subsists on materials stored within its body, including glycogen in the liver. In some Southern states many amphibians are active at all seasons, but in hot dry areas some estivate during the summer. In temperate zones most species have a definite time of spring or fall emergence, triggered by temperature and/or rainfall.

Amphibians resist desiccation by behavioral, structural, and physiologic means. Cutaneous respiration limits the extent to which the skin can become impervious, and water is usually readily absorbed and lost through the skin. However, some amphibians, especially terrestrial species, have a jellylike acid mucopolysaccharide layer in the dermis that strongly resists water loss and is able to hold water absorbed by the skin and release it as needed to the tissues. Most amphibians, even when severely dehydrated, can restore body water within a few hours by contacting a moist surface. Toads have a blotterlike "seat patch" of thin skin specialized for this purpose. On wet surfaces some species move water quickly by capillary action to upper skin surfaces for absorption and respiration along skin furrows (costal grooves of salamanders and wrinkles in the skin of toads).

Some physiologic mechanisms are (1) release of pituitary secretion (arginine vasotocin), which reduces the amount of water that leaves the kidneys (antidiuretic effect) and enhances water absorption by the skin and resorption of stored water from the bladder; (2) increased retention of urea, which elevates tissue osmolarity ("saltiness") and promotes water uptake from and reduces water loss to the soil; and (3) increased tissue tolerance to water loss.

Behavioral adjustments include (1) reduction of exposed body surfaces—folding the limbs, closing the eyes, lying flat, coiling the body and tail; (2) retreating to moist locations in crevices, decayed wood, or animal burrows; (3) burrowing; and (4) remaining inactive during dry weather. In dry environments many species are active at night when air humidity rises, and some, during times of drought, seal themselves into self-constructed cavities underground. The African bullfrog (*Pyxicephalus adspersus*) encloses itself in a water-conserving, cellophanelike mucous envelope, secreted by its skin glands. Other species produce "cocoons" of dead epidermis that slow evaporation.

31-20 Voice and hearing   Vocal cords in the larynx of frogs and toads serve to make the familiar calls, distinctive for each species, that bring the sexes together for mating—chiefly in the spring. The calls are given by the males. Some species have resonating pouches in the throat that amplify the sounds. Salamanders lack vocal cords, but a few kinds make faint squeaks. *Dicamptodon*, the one exception having vocal cords, makes a growling or barking sound.

Frogs have a two-channel hearing system: one channel for lower frequencies (less than 1 kHz) and one for higher frequencies (greater than 1 kHz). The vocalizations of a given species contain a large amount of energy in the higher frequency range. An ability to diminish the perception of environmental "noise" (generally less than 1 kHz) may help to improve recognition of vocalizations during reproductive chorusing. The mechanism is as follows (Fig. 31-11): The columella transmits vibrations of the tympanum to the oval membrane and thence to the inner ear. Loosely coupled to its base is another ear ossicle, the operculum, found only in amphibians. A muscle from the suprascapula inserts on the operculum. Upon contraction it locks the latter to the columella, thereby increasing the inertia of the sound-transmitting apparatus. This creates a mechanically more efficient system for responding to low frequencies such as environmental sounds. When relaxed, the operculum is uncoupled from the columella. This produces a system better able to respond to the higher frequencies of the vocal signals. At the same time it reduces the ability of the animal to perceive lower frequencies. Terrestrial sound communication

Figure 31-11 Right middle ear of the Pacific treefrog (*Hyla regilla*), showing the structures which permit a frog to focus its hearing either on general environmental sounds or on the calls of its species (see Sec. 31-20). (*Courtesy of Eric Lombard and Ian Straughan.*)

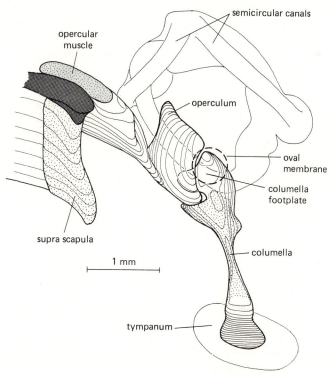

has yet to be convincingly demonstrated in salamanders. Their ear structure appears to be suitable only for low-frequency sound transmission.

31-21 Migration and homing Many frogs and salamanders regularly travel considerable distances to water to breed and then return to land. Some move between familiar sites, engaging in true migration. Movement of newts (*Notophthalmus*) to aquatic breeding sites is promoted by the pituitary hormone prolactin.

In experiments, treefrogs, toads, newts, and other salamanders transported into unfamiliar territory have returned to their home sites with surprising accuracy. The red-bellied newt (*Taricha rivularis*) has done so over hilly country in California from a distance of 8 km (5 mi). Celestial and olfactory cues appear to be used.

The home shore is important in the lives of most amphibians, whether they are completely aquatic or amphibious. They escape from danger by swimming to or from shore, the direction depending on the species, the stage in the life cycle, and the season. Cricket frogs (*Acris gryllus*), when thrown into the water, swim toward shore. Tadpoles swim toward deeper water when frightened but toward land when ready to transform. The axis of movement (called the Y axis) is roughly at a right angle to the general trend of the shoreline (X axis). Y-axis orientation appears to depend primarily on the detection of celestial cues provided by the sun, moon, or stars, familiarity with an area of shore, and a time sense phased to local time. Extraretinal photoreceptors (ERPs; Sec. 9-17), responsive to polarized light, appear to be the receptors used in determining both compass direction and time of day. In general, blinding has not prevented Y-axis orientation, but blocking of ERPs has.

31-22 Food Adult amphibians and the larvae of salamanders eat live moving animals, such as insects, worms, and small mollusks. Large aquatic species also take small fishes; the bullfrog sometimes catches small fishes, birds, or mammals; and large amphibians will devour small individuals of their own or other species. Small conical teeth are used for

holding and crushing the prey. Toads (*Bufo*), however, are toothless. In most terrestrial species the prey is caught by sudden protrusion of the sticky tongue, which in frogs and toads is flipped out hind part foremost (Fig. 31-12). Some terrestrial salamanders, which have the tongue pad mounted at the end of an elongate hyoid skeleton, are able to fire the tongue in projectilelike fashion, and with great accuracy, a third of their body length. Tadpoles feed mainly on algae and microorganisms suspended in the water. The particulate matter is entrapped in mucous strands produced by a filtering mechanism in the throat. Specialized filter feeders, such as the tadpole of the clawed frog (*Xenopus*), can remove particles down to 1.1 μm in diameter.

31-23 Predators The acrid secretions of the warts on toads protect these animals from many predators but not all. Raccoons and skunks roll toads underfoot to get rid of these secretions before eating them. Puppies soon learn to leave toads alone, and adult dogs rarely attack them. For the same reason, newts are seldom eaten by other animals. Newt poison, **tarichatoxin,** blocks nerve transmission and is capable of killing large animals. Highly poisonous amphibians are often warningly colored. Poison glands are usually concentrated where they are most effective in defense—in the parotid glands at the back of the head of toads (*Bufo*), in the dorsolateral folds of frogs (*Rana*), and on the dorsum of the tail in some salamanders. Amphibian larvae are the prey of large water bugs, beetles, dragonfly nymphs, fish, snakes, birds, and adult amphibians.

31-24 Reproduction Mature females and males differ in most species of frogs and toads. Males have heavier arm muscles, shorter but stouter inner fin-

Figure 31-12 Tongues of a plethodontid salamander and a frog as extended to capture prey.

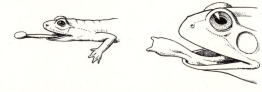

A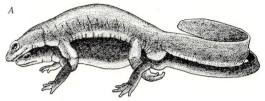

B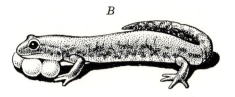

Figure 31-13 *A.* Amplexus in California newts (*Taricha torosa*, family SALAMANDRIDAE), male above female. Mating occurs in water. The male induces receptivity in the female by rubbing her snout with secretion from courtship glands on his throat and by stroking her vent with his toes. Back-and-forth movements rub the lining of his swollen vent over the rough skin of the female's tail base. Spermatophore pickup occurs as described in Sec. 31-24. *B.* Ensatina (*Ensatina eschscholtzi*, family PLETHODONTIDAE), a terrestrial salamander, brooding her eggs.

gers, and roughened "nuptial pads" on the fingers. Male toads and treefrogs have a median resonating pouch on the throat, and there are paired pouches in some frogs. The eardrum is larger in the male bullfrog and green frog, and the two sexes of some toads differ in coloration. Endocrine secretions influence the development of some of these characteristics. Implants of testes into female toads produce thick nuptial pads such as occur in males.

Most amphibians mate in water, where their eggs are deposited and hatch and where the resulting larvae live and grow until they metamorphose into the adult stages. Each species has a characteristic type of breeding place such as a large quiet lake or pond, a stream, or a transient pool; some breed on land. Male toads and frogs, upon entering the water, begin croaking to attract females. As each "ripe" female enters, she is clasped by a male, who clings on her back, a sexual embrace known as **amplexus.** His enlarged "thumb" bases with their roughened nuptial pads are pressed firmly into her armpits. As she extrudes her eggs, the clasping male discharges sperm, or milt, over them to effect fertilization (Fig. 31-15). Many aquatic and terrestrial salamanders have elaborate courting performances in which the male noses about or may mount the female (Fig. 31-13), but he eventually deposits one or more gelatinous spermatophores on the bottom of a stream or pond or on the ground. (The spermatophore is a sperm packet on a gelatinous base.) He then leads the female over the spermatophore and she takes it into her cloaca, where the spermatozoa are stored in the seminal receptacle, later to fertilize her eggs internally before they are laid.

Eggs of amphibians are covered with one or more gelatinous coatings that protect against mechanical

injury, pathogenic organisms, and predators. After the eggs are laid, the jelly coats swell and may adhere to plant stems or other objects. The form of the egg mass and jelly coats are distinctive for each species (Fig. 31-14). Toad eggs are usually in long strings, those of frogs in tapiocalike masses, and those of some aquatic salamanders in small clumps; a few species deposit the eggs individually. Eggs of land salamanders (PLETHODONTIDAE) are often attached by a stalk to some object. Oviparous caecilians lay eggs in burrows, and the female coils around her egg mass. A small frog (*Sminthillus*) of Cuba may lay but one egg; many land salamanders deposit a few to about two dozen; small frogs and treefrogs produce up to 1000 and the bullfrog up to 25,000; and one large toad (*Bufo marinus*) may lay up to 32,000 eggs at one time. The time and rate of development vary widely—scarcely a month for spadefoot toads (*Scaphiopus*) but for the bullfrog 2 or 3 years from egg to young frog in northern regions.

Salamander larvae (Fig. 31-16) resemble their parents in general form, having limbs and feet early in

Figure 31-14 Some amphibian eggs and their gelatinous coverings; diagrammatic.

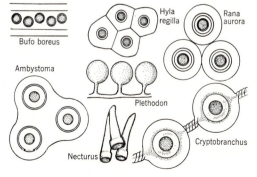

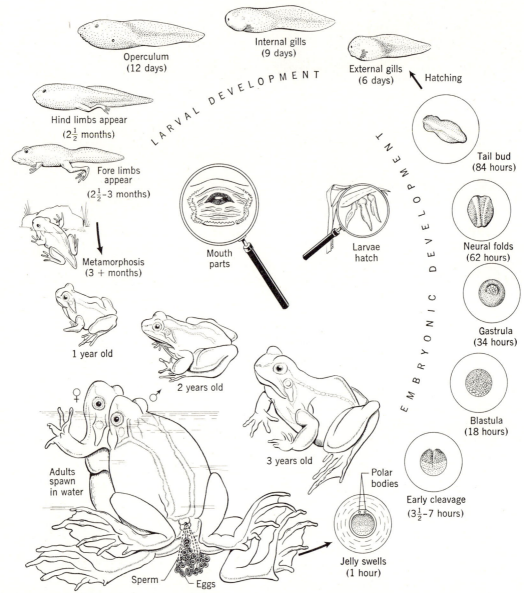

Figure 31-15 Life cycle of a frog. Magnified figures show newly hatched tadpoles clinging to vegetation by their adhesive organs and the face of an older larva with its dark horny jaws and minute labial teeth. (Compare Fig. 31-10.) (*Partly after Rosel, 1758.*)

life and a mouth that is adapted for feeding on animals. At metamorphosis their larval characteristics of gills, gill slits, and tail fins are usually lost and the hyoid apparatus is modified to support the tongue.

Larvae of toads and frogs are the familiar tadpoles (polliwogs) with ovoid head and body and a long tail (Fig. 31-15). The external gills are soon replaced by internal gills, under a delicate membrane (operculum). The limbs develop late in larval life, the forelimbs within the opercular chamber. The mouth typically has horny mandibles and rows of comblike labial teeth (Fig. 31-15) used to bite and scrape food materials (algae, etc.) from objects in the water. The intestine is long, slender, and spirally coiled. Meta-

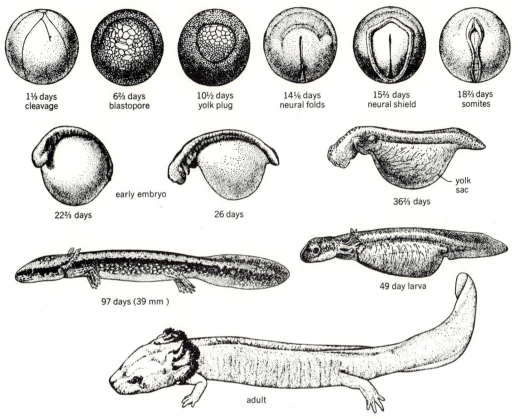

1⅓ days
cleavage

6⅔ days
blastopore

10½ days
yolk plug

14⅙ days
neural folds

15⅔ days
neural shield

18⅔ days
somites

early embryo

22⅔ days

26 days

yolk
sac

36⅔ days

97 days (39 mm)

49 day larva

adult

Figure 31-16 Mud puppy (*Necturus maculosus*, family PROTEIDAE): development and adult. (*Rearranged from Noble, Biology of the Amphibia.*)

morphosis usually involves (1) growth of a wide mouth and loss of the horny jaws; (2) loss of the gills, closure of the gill slits, and development of lungs; (3) development of a protrusible tongue; (4) emergence of the forelegs from the opercular chamber; (5) reduction in length of the intestine from the long (herbivorous) type to the short (carnivorous) form of the frog; and (6) resorption of the tail and median fins. There are also biochemical changes: (1) excretion of ammonia shifts more to the less toxic urea as transformation approaches; (2) the proportion of globulin to albumin in the blood shifts toward the latter, conserving water in the blood by preventing a drop in osmotic pressure; (3) the capacity of the blood to release oxygen quickly to the tissues increases as life becomes more active; (4) in carnivorous forms the activity of pepsin and trypsin, the enzymes of protein digestion, increases; and (5) enzymes released

from lysosomes in cells of the tail cause its resorption. During metamorphosis the young inhabit shallow water where both gill and lung respiration are possible and where insect food may be captured. Later they hide in crevices to avoid desiccation. Thousands may appear suddenly during a shower, thus accounting for the age-old belief of an occasional "rain of toads."

Fertilization is external in the CRYPTOBRANCHIDAE and HYNOBIIDAE and in most frogs and toads but is internal in the tailed frog (*Ascaphus*), most salamanders, and all caecilians. In copulation the tailed frog uses the "tail" and caecilians an eversible copulatory organ. Eggs of some land salamanders (PLETHODONTIDAE) are deposited in damp cavities; some hatch as larvae that have an aquatic stage, but most complete development within the egg to emerge as miniature adults, as do certain frogs. The embryos have broad, thin gills that serve for air breathing while in the egg. The many species of salamanders, frogs, and caecilians that have evolved, independently, complete development of eggs on land show the strong selection for terrestrial life among amphibians. Such a development long ago led to the evolution of the amniote eggs of higher vertebrates. Trends toward terrestrialism are seen even within closely related amphibian groups, which may contain a series of stages bridging the transition between aquatic and terrestrial habitats (leptodactylid frogs). The diversity in modes of reproduction is perhaps excelled only by that of the bony fishes. The male obstetrical toad (*Alytes*) loops the egg strings about the base of his hind legs until the larvae are ready for release into water. The female tongueless toad (*Pipa*) carries her eggs in separate dermal pockets on her back, and the marsupial frogs (Fig. 31-17) carry theirs in a broad pouch on the back. Male phyllobatid frogs carry the tadpoles on their backs (Fig. 31-20). The eggs of Darwin's frog (*Rhinoderma darwini*) are carried throughout their development in the vocal pouch of the male, and *Rheobatrachus silus*, a stream-dwelling Australian frog, transports and broods its larvae and juveniles in the stomach of the female! In the European salamander (*Salamandra salamandra*) the young are usually born as gilled larvae which are released into water after 10 to 12 months of development, but in the cold climates of the mountains of Western Europe they may be almost or completely metamorphosed at birth.

Some salamanders that have both an aquatic larval and terrestrial adult stage in the life cycle sometimes produce perennial larvae that reach adult size and persist indefinitely as larvae, a condition called **neoteny.** (The term is also applied to retarded development of structures.) If such larvae breed, they are called **paedogenic.** Paedogenesis occurs in ambystomatids, the tiger salamander (*Ambystoma tigrinum*) in the Rocky Mountains and elsewhere, the northwestern salamander (*A. gracile*), Cope's salamander (*Dicamptodon copei*) in western Washington, and others and in a plethodontid, the many-ribbed salamander (*Eurycea multiplicata*), in central United States. Paedogenesis may be considered as stopgap reproduction which maintains a population when conditions are unfavorable for transformation. Other salamanders are paedomorphic (having larval characteristics) but never transform—the mud puppy (*Necturus*), the European olm (*Proteus*), and the mud eel (*Siren*).

Some small amphibians may breed at 2 years of age, but larger forms mature more slowly. The length of life is known only for specimens in zoos; a few records are giant salamander, 55 years; European toad, 36 years; tiger salamander, 25 years as a larva and 11 years as an adult. Some red-bellied newts (*Taricha rivularis*) are known to have lived at least 17 years in the wild.

Figure 31-17 A tropical marsupial treefrog (*Gastrotheca marsupiata,* family HYLIDAE) with brood pouch cut open to show the eggs. (*After Noble, Biology of the Amphibia.*)

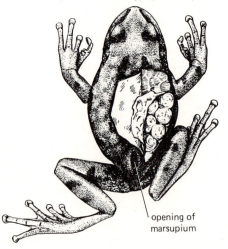

opening of
marsupium

31-25 Relation to man Amphibians furnish art motifs and play a role in the religions of primitive peoples. The American Indian medicine men used parts of frogs and toads in their magic. Toads, long used in Chinese medicine, may have some therapeutic value because of the digitalislike secretions in the skin.

Mark Twain's humorous story "The Celebrated Jumping Frog of Calaveras County" is now the basis for an annual jumping-frog contest at Angels Camp, Calif., where frogs from all over the United States are entered.

Frogs are used for elementary courses in biology, for fish bait, and for human food and have been used for human pregnancy tests. Many thousands are caught annually in the United States and Mexico. They are taken by hand, with nets, with spears, or by shooting. Axolotls of lakes near Mexico City are also used as food. The marine toad *Bufo marinus* is raised commercially for teaching and research, but most "frog farms" have not proved commercially practicable. The long developmental period of larvae and adults and low value of the final product often preclude financial success. The large marine toad (*Bufo marinus*) has been introduced in sugarcane fields in many countries to control insects.

The study of amphibians has contributed greatly to basic research in vertebrate evolution, physiology, and pharmacology. Their jelly-coated eggs, with an unobstructed view of embryonic growth, have made possible great advances in vertebrate embryology.

31-26 Fossil amphibians The earliest recognized amphibians (ichthyostegans; †*Ichthyostega* and others) are in Devonian rocks of Greenland and were of some size, as the skulls are 15 cm (6 in) long. The limbs were well developed, and there were opercular vestiges and a tail fin supported by bony fin rays. They probably derived from crossopterygian fishes, had both aquatic and aerial respiration, and had fins with skeletal supports from which the tetrapod limb could be derived. The Devonian was evidently a time of seasonal droughts, when the chances of survival were greater for an animal that could leave a drying pool to travel over land and seek water elsewhere. A less catastrophic view holds that terrestrial vertebrates evolved in moist environments where conditions were favorable for a gradual transition to life on land. Reduced competition and less predation in the land habitat, a new food supply, and higher environmental temperatures may have been important in this process. Some investigators think the primordial amphibian spent nights on land, away from aquatic predators, and days in the water, where it obtained most of its food. By Carboniferous time, with its coal swamps, there were two subclasses and some six orders of ancient amphibians. The skull became flattened and roofed with bone, and many had external armor of bony plates, possibly as pro-

tection against large eurypterids (arthropods) then in fresh waters. These amphibians were from a few centimeters to 4.6 m (15 ft) in length, some aquatic and some terrestrial. Some of the last types (Triassic) were permanently aquatic, and several had degenerated to become limbless, like eels or snakes. Salamanders appeared in the late Jurassic and were distinct from the earliest froglike forms, which appeared in the Triassic. Both show simplifying trends in the skull by loss of bones. There is as yet no agreement as to whether the three Recent groups of amphibians (frogs, salamanders, and caecilians) arose from one, two, or three separate ancestral lines.

Classification

Class Amphibia (Batrachia).

Amphibians. Living forms with moist glandular skin, no scales except in caecilians; typically 2 pairs of limbs (no paired fins); 2 nostrils connecting to mouth cavity, skull with 2 occipital condyles; heart 3-chambered; respiration by gills, lungs, skin, or lining of mouth cavity; eggs with gelatinous coverings, usually laid in water; larvae usually aquatic; adults in water or in moist places on land; Devonian to Recent; over 3400 living species.

Subclass 1. †Labyrinthodontia. Two sets of bony arch structures, intercentrum and pleurocentrum, in central region of vertebrae below each neural arch; not froglike; tail present, often long; legs all of about same size; cranium and face completely roofed with bony plates; often armored ventrally with overlapping scales, some also with dorsal scales.

Order 1. †Ichthyostegalia. Vertebrae resemble those of temnospondyls; caudal fin partly supported by dermal rays; primarily aquatic; Late Devonian to Early Carboniferous. †*Icthyostega*, †*Elpistostege*.

Order 2. †Temnospondyli. Most with pleurocentra of vertebrae as small paired blocks, typically laterodorsal, but in later forms reduced or absent and intercentrum alone present below neural arches; vomers broad; Late Devonian to Triassic. †*Eryops*, to 2.5 m (8 ft) long, body narrow; †*Mastodonsaurus*, large, skull 90 cm (3 ft) long.

Order 3. †Anthracosauria. Pleurocentra complete disks enclosing notochord; intercentra disk-shaped or reduced to small ventral wedges; vomers narrow; Upper Carboniferous to Upper Permian. †*Eogyrinus*, to 4.5 m (15 ft), limbs small; †*Seymouria*, to 50 cm (20 in) long, with ribs, tail short, general structure intermediate between amphibians and reptiles.

Subclass 2. Lepospondyli.   Centrum of vertebrae single, often spool-shaped, with longitudinal opening for persistent notochord.

Order 4. †Nectridea.   Tail with expanded neural and haemal processes; Carboniferous to Lower Permian. †*Sauropleura*, body long, limbs small or none, skull pointed. †*Diplocaulus*, skull triangular, "horned," limbs and jaws small.

Order 5. †Aistopoda.   No limbs; body long, snakelike, ribs 100 +; Carboniferous and Permian. †*Ophiderpeton*, †*Phlegethontia*.

Order 6. †Microsauria.   Small, body slender, limbs short; skull lengthened postorbitally; Carboniferous to Permian. †*Euryodus*, †*Microbrachis*, †*Lysorophus*.

(Note: The following three living subclasses of amphibians are frequently combined in the subclass Lissamphibia, not recognized here.)

Subclass 3. Salientia.   Centra of vertebrae reduced or absent, replaced by downgrowths of neural arches; skull bones reduced; trunk short; elongated ilia; hind legs usually longer than forelegs.

Order 7. †Proanura.   †PROTOBATRACHIDAE.   Early frogs. Precaudal vertebrae 16, caudal 3 or 4; no urostyle; hind legs only slightly enlarged; Triassic (Madagascar). †*Triadobatrachus* (*Protobatrachus*).

Order 8. Anura.   Toads and frogs. No tail; skull thin, no solid roof, much reduced, few bones; vertebrae few, the last a slender urostyle formed by fusion of vertebrae; ribs reduced or none; hind legs usually enlarged for leaping, webbed between toes; considerable cartilage in skeleton; egg deposition and fertilization usually external by clasped pairs of adults (Fig. 31-19); larva (tadpole) of fused ovoid head + body and long tail with median fins, no true teeth, usually aquatic; metamorphosis conspicuous; about 2900 species; Jurassic to Recent.

[Note: In the first five families listed, with one exception, the posterior portion of the halves of the pectoral girdle move freely over one another and are arciferal, and males typically clasp females about the waist when mating (pelvic amplexus).]

LEIOPELMATIDAE (Ascaphidae).   Ribbed frogs. Vertebrae connected by undivided intervertebral disks, not truly amphicoelous; presacral vertebrae 9; free ribs in adults; relict "tail" muscles; only 2 genera: *Leiopelma*, 3 species, only amphibians native to New Zealand; direct development of eggs; *Ascaphus truei* (Fig. 31-18), tailed frog, northwestern California to southwestern British Columbia and Montana, in or near cool mountain streams; male has taillike copulatory organ used for internal fertilization (unique in ANURA); tadpole with single midventral spiracle and large suctorial mouth for adhering to stones in swift waters, many labial tooth rows. †*Vieraella*, Early Jurassic (Argentina). Earliest known frog.

DISCOGLOSSIDAE. Disk-tongued toads. Resemble leiopelmatids but no tail muscles or cloacal appendage; vertebrae opisthocoelous; tongue and eyelids present; tadpoles with midventral spiracle, well-developed mandibles, and labial tooth rows; North Africa, Europe, Asia, Philippines; Cretaceous. *Discoglossus*; *Bombina*, fire-bellied toads, Eurasia, black and bright red or yellow below; *Alytes obstetricans*, midwife toad, Europe, eggs looped around base of hind legs of male, then hatching tadpoles deposited in water.

†PALAEOBATRACHIDAE.   Vertebrae procoelous; urostyle with double condyle. †*Palaeobatrachus*, Late Jurassic to Pliocene, Europe.

PIPIDAE. Tongueless frogs. No tongue; feet fully

Figure 31-18 Male tailed frog (*Ascaphus truei***, family LEIO-PELMATIDAE) of northwestern United States.**   Head-and-body length to 5 cm (2 in). Enlarged inset shows cloacal appendage of male in ventral view. (*After Noble, Biology of the Amphibia.*)

webbed; strictly aquatic; tadpoles with a spiracle on each side of body, mouth slitlike, with narrow lips, filter feeders; South America, Africa; Cretaceous. *Pipa pipa*, Surinam toad, northern South America, no eyelids, front toes end in groups of papillae, eggs in separate pockets on female's back, hatched as small young; *P. parva*, young leave pockets in tadpole stage; *Xenopus*, clawed frog, Africa, 3 inner hind toes with black horny claws, eggs many, laid singly on plant, adults used in pregnancy tests.

RHINOPHRYNIDAE. Mexican burrowing toad, *Rhinophrynus dorsalis*, only living species in family; Middle America; Eocene–Oligocene. Tongue free, rather than attached in front, protrusible; tadpole resembles that of pipids; adults apparently feed on termites.

PELOBATIDAE. Spadefoot toads. Vertebrae procoelous or opisthocoelous; no ribs; teeth in upper jaw; horny sharp-edged spade on inner edge of hind foot; pupil vertically elliptical; tadpole resembles that of leptodactylids and "higher" frogs. *Pelobates*, Europe and North Africa; *Megophrys*, horned-nosed frogs, southeastern Asia; *Scaphiopus*, spadefoot toads, North America; nocturnal, often hide in burrows by day, seldom seen except when spawning in shallow pools after spring or summer rains; development rapid, egg to young toad in 15 to 30 days.

MICROHYLIDAE (Brevicipitidae). Microhylids. Extremely diverse; vertebrae mostly procoelous; firmisternal; tadpoles with midventral spiracle and lips with folds, more elaborate than in pipids, filter feeding common; many species have direct development, metamorphosing within egg. New World, Africa, Asia, Australia; Miocene. Arboreal to burrowing, the latter stocky with small pointed heads and slender toes. *Gastrophryne carolinensis*, eastern narrow-mouth toad, southern Maryland and Florida to Texas; nocturnal, hides in ground by day.

[*Note: Remaining families except leptodactylids, some bufonids, hylids, and centrolenids typically are firmisternal, having the two halves of the pectoral girdle firmly joined medially; all typically have tadpoles with one spiracle on the left side, horny mandibles, and rows of comblike labial teeth; males clasp females around the chest when mating* (pectoral embrace).]

LEPTODACTYLIDAE. Southern frogs. Resemble bufonids but maxillary teeth usually present and no Bidder's organ; tadpoles with single spiracle on left side; mouth typically with mandibles and labial teeth. Southern United States to southern South America, West Indies. Some 625 species and some 40 genera. *Leptodactylus*, South America to Mexico and Texas, eggs in frothy nests near water; *Eleutherodactylus*, with some 340 species, includes barking frog, Central America, West Indies, 1 species in extreme southern United States, eggs laid on moist land, no tadpole stage.

MYOBATRACHIDAE. Australian frogs. Resemble leptodactylids; confined to Australia. Habits vary, aquatic or terrestrial; development of eggs and larvae in water or direct development on land; some build foam nests in burrows and some brood eggs and young in stomach or inguinal pouch; 24 genera and about 100 species.

BUFONIDAE. True toads. No teeth; males usually have Bidder's organs, a pair of structures anterior to the testes which, following removal of the latter, may develop into functional ovaries; often a large parotid gland behind each eye; most with aquatic larvae; virtually worldwide, not native to Australo-Papuan region; Paleocene. *Bufo*, common toads, widely distributed; skin rough, many warts (Fig. 31-19); terrestrial and nocturnal, hiding under logs, stones, etc., or in burrows by day; enter water briefly to spawn; eggs usually in gelatinous strings; some 200 species, 18 in United States. *B. americanus*, American toad, Hudson Bay to Louisiana and west to Missouri River; *B. boreas*, western toad, Rocky Mountains and Pacific states to Alaska. Other genera, apparently representing separate bufonid radiations in southeastern Asia, Africa, and the Americas, are sometimes placed in other families; among such genera are *Ansonia* (about 15 species), *Nectophrynoides* (5 species; the only live-bearing anurans), and *Atelopus* (40 species).

BRACHYCEPHALIDAE. Related to bufonids; halves of pectoral girdle more or less fused ventrally; no Bidder's organ. One species, *Brachycephalus ephippium*, golden frog, less than 2 cm ($\frac{3}{4}$ in) long; southeastern Brazil.

RHINODERMATIDAE. Darwin's frog. Resemble leptodactylids; tadpoles carried in vocal pouch of male; southern South America.

Figure 31-19 Mated pair of Yosemite toads (*Bufo canorus*). These exhibit the greatest sexual difference in color of any amphibians in the United States: male (above) olive green, female black-and-white. (*Photo by T. I. Storer.*)

HYLIDAE. Treefrogs. Many under 5 cm (2 in), but some to 14 cm (5½ in), teeth in upper jaw; terminal bone of each digit claw-shaped; many with expanded "adhesive" disks on toes used to climb trees, rocks, etc.; an intercalary cartilage in the digits; voice often loud; eggs usually in water; some burrow, a few are highly aquatic; widespread but absent in Africa south of Sahara; common in American tropics; some 80 species, many in genus *Hyla* (color plate). *Hyla versicolor,* gray treefrog, southern Canada to Texas and eastward; *H. regilla,* Pacific treefrog, Lower California to British Columbia; *H. crucifer,* spring peeper, southern Canada to Texas and eastward; *Acris crepitans,* cricket frog, South Dakota and New Mexico eastward; *Gastrotheca,* marsupial frogs, South America, egg mass carried on back in skin sac, whence young frogs hatch (Fig. 31-17).

CENTROLENIDAE. Leaf frogs. Resemble hylids; most are small, green above, venter transparent; eggs in masses on leaves above streams; hatching tadpoles fall into water; tropical America. *Centrolene.*

DENDROBATIDAE. Arrow poison frogs. Have features in common with leptodactylids; maxillary teeth often present; toes may have dermal scutes; small, brightly colored, diurnal; toxic skin secretion of some used to poison arrow and dart points; tadpoles transported to water on back of parent (Fig. 31-20). New World tropics, Nicaragua south. *Dendrobates, Phyllobates.*

RANIDAE. True frogs. Teeth in upper jaw; eggs of species in United States usually in tapiocalike masses; worldwide except southern South America and most of Australia; Oligocene. Only *Rana,* with 19 species, in United States (Fig. 31-1), a common species being *R. catesbeiana,* bullfrog, southern Canada

to Texas and eastward, introduced in Western states; *R. clamitans,* green frog, southern Canada to Texas and eastward; *R. palustris,* pickerel frog, Arkansas to southern Canada and eastward; *R. pipiens,* northern leopard frog or grass frog, North America chiefly east of the Sierra-Cascade mountains to East Coast and Canada to New Mexico; *R. sylvatica,* wood frog, South Carolina to Nova Scotia and Alaska; *R. aurora,* red-legged frog, British Columbia to Lower California.

RHACOPHORIDAE. Old World treefrogs. Resemble ranids but intercalary cartilage in digits and toe tips specialized for climbing; Africa, southeastern Asia; mostly arboreal. *Rhacophorus,* eggs usually laid in gelatinous foam over or near water, whence tadpoles drop or are washed into water; *R. nigropalmatus,* "flying" frog, glides or parachutes on widely spread webbed feet.

SOOGLOSSIDAE. Seychelles frogs. Resemble ranids; confined to Seychelles Islands, Indian Ocean; tadpoles transported on back of adult. *Sooglossus.*

Subclass 4. Urodela. Lepospondylous vertebral centra formed by direct deposition of bone around notochord; no fusion of frontal and parietal; trunk not shortened; no elongation of ilia; hind legs and forelegs about same length.

Order 9. Caudata. Includes salamanders, newts. Head, trunk, and tail usually distinct; limbs of about equal size; limb girdles largely cartilaginous; no dermal shoulder girdle; larvae resemble adults in form and have teeth in both jaws; Upper Jurassic to Recent; about 325 species.

Suborder 1. Cryptobranchoidea. No gills in adult; angular and prearticular of lower jaw separate; no spermatophores, fertilization external; eggs in gelatinous sacs; Upper Cretaceous to Recent.

HYNOBIIDAE. Asiatic salamanders. Small; with eyelids; vomerine teeth in V or M shape behind nares; fertilization external; eggs laid in sacs; metamorphosis complete. About 30 species, northern Asia, Kamchatka, Japan, and Taiwan to Hupeh and Ural Mountains; Paleocene (?) to Recent. *Hynobius; Ranodon,* lives in brooks; *Onychodactylus,* lungless, in mountain brooks, larvae with clawed digits.

CRYPTOBRANCHIDAE. Giant salamanders and hellbenders. Body depressed, skin soft, flabby, with fleshy folds at sides; no eyelids; vomerine teeth in arc parallel to maxillary teeth; permanently aquatic; 3 species; Oligocene to Recent. *Cryptobranchus alleganiensis,* hellbender (Fig. 31-21), Mississippi to Ohio and New York, in running water with rocks, snags, or other cover; to 74

Figure 31-20 Tadpole-carrying frog (*Colostethus punctatus,* family DENDROBATIDAE); the male serves as the tadpole nurse (× 2).

Figure 31-21 The hellbender (*Cryptobranchus alleganiensis,* suborder CRYPTOBRANCHOIDEA); length to 73 cm (29 in). (*After Noble, Biology of the Amphibia.*)

cm (29 in) long; eats small fishes and invertebrates; fertilization external; in late summer mates and lays to as many as 450 eggs in rosarylike strings; larvae hatch in November, require several years to mature. *Andrias* (*Megalobatrachus*)*japonicus,* giant salamander of Japan, and *A. davidianus* of eastern China, in cold brooks; *A. japonicus* to 175 cm (69 in) long; †*A. scheuchzeri,* Miocene in Europe, first referred to several centuries ago as "*homo diluvii testis*" (human witness of the deluge).

Suborder 2. Sirenoidea. Permanent larvae, few adult characteristics; body slender; no eyelids or hind legs, front legs small; gills persistent; jaws with horny covering; no cloacal glands; fertilization probably external; strictly aquatic; Late Cretaceous to Recent.

SIRENIDAE. Sirens. Three species. Late Cretaceous to Recent. *Siren* (2 species), Virgina and southern Michigan to Florida and Texas, in muddy ditches and ponds; *S. lacertina,* greater siren, grows to 96 cm (38 in), 4 toes, 3 pairs of gill slits; *Pseudobranchus striatus,* dwarf siren, South Carolina through Florida, to 25 cm (10 in) long, 3 toes, 1 pair of gill glits, burrows. Courtship and mating of sirenids unknown.

(*Note: In salamanders of remaining families the male produces spermatophores; the female has a seminal receptacle, and fertilization is internal.*)

Suborder 3. Salamandroidea. Tooth-bearing extensions of prevomers along parasphenoids; Paleocene to Recent.

PROTEIDAE. Mud puppies and olms. Body depressed, tail with fin; permanent larvae, adults with 3 pairs of gills, 2 pairs of gill slits; lungs present, no eyelids; teeth along jaws, 2 rows in upper, 1 in lower; aquatic bottom dwellers; 6 spe-

cies; Miocene to Recent. *Necturus,* mud puppies; eastern United States. *Necturus maculosus,* mud puppy, or waterdog (Fig. 31-16), Manitoba to Louisiana; to over 33 cm (13 in) long, gray or rusty brown with blackish spots; eats small fishes and invertebrates; mates in autumn, when females take up spermatophores deposited by males; eggs laid in May or June, 18 to 180 in "nests," attached individually by jelly stalks to undersides of stones and guarded by female; hatch in 5 to 9 weeks, larvae about 2.5 cm (1 in) long; growth slow, about 6 years to full size; *N. punctatus,* dwarf waterdog, coastal plain from Virginia to Alabama; *Proteus anguineus,* olm of southeastern Europe, in waters of deep caves, unpigmented, eyes under skin.

SALAMANDRIDAE. Salamanders and newts. Teeth in roof of mouth behind nares in 2 long rows, diverging posteriorly; adults with lungs but no gills; eyelids present; vertebrae usually opisthocoelous; Upper Cretaceous, Paleocene to Recent. Europe, North Africa, eastern Asia, and North America; dominant family of Old World salamanders. *Notophthalmus viridescens,* red-spotted newt, or eft, Atlantic Coast to southern Canada and Texas, with reddish spots along sides, to 12.5 cm (5 in) long, eggs single, in water; *Taricha* (*Triturus*) *torosa* and others, Pacific Coast newts, or waterdogs, California to southern Alaska, brown above and orange beneath, to 23 cm (9 in) long, eggs in small masses or laid singly in water; *Triturus cristatus,* European crested newt, male brilliant, with dorsal crest in breeding season; *Salamandra salamandra,* European fire salamander, black with yellow spots, to 15 cm (6 in) long, adults terrestrial; *S. atra,* European alpine salamander, live-bearing.

AMPHIUMIDAE. Congo eels. Body cylindrical; no eyelids; adults with lungs and 1 pair of gill clefts but no gills; limbs diminutive, toes 1, 2, or

3; 3 species; southeastern United States; *Amphiuma means,* two-toed amphiuma. Virginia to Louisiana and Florida, in streams, muddy pools, swamps, and rice fields, to over 100 cm (40 in) long, feeds on mollusks, crayfishes, and small fishes; Late Cretaceous to Recent.

Suborder 4. Ambystomatoidea. Prevomers short, lack posterior processes; teeth across rear margins of vomers, angular and prearticular of lower jaw fused; eyelids and lungs present; adults usually on land; Paleocene to Recent.

AMBYSTOMATIDAE. Mole salamanders, or ambystomatids. Teeth in transverse row behind nares, across posterior margins of vomers, none on palatine bones; with eyelids; vertebrae amphicoelous; adults usually terrestrial; 35 species, North America south to southern margin of Mexican Plateau; Miocene to Recent. *Ambystoma,* 28 species; *A. tigrinum,* tiger salamander (Figs. 31-1, 13-19), southern Canada to southern Mexico, adults to about 40 cm (14 in), most about 20 cm (8 in) long, larvae often neotenic as axolotls; *A. opacum,* marbled salamander, Eastern states, eggs laid in moist places; *Dicamptodon ensatus,* Pacific giant salamander, Pacific Coast, to 30 cm (12 in) long, larvae may be neotenic; *Rhyacotriton olympicus,* Olympic salamander, northwestern California to Washington, in cold mountain streams, lungs only 6 mm (¼ in) long; *Ambystoma (Bathysiredon) dumerili,* paedogenic, never known to metamorphose, in Mexican lakes.

PLETHODONTIDAE. Lungless salamanders (color plate). Length 2.5 to 30 cm (1 to 12 in), most under 15 cm (6 in); a minute nasolabial groove from each nostril to upper lip; vomerine teeth usually in two arched series behind nares, continuing posteriorly between orbits, where they may form isolated single or paired parasphenoid patch; eyelids present; no lungs; skin delicate; most adults terrestrial in moist places where eggs are deposited, guarded by female, and young emerge with no larval stage; few in cold brooks with aquatic larvae; Upper Cretaceous to Recent. Around 225 species, 150 of them tropical, southern Canada to Bolivia and Lower California, none in Great Basin or Rocky Mountains except extreme southern part; two species in southern Europe, *Hydromantes; Pseudotriton ruber,* red salamander; *Desmognathus fuscus,* dusky salamander; *Plethodon glutinosus,* slimy salamander; *Aneides lugubris,* arboreal salamander of California; *Eurycea bislineata,* two-lined salamander; *Typhlomolge rathbuni,* blind, nearly colorless, permanently aquatic, paedomorphic, in underground waters of Balcones

Escarpment, central Texas; *Bolitoglossa,* largest genus (over 50 species) with many arboreal, Mexico to Bolivia; *Thorius,* Mexico, smallest living salamanders.

Subclass 5. Gymnophiona (Apoda).

Order 10. Gymnophiona. Caecilians. Body slender, elongate, no limbs or limb girdles; skull compact, roofed with bone, modified for burrowing; vertebrae many; ribs long; skin smooth, with transverse furrows (annuli) and both slime glands and "squirt glands" (discharging an irritating fluid); some with dermal scales recessed in annular folds; a small protrusible tentacle between eye and nostril; eyes lidless, degenerate, beneath skin or skull bones; tail short, anus near or at end of body; protrusible copulatory organ; development of fetuses of viviparous species depends on ingesting the secretion of epithelial cells scraped from maternal oviducal lining with specialized fetal teeth; tropical; Paleocene fossil (single vertebra) from Brazil; about 165 species.

CAECILIIDAE. India to Phillipines, Borneo, Java, and Seychelles; east and west Africa; Mexico to Argentina; live and burrow in moist ground; feed on invertebrates; some genera and species deposit their eggs in moist ground, others viviparous; some egg-laying species complete larval development in egg, others have aquatic larvae. *Gymnopis,* viviparous; *Hypogeophis,* direct development.

TYPHLONECTIDAE. Northern South America; aquatic; *Typhlonectes,* aquatic throughout life and viviparous, producing up to 9 embryos that grow to 150 mm (6 in) long before birth.

ICHTHYOPHIIDAE. Southeastern Asia, northern South America; oviparous; free-living larvae; *Ichthyophis* (Fig. 31-1).

SCOLECOMORPHIDAE. East Africa; viviparous; *Scolecomorphus.*

RHINATREMATIDAE. Northern South America; oviparous; free-living larvae; *Rhinatrema.*

References

Articles on amphibians appear in many periodicals, especially in the quarterly *Copeia* (published by the American Society of Ichthyologists and Herpetologists), in *Herpetologica* (Herpetologists League), *Journal of Herpetology* (Society for the Study of Amphibians and Reptiles), and *British Journal of Herpetology* (British Society of Herpetologists). For many states and Canadian provinces there are state or local lists giving the distribution and natural history of amphibians occurring in those areas. Individual accounts of American species of amphibians and reptiles are published in *The Catalogue of American*

Amphibians and Reptiles, issued by the Catalogue Committee of the Society for the Study of Amphibians and Reptiles, American Museum of Natural History, New York, N.Y. The accounts provide a concise description, pertinent literature, distribution map, and other information for each species. Over 200 accounts, mostly of North American species, have been published.

Bishop, S. C. 1943. Handbook of salamanders of the United States, of Canada and of Lower California. Ithaca, N.Y., Comstock Publishing Associates, a division of Cornell University Press. xiv + 555 pp. *Excellent natural history accounts but taxonomy out of date.*

Blair, W. F. (editor). 1972. Evolution in the genus *Bufo.* Austin, University of Texas Press. viii + 459 pp. *Evolutionary relationships among toads assessed by a variety of methods.*

Cochran, D. M. 1961. Living amphibians of the world. Garden City, N.Y., Doubleday & Co., Inc. 199 pp., 220 figs., incl. 77 in color.

————, and **C. J. Goin.** 1970. The new field book of reptiles and amphibians. New York, G. P. Putnam's Sons. xxii + 359 pp., 16 pls., 100 figs. *Identification and habits of all species in the United States.*

Conant, R. 1975. A field guide to reptiles and amphibians of eastern and central North America. Boston, Houghton Mifflin Company. xviii + 429 pp., 32 color pls., 105 figs. *Identification, habits in brief, distribution with maps; covers the United States and Canada east of the 100th meridian.*

Goin, C. J., O. B. Goin, and **G. R. Zug.** 1978. Introduction to herpetology, 3d ed. San Francisco, W. H. Freeman and Company. xi + 378 pp. *An introductory text for the study of amphibians and reptiles.*

Lofts, B. 1974. Physiology of the Amphibia, Vol. II. New York, Academic Press, Inc. xi + 592 pp., 114 figs. *Focuses primarily on the physiologic and environmental processes associated with amphibian reproduction.*

Moore, J. A. (editor). 1964. Physiology of the Amphibia, Vol. I. New York, Academic Press, Inc. xii + 654 pp., few figs. *Digestion, respiration, metabolism, and the usefulness of amphibians for elucidating the processes of development, metamorphosis, and regeneration.*

Muntz, W. R. A. 1964. Vision in frogs. *Sci. Am.,* vol. 210, no. 3, pp. 110–119. *The frog's response to blue light and the analyzing role of the retina.*

Noble, G. K. 1931. Biology of the Amphibia. New York, McGraw-Hill Book Company. Reprinted 1955. Dover Publications, Inc. xiii + 577 pp., 174 figs. *Structure, function, life histories, and classification.*

Porter, K. R. 1972. Herpetology. Philadelphia, W. B. Saunders Company. xi + 524 pp. *Textbook in herpetology; structure, function, origin, relationships, distribution, and natural history of amphibians and reptiles.*

Stebbins, R. C. 1966. A field guide to western reptiles and amphibians. Boston, Houghton Mifflin Company. xiv + 270 pp., 39 pl., 39 figs. *Identification, habits, distribution with maps.*

Twitty, V. C. 1966. Of scientists and salamanders. San Francisco, W. H. Freeman and Company. ix + 178 pp., 59 figs. *Autobiography of a scientist and the major role of salamanders in experimental and field research.*

Vial, J. L. 1973. Evolutionary biology of the anurans. Columbia, University of Missouri Press, xii + 470 pp.

Wake, M. H. 1977. The reproductive biology of caecilians: an evolutionary perspective. In The reproductive biology of amphibians. Taylor, D. H. and S. I. Guttman, editors. New York, Plenum Publishing Corp. 475 pp.

Wright, A. A., and **A. H. Wright.** 1949. Handbook of frogs and toads of the United States and Canada. 3d ed. Ithaca, N.Y., Comstock Publishing Associates, a division of Cornell University Press. xi + 652 pp., 163 figs. *Keys and descriptions.*

CLASS REPTILIA: REPTILES

The class REPTILIA includes the lizards and snakes (order SQUAMATA), the turtles and tortoises (order CHELONIA), the crocodiles and alligators (order CROCODILIA), and the New Zealand tuatara (*Sphenodon punctatus*, order RHYNCHOCEPHALIA). (Fig. 32-1). These are only four of the some sixteen known orders that flourished during Mesozoic time, the Age of Reptiles, when they were the dominant animals. The class name refers to the mode of travel (L. *reptum*, creep), and the study of reptiles (and often including amphibians) is called herpetology (Gr. *herpeton*, reptile).

32-1 Characteristics

1 Body covered with dry cornified skin (not slimy), usually with ectodermal scales or scutes; few or no skin glands.

2 Two pairs of limbs, each typically with five toes ending in horny claws and suited to running, crawling, or climbing; limbs paddlelike in marine turtles, reduced in some lizards, and absent in a few other lizards and in all snakes (vestiges in boas, etc.).

3 Skeleton well ossified; skull with one occipital condyle.

4 Heart incompletely four-chambered, two atria and a partly divided ventricle (ventricles separate in crocodilians); one pair of aortic arches; red blood corpuscles nucleated, biconvex, and oval.

5 Respiration by lungs; gills absent; pharyngeal and perhaps cloacal respiration in some aquatic turtles.

6 Twelve pairs of cranial nerves.

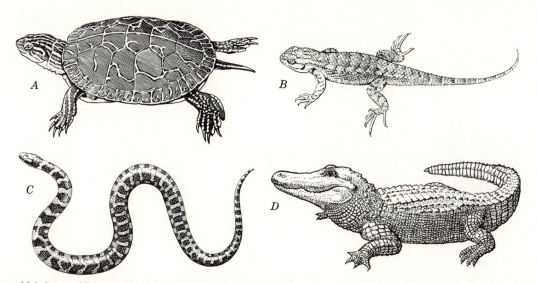

Figure 32-1 Types of living reptiles (class REPTILIA); all reduced, not to scale. *A*. Painted turtle (*Chrysemys*, order CHELONIA). *B*. Lizard (*Sceloporus*), and *C*. water snake (*Natrix*; both, order SQUAMATA). *D*. Alligator (*Alligator*; order CROCODILIA). (*A, B, C from Stebbins, Amphibians and reptiles of western North America; D after Palmer, Field book of natural history.*)

7 Excretion by metanephric kidneys; nitrogenous wastes chiefly uric acid in lizards, snakes, and terrestrial chelonians.

8 Body temperature variable (ectothermal).

9 Fertilization internal, usually by copulatory organ; eggs large, with much yolk, in leathery or limy shells; usually laid, but retained for development by some lizards and snakes.

10 Cleavage meroblastic; embryonic membranes (amnion, chorion, yolk sac, and allantois) present during development; young when hatched (or born) resemble adults; no metamorphosis.

The reptiles show advance over the amphibians in having (1) a dry scaly body covering adapted to life away from water, (2) limbs suited for rapid locomotion, (3) further separation of the oxygenated and unoxygenated blood in the heart, (4) complete ossification of the skeleton, and (5) eggs suited for development on land, with membranes and shells to protect the embryo. Reptiles lack the insulated body covering, internally regulated body temperature, and some other features of birds and mammals.

32-2 Evolution Reptiles were the first group of vertebrates to adapt to life in dry places on land.

Some of their characteristics which make this possible are as follows:

1 Increased resistance of the skin to water loss (as compared with amphibians). Skin glands, important in amphibian cutaneous respiration, decreased in importance as lungs assumed the primary role in respiration and poison glands gave way to other methods of defense.

2 Thickening and cornification of the skin with abandonment of its respiratory role. The skin thus afforded increasing protection against the hazards of abrasion and water loss in dry environments. However, small areas of thin skin have remained between reptilian scales, providing flexibility on parts that would otherwise have become a rigid dermal armor.

3 Claws which protect the toe tips and aid locomotion on rough surfaces.

4 A copulatory organ (absent or uncommon in lower vertebrates) for direct transfer of sperm to the reproductive tract of the female.

5 Eggs with shells resistant to water loss and containing a fluid-filled chamber, the amniotic cavity, which protects the developing embryo against desiccation and mechanical injury (Sec. 32-26).

6 Reduction in urinary water loss through a shift toward production of hypertonic urine and excretion (chiefly in arid-land species) of nitrogenous wastes as uric acid.

7 A general increase and narrowing of the range in temperatures selected for activity, permitting greater exploitation of warmer parts of the thermal environment; probable attainment of endothermy in some fossil groups, including some dinosaurs.

8 Action of the tongue in transporting chemical sense data from the external environment to the Jacobson organs in the absence of water transport as found in amphibians (Sec. 31-2).

Although reptiles have entered all major habitats, the above characteristics appear to be primarily responsible for their abundance in arid and semiarid regions. In many desert regions they are the dominant diurnal vertebrates, and the snakes, slender in form and with well-developed powers of chemoreception, have become the most important subterranean vertebrate predators in such arid lands.

32-3 Size Fossil reptiles ranged in size from the small lizards to the great sauropod dinosaurs, over 24 m (80 ft) long, the largest terrestrial animals ever known. Of living reptiles, size extremes are the reticulated python of Southeast Asia, to 10 m (33 ft); the South American anaconda, to over 9 m (30 ft), perhaps 11 m (37 ft); the Komodo lizard (*Varanus komodoensis*) to 3 m (10 ft); a slimmer relative, Salvador's monitor (*V. salvadorii*) perhaps to 5 m (15 ft); and the marine leatherback turtle, which may exceed 2 m (6 ft) and a weight of 544 kg (1200 lb). Some land tortoises of the Galápagos Islands exceed 120 cm (4 ft) in length and formerly reached weights of over 180 kg (400 lb). Large crocodiles average 2.5 to 4.5 m (8 to 15 ft), with the Orinoco crocodile recorded at 7 m (23 ft). Some burrowing blind snakes (*Leptotyphlops* and *Typhlops*) are around 15 cm (6 in) long, and a lizard, a gecko (*Sphaerodactylus*) of the West Indies, is about 5 cm (2 in) long. Most North American snakes are 25 to 150 cm (10 to 60 in) long, and the lizards usually under 30 cm (12 in).

Structure of a reptile: The alligator

32-4 External features The **body** comprises a distinct head, neck, trunk, and tail; each of the short **limbs** bears toes tipped with horny claws and with webs between. The long **mouth** is margined with conical teeth, set in sockets. Near the tip of the snout are two small valvular **nostrils.** The **eyes** are large and lateral, with upper and lower eyelids, and a transparent **nictitating membrane** moves backward beneath the lids. The **ear** opening is behind the eye under a movable flap. The **vent** is a longitudinal slit, behind the bases of the hind limbs.

32-5 Body covering The tough leathery **skin** is sculptured into rectangular **horny scales** (or scutes) over most of the trunk and tail. The scales are generally in transverse and lengthwise rows, with furrows of softer skin between. The cornified exterior wears and is replaced by additional cornified layers from the epidermis beneath. Adults have separate bony **dermal plates** (*osteoderms*) under the dorsal scales from neck to tail; these are rectangular or oval, often pitted, and some have a median keel. Some species also have osteoderms on the belly. There are two pairs of epidermal **musk glands,** one pair opening on each side of the undersurface of the jaw and the other pair within the cloaca.

32-6 Skeleton (Fig. 32-2) The massive **skull** includes a long snout, and the bones are usually pitted in old adults. The long lower jaw articulates at each side of the posterior margin of the skull on a fixed quadrate bone. Ventrally on the cranium is the long **hard palate,** above which are the respiratory passages.

The **vertebral column** comprises five types of vertebrae: 9 cervical, 10 thoracic, 5 lumbar, 2 sacral, and about 39 caudal. On the cervical vertebrae are short, free **cervical ribs;** the thoracic vertebrae and sternum are connected by **thoracic ribs,** with cartilaginous ventral extensions, and between the sternum and pubic bones are 7 pairs of V-shaped **abdominal ribs** (*gastralia*) held in a longitudinal series by ligaments.

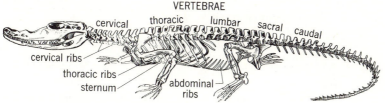

Figure 32-2 Skeleton of a crocodile. (*Adapted from Claus.*)

32-7 Muscular system (Fig. 32-3) As compared with a frog, the alligator has more diversity in its muscles, in keeping with the greater variety of its bodily movements both on land and in water. The muscles of the head, neck, and limbs are well differentiated, though of less bulk than those of a mammal. The segmental muscles of the vertebral column, ribs, and tail are conspicuous.

32-8 Digestive system The large **mouth** (Fig. 32-4) can open widely and has stout **teeth** used for offense and defense and also to grasp and twist large items of prey. A flat **tongue** lies in the floor of the mouth cavity but cannot be protruded. On the hind margin of the tongue is a transverse fold opposite a similar one on the palate; when pressed together, these shut off the mouth cavity from the pharynx, so that an alligator in water may open its mouth without having water enter the lungs. Beyond the short

pharynx is the **esophagus,** a slender tube leading to the **stomach.** The latter comprises a large spherical or fundus region and a smaller pyloric portion on the right side. This connects to the coiled **small intestine,** which joins the larger **rectum** leading to the **cloaca** and **vent.** The **liver,** of two lobes, lies anterior to the stomach, and the **pancreas** is in the first or duodenal loop of the intestine, the ducts of both enter the forepart of the intestine. The digestive, excretory, and reproductive systems terminate in the cloaca.

32-9 Circulatory system The **heart** lies in the anteroventral part of the thorax; it comprises a small **sinus venosus,** two **atria,** and two **ventricles.** The ventricles are completely separated in crocodilians but incompletely so in other reptiles. Blood from the veins passes in turn through the (1) sinus venosus, (2) right atrium, (3) right ventricle, (4) pulmonary

Figure 32-3 Muscles of a lizard (*Lacerta*). Behind the ''arm'' successive muscle layers are removed to reveal those beneath: 1, outermost layer; 6, innermost layer. (*After Nierstrasz & Hirsch.*)

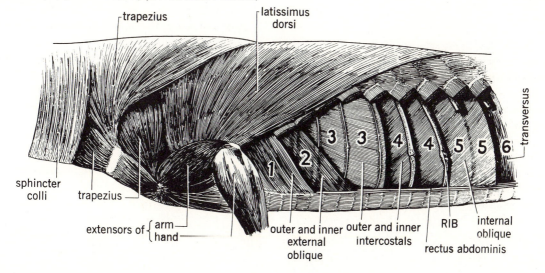

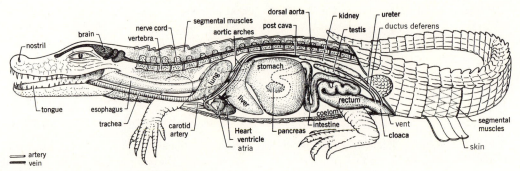

Figure 32-4 American alligator: general structure.

artery to each lung, (5) pulmonary veins from lungs, to the (6) left atrium, and (7) left ventricle. It emerges in a pair of **aortic arches** that pass dorsally around the esophagus; from the base of the right arch two **carotid arteries** lead to the neck and head, and a **subclavian artery** to each forelimb. The two aortic arches join as a **dorsal aorta** that distributes blood to organs in the body cavity and to the hind limbs and tail. Venous blood is collected (1) by an **anterior vena cava** on each side from the head, neck, and forelimb; (2) by a single middorsal **posterior vena cava** from the reproductive organs and kidneys; (3) by a **hepatic portal vein** from the digestive tract that breaks up into capillaries in the liver and collects as a short **hepatic vein;** and (4) by an **epigastric vein** on each side of the abdominal cavity from the posterior limbs, tail, and body. All these veins empty into the sinus venosus.

32-10 Respiratory system Air enters the **nostrils** (external nares) and passes above the hard palate to the **internal nares** behind the **palatine valve** (*velum*) and thence through the **glottis** in the larynx, just behind the tongue. The **larynx** is supported by several cartilages and contains the paired **vocal cords;** it connects to the tubular **trachea,** which is reinforced with rings of cartilage. The trachea extends to the forepart of the thorax and divides into two short **bronchi,** one to each lung. The **lungs** contain higher interior partitions than the frog and are spongier.

32-11 Excretory system The two flat lobular **kidneys** lie in the posterior dorsal part of the body cav-

ity; a **ureter** from each extends back to the side of the cloaca.

32-12 Nervous system and sense organs The **brain** (Fig. 9-6) has two long olfactory lobes connected to the large cerebral hemispheres; behind the latter are two oval optic lobes. Next is the median pear-shaped cerebellum, which is larger than in amphibians. The medulla oblongata is spread laterally below the cerebellum, then narrows to the spinal **nerve cord.** Ventrally, between the bases of the cerebral hemispheres, are the optic tracts and optic nerves, followed by the infundibulum and the hypophysis. There are 12 pairs of **cranial nerves** and paired **spinal nerves** to each body somite.

There are **taste buds** on the tongue and **olfactory cells** in the nose. The eyes have **lachrymal glands** that keep the cornea or surface of the eyeball moist when out of water. The ears are of the type characteristic of land vertebrates. Each ear has a short **external auditory canal** under a flap of skin, with a **tympanic membrane** at the inner end, a tympanic cavity, or middle ear, housing the one ear bone, or stapes, and an inner ear containing three semicircular canals and the organ of hearing. From each tympanic cavity two **Eustachian tubes** lead medially to a common opening on the roof of the pharynx behind the internal nares. In other vertebrates the tubes open separately on each side of the pharynx.

32-13 Reproductive system The paired gonads and ducts are much alike in young of the two sexes. In mature males two roundish **testes** lie near the ventromedial borders of the kidneys; a **ductus deferens** from each passes back to enter the cloaca just anterior to the ureter and joins the single median **penis** on the ventral floor of the cloaca. In adult females two **ovaries** are similarly attached near the kidneys. Near the anterior end of each kidney is the open funnel of an **oviduct,** and the latter runs back to the cloaca. Eggs form in the ovaries and pass into the funnels; in the oviducts, each is fertilized and covered with albumen, shell membranes, and a shell before being laid.

32-14 Natural history Many crocodilians, including the American alligator, live in swamps or rivers; some inhabit seacoasts. They dig burrows in stream banks for shelter and eat various kinds of animals—insects, mollusks, and vertebrates including dogs and hogs, but few attack humans. The female American alligator builds a nest of decaying vegetation and lays 30 to 60 hard-shelled eggs, which are incubated by decay in the nest for about 60 days. The young are 22 to 24 cm (8½ to 9½ in) long at hatching and grow 30 cm (12 in) or so per year. At 10 years a male is about 280 cm (110 in) long and weighs 113 kg (250 lb), a female 220 cm and 51 kg (87 in and 113 lb).

Structure of other reptiles

32-15 Body covering (Fig. 32-5) Most reptiles have a nonglandular skin consisting of a stratified epidermis and underlying dermis. The germinative layer of the epidermis produces cells that pass to the surface, become cornified, and provide the exterior covering.

The epidermis develops an outer hard layer of beta keratin and an inner softer one of alpha keratin. The alpha layer predominates in the hinge region of scales and permits the skin to stretch. The scale surface often has microornamentation of ridges, tubercles, or spines which scatters light and reduces the amount of radiation deeply penetrating the body. Such sculpturings in geckos and anoles have apparently given rise to the digital setae (p. 736) which cover the undersurfaces of their expanded toe tips and aid these animals in clinging to vertical (or even inverted) surfaces. The epidermal cells are bound to one another by many fine intercellular "bridges," which resist mechanical displacement. The dermis is chiefly of connective tissue and contains pigment cells, blood vessels, and nerves; in some species it includes dermal bones (*osteoderms*). The connective tissue fibers are in definite layers nearly parallel to the surface, those of each layer approximately at right angles to fibers immediately above and below and "on the bias" at about 45° to the body axis. Small fibers through the dermis and into the epidermis bind the whole skin together. This structural arrangement provides mechanical strength, as in a woven fabric, yet is elastic so that the skin can stretch, as when a snake is swallowing large food items or a lizard puffs up in its courting performance. The pattern of scales is essentially constant on each species, and the form and arrangement of the scales are useful in the classification of reptiles.

Figure 32-5 Generalized body scales of squamate reptile. The alpha and beta layers, produced by the stratum germinativum, are horny (keratinized).

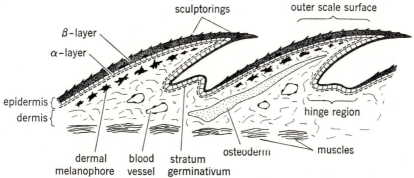

The outer cornified epidermis is molted at intervals by lizards and snakes. Prior to a molt the germinative layer produces a new cuticle beneath the older one. The latter then loosens by cell dissolution at its base, with some production of moisture between new and old layers. In snakes, loosening of the cuticle over the eye may temporarily impair the animal's vision, whence an old belief that "snakes go blind in August," when many molt. The slough of snakes and of small-limbed or limbless lizards comes off in one piece, but some lizards molt the cuticle in pieces. The rattle of a rattlesnake results from retention of the heavier cornified covering at the end of the tail in successive molts (Fig. 32-14). Some turtles and crocodilians apparently shed the outer layer of their scutes as single pieces. In some turtles the outer surface apparently gradually wears away.

32-16 Coloration Some lizards and many snakes have color patterns of brilliance and beauty—stripes or bands of alternating colors, spots, and diamond or rectangular markings. Their pigment cells (chromatophores) resemble those of amphibians (Sec. 31-5, Fig. 31-4). In the green anole (*Anolis carolinensis*), which can change rapidly from green to brown, yellow chromatophores (xanthophores) lie just beneath the epidermis. Under them are whitish iridophores and, still deeper, large branched melanophores. Branches of the melanophores extend among and around the iridophores and between the xanthophores and the epidermis. When melanin migrates into the melanophore branches, it more or less conceals the other chromatophores. In the green phase melanin is concentrated, but in the dark phase it is dispersed and the lizard is brown.

Many lizards can change color. Some species pass from the light to the dark phase in 1 to 10 min in response to light and temperature changes, excitement, or other factors. Many assume the dark phase when cold and in darkness, but some are pale on warm nights. Excitement or fright may cause paling, darkening, or both, depending on the species. Color changes are under hormonal and nervous control. Responses to light may be initiated through the eyes and brain or directly by the skin. Release of **intermedin** by the pituitary, which disperses melanin, is triggered by nervous impulses from eye and brain. Release of epinephrine, which concentrates melanin and causes the skin to pale, is under both hormonal and nervous control. The capacity for color change may aid some lizards in concealing themselves by increasing their versatility in background matching. It may also help some to control body temperature more effectively by influencing the rate of dermal absorption and reflection of sunlight. Most other reptiles have little ability to change color.

The black peritoneums of many diurnal desert lizards seem to function primarily in absorbing mutation-inducing ultraviolet light.

32-17 Turtles The body is encased in an oval shell, a layer of platelike bones in definite pattern and sutured to one another, over which there is a covering of cornified scutes, also of regular arrangement (Fig. 32-6). The dorsal convex portion is the **carapace,** and the flatter ventral part the **plastron.** The thoracic vertebrae and ribs are usually consolidated with the bony carapace. Soft-shelled turtles have leathery integument, not divided into scutes, and the shell is poorly ossified. The head, tail, and limbs of turtles protrude between the two parts of the shell and in most species can be drawn completely within the margins for safely. Their jaws lack teeth but bear stout cornified sheaths serving to cut, tear, and crush their food. The toes end in horny claws that are useful in crawling or digging. In terrestrial tortoises the feet are stumpy, and in marine turtles the limbs are flipperlike, for swimming. The excretory system includes a bladder; the urine is fluid and

Figure 32-6 The shell of a turtle (*Chrysemys*), showing the arrangement of external horny scutes over the bony plates beneath.

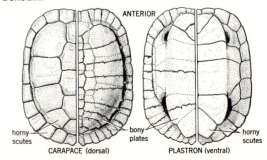

CARAPACE (dorsal)   PLASTRON (ventral)

contains urea and uric acid. The male has an erectile penis on the ventral wall of the cloaca. Connected to the cloaca of aquatic turtles (not including marine turtles and soft-shells) are thin-walled sacs that perhaps serve as "cloacal gills" when the animals are submerged.

32-18 Sphenodon (Fig. 32-17) The tuatara, only living member of the order RHYNCHOCEPHALIA, is restricted to New Zealand. It is lizardlike in superficial appearance but has various primitive or conservative characters including (1) two temporal arches in the skull, (2) a widely roofed mouth, (3) firm anchorage of the quadrate, (4) teeth firmly fixed on rims of jaws, (5) a median parietal eye, (6) persistent abdominal ribs, (7) absence of a copulatory organ, and (8) low thermal preferences—4 to 28°C, average 18°C (39 to 82°F; average 64°F)—in laboratory experiments.

32-19 Lizards The body shape is varied; many species are slender, some compressed laterally and some flattened dorsoventrally, such as the horned lizards. The limbs may be long or short, stout or delicate; they are reduced on some and entirely lacking in a number of limbless lizards ("glass snake", legless lizards). Virtually all the limbless forms live in the soil, through which they usually progress by wriggling the body from side to side. In some rapidly running lizards the long slender tail serves as a counterbalance, but in sluggish forms it may be short and stumpy. The tail vertebrae in many species are incompletely ossified in a zone midway across each vertebra (Fig. 32-7). If the tail is seized, the vertebrae separate at one of these "breaking points," and the animal scuttles to freedom; in time the lost part is regenerated.

The skin is usually flexible, is loosely attached to the body, and contains many scales arranged in rows, longitudinal, transverse, or diagonal. On most lizards the scales overlap, like shingles on a roof. Individual scales are variously smooth, pitted, or have a longitudinal keel. An external eardrum is usually evident on each side of the head.

The tongue may be only slightly mobile or freely extensible; in chameleons it can be "shot" several

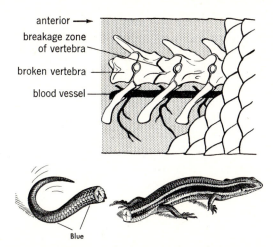

Figure 32-7 Side view of tail vertebrae of western skink (*Eumeces skiltonianus*), showing breakage zones; severed tail moves rapidly. (*From Stebbins, Reptiles and amphibians of the San Francisco Bay region, University of California Press.*)

inches beyond the snout to capture insects on the mucus-coated tip. The jaws are variously provided with teeth, usually short and alike but differentiated as to form in the Old World AGAMIDAE and in some teiids and skinks. A bladder is present in lizards, but the excretory wastes are semisolid, as in birds and most reptiles, being passed from the cloaca as whitish material (urates) with the feces. The male has two hemipenes in the base of the tail. At copulation one or both are everted, but usually only one is used. Seminal fluid passes in a groove on the hemipenis into the cloaca of the female. Snakes also have hemipenes, a character uniting the two groups.

Many diurnal lizards (agamids, iguanids, etc.) display in defense of territory or when fighting, courting, or in other social interactions. Display behavior is especially prevalent in adult males, and they are usually more brightly colored than females. Displays include attention getting movements such as "push-ups," flattening of the sides, bobbing the head, and lowering the throat skin. Many of the movements reveal otherwise hidden colors. Each species has an innate and distinctive **display action pattern** (DAP), differing in amplitude and configuration of bobs, their cadence, and their duration. The DAP is important in species recognition and communication and is distinctive, like the song of a bird, serving a comparable function in these largely voiceless animals (geckos are an exception). Individual differences superimposed on the species DAP appear to aid individual recognition. The DAP functions to space individuals in relation to available food, shelter, and mates and to position individuals within peck orders. It signals information on an individual's physiologic state and intent. In nocturnal species (most geckos) or those species that are active where visibility is poor (skinks, etc.), bright colors and display are usually less evident, and voice (geckos) or chemoreception may assume the dominant role in social interactions.

32-20 Snakes Loss of appendages and elongation of the body have occurred as a parallel development in several groups of vertebrates, including the eels and morays among fishes, the caecilians among amphibians, and the worm lizards (AMPHISBAENIDAE), legless lizards (ANNIELLIDAE), "glass snake" and "slowworm" (ANGUIDAE), and other lizards, besides all the snakes. Vestiges of the pelvic girdle and hind limbs persist in the boas, pythons, etc. (Fig. 32-11). The snakes also lack limb girdles, a sternum, eyelids, external ear openings, and a bladder, and they show many specializations in form and function of other parts of the body.

As in lizards, the skin of snakes bears rows of scales, either smooth or keeled (Fig. 32-8). The ven-

tral surface of nearly all snakes has one row of large transverse scales (ventrals, or gastrosteges) from chin to vent, and either one or two rows (caudals, or urosteges) on the tail. Blind snakes, some sea snakes and a few others lack enlarged ventral scales.

The eyes are covered with transparent cuticle; on blind snakes the vestigial eyes are beneath scales. There is no external ear membrane or opening. The skull is delicate, and several of the bones can move upon one another. Teeth, having a backward slant, are present on the jaws and on bones in the roof of the mouth; they serve to hold food being swallowed. Poisonous snakes have a pair of specialized teeth, or fangs, on the two maxillary bones of the upper jaw that conduct the venom used to kill prey; these are fixed in the cobra and its allies and in sea snakes but folded back, when not in used, in rattlesnakes and other vipers. The fangs are hollow. In back-fanged snakes (African boomslang, American lyre snakes, etc.) the fangs are rigid and grooved. Venom is injected by capillary action and sometimes by chewing. The serpent tongue is narrow, flexible, and ribbonlike, with a forked tip, and can protrude through a notch in the upper jaw when the mouth is closed. Odors from food, mates, predators, and perhaps rivals are brought by the tongue to the **Jacobson organs,** two small sensory chambers with ducts opening far forward on the palate. These organs enable a rattlesnake to detect the presence of a king snake and respond by defensive movements. The digestive tract is essentially a straight tube, from mouth to vent; virtually all the other internal organs are elongated; and the left lung is usually vestigial. Hemipenes are present in the male, as in lizards.

In long snakes the vertebrae number 200 to 400 (Fig. 32-9). The many segmental body muscles are slender, connecting vertebrae to vertebrae, vertebrae to ribs, ribs to ribs, ribs to skin, and skin to skin. Many extend from one body segment to the next, but others connect by tendons between segments far removed from one another. This arrangement makes possible the graceful sinuous movements of a snake.

Four types of snake locomotion are recognized:

1 Horizontal (or lateral) **undulatory locomotion** is commonly used when a snake is crawling at moderate to high speed. The body is formed into S-

Figure 32-8 Scales of snakes. *A*. Head of a ringneck snake. *B*. Lower surface of a snake in the vent region showing the enlarged ventral scales on the body and tail. *C*. Snake body with smooth scales. *D*. Snake body with keeled scales. (*Adapted from Blanchard, 1924.*)

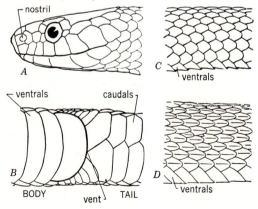

A

B

C
ventrals

D
ventrals

ventrals   caudals

BODY   vent   TAIL

nostril

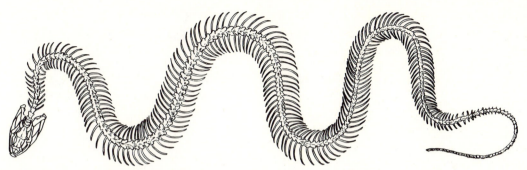

Figure 32-9 Skeleton of a snake, showing the loss of limbs, limb girdles, and sternum and the greatly increased number of vertebrae and ribs, all virtually alike in structure.

shaped curves, and the animal propels itself forward by horizontal and backward pressure exerted by its body loops against irregularities of the substratum. The places where pressure is applied may be indicated by ridges of soil at the rear of each body loop. The slick ventral surface of the snake reduces friction with the ground.

2 **Rectilinear locomotion**   (or "caterpillar" action) may be employed in stalking prey or moving cautiously. The snake crawls slowly, usually with the body held nearly straight. The loosely attached belly skin is advanced in undulating waves that seem to pass from tail to head. In lateral aspect low undulations appear along the ventral surface as the ventral scales are advanced in a progressive series, pressed against the ground, and the body is shifted forward within the skin. In contrast with horizontal undulatory locomotion, friction between the ventral surface of the snake and the substratum is essential.

3 **Concertina locomotion**   involves extending or retracting the body from one or more "friction" or anchor points. It may be used on a flat surface, in crawling through a tunnel, or in climbing. The accordionlike movements of the body, the basis for the name, are perhaps best seen when concertina is used to ascend a crevice. The snake anchors the anterior part of its body by pressing against nearby surfaces with body loops. It then draws up its hind parts, anchoring them in the same way. The foreparts are now released and extended (opening of the bellows). By alternating the two actions, the snake progresses upward. Sometimes

concertina and horizontal undulatory locomotion are combined.

4 **"Sidewinding"**   (Fig. 32-10) is a special form of locomotion found in several African and western Asian vipers and in the North American rattlesnake, the sidewinder. All are arid-lands species. Sidewinding appears to be an adaptation for moving rapidly over smooth surfaces largely free of impediments, such as sand or hardpan. It prevents slippage and conserves energy. The speed achieved aids in escaping enemies. Sidewinding has been considered a modified form of horizontal undulatory locomotion. The body loops on one side are lifted slightly off the ground, and the thrust exerted by those on the opposite side propels the snake sideways. The propulsive thrust is no longer exerted horizontally but at an angle, obliquely into the surface. This inclined thrust compensates for the lack of surface irregularities required as purchase points for horizontal undulatory locomotion. Alternatively, sidewinding has been considered a special case of concertina locomotion.

Snakes neither chew nor tear their food but swallow it whole. They can take prey greater in diameter than their bodies (Figs. 32-11, 32-12) by reason of a series of adaptive modifications, which include (1) joining of the mandibles anteriorly with flexible tissue, (2) loose attachment of the quadrate bone on either side to both the skull and mandible, and (3) movement of bones of the palate. In consequence of these three features, the mouth can be widely dis-

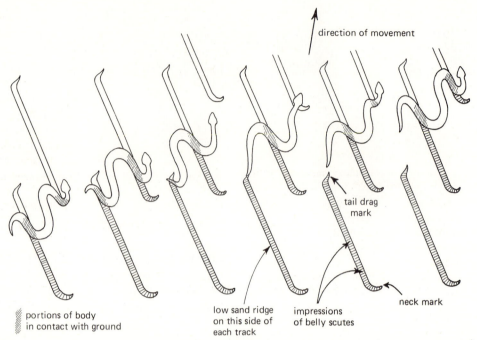

direction of movement

tail drag
mark

neck mark

portions of body
in contact with ground

low sand ridge
on this side of
each track

impressions
of belly scutes

Figure 32-10 Locomotion of a sidewinder (*Crotalus cerastes*) on smooth sand. The body moves sideways and the head points in the direction of locomotion. Much of the body is elevated above the ground as the snake crawls. The snake's body presses backward, building up a low ridge of the sand on the posterior side of each track. The track is a series of separate, reversed J-shaped marks that bear impressions of the belly scutes, indicating absence of slippage. The hook of the J is made by the neck. Note tail-drag mark at other end of track.

Figure 32-11 Skeletal features of a python. *Above.* Skull with large teeth and quadrate articulation of lower jaw that permits distension of the mouth when swallowing. *Left.* External appearance of pelvic region, with black horny caps on vestigial hind limbs. *Right.* Vestigial pelvic girdle and limb bones. (*From Guide to reptile gallery, British Museum.*)

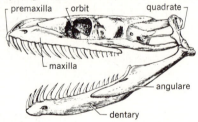

premaxilla   orbit   quadrate

maxilla

angulare

dentary

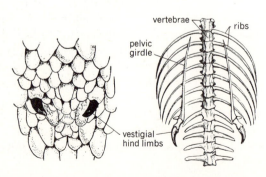

vertebrae   ribs

pelvic
girdle

vestigial
hind limbs

tended. Other modifications are (4) the slender backward-pointing teeth on the jaws and palate, which prevent the food from slipping forward, once swallowing has begun; (5) absence of a sternum (breastbone) and ribs free of any bony articulation ventrally, so that the body wall may be dilated; (6) presence of soft, elastic skin between scales on the

Figure 32-12 King snake (at right) swallowing a rattlesnake that it had killed by constriction. The skin becomes stretched, separating the rows of scales, as the large prey is swallowed. The glottis opening, just inside the lower jaw, permits respiration to continue. (*Photo by Karl S. Hazeltine.*)

back and sides of body, permitting wide distension; (7) thin and easily stretched walls of esophagus and stomach; and (8) placement of the glottis far forward, between the jaws and just behind the sheath for the slender tongue, which permits respiration during the swallowing of food. During swallowing the glottis may be projected forward to aid breathing, and the flexible jaws are advanced alternately in a ''walking'' movement over the prey.

Most snakes are seclusive. Species and individual recognition in most probably depends more on chemoreception (Jacobson organs and olfactory sense) than on vision. There may be little or no difference in coloration between the sexes, and there is a notable lack of elaborate display action patterns like those of lizards (Sec. 32-19). Fighting between males occasionally occurs, perhaps over territory or mates. The contestants engage in a kind of wrestling match, their foreparts entwined and elevated well above the ground (Fig. 32-13).

Natural history

32-21 Distribution Most reptiles, both species and individuals, live in tropical and subtropical regions; their numbers decline rapidly toward the poles and in high altitudes. Thus Louisiana has 22 kinds of turtles, 13 lizards, 39 snakes, and the alliga-

Figure 32-13 Combat in male rattlesnake. The contestants attempt to throw one another to the ground. Fights appear to be a form of territorial behavior.

tor, but northern Alberta has only 1, a garter snake. Turtles and snakes are more numerous in humid regions such as the Southeastern states, and lizards are in arid territory like that of the Southwest.

Reptiles occupy a wide variety of habitats. Large pythons and boas dwell in the tropics, the crocodilians in swamps or rivers or along seacoasts, the biggest turtles in the ocean, and the giant land tortoises on arid oceanic islands. Most lizards and snakes are terrestrial, but some climb rocks and trees. The racers and whipsnakes often ascend bushes and trees in search of food, and some tropical serpents are predominantly arboreal. Small geckos and other lizards seek shelter in crevices of rocks, trees, or buildings, snakes often use rodent burrows, and some lizards and snakes burrow into sand, the snakes using an enlarged or upturned (rostral) plate on the snout. Most turtles live in and about water, but box turtles inhabit the open floor of forests, and the land tortoises inhabit dry regions. The sea snakes live in tropical oceans.

32-22 Activity Being without the well-developed internal mechanisms of heat production and control of body temperature found in birds and mammals, reptiles are influenced markedly by the temperature of their environments. They regulate their body temperatures chiefly by actively seeking suitable environmental temperatures, but in addition, use some or all of the following methods: (1) varying the amount of skin surface exposed to the sun or in contact with warm or cool places, flattening the body and tilting it toward the sun, or compressing and aligning it with the sun's rays; (2) delaying heat gain through panting (usually only at critical thermal levels); (3) accelerating heat gain during warming by peripheral vasodilation and retarding loss during cooling by vasoconstriction; (4) having a faster heartbeat at a given body temperature during warming than during cooling, thereby accelerating heat gain and retarding heat loss; (5) retention of small amounts of metabolic heat.

Many reptiles maintain their body temperatures within rather well-defined limits when sufficient heat is available. Within this thermal range they carry on their normal life activities such as foraging and mating.

In the tropics reptiles can be active at all seasons, but in temperate regions only in the warmer months. On deserts or semideserts they are abroad on warm days in spring and sometimes in autumn, but in summer many species avoid excessive midday heat by appearing only in early morning or late in the day. A few lizards and some desert snakes are always nocturnal, and other snakes become night prowlers during hot weather. Snakes and lizards of the Southwest are active between environmental extremes of 16 to 20°C (61 to 68°F) and 35 to 41°C (95 to 106°F) or higher. Usually each emerges and "warms up" by exposure to the sun before moving about. Many lizards are most agile when the body temperature is only 5 to 8°C below the critical limit (about 45°C; 113°F), at which death ensues quickly from overheating.

In winter reptiles are dormant for a period depending on the length and severity of the cold season. Lizards and snakes hibernate in crevices or holes in the ground, rattlesnakes and some other snakes may den up in groups in caves or large burrows, and some freshwater turtles go to the bottoms of ponds.

32-23 Hearing Although it has long been known that crocodilians and geckos communicate vocally, sound detection by other reptiles has been in doubt. Experiments now suggest that most reptiles hear. Sensitivities to the following sound-frequency ranges have been reported: crocodilians 20 to 3000 Hz, lizards 100 to 10,000 Hz (some to 19,000), snakes 100 to 700 Hz, and turtles up to 3000 Hz (for comparison, the human range is 15 to 20,000 Hz). Snakes lack an eardrum and middle ear cavity, but the sound-transmitting columella is present and extends from the quadrate bone that suspends the lower jaw to the inner ear. The head skin and underlying loosely articulated quadrate evidently act as an eardrum. Breaking the latter results in some loss of auditory sensitivity. Crocodilians and geckos produce territorial and mating sounds, and crocodilians respond defensively to the cries of their young. However, for most reptiles the function in nature of their demonstrated hearing capability remains unknown. It is of interest that sounds expected to be of significance to reptiles (the stealthy approach of predators, sound of prey, etc.) often have frequencies that correspond with their range of auditory sensitivity. Although most reptiles appear to be voiceless, snakes and some lizards hiss, and during mating and fighting scuffling and breathing sounds may be heard. Many reptiles are sensitive to vibrations of the substratum. Snakes, with so much of their body in contact with the ground, are notable in this respect. Such sensitivity may be one reason these animals are so seldom encountered; they may retreat from the path of an approaching large animal long before it arrives.

32-24 Food Most reptiles feed chiefly on animals, and a number of species scavenge; land tortoises, large and small, some turtles, and some lizards eat vegetation. The large crested Galápagos marine iguana (*Amblyrhynchus cristatus*) enters the sea to browse on seaweed (marine algae), its chief food.

Lizards and small snakes take insects and other small invertebrates; small turtles use aquatic invertebrates; and large lizards, turtles, snakes, and crocodilians eat various vertebrates from fishes to mammals (Fig. 32-12). Snakes feed strictly on animals, and many take other reptiles, including snakes. King snakes often feed on venomous snakes (including rattlers) and are immune to the venom. Many predatory reptiles detect their prey by sight, and movement of the prey is required to trigger attack. The total intake of food is small in amount as compared with that necessary for birds and mammals. Reptiles eat more and digestion is faster at higher temperatures. Small prey is simply seized with the jaws and swallowed, being killed by crushing action of the jaws, by suffocation, or by the digestive juices. Racers and whipsnakes overcome larger prey by holding it in their jaws or pressing it against the ground. Boas, king snakes, and others encircle the prey quickly with coils of the body, then constrict the victim until death results from asphyxiation. Poisonous snakes inject venom, and the victim usually dies quickly.

Reptiles exposed to high salt intake have special glands that aid in the elimination of the excess. Eye glands in marine turtles expel salt by tearing, and nasal glands in lizards excrete it into the nasal passages. Nasal salt glands have been found in the marine iguana, which feeds on salty marine vegetation, and in many lizards of arid lands that feed on insects and plants high in salt content (sodium, potassium, etc.). Since the nasal glands can excrete salt at high concentration (some 50 percent above that of seawater) with minimal loss of water, they are of special importance in dry environments. The salt is eliminated by snorting and in the marine iguana may be expelled audibly for a distance of over a meter.

32-25 Life span In captivity, several tortoises have survived beyond 100 years, various turtles for 20 to 90 years, occasional crocodiles and large snakes for 25 to 40 years, and smaller species for 10 to 20 years. In former times some free-living tortoises, crocodilians, and the tuatara probably lived to over 100 years, but few do so now because of human disturbances. By contrast, in marked wild populations of some small lizards (*Anolis, Uta*), over 90 percent of those marked disappear within a year, and few live more than a year after reaching sexual maturity.

32-26 Reproduction Reptiles are intermediate between the anamniotes (fishes and amphibians) and the mammals in manner of reproduction. Fertilization is internal, yet most species deposit their eggs for development outside the body. Internal development among land vertebrates may have begun as a regular phenomenon in reptiles, since some strictly oviparous turtles and snakes retain their eggs temporarily when conditions for deposition are unfavorable. When this happens in the European ring snake (*Tropidonotus*), the embryos develop so that after laying only 3 weeks (instead of the usual 7 or 8) ensue until hatching. Internal development is the rule for vipers, rattlesnakes, water and garter snakes, and sea snakes. In certain genera of lizards, some species lay eggs and others are live-bearing; horned lizards are oviparous except one species, but several spiny lizards (*Sceloporus*) are live-bearing.

In adaptation to life on land, reptile eggs resem-

ble somewhat those of birds, being covered by a tough shell with a shell membrane inside. The shells are hardened by limy salts in crocodilians, some turtles, and most geckos. The shape is usually a long oval but spherical in land tortoises and sea turtles and typically unpigmented. Much yolk is provided to nourish the embryo, so that the egg is often large in proportion to the size of the female. During development the embryo is surrounded by embryonic membranes, the **amnion, chorion,** and **allantois** (Fig. 10-14); these "new" features of vertebrates, first occurring in reptiles, are an adaptation to protect the delicate embryo against desiccation and physical shock during development. The allantois and chorion serve in respiration. The tip of the upper mandible develops an egg tooth, calcareous in snakes and lizards but horny in others. It serves to cut the egg membrane and shell at hatching, then drops off. The young upon emerging usually resemble their parents and become independent at once.

The number of eggs produced annually by a female varies from nearly 400 in a sea turtle (*Caretta*) to a single egg in the house gecko (*Sphaerodactylus*). Small pond turtles lay 5 to 11 eggs, the lesser snakes and lizards about 10 to 20, and the American alligator 30 to 60; and garter snakes have produced over 70 living young in a single brood. The eggs are deposited in natural cavities beneath rocks or within logs, under vegetable debris, or in earth or sand. Female turtles seek sandbanks or hill slopes to excavate nest cavities by digging with the hind feet, later covering the eggs. The duration of development in different reptiles is from a few weeks to several months; the New Zealand tuatara is unique in requiring about 13 months.

In ovoviviparous reptiles that retain large yolked eggs in the oviduct for development, the "shell" is but a thin membrane. When development is completed, such eggs are laid and the young hatch at once. Among viviparous forms, blood vessels of the yolk sac and chorion lie close to maternal vessels on the inner surface of the oviduct ("uterus"), providing respiration, and in some species perhaps nutrients for the embryo, in an arrangement functionally equivalent to the placenta in mammals (Sec. 10-20).

Populations of females, reproducing parthenogenetically, occur in rock lizards (*Lacerta*) of the Caucasus, whiptailed lizards (*Cnemidophorus*) in the United States and Mexico, and a few other lizards.

32-27 Relation to humans   Many kinds of snakes and lizards feed upon harmful rodents and insects, but some snakes prey on the eggs of game and songbirds in season.

Skins of crocodiles and alligators have long been used for fancy leather, and in recent years those of large snakes and lizards have been made into shoes, purses, and similar articles. Millions of skins have been taken for such purposes. Fortunately, steps are being taken to reduce such exploitation. The IUCN (International Union for the Conservation of Nature) has declared crocodilians *threatened* and the U.S. Bureau of Sport Fisheries and Wildlife has classified the American alligator as an *endangered* species. The alligator is protected by law in all states, and Florida protects the American crocodile. The Bureau bans importation of wildlife threatened with extinction, including, among reptiles, certain crocodilians, turtles, lizards, and the tuatara. Turtles are useful chiefly for food. Flesh of the green turtle, both fresh and dried, has been in such demand that the fisheries for it on tropical coasts have been greatly depleted. The Galápagos giant land tortoises, once enormously abundant, were killed by Pacific Ocean mariners for three centuries as a source of fresh meat. During the whaling period from 50 to 300 were removed by each ship, and an estimated 100,000 were taken during the nineteenth century. Feral animals have also taken their toll. Rats dig into tortoise nests and eat the eggs. Pigs eat the young, and goats and burros compete with the tortoises for plant foods. Some varieties have been exterminated, and the reduced populations of others are now protected. Eggs and young of some of the threatened forms are now being reared in captivity, the young tortoises being released only after they reach "pig-proof" size. The smaller turtles of fresh waters and brackish waters in the United States also serve as food, especially the snapper, diamondback terrapin, and soft-shelled turtles. Other small turtles are used in biologic laboratories. The real tortoiseshell for combs and other fancy articles is obtained from the marine hawksbill turtle.

Snakes have long been objects of fear and superstition. They are venerated and worshiped by many primitive peoples. Hopi Indians in northern Arizona have an annual ceremonial, part of which is a snake dance, during which live snakes, including rattlesnakes, are carried in the mouths of certain priests. Many erroneous beliefs about snakes still persist among people in civilized countries, despite the amount now known about the structure and biology of snakes and their place in nature as one group of predaceous animals.

32-28 Venomous reptiles   The two species of *Heloderma,* one of which is the Gila monster of the Southwest, are the only venomous reptiles other than some snakes. Their venom is as poisonous as that of some rattlesnakes, but the mechanism for venom transfer is poorly developed, and only rarely, if ever, are human beings injured or killed by the bite alone. Venomous snakes occur on all the continents and on most large islands except Madagascar, Ireland, and New Zealand. Venom serves for the capture of small prey and is used defensively against large animals and humans. Among the crowded populations of India, snake poisoning causes an estimated 15,000 or more human deaths each year, and many cattle are killed by snakes. Accurate information on this subject, however, is not available. In the United States several hundred persons are bitten annually, and a number of deaths result.

In North America, north of Mexico, the snakes poisonous to humans include two coral snakes, the moccasin, the copperhead, and 15 species of rattlesnakes. Coral snakes have a small head and eye, an elliptical pupil, and the body marked with transverse rings of red, black, and yellow, each black ring being bordered by yellow. All other dangerous snakes in this region are pit vipers, with wide heads and vertically elliptical pupils. The pit, between the nostril and eye, on each side, is a thermosensory organ enabling the snake to detect the presence of warm-bodied prey and to strike accurately when visibility is poor. (Structures of similar function are present on the labial scales of boas and pythons.) The tail on the moccasin and copperhead is slendertipped; on all but one species of rattlesnake it ends with the characteristic rattle (Fig. 32-14B,C). The Eastern coral snake occurs from states of the Mexican border and Gulf north to Arkansas and North Carolina; the copperhead from Massachusetts to Illinois and southward; and the moccasin from Virginia and Illinois southward. Rattlesnakes are present from southern Can-

Figure 32-14 Rattlesnake. *A*. Mechanism of the head used in striking. The sphenopterygoid muscle (1) contracts to rock the fang forward; after the fang enters the victim, the external pterygoid (2) and sphenopalatine (3) contract to draw the fang more deeply; then the anterior temporal muscle (4) draws up the lower jaw and compresses the poison gland to force venom through the duct and fang into the wound. All head and neck muscles relax for withdrawal. The whole process requires but an instant. *B, C*. The dry cornified rattle in external view and sagittal section to show how the segments are held loosely together. (*After Grinnell and Storer, Animal life in the Yosemite.*)

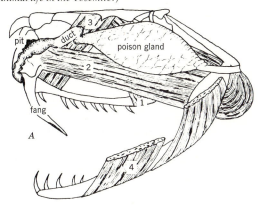

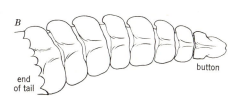

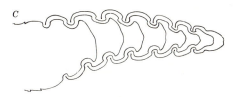

ada southward, being more numerous as to species in the arid Southwest.

The **venom** is secreted in a pair of glands, one on either side of the upper jaw, each connected by a duct to a fang. Reserve fangs develop behind the functional pair to replace the latter when lost (as with teeth in all snakes). Venom is a complex protein substance having various physiologic effects, and each kind of venom has distinctive characteristics and different toxicity. The neurotoxic venom of cobras and related snakes (ELAPIDAE) affects the nervous system, particularly the respiratory center, and death results from asphyxiation; the hemotoxic venom of rattlesnakes and vipers (VIPERIDAE) usually affects the circulatory system, breaking down capillary walls and destroying the blood cells.

Rattlesnakes and other pit vipers strike from a nearby horizontal S-shaped position of the body; the fangs are buried in the victim, the snake's lower jaw clamps upward, muscles over each poison gland force the venom through the hollow fangs into the flesh, and then the head is withdrawn—all with amazing rapidity (Fig. 32-14A). The venom of vipers and rattlesnakes usually travels slowly in the lymphatic circulation.

Efforts to treat victims of snakebite date from antiquity, and many irrational means, chemical and others, have been tried. In 1887 Sewell discovered that repeated injection of minute sublethal doses of rattlesnake venom into pigeons caused development of an **antibody** (Sec. 6-7). For snakebite treatment, horses are immunized to snake venom, then the blood serum is withdrawn to serve as **antivenin,** which is injected into a victim of snakebite. Each kind of venom has distinctive qualities, and no single antivenin will serve against all snakes. Antivenins for snakebite are now generally available.

Some aboriginal peoples use snake venom to prepare poisoned arrows for killing game and for warfare. An extract of water moccasin venom has been used by physicians to check persistent bleeding in some diseases, and cobra venom has been employed to relieve severe pain in human cases of inoperable cancer.

32-29 Fossil reptiles The living reptiles, although numerous, are mostly small and inconspicuous. During the Mesozoic era, or Age of Reptiles (Triassic to Upper Cretaceous), however, they were the dominant vertebrates and occupied most animal habitats then available, from semideserts and dry uplands through marshes and swamps to the open ocean. They varied in size from small to large, some far exceeding elephants in length and weight, and were greatly diversified as to structure and habits.

The most primitive reptiles (†Cotylosauria) closely resembled early amphibians. During the Permian, reptiles became numerous and began to radiate in both structure and manner of life. By the end of that period they had largely replaced the amphibians, by the end of Triassic time virtually all major groups of reptiles had appeared, and during the Jurassic and Cretaceous these attained a climax as to numbers of species and individuals and also in diversity of form and manner of life. Then, with the end of the Cretaceous, this great reptilian host disappeared, and only 4 orders survived.

The most important evolutionary achievement of the reptiles was to become adapted for terrestrial life apart from water. The acquisition of a dry cornified skin that reduced loss of moisture from the body and the production of eggs capable of development on land were significant in this adaptation. The small primitive reptiles had stocky to rather slender bodies and tails and four short five-toed limbs spread widely out at the sides of the body. From this generalized form some of the lines of radiation or specialization were (1) increase in size, as to the huge proportions of the brontosaurs; (2) acquisition of defensive armor, including plates in the skin and horns or spines on the head, as by some dinosaurs; (3) shift of limbs to beneath the body, moving forward and backward in a vertical plane; (4) lightened build, for rapid running on four legs or two, as with some ostrichlike dinosaurs; (5) adaptation to flight, by increased length of forelimbs (and tail) and development of flight membranes of skin by the pterodactyls; (6) adaptation to strictly aquatic life, with paddlelike limbs and a spindle-shaped body as in the ichthyosaurs, plesiosaurs, and mosasaurs; and (7) perhaps the evolution of endothermy in certain dinosaurs.

†*Brontosaurus* (Fig. 32-15) attained a length of 23 m (75 ft) and †*Diplodocus* 27 m (87 ft), the largest terrestrial animals ever known. They may have weighed 23,000 to 32,000 kg (25 to 35 tons) and used more than 230 kg ($\frac{1}{4}$ ton) of plant food per day. †*Stegosaurus* (Fig. 32-15) had a series of enormous bony plates in a dorsal crest extending from the neck to the end of the tail. †*Triceratops* bore a skull scarcely equaled in size and weight by any animal but the modern whales, with a sharp cutting beak on the upper jaw, a short, stout horn on the nose, and a pair of long forward-pointing horns on the top of the head. The brains of these huge dinosaurs were proportionately small, that of the 5.5-m (18-ft) †*Stegosaurus* having an estimated weight of only 71 g (2.5 oz), but it had enlarged ganglia on the nerve cord near the pectoral and pelvic girdles.

The flying reptiles, or pterodactyls (Figs. 32-15, 13-1) were of light build. The head was long and the fourth "finger" greatly lengthened as support for the wing. The hind limbs also helped to support the flight membrane, somewhat as in modern bats. One form (†*Pteranodon*) had a wingspread of 7.5 m (25 ft). The marine ichthyosaurs were whalelike but had four (instead of two) paddlelike limbs used for steering and a large tail fin. The Mesozoic reptiles included both herbivorous and carnivorous species. Nests of dinosaur eggs (†*Protoceratops*) discovered in Mongolia prove that some ancient reptiles laid eggs in nests, but the marine ichthyosaurs were live-bearing.

Remains of fossil reptiles have been found on all continents and are abundant in some rocks of the Western states, especially in Wyoming, Utah, and Colorado. Footprints of Mesozoic reptiles, made on the shores of ancient waters, are found in sandstones in several places in the United States.

There has been much speculation as to why the ancient reptiles disappeared so suddenly, geologically speaking. One theory suggests that the earliest mammals preyed upon their eggs; another and more plausible explanation is that change in climate, as by alteration of temperature or moisture relations, affected adversely either the reptiles themselves or their habitats, or both. Disappearance of the late Mesozoic reptiles evidently gave opportunity for primitive mammals, then small in size, to begin the spectacular development that characterized the Tertiary era.

AIR

Pteranodon-spread 7.6 (25)

Brontosaurus-23 (75)

Tyrannosaurus-15 (50)

LAND

Stegosaurus-6 (20)

Triceratops-6 (20)

Trachydon-9 (30)

Ornithomimus-2 (7)

Euoplocephalus (Ankylosaurus) 7.6 (25)

SEA

Elasmosaurus-to 15 (50)

Tylosaurus-6+ (20+)

Archelon-3.6 (12)

Ophthalmosaurus-2 (7)

Figure 32-15 Restorations of Mesozoic reptiles.   †*Brontosaurus* and †*Stegosaurus* Jurassic, all others Cretaceous; four at bottom are marine. Numbers are total length in meters and feet (latter in parentheses). (*After various authors.*)

Classification

Class Reptilia

Reptiles (characteristics of living forms). Skin nonglandular (a few exceptions), horny, usually with scales or scutes; limbs typically 4, each with 5 clawed toes but reduced or absent in some; skeleton bony, 1 occipital condyle; Permian to Recent; terrestrial, freshwater, or marine, mostly in tropics and warm temperate regions; about 6500 species.

Subclass 1. Anapsida. Skull roof solid, no opening behind eye.

Order 1. †Cotylosauria. Stem reptiles. Earliest resemble labyrinthodont amphibians. Permian, †*Captorhinus*; Triassic, †*Scutosaurus*, ox-sized; †*Hypsognathus*.

Order 2. †Mesosauria. Small, slender, aquatic. Lower Permian. †*Mesosaurus.*

Order 3. Chelonia (*Testudinata*). Turtles, terrapins, and tortoises. Body broad, encased in a firm "shell" of rounded dorsal carapace and flat ventral plastron, joined at sides and covered by polygonal scutes or leathery skin; no teeth, jaws with horny sheaths; quadrate not movable; thoracic vertebrae and ribs usually fused to carapace; vent a lengthwise slit; unpaired copulatory organ, everted from floor of cloaca; oviparous, eggs laid in cavities (nests) in ground, dug and covered by females; Jurassic to Recent; worldwide except New Zealand and western South America; in fresh waters or salt waters or on land; about 250 species (Fig. 32-16).

Suborder 1. Cryptodira. Neck bends in vertical S curve if retractile; pelvis not fused to plastron.

Superfamily 1. Testudinoidea. Shell usually complete, with scutes; limbs with claws, not flipperlike.

Dermatemydidae. American river turtles. Flat, tail short, aquatic; 1 species, *Dermatemys mawii*, Vera Cruz to Honduras.

Figure 32-16 Turtles (order Chelonia); all reduced, not to scale. *A*. Green turtle (*Chelonia mydas*), marine, shell length to 152 cm (60 in). *B*. Snapping turtle (*Chelydra serpentina*), in fresh waters, shell to 46 cm (18 in) long. *C*. Desert tortoise (*Gopherus agassizi*), to 36 cm (14 in) long, on land in arid Southwest. *D*. Soft-shelled turtle (*Trionyx spiniferus*), in fresh waters of central and eastern United States, to 46 cm (18 in) long. *B*, *C*, and *D* are young individuals. (*From Stebbins, Amphibians and reptiles of western North America.*)

CHELYDRIDAE. Snapping turtles. Plastron small, cross-shaped; much of head, neck, and limbs exposed from below when drawn beneath shell; beak hooked, tail long. *Chelydra serpentina,* common snapper, Gulf Coast to southern Canada, west to Colorado, shell to 45 cm (18 in) long (Fig. 32-16*B*); *Macrochelys temminckii,* alligator snapper, Illinois, Georgia, and southward, length to 66 cm (26 in), weight usually under 68 kg (150 lb), record 91+ kg (200 lb).

STAUROTYPIDAE. Small, shell flat. *Staurotypus,* Central America.

KINOSTERNIDAE. Musk turtles. Plastron well developed, both ends hinged. *Sternotherus,* musk turtle, Gulf states to Canadian border, head large, plastron reduced; *Kinosternon,* mud turtles, Eastern states, to lower Colorado River and into Mexico to South America, in shallow waters.

PLATYSTERNIDAE. Flat, head large, mandible hooked, tail long. *Platysternon megacephalum,* southeastern Asia, in mountain streams.

EMYDIDAE. Common turtles, terrapins, and box turtles. Head can be drawn into shell, middle front toe with 3 phalanges, toes webbed at least in part; Eocene to Recent; worldwide except Australian region; includes most chelonians. *Clemmys, Emys,* pond turtles, Europe, Asia, Africa, and North America; *Terrapene,* box turtles, Mexico and central United States to Atlantic Coast, plastron with transverse hinge; *Malaclemys terrapin,* diamondback terrapins, Massachusetts to Texas, in salt water and brackish coastal waters, esteemed for eating, but popularity has waned; *Graptemys,* map turtles, Florida to Texas and South Dakota, in fresh waters; *Chrysemys,* sliders and cooters, South Central and Eastern states; *Chrysemys picta,* painted turtle, Atlantic Coast to western Washington, in clear waters (Fig. 32-1); *Deirochelys,* chicken turtle, western Virginia to Texas.

TESTUDINIDAE. Land tortoises. Toes not webbed, hind legs cylindrical; shell high. *Geochelone (Testudo),* giant land tortoises, Galápagos Islands, Pacific Ocean, and Aldabra Island, Indian Ocean, some to 1.2 m (4 ft) long; *Gopherus,* small land tortoises, take shelter in self-made burrows, Florida to Texas, southern Nevada and California into Mexico (Fig. 32-16*C*); other species in southern Europe, Africa, Madagascar, and southeastern Asia.

Superfamily 2. CHELONIOIDEA. Family CHELO-NIIDAE. Sea turtles. Limbs flipperlike; shell shields smooth; upper Cretaceous to Recent; tropical and subtropical seas, occasional on coasts of United States; usually come ashore only to lay eggs. *Chelonia,* green turtle, horny scutes large, adults 0.9 to 1.5 m (3 to 5 ft) and over 90 kg (200 lb), prized for food (Fig. 32-16*A*); *Eretmochelys,* hawksbill, upper jaw hooked, adults to 56 cm and 45 kg (22 in and 100 lb), dorsal scutes are "tortoiseshell" of commerce; *Caretta,* loggerhead, big-headed, adults 1.2 m and 136 kg (4+ ft and 300 lb); *Lepidochelys,* ridley, to 63 cm (25 in) long, gray or olive.

Family DERMOCHELYIDAE. Leatherback turtles. *Dermochelys coriacea,* leatherback turtle, skin smooth, scaleless, black, seven ridges along back, adults to 2.4 m and 725 kg (8 ft and 1,600 lb), in tropical oceans, occasionally along coasts of United States.

Superfamily 3. TRIONYCHOIDEA. Family TRIONY-CHIDAE. Soft-shelled turtles. Shell reduced, covered by leathery skin. *Trionyx,* Gulf states to St. Lawrence River, Montana and lower Colorado River, *T. spiniferus,* most common, *T. ferox,* both to 45 cm (18 in) long (Fig. 32-16*D*).

Family CARETTOCHELYIDAE. Pitted-shell turtles. Shell soft, limbs paddleshaped. New Guinea. *Carettochelys insculpta,* highly aquatic.

Suborder 2. Pleurodira. Side-neck turtles. Neck folds sideways under shell; pelvis fused to shell; restricted to southern continents.

PELOMEDUSIDAE. Hidden-neck turtles. *Pelomedusa,* Africa; *Podocnemis,* tartaruga, South America and Madagascar.

CHELIDAE. Snake-neck turtles. *Chelus fimbriata,* matamata, South America; *Chelodina,* Australia.

Subclass 2. †Synapsida. Primitive skull roof with lower opening behind eye bounded above by postorbital and squamosal.

Order 4. †Pelycosauria. Limbs spread, some with long dorsal spines (sail lizards). Upper Carboniferous and Permian. †*Varanosaurus;* †*Dimetrodon.*

Order 5. †Therapsida. Mammallike reptiles. Some large and heavy. Permian into Triassic, few Jurassic. †*Moschops;* †*Dicynodon;* †*Cynognathus.*

Subclass 3. †Euryapsida. Skull roof with upper opening behind eye bounded below by postorbital and squamosal.

Order 6. †Araeoscelidia (Protorosauria). Small lizardlike. Permian to Triassic. †*Araeoscelis;* †*Protorosaurus.*

Order 7. †Sauropterygia. Includes plesiosaurs. Slender, long, paddle limbs, marine. Triassic to Cretaceous. *†Plesiosaurus; †Elasmosaurus.*

Subclass 4. †Ichthyopterygia. Skull roof with upper opening behind eye bounded below by postfrontal and supratemporal.

Order 8. †Ichthyosauria. Ichthyosaurs. Size medium, some porpoiselike, limbs as paddles. Triassic to Cretaceous. *†Ichthyosaurus; †Ophthalmosaurus.*

Subclass 5. Lepidosauria. Primitive diapsids and their descendants; skull roof with 2 openings behind eye separated by a bar formed by postorbital and squamosal; no anteorbital opening or depression.

Order 9. †Eosuchia. Ancient two-arched reptiles. Permian. *†Youngina.*

Order 10. Rhynchocephalia. Living representative lizardlike; scales granular; a middorsal row of low spines; quadrate immovable; mandibles joined by ligament; vertebrae amphicoelous; abdominal ribs present; anal opening transverse; no copulatory organ; Triassic to Cretaceous, and Recent; New Zealand. One living species, *Sphenodon punctatus*, tuatara (Fig. 32-17), length to 76 cm (30 in); lives on land and burrows; eats insects, mollusks, or small vertebrates; eggs about 10, with hard white shells, laid in holes in ground, require about 13 months to hatch.

Order 11. Squamata. Lizards and snakes. Skin with horny epidermal scales or shields; quadrate bone movable; vertebrae usually procoelous; copulatory organ (hemipenes) double and eversible; anal opening transverse.

Suborder 1. Sauria (Lacertilia). Lizards. Body usually slender; limbs typically 4, sometimes reduced or absent; mandibles fused anteriorly; pterygoid in contact with quadrate; eyelids usually movable; pectoral girdle well developed (or vestigial); ribs connected to sternum, enclosing thoracic cavity; tongue usually entire; bladder ordinarily present; length usually under 30 cm (12 in), a few over 2 m (6 ft); mostly oviparous, some ovoviviparous and viviparous, food chiefly insects and small invertebrates; some herbivorous; tropics and temperate regions, on continents and islands; about 3700 species.

(*Note: Of the 26 to 30 families, several entirely fossil, some are omitted here.*)

GEKKONIDAE. Geckos. Small; toes often with rounded adhesive plates or pads for climbing; undersides of pads contain many hairlike, often branched digital setae or microvilli with spatulate tips; microvilli to 100 μm long; adhesion of the pads results apparently from frictional forces generated by the many setae; eyes usually lack movable lids; vertebrae amphicoelous; tongue protrusible; many can produce sounds; eggs 1 or 2, usually with calcified shells; worldwide, in warm regions; some climb rocks, trees, and buildings, others live on sandy deserts; over 650 species. *Phyllodactylus* and *Coleonyx*, native to southwestern United States and Mexico (Fig. 32-18A,B); *Hemidactylus* and *Sphaerodactylus*, Key West and southern tip of Florida, former introduced from Old World.

Figure 32-17 Tuatara (*Sphenodon punctatus*); length to 76 cm (30 in). Native to New Zealand and only living member of the order RHYNCHOCEPHALIA. (*Photo by T. I. Storer, live specimen at California Academy of Sciences.*)

Figure 32-18 Geckos of western North America. *A*. Leaf-toed gecko (*Phyllodactylus xanti*), southern California and Lower California. Length to 11 cm (4 in). *B*. Banded gecko (*Coleonyx variegatus*), southwestern United States and northwestern Mexico. Length to 22 cm (8½ in). (*From Stebbins, A field guide to Western reptiles and amphibians.*)

PYGOPODIDAE. Flapfooted lizards. Body snake-like; no forelimbs; hind limbs reduced to flat flaps of skin; scales rounded; tail long and fragile; no eyelids; Australia, Tasmania, New Guinea; about 30 species. *Pygopus.*

IGUANIDAE. New World lizards. Limbs normal; teeth usually alike (homodont) and fixed to sides of jaws (pleurodont); tongue fleshy, not protractile; eyelids complete; chiefly in the Americas; includes most North American lizards, a majority of them west of the Mississippi River, about 625 species. *Anolis carolinensis,* green anoles, North Carolina to Texas, skin color changes readily, green to brown, green on leaves, etc.; *Sceloporus* (Fig. 32-1), spiny lizards, on rocks, trees, or ground; *S. undulatus,* in Central and Eastern states, some 50 species, many in Mexico; *Phrynosoma,* horned lizards (Fig. 32-19), body broad, flat, and spiny, enlarged ''horns'' (scales) on head, southwestern Canada and Kansas to California and Mexico, live on dry ground and sand; *Uta stansburiana,* side-blotched lizards, western United States and Mexico; *Crotaphytus,* collared lizards (Fig. 32-20); *Gambelia,* leopard lizards; *Sauromalus,* chuckwallas; *Callisaurus,* zebra-tailed lizard; *Iguana, Cyclura,* iguanas, Mexico and southward, some to 2 m (6 ft) long, body compressed, tail long, a middorsal crest of lance-like spines; active in running and climbing; feed on leaves, fruits, some insects, and small vertebrates; used for food in tropics.

AGAMIDAE. Old World lizards. Limbs normal; teeth usually differentiated (heterodont) and attached to edges of jaws (acrodont); tongue short, thick; some with throat fan or dorsal spines; southern Europe, Africa, Asia, and Australia; about 325 species. *Agama; Draco,* flying dragons, long ribs covered by membrane used to volplane from a height; *Moloch horridus,* Australia, somewhat like a horned lizard, feeds on ants.

CHAMAELEONIDAE. Chameleons. Head angular; body high and narrow; tail prehensile; toes

Figure 32-19 Coast horned lizard (*Phrynosoma coronatum*); length to 14 cm (5½ in). Seven species of horned lizards occur on dry ground and sand in the arid western states. (*From Stebbins, Amphibians and reptiles of western North America.*)

Figure 32-20 Collared lizards (*Crotaphytus collaris*); length to 33 cm (13 in). The sexes differ in coloration. Live on rocks and open ground in Central and Western states. (*From Van Denburgh.*)

opposed (2 versus 3) for grasping; tip of tongue expanded and mucus-coated, can be "shot" several centimeters beyond head; eyes large, move independently; eyelids fused over eyeball, with small central circular aperture; teeth acrodont; lungs with air sacs; Africa, Madagascar, and India, about 100 species. *Chamaeleo*, chameleon, changes color readily (Fig. 32-21).

XANTUSIIDAE. Night lizards. Small; dorsal scales granular, ventral scales squarish and nonimbricate; eyes without movable lids; femoral

Figure 32-21 African grass chameleon (*Chamaeleo bitaeniatus*); length to 11 cm (4½ in). Chameleons are native to warm regions of the Old World. They are highly specialized, having a compressed body, prehensile tail, grasping toes, and protrusible tongue.

A

pores present; southwestern United States to Central America, and Cuba, 17 species. *Xantusia*.

TEIIDAE. New World lizards. Slender; limbs reduced in some; temporal fossa not roofed; ventral scales usually squarish and nonimbricate; tongue long and bifid, with scalelike papillae; the Americas and West Indies, about 200 species. *Cnemidophorus*, whiptailed lizards, some 32 species, 9 of which are known to be parthenogenetic. *Ameiva*; *Tupinambis*, tegu, to over 90 cm (3 ft) long.

SCINCIDAE. Skinks. Limbs or toes often reduced, sometimes absent; scales commonly smooth and round-margined (cycloid), with dermal ossifications beneath; tongue with scalelike papillae, tip feebly indented; oviparous or viviparous, some with well-developed placentation; Old World tropics, Australia, and Pacific islands, few in the Americas, over 1000 species. *Eumeces fasciatus* and *E. skiltonianus*, blue-tailed skinks, in United States.

CORDYLIDAE. Africa and Madagascar. *Chamaesaura*, limbs much reduced; *Cordylus* (*Zonura*), tail with whorls of sharp spines.

LACERTIDAE. Old World lizards. Limbs normal; posttemporal fossa roofed; intermaxillary unpaired; Old World counterpart of TEIIDAE. Europe, Asia, and Africa, about 200 species. *Lacerta*.

ANGUIDAE. Includes alligator lizards. Body slender, often a deep fold in skin along each side; limbs small or none; tail long, fragile, regenerates readily; tongue delicate, retracts into thick base; scales usually rectangular over thin bony plates; southwestern United States into South America, Europe, Northern Africa, and India, about 70 species. *Gerrhonotus*, in United States and Mexico, limbs small; *Ophisaurus attenuatus*, slender glass lizard, Wisconsin to Texas and eastward, limbless, the very fragile tail about two-thirds of total length; *O. apus*, Europe; *Anguis fragilis*, slowworm, or blindworm, Europe, western Asia, and northern Africa.

ANNIELLIDAE. California legless lizards. Slender, limbless; eyes small and no ear openings; scales smooth, soft; subterranean; California and Lower California. *Anniella* (2 species).

XENOSAURIDAE. Back with 2 lines of large scales, others small; *Xenosaurus,* central Mexico to Guatemala; *Shinisaurus,* southern China, like small crocodile, eats tadpoles and fish.

HELODERMATIDAE. Beaded (poisonous) lizards. Trunk and tail stout, rounded, limbs short; dorsal scales beadlike, over bony tubercles (osteoderms); ventral scales flat; tongue fleshy, protrusible; teeth bluntly fanglike, grooved, with poison glands opening on outer "gum" of lower jaw; venom potent, bite fatal to small animals, rarely to humans; 2 species, the only venomous lizards. *Heloderma suspectum,* Gila monster, Arizona and Mexico to southwestern Utah and southern Nevada, pink and black to 48 cm (19 in) in total length (Fig. 32-22); *H. horridum,* beaded lizard, Mexico and Central America, yellow and much black.

VARANIDAE. Monitors. Large trunk and limbs stout, neck and tail long; scales small, smooth; teeth large, pointed, and pleurodont; tongue long, smooth, deeply bifid, and protrusible; Africa, southern Asia, and islands southeast to Australia. *Varanus; V. salvator,* Ceylon and Malaya, to 2.5 m (8 ft); *V. komodoensis,* to 3 m (10 ft); *V. acanthurus,* to about 45 cm (1½ ft) long.

LANTHONOTIDAE. Face short, brain region longer; limbs short; scales alligatorlike. One species. *Lanthonotus borneensis,* earless monitor, Borneo. Close to ancestors of snakes.

†MOSASAURIDAE. Mosasaurs. Body and tail slender; limb girdles small; limbs paddlelike. Cretaceous, marine, some to 6 m (20 ft) long. †*Mosasaurus,* †*Tylosaurus* (Fig. 32-15).

Suborder 2. Serpentes (Ophidia). Snakes. No limbs, feet, ear openings, sternum, or urinary bladder; mandibles joined anteriorly by ligament; eyes capable of little movement, each covered by transparent scale (speculum), no lids; tongue slender, bifid, and protrusible; left lung usually vestigial; teeth slender and conical, usually on jaws and on roof of mouth (palatine and pterygoid bones); about 2500 species.

TYPHLOPIDAE. Blind snakes. Eyes vestigial, under opaque scales; teeth only on transversely placed maxillary bones, none on lower jaw; scales thin, overlapping, cycloid, in multiple rows over whole body; tropics and subtropics of both hemispheres, subterranean, about 150 species. *Typhlops.*

LEPTOTYPHLOPIDAE. Blind snakes. Resemble TYPHLOPIDAE, but teeth only on lower jaw; vestiges of femur and all pelvic bones present; southwestern United States to Brazil, the West Indies, Africa, and southeastern Asia; subterranean; around 65 species. *Leptotyphlops.*

(*Note: The following snakes have the eyes exposed and functional and one row of large transverse scales, the ventrals, along the belly.*)

BOIDAE. Includes pythons, boas. With vestiges of pelvic girdle and hind limbs as 2 spurs, latter usually at sides of vent (Fig. 32-11); nearly worldwide; variously on ground, in trees of tropical jungles, or in dry, rocky, or sandy places; includes largest living serpents, also smaller ones, about 80 species. *Python reticulatus,* reticulated python of Malay region, to 10 m (33 ft) long; *P. molurus,* Indian python, to 7.5 m (25 ft); *Eunectes murinus,* anaconda of South America, to 9 m (30 ft); *Charina bottae,* rubber boa, south-

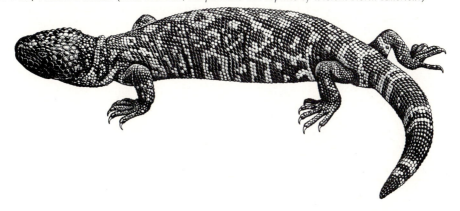

Figure 32-22 The Gila monster (*Heloderma suspectum*) of the American Southwest; length to 48 cm (19 in). This and a related species are the only venomous lizards. (*From Stebbins, Amphibians and reptiles of western North America.*)

Figure 32-23 Rubber boa (*Charina bottae*), a small boa of the western United States; length around 76 cm (30 in). (*Photo by T. I. Storer.*)

Figure 32-25 The gopher snake, or bull snake (*Pituophis melanoleucus*); length around 240 cm (8 ft). (*Photo by T. I. Storer.*)

western Canada to southern California, mostly under 50 cm (20 in) long (Fig. 32-23).

ANILIIDAE (Llysiidae).   Vestiges of pelvic girdles and hind limbs, the latter as spurs beside vent; head small; cranium firm; tail short; scales smooth, shiny, ventrals small; mostly burrowing; 11 species. *Anilius,* northeastern South America; *Cylindrophus,* southeastern Asia.

UROPELTIDAE. Shield-tailed snakes, or roughtails. Head small and pointed; eyes small; tail short, often ending in flat disk or ridge; scales smooth and shiny; no supratemporal bone; no vestiges of hind limbs or pelvic girdle; livebearing; Sri Lanka and southern India. *Uropeltis.*

XENOPELTIDAE. Dentary bone loosely attached to articular; scales smooth, iridescent; southeastern Asia. One species, *Xenopeltis unicolor,* sunbeam snake.

COLUBRIDAE. Common snakes. Facial bones movable; squamosals loosely attached to skull; maxillaries horizontal, forming most of upper jaw; both jaws with teeth but no elaborate fangs; 6 subfamilies, some 1500 species. Subfamily COLUBRINAE (and others), harmless, all teeth solid, none grooved; mostly terrestrial and oviparous; nearly worldwide; includes most living snakes (Figs. 32-24, 32-25). *Natrix,* water snakes,

Figure 32-24 A garter snake (*Thamnophis*); length mostly 60 to 120 cm (2 to 4 ft). (*Photo by T. I. Storer.*)

and *Thamnophis,* garter snakes, aquatic in varying degree, feed on frogs, fishes, etc.; *Coluber* and *Masticophis,* racers and whipsnakes, speediest snakes in United States; *C. constrictor,* racer, climbs readily, preys on small vertebrates including birds; *Pituophis,* bull snakes and gopher snakes, terrestrial, eat many rodents; *Lampropeltis,* milk and kingsnakes, feed in part on reptiles, including rattlesnakes (Fig. 32-12); *Tantilla,* blackheaded snakes, American Southwest, feed on land invertebrates. Subfamily DASYPELTINAE, *Dasypeltis,* central and south Africa, ventral spines on anterior vertebrae used to crack shells of bird eggs. Subfamily DIPSADINAE, *Dipsas,* Mexico to Argentina, teeth long, slender, feed on snails and slugs. Subfamily HOMALOPSINAE, *Enhydris,* southeastern Asia to Australia, and *Erpeton,* tentacled snake, with two scaly appendages on snout, southeastern Asia; grooved rear fangs, feed mostly on fishes, not dangerous to humans.

ELAPIDAE. Coral snakes, cobras, kraits. A pair (or more) of short, rigid, erect, venom-conducting teeth (fangs), grooved or with a longitudinal canal, on maxilla in anterior part of upper jaw (Proteroglypha); tail rounded, tapered, not compressed; venom powerful in many, dangerous to humans; worldwide, except northern North America and northern Eurasia; some 200 species. *Naja,* cobras (Fig. 32-26); *Bungarus,* kraits; *Denisonia; Micrurus,* coral snakes, colored with bands of black, red, and yellow, slow to bite and thus often erroneously considered harmless but strong venom causes occasional human deaths; *M. fulvius,* North Carolina to Gulf states; *M. euryxanthus,* southern Arizona and southwestern New Mexico southward.

HYDROPHIIDAE. Sea snakes. Fangs as in ELAPIDAE; tail compressed, used for swimming; veno-

Figure 32-26 Oriental common cobra (*Naja naja*) with the hood expanded and in the striking position; length to about 200 cm (80 in); native to the Old World and highly venomous.

Figure 32-27 Western rattlesnake (*Crotalus viridis*), in the striking position and with the rattle elevated; dangerously venomous. (*Photo by T. I. Storer.*)

mous and dangerous to humans; live-bearing; in tropical seas; about 50 species. *Hydrophis; Pelamis* (*Hydrus*) *platurus,* along Pacific Coast from southern Mexico to northern South America, west to east African coast, feeds on small fishes.

VIPERIDAE. Paired erectile fangs in front of upper jaw, one on each maxillary bone and folded backward when not in use (Solenoglypha); maxillary short, thick, and movable in vertical plane; facial bones movable; venomous; usually viviparous; about 200 species. Subfamily VIPERINAE. Old World vipers. No pit between nostril and eye. Eurasia, Africa, and southern India. *Vipera,* vipers; *Cerastes cerastes,* horned viper, sidewinds; *Bitis,* puff adders. Subfamily CROTALINAE. Pit vipers, rattlesnakes. A pitlike depression on each side of upper jaw between nostril and eye; some with horny rattle at end of tail; the Americas to Philippines, west through Asia (*Trimeresurus*) to Caspian Sea (*Agkistrodon*). Tail without rattle: *A. piscivorus,* water moccasin, Virginia and Illinois southward, in swampy places; *A. contortrix,* copperhead, Massachusetts and Illinois to Gulf Coast, on drier land; *Bothrops atrox,* fer-de-lance, tropical Americas and West

Indies; *Lachesis muta,* bushmaster, Costa Rica to Peru and Brazil. Tail with rattle of cornified epidermis: *Sistrurus,* pigmy rattlesnakes, and *Crotalus,* rattlesnakes; about 30 species in Western Hemisphere, 15 in United States (Figs. 32-13, 32-27); *C. horridus,* timber or banded rattlesnake, Maine to Florida and west to Iowa and Texas; *C. viridis,* western rattlesnake, Great Plains to Pacific Coast; *C. atrox,* western diamondback rattlesnake, Texas to southeastern California and south into Mexico; *C. adamanteus,* eastern diamondback rattlesnake, North Carolina and Louisiana to Florida, to over 2 m (7 ft) long; *C. cerastes,* sidewinder, or horned rattlesnake, Southwestern deserts, travels sideways with S-shaped loops of body.

Suborder 3. Amphisbaenia (*Annulata*). Worm lizards. Skull bones solidly united; highly adapted for burrowing.

AMPHISBAENIDAE. Worm lizards. Body wormlike; limbs and girdles vestigial or absent; tail short, blunt; skin soft, folded into numerous rings; eyes and ear openings concealed; subterranean; tropical America, the West Indies, and Africa, about 125 species. *Bipes biporus,* southern Lower California, forelimbs molelike, each with 5 toes; *Rhineura floridiana,* in Florida, no limbs.

Subclass 6. Archosauria.

More advanced diapsids, many of which tended toward bipedal locomotion; usually an anteorbital space in skull.

Order 12. †*Thecodontia.* Dinosaur ancestors. Teeth in sockets. †*Saltoposuchus,* bipedal; †*Phytosaurus,* phytosaurs, crocodilelike.

Order 13. Crocodilia. Alligators, caimans, gavials, and crocodiles. Body long; head large and long, jaws powerful, with numerous bluntly conical teeth; 4 short limbs ending in clawed toes with webs between; ribs connected to sternum; tail long, heavy, and compressed; skin thick and leathery, with horny scutes, those on back and belly rectangular and reinforced beneath by dermal bones; ear opening protected by flap of skin; tongue not protrusible; heart 4-chambered, with separate ventricles; no bladder; oviparous; eggs deposited in "nests" of decaying vegetation; Triassic to Recent; tropics of New and Old Worlds and Australia, and Gulf states; 23 species.

Crocodylidae. *Crocodylus acutus,* American crocodile, southern Florida and Greater Antilles to Ecuador and Colombia, snout narrow; *C. niloticus,* Africa and River Nile, dangerous to humans, some deaths; *Alligator mississippiensis,* American alligator, North Carolina to Florida and Rio Grande in Texas (Fig. 32-1); *Gavialis gangeticus,* India and Burma; snout long, slender, sharply distinct from rest of skull, length to 6.5 m (21 ft); feeds on fish.

Order 14. †Saurischia. Dinosaurs. Ischium and pubis diverge.

Suborder 1. †Theropoda. Bipedal, carnivorous; Triassic to Cretaceous. †*Ornithomimus,* ostrichlike; †*Tyrannosaurus,* to about 9 m (30 ft) long.

Suborder 2. †Sauropodomorpha. Sauropods. Huge, 4-footed, herbivorous; Triassic to Cretaceous; †*Brontosaurus;* †*Diplodocus,* to 26.5 m (87 ft) long.

Order 15. †Ornithischia. Dinosaurs. Pelvis birdlike, ischium and pubis together.

Suborder 1. †Ornithopoda. Duck-billed dinosaurs. Two- or four-footed, in water margins; Jurassic to Cretaceous. †*Camptosaurus,* †*Iguanodon.*

Suborder 2. †Stegosauria. Armored; Jurassic. †*Stegosaurus.*

Suborder 3. †Ankylosauria. Armored; Cretaceous. †*Euoplocephalus.*

Suborder 4. †Ceratopsia. Horned; Cretaceous. †*Triceratops.*

Order 16. †Pterosauria. Flying reptiles, or pterosaurs. Forelimb with wing membrane. †*Rhamphor-* *hynchus,* Jurassic, tail long; †*Pteranodon,* Cretaceous, no tail; a pterosaur recently found in Texas estimated to have had a wing span of over 15 m (50 ft).

References

(See also works by Conant, Goin et al., Oliver, Schmidt, and Stebbins and references to journals in Chap. 31.)

Bakker, R. T. 1975. Dinosaur renaissance. *Sci. Am.,* vol. 232, no. 4, pp. 58–78. *A case is made for "warm-bloodedness" of dinosaurs; their descendants are birds.* (See also Hopson, J. A. 1976. Review. Hot- or cold-blooded dinosaurs? *Paleobiology,* vol. 2, no. 3, pp. 271–275.)

Bellairs, Angus. 1970. The life of reptiles. New York, Universe Books. Vol. 1, xii + 282 pp. Vol. 2, 307 pp.

Blair, W. F., and others. 1968. Vertebrates of the United States. 2d ed. New York, McGraw-Hill Book Company. ix + 616 pp. *A complete survey of the vertebrates of the United States.*

Bogert, C. M., and **R. M. del Campo.** 1956. The Gila monster and its allies. *American Museum of Natural History Bulletin,* vol. 109, pt. 1, pp. 1–238, 20 pls., 35 figs.

Carr, Archie. 1952. Handbook of turtles of the United States, Canada, and Baja California. Ithaca, N.Y., Cornell University Press. xv + 542 pp., 82 pls., 37 figs. *Many detailed accounts, structure, distribution, life history.*

———. 1965. The navigation of the green turtle. *Sci. Am.,* vol. 212, no. 5, pp. 78–86. *Migration of this marine turtle covers great distances; it is able to find an island 1400 mi at sea.*

———, and the editors of *Life.* 1963. The reptiles. New York, Life Nature Library, Time, Inc. 192 pp. *Many color figs. Popular.*

Cloudsley-Thompson, J. L. 1971. The temperature and water relations of reptiles. Watford, England, Merrow Publishing Co. Ltd. vi + 159 pp.

Colbert, E. H. 1961. Dinosaurs: Their discovery and their world. New York, E. P. Dutton & Co., Inc. xiv + 300 pp., 100 pls., 150 figs.

———. 1965. The age of reptiles. New York, W. W. Norton & Company, Inc. 228 pp., 20 pls., 67 figs. *The prehistoric world of reptiles.*

Cowles, R. B. 1965. Hyperthermia, aspermia, mutation rates and evolution. *Quart. Rev. Biol,* vol. 40, no. 4, pp. 341–367.

Dawbin, W. H. 1962. The tuatara in its natural habitat. *Endeavour,* vol. 21, no. 81, pp. 16–24. 20 figs. incl. 18 in color. *Habits in natural surroundings.*

Ernst, C. H., and **R. W. Barbour.** 1972. Turtles of the United States. Lexington, The University of Kentucky Press. x + 347 pp.

Gamow, R. T., and **J. F. Harris.** 1973. The infrared receptors of snakes. *Sci. Am.,* vol. 228, no. 5, pp. 94–100. *Heat reception by the labial and loreal pits of snakes, used in detecting warm-blooded prey.*

Gans, Carl (editor). 1969–1977. Biology of the Reptilia. Academic Press, Inc. 1969. Vol. 1. A. Bellairs and T. S. Parsons (coeditors). Morphology A: Origin of reptiles, dentition and osteology. xv + 373 pp. 1970. Vol. 2. Carl Gans and T. S. Parsons (coeditors). Morphology B: Sense organs. xiii + 374 pp. 1970. Vol. 3. Carl Gans and T. S. Parsons (coeditors). Morphology C: Blood, endocrines, xiv + 385 pp. 1973. Vol. 4. Carl Gans and T. S. Parsons (coeditors). Morphology D: Bones and muscles. xii + 539 pp. 1976. Vol. 5. Carl Gans and W. R. Dawson (coeditors). Physiology A. xv + 556 pp. 1977. Vol. 6. Carl Gans and T. S. Parsons (coeditors). Morphology E: Urogenital, digestive and lymphatic systems. xiii + 505 pp.

———. 1970. How snakes move. *Sci. Am.,* vol. 222, no. 6, pp. 82–96. *In their chief mode of locomotion, snakes push sideways rather than downward.*

Grassé, P.-P. (editor). 1970. Traité de zoologie. Vol. 14. Reptiles: Charactères généraux et anatomie. Paris, Masson et Cie. xliv + 680 pp., illus.

Guggisberg, C. A. W. 1972. Crocodiles: Their natural history, folklore, and conservation. Harrisburg, Pa., Stackpole Books. x + 195 pp.

Huey, R. B., and **M. Slatkin.** 1976. Cost and benefits of lizard thermoregulation. *Quart. Rev. Biol.,* vol. 51, no. 3, pp. 363–384.

Klauber, L. M. 1972. Rattlesnakes: Their habits, life histories, and influence on mankind. Berkeley, University of California Press. 2 vols., xxx + 1533 pp.

Mertens, Roberts. 1960. The world of amphibians and reptiles. New York, McGraw-Hill Book Company. 207 pp., 80 pls. (black and white), 16 pls. (color). *Popular.*

Milstead, W. W. 1967. Lizard ecology: A symposium. University of Missouri at Kansas City, June 13–15, 1965. Columbia, University of Missouri Press. ix + 300 pp.

Minton, S. A., Jr., and **M. Minton.** 1969. Venomous reptiles. New York: Charles Scribner's Sons. 274 pp. *Poisonous snakes and their role in human culture.*

Oliver, J. A. 1958. Snakes in fact and fiction. New York, The Macmillan Company. xv + 199 pp., illus.

Ostrom, J. H. 1978. Startling finds prompt a new look at dinosaurs. *National Geographic Magazine,* vol. 154, no. 2, pp. 152–185.

Parker, H. W. 1965. Natural history of snakes. London, British Museum of Natural History. v + 95 pp., 6 pls., 18 figs. *A brief guide to their biology.*

Pooley, A. C., and **C. Gans.** 1976. The Nile crocodile. *Sci. Am.,* vol. 234, no. 4, pp. 114–124. *This large carnivore takes care of its young.*

Pope, C. H. 1937. Snakes alive and how they live. New York, The Viking Press, Inc. xii + 238 pp., illus.

———. 1939. Turtles of the United States and Canada. New York, Alfred A. Knopf, Inc. xviii + 343 pp., 99 illus.

———. 1955. The reptile world: A natural history of the snakes, lizards, turtles, and crocodilians. New York, Alfred A. Knopf, Inc. xxv + 325 + xiii pp., 212 figs. (photos).

Porter, K. 1972. Herpetology, Philadelphia, W. B. Saunders Co. xi + 524 pp. *An introductory text.*

Schmidt, K. P., and **R. F. Inger.** 1957. Living reptiles of the world. Garden City, N.Y., Hanover House, Doubleday & Company, Inc. 287 pp. *A well-illustrated popular survey of reptile families.*

Smith, H. M. 1946. Handbook of lizards (of the United States and Canada). Ithaca, N.Y., Comstock Publishing Associates, a division of Cornell University Press. xxi + 557 pp., 135 pls., 135 figs., 41 maps.

Van Riper, W. 1953. How a rattlesnake strikes. *Sci. Am.,* vol. 189, no. 4, pp. 100–102. *In striking, a rattlesnake may either stab or bite.*

Wright, A. H., and **A. A. Wright.** 1957; 1962. Handbook of snakes of the United States and Canada. Ithaca, N.Y., Comstock Publishing Associates, a division of Cornell University Press. Vols. I and II, xviii + 1133 pp., 305 figs., 68 maps. Vol. III, Bibliography. Ann Arbor, Mich., J. W. Edwards, Publisher, Incorporated. 179 pp. *About 3500 titles.*

33

CLASS AVES: BIRDS

Birds are the best known and most easily recognized of all animals, because they are common, active by day, and easily seen. They are unique in having feathers that cover and insulate their bodies, making possible a closely regulated body temperature and aiding in flight (Fig. 33-1); no other animals possess feathers. The ability to fly enables birds to occupy some habitats denied to other animals. The distinctive coloration and voices of birds appeal to human eyes and ears, and many bird species are of economic importance because of their food habits. Certain kinds are hunted as game, and both domesticated and wild species contribute to the human food supply. The ancient classical names for birds (L. *aves*, Gr. *ornis*) are perpetuated in the name of the class and the term **ornithology,** the study of birds.

33-1 Characteristics

1 Body covered with feathers.

2 Two pairs of limbs; anterior pair usually modified as wings for flight; posterior pair adapted for perching, walking, or swimming (with lobes or webs); each foot usually with four toes; shanks and toes sheathed with scales of cornified skin.

3 Skeleton light, strong, fully ossified; many bones fused, providing rigidity; mouth a projecting beak or bill, with horny sheath, no teeth in living birds; skull with one occipital condyle; neck usually long and flexible; pelvis fused to numerous vertebrae but open ventrally; sternum enlarged, usually with median keel; tail vertebrae few, compressed posteriorly.

Figure 33-1 Class Aves: the American robin (*Turdus migratorius*)**, a representative bird.** The contour feathers insulate the body against loss of heat and form a smooth streamlined exterior surface; the large feathers of the wings and tail provide extended surfaces for flight. The beak, or bill, is of bone, with a smooth cornified covering, and serves as both mouth and hands. The slender lower legs, or tarsi, and the feet also have a cornified covering.

4 Heart with four pumping chambers (two atria, two separate ventricles); only the right aortic (systemic) arch persists; red blood corpuscles nucleated, oval, and biconvex.

5 Respiration by compact, highly efficient lungs attached to ribs and connected to thin-walled air sacs extending between internal organs and into parts of skeleton; voice box (syrinx) at base of trachea.

6 Twelve pairs of cranial nerves.

7 Excretion by metanephric kidneys; uric acid chief nitrogenous waste; urine semisolid; no urinary bladder (except in rheas and ostriches); a renal portal system.

8 Body temperature essentially constant (endothermal).

9 Fertilization internal; females usually with only left ovary and oviduct; eggs with large yolk, covered with hard limy shell, incubated externally; segmentation meroblastic; extraembryonic membranes (amnion, chorion, yolk sac, and allantois) present during development within the egg; young at hatching cared for by parents.

33-2 Evolution Birds appear to have been derived from rather slim, long-tailed, bipedal reptiles (Sec. 33-25). These animals probably ran rapidly on their hind legs, with forelimbs raised, free to develop into wings. Modern bipedal lizards use the tail as a counterweight in turning and balancing. Thus the proavian tail may have been preadapted for steering and flight.

The selective factors in the evolution of feathers are unclear. One theory holds that movable featherlike scales appeared before endothermy and were unrelated, at first, to flight. They formed an insulated covering that prolonged ectothermal activity in warm sunshine by slowing rates of heating and cooling. Their use in retarding loss of internally generated body heat came later. Alternately, in a postulated cooler environment, they may have evolved along with endothermy as a mechanism to conserve body heat.

Gliding must have preceded powered flight, and the proavian may have at first glided from heights (slopes, vegetation) or when running at high speed. The selective advantages of increased speed and penetration of the little-occupied aerial niche appears to have set the course of avian evolution. The great activity and energy demands of flight required endothermy, evolved independent of that of mammals and pushed to the highest level among vertebrates. This permitted expansion of birds into areas of thermal extremes, and their powers of flight made possible temporary use of such areas when conditions were most favorable. Many species are able to occupy polar and high mountain areas and hot deserts.

Birds inherited from reptiles several features that contributed to their success as fliers by reducing weight. Eggs develop fully outside the body, and nitrogenous urinary wastes are excreted without the burden of abundant watery urine. Further weight reductions have been achieved by loss of the bladder and lightening of the skeleton.

Visceral changes relating to endothermy include a four-chambered heart, complete separation of venous and arterial circulation, and improvement in respiration. Internal air sacs, opening to the outside via the respiratory tract, aid breathing and help to dissipate the heat generated by elevated metabolism.

Flight requires a compact, streamlined, rigid body, achieved in birds by fusion, loss, and reinforcement of bones. Many changes have occurred in the skeleton to lighten the total body mass. The legs are positioned beneath the body and are retractible among the belly feathers.

Great visual acuity and rapid accommodation are required of a flying animal, and sight is a primary sense in birds. High mobility and the need to communicate over great distances promoted elaboration of the voice (poorly developed in reptiles) and hearing. Chemoreception, important in lower vertebrates, declined, including the Jacobson organ sense.

Parental care of eggs and young in general is far more advanced than in ectotherms, but no birds are viviparous.

33-3 Size The largest living birds include the ostrich of Africa, which stands fully 2 m (7 ft) tall and weighs up to 136 kg (300 lb), and the great condors of the Americas, with wingspreads up to 3 m (10 ft); the smallest is Helena's hummingbird of Cuba, less than 6 cm ($2\frac{1}{4}$ in) long and weighing less than 3 g ($\frac{1}{10}$ oz); no bird, living or fossil, approaches the largest fishes or mammals in size.

Structure of a bird: The domestic fowl

33-4 External features The chicken or fowl has a distinct **head,** a long flexible **neck,** and a stout spindle-shaped **body,** or trunk. The two forelimbs, or **wings,** are attached high on the back and have long flight feathers (remiges); the wings are deftly folded in Z-shape at rest. On each **hind limb** the two upper segments are muscular; the slender lower leg, or **shank,** contains tendons but little muscle and is sheathed with cornified scales, as are the four **toes,** which end in horny **claws.** The short **tail** bears a fanlike group of long tail feathers (rectrices).

The mouth is extended as a pointed bill, or **beak,** with horny covering. On the upper mandible are two slitlike **nostrils.** The **eyes** are large and lateral, each with an upper and lower **eyelid;** beneath these is the membranous **nictitating membrane,** which can be drawn independently across the eyeball from the anterior corner. Below and behind each eye is an **ear opening,** hidden under special feathers. The fleshy median **comb** and lateral **wattles** on the head and the cornified spurs on the legs are peculiar to the chicken, pheasant, and a few other birds. Below the base of the tail is the **vent.**

33-5 Body Covering The soft, flexible **skin** is loosely attached to the muscles beneath. It lacks glands save for the **preen** (or uropygial) **gland** above the base of the tail, which secretes an oily substance for waterproofing the feathers and keeping the bill and perhaps feathers from becoming brittle. The **feathers** grow from follicles in the skin.

33-6 Feathers These distinctive epidermal structures provide a lightweight, flexible, but resistant body covering, with innumerable dead air spaces providing insulation; they protect the skin from wear, and the thin, flat, and overlapping wing and tail feathers form surfaces to support the bird in flight.

Growth of a feather begins, like the scale of a reptile, as a local **dermal papilla** thrusting up the overlying epidermis (Fig. 33-2). The base of this **feather bud** sinks into a circular depression, the future **follicle,** which will hold the feather in the skin. The outermost epidermal cells on the bud become a smooth cornified sheath (*periderm*) within which other epidermal layers are arranged in parallel ribs, a larger median one forming the future shaft and the others producing the barbs. The central soft dermal pulp (original dermal papilla) contains blood vessels and is wholly nutritive, drying upon completion of growth, so the feather is purely an epidermal structure. Pigment for coloration is deposited in the epidermal cells during growth in the follicle but not thereafter. When growth is completed, the sheath breaks and crumbles away or is removed by preening; then the feather spreads to its completed shape.

1 Contour feathers. These provide the external covering and establish the contour of the bird's body, including the enlarged **flight feathers** of the wings and tail; several thousand are present on a chicken. Each consists of a flattish **vane,** supported by the central **shaft** (rachis), which is an extension of the hollow **quill** (calamus) attaching to the follicle. Each half of the vane is composed of many narrow, parallel, closely spaced **barbs** joining the

sides of the shaft. On the proximal and distal side of each barb are numerous smaller parallel **barbules** provided with minute **barbicels,** or **hooklets,** serving to hold opposing rows of barbules loosely together (Fig. 33-3). Many body feathers have a secondary shaft, or aftershaft, and vane attached to the junction of the principal shaft and quill. Smooth and striated muscles and elastic fibers in the skin enable a bird to ruffle or raise its feathers away from the body, to facilitate their rearrangement when bathing and preening, and to increase the insulation value of the feather cover-

Figure 33-3 *Above.* Four types of feathers. *Below.* Stereogram of the parts in a contour feather; asterisks indicate two proximal barbules cut to show curved edge along which the hooklets slide to make the feather flexible.

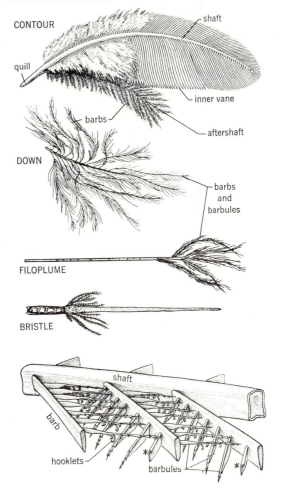

Figure 33-2 Development of a contour feather. *A*, *B*. Early stages, as seen in section. *C*. Stereogram of later stage.

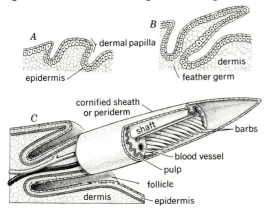

ing during cold weather. When a feather vane is parted by mechanical impact, the separated barbs are rejoined by preening.

2 **Down feathers.** Young chicks and many other birds at hatching are covered with soft downy plumage, providing excellent insulation. A down feather has a short quill, a reduced shaft, and long flexible barbs with short barbules and no hooks. Down is also present beneath the contour feathers on ducks, many other water birds, and some land birds.

3 **Filoplumes.** These are fine hairlike shafts with a few barbs and barbules at their tips, sparsely distributed over the body, as seen on a plucked fowl. They grow in clusters near the follicles of some contour feathers, where they may have a sensory function in controlling feather movements.

4 **Bristles.** These hairlike modified feathers, each with a short quill and slender shaft with a few or no barbs at the base, are seen about the nostrils, where they filter out dust, and around the mouths of flycatchers and nightjars, where they function as a tactile insect trap.

5 **Powder-down feathers** (absent in chicken). The barbs at the tips of these downlike feathers disintegrate as they grow, forming a fine powder that waterproofs the feathers; the powder is distributed by fluffing or preening. Powder-down feathers are found in herons, bitterns, hawks, parrots, etc.

The varied coloration of feathers results from pigments (melanins and lipochromes) deposited during their growth and from structural characteristics which cause reflection and refraction of certain wavelengths (**structural colors**). Melanins produce black, brown, yellowish, and reddish colors, while lipochromes (carotenoids and porphyrins) produce red, violet, yellow, orange, green, and brown. Iridescent colors and most blues are structural. Minute particles (under 0.6 μm) in feathers cause a Tyndall blue light-scattering effect like that resulting from particles suspended in the sky or ocean. Iridescent colors, as seen on the neck feathers of some poultry, on hummingbirds, and in oil films and soap bubbles result from interference phenomena. The colors seen

depend on the thickness of the film and the angle of viewing. Some reflected wavelengths are reinforced (those seen), others are damped out. Platelets and air bubbles in feathers are responsible for such iridescent colors in birds. A backdrop of melanin enhances a structural color by absorbing other wavelengths. A jay feather looks dark when viewed with transmitted light or when its structure is damaged by pounding. Some green colors result from yellow pigment combined with structural blue. A feather once grown changes color only by wear, fading, or other discoloration.

With the exception of ostriches, penguins, and a few other completely feathered birds, feathers grow only on certain areas or feather tracts (*pterylae*) of the skin, between which are bare spaces (*apteria*), as can be seen upon parting the feathers of a bird (Fig. 33-4). The assemblage of feathers on a bird at any one time is called **plumage,** and the process of feather replacement is known as **molt.** The molt is usually an orderly and gradual process, so that portions of the body are not left bare; the large feathers on the wings and tail are molted in symmetrical pairs so that flight is not hindered. Ducks molt all wing feathers simultaneously and are flightless until replacements grow out.

33-7 Skeleton The bird skeleton (Fig. 33-5) is light and delicate compared with that of most mammals; many of the bones contain air cavities to lessen

Figure 33-4 Feather tracts of the domestic fowl. (*After Nitzch.*)

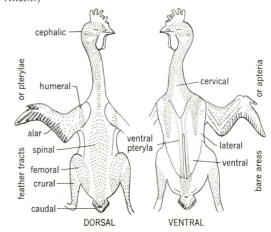

DORSAL   VENTRAL

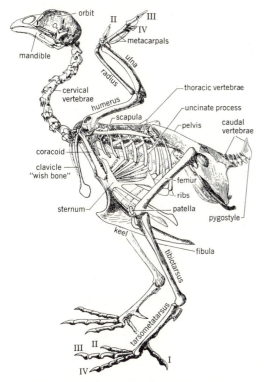

Figure 33-5 Skeleton of the domestic fowl. (*Adapted from Elenberger and Baum.*)

the weight and a strutwork of bony braces that provides strength. The skeleton is modified in relation to flight, bipedal locomotion, and the laying of large eggs with hard shells. The bones of the cranium are separate in young birds but fused in adults except for a nasofrontal hinge that permits slight movement (pronounced in parrots) of the upper jaw in many species. The **brain case** is rounded, the **orbits** sheltering the eyes are large, and the **jaws** (premaxillae + maxillae, and mandibles) project forward as the bony **beak.** The lower jaw hinges on the movable **quadrates** connecting to the squamosals. The skull articulates on a single **occipital condyle** with the first neck vertebra.

The neck contains about 16 **cervical vertebrae,** each with saddle-shaped bearing surfaces that permit free movements in feeding, preening, and other activities. The **trunk vertebrae** are closely fitted together; those of the thorax have rib articulations laterally, and the remainder are fused into a solid **synsacrum** to which the pelvis attaches. The synsacrum

contributes to the formation of the compact body mass of the bird, important in streamlining, control of the wings and tail, and transmission of force to the body from the catapult action of the legs. No lumbar region is evident. The four free **caudal vertebrae** and the compressed terminal **pygostyle** (five or six fused vertebrae) serve in movements of the tail feathers.

The **bony thorax** protects the internal organs and provides a rigid support for the flight mechanism, yet is capable of slight expansion and contraction for respiration. It consists of (1) the **vertebrae** dorsally, (2) the flat **ribs** laterally, and (3) the breastbone, or **sternum,** ventrally, with a prominent median **keel** below providing a broad area of attachment for the powerful flight muscles. Each thoracic rib has a distinct vertebral and a sternal part, the two joining nearly at a right angle; the first four or five thoracic ribs each has posteriorly a flat **uncinate process** overlapping the rib behind to strengthen the thorax.

The **pectoral girdle** on each side consists of (1) the swordlike **scapula** (shoulder blade) lying parallel to the vertebrae and over the ribs, (2) the **coracoid,** a stout stay between the scapula and sternum that braces against the powerful down strokes of the wings, and (3) the **clavicle,** hanging vertically from the scapula; the two clavicles are fused at their ventral ends to form the V-shaped **furcula,** or "wishbone," attaching to the sternum. The three bones meet dorsally on either side to form a circular canal as a pulley for the tendon of the supracoracoideus muscle that lifts the wing.

Each forelimb attaches high on the dorsal surface, the **humerus** pivoting in the glenoid fossa on the coracoid. The "forearm," attached by a hinge joint to the humerus, contains the **radius** and **ulna,** as in other land vertebrates, but other wing bones are modified to provide stable supports for the flight feathers. Only two **carpals** and three **digits** (II, III, IV) are present; the other carpal bones are fused to the three metacarpals to form the **carpometacarpus.** In the "hand," the anterior (second) digit bears the alula (Fig. 33-6), the next is longest and of two segments, and the innermost (fourth) has a single segment. The outermost and principal flight feathers, or **primaries,** are supported on digits III and IV and the carpometacarpus, the **secondaries** on the ulna, and the **tertiaries** on the humerus. Quills of the flight

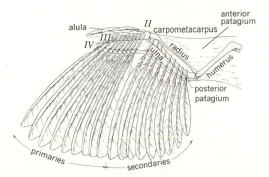

Figure 33-6 Spread left wing of a white-crowned sparrow from the dorsal side, dissected to show how the flight feathers are supported, primaries on the "hand" and secondaries on the "forearm," by the bones and also the membranous patagium. II–IV, digits.

feathers are kept from twisting during flight by connective tissue in extensions of the skin (*posterior patagium*, Fig. 33-6). The hinged elbow helps prevent twisting of the wing during flight.

The **pelvic girdle** is a broad, thin saddle firmly united to the synsacrum but widely open ventrally, permitting easy passage of large eggs in the female. Each half consists of the large **ilium** anteriorly, the **ischium** posteriorly, and the slender **pubis** ventrally; where the three meet, a socket, the **acetabulum,** receives the head of the thighbone, or femur.

Each leg consists of (1) the **femur,** or thighbone; (2) the long and triangular-headed **tibiotarsus,** which is paralleled by the slender and often incomplete **fibula;** (3) the fused **tarsometatarsus,** or shank; (4) a knee bone, or **patella,** held in ligaments before the femur-tibiotarsal joint; and (5) the four **toes,** three in front and one behind, each consisting of several bones. Of the tarsals, or ankle bones, seen in other vertebrates, the proximal row in birds is fused to the tibia and the distal row to the metatarsus, the latter consisting of three fused metatarsals, distinguishable only at the ends of the bone.

33-8 Muscular system   In the bodies of most lower vertebrates, segmental muscles predominate over nonsegmental ones, but the reverse is true among birds and mammals, where the limb muscles are enlarged for rapid activity. Movement of the wings in flight is due chiefly to the large pectoral muscles of the breast, the "white meat" of chickens and turkeys, a major fraction of the entire muscula-

ture (Fig. 33-7*B*). At either side, the **pectoralis** originates on the outer part of the sternal keel and inserts on the ventrolateral head of the humerus; its contraction moves the wing downward and lifts the bird's body in flight. In other land vertebrates the forelimb is raised by muscles on the dorsal surface, but in birds such movement is due also to a ventral muscle, the **supracoracoideus.** This originates on the keel (beneath the pectoralis) and tapers to a strong tendon passing upward to insert on the dorsoposterior surface of the humerus. Both these muscles are symmetrically paired and in turn exert an equal and opposite pull on the thin keel of the sternum.

The muscles of the femur and tibiotarsus are the principal ones used for running and perching. The shanks and feet contain little muscle, an adaptation that provides streamlining and prevents loss of heat from these unfeathered parts. The toes are moved by tendons connected to muscles in the upper segments of the legs. The tendons move through spaces lubricated by fluid, and their action on the toes is directed through loops of tendon (Fig. 33-7*A*).

33-9 Digestive system   The **tongue** is small and pointed, with a horny covering. The **mouth cavity** is roofed with long palatal folds; a short **pharynx** fol-

Figure 33-7 *A*. Perching mechanism of a bird. From muscles on the thigh and lower leg, tendons pass behind the "heel" (tarsal joint) and shank, then through an annular ligament beneath the foot to insert separately on the toes. When a bird squats down for resting or sleeping, this arrangement flexes the toes and holds the bird firmly to its perch. (*Adapted from Wolcott.*) *B*. Diagram of the pectoral muscles (pectoralis and supracoracoideus) that move the bird's wings in flight.

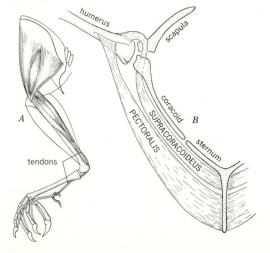

lows; then the tubular and muscular **esophagus** extends to the base of the neck, where it dilates into a large soft-walled **crop,** in which food is stored and moistened. The stomach comprises a soft anterior **proventriculus,** with thick walls secreting the gastric juices, and the disk-shaped **ventriculus,** or **gizzard,** with walls of thick, dense musculature, lined internally by hardened epithelial secretion. Here the food is ground up by action of the muscular walls, aided by bits of gravel or other hard particles swallowed for the purpose—these are, functionally, the "teeth." The **intestine** is slender, with several coils, and leads to the larger **rectum;** at the junction are two slender

caeca, or blind pouches, in which bacterial decomposition of fibrous foods occurs. Beyond is the dilated **cloaca,** the common exit for undigested food wastes and materials from the excretory and reproductive organs, ending with the **vent.** Dorsally, in the young, the cloacal wall bears a small outgrowth, the bursa of Fabricius, a lymphatic organ which forms antibodies that protect against infections. The structure is useful for age determination. The large reddish **liver** is bilobed, with a gallbladder and two bile ducts. The **pancreas** usually has three ducts; all these ducts discharge into the anterior loop of the intestine (Fig. 33-8).

Figure 33-8 The domestic fowl; internal structure. The two caeca are cut off.

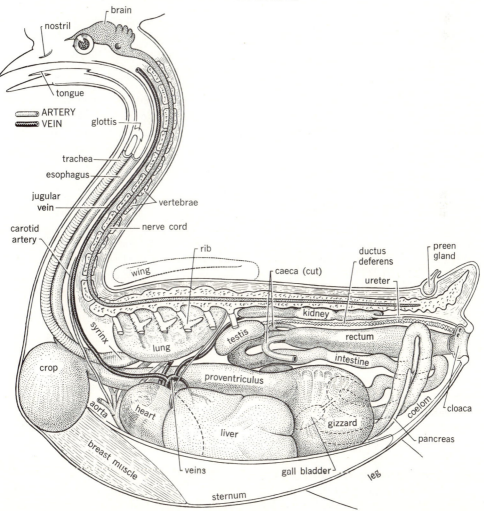

33-10 Circulatory system The bird **heart** has two thin-walled atria and two distinct thick-walled ventricles, separating completely the venous and arterial bloodstreams. This is a major factor in endothermy. The sinus venosus is incorporated into the right atrium. Blood from the two precaval veins and one postcaval enters the right atrium, passes to the right ventricle, and thence by the pulmonary artery to capillaries in the lungs for oxygenation. It all returns in the pulmonary vein to the left atrium, then the left ventricle, and into the single **right aortic arch.** The latter gives off two innominate arteries, each with three large branches, the carotid to the head and neck, brachial to the wing, and pectoral to the breast muscles of flight. The arch continues posteriorly as the **dorsal aorta,** serving the remainder of the body. The venous system retains some reptilian features. Two precavals collect anteriorly, and the single postcaval vein is formed of two large iliac veins draining from the hind limbs and body. A **hepatic portal system** is present as in other vertebrates, but the renal portal system in the kidneys is reduced. A special adaptation is the cross connection in the jugular veins below the head that prevents stoppage of circulation if one vein is compressed by movements of the head or neck. The small rounded **spleen** lies near the stomach. Behind the pericardial sac containing the heart, a delicate **oblique septum** separates the heart and lungs from the other viscera.

33-11 Respiratory system (Figs. 33-9, 33-10) The **nostrils** connect to **internal nares** above the mouth cavity. The slitlike **glottis** in the floor of the pharynx opens into the long, flexible **trachea,** which is reinforced by hooplike cartilages, partly calcified. The trachea continues to the **syrinx** (voice box), around which are the vocal muscles. From the syrinx a **bronchus** leads to each **lung.**

A bird's lung is perhaps more efficient than that of other vertebrates because the inhaled air passes through it instead of in and then out in a reverse flow. Air-flow reversal results in considerable mixing of fresh and residual air (Fig. 33-10). There are many small interconnecting canals (*air capillaries*) opening into larger ones, the **parabronchi.** Gas exchange takes place in the walls of the air capillaries.

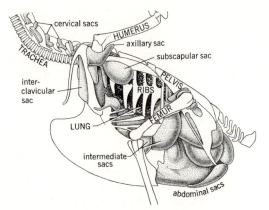

Figure 33-9 Respiratory system of the pigeon. The air sacs are stippled, and bones are outlined; attachments of the lungs to the ribs are indicated. (*After Muller, 1910.*)

The parabronchi communicate with the bronchi and with the air sacs, which extend between organs in the body cavity. Air is inhaled by movements of abdominal and costal muscles, the sternum moving downward and the ribs bowing laterally, and is expelled by contraction of the thorax. The precise path of air flow within the bird's lung is not fully understood, but the following steps appear to be involved. On inhalation a bolus of air enters the bronchus and mostly bypasses the contracted lung, entering the posterior air sacs. On exhalation it flows into and expands the lung. Upon the next inhalation, as a second bolus of air fills the posterior sacs, the first passes from the now shrinking lung into the anterior air sacs. Then with exhalation, that in the anterior sacs is expelled into the bronchus and to the outside and the second bolus of air from the lungs enters the anterior sacs. In this way a one-way movement of air is maintained through the lung. Its direction is opposite to that of pulmonary blood flow, creating a countercurrent effect (see Sec. 30-9). This may explain why many birds are so efficient at extracting oxygen and flying at high altitudes.

The songs and calls of birds are produced by air forced across membranes in the walls of the syrinx, which vibrate and can be varied in tension to give notes of different pitch (Fig. 33-15).

33-12 Excretory system The paired **kidneys** are dark brown, three-lobed structures attached dorsally

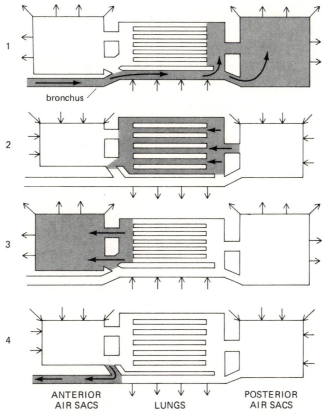

bronchus

ANTERIOR
AIR SACS   LUNGS   POSTERIOR
AIR SACS

Figure 33-10 Movement of air through the lungs of a bird. A bolus of air (shaded) passes through the respiratory tract in the course of two cycles of inhalation and exhalation; 1 and 3, inhalation, 2 and 4, exhalation (see Sec. 33-11). (*After Welty, 1975, as redrawn from Schmidt-Nielson, 1971.*)

under the pelvis. From each, a slender **ureter** extends posteriorly to the dorsal wall of the cloaca. Bird urine contains chiefly minute spherical bodies (2 to 10 μm) of uric acid and urate salts. This appears as whitish material associated with the feces.

33-13 Nervous system and sense organs The bird **brain** is proportionately larger than that of a reptile and is short and broad. The olfactory lobes are small (poor sense of smell in most birds), the cerebral hemispheres are large and smooth, and the optic lobes on the midbrain are conspicuously developed (keen sight). The cerebellum has increased surface with many superficial folds (many activities of coordination). The **nerve cord** and paired **spinal nerves** are essentially as in other vertebrates. The thoracic, or

brachial, plexus, serving the great muscles of flight, is especially large.

The eye is proportionately larger than in other vertebrates, and the sense of **sight** phenomenally keen. Birds have more photoreceptors per unit of retina than other vertebrates; in the region of the fovea in some hawks, eight times that of humans. Color vision in diurnal species is acute; the retina contains many cones provided with image-sharpening oil droplets (Sec. 9-16). The eyes of nocturnal species (owls and nightjars) are large and contain numerous rods. Accommodation, or adjustment of focus to objects at different distances, is very rapid, as required for the quick changes from near to distant vision during flight and in other activities. In close vision the lens is rounded, in most birds evidently by a lens-squeezing action of the ciliary muscles between the sclerotic ring and lens, as in lizards. The sclerotic ring is a group of small shinglelike bones that reinforce and stiffen the front of the eyeball surrounding the cornea.

Birds, like reptiles, have a vascular body, the **pecten,** that projects from the back of the eye into the vitreous humor. In both groups the retina lacks blood vessels, and the pecten is thought to supply the eye with oxygen and nourishment and to remove wastes.

The jerky back-and-forth movement of the head of fowl, coots, and other birds in motion gives them brief intervals of optical fixation on their surroundings, useful in judging distance and detecting moving objects.

The sense of **hearing** is highly developed. A short **external auditory canal** at the posterolateral angle of the head leads to the **tympanic membrane,** or eardrum. From the latter a bone, the **columella auris,** transmits sound waves across the middle ear cavity to the oval membrane of the inner ear, as with reptiles and amphibians. From the **middle ear** on each side, a **Eustachian tube** leads to the pharynx, the two having a common opening at the back of the palate. The **cochlea** of the inner ear is a short, blind-ended tube, larger than in reptiles but less developed than in mammals. However, birds have about 10 times more hair cells per unit of cochlear length than mammals. Their frequency range is about 40 to 29,000 Hz. They are far better than humans at hearing as separate sounds notes that are received in rapid sequence. The oilbird (*Steatornis*) uses rapid vocal pulses (batlike echolocation) to avoid obstructions in caves.

The senses of smell and taste are generally poorly developed in birds. Exceptions are the kiwi, a primitive nocturnal species that probes for earthworms, and the turkey vulture (*Cathartes aura*), now known to use olfaction in locating carrion. Albatrosses, shearwaters, and petrels have relatively large olfactory lobes. There are taste buds on parts of the mouth cavity and tongue in various birds. Tactile receptors on the tongues and/or beaks of woodpeckers, ducks, and mud-probing species aid in detecting unseen food. In general, however, recognition of food depends primarily on sight.

33-14 Endocrine glands The pituitary, thyroid, pancreatic islets, and adrenals are, in structure and function, like those of other vertebrates. Endocrine secretions of the **gonads** regulate the secondary sexual characters, especially those of plumage. In many species of birds the males and females differ in color and adornment of plumage; in chickens and turkeys the wattles and comb are unlike in the two sexes. Castration in a hen produces a **poulard,** a female with small comb, spurs, and malelike plumage. A poulard with an implanted testis resembles a cock. Castration in a male chicken results in a **capon,** with some female feathers, little development of comb and wattles, and loss of pugnacity and ability to crow. Implanting an ovary in a capon usually is followed by assumption of female plumage.

33-15 Reproductive system (Fig. 33-11). In the male the two oval, whitish **testes** are attached near the anterior end of the kidneys. From each testis a much convoluted **ductus deferens** extends posteriorly parallel to the ureter; in many birds it is dilated as a **seminal vesicle** just before entering the cloaca on a papilla close to the urinary opening. In the cloaca of ducks, geese, tinamous, and ostriches there is a median erectile **penis,** as in the alligator; this is present only as a vestige in the chick for a few days after hatching, when it provides a means of determining the sex of downy young. The testes become enormously enlarged in the breeding season. Sperm developed in the testes is accumulated in the

Figure 33-11 Urogenital systems of the domestic fowl, in ventral view. Cloaca opened to show exits of the ducts.

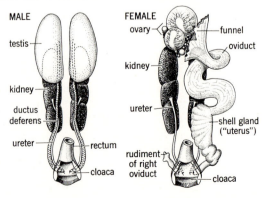

MALE
testis
kidney
ductus deferens
ureter
rectum
cloaca

FEMALE
ovary
funnel
oviduct
kidney
ureter
shell gland ("uterus")
rudiment of right oviduct
cloaca

ductus deferens, to be transferred by cloacal contact from the cloaca of the male to that of the female during copulation.

The **female** reproductive system usually develops only on the left side. The **ovary** is near the left kidney, and close by is the open expanded funnel, or **infundibulum,** of the **oviduct;** the duct extends posteriorly to the cloaca. In nonlaying birds the ovary is small, containing minute eggs, and the duct is also small, but in the season of egg laying both are much enlarged. Each ovum receives its full quota of yolk before being released. A mature ovum escapes from its ovarian follicle into the abdominal cavity, is grasped by active gaping movements of the funnel, and enters the oviduct, moving slowly down the oviduct by the action of muscles in the walls and bands of cilia on the inner surface. Fertilization probably occurs in the upper oviduct; albumen (egg white) is added by glands in the middle portion; and the shell membranes and shell are secreted by glands in the posterior part ("uterus"), after which the egg is ready to be laid. A small nonfunctioning right ovary is commonly present and may enlarge to function if the left one is removed.

Structure of other birds

33-16 Adaptive features As a consequence of the special requirements of flight, birds as a group are more like one another than are members of other vertebrate classes. The external form is usually spindle-shaped (streamlined), offering minimum resistance to the air during flight or to water in diving.

The **coloration** is often varied and striking. A few birds are nearly all of one color (black crows, some blackbirds, white egrets), but the feathers of most species are marked with spots, stripes, or bars. The pattern often blends well with that of the environment, and such **concealing coloration** renders the bird less visible; the effect is enhanced by **countershading,** the upper parts which receive stronger lighting being darker than the undersurface. Birds of arid regions tend to be pale-colored and those of humid places darker (Gloger's rule). Males are often more brilliant than females and marked differently. The colorful plumage of the male functions in identification and threat display, in defense of territory, stimulates sexual behavior in the female, and may divert attack from the nest and young. In some species both sexes acquire a **nuptial plumage** by a spring or prenuptial molt and return to duller colors for the winter by a postnuptial molt in late summer.

The **beak** is at once a mouth and hands, serving to obtain and "handle" food, to preen the feathers, to gather and arrange nest materials, and for other purposes, including defense (Fig. 33-12). The cornified covering grows continually to replace that lost by wear. The form of beak usually indicates the food habits of a bird, being slender in species that probe into crevices or capture insects; stouter but still elongate in woodpeckers, which dig in wood; wide but delicate in swallows and flycatchers, which capture living insects in flight; stout and conical in seed-eating finches and sparrows; sharp-edged with hooked tip in flesh-eating hawks, owls, and shrikes; and wide, with serrated margins, in ducks that sieve

Figure 33-12 Some types of beaks in birds.

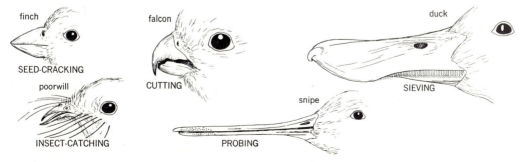

small materials from water. The **tongue** in most birds cannot be extended, but in woodpeckers it can be protruded beyond the bill for extracting insects in wood, and in hummingbirds and sunbirds it is an extensible probe for obtaining nectar from flowers. The two long hyoid bones that support the tongue in such birds extend in muscular sheaths over the back and top of the head and, in some woodpeckers, terminate in the nasal cavity.

The **wings** (Fig. 33-14) are shaped as airfoils providing lift in flight, and various tilting and banking movements aid in altering the manner of flight. The penguins, auks, and some other waterfowl use their wings to "fly" under water. The ostrich, kiwi, and a few other birds have degenerate wings and cannot fly. The **tail** serves as a rudder and brake in flight, as a counterbalance in perching, and for display in courting by the males of many species. Creepers, woodpeckers, and some others that forage on vertical surfaces have tail feathers with stiffened shafts that help to support the body when these birds are at work. The **feet** serve variously for running or climbing, for support of the body at rest, for arranging nest materials, and in some species for handling food and for offense and defense (Fig. 33-13).

The **crop** is used by some birds to carry food for the young, which are fed either by the parent's regurgitation or by the young bird putting its head down the parent's gullet. In adult pigeons and doves, while rearing young, the epithelial lining sloughs off as "pigeon milk," used to nourish squabs in the

nest. The milk closely resembles that of mammals in its composition, formation, and hormonal (prolactin) control. In some flesh-eating birds, such as the kingfisher, there is no distinct crop, and the stomach is a thin-walled distensible sac.

In males of some passerine birds, during the breeding season, the lower end of the ductus deferens is swollen, causing the cloaca to protrude and indicating the sex of the bird.

Natural history

33-17 Distribution Birds occupy all continents, the seas, and most islands, penetrating the Arctic to beyond 80°N and the Antarctic, and live from sea level to above timberline on mountains, even above 6000 m (20,000 ft) on Mount Everest. Despite the ability to fly, they conform to the laws of animal distribution, each species occupying a definite geographic range and particular kind of habitat. Albatrosses and petrels live on the open ocean except when nesting; gulls, auks, and murres occur along seacoasts; shorebirds, except when nesting, inhabit ocean beaches and borders of inland waters; ducks are found in marshes and on open fresh water and salt water; larks and certain sparrows dwell in grasslands; a great variety of birds inhabit thickets; and bustards, ostriches, and other long-legged runners prefer open areas. Many birds utilize trees: the woodpeckers,

Figure 33-13 Some types of feet in birds.

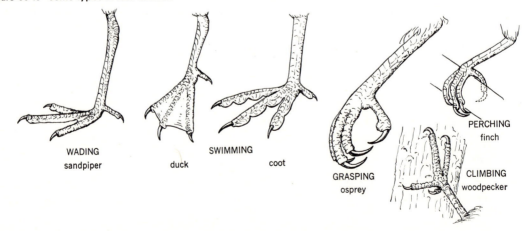

WADING
sandpiper

SWIMMING
duck

coot

GRASPING
osprey

PERCHING
finch

CLIMBING
woodpecker

nuthatches, and creepers forage on trunks and large branches, the warblers, chickadees, and others in foliage; and the flycatchers sit on tree perches to watch for passing insects, as do hawks and owls for their larger prey. Swallows and swifts capture their insect food in the air, and some hawks soar when hunting. Owls, woodpeckers, and others use cavities for nesting and sleeping, but no bird is strictly subterranean. Insect eaters and birds of prey usually live and hunt alone; quail, ducks, robins, and juncos scatter in pairs to nest but flock at other seasons; and some seabirds, pigeons, and blackbirds are almost always in flocks.

In polar regions the species are few but individuals numerous. In temperate lands, 150 to 200 kinds of birds may occur in a locality at various seasons. Tropical regions tend to have many species. Birds sometimes occur in enormous aggregations, such as "rafts" of 100,000 ducks and "clouds" of blackbirds in rice fields; the now extinct passenger pigeon once was said to "darken the sky" with its numbers in Eastern states.

33-18 Flight The spindle-shaped streamlined body results in minimum air resistance to forward movement, and the shape of the wings and manner of moving them result in propulsion with limited expenditure of energy. The wing is an airfoil, with leading edge thicker, trailing edge thin, upper surface slightly convex, and lower concave (Fig. 33-14). This form efficiently parts the air ahead and leaves a minimum of turbulent backwash behind. Air passing over the curved upper surface travels farther and thus faster than that below and exerts less pressure

on the wing surface. The relatively greater pressure on the underside of the wing, combined with slight negative pressure or partial vacuum over the upper surface, results in lift. If the tilt of the wing (angle of attack) is increased, the lift is also increased, but if it is too great, smooth airflow over the wing is disrupted (the flow "breaks away"), there is turbulence behind, the lifting effect is lost, and stalling results. Turbulence is caused by the tendency at slow speeds for the air on the underside of the wing to swirl upward into the low-pressure region over the wing. At slow speeds, as when landing or taking off, toward the end of a glide, or during labored flight, birds may increase the angle of attack. In many species stalling is avoided and lift increased by elevation of the **alula** (Fig. 33-6). The alula acts like an airplane wing slot. It directs a stream of rapidly moving air close over the upper surface of the wing, helping to prevent break-away turbulence. Turbulence tends to be greatest toward the tips of wings, and its effect is reduced by spreading the primaries, thereby creating a series of slots, or by separating the turbulent regions by wing elongation as in long-winged soaring birds.

Unlike the fixed wing of an airplane, that of a bird is not only hinged for down and up movement but its outline can be altered by muscular action, thereby changing the amount of planing surface. The flight feathers can be spread, rotated, or drawn together. The shape of the wing is related to the type of flight, short and broad in birds inhabiting shrubbery but long and often narrow in those which fly or soar in open air. The former have a low **aspect ratio** (ratio of wing length to width), the latter a high aspect ratio.

Gliding is the simplest type of flight, with wings

Figure 33-14 Bird flight. *A.* Air flow over a wing. *B.* Soaring on upwelling air. *C.* Dynamic soaring in wind over ocean (albatross or frigate bird).

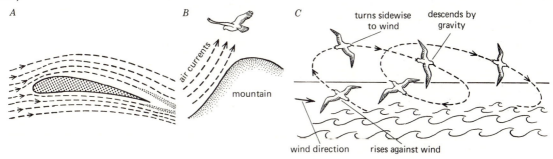

spread but motionless as in ducks coming down on water and quail or other birds alighting on land. It requires no propulsion on the part of the bird. The bird makes use of previous momentum generated in flight and/or the pull of gravity to overcome the air resistance to its forward passage.

Soaring on spread wings, with minor adjustments in wing outline or position, is used by gulls, certain hawks, and others, taking advantage of air currents. The birds maintain or increase their altitude without flapping their wings. Warm air rising from sun-heated soil or rocks provides upward thermals in which birds often circle to maintain their position. Winds deflected upward by mountain slopes are also used. In the usual horizontal winds a soaring bird may face into the wind, remaining stationary or moving slowly forward, or it may glide with the wind, turning about occasionally into the wind to regain altitude.

Albatrosses and other soaring seabirds make use of a wind-velocity gradient near the ocean's surface. Because of friction at the surface, the air moves more slowly there than higher up. The birds glide downward with the wind into the slower-moving air. As they gain momentum they bank into the wind and rise with increasing air speed (as they enter the faster-moving air layers) to near their original height, and the process is then repeated (Fig. 33-14).

In flapping propulsive flight the wings are fully spread and swept downward and *forward* (action of pectoralis muscle), then partly folded for the recovery stroke upward and *backward*. Most of the propulsion is achieved with the "hand" and its primary feathers, on the down stroke. The flexible rear edge of the wing bends upward, giving the wing **pitch** like a propeller blade. The individual primary feathers, with their flexible rear margins, do likewise. On the recovery stroke the primaries tend to separate, reducing air resistance.

There is great variety in the flight of different species, from the duck hawk that dives on partly folded wings to strike its prey to the hummingbird that can fly forward or backward and hover, at 30 to 50 wing beats per second.

33-19 Activity Most birds are active at all seasons, by reason of being endothermal, but the poor-will is known to become torpid and hibernate in winter. Diurnal species are active from dawn to dusk, and owls, nightjars, and others feed at night. Diurnal birds often sleep at night with the head turned back under feathers over the wing. This keeps the beak warm and the bird breathes warm air. Some water birds may sleep by day, floating and dragging a foot, which perhaps minimizes drifting.

Owls have velvety feathers and are able to descend on their prey in silence. Their night vision and hearing are acute. Barn owls are able to strike accurately in total darkness but have difficulty if the heart-shaped facial disk is clipped away. The disk apparently focuses sound waves. Sound localization in owls also appears to be aided by asymmetry in size and location of the ear openings in many species.

A bird's body temperature is closely regulated and usually above that of the environment. The daytime temperature of most adults is about 40 to 42°C (105 to 108°F), varying with the species. This is higher than that found in other vertebrate groups, with the exception of some lizards, and is related to the high metabolic rate and active life of birds dictated by the demands of flight. Evaporative cooling by means of panting helps to prevent overheating. Poultry and passerine birds have a daily rhythm of 41 to 42°C (107 to 108°F) by day to 40°C (104°F) at night; in some swifts and hummingbirds the temperature drops markedly at night. The young of house wrens and others are "reptilian" at birth, assuming the temperature of the environment, but their heat regulation develops within a few days. Young megapodes, on the other hand, are fully endothermic upon emergence.

The ability to fly gives greater opportunity to seek food and escape enemies and makes long migrations possible. Quail have a short direct flight; others can remain awing for longer periods; and the swallows, swifts, terns, and some hawks are in the air much of the time. The speed of flight varies from 30 to 80 km/h (20 to 50 mi/h) or more. Diving falcons may exceed over 200 km/h (125 mi/h).

33-20 Voice Most birds utter calls and songs. Calls generally are brief, relatively simple, stereotyped sounds that influence daily maintenance behavior—feeding, the interaction between parents and young, movements (migration), avoiding danger, and flocking. Songs tend to be more complex than calls, are mostly uttered by males, are strongly influenced by the endocrine changes of the reproductive cycle, and are related to reproduction, including establishment and defense of territory, attracting a mate, maintaining the pair bond, and synchronizing the male and female reproductive cycles. Songs are usually susceptible to some modification by imprinting (Sec. 9-10) in early stages. Some species have only fixed calls, but most songbirds (order Passeriformes, suborder Passeres) have definite songs. Jackdaws, crows, and jays use various notes to convey different information to others of their species, and in some cases the information is passed from generation to generation, as with human traditions. Parrots, mynas, magpies, and mockingbirds have the power of song mimicry. Some notes are used at all times of year, some only during the nesting season, and a few only during migration.

Song pitch and volume are often related to communication distance and openness of habitat. Dense vegetation greatly impedes transmission of high-pitched sounds. In a study of tropical African birds, Chapuis found that species in such habitats tended to have lower-pitched voices with fewer overtones than those in more open environments.

Song patterns within a species may vary geographically, and distinct **dialects** may be present in different parts of the species range. Imprinting on dialect and habitat probably reduces mixing of populations. Only birds and humans are known to transmit dialects.

The voice box of birds is the syrinx (absent in the ostrich and vultures). It varies in complexity. In simple form it may be an enlarged resonating chamber without special muscles or membranes at the base of the trachea (male ducks) or may contain a thin membrane, stretched across the open ends of the tracheal rings, that vibrates in the stream of expelled air. In more complex form (herring gull, Fig. 33-15), a thin, clear internal tympanic membrane is stretched across the open ends of the upper bronchial rings in the syrinx, forming a "window" in the wall at the base of each bronchus. Pressure is increased in the interclavicular air sac surrounding the syrinx, and the sac presses against the "windows," causing them to bow inward. Air from the lungs vibrates the tympanic membranes as it passes through the constrictions so produced. Muscles on the outer walls of the syrinx change the tension on the membranes and, in some species (certain passerines with complex syringeal muscles), can stretch them in almost any direction. The two halves of the syrinx can operate independently, so two notes or phrases can be produced simultaneously. The notable carrying power of some bird songs results from resonance in the air sacs.

Bird song is analyzed by use of the sonograph, which makes a tracing on paper (sonogram) of frequency plotted against time. Much information in bird song is not evident to human ears, but the song

Figure 33-15 Syrinx of herring gull. *A*. Dorsal view, showing internal tympanic membranes that provide the source of sound. *B*. Ventral view, showing attachment of bronchotrachealis muscles to movable second bronchial half rings that change tension on the tympanic membranes. (*After Rüppell.*)

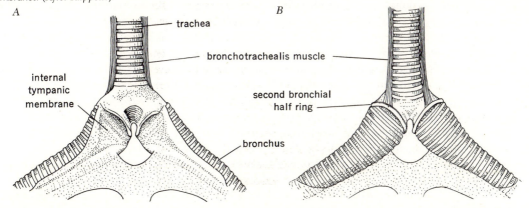

A
trachea
bronchotrachealis muscle
internal tympanic membrane
second bronchial half ring
bronchus
B

is the means by which birds recognize one another, both as to species and individually. Human and avian frequency discrimination seems to be about the same, but time discrimination is far less in humans. Sounds that are buzzlike to us may be a series of distinct notes to a bird.

More refined records of bird song can be obtained with the oscilloscope, which provides detailed information on amplitude modulations. In some birds such modulations may be of greater importance than the frequency or pitch at which phrases are sung.

33-21 Migration Some birds, such as the bobwhite, are strictly resident, but many species **migrate,** or shift regularly from one region to another with the change of seasons. Both the summer and winter ranges of species are well defined. Results from marking individual birds with numbered leg bands show that many return to places previously occupied. Most migration is north and south, or **latitudinal;** birds move into the wide land masses of the north temperate and subarctic regions, where there are habitats for feeding and nesting during the warmer months, and then retire south for the winter. A lesser and opposite movement occurs in the Southern Hemisphere, where the seasons are reversed. Some other birds perform **altitudinal migrations** into mountainous regions for the summer and return to the lowlands to winter; this occurs in the Rocky Mountains and the Cascade–Sierra Nevada systems of western North America.

Most species use established routes for migration and travel more or less on schedule, arriving and disappearing regularly "according to the calendar" (Fig. 33-16). Some birds migrate close to the earth, others up to 900 to 1500 m (3000 to 5000 ft) but rarely higher. Although individuals may fly at 50 to 80 km/h (30 to 50 mi/h), they stop to feed, are passed by others, then go on; hence the "migration front" progresses rather slowly, averaging about 40 km (25 mi)/day.

Migration, breeding, and molt are phases in the annual cycle of birds, all regulated by the neuroendocrine system. Usually, prior to migration, fat reserves, not present at other times, are accumulated rapidly for extra fuel during long flights. Also, many strictly diurnal birds become nocturnal during mi-

gration. Prior to migration a restlessness (*Zugunruhe*) appears, and birds begin to move in the direction of their migratory goal. Although *Zugunruhe* appears to be innate, its timing is influenced by environmental stimuli, acting through the bird's neuroendocrine system. In Temperate Zone species the most important stimulus is changing day length, but weather conditions may also influence departure times. Experiments on the white-throated sparrow have shown that corticosterone from the adrenals and pituitary prolactin interact to bring about the fattening, gonadal growth, and restlessness associated with the onset of migration, and that timing of the daily fluctuation in blood concentration of these two hormones is critical in initiating the premigratory changes. The biological clock involved may be the pineal gland. The pineal in birds has been shown to exert control over their daily (circadian) locomotor activity (Sec. 8-13). It is responsive to photoperiodic stimulation (daily light changes) and perhaps acts on the hypothalamus or pituitary, promoting release of corticosterone and prolactin.

In migrating, some birds follow obvious landmarks—coasts, rivers, mountain ranges—but others pass over seas or lands with few directional features. Many quickly learn and remember topographic features. Furthermore, in some young birds that migrate independently of their parents, recognition of major landmarks may be innate. Imprinting during early development enables some birds to recognize the region of their birth upon their return. In many species migration in daytime appears to be guided by the position (azimuth) of the sun and at night by the stellar axis of rotation. Heavy cloud cover or fog interferes with orientation. Use of celestial cues usually requires a time sense (biologic clock), well developed in birds. However, in the indigo bunting it appears that directional information is obtained from star patterns and a biologic clock is not required. Young buntings first develop a north–south reference axis based on celestial rotation. They then learn the pattern of stars around the northern celestial pole. After that, they can orient on star patterns alone, and celestial rotation becomes a redundant clue. In some species sensitivity to the earth's magnetic field aids orientation, as has been shown by distortions created through use of head magnets and artificial magnetic fields. Much remains to be learned

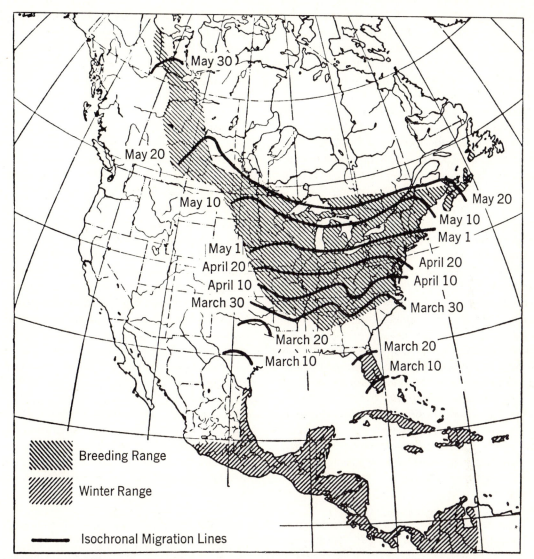

Figure 33-16 Summer (breeding) and winter ranges of the black-and-white warbler (*Mniotilta varia*), a migratory species. The advance during the spring migration is shown by isochronal lines that connect points of arrival on certain dates. (*From U.S. Fish and Wildlife Service.*)

about the biologic-clock navigational mechanism of birds, and no doubt differences will be found among species, having been impressed on the nervous system through countless generations of natural selection.

33-22 Food The high and regulated body temperature of birds, their great activity, and their light weight make for a large requirement of food having high energy value. Blood sugar levels in birds are usually about twice that in mammals. Since few have more than a limited fat storage, a bird cannot long survive without food. Quail and grouse take much leafy vegetation, but most species use concentrated materials such as seeds or fruits and various animals, including worms, arthropods, mollusks, and vertebrates. Small flycatchers, warblers, and vireos eat little but insects; fish are the staple diet of loons, cormorants, pelicans, terns, and kingfishers; frogs are eaten by herons; both snakes and lizards are captured by some hawks; the bird hawks (e.g., the sharp-shinned and the duck hawk) feed chiefly on birds; and rodents and rabbits bulk large in the diet of many hawks and owls. Vultures feed only on carrion, and ravens and some hawks also use such food.

Nestlings may eat more than their own weight in a day, and adults of many small species need one-quarter to one-half their weight daily. The crop and stomach may be filled twice a day or more often; the growing young of some species are fed several hundred times a day by their parents.

The diminutive hummingbirds that fly and hover through much of the day have an enormously high rate of metabolism (four times that of mice). This need is met by frequent feeding on nectar (largely sugar solution), together with some insects. To survive the night, they feed most heavily at dusk, sleep briefly, then become torpid for most of the hours of darkness, when their metabolism sinks to about 10 or 5 percent of that in daytime.

Most nonmigratory bird species fit Bergmann's rule (Sec. 13-27). Their body size averages larger at high latitudes. This is a reflection of endothermy and the high metabolic rate of birds.

Water requirements differ widely. Many species can live only where drinking water is readily accessible. Various finches require 20 to 60 percent of their body weight of water daily; some need less, the California quail only 5 percent. The Australian zebra finch has survived for $1\frac{1}{2}$ years on dry seeds. Marine birds (gulls, petrels, etc.) have special salt-excreting nasal glands which dispose of the excess of salts taken in their diet or in drinking seawater (see Sec. 7-17).

33-23 Reproduction Fertilization is always internal, and all birds lay eggs with much yolk and a hard limy shell, which must be warmed, or **incubated,** for growth of the embryo (Fig. 33-17). In all except the mound builders (megapodes) (p. 768), either one or both parents sit upon the eggs to supply the necessary heat. The young of chickens, quail, ducks, shorebirds, and others are **precocial,** being well formed, fully covered with down, and able to run about at once when hatched, whereas those of songbirds, woodpeckers, pigeons, and others are "blind" (eyes closed), naked, and helpless when hatched and must be fed and cared for in the nest; these are termed **altricial** (Fig. 33-18). Their mouth lining and borders of the mouth are often colorful, providing a feeding target for the adults when the young gape for food. In precocial and some altricial species, hatchlings quickly learn to respond to the visual image and sounds of the parent. In some species response to parental sounds may begin even before hatching. Such rapid learning, called **imprinting** (Sec. 9-10), is particularly important in precocial species to ensure that the young birds will

Figure 33-17 Some types of bird nests. *Left.* Killdeer, eggs laid on ground. *Center.* California quail, eggs in a lined depression. *Right.* Cliff swallow, nests formed of mud pellets and closely placed in a dense colony on a cliff. (*Photographs by T. I. Storer.*)

follow and respond to their parents. Later imprinting establishes sex recognition.

Each species has a characteristic season of reproduction, usually a few weeks in spring or summer—horned owls in February or March, goldfinches in July, and so on. Breeding activities follow increase in the size and functioning of the gonads and are controlled by hormones. Gonadal enlargement is effected through the pituitary, which, in turn, is affected indirectly by light. Experimental exposure of birds in winter to an increased "length of day," even with a 40-watt electric light, will cause enlargement of the gonads, song, mating behavior, and even egg production long before the normal season of reproduction. Poultry farms use electric lights in laying houses during winter to increase egg production, but the hens are shorter-lived.

The breeding season for many species begins with the males uttering their characteristic songs at frequent intervals, and many go through **courting performances** in the presence of females. The sights, sounds, and contacts involved in courtship reinforce the pair bond and promote sexual readiness. Among many small land birds each male takes up and defends a **territory** (Sec. 12-8) suitable to the needs of a pair for the period necessary to rear a brood. Other males of the same species, occasionally members of other species, and predators are not tolerated. When a female joins a male so established, the pair proceeds with nest construction, mating, egg laying, incubation, and care of the young. In some species both sexes participate fully, whereas in others one sex performs most of the duties. Colonial waterbirds (murres, cormorants, gulls) on rocky sea cliffs and small islands defend small nesting territories, since suitable places are scarce. Brood production usually leaves the parents with worn plumage, which is renewed at the postnuptial molt.

Some waterfowl and a few land birds lay on bare rocks or ground, but most species construct a **nest** to hold the eggs and shelter the young (Fig. 33-17). This may be a mere depression in soil or gravel (killdeer), a few bits of vegetation (gulls), a loose framework of twigs (doves), or a cuplike structure woven of grasses or other plant materials (many songbirds). Most kingfishers and some swallows dig nest tunnels in stream banks, woodpeckers usually excavate

gourd-shaped nest cavities in trees, and some swallows build nests of mud. Bluebirds, titmice, and others use natural or artificial cavities, including woodpecker holes and nest boxes of human construction.

The average number of **eggs** laid by a female in one clutch is smallest in birds nesting in safe locations and largest among ground nesters. In general, Temperate Zone species have larger clutches than tropical species. The murre and the band-tailed pigeon lay but one egg, the mourning dove two, large hawks two or three, most small land birds three to five, and the bobwhite and California quail average 14 to a set. Some birds are **determinant layers,** depositing a clutch of fixed size (doves, many passerines) and others are **indeterminant layers,** capable of depositing additional eggs to make up losses (some woodpeckers, ducks, and galliforms). The **shells** are white in owls and woodpeckers, plain blue in the robin, and of a common ground color overlaid with darker spots, streaks, or blotches in many species. Shorebirds and some others nesting on bare places have eggs that are **protectively colored,** closely resembling their average surroundings. The usual ovoid, or "egg," shape permits the eggs to be closely placed and covered by the parent during incubation. **Incubation** for many small land birds requires about 14 days, for the domestic fowl 21 days, for pheasants 21 to 26 days, for many ducks and the larger hawks 28 days, and for the ostrich 42 to 60 days. Young of small altricial species require about a week after hatching before they leave the nest.

All young birds (except megapodes) require **care after hatching.** This includes feeding, guarding, and brooding (protection against chilling, wetting, or undue heat of the sun) (Fig. 33-18). The young are attended and defended by the adults for some time after leaving the nest. Family groups form the nuclei for winter flocks in gregarious birds such as blackbirds or quail; among solitary forms, such as hawks and insect eaters, the young scatter or are driven away by the parents, and all take up independent life.

33 24 Relation to humans Some people use wild birds for food and garments. Ducks, quail, herons,

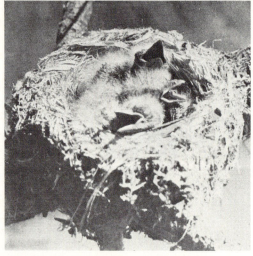

Figure 33-18 Young birds. *Above*. Precocial young killdeer hatched with a full covering of down, eyes open, and able to get about at once. *Below*. Helpless (altricial) young of the wood pewee still in the nest some days after hatching; the down has grown out, the contour feathers are just appearing, the eyes are still closed, and the young must be fed by parents. The killdeer shows white egg tooth on tip of bill; the leg bears a bird band for identification. (*Photographs by T. I. Storer.*)

and other birds down to small finches have been sought as game in Europe for centuries, and today small songbirds are regularly marketed in some Mediterranean countries. Early settlers in North America used game birds for food, and later enormous numbers of wild ducks, geese, prairie

chickens, pigeons, and shorebirds were sold in city markets, but wild populations cannot withstand such continued slaughter. The great auk and passenger pigeon became extinct, and many others were greatly reduced. Ducks, geese, quail, and other game birds are now sought by several million hunters every year in the United States, under legal restrictions as to season, bag limit, and manner of capture. Means are being developed to increase the supply by game management. Bird feathers of distinctive color or shape have long been used for human adornment.

Chlorinated hydrocarbon insecticides are now a serious threat to bird life, particularly in species at the top of food chains (see Secs. 12-6, 12-7, and 36-8).

Many small birds consume harmful insects and weed seeds, and they are one agency in reducing such pests. Most hawks and owls feed upon rodents that are injurious to crops; a few may destroy poultry, game birds, and songbirds, but their beneficial effects usually outweigh their harm. Scavenging species dispose of carrion. Nectar eaters (about one-fifth of all bird species) are important pollinators of flowering plants. Certain birds sometimes damage crops by eating newly planted seeds or young plants or the mature seeds, fruits, and berries, or they may be agents in the spread of weeds. Control then may be necessary. Human crop monocultures favor explosive growth of pests, including certain birds such as the starling.

A few birds are vectors of diseases that can be transmitted to humans (encephalitis, ornithosis), and humans may become infected with a fungal lung disease (histoplasmosis) in areas where large numbers of birds roost. The fungus is carried in dust from bird feces.

Bird study with field glasses or camera is enjoyed by many people, and there are numerous organizations devoted to this subject; the American Ornithologists' Union and the National Audubon Society are two of the largest in North America.

For centuries humans have kept wild birds as pets in cages and aviaries. Until recently this traffic depended upon capture of adults or young birds in the wild, but aviculturists now breed many species in captivity. Populations of some rare and endangered wild birds are now being increased by breeding of captives.

The greatest economic contribution of birds to human welfare is from the guano birds of offshore islands (cormorants and others), whose feces are gathered for fertilizer, and from domesticated species, the barnyard fowls, or poultry—chickens, turkeys, ducks, and geese—that serve for food, supply eggs, and provide feathers for a variety of uses. Many other birds have been domesticated for aesthetic reasons.

33-25 Fossil birds Fossil remains of birds are scarcer than those of some other land vertebrates, because their delicate skeletons are less likely to be preserved. Birds are derived from reptiles, probably from small, primitive bipedal thecodonts, which were ancestral also to dinosaurs, or from dinosaurs themselves.

The oldest bird relics are in slabs of Upper Jurassic slate from near Solnhofen in Germany, containing impressions of †*Archaeopteryx* (Fig. 33-19). The size is that of a large pigeon, with a small head, slender jaws with teeth but no horny sheath, the wing with long flight feathers and three clawed digits, and the tail of 21 vertebrae with a pair of lateral feathers on each segment. The body and thighs were feathered, and there was a basal ruff on the otherwise bare neck and head. Perhaps only gliding flight was possible.

In Cretaceous time there were toothed birds in Kansas and Montana; †*Hesperornis*, about 1.5 m (5 ft) long, was flightless and aquatic. †*Ichthyornis* was a ternlike toothed bird. From the Tertiary (Eocene) onward, the birds lack teeth and become progressively more modern in structure and appearance. The terrestrial †*Diatryma* of the Eocene in Wyoming was about 2 m (7 ft) tall, with a huge beak but vestigial wings. A giant "bony toothed" seabird (†*Osteodontornis*) from the Miocene of California had an estimated total wingspread of 4.5 m (15 ft). The oldest known perching birds (passeriforms) go back to the Eocene. Bird remains are common in the Pleistocene of California, Florida, and elsewhere. The asphalt pits of Rancho La Brea in Los Angeles, Calif., have yielded over 130 species, including †*Teratornis merriami*, a giant condor larger than any now existing. A large member of this genus found in Nevada had a

wingspan estimated at 5 m (over 15 ft). About 1700 kinds of fossil birds have been found, some 900 of which are extinct.

Classification

Class Aves.

Birds. Body covered with feathers; forelimbs usually modified as wings and adapted for flight; hind limbs for walking, perching, or swimming, usually 4 toes, never more; mouth extended as a beak, no teeth (in living birds); skull with 1 occipital condyle; pelvis fused to numerous vertebrae, forming a synsacrum, open ventrally; heart 4-chambered; lungs compact, with air sacs; voice box at base of trachea; usually no urinary bladder; body temperature closely regulated; oviparous; Upper Jurassic to Recent; 8700 species.

Subclass 1. †Archaeornithes. Ancestral birds, or "lizard birds." Both jaws with teeth in sockets (alveoli); wing broadly feathered, digits 3, each with claw; tail of more than 13 vertebrae with pairs of lateral feathers; no pygostyle; Upper Jurassic. One species, †*Archaeopteryx lithographica*, pigeon-sized, 4 specimens (Fig. 33-19).

Subclass 2. Neornithes. True birds. Metacarpals fused, second finger longest; tail vertebrae 13 or fewer, usually with pygostyle; sternum keeled or flat; Cretaceous to Recent.

Figure 33-19 Fossil birds. *Left.* Restoration of the Jurassic lizard-bird (†*Archaeopteryx*). *Right.* Skeleton of Cretaceous toothed bird (†*Hesperornis*). *(After Marsh.)*

Superorder A. †Odontognathae. New World toothed birds.

Order 1. Hesperornithiformes. Clavicle not fused; sternum lacks keel; wing of vestigial humerus only; both jaws with teeth; Upper Cretaceous. *†Hesperornis,* flightless, specialized for swimming, to nearly 1.5 m (5 ft) long.

Order 2. Ichthyornithiformes. Clavicles fused; sternum keeled; wing well developed; toothed jaws. Upper Cretaceous. *Ichthyornis,* ternlike.

Superorder B. Neognathae. Typical birds (toothless).

Order 1. Tinamiformes. Tinamous. Wings short, rounded, functional, but flight weak; sternum keeled; tail very short, pygostyle reduced; eggshells with high gloss; southern Mexico to southern South America. *Tinamus, Rhynchotus,* tinamous, size medium, flight direct, but usually run; used as game birds. 60 cm (24 in).

Order 2. Rheiformes. Rheas. Flightless; terrestrial; wings reduced; coracoid and scapula small; sternum unkeeled; 3 front toes on each foot; head and neck partly feathered; feathers lack aftershafts; precocial young. Miocene to Recent; South America. *Rhea,* to 122 cm (48 in) tall, on open lands.

Order 3. Struthioniformes. Ostriches. Flightless walking birds; wings reduced; coracoid and scapula small, usually fused; sternum lacks keel; pubic symphysis present (unique in birds); pygostyle very small; only 2 toes (third and fourth) on each foot; head, neck, and legs sparsely feathered; feathers without aftershaft; Miocene to Recent; Africa and Arabia. *Struthio camelus,* ostrich, largest living bird, 2.2 m (over 7 ft) tall and weighs to 135 kg (300 lb); inhabits arid lands in flocks of 3 to 20; omnivorous; at nesting a male defends 4 or 5 females that lay up to 30 eggs in one nest, male usually incubating at night, female by day; precocial young; feathers of adults formerly widely used by women as decorative plumes, plucked from living birds on ostrich farms in Africa.

Order 4. Casuariiformes. Cassowaries and emus. Flightless; wings reduced; coracoid and scapula small; terrestrial; sternum unkeeled; 3 front toes on each foot; neck and body densely feathered; feathers with aftershaft nearly equal to shaft; precocial young; Pliocene to Recent. Australia, New Guinea: *Dromaeus,* emu, to 1.5 m (5 ft) tall, in sparsely wooded regions. New Guinea and nearby islands: *Casuarius,* cassowary, to 1.5 m (5 ft) tall, bare skin on head and neck, helmetlike casque on head, in forest, nocturnal.

Order 5. †Aepyornithiformes. Elephant birds. Flightless; terrestrial; sternum short, broad, and unkeeled; wings vestigial; 4 toes; Recent, but extinct for several centuries; Madagascar. *Aepyornis,* 9 species, about 3 m (10 ft) tall; eggs to 24 × 33 cm (9½ × 13 in), largest known animal eggs.

Order 6. Dinornithiformes. Moas and kiwis.

Suborder A. †Dinornithes. Moas. Flightless; terrestrial; sternum reduced and unkeeled; coracoid, scapula, and wing bones reduced or absent; hind limbs massive, 3 or 4 toes; feathers with large aftershaft; Recent, but extinct for several centuries; New Zealand; over 3 m (9½ ft) tall and about 450 kg (1000 lb); eggs to 14 × 19.5 cm (5½ × 7¾ in). *Dinornis,* 8 species; *Anomalopteryx,* 20 species.

Suborder B. Apteryges. Kiwis. Flightless; terrestrial; bill long and slender with nostrils at tip; wing degenerate (humerus vestigial, only one digit, no flight feathers); sternum unkeeled; 4 toes; body plumage fluffy, hairlike, without aftershafts; New Zealand. *Apteryx,* kiwi (Fig. 33-20), 3 species, mostly under 60 cm (24 in) long; nocturnal; omnivorous; nest in burrows, eggs 1 or 2, about 8 × 13 cm (3 × 5 in).

Order 7. Podicipediformes. Grebes. Tail a tuft of downy feathers; legs far back on body; patella large; tarsus compressed; feet lobed; on quiet fresh water, some on bays or ocean margins; adept at diving "at flash of a gun"; prococial young; food small aquatic animals, some plant materials. *Podiceps auritas,* horned grebe; *Podilymbus podiceps,* pied-billed grebe, or dab-chick.

Order 8. Sphenisciformes. Penguins. Flightless; forelimbs ("wings") paddlelike, bones much compressed; tarsometatarsus incompletely fused; toes 4 (first small), all directed forward and outward, feet webbed; feathers small, scalelike, dense over entire body (no apteria); thick layer of fat under skin; stand erect on tarsi; dive readily and swim by strokes of forelimbs; nest in colonies on rocky islands or ice; southern oceans north to Galápagos Islands; gregarious; 17 species (Fig. 33-21). *Aptenodytes forsteri,* emperor penguin, shores of Antarctica, to 122 cm (48 in) tall; other species smaller.

Figure 33-20 Kiwi (*Apteryx*, order DINORNITHIFORMES), wings vestigial, plumage hairlike; native to New Zealand. (*Photo by San Diego Zoological Society.*)

Figure 33-21 Penguins (order SPHENISCIFORMES). The paddle-like wings serve for swimming but not for flight; native to the Southern Hemisphere. (*Photo by National Zoological Park.*)

Order 9. Procellariiformes. Albatrosses, fulmars, and petrels. Nostrils tubular; horny sheath of bill compounded of several plates; large nasal salt glands; hind toe vestigial or none (diving petrels); plumage compact, "oily" in texture; wings long, narrow; young downy when hatched; strictly oceanic, nest usually on islands.

DIOMEDEIDAE. Albatrosses. Mostly large and on southern oceans, often follow ships at sea. *Diomedea exulans,* wandering albatross, wingspread to 3.5 m (over 10 ft).

PROCELLARIIDAE. Gull-sized birds of open sea. *Puffinus,* shearwaters.

HYDROBATIDAE. *Oceanodroma,* stormpetrels, or Mother Carey's chickens, small, nest in crevices or burrows.

Order 10. Pelecaniformes. Includes pelicans, cormorants, boobies, gannets. All 4 toes included in footweb; nostrils vestigial or absent; a gular pouch on throat (except tropic birds); aquatic; fish-eating; colonial nesting common; young naked at hatching.

PELECANIDAE. Pelicans. Body heavy, bill to 46 cm (18 in) long. *Pelecanus erythrorhynchos,* white pelican, on inland fresh waters of western and central North America; pouch used to scoop fish from water; *P. occidentalis,* brown pelican, coastwise in Gulf states and California to South America.

PHALACROCORACIDAE. Cormorants. *Phalacrocorax,* many species, virtually worldwide, on both coasts of North America, nests on sea cliffs, islands, and some inland waters and in trees; slender-bodied with small pouch, fish by diving; captives (with ring on neck) used by Orientals to catch fish.

SULIDAE. *Sula,* booby, tropic birds; *Morus bassanus,* gannet.

FREGATIDAE. *Fregata,* frigate birds, or man-o'-war birds, swift predators of tropical seas.

Order 11. Anseriformes. Ducks, geese, and swans. Bill broadened, covered with soft cornified epidermis containing numerous tactile nerve endings in sense pits, with harder "nail," or cap, at tip; margins of bill with many transverse horny ridges (lamellae); tongue fleshy; legs short; feet webbed; tail usually short, of many feathers; nests lined with down; eggs plain, without spots; young downy and precocial when hatched; worldwide; over 200 species; the waterfowl of hunters, prized for sport and eating; many kept as captives (Fig. 33-22).

ANHIMIDAE. Screamers. In South America, bill slender, feet not webbed. *Anhima (Palamedea).*

ANATIDAE. Waterfowl. Surface ducks (tip up, do not dive in feeding): *Anas platyrhynchos,* mallard, or greenhead; *Anas acuta,* pintail, or sprig; shoveler, teal, wood duck. Diving ducks (more often on salt

Figure 33-22 **Order ANSERIFORMES.** *A.·* Canada goose (*Branta canadensis*). *B.* Mallard (*Anas platyrhynchos*). *C.* Common merganser (*Mergus merganser*).

Figure 33-23 Order CICONIIFORMES: great blue heron (*Ardea herodias*); height about 120 cm (4 ft).

waters): *Aythya valisineria,* canvasback; *A. americana,* redhead; scaup, goldeneye, eider, scoter. Geese (neck longer): *Branta canadensis,* Canada goose, several subspecies; *Chen hyperborea,* snow goose, *Anser albifrons,* white-fronted goose. Swan (neck very long): *Olor columbianus,* whistling swan, winters in United States.

Order 12. Ciconiiformes. Herons, storks, ibises, and flamingos. Long-necked and long-legged wading birds, with decorative plumes (herons), bare areas on head (storks), or bill abruptly decurved at middle (flamingos); little or no web between toes (except flamingos); hatchlings downy (except storks); chiefly tropical and subtropical, usually about water; food chiefly fish and other aquatic animals; many nest in colonies in trees.

ARDEIDAE. Herons and bitterns. *Ardea herodias,* great blue heron (Fig. 33-23); *Butorides virescens,* green heron; *Nycticorax,* night heron; *Casmerodius albus,* common egret, and *Egretta thula,* snowy egret, both with pure white plumage and nuptial plumes (aigrettes) on back in nesting season.

PHOENICOPTERIDAE. *Phoenicopterus ruber,* greater flamingo, Brazil to Florida Keys, mandibles bent, forming box with serrated margins, birds feed with head down, sieving plankton from water; nest in dense colonies, eggs on cylindrical mounds of mud. Sometimes placed in separate order.

Order 13. Falconiformes. Vultures, kites, hawks, falcons, and eagles (Fig. 33-24). Bill stout, hooked at tip, with soft naked skin (cere) at base; mandibles sharp-edged; feet usually with talons adapted for grasping, with sharp curved claws; predaceous, active by day; flight strong, rapid in some; food variously of vertebrates and smaller animals. Carrion feeders: *Cathartes aura,* turkey vulture, or buzzard, with naked red head; *Coragyps atratus,* black

vulture, chiefly in southeastern United States and south to Argentina, head black; *Gymnogyps californianus,* California condor, head yellow, wingspread to 2.7 m (9 ft), local in southern California. Bird hunters: *Accipiter cooperii,* Cooper's hawk; *A. gentilis,* goshawk. Broad-winged soaring types, feed largely on rodents: *Buteo jamaicensis,* red-tailed hawk; *Aquila chrysaetos,* golden eagle. Fish-eating: *Haliaeetus leucocephalus,* bald eagle, the United States emblem; *Pandion haliaetus,* osprey. Narrow-winged speedy hunters, primarily of birds in open country: *Falco peregrinus,* peregrine falcon (duck hawk), used in falconry; *F. sparverius,* American kestrel (sparrow hawk), feeds largely on insects.

Order 14. Galliformes. Includes grouse, quail, pheasants, turkeys. Bill short; feathers with aftershaft; feet usually adapted for scratching and running; young downy at hatching and precocial; the upland game birds of hunters, several species domesticated, many gregarious ground dwellers where most of them nest; food chiefly of plant materials (Fig. 33-25).

OPISTHOCOMIDAE. *Opisthocomus hoazin,* hoatzin, South America, inhabits trees and feeds on leaves; young with claws on first 2 digits of wing, used for climbing until wing feathers are grown.

MEGAPODIIDAE. Megapodes, or mound builders, Australia and Pacific Islands; eggs buried in soil and/or plant debris; young precocial, able to fly soon

Figure 33-24 Order FALCONIFORMES. *A.* Turkey vulture, spread to 2 m (6½ ft). *B.* Red-tailed hawk, spread 120 to 130 cm (4 to 4½ ft) *C.* American kestrel, length 23 to 30 cm (9 to 12 in). *D.* Osprey, spread to 2 m (6½ ft).

after hatching; *Leipoa ocellata,* mallee fowl, South Australia, temperature of egg site controlled by removing or adding covering materials with their large feet; *Alectura lathami,* brush turkey, eastern coast of Australia.

PHASIANIDAE. Grouse, quails, pheasants, and turkeys. *Bonasa umbellus,* ruffed grouse, in broad-leaved woods; *Dendragapus,* blue grouse, in conifers; *Lagopus,* ptarmigan, north to Arctic, turns white in winter; *Tympanuchus,* prairie chicken; *Centrocercus urophasianus,* sage grouse, in sagebrush of West; *Perdix perdix,* European or Hungarian partridge, introduced successfully in northwestern prairie regions of United States and Canada; *Colinus virginianus,* bobwhite, or eastern quail; *Lophortyx californicus,* California (or valley) quail of Pacific Coast. Native to Asia, widely introduced: *Phasianus colchicus,* ring-necked pheasant, in many states; *Pavo cristatus,* peacock, or peafowl, native of India;

Figure 33-25 Order GALLIFORMES. Two important upland game birds of North America. Bobwhite quail (*Colinus virginianus*), length 25 cm (10 in). Ruffed grouse (*Bonasa umbellus*), length 40 to 48 cm (16 to 19 in). (*From Pennsylvania Game News.*)

Gallus, jungle fowl and domestic chickens; *Meleagris gallopavo,* wild turkey, eastern North America to Mexican plateau, also domesticated.

Order 15. Gruiformes. Includes cranes, rails, coots. Feathers with aftershaft; either of large size and strong flight, inhabiting open marshes or prairies (cranes), or of medium to small size, of weak flight (occasionally flightless), and shy inhabitants of marshes (rails); young precocial.

RALLIDAE. Includes rails, coots. Size of chickens or smaller, neck and legs moderate; worldwide; *Rallus,* rails, body compressed, live in marshes, slip readily through vegetation; *Fulica americana,* American coot, or mud hen, with lobate toes, common on ponds, diving or swimming, over much of North America; *Gallinula,* gallinule, toes elongate, used to walk on lily pads in ponds.

GRUIDAE. Cranes. Some bare skin about head; legs and neck long; all continents except South America. *Grus canadensis,* sandhill crane, tall, gray-colored, nests in temperate North America, often flocks on marshes in winter.

†DIATRYMIDAE. Large, flightless, wings atrophied; bill huge; 4 toes on each foot. Eocene of North America and Europe. †*Diatryma.*

Order 16. Charadriiformes. Includes shorebirds or waders, gulls, auks. Toes usually webbed at least at base; plumage dense and firm; more or less long-legged (shorebirds), strong-winged (gulls), or with only 3 toes and legs far back on body (auks); eggs heavily spotted; most species with precocial young.

CHARADRIIDAE, SCOLOPACIDAE, etc. Shorebirds. On seabeaches or inland watery flats—the strand probing for food in sand, mud, or shallow water;

Figure 33-26 Killdeer (*Charadrius vociferus,*** order CHARA-DRIIFORMES); length 28 cm (11 in).** A common shorebird, often along inland waters. (*After Hoffman, Birds of the Pacific states, Houghton Mifflin Company.*)

Figure 33-27 Loon (*Gavia immer,*** order GAVIIFORMES); length 71 to 91 cm (28 to 36 in); highly aquatic; native to Northern Hemisphere.**

many nest in Arctic regions and migrate far south for winter; formerly hunted for market. *Charadrius vociferus,* killdeer (Fig. 33-26), often inland; *Philohela minor,* American woodcock, in woods of Eastern states; *Gallinago,* snipe, in marshy places; *Calidris,* sandpiper.

LARIDAE. Gulls and terns. On seashores and inland waters; worldwide; long-winged and commonly gray-backed, white beneath. *Sterna,* terns, fish actively over water, plunging to capture prey; *Larus,* gulls, feed on aquatic animals in surface waters or on shore and on garbage about harbors.

ALCIDAE. Includes auks, murres, puffins. Flight in air swift, direct; swim and dive readily, "fly" under water; nest in dense colonies on sea cliffs or in burrows, each pair with 1 (or 2) eggs; on seacoasts of Northern Hemisphere. *Pinguinus (Plautus) impennis,* great auk, or garefowl, 61 cm (24 in) tall, North Atlantic coasts, south to Massachusetts, extinct since 1844; *Uria aalge,* common murre, eggs earlier collected for market; *Fratercula,* puffin.

Order 17. Gaviiformes.   Loons. Legs short, at end of body; toes fully webbed; patella reduced; tail of 18 to 20 short stiff feathers; flight swift, direct, adept at diving; precocial young; feed on fish; northern part of Northern Hemisphere; 4 species. *Gavia immer,* common loon (Fig. 33-27).

Order 18. Columbiformes.   Pigeons and doves. Bill usually short and slender, with thick soft skin (cere) at base; tarsus usually shorter than toes; crop large, producing "pigeon milk" to feed small young; eggs unmarked, usually white; young naked; worldwide.

COLUMBIDAE. *Columba fasciata,* band-tailed pigeon, forested portions of Pacific Coast; *Ectopistes*

migratorius, passenger pigeon, formerly in forested regions of eastern North America, killed extensively for market, extinct since about 1900; *Zenaidura macroura,* mourning dove, over whole of United States (Fig. 33-28*A*).

Order 19. Psittaciformes.   Parrots, lories. Beak stout, narrow, sharp-edged and hooked at tip; upper mandible highly movable on frontal bone of skull; bill with soft cere, often feathered; toes 2 in front and 2 behind, outer hind toe not reversible; feet adapted for grasping; plumage brilliant, green, blue, yellow, or red; in forests of tropics and subtropics; many gregarious and with loud voices; food chiefly fruits. *Rhynchopsitta pachyrhyncha,* thick-billed parrot, Mexico north to southern Arizona; *Conuropsis carolinensis,* Carolina paroquet, small, formerly to Wisconsin and Colorado, extinct.

Figure 33-28   *A.* Mourning dove (*Zenaidura macroura,* order COLUMBIFORMES), length about 30 cm (12 in). *B.* Common nighthawk (*Chordeiles minor,* order CAPRIMULGIFORMES), length 21 cm (8½ in).

A   *B*

Order 20. Cuculiformes. Cuckoos, roadrunners, plantain eaters, and touracos. Worldwide. Toes 2 in front and 2 behind, outer hind toe reversible; feet not adapted for grasping; tail long; bill moderate; many Old World cuckoos "parasitic," the female laying egg in nest of other small birds for incubation and rearing; young cuckoo usually "elbows out" rightful owners; two North American cuckoos, *Coccyzus,* not usually parasitic, and *Geococcyx californianus,* road-runner, Kansas to Texas and California, terrestrial, legs and tail long.

Order 21. Strigiformes. Owls. Worldwide. Head large and rounded; eyes large and directed forward, each in a disk of radial feathers, ear openings large, often with flaplike cover, sometimes asymmetrical in size and position; beak short, feet adapted for grasping; claws sharp; plumage soft-textured and lax; eggs white; young downy at hatching; active chiefly by night; hide in retreats by day; food is land vertebrates, especially mammals, some birds, and arthropods. *Tyto alba,* barn owl (Fig. 33-29); *Bubo virginianus,* great horned owl; *Otus asio,* screech owl; *Nyctea scandiaca,* snowy owl in Arctic, sometimes invades Northern states in winter; *Speotyto cunicularia,* burrowing owl, inhabits burrows in ground dug by prairie dogs, ground squirrels, etc., or sometimes digs own burrow.

Order 22. Caprimulgiformes. Goatsuckers, nightjars, oilbirds. Beak small and delicate, but mouth wide and margined with long bristlelike feathers (not in nighthawks); legs and feet small and weak, not adapted for grasping; plumage soft and lax, active mostly at dusk and by night, food is night-flying insects captured in air; North American forms do not build nest. *Caprimulgus vociferus,*

whippoorwill, woods of Eastern states, named for its call; *Phalaenoptilus nuttalii,* poorwill, brush areas of Western states, two-syllabled call; both forage near ground; *Chordeiles minor,* common nighthawk, usually forages high in air (Fig. 33-28B.)

Order 23. Apodiformes. Swifts and humming-birds. Size usually small; legs very short and feet very small; wings pointed; beak small and weak (swifts) or slender with long tubular tongue (hummingbirds); eggs white; active by day, food captured on the wing; mostly tropical.

APODIDAE. Swifts. Nests in rock cavities or trees held together with salivary secretion used in Orient for bird's-nest soup; about 70 species, 4 in United States. *Chaetura pelagica,* chimney swift, eastern North America.

TROCHILIDAE. Hummingbirds (Fig. 33-30A). Only in Western Hemisphere, over 300 species, mostly in tropics, 15 in United States; length 5.5 to 21 cm ($2\frac{1}{4}$ to $8\frac{1}{2}$ in); plumage brilliant, iridescent especially on head and neck of males; food of nectar, small insects, and spiders taken from blossoms by the tubular protrusible tongue and needlelike beak; nests small, exquisitely felted of mosses, plant down, and spider web; eggs 2, young naked, fed by regurgitation. *Archilochus colubris,* ruby-throated hummingbird, Alberta to central Texas and eastward; others chiefly in California and far Western states; *Selasphorus rufus,* rufous hummingbird, to 61°N in Alaska for nesting.

Order 24. Coliiformes. Colies, mousebirds. Small, passerinelike; first and fourth toes reversible; tail very long; gregarious; Africa. *Colius.*

Order 25. Trogoniformes. Trogons. Tropical. Beak short and stout, with bristles at base; feet small

Figure 33-29 Order STRIGIFORMES. Great horned owl (*Bubo virginianus*); length 46 to 58 cm (18 to 23 in). Barn owl (*Tyto alba*); length 38 to 53 cm (15 to 21 in). (*After Grinnell and Storer, Animal life in the Yosemite.*)

Figure 33-30 A. Hummingbird (*Stellula,* order APODIFORMES), length to 9 cm ($3\frac{1}{2}$ in). B. Belted kingfisher (*Megaceryle alcyon,* order CORACIIFORMES), length 28 to 37 cm (11 to $14\frac{1}{2}$ in).

and weak; plumage brilliant, often green, but soft and lax; nest in cavities; *Trogon elegans,* coppery-tailed trogon, Mexico and southern Arizona; *Pharomacrus moccino,* quetzal, Central America, one of the most beautiful of birds.

Order 26. Coraciiformes.   Kingfishers, hornbills. Third and fourth toes fused at base; beak strong; includes also motmots, bee eaters, rollers, and hoopoes, mostly in tropics. *Megaceryle alcyon,* belted kingfisher (Fig. 33-30B), near fresh waters over much of North America, captures small fishes and frogs by diving from a perch; excavates horizontal tunnel in stream bank for nest.

Order 27. Piciformes.   Woodpeckers, jacamars, puffbirds, barbets, honeyguides, and toucans.
PICIDAE. Woodpeckers. Worldwide except Australia. Tail feathers stiff with pointed tips; beak stout, awllike; tongue roughened or with barbs near tip, and protrusible; toes 2 in front and 2 (or 1) behind, not reversible; mostly in forests, cling to tree trunks, dig insects and larvae out of wood, and excavate nest cavities in trees; some on cactus in southwestern deserts. *Dendrocopos (Dryobates) villosus,* hairy woodpecker; *D. pubescens,* downy woodpecker; *Sphyrapicus,* sapsuckers, eat cambium and sap of trees and some insects (ants); *Colaptes,* flickers, forages also on ground; *Dryocopus,* pileated woodpecker (Fig. 33-31); *Melanerpes formicivorus,* acorn woodpecker, stores many acorns individually in holes drilled in trees.

Order 28. Passeriformes.   Perching birds. Toes 3 in front and 1 behind, adapted for perching on twigs and stems and neither reversible nor united; wing with 9 or 10 primaries; includes great majority of all known birds, 4 suborders, 69 families, about 5100 species.
Families in North America, north of Mexico:

1. *Suborder Tyranni*
COTINGIDAE: cotingas
TYRANNIDAE: tyrant flycatchers

2. *Suborder Passeres.*   Songbirds (52 families), so named because of structure of syrinx; many produce beautiful songs, others are poor songsters, virtually all have call notes. Size from that of raven and crow to small chicadee; in all types of habitats, some strictly terrestrial, a majority in trees and shrubs: the ouzels or dippers (*Cinclus*) swim and dive in cold freshwater streams, and flycatchers and swallows capture insects in flight (Fig. 33-32). Usually build nests, some elaborate, eggs 3 to 8, variously colored, young naked and

Figure 33-31   Pileated woodpecker (*Dryocopus pileatus,* order PICIFORMES), length 38 to 48 cm (15 to 19 in), one of the largest of North American woodpeckers, in forested regions.

blind at hatching, require feeding and care by parents before becoming independent. Food varied, chiefly of small size, many strictly insectivorous, some exclusively seed eaters, others have mixed diet. Common examples: *Corvus,* crow, raven; *Melospiza,* song sparrow; *Dendroica,* wood warblers; *Turdus,* American robin (Fig. 33-1).

ALAUDIDAE: larks
HIRUNDINIDAE: swallows
LANIIDAE: shrikes (butcherbirds)
BOMBYCILLIDAE: waxwings
PTILOGONATIDAE: silky flycatchers (phainopepla)
MOTACILLIDAE: pipits, wagtails
CINCLIDAE: dippers (water ouzels)
TROGLODYTIDAE: wrens
MIMIDAE: thrashers, mockingbirds, catbirds
TURDIDAE: thrushes, robins, bluebirds
SYLVIIDAE: kinglets, gnatcatchers
CHAMAEIDAE: wrentits
CERTHIIDAE: creepers
SITTIDAE: nuthatches
PARIDAE: chickadees, titmice, verdins
CORVIDAE: crows, magpies, jays

Figure 33-32 Order PASSERIFORMES: common small North American birds. *A.* Olive-sided flycatcher (*Nuttallornis borealis*), length 19 cm (7½ in). *B.* Song sparrow (*Melospiza melodia*), length 17 cm (6½ in). *C.* Barn swallow (*Hirundo rustica*), length 20 cm (7¾ in). *D.* House wren (*Troglodytes aedon*), length 13 cm (5¼ in).

STURNIDAE: European starling and crested myna (introduced)

FRINGILLIDAE: grosbeaks, finches, sparrows, towhees, buntings

VIREONIDAE: vireos

PARULIDAE: wood warblers

THRAUPIDAE: tanagers, buntings

ICTERIDAE: blackbirds, orioles, meadowlarks

References

Special periodicals that deal with birds include *The Auk* (published by the American Ornithologists' Union), chiefly on North American birds; *The Condor* (Cooper Ornithological Society), especially for western North America; *The Wilson Bulletin* (Wilson Ornithological Club), especially for the Middle West; *Audubon Magazine* (National Audubon Society), popular. The scientific and common name and geographic range for each bird species and subspecies north of Mexico is given in the 1957 *Checklist of North American birds*, 5th ed., American Ornithologist's Union, xii + 691 pp. Two supplements have been published in *The Auk:* vol. 90, no. 2, pp. 411–419 and vol. 93, no. 4 pp. 875–879.

Bellrose, F. C. 1967. Orientation in waterfowl migration. In R. M. Storm (editor), Animal orientation and navigation. Corvallis, Oregon State University Press.

Bent, A. C. 1919–1968. Life histories of North American birds. U.S. National Museum Bulletins 107–237. Reprinted by Dover Publications, Inc. *Detailed account of habits, migration, and nesting for each species; many illustrations.*

Cone, C. D., Jr. 1962. The soaring flight of birds. *Sci. Am.*, vol. 206, no. 4, pp. 130–140. *Soaring flight has revealed much about the nature of air currents.*

Emlen, J. T., and R. L. Penney. 1966. The navigation of penguins. *Sci. Am.*, vol. 215, no. 4, pp. 104–113. *A study of navigation of the Adelie penguin, a bird that travels hundreds of miles over nearly featureless land and water and returns to its nest.*

Emlen, S. T. 1975. The stellar-orientation system of a migratory bird. *Sci. Am.*, vol. 233, no. 2, pp. 102–111. *The indigo bunting orients itself by the stars but needs other clues for long-distance migration.*

Farner, D. S., J. R. King, and K. C. Parkes (editors). 1971–1974. Avian biology. New York, Academic Press, Inc. 1971. Vol. 1, xix + 586 pp. 1972. Vol. 2, xxiii + 612 pp. 1973. Vol. 3, xx + 573 pp. 1974. Vol. 4, xxii + 504 pp.

Frings, H., and M. Frings. 1959. The language of crows. *Sci. Am.*, vol. 201, no. 5, pp. 119–131. *Some cosmopolitan crows can understand the "languages" of related species.*

Gilliard, E. T. 1958. Living birds of the world. New York, Doubleday & Company, Inc. 400 pp. *Popular general survey with many colored illustrations.*

Greenewalt, C. H. 1969. How birds sing. *Sci. Am.*, vol. 221, no. 5, pp. 126–139. *The mechanism is different from that of a wind instrument or the voice.*

Griffin, D. R. 1964. Bird migration. Garden City, New York, The Natural History Press. xv + 180 pp. Paper, Dover Publications Inc., 1974.

Hess, E. H. 1972. "Imprinting" in a natural laboratory. *Sci. Am.*, vol. 227, no. 2, pp. 24–31. *Imprinting, by which newly hatched birds rapidly form a permanent bond to the parent, begins prior to hatching in mallards. Parent and prehatching young respond to each other's vocalizations.*

Keeton, W. T. 1974. The mystery of pigeon homing. *Sci. Am.*, vol. 231, no. 6, pp. 96–107. *Pigeons appear to have more than one compass system for determining direction.*

Konishi, M. 1973. How the owl tracks its prey. *Am. Sci.*, vol. 61, no. 4, pp. 414–424.

Lack, David. 1966. Population studies of birds. New York, Oxford University Press. 341 pp.

Lorenz, K. Z. 1952. King Solomon's ring. New York, Thomas Y. Crowell Company. 202 pp. *A delightful account of animal behavior, including much information on birds.*

Marshall, A. J. (editor). 1960–1961. Biology and comparative physiology of birds. New York, Academic Press, Inc. 2 vols. xxii + 986 pp., illus.; 26 chapters, each by specialists, many references.

Matthews, G. V. T. 1968. Bird navigation. 2d ed. London, Cambridge·University Press. x + 197 pp., 40 figs.

Peakall, D. B. 1970. Pesticides and the reproduction of birds. *Sci. Am.*, vol. 222, no. 4, pp. 72–78. *DDT and other chlorinated hydrocarbon pesticides ingested by flesh-eating birds (hawks and pelicans) cause disturbed reproduction and eggs with shells too fragile to survive.*

Peterson, R. T. 1947. A field guide to the birds [of eastern North America]. Boston, Houghton Mifflin Company. xxvi + 290 pp., 60 pls. (36 color), 25 figs.

———. 1961. A field guide to western birds [and Hawaii]. 2d ed. Boston, Houghton Mifflin Company. xxvi + 366 pp., 60 pls. (36 color), some text figs.

Pettingill, O. S. 1970. Ornithology in laboratory and field. 4th ed. Minneapolis, Burgess Publishing Company. xvii + 524 pp.

Schmidt-Nielsen, K. 1959. Salt glands. *Sci. Am.*, vol. 200, no. 1, pp. 109–116. *A special organ which eliminates salt enables marine birds to meet their fluid needs by drinking seawater.*

———. 1971. How birds breathe. *Sci. Am.*, vol. 225, no. 6, pp. 72–79. *In avian respiration air flows not only through the lungs but through air sacs and bones.*

Smith, N. G. 1967. Visual isolation in gulls. *Sci. Am.*, vol. 217, no. 4, pp. 94–102. *Experiments with color of the fleshy eye ring in gulls shows that species recognize one another by subtle visual signals.*

Stittner, L. J., and **K. A. Matyniak.** 1968. The brain of birds. *Sci. Am.*, vol. 218, no. 6, pp. 64–76. *Intelligent behavior in birds appears to depend on a different part of the brain than in mammals.*

Sturkie, P. D. 1976. Avian physiology. 3d ed. New York, Springer-Verlag New York Inc. xiii + 400 pp. *Chiefly for advanced students.*

Taylor, T. G. 1970. How an eggshell is made. *Sci. Am.*, vol. 222, no. 3, pp. 88–95. *The chief constituent of eggshell, crystalline calcium carbonate, comes partly from a hen's bones; when necessary she can mobilize 10 percent of her bone for this purpose in a day.*

Thorpe, W. H. 1956. The language of birds. *Sci. Am.*, vol. 195, no. 4, pp. 128–138. *Songs and calls of birds make up a complex communications system, with the various sounds suited to various types of message.*

Tinbergen, Niko. 1953. Social behavior in animals. London, Methuen & Co., Ltd. xi + 150 pp., 8 pls., 67 figs. *Includes excellent information on bird behavior.*

———. 1971. The herring gull's world. rev. ed. New York, Harper Torchbooks, Harper and Row Publishers. xvi + 255 pp., paper. *An excellent account of bird behavior.*

Tucker, V. A. 1969. The energetics of bird flight. *Sci. Am.*, vol. 220, no. 5, pp. 70–78. *How flying birds husband their fuel.*

Van Tyne, J., and **A. J. Berger.** 1976. Fundamentals of ornithology. 2d ed. New York, John Wiley & Sons, Inc. xviii + 808 pp., illus. *Structure, habits, summaries of bird families, references.*

Wetty, C. 1957. The geography of birds. *Sci. Am.*, vol. 197, no. 1, pp. 118–128. *Few birds are cosmopolitan, most species have provincial abodes.*

———. 1975. The life of birds. 2d ed. Philadelphia, W. B. Saunders Company. xv + 623 pp., illus. *Structure, function, behavior, ecology, flight, migration, distribution, evolution; many references.*

34

CLASS MAMMALIA: MAMMALS

The mammals are the "highest" group in the Animal Kingdom. They include the moles, bats, rodents, cats, monkeys, whales, horses, deer, humans, and other living forms, besides a host of extinct species and orders. All are more or less covered with hair or fur and are warm-blooded. The distinctive term *mammal* refers to the mammary glands on females that supply milk for suckling the young. Parental care is most highly developed in this class and reaches its climax in the human species (Chap. 35). Various mammals live in all sorts of habitats from polar regions to the tropics and from the sea to the densest forests and driest deserts. Many are of retiring habits or are nocturnal so that they are seldom seen, but together with the arthropods, they dominate terrestrial habitats in the present-day world. Certain wild species are hunted as game and others

for their fur. Some rodents and flesh eaters damage crops and livestock, and a few species are reservoirs of disease (Fig. 34-1). The domestic mammals provide humans with food, clothing, and transportation.

34-1 Characteristics

1 Body usually covered with hair (scant on some), which is molted periodically; skin with many glands (sebaceous, sweat, scent, and mammary).

2 Skull with two occipital condyles; each half of lower jaw consisting of a single bone (dentary); three ear bones; nasal region usually slender; mouth with teeth (rarely absent) in sockets, on

Figure 34-1 Class MAMMALIA: the Norway or house rat (*Rattus norvegicus*), a representative mammal. Body covered with hair, nose bearing sensory vibrissae, lips soft and movable, external ears thin, feet with fleshy sole pads, toes with claws, and slender tail serving mainly for balancing.

both jaws, and differentiated in relation to food habits.

3 External ear opening usually surrounded by a fleshy pinna. Tongue usually mobile; eyes with movable lids.

4 Vertebral column with five well-differentiated regions: cervical, thoracic, lumbar, sacral, and caudal.

5 Four limbs (cetaceans and sirenians lack hind limbs); each foot with five (or fewer) toes and variously adapted for walking, running, climbing, burrowing, swimming, or flying; toes with horny claws, nails, or hoofs and often fleshy pads.

6 Heart completely four-chambered (two atria, two distinct ventricles); only the left aortic arch persists; red blood cells nonnucleated, usually biconcave disks.

7 Respiration only by lungs; larynx with vocal cords (except in giraffes); a complete muscular diaphragm separating lungs and heart from abdominal cavity.

8 Twelve pairs of cranial nerves; brain highly developed, both cerebrum and cerebellum large, the former having development of the corpus callosum and an expanded neopallium (Sec. 9-6).

9 Endothermal (homoiothermal).

10 Male with copulatory organ (penis); testes commonly in a scrotum external to abdomen; fertilization internal; eggs usually minute, without shells, and retained in uterus (modified oviduct) of female for development; embryonic membranes (amnion, chorion, and allantois) present; usually with a placenta affixing embryo to uterus for nutrition, respiration, and waste removal; young nourished after birth by milk secreted from mammary glands of female.

The insulated body covering (hair and subcutaneous fat) and the complete separation of venous and arterial blood in the heart make precise internal regulation of body temperature possible. With this go a high rate of metabolism and a consequent large requirement for food. The teeth are usually conspicuous and differentiated. The senses of sight, hearing, and smell are highly developed. The large cerebellum and cerebrum provide for a high degree of coordination in all activities and for learning and a retentive memory.

34-2 Evolutionary advances of mammals The presence of an outer covering of hair with associated numerous skin glands gave mammals distinct advantages over their reptilian ancestors. The hair served to insulate against both heat and cold, thus helping maintain a constant internal temperature. The associated sebaceous glands secreted substances which lubricated the hair and waterproofed it, while sweat glands secreted water and wastes, aiding both excretion and temperature regulation.

Ability to maintain a constant internal temperature not only allowed mammals to penetrate many habitats but made possible the development of more efficient and sensitive physiologic mechanisms within the organism. The great enlargement of the brain in mammals, especially the cerebrum and cerebellum, coupled with the constant internal temperature and efficient, fast-acting physiologic mechanisms, permitted mammals to become alert and intelligent organisms with the potential for fast movement.

Quick response and movement, hallmarks of mammals, require efficient limbs. To this end, the legs and arms, originally jutting out in reptiles and amphibians, were brought into the center so that the body was centered over them and not between them. In this position they not only support the animal better but can more efficiently and quickly propel it forward.

Reproductive success was enhanced in the advanced mammals by eliminating egg laying and retaining the young in the reproductive tract of the female for a longer period of development, thus allowing a safer maturation period. Although this reproductive method usually meant that fewer young could be produced at one time, this was offset by the increased chances of survival of each.

Finally, all senses were highly developed, especially that of smell, absent or poorly developed in birds, enabling mammals to obtain maximum information concerning the world about them.

The unique combination of the above evolutionary advances has made the mammals highly successful animals in all habitats.

34-3 Size The smallest mammals are shrews, bats, and mice, less than 5 cm (2 in) in head-and-body length and weighing only a few grams. Others range variously in size up to the elephants and large whales; the blue whale (*Balaenoptera musculus*), which grows to 30 m (100 ft) long and a weight of nearly 120,000 kg (130 tons), is the largest animal ever known.

Structure of a mammal: the domestic cat

The cat (*Felis catus*) is a member of the order CARNIVORA (L. *caro*, flesh + *voro*, devour) that feeds typically on small mammals and birds. Its forward-directed eyes, keen senses, and silent tread on padded feet make it an effective hunter. The lithe body and sharp claws enable it to pounce upon and grasp prey easily, and the sharp, shearing teeth facilitate the cutting of flesh.

34-4 External features The body is densely covered with **hair,** or fur, and consists of a rounded **head,** short **neck,** narrow **trunk,** and long flexible **tail.** Each **forelimb** has five **toes** provided with fleshy **pads** and curved retractile **claws.** The **hind limbs** are stouter, providing the principal power in locomotion; each has four toes and claws. A cat walks on its toes and hence is **digitigrade** (L. *digitus*, finger + *gradior*, to step; Fig. 34-6).

The tip of the nose has two narrow **nostrils,** and the **mouth** is bordered by thin, fleshy **lips.** Each eye has two fleshy **eyelids,** margined with fine hairs, or **eyelashes;** and beneath the lids is a transparent **nictitating membrane** that moves across the eyeball from the inner angle. About the eyes and nose are long sensory hairs, the **vibrissae,** or whiskers. Behind the eyes are two thin, fleshy **external ears** (*pinnae*), each with an external auditory canal leading into the head. The trunk comprises a narrow chest, or **thorax,** sheltered by the ribs, and the broader **abdomen** posteriorly; along the ventral surface on the female are four or five pairs of small elevated **teats,** or **mammae,** through which the milk, or mammary glands open. The **anus** is ventral to the base of the tail, and close below is the **urogenital opening.** In males the **scrotum,** containing the testes, hangs beneath the anus.

34-5 Body covering The **skin** is soft and thin except for the thickened cornified pads of the feet. It is covered densely with closely spaced **hairs,** and all the hairs collectively are termed the coat, or **pelage.**

Each hair (Fig. 4-1) grows from a **hair papilla** at the base of a deep tubular pit, or **hair follicle,** that is embedded in the skin and lined by epidermis. Each follicle is joined to a small **sebaceous gland** that provides an oily secretion (sebum) to lubricate the hair. The follicles lie slantingly in the skin, and to each is attached an **erector muscle** by which the hair may be made to "stand on end" when the cat is cold, unduly stressed, or angry. Hair, when grown, is a nonliving epidermal product subject to wear and fading and therefore is periodically renewed by a molt, usually in spring and again in autumn in the Temperate

Zones. The hairs drop out individually, and new ones grow from the same follicles. An individual hair (Fig. 34-9) is covered by **cuticular scales** over a tubular **cortex** that surrounds an inner **medulla.** The coloration of hair and hence of the pelage is dependent upon the kind of pigment granules present in the cortex—black, brown, red, or yellow. The blue color of some cats is due to the combined effect of pigment and of light-interference phenomena in the cortex and cuticle. White cats lack pigment.

34-6 Skeleton The skeleton is largely of bone, with cartilage over joint surfaces, on parts of the ribs, and in a few other places (Fig. 34-2). Besides the cartilage and membrane bones (Chap. 3), certain tendons contain ossifications known as **sesamoid bones;** the most conspicuous of these is the kneecap, or **patella,** but others occur on the feet. The rounded **skull** is a hard case with all the bones closely united by irregular **sutures** that may be obliterated in later life. The facial region contains, dorsally, the **nostrils,** and the large **orbits** that shelter the eyes, and ventrally there is a flat **palate** margined by the teeth of

Figure 34-2 Skeleton of the domestic cat.

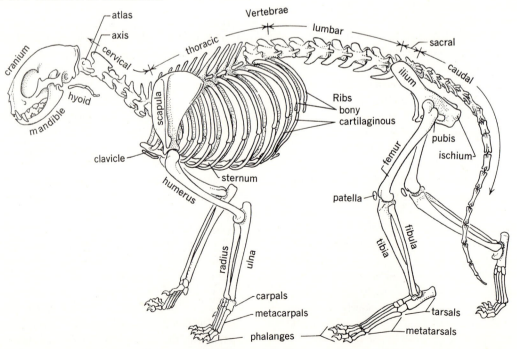

the upper jaw. Outside of each orbit is a conspicuous horizontal bar, the **zygomatic arch.** On the posterior surface is the large **foramen magnum** through which the spinal cord connects to the brain, and at either side of this is a rounded **occipital condyle** by which the skull articulates to the first vertebra, or atlas. The **lower jaw,** which also bears teeth, consists of a single bone on each side that articulates to the squamosal bone of the cranium.

The **vertebral,** or **spinal, column** forms a flexible support for the body and shelters the spinal cord; adjacent vertebrae are separated by **intervertebral disks** of dense fibrocartilage. the column comprises five regions: (1) the short neck of 7 **cervical vertebrae,** (2) the 13 **thoracic vertebrae** on which the movable ribs articulate, (3) the 7 **lumbar vertebrae** of the lower back, (4) the 3 **sacral vertebrae** which are fused for attachment of the pelvic girdle, and (5) the 16 to 20 slender **caudal vertebrae** in the tapered tail. The 13 pairs of **ribs** and the slender midventral **sternum** form a flexible "thoracic basket" that protects vital organs within and also performs respiratory movements.

The **pectoral girdle** attaches by muscles to the thorax and supports the forelimbs. On each side it comprises a flat triangular shoulder blade, or **scapula,** which receives the head of the humerus. A delicate collarbone, or **clavicle,** "floats" in nearby muscles. The **forelimb** comprises a humerus, distinct radius and ulna, seven carpal bones, five metacarpals (innermost short), and the phalanges of the toes. The **pelvic girdle** attaches rigidly to the sacrum; each half (the innominate bone, or hipbone) consists of an anterior dorsal **ilium,** a posterior **ischium** with a tuberosity on which the cat sits, and the ventral **pubis.** At the junction of the three bones is a cuplike socket, or **acetabulum,** in which the head of the femur articulates. The two pubes and ischia join in symphyses ventrally below the vertebrae. The bones of each **hind limb** are the femur, separate tibia and fibula, seven tarsals of the ankle, four long metatarsals (and a vestige of the innermost, or first), and the phalanges.

34-7 Muscular system   (Fig. 34-3) As compared with the lower vertebrates, mammals have a lesser bulk of segmental muscles on the vertebrae and ribs

Figure 34-3 Superficial muscles and salivary glands of the domestic cat; some of the muscles are cut to show others beneath.

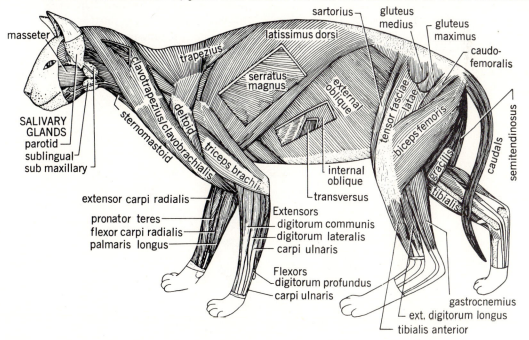

and more elaborately developed muscles on the head, neck, and limbs. The cat has facial muscles that permit some degree of "expression" in relation to emotional states. The limbs have been rotated so that the elbow is posterior and the knee is anterior and both limbs project ventrally; this is quite unlike the laterally projecting limbs of amphibians and reptiles. The cat's ability to "land on all fours" when dropped results in part from the freedom in its skeletal articulations and the diversity of muscular movements that it can make. A distinctive mammalian characteristic is the dome-shaped transverse muscular partition, the **diaphragm,** covered by peritoneum, that separates the coelom into an anterior **thoracic cavity** containing the heart and lungs and the posterior **abdominal cavity** which contains the other viscera.

34-8 Teeth Mammalian teeth are fixed in sockets and of definite number. In various mammals the teeth are specialized in form and function according to the kind of food used, those of the cat being adapted for cutting and tearing flesh. On each tooth (Fig. 34-13) the exposed part, or **crown,** is covered by hard white **enamel** over a bonelike **dentine** that contains a pulp cavity. The base, or **root,** of the tooth, below the gums, is fixed by bony **cement** in a socket, or **alveolus,** in the jaw. Like most mammals, the cat

has four kinds of teeth: short **incisors** at the anterior ends of the jaws, to cut or scrape off food; slender stabbing **canines** used to seize or kill prey and in fighting; and angular **premolars** and **molars** for shearing and crushing the food (Fig. 34-12). Incisors, canines, and premolars constitute the milk teeth of a kitten; these are later replaced and the molars, having no "milk" predecessors, are added. The teeth are alike on the two sides but differ in the upper and lower jaws. All the teeth collectively form the **dentition,** and their number and kind are expressed in a dental formula that indicates those of the upper and lower jaws on one side thus:

For a kitten,

$$i\,\frac{3}{3},\,c\,\frac{1}{1},\,p\,\frac{3}{2} \times 2 = 26$$

For an adult cat,

$$I\,\frac{3}{3},\,C\,\frac{1}{1},\,P\,\frac{3}{2},\,M\,\frac{1}{1} \times 2 = 30$$

34-9 Digestive system (Fig. 34-4) The mouth cavity is margined by the thin, soft **lips,** within which are fleshy **gums** around the bases of the **teeth.** The lips and cavity are lined with a soft mucous membrane. The **tongue** is a flexible muscular organ,

Figure 34-4 The cat: internal structure.

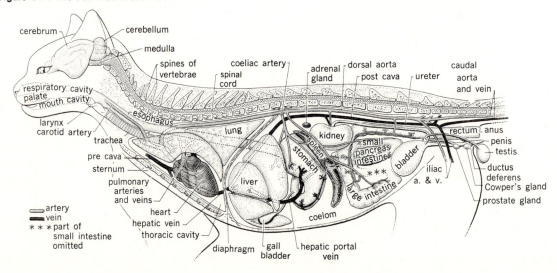

cerebrum — cerebellum — medulla — spines of vertebrae — coeliac artery — spinal cord — adrenal gland — dorsal aorta — post cava — ureter — caudal aorta and vein

respiratory cavity — palate — mouth cavity — esophagus — larynx — carotid artery — trachea — pre cava — sternum — pulmonary arteries and veins — lung — liver — stomach — spleen — kidney — pancreas — small intestine — large intestine — bladder — iliac a. & v. — rectum — anus — penis — testis — ductus deferens — Cowper's gland — prostate gland

heart — hepatic vein — thoracic cavity — coelom — diaphragm — gall bladder — hepatic portal vein

artery — vein — *** part of small intestine omitted

attached ventrally and supported by the hyoid bones; its rough and cornified upper surface contains four kinds of **papillae** and microscopic **taste buds.** The mouth cavity is roofed by the **palate;** the anterior hard palate is formed of bone and crossed by fleshy ridges that help in holding food; the short, fleshy soft palate behind closes the respiratory passage above during swallowing. At either side of the soft palate is a reddish **tonsil** of lymphoid tissue. Four pairs of **salivary glands** pour watery and mucous secretions through ducts into the mouth to moisten the food; these are (1) the parotids, below the ears; (2) the submaxillaries, behind the lower jaws; (3) the sublinguals, near the preceding; and (4) the infraorbitals, below the eyes.

The **pharynx** is the cavity behind the mouth where respiratory and food paths cross. Into its dorsal surface open the **internal nares** from the nasal cavity above the palate, and behind these are two slitlike openings of the **Eustachian tubes.** Beyond the tongue and ventral in the pharynx is the respiratory opening, or **glottis;** when food is passing, this is covered by an anterior flap, the **epiglottis.** The **esophagus** is a narrow muscular tube extending from the back of the mouth through the thorax to the enlarged **stomach** behind the diaphragm. The stomach connects, through a pyloric valve, to the slender and coiled **small intestine.** To the anterior portion of the latter are joined ducts of the digestive glands, the large brownish **liver** of several lobes, and the small whitish **pancreas.** At the junction of the small and large intestine there is a short blind sac, or **caecum.** The **large intestine** extends up the right side and across the abdomen, then descends on the left to a short muscular **rectum** opening at the **anus.** Most mammals have no cloaca; the digestive, excretory, and reproductive systems have separate external openings.

34-10 Circulatory system The red blood corpuscles of mammals are unlike those of all lower vertebrates in being biconcave disks (spherical in camels) and nonnucleated. The **heart,** in the thoracic cavity, is enclosed in a delicate pericardial sac. It is completely four-chambered, as in birds, with two atria and two thick muscular ventricles. The course of blood through the heart and lungs is the same in the cat and the bird, but in the mammal it leaves the left ventricle of the heart through a **left aortic arch.** Shortly the arch gives off an **innominate artery** (relict of the right arch), whence the two **common carotid arteries** arise, and the innominate continues as the **subclavian** artery to the right forelimb. The arch gives off a left subclavian and then turns as the **dorsal aorta,** which extends posteriorly. The latter gives branches to the internal organs, body wall, and posterior limbs, then continues as the **caudal artery** of the tail.

The **venous system** includes paired **jugular veins** from the head and neck and **subclavian veins** from the forelimbs; these all join as a **precaval vein,** which enters the right atrium. The **postcaval vein** returns blood from the tail, hind limbs, kidneys, gonads, and dorsal muscles. From the digestive organs a **hepatic portal system** carries blood to the liver by means of the **hepatic portal vein. Hepatic veins** drain the liver and join the postcaval vein as the latter enters the right atrium. There is no renal portal system, but each kidney is drained by a **renal vein** which enters the postcaval vein. The **spleen** is a slender dark mass behind the stomach.

34-11 Respiratory system Air enters the **nostrils** to pass above the palate through a maze of coiled turbinate bones covered by mucous epithelium, where it is cleaned, moistened, and warmed. Behind the soft palate the air crosses the pharynx to enter the **glottis.** This is the opening in the voice box, or **larynx** (Adam's apple in humans), a framework of several cartilages that contain the **vocal cords** by which the cat's calls and squalls are produced. From the larynx the air passes down the flexible windpipe, or **trachea,** which is reinforced against collapse by C-shaped cartilages. The trachea continues into the thorax to divide into two **bronchi.** These distribute air through subdividing branches that terminate in the microscopic **alveoli** of the lungs. The alveoli are surrounded by pulmonary capillaries in which the O_2–CO_2 exchange of external respiration occurs. The lungs are spongy elastic structures, each of three lobes (Fig. 7 3). The exterior of the lungs and interior of the thorax (pleural cavity) are lined by smooth peritoneum, the **pleura.** The mechanics of breathing are described in Chap. 7 (Fig. 7-4 and Sec. 7-8).

34-12 Excretory system   The two **kidneys** lie in the lumbar region above the peritoneum. The liquid urine passes from each kidney down a duct, the **ureter,** to be stored in the distensible **bladder** that lies midventrally below the rectum. At intervals the muscular walls of the bladder are voluntarily contracted to force urine out through the single **urethra.** In a female this empties at the **vestibule,** the site where the reproductive and urinary tracts meet. In a male the urethra traverses the penis.

34-13 Nervous system and sense organs   The **brain** is proportionately larger than in terrestrial nonmammalian vertebrates. The **olfactory lobes** and brainstem are covered by the greatly enlarged **cerebral hemispheres,** whose increased size is mainly due to the increased development of the **neopallium.** The exterior of the neopallium is convoluted with elevations (*gyri*) separated by furrows (*sulci*); this increase of cerebral surface area is in keeping with the greater degree of intelligence displayed by the cat. The two hemispheres are joined internally by a transverse band of fibers, the **corpus callosum,** peculiar to mammals, which functions as a communication network between the two halves of the cerebrum. The **cerebellum** likewise is large and conspicuously folded, being formed of a median lobe and two lateral lobes (Fig. 9-3). Its greater development is related to the fine coordination in the cat's activities. There are 12 pairs of **cranial nerves,** and from the **spinal cord** a pair of **spinal nerves** passes to each body somite. The trunks of the **sympathetic system** lie close below the vertebrae.

The organs of taste, smell, sight, and hearing in the cat are essentially like those of humans as to location and function (see Chap. 9), but the eyes are nocturnal in adaptation. Sound waves collected by the movable external ears (pinnae) pass into the external auditory canal that leads to the eardrum, or tympanic membrane. The middle ear of a mammal has three auditory ossicles (malleus, incus, and stapes) that transfer vibrations to the inner ear—unlike the ears of birds, reptiles, and amphibians,

which have only a single bone (columella)—and to the cochlea, which is spirally coiled. The malleus and the incus were derived, respectively, by evolutionary change of function from the reptilian articular bone of the lower jaw and the quadrate bone of the skull (upon which the jaw articulates).

34-14 Endocrine glands   In the cat these comprise the pituitary, thyroid, parathyroids, adrenals, islets of Langerhans, and gonads. Their functions are much as in other mammals (see Chap. 8).

34-15 Reproductive system   (Fig. 34-5). In a male the two **testes** lie within the **scrotum,** a skin-covered double extension of the abdominal cavity suspended somewhat anterior to and below the anus. From each testis the spermatozoa are gathered in a network of minute tubules in the **epididymis** to enter the sperm duct, or **ductus deferens.** This, together with blood vessels and nerves, constitutes a **spermatic cord** that enters the abdomen through a small **inguinal canal.** The two ducti deferentia enter the base of the urethra, which is a common urogenital canal, through the male copulatory organ, or **penis,** that serves to transfer sperm into the vagina of a female during copulation. Two small accessory glands, the **prostate gland** around the base of the urethra and the Cowper's or **bulbourethral glands** posteriorly, provide secretions that aid in the transfer of sperm.

The female has, behind the kidneys, two small **ovaries.** Lateral to each ovary is the funnel, or **os-**

Figure 34-5   The cat: urogenital organs in ventral view; bladder turned aside.

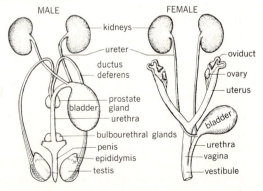

tium, joining a slender coiled oviduct. The latter is continued backward as a thick-walled **horn of the uterus.** Posteriorly, the two horns unite medially as the **body of the uterus,** from which the **vagina** extends between the bladder and rectum to the **urogenital opening.** Ventral in the latter is a minute rod, the **clitoris,** corresponding to the penis of the male.

In reproduction, ova are discharged by the Graafian follicles and enter the ostia. In the oviducts they are fertilized by sperm that migrate up from the vagina after copulation. The fertilized ova settle separately against the inner wall of the uterine horns and become implanted. The fetal membranes of the developing embryo plus the associated maternal tissues form a **placenta** (Fig. 10-15) through which the embryo receives nourishment and oxygen and disposes of wastes by way of the maternal blood circulation. The **period of gestation,** from fertilization until birth, is 60 days.

Structure of other mammals

34-16 External features The many kinds of mammals differ in size, form, proportions, nature of the pelage, and coloration. Swift-running species have narrow bodies and long limbs, large sedentary species are heavy in all respects, and the whales, seals, and others that swim have torpedo-shaped bodies. The mammalian **head** is proportionately large, because of the greater size of the brain, and usually has a long snout. The **eyes** of rodents, hoofed animals, and other plant feeders are lateral (to watch on both sides for enemies); those of primates, bats, and carnivores are directed forward to provide binocular vision, important for depth perception. The **external ears** (pinnae) are large and mobile in the grazing deer, horses, and hares, reduced in the burrowing ground squirrels, moles, and pocket gophers, and reduced or absent in seals and other swimming mammals. The **neck** is long in deer, horses, and giraffes but short in burrowing types and not evident in whales. The **trunk** is cylindrical in the agile weasels, casklike in the ponderous elephants, compressed in deer and other speedy runners, and depressed in the burrowing gophers and moles. Mammalian **tails** are of diverse form and serve variously—being brushy "flyswatters" in hoofed mammals, stout for support and balance in kangaroos, flat as rudders in whales, beavers, and muskrats, and prehensile for grasping in opossums and some monkeys.

The **limbs** are slender and tapered in the agile deer and antelopes, huge and stumplike in the elephants and hippopotami, short and with broad palms in the burrowing moles, and paddlelike in the seals and whales. Kangaroos, jerboas, and kangaroo rats have long hind legs, feet, and tail that serve for their jumping gait. Bats have long delicate forelimbs and fingers to support their thin flight membranes, or wings. Other "flying" mammals—squirrels, lemurs, and marsupials—have normal limbs with lateral extensions of skin along the body by which they merely glide down from a height.

The mammalian **foot** (Fig. 34-6) ends typically in

Figure 34-6 Mammalian feet (left hind limb). *A.* Human, generalized with five toes, walks on entire foot. *B.* Dog, with four toes, "heel" raised; walks on fleshy pads under toes. *C.* Horse, with only one toe (third), "heel" raised; walks on cornified hoof over tip of toe.

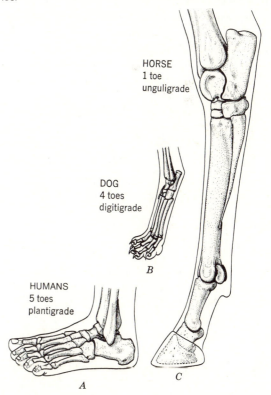

HORSE
1 toe
unguligrade

DOG
4 toes
digitigrade

B

HUMANS
5 toes
plantigrade

A

C

five toes; hence in this respect, humans and other primates are generalized. Narrowing and lengthening of the foot and reduction in the number of toes occur particularly in running mammals, the horse (Fig. 13-13) being an extreme case. Cattle, sheep, and deer are cloven-hoofed, the third and fourth toes persist, and the second and fifth remain as vestigial **dewclaws.**

34-17 Body covering The **skin** produces hair and various horny or cornified structures and contains many glands. Where subjected to heavy wear it forms a dense cornified coating such as the calluses on the human palms or soles and the foot pads of bears, mice, etc. The claws, nails, and hoofs of various mammals and the horns of cattle, sheep, and Old World antelope are also cornified, resistant to wear and chemical disintegration. Growth in all of them is continuous from the base to compensate for external wear. Horns are supported by a bony core. Antlers of deer are annual growths of dense connective tissue that later become calcified (Figs. 34-7, 34-8).

The **pelage,** or coat of hair (Fig. 34-9), varies in length, density, texture, and color in different species. It is heaviest on Arctic mammals but thin and short on tropical species. The elephant, rhinoceros, and hippopotamus are, like humans, sparsely haired. Whales and manatees are naked save for a few bristles about the lips. The coat on many mammals is differentiated into a dense fine **underfur** for body insulation and a lesser number of heavier long **guard hairs** that protect against wear. About the nose and eyes of carnivores and rodents are long sensory **vibrissae,** with the base of each hair surrounded by sensory nerve fibers. As the animal moves about, these receive touch stimuli. A rat deprived of its vibrissae does not run freely and behaves abnormally. Spiny anteaters, hedgehogs, and porcupines are covered by sharp pointed quills (modified hairs). Scales, with hairs between, occur on the tails of beavers, muskrats, and many rats and mice; the pangolins (*Manis*) are entirely covered by scales. The armadillo has a jointed armor of epidermal scutes over bony plates and a few hairs.

The pelage is replaced gradually by a periodic **molt,** usually in autumn so that a new coat is provided for winter. Most mammals that inhabit the

Figure 34-7 Structure and growth stages of horns and antlers; diagrammatic sections. *A.* Pronghorn antelope. *B–D.* Horns of cow or sheep. *E–I.* Antlers of deer or elk. Compare Fig. 34-8.

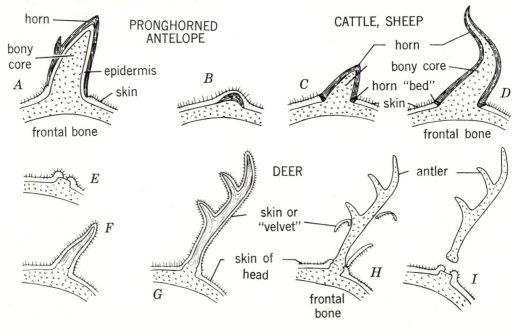

Figure 34-8 Horns and antlers. *A*. The mountain sheep has thick, permanent, hollow horns (of keratin), each over a bony core. *B*. The pronghorn antelope has thin hollow horns over bony cores, and the horns are molted annually. *C*. The mule deer has solid calcified antlers that are shed and grow anew each year.

cooler (temperate) regions of the world also have a spring molt that yields a shorter, sparser summer pelage. In northern regions the weasels and hares have a brown summer coat that is replaced by one of white for the period roughly corresponding to the season of snow. Each hair usually grows to a definite length and stops, but on the human scalp and the manes and tails of horses the growth is continuous. The various **color patterns** of mammals results from differences in pigmentation of the hair. Some coats are virtually solid color, but stripes, spots, bars, or other markings are common.

Whales and seals have a thick layer of subcutaneous fat (blubber) that insulates the body against loss

Figure 34-9 Structure of mammalian hairs. (*After Hausman.*)

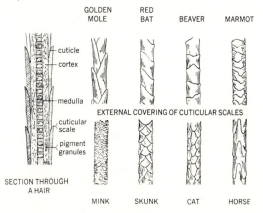

of heat in water; blubber is the source of oil in the whaling industry.

34-18 Glands The superficial **glands** include (1) sebaceous glands, as described earlier for the cat; (2) scent glands of various types; (3) mammary glands that produce milk; (4) sweat glands on the horse and human; and (5) lachrymal (tear) glands that moisten and cleanse the surface of the eye. **Scent glands** are variously placed: suborbital and metatarsal on deer, middorsal on peccaries and some squirrels, between the toes and on the tail in the dogs, between the toes of sheep, and about the anal region in rabbits, muskrats, beavers, skunks, weasels, and others. The secretions may be abundant and odoriferous (skunks) or scant and delicate (squirrels). They serve variously (1) to mark individual territories by an odor, or scent track, for animals of the same or other species, (2) to bring the sexes together for mating, and (3) as a means of defense. Such extensive use of scent is related to the keen sense of smell in mammals. **Mammary glands** (Fig. 34-10) usually occur only on females but are present, albeit reduced and nonfunctional, on males in primates and some other mammals. They become activated late in pregnancy by action of a hormone (prolactin from the anterior pituitary), to provide milk during early growth of the young, and then recede (Sec. 8-12). The nipples are usually more numerous than the average number of

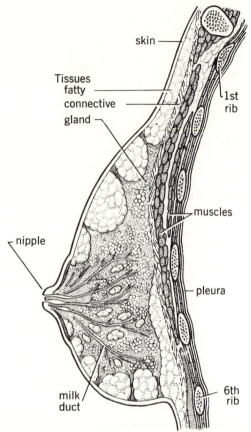

skin

Tissues
fatty
connective
gland

1st
rib

nipple

muscles

pleura

milk
duct

6th
rib

Figure 34-10 Human mammary gland in sagittal section; nonlactating. (*After Testut.*)

young in a litter—2 in the human female, mare, and ewe; 4 in cattle and some rodents; and 10 to 12 in swine. **Sweat glands** produce a watery secretion that, by evaporation from the surface of the skin, serves to cool the body. They cover the body of humans, horses, and some other mammals or may be localized on the feet or other regions, as in many rodents, sheep, cattle, and dogs, or absent or nearly so in whales and seals.

34-19 Dentition Mammalian **teeth,** unlike those in most lower vertebrates, are differentiated in form and function (Figs. 34-11 to 34-13); in each mammalian order they are specialized for the kind of food used. The teeth are narrowly pointed in moles and bats, which feed on insects, and sharp for shearing

and piercing in flesh eaters. Cheek teeth are flattened and have low cusps for crushing various foods in squirrels, pigs, and human beings and of rasplike structure with many enamel ridges for grinding green vegetation in hoofed animals and many rodents. The incisors of rodents have enamel only on the anterior surface; the softer dentine behind wears away more rapidly, so these teeth are always of chisellike form for gnawing. Teeth may be low-crowned as in cats or high-crowned as in the cheek teeth of horses; they may have distinct roots in which growth soon ceases or may be rootless and grow from a persistent pulp, as with the incisors of rodents. Among higher mammals a full dentition never exceeds 44 teeth, and many species have fewer; some representative dental formulas are:

Mole and pig:

$$I\frac{3}{3}, C\frac{1}{1}, P\frac{4}{4}, M\frac{3}{3} \times 2 = 44$$

House mouse and house rat:

$$I\frac{1}{1}, C\frac{0}{0}, P\frac{0}{0}, M\frac{3}{3} \times 2 = 16$$

The tusks of pigs are enlarged canines and those of elephants are upper incisors. Adult monotremes, some edentates, and baleen whales lack teeth entirely.

34-20 Digestive tract The intestine is short in species that consume such concentrated foods as insects or flesh but long in rodents and hoofed mammals which eat grass or leafy vegetation. All the latter have a large **caecum** that provides additional space for such bulky food while it is undergoing slow digestion. In the cattle, deer, antelopes, camels, etc., the ruminants which "chew the cud," the stomach is of four compartments (Fig. 34-14). Their food is gathered rapidly, mixed with saliva, chewed slightly, and passed into the rumen. Then the animal seeks a safe quiet place to ruminate. In small masses, or cuds, the food is regurgitated, chewed thoroughly, and reswallowed. It then passes in turn through the other compartments, the first three of which are specialized parts of the esophagus and have cornified

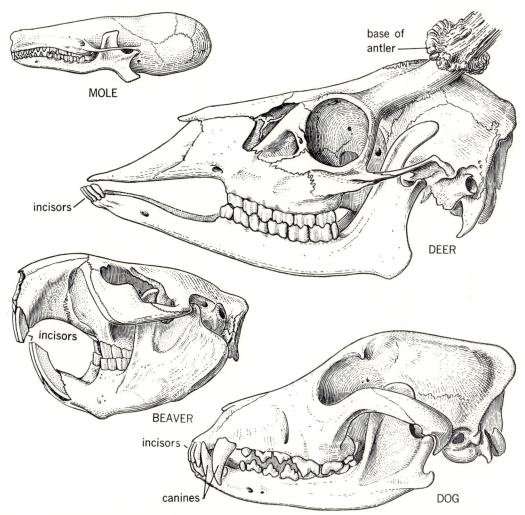

base of
antler

MOLE

incisors

DEER

incisors

BEAVER

incisors

canines

DOG

Figure 34-11 Some types of mammalian skulls and teeth; not to same scale. Mole (order INSECTIVORA): teeth fine and conical for grasping insects and worms. Mule deer (order ARTIODACTYLA): lower incisor teeth close against fleshy pad on the upper jaw to nip off vegetation, which is ground between the enamel-ridged cheek teeth (premolars and molars); no canines. Beaver (order RODENTIA): two pairs of chisellike incisors used for gnawing; no canines; premolars and molars cross-ridged with enamel to grind food. Dog (order CARNIVORA): incisors small for nipping, canines large for stabbing, and cheek teeth differentiated for shearing and crushing.

linings with ridges or folds that subject it to further abrasion.

Natural history

34-21 Distribution Mammals inhabit virtually all parts of the earth—land, water, and air. The walrus and some seals live in the Arctic or Antarctic seas and on ice, other seals and sea lions are native to tem-perate ocean shores, and whales and porpoises inhabit the open sea. The beaver, muskrat, mink, and otter inhabit fresh waters. Grasslands, brushlands, and forests are the homes of many rodents, carnivores, and hoofed animals. Still other mammals live in dense tropical jungles, on the treeless Arctic tundra, and in desert regions. Tree squirrels, lemurs, many monkeys, and some small carnivores live chiefly in trees, moles and pocket gophers are strict inhabitants of the soil, and the insect-eating bats

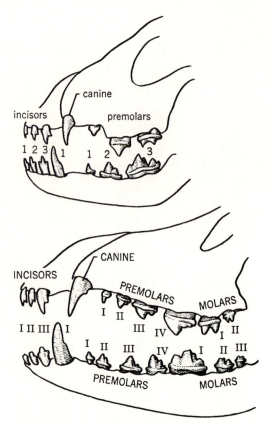

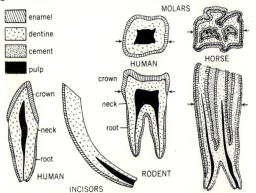

Figure 34-12 The two dentitions of the dog. *Above.* The temporary or milk teeth of a puppy, all of which are later replaced. *Below.* The adult (or permanent) teeth.

Figure 34-13 Structure of mammalian teeth in diagrammatic sections. On the molar teeth the arrows indicate the plane of section for the diagram above or below. Note the large pulp opening in the rodent incisor. This ensures continuous growth of this tooth.

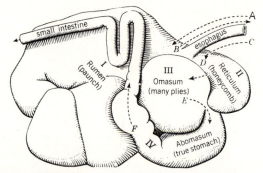

Figure 34-14 The four-compartment stomach of the cow, a ruminant animal. Grasses and similar roughage pass down (*A*) the esophagus to the rumen, return (*B*) to the mouth for rechewing or rumination, and then pass (*C*) into the reticulum. Grain and other concentrates go at once into the reticulum. The food while being digested is moved successively to the (*D*) omasum, (*E*) abomasum, and (*F*) small intestine.

find their forage at night in the air. Different mammals range in altitude from sea level to high mountains, where some species such as the mountain sheep and pikas live above timberline. Each kind of mammal has a definite geographic and ecologic range. The habitat limitations may be narrow, as with a beaver, which requires the bark of broad-leaved trees as food and quiet fresh water for shelter, or wider, as with Norway rats, which accommodate themselves to various environments.

34-22 Populations Many mammals are nocturnal or hidden and escape human notice; their presence is revealed by tracks, droppings, or other sign, but trapping or hunting is necessary to determine their numbers. The mountain lion, mink, and many rodents live solitarily. Wolves and hyenas may hunt in packs, the prairie dog (*Cynomys*) and some western American ground squirrels inhabit colonial burrows, and buffaloes, antelopes, fur seals, and sea lions often live in herds. Some primates live in social groups.

The population density varies with available cover and food. Small shrews and mice may number 50 to 100 per acre and large ground squirrels from 2 to 10 per acre; American deer average 10 to 40 acres per animal, and the black bear up to a township (93 km², or 36 mi²) for each individual. Many seals and sea lions have large areas of the sea for foraging but often assemble in herds on the shore. The number of species is least in polar regions and greatest in the tropics.

In many species the general population is rather constant unless modified by drought, flood, food shortage, human interference, or similar factors. It reaches an annual peak when the young appear and then declines until completion of the next breeding season. Cyclic fluctuations in population occur with some meadow voles (*Microtus*), lemmings (*Lemmus, Dicrostonyx*), snowshoe hares, arctic foxes, and other species in the Northern states, Canada, Alaska, and northern Europe. The cycle is about 4 years in lemmings and about 10 or 11 years in arctic foxes and hares. Meadow mice and house mice (*Mus musculus*) occasionally irrupt, with a great local increase in numbers for a short time.

34-23 Nests and shelters   Many mammals use shelters where they rest, sleep, avoid inclement weather, and rear their young. Natural crevices amid rocks serve wood rats, pikas, and some carnivores. Holes in trees are used by mice, chipmunks, raccoons, opossums, and squirrels. Bats resort to caves, trees, or buildings. Tree squirrels, some wood rats, and certain mice build nests amid tree foliage. Ground squirrels, badgers, coyotes, and skunks dig burrows in the ground. Moles and pocket gophers spend their lives in elaborate systems of tunnels beneath the ground surface. Beavers construct dams to form ponds where they can float logs and build moundlike houses of sticks and mud for shelter. In regions of heavy snow, deer and moose trample out winter yards where they can move about and forage. Many of the small rodents shred plant fibers lengthwise to form a warm, soft nest.

34-24 Voice   Many mammals use their voices often to express "emotions" and transfer information between individuals. The calls and notes serve (1) to warn of danger, (2) to intimidate enemies, (3) to assemble gregarious species, (4) to bring the sexes together for mating, and (5) to locate parents or young. The "language" is usually of stereotyped calls but is more varied among the primates, leading toward articulate human speech. Small bats emit short bursts of ultrasonic notes that, echoing from nearby objects, serve to guide them in flight and in capturing prey (10 to 60 bursts per second each of 0.005-second duration, frequency 50,000 cycles per second). Some shrews do likewise (28,000 to 60,000 cycles per second). Porpoises emit series of clicking notes thought to serve in echolocation of nearby animals and high-pitched whistles that may be for communication with other porpoises. Some large whales produce sounds (20 cycles per second) that can be detected several kilometers away.

34-25 Food   Hoofed mammals and most rodents (along with insects) are primary consumers (converters of vegetation), since they subsist on plant materials; in turn they serve as food for flesh eaters. The **herbivorous** mammals include horses, cattle, bison, and others, which **graze** on grasses and forbs, and deer, goats, elephants, and giraffes, which **browse** on leaves and twigs of shrubs and trees. Rabbits eat grasses, forbs, and bark; squirrels and chipmunks consume many seeds; tree squirrels eat nuts; beavers feed on the inner bark, or cambium, of willow and aspen, and porcupines take that of coniferous trees. **Omnivorous** mammals are those which take both plant and animal materials; examples are the house rat, raccoon (Fig. 34-15), bear, pig, and

Figure 34-15 Proportions of different foods taken by an omnivorous mammal, the raccoon (*Procyon*), in fall and winter. (*From Hamilton, American mammals.*)

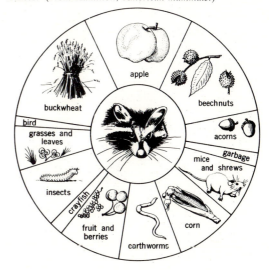

apple

beechnuts

buckwheat

bird

grasses and leaves

acorns

garbage

mice and shrews

insects

crayfish

corn

fruit and berries

earthworms

human. The **carnivorous** mammals, or flesh eaters, include the cats, weasels, and mink. Seals and toothed whales are predominantly fish-eating, or **piscivorous.** Many small bats are exclusively **insectivorous,** and moles, shrews, skunks, and others subsist largely on insects. Other bats are **frugivorous** (fruit, nectar, and pollen feeders), and the vampire bat of the New World tropics is **sanguivorous,** subsisting entirely on blood. Small mammals have disproportionately large food requirements; some mice eat nearly their own weight in food daily, and a shrew has been known to consume its own weight in insects during 24 h. Rodents and deer may forage on farm crops, and coyotes sometimes kill sheep, but the beneficial role of predators exceeds the harm they do.

Mammals overcome a seasonal scarcity of food in several ways. The buffalos formerly on the Great Plains performed a regular **migration** to northern grasslands for the summer and retired to others farther south for the winter. Some western deer migrate up the Sierra-Cascade and Rocky Mountains for summer but winter in the foothills when deep snow blankets the higher ranges. Herds of antelope and zebras in Africa shift about in various ways, depending on the supplies of forage grasses, forbs, or shrubs to which each is adapted. Some bats, like insectivorous birds, leave their northern summer habitats when insects become scarce and migrate southward to winter.

Squirrels, chipmunks, and kangaroo rats gather seeds in their cheek pouches and **store** them by burying them in the ground. The pika (*Ochotona*) cuts and cures various plants as "hay" during the summer; this material is put into dry, well-aired spaces ("barns") among the rocks of its high mountain habitat, to be eaten when the region is blanketed with snow.

In many mammals the **body temperature** is closely regulated; some bats, rodents, and lower mammals have less stable temperatures. Bats roosting in cool caves by day must "warm up" before flying out to forage.

Winter sleep, or **hibernation,** is practiced by ground squirrels, chipmunks, marmots, some other rodents, and Temperate Zone insectivores and bats to tide over winter food scarcity. During hibernation all bodily metabolism drops to a low level; the respiration and heartbeat are slowed, and the body temperature is far reduced. Excess fat is accumulated prior to hibernation and serves as bodily fuel through the dormant period. Some northern ground squirrels hibernate for about two-thirds of the year. Bears have a winter dormancy that varies with latitude. Some mice go through short **torpor** stages during hot or cold spells or other stressful periods. The animals are inactive, but physiologic processes are not shut down to the extent seen during hibernation. Other species such as raccoons, tree squirrels, opossums, and skunks may sleep for several days during more severe winter periods.

34-26 Predators Herbivorous mammals fall prey to various flesh eaters according to size. Deer are eaten by tigers; antelopes and zebras by African lions; American deer by mountain lions, wolves, and coyotes; and so on, down to mice eaten by weasels. Hawks, owls, and large snakes prey on small rodents and rabbits. Humans are a major enemy. They hunt game species, trap furbearers, and kill the predators that prey on domestic livestock and rodents that forage on crops. They also alter vast land areas which can then no longer support wild mammal populations. Mammals are hosts to various parasites and diseases (Table 25-1) that may reduce their vitality or cause death.

34-27 Reproduction Fertilization is always internal, and the young are nourished by milk after birth. The lowest mammals, the duckbilled platypus and spiny anteaters (Monotremata), lay reptilelike eggs about 1.3 to 2.5 cm ($\frac{1}{2}$ to 1 in) in diameter, which are incubated. Opossums, kangaroos, and other Marsupialia have minute eggs that develop for some weeks in the uterus; then the immature "fetuses" crawl out to the ventral pouch, or marsupium, and onto the abdomen of the female, where they attach firmly to the mammary nipples for nutrition and remain until their development is complete. In all higher mammals the minute eggs are retained longer in the uterus, which becomes modified for their reception and nourishment by development of a placenta (Figs. 10-15, 34-16).

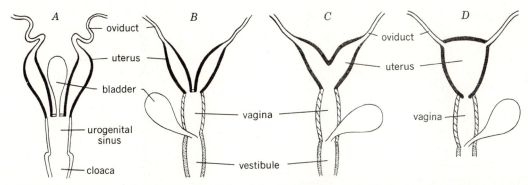

Figure 34-16 Types of uteri. *A.* Duck-billed platypus (MONOTREMATA): separate oviducts (uteri) enter a urogenital sinus joined to the cloaca. *B.* Rabbit (LAGOMORPHA): uterus duplex, of separate "horns" joined to a common vagina. *C.* Cow (ARTIODACTYLA): uterus bicornate, horns partly fused. *D.* Human (PRIMATES): uterus simplex, with a common central chamber and no horns.

Most mammals have rather definite mating seasons, often in winter or spring. Many produce but one brood a year, some have more than one, and the alien house mouse and rats may breed at any season. In rodents, the testes enlarge and descend into the scrotum to become functional during the mating season, then decrease and return into the abdomen. Many mammals fail to breed if the testes do not descend into the scrotum, a condition known as **cryptorchidism,** for no viable sperm are produced in the higher temperature of the abdomen. Male elk bugle and become more pugnacious during the mating period, and other mammals have special mating calls between the sexes, as with domestic cats and cattle.

Female mammals experience a recurrent **estrous cycle,** marked by cellular changes in the uterus and vagina and by differences in behavior. The successive stages are termed diestrus (quiescent period), proestrus (preparation for mating), estrus, or "heat" (acceptance of male), and metestrus (regressive changes). Discharge of ova from the ovary occurs usually late in estrus or soon thereafter. The female rat is in heat for a few hours about every 4 days, the cow has recurrent periods lasting about 24 h every 18 to 21 days, and the female dog is in estrus for 6 to 12 days about every 6 months. Female rabbits and ferrets will usually breed at any time. Estrus, pregnancy, and lactation are regulated by hormones of the hypothalamus, pituitary gland, and ovaries; environmental factors such as the length of day and temperature affect pituitary activity.

The relations between the sexes during the repro-

ductive season are various. In some carnivores the male remains with the female and helps to gather food until the young are weaned. Many other mammals are **promiscuous** in mating, that is, more than one male copulates with a female during her receptive period. **Polygamy,** the mating of one male with several or many females, is the habit of sea lions, fur seals, cattle, wild horses, elk, and many others. The period of **gestation,** or pregnancy, usually varies with size.

The number of young produced at one time is usually inversely proportional to size: elephant and horse, one; sheep and deer, one or two; carnivores, three to five; rodents, two or three to eight or nine. The black rat averages six and the Norway rat about eight, but females of the latter may have up to 15 embryos at one pregnancy. Large mammals usually produce but one young per year, but smaller species with short gestation time and multiple broods are more prolific. A pair of captive meadow mice has produced 13 to 17 litters, totaling 78 to 83 young, in 1 year.

The **precocial** young of hares and jackrabbits (*Lepus*), deer, and domestic livestock are fully haired at birth, have their eyes open, and are able to move about at once (Fig. 34-17). In contrast, the **altricial** young of domestic rabbits (*Oryctolagus*) and cottontails (*Sylvilagus*), most rodents, and carnivores are naked, blind, and helpless when born and require extensive development in a nest before becoming independent. The young of all mammals receive parental care before they become independent.

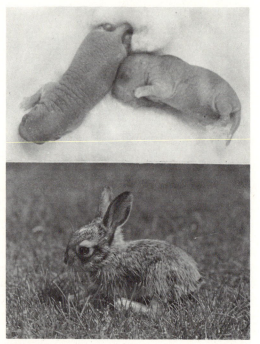

Figure 34-17 Young mammals. *Above.* Rats, born naked, blind, and helpless; require nursing and care in a nest before becoming independent. *Below.* Jackrabbit, fully haired and with eyes open at birth, able to move about at once. (*Photos by T. I. Storer.*)

Some small mice may breed a few weeks after birth, whereas the larger land mammals require several years before sexual maturity; large whales, however, breed in the second or third year of life.

The average span of life is scarcely a year in small rodents and shrews and is generally proportional to size in the larger species; yet few mammals, even among captives, attain the average lifespan of humans. The Indian elephant, commonly seen in zoos and the longest-lived of nonhuman mammals, reaches an age approaching 70 years.

34-28 Relation to humans Mammals have served or harmed humans in various ways from the earliest times. Humans depend on some mammals for food, clothing, and other necessities, but others have been a physical menace or a source of disease, and some injure human possessions. Early cave paintings by primitive peoples in Europe depict the chase for food; the great interest in hunting by all peoples demonstrates this continuing relation.

Eskimos in the Arctic use whale and seal blubber as fuel oil, caribou skins for clothing, seal hides for footwear, and skins of other species for boats. The Plains Indians were supported largely by the buffalo, using its flesh for food, its hides for clothing and tepees (tents), and its sinews for sewing. Civilized races in northern countries have used animal pelts for clothing for many centuries, and much of the early exploration of North America was by trappers in search of furs. Today, the demand for fur garments supports a large industry. Dense natural furs of soft texture and pleasing colors, such as marten, ermine, fur seal, and beaver, bring high prices. Millions of muskrats, skunks, and opossums are still taken in the United States, and lesser numbers of other furbearers. The rearing of black foxes and mink on fur farms has become successful. Pelts of rabbits, house cats, muskrats, and others are dyed, clipped, or otherwise altered to resemble the more expensive furs.

Civilized people hunt now for pleasure and recreation. The moose, elk, deer, mountain sheep, and bears are "big game," sought for their heads, hides, and meat; lesser forms such as raccoons, opossums, rabbits, and tree squirrels are pursued by thousands of hunters.

Live mammals, especially the larger or rarer kinds, are exhibited in many zoological gardens, and in national parks they are protected so that visitors may see them at close range.

Harmful mammals are those which damage crops or domestic animals, those which harbor disease, and the large kinds which are dangerous to humans themselves. Rabbits, woodchucks, and some mice will forage in vegetable patches, gardens, and fields and may gnaw the bark of trees. In the Western states ground squirrels damage grain and other crops, and pocket gophers eat roots of plants in gardens and fields. The alien house rats and mice, spread by shipping and commerce to many lands, do much damage. They gnaw into buildings, consume all kinds of foodstuffs, and destroy much property. The carnivores, or predatory mammals, in nature feed on other animals appropriate to their size, from rodents to deer. Wolves, coyotes, bobcats, and bears also prey at times on domestic cattle, sheep, hogs, and poultry. Since the beginning of agriculture humans have endeavored to control the damage by rodents to their crops and property and by predators to their livestock.

Important diseases in mammals that may be transmitted to humans are plague, typhus, and rat-bite fever in Norway rats; plague, relapsing fever, spotted fever, and tularemia in various wild rodents; trichina in rats, pigs, and cats; and rabies in bats and various carnivores.

34-29 Domesticated mammals   Various mammals have been domesticated—tamed and reared in captivity—to serve human needs. This began many centuries ago in Asia (Sec. 35-22), where many of the wild stocks are native. The horse, ox, and camel serve for transport and draft; the cattle, sheep, and goats provide meat and milk; the pig and rabbit afford meat; and all yield hides to tan into leather or fur. The wool and hair of sheep and goats are spun into yarn or thread and woven into fabrics. The dog is a guardian and friend, is used in hunting, sometimes is a beast of burden, and is eaten by some primitive peoples.

Many special breeds of horses, cattle, sheep, pigs, and dogs have been developed, to suit particular needs, by artificial selection in breeding. Some of these differ greatly from their wild ancestors in size, conformation, color, reproductive ability, and other features, although no new species have been created. Examples of such breeds among cattle are the Hereford (beef), Holstein-Friesian (heavy milk production), Jersey (high butterfat content of milk), and Brown Swiss (of triple purpose: meat, milk, and draft). Domestic mammals and birds provide most of the protein for human food, aside from fish, and also many other necessities of modern life. The livestock industry breeds and rears domesticated mammals and birds, and the packing industry slaughters these animals, distributes meat products, and recovers useful by-products.

34-30 Fossil mammals   The mammals probably arose in Late Triassic time from the cynodont reptiles, which had differentiated teeth. The earliest forms were small, and fossils are few. Five extinct orders are known from teeth and jaws in the Jurassic. Later stocks increased in size and became varied in structure. Marsupials and shrewlike insectivores appeared in the Cretaceous; from the latter group the higher placental mammals are considered to have arisen. With the close of the Cretaceous the "ruling reptiles" (dinosaurs) disappeared and the mammals proliferated. First came a series of ungainly archaic stocks (orders †Taeniodonta, †Creodonta, †Condylarthra, †Amblypoda), then modern types appeared. The Cenozoic era (Paleocene to Pleistocene) is called the Age of Mammals, because members of this class were dominant throughout. About half the known orders consist entirely of extinct forms, and all the others contain, besides the living representatives, some or many that have died out.

Some important structural features of mammals, as compared with reptiles, are (1) two occipital condyles instead of one; (2) each ramus of the lower jaw of one bone, not several; (3) articulation of the jaw directly on the skull, not through the quadrate; (4) teeth usually differentiated and of only two sets (milk, permanent), instead of alike with many replacements; and (5) centra of vertebrae and long bones ossifying from three centers instead of one.

The factors that led to development of the mammals have been inferred to be the increasing aridity from Permian time onward, which placed a premium on speed of locomotion, and the colder glacial climates, in which animals with regulated body temperature that progressed with their bodies off the ground would be more likely to survive.

Some early mammals were exceedingly abundant, to judge from the quantities of their teeth and skeletons that have been found. Fossil relics of carnivores, whales, elephants, horses, camels, and others are sufficiently numerous to enable paleontologists to trace their probable evolution in some detail. The phylogeny of some groups can be traced through several geologic periods in one continent. In other cases a stock suddenly appears in the rocks and provides little or no clue to its immediate ancestry or probable path of migration.

Classification

Class Mammalia.

Mammals. Body usually covered with hair; skin with various glands; skull with 2 occipital condyles; jaws usually with differentiated teeth, in sockets;

limbs adapted for walking, climbing, burrowing, swimming, or flying; toes with claws, nails, or hoofs; heart 4-chambered, with left aortic arch only; lungs large, elastic; a diaphragm between thoracic and abdominal cavities; male with a penis; fertilization internal; eggs small or minute, usually retained in uterus for development; females with mammary glands that secrete milk to nourish the young; body temperature regulated (homoiothermal); Triassic to Recent; 4060 living species (many subspecies) and numerous fossil forms.

Subclass 1. Prototheria

Infraclass 1. †Eotheria

Order 1. †Triconodonta. Carnivores, cat-sized; I 4, C 1, P 3-4, M 4-5, with three subequal cusps in a line; Upper Triassic to Lower Cretaceous. †*Eozostrodon*, †*Triconodon*.

Order 2. †Docodonta. Probably omnivores, cat-sized; I 3?, C 1, P-M 10?; upper molars figure-eight-shaped; lower molars quadrate; Upper Jurassic. †*Docodon*, †*Haldanodon*.

Infraclass 2. Ornithodelphia

Order 3. Monotremata. Monotremes, or egg-laying mammals. Coracoid and precoracoid present; no external pinna on ear; teeth only in young, adults with horny beak; cloaca present; ureters open in dorsal wall of urogenital passage; testes abdominal; penis conducts only sperm; oviducts distinct, entering cloaca separately; no uterus or vagina; mammary glands without nipples; females oviparous; Australian region. *Ornithorhynchus anatinus*, duck-billed platypus (Fig. 34-18), eastern Australia and Tasmania, 45 to 50 cm (18 to 20 in) long; ducklike bill, internal cheek pouches, feet broadly webbed, tail

Figure 34-18 An egg-laying mammal, the duck-billed platypus (*Ornithorhynchus anatinus*, order MONOTREMATA) of Australia. (*After Vogt and Specht, The natural history of animals, Blackie & Son, Ltd.*)

flat and beaverlike, male with roosterlike spur on heel connected to poison gland; aquatic, burrows to 12 m (40 ft) long in river banks; food freshwater invertebrates, carried in cheek pouches; during spring (August to December) female makes nest of roots, leaves, etc., in burrow and lays 1 to 3 eggs, each about 1.3 by 2 cm ($\frac{1}{2}$ by $\frac{3}{4}$ in), with strong flexible white shell; probably incubated; young about 2.5 cm (1 in) long when hatched, nurse by lapping up milk secreted by scattered mammary glands on abdomen of female. *Tachyglossus*, *Zaglossus*, spiny anteaters, Australia, Tasmania, and New Guinea. *T. aculeatus*, of Australia, 43 cm (17 in) long, beak elongate, cylindrical, tongue extensile, body covered with coarse hairs and spines, limbs short, toes 3 to 5 with claws; nocturnal, feeds on ants found under stones; female lays 1 egg, probably carried in pouch (marsupium) on abdomen.

Infraclass 3. †Allotheria

Order 4. †Multituberculata. Herbivorous; analogous to placental rodents in size and function; highly specialized and varied in form; skull stout, beaverlike; incisors 1+/1, no canines, molars with 2 or 3 rows of tubercles; Upper Jurassic to Eocene. †*Taeniolabis*, †*Ptilodus*.

Subclass 2. Theria Marsupials and placental mammals. Ear usually with external pinna; teeth usually in both young and adults, differentiated as to form and function; cloaca usually absent; ureters open into base of bladder; testes of males usually in a scrotal sac, ducti deferentia and bladder opening through a common urethra in penis; females with each oviduct differentiated into upper fallopian and lower uterine portions, both opening into a distinct vagina; mammary glands with nipples; females produce living young (viviparous).

Infraclass 1. †Pantotheria

Order 5. †Symmetrodonta. Carnivores, moderate-sized; I 3, C 1, P 3?, M 7; molars symmetrically triangular in shape with 1 large and 2 small cusps; Upper Triassic to Lower Cretaceous. †*Kuehneotherium*; †*Eurylambda*.

Order 6. †Pantotheria (Trituberculata). Carnivores; small-sized; only jaws known; I 4, C 1, P 4, M to 7 or 8; molars asymmetrically triangular in shape with 3 major cusps; Middle Jurassic to Lower Cretaceous. †*Amphitherium*; †*Melanodon*.

Infraclass 2. Metatheria

Order 7. Marsupialia. Marsupials, or pouched mammals. Epipubic (marsupial) bones usually present; incisor teeth not equal in the 2 jaws; female usually with marsupium (ventral pouch) or marsupial folds surrounding nipples on abdomen; uterus and vagina double; usually no placenta; eggs fertilized internally and begin development in uterus, but after a few weeks the premature fetuses leave and crawl to marsupium, where each attaches closely by its mouth to a mammary nipple and remains (as a "mammary fetus") until fully formed; thereafter small young retreat to marsupium for shelter; Cretaceous to Recent.

(*Note: All living marsupials except* DIDELPHIDAE *and* CAENOLESTIDAE *are restricted to The Australian region; where they are the dominant mammals and have radiated in many diverse ecologic niches; many have forelimbs and hind limbs of about equal length, but others have long hind legs as in kangaroos; they range in size from small mice to large human-size kangaroos.*)

DIDELPHIDAE. American opossums. Muzzle long, naked; tail bare, prehensile. *Chironectes,* water opossum, Central and South America; *Didelphis virginiana,* opossum (Fig. 34-19), southern New England and Wisconsin to Gulf states, Texas, and Mexico, introduced into California; tail scaly and prehensile; hides by day in hollow tree near water; omnivorous. Up to 22 embryos, mammae only 13; fetuses when 12¾ days old (11 mm long) crawl to marsupium and remain 50 to 60 days, then become free. Adults feign death or "play possum," when disturbed; favored for hunting and eating in Southern states.

DASYURIDAE. Limbs alike; carnivorous or insectivorous. *Thylacinus,* marsupial wolf, size of large dog; *Sarcophilus,* Tasmanian devil; *Dasyurus,* tiger

cat; *Phascogale,* rat-size; *Sminthopsis,* pouched mouse, 7.5 to 12.5 cm (3 to 5 in) long.

NOTORYCTIDAE. Foreclaws enlarged; subterranean, insectivorous. *Notoryctes,* marsupial mole, 16 to 20 cm (6½ to 8 in) long.

PERAMELIDAE. Terrestrial and burrowing, largely insectivorous. *Perameles,* bandicoot.

CAENOLESTIDAE. Ratlike or shrewlike in appearance; insectivorous, small-sized; 3 genera, 7 species, in Andes of South America; *Caenolestes.*

PHALANGERIDAE. Claws sharp, hallux opposed; arboreal, herbivorous. *Phalanger,* cuscus; *Petaurus,* flying phalanger (like flying squirrel); *Phascolarctus,* koala, or "teddy bear."

PHASCOLOMYIDAE. Terrestrial and burrowing, herbivorous. *Vombatus,* wombat.

MACROPODIDAE. Hind legs, hind feet, and tail long; habitat various, herbivorous. *Megaleia* and *Macropus,* kangaroos, open forest or plains (Fig. 34-20); *Petrogale,* rock wallaby, about rocks; *Lagorchestes,* hare wallaby, like hare or jackrabbit; *Dendrolagus,* tree kangaroo, arboreal.

Infraclass 3. Eutheria. Placental mammals. No marsupial bones or pouch; vagina single; fetus developed entirely within body of female, attached by a placenta to wall of uterus, and with growth of chorion.

Order 8. Insectivora. Moles, shrews. Size small; snout usually long and tapered; feet usually 5-toed,

Figure 34-20 A marsupial mammal (order MARSUPIALIA) **of Australia; female kangaroo with young in the pouch; long tail of female not shown.** (*Photo by San Diego Zoological Society.*)

Figure 34-19 Opossum (*Didelphis virginiana,* **order** MARSUPIALIA)**; head-and-body to 50 cm (20 in) long, tail 33 cm (13 in).** (*Photo by D. R. Dickey.*)

with claws, inner toes not opposable; teeth sharp-pointed; Cretaceous to Recent; Northern Hemisphere, West Indies, and Africa. Eight recent families, 77 recent genera, about 400 species.

Suborder 1. Lipotyphla. Auditory bulla unossified, intestinal caecum absent.

ERINACEIDAE. Hedgehogs. Back and sides spiny, roll into ball for protection. *Erinaceus,* Europe, Asia, North Africa.

SORICIDAE. Shrews. Mouselike, eyes not covered, feet and tail normal; pelage soft, short; forage on surface of ground and in animal burrows, feeding voraciously on insects, small invertebrates, and some rodents; *Sorex* (Fig. 34-21*A*); *Echinosorex albus,* Borneo, largest, to 653 mm (26 in) long.

TALPIDAE. Moles, eyes covered, palms enlarged, foreclaws large; pelage short, velvety; most species strictly subterranean, in mole runs at ground surface and in deeper tunnels from which earth is pushed out on surface as molehills; food chiefly insects and worms, some sprouted seeds; *Scalopus, Scapanus* (Fig. 34-21*B*).

Suborder 2. Menotyphla. Elephant shrews. Auditory bulla ossified, intestinal caecum present.

MACROSCELIDIDAE. Elephant shrews. Snout trunklike, hind legs longer, gait leaping, partly diurnal, resemble tree shrews (TUPAIIDAE). *Macroscelides,* Africa.

Order 9. Dermoptera. Flying lemurs. Resemble flying squirrel in appearance; the four limbs equal and included with tail in a wide, thin, fur-covered "parachute" (patagium); incisor teeth 2/3. South-

eastern Asia and adjacent islands. One living family and genus, *Cynocephalus* (*Galeopithicus*), 2 species, nocturnal, in trees, glide on extended membranes; feed on leaves and fruits.

Order 10. Chiroptera. Bats. Flying mammals; size small; forelimbs and 2d to 5th digits long, supporting thin integumental flight membrane, or wing, that includes hind limbs (and tail in some); only 1st digit (2d also in fruit bats) of forelimb with claw; hind feet small, with sharp curved claws; teeth sharp; mostly nocturnal, capable of true flight; Paleocene to Recent; worldwide; 16 families in 2 suborders.

Suborder 1. Megachiroptera. Fruit bats, or flying foxes. Sleep by day on tree branches, hanging head downward with wings folded cloaklike around body; feed on fruits, sometimes damaging orchards; Africa and southern Asia to Australia; often in great flocks. *Pteropus giganteus,* of Java, largest species, head and body about 30 cm (12 in), wingspread to 1.5 m (5 ft); one family, 38 genera, and 154 species.

Suborder 2. Microchiroptera. Insectivorous bats. Hang by hind claws head downward during day, in crevices of rocks or trees or in caves or buildings; some gregarious, others solitary; forage independently at night (Fig. 34-21*C*), in open air or beneath trees; feed on flying insects; some infected with rabies. *Myotis, Eptesicus,* brown bats; *Tadarida brasiliensis,* free-tailed bat, roosts in companies of a dozen to millions, as in Carlsbad Caverns, New Mexico, the accumulated drop-

Figure 34-21 Order INSECTIVORA. *A.* Shrew (*Sorex*), length to 13 cm (5 in). *B.* Mole (*Scapanus*), length to 15 cm (6 in). *C.* Order CHIROPTERA. Big-eared bat (*Plecotus*), wingspread to 30 cm (12 in).

pings (guano) under roosts in dry caves gathered for fertilizer; *Desmodus*, vampire bat, in American tropics, with enlarged incisors and canine teeth, makes an incision on the skin of horses, cattle, and occasionally humans and laps up the streaming blood as food; transmits rabies and a trypanosome disease of horses; some tropical forms are fruit and nectar feeders; 15 Recent families, about 135 genera, and 713 species.

Order 11. Primates. Lemurs, monkeys, apes, and humans. Head turns readily on neck; limb bones separate, freely jointed; hands and feet often enlarged, each with 5 distinct digits; innermost toes and thumb usually opposable; eyes usually directed forward, orbit ringed by bone; molars usually tuberculate; food largely fruit and seeds, with some small animals; Paleocene to Recent, chiefly tropical and subtropical, most species arboreal.

Suborder 1. Prosimii (Lemuroidea). Head usually with snout; some toes with claws, others with flat nails; tail long (rarely absent), never prehensile; fur usually thick, woolly; solitary; mostly nocturnal.

Infraorder 1. Lemuriformes. Orbits lateral; 2d hind toe with claw.
TUPAIIDAE. Tree shrews. Squirrellike, small; all toes with claws, feet not used for grasping; tail usually bushy; 38 teeth; diurnal. *Tupaia*, India, East Indies, Philippines, and China.
LEMURIDAE. Lemurs. Usually a long tail; fur woolly; 2d toe with sharp nail; only in Madagascar (5 genera). *Lemur.*
INDRIDAE. Woolly lemurs. Web of skin unites digits of foot; Madagascar (3 genera). *Indris.*
DAUBENTONIIDAE. Aye-aye. Rodentlike skull with constantly growing incisors and no canines; Madagascar. *Daubentonia.*

Infraorder 2. Lorisiformes. LORISIDAE. Face short, eyes large, orbits directed forward, *Nycticebus*, slow loris, southern Asia to Phillippines, no external tail; *Galago*, bush baby, southern and equatorial Africa, tail bushy.

Infraorder 3. Tarsiiformes. TARSIIDAE. Tarsiers. Eyes large, protruding; ears large, thin; tarsal region of foot long, 2d and 3d toes with claws, tips of all digits with rounded pads; tail long, ratlike; Philippines and Indo-Australian islands. *Tarsius* (1 or more species), size of red squirrel, solitary, arboreal, nocturnal, agile, jumps like a frog; food chiefly insects and lizards.

Suborder 2. Anthropoidea. Monkeys, apes, humans. Cranium enlarged, cerebral hemispheres extend over cerebellum; eyes directed forward, vision binocular; nostrils ringed by bare skin; upper lip freely protrusible; facial muscles permit emotional expression; external ears reduced, often lie against head; incisors broad; posture more or less upright; forelimbs usually longer than hind; digits with flat nails (except CALLITHRICIDAE); tree-inhabiting or terrestrial; diurnal; often social; Oligocene to Recent.

Superfamily 1. CEBOIDEA (PLATYRRHINA). New World monkeys. Nostrils diverge anteriorly; no cheek pouches; premolars 3; Central and South America.
CEBIDAE. Tail usually furry, prehensile in some. *Aotus*, douracouli, nocturnal; *Alouatta*, howler, hyoid of males enlarged, voice loud; *Cebus*, capuchin; *Saimiri*, squirrel monkey; *Ateles*, spider monkey, very slender.
CALLITHRICIDAE. Marmosets. Squirrel-sized; head often tufted; tail not prehensile. *Callithrix.*

Superfamily 2. CERCOPITHECOIDEA (CATARRHINI). Old World monkeys. Nostyrils parallel, directed downward; often with internal cheek pouches; premolars 2; tail often long, usually sparsely haired, never prehensile; ischial tuberosities on buttocks with exposed, calloused skin. *Macaca*, macaques, or rhesus monkeys, Asia (1 species at Gibraltar), used as pets and in biological research; *Papio*, baboon and mandrill, Africa, head stout, dog-like, tail short, disposition savage, largely terrestrial; *Cercopithecus*, guenon and other common African monkeys; *Presbytis*, langur, India, southern China, Malaya, East Indies, includes sacred hanuman of Hindus; *Nasalis*, proboscis monkey, Borneo.

Superfamily 3. HOMINOIDEA. Anthropoid apes and humans. No tail or cheek pouches.
PONGIDAE. Anthropoid apes. *Hylobates*, gibbon, and *Symphalangus*, siamang, southeastern Asia and Malay Archipelago, body slender, limbs long, height to 1 m (3 ft), chiefly arboreal, travel easily through trees but walk slowly on ground, using only the feet; voice powerful, often used; diet omnivorous. *Pongo pygmaeus*, orangutan, Sumatra and Borneo, in swampy forests; skull high-topped, face flattish, brain humanoid; to 140 cm (54 in) tall; hair long, lax, and reddish, old males often with beard; lives in trees and constructs crude nests; diet of plant materials. *Gorilla gorilla* (Fig. 34-22), gorilla, west and east central Africa, in forests; body and limbs heavy, males to

Figure 34-22   Mountain gorilla (*Gorilla gorilla*, order PRI-MATES).   Distinctive features are dome-shaped cranium, small ears, heavy brow ridges, long arms, and both hand and foot with opposable first digit. (*Photo by T. I. Storer.*)

Figure 34-23   Chimpanzee (*Pan troglodytes*, order PRI-MATES).   Conspicuous are the flat cranium, large ears, moderate brow ridges, long arms, and opposable first digit on each foot. (*Photo by San Diego Zoological Society.*)

170 cm (65 in) high and over 225 kg (500 lb); hair and skin black; walks on soles of feet aided by knuckles of hands, with body inclined; lives chiefly on ground, in family groups; food of plant materials. *Pan troglodytes*, chimpanzee (Fig. 34-23), west Africa, in forests, 150 cm (54 in) tall, weight to 68 kg (150 lb); head rounded, ears large, forearms short, feeds on plant materials; resembles human beings more than other anthropoids; in captivity can be trained to perform many acts.

HOMINIDAE. *Homo sapiens*, human beings (Chap. 35).

Order 12. †Tillodontia.   Skull long, brain case small; 2 pairs of rootless incisors, 2d pair small; canines minute, molars low-crowned; feet plantigrade, 5-toed, claws large; Paleocene-Eocene. †*Tillotherium,* size of a bear.

Order 13. †Taeniodonta.   Early forms small, like insectivores, skull slender, teeth rooted, enamel-covered; later types larger, skull short and deep, 1 pair of rodentlike incisors, canines large, molars peglike, no roots, scant enamel; Paleocene-Eocene. †*Conorytes,* †*Stylinodon.*

Order 14. Edentata (Xenarthra).   Includes sloths. Teeth reduced to molars in forepart of jaws, or none, no enamel; toes clawed; Eocene to Recent. Includes large fossil ground sloths and glyptodonts; Paleocene to Recent.

MYRMECOPHAGIDAE. Anteaters. No teeth; head and neck long, slender, tapered; mouth tubular, tongue long, protrusible, sticky; hair long; forefeet with stout curved claws used to open ant and termite nests; Central and South America. *Myrmecophaga,* giant anteater.

BRADYPODIDAE. Arboreal sloths (Fig. 34-24). Cheek teeth present (5/4–5); all feet with long curved claws by which sloth hangs upside down on tree branches; hair long, slopes dorsally; food of leaves of certain trees. In tropical America. *Bradypus* (3-toed), *Choloepus* (2-toed).

DASYPODIDAE. Armadillos. Teeth present, variable; a dorsal horny protective shell over bony plates in the skin, commonly divided by transverse furrows of softer skin so that animal can curl up when disturbed. *Dasypus novemcinctus*, nine-banded armadillo, southern Kansas to South America; others in South America are hosts of *Triatoma* bugs that carry Chagas' disease.

Figure 34-24 Captive three-toed sloths (*Bradypus,*** order EDENTATA).** In nature they hang inverted by the long curved claws. The slope of hair is reversed from that of most mammals, thus shedding rain more effectively. (*Photo by T. I. Storer.*)

Figure 34-25 Order LAGOMORPHA. *A.* Pika (coney) (*Ochotona*), length about 15 cm (6 in), lives in rock slides of Western mountains. *B.* Jackrabbit (*Lepus californicus*), head and body 45 to 48 cm (18 to 19 in), ears 14 to 17 cm (5½ to 6½ in), a hare of open grasslands.

Order 15. Pholidota. Body covered by large overlapping horny plates with sparse hair between; no teeth, tongue slender and used to capture insects. *Manis,* pangolin or scaly anteater, Africa and southeastern Asia.

Order 16. Lagomorpha. Pikas, hares, and rabbits. Size moderate to small; toes with claws; tail stubby; incisors chisellike, grow continually, 2/1 × 2 (2d upper incisors small, behind 1st); no canines; premolars 3/2; molars unrooted; total cheek teeth to 6/5, those of upper jaw wider; palate broad; jaw motion only lateral; elbow joint nonrotating; Eocene to Recent; more than 300 species and subspecies; food of leaves, stems, and bark.

OCHOTONIDAE. Pikas, or coneys (not Hyracoidea). About 15 cm (6 in) long, ears rounded, legs equal, tail vestigial. *Ochotona* (Fig. 34-25*A*), California and Colorado to Alaska and Asia, at high altitudes; live about rock piles and gather plants for "winter hay."

LEPORIDAE. Hares and rabbits. Ears long, hind legs long for jumping, feed on leaves and stems of plants. *Lepus americanus,* varying hare, or snowshoe rabbit, turns white in winter; *L. californicus,* etc., jackrabbit (Fig. 34-25*B*), arid Western states; both surface dwellers, young fully haired when born; *Sylvilagus,* American cottontail, lives about thickets, young born naked and blind; *Oryctolagus cuniculus,* European gray rabbit, source of all domestic breeds, introduced and widespread in Australia and New Zealand.

Order 17. Rodentia. Rodents, or gnawing mammals. Usually small; limbs usually with 5 toes and claws; incisors 1/1 × 2, exposed, chisellike, rootless, grow continually; no canines; a gap between incisors and cheek teeth; premolars 2/1 or fewer; molars 3/3 (rarely 2/2); upper and lower cheek teeth about equal-sized; palate narrow; jaw motion both back-and-forth and lateral; elbow joint rotates; Paleocene to Recent; worldwide, on all continents and many islands, from sea level to above 5800 m (19,000 ft) in Himalayas, from dry deserts to rain forests, some in swamps and fresh waters, none marine; includes majority of living mammals, about 1687 species; size varies from mice only 5 cm (2 in) long to capybara 120 cm (48 in) long, but few exceed 30 cm (12 in); food chiefly leaves, stems, seeds, or roots, some partly or largely insectivorous; smaller species often abundant and with high reproductive potential; serve as staple food for many carnivorous mammals, birds, and reptiles; of 34 Recent families, a few are listed here.

Suborder 1. Sciuromorpha. Squirrels. Infraorbital canal small but not slitlike.

Sciuridae. Squirrels. Mostly diurnal, feed chiefly on seeds and nuts. *Sciurus,* tree squirrels (red, gray, fox, etc.), in forested regions, often sought by hunters, (Fig. 34-26*B*); *Spermophilus* (*Citellus*), ground squirrels, on open lands of Western states, make burrows in ground, feed on seeds and grasses, some damage crops, a few carry diseases transmissible to humans; *Tamias, Eutamias,* chipmunks, about logs, brush, and rocks; *Marmota,* woodchuck, groundhog, or marmot, in meadowlands, feeds on grasses; *Glaucomys,* flying squirrel, with broad membranes along sides of body, in forests, nocturnal (Fig. 34-26*A*).

Geomyidae. Pocket gophers. With external fur-lined cheek pouches to carry food; strictly subterranean in self-constructed burrows; vegetarian; southern and western North America. *Geomys, Thomomys.*

Heteromyidae. Pouched mice. With fur-lined external cheek pouches; forage on surface of ground at night, chiefly on seeds; in arid southwestern North America. *Dipodomys,* kangaroo rat (Fig. 34-26*C*); *Perognathus,* pocket mouse.

Castoridae. Beavers. *Castor,* North America and Europe, head and body to 75 cm (30 in) long; tail flat, oval, scaly; body pelage soft, a desirable fur; builds dams and ponds, to float logs cut for food and to contain houses for shelter; feeds on cambium of aspen, willow, etc. (Fig. 34-26*D*).

Suborder 2. Myomorpha. Mice. Infraorbital canal slitlike.

Cricetidae. Rats and mice. *Peromyscus,* white-footed mice, widespread over North America; *Sigmodon,* cotton rats; *Neotoma,* wood rats or pack rats; *Lemmus,* lemming, in Arctic tundra; *Ondatra zibethicus,* muskrat, in marshes, pelt valuable for fur; *Microtus,* meadow mice or voles, in grasslands.

Muridae. Old World rats and mice. Include 3 species spread by commerce and shipping to all civilized countries, common in much of North America, damage property and food supplies and carry various diseases, including plague, typhus, and rat-bite fever. *Rattus norvegicus,* Norway rat (Fig. 34-1); *R. rattus,* black and roof rats; *Mus musculus,* house mouse.

Figure 34-26   Order Rodentia: **some specialized representatives (compare Fig. 34-1).** *A.* Flying squirrel (*Glaucomys*), head and body 14 to 17 cm (5½ to 6½ in), with membranes between legs for gliding in air. *B.* Red squirrel (*Tamiasciurus*), head and body 18 to 20 cm (7 to 8 in), claws sharp for climbing trees, tail a bushy counterbalance. *C.* Kangaroo rat (*Dipodomys*), head and body 10 to 12 cm (4 to 5 in), tail 15 to 18 cm (6 to 7 in) long, jumps on long hind legs. *D.* Beaver (*Castor canadensis*), head and body 61 to 77 cm (24 to 31 in), tail 28 to 40 cm (11 to 16 in), aquatic. *E.* American porcupine (*Erethizon dorsatum*), head and body 50 to 68 cm (20 to 27 in), tail 12 to 30 cm (5 to 12 in), armed with spines, tree-climbing.

Suborder 3. Hystricomorpha. Porcupines, cavies. Infraorbital canal very large.

ERETHIZONTIDAE. Porcupines. *Erethizon,* protectively covered with slender, pointed quills (modified hairs); feeds on cambium of coniferous trees (Fig. 34-26E).

CAVIIDAE. Tail short or none. South America. *Cavia,* guinea pig, about 20 cm (8 in) long, domesticated for pets and research.

Order 18. Cetacea. Whales, dolphins, and porpoises. Size medium to very large; body usually torpedo-shaped; head long, often pointed, joined directly to body (no neck region); some with a fleshy dorsal "fin"; forelimbs (flippers) broad and paddle-like, digits embedded, no claws; no hind limbs; tail long, ending in two broad transverse fleshy flukes and notched in midline; teeth alike when present; if lacking, then plates of whalebone (baleen); nostrils on top of head; ear openings minute; body surface smooth, no hairs save for few on muzzle; no skin glands except mammary and conjunctival glands; a thick layer of fat (blubber) under skin affording insulation; stomach complex; Eocene to Recent; oceanic, over the world, always in water, die from collapse of lungs if stranded on shore; whales may dive to 1100 m (3600 ft) when wounded and can remain submerged many minutes without breathing; upon reaching surface, warm moist air is blown from lungs, forming a spout as condensed by cooler air over ocean; mate and bear young in the sea, sometimes in shallow water, young large at birth and suckled like those of other mammals (Fig. 34-27).

Suborder 1. †Archaeoceti. Zeuglodonts. Eocene to Oligocene.

Suborder 2. Odontoceti. Toothed whales. Teeth 2 to 40 in various species; 1 nostril. *Physeter catodon,* sperm whale, to 18 m (60 ft) long, head squarish, about one-third total length; food large squids and some fishes; a large reservoir in head yields sperm oil, a fine lubricant; ambergris, formed in the stomach, is used in perfume. *Delphinus delphis,* common dolphin, to 2 m (7 ft) long, and *Phocaena phocaena,* harbor porpoise, both feed on fishes; 4 species of dolphin native to rivers in China, India, and South America; *Orcinus,* killer whale, predatory cetacean, attacks large fishes, seals, porpoises, and even big whales; *Monodon monoceros,* narwhal, of Arctic, with but 2 teeth, of which 1 in male is a slender twisted tusk 2.5 to 3 m (8 to 9 ft) long, involved in "unicorn" of European fables; 7 Recent families, 33 genera, and 74 species.

Suborder 3. Mysticeti. No teeth; 2 nostrils; mouth with many parallel horny plates of whalebone on sides of upper jaw, used to strain small animals from water; food chiefly krill [*Euphausia,* a small crustacean about 5 cm (2 in) long, often enormously abundant] *Balaena,* right whale, no dorsal fin; *Balaenoptera (Sibbaldia) musculus,* blue whale, with fin on back and lengthwise grooves on throat; the largest living animal, to about 32 m (105 ft) long, young 7 m (23 ft) long at birth, 16 m (52 ft) at weaning, and 22.5 to 23.5 m (74 to 77 ft)

Figure 34-27 Humpback whale (*Megaptera novaeangliae,* order CETACEA), length to 15 m (50 ft), female suckling young. (*After Scammon, Marine mammals, 1874.*)

long when sexually mature in second winter of life; *Eschrichtius robustus,* California gray whale, along Pacific Coast; 3 Recent families, 5 genera, 10 species.

Order 19. Carnivora.   Carnivores. Small to large; toes usually 5 (at least 4), all with claws; limbs mobile, radius and ulna, tibia and fibula complete and separate; incisors small, usually 3/3; canines, 1/1, as slender "fangs"; uterus bihorned; placenta zonary.

Suborder 1. †*Creodonta.*   Paleocene to Miocene.

Suborder 2. Fissipedia.   Includes dogs, cats, bears. Toes separate; Eocene to Recent; native to all continents except Australia and New Zealand, where some now introduced; flesh eaters or predatory animals, but some eat plant materials; pelts of many are valuable furs; Paleocene to Recent.

CANIDAE. *Canis lupus,* wolf, once common in Europe and America, now much reduced; *C. latrans,* coyote, or prairie wolf, usually under 14 kg (30 lb), open regions of arid West but has spread to Alaska and eastern part of North America; eats rodents, deer, antelope, domestic sheep, and some plant material; *Canis,* domestic dogs, derived in Old World; *Urocyon,* gray fox (Fig. 34-28*A*), in brushlands; *Vulpes,* red fox, chiefly in woodlands, domesticated "black" and "silver" phases now reared commonly on fox farms.

URSIDAE. Bears. *Thalarctos maritimus,* polar bear, circumpolar in Arctic regions to southern limits of pack ice, pelage white and dense, feet fully furred, teeth pointed, food seals and fishes; *Ursus (Euarctos) americanus,* black bear (also with cinnamon phase), in forests of North America from subarctic regions to Mexico, climbs readily, adults sometimes to 275 kg (600 lb); sleeps in midwinter, when young are born; omnivorous; *Ursus arctos,* brown bear of Eurasia and western North America, including grizzly bear, large, some Alaskan individuals weighing over 545 kg (1200 lb), Mexico to Arctic, in open and brushy areas, terrestrial, mostly exterminated in United States.

PROCYONIDAE. *Procyon lotor,* raccoon (Fig. 34-15), over most of United States, especially near streams, tail black-and-white–banded, food, frogs, fishes, crayfishes, fruits; *Ailurus,* lesser panda, in Himalayas; *Ailuropoda,* giant panda, mountains of western China, feeds on bamboo.

MUSTELIDAE. Weasels, otters, etc. Many strictly predatory, eat birds, mammals, or fishes. *Martes,* marten, fisher, in woods; *Gulo,* wolverine; *Mustela,* weasels (Fig. 34-28*B*), stoats, and ferrets, live in burrows; *Taxidea,* badger (Fig. 34-28*C*), on open lands; *Mephitis,* striped skunk (Fig. 34-28*D*), and *Spilogale,* spotted skunk, both with nauseating secretion (containing mercaptans), eat small vertebrates, insects, and fruits; *Mustela vison,* mink, and *Lutra,* river otter, both aquatic; *Enhydra lutris,* sea otter, California to Alaska, lives in ocean surf, feeds on marine invertebrates.

VIVERRIDAE. Mostly small, body slender. Africa and southern Asia. *Herpestes mungo,* com-

Figure 34-28   Order CARNIVORA: land carnivores.   *A.* Gray fox (*Urocyon cinereoargenteus*), head and body 56 to 69 cm (22 to 27 in). *B.* Longtail weasel (*Mustela frenata*), head and body 23 to 27 cm (9 to 10½ in). *C.* Badger (*Taxidea taxus*), head and body 60 to 75 cm (24 to 30 in). *D.* Striped skunk (*Mephitis mephitis*), head and body 30 to 43 cm (12 to 17 in), tail 25 to 33 cm (10 to 13 in). *E.* Bobcat (*Lynx rufus*), head and body 65 to 75 cm (25 to 30 in), tail 10 to 15 cm (4 to 6 in).

mon mongoose, of India, introduced into Jamaica and Hawaii; *Viverra,* civet, scent glands used to make perfume.

HYAENIDAE. Hyenas. *Hyaena,* Africa and southern Asia, doglike in size and form but with hind limbs shorter than forelimbs, jaws powerful, teeth large.

FELIDAE. Cats. *Leo leo,* lion, of Africa, and *L. tigris,* tiger, eastern Asia and India, both large and dangerous to humans; *Felis catus,* house cat, widely feral in United States; *F. concolor,* mountain lion (also cougar, puma, or panther), the Americas, but exterminated in many Eastern states; to about 2 m (7 ft) and 72 kg (160 lb), food deer, sometimes colts and livestock, virtually never attacks humans; *Lynx canadensis,* Canada lynx, and *L. rufus,* bobcat (Fig. 34-28*E*), both with short tails.

Order 20. Pinnipedia. Seals and sea lions. Size medium to large, males usually much bigger than females; body spindle-shaped, limbs formed as flippers or paddles for swimming, with toes included in webs; tail very short; Miocene to Recent; oceans and seacoasts, gregarious; food chiefly fishes.

OTARIIDAE. Eared seals. Hind feet can turn; able to travel on rocky or sandy shores. *Callorhinus ursinus,* northern fur seal, adults haul out on rocky

shores of Pribilof Islands in Bering Sea through summer, in assemblages (rookeries) numbering thousands. An adult male (bull) controls a group (harem) of up to 60 females (cows); each female usually produces 1 young (pup), nurses it on land, later breeds; at end of summer entire population (herd) puts to sea for ensuing 9 months; summer population about 2,600,000, under supervision of United States government; more than 60,000 three-year males (bachelors) taken annually to provide sealskins of commerce; *Eumetopias jubata,* Steller's sea lion (Fig. 34-29), California to Alaska; *Zalophus californianus,* California sea lion, smaller, central California southward.

ODOBENIDAE. Walruses. Skin thick, sparsely haired; 2 upper canine teeth form tusks up to 1 m (3 ft) long, used to dig mollusks and crustaceans on sea bottom and used by humans as ivory. *Odobenus,* in Arctic, to 3 m (10 ft) long.

PHOCIDAE. Earless, or hair seals. Hind feet cannot turn forward, hence less agile on land. *Phoca vitulina,* harbor seal, common on coasts and bays of North America and elsewhere; *Mirounga,* elephant seal (Fig. 34-29), to 5.5 m (18 ft) long, California coastal islands and southward.

Order 21. †Condylarthra. Primitive ungulates, some with canines; cheek teeth with pointed cusps;

Figure 34-29 Order PINNIPEDIA: marine carnivores. *In front.* Immature elephant seals (*Mirounga angustirostris*). *Beyond.* Steller's sea lions (*Eumetopias jubata*), young and adult males or bulls. (*Photo by Robert T. Orr, Ano Nuevo Island near San Francisco.*)

Paleocene-Eocene. Some small-bodied (†*Hyopsodus*), others over 1.5 m (5 ft) long, carnivorelike, tail long, toes 5 (†*Phenacodus*).

Order 22. †Litopterna. Of ungulate form, some to size of camel; toes 3 or 1, with hoofs; upper incisors present; cheek teeth well developed, with folded enamel; Paleocene-Pleistocene, South America. †*Thoatherium*, horselike; †*Macrauchenia*, camellike.

Order 23. †Notoungulata. Form varied, rabbit-sized to 6 m (20 ft) long; some heavy-bodied; digitigrade, mostly 3-toed, with claws, some hoofed; skull short, broad; teeth varied, incisors large in some, upper molars incurved; Paleocene-Pleistocene, mostly in South America. †*Palaeostylops*, Eocene, Asia; †*Toxodon*, †*Typotherium*.

Order 24. †Astrapotheria. Small to large, hind quarters weak; no upper incisors, canines and molars large; toes 5, digitigrade; Paleocene-Miocene, South America. †*Astrapotherium*.

Order 25. Tubulidentata. Aardvark. Body stout, somewhat piglike, sparsely haired; ears and snout long; mouth tubular, tongue slender, protrusible; milk teeth many, permanent teeth fewer (no incisors or canines), unrooted, no enamel; toes 4 or 5, with heavy claws; Eocene to Recent; Africa. *Orycteropus afer*, nocturnal, ants and termites dug from ground nests and captured with sticky tongue.

Order 26. †Pantodonta (Amblypoda). Medium-sized, limbs short, feet broad, hoofed (some clawed); canines large; cheek teeth with looped ridges; Paleocene-Oligocene. †*Pantolambda*, †*Coryphodon*.

Order 27. †Dinocerata (Uintatheria). Size and form of rhinoceros; skull to 75 cm (30 in) long, with hornlike bony growths; canines large, upper molars double-crested; Paleocene to Eocene. †*Uintatherium*.

Order 28. †Pyrotheria. Elephantlike in size and form; incisors chisellike tusks; Paleocene-Oligocene; South America. †*Pyrotherium*.

Order 29. Proboscidea. Elephants. Massive; head large, ears broad and flat, neck short, body huge, legs pillarlike, skin thick (pachyderm), loose, and sparsely haired; nose and upper lip a long, flexible muscular proboscis, containing nasal passages, with nostrils at tip; 2 upper incisors elongated as tusks; each molar tooth with many transverse rows of enamel on exposed grinding surface, only 1 tooth (or 2) functional at a time in each side of jaw; feet

clublike, toes 5, 3, (or 4), each with small naillike hoof, weight borne on elastic pad behind toes. Eocene to Recent. Africa and southeastern Asia; inhabit forests and tall grass; gregarious, in herds of 10 to 100; feed on trees, grasses, and bamboos, adults eating 275 to 325 kg (600 to 700 lb) daily; gestation period about 20 months, young about 1 m (3 ft) tall, weighing 90 kg (200 lb) at birth; maximum recorded age about 50 years.

ELEPHANTIDAE. *Elephas maximus*, Indian elephant, India to Sri Lanka and Borneo, tusks to 2.5 m (9 ft) long; used for transport and in circuses; *Loxodonta africana*, African elephant (Fig. 34-30), to over 3 m (10 ft) in height, ears huge; tamed for transport in war and pageants by Carthaginians and Romans but not often since.

Order 30. †Embrithopoda. Size of rhinoceros, forelimbs shorter than hind, toes 5, spreading; a huge pair of horns on nasal bones and small pair on frontals; teeth equal-sized; Oligocene; Africa. †*Arsinotherium*.

Order 31. Hyracoidea. Hyraxes, coneys. Small; 4 toes on forelimb, 3 on hind; digits with small hoofs, except 2d clawed toe; ears and tail short; incisors 1/2; no canines; superficially like guinea pigs but related to PROBOSCIDEA; Oligocene to Recent; Africa and Syria. *Procavia* (*Hyrax*), diurnal in rocky areas; *Dendrohyrax*, in tree cavities, nocturnal.

Order 32. Sirenia. Manatees, or sea cows. Large, body spindle-shaped; forelimbs paddlelike, no hind limbs; tail with lateral flukes, not notched; muzzle blunt, mouth small, lips fleshy; no external ears; teeth with enamel; hairs few, scattered; stomach complex; Paleocene to Recent; tropical and subtropical seas and rivers; herbivorous. Two families, 3 gen-

Figure 34-30 African elephant (*Loxodonta africana*, **order PROBOSCIDEA**). (*Photo by R. L. Usinger.*)

era; *Trichechus,* manatees, in warm rivers of Florida, West Indies, Brazil, and west Africa; *T. latirostris,* Florida manatee, to 3.5 (12 ft) long; *Dugong dugon,* dugong, Red Sea and coasts of India, New Guinea, and Australia; *Hydrodamalis stelleri,* Steller's sea cow, large, formerly on shores of the Komandorskie Islands in the North Pacific Ocean, exterminated by hunters in 1768.

Order 33. Perissodactyla. Odd-toed hoofed mammals. Large; legs long; foot with odd number of toes, each sheathed in cornified hoof, functional axis of leg passing through middle (3d) toe; stomach simple; Eocene to Recent; Eurasia, Africa, and tropical America.

EQUIDAE. Horses, asses, and zebras. One functional digit and hoof on each leg (in Recent forms); inhabit open plains or deserts, diet of grass. *Equus caballus,* horse, principal means of human transport and work for centuries, with 50 or 60 domestic races, from Shetland pony 105 cm (42 in) high at shoulder and weighing scarcely 135 kg (300 lb) to Shire and Percheron stallions 185 cm (6 ft) high, weighing to 1180 kg (2600 lb) and more; *E. asinus,* ass, adapted to desert life, domesticated as donkey, or burro, for riding and transport; the mule is an F_1 hybrid of male ass (jack) and female horse (mare), with stamina of ass and size of horse; *E. burchelli,* zebra (Fig. 34-31), native to open lands of Africa, seldom domesticated.

TAPIRIDAE. Tapirs. Toes 4 on front feet, 3 on hind feet; in tropical forests. *Tapirus,* 4 species, Central and South America, Malay Peninsula, and Sumatra.

RHINOCEROTIDAE. Rhinoceroses. Large, heavy-bodied, hide thick and armorlike; 1 or 2 median horns on snout; toes 3; in woody or grassy areas *Diceros, Ceratotherium* in Africa south of Sahara; *Rhinoceros* and *Didermocerus* in southeastern Asia.

Figure 34-31 Zebra (*Equus burchelli,* order PERISSODACTYLA); **about 120 cm (48 in) high at shoulder; native to Africa.** (*Photo by T. I. Storer.*)

Figure 34-32 Hippopotamus (*Hippopotamus amphibius,* order ARTIODACTYLA); **length to 3.5 m (12 ft).** (*Photo by National Zoological Park.*)

Order 34. Artiodactyla. Even-toed hoofed mammals. Size various; legs usually long; 2 (rarely 4) functional toes on each foot, each usually sheathed in cornified hoof; axis of leg between toes; many with antlers or horns on head; all but pigs with reduced dentition, most with a 4-compartment stomach, and ruminate, or "chew the cud"; Eocene to Recent; all continents except Australia. Size from antelope (dik-diks) no bigger than a hare to heavy buffalo and hippopotamus; some domesticated; 3 suborders and 25 families, 16 families extinct, others all with fossil species.

Suborder 1. Suiformes. Pigs and allies. No horns or antlers, teeth 38 to 44, canines enlarged as curved tusks.

SUIDAE. Old World pigs. Many in southern Asia; *Sus scrofa,* European wild boar, source of most domestic pigs; *Phacochoerus,* warthog, in Africa.

TAYASSUIDAE. Peccaries, or New World pigs. Patagonia to southern United States; *Tayassu tajacu,* southern Texas and Arizona into South America.

HIPPOPOTAMIDAE. Hippopotamuses ("horse of the river"). Body and legs very stout, skin heavy but scantily haired, adept at swimming, feed mostly on land plants. *Hippopotamus amphibius,* Nile River and south in Africa (Fig. 34-32), to 4 m (12 ft) long; *Choeropsis liberiensis,* pigmy hippopotamus, in west Africa, smaller.

Suborder 2. Tylopoda. CAMELIDAE. Camels, dromedaries, llamas, alpacas. Feet soft, bearing nails, not hooves; 1 pair of upper incisors; stomach 3-parted, chew cud. *Camelus,* camel and dromedary (Fig. 34-33), Africa and Asia, domesticated for riding and transport; *Llama,* llama and alpaca, western South America, provide transport, flesh, hides, and wool.

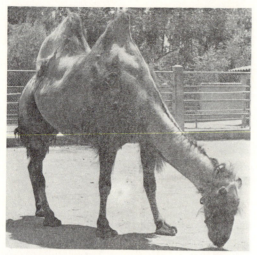

Figure 34-33 Bactrian, or two-humped, camel (*Camelus bactrianus,* order ARTIODACTYLA), of Asia. Most of the long, dense winter hair has molted, and the summer coat is short. (*Photo by T. I. Storer.*)

Suborder 3. Ruminantia. Ruminants. Feet with hoofs; chew cud.

Infraorder 1. Tragulina. TRAGULIDAE. Chevrotains, or mouse deer. Intermediate between pigs, camels, and deer; no horns; teeth 34; upper canines present, larger in males; stomach 3-parted; some species the size of large rodents. *Tragulus,* southeastern Asia; *Hyemoschus,* West Africa.

Infraorder 2. Pecora. True ruminants. Horns or antlers usually present, at least in males; teeth 32 to 34; 4 pairs of incisorlike teeth at front of lower jaw, the fourth pair being modified canines; no upper incisors, usually no upper canines; stomach 4-parted (Fig. 34-14).

CERVIDAE. Deer, etc. Head of male with pair of solid bony antlers, shed and grown anew each year. *Rangifer,* reindeer and caribou, in northern regions, antlers in both sexes; *Alces alces,* moose, Alaska to Maine, North Dakota, and Yellowstone Park, adult male (bull) to 2 m (6 ft) at shoulder and 400 to 625 kg (900 to 1400 lb), enters water, feeds on aquatic plants; *Cervus canadensis,* American elk (wapiti), formerly over much of United States except deserts, now restricted to mountains of the West, males to 1.5 m (5 ft) high and 225 kg (500 lb); *C. elaphus,* European red deer; *Odocoileus,* American deer, males of larger races to 1 m (42 in) high and occasionally 135 kg (300 lb); *O. virginianus,* whitetail deer, Atlantic Coast to eastern Oregon; *O. hemionus,* mule or blacktail deer, western North America (Fig. 34-8C).

GIRAFFIDAE. Giraffes, okapis. Neck and legs very long; Africa. *Giraffa camelopardalis,* giraffe (Fig. 34-34), 5.5 to 6 m (18 to 20 ft) tall, head with 2 to 6 skin-covered "horns"; browses on foliage of mimosa and other trees; *Okapia,* okapi, smaller, neck shorter.

ANTILOCAPRIDAE. One living species, *Antilocapra americana,* pronghorn antelope (Fig. 34-8B), on treeless plains and deserts from Great Plains to California, now reduced and abundant only in certain areas; adults 90 cm (36 in) high, weight to 55 kg (125 lb), both sexes with pair of permanent bony horn cores on head, on which compressed black horns with lateral prongs grow and are shed annually.

BOVIDAE. Hollow-horned ruminants. Horns hollow, paired, unbranched, composed of keratin, growing slowly and continuously from base, over bony cores on frontal bones, usually in both sexes, larger in males; more than 50 genera in Northern Hemisphere, southeastern Asia and Africa. *Ovis canadensis,* mountain sheep, or bighorn (Fig. 34-8A), Mexico, Rocky Mountains, and California to Alaska, on rugged mountains; males (rams) to 100 cm (40 in) high and 135 kg (300 lb), with horns to 100 cm (40 in) long; *Oreamnos americanus,* mountain goat, Idaho and Washington to Alaska, coat long, shaggy, and white, horns short, black; *Bison bison* (Fig. 34-35), American buffalo, or bison, formerly Texas and Great Plains to southern Canada, with hump over shoulder and long crinkly hair on foreparts; adult bulls to 175 cm (70 in) high and 800 kg (1800 lb), but cows smaller; millions formerly on flatlands but slaughtered for meat and hides, now some on reservations and in parks; *Bison bonasus,* Euro-

Figure 34-34 Giraffe (*Giraffa camelopardalis,* order ARTIODACTYLA), inhabits tree-studded grasslands of east Africa. (*Photo by Stuart MacKay.*)

Figure 34-35 American bison, or buffalo (*Bison bison,* order ARTIODACTYLA), in full winter coat. (*Photo by National Zoological Park.*)

pean wisent, nearly extinct; domestic cattle, sheep, and goats (Sec. 34-29); *Bos gaurus,* gaur, India and Burma; *Bos banteng,* banteng, southeastern Asia, Borneo, and Java; *Bubalus bubalis,* water buffalo or carabao, native of India, widely domesticated, Italy to Pacific Islands; *Ovibos moschatus,* musk ox, on treeless Arctic lands from Hudson Bay to 83°N, to 130 cm (52 in) high and over 225 kg (500 lb); African and Asian antelopes, many species of various sizes.

References

The *Journal of Mammalogy* (quarterly, since 1919, published by the American Society of Mammalogists) is the principal periodical in English. Museums, universities, etc., publish state lists or books on mammals. The U.S. Departments of Agriculture and the Interior have many circulars and bulletins on the economic and other relations of mammals. *The North American Fauna* includes accounts of mammals of various states and regions. Information on game and furbearers may be found in books on hunting, shooting, and trapping.

Anderson, Sydney, and **J. K. Jones, Jr.** (editors). 1967. Recent mammals of the world: A synopsis of families. New York, The Ronald Press Company, viii + 453 pp., illus.

Audubon, John James, and **J. Bachman.** 1851–1854. The quadrupeds of North America. 3 vols. New York, V. G. Audubon. 155 pls. and text. First issued (1845–1848) as "The Viviparous Quadrupeds of North America" in 3 folio vols. containing 150 plates and a separate text (1846–1853) of 3 quarto vols.; combined as "The Quadrupeds . . ." of 3 royal octavo vols. *The first elaborate work on mammals of North America.*

Barbour, R. W., and **W. H. Davis.** 1969. Bats of America. Lexington, University of Kentucky Press. 286 pp., illus.

Bourlière, François. 1970. The natural history of mammals. 3d ed. New York, Alfred A. Knopf, Inc. xxi + 387 + xi pp., illus.

Burt, W. H., and **R. P. Grossenheider.** 1976. A field guide to the mammals [north of Mexico]. 3d ed. Boston, Houghton Mifflin Company. 289 pp., many figs. (some colored) and maps.

Cahalan, V. H. 1947. Mammals of North America. New York, The Macmillan Company. x + 682 pp., illus.

Cockrum, E. L. 1962. Introduction to mammalogy. New York, The Ronald Press Company. viii + 455 pp., illus.

Davis, D. E., and **F. B. Golley.** 1963. Principles in mammalogy. New York, Reinhold Publishing Corporation. xiii + 335 pp., illus. *Activity, behavior, reproduction, populations, adaptations.*

Ewer, R. F. 1968. Ethology of mammals. New York, Plenum Press, Plenum Publishing Corporation. xiv + 418 pp., 12 figs., 8 pls.

Flower, W. H., and **R. Lydekker.** 1891. An introduction to the study of mammals living and extinct. London, A. & C. Black, Ltd. xvi + 763 pp., 357 figs.

Glass, B. P. 1973. A key to the skulls of North American mammals. Stillwater, Okla., Oklahoma State University Press. iii + 59 pp., illus.

Grassé, Pierre-P. (editor). 1955. Mammifères. In Traité de zoologie. Paris, Masson et Cie. Vol. 17 (2 pts.), 2300 pp., 2106 figs.

Grzimek, H. C. B. (editor). 1972–1975. Grzimek's animal life encyclopedia. Vols. 10–13. Mammals I–IV. New York, Van Nostrand-Reinhold.

Gunderson, H. L. 1976. Mammalogy. New York, McGraw-Hill Book Company. viii + 438 pp., illus.

Hall, E. R., and **K. R. Kelson.** 1959. The mammals of North America. New York, The Ronald Press Company. 2 vols. xxx + 1083 + 79 pp., 553 figs. (skulls), 500 maps (distribution), 186 sketches of generic types. *Keys, descriptions), ranges, references.*

Hamilton, W. J., Jr. 1939. American mammals: Their lives, habits, and economic relations. New York, McGraw-Hill Book Company. xii + 434 pp., 92 illus.

———. 1943. The mammals of eastern United States. Ithaca, N.Y., Comstock Publishing Associates, a division of Cornell University Press. 7 + 432 pp., 183 figs. Reprint, 1963. New York, Hafner Publishing Company, Inc.

Kükenthal, Willy, and **Thilo Krumbach.** 1955–1969. Handbuch der Zoologie. Vol. 8. Mammalia, pts. 1 to 45. Berlin, Walter de Gruyter & Co. 4359 pp., illus.

Matthews, H. L. The life of mammals. 1969. Vol. I. 340 pp., illus. 1971. Vol. II. 400 pp. illus., 1971. New York, Universe Books.

Miller, G. S., Jr., and **Remington Kellogg.** 1955. List of North American Recent mammals. U.S. National Museum Bulletin 205. xii + 954 pp. *Scientific name and geographic range of each species and subspecies.*

Napier, J. R., and **P. H. Napier.** 1967. A handbook of living primates. New York, Academic Press, Inc. xiv + 456 pp., 114 pls., 10 figs.

Orr, R. T. 1976. Vertebrate Biology. 4th ed. Philadelphia, W. B. Saunders Company. viii + 472 pp., illus.

Palmer, R. S. 1954. The mammal guide . . . North America north of Mexico. Garden City. N.Y., Doubleday & Co., Inc. 384 pp., 40 col. pls., 37 figs., 145 maps. *Identification, biology, and economic status.*

Seton, E. T. 1925–1928. Lives of game animals. Garden City, N.Y., Doubleday & Co., Inc. 4 vols., Illus. *Extensive account of hoofed mammals, carnivores, and some rodents.*

Simpson, G. G. 1945. The principles of classification and a classification of mammals. *American Museum of Natural History Bulletin 85.* xvi + 350 pp. *Technical; both fossil and Recent groups down to genera.*

Slijper, E. J. 1962. Whales, New York, Hillary House Publishers Ltd. 475 pp., illus.

Vaughn, T. A. 1972. Mammalogy. Philadelphia, W. B. Saunders Co. vi + 463 pp., illus.

Walker, E. P. and others. 1975. Mammals of the world. 3d ed. 2 vols. Baltimore, The Johns Hopkins Press. xlvii + 1500 pp., illus.

Young J. Z. 1975. The life of mammals. 2d ed. New York, Oxford University Press. xv + 528 pp., many figs. *Structure and function of organs and systems.*

35

THE HUMAN ANIMAL

35-1 The place of humans in nature   The derivation and position of the human species, *Homo sapiens,* in the realm of living things has long been a subject of great interest. Primitive peoples have myths that imply their ancestry from animals, or inanimate materials, or other sources. To other peoples, humans are the creation of a supernatural power. Biologists have determined the origin and position of humans using knowledge of the structure and physiology of the human body, its embryonic development, and the historic, prehistoric, and fossil records. On this basis, the human species belongs to the following groups:

1 Animal Kingdom, in requiring complex organic food, etc.
2 Phylum CHORDATA, in having a notochord and gill arches and pouches during embryonic development and a dorsal tubular brain and nerve cord throughout life
3 Subphylum VERTEBRATA, in having a cranium and segmented spinal column
4 Superclass TETRAPODA, in having two pairs of limbs
5 Class MAMMALIA, in having hair and mammary glands
6 Order PRIMATES, in having four generalized limbs each with five digits bearing nails
7 Superfamily HOMINOIDEA, in lacking cheek pouches and a tail
8 Family HOMINIDAE (as distinguished from the family PONGIDAE, or anthropoid apes); in the possession of many features, including:
   a Brain with far greater functional ability and of larger size (minimum human, 1000 cm³;

maximum gorilla, 650 cm³); brain case larger than face

b   Face flatter and more vertical, brow ridges reduced, lower jaw less protruding, teeth more evenly sized

c   Hair long and of continuous growth on head but sparse and short on body

d   Hands more generalized, thumbs better developed; legs 30 percent longer than arms, and straight; big toe not opposable to other toes

e   Skeleton and soft parts of different configuration and proportions; body with subcutaneous fat

f   Arcuate (bow-shaped) tooth row, small canines, biscuspid premolars

g   Prolonged infancy and skeletal maturation

The human species habitually walks erect on two feet, is terrestrial in habits, highly gregarious, and omnivorous in diet, and commonly uses cooked food. By contrast, our closest relatives, the anthropoid apes (gorilla, chimpanzee, orangutan, gibbon) are semierect in posture and literally "four-handed." The gorilla and chimpanzee walk on their knuckles, a unique method of locomotion. The gorilla is a ground-dwelling vegetarian that seldom eats fruit and only occasionally climbs. The chimpanzee is both arboreal and terrestrial, feeding on fruit and some animal matter. Both the orangutan and gibbon are arboreal vegetarians and often eat fruit. In general the anthropoid apes are not as strongly gregarious as humans, but recent studies of gorilla and chimpanzee bands reveal considerable cohesiveness and social organization.

Studies of macromolecules, serum albumins, transferrins, hemoglobins, and DNA of Old World monkeys, anthropoid apes, and humans reveal a much closer relationship between humans and the anthropoid apes than between monkeys and apes. Divergence of the human lineage from that leading to African apes may have occurred 5 to 15 million years ago; biochemical evidence suggests the former date.

Human beings surpass all other living organisms in many functional abilities such as (1) the construction and use of tools; (2) modification of the environment to their own advantage, including the production of food; (3) possession of articulate speech and

language; (4) organization of a complex social life with cooperative effort; and (5) formation of abstract concepts. These are possible by reason of having generalized hands but more importantly by the superior development of the human brain in size, structure, and functional capacity. By the use of articulate speech, language, writing, and records, humans are able to assemble and transmit accumulated knowledge to succeeding generations. Material, social, cultural, and ethical advances have thus resulted. In most other animals, transmission of abilities between generations is limited to such instincts and reflexes as are fixed by heredity.

Structure and physiology

35-2 Size   Most human adults are from about 142 to 198 cm (56 to 78 in) tall, although occasional midgets are only 46 to 91 cm (18 to 36 in) high and exceptional giants grow to 2.7 m (9 ft) because of excessive pituitary function. Races vary in average height from 129 cm (51 in; African pigmies) to 183 cm (72 in; Patagonian and Australian natives). Peoples lacking suitable food or living in extreme climates tend to be small; active races in temperate lands and with better food are taller. Recent generations in the United States have shown a definite increase in average height, which now appears to have stabilized, probably due to improved nutrition and control of disease. In most races males are taller than females. The weight varies with race, sex, height, and age.

35-3 Body covering[1]   Human skin is thin and delicate, except for calluses (functional adaptations) on the palms and soles, especially among those who go barefoot. The skin coloration results from the presence of blood capillaries containing red cells (with hemoglobin) in the dermis and also of pigment, including melanin (black) and carotene (yellow). The white, yellow, brown, and black races differ in the amounts of melanin pigment in the skin. Blonds of the white race have less melanin than bru-

---

[1] *Many features of human structure and physiology are described in Chaps. 4 to 10 and some aspects of human heredity in Chap. 11.*

nets, and women usually less than men. The yellow tinge of the palms, soles, and eyelids is ascribed to carotene. Pigmentation is commonly more pronounced in races exposed to strong sunlight. The tanning, or darkening, of the skin in white persons is a defensive mechanism against undue amounts of ultraviolet radiation.

The human skin (Fig. 4-1) differs from that of most mammals in having many sweat glands that produce watery perspiration, containing salts and excretory products. Evaporation of perspiration aids in cooling and regulating the body temperature.

35-4 Hair Human hair is mammalian as to structure and manner of growth but differs (1) in being long or dense chiefly on the head; (2) in amount, distribution, and form at various ages of an individual; (3) between the sexes as to amount and length; and (4) in amount and structure between races of humans.

Long hair of almost continuous growth on the scalp is a distinctive human characteristic. The eyebrows and eyelids (lashes), external ear canals, and nostrils all bear specialized hairs. The body elsewhere has a scant covering, and the lips, soles, palms, outer (extensor) surface on the terminal segment of each finger, and parts of the genitalia are hairless.

The human fetus at about the twentieth week is covered with fine soft hair (*lanugo*) which is lost at or before birth. The newborn usually has dark scalp hair that later may change color. Body hair is scant during childhood but begins to grow at puberty (14 ± years), especially in the armpits (*axillae*) and in the pubic region. As boys become men, the beard and chest hairs grow. Dense facial hair is characteristic of males after puberty; women have varying small amounts, and there are occasional "bearded ladies."

Further changes occur later in life, when the hair gradually becomes gray and then white through failure of the hair papillae to produce pigment—but no one's hair ever "turns white overnight," despite popular belief. The head hair usually becomes sparse with advancing years, and definite patterns then develop, leading to graying of the scalp and beard and, in men, and occasionally women, to baldness. These are probably hereditary changes, although pathologic ones also occur; genes for baldness are probably transmitted by, but not often expressed in, females.

Differences in human hair are exemplified by the straight black hair of Orientals and American Indians, the kinky hair of Negroes, and the yellowish hair of Scandinavians. In cross section, the hair varies from nearly round in Mongolians to elliptical in kinky-haired Hottentots.

All human hairs are replaced at intervals, those of the eyebrows and eyelashes every few months and scalp hairs at from a few months to 4 or 5 years. There is evidence of slight spring and autumn molts in human beings.

35-5 Skeleton and teeth The human cranium is a rounded bony box, lacking the sagittal crest of many other mammals. The occipital condyles are ventral rather than posterior, in keeping with the upright posture, the brain case is larger, the nasal region is short, and the lower jaw is U-shaped rather than V-shaped. The skull increases in size for about seven years and again at puberty, and final closure of the sutures may be delayed until old age. Other notable features of the human skeleton are the absence of a tail and the complete and generalized skeleton of the limbs, hands, and feet.

The human teeth are less conspicuously differentiated than the teeth of other mammals and are of nearly equal height (Fig. 35-1). The permanent, or

Figure 35-1 Change in outline and relative shape of the human mandible with age. (*Altered from Heitzmann.*)

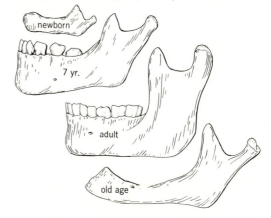

adult, dentition includes 28 to 32 teeth, depending on how many of the third molars (wisdom teeth) appear, these being irregular as to time and manner of eruption.

35-6 Muscular system
The five hundred or more muscles of the human body are much like those of other mammals. The names for muscles that were first applied by early anatomists to the human subject serve in the main for other mammals whose muscles may display minor differences as to origin, insertion, and size. Most of the peculiarities of human muscles relate to the upright posture, the support of the body and walking on the "hind" limbs, and the greater flexibility of the arms, hands, and fingers. Of special interest are the many facial muscles that make possible the expression of pleasure, anger, and other emotions. Humans and other primates have the greatest development of these muscles; a few carnivores, such as the cat, may indicate pleasure and anger, but in all vertebrates below the mammals the face is a rigid mask.

35-7 Internal organs
In the abdominal cavity the organs are supported by mesenteries attached to the posterior wall and by a large sheet (greater omentum) hanging apronlike between the intestines and the inner front wall of the abdomen. It is composed of four layers of peritoneum and contains deposits of fat, serving to cushion the internal organs and protect them against loss of heat.

35-8 Nervous system
The human brain (Fig. 9-7) is notable for its proportionately large size and the great development of the cerebral hemispheres (gray matter), which have a much convoluted surface and are so enlarged that they almost completely cover the other parts of the brain.

The average human brain weighs about 1250 (women) to 1350 g (men), and the volume is 1200 to 1500 cm³. Brain size and mental ability are not closely correlated. The human brain far surpasses that of other animals in proportionate size as well as functional development. Thus a dog, gorilla, and human of about equal weights have brains of about 135, 430, and 1350 g, respectively.

35-9 Distribution
Human populations range from the Arctic to islands beyond the southern continents, from sea level to over 7000 m (20,000 ft) in some mountain regions, and from moist tropical jungles to the driest of deserts. The local races in these various habitats differ in stature, configuration, and coloration, in mode of life, in foods and shelters utilized, and in social organization and customs; but in recent times these differences have been declining.

35-10 Numbers
Precise information on changes in human numbers is lacking, but the major trend is clear. After a long period of gradual increase, population growth began to accelerate with the beginning of the Christian era and has now become explosive. World population now stands at over 4 billion and is currently estimated to be increasing at the rate of 70 to 80 million annually. The figures may be conservative, because population size is often underestimated. Increase may have been in surges, as it was in later stages when invention and discovery improved human chances of survival and capacity to exploit the environment. Major spurts probably followed the development of tools and language and are known to have accompanied the development of agriculture, industrialization, sanitation and medicine, and modern science.

In the absence of imports the ultimate human population of any region depends upon soil fertility, water supply, and climate. Some local human populations in Asia Minor, North Africa, and parts of the United States have decreased because of depletion of soil fertility, loss of top soil by erosion, and increasing aridity.

35-11 Factors regulating human populations
Humans surpass animals in being able to control their environment in some degree, by (1) **food production** (agriculture) and **food storage** (warehouses, canning, refrigeration) against seasonal shortage and deterioration; (2) improved **shelter** for adults and offspring against unfavorable weather and other threats (housing, flood control, police protection, national defense); (3) **destruction of animal "ene-

mies'' (wolves, rats and mice, household insects); (4) **reducing competition** from other animals (fencing crops and livestock; shooting, trapping, or poisoning species that attack human food and crops and domesticated livestock) and unwanted plants (use of herbicides); and (5) **control of diseases** that attack humans (sanitation, immunization). By these means, modern humans in civilized countries have increased in numbers over the population formerly possible.

Human beings are reduced in numbers by several agencies, some of them the same as those affecting wild animals; these include (1) **food shortage** or famine due to crop failure, as from unfavorable weather, especially in densely populated lands such as China, India, and parts of Africa where food reserves are periodically scant or unequally distributed, and malnutrition (improper diet) leading to physical impairment with greater susceptibility to disease and death; (2) **inadequate shelter,** especially during the cold of winter outside the tropics; (3) **calamities** such as floods, hurricanes, and volcanic eruptions; (4) **enemies** such as large predatory animals (crocodilians, sharks, wolves, lions, tigers) and venomous snakes, which cause many deaths in some tropical areas; (5)

warfare, a form of competition between races or nations for land, natural resources, trade routes, or other desired features of human environment (war and slavery have long been major factors in depopulation); and (6) **disease,** uncontrolled as among primitive peoples, the result of neglect among "civilized" peoples, or uncontrollable as with epidemic influenza.

Where many diseases are under control by public health measures in civilized countries, fewer individuals die in childhood or early life. Proportionately more die from diseases of later life or from old age (senility). Explosions, fire, earthquake, shipwreck, and transportation accidents kill many in modern civilizations. In recent years pollution and the stresses of crowding have been contributing to mortality, but the extent of their impact has not been adequately determined.

35-12 Reproduction   During childhood secondary sexual characteristics are usually undeveloped. When an individual reaches about 14 (10 to 17) years of age, conspicuous changes take place that lead toward sexual maturity and the ability to reproduce.

Figure 35-2 The male urogenital system. *Left.* In perspective. *Right.* Median section.

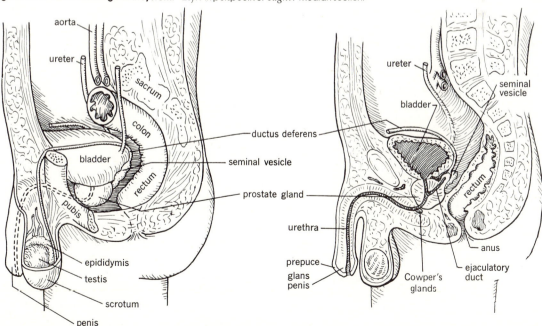

This time is known as **puberty,** when many of the characteristics that differentiate the sexes gradually appear. In males the beard and body hair begin growth, the shoulders broaden, the voice becomes deeper-pitched, and more attention is paid to the opposite sex. In females the mammary glands and hips enlarge, subcutaneous fat is deposited, the reproductive tract begins a series of cyclic changes, and the attentions of men become of interest.

Healthy females from puberty onward experience a recurrent discharge of blood, mucus, and epithelial cells from the lining of the uterus, termed **menstruation** (L. *menses,* months); on the average, this occurs every 28 days (1 lunar month) and lasts 4 or 5 days. Menstruation is regulated by endocrine secretions (pituitary gonadotropins and estrogen and progesterone of the ovary) and prepares the uterus for implantation of an ovum. It usually ceases during pregnancy and lactation and stops if the ovaries are removed. Menstruation gradually stops at about age 45 to 50, and its cessation (the menopause) marks the end of reproductive ability. This *change of life* is a time of physiologic and emotional distress for many women. There is no truly comparable sexual cycle in the male, although there are diurnal variations and perhaps a menopause-type response in some male vertebrates.

Each ovary in a young female contains many thousands of immature ova, the numbers decreasing with age. One enlarged mature ovum (occasionally more) is released to pass down the oviduct some days after each menstruation, possibly 400 ova in all during the life of an average woman.

Exceptional human females only 8 or 9 years old have borne children, but there is usually a time after menstruation begins (the menarche), termed *adolescent sterility,* before pregnancy occurs in any considerable percentage of young women. Age at menarche had decreased worldwide but has now stablized. The time at which a young male becomes capable of procreation varies with individuals, and the same is true for cessation of potency, as some men of advanced age have become fathers.

If an ovum is fertilized, it becomes implanted in the mucosa of the uterus, begins embryonic development, and is surrounded by embryonic membranes. Shortly a placenta develops by which the maternal and embryonic circulations are brought close together for transfer of nutrients, respiration, and removal of wastes during pregnancy (Fig. 35-4). The period from conception to birth in the human embryo normally requires about 280 days. In the vast majority of cases a single infant is produced. Twins occur about once in 100 births, triplets once in

Figure 35-3 The female urogenital system. *Left.* In perspective. *Right.* Median section.

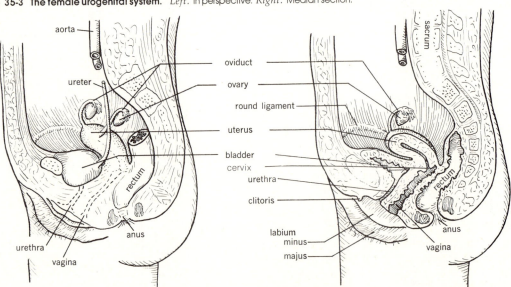

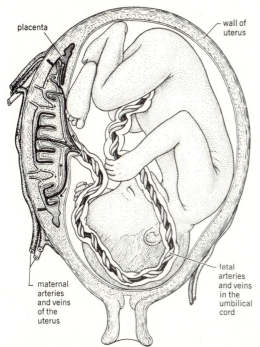

placenta

wall of uterus

maternal arteries and veins of the uterus

fetal arteries and veins in the umbilical cord

Figure 35-4 Section of the human uterus with a fetus connected by the umbilical cord to the placenta; maternal blood circulation indicated by broken arrows, embryonic circulation by solid arrows. (*Adapted from Ahfeld and Spanner.*)

10,000, and quadruplets once in 600,000; some families produce a greater percentage of twins. Quintuplets and even more numerous multiple births also occasionally occur. The average infant weighs about 3000 g (7 lb) at birth but may be from less than 2300 to 6000 g (less than 5 to 13 lb) or more.

The sex ratio at birth is about 105 to 106 males per 100 females; it is even more disproportionate before birth. A higher mortality rate occurs in males, before birth and during the early years of life, so that from about age 40 to 80 females decidedly outnumber males.

35-13 Hormonal control of reproduction Endocrine events during ovulation and during the uterine changes of the menstrual cycle are as follows (Fig. 35-5): During the **follicular phase** the follicle and ovum mature and the encircling follicular cells proliferate under the action of the **follicle-stimulating hormone** (FSH) and the **luteinizing hormone** (LH) of the

pituitary and also through increased production of the hormone **estrogen** secreted by the follicular cells themselves as the follicle enlarges. Maturation requires about 14 days. Under the influence of FSH and LH, estrogen secretion by the follicular cells and other ovarian cell types greatly increases during the latter part of the follicular phase (Fig. 35-5). Estrogen begins the preparation of the uterine lining for gestation. The lining proliferates and thickens and becomes more vascular. The mucus of the cervix (Fig. 35-3) becomes more fluid and copious in preparation for the entry of sperm. What causes ovulation? It is associated with an abrupt and brief outpouring of large amounts of LH. The rapidly rising level of estrogen during the last half of the follicular phase ap-

Figure 35-5 Human ovulation and menstrual cycle. *A.* Changes in the uterine mucosa; period of menstruation (lasting 3 to 5 days) indicated by horizontal black bar. Accompanying follicular events 1 to 3 are enlargement of a follicle (1), ovulation (2), and formation and degeneration of the corpus luteum (3). *B.* Changes in blood levels of hormones during the 28-day cycle.

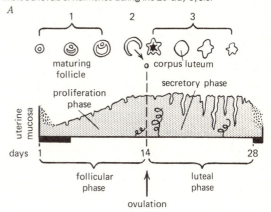

A

1   2   3

maturing follicle   corpus luteum

secretory phase

proliferation phase

uterine mucosa

days 1   14   28

follicular phase   luteal phase

ovulation

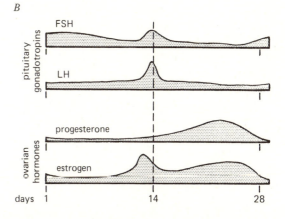

B

FSH

LH

pituitary gonadotropins

progesterone

estrogen

ovarian hormones

days 1   14   28

pears to cause this LH surge (Fig. 35-5). The ovum bursts out of the follicle and through the wall of the ovary. Ovulation occurs about 2 days after the peak in the plasma estrogen and about 1 day after the peak in plasma LH.

Ovulation initiates the **luteal phase** of the cycle. Follicular cells of the collapsed follicle (3 in Fig. 35-5*A*) proliferate and enlarge and accumulate lipid to form a structure called the **corpus luteum** (yellow body). It secretes both estrogen and the hormone **progesterone.** LH is required for luteum activity. The estrogen and progesterone produced inhibit release of FSH and LH, maintaining them at low blood levels and thereby preventing maturation of other ova. Progesterone seems to prevent any additional surges of ovulation-promoting LH. Progesterone, with a minor contribution from estrogen, induces the secretory phase of the uterine cycle and stimulates the uterus to make final preparations for gestation. It causes the cervical mucus to become more viscous, a change perhaps important in blocking bacterial entry into the uterus during embryonic growth. If implantation of the embryo occurs, the corpus luteum continues its hormonal activity and maintains the active state of the uterine lining. Its persistence during pregnancy is due at least in part to a placental hormone, **chorionic gonadotropin** (CG), whose production begins in the very early stages of implantation. By about the third month of pregnancy the placenta produces enough progesterone and estrogen so that it can take over the role of the corpus luteum, and CG production falls off rapidly. If fertilization does not occur, the corpus luteum degenerates after about 10 or 12 days, and blood levels of progesterone and estrogen drop. The uterine lining is no longer maintained in its receptive state and menstrual sloughing occurs. The drop in progesterone and estrogen levels, in negative-feedback fashion, removes the inhibition on FSH production and the cycle resumes.

The menstrual cycle is characteristic of primates only. Other mammals have an **estrous** cycle in which the female is receptive to the male usually only during a period of *heat* and the uterine lining is not sloughed. Hormonal regulation, however, is similar throughout the mammals.

Males also produce FSH and LH. The former stimulates spermatogenesis and the latter, testosterone production (Sec. 8-11). Since testosterone is required for spermatogenesis, LH is involved in this process as well. Testosterone exerts negative-feedback inhibition on LH secretion via both the pituitary and the hypothalamus. The endocrine control of reproduction in the male proceeds at a rather fixed and continuous rate, in contrast to the cyclic control in the female.

35-14 Growth The human individual has the longest period of development after birth of any living creature, and its growth curve (age to weight) differs both quantitatively and qualitatively from that of other animals (Figs. 35-6, 35-7). There is accelerated growth in the prenatal and adolescent periods but retardation in the juvenile and postadolescent stages. Little or no living tissue is added after the latter; any subsequent changes in weight is due chiefly to the addition of fat, dehydration of tissues, and further mineralization of bone.

35-15 Length of life The extreme limit of human life is about 100 years. Humans are among the longest-lived animals. A few people live well over 100 years, but accounts of persons living much over a century often are undocumented. The absolu human lifespan appears not to have changed materially within historic times. The Biblical "three score years and ten" thus remains a fairly good approximation.

The **mean length of life** is the average number of years lived by all persons born in a given period. It is least among aboriginal and tropical native populations. In white populations of the North Temperate region it has increased during the twentieth century with improved sanitation, control of communicable diseases, corrective surgery, and better nutrition. For white persons in the United States the average life expectancy at birth in 1900 was about 49 years; by 1954 it was 67 for males and 74 for females; in 1965, 67 for males and 74 for females; and in 1973, 68 for males and 76 for females. The effect of increased mean length of life has been a general aging

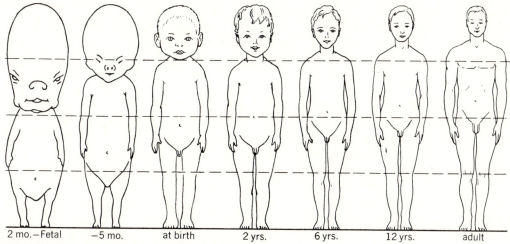

Figure 35-6 Changes in proportions of the human body with age. (*After Stratz.*)

| 2 mo.–Fetal | –5 mo. | at birth | 2 yrs. | 6 yrs. | 12 yrs. | adult |

of the population. Thus persons aged 65 and over represented about 4 percent of the population of the United States in 1900, 6 percent in 1935, 9 percent in 1962 and increased to about 11 percent in 1978.

In ancient Egypt and Rome the **average expectation of life** was probably never over 30 years, whereas in the United States it is now about 68 for males and 76 for females. This has resulted from great reductions in infant mortality and the other factors just mentioned.

History of the human species

The science of anthropology deals with the study of human and related hominoid species, living, prehistoric, and fossil; that of archaeology, with past human artifacts and works.

35-16 Sources   Remains of early humanids comprise a limited number of skulls, often incomplete,

Figure 35-7 Comparison of growth curve for weight in humans and mammals (guinea pig and cow). (*After Brody, 1927.*)

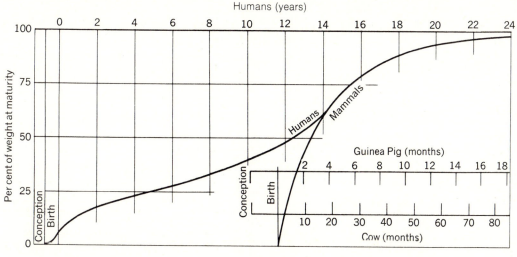

and some other bones. Complete skeletons and skulls of later periods, after burial of the dead became a practice, are more numerous. The works of prehistoric peoples provide much material that permits inferences about their manner of life and activities. The making of stone implements by *Australopithecus* probably began at least 3 million years ago, and such toolmaking has been continued into historical times by some living races. Many of these tools (artifacts) have been found on the ground or below the surface in association with ancient hearths, skeletal remains, and debris of early camp sites and communities. From the changes in kinds of tools with the passage of time, paleoanthropologists and archaeologists have established a sequence of cultural periods (Paleolithic, etc.) and stages of culture (Acheulean . . . Magdalenian) as a guide in dating artifacts obtained from various sites (Figs. 35-8, 35-9).

Excavations of prehistoric habitations from European caves and rock shelters occupied during the last Ice Age and on to settled communities in the Middle East, India, and China have revealed much of the beginnings of civilization. Some Old World sites show a vertical succession of towns, each built on relics of its predecessor, through several millennia. The beginnings of food production, writing, mathematics, and some religious practices thus have been dated by ^{14}C (carbon 14) and other methods. In the New World the entire record is shorter, tools and occupied sites are far fewer, and early skulls or skeletons are rare. Knowledge of human prehistory is advancing rapidly as increasing numbers of trained investigators go afield to "dig" and return to study and interpret their findings.

35-17 Human characteristics   In physical features the human line shows a series of progressive changes, leading it to differ from the anthropoids. The face becomes flattened, without a muzzle, the brow ridges decline and often disappear, and the cranium rises well above the orbits to accommodate a larger brain. At the rear the skull becomes rounded, and the foramen magnum and occipital condyles move ventrally to meet the upright vertebral column.

Figure 35-8 Time scales in the history of the human species. In *A* and *B* years are approximate, before the present time. Types of culture are shown for western Europe, limits somewhat uncertain. *Australopithecus*, with Olduvan culture of crude stone tools (not shown), was replaced by *Homo erectus* about 1 million years ago.

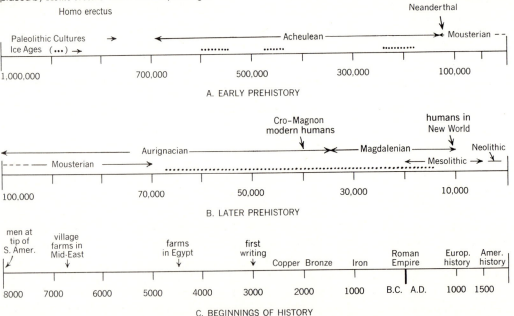

A mastoid process develops below the ear region. The teeth become smaller in size and of equal height, without large canines, and are set in a U-shaped arc. The arms and fingers become proportionately shorter, the foot nongrasping, and the toes in line and the heel bone lengthens, improving muscle insertion for upright posture, walking, and running. Slight curvature of the vertebral column places the head and trunk directly above the pelvis, centering the weight over the legs. The blade (ilium) of the pelvis becomes wider than long, affording broader attachment for the big gluteal muscles involved in balance. The entire pelvis becomes basketlike, supporting the viscera now positioned above it.

Sounds made by great apes indicate some desires and emotions but have limited use in denoting objects. One great evolutionary advance was to progress from sounds to words—symbols to indicate things or ideas. Speech makes it possible for experience to become cumulative and benefit later generations of individuals. The anthropoid-prehuman transition also included descending from the trees to the ground—probably grassland—and standing upright. Erect posture aided use of tools (Fig. 35-9), seeing and avoiding danger, and pursuit of prey. Increased brain development accompanied involvement with tools and hunting. Tool use, brain development, and bipedalism interacted to stimulate evolutionary advance. Toolmaking is a farsighted endeavor, involving time and labor to make devices that later will enable tasks to be performed more efficiently. Increase in size and functional complexity of the brain accompanied these advances in human activities.

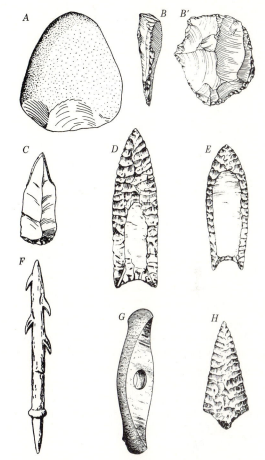

Figure 35-9 Some tools of prehistoric humanids; not to scale. *A.* Hand ax or chopper. *B, B′.* Scraper, edge and side (Levalloisian). *C.* End scraper (Aurignacian). *D.* Clovis projectile point. *E.* Folsom projectile point. *F.* Bone harpoon (Magdalenian). *G.* Neolithic polished stone ax head. *H.* Arrowhead (Gypsum Cave). *A, B, B′, C, F,* and *G* are Old World; *D, E,* and *H,* New World. (*A, B,* after R. J. Braidwood, 1961, *Prehistoric men,* 5th ed., *Chicago Natural History Museum; C, F,* after Miles Burkitt, 1963, *The Old Stone Age,* 4th ed., *Atheneum Publishers; D, E, H* after J. M. Wormington, 1957, *Ancient man in North America,* 4th ed., *Denver Museum of Natural History.*)

35-18 Fossil and subfossil record Primate beginnings date from the Paleocene, and the earliest small anthropoids (†*Parapithecus,* †*Propliopithecus,* †*Aegyptopithecus*) are represented by jaws and skull and limb bones from the Oligocene of Egypt. During Miocene time anthropoids in Africa and Europe developed considerable diversity, some (†*Proconsul*) leading toward the great apes and others having prehuman features in the face, jaws, teeth, and limbs. †*Dryopithecus,* the size of a chimpanzee, is close to the point of divergence. †*Ramapithecus,* a humanoid primate, may be on the direct line leading to the human species.

In dealing with early human ancestry, some anthropologists are prone to make a new genus and species for each new kind found. (A similar trend occurred early in systematic zoology.) Other investigators employ a broader grouping with fewer names, making it easier to comprehend relationships; this pattern is followed here. The place and time when the modern human, *Homo sapiens*, first appeared are controversial; the Neanderthals are estimated to have appeared about 100,000 years ago. Skeletons of modern types from late prehistoric time are fairly common from Western Europe and North Africa to Mesopotamia. Future discoveries will modify and improve our ideas of human evolution. The principal fossils now in the record (Fig. 35-10) are as follows:

Australopithecines These have been found in South Africa, equatorial eastern Africa, and Java (?). *Australopithecus africanus* (3 million to 1 million years ago) probably was 150 cm (5 ft) tall, weighed 27 to 32 kg (60 to 70 lb), was slender and lithe, and evolved rapidly into larger, increasingly humanoid forms. *A. africanus* used simple tools of stone and perhaps bone and wood. *A. robustus* (*Paranthropus*) (about 3 to $\frac{1}{2}$ million years ago) overlapped *A. africanus*; this form was not in the direct line of human evolution, probably was about 150 cm (5 ft) tall, weighed 59 to 68 kg (130 to 150 lb), was gorillalike, perhaps a knuckle walker and vegetarian, and probably used tools. The australopithecines changed little and may have been eliminated by *Homo erectus*. *Homo habilis* may have been a transitional form between *A. africanus* and *H. erectus*.

Homo erectus (about 1 million to 100,000 years ago). Most specimens are about $\frac{1}{2}$ million years old; *H. erectus* probably was ancestral to *H. sapiens*. Specimens have been found in Europe, southeastern Asia, and Africa. *H. erectus* was about 150 cm (5 ft) tall, slim-legged, a better walker than the australopithecines, with a larger brain of 775 to 1300 cm³ (*H. sapiens*, 1200 to 1500 cm³). *H. erectus* includes the Dubois Java man (*Pithecanthropus erectus*) and Black's Peking man (*P. pekinensis*). Their tools were more elaborate than those of the australopithecines and included hand axes; hearths and charred bones show the use of fire. There is evidence for cooperative hunting and stampeding of game. Probably they followed game and returned to settled sites. They hunted deer, horse, rhinoceros, and elephant; some crushed skulls suggest cannibalism.

Homo sapiens neanderthalensis. This is the Neanderthal, of the Late Pleistocene, before and during the last Ice Age, about 100,000 to 35,000 years ago, who disappeared abruptly from the fossil record. Specimens have been found from Turkestan and Iran to western Europe and North Africa and in southern Asia (Solo man) and southern Africa (Rhodesian man); this subspecies was well established in Europe 75,000 years ago. Males were about 150 to 170 cm (5 to 5$\frac{1}{2}$ ft) tall, females shorter; they were

Figure 35-10 Skulls of prehistoric humanid species and races. *Above. Australopithecus* and *Homo erectus* partly restored. *Below.* Successive forms of *H. sapiens.* (*After various authors.*)

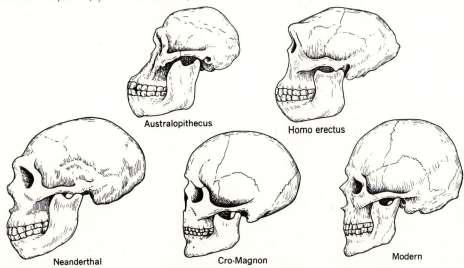

Australopithecus

Homo erectus

Neanderthal

Cro-Magnon

Modern

stocky, heavy-boned, and powerful, with short fingers and toes and a massive skull that was long, flat-topped, and beetle-browed. Their characteristics were extreme in western Europe toward the end of their recorded residency there. In the eastern part of their range they were taller, less robust, with a less rugged skull; the eastern form may have given rise to modern humans. They used tools extensively (Mousterian culture). They ate rodents, deer, horse, reindeer, chamois, and ibex; there is some evidence for cannibalism. They used fire, wore clothing of skins and lived in caves and in the open. They gave the first indications of social and religious awareness, with burial of the dead, often in sleeplike poses, with food and tools and decorated implements.

Homo sapiens sapiens. This is Cro-Magnon man (37,000 to 15,000 years ago). Specimens from Europe are considered to be clearly *sapiens.* Similar populations existed in Asia and Africa. Males were about 170 cm (5 ft 10 in) tall, with a strong modern face, no prominent brow ridges, large eyes, conspicuous chin, a nose with a high bridge, occiput rounded, and brain size modern. Their tools were advanced, including delicately flaked stone implements and carved bone. Their clothing was of skins. They used fire. Their aesthetic and religious senses were well developed; they made cave paintings, mostly of animals hunted, sequestered in places perhaps representing shrines, and sculpture (female figurines). There is evidence of totems, taboos, sorcery, and perhaps some ritualistic cannibalism. The cave bear apparently was revered. They hunted horse, reindeer, deer, ibex, bison, wild cattle, woolly rhinoceros, and mammoth and had the beginnings of settled life. All populations after this time are considered to be *H. sapiens sapiens.*

35-19 Modern humans After the Old Stone Age (Paleolithic) there was a brief stage termed Mesolithic with very small flints (Fig. 35-9), and then with the waning of the glacial cold came the New Stone Age (Neolithic), in which stone implements were refined and some were polished. In time people began to plant cereals and to domesticate livestock,

became food producers instead of food gatherers. They made pottery, learned to weave textiles, to construct dwellings, and to live in organized communities. In turn came the Ages of Copper, Bronze, and Iron. Written history begins at various times in different portions of the Middle East, Egypt, Europe, and Asia. Thenceforth the human species has progressed to the Age of Steel and to the metal alloys and plastics of the present day.

35-20 Humans in the New World The route and time of arrival of humans in the Western Hemisphere are unknown.

Anthropologists believe the most likely route was across the Bering Strait, where only 48 km (30 mi) of shallow waters separate Siberia from Alaska. Ice sometimes connects the two in winter, and a land bridge may have joined the two continents there during the last Ice Age. In that period there was a corridor free of glaciers through central Alaska and east of the northern Rockies, a route along which early hunting people could have traveled and found large game animals. A few cultural hints suggest more recent occasional migration from islands in the South Pacific to Central or South America.

The first inhabitants were long-headed (dolichocephalic), and people of this type have persisted to the present. Some later populations, including some modern American Indians, were round-headed (brachycephalic).

Evidences of early human activities in the New World are primarily finely shaped, lancelike projectile points made by pressure flaking. Clovis points (Fig. 35-9D), some associated with mammoth remains, have been found distributed across the United States and northern Mexico, indicating the existence of a hunting culture (the Llano) between 11,000 and 11,500 years ago. It is estimated to have spanned about 500 years. Mammoth hunting then gave way rapidly to the hunting of bison (*Bison antiquus*) and use of Folsom projectile points (Fig. 35-9E). This culture existed between 10,000 and 11,000 years ago, lasting about 1000 years. The cultural shift correlates with the rather sudden disappearance of mammoths. Fluted fishtail points of the older Llano culture have been traced into South America. Patagonian points are somewhat over 10,000 years old, the oldest reliably dated industry of projectile points in South America.

Although the paleohunters seem to have concentrated on certain large mammals such as mammoths and bison, they undoubtedly sought a variety of game, including mastodons, camels, horses, and other members of the Pleistocene megafauna. The large mammals disappeared about 8000 to 7000 years ago, leaving only the plains bison that survives today on reservations.

Extinction may have resulted from changes in climate, or ecologic conditions, or hunting pressure. Thereafter in the West deer, rabbits, and other small game together with seeds, nuts, and berries became the food of the natives. Cultivation of primitive maize (corn) began about 2500 B.C. in parts of Arizona and New Mexico, to which later beans, squash, and other crops were added, all perhaps from Middle America. In lesser degree and later, agriculture spread to eastern North America. In Middle and South America the beginnings of agriculture date from about the same time. In the eastern woodlands of the United States the natives had access to deer and small game, many kinds of fishes and shellfish, and various crops of seeds, berries, nuts, and acorns. Peoples living along the ocean coasts, east and west, subsisted on the rich marine fauna of fishes, seabirds, mammals, and shellfish, as evidenced by accumulated remains in many shell mounds. Natives of the Americas, when first studied by anthropologists, had become diversified into a wide variety of tribes differing in physical features, language, and cultural patterns.

Two large New World culture centers began in pre-Christian time, the Mayas in Central America and the Incas in the Andean highlands of South America. The Mayas were in the Yucatan Peninsula, Guatemala, and bordering areas. Their earliest settled communities, from possibly 1000 B.C. to about A.D. 200, contained walled and thatched houses; the people grew corn and probably other crops, made tools of obsidian, and produced baskets, mats, and textiles. Then from about 300 to 900 A.D. a high civilization flourished, with hieroglyphic writing, arithmetic with the concept of zero, an accurate calendar, sculpture, and ceremonial buildings of stone (a pyra-

mid at Tical was 70 m (230 ft) high). This civilization suddenly collapsed, to be followed by another, which lasted until the Spanish Conquest. These people had the domestic dog. Their crop plants, all from the New World, included corn, beans, and squash; manioc, or cassava; sweet potato, peppers and tomatoes, peanuts, cotton, and tobacco.

The Incas in Peru are known from archaeologic materials other than writing or a recorded calendar, which they lacked. Their beginnings (about 1200 B.C.) show corn and manioc as crops, ceramics, goldwork, weaving, and fairly elaborate stone buildings. They had domesticated the llama. They progressed through several stages to the Inca Empire (about 1438–1532), described in Spanish documents. It embraced an area about 2220 by 815 km (2000 by 500 mi), with narrow roads connecting towns that contained houses and temples of fitted stone construction. Social, political, and religious organization was complex.

35-21 Shelter   The anthropoid apes live in tropical forests and have little need of shelter; at most, they construct crude nests of leaves and branches in trees. Humans, long dwellers on the earth both in and beyond the tropics, require protection against weather and enemies. During the glacial period, early humans in Europe lived in the mouths of caves or under rock shelters. The Plains Indian hunters of North America, who followed buffalo and other game, made conical shelters (tepees) of animal skins supported on poles; nomad tribesmen with herds of sheep and goats in Cental Asia today use crude tents of skins, felt, or woven fabrics of animal fibers. The oldest fixed human shelters in Egypt were of wattle (woven sticks). Sun-dried (adobe) bricks later served as building material in the eastern Mediterranean countries and in pueblos of the American Southwest. Rough (unhewn) stones were first employed to form burial tombs in the Mediterranean region, in England, and in Scandinavia. Structures wholly of wood were built in forested regions, as by the *Danube people,* who moved up the river of that name about 2600 B.C.; such structures are made by Indians of the Pacific Northwest today. Humans have continued to use all these materials, improving their work with wood and stone by invention and development of iron and steel tools. To these have been added, in the past century, concrete (with reinforcement) and iron or steel for construction.

35-22 Food Humans were first **food gatherers,** either hunting wild animals for their flesh and marrow (using the hides for clothing) or gathering wild roots, seeds, and fruits, as did American Indians in recent centuries. Later they became **food producers,** when flesh-producing animals were domesticated and reared in captivity and seeds were planted and tended to produce more abundant harvests. In Arizona these two stages existed side by side up to the last century with the hunting Apaches and the agricultural Papagos. Conscious food production is an attribute solely of the later human populations. The human species and the cultivated cereals now exist symbiotically, as neither can survive without the other. The only instinctive food producers among animals are a few specialized fungus-growing ants and termites and some ants that colonize aphids (*ant cows*) on plants to afford honeydew for the ants.

Domestication of animals (Secs. 33-24, 34-29) to furnish meat for food, hides for clothing, and fibers for weaving and for use in draft and transport began long ago. Mesolithic people in Europe after the Glacial period had the first domestic dogs, 7500 B.C. or earlier, and American Indians all had dogs. Horses for transport and cattle for milk production are on record in Asia Minor about 3000 B.C. Sheep were herded in Persia by 6000 B.C. Asses were under domestication in predynastic Egypt. The gathering and planting of seeds of native wheat and barley began long ago, as did the use of scythes of wood with blades of flint. Storage of harvested grain in pits lined with mud and straw began in Egypt between 5000 and 4000 B.C. Of all living creatures, only humans use cooked food.

35-23 Implements Use of fire is associated with Peking man of the early Pleistocene; later its use spread to all peoples. The spear is known from Mesolithic time and probably was used much earlier by the Neanderthals. A progressive series in the development of tools (artifacts) can be traced from association with older human relics down to the present day (Fig. 35-9). At first, stones of convenient size and shape (eoliths) probably were selected as tools. These were followed by crudely chipped chopping tools such as were made by early humans and the australopithecines nearly $2\frac{1}{2}$ million years ago. Crude implements were formed from flint or other rocks with a glassy texture by striking off chips with another stone (percussion method of Paleolithic time, the Old Stone Age). The wedge-shaped fist or hand ax was one of the earliest tools thus produced. Later the material was shaped more accurately by the removal of small bits by pressure with a piece of bone, wood, or antler (pressure method of the later Paleolithic). Still later, such implements were smoothed by grinding on some abrasive stone (polished implements of the Neolithic or New Stone Age). Meanwhile, some prehistoric races made awls, needles, and fishhooks of bone, antler, or mollusk shell. Native copper was discovered and first forged into simple tools about 4000 B.C. Soon the art of smelting copper from ores was developed, and the practice spread widely; axes, chisels, other tools, and weapons were made by casting. Later small amounts of tin were added to the copper to make bronze, an alloy of lower melting point yet of superior hardness and better suited to making edge tools. In turn, there followed the recovery of iron by smelting (1000 B.C.) and much later the production of steel by adding a trace of carbon to the iron. The special alloys of steel and other metals now used are a further development of this ancient practice.

Other early inventions of fundamental significance in human social development were (1) drawing as first practiced on Paleolithic cave walls in France and Spain; (2) pictorial records (hieroglyphics) leading to phonetic signs and writing (Egypt, about 3500 B.C.); (3) baking of clay to make pottery; and (4) clay tablets and then papyrus sheets on which written records were preserved. In the realm of mechanics, the wheel for carts and much later for machines and the arch in stone construction were great advances, as were the boat and sail for water transport. Development of counting and computation systems led to means for calculating the movements of heavenly bodies, determination of the length of the year, the size of the earth, etc., and slowly paved the way for the scientific research and technical knowledge of the present day.

35-24 Energy supplies During the past 500 million years, a fraction of the organisms living on the earth have been entombed in sediments under conditions that preserved some of their energy content derived from solar radiation. Coal, oil, shale, petroleum, and natural gas are rich in such *fossil energy*.

Through its long early history, the human species existed almost entirely on energy from food, perhaps 2000 calories a day, with a little added heat later from wood fires. In the ecologic complex of nature humans were merely one animal species competing for the immediate products of solar energy. Civilization is essentially a means for using more energy, both current and fossil supplies. It has upset the ecologic balance to favor increase in numbers of the human species at the expense of the environment, and this process goes on at an ever-accelerated rate. Wind and water power are derived from replaceable resources, and wood for fuel may be replaceable when human populations are small. Beasts of burden are sustained by the recurrent supplies of solar energy in their plant foods, but with some eventual decline in fertility of crop lands. Modern civilization, however, with its enlarged human populations and complex activities, depends largely on "mining" the fossil energy from past geologic periods, a nonrenewable resource.

35-25 The future of the human species Against human ingenuity in devising machines to multiply the results of human labor must be set the effects of increasing the human population and of the consequent greater struggle for lands (territory) and the natural resources required in modern life. The human species has already modified the environment profoundly by agriculture, drainage, irrigation, deforestation, planting, and the construction of huge cities with manifold biologic requirements (Chap. 36). Unwise use of land has destroyed the productive capacity of many areas by loss of topsoil through erosion by water and wind. Commerce and agriculture have spread pests, parasites, and diseases to countries and continents where these were not native and where they levy on the human population, its domestic animals, and its agriculture. Efforts to alleviate sickness and suffering by applying the discoveries of modern medicine have offset some of the effects of natural selection that formerly eliminated less fit individuals from the population.

After the distant times when early humans lived much like any other omnivorous mammals, their descendants developed the fine arts, as demonstrated in the relics of pottery and buildings found in the Middle East, Egypt, and Middle America. Yet a gory interest in human and animal sacrifice extended throughout much of human development, as exemplified in the Roman and Aztec empires, and it is covered only by a thin veneer of civilization today. In the more advanced countries much of the population now shows a general respect for law and order, but when passions are aroused in cases of mob violence or when nations seek war as a means of settling differences, the basic barbarism flares anew. In the face of the current worldwide population explosion, it remains to be seen whether the conference table rather than a resort to primitive instincts will be employed to adjust disparities in allocation of land and food supplies and in the development of governments in newly established nations.

Classification

All living human beings belong to the one highly variable species *Homo sapiens*. They are divided into numerous races, all able to produce fertile offspring when intercrossed. A **race** is a stock of common ancestry and physical characteristics, whereas a **people** is the assemblage of individuals occupying a given area; the two terms are not synonymous. The culture and languages of a people are of sociologic interest but do not necessarily indicate its origin, as some habits and inventions are borrowed from other peoples and some are independently developed. The human races have probably always been variable and unclearly separated from one another, just as they are in other species of plants and animals. In the past few thousand years migrations, war, conquest, and slavery have probably made them even less distinct. Their number is uncertain, classification is difficult, and anthropologists disagree as to how they should be divided. Characters that have been used in classification are (1) hair form—straight, smooth, wavy, or woolly; (2) pigmentation—skin, hair, and eye color; (3) hair abundance and distribution; (4) form of head as shown by various measurements; (5) facial features—general shape, form of forehead, shape and proportions of nose, lips, jaw, and eyes; (6) stature, or average height of adult males; and (7) gene frequencies (blood groups, enzymes, etc.; see Sec. 6-8).

References

Barnett, L., and others. 1961. The epic of man. New York, Time. Inc. 307 pp., Many colored illus. *Popular, pictorial.*

Brues, A. M. 1977. People and races. New York, The Macmillan Company. vi + 336 pp., 60 figs.

Campbell, B. G. 1976. Humankind emerging. Boston, Little, Brown and Company. xii + 463 pp.

Clark, J. D. 1970. The prehistory of Africa. Southhampton, Thames & Hudson. 302 pp., 48 pls., 72 figs.

Clark, J. G. D. 1952. Prehistoric Europe: The economic basis. New York, Philosophical Library, Inc. xix + 349 pp., 16 pls., 180 figs. *Discusses hunting, fishing, food gathering, etc.*

_____ and **S. Piggot.** 1967. Prehistoric societies. New York, Alfred A. Knopf, Inc. 357 pp., 95 figs.

Clark, W. E. LeG. 1960. The antecedents of man and introduction to the evolution of the Primates. 2d ed. Edinburgh, Edinburgh University Press. vii + 374 pp., 152 figs.

_____. 1962. History of the Primates: An introduction to the study of fossil man. 8th ed. London, British Museum (Natural History). 5 + 119 pp., 40 figs.

Day, M. H. 1965. Guide to fossil man. Cleveland, The World Publishing Company. 289., 94 figs.

Dobzhansky, T. 1955. Evolution, genetics, and man. New York, John Wiley & Sons, Inc. 8 + 398 pp., illus. Science paper edition, 1963.

_____. 1962. Mankind evolving. New Haven, Conn., Yale University Press. ix + 381 pp., 10 figs.

Eckhardt, R. B. 1972. Population genetics and human origins. *Sci. Am.,* vol. 226, no. 1, pp. 94–103. *When genetic variability is considered, it appears impossible to decide which of the fossil apes is our hominid ancestor.*

Garn, S. M. 1971. Human races. 3d ed. Springfield, Ill., Charles C Thomas, Publisher. xiv + 196 pp.; 28 figs.

Graham, J. A. and **R. F. Heizer.** 1967. Man's antiquity in North America: Views and facts. *Quaternaria,* vol. 9, pp. 225–235.

Howell, F. C., and The Editors of Time-Life. 1965. Early man. New York, Time-Life Books. 200 pp., many illus. *Fossil hominids and their tools.*

Howells, W. W. 1960. The distribution of man. *Sci. Am.,* vol. 203, no. 3, pp. 112–127. *Humans originated in the Old World but have since become the most widely distributed of all animal species; they have differentiated into three principal strains.*

_____. 1967. Mankind in the making. (Rev. ed.) Garden City, New York, Doubleday & Company, Inc. 384 pp.

Hulse, F. S. 1971. The human species: An introduction to physical anthropology. 2d ed. New York, Random House, Inc. xvi + 524 pp.

Luckett, W. P., and **F. S. Szalay** (editors). 1975. Phylogeny of the primates: A multidisciplinary approach. New York, Plenum Press. xvi + 483 pp.

Macgowan, Kenneth, and **J. A. Hester, Jr.** 1962. Early man in the New World. Rev. ed. Garden City, N.Y. Doubleday & Company, Inc. xxiii + 330 pp., illus.

Scientific American. 1960. The human species. Vol. 103, no. 3, *Entire issue devoted to the human being.*

Simons, E. L. 1964. The early relatives of man. *Sci. Am.,* vol. 211, no. 1, pp. 50–62. *New evidence singles out the ape stock from which the human line arose.*

_____. 1967. The earliest apes. *Sci. Am.,* vol. 217, no. 6, pp. 28–35. *A precursor of apes and humans found in Egypt.*

_____. 1977. Ramapithicus. *Sci. Am.,* vol. 236, no. 5, pp. 28–35. *This Miocene primate is the earliest distinctively humanoid member of the human family tree.*

Stewart, T. D. 1973. The people of America. New York, Charles Scribner's Sons. x + 261 pp.

36
POPULATION AND ENVIRONMENT

The human population of the world was about $\frac{1}{2}$ billion early in the 1700s and only twice that number in the mid-1800s. By 1950 it had trebled, to about 3 billion, and now stands at over 4 billion. It grows at the rate of about 70 to 80 million yearly. Such enormous numbers far exceed the long-term carrying capacity of the world as a human habitat.[1] Current food production, even if expanded to capacity, is

[1] *Human populations, like populations of other organisms, have probably always tended to breed up to the carrying capacity of the environment (Sec. 12-14). In this sense high human population densities are not something new. Except for occasional periods of recovery from population setbacks, times of expansion into favorable new environments, and periods following the development of new techniques, people have been about as numerous as the technology and conditions of life at any given time permitted.*

insufficient to give all an adequate diet. The depletion of natural resources is accelerating, and the quality of the environment is declining.

Thomas R. Malthus (1766–1834) in his *Essay on the Principles of Population* (1798) stated that population increases in geometric ratio (Sec. 12-14) but food in arithmetic ratio, that the means of life deteriorate and become inadequate unless population increase is checked, as by famine, disease, or war. Thereafter, a few perceptive persons wrote of the evils of excess numbers and depletion, but they were largely ignored. Since World War II increasing numbers of observers have emphasized the crash course on which civilization is headed, until now there is rapidly mounting concern and ecology is a household word. How bad is the damage? Where is our economy headed? What changes and improvements are possible? The problems are not all new; some began long ago, but all now are pressing. This chapter traces the origins of some difficulties and mentions means toward alleviation.

36-1 Population   Unlike most animals, humans have no seasonal restraint in sexual activity or birth of young. One female may bear a dozen or more children, and a polygamous male can sire far more. The human reproductive potential is high, despite a 9-month pregnancy and successive generations at only 20 to 25 years. Throughout history, as food supplies were increased by improvements in agriculture, human populations grew correspondingly. More efficient farming practices, including mechanization, together with genetic improvement of crops, have resulted in a relatively enormous yield. A second factor for population increase in the past century has been health improvement by sanitary measures and medical care.

In the past, inherent human fertility was offset by deaths from famine, pestilence, and war. Crop failure followed by famine goes back to the beginnings of agriculture. Plague, the Black Death, killed 25 million people in the fourteenth century and lesser numbers in later years and still occurs. Today there is greater crop production, food may be shipped to an area threatened by famine, and many ailments have been conquered, but there are already too many people to be adequately supported by the finite resources of the earth. Projected increases will worsen the situation. All the problems discussed in this chapter are compounded by population growth.

The only rational means to lessen population pressure is to reduce births. Contraceptives, medicinal or mechanical, are useful for this purpose. No known method is infallible. The ideal contraceptive would be easily administered, reliable, long-lasting but reversible when desired, without undesirable side effects, and inexpensive. It has yet to be developed.

36-2 Food   About half the world population is in Asia, Africa, and Latin America, in countries with scant industry, mostly subsistence agriculture, low per capita income, many illiterate people, and a high birth rate. About one-fifth of the people of the world are undernourished, and three-fifths suffer protein shortage. Kwashiorkor, a caloric and protein deficiency disease in children, is now the world's major health problem. Throughout the world some 10,000 deaths occur daily from malnutrition. The grain-rich nations (United States, Canada, Australia, Argentina) have sold or given surplus cereals to needy countries, but this cannot continue. Within a decade or two the United States may have no wheat for export. World population is expected to exceed 6 billion by the year 2000. To feed people at that time even at present inadequate levels would require a 40 to 50 percent increase in food production; yet the expected increase in arable land is estimated to be 5 percent. Particularly hazardous in the mechanized West is the dependence of agriculture on a finite and diminishing resource with which to plant and harvest—petroleum. More famines are likely.

World food supply may be somewhat expanded by selecting improved crop plants with higher yields and enlarging the acreage of land under cultivation. Unfortunately, the effects of even a moderately larger supply will quickly be offset by the multitude of new mouths being added by the population explosion. Other means have been suggested for meeting the food crisis, but some are untested and others require much new capital and time not available to effect a solution. These include greater use of fertilizers, growing algae on sewage, culturing microorganisms, deriving synthetic products from petroleum, more widespread crop irrigation from desalinated seawater, clearing tropical forests for farms, and seeking greater yields from the oceans. The last two will be discussed here.

The vast oceans (71 percent of the world area) cannot, as once was believed, yield an inexhaustible supply of human food. Indeed, marine fisheries now are close to maximum sustained yield. Open, or oceanic, waters, 90 percent of the total, have an estimated gross annual production of only 1.6 million tons of fish. Coastal waters along the continental shelf (area 9.9 percent) yield 120 million tons. An equal amount derives from small local upwelling areas (only 0.1 percent), very rich in nutrients, along some western coasts (California, Peru, West Africa). Of the total 241.6 million, about half goes to natural mortality—other fishes and marine birds and mammals. Humans now harvest about 75 million tons annually, and 100 million tons is probably the safe limit for continuing yield. Some Atlantic coastal areas already are overfished. The whaling industry may soon end because of overkill. Pollution is also a problem both in the open ocean and close to shore. It has apparently already destroyed the California mackerel fishery and perhaps is threatening the sardine. Marine life in the Baltic Sea is endangered because contaminants have lowered the oxygen content to nearly the minimum required to sustain life. Salt marshes, mud flats, and below-tide shallows are important spawning grounds. Because they are close to sources of contamination and subject to landfills, they are particularly susceptible to damage. Their deterioration can affect marine life well beyond their borders. "Farming" the sea, even coastal waters, is not yet economical on a large scale. Estuarine fishponds along Asiatic coasts provide some animal protein for local residents, but the world total is not great. Straining out marine plankton for human food has been suggested, but this would require great amounts of energy, and the concentrate might be unpalatable; its removal could cause damage to marine ecosystems. The oceans cannot relieve all hunger on the land.

Some persons think that the tropics hold the answer for greatly expanded food production. Tropical forest is a many-layered canopy of evergreen growth, often dense, subjected to heavy rainfall and high, almost uniform, temperatures. Leaf fall is continuous but often slow, and the organic litter on the forest floor oxidizes rapidly. Common tropical soils, called laterite, or latisols, are reddish; they contain a porous clay high in aluminum and iron hydroxides with varying amounts of sand or silt. They are usually low in humus and nutrients and are easily leached by heavy rains. When undisturbed, the forest is a protective cover where nutrients cycle rapidly from plant to soil and back.

When large areas are cleared, the soil overheats and leaching accelerates. Nutrients are lost quickly. Oxides of iron and aluminum increase, and if exposure is prolonged, some soils harden, a process called **laterization.** When this is severe, a slaglike or bricklike pavement forms and there is permanent damage.

Native tropical peoples clear a small area and grow crops for 1 to 3 years, then move to another site. The cleared area quickly reverts to forest, soil fertility returns in time, and after some years the place can be reused; thus laterization is avoided. Now, however, it is becoming increasingly difficult, because of population growth, to allow the land to lie fallow.

Modern land-clearing machinery can remove large areas of jungle for farming, but where this has been tried, the results have often been disastrous because of soil deterioration. Chemical clearing with defoliants (2,4,5-T, etc.), as done in Vietnam, has essentially sterilized some sites for 6 years. In some areas tree crops that simulate the natural forest cover probably are the best to grow, but human carrying capacity is low. Research in tropical agriculture for food production is greatly needed.

36-3 Debris   Few wild animals accumulate enough food discards to mark their dens for long—but humans often "foul their lair." Indeed, our knowledge of prehistoric humans derives from relics at ancient homesites: skeletal parts of food animals, tools made from stone, and occasional human burials. The kitchen midden, which was a combined living site, garbage dump, and cemetery, is conspicuous evidence of early waste accumulation. Ancient Rome, from 264 to 69 B.C., had a million or so inhabitants, probably fed wet garbage to hogs and fowls, and discarded innumerable oxcarts of broken ceramics, bricks, and scraps of stone. For twenty centuries all other communities discarded refuse in increasing amounts; lately many new materials have been added. In the industrialized world debris is now prodigious. A United States garbage dump with discarded refrigerators, TV sets, scrap metals, and other salvageable materials would not be tolerated in a country less prodigal with its wastes.

36-4 Soil deterioration and loss   Throughout geologic time, rock masses have been elevated, then eroded by wind, water, and other agencies into soil that later was washed down streams to the sea. When humans began to clear and cultivate, the fertile topsoil that had been formed by weathering and the action of organisms for millenia was washed away at an increasing rate. Removing forests, burning off brush cover, plowing up the grass on natural range areas, and overgrazing aggravated the situation, with subsequent decline in fertility. In other areas, heavily irrigated to produce crops and not properly drained, underground accumulations of alkali were brought to the surface, degrading or ruining productivity. Corrective chemical treatments are costly and not always successful. Heavy use of inorganic fertilizers may lead to compaction of some soils in default of efforts to maintain humus and restore fertility with manure and other organic materials. Chemical control of insects or fungus pests of crops may permanently damage the topsoil. Repeated application of arsenicals to apple trees for codling moth control in eastern Washington so loaded the soil with those chemicals that replacement trees failed to grow. With the population explosion, much land now is being permanently covered by urban and suburban housing, paved streets, and great highways. In California an estimated 50,000 to 150,000 acres of land has been lost thus during the past quarter century. Lesser expansions in many other states and developed countries have had similar results.

In affluent countries, especially those with extensive public lands such as the United States, Canada, and Australia, a new threat to the soil has emerged—recreational off-road driving. There are now over 12 million off-road vehicles (ORVs)—motorcycles, minibikes, four-wheel-drive vehicles, etc.—operating in the United States alone. In less than a decade, ORV recreation has become a substantial political and economic force. Since a single motorcycle driven off-road for 20 miles impacts an acre of land and an average four-wheel-drive vehicle does the same in 6 miles of travel, the total effect of such large numbers of machines used repeatedly is great. So far ORV damage is greatest in arid parts of the Southwest, on desert lands with fragile ecology, slow to recover from mechanical disturbances. However, there are growing signs of significant and lasting damage in other environments, from forest trails and meadows to beach dunes, grasslands, and chaparral. ORVs greatly accelerate erosion, particularly on slopes, and they cause degradation of stabilized fertile soils in catchment areas by burial with less fertile upland materials. Acceleration of soil erosion is caused by loss of protective plant cover and the destruction of chemical and biological crusts (lichens, fungi, etc.) that seal the surface against the action of wind and water and by mechanical dislodgment of the soil, especially on slopes, from the shearing action of the vehicle wheels. On some soils ORVs cause severe compaction, which interferes with plant root growth, burrow construction by animals, penetration of water, and stability of the soil temperature. In addition to soil damage, in areas of persistent use a marked decline in plant and animal species diversity and population densities has been recorded (Fig. 36-1). The disturbances created by ORVs favor the intrusion of weeds. Russian thistle is spreading in the California desert from focal points of off-road driving and has been found growing even in single motorcycle tracks. This aggressive plant has already caused marked deterioration and simplification of the natural desert ecosystem of the Coachella Valley of California, where it has evidently spread from areas where the land was cleared for agriculture, housing, and roads.

36-5 Water pollution Most cities throughout the world use the nearest bay, lake, or river as the civic outlet, and many still release raw sewage. Such sewage carries the threat of typhoid and hepatitis infections. In some cases the same waters may be used as the water supply. A water molecule in the Mississippi River may make many diversions en route: filter plant—human use—sewage plant—discharge into the river. In the Orient, "night soil" (feces) serves to fertilize vegetable gardens and fish ponds, and contaminants may reach water supplies. Where untreated urine and feces contaminate waters, internal parasites spread readily.

Early American factories were built beside streams for power from waterwheels. Production wastes were dumped into the water from sawmills, paper mills, metal industries, and others. Conservationists have since forced cleanup of some pollutions, but others persist.

Figure 36-1 Environmental deterioration at Dove Springs in the Mojave Desert, California, caused by off-road-vehicle recreation. The nearly complete loss of plant cover occurred in a period of four years and has been accompanied by a marked decline in the fauna. Chances for recovery are poor. Photographs taken in 1968 (top), 1970 (bottom), and 1972 (p. 831). (*Courtesy of the U.S. Bureau of Land Management in California.*)

Pesticides are leached from the land and carried into streams; these go to the oceans, where they are now widespread, affecting various kinds of animals. They also adsorb to airborne particles during spraying and fall out from the atmosphere. Plastic objects discarded inland or elsewhere accumulate on the bottom in coastal waters and do not disintegrate. Petroleum is spilled in ever-increasing amounts by damaged or sinking tanker ships and from oil wells on the continental shelf. Crude oil ruins many beaches, fouls small craft, and coats and kills innumerable seabirds, mammals, and other marine life. Residual oils may float as amorphous soft masses or sink as a viscous blanket covering the bottom and destroying some of the benthos organisms.

Many fresh waters undergo artificial enrichment, or eutrophication (Gr. *eu*, well + *trophos,* to feed). Raw or treated sewage, runoff from farms, and the wastes of industry yield an oversupply of inorganic nitrates (a common source is farm fertilizers), phosphates (mainly from detergents), and other compounds. Growth of certain aquatic plants is stimulated, and "blooms" of algae are common. When the plants die, decomposition robs the water of oxygen, while the nitrates and phosphates go into the bottom muck, later to promote further growth. The aquatic community is markedly changed. Fishes and other organisms with high oxygen needs die and those requiring less oxygen increase. Lake Erie (9600 mi²) is almost dead biologically from excessive pollution. Even a clear mountain lake, if oversupplied with nutrients, may become choked with vegetation. At present rates of pollution, oxygen supplies in most major rivers in the United States may near exhaustion within a few decades. Eutrophication taints drinking water with unpleasant odors and taste and affects the composition of aquatic life and the human food supply.

Much improvement is possible. Lake Washington at Seattle became polluted and turbid, unattractive, and unsuited for swimmers. Residents of bordering communities decided cleanup was imperative. They voted sewage bonds, modern treatment plants were built, and the waters again are clear and safe.

Some sewage plants compress, dry, and sterilize organic residues as fertilizer to replace chemical types. Laws against dumping garbage or factory wastes into any waters, when adequately enforced, can do much to restore our lakes, streams, and ocean shores. Strict regulation of petroleum traffic on the oceans and inland waters could greatly reduce the damage.

36-6 Air pollution Before humans appeared, air over the earth received ash, smoke, and gases from volcanoes and from lightning fires in grasslands and forests. Prehistoric peoples burned wood for fuel, and some set fire to brushy or grassy areas to aid food production and gathering, a process that continues today in many parts of the world. All civilizations until the Renaissance used wood for cooking and heating with scant effect on the air or humans. When soft coal, with high sulfur content, became a major fuel, soot darkened the skies and buildings (also trees, Sec. 13-24), especially during the Industrial Revolution of Europe and eastern North America. Relief came in regions where soft coal was outlawed as fuel and replaced by gas. Petroleum fuels also reduced smoke but released new invisible wastes. As motor vehicles and later airplanes increased in the middle decades of the twentieth century, their exhausts added enormously to the contamination. At the same time industrial output, burgeoning in amount and variety—including plastics and other entirely new products—has added quantities of aerial wastes of unknown potential. The end result is smog (literally, smoke plus fog), a visible mixture of fine particles with various invisible chemical additives, which now blankets many cities of the world and is spreading. It darkens the sky, irritates the eyes and nose, and produces respiratory troubles.

Fires of wood or trash yield carbon particles (C) as smoke or ash, carbon dioxide (CO_2), and carbon monoxide (CO). Burning soft coal yields much soot (C), the gases just named, and sulfur dioxide (SO_2). Under most air conditions sulfur dioxide produces sulfuric acid (H_2SO_4), which attacks stone buildings and metal. Motor vehicle and airplane exhausts contain oxides of nitrogen (e.g., NO_2) and lead derived from leaded gasolines. Pesticides when sprayed become airborne as they adsorb to particles in the atmosphere. Polychlorinated biphenyls (PCBs), related to persistent pesticides, enter the air as industrial wastes, by the burning of plastics, and by wear of automobile tires. Hydrocarbons and other organic residues also are common in city smog.

Wood smoke, at levels formerly experienced, was probably merely offensive, but modern air pollutants seriously affect humans. Some corrode paints, metals, and clothing. Others jeopardize the health of humans, animals, and plants. Ponderosa pine forests in mountains of southern California, 60 miles from Los Angeles, are dying from effects of smog. Many crops fail to grow and produce in smoggy air; worldwide losses of $500 million annually are attributed to this cause. The long-term effects of many pollutants are unknown, but statistics suggest they are important in heightening the effects of certain diseases. Carbon monoxide combines readily with hemoglobin in the blood, reducing its oxygen-carrying capacity, thereby increasing the work load of the heart. Joined with respiratory difficulties, it may increase heart failure in urban areas. Sulfur dioxide is a strong irritant for respiratory tissues. Nitrogen oxides have effects similar to carbon monoxide. PCBs, like the persistent pesticides, now occur in the milk of nursing mothers and are implicated in reproductive failure in birds. The health of entire human populations may be declining. Daily irritation of the eyes and nose is common. Difficulties are pronounced in children, the aged, and the chronically ill. Respiratory ailments such as asthma, bronchitis, and emphysema may be induced or severely aggravated. Persons with these ailments often receive medical advice to move to less contaminated environments. Some pollutants are cancer-producing in laboratory animals.

There are several steps to lessen air pollution. Prohibition of trash burning except in high-temperature incinerators will reduce much particulate ash production. Internal combustion engines, piston or jet, must be redesigned to yield fewer contaminants. Fuel-injection engine systems can reduce undesirable contaminants. Electric power for automobiles is possible and has fewer undesirable by-products. Prohibition of use of high-sulfur fuels and changing to wind, geothermal, and solar power for electric generation would reduce pollutants. The current trend favors atomic power, but present nuclear technology risks increases in exposure to high-energy radiation and presents severe problems of atomic waste disposal and containment of use for safe and peaceful purposes. Citizens' groups can pressure government to insist that nuclear technology have adequate safeguards before further expansion and that other industries and refineries redesign their operations so that harmful wastes are recycled or reduced to nonobjectionable end products. Furthermore, efforts can be made by everyone to reduce unnecessary and wasteful burning of fossil fuels.

36-7 Climatic effects   Carbon dioxide and fine particulate matter are increasing in the atmosphere. An 18 percent increase in carbon dioxide from the burning of fossil fuels and the destruction of forests is projected for the year 2000. An effect on world temperature is probable through a shift in the balance between atmospheric absorption and reflection of sunlight. The direction of the shift, however, cannot be predicted. Of the radiant energy from the sun, much (60 percent) is reflected back by clouds. That which reaches and warms the earth is mostly of short wavelengths—visible and near ultraviolet—but that reradiated as the surface warms is of longer infrared rays. The latter tend to be trapped in the atmosphere by carbon dioxide and water vapor, which absorb outbound infrared and reradiate some of it back toward the surface. This is known as the **greenhouse effect.** In a greenhouse, the glass covering is transparent to short wavelengths but absorbs and reradiates the longer ones, so that the interior becomes hotter than outside air. By-products from burning fossil fuels could add to the greenhouse effect, whereas jet trails in the stratosphere might increase cloud cover, intercepting more solar energy. Atmospheric reflectance is augmented by that from the earth's surface, which increases with spread of deserts and deforestation. There is a further factor that may affect world climate—atmospheric accumulation of waste heat from energy used by humans. Already, weather anomalies occur in "heat islands" around large cities. A thermal shift in atmospheric temperature of only a few degrees could change con-

vection currents and affect climatic zones and the polar caps. There would be a marked effect on the earth's biota. Despite the uncertainties surrounding unintentional human effects on the climate, millions of dollars are being spent on intentional weather modification with little effort to determine possible adverse effects.

36-8 Pesticides   Virtually every wild plant species in nature is fed on by one or more kinds of insects and subject to attacks by bacterial, fungus, and virus diseases. Damage is usually inconspicuous because the plants are scattered and adapted to withstand attack, and a great variety of other organisms help keep the pests in check. In farming, thousands or millions of one crop species are planted close together. They are often specialized for a narrow set of conditions and little protected by other organisms. Under these conditions pest insects or plant pathogens can multiply and effect much economic damage—even total crop destruction. Pest control is essential for successful production. The ideal pesticide would kill only the attackers, leave no toxic residues on the crop, and be biologically degradable (reduced to harmless residues); until recently no such material was available. Now pesticides made of insect hormones give promise that this ideal may be achieved. Before the 1940s, pyrethrum, rotenone, and nicotine, all derived from plants, found limited use in home flower gardens. They all degraded quickly. Farm crop protection relied on several inorganic salts of arsenic, copper, lead, and zinc plus sulfur, lime-sulfur mixtures, and some oil sprays.

After World War II came the chlorinated hydrocarbon insecticides (DDT, dieldrin, aldrin, lindane, chlordane, etc.) and herbicides (2,4-D, 2,4,5-T). DDT deloused huge human populations to control typhus in war-torn Mediterranean countries and markedly checked malaria by killing mosquitos in many tropical countries and in parts of the United States. It virtually rid dairy barns of flies (until DDT-resistant flies developed). Soon the chlorinated hydrocarbons became the dominant pesticides for farms and home gardens. These synthetic chemicals were produced in enormous quantities and as dust or sprays were spread widely, often overabundantly, by airplanes.

Unfortunately, these insecticides caused destruction of beneficial insects and other animal life that went unnoticed or was disregarded. Elm trees in North Central and Eastern states were drenched annually with enormous amounts of DDT in efforts to control the beetle carrying the virus of Dutch elm disease. Losses in songbirds became evident as insect-feeding tree foragers became doused with the poison and died. Earthworms ingested fallen leaves covered with DDT spray and concentrated the poison in their bodies. When robins ate the worms, they developed DDT tremors and died. Amphibians and small fishes perished in streams and ponds showered with this chemical.

Evidence to date indicates that chlorinated hydrocarbons degrade very slowly. Of the millions of pounds sprayed or dusted on land, much has since washed into lakes and streams and into the oceans. Airborne during spraying, these pesticides adsorb to dust particles in the air and are carried by winds to settle out or be rained down many miles from their source. DDT sprayed over Texas cotton fields has been traced in Texas dust to the soils of Ohio. DDT is now widespread in the world's oceans. Experiments have shown that marine phytoplankton (algae, diatoms, etc.) are adversely affected by a few parts per billion. Because of the persistent character of the poison, it accumulates in the higher levels of the food chain, in the carnivorous animals.

Evidence accumulated since 1967 shows an increasing number of bird species suffering nesting failures because of eggshell thinning and breakage and mortality of chicks. There have been severe population declines in the peregrine falcon, bald eagle, and brown pelican, the first two of which are threatened with possible extinction. Eggshell thinning and breakage are associated with high residues of DDE (DDT derivative) and other chlorinated compounds in eggs and adults. Thin-shelled eggs have been produced experimentally in captive sparrow hawks and mallards by feeding them with sublethal amounts of DDT or DDE. Thin-shelled eggs have been found in the nests of at least 12 additional species of raptors and fish-eating birds high in the food chain, including the osprey, white-faced ibis, and several species of herons and egrets. Increasing reproductive failures can be expected in these groups of predaceous birds unless measures are sustained to prevent environmental contamination by the pesticides responsible. As animals high in the food chain, humans accumulate DDT, but its effects on humans are unclear. In people sampled in 1964, the concentration of DDT and toxic breakdown products in body fat was found to be 26 parts per million (ppm) in India, 19 ppm in Israel, and 7.6 ppm in the United States.

36-9 Heavy-metal poisoning The heavy metals, mercury and lead, are increasing as environmental pollutants and are undergoing bioconcentration in food chains. Mercury is discharged into lakes and streams by chemical plants and industries that produce chlorine, paper, electrical appliances, batteries, etc. It was thought to settle harmlessly into bottom mud and to remain inert. Recent studies show that actually anaerobic bacteria in mud convert it to volatile form, dimethyl mercury, which enters water and the food chain or is directly absorbed through the gills of fish. Interstate transport of some fish shipments has been banned because of high concentrations of mercury in fish destined for human consumption. Mercury has been implicated in fish die-offs and the death of bald eagles. In humans, inorganic mercury causes muscular tremors, depression, and kidney damage; methyl mercury damages the brain.

Sources of lead pollution include burning of antiknock gasolines, smelting plants, pesticides, soldered cans, and weathering of lead-base paints. In recent decades the automobile has been a major contributor. Lead is contaminating air, water, and food. Exposure through the digestive tract and lungs is chronic and most severe in urban areas. Lead is a cumulative poison and can cause debility, tissue lesions, and death. Snow taken at successively deeper levels in the Greenland ice cap indicates exponential increase in atmospheric lead from the mid-1700s—a reflection of increasing industrial use and release from automobile exhausts.

36-10 Depletion of resources Sparse aboriginal populations scarcely affected the environment or its resources. A minimum of wood served as fuel; native plants afforded food, fiber, materials for weapons or other needs; and a moderate levy on local animals provided meat and pelts. All these supplies were renewable. The rise of modern civilization has resulted in an ever-increasing use of nonrenewable resources—supplies of metals and of the fossil fuels coal, natural gas, and petroleum. At the present rate of mining and use, and given present price structures, it has been predicted that supplies of copper, lead, zinc, tin, silver, tungsten, and crude oil will be exhausted by A.D. 2000, some before. Stocks of coal, lignite, good iron ore, and chromium will last longer. Important recent discoveries of oil on the Alaskan north slope and elsewhere and extensive iron deposits in western Australia will help. Oil and copper are especially important. Oil powers the agriculture of industrialized societies and makes possible the mass food production required to maintain present population densities. Copper is important to electrification. If metal supplies are depleted, many amenities of life in the Western world may disappear—household appliances, electrical equipment, and others. Means to conserve and recycle the critical metals are imperative. No longer can they go to the garbage dump. Most undeveloped nations hope to emulate the material standards of the industrialized ones. If their desires do not change and the industrialized world continues its present course, vast quantities of common rock may have to be processed to obtain many of the necessary minerals. This cannot be done without greatly increasing air and water pollution.

The situation on energy supplies is equally unbalanced and dismal. Wood, coal, petroleum, natural gas, and atomic, geothermal, tidal, and solar energy are the sources. Only the first and the last three are renewable. The fossil fuels are finite in amount. Crude oil and gasoline may be in short supply within a decade or two. Atomic energy depends on the mineral uranium and on plutonium produced by atomic reactors. In producing energy, a nuclear reactor releases enormous amounts of heat. This is removed by circulating water and discharged into a stream, lake, or ocean margin. The local aquatic life may be altered, damaged, or killed by the extra heat. There are also problems of disposing of radioactive wastes, which must be sealed off from the environment for

many years, and of preventing release of damaging amounts of radiation caused by reactor malfunction. Like many other environmental pollutants, radioactive substances undergo bioconcentration in food chains. Because of problems of pollution, atomic energy may have only a minor role in world needs. The extent to which it can be employed on farms as a replacement for petroleum is problematic. Harnessing solar energy is a most promising possibility but as yet has not been achieved on a large scale.

In less industrialized lands cooking is simple, mostly with wood (cow dung in southern Asia), and little fuel is used for heating. Elsewhere the amount burned somewhat reflects the economy of a nation. Those low in resources are sparing of cooking fuel. The larger cities and towns use vastly more per capita for cooking, heating, and lighting. The extreme is in the United States, where 5 percent of the world's population expends 30 percent of the energy supply. This goes for overwarmed houses in winter, air conditioning in summer, much illumination both indoors and out, and heavy fuel demands for transportation and recreation. The long lines of automobiles with one person per car going to and from work may soon be a part of history. A generation ago electric streetcars, urban and suburban, carried commuting workers with far less expenditure of energy and little atmospheric contamination. Electric rapid transit can do much to reduce energy demands and clear the air in areas where water power can be used to generate electricity.

36-11 Changes in wildlife   As human population increases and technological developments attempt to keep pace, wild plants and animals and their habitats disappear. Forests and woodlands are removed, marshes are drained and filled, rivers are blocked by dams, and great areas are plowed and planted to produce cereals and other crops. Other areas are used to produce domestic livestock in quantity. Each alteration changes the conditions of various native species for successful life. Extinction of bird and mammal species has closely followed the curve of human population growth over the last three centuries. Many other species have also disappeared.

Extinction records only obvious damage. There is

great piecemeal reduction in species populations that perturbs the function of ecosystems and deprives us of sources of knowledge and enjoyment. Often destruction is unwitting, and it may have far-reaching effects. A seemingly insignificant organism may be highly important. For example, for a thousand years people have destroyed the nests of ants (species of *Formica*) in the forests of Europe to obtain ant pupae to feed chickens, pet birds, and fishes and adult ants for formic acid to treat rheumatism and arthritis. A decline in the vigor of forest trees became evident. Growth of the trees was less exuberant and the attacks of pests more severe. Ants aid the forests in many ways. Their burrows open the soil to air and water. Excavated soil buries leaves and speeds formation of humus, and wastes and remains of ants and insects they capture add nitrogen. Tree growth is stimulated near nests. The ants transport seeds and aid the dispersal of forest trees. When caterpillars increase, they help destroy them. European forests have thrived on soil capital contributed to by ants over thousands of years. Forests are vital to the economy of Europe, and the ant is an important part of the forest ecosystem upon which humans depend.

The trend throughout evolution has been to increase organic diversity and to produce complex, integrated, stable ecosystems. We humans are reversing this trend. We have steadily diverted increasing amounts of energy into our agrarian-industrial system. The result has been reduction in organic variety and loss of stability of ecosystems. Smaller amounts of energy fixed by ecosystems are now readily available, and our waste products increasingly interfere with our efforts to divert energy to ourselves. Establishing wilderness areas and nature reserves is an attempt to protect natural environments for their scientific, economic, and aesthetic values. This practice is widespread in the United States, but many countries, long overpopulated, have no wild lands to conserve. In some few regions the original conditions still exist; in others, partially altered, the primeval can be visualized; but in many places, long inhabited and modified by man, no inference is possible.

How important are natural surroundings and solitude? Urbanization is only a few thousand years old; the giant city (megalopolis), only a few decades; but the life of the human being as a hunter and food gatherer, a member of a small band, goes back several million years. Evolution has prepared humans for acts of great exertion and the cooperation necessary to kill dangerous animals. Our ancestors pursued the Pleistocene megafauna for thousands of years. What are the effects of crowding, impersonality, lack of identity, and physical restriction that are the lot of many in large cities? Crowding brings frequent contact. Emotions are aroused but often suppressed and seldom followed by action. A wise course would avoid the megalopolis and would keep population densities low to provide natural surroundings in and near cities and rapid public transit to them.

The conservation of nature and our own self-protection require that we go beyond merely setting up reserves to save remnants of natural systems. We must greatly reduce our impact on natural environments everywhere. The human ecosystem must achieve a steady state, with a balance between energy input and outflow, based on the use of renewable resources. For the system to be long-lasting, it will probably have to exist at a much lower level of energy flux than at present. To achieve this, world population must be greatly reduced and cultural patterns significantly altered. The prevalent view that equates growth with progress will have to be changed.

36-12 Limits to growth On a finite earth, no species can expand its population indefinitely. In nature, as a species' numbers increase, environmental resistance (Sec. 12-14) mounts and population growth finally stops; the population then achieves a sustainable level. An equilibrium tends to be maintained between population size and available environmental resources. Humans are subject to many of the same controls, although our present success in supporting population growth through technological advances (discovery and invention) seems to suggest otherwise. Actually, in the long term, we have only temporarily delayed the action of balancing mortality. Many of our socioeconomic and environmental problems (most of them chronic) appear to be related to the birth-death imbalance.

Much of the expansion of human technology has been fostered by population growth, although this has often not been recognized. Technology has repeatedly blunted the action of the natural checks and has often created temporary material abundance, only to be surpassed again by population increase. Success in maintaining a technological lead over population has been especially notable in industrialized societies. It is therefore not surprising that the Western world embraces a "growth ethic" which regards technological advance, increasing numbers of consumers, and growth of the gross national product (GNP) as measures of economic health. However, at present population levels and rates of growth even the most sophisticated technology cannot keep ahead indefinitely. There are limits to critical earth resources. It is now evident that a technological lead cannot be maintained much longer even in the most industrialized societies (Fig. 36-2). Furthermore, technology has failed to solve many long-term human problems and it has added new ones. All soci-

eties today, in varying degree, are experiencing poverty, environmental degradation, unemployment, alienation of youth, and inflation and other economic disruptions. Although the problems are readily perceived, their origins and interrelationships are not, and effective solutions have not been devised.

In 1968, The Club of Rome, an international group of scientists, educators, industrialists, and national leaders and civil servants, met to consider the "predicament of mankind." The group felt that human problems had to be addressed as a whole rather than piecemeal. The meeting led to a study of five interacting factors identified as limiting human expansion on earth: population growth, agricultural production, natural resources, industrial production, and pollution. An MIT team, using computer-simulated changing patterns of growth and interaction among these factors, reached the following conclusions (D. H. Meadows et al., 1972, *The Limits to Growth*): "1. If the present growth trends in world

Figure 36-2 The behavior of five factors considered of major importance in governing the limits of human expansion on earth are shown. The "standard run" (*A*) is a projection of current trends; the "stabilized" world model (*B*) shows the expected effect of adjustments to achieve a steady state. The curves in *A*. include projections of historical values from 1900 to 1970 and assume no major change in the physical, social, or economic relationships that have produced the present world system. Population and industrial output grow exponentially until declining resources halt industrial growth. Population and pollution continue to increase for a time, but population growth is finally stopped by a rise in the death rate due to decreased food and medical services. The stabilized world model *B* is achieved by holding population constant (equating birth and death rates), restricting capital growth (requiring that capital investment equal depreciation), and instituting technological policies of resource cycling, pollution control, increased lifetime of all forms of capital, and methods to restore eroded and infertile soils. Emphasis is on food and services rather than industrial production. (*From Donella H. Meadows, Dennis L. Meadows, Jørgen Randers, and William W. Behrens III, The Limits to Growth: A report for The Club of Rome's Project on the Predicament of Mankind, 2d ed. A Potomac Associates book, New York, Universe Books, 1974.*)

A. WORLD MODEL-STANDARD RUN

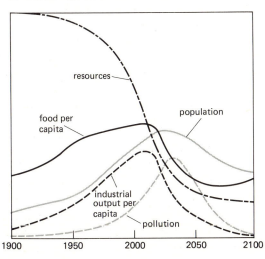

B. STABILIZED WORLD MODEL

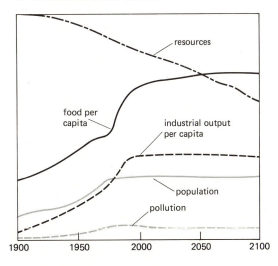

population, industrialization, pollution, food production, and resource depletion continue unchanged, the limits to growth on this planet will be reached sometime within the next one hundred years. The most probable result will be a rather sudden and uncontrollable decline in both population and industrial capacity. 2. It is possible to alter these growth trends and to establish a condition of ecological and economic stability that is sustainable far into the future. The state of global equilibrium could be designed so that the basic material needs of each person on earth are satisfied and each person has an equal opportunity to realize his individual human potential. 3. If the world's people decide to strive for this second outcome rather than the first, the sooner they begin working to attain it, the greater will be their chances of success."

These findings deserve the most serious attention of the world's leadership and all persons concerned with the present and future well-being of humanity. They argue strongly and urgently for a transition from growth to global equilibrium, a human steady state, in which all the above factors are maintained in equilibrium. Such a state does not portend stagnation or the end of progress. As Meadows et al. (1972) point out, "Population and capital are the only quantities that need be constant in the equilibrium state. Any human activity that does not require a large flow of irreplaceable resources or produce severe environmental degradation might continue to grow indefinitely. In particular, those pursuits that many people would list as the most desirable and satisfying activities of man—education, art, music, religion, basic scientific research, athletics, and social interactions—could flourish."

36-13 A conservation ethic   The shift to equilibrium must be accompanied by a conservation ethic. Ecologically, according to Aldo Leopold, this is a limitation on freedom of action in the human struggle for existence. The conservation ethic recognizes that other organisms, whether we regard them as beneficial or not, have a right to exist and that we have a moral obligation to protect them, at some self-sacrifice if necessary. The ethic should be embraced whether or not one cares about other life—as

a matter of survival. In view of our reproductive success and ability to modify the earth's environments, continued unlimited pursuit of immediate human interests and needs at the expense of other organisms will ultimately result in the overpowering of natural ecosystems and widespread major disruption of the biosphere. Our own existence will then be in jeopardy. The basis for such an ethic would include the following principles:

1   The earth is a closed system with finite supplies of air, water, food, and capacity for wastes.

2   Soil in good condition is essential for all terrestrial life.

3   Principles of ecology apply to management of nature and to human beings.

4   Other organisms have a right to exist; humans are part of nature, not the conquerors of other living things.

5   Diversity in natural ecosystems provides stability and adaptability to environmental change.

6   Major alterations in the balances of nature may prejudice the affairs of both humans and other organisms.

7   Low human population density reduces competition, protects organic diversity, lessens pollution, and lowers demands on dwindling resources.

8   A high standard of living must include clean air, water, and food, pleasant surroundings, and tranquillity.

9   Education should develop a sound appreciation of nature.

References

Börgstrom, G. 1969. Too many: A study of earth's biological limitations. Toronto, Collier-Macmillan Canada Ltd. xiii + 368 pp., 54 figs. *An analysis of the world food situation and prospects for increasing food production.*

Boulding, K. E. 1966. The economics of the coming spaceship earth. In H. Jarrett (editor). Environmental quality in a growing economy. Baltimore, The Johns Hopkins Press.

Brown, Harrison. 1954. The challenge of man's future. New York, The Viking Press, Inc. xii + 290 pp., illus. *Classic work on population and resources.*

Chisolm, J. J., Jr. 1971. Lead poisoning. *Sci. Am.*, vol. 224, no. 2, pp. 15–23. *A natural substance humans concentrate in their immediate environment that can have deleterious effects on their health.*

Clark, J. R. 1969. Thermal pollution and aquatic life. *Sci. Am.*, vol. 220, no. 3, pp. 19–27. *Use of river and lake waters for industrial cooling presents a threat to fish and other organisms.*

Cloud, P. E., Jr. (editor). 1969. Resources and man. San Francisco, W. H. Freeman and Company. xi + 259 pp., illus.

Commoner, B. 1971. The closing circle. New York, Alfred A. Knopf, Inc. 326 pp. *The logic of ecology illuminates many of the troubles which afflict the earth and its inhabitants.*

Cowles, R. B. 1959. Zulu journal: Field notes of a naturalist in South Africa. Berkeley, University of California Press. xiii + 267 pp., illus. *The human impact on the South African environment.*

Dasmann, R. F. 1972. Environmental conservation. 3d ed. New York, John Wiley & Sons, Inc. x + 473 pp.

Edwards, C. A. 1969. Soil pollutants and soil animals. *Sci. Am.*, vol. 220, no. 4, pp. 88–99. *Pesticides in forest and woodland soils may slow down the process of soil formation and maintenance of soil fertility.*

Ehrlich, Paul. 1968. The population bomb. New York, Ballantine Books, Inc. 223 pp.

————— and **A. H. Ehrlich.** 1972. Population, resources, environment: Issues in human ecology. 2d ed. San Francisco, W. H. Freeman and Company. 509 pp. *Comprehensive, detailed analysis of worldwide crisis of overpopulation and resulting demands on food, resources, and environment; a broad ecologic approach; extensive bibliography.*

Eliassen, R. 1952. Stream pollution. *Sci. Am.*, vol. 186, no. 3, pp. 17–21. *Modern methods of waste treatment can help solve the problem of water pollution.*

Goldwater, L. J. 1971. Mercury in the environment. *Sci. Am.*, vol. 224, no. 5, pp. 15–21. *Is the concentration by industrial and biological processes endangering animals and humans?*

Haggen-Smit, A. J. 1964. The control of air pollution. *Sci. Am.*, vol. 210, no. 1, pp. 24–31.

Hardin, G. 1968. The tragedy of the commons. *Science*, vol. 162, no. 3859, pp. 1243–1248.

Leopold, Aldo. 1966. A Sand County almanac, with essays on conservation from Round River. New York, Ballantine Books, Inc. (First published by Oxford University Press in 1949.) xix + 295 pp., illus., paper. *An eloquent early voice on the need for a conservation ethic.*

Marshall, A. J. (editor). 1966. The great extermination. London, William Heinemann, Ltd. 221 pp. *Destruction of the natural environment in Australia.*

McDermott, W. 1961. Air pollution and public health. *Sci. Am.*, vol. 205, no. 4, pp. 49–57.

Meadows, D. H., and others. 1974. The limits to growth. 2d ed. New York, Universe Books. 207 pp.

Newell, R. E. 1971. The global circulation of atmospheric pollutants. *Sci. Am.*, vol. 224, no. 1, pp. 32–42. *A study of how atmospheric pollutants may travel and how they might affect world climate.*

Population Bulletin. A publication of the Population Reference Bureau, Inc. Washington, D.C. *Analyzes and interprets local, national, and international information concerning human population and related topics.*

Rudd, R. L. 1964. Pesticides and the living landscape. Madison, The University of Wisconsin Press. xiv + 320 pp.

Scientific American, The origin and characteristics of cities. 1965. Vol. 213, no. 3.

————— . 1970. Processes in the biosphere. Vol. 223, no. 3. *Discusses cycles in energy, water, oxygen, carbon, nitrogen and minerals and production of human food, energy, and materials.*

————— . 1971. Energy and power. Vol. 225, no. 3. *The role of energy in human life, past, present, and future.*

————— . 1974. The human population. Vol. 231, no. 3

————— . 1976. Food and Agriculture. Vol. 235, no. 3. *Entire issue devoted to food, agriculture, and human nutrition.*

Sears, P. B. 1937. This is our world. Norman, University of Oklahoma Press. xi + 292 pp. *One of the first comprehensive discussions of human beings and their environment.*

Shepard, Paul, and **David McKinley** (editors). 1969. The subversive science. Boston, Houghton Mifflin Company. x + 453 pp. *Essays toward an ecology of human beings.*

Spengler, J. J. 1960. Population and world economic development. *Science*, vol. 131, no. 3412, pp. 1497–1502. *In a finite world, population growth, before it is finally halted, entails diverse costs.*

Vogt, William. 1948. Road to survival. 2d ed. New York, William Sloane Associates. xvi + 335 pp. *An early warning; outlines course that must be followed to find a harmonious adjustment between humans and nature.*

White, Lynn. 1967. The historical roots of our ecologic crisis. *Science*, vol. 155, no. 3767, pp. 1203–1207.

Wilshire, H. G., and others. 1977. Impacts and management of off-road vehicles. The Geological Society of America. Report of the committee on environment and public policy. May. 8 pp.

Woodwell, G. M. 1967. Toxic substances and ecological cycles. *Sci. Am.*, vol. 216, no. 3, pp. 24–31. *Natural meteorological and biological cycles can distribute and* concentrate toxic substances such as pesticides and radioactive elements to dangerous levels.

———. 1978. The carbon dioxide question. *Sci. Am.,* vol. 238, no. 1, pp. 34–43. *Human activities are increasing carbon dioxide in the atmosphere; will enough carbon be stored in forests and the ocean to avert a major change in climate?*

———, **P. P. Craig,** and **H. A. Johnson.** 1971. DDT in the biosphere—where does it go? *Science,* vol. 174, no. 4014, pp. 1101–1107.

Ziswiler, Vinzenz. 1967. Extinct and vanishing animals. The Heidelberg Science Library, vol. 2. New York, Springer-Verlag New York, Inc. x + 133 pp., 74 figs. Revised English edition by F. Bunnell and P. Bunnell. *Human destruction of animal life and the relationship between human population growth and the decline and extinction of wild animals.*

GLOSSARY

L. = Latin; Gr. = Greek; dim. = diminutive.

Definitions of terms not included in the glossary can be found by referring to the index for the pages on which they occur in the text.

Pronunciation is indicated by accent-marks and the division of words into syllables is indicated by hyphens. In longer words a double accent-mark ["] is sometimes used for the syllable that is stressed most heavily; in such cases ['] indicates a syllable which bears only secondary stress. In syllables marked by either ['] or ["], the pronunciation of vowels that might be mispronounced can be told by a simple rule: the vowel of a syllable is short if the syllable is divided after a consonant, but long if it is not. Thus **pe-lag'-ic** represents the pronunciation [pa-**ladj**-ik] but **ra'-mus** represents [**ray**-mus].

Latin and Greek plurals are usually formed as follows: sing. *us* to plural *i* (nucleus, nuclei); *a* to *ae* (larva, larvae); *um* to *a* (cilium, cilia); *on* to *a* (pleuron, pleura); *is* to *es* (testis, testes). A few exceptions are genus, genera; species, species; vas, vasa.

ab-do'-men The major body division posterior to the thorax; in vertebrates, that region containing most of the digestive tract; in mammals, the region between the diaphragm and the pelvis. (L.)

ab-duc'-tor A muscle that draws a part away from the axis of the body or a limb or separates two parts. (L. *ab,* away + *duco,* lead)

ab-o'-ral Opposite the mouth. (L. *ab,* from + *os,* mouth)

ab-sorp'-tion The selective taking up of fluids or other substances (digested food, oxygen, etc.) by cells or tissues. (L. *ab,* from + *sorbeo,* suck in)

ac-cli'-ma-tize To become habituated to an environment not native. (L. *ad,* to + *clima,* climate)

ac'-i-nus A small terminal sac in a lung or multicellular gland. (L., grape)

a-coe'-lom-ate Lacking a coelom. (Gr. *a,* without + *koiloma,* cavity, from *koilos,* hollow)

acquired character One that originates during the life of an individual through environmental or functional causes.

ad'-ap-ta"-tion The fitness of a structure, function, or entire organism for a particular environment; the process of becoming so fitted. (L. *ad*, to + *apto*, fit)

ad-duc'-tor A muscle that draws a part toward the median axis or that draws parts together. (L. *ad*, to + *duco*, lead)

ad'-i-pose Pertaining to fat or the tissue in which fat is stored. (L. *adeps*, fat)

ad-o'-ral Near the mouth. (L. *ad*, toward + *oris*, mouth)

ad-sorp'-tion Adhesion of an extremely thin layer of gas molecules, dissolved substance, or liquid to a solid surface. Compare *absorption*. (L. *ad*, to + *sorbeo*, suck in)

aes'-ti-vate To pass the summer in a quiet, torpid condition. (L. *aestas*, summer)

af'-fer-ent Carrying to or toward a certain region; a vessel or nerve which transmits blood or impulses toward a given position. (L. *ad*, to + *fero*, bear)

al'-bin-ism Lack of pigment that is normally present. (Spanish, *albino*, from L. *albus*, white)

al'i-men"-ta-ry Pertaining to food, digestion, or the digestive tract. (L. *alimentum*, food)

al-lan'-to-is An extraembryonic membrane outpocketed from the hindgut and serving for respiration and excretion in embryos of reptiles and birds; becomes part of umbilical cord and unites with the chorion to form the placenta in mammals. (Gr. *allas*, stem + *eidos*, form)

al-lele' One of the alternative forms of a gene, having the same locus in homologous chromosomes; also, the alternative forms of a Mendelian character; allelomorph. (Gr. *allelon*, of one another)

alternation of generations See Metagenesis.

al-ve'-o-lus A small cavity or pit; a tooth socket; a minute terminal air sac in a lung; a terminal unit in an alveolar gland; one droplet in an emulsion. (L., a small cavity)

a-mi'-no acid Any organic compound containing an amino group (NH_2) and a carboxyl group (COOH); amino acids are the structural subunits of proteins.

am'-ni-on The innermost extraembryonic membrane, filled with watery amniotic fluid, that encloses the developing embryo of a reptile, bird, or mammal; a similar single membrane around the insect embryo. (Gr., fetal membrane)

am'-ni-ote A vertebrate that develops an amnion during its embryonic stage—the reptiles, birds and mammals. (L. *Amniota*)

a-moe'-boid Like an amoeba; pertaining to cell movements (pseudopodia projection) resembling those of an amoeba. (Gr. *amoibe*, change)

am-phib'-i-ous Capable of either terrestrial or aquatic life, as a frog. (Gr. *amphi*, on both sides + *bios*, life)

am'-phi-coe"-lous Concave at both ends, as the centrum of some vertebrae. (Gr. *amphi*, on both sides + *koilos*, hollow)

am-pul'-la A small bladderlike enlargement; the dilated portion of a canal or duct (e.g., the semicircular canals of the ear). (L., flask)

an-ab'-o-lism Constructive or synthetic chemical reactions of metabolism, from digestion to assimilation. (Gr. *ana*, up + *ballo*, throw)

a-nal"-o-gous Corresponding in function but not in origin, development, or structure. (Gr. *analogia*, correspondence, ratio)

an-as'-to-mo"-sis A communication or joining as of two or more arteries, veins, or other vessels. (Gr. creation of an opening, from *anastomoö*, "mouth up," make a mouth, from *ana*, up + *stoma*, mouth)

an-ten'-na A projecting sensory appendage, especially on arthropods, not concerned with light perception or sight. (Gr. *ana*, up + *teino*, stretch)

an-te'-ri-or The forward-moving or head end of an animal, or toward that end. (L., from *ante*, before)

an'-ti-mere One of the several similar or equivalent parts into which a radially symmetrical animal may be divided. (Gr. *anti*, against + *meros*, part)

a'-nus The posterior opening of the digestive tract. (L.)

a-or'-ta A large main artery connected to the heart. (Gr. *aorta*, artery)

a-or'-tic arch A large artery arising from the heart in vertebrates; one of paired arteries connecting the ventral aorta and dorsal aorta in the region of the pharynx or gills.

ap'-i-cal At the apex or top, as of a conical structure. (L. *apex*, peak)

ap-pend'-age A movable projecting part on a metazoan body (L. *ad*, to + *pendeo*, hang)

aq-uat'-ic Pertaining to or living in water. (L. *aqua*, water)

ar-bo'-re-al Pertaining to or living in trees, as tree-inhabiting animals. (L. *arbor*, tree)

arch-en'-ter-on The primitive digestive cavity of a metazoan embryo, formed by gastrulation. (Gr. *archo*, govern, be chief + *enteron*, intestine)

ar'-te-ry   A tubular vessel conveying blood away from the heart. (Gr. *arteria*, artery)

ar-tic'-u-late   To attach by a joint. (L. *articulus*, joint)

a-sex'-u-al   Not related to sex; not involving gametes or fusion of their nuclei. (Gr. *a*, without + L. *sexus*, sex)

as-sim'-i-la"-tion   Incorporation of digested nutriment, after absorption, into living protoplasm. (L. *ad*, to + *similis*, like)

a'-sym-met"-ri-cal   Without symmetry.

a'-tri-um   An outer cavity or chamber; a chamber of the heart receiving blood from veins. (L. *atrium*, chamber)

au"-di-to'-ry   Pertaining to the organ or sense of hearing. (L. *audio*, hear)

au'-ri-cle   The external pinna of the ear in mammals; also a chamber of the heart that receives blood from the veins; a synonym for atrium. (L. *auricula*, dim. of *auris*, ear)

au'-to-some   Any ordinary chromosome as contrasted with a sex chromosome. (Gr. *autos*, self + *soma*, body)

au-tot'-o-my   The automatic "voluntary" breaking off of a part by an animal. (Gr. *autos*, self + *temno*, to cut)

au'-to-tro"-phic   Capable of synthesizing organic compounds directly from inorganic materials. (Gr. *autos*, self + *trepho*, to feed)

au'-to-tro"-phic nu-tri'tion   That process by which an organism manufactures its own food from inorganic compounds, as in a plant.

ax'-i-al skel'-e-ton   That part of the vertebrate skeleton on the axis of the body—skull, vertebrae, ribs, and sternum.

ax'-is   A line of reference or one about which parts are arranged symmetrically. (L. *axis*, axle)

ax'-on, ax'-one   The fiber of a nerve cell that conducts impulses away from the cell body of which it is a part. (Gr. *axon*, axis)

bi-lat'-er-al sym'-me-try   Symmetry of such a kind that a body or part can be divided by one median plane into equivalent right and left halves, each a mirror image of the other.

bi-o-chem-is-try   The study of the chemical substances and reactions which occur in living organisms. (Gr. *bios*, life + chemistry, from Gr. *chemeia*, alchemy, or from *chyma*, fluid)

bi-ra-di-al sym-metry   Symmetry in which the organism has two planes on which it may be divided into mirror images.

bi-ra'-mous   Consisting of or possessing two branches, as a crustacean appendage. (L. *bis*, twice + *ramus*, branch)

blad'-der   A thin-walled sac, or bag, that contains fluid or gas.

blas'-to-disc   The germinal area on a large-yolked egg that gives rise to the embryo. (Gr. *blastos*, germ + *diskos*, platter)

blas'-to-mere   One of the early cells formed by the division of an ovum. (Gr. *blastos*, germ + *meros*, part)

blas'-to-pore   The mouthlike opening of a gastrula. (Gr. *blastos*, germ + *poros*, passage)

blas'-tu-la   Early stage of an embryo, usually a hollow sphere of cells. (Gr. dim. of *blastos*, germ)

blood   The fluid that circulates in the vascular system of many animals.

bod'-y cav'-i-ty   The cavity between the body wall and internal organs of an animal. See *coelom*.

bra'-chi-al   Refering to the forelimb or pectoral appendage. (L. *brachium*, forearm)

bran'-chi-al   Referring to gills. (Gr. *branchia*, gills)

bron'-chus   Either of the two main divisions of the trachea conveying air into the lungs of vertebrates. (Gr. *bronchos*, windpipe)

buc'-cal   Pertaining to the mouth or cheek. (L. *bucca*, cheek)

bud   Part of an organism that grows out to produce a new individual or new part.

buff-er   A solution constituted to minimize changes in pH which would otherwise occur upon the addition of other chemicals.

bur'-sa   A pouch, or sac, such as the bursa of a joint. (Gr. hide or skin)

cae'-cum, pl. **cae'-ca**   A pouch or blind saclike extension on the digestive tract. (L. *caecus*, blind)

cal-car'-e-ous   Composed of or containing calcium carbonate ($CaCO_3$); limy. (L. *calx*, lime)

cal'-or-ie, cal'-o-ry   Unit of heat. Small calorie (cal), amount of heat needed to raise 1 gram of water 1°C (at 15°C); large calorie (kcal), amount necessary to raise 1 kilogram (2.2 pounds) of water 1°C. (L. *calor*, heat)

cap'-il-la-ry   A minute blood vessel with walls composed of a single layer of endothelium through which diffusion may occur; commonly in a connecting network between arteries and veins. (L. *capillus*, hair)

car'-a-pace   The hard dorsal shell of turtles and crustaceans. (Spanish *carapacho*)

car'-di-ac   Pertaining to or near the heart. (Gr. *kardia*, heart)

car-niv'-o-rous   Eating or living on flesh of other animals. (L. *caro*, flesh + *voro*, eat)

cas-tra'-tion Removal of the gonads, or sex glands. (L. *castro*, to castrate)

ca-tab'-o-lism The breaking down of complex chemical substances in the cell. (Gr. *kata*, down + *ballo*, throw)

cau'-dal Pertaining to the tail or posterior part of the body. (L. *cauda*, tail)

cell The basic structural and functional unit of life consisting of a complex of organelles and bounded by a unit membrane; usually microscopic in size. (L. *cella*, small room)

cel'-lu-lar Pertaining to or consisting of cells. (L. dim. of *cella*, small room)

cel'-lu-lose The carbohydrate forming the wall of plant cells; also in the mantle of tunicates.

cen'-trum The body of a vertebra, to which the arches and transverse process attach. (L., center)

ce-phal'-ic Pertaining to or toward the head. (Gr. *kephale*, head)

ceph'-a-lo-tho"rax A body division with the head and thorax combined. (Gr. *kephale*, head + *thorax*, chest)

cer'-e-bel"lum The anterior development from the hindbrain which controls muscular coordination. (L. dim. of *cerebrum*, brain)

cer'-e-bral or **ce-re'-bral** Of or pertaining to the brain as a whole or the anterior dorsal (cerebral) hemispheres; also to the anterior brainlike nerve ganglia of various invertebrates. (L. *cerebrum*, brain)

ce-re'-brum or **ce'-re-brum** The dorsal anterior part of the vertebrate forebrain, consisting of two hemispherical masses. (L., brain)

cer'-vi-cal Pertaining to a neck. (L. *cervix*, neck)

chae'-ta See Seta.

char'-ac-ter, characteristic A distinguishing feature, trait, or property of an organism. (Gr.)

che-lic'-e-ra One of the most anterior pair of appendages on arachnoids such as spiders, scorpions, and the horseshoe crab. (Gr. *chele*, claw + *keras*, horn)

che'-li-ped" The first thoracic appendage (pincer) of crayfish and related crustaceans. (Gr. *chele*, claw + L. *pes*, foot)

chi'-tin The protein-carbohydrate that forms the exoskeleton in arthropods and some other animals. (Gr. *chiton*, tunic)

chlo'-ro-phyll The green pigment of plants and certain protists and monerans (some bacteria, and blue-green algae), involved in photosynthesis. (Gr. *chloros*, green + *phyllon*, leaf)

chon'-dro-cra"-ni-um The cartilaginous skull of cyclostomes and elasmobranchs; also that part of the embryonic skull in higher vertebrates first formed as cartilage. (Gr. *chondros*, cartilage + *kranion*, skull)

Chor-da'-ta The phylum of animals with a notochord, persistent or transient; includes the vertebrates, amphioxus, and tunicates; the chordates. (L. *chorda*, cord or string, from Gr. *chorde*, gut, sausage)

cho'-ri-on The outer double extraembryonic membrane surrounding the embryo of a reptile, bird, or mammal, in mammals uniting with the allantois to form the placenta; the outer membrane of an insect egg. (Gr., membrane, especially of a fetus or an egg)

chro'-ma-tid One of the two identical longitudinal halves into which a chromosome divides during cell division and which in turn develops into a complete new chromosome. (Gr. *chroma*, color + *id*, particle)

chro'-ma-tin The staining substance in a cell nucleus, conspicuous in the nuclear network and in the chromosomes at mitosis containing DNA and protein. (Gr. *chroma*, color)

chro-mat'-o-phore A pigment cell containing granules of coloring material and responsible for color markings on many animals. (Gr. *chroma*, color + *phero*, bear)

chro'-mo-mere An individual chromatin granule in a chromosome. (Gr. *chroma*, color + *meros*, part)

chro'-mo-somes Characteristic deeply staining bodies, formed of chromatin in the nucleus of a cell during mitosis, that bear the genes or determiners of heredity. (Gr. *chroma*, color + *soma*, body)

cil'-i-um, pl. **cil'-i-a** A microscopic hairlike process attached to a free cell surface; usually numerous, often arranged in rows, and capable of vibration. (L., eyelid)

cir'-rus, pl. **cir'-ri** A small, slender, and usually flexible structure or appendage. (L., tuft of hair)

cleav'-age The division or splitting of an egg cell into many cells.

clo-a'-ca The terminal portion of the digestive tract in many insects; the common passage from the digestive, excretory, and reproductive organs in various lower vertebrates. (L., sewer)

co-coon' A protective case or covering about a mass of eggs, a larva or pupa, or even an adult animal.

coe'-lom The body cavity or space between the body wall and internal organs in many metazoan animals, lined with mesoderm. (Gr. *kiolama*, cavity, from *koilos*, hollow)

coe-lom″-o-duct A duct, derived from mesoderm, that conveys gametes or excretory products (or both) from the coelom to the exterior. (Gr. *koilama*, cavity + L. *duco*, to lead)

col′-o-ny A group of organisms of the same species living together. (L. *colonus*, farmer)

com-men′-sal-ism The association of two or more individuals of different species in which one or more is benefited and the others are not harmed. (L. *con*, together + *mensa*, table)

com-mu′-ni-ty A group of organisms of one or more species living together and related by environmental requirements. (L. *communis*, common)

com-pressed′ Reduced in breadth; flattened laterally. (L. *con*, together + *premo*, press)

con-ver′-gent Approaching each other or tending toward a common point. (L. *cum*, together + *vergo*, incline)

con′-vo-lut″-ed Coiled or twisted. (L. *convolvo*, roll together)

cop′-u-la″-tion Sexual union. (L. *copulo*, join together)

co′-ri-um The layer of skin lying beneath the epidermis; the dermis. (L., skin, hide)

cor′-ne-a The outer transparent coat of the eye. (L. *corneus*, horny)

cor′-pus-cle A small or minute structure or a cell, free or attached, as a blood corpuscle or bone corpuscle. (L., *corpusculum*, dim. of *corpus*, body)

cor′-tex The outer or covering layer of a structure. (L., rind, bark)

cra′-ni-al Of or pertaining to the skull or brain, as a cranial nerve. (Gr. *kranion*, skull)

cra′-ni-um The skull, specifically the braincase. (Gr. *kranion*, skull)

crop A thin-walled and expanded portion of the digestive tract, primarily for food storage.

cross-fer′-til-i-za″-tion Union of an egg cell from one individual with a sperm cell from another; opposite of self-fertilization.

cu-ta′-ne-ous Pertaining to the skin. (L. *cutis*, skin)

cu′-ti-cle A thin, noncellular external covering on an organism. (L. dim. of *cutis*, skin)

cyst A resistant protective covering formed about a protozoan or other small organism during unfavorable conditions or reproduction; a small sac or capsule, usually containing a liquid. (Gr. *kystis*, bladder)

cy′-to-kin-e″-sis That part of the cell cycle in which the cytoplasm divides. (Gr. *kytos*, hollow + *kinesis*, movement)

cy′-to-plasm That part of a cell outside the nucleus and within the cell membrane. (Gr. *kytos*, hollow + *plasma*, form)

cy′-to-some The cell body inside the plasma membrane. (Gr. *kytos*, hollow + *soma*, body)

def′-e-cate To discharge food residues (feces) through the anus. (L. *de*, from + *faex*, dung)

den′-drite A branching process on a nerve cell that conducts impulses to the cell body. (Gr. *dendron*, tree)

de-pressed′ Flattened vertically, from above. (L. *de*, down + *premo*, press)

der′-mal Pertaining to the skin, especially the inner layers of vertebrate skin. (Gr. *derma*, skin)

der′-mis The deeper portion of the skin, or "true" skin, beneath the epidermis in a vertebrate, derived from mesoderm. The dermis contains blood vessels, and nerves, in addition to connective tissue. (Gr. *derma*, skin)

determinate development An embryonic development pattern in which the destiny of individual blastomeres to produce specific body parts of the adult is established early in cleavage (development), so that if the individual blastomeres are separated, each will give rise to only a part of the larva or adult; characteristic of protostomes.

deu″-ter-o-stome′ One of two major lines of evolution in the Animal Kingdom, characterized by radial indeterminate cleavage, mouth arising distant from the blastopore, and enterocoel formation. (Gr. *deuteros*, second + *stoma*, mouth)

di′-a-phragm A dividing membrane, as the diaphragm of the ear; the fibro-muscular partition between the thoracic and abdominal cavities in mammals used in respiration. (Gr. *dia*, through + *phragma*, fence)

di-ges′-tion The process of preparing food for absorption and assimilation. (L. *digero*, divide, dissolve)

dig′-it A finger or toe; one of the terminal divisions of a limb in tetrapods. (L. *digitus*, finger)

dig′-i-ti-grade Walking on the toes (e.g., dogs and cats). (L. *digitus*, finger + *gradior*, to step)

di-mor′-phism Existing under two distinct forms. (Gr. *di*, two + *morphe*, form)

di-oe′-cious Male and female organs located in separate individuals; gonochoristic; opposite of monoecious. (Gr. *di*, two + *oikos*, house)

dip′-lo-blas″-tic Derived from two embryonic germ layers, ectoderm and endoderm. (Gr. *diplous*, double + *blastos*, germ)

dip′-loid The dual or somatic number of chromosomes (2*n*), twice the number found in the matured germ cells of an organism. (Gr. *diplous*, double + *eidos*, form)

dis-sim'-i-la"tion The chemical disintegration of protoplasm, usually by oxidation, with release of energy; catabolism. (L. *dissimilis*, different)

dis'-tal Away from the point of attachment or place of reference. (L. *disto*, stand apart)

di-ur'-nal Pertaining to the daytime, as opposed to nocturnal. (L. *dies,* day)

di-ver'-gent Going farther apart; separating from a common source. (L. *dis*, apart + *vergo*, incline)

DNA Abbreviation for *deoxyribonucleic acid.*

DNA po-ly"-mer-ase' An enzyme that affects DNA synthesis.

dominant character A character from one parent that manifests itself in offspring to the exclusion of a contrasted (recessive) character from the other parent.

dor'-sal Toward or pertaining to the back or upper surface. (L. *dorsum*, back)

duct A tube by which a liquid or other product of metabolism is conveyed, as a secretion from a gland; usually opening on a surface or in a larger compartment (L. *duco*, lead).

duct'-less gland A gland that elaborates and secretes a hormone, or internal secretion, directly into the bloodstream; an endocrine gland.

duc'-tus de'-fer-ens The sperm duct from the efferent ductules to the cloaca or ejaculatory duct. (L. *deferens*, carrying out)

ec'-dy-sis The process of shedding the exoskeleton in arthropods or the skin in some vertebrates.

e-col'-o-gy The study of interrelationships between organisms and their environments. (Gr. *oikos*, home + *logos*, knowledge)

ec'-to-derm The outer germ layer or cell layer of an early embryo; contributes importantly to the formation of the skin, sense organs, and nervous system. (Gr. *ektos*, outside + *derma*, skin)

ec'-to-par"-a-site A parasite that lives on the exterior of its host. (Gr. *ektos*, outside + parasite)

ec'-to-ther"-mal Deriving body heat from the environment; characteristic of all animals but birds and mammals. (Gr. *ektos*, to or on the outside + thermal)

ef-fec'-tor An organ, tissue, or cell that can react to stimuli (e.g., muscles and glands which transform motor impulses into motor action). (L. *efficio*, effect, bring to pass)

ef'-fer-ent Leading away from a given point of reference, as an efferent artery or nerve. (L. *ex*, out + *fero*, carry)

ef'-fer-ent duc'-tules Short ducts carrying sperm from the testis to the ductus deferens. (L. *ex*, out + *fero*, carry + *duco*, lead)

e-gest' To discharge unabsorbed food or residues from the digestive tract. (L. *e*, out + *gero*, bear)

egg A female germ cell produced by a functionally female organism; an ovum.

em'-bry-o A newly forming young animal in the stages of development before hatching or birth. (Gr.)

em'-bry-og"-e-ny The process of development of the embryo. (Gr. *embryon* + *genesis*, generation)

embryonic membranes Cellular membranes formed as part of an embryo during its development and necessary for its metabolism and protection; the amnion, chorion, and allantois of reptiles, birds, and mammals; some also in insects.

en-am'-el The dense whitish covering on teeth of vertebrates, the hardest substance produced by animal bodies.

en'-do-crine Pertaining to a ductless gland; also its secretion, or hormone, which is carried in the blood or lymph and influences or regulates other organs in the body. (Gr. *endon*, within + *krino*, separate)

en'-do-derm, entoderm The layer or group of cells lining the primitive gut, or gastrocoel, in an early embryo, beginning in the gastrula stage. (Gr. *endon*, within + *derma*, skin)

en-dog'-e-nous Growing or originating from within, as opposed to exogenous. (Gr. *endon*, within + *gigno*, be born)

en'-do-par"-a-site A parasite that lives within its host. (Gr. *endon*, within + parasite from Gr. *para*, beside + *sitos*, food)

en'-do-skel"-e-ton An internal supporting framework or structure. (Gr. *endon*, within + skeleton)

en'-do-style The ventral ciliated groove in the pharynx of tunicates, amphioxus, and larval lampreys, used in food getting; homologous with the thyroid gland of vertebrates. (Gr. *endon*, within + *stylos*, column)

en'-do-the"-li-um Layer of squamous cells lining the inner surface of the blood and lymph vessels and heart of vertebrates. (Gr. *endon*, within + *thele*, nipple)

en'-do-ther"-mal Generating and maintaining body heat metabolically; characteristic of birds and mammals. (Gr. *endo*, within + thermal)

en'-ter-o-coel A type of coelom that originates as an outpocketing of the archenteron.

en'-ter-on The digestive cavity, especially that part lined by endoderm. (Gr., intestine)

environment   The aggregate of conditions surrounding an organism.

en'-zyme   A complex protein produced by living cells that in minute amount accelerates specific chemical transformation such as hydrolysis, oxidation, or reduction but that is not used up in the process; a catalyst. (Gr. *en*, in + *zyme*, leaven)

ep'-i-der"-mis   A layer of cells (sometimes stratified) covering an external surface; the ectodermal portion of the skin of most animals; secretes cuticle on arthropods and some other animals. (Gr. *epi*, upon + *derma*, skin)

ep'-i-did"-y-mis   A coiled structure containing the efferent tubules of the mammalian testis. (Gr. *epi*, upon + *didymos*, testicle)

e-piph'-y-sis   The end or other external part of a bone that ossifies separately; also the pineal body, a dorsal outgrowth on the diencephalon of the vertebrate brain. (Gr. *epi*, upon + *phyto*, to grow)

ep'-i-the"-li-um   A layer (or layers) of cells covering a surface or lining a cavity. (Gr. *epi*, upon + *thele*, nipple)

e-ryth'-ro-cyte   A red blood cell or corpuscle; characteristic of vertebrates. (Gr. *erthros*, red + *kytos*, hollow vessel)

e-soph'-a-gus   That part of the digestive tract between the pharynx and stomach. (Gr.)

eu-kar"-y-ote   An organism in which the genetic material is bounded by a membrane system separating it from the remainder of the cell. (Gr. *eu*, well + *karyote*, given a nucleus, from Gr. *karyon*, nut, kernel, kernellike thing)

eu-sta'-chi-an tube   The passage between the pharynx and middle ear in land vertebrates. (Eustachio, an Italian anatomist)

e-vag'-i-na"-tion   An outpocketing from a hollow structure. (L. *e*, out, from + *vagina*, a sheath)

evolution   The process by which living organisms have come to be what they are, structurally and functionally, complex forms being derived from simpler forms; hence, descent with modification. (L. *evolvo*, unroll, unfold)

ex-cre'-tion   Discharge of metabolic wastes by an organism; also, the substances discharged. (L. *excerno*, separate, secrete)

ex'-o-skel"-e-ton   An external supporting structure or covering. (Gr. *exo*, outside + skeleton)

F₁, F₂, etc.   Abbreviations for first filial, second filial, etc., indicating the successive generations following cross-breeding.

fac'-tor   An agent or cause; in genetics, a specific germinal cause of a hereditary character; same as gene.

fas'-cia   A fibrous sheet of connective tissue that covers, supports, or binds together muscles and other organs. (L., a band)

fau'-na   All the animal life in a given region or period of time.

fe'-ces   Excrement; unabsorbed or indigestible food residues discharged from the digestive tract as waste. (L., dregs)

fer'-til-i-za"-tion   Union of two gametes (egg and sperm) to form a zygote and initiate the development of an embryo. (L. *fertilis*, fruitful, from *fero*, bear)

fe'-tus   The later developmental stages of an embryo while within the egg or uterus, after the third month in humans. (L., offspring)

fi'-ber   A delicate threadlike part in a tissue; also, a threadlike cell (muscle fiber) or cell process (nerve fiber). (L. *fibra*)

fi'-bril   A small fiber. (L. dim. of *fibra*, fiber)

fin   An extension of the body on an aquatic animal used in locomotion or steering.

fis'-sion   Asexual reproduction by division into two or more parts, usually equivalent. (L. *findo*, split)

fla-gel'-lum   A long lash or threadlike extension capable of vibration; e.g. on flagellate protozoans and on collar cells of sponges. (L., little whip)

flame cell   A type of hollow terminal excretory cell in certain invertebrates that contains a beating (flamelike) group of cilia.

foetus   See Fetus.

fol'-li-cle   A minute cellular sac or covering. (L. dim. of *follis*, bag)

food vac'-u-ole   An intracellular digestive organelle.

fo-ra'-men   An opening or perforation through a bone, membrane, or partition. (L. *foro*, bore)

fos'-sil   Any relic of an organism buried in the earth or rocks by natural causes in past geologic time. (L. *fadio*, dig)

free-liv'-ing   Not attached or parasitic; capable of independent movement and existence. Compare *sessile*.

fron'-tal   Of or pertaining to the front, or forehead; also a plane or section parallel to the main body axis and at right angles to the sagittal plane. (L. *frons*, the brow)

func'-tion   The activity or action of any part of an organism. (L. *functio*, perform)

ga-mete'   A mature reproductive or germ cell, either male (sperm) or female (ovum). (Gr. *gametes*, husband, from *gamos*, marriage)

ga-me'-to-gen"-e-sis   The process of formation of

mature germ cells, or gametes; maturation. (Gr. *gametes* + *genesis*, birth)

gan'gli-on, pl. **gan'gli-a** A group or concentration of nerve cell bodies usually located outside the central nervous system in vertebrates but often within the central nervous system in invertebrates.

gas'-tro-coel The primitive digestive cavity of a metazoan embryo, formed by gastrulation. (Gr. *gaster*, stomach + *koilos*, hollow)

gas'-tro-der"-mis Lining of the digestive cavity in cnidarians. (Gr. *gaster*, stomach + *derma*, skin)

gas'-tro-vas"-cu-lar Serving for both digestion and circulation.

gas'-tru-la Early stage in development when the embryo is saclike and usually consists of two germ layers (ectoderm and endoderm). (dim. from Gr. *gaster*, stomach)

gas'-tru-la"-tion The process during embryonic development involving cellular movements which result in the formulation of a gastrula.

gene The unit of inheritance located in a chromosome, which is transmitted from one generation to another in the gametes and controls the development of a character in the new individual; the factor, or hereditary determiner. (Gr. stem *gen*, from *gignomai*, be born)

gen'-i-tal Referring to reproductive organs or the process of generation. (L. *gigno*, beget)

gen'-ome The genes of a haploid set of chromosomes; also all the genes of an individual or population. (Gr. stem *gen*, from *gignomai*, be born + L. *oma*, group)

gen'-o-type The internal genetic or hereditary constitution of an organism without regard to its external appearance. Compare *phenotype*. (Gr. *genos*, race + *typos*, impression, form)

germ cell A reproductive cell or gamete in a multicellular organism (spermatozoon or ovum).

germ layer One of the (two or three) fundamental cell layers (ecto-, endo-, mesoderm) in an early embryo of a multicellular animal, from which tissues and organs of the adult are formed.

germ plasm The material basis of inheritance; the gametes and the cells and tissues from which they form, considered as a unit.

gill An organ for respiration in aquatic organisms.

giz'-zard A heavily muscled portion of the digestive tract found in earthworms, insects, and birds, used for grinding food.

gland An organ of secretion or excretion. (L. *glans*, acorn)

glo-mer'-u-lus A small rounded clump of vessels; the knot of capillaries in a renal corpuscle. (L. dim. of *glomus*, ball)

glot'-tis The opening from the pharynx into the trachea. (Gr. *glotta*, tongue)

gly'-co-gen A carbohydrate (polysaccharide) stored in the muscles and liver. (Gr. *glykys* sweet + *gen*, come into being)

go'-nad A reproductive organ (ovary, testis) in which gametes (ova or sperm) are produced. (Gr. *gonos*, generation, seed)

gre-gar'-i-ous Habitually living in groups, flocks, etc., of numerous individuals. (L., *grex*, herd)

gyn-an'-dro-morph An individual in a dioecious species having one part of the body female and another part male in constitution. (Gr. *gyne*, woman + *aner*, man)

hab'-i-tat The natural or usual dwelling place of an individual or group of organisms. (L. *habitus*, condition)

hair A slender filamentous growth on the skin of mammals and on the exposed surfaces of some arthropods.

hap'-loid The single, or halved, number of chromosomes (*n*) as found in matured germ cells. (Gr. *haplous*, single + *eidos*, form)

he'-mal Pertaining to the blood or the blood-vascular system. (Gr. *aima*, blood)

he"-mo-coel' Portion of a body cavity reduced in size and functioning as part of a blood-vascular system. (Gr. *haima*, blood + *koilos*, hollow)

hem"-o-glo'-bin The coloring matter of red corpuscles in vertebrate blood and of blood plasma in some invertebrates; a protein pigment containing iron that combines with and transports oxygen to the tissues. (Gr. *haima*, blood + L. *globus*, ball, round object, + L. *inus*, of, belonging to)

he-pat'-ic Pertaining to the liver. (Gr. *hepar*, liver)

he-pat'-ic por'-tal sys'-tem A system of veins leading from the digestive tract to capillaries (sinusoids) in the liver of a vertebrate.

her-biv'-o-rous Feeding only or chiefly on herbs, grasses, or other vegetable matter. (L. *herba*, grass + *voro*, devour).

he-red"-i-tar'y Passing by inheritance from one generation to another. (L. *heres*, heir)

he-red'-i-ty The transmission of characters, physical and other, from parent to offspring. (L. *heres*, heir)

her-maph'-ro-dite An animal with both male and female reproductive organs. (Gr. *Hermes* + *Aphrodite*)

het'-er-o-zy"-gote An individual produced by union of two germ cells that contain unlike genes for a given character, either both genes of an allelomorphic pair or two different genes of an allelomorphic series. Compare *homozygote*. (Gr. *heteros*, another + *zygon*, yoke)

hi'-ber-nate To pass the winter in an inactive or torpid condition. (L. *hiberno*, pass the winter)

hol'-o-blas"-tic Cleavage in which an entire egg cell divides. (Gr. *holos*, whole + *blastos*, germ)

hol'-o-phyt"-ic nu-tri'-tion Nutrition involving photosynthesis of simple inorganic chemical substances, as in green plants and some flagellate protozoans. (Gr. *holos*, whole + *phyton*, plant)

hol'o-zo"-ic nu-tri'-tion Nutrition requiring complex organic foodstuffs, and characteristic of most animals. (Gr. *holos*, whole + *zoön*, animal)

ho'-moi-o-ther"-mal Having constant internal temperature, often maintained above that of the environment; characteristic of birds and mammals. (Gr. *homoios* like + thermal)

ho-mol'-o-gous Corresponding in structure and origin but not necessarily in function or appearance. (Gr. *homos*, same + *lego*, speak)

ho-mol'-o-gous chro'-mo-somes A pair of chromosomes having relatively similar structure and function and a common origin, one from each parent. (Gr. *homologos*, agreeing)

ho-mol'-o-gy Fundamental similarity; structural likeness of an organ or part in one kind of animal with the comparable unit in another resulting from descent from a common ancestry. Compare *analogy*. (Gr. *homologia*, agreement)

ho'-mo-zy"-gote An individual produced by union of two germ cells that contain like genes for a given character. Compare *heterozygote*. (Gr. *homos*, like + *zygon*, yoke)

hor'-mone A chemical regulator or coordinator secreted by cells or ductless glands and carried in the bloodstream to act on cells in other parts of the body. (Gr. *hormao*, urge on, spur)

host An organism that harbors another organism as a parasite. (from Old French, from L. *hospes*, host)

hy'-a-line Glassy or semitransparent. (Gr. *hyalos*, glass)

hy'-brid The offspring of two parents that differ in one or more heritable characters; a heterozygote. (L. *hybrida*, mongrel)

hy-drol'-y-sis Alteration of a chemical compound through the action of water.

hy-per'-tro-phy Abnormal increase or overgrowth in the size of a cell, tissue, or organ. (Gr. *hyper*, over + *tropho*, nourish)

hy'-po-thal"-a-mus A region of the forebrain which coordinates such physiological activities as water balance, temperature, reproduction, and metabolism.

in-breed To mate related animals or plants.

indeterminant development A development pattern in which the destiny of individual blastomeres remains unfixed until late in development, so that if individual blastomeres are separated, each will give rise to a whole larva or adult; characteristic of deuterostomes.

in-gest' To take food into the digestive tract. (L. *in* + *gero*, bear)

in-her'i-tance The sum of all characters that are transmitted by the germ cells from generation to generation. (L. *in*, in + *heres*, heir)

in'-stinct An inherited response or pattern of behavior, invoked by a certain stimulus and often of complex nature, combining associated reflex acts and leading to a particular end.

in-teg'-u-ment An outer covering, especially the skin of a vertebrate and its derivatives. (L. *intego*, to cover)

in'-ter-cel"-lu-lar Between or among cells. (L. *inter*, between, among + cellular)

in-tes'-tine That part of the digestive or alimentary canal between the stomach and anus (or cloaca). (L. *intus*, inside)

in'-tra-cel"-lu-lar Within a cell or cells. (L. *intra*, within + cellular)

in'-tus-sus-cep"-tion Growing by addition from within. (L. *intus*, inside + *suscipio*, undertake, begin, from *sub*, under + *capio*, take)

in-vag'-i-na"-tion An inpocketing or folding in, as of the vegetal pole of a blastula to form the gastrula. (L. *in* + *vagina*, a sheath)

in-ver'-te-brate Any animal without a dorsal column of vertebrae.

joint A place of union between two separate bones or other hardened structures. (L. *junctus*, from *jungo*, join)

kar'-yo-type The total of characteristics, including size, shape, and number of the set of chromosomes of a somatic cell. (Gr. *karyon*, nut, kernel, kernellike thing, hence, in the usage of scientific vocabulary, nucleus)

ker'-a-tin A water insoluable protein present in epidermal tissues and modified into hard structures such as hair, horns, feathers, and nails. (Gr. *keras*, horn)

la'-bi-al Pertaining to the lips.

la-mel'-la A thin, sheetlike layer. (L. *lamina*, plate)

lar'-va The early and usually active feeding stage of

an animal, after the embryo and unlike the adult. (L., mask)

leu'-ko-cyte A white blood cell or corpuscle. (Gr. *leukos*, white + *kytos*, hollow vessel)

li'-my Calcareous; containing calcium salts, especially $CaCo_3$.

lin'-gual Pertaining to the tongue. (L., tongue)

link'-age Inheritance of characters in groups, probably because their genes lie in the same chromosome.

loph'-o-phore A ridge about the mouth region bearing tentacles in some invertebrates. (Gr. *lophos*, crest + *phero*, bear)

lu'-men The cavity in a gland, duct, vessel, or organ. (L. light)

lu'-mi-nes"-cence Emission of light as a result of chemical reactions within cells. (L. *lumen*, light)

lung An organ for breathing air.

lymph Colorless blood fluid (without red blood cells) found among tissues and in lymph capillaries or vessels, (L. *lympha*, water)

lym-phat'-ic sys'-tem A system of delicate vessels in vertebrates that lead from spaces between tissues to large veins entering the heart; part of the circulatory system.

lymph'-o-cyte A white blood cell with one large undivided and nongranular nucleus; present in blood and lymph vessels. (L. *lympha*, water + Gr. *kytos*, hollow vessel)

man'-di-ble The lower jaw of a vertebrate; either jaw of an arthropod. (L. *mandibula*, jaw)

ma-rine' Pertaining to or inhabiting the sea, ocean, or other salt waters. (L. *mare*, sea)

ma'-trix Intercellular substance, as in connective tissues, cartilage, etc. (L., womb)

mat'-u-ra"-tion Final stages in preparation of sex cells for mating, with segregation of homologous chromosomes so that each cell or gamete contains half the usual (diploid) number.

me-du'-sa A free-swimming organism (cnidarian) with gelatinous body of bell or umbrella shape, margined with tentacles and having the mouth centered on a projection in the concave surface. (Gr. myth, Gorgon with snakes for hair)

mei-o'-sis Nuclear changes in maturation of germ cells, resulting in cells containing the haploid number of chromosomes. (Gr. dim., decrease, from *meion*, less)

mem'-brane A thin soft sheet of cells or of material secreted by cells. (L. *membrana*, skin, parchment)

mer-is'-tic Pertaining to features of an organism which can be counted. (Gr. *meros*, part)

mer'-o-blas"-tic Cleavage of an egg in which only part of the protoplasm divides, leaving the yolk undivided; characteristic of eggs with much yolk. (Gr. *meros*, part + *blastos*, germ)

mes'-en-chyme The gelatinous region between the outer and inner epithelia that contains amoeboid or other cells in lower invertebrates. Also part of the mesoderm in a vertebrate embryo that produces connective and circulatory tissues. (Gr. *mesos*, middle + *chymos*, fluid)

mes'-en-ter-y The sheet of tissue that suspends organs in the body cavity and is continuous with the peritoneum lining that cavity. (Gr. *mesos*, middle + *enteron*, intestine)

mes'-o-derm The embryonic cells or cell layers between ectoderm and endoderm. (Gr. *mesos*, middle + *derma*, skin)

mes'-o-gle"-a The gelatinous filling between the outer and inner cell layers of a two-layered animal such as a jellyfish. (Gr. *mesos*, middle + *gloios*, glutinous)

me-tab'-o-lism The sum of the constructive and destructive processes (anabolism and catabolism), physical and chemical, that occur in living organisms. (Gr. *metabolos*, changeable, from *meta*, after, with + *ballo*, throw)

met'-a-gen"-e-sis Alternation of sexual and asexual reproduction in the life cycle of certain animals; alternation of generations. (Gr. *meta*, after, with + *genesis*, origin)

met'-a-mere Any one of a series of homologous parts in the body, as with annelids, arthropods, or chordates; a somite. (Gr. *meta*, after, with + *meros*, part)

me-tam'-er-ism Segmental repetition of homologous parts (metameres).

met'a-mor"-pho-sis Marked change in form from one stage of development to another, as of a larva to an adult. (Gr., from *meta*, after, with + *morphe*, shape)

met'-a-ne-phrid"-i-um A tubular excretory organ, the open inner end draining from the coelom and the outer discharging to the exterior, as in the earthworm. (Gr. *meta*, after + *nephros*, kidney)

Met'-a-zo"-a Multicellular animals with cells usually arranged into tissues; includes all animals above the sponges. (Gr. *meta*, after, with + *zoön*, animal)

mi"-cro-me'-ter The unit of microscopic measurement, 1/1000 of a millimeter, represented by μm (Greek letter *mu* + *m*); formerly *micron*. (Gr. *mikros*, small)

mi"-cro-tu'-bule A small tube; groups of such tubules make up many intracellular "fibers." Microtubules are basic components of centrioles, basal bodies, cilia, and flagella. (Gr. *mikros*, small + L. *tubules*, dim. of *tubus*, tube).

mi-to'-sis Indirect cell division, characterized by the appearance of a fibrous spindle and a definite number of chromosomes, which split longitudinally to form two equal sets of daughter chromosomes; the latter diverge to opposite poles of the spindle to become parts of the two new nuclei. (Gr. *mitos*, thread)

mo'-lar A posterior permanent tooth of a mammal. (L. *molo*, grind)

molt To cast off an outer covering such as cuticle, scales, feathers, or hair.

Monera A kingdom of organisms defined by having prokaryotic cells. (Gr. *monos*, single)

mo-noe'-cious Having both male and female gonads in the same individual; hermaphroditic. (Gr. *monos*, single + *oikos*, house)

mon'-o-hy"-brid The offspring of parents differing in one character. (Gr. *monos*, single + *hybrida*, mongrel)

mon'-o-phy-let"-tic Originating from a single known evolutionary derivation. (Gr. *monos*, single + *phyle*, tribe)

mu'-cous Secreting mucus or similar substance. (L. *mucosus*, from *mucus*)

mu-ta'-tion Abrupt, stable, and heritable modification of a character; also the change in a gene responsible for it. (L. *mutatus*, changed)

mu'-tu-al-ism Jointly beneficial association between individuals of two different species. (L. *mutuus*, exchanged)

my"-o-fi'-bril A small, longitudinal contractile filament within a muscle cell or fiber; usually large numbers are present within each muscle cell. (Gr. *mys*, muscle + L. dim. of *fibra*, fiber)

my'-o-mere A muscle segment or somite. (Gr. *mys*, muscle + *meros*, part)

na'-ris, pl. na'-res The opening of the air passages, both internal and external, in the head of a vertebrate. (L., nostril)

na'-sal Pertaining to the nose. (L. *nasus*, nose)

natural selection The elimination of less fit individuals in the struggle to live.

ne"-o-te'-ny Condition of having the period of immaturity or larval form indefinitely prolonged, as in the axolotl. (Gr. *neos*, new + *teinein*, extend)

ne-phrid'-i-um A tubular excretory organ found in mollusks, annelids, arthropods, and other invertebrates. (Gr. *nephros*, kidney)

neph'-ro-stome The ciliated entrance from the coelomic cavity into a nephridium, or kidney tubule. (Gr. *nephros*, kidney + *stoma*, mouth)

nerve A bundle of nerve fibers lying outside the central nervous system, most containing both afferent and efferent neurons. (L. *nervus*, sinew)

nerve cord A compact cord, composed of neurons and usually with ganglia, forming part of a central nervous system.

neu'-ron A nerve cell with cytoplasmic extensions (dendrites, axon) over which nerve impulses pass; the structural unit of the nervous system.

noc-tur'-nal Active at night. (L. *nocturnus*, nightly)

no'-to-chord The cellular axial support in the early embryo of all chordates, found ventral to the nerve cord and in the antero-posterior axis; later surrounded or supplanted by the vertebrae in most vertebrates. (Gr. *notos*, the back + *chorde*, string)

nu-cle'-o-lus An oval mass within the nucleus of most cells; responsible for the synthesis of ribosomes; disappears during mitosis. (L. dim. of *nucleus*)

nu'-cle-us A differentiated structure of specialized protoplasm within a cell, refractile and with deeply staining chromatin, that controls metabolic activities and contains the hereditary material in cells of all organisms except the monerans (prokaryotes). (L. dim. of *nux*, nut)

o-cel'-lus A small, simple eye as on many invertebrates. (L. dim. of *oculus*, eye)

oc'-u-lar Pertaining to the eye. (L. *oculus*, eye)

ol-fac'-to-ry Pertaining to the sense of smell. (L. *olfacto*, smell)

om-niv'-o-rous Eating all kinds of food; feeding upon both plants and animals. (L. *omnis*, all + *voro*, eat)

on-tog'-e-ny The developmental history of an individual. (Gr. *on*, being + *gen*, become)

o'-o-sperm A fertilized egg; a zygote. (Gr. *oön*, egg + *sperma*, seed)

o-per'-cu-lum The plate covering the gills of a bony fish; also the plate serving to cover the opening of some snail shells. (L.)

oph-thal'-mic Pertaining to the eye. (Gr. *ophthalmos*, eye)

op'-tic Pertaining to the eye or sense of sight. (Gr. *optos*, seen)

o'-ral Pertaining to or near the mouth. (L. *os*, mouth)

or'-bit The eye socket. (L. *orbis*, circle)

or'-gan Any part of an animal performing some definite function; a group of cells or tissues acting as a unit for some special purpose. (Gr. *organon,* instrument)

or-gan-elle A specialized part in a protozoan that performs some special function (like an organ in a metazoan); an intracellular structure, mitochondrium, ribosome, etc. (Gr. dim. of *organon,* instrument)

or'-gan-ism A single plant or animal; one that functions as a unit.

os-mo'-sis Diffusion through a differentially permeable membrane. (Gr. *osmos,* pushing)

os'-ti-um, pl. **os'-ti-a** An opening to a passage, usually guarded by a valve or circular muscle. (L. *os,* mouth)

o'-tic Pertaining to the ear. (Gr. *otikos,* pertaining to the ear)

o'-to-lith A concretion of calcium salts in the inner ear of vertebrates or in the auditory organ of some invertebrates. (Gr. *otikos,* pertaining to the ear + *lithos,* stone)

o'-va-ry Female gonad in which the egg cells multiply and develop. (L. *ovum,* egg)

o'-vi-duct The tube in which eggs are conveyed from the ovary to the uterus or to the exterior. (L. *ovum,* egg + *duco,* lead)

o-vip'-a-rous Egg-laying; producing eggs that hatch outside the mother's body. (L. *ovum,* egg + *pario,* produce)

o'-vi-pos"-i-tor Organ consisting of paired abdominal appendages variously modified to lay eggs, sometimes slender for boring in wood or sharp and barbed as a sting. (L. *ovum,* egg + *pono,* to place)

o'-vo-vi-vip"-a-rous Producing eggs that are incubated and hatched within the parent's body, as with some fishes, reptiles, and invertebrates. (L. *ovum,* egg + *vivus,* alive + *pario,* produce)

o'-vum An egg, the sex cell of a female. (L.)

ox'-i-da"-tion The addition of oxygen, the removal of hydrogen, or the removal of electrons from a compound or element.

P The first parental generation—parents of an individual of the F_1 generation.

pae"-do-gen'-e-sis Reproduction by larvae or other preadult forms. (G. *pais,* child + *genesis,* birth)

pae"-do-mor'phic Having the form of a larva or other preadult stage. (Gr. *pais,* child + *morphe,* form)

palp, palpus A projecting part or appendage, often sensory, on the head or near the mouth in some invertebrates. (L. *palpo,* stroke)

pa-pil'-la Any nipplelike structure, large or small. (L., nipple)

par'-a-site An organism that lives on or in another more or less at the expense of the latter (the host). (Gr. *para,* beside + *sitos,* food)

pa-ren'-chy-ma Soft cellular substance filling space between organs. (Gr. *para,* beside + *en,* in + *chyma,* fluid)

par'-the-no-gen"-e-sis Development of a new individual from an unfertilized egg, as in rotifers, plant lice, drone bees, certain lizards etc. (Gr. *parthenos,* virgin + *genesis,* birth)

path'-o-gen"-ic Causing or productive of disease. (Gr. *pathos,* suffering + *genesis,* birth)

pec'-to-ral Pertaining to the upper thoracic region, or breast. (L. *pectus,* chest)

pe-lag'-ic Pertaining to the open sea, away from the shore. (Gr. *pelagos,* open sea)

pel'-vic Pertaining to the posterior girdle and paired appendages of vertebrates; the posterior abdominal region of a mammal. (L. *pelvis,* a basin)

pe'-nis The copulatory organ of a male for conveying sperm to the genital tract of a female. (L.)

pen'-ta-dac"-tyl Having five fingers, toes, or digits. (Gr. *pente,* five + *daktylos,* finger)

per'-i-car"-di-um The cavity enclosing the heart; also the membranes lining the cavity and covering the heart. (Gr. *peri,* around + *kardia,* heart)

pe-riph'-er-al To or toward the surface, away from the center. (Gr. *peri,* around + *phero,* to bear)

per'-i-stal"-sis Rhythmic involuntary muscular contractions passing along a hollow organ, especially of the digestive tract. (Gr. *peri,* around + *stalsis,* constriction)

per'-i-to-ne"-um The thin serous membrane (mesodermal) that lines the body cavity and covers the organs therein in many animals. (Gr. *peri,* around + *teino,* stretch)

pH (potential of hydrogen) A symbol used (with a number) to express acidity or alkalinity. It is a measure of hydrogen-ion concentration in a given solution. The pH scale in common use ranges from 0 to 14, with pH 7 as neutral, 6 to 0 increasingly acid, and 8 to 14 increasingly alkaline. The lower the pH value the more acid or hydrogen ions in the solution.

phag'-o-cyte A white blood cell that engulfs and digests bacteria and other foreign materials. (Gr. *phagein,* to eat + *kytos,* hollow vessel)

phar'-ynx The region of the digestive tract between the mouth cavity and the esophagus; often muscular and sometimes with teeth in invertebrates; the gill region of many aquatic vertebrates. (Gr., throat)

phe'-no-type The external appearance of an individual without regard to its genetic or hereditary constitution. Compare *genotype*. (Gr. *phaino*, show + *typos*, impression, type)

pher'-o-mone A chemical signal transmitted between members of the same species which may influence development or behavior. (Gr. *pherein*, to carry + *horman*, to excite)

pho'-to-per"-iod The length of daily exposure of a plant or animal to light, considered especially with reference to the effect of light on reproduction, growth, and development. (Gr. *phos*, light + *periodos*, cycle)

pho'-to-syn"-the-sis The formation of carbohydrates from carbon dioxide and water by the chlorophyll in green plants or flagellate protozoans in the presence of light. (Gr. *phos*, light + *synthesis*, place together)

phy-log'-e-ny The evolutionary history of a species or higher group. (Gr. *phylon*, race + *gen*, stem from *gignomai*, be born, become)

pin'-na A wing or fin; also the projecting part of the external ear in mammals. (L., feather)

pla-cen'-ta The organ by which the fetus (embryo) of higher mammals is attached in the uterus of the mother and through which diffusible substances (oxygen, nutrients, wastes) pass for the metabolism of the fetus. See Chorion. (L., flat cake)

plan'-ti-grade Walking on the whole sole of the foot, as a human or a bear. (L. *planta*, sole + *gradior*, walk)

plas'-ma The fluid portion of blood or lymph. (Gr., a thing molded)

pleu'-ra The membrane covering the lung and lining the inner wall of the thorax. (Gr., rib, side)

pleu'-ron The lateral plate on either side of a somite in arthropods. (Gr., rib, side)

plex'-us A network of interlaced nerves or blood vessels. (L., a plaiting)

poi'-ki-lother"-mal Having varying body temperature; characteristic of all animals but birds and mammals. (Gr. *poikilos*, variegated + *therme*, heat)

pol-y-mer A chemical compound made up of repeating units of simple molecules (monomers). (Gr. *polys*, many + *meros*, part)

pol'-y-mor"-phism Of two or more forms; existence of individuals of more than one form in a species. (Gr. *polys*, many + *morpho*, form)

pol'-yp An organism (cnidarian) typically with a body wall of two layers and tubular in form, one end being closed and attached and the other with a central mouth usually surrounded by tentacles. (Gr. *polypous*, many-footed)

pol'-y-phy-let"-ic From more than one known evolutionary derivation. (Gr. *poly*, many + *phyle*, tribe)

pol'-yp-loid Having three or more times the haploid number of chromosomes.

population The aggregate of individuals of a species, race, or variety that inhabit a particular locality or region. (L. *populus*, people, nation)

por'-tal vein A vein that begins and ends in a capillary bed. (L. *porta*, gate)

pos-te'-ri-or The hinder part or toward the hinder (tail) end, away from the head; opposite of anterior. (L., following)

pred'-a-tor An animal that captures or preys upon other animals for its food. (L. *praeda*, booty)

pre-hen'-sile Adapted for grasping or holding. (L. *prehendo*, seize)

prim'-i-tive Not specialized; the early or beginning type or stage. (L. *primus*, first)

proc'-to-de"-um The ectoderm-lined terminal part of the digestive tract near the anus. (Gr. *proktos*, anus + *hodos*, a road)

pro-kar-y-ote An organism which lacks a membrane around the genetic material. (Gr. *pro*, rudimentarily, primitively + *karyote*, given a nucleus, from *karyon*, nut, kernel, kernellike thing)

pro-sto'-mi-um The preoral segment in annelids. (Gr. *pro*, before + *stoma*, mouth)

pro-tan'-dry Production of sperm and later of eggs by the same gonad. (Gr. *protos*, first + *aner*, man)

Pro-tis'-ta A kingdom of organisms which includes the eukaryotic unicellular organisms. (Gr., the first ones)

pro'-to-ne-phrid"-i-um An invertebrate excretory organ (of one or more cells), the inner end closed and either branched or with one terminal cell (solenocyte). (Gr. *protos*, first + *nephros*, kidney)

pro'-to-plasm Older term for the colloidal complex of chemicals comprising the living substance of a cell. (Gr. *protos*, first + *plasma*, form)

pro"-to'-stome One of two major lines of evolution in the Animal Kingdom, characterized by spiral determinate cleavage, mouth arising from the blastopore, and schizocoel formation. (Gr. *protos*, first + *stoma*, mouth)

prox'-i-mal Toward or nearer the place of attachment or reference of the center of the body; opposite of distal. (L. *proximus*, nearest)

pseu'-do-coel A body cavity not lined with peritoneum and not part of a blood-vascular system, as in nematodes and some other invertebrates. Compare *coelom, hemocoel*. (Gr. *pseudes*, false + *koilos*, hollow)

pseu'-do-po"-di-um A flowing extension of protoplasm used in locomotion or feeding by a cell or protozoan. (Gr. *pseudes*, false + *podion*, foot)

pul'-mo-na-ry Pertaining to the lungs. (L. *pulmo*, lungs)

ra'-di-al cleav'-age A pattern of division of cells in a zygote in which successive sets of blastomeres lie directly above the set below them.

ra'-di-al sym'-me-try Having similar parts arranged around a common central axis, as in a starfish.

ra'-mus A branch or outgrowth of a structure. (L., branch)

Re'-cent The present, or Holocene, epoch in geology. Compare *fossil*. (L. *recens*, fresh)

re-cep'-tor A free nerve ending or sense organ capable of receiving and transforming certain environmental stimuli into sensory nerve impulses; also, a chemical specialized for reception of stimuli such as the hormone receptors (protein) in certain cells. (L., receiver)

recessive character A character from one parent that remains undeveloped in offspring when associated with the corresponding dominant character from the other parent. (L. *recessus*, a going back)

rec'-tum The terminal enlarged portion of the digestive tract. (L. *rectus*, straight)

re-duc'-tion di-vi'-sion That division of the maturing germ cells by which the somatic, or diploid, number of chromosomes becomes reduced to the haploid number.

re'-flex ac'-tion Action resulting from an afferent sensory impulse on a nerve center and its reflection as an efferent motor impulse, independent of higher nerve centers or the brain; an automatic response to a stimulus. (L. *re*, back + *flecto*, bend)

re-gen'-er-a''-tion Replacement or repair of parts lost through mutilation or otherwise.

re'-nal Pertaining to a kidney. (L. *renes*, kidneys)

re'-pro-duc''-tion The maintenance of a species from generation to generation.

res'-pi-ra''-tion Two meanings: (1) process by which cells oxidize nutrient molecules and obtain energy; (2) breathing. (L. *re*, back + *spiro*, breathe)

ret'-i-na The cell layer of an eye containing the receptors of light impulses. (L. *rete*, net)

re-ver'-sion The reappearance of ancestral traits that have been in abeyance for one or more generations. (L. *re*, back + *verto*, turn)

ros'-trum A projecting snout or similar process on the head. (L., beak)

ru'-di-men''-ta-ry Incompletely developed or having no function. Compare *vestigial*. (L. *rudis*, unwrought)

ru'-mi-nant A herbivorous land mammal that chews a cud, as a cow or deer. (L. *rumen*, throat)

sa'-crum The posterior part of the vertebral column that is attached to the pelvic girdle. (L. from *sacer*, sacred, offered in sacrifice)

sag'-it-tal Of or pertaining to the median anteroposterior plane in a bilaterally symmetrical animal or a section parallel to that plane. (L. *sagitta*, arrow)

sal'-i-va-ry Pertaining to the glands of the mouth that secrete saliva. (L. *saliva*, spittle)

sap'-ro-phyte An organism that lives upon dead organic matter. (Gr. *sapros*, rotten + *phyton*, plant)

sar''-co-mere The region between the Z bands of a myofibril. (Myofibrils are contractile filaments within muscle cell.) (Gr. *sarx*, flesh + *meros*, part)

scan-so'-ri-al Pertaining to or adapted for climbing. (L. *scando*, climb)

schiz'-o-coel Type of coelom which arises as a split in the embryonic mesoderm. (Gr. *schizo*, split + *koilos*, hollow)

schi-zog'-o-ny Multiple asexual fission in Protozoa. (Gr. *schizo*, split + *gonos*, generation, seed)

secondary sexual characters Those characters which distinguish one sex from the other but do not function directly in reproduction.

se-cre'-tion A useful substance produced in the body by a cell or multicellular gland; also the process of its production and passage. Compare Excretion. (L. *secretus*, separated)

sed'-en-ta-ry Remaining in one place. (L. *sedeo*, sit)

sed'-i-men''ta-ry In geology, rocks formed of calcium carbonates, clay, mud, sand, or gravel, deposited in water or depressions on land and cemented or pressed into solid form; fossils occur in such rocks. (L. *sedimentum*, settling)

seg'-ment A part that is marked off or separate from others; any of the several serial divisions of a body or an appendage. Compare Somite. (L. *segmentum*)

sem'-i-nal Pertaining to structures or fluid containing spermatozoa (semen). (L. *semen*, seed)

sense or'-gan An organ containing a part sensitive to a particular kind of stimulus.

sep'-tum A dividing wall, or partition, between two cavities or structures. (L. *sepes*, fence)

se'-rous Secreting watery colorless serum, as a gland or serous membrane. (L., milky fluid)

se'-rum The fluid of blood that separates from a clot and contains no cells or fibrin. (L.)

ses'-sile Permanently fixed, sedentary, not free-moving. (L. *sedeo*, sit)

se'-ta A bristle or slender, stiff bristlelike structure. (L., bristle)

sex chro'-mo-somes Special chromosomes different in males and females and concerned in the determination of sex; the X and Y chromosomes.

sex-limited character A character belonging to one sex only; commonly a secondary sexual character.

sex-linked character A character the gene of which is located in the sex chromosome.

sexual union Temporary connection of a male and female for transfer of sperm into the female's reproductive tract.

si-lic'-eous Containing silica or silicon dioxide (SiO_2). (L. *silex*, flint)

si'-nus A cavity in a bone, a tissue space, or an enlargement in a blood vessel. (L., fold, hollow)

skel'-e-ton The hardened framework of an animal body serving for support and to protect the soft parts; it may be external or internal and either solid or jointed. (Gr. *skeletos*, dried up)

sol'-i-ta-ry Living alone, not in colonies or groups. (L. *solus*, alone)

sol'-ute A substance that will dissolve or go into solution, as salt in water. (L. *solvo*, let loose, dissolve)

so-lu'-tion A single homogeneous phase of matter (gas, liquid, solid) that is a mixture in which the components are equally distributed. (L. *solvo*, let loose, dissolve)

sol'-vent A fluid capable of dissolving substances. (L. *solvo*, let loose, dissolve)

so'-ma The body except for its germ cells. (Gr.)

so-mat'-ic Pertaining to the body or body cells, as contrasted with germ cells. (Gr. *soma*, body)

so'-mite A serial segment or homologous part of the body; a metamere. (Gr. *soma*, body)

spe'-cial-ized Not generalized; adapted by structure or function for a particular purpose or mode of life. (L. *species*, kind)

spe'-ci-a"-tion The formation of species by evolutionary processes. (L. *species*, kind)

spe'-cies, pl. **species** The taxonomic unit in classification of animals or plants. (L., kind)

sperm See Spermatozoa.

sper'-mat-o-phore A capsule or packet of sperm extruded by the male of some species and trans-ferred to the female. (Gr. *sperma*, seed + *phero*, to bear)

sper'-ma-to-zo"-a The matured and functional male sex cells or male gametes. (Gr. *sperma*, seed + *zoön*, animal)

spir'-a-cle In insects, an external opening to the tracheal or respiratory system; in cartilaginous fishes, the modified first gill slit through which water enters the pharynx; in tadpoles, the opening through which water exits from the pharynx. (L. *spiraculum*, air hole, from *spiro*, breathe)

spi'-ral cleav'-age A pattern of division of cells in a zygote in which successive sets of blastomeres lie above the furrows of the set below them, thus creating a spiral pattern. (L. *spira*, coil)

spore A cell in a resistant covering, capable of developing independently into a new individual. (Gr. *spora*, seed)

spo-rog'-o-ny The division of a zygote or oöcyst into spores. (Gr. *spora*, seed + *gonos*, generation, seed)

stat'-o-cyst An organ of equilibrium in some invertebrates. (Gr. *statos*, standing + *kystis*, bladder)

stat'-o-lith A calcareous granule or other particle in a statocyst.

stim'-u-lus A change in the external or internal environment capable of influencing some activity in an organism or its parts.

sto'-mo-de"-um The ectoderm-lined portion of the mouth cavity. (Gr. *stoma*, mouth + *hodos*, a road)

strat'-i-fied Having a series of layers, one above another. (L. *stratum*, covering)

stra'-tum A layer or sheet of tissue (anatomy); a layer or sheet of sedimentary rock (geology). (L., covering)

sub'-strate The compound or material upon which an enzyme acts to produce a product. (L. *sub*, under + *stratum*, covering, coat, layer)

su'-ture Line of junction between two bones or between two parts of an exoskeleton. (L. *suo*, sew)

sym'-bi-o"-sis Interrelation between two organisms of different species; see Commensalism, Mutualism, Parasite. (Gr. *syn*, together + *bios*, life)

sym'-phy-sis A union between two parts. (Gr. *syn*, together + *phyein*, grow)

syn'-apse The contact of one nerve cell with another, across which action potentials are transmitted by neurotransmitter chemicals. (Gr. *syn*, together + *hapto*, unite)

syn-ap'-sis Temporary union of the chromosomes

in pairs preliminary to the first maturation division. (Gr. *syn*, together + *hapto*, unite)

syn-cyt'-i-um A mass or layer of protoplasm containing several or many nuclei not separated by cell membranes. (Gr. *syn*, together + *kytos*, cell)

sys-tem'-ic circulation Portion of the circulatory system not directly involved in respiration. (Gr. *systema*, organization, from *syn*, together + *histemi*, stand)

tac'-tile Pertaining to the organs or the sense of touch. (L. *tactus*, touch)

tag'-ma-ti-za"-tion In metameric animals, the process of organizing groups of segments into units specialized to perform certain functions and anatomically differentiated from other segments. (Gr. *tagma*, order, rank)

ten'-don Fibrous connective tissue serving for the attachment of a muscle, usually to a bone. (L. *tendo*, to stretch)

ten'-ta-cle An elongate flexible appendage usually near the mouth. (L. *tento* from *teneo*, hold)

ter-res'-tri-al Belonging to or living on the ground or earth. (L. *terra*, earth)

tes'-tis, pl. **tes'-tes** The male reproductive gland or gonad, in which spermatozoa are formed. (L.)

tet'-ra-pod A four-legged vertebrate—the amphibians, reptiles, birds, and mammals. (Gr. *tetra*, four + *pous*, foot)

ther'-mo-dy-nam"-ics That branch of physics which deals with energy relationships, particularly the conversion from one form to another. (Gr. *thermos*, heat + *dynamis*, power)

tho'-rax The major division of an animal just behind or below the head; in land vertebrates, the part enclosed by the ribs. (Gr.)

tis'sue A layer or group of cells in an organ or body part having essentially the same structure and function; e.g., bone tissue, nervous tissue, etc. (L. *texo*, weave)

tor'-sion In gastropod mollusks, the process of rotating the visceral mass 180° counterclockwise with respect to the head-foot.

tra'-che-a An air tube; the windpipe of land vertebrates, from the glottis to the lungs; part of the respiratory system in insects and other arthropods. (Gr. *trachys*, rough)

trip'-lo-blas"-tic Derived from three embryonic germ layers—ectoderm, endoderm, and mesoderm. (Gr. *triplous*, threefold + *blastos*, germ)

tro'-cho-phore An invertebrate larva, commonly pear-shaped and with an equatorial band of cilia. (Gr. *trochos*, wheel + *phoros*, bearing)

tym'-pa-num A vibrating membrane involved in hearing; the eardrum, or tympanic membrane. (Gr. *tympanon*, drum)

um-bil'-i-cal cord The cord containing blood vessels supported by connective tissue that unites the embryo or fetus of a mammal with the mother during development in the uterus. (L. *umbilicus*, navel)

un-guic'-u-late Having claws, as a cat. (L. *unguis*, claw)

un'-gu-late Having hoofs, as a deer or horse. (L. *ungula*, hoof)

un'-gu-li-grade' Walking or adapted for walking on hoofs. (L. *ungula*, hoof + *gradior*, to step)

u-nit char'-ac-ter A trait that behaves more or less as a unit in heredity and may be inherited independently of other traits.

u-re'-ter The duct carrying urine from the kidney to the urinary bladder or to the cloaca. (Gr. from *ouron*, urine)

u-re'-thra The duct by which urine is discharged from the bladder to the outside in mammals; joined by the ductus deferens in the male. (Gr. from *ouron*, urine)

u'-ri-no-gen"-i-tal See Urogenital.

u'-ro-gen"-i-tal Pertaining to the excretory and reproductive organs and functions. (L. *urina*, urine + *genitalis*, genital)

u'-ter-us The enlarged posterior portion of an oviduct; that segment of the female reproductive tract in which the developing embryo is retained in viviparous and ovoviviparous species. (L., womb)

vac'-u-ole A minute cavity within a cell, usually filled with some liquid product of protoplasmic activity. (L. dim. of *vacuus*, empty)

va-gi'-na The terminal portion of the female reproductive tract, which receives the copulatory organ of a male in mating. (L., sheath)

valve In animals, any structure that limits or closes an opening; some examples are the thin folds in veins, lymph vessels, or heart or the circular muscles about a tubular exit; also either external shell of a bivalve mollusk or brachiopod or some crustaceans.

vas, pl. **va'-sa** A small tubular vessel, or duct. (L., vessel)

vas'-cu-lar Pertaining to vessels or ducts for conveying or circulating blood or lymph. (L. dim of *vas*, vessel)

vein A relatively thin-walled vessel carrying blood from capillaries toward the heart. (L. *vena*, vein)

vent The external opening of the cloaca or the intestine on the surface of the body, especially in nonmammals, as birds, fishes, and reptiles.

ven'-tral Toward the lower side or belly; away from the back; opposite of dorsal. (L. *venter*, belly)

ven'-tri-cle A cavity in the brain of a vertebrate; also, a chamber of the heart that receives blood from the atria. (L. dim. of *venter*, belly)

ver'-te-bra One of the segmental structural units of the axial skeleton or spinal column in a vertebrate. (L., joint)

ver'-te-brate An animal having a segmental backbone, or vertebral column; the cyclostomes to the mammals. (L. *vertebratus*, jointed)

ves'-sel A tubular structure that conveys fluid, especially blood or lymph. (L. *vascellum*, dim. of *vas*, vessel)

ves-tig'-ial Small or degenerate but representing a structure that formerly was more fully developed or functional. Compare Rudimentary. (L., *vestigium*, footstep, a trace)

vil'-lus, pl. **vil'-li** A minute fingerlike projection, especially one of those of the intestinal lining of vertebrates. (L. *villus*, shaggy hair)

vis'-cer-a The organs within the cranium, thorax, and abdomen, especially the latter. (L., internal organs)

vis'-cer-al skel'-e-ton The supporting framework of the jaws and gill arches and their derivatives in vertebrates.

vis'-u-al Pertaining to sight. (L. *visus*, sight)

vi'-ta-min An organic substance that is an essential food factor needed in minute amounts for normal growth and function. (L. *vita*, life + *amin*, a chemical radical)

vi-vip'-a-rous Producing living young that develop from eggs retained within the mother's body and nourished by her bloodstream, as with most mammals. (L. *vivo*, live + *pario*, to bear)

X, Y chromosomes Chromosomes associated with sex in many animals.

yolk Nutritive materials (fats and others) stored within or with an egg for nourishment of the future embryo.

zy'-gote A fertilized egg (diploid cell) resulting from the union of two gametes of opposite kind, ovum and sperm. (Gr. *zygon*, yoke)

INDEX

Page numbers in **boldface** indicate pages on which illustrations occur. Scientific names of genera and species are in *italics* and those of higher groups are in small CAPITALS. Names of fossil species or groups are preceded by an asterisk (*). This index omits incidental references to common structures such as stomach, heart, etc., but contains entries for pages where these are described in detail. Such items may be sought under entries to Digestive system, Circulatory system, etc., or to an animal type such as Cat, Crayfish, etc.

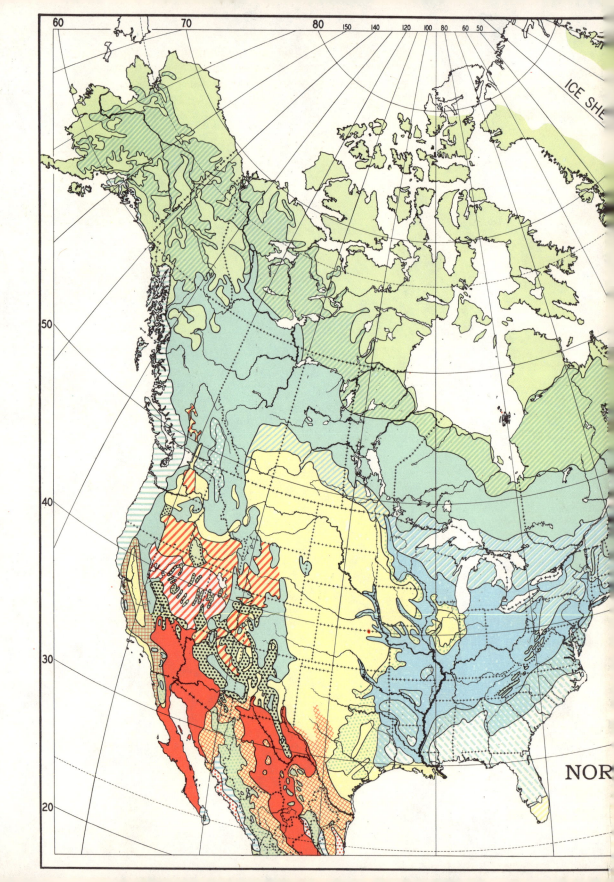

60 70 80 150 140 120 100 80 60 50

ICE SHE

50

40

30

20

NOR